Comprehensive Guide to Hallucinogenic Plants

Hallucinogens have been traditionally used to encourage spiritual growth, heighten perception, inspire personal development, or expand reality. *Comprehensive Guide to Hallucinogenic Plants* focuses on ethnobotanical aspects of hallucinogenic plant species, featuring the history on how they were used in ancient societies, identifying chemical compounds, and explaining modern medicinal uses, as well as conservation initiatives. The book emphasizes the importance of understanding the cultural, countrywide, environmental, and scientific importance of these medicinal plants.

Some of the 50 plants covered in this work include: ayahuasca, ginger, kanna, dream herb, iboga, peyote, canary broom, coral tree, catnip, wild rue, kava, mandrake, and golden angel's trumpet. Each chapter includes information on historical plant use and identification of chemical compounds, and explains modern medicinal uses. The text highlights the importance of studying, evaluating, and utilizing these plants not in isolation, but from a global perspective.

Comprehensive Guide to Hallucinogenic Plants appeals to plant scientists, botanists, ethnobotanists, pharmacologists, and those with an interest in alternative or herbal medicine.

Comprehensive Guide to Hallucinogenic Plants

Edited by
Noureddine Chaachouay, Abdelhamid Azeroual,
and Lahcen Zidane

CRC Press
Taylor & Francis Group
Boca Raton London New York

CRC Press is an imprint of the
Taylor & Francis Group, an **informa** business

Designed cover image: Shutterstock

First edition published 2025
by CRC Press
2385 NW Executive Center Drive, Suite 320, Boca Raton FL 33431

and by CRC Press
4 Park Square, Milton Park, Abingdon, Oxon, OX14 4RN

CRC Press is an imprint of Taylor & Francis Group, LLC

Library of Congress Cataloging-in-Publication Data
Names: Chaachouay, Noureddine, editor. | Azeroual, Abdelhamid, editor. | Zidane, Lahcen, editor.
Title: Comprehensive guide to hallucinogenic plants / edited by Noureddine
Chaachouay, Abdelhamid Azeroual, and Lahcen Zidane Other titles: Exploring medicinal plants.
Description: First edition | Boca Raton, FL : CRC Press, 2025 | Includes bibliographical references and index |
Summary: "Hallucinogens have been utilized to encourage spiritual growth, heighten perception, inspire personal development, or expand reality. Comprehensive Guide to Hallucinogenic Plants discusses the ethnobotany of hallucinogenic plant species, featuring history on how they were used in ancient societies, identifying chemical compounds, and explaining modern medicinal uses and conservation initiatives. This book examines the indigenous medicinal practices of many countries and civilizations, and provides comprehensive and up-to-date information on recent investigations and trends in medicinal plants and aromatics.
Additionally, it emphasizes the importance of understanding their cultural, country-wide, environmental, and scientific factors. It highlights that studying, evaluating, and utilizing these plants cannot be done effectively in isolation but requires a global perspective.
Approximately 120 hallucinogenic plants have been identified globally, although 3,000 plant species have been routinely ingested by humans at some point in history, and only 150 are significant enough to join international trade. Some of the 50 plants covered include Ayahuasca, ginger, kanna, dream herb, iboga, peyote, canary broom, coral tree, catnip, wild rue, kava, mandrake and golden's angel trumpet. Each chapter follows the structure of including information on historical plant use, identification of chemical compounds, and an explanation of the modern medicinal uses. Comprehensive Guide to Hallucinogenic Plants will appeal to plant scientists, botanists, ethnobotanists, pharmacologists, and those with an interest in alternative medicine"– Provided by publisher
Identifiers: LCCN 2024009960 (print) | LCCN 2024009961 (ebook) |
ISBN 9781032591803 (hardback) | ISBN 9781032607320 (paperback) |
ISBN 9781003460336 (ebook)
Subjects: LCSH: Hallucinogenic plants. | Ethnobotany. | Materia medica, Vegetable.
Classification: LCC SB293 .C56 2025 (print) | LCC SB293 (ebook) | DDC 633.8/8–dc23/eng/20240709
LC record available at https://lccn.loc.gov/2024009960
LC ebook record available at https://lccn.loc.gov/2024009961

ISBN: 978-1-032-59180-3 (hbk)
ISBN: 978-1-032-60732-0 (pbk)
ISBN: 978-1-003-46033-6 (ebk)

DOI: 10.1201/9781003460336

Typeset in Times
by SPi Technologies India Pvt Ltd (Straive)

This book is dedicated to

All our families, Chaachouay, Azeroual, and Zidane*, for their unwavering support and endless love. Your encouragement has been the guiding light through the journey of writing this book. This work is a tribute to your daily strength and inspiration.*

Our esteemed professors and cherished friends*, whose wisdom, guidance, and companionship have been the cornerstones of our journey. Your insights have not only shaped this book but have profoundly influenced our personal and academic growth. This work is a testament to the invaluable lessons you've imparted and the enduring bonds we've formed.*

Contents

About the Editors

Noureddine Chaachouay, a professor at the Ecole Normale Supérieure (ENS) of Berrechid, Hassan First University in Settat, Morocco, has had a distinguished academic and professional journey. His educational background is deeply rooted in the sciences, particularly botany and life sciences. He achieved his doctorate in the sciences in 2020 from the Faculty of Sciences at Ibn Tofail University, Kenitra, with his thesis focusing on the floristic and ethnomedicinal study of aromatic and medicinal plants in the Rif region of Northern Morocco. Before his current role as an assistant professor, which he has held since 2020, he spent five years teaching secondary school in life and earth sciences at the Provincial Directorate of Education in Salé, Morocco. His credentials, which include a Certificate of Pedagogical Qualification for Secondary Education and a Certificate of Recognition of Success in the Professional Aptitude Examination for Secondary Education, further demonstrate his dedication to education. His technical skills are grounded in his diploma in systems and network computing. As a researcher, he is keenly interested in studying the use of plants in various areas, such as alternative medicine, pharmacy, cosmetics, clothing, rituals, flavors, fragrances, and dyes. In recognition of his contributions to scientific research, Pr. Noureddine was awarded the prestigious Hassan First University Prize for Scientific Research in 2023, underscoring his status as a "Young Researcher". This accolade is a testament to his commitment to advancing knowledge in his field. His academic and professional experiences reflect a deep commitment to education and research, particularly in botany, climate change, and sustainable development.

Abdelhamid Azeroual is a professor at the Ecole Normale Supérieure (ENS) of Berrechid, Hassan First University of Settat, Morocco. He earned his PhD from Mohammed V-Agdal University in Rabat, Morocco. He has over 20 years of experience in research, teaching, and administration. As part of his professional activities, he has received several scientific research grants (including the German DAAD) and participated in several European Tempus programs. As a permanent member of the Agri-Food and Health research laboratory, he first developed his research on the ichthyofauna of the continental waters of Morocco, conducting the very first assessment of North African freshwater fish in 2007. He later extended his area of research to include the field of aromatic and medicinal plants. Since then, he has successfully published numerous papers in reputed and internationally recognized journals and several book chapters for internationally renowned publishers.

Lahcen Zidane has been a professor since 1997 in the Department of Biology, Faculty of Science, Ibn Tofail University, Kenitra, Morocco. His work focuses primarily on plant species and agro-industrial production. This indicates a strong focus on applied biological sciences, especially in areas that intersect with botany, agriculture, and biotechnology. In addition to teaching, Prof. Zidane is interested in research on the use of medicinal and aromatic plants by the local communities of Morocco. In addition, he is also the owner of a project in Morocco focused on the inventory and preservation of plant species at the national level. This project aims to establish the cataloging and mapping of different plant species present and the identification of threatened or endangered species. Given his position and area of expertise, Prof. Zidane is involved in teaching, mentoring students, and contributing to the academic community at the university. His role as a professor at a higher education institution confirms his commitment to educating the next generation of scientists and researchers in plant biology.

Contributors

Abdullah Abdullah
Faculty of Biology
Ludwig-Maximilians-University of Munich (LMU)
Planegg-Martinsried, Germany

Mushtaq Ahmad
Department of Plant Sciences
Plant Systematics and Biodiversity Lab Quaid-i-Azam University
Islamabad Pakistan
Pakistan Academy of Sciences
Islamabad, Pakistan

Ahmad Ali
Department of Life Sciences
University of Mumbai, Mumbai
Maharashtra, India

Ehsan Ali
Department of Forestry and Range Management
Faculty of Agricultural Sciences & Technology
Bahauddin Zakariya University
Multan, Pakistan

Volkan Altay
Hatay Mustafa Kemal University
Faculty of Arts & Sciences
Department of Biology
Hatay, Turkey

Nimrah Ameen
Department of Botany
University of Faisalabad
Pakistan

Mohd Kafeel Ahmad Ansari
Department of Biology
Faculty of Natural Sciences
University of Guyana Georgetown
Guyana, South America

Wajiha Zaka Ansari
Department of Botany
University of Mianwali
Mianwali, Pakistan

Godwin Anywar
Department of Plant Sciences
Microbiology & Biotechnology
College of Natural Sciences
Makerere University
Kampala, Uganda

Yasir Arafat
State Key Laboratory of Desert and Oasis and Ecology
Xinjiang Institute of Ecology and Geography
Chinese Academy of Sciences
Urumqi, PR China

Everaldo Attard
Division of Rural Sciences and Food Systems
Institute of Earth Systems
University of Malta
Msida, Malta

Tamara Attard
Department of Pharmacy
Faculty of Medicine and Surgery
University of Malta
Msida, Malta

Abdelhamid Azeroual
Agri-Food and Health Laboratory (AFHL)
ENS of Berrechid, Hassan First University of Settat
Morocco

Muhammad Farooq Azhar
Department of Forestry and Range Management
Faculty of Agricultural Sciences & Technology
Bahauddin Zakariya University
Multan, Pakistan

Aziz Baidani
Agri-Food and Health Laboratory (AFHL)
FST of Settat, Hassan First University of Settat
Morocco

Viktória L. Balázs
Department of Pharmacognosy
Faculty of Pharmacy
University of Pécs
Hungary

Shehar Bano
Atta-ur-Rahman School of Applied Biosciences
National University of Sciences and Technology
Islamabad, Pakistan

Bouchaib Bencharki
Agri-Food and Health Laboratory (AFHL)
FST of Settat, Hassan First University of Settat
Morocco

Ouafae Benkhnigue
Department of Botany and Plant Ecology
Scientific Institute, Mohammed V University
Rabat, Morocco

Fathia Benlarbi
Laboratory for Exploration and Valorization of Steppe Ecosystems (EVES)
Department of Biology
Faculty of Natural Sciences and Life
University of Djelfa
Djelfa, Algeria

Noureddine Chaachouay
Agri-Food and Health Laboratory (AFHL)
ENS of Berrechid, Hassan First University of Settat
Morocco

İlker Çinbilgel
Department of Tourism Guidance
Manavgat Faculty of Tourism
Akdeniz University
Antalya, Turkey

Eszter Csikós
Department of Pharmacognosy
Faculty of Pharmacy
University of Pécs
Hungary

Hussaini Danlami
Department of Biological Sciences
Faculty of Life Sciences
Kaduna State University
Kaduna

Gopal Dixit
Department of Botany
Upadhi PG College (MJP Rohilkhand University)
Pilibhit, India

Abdulrahman Mahmoud Dogara
Biology Education Department
Faculty of Education
Tishk International University-Erbil
Kurdistan Region, Iraq

Allal Douira
Plant, Animal Productions and Agro-industry Laboratory
Department of Biology
Faculty of Sciences
Ibn Tofail University
Kenitra, Morocco

Abdelkarim El Djiri
Agri-Food and Health Laboratory (AFHL)
ENS of Berrechid, Hassan First University of Settat
Morocco

Omayma A. Eldahshan
Department of Pharmacognosy
Faculty of Pharmacy
Ain Shams University
Abbassia, Egypt

Ahmed M. Elissawy
Department of Pharmacognosy
Faculty of Pharmacy
Ain Shams University
Abbassia, Egypt

Kakudidi Esezah
Department of Plant Sciences, Microbiology & Biotechnology
College of Natural Sciences
Makerere University
Kampala, Uganda

Abdel Khalid Essamadi
Biochemistry, Neuroscience, Natural Resources, and Environment Laboratory (BNREL)
FST of Settat
Hassan First University of Settat
Settat, Morocco

Ágnes Farkas
Department of Pharmacognosy
Faculty of Pharmacy
University of Pécs
Pécs, Hungary

Anam Fatima
Department of Botany
University of Mianwali
Mianwali, Pakistan

Anywar Godwin
Department of Plant Sciences, Microbiology & Biotechnology
College of Natural Sciences
Makerere University
Kampala, Uganda

Sefa Gözcü
Department of Pharmacognosy
Faculty of Pharmacy
Erzincan Binali Yildirim University
Erzincan, Turkey

Alvina Gul
Department of Plant Biotechnology
Atta-Ur-Rehman School of Applied Biosciences (ASAB)
National University of Sciences and Technology (NUST)
Islamabad, Pakistan

Jamaa Habchaoui
Laboratory of Animal Plant Production and Agro-Industry
Department of Biology
Faculty of Sciences
Ibn Tofail University
Kenitra, Morocco

Bawale Sani Halliru
Department of Biological Sciences
Al-Qalam University
Katsina, Nigeria

Harmand A. Hama
Biology Education Department
Faculty of Education
Tishk International University-Erbil
Kurdistan Region, Iraq

Aneeqa Hanif
Atta-ur-Rahman School of Applied Biosciences
National University of Sciences and Technology
Islamabad, Pakistan

Daniyal Hussain
Department of Forestry and Range Management
Faculty of Agricultural Sciences & Technology
Bahauddin Zakariya University
Multan, Pakistan

Talha Imtiaz
Department of Forestry and Range Management
Faculty of Agricultural Sciences & Technology
Bahauddin Zakariya University
Multan, Pakistan

Muhammad Ishaq
Office of Research, Innovation & Commercialization
PMAS-Arid Agriculture University Rawalpindi
Rawalpindi, Pakistan

Bushra Jabeen
Directorate of Hemp Project
Pir Mehr Ali Shah Arid Agriculture University
Rawalpindi, Pakistan

Ssenku Jamilu
Department of Plant Sciences
Microbiology & Biotechnology, College of Natural Sciences
Makerere University
Kampala, Uganda

Frank Kalungi
Department of Plant Sciences
Microbiology and Biotechnology
Makerere University
Kampala, Uganda

Abderrahmaine Katiri
Laboratory of Biotechnologies and Valorization of Natural Resources
Ibn Zohr University, Faculty of Sciences
Agadir, Morocco

Mustafa Keskin
Institute of Pure and Applied Sciences
Biology Program, Göztepe
Marmara University
İstanbul, Turkey

Isah Labaran
Department of Sciences Laboratory Technology
School of Science
Federal polytechnic Mubi
Adamawa state, Nigeria

Ume Laila
Department of Botany
University of the Punjab
Lahore, Pakistan

Saba Latif
Department of Botany
Pir Mehr Ali Shah Arid Agriculture University
Rawalpindi, Pakistan

Aisha Abdullahi Mahmud
Department of Plant Science and Biotechnology
Federal University Dutsin-Ma
Katsina State, Nigeria

Salman Majeed
Department of Botany
University of Mianwali
Mianwali, Pakistan
Department of Plant Sciences
Plant Systematics and Biodiversity Lab Quaid-i-Azam University
Islamabad, Pakistan

Ruvimbo Jessy Mapaya
Department of Applied Biosciences and Biotechnology
Midlands State University
Senga, Gweru, Zimbabwe

Muhammad Maqsood
Department of Botany
PMAS-Arid Agriculture University Rawalpindi
Rawalpindi, Pakistan

Alfred Maroyi
Department of Botany
University of Fort Hare
Alice, South Africa

Zia ur Rehman Mashwani
Department of Biological Sciences
University of Veterinary & Animal Sciences
Lahore, Pakistan

Murad Muhammad
State Key Laboratory of Desert and Oasis and Ecology
Xinjiang Institute of Ecology and Geography
Chinese Academy of Sciences
Urumqi, PR China
University of Chinese Academy of Sciences

Beijing, PR China

Emmanuel Muhumuza
Department of Plant Sciences
Microbiology & Biotechnology
College of Natural Sciences
Makerere University
Kampala, Uganda

Mehmooda Munazir
Department of Botany
Government College Women University
Sialkot, Pakistan

Mubashrah Munir
Department of Biological Sciences
University of Veterinary & Animal Sciences
Lahore, Pakistan

Lilla Nagy-Radványi
Department of Pharmacognosy
Faculty of Pharmacy
University of Pécs
Pécs, Hungary

Samir M. Osman
Department of Pharmacognosy
Faculty of Pharmacy
October 6 University
6 October City, Egypt

Munir Ozturk
Department of Botany and Centre for Environmental Studies
Ege University
Izmir, Turkey

Tugume Patience
Department of Plant Sciences
Microbiology & Biotechnology
College of Natural Sciences
Makerere University
Kampala, Uganda

Priyatama Vijaysing Powar
Assistant Professor
Dr. D. Y. Patil College of Pharmacy
Akurdi, India

Rahmatullah Qureshi
Department of Botany
Pir Mehr Ali Shah Arid Agriculture University
Rawalpindi, Pakistan
Directorate of Hemp Project
Pir Mehr Ali Shah Arid Agriculture University
Rawalpindi, Pakistan

Khola Rafique
Pest Warning and Quality Control of Pesticides
Agriculture Department
Punjab, Pakistan

Sabith Rehman
Department of Botany
Hazara University Mansehra
Khyber Pakhtunkhwa, Pakistan

Sehrish Sadia
Department of Biological Sciences
University of Veterinary & Animal Sciences
Lahore, Pakistan

Ojelel Samuel
Department of Plant Sciences
Microbiology & Biotechnology
College of Natural Sciences
Makerere University
Kampala, Uganda

Syed Sabir Hussain Shah
Department of Botany
Pir Mehr Ali Shah Arid Agriculture University
Rawalpindi, Pakistan

Ayesha Siddiqa
Directorate of Hemp Project
Pir Mehr Ali Shah Arid Agriculture University
Rawalpindi, Pakistan

Iqra Siddique
Department of Plant Biotechnology
Atta-Ur-Rehman School of Applied Biosciences (ASAB)
National University of Sciences and Technology (NUST)
Islamabad, Pakistan

Abdel Nasser B. Singab
Department of Pharmacognosy
Faculty of Pharmacy
Ain Shams University
Abbassia, Egypt

Ahmet Zafer Tel
Department of Agricultural Biotechnology
Faculty of Agriculture
Iğdır University
Iğdır, Turkey

Khadija Temmante
Agri-Food and Health Laboratory (AFHL)
ENS of Berrechid, Hassan First University
Settat
Morocco

Abdul Wahab
Shanghai Center for Plant Stress Biology
CAS Center for Excellence in Molecular Plant Sciences
Chinese Academy of Sciences
Shanghai, China
University of Chinese Academy of Sciences
Beijing, PR China

Belkıs Muca Yiğit
Department of Organic Agriculture Management
Faculty of Applied Sciences
Iğdır University
Iğdır, Turkey

Muhammad Younas
Department of Botany
PMAS-Arid Agriculture University Rawalpindi
Rawalpindi, Pakistan

Mai M. Younis
Department of Pharmacognosy
Faculty of Pharmacy
Ain Shams University
Abbassia, Egypt

Muhammad Zafar
Department of Plant Sciences
Plant Systematics and Biodiversity Lab Quaid-i-Azam University
Islamabad, Pakistan

Nida Zahra
Department of Botany
University of Mianwali
Mianwali, Pakistan

Gokhan Zengin
Department of Biology
Science Faculty
Selcuk University
Konya, Turkey

Lahcen Zidane
Plant, Animal Productions and Agro-industry Laboratory
Department of Biology
Faculty of Sciences
Ibn Tofail University
Kenitra, Morocco

Preface

The strong urge that drives man to leave the humdrum of daily existence has led him to intuitively explore unexpected substances. It has done so even when nature has been most stingy in supply and the goods seem unable to hold the attributes allowing him to fulfill this need. In recent years, hallucinogens have been utilized by some who think they encourage spiritual growth, heighten perception, inspire personal development, and expand reality. Depending on their chemical nature, naturally occurring hallucinogens are employed in various methods. Most of these hallucinogens are alkaloids, a class of over 5,000 different biochemical molecules responsible for the biological activity of the majority of toxic and medicinal plants. As shown by the number of traditional treatments, they may be absorbed by the human body in various ways. These hallucinogens are only present in varying concentrations in the most advanced blooming plants in the plant world. So far approximately 120 hallucinogenic plants have been identified globally. Perhaps 3,000 plant species were routinely ingested by humans at some point in history, but just 150 taxa are still significant enough to join international trade today. Ancient man was familiar with several of these plants and, due to their properties, attributed magical or mystical significance to them, using them in religious ceremonies. The pharmacological effect of hallucinogens is attributable to a restricted number of chemical components. While modern science has been able to effectively reproduce these chemicals in most instances or even change their molecular structures to generate unique synthetic versions, practically all hallucinogens have their beginnings in plants. This book covers several subjects, such as botany, geographic distribution, habitat and ecology, traditional and hallucinogenic usage, phytochemicals, toxicity, clinical effects, and pharmacological uses of aromatic and medicinal plants. Considering that plants are sometimes geographically isolated, this text also examines the indigenous medicinal practices of many countries and civilizations. This book aims to provide comprehensive and up-to-date information on recent investigations and trends in medicinal plants and aromatics. It is specifically designed to meet the needs of students and researchers interested in medicinal plants, including hallucinogenic varieties. Additionally, it seeks to generate general interest in the study of medicinal plants by emphasizing the importance of understanding their cultural, countrywide, environmental, and scientific factors. It highlights that studying, evaluating, and utilizing these plants cannot be done effectively in isolation, but requires a global perspective.

Noureddine Chaachouay
Abdelhamid Azeroual
Lahcen Zidane

Acknowledgments

We express our profound gratitude to all the esteemed authors and co-authors who contributed chapters and generously dedicated their time and skills to this edited publication. We thank Randy Brehm, Tom Connelly, and all Taylor & Francis Group, LLC/CRC Press partners for their enthusiastic collaboration. In addition, we would like to express our gratitude to the reviewers who contributed to enhancing the chapters. We sincerely express our thanks and appreciation to all who have helped, either directly or indirectly, in producing this work. We encourage feedback and constructive criticism from subject-matter experts and readers.

1 Chambá (*Justicia pectoralis* Jacq. Acanthaceae)

Noureddine Chaachouay and Lahcen Zidane

1.1 INTRODUCTION

Employing plants for medicinal reasons, either in the management or prevention of illnesses, is one of the most ancient traditions in human medicine, and several traditional medical treatments have been documented using various plant species (Houghton, 1995). Recent technological advancements have successfully addressed multiple challenges associated with discovering new drugs derived from natural sources (Atanasov et al., 2021; Cragg & Newman, 2001). These include improvements in chemical pharmaceuticals and pharmacological studies and integrating metabolomics and genomics approaches to enhance traditional natural product study methods (Sumner et al., 2015). This has facilitated the development of secure, effective, and high-quality herbal medications, which excel in the global pharmaceutical industry.

Justicia pectoralis Jacq. 'fresh cut' is a tropical plant classified under the Acanthaceae family. It is indigenous to Central and South America and is present in several countries (Leal et al., 2017). The plant is renowned for its conventional therapeutic use and is sometimes planted for decorative reasons. *J. pectoralis* is often found in tropical and subtropical areas, flourishing in warm and humid conditions (Oliveira & Andrade, 2000; Benlarbi et al., 2023). It often thrives in unobstructed spaces beside highways, disrupted environments, and the lower layer of woods (Roersch, 2018; Chaachouay et al., 2022a, 2022b, 2022c). The plant has a preference for well-drained soils and possesses the ability to endure various soil types. *J. pectoralis*, commonly called chambá in Brazil, is extensively used in traditional northern and northeastern medicine to manage coughs, asthma, and bronchitis (Leal et al., 2017; Locklear et al., 2010; Benkhnigue et al., 2022).

Much research suggests that the aerial components of the plant are used to alleviate symptoms of insomnia, anxiousness, hypotension, and epilepsy (Chanfrau et al., 2013). Nevertheless, only a few authors have documented its antineoplastic properties. In contrast, most research on this plant has focused on its efficacy in treating respiratory tract diseases and its potential as an anti-inflammatory agent (Lino et al., 1997). These investigations include clinical trials using *J. pectoralis* syrups to treat individuals with asthma, specifically targeting inflammation. Phytochemical analysis of *J. pectoralis* identified many substances, such as coumarin, Umbelliferone (UMB), ortho-methoxylated glycosylflavones, and justicidin B (de Lima et al., 2022; Leal et al., 2017). The hydroalcoholic extract of 'chambá' underwent a preclinical toxicological study, which revealed that the plant has a low level of toxicity (Guimarães et al., 2023).

Furthermore, the plant extract and its chemical components, coumarin and UMB, have been shown to possess antioxidant, anti-inflammatory, antinociceptive, and muscle-relaxant properties. Recent studies have shown that they also have central effects in rats (Arteaga et al., 2011; Carneiro et al., 2023; de Oliveira et al., 2023; Chaachouay et al., 2019a, 2019b). Nevertheless, more comprehensive morpho-anatomical studies and phytochemical analyses on *J. pectoralis* are needed, which are imperative for ensuring the efficacy and safety of this plant as an herbal remedy. This chapter thoroughly evaluates the existing scientific literature on the beneficial impacts of *J. pectoralis* on health. This chapter covers traditional applications, taxonomy, habitat and geographical distribution, ecology, phytochemistry, pharmacological activities, toxicity, and clinical uses of *J. pectoralis.*

1.2 SCIENTIFIC CLASSIFICATION

Justicia pectoralis Jacq. is classified as a species within the genus *Justicia*, which comprises an estimated 1163 species and is a member of the Acanthaceae family. *Justicia pectoralis* Jacq. was first described by Nikolaus Joseph von Jacquin in 1760, and he published more information in 1763. 1958, Emery Clarence Leonard reported a distinct and identifiable variation called *Justicia pectoralis* var. *Stenophylla.* This species is only distributed in South America, while *Justicia pectoralis* Jacq. has a broader distribution range that includes Central America and the Caribbean.

Domain: Eukaryota
Kingdom: Plantae
Phylum: Spermatophyta
Subphylum: Angiospermae
Class: Dicotyledonae
Order: Scrophulariales
Family: Acanthaceae

DOI: 10.1201/9781003460336-1

Genus: *Justicia*
Species: *Justicia pectoralis* Jacq.

1.3 VERNACULAR NAMES

Arabic: Adalah Al-Sadr; **English**: Carpenter's, fresh cut, grassdeath-angel; **Spanish**: Carpintero, cariñito, hierba tila, tila, tilo, hierba de carpintero, té criollo; **French**: Herbe à charpentiers; **Portuguese**: Anamú, Takurushni; **German**: Justicia; **Italian**: Giustizia; **Dutch**: Justicia; **Russian**: Жустиция пекторалис (Zhustitsiya pektoralis); **Chinese**: 胸型爵床 (Xiōng xíng juéchuáng); **Hindi**: जस्टीशिया पेक्टोरालिस जैक (Jastiśiyā pekṭorālis jaiḳ); **Korean**: 저스티시아 펙토랄리스 자크 (Jeosṭisia Pektoraliseu Jak); **Greek**: Δικαιοσύνη πεκτοράλις (Dikaiosýni pektorális); **Swahili**: Haki ya kifua Jacq.; **Thai**: ยูสติเชีย เป็คทอราลิส จัค (Yūs ṭi cīyā Pĕk t^{h}xrā lịs Cạk); **Hebrew**: ג'וסטיסיה פקטורליס יאק (Jusṭisya Pekṭorlis Yak); **Filipino**: Katarungan ng Dibdib Jacq.; **Bengali**: জাস্টিসিয়া পেক্টোরালিস জ্যাক (Jāsṭisīyā pekṭōrālis jyāk); **Malay**: Keadilan pectoralis Jacq.; **Malayalam**: ജസ്റ്റിഷ്യ് പെക്ടൊറാലിസ് ജാക്ക് (Jasṟṟīṣya pekṭoṟālis jākkŭ); **Urdu**: جسٹسیا پییکٹورالس جیک (Jasṭsiyā pekṭorālis jaiḳ); **Persian (Farsi)**: عدالت پکتورالیس جاک (Adalat Pectoralis Jacq.)

1.4 ETYMOLOGY AND SYNONYMS

The genus *Justicia* is named in honor of Giacomo Justice, a Scottish farmer who authored the Scottish Gardener in 1754 and the English Gardener in 1767 (Carneiro et al., 2023). The particular epithet pectoralis is derived from the Latin word 'pectoralis', which means 'of the breast', originating from the Latin phrase pectus (genitive pectoris), meaning 'breast'. The etymology of pectus is still being determined. The terms are synonymous with *J. pectoralis*:

Dianthera pectoralis (Jacq.) J.F.Gmel.;
Dianthera pectoralis (Jacq.) Murray;
Ecbolium pectorale (Jacq.) Kuntze;
Justicia stuebelii Lindau;
Leptostachya pectoralis (Jacq.) Nees & Mart.;
Psacadocalymma pectorale (Jacq.) Bremek.;
Rhytiglossa pectoralis (Jacq.) Nees;
Stethoma pectoralis (Jacq.) Raf.

1.5 BOTANICAL DESCRIPTION

J. pectoralis is an herb that grows upright at an average height of 30–40 cm. The plant stems are weak and may either be erect or ascending (Figure 1.1). They frequently root at the lower nodes and have a sub-quadrangular shape with grooves. The stems are smooth or somewhat hairy, with the hairs pointing backward.

The leaves are narrow to broadly lanceolate, measuring 5–11 × 0.5–1.8 cm. They have an acuminate apex and an acute to obtuse base. The cystoliths are dense and prominent on the upper surface of the leaves but not visible on the lower surface. The upper surface of the leaves is glabrous to strigose along the costa, while the lower surface is glabrous. The leaf margins are entirely undulatory shaped (Roersch, 2018). The petioles are short, approximately 5 mm long, and have a strigose upper surface. The inflorescences are arranged in loose, spike-like clusters at the end of the stem. They can grow up to 25 cm long and 15 cm wide. The lower sections of the groups are about 2 cm long and covered in tiny glandular hairs (Guimarães et al., 2023). The bracts and bracteoles are slender, measuring 2 mm long and 1 mm wide at the base. They are covered in fine hairs and have a few glandular hairs (Oliveira & Andrade, 2000). The flowers have a calyx with five narrow and subulate segments, measuring up to 2 mm long and 0.5 mm wide at the base. The calyx is glandular and tuberculous. The corollas are white or purple, measuring 8–12 mm long. They are tuberculous on the outside and have a transversely plicate throat, often spotted with dark purple (de Oliveira et al., 2023). The corolla tube is narrowly funnel-shaped, 1.25 mm wide at the base and 2.5 mm wide at the mouth. The upper lip of the corolla is erect and triangular, measuring 3.5 mm long and 3 mm wide, with an acute apex. The lower lip is spreading, measuring up to 5 mm long, and has three lobes. The lobes are 1 mm long, with the middle lobe 2 mm wide and the lateral lobes 1.5 mm wide. All lobes have obtuse apices. The stamens extend to the tip of the upper lip, with slightly superposed anther cells that are apiculate at the base (Chanfrau et al., 2013). The filaments are glabrous. The capsules are slender and club-shaped, measuring up to 8 mm in length and 2 mm in width. They are covered with fine hairs. The seeds are round, flattened, and have a diameter of 1.5 mm (Figure 1.2).

1.6 ORIGIN AND GEOGRAPHIC DISTRIBUTION

Since its description in 1760 by NJ Jacquin based on specimens taken in the Caribbean islands of Dominica, it has been proposed that the natural distribution of the species *J. pectoralis* encompasses tropical America (such as Belize, Bolivia, Brazil, Costa Rica, Colombia, Ecuador, Guatemala, Honduras, Mexico, Nicaragua, Panama, Peru, Venezuela, and others) and the Caribbean islands. Nevertheless, several writers argue that its origin is exclusively confined to tropical America and subsequently introduced to the Caribbean islands (Roersch, 2018; UK, 2014). Nowadays, *J. pectoralis* is only found in tropical America, which is regarded as its indigenous distribution area, as well as in the Caribbean. Regarding the Caribbean, more than the available data is needed to determine if this species is indigenous or foreign. Nevertheless, if the species were transported into the Caribbean islands, its introduction would have had

FIGURE 1.1 Aerial parts of *J. pectoralis* (leaves and flowers).

FIGURE 1.2 Young seedlings of *J. pectoralis*.

to occur in the early stages of European colonization (Oliveira & Andrade, 2000). The likelihood of the opening of *J. pectoralis* is modest. This species, which tends to grow like a weed, reproduces via seeds and vegetative means by forming roots at its nodes. It is often cultivated in gardens and residential yards. Therefore, *J. pectoralis* can expand its distribution range well beyond its current extent (Leal et al., 2017).

1.7 HABITAT AND ECOLOGY

J. pectoralis has a wide distribution encompassing several countries, such as Brazil, Mexico, Venezuela, Colombia, Puerto Rico, and other Central America and the Caribbean areas. This herbaceous plant is very adaptable and may be found in a wide range of environments, including open spaces, road edges, woodland edges, and regions that have been disrupted (UK, 2014). Moreover, it thrives in grasslands and savannas, demonstrating its capacity to adapt to many ecological habitats. *J. pectoralis* is a common plant that thrives along highways and abandoned areas (Guimarães et al., 2023). It may be found at various altitudes ranging from sea level to 1400 m in tropical regions of America and the Caribbean. Inside Central and South America, this plant thrives in low-lying areas inside the lower layer of damp to saturated woods, as well as beside streams and on the banks of rivers (Cassola et al., 2019; de Lima et al., 2022). The leaves of this plant are organized in an opposing pattern and have a simple and elliptical shape. They have a noticeable light green color and provide a pleasant fragrance when crushed. The plant's reproductive cycle entails the generation of tubular, two-lipped flowers organized in spikes, a characteristic shared by plants in the Acanthaceae family. *J. pectoralis* is a plant collected in its original form for local use as a medicinal substance (Kujawska et al., 2023). This plant is often seen growing beside highways and in areas not planted, typically at heights around sea level. This fragrant plant may be reproduced by seeds or by splitting rooted stems. It thrives best in well-drained and moist soils. Pollination, which insects like bees and butterflies most likely aid, plays a crucial role in the plant's seed production (Aoyama & Indriunas, 2015; Arteaga et al., 2019). These seeds are distributed through many mechanisms, including wind and animals. Its remarkable adaptation to various climatic circumstances and soil types is shown in its capacity to flourish in disturbed regions and beside road networks, highlighting its endurance in the face of human-altered environments (Vieira et al., 2019; Chaachouay et al., 2020a, 2020b).

1.8 ETHNOBOTANY

J. pectoralis is a medicinal herb that has been traditionally used for a significant period in South and Central America. *J. pectoralis* is often referred to as tilo in Latin America and Cuba. The term used for this in Haiti is chapantye, whereas, in Dominica and Martinique, it is referred to as Zeb chapantyè. Additional well-known aliases include freshcut, chambá, carpintero ('carpenter'), té criollo ('Criollo tea'), curia, the angel of death, masha-hari, or 'piri piri'. The plant has leaves and stems that contain coumarin, an anticoagulant compound that functions as a glucose metabolism regulator due to luteolin (Locklear et al., 2010). It is used in traditional medicine in South America and added to *Virola* cigarettes, mainly because of its fragrance, which is like that of vanilla. Cubans in South Florida cultivate this plant, initially used by the indigenous populations of Eastern Colombia and the Amazon basin. Waika communities from different groups desiccate and grind the leaves, combining them with the powdered 'resin' derived from *Virola thetodora* to produce a hallucinogenic snuff (Leite & de Oliveira, 2017; Chaachouay et al., 2021a, 2021b, 2021c; Orch et al., 2021). The Peruvian Amazon extracts comprise betaine, coumarin, and Umbelliferone, which were grown and harvested from the region (Leal et al., 2017). However, in animal trials, these extracts did not exhibit hallucinogenic properties, nor did they generate synergistic effects similar to those caused by tryptamines. Herb vendors in Cuba and Panama offer bundles of this botanical specimen. A decoction in Panama is used to alleviate stomach sickness and leg pain (de Oliveira et al., 2023). In Mexican folk medicine, the plant known locally as turbo is administered to treat elevated body temperatures. Barbadians and Puerto Ricans use sweet herbs as a remedy to promote the expulsion of mucus from the respiratory system (Fonseca et al., 2010). In Guadeloupe and Martinique, a decoction is used as a digestive aid, while the infusion of the plant combined with frying oil serves as a treatment for chest afflictions. Occasionally, dampened foliage is used as an exterior dressing for wounds. In Brazil, the leaves of *J. pectoralis* are often used in homemade remedies to treat respiratory tract ailments, including cough, bronchitis, and asthma (Carvalho et al., 2021; Guimarães et al., 2023). The syrup is extensively manufactured and supplied under the public phytotherapy program 'Farmácias Vivas'. In Trinidad, a decoction treats coughs, chest colds, influenza, fever, and vomiting. A traditional Jamaican cure for infantile colic and TB combines *J. pectoralis* and Cuscuta Americana with orange and sugar.

1.9 HALLUCINOGENIC EFFECTS

Hallucinogens are widely utilized in cultures that ascribe positive connotations to the altered state of consciousness they induce, such as providing access to the spiritual realm (De Rios et al., 2002; Dupuis, 2022). The widespread use of *J. pectoralis*-based hallucinogenic snuffs in the Amazonian areas has been documented during the past century. The bark resin of this plant was reported as one of the sources of these snuffs, and their uses are now well-established (Schultes, 1969, Chaachouay & Zidane, 2021). The primary benefit of *J. pectoralis* leaves is that they are used as an ingredient in psychedelic snuffs made from the dried resin of DMT-containing plants from the *Anadenanthera* and *Virola* genera. In certain regions along the Caribbean coast of Central America, it is regarded as possessing aphrodisiac qualities (Chanfrau et al., 2013). Before use, the whole plant is desiccated under sunlight and then pulverized into a fine powder. Significantly, the powder demonstrates efficacy only when administered via the nasal membranes, such as snuff, or when smoked in combination with *Cannabis sativa*. *Justicia pectoralis* var. *stenophylla*, a kind of plant found in South America, is used for its ability to induce psychedelic experiences or as an ingredient in ayahuasca (*Banisteriopsis caapi*). The plant contains tryptamines that have hallucinogenic properties. Tilo tea, known for its anti-inflammatory and relaxing properties, is often available in supermarkets (Brewer-Carias & Steyermark, 1976; Lima et al., 2020).

The desiccated leaves possess a fragrant perfume. Several indigenous groups in the Amazon area use these methods. The Waika or Yanomamö tribe uses *J. pectoralis* leaves and *Virola* resin to produce a kind of snuff known as machohara. According to popular belief, alone, either of the two substances may be inhaled to produce slight hallucinations. However, when combined, the effects are enhanced, increasing the potency (Leal et al., 2017). It is possible that *J. pectoralis* was used as a powdered tobacco product in ancient Mexico. According to Shipibo, the plant stimulates the work ethic of individuals and enhances their chances of success in fishing (Roersch, 2018). To get these advantages, an individual must consume a concoction made from the leaves.

1.10 PHYTOCHEMISTRY

Initial phytochemical investigations conducted on the *J. pectoralis* extract using general identification procedures or thin-layer chromatography demonstrated the existence of coumarins, flavonoids, steroids, triterpenoids, and alkaloids (Guimarães et al., 2023; Vieira et al., 2019). Research reports the presence of alkaloids in *J. pectoralis* at low levels, which may explain the plant's hallucinogenic properties and its use for this purpose by indigenous groups in South America. Nevertheless, other writers deny the existence of alkaloids and argue that using these substances in ritual preparations is only for enhancing the taste, attributed to the delightful fragrance of coumarins (de Sousa Nunes et al., 2018). The aerial component of *J. pectoralis* was analyzed using chromatographic and spectrometric techniques, resulting in the identification and characterization of many compounds. These compounds include coumarin derivatives such as 1,2-benzopyrone-coumarin, dihydrocoumarin, 7-hydroxydocumarine, and Umbelliferone (Carvalho et al., 2021; Fonseca et al., 2010; Roersch, 2018).

Additionally, it has been documented that ortho-hydroxytranscinnamic acetylated acid, ortho-hydroxy-dihydro cinnamic acetylated acid, betaine, and justice in B exist, as well as o-methoxylated C-glycosyl flavones including swertisin, 2-O-rhamnosyl-eswertisin, eswertiajaponin, and 2-O-ramnosileswertiajaponin. Also, derivatives of phenylpropionic acid were detected and characterized. According to some publications, *J. pectoralis* is believed to contain essential oils (Leal et al., 2017). Essential oils are volatile substances that have a role in the distinct scent of plants. Essential oils typically consist of a combination of terpenes and terpenoids, but their specific composition may vary. Another research indicates the existence of alkaloids in small amounts in *J. pectoralis*, which likely explains the plant's hallucinogenic properties and its use for this purpose by indigenous tribes in South America (Arteaga et al., 2019; de Oliveira et al., 2023). Nevertheless, other writers deny the existence of alkaloids and argue that using these substances in ritual preparations is only for enhancing the taste, attributed to the delightful fragrance of coumarins.

TABLE 1.1
Phytochemicals Present in Different Parts of *J. pectoralis*

Chemical Compounds	Sub Class	References
Coumarins	1,2-benzopyrone-coumarin, dihydro coumarin, 7-hydroxydocumarine, Umbelliferone, phenyl propionic acid, ortho-hydroxytranscinnamic acetylated acid, ortho-hydroxy-dihydro cinnamic acetylated acid, betaine, justicidin B, o-methoxylated C-glycosyl flavones, 2-O-rhamnosyl-eswertisin, eswertiajaponin, 2-O-ramnosileswertiajaponin	Carneiro et al. (2023), de Lima et al. (2022), Leal et al. (2017), Lima et al. (2020), de Oliveira et al. (2023)
Flavonoids	O-methylated, 2′-Orhamnosylswertisin, 2′-O-rhamnosylswertiajaponin, Swertisin, swertiajaponin, C-glycosyl flavones, justicidin B	Guimarães et al. (2023), Vieira et al. (2019)
Other Compounds	2-methyl-2-pentanol, 3-hexanol, 3-methyl-3-pentanol, 3-hexanone, 2-hexanone, 2-hexanol, Cis (Z)-2-hexenal, Trans (E)-2-hexenal, Cis (Z)-3-hexenol, Cis (Z)-2-hexenol, Cyclohexanol, Heptanal, Benzaldehíde, 1-octen-3-ol, 1-octen-3-ona, 3-octanol, Octanal, Nonanal, Phytol, Undecanal, Dodecanal, Ethyl decanoate, (E)-2-Decenal, Safranal, 2-Decanone, Phenylacetaldehyde, Limonene, 2-Furfural, Benzaldehyde,	de Sousa Nunes et al. (2018), Locklear (2008), Vieira et al. (2019)

1.11 PHARMACOLOGY

1.11.1 Anti-Inflammatory Effects

In the formalin model of nociception in mice, the antinociceptive effects of the hydroalcoholic extract derived from *J. pectoralis* leaves and its isolated compounds were investigated. It was observed that the opioid system does not play a role in these effects, as the administration of naloxone, an opioid antagonist, did not reverse the antinociceptive effects (Lino et al., 1997). Nevertheless, the administration of L-arginine before therapy counteracted the impact of UMB, indicating the involvement of the nitric oxide pathway in this phenomenon (Locklear et al., 2010). Furthermore, hydroalcoholic extract exhibited bronchodilation effects in the trachea of guinea pigs, supporting the traditional usage of *J. pectoralis* in treating respiratory conditions, including bronchitis and asthma (Moura et al., 2017). A study assessed the biological impact of hydroalcoholic extracts from coumarin plants, commonly used in Brazil's Northeast region to treat respiratory disorders (de Sousa Nunes et al., 2018). The researchers observed that when administered to rats at 200 and 400 mg/kg and 50 mg/kg (i.p.), the extracts exhibited antinociceptive and antidematogenic effects. Research has shown that the standardized dry extract of *J. pectoralis*, with concentrations of 8.84 mg/g (CM) and 1.27 mg/g (UMB), decreased paw edema generated by carrageenan by up to 38% and capsaicin-induced nociception in mice by 79.6% at doses of 100, 200, and 400 mg/kg (Lino et al., 1997).

1.11.2 Antibacterial Effect

An investigation assessed the antibacterial efficacy of medicinal herbs commonly used in Trinidad and other Caribbean islands. The study discovered that the petroleum ether fraction derived from the aerial parts of *J. pectoralis* var. *stenophylla* exhibited significant antibacterial effects against *E. coli*, *P. aeruginosa*, *S. epidermidis*, *E. faecalis*, and *S. typhimurium* (Guimarães et al., 2023; de Oliveira et al., 2023). Furthermore, the ethyl acetate fraction inhibited the growth of these microorganisms and demonstrated inhibitory effects on *S. aureus*. The same investigation showed that both fractions, when present at a concentration of 0.5 mg/ml, have a toxic impact on stage IV larvae of *A. aegypti*. This resulted in the death of the larvae after 2 and 9 days, respectively. In a separate investigation, an unstandardized aqueous extract of *J. pectoralis* was shown to have no antibacterial effects at any of the tested doses (25, 50, and 100 mg/ml) against *E. coli* and *K. pneumonia* (Arteaga et al., 2011; de Sousa Nunes et al., 2018).

1.11.3 Anxiolytic and Anti-Asthmatic Effects

Studies have shown that *J. pectoralis* has anti-inflammatory, anxiolytic, and anti-asthmatic actions and hormonal and gastroprotective properties. The methanolic extracts derived from the aboveground sections of the plant caused a reduction in the binding strength of estradiol and progesterone to their receptors in human breast cancer cells when tested in a laboratory setting (Leal et al., 2017; Moura et al., 2017). Additionally, these extracts increased the activity of genes sensitive to estrogen in human osteoblastic osteosarcoma cells (de Sousa Nunes et al., 2018). Furthermore, they were shown to suppress the activity of recombinant human COX-2. These effects explain the conventional use of the plant for alleviating symptoms associated with premenstrual syndrome and menopause, as well as its anti-inflammatory properties. Recent research found that administering the standardized dry extract effectively reduced the development of stomach ulcers caused by ethanol (Carvalho et al., 2021).

1.11.4 Anti-Asthmatic Properties

A research group recently assessed the anti-asthmatic properties of the standardized hydroalcoholic extract of *J. pectoralis* in rats exposed to ovalbumin (OVA), an experimental model that mimics clinical asthma (Leal et al., 2017). The section contains bioactive markers, specifically coumarin (1.5 mg/ml) and Umbelliferone (0.17 mg/ml). When administered orally, the standardized extract decreased the hyperresponsiveness in trachea preparations that were challenged with OVA and stimulated with either potassium chloride) or acetylcholine. The observed effects on the airways of rats may be attributed to the anti-inflammatory properties of the substance, as shown by its considerable inhibition of the elevation of TNF-α and IL-1β levels, which are pro-inflammatory cytokines, in the bronchoalveolar lavage of rats challenged with OVA (Moura et al., 2017). Furthermore, the *J. pectoralis* extract seems to modulate the gene expression of TRPC (canonical transient receptor potential) proteins in the lung tissue of animals challenged with OVA. The TRPC channels are non-selective and allow the passage of Ca^{2+}. They regulate many processes, including smooth muscle activity, endothelial cell function, and cell cycle regulation (de Sousa Nunes et al., 2018).

1.11.5 Effects on Central Nervous System

The aqueous extract derived from the leaves of *J. pectoralis* was shown to decrease aggressive behavior and exploratory activity in rats when its impact on the central nervous system was assessed (Venâncio et al., 2011). It inhibited the stimulation caused by phencyclidine, an NMDA receptor antagonist. Nevertheless, the plant was ineffective in stopping seizures induced by GABAergic antagonists (picrotoxin and pentilenotetrazole), unlike diazepam, which exhibited neuroprotection (Chanfrau et al., 2013). This indicates that the plant's anxiolytic effects do not operate via a benzodiazepine-mediated mechanism. In addition, it was shown that the intraperitoneal injection of the plant's aqueous extract could not counteract the effects of apomorphine, a dopaminergic agonist (Venâncioa et al., 2019). This indicates a lack of antidopaminergic action comparable to standard neuroleptics. A study group recently studied the impact of coumarin, a critical active compound found in *J. pectoralis*, on the central nervous system. Research provided evidence of the sedative, anxiogenic, and depressant-like properties of coumarin.

1.11.6 Antioxidant Properties

An investigation demonstrated the antioxidant properties of *J. pectoralis* dry extract (at concentrations ranging from 0.0138 to 2.97 mg/ml) in preventing lipid peroxidation in rat brain homogenates (Arteaga et al., 2011; Cassola et al., 2019; Chanfrau et al., 2013; Guimarães et al., 2023). In a streptozotocin-induced model of diabetes, rats exhibited restored lipid peroxidation levels after the oral administration of Umbelliferone at a dosage of 30 mg/kg. This indicates a potential impact on the β cells of the pancreas (de Oliveira et al., 2023).

1.11.7 Anticonvulsant Properties

An experiment showcased the anticonvulsant properties of the SDEJP in mice, revealing its efficacy in countering seizures caused by picrotoxin, strychnine, electroshock, or pilocarpine (Carneiro et al., 2023). In seizures caused in the pilocarpine model, the levels of inhibitory amino acids (GABA, glycine, and taurine) were elevated. In contrast, the levels of excitatory amino acids (glutamate and aspartate) were reduced (Leal et al., 2017). Neuroprotection was detected in this research, as shown by decreased neuronal damage in the CA1 and CA3 regions of the hippocampus (Locklear et al., 2010).

1.11.8 Smooth Muscle Effects

An investigation conducted in the laboratory assessed the calming impact of a hydroalcoholic extract derived from the leaves of *J. pectoralis*, as well as coumarin and other coumarin plants (*Pterodon polygaliflorus*, *Amburana cearenses*, *Eclipta alba*, and *Hybanthus ipacacuanha*), on the guinea-pig trachea that had been contracted with carbachol (Leal et al., 2017; Moura et al., 2017). The hydroalcoholic extract elicited a relaxing effect that varied in intensity depending on the dose, with an EC50 value of 1.5 ± 0.18 mg/ml. This EC50 value is twice as high as that of CM (0.08 ± 0.01 mg/ml), which is one of the active compounds found in the plant (Locklear, 2008).

1.12 CLINICAL EFFECTS

A pilot clinical trial investigated the effects of administering *J. pectoralis* and *Plectranthus amboinicus* syrup to patients with mild or moderate asthma. The trial included 37 participants given a 5 ml syrup dose thrice daily for two consecutive weeks. The results showed a reduction in airway obstruction, as assessed by respiratory parameters, including forced expiratory volume in the first second (FEV1), forced vital capacity (FVC), and maximum expiratory flow volume (MEFV) (Locklear et al., 2010). Furthermore, there was an evident improvement in sputum quality and a lack of any adverse reactions. A clinical trial involving 21 asthmatic patients aged 6–12 years showed that administering *J. pectoralis* syrup for two weeks improved symptoms and reduced obstruction in patients with intermittent, mild, or persistent moderate asthma. This trial revealed that the herbal medicine has a bronchodilator effect (Carvalho et al., 2021; Venâncio et al., 2011).

In a double-masked placebo-controlled clinical trial, researchers conducted a study to evaluate the therapeutic effects of a standardized syrup containing leaves of *J. pectoralis*, *Plectranthus amboinicus*, and *Mentha arvenis*. The syrup, prepared by the Pharmacy School of the Federal University of Ceará, was administered to 35 patients with mild asthma as a complementary therapy. The patients were instructed to take 20 ml of syrup thrice daily for 14 consecutive days. However, the study found that this treatment did not change lung test parameters, including FVC, FEV1, FEV1/FVC ratio, and forced expiratory flow (FEF) between 25 and 75% of FVC (Venâncioa et al., 2019). Nevertheless, there was a notable improvement in the quality of life, as assessed by the outcomes derived from the standardized Asthma Quality of Life Questionnaire – AQLQ(S), without inducing any toxic or detrimental consequences. The disparities in the therapeutic effectiveness of *J. pectoralis* syrup are likely due to many factors, such as variances in the syrup's composition (whether it contains just *J. pectoralis* or is combined with other species) and maybe inadequate analytical characterization of the herbal medication (Nascimento et al., 2021).

1.13 TOXICITY

Preclinical assessment of the toxicity of natural products is crucial to substantiate their safety for subsequent clinical investigations. An investigation found that the lethal dosage (LD50) of the non-standardized hydroalcoholic extract derived from the leaves of *J. pectoralis* in rats was 3.0 ± 0.2 g/kg. Nevertheless, the animals did not experience any fatalities when the extract was delivered orally in a single dosage (de Oliveira et al., 2023). Administering a plant extract to rats at a dosage of 400 mg/kg daily for 30 days resulted in changes in their blood composition and biochemistry. These changes included an elevation in hematocrit and alkaline phosphatase levels and a decrease in hemoglobin concentration. This investigation demonstrated that the oral administration of a non-standard hydroalcoholic extract of *J. pectoralis* at a single dosage of 2 g/kg did not result in any deaths or anatomopathological changes (Lagarto et al., 2009). The extract is categorized as non-toxic according to the Global Harmonized System of Classification and Labeling of Chemicals. Another investigation assessed the harmfulness of the unconventional hydroalcoholic extract from *J. pectoralis* on *Artemia salina* L. It established its fatal concentration as 60.14 μg/ml (Carvalho et al., 2021). An experiment showed that the standardized dry extract (ESJP: 10–100 μg/ml) did not have a significant impact on the survival of human neutrophils, as measured by the activity of lactate dehydrogenase (LDH) and the MTT (3-(4,5-dimethylthiazol-2-yl)-2,5-diphenyltetrazolium bromide) test. Furthermore, the unconventional water-based extract of *J. pectoralis* did not harm the genetic material of rodent bone marrow cells (de Sousa Nunes et al., 2018; Guimarães et al., 2023; Roersch, 2018).

1.14 CONCLUSION

J. pectoralis is highly regarded in Central and South American traditional medicinal practices. Indigenous societies have used the herb for several therapeutic applications, including treating inflammatory illnesses, respiratory disorders, and fever. The substance is recognized for its diverse bioactive components, including alkaloids, coumarins, flavonoids, and terpenoids. The plant's chemical components contribute to its pharmacological effects and may be accountable for its traditional use in the treatment of many health disorders. *J. pectoralis* has pharmacological actions that include anti-inflammatory, antibacterial, and antispasmodic effects. Investigation into the pharmacology of the plant provides a scientific foundation for its customary use and can facilitate the creation of novel medicinal substances. Although *J. pectoralis* displays pharmacological properties, it is crucial to consider the possibility of toxicity. Research on the plant's toxicity aids in establishing criteria for safe consumption, assuring that its medical applications do not present any hazards to human health. However, it is crucial to use contemporary pharmaceutical technology and analytical techniques to research *J. pectoralis* further and guarantee its quality. Furthermore, it is essential to establish a robust partnership between preclinical and clinical investigations to advance the development of an herbal remedy derived from the aerial components of *J. pectoralis*. This extra study on *J. pectoralis* will have a noteworthy socioeconomic effect, permitting the creation of a novel treatment for this species.

LIST OF ABBREVIATIONS

DMT	Dimethyltryptamine
FEV1	Volume in the First Second
FVC	Forced Vital Capacity
GABA	Gamma-Aminobutyric Acid
IL	Interleukin
J. pectoralis	*Justicia pectoralis* Jacq.
MEFV	Maximum Expiratory Flow Volume
NMDA	N- methyl-D-aspartate
OVA	Ovalbumin
TB	Tuberculosis
TNF	Tumor Necrosis Factor
TRPC	Canonical Transient Receptor Potential
UMB	Umbelliferone

REFERENCES

Aoyama, E. M., & Indriunas, A. (2015). Use of anatomy for the identification of vegetal drugs from two species of Justicia (Acanthaceae) marketed as 'anador'. *Scientia Plena, 11*(7), 7–7.

Arteaga, J. J. M., Perea, E. M., & Varón, E. Y. (2011). Influencia del clima en la composición química y la actividad antioxidante de Justicia pectorales Jacq. *Revista Cubana de Farmacia, 45*(1), 88–100.

Arteaga, S., Yabor, L., Torres, J., Solbes, E., Muñoz, E., Díez, M. J., Vicente, O., & Boscaiu, M. (2019). Morphological and agronomic characterization of Spanish landraces of *Phaseolus vulgaris* L. *Agriculture, 9*(7), 149.

Atanasov, A. G., Zotchev, S. B., Dirsch, V. M., & Supuran, C. T. (2021). Natural products in drug discovery: Advances and opportunities. *Nature Reviews Drug Discovery, 20*(3), 200–216.

Benlarbi, F., Mimoune, N., Chaachouay, N., Souttou, K., Saidi, R., Mokhtar, M.R., Kaidi, R., & Benaissa, M. T. (2023). Ethnobotanical survey of the traditional antiparasitic use of medicinal plants in humans and animals in Laghouat (Southern Algeria). *Veterinary World, 16*(2), 357–368. https://doi.org/10.14202/vetworld.2023.357-368

Benkhnigue, O., Chaachouay, N., Khamar, H., El Azzouzi, F., Douira, A, & Zidane, L. (2022). Ethnobotanical and ethnopharmacological study of medicinal plants used in the treatment of anemia in the region of Haouz-Rehamna (Morocco). *Journal of Pharmacy & Pharmacognosy Research, 10*(2), 279–302. https://doi.org/10.56499/jppres21.1196_10.2.279

Brewer-Carias, C., & Steyermark, J. A. (1976). Hallucinogenic snuff drugs of the Yanomamo Caburiwe-Teri in the Cauaburi river, Brazil. *Economic Botany, 30*, 57–66. https://doi.org/10.1007/BF02866785

Carneiro, M. R. B., Sallum, L. O., Martins, J. L. R., Peixoto, J. de C., Napolitano, H. B., & Rosseto, L. P. (2023). Overview of the *Justicia* genus: Insights into its chemical diversity and biological potential. *Molecules, 28*(3), 1190.

Chaachouay, N., Benkhnigue, O., Fadli, M., Ibaoui, H., & Zidane, L. (2019a). Ethnobotanical and ethnopharmacological studies of medicinal and aromatic plants used in the treatment of metabolic diseases in the Moroccan Rif. *Heliyon, 5*(10), e02191. https://doi.org/10.1080/10477845.2021.1942433

Chaachouay, N., Benkhnigue, O., Fadli, M., Ibaoui, H., El Ayadi, R., & Zidane, L. (2019b). Ethnobotanical and ethnopharmacological study of medicinal and aromatic plants used in the treatment of respiratory system disorders in the Moroccan Rif. *Ethnobotany Research and Applications, 18*(22), 1–16.

Chaachouay, N., Douira, A., Hassikou, R., Brhadda, N., Dahmani, J., Belahbib, N., Ziri, R., & Zidane, L. (2020a). Ethnobotanical and ethnopharmacological study of medicinal and aromatic plants used in the treatment of respiratory system disorders in the Moroccan Rif. PhDThesis, Département de Biologie - Université Ibn Tofail - Kénitra, Maroc, https://tel.archives-ouvertes.fr/tel-03376377

Chaachouay, N., Benkhnigue, O., & Zidane, L. (2020b). Ethnobotanical study aimed at investigating the use of medicinal plants to treat nervous system diseases in the Rif of Morocco. *Journal of chiropractic medicine, 19*(1), 70–81. https://doi.org/10.1016/j.jcm.2020.02.004

Chaachouay, N., & Zidane, L. (2021). The Symbolic Efficacy of Plants in Rituals and Socio-Religious Ceremonies in Morocco, Northwest of Africa. *Journal of Religious & Theological Information, 21*(1–2), 34–53. https://doi.org/10.1080/10477845.2021.1942433

Chaachouay, N., Benkhnigue, O., Douira, A., & Zidane, L. (2021a). Poisonous medicinal plants used in the popular pharmacopoeia of the Rif, northern Morocco. *Toxicon*, 189, 24–32. https://doi.org/10.1016/j.toxicon.2020.10.028

Chaachouay, N., Douira, A., & Zidane, L. (2021b). COVID-19, prevention and treatment with herbal medicine in the herbal markets of Salé Prefecture, North-Western Morocco. *European Journal of Integrative Medicine*, 42, 101285. https://doi.org/10.1016/j.eujim.2021.101285

Chaachouay, N., Douira, A., & Zidane, L. (2021c). Herbal Medicine Used in the Treatment of Human Diseases in the Rif, Northern Morocco. *Arabian Journal for Science and Engineering*, 47, 131–153. https://doi.org/10.1007/s13369-021-05501-1

Chaachouay, N., Benkhnigue, O., & Zidane, L. (2022a). Ethnobotanical and Ethnomedicinal study of medicinal and aromatic plants used against dermatological diseases by the people of Rif, Morocco. *Journal of Herbal Medicine*, 32, 100542. https://doi.org/10.1016/j.hermed.2022.100542

Chaachouay, N., Azeroual, A., Bencharki, B., Douira, A., & Zidane, L. (2022b). Ethnoveterinary medicines plants for animal therapy in the Rif, North of Morocco. *South African Journal of Botany*, *147*, 176–191. https://doi.org/10.1016/j.sajb.2021.12.037

Chaachouay, N., Azeroual, A., Douira, A., & Zidane, L. (2022c). Ethnoveterinary practices of medicinal plants among the Zemmour and Zayane tribes, Middle Atlas, Morocco. *South African Journal of Botany*, *151*, 826–840. https://doi.org/10.1016/j.sajb.2022.11.009

Carvalho, M., Caixeta, G. A. B., Lima, A. R. S., Teófilo, M. N. G., de Melo Cruvinel, W., Gomes, C. M., Fleury, L. F. F., de Paula, J. A. M., & Amaral, V. C. S. (2021). Assessing the safety of using the dry extract of *Justicia pectoralis* Jacq. (Acanthaceae) during pregnancy of Wistar rats. *Journal of Ethnopharmacology*, *268*, 113618.

Cassola, F., da Silva, M. H. R., Borghi, A. A., Lusa, M. G., Sawaya, A. C. H. F., Garcia, V. L., & Mayer, J. L. S. (2019). Morphoanatomical characteristics, chemical profiles, and antioxidant activity of three species of *Justicia L.* (Acanthaceae) under different growth conditions. *Industrial Crops and Products*, *131*, 257–265.

Chanfrau, J. E. R., Hernández, O. D. L., Figueredo, Y. N., Ferradá, C. R., & Mendoza, A. N. (2013). Obtención de extractos secos a partir de extractos acuosos de *Justicia pectoralis* (tilo). *Revista Cubana de Plantas Medicinales*, *18*(4), 543–554.

Cragg, G. M., & Newman, D. J. (2001). Natural product drug discovery in the next millennium. *Pharmaceutical Biology*, *39*(supl), 8–17.

de Lima, P. Z., de Freitas Morel, L. J., Pinto, G. H. T., Contini, S. H. T., Crevelin, E. J., Do Nascimento Júnior, J. E., De Castro França, S., Bertoni, B. W., & Pereira, A. M. S. (2022). Variations in biomass and coumarin content of *Justicia pectoralis* Jacq.: Influence of season, harvest frequency and shade level. *Biochemical Systematics and Ecology*, *100*, 104374.

Oliveira, L. M. N. de, Bastos, M. Do S. R., Lima, A. C. S. de, Figueiredo, E. A. T. de, Dias, F. G. B., Silva, L. M. R. da, Benevides, S. D., Ribeiro, P. R. V., Sabino, L. B. de S., & Figueiredo, R. W. de. (2023). Chemical characterization, cytotoxicity, antimicrobial and antioxidant potential of *Justicia pectoralis* Jacq and *Croton jacobinensis* Baill extracts. *Brazilian Journal of Pharmaceutical Sciences*, *59*, e21224.

De Rios, M. D., Grob, C. S., & Baker, J. R. (2002). Hallucinogens and redemption. *Journal of Psychoactive Drugs*, *34*(3), 239–248.

de Sousa Nunes, T. R., Cordeiro, M. F., Beserra, F. G., De Souza, M. L., Da Silva, W. A. V., Ferreira, M. R. A., Soares, L. A. L., Costa-Junior, S. D., Cavalcanti, I. M. F., & da Rocha Pitta, M. G. (2018). Organic extract of *Justicia pectoralis* Jacq. Leaf inhibits interferon-γ secretion and has bacteriostatic activity against *Acinetobacter baumannii* and *Klebsiella pneumoniae*. *Evidence-based Complementary and Alternative Medicine: eCAM*, 2018. https://www.ncbi.nlm.nih.gov/pmc/articles/PMC6126107

Dupuis, D. (2022). The socialization of hallucinations: Cultural priors, social interactions, and contextual factors in the use of psychedelics. *Transcultural Psychiatry*, *59*(5), 625–637. https://doi.org/10.1177/13634615211036388

Fonseca, F. N., Silva, A. H., & Leal, L. K. (2010). *Justicia pectoralis* Jacq:, Acanthaceae: Preparation and characterisation of the plant drug including chromatographic analysis by HPLC-PDA. *Revista Brasileira de Farmacognosia*, *20*, 871–877.

Guimarães, T. L. F., da Silva, L. M. R., Muniz, C. R., Vieira, Í. G. P., de Oliveira, L. S., Dias, F. G. B., de Brito Lima, C., da Costa Gonzaga, M. L., Lima, J. da S. S., & de Sousa Sabino, L. B. (2023). Potential of chambá (*Justicia pectoralis* Jacq.) leaves extracts as a source of bioactive compounds and natural antimicrobial agent. *Food Chemistry Advances*, *3*, 100367.

Houghton, P. J. (1995). The role of plants in traditional medicine and current therapy. The *Journal of Alternative and Complementary Medicine*, *1*(2), 131–143.

Kujawska, M., Zamudio, F., Castillo, J. A., & Sosnowska, J. (2023). The relation between Ashaninka Amazonian Society and cultivated acanthaceae plants. *Economic Botany*, *77*(4), 372–409. https://doi.org/10.1007/s12231-023-09585-8

Lagarto, A., Bueno, V., Guerra, I., Valdés, O., Gabilondo, T., & Rodríguez, J. (2009). Acute and subchronic oral toxicities of *Justicia pectoralis* J. Extract in Wistar rats. *The Open Natural Products Journal*, *2*(1). https://benthamopen.com/ABSTRACT/TONPJ-2-53

Leal, L. K. A. M., Silva, A. H., & Viana, G. S. de B. (2017). *Justicia pectoralis*, a coumarin medicinal plant have potential for the development of antiasthmatic drugs? *Revista Brasileira de Farmacognosia*, *27*, 794–802.

Leite, M. N., & de Oliveira, P. C. (2017). Tapajós river communities in savannah ecosystems: Ethnobotanical profile and diversity of traditional knowledge. *International Journal of Botany Studies* 2, 63–70 Consulté 2 janvier 2024, à l'adresse https://www.academia.edu/download/58215472/2-5-13-798.pdf

Lima, A. R. S., Lôbo, L. A. C., Costa, T. L. da C., Peixoto, J. de C., Borges, L. L., Amaral, V. C. de S., & Paula, J. A. M. de. (2020). Eco-friendly extraction and simultaneous determination of two coumarins in *Justicia pectoralis* (Acanthaceae). Rodriguésia, 71. https://www.scielo.br/j/rod/a/r875nqZRtrnwcWRvq36YzHg/?lang=en&format=html

Lino, C. S., Taveira, M. L., Viana, G. S. B., & Matos, F. J. A. (1997). Analgesic and antiinflammatory activities of *Justicia pectoralis* Jacq and its main constituents: Coumarin and umbelliferone. *Phytotherapy Research*, *11*(3), 211–215. https://doi.org/10.1002/(SICI)1099-1573(199705)11:3<211::AID-PTR72>3.0.CO;2-W

Locklear, T. D. (2008). *Biologically active compounds from Justicia pectoralis: Significance for the treatment of dysmenorrhea*. University of Illinois at Chicago, Health Sciences Center. https://search.proquest.com/openview/e891dad92bf7edd57099372dd5ab30f5/1?pq-origsite=gscholar&cbl=18750

Locklear, T. D., Huang, Y., Frasor, J., Doyle, B. J., Perez, A., Gomez-Laurito, J., & Mahady, G. B. (2010). Estrogenic and progestagenic effects of extracts of *Justicia pectoralis* Jacq., an herbal medicine from Costa Rica used for the treatment of menopause and PMS. *Maturitas*, *66*(3), 315–322.

Moura, C. T., Batista-Lima, F. J., Brito, T. S., Silva, A. A., Ferreira, L. C., Roque, C. R., Aragão, K. S., Havt, A., Fonseca, F. N., & Leal, L. K. (2017). Inhibitory effects of a standardized extract of *Justicia pectoralis* in an experimental rat model of airway hyper-responsiveness. *Journal of Pharmacy and Pharmacology*, *69*(6), 722–732.

Nascimento, W. de M. C., Arcanjo, F. P. N., Bandeira, M. A. M., Prado, B. L., Fernandes, M. E. F., Brito, E. C. O., Monte, B. F., Nascimento, A. A. do, Oliveira, A. F. Q. de, Portela, L. C., & Arcanjo, C. C. (2021). Chamb (*Justicia pectoralis*) syrup for the treatment of cough and respiratory symptoms: A randomized clinical trial. *Journal of Medicinal Plants Research*, *15*(3), 118–124. https://doi.org/10.5897/JMPR2020.7047

Oliveira, A. F. M., & Andrade, L. (2000). Morphological characterization of *Justicia pectoralis* Jacq. And *J. gendarussa* Burm. F.(Acanthaceae). *Acta Amazonica*, *30*(4), 569–578.

Orch, H., Chaachouay, N., Douiri, E. M., Faiz, N., Zidane, L., & Douira, A. (2021). Use of medicinal plants in dermato-cosmetology: An ethnobotanical study among the population of Izarène. *Jordan Journal of Pharmaceutical Sciences*, *14*(3), 323–340.

Roersch, C. (2018). *Justicia pectoralis* Jacq. In U. P. Albuquerque, U. Patil, & Á. Máthé (Éds.), *Medicinal and aromatic plants of South America* (Vol. 5, pp. 251–264). Springer Netherlands. https://doi.org/10.1007/978-94-024-1552-0_22

Schultes, R. E. (1969). Hallucinogens of plant origin: Interdisciplinary studies of plants sacred in primitive cultures yield results of academic and practical interest. *Science*, *163*(3864), 245–254.

Sumner, L. W., Lei, Z., Nikolau, B. J., & Saito, K. (2015). Modern plant metabolomics: Advanced natural product gene discoveries, improved technologies, and future prospects. *Natural Product Reports*, *32*(2), 212–229.

Uk, C. (2014). *Justicia pectoralis* (Jacq.), freshcut.[invasive species]. AQB ISC record. https://www.cabdirect.org/cabdirect/abstract/20107601611

Venâncio, E. T., Rocha, N. F. M., Rios, E. R. V., Feitosa, M. L., Linhares, M. I., Melo, F. H. C., Matias, M. S., Fonseca, F. N., Sousa, F. C. F., Leal, L. K. A. M., & Fonteles, M. M. F. (2011). Anxiolytic-like effects of standardized extract of *Justicia pectoralis* (SEJP) in mice: Involvement of GABA/benzodiazepine in receptor. *Phytotherapy Research*, *25*(3), 444–450. https://doi.org/10.1002/ptr.3274

Venâncioa, E. T., Souza, A. G. de, Limaa, K. A. de, Carvalhoa, M. A. J. de, Souzaa, D. A. A. de, Oliveiraa, J. V. S., Leala, L. K. A. M., & Fontelesa, M. M. De F. (2019). Evaluation of monoaminergic levels after acute administration of standardized extract of *Justicia pectoralis* in mice. *Revista Cubana de Farmacia*, *51*(4), Article 4. https://revfarmacia.sld.cu/index.php/far/article/view/128

Vieira, R. M., Vieira, A. S., Rezende, C. F. A., & Peixoto, J. De C. (2019). Growth, nutrient accumulation and phytochemical prospection of *Justicia pectoralis* Jacq as a function of fertilization type. *Revista Brasileira de Agropecuária Sustentável*, *9*(4), 27–33.

2 Sweet Flag (*Acorus calamus* L. Acoraceae)

Mustafa Keskin, Volkan Altay, and Munir Ozturk

2.1 INTRODUCTION

Acorus calamus has a long history of cultural, medicinal, and culinary uses worldwide. It is distributed on natural wetlands but is at the same time cultivated in many parts of Eurasia and North America (Motley, 1994; Sharma et al., 2014; Shu et al., 2018).

It is among one of the plants that has become an important symbol in the Dragon Boat Festival, a festival traditionally celebrated throughout history for 4,000–5,000 years by communities living in China; the *Acorus calamus festival* (Shu et al., 2018). In the northeastern areas of the Indian state of Assam, local people traditionally believed that *A. calamus* was a supernatural plant and that spirits and demons could not enter it due to its scent. There has been a common habit of placing the rhizomes of this plant in the four corners of the altar where they worship or at the entrance of a ritual area (Sharma and Pegu, 2011; Shu et al., 2018).

While the genus *Acorus* was previously included in the Araceae family, it has now been included in the Acoraceae family under the order Acorales (Grayum, 1990; ITIS, 2023). The systematic phylogenetic analyses reveal that monocotyledons are considered to be a member of the oldest existing lineage (Duvall et al., 1993). The name *Acorus* is derived from the Greek word 'acoron', used by Dioscorides, and 'coreon', meaning pupil of the eye. The derivation of these words was pioneered by the use of this plant in traditional medicine to treat eye inflammation (Motley, 1994; Singh et al., 2011; Shu et al., 2018).

The aim here is to present a detailed account of various features of the sweet flag plant, including its botanical aspects with vernacular names, origin and geographical distribution, ecological features, traditional uses, chemical composition, and some pharmacological effects.

2.2 SCIENTIFIC CLASSIFICATION

The latest updated classification of this species according to the Integrated Taxonomic Information System (ITIS) (2023) is presented next (Table 2.1):.

2.3 VERNACULAR NAMES

A. calamus, widely named as 'sweet flag' in English, is known by different local names in different countries. In Turkiye locals call it 'Eğirotu'; in Chinese 'Chāng pu, Pai ch'ang-p'u, Shui ch'ang-p'u'; in India as 'Dlingo'; in France as 'Acore aromatique, Acore calame, Acore odorant, Acore vrai, Canne aromatique, Jonc odorant, Lis des marais, Roseau odorant'; in Germany as 'Arzneikalmus, Deutscher Ingwer, Kalmus'; in Spain as 'Acoro, Calamis, Cálamo aquático, Cálamo aromático'; in Italy, it has different local names such as 'Acoro aromatico, Acoro vero, Calamo aromatico'; in Far Eastern countries, especially in Japan as 'Shôbu' and in Korea as 'Changpo' (Akbar, 2020).

2.4 BOTANICAL DESCRIPTION

A. calamus is a fragrant, stout, perennial herbaceous plant with branched and creeping rhizomes, up to 150 cm tall, growing on the edges of fresh waters. The rhizome is up to 3 cm thick, more or less spongy due to the large number of intercellular spaces. The stems are stout, erect, simple; leaves erect, long, linear, 0.7–2 cm wide. The flowers are small, many together, continuous, 6–8 cm long and 1 cm wide, gathered into a dense and cylindrical spadix. It has an erect, leafy, elongated spathe that extends singly at the top of the stem and as a continuation of the stem. Flowers are hermaphrodites, with six segmented perigons, the segments being free and greenish; there are six stamens with a three-eyed ovary, and many ovules. The fruit is a red bacca with 2–5 seeds (Baytop, 1979).

2.5 ORIGIN AND GEOGRAPHICAL DISTRIBUTION

The natural distribution of *A. calamus* reveals that it is a semi-aquatic and perennial rhizomatous geophyte distributed in the temperate and semi-temperate regions of Asia, Eurasia, America, and Southeast Asia (Jeon et al., 2013). The natural distribution areas for *A. calamus* have been outlined mainly as; Bangladesh, Borneo, Cambodia, China, Himalayas, India, Iran, Japan, Kazakhstan, Kirghizistan, Korea, Mongolia, Myanmar, Nepal, North America, Philippines, Sri Lanka, Thailand, Uzbekistan and Vietnam (www.powo.science.kew.org). It grows widely in Turkiye alongside the aquatic habitats (Secmen and Leblebici, 2008).

2.6 HABITAT AND ECOLOGY

In terms of habitat preferences, this species is generally spread on wetlands, alongside the pond edges and shallow

DOI: 10.1201/9781003460336-2

TABLE 2.1
Hierarchical Taxonomy of *Acorus calamus* L.

Kingdom	Plantae
Subkingdom	Viridiplantae
Infrakingdom	Streptophyta
Superdivision	Embryophyta
Division	Tracheophyta
Subdivision	Spermatophytina
Class	Magnoliopsida
Superorder	Lilianae
Order	Acorales
Family	Acoraceae
Genus	*Acorus* L.
Species	*Acorus calamus* L.

waters of the rivers, lake shores and swampy areas in temperate and subtropical regions between sea level up to 2600 meters (Secmen and Leblebici, 2008; Cao et al., 2015; Shu et al., 2018). It is a semi-aquatic and perennial rhizomatous geophyte and its morpho-anatomical structure allows it to easily develop alongside such habitats. The rhizomes of the species have the feature of growing ventrally in existing habitats, allowing the plant to be fixed in lower layer depths of the sediment (Dykyjová, 1980; Weber and Brändle, 1996; Jeon et al., 2013). In addition, these plants cluster in small and dense colonies alongside shallow coastal areas with high trophic status, possess relatively short growing seasons, and are sensitive to water fluctuations which have given these plants an important advantage in their ecological adaptations (Henjý and Husák, 1978; Dykyjová, 1980; Monk et al., 1984; Crawford and Brändle, 1996; Weber and Brändle, 1996; Jeon et al., 2013). The morpho-anatomical features of this species have given this plant an ecological resilience to floods, as such they are able to live underwater for a long time and show good development (Weber and Brändle, 1996; Cao et al., 2015). In many countries, this species has the advantage of being one of the preferred species in protecting and purifying the aquatic environment of wetlands and in wetland restoration, in the light of its important ecological features (Cao, 2010; Cao et al., 2015).

2.7 ETHNOBOTANY

This plant has a long history of use in traditional medicine especially in China and India (Rajput et al., 2014; Shu et al., 2018; Chaachouay et al., 2021a). The plant has been used in the Asian continent, especially in Chinese traditional medicine, to reduce swelling and relieve constipation. It has also been reported that the rhizomes of this plant have been potentially used in Indian traditional medicine as an aphrodisiac, laxative, diuretic, emmenagogue, antispasmodic, carminative, stimulant, bitter tonic, emetic, expectorant, and anthelmintic (Rajput et al., 2014). *A. calamus* is well known to many Indigenous people on the American continent and they have used it as an anesthetic for toothaches and headaches. It has also been reported to have been used by the ancient Greeks and has been included in their traditional medicine, similar observations are reported from other European cultures as well (Rajput et al., 2014). At present this plant is generally used in the treatment of some disorders especially loss of appetite, colic, cramps, diarrhea, digestive disorders, bronchitis, chest pain, bloating, gas, indigestion, nervous disorders, rheumatism, sedative and vascular disorders (Bangasen, 1984; Kirtikar and Basu, 1987; Chunekar and Pandey, 1998; Rajput et al., 2014; Orch et al., 2021).

2.8 PHARMACOLOGY

Very recently a phytochemical review study has been published which has revealed that the main groups of compounds isolated from the rhizomes and leaves of *A. calamus* include alkaloids, phenylpropanoids, monoterpenes, sesquiterpenoids, sterols, triterpene glycosides and triterpenoid saponins (Sharma et al., 2020). Among these chemical compounds, the main effective compounds are mentioned to be asarone and eugenol compounds from the phenylpropanoids chemical compounds group, as well as sesquiterpenoids (Sharma et al., 2020). The chemical compound groups isolated from the *A. calamus* are presented in detail below (Sharma et al., 2020; Masoodi and Rehman, 2022):

Alkaloids: Pyrimidin-2-one,4-[N-methylureido]-1-[4methyl amino carbonloxy methy]; Trimethoxy-amphetamine, 2,3,5.

Amino acids: α-alanine; Arginine; Asparagine; Aspartic acid; Glutamic acid; Lysine; Norvaline; Phenylalanine; Proline; Threonine; Tryptophan; Tyrosine.

Fatty acids: Myristic acid; Palmitic acid; Palmitoleic acid; Stearic acid.

Monoterpenes: 2-hexenal; α-Acoradiene; α-Pinene; β-Acoradiene; β-Pinene; o-Cymol; p-Cymene; γ-Terpinene; α-Terpineol; α-Terpinolene; Anethole; Camphene; Camphor; Cis-linaloloxide; Cis-sabinene hydrate; Hexane; Isoaromadendrene epoxide; Isoborneol; Limonene; Linalool; Sabinene; Terpinen-4-ol; Thymol; Trans-linalol oxide.

Phenylpropanoids: 1(5),3-aromadenedradiene; 1,8-cineole; 2-cyclohexane-1-one; 2,3-dihydro-4,5,7-trimethoxy-1-ethyl-2-methyl-3 (2,4,5-trimethoxyphenyl) indene; 4,5-dehydro-isolongifolene; 5,8-dimethyl isoquinoline; 5-n-butyltetraline; α-Asarone; α-Calacorene; β-Asarone; γ-Asarone; γ-Cadinene; Acorin; Aristolene; Asaronaldehyde; Azulene; Calamol; Calarene; Carvacrol; Cinnamaldehyde; Cis-methyl isoeugenol; Cyclohexanone; Dipentene; (E)-caryophyllene; (E)-methyl isoeugenol; (E)-isoeugenol acetate; Estragole; Eugenyl acetate; Eugenol; Eugenol methyl ether, Euasarone; Isoasarone; Isoelemicin; Isoeugenol; Isohomogenol; Longifolene; Methyl eugenol; Methyl isoeugenol; Naphthalene; Safrole; Terpinolene; (Z)-asarone; Z-3-(2,4,5-trimethoxyphenyl)-2-propenal; Zingiberene.

Sesquiterpenoids: 1,4-(cis)-1,7-(trans)-acorenone; 1,4-(trans)1,7(trans)-acorenone; 1β,5α-Guaiane-4β,10α-diol-6-one; 2,6 diepishyobunone; 2-acetoxyacorenone; 2-hydroxyacorenone; 4α, 7-Methano-4α-naphth[1,8a-b] oxirene; 4-(2-formyl-5-methoxymethyl pyrrol-1-yl) butyric acid methyl ester; 7-tetracycloundecanol, 4,4,11,11-tetramethyl; α-Bisabolol; α-Cadinol; α-Cedrene; α-Curcumene; α-Gurjunene; α-Selinene; β-Cedrene; β-Caryophyllene; β-Elemene; β-Gurjunene; δ-Cadinene; Acoramol; Acoramone; Acorenone; Acorone; Acoric acid; Acortatarin A; Asaronaldehyde; Calamene; Calamenenol; Calameone; Calamusin D; Calamusenone; Caryophyllene oxide; Cedrol; Dehydroxyiso-calamendiol; Dioxosarcoguaiacol; Elemol; Epishyobunone; Humulene epoxide II; Humulene oxide II; Isocalamendiol; Isoshyobunone; N-transferuloyl tyramine; Neoacorane A; Preisocalamendiol; Shyobunone; Spathulenol; Tatanan A; Tatarinoid A; Tatarinoid B; Valencene; Viridiflorene; Vulgarol B.

Steroids/Sterols: β-daucosterol.

Triterpenoid saponins: 1β,2α,3β, 19α-Tetrahydroxyurs-12-en-28-oic acid-28-O-{(β-D-glucopyranosyl (1-2)}-β-D galactopyranoside 3-β; 22-α-24,29-Tetrahydroxyolean-12-en-3-O-(β-Darabinosyl(1,3)}-β-D-arabinopyranoside.

Triterpene glycoside: 22-[(6-deoxy-α-L-rhamnopyranosyl) oxy]-3,23-dihydroxy-, methyl ester, (3β,4β,20α,22β).

Xanthone glycosides: 4,5,8-trimethoxy-xanthone-2-O-β-D-glucopyranosyl(1-2)-O-β-D-galactopyranoside.

Clinical experimental studies on the rhizomes of the *A. calamus* and essential oils obtained from this plant are reported to show important pharmacological activities such as immunomodulatory, antidiabetic, antitumor and antimicrobial (Rajput et al., 2014).

2.9 MEDICINAL EFFECTS

It is one of the plants popularly used in the Asian continent, especially in Ayurveda and Unani medicine systems. In particular, the 'Brahmi Bati (Budhivardhar)' drug known in Ayurvedic medicine, and 'Ma'jun Baladur' drug known in Unani medicine, contain components of this plant and are widely prescribed (Jadhav, 1994; Rajput et al., 2014). In the traditional Ayurvedic medical system, the rhizomes of *A. calamus* are used in the treatment of mental disorders such as schizophrenia, psychoneurosis, insomnia, hysteria, epilepsy and memory loss (Nadkarni, 1998; Prajapati et al., 2003; Shah, 2005; Rajput et al., 2014; Shu et al., 2018). A mixture of honey, water, and a paste of rhizomes of *A. calamus* is given to the newborn child by local people in India by applying it on the tongue in the form of a paste for proper brain development, speech ability, better visual power and increased seminal power (Jadhav, 1994; Rajput et al., 2014). In addition, it has a nervous system stimulating feature to get rid of depression. It has been reported that it may be useful in improving speech in stuttering and other disorders (Jadhav, 1994; Rajput et al., 2014; Chaachouay et al., 2021b).

Schultes has reported in 1976 that Northern Canadian Indians traditionally chewed the rootstock parts of the *A. calamus* as a hallucinogen for stimulant purposes. It has even been reported that it can cause strong visual hallucinations in excessive doses. The same author has emphasized that these intoxicating properties may be caused by phytochemicals related to α-asaron and ß-asaron (Schultes 1976).

In some of the pharmacological experimental studies conducted to date, it has been reported that the essential oils present in the ethanol extract obtained from the rhizomes of the *A. calamus* exhibit hypnosis-enhancing and show hypotensive activities (Dandiya et al., 1958, 1959; Dandiya and Cullumbine, 1959; Akbar, 2020). Water-soluble alcohol extract powder obtained from the dried rhizomes of *A. calamus* has been reported to have a sedative effect in experimental mice. It has also been reported to antagonize spontaneous motor activity and amphetamine-induced hyperactivity (Panchal et al., 1989; Akbar, 2020). In some experimental studies conducted on rats, cats, and dogs, extracts from this plant have been reported to produce sedative effects (Dhalla and Bhattacharya, 1968; Akbar, 2020).

In some pharmacological experimental studies, different extracts obtained from the rhizomes of this plant are reported to prevent cognitive impairment in rats, as well as memory impairment caused by noise stress (Sundaramahalingam et al., 2013; Reddy et al., 2015). Some of the studies have revealed that work done on the existing phytochemicals of this plant, asarone and b-asarones, can prolong pentobarbital, hexabarbital, and ethanol-induced hypnosis in mice (Dandiya and Sharma, 1962; Menon and Dandiya, 1963; Akbar, 2020). Similarly, it has also been reported that asaron could potentially reduce spontaneous motor activity and improve anxiety in experimental rats without affecting perception and without affecting the 5-HT content in the brain (Dandiya and Menon, 1964a, 1964b; Akbar, 2020). Asarone has been shown to antagonize mescaline-induced hyperactivity and hallucinogenic effects in experimental rats and has protective potential in dextroamphetamine-treated mass experimental mice (Dandiya and Menon, 1965).

2.10 CONCLUSION

Sweet flag has a long history of traditional use. It is a natural genetic resource with much scientific evidence supporting its therapeutic potential through pharmacological and clinical studies. Due to its different chemical composition and some important pharmacological effects, it appears as an indispensable potential candidate for the development of innovative treatments for various neurodegenerative disorders. Therefore, more research is needed to fully reveal its therapeutic potential and ensure its safe use in modern medicine, so that it can be made available to all humans on a global scale. The populations of the plant are facing serious threat due to their habitat preferences. There is a need for their conservation for future evaluations.

LIST OF ABBREVIATIONS

A. calamus	*Acorus calamus*
ITIS	Integrated Taxonomic Information System

REFERENCES

Akbar, S. 2020. *Handbook of 200 medicinal plants: A comprehensive review of their traditional medical uses and scientific justifications*. Cham: Springer. https://doi.org/10.1007/978-3-030-16807-0_10.

Bangasen, 1984. *Bangasen* (1st ed.). India: Chaukhambha Sanskrit Bharati.

Baytop, A. 1979. Türkiye'de *Acorus calamus*. *Ankara Ecz. Fak. Mec.*, *9*, 12–17.

Cao, Y., Guo, Z., Yang, Y., Wang, G., & Xie, Z. 2015. The ecological amplitude of *Acorus calamus* young shoots under water level gradient. *Polish Journal of Ecology*, *63*(4), 585–592.

Cao, Y.M. 2010. Urban sewage treatment performance of *calamus* artificial wetlands filled with soil and/or coal ash. *Environ - Poll. Control*, *32*, 54–58.

Chaachouay, N., Douira, A., & Zidane, L. 2021a. Herbal Medicine Used in the Treatment of Human Diseases in the Rif, Northern Morocco. *Arabian Journal for Science and Engineering*, 47, 131–153. https://doi.org/10.1007/s13369-021-05501-1

Chaachouay, N., Douira, A., & Zidane, L. 2021b. COVID-19, prevention and treatment with herbal medicine in the herbal markets of Salé Prefecture, North-Western Morocco. *European Journal of Integrative Medicine*, 42, 101285. https://doi.org/10.1016/j.eujim.2021.101285

Chunekar, K.C., & Pandey, G.S. 1998. *Bhavprakash Nighantu* (1st ed.). India: ChaukhambhaBharati Acadamy.

Crawford, R.M.M., & Brändle, R. 1996. Oxygen deprivation stress in a changing environment. *Journal of Experimental Botany*, *47*, 145–159.

Dandiya, P.C., Baxter, R.M., & Cullumbine, H. 1958. Studies on *Acorus calamus*. I: A phytochemical investigation. *Canadian Pharmacists Journal*, *91*, 607–610.

Dandiya, P.C., & Cullumbine, H. 1959. Studies on *Acorus calamus*. III: some pharmacological properties of the volatile oil. *Journal of Pharmacology and Experimental Therapeutics*, *125*, 353–359.

Dandiya, P.C., Cullumbine, H., and Sellers, E.A. 1959. Studies on *Acorus calamus*. IV. Investigations on mechanism of action in mice. *Journal of Pharmacology and Experimental Therapeutics*, *126*, 334–337.

Dandiya, P.C., & Menon, M.K. 1964a. Some more pharmacological actions of asarone, the active principle of *Acorus calamus*. *Indian Journal of Physiology and Pharmacology*, *8*, 44.

Dandiya, P.C., & Menon, M.K. 1964b. Actions of asarone on behaviour, stress and hyperpyrexia and its interaction with central stimulants. *Journal of Pharmacology and Experimental Therapeutics*, *145*, 42.

Dandiya, P.C., & Menon, M.K. 1965. Interaction of asarone with mescaline, amphetamine and tremonine. *Life Sci.*, *4*, 1635.

Dandiya, P.C., & Sharma, J.D. 1962. Studies on *Acorus calamus* Part I. Pharmacological actions of asarone and b-asarone on central nervous system. *Indian Journal of Medical Research*, *50*, 46–60.

Dhalla, N.S., & Bhattacharya, I.S. 1968. Further studies on neuropharmacological actions of *Acorus* oil. *Arch Int Pharmacodyn.*, *172*, 356.

Duvall, M.R., Learn, G.H. Jr., Eguiarte, L.E., and Clegg, M.T. 1993. Phylogenetic analysis of rbcL sequences identifies *Acorus calamus* as the primal extant monocotyledon. *Proceedings of the National Academy of Sciences of the United States of America*, *90*(10), 4641–4644.

Dykyjová, D. 1980. Production ecology of *Acorus calamus*. *Folia Geobot Phytotaxon*, *15*, 29–57.

Grayum, M.H. 1990. Evolution and phylogeny of the Araceae. *Annals of the Missouri Botanical Garden*, *1990*, 628–697.

Henjý, S., & Husák, S. 1978. Higher plant communities. In D. Dykyjová, & J. Květ (Eds.), *Pond littoral ecosystems-structure and functioning* (pp. 23–64). New York: Springer-Verlag.

ITIS. 2023. Integrated taxonomic information system. *Acorus calamus* L. Taxonomic Serial No.: 564989. www.itis.gov, CC0.

Jadhav, T. 1994. *Sushrut Samhita* (1st ed.). India: Chaukhambha Surbharati Acadamy.

Jeon, S.H., Kim, H., Nam, J.M., & Kim, J.G. 2013. Habitat characteristics of sweet flag (*Acorus calamus*) and their relationships with sweet flag biomass. *Landscape and Ecological Engineering*, *9*, 67–75.

Kirtikar, K.R., & Basu, B.D. 1987. *Indian medicinal plants* (4th ed.). India: Mahendra Pal SingPublishers.

Masoodi, M.H., & Rehman, M.U. (Eds.). 2022. *Edible plants in health and diseases: Volume II: Phytochemical and pharmacological properties*. Singapore: Springer Nature.

Menon, M.K., & Dandiya, P.C. 1963. Tranquillizing properties of asarone and a-asarone, active constituents of *A. calamus*. *Indian Journal of Physiology and Pharmacology*, *7*, 14.

Monk, L.S., Crawford, R.M.M., & Braendle, R. 1984. Fermentation rates and ethanol accumulation in relation to flooding tolerance in rhizomes of monocotyledonous species. *Journal of Experimental Botany*, *35*, 738–745.

Motley, T.J. 1994. The ethnobotany of sweet flag, *Acorus calamus* (Araceae). *Economic Botany*, *48*(4), 397–412.

Nadkarni, K.M. 1998. *Indian Materia Medica* (1st ed.). India: Popular Prakashan.

Orch, H., Chaachouay, N., Douiri, E. M., Faiz, N., Zidane, L., & Douira, A. (2021). Use of medicinal plants in dermato-cosmetology: An ethnobotanical study among the population of Izarène. *Jordan Journal of Pharmaceutical Sciences*, *14*(3), 323–340.

Panchal, G.M., Venkatakrishna-Bhatt, H., Doctor, R.B., & Vajpayee, S. 1989. Pharmacology of *Acorus calamus* L. *Indian Journal of Experimental Biology*, *27*, 561–567.

Prajapati, N.D., Purohit, S.S., Sharma, D.D., & Tarun, K. 2003. *A handbook of medicinal plants* (Section II, 1st ed.). India: Agrobiaos.

Rajput, S.B., Tonge, M.B., & Karuppayil, S.M. 2014. An overview on traditional uses and pharmacological profile of *Acorus calamus* Linn. (Sweet flag) and other *Acorus* species. *Phytomedicine*, *21*(3), 268–276.

Reddy, S., Rao, G., Shetty, B., & Hn, G. 2015. Effects of *Acorus calamus* rhizome extract on the neuromodulatory system in restraint stress male rats. *Turkish Neurosurgery*, *25*, 425–431.

Schultes, R.E. 1976. *Hallucinogenic Plants*. New York: Golden Press; Western Publishing Company, Inc.

Secmen, O., & Leblebici, E. 2008. *Wetland of Turkiye and their Plant Cover*. Turkiye: Ege University Press, Second Printing, Bornova-İzmir.

Shah, N.C. 2005. *Bharat Bhaishajya Ratnakar*. India: B. Jain Publishers Pvt. Ltd.

Sharma, U.K., & Pegu, S. 2011. Ethnobotany of religious and supernatural beliefs of the missing tribes of Assam with special reference to the 'Dobur Uie'. *Journal of Ethnobiology and Ethnomedicine*, *7*, 16.

Sharma, V., Sharma, R., Gautam, D.S., Kuca, K., Nepovimova, E., & Martins, N. 2020. Role of Vacha (*Acorus calamus* Linn.) in neurological and metabolic disorders: evidence from ethnopharmacology, phytochemistry, pharmacology and clinical study. *Journal of Clinical Medicine*, *9*(4), 1176.

Sharma, V., Singh, I., & Chaudhary, P. 2014. *Acorus calamus* (the healing plant): A review on its medicinal potential, micropropagation and conservation. *Natural Product Research*, *28*, 1454–1466.

Shu, H., Zhang, S., Lei, Q., Zhou, J., Ji, Y., Luo, B., ... Long, C. 2018. Ethnobotany of *Acorus* in China. *Acta Societatis Botanicorum Poloniae*, *87*(2), 1–14.

Singh, R., Sharma, P.K., & Malviya, R. 2011. Pharmacological properties and ayurvedic value of Indian buch plant (*Acorus calamus*): A short review. *Advances in Biological Research (Rennes)*, *5*(3), 145–154.

Sundaramahalingam, M., Ramasundaram, S., Rathinasamy, S.D., Natarajan, R.P., & Somasundaram, T. 2013. Role of *Acorus calamus* and alpha-asarone on hippocampal dependent-memory in noise stress exposed rats. *Pakistan Journal of Biological Sciences*, *16*, 770–778.

Weber, M., & Brändle, R. 1996. Some aspects of the extreme anoxia tolerance of the sweet flag *Acorus calamus* L. *Folia Geobot Phytotaxon*, *31*, 37–46. www.powo.science.kew.org. Accesses on August 25, 2023.

3 Kanna (*Sceletium tortuosum* (L.) N.E. Br. Aizoaceae)

Aneeqa Hanif and Alvina Gul

3.1 INTRODUCTION

South African native *Sceletium tortuosum*, sometimes known as 'Kanna' or 'kougoed', is a succulent medicinal plant. Common uses for the plant's aerial portions include chewing, drinking tea or tinctures, and occasionally smoking. The plant's fresh juice is given to young children to help them go asleep, and it is chewed for the alleviation of hunger, toothaches, and abdominal pain. It is currently prescribed commercially to treat a variety of problems of the central nervous system, such as stress, depression, and anxiety. Due to its potential to treat neurological illnesses and neurodegenerative diseases, this plant species has attracted particular scientific interest (Olatunji et al., 2022; Chaachouay et al., 2019). Kanna has also been identified as a prospective therapy for neurological, psychiatric, and inflammatory illnesses ranging from Alzheimer's disease to drug addiction to human immunodeficiency virus (HIV) (Sreekissoon et al., 2021). Kanna, as well as other species, have emerged as desirable commodities in the commercialization of South African medicinal plants. Various versions of the plant are currently available for purchase; for example, tea bags, which are frequently combined with Red Bush Tea or Honeybush tea, may be found in South African supermarkets. Extracts of the plant are available in raw powdered plant material, tablets, and capsules, which are regularly traded on the Internet, and people utilize it to improve their well-being and relieve stress. Furthermore, *S. tortuosum* plants are a safe, effective, and inexpensive natural remedy for veterinary and pharmaceutical reasons (Manganyi et al., 2021).

3.2 SCIENTIFIC CLASSIFICATION

Kingdom:	Plantae
Clade:	Tracheophytes
Clade:	Angiosperms
Clade:	Eudicots
Order:	Caryophyllales
Family:	Aizoaceae
Genus:	*Sceletium*
Species:	*Sceletium tortuosum* (L.) N.E. Br.

3.3 VERNACULAR NAMES

Sceletium tortuosum, sometimes known as 'Kanna' or 'kougoed'.

3.4 BOTANICAL DESCRIPTION

Kanna (*Sceletium tortuosum*) is a species of *Sceletium*. According to the World Flora Online (WFO), the genus presently has 24 species, comprising intraspecific names, but only 20 correspond to approved names. The *Sceletium tortuosum* species is the most popular and often used species in the *Sceletium* genus. Succulents include *Sceletium tortuosum*. The word 'sceltium' is derived from the Latin word 'sceletus', which describes the leaf veins that persist in dry leaves as observable skeleton-like structures. A specific word with the meaning of 'twisted' or 'tortuous' is 'tortuosum'. A creeping or climbing plant called *Sceletium tortuosum* has thin branches that, as they become older, thicken and become just a little bit woody (Olatunji et al., 2022). The leaves are flat and succulent in the dry summer, connate at the base, and skeletonize around the vegetative buds. The majority of the time, the flowers are solitary, sessile, 20–40 mm in diameter, and almost generally yellow, however they occasionally are white or pink as well. The fruit capsules are exceedingly light, 4- to 5- (or perhaps 6-) locular, and break off to be carried by the wind. The D-shaped seeds have a diameter of 1.5–2 mm and are either black or brown. Because of the species' and its populations' wide geographic range, there are a number of regional variants that were formerly believed to belong to different species (Brendler et al., 2021).

Despite sharing many morphological traits with other species in the genus, *S. tortuosum* is typically identified by contrasting flower, seed, fruit, and vegetative characteristics. Only two species (*S. tortuosum* and *S. crassicaule*) in the genus generate the four main mesembrine-type alkaloids that are responsible for their psychedelic characteristics (Shikanga et al., 2013).

3.5 ORIGIN AND GEOGRAPHICAL DISTRIBUTION

Kanna (*Sceletium tortuosum*), a South African endemic plant, has a long history of traditional use as a masticatory and medicine by San and Khoikhoi people, and later as a psychoactive in tincture form by European colonial farmers (Terburg et al., 2013). *Sceletium tortuosum* has long been used by hunter-gatherers and pastoralists to increase endurance during hunting expeditions and to cope with the stress of living in the harsh arid environments of Bushmanland, Namaqualand, and the Karoo. The

DOI: 10.1201/9781003460336-3

indigenous Nama name for *Sceletium tortuosum* is Kanna or Channa, whereas the /Xam San people called it Kwain. *Sceletium tortuosum* was known as Kaauwgoed or Kauwgood by the Dutch colonists, which translates as 'chewing stuff' (Dimpfel et al., 2016).

3.6 HABITAT AND ECOLOGY

The genus is found in the south-western areas of South Africa, from Namaqualand to Montagu and Aberdeen, and is most usually found in quartz patches. It prefers arid areas and can be seen growing behind shrubs in partial shade (Appleton, 2021).

3.7 ETHNOBOTANY

Sceletium tortuosum, also known as 'kougoed' or 'Kanna', has been used by South African pastoralists and hunter-gatherers as a mood-altering narcotic since prehistoric times (Shikanga et al., 2011). The first known recorded account of the use of the herb, known as Kanna, was penned by Jan van Riebeeck in 1662. Farmers in the Cape utilized *S. tortuosum* as a narcotic and as a sedative in the form of a decoction or tincture, according to Watt and Breyer-Brandwijk (1962). After fermentation, it is reported that the narcotic impact becomes significantly stronger (Chen and Viljoen, 2019). During hunting trips, the leaf is also chewed to treat toothache and gastrointestinal pain, as well as to reduce appetite and thirst (Gericke, 2020). *S. tortuosum* is commonly chewed by pregnant women to relieve constipation, indigestion, and nausea. An infusion of *S. tortuosum* is administered to new mothers in order to contract the uterus, evict any residual afterbirth, and relieve stomach discomfort. The liquid from dried *Sceletium* was traditionally prepared by chewing and swallowing, but it has also been turned into tablets, gel caps, teas, and tinctures. Additionally, it is smoked and snuffed. Additionally, oil made from fried *S. tortuosum* and the last segment of a sheep's tail is administered to infants to relieve colic (Olatunji et al., 2022; Chaachouay et al., 2021).

Sceletium tortuosum is one of the plants that has long been believed to have sedative and anti-anxiety qualities. Anecdotal evidence of the usefulness of this plant, which grows primarily in the eastern and western areas of South Africa and is commonly known as 'kougoed' or 'Kanna' (Smith, 2011). The succulent sub-shrub *Sceletium tortuosum*, which is native to South Africa, is extremely valuable for its traditional applications as a masticant and medication (Shikanga et al., 2012; Chaachouay et al., 2021b). From the plant, a traditional remedy known as 'Kougoed' is made, which is applied to cases of intoxication (Manganyi et al., 2018). The Khoisan people of Southern Africa have traditionally used *Sceletium tortuosum*, among other *Sceletium* species, to treat toothaches and stomach pain, and commercial availability of this herbal supplement has significantly increased as a result of anecdotal claims of its mood-elevating and anxiolytic properties (Bennett and Smith, 2018).

3.8 PHARMACOLOGY

Biomarkers or pharmacologically active substances are commonly utilized as quality control criteria in traditional medications (Shikanga et al., 2012). Mesembrine alkaloids are the major active components of the South African medicinal plant *Sceletium tortuosum*, and it is utilized as the plant's dried or fermented aerial material, known as Kanna (Chiu et al., 2014). The main active ingredients are a group of mesembrine alkaloids, specifically mesembrine and mesembrenone. They are serotonin reuptake inhibitors, which explains why the herb has traditionally been used as an antidepressant (Krstenansky, 2017).

Sceletium is the only known genus with species that contain reasonably high quantities of mesembrine. Other similar plants or genera, such as *Narcissus*, contain substantially less alkaloid, if any at all (Krstenansky, 2017). *M. tortuosum*'s alkaloid chemistry in wild populations varies greatly in terms of both alkaloid concentration and relative alkaloid composition. The total alkaloid content of wild *M. tortuosum* can range from 0.05% to nearly 3.0% of dry weight in outliers. The average total alkaloid content of traditionally preferred wild *M. tortuosum* types ranges from 0.2% for 'mak' (mild) to 2% for 'trek' (strong). In-depth research revealed that the genus Sceletium generates indole alkaloids such as mesembrenol, mesembranol, mesembrine, and mesembranone, the chemical formulae for which are listed in U.S. Pat. No. 6288104 (Brendler et al., 2010). Figure 3.1 illustrated the chemical structures of *Sceletium tortuosum*'s alkaloids. Further investigation revealed that the only alkaloid found in the leaves of *S. tortuosum* was mesembrine (Manganyi et al., 2021). Roscher et al. (2012) carried out a forensic analysis of mesembrine alkaloids using capillary electrophoresis mass spectrometry (CEMS) and compared chemotypically a number of *S. tortuosum* extracts from various vendors, self-fermented samples, and consumables (Roscher et al., 2012). Patnala and Kanfer in 2013 discovered significant chemotaxonomic variation among *Sceletium* species in terms of their mesembrine-type alkaloids. They created a quality control system to make sure that the right *Sceletium* species are used in commercial products after discovering from their research that not all *Sceletium* species possess mesembrine-type alkaloids (Patnala and Kanfer, 2013).

S. tortuosum has a long history of usage as an anxiolytic and antidepressant, and its extract fraction has antidepressant-like effects in a mouse model of depression (Carpenter et al., 2016).

3.9 MEDICINAL EFFECTS

Despite the plant's long recorded history of use as a masticatory, health tea, and herbal medicine, there have been few reports on its pharmacological activity. Its reported efficacy in cases of depression has been connected to inhibitory effects on 5-HT reuptake, yet these effects have not been well explored (Gericke and Viljoen, 2008). Mesembrine,

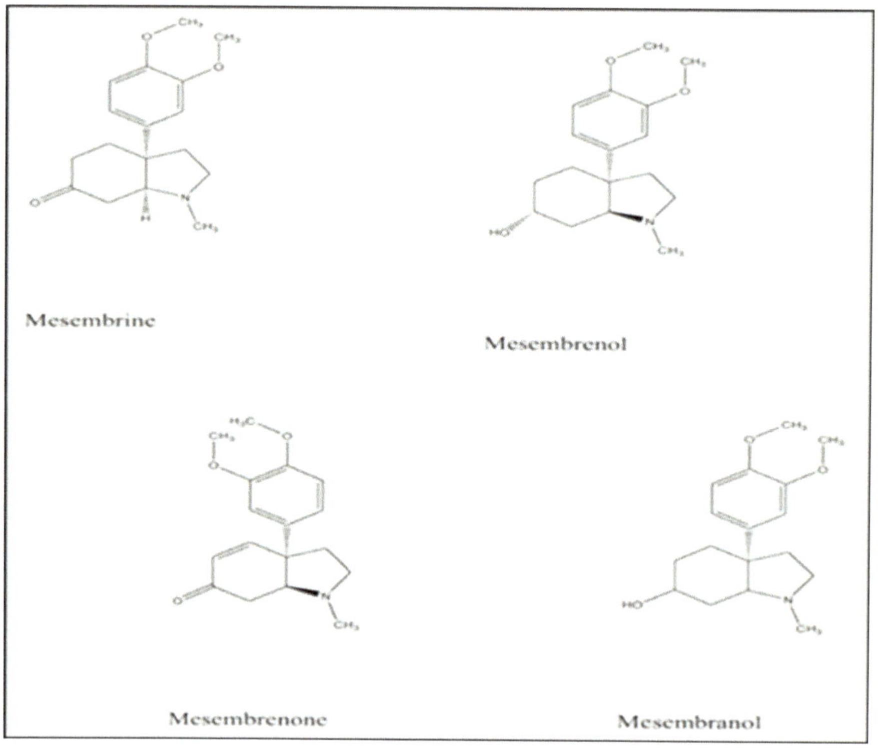

FIGURE 3.1 Chemical structures of *Sceletium tortuosum* alkaloids (Harvey et al., 2011).

isolated from *Sceletium tortuosum*, was found to be a powerful inhibitor of 5-hydroxytryptamine (5-HT) reuptake, which was verified with synthetic mesembrine (Gericke and Van Wyk, 2001). Mesembrine and its synthetic equivalents have been demonstrated to suppress the activity of phosphodiesterase 4 (PDE4). Mesembrine was a very weak PDE4 inhibitor, but certain synthesized counterparts were much more potent (Napoletano et al., 2001). The therapeutic benefits of *Sceletium tortuosum* extract (Zembrin®) are most likely due to their inhibitory effects on 5-HT absorption pathways and PDE4 activity. Selective serotonin reuptake inhibitors are well-known antidepressants, while PDE4 inhibitors have also gotten a lot of interest as possible antidepressants, despite the fact that the pharmaceutical compounds tested clinically showed dose-limiting adverse effects such as nausea and emesis (Kanes et al., 2007). As a result, the discovery that the extract *Sceletium tortuosum* Zembrin® was a potent inhibitor of both 5-HT absorption and PDE4 activity may have far-reaching implications for human health (Cashman et al., 2009). Figure 3.2 elaborate the potential benefits of Kanna (*Sceletium tortuosum*).

Because of its clinical promise for treating anxiety and depression, lowering stress in healthy individuals, and enhancing cognitive capacities, *Sceletium tortuosum* is currently receiving a lot of scientific interest. The phytochemical components of this plant, known as mesembrine-type alkaloids, are responsible for these therapeutic benefits (Olatunji et al., 2022). To treat a variety of central nervous system (CNS) issues such stress, sadness, and anxiety, it is currently used commercially. This plant species has attracted the attention of scientists due to its efficacy in treating neurological illnesses and neurodegenerative diseases (Van Wyk and Gericke, 2000; Yin et al., 2019). The psychoactive properties of the mesembrine-type alkaloids have been related to their ability to act as phosphodiesterase-4 (PDE4) and serotonin reuptake inhibitors (SRIs), which are responsible for modulating intracellular messengers and synaptic and neuronal serotonin levels, respectively (Hoffman et al., 2020). *Sceletium* extract has been demonstrated to be effective as a monoamine-releasing agent, a selective serotonin reuptake inhibitor, and a glucocorticoid production inhibitor by inhibiting particular adrenal p450 enzymes in the context of stress (Coetzee et al., 2016; Swart and Smith, 2016).

3.10 CONCLUSION

Plant-based medication and/or treatments are now a global expanding trend, increasing the demand for natural bioactive substances derived from natural sources. As a result, experts anticipate enormous problems in numerous parts of plant-based therapy. Although South Africa has recognized African traditional medicine as a component of the public health care system, there are significant problems in plant-based medicine quality, standardization, efficacy, authenticity, and safety. Kanna is widely regarded as safe, yet, more research is required to completely comprehend its long-term effects, potential drug interactions, and appropriate dosing regimes. Furthermore, sustainable farming practices and conservation initiatives are critical to the survival of this culturally significant plant. To summarize, the Kanna plant is a botanical treasure with a long history of traditional use as well as a growing body of scientific

FIGURE 3.2 Potential health benefits of Kanna (*Sceletium tortuosum*).

evidence supporting its therapeutic potential. Because of its distinct chemical composition and pharmacological actions, it is a prospective candidate for the development of innovative treatments for a variety of mental health illnesses and neurodegenerative ailments. However, further research is needed to fully realize its therapeutic potential and ensure its safe usage in modern medicine.

LIST OF ABBREVIATIONS

5-HT 5-hydroxytryptamine
CEMS Capillary electrophoresis mass spectrometry
CNS Central nervous system
HIV human immunodeficiency virus
PDE4 Phosphodiesterase 4
SRIs Serotonin reuptake inhibitors
WFO World Flora Online

ACKNOWLEDGMENTS

Our special thanks go to Prof. Dr. Munir Ozturk and Prof. Dr. Volkan Altay for critical evaluation of this chapter.

REFERENCES

Appleton, J. 2021. *Sceletium tortuosum*: A South African plant for mood and stress [WWW Document]. Naturop. Dr. News Rev. URL. https://ndnr.com/botanical-medicine/sceletium-tortuosum-a-south-african-plant-for-mood-and-stress/. accessed, 3, 21.

Bennett, A.C., and Smith, C. 2018. Immunomodulatory effects of *Sceletium tortuosum* (Trimesemine™) elucidated in vitro: Implications for chronic disease. *Journal of Ethnopharmacology*, 214: 134–140.

Brendler, T., Brinckmann, J.A., Feiter, U., Gericke, N., Lang, L., Pozharitskaya, O.N., … Van Wyk, B.E. 2021. Sceletium for managing anxiety, depression and cognitive impairment: A traditional herbal medicine in modern-day regulatory systems. *Current Neuropharmacology*, 19(9): 1384.

Brendler, T., Eloff, J.N., Gurib-Fakim, A., and Phillips, L.D. 2010. African herbal pharmacopoeia; Association for African Medicinal Plants Standards (AAMPS), Mauritius.

Carpenter, J.M., Jourdan, M.K., Fountain, E.M., Ali, Z., Abe, N., Khan, I.A., and Sufka, K.J. 2016. The effects of *Sceletium tortuosum* (L.) NE Br. extract fraction in the chick anxiety-depression model. *Journal of Ethnopharmacology*, 193: 329–332.

Cashman, J.R., Voelker, T., Johnson, R., and Janowsky, A. 2009. Stereoselective inhibition of serotonin re-uptake and phosphodiesterase by dual inhibitors as potential agents for depression. *Bioorganic & Medicinal Chemistry*, 17(1): 337–343.

Chaachouay, N., Benkhnigue, O., Fadli, M., Ibaoui, H., & Zidane, L. 2019. Ethnobotanical and ethnopharmacological studies of medicinal and aromatic plants used in the treatment of metabolic diseases in the Moroccan Rif. *Heliyon*, *5*(10), e02191. https://doi.org/10.1080/10477845.2021.1942433

Chaachouay, N., Douira, A., & Zidane, L. 2021. Herbal medicine used in the treatment of human diseases in the Rif, Northern Morocco. *Arabian Journal for Science and Engineering*, 47, 131–153. https://doi.org/10.1007/s13369-021-05501-1

Chaachouay, N., Douira, A., & Zidane, L. 2021b. COVID-19, prevention and treatment with herbal medicine in the herbal markets of Salé Prefecture, North-Western Morocco. *European Journal of Integrative Medicine*, 42, 101285. https://doi.org/10.1016/j.eujim.2021.101285

Chen, W., and Viljoen, A.M. 2019. To ferment or not to ferment *Sceletium tortuosum*-Do our ancestors hold the answer? *South African Journal of Botany*, 122: 543–546.

Chiu, S., Gericke, N., Farina-Woodbury, M., Badmaev, V., Raheb, H., Terpstra, K., … Goble, L. 2014. Proof-of-concept randomized controlled study of cognition effects of the proprietary extract *Sceletium tortuosum* (Zembrin) targeting phosphodiesterase-4 in cognitively healthy subjects: implications for Alzheimer's dementia. *Evidence-based Complementary and Alternative Medicine*, 2014: 2014.

Coetzee, D.D., López, V., and Smith, C. 2016. High-mesembrine Sceletium extract (Trimesemine™) is a monoamine releasing agent, rather than only a selective serotonin reuptake inhibitor. *Journal of Ethnopharmacology*, 177: 111–116.

Dimpfel, W., Schombert, L., and Gericke, N. 2016. Electropharmacogram of *Sceletium tortuosum* extract based on spectral local field power in conscious freely moving rats. *Journal of Ethnopharmacology*, 177: 140–147.

Gericke, J. 2020. Evaluating the antidepressant-like properties of *Sceletium tortuosum*, alone and as adjunctive treatment. *Doctoral dissertation*, North-West University, South-Africa.

Gericke, N., and Viljoen, A.M. 2008. *Sceletium*-a review update. *Journal of Ethnopharmacology*, 119(3): 653–663.

Gericke, N.P., and Van Wyk, B.E. 2001. U.S. Patent No. 6,288,104. Washington, DC: U.S. Patent and Trademark Office.

Harvey, A.L., Young, L.C., Viljoen, A.M., and Gericke, N.P. 2011. Pharmacological actions of the South African medicinal and functional food plant *Sceletium tortuosum* and its principal alkaloids. *Journal of Ethnopharmacology*, 137(3): 1124–1129.

Hoffman, J.R., Markus, I., Dubnov-Raz, G., and Gepner, Y. 2020. Ergogenic effects of 8 days of *Sceletium tortuosum* supplementation on mood, visual tracking, and reaction in recreationally trained men and women. *The Journal of Strength & Conditioning Research*, 34(9): 2476–2481.

Kanes, S.J., Tokarczyk, J., Siegel, S.J., Bilker, W., Abel, T., and Kelly, M.P. 2007. Rolipram: A specific phosphodiesterase 4 inhibitor with potential antipsychotic activity. *Neuroscience*, 144(1): 239–246.

Krstenansky, J.L. 2017. Mesembrine alkaloids: Review of their occurrence, chemistry, and pharmacology. *Journal of Ethnopharmacology*, 195: 10–19.

Manganyi, M.C., Bezuidenhout, C.C., Regnier, T., and Ateba, C.N. 2021. A chewable cure 'kanna': Biological and pharmaceutical properties of *Sceletium tortuosum*. *Molecules*, 26(9): 2557.

Manganyi, M.C., Regnier, T., Kumar, A., Bezuidenhout, C.C., and Ateba, C.N. 2018. Phylogenetic analysis and diversity of novel endophytic fungi isolated from medicinal plant *Sceletium tortuosum*. *Phytochemistry Letters*, 27: 36–43.

Napoletano, M., Fraire, C., Santangelo, F., and Moriggi, E. 2001. Mesembrine is an inhibitor of PDE4 that follows structure-activity relationship of rolipram. *Science Direct Working Paper*, 2001: 4.

Olatunji, T.L., Siebert, F., Adetunji, A.E., Harvey, B.H., Gericke, J., Hamman, J.H., and Van der Kooy, F. 2022. *Sceletium tortuosum*: A review on its phytochemistry, pharmacokinetics, biological, pre-clinical and clinical activities. *Journal of Ethnopharmacology*, 287: 114711.

Patnala, S., and Kanfer, I. 2013. Chemotaxonomic studies of mesembrine-type alkaloids in *Sceletium* plant species. *South African Journal of Science*, 109(3): 1–5.

Roscher, J., Posch, T.N., Pütz, M., and Huhn, C. 2012. Forensic analysis of mesembrine alkaloids in *Sceletium tortuosum* by nonaqueous capillary electrophoresis mass spectrometry. *Electrophoresis*, 33(11): 1567–1570.

Shikanga, E.A., Viljoen, A., Combrinck, S., and Marston, A. 2011. Isolation of *Sceletium* alkaloids by high-speed countercurrent chromatography. *Phytochemistry Letters*, 4(2): 190–193.

Shikanga, E.A., Viljoen, A.M., Combrinck, S., Marston, A., and Gericke, N. 2012. The chemotypic variation of *Sceletium tortuosum* alkaloids and commercial product formulations. *Biochemical Systematics and Ecology*, 44: 364–373.

Shikanga, E.A., Viljoen, A.M., Vermaak, I., and Combrinck, S. 2013. A novel approach in herbal quality control using hyperspectral imaging: discriminating between *Sceletium tortuosum* and *Sceletium crassicaule*. *Phytochemical Analysis*, 24(6): 550–555.

Smith, C. 2011. The effects of *Sceletium tortuosum* in an in vivo model of psychological stress. *Journal of Ethnopharmacology*, 133(1): 31–36.

Sreekissoon, A., Plačková, L., Doležal, K., Finnie, J.F., and Van Staden, J. 2021. In vitro and ex vivo vegetative propagation and cytokinin profiles of *Sceletium tortuosum* (L.) NE Br.: A South African medicinal plant. *Plant Cell, Tissue and Organ Culture (PCTOC)*, 145: 191–202.

Swart, A.C., and Smith, C. 2016. Modulation of glucocorticoid, mineralocorticoid and androgen production in H295 cells by Trimesemine™, a mesembrine-rich Sceletium extract. *Journal of Ethnopharmacology*, 177: 35–45.

Terburg, D., Syal, S., Rosenberger, L.A., Heany, S., Phillips, N., Gericke, N., ... and Van Honk, J. 2013. Acute effects of *Sceletium tortuosum* (Zembrin), a dual 5-HT reuptake and PDE4 inhibitor, in the human amygdala and its connection to the hypothalamus. *Neuropsychopharmacology*, 38(13): 2708–2716.

Van Wyk, B.-E., and Gericke, N. 2000. *People's plants: A guide to useful plants of Southern Africa*. Briza Publications.

Watt, J.M., and Breyer-Brandwijk, M.G. 1962. *The medicinal and poisonous plants of Southern and Eastern Africa being an account of their medicinal and other uses, chemical composition, pharmacological effects and toxicology in man and animal*. E. & S. Livingstone Ltd. Press, UK.

Yin, H., Ali, Z., Ding, Y., Wang, Y.H., Cunningham, M.J., Ibrahim, M.A., ... and Khan, I.A. 2019. Sceletorines A and B, two minor novel dimeric alkaloids of *Mesembryanthemum tortuosum* (synonym *Sceletium tortuosum*). *Phytochemistry Letters*, 31: 78–84.

4 Pancratium Lily (*Pancratium trianthum* Herb. Amaryllidaceae)

Munir Ozturk, Mustafa Keskin, and Volkan Altay

4.1 INTRODUCTION

Pancratium genus is naturally distributed in the world, especially in Africa, the Canary Islands, West Africa, the Mediterranean, Namibia, and tropical Asia, and includes important bulbous geophytes (De Castro et al. 2012; Chaachouay et al., 2021a). The genus name '*Pancratium*' was first described by Linnaeus (1753) being of Greek origin, meaning 'all strength' alluding to the medicinal characteristics of the bulbs of these plants. It is globally represented by 21 species (WFO 2021; Prameela et al., 2022). One of these species is the *Pancratium trianthum*, generally known as 'Pancratium lily' in African countries.

This chapter presents data on various features of the *P. trianthum*, covering its botanical aspects with vernacular names in some African countries, geographical distribution, ecological features, traditional uses, and some pharmacological effects.

4.2 SCIENTIFIC CLASSIFICATION

The latest updated classification of this species according to the United States Department of Agriculture Natural Resources Conservation Service (USDA) (2023) is presented below:

Kingdom	Plantae
Subkingdom	Tracheobionta
Superdivision	Spermatophyta
Division	Magnoliophyta
Class	Liliopsida
Subclass	Liliidae
Order	Liliales
Family	Liliaceae
Genus	*Pancratium* L.
Species	*Pancratium trianthum* Herb.

4.3 VERNACULAR NAMES

The *P. trianthum*, widely named as 'Pancratium lily' in English, is known by different local names in different African countries. In Senegal locals call it 'Fula-Pulaar nagada'; in Guinea 'Manding-Maninka boa'; in Mali as 'Songhai'; and in Nigeria as 'Hausa hátsín mánoòmaá' (Burkill, 1985).

4.4 BOTANICAL DESCRIPTION

P. trianthum is 40–60 cm tall glabrous herbaceous perennial with a solitary or gregarious bulb, globose or subglobose to ovoid, 3–6 cm long, 2.5–5 cm in diameter; the neck is (8-)10–13 cm long; dark or blackish-brown tunic; leaves (3-)5–10(-15) developing with or after the flowers, glaucous, linear to lanceolate-linear, (6.5)10–30 cm long, 0.3–1.0 cm diameter, broadest at middle or slightly below, erect, slightly twisted, gradually tapering above into an acute apex, glabrous near the base; base long attenuate, without scarious-membranous sheath. Aerial scape solitary, solid, compressed, glaucous, (6-)10–25 cm long, dilated at the glabrous base (0.2–0.6 cm diameter); spathe 1-valved, lanceolate or ovate-lanceolate, and with ± contorted margins at the base, tapering above into an acute apex; the flowers (1-) 2–5 are large-sized, (18-)20–24 cm long, 4–6 times as long as the spathe, fragrant, subsessile, or shortly pedicelled; pedicel 0.2–0.8 (–1.0) cm, shorter than the ovary; perianth white, 17–22 cm long; perianth-tube glaucous, 11–16 cm long, 2–3 times as long as the perianth segments, white, with a narrow green mid-stripe outside, 5–7 cm long, narrowly spathulate-oblong to lanceolate-linear, gradually tapering above to a mucronate apex, inner linear-lanceolate, acute or mucronulate; perianth segments 1.5–2 times as long as the corona; corona (2.5-)3–4 × 2–3 cm, margin 12-teethed; teeth triangular-acuminate (subulate), 7–10 mm long, shorter than the free filaments; free part of filament glabrous, 10–15(–20) mm long; anthers yellow, 6–10 mm long, 1.0–1.5 mm diameter; ovary glaucous, glabrous, narrowly ellipsoid, 2–3 cm long, 0.5–1.0 cm diam.; style glaucous, filiform, (18-)20–22 cm long; stigma glaucous, obscurely 3-lobed, slightly shorter than the perianth segments (about 0.7–1.0 cm shorter); capsule ellipsoidal or cylindrical, c. 3 × 2 cm (from type spec.); seeds not seen (Azza et al., 2011).

4.5 ORIGIN AND GEOGRAPHICAL DISTRIBUTION

The natural distribution of *P. trianthum* reveals that it is a bulbous geophyte distributed in the seasonally dry tropical biome regions of Sahara to West and North Tropical Africa. The natural distribution areas are outlined mainly as; Algeria, Benin, Burkina, Cameroon, Central African Repu, Chad, Egypt, Gambia, Ghana, Ivory Coast, Mali,

DOI: 10.1201/9781003460336-4

Mauritania, Morocco, Niger, Nigeria, Senegal, Sierra Leone, Tunisia, and Morocco (www.powo.science.kew.org; Azza et al., 2011).

4.6 HABITAT AND ECOLOGY

Generally, this bulbous geophyte species is native to northern and western tropical Africa, where its habitat preferences are light soil pockets between rocks. From an ecological perspective, it develops both in seasonally dry tropical biome regions and in the hard terrains of savannahs (De Castro et al., 2012). Its sister species, *P. maritimum*, is a psammophyllous and covers an ecologically interesting habitat, the Mediterranean coasts and their sandy beaches. Much work has been done on this plant in the Mediterranean and great efforts have been made to prevent it from going extinct. The species too has not been evaluated much from medicinal point of view (Altay and Ozturk, 2012; Ozturk et al., 2014).

4.7 ETHNOBOTANY

The fact that this plant is used by local people in some African countries has made it known in the African traditional medicine system (Chaachouay et al., 2021b; Orch et al., 2021). In Senegal and Sudan, the plant is only used externally due to its toxic features (Kerharo and Adam, 1974; Ka et al., 2021a). However, it is thought to be potentially edible by local people in Nigeria and Western Sahara (Kerharo and Bouquet, 1950; Monteil, 1953; Ka et al., 2021a). In traditional medicine, extracts of this plant are generally used as anti-epileptic, anti-fungicidal, antiseptic, anti-dermatosis, anti-edema, psychotropic, soothing irritation, and wound healing (Schultes, 1978; Ba, 2020; Ka et al., 2021a). A recent study reports that the bulbs of this plant were used in the treatment of cough and asthma in the traditional folk medicine of Southern Benin (Davo et al., 2021). The preparations prepared from this plant are also reported to be used in traditional medicine in the treatment of central nervous system disorders (Ka et al., 2021a). Interestingly, there are also records of its use as a psychoactive agent by applying the bulb of this plant over cuts on the head and repeatedly rubbing the cuts on the head to cause visual hallucinations (www.zauberpilz.com/golden; Schultes 1977).

4.8 PHARMACOLOGY

The studies undertaken on the determination of phytochemical compounds present in the bulbs of this plant have revealed that these especially include; alkaloids, flavonoids, polyphenols, and saponoside (Kerharo and Adam, 1974; Davo et al., 2021). Frederik and Muraveva (1983) have identified galantamine, hippeastrin, hordenine, lycorine, pancratin, tazettin, triantini, and trisphaeridine compounds in all parts of this plant as, all being noted as Amaryllidaceae alkaloids. Recently, much attention has been paid to these alkaloids.

Among Amaryllidaceae-type alkaloids detected in the *P. trianthum*, lycorine has proven to be cytotoxic in terms of bioactivity (Kerharo and Adam, 1974; Ka et al., 2021a). In some pharmacological experimental studies, the alkaloid lycorine has been reported to exhibit potent antitumor bioactivity (Lamoral-Theys et al., 2009, 2010). In a recent study, it was found that lycorine, one of the alkaloids present in the extracts obtained from this plant, showed cytotoxic activity in hepatocarcinoma Huh7 and monocytic leukemia THP-1 cells, depending on the concentration (Ka et al., 2021a).

In many antibacterial studies conducted on the Gram-positive *S. aureus* strain, it has been reported that the crine and lycorine alkaloids present in the extracts obtained from this plant can strongly inhibit this bacterium (Adesanya et al., 1992; Evidente et al., 2004; Tan et al., 2011). A similar situation was also supported by the study by Ka et al. (2021a). It has also been emphasized that lycorine alkaloid can strongly inhibit HIV-1 and flaviviruses in vitro and in vivo modeling pharmacological studies (Szlavik et al., 2004; Zou et al., 2009; Peng et al., 2014; Wang et al., 2014; Chen et al., 2020; Ka et al., 2021a, 2021b).

4.9 MEDICINAL EFFECTS

AchE has the potential to increase both the level and duration of neurotransmitter action by preventing the cholinesterase enzyme from breaking down ACh (Colovic et al., 2013). It has also been reported that galantamine and lycorine alkaloids may also potentially exhibit AchE inhibitory bioactivity (Lopez et al., 2002; Houghton et al., 2006; Ka et al., 2021a). In African traditional medicine, although preparations prepared from this plant are used in the treatment of central nervous system disorders and psychotropics, there are very few reports on this subject. More clinical and pharmacological studies are needed to better elucidate this issue.

4.10 CONCLUSION

The Pancratium lily plant occupies an important place among the plants used in African traditional medicine. Although there are limited pharmacological and clinical studies, it is a natural genetic resource with therapeutic potential, mostly with cytotoxic, antibacterial, and antiviral activities. Keeping in view their different chemical composition, Amaryllidaceae-type alkaloids, in particular, are promising as potential candidates in the development of some innovative treatments. Therefore, detailed research studies should be conducted in order to make it available to people at a global level, especially in Africa. There is a need to fully reveal its therapeutic potential in neurodegenerative disorders and other potential diseases and to ensure its safe use in modern medicine.

LIST OF ABBREVIATIONS

AchE	anti-cholinesterases
P. trianthum	*Pancratium trianthum*
USDA	United States Department of Agriculture Natural Resources Conservation Service

REFERENCES

Adesanya, S., Olugbade, T., Odebiyl, O., and Aladesanmi, J. 1992. Antibacterial alkaloids in *Crinum jagus*. *International Journal of Pharmacognosy*, 30: 303–307.

Altay, V., and Ozturk, M. 2012. Land degradation and halophytic plant diversity of Milleyha Wetland Ecosystem (Samandağ-Hatay), Turkey. *Pakistan Journal of Botany*, 44: 37–50.

Azza, E.H., Abd El-Ghani, M., Wafaa, A., and Hassan, R. 2011. Systematic revision of the genus *Pancratium* L.(Amaryllidaceae) in Egypt with a new addition. *Notulae Scientia Biologicae*, 3(2): 24–38.

Ba, D. 2020. Le *Pancratium trianthum*. http://seyilaabe-htkm.blogspot.com/2015/10. Accessed on 25 November 2020.

Burkill, H.M. 1985. The useful plants of west tropical Africa. *Global Plants (JSTOR)*, 1985: 1.

Chaachouay, N., Douira, A., & Zidane, L. 2021a. Herbal Medicine Used in the Treatment of Human Diseases in the Rif, Northern Morocco. *Arabian Journal for Science and Engineering*, 47, 131–153. https://doi.org/10.1007/s13369-021-05501-1

Chaachouay, N., Douira, A., & Zidane, L. 2021b. COVID-19, prevention and treatment with herbal medicine in the herbal markets of Salé Prefecture, North-Western Morocco. *European Journal of Integrative Medicine*, 42, 101285. https://doi.org/10.1016/j.eujim.2021.101285

Chen, H., Lao, Z., Xu, J., Li, Z., Long, H., Li, D., Lin, L., Liu, X., Yu, L., Liu, W. et al. 2020. Antiviral activity of lycorine against Zika virus in vivo and in vitro. *Virology*, 546: 88–97.

Colovic, M.B., Krstic, D.Z., Lazarevic-Pasti, T.D., Bondzic, A.M., and Vasic, V.M. 2013. Acetylcholinesterase inhibitors: pharmacology and toxicology. *Current Neuropharmacology*, 11(3): 315–335.

Davo, A., Assogba, M.F., Aïkpe, J.A., Ahoton, D., Kinsou, L.D., Ladekan, E.Y., and Gbenou, J.D. 2021. Sub-acute oral toxicity of *Pancratium trianthum* Herb.(Amaryllidaceae) bulb's aqueous and ethanolic extracts in wistar rat. *American Journal of Applied Chemistry*, 9(5): 164–170.

De Castro, O., Brullo, S., Colombo, P., Jury, S., De Luca, P., and Di Maio, A. 2012. Phylogenetic and biogeographical inferences for *Pancratium* (Amaryllidaceae), with an emphasis on the Mediterranean species based on plastid sequence data. *Botanical Journal of the Linnean Society*, 170(1): 12–28.

Evidente, A., Andolfi, A., Abou-Donia, A.H., Touema, S.M., Hammoda, H.M., Shawky, E., and Motta, A. 2004. Amarbellisine, a lycorine-type alkaloid from *Amaryllis belladonna* L. growing in Egypt. *Phytochemistry*, 65: 2113–2118.

Frederik, D.M., and Muraveva, D. 1983. Alkaloid composition of *Pancratium trianthum*. *Chemistry of Natural Compounds*, 18: 510.

Houghton, P.J., Ren, Y., and Howes, M.-J. 2006. Acetylcholinesterase inhibitors from plants and fungi. *Natural Product Reports*, 23: 181–199.

Ka, S., Mérindol, N., Seck, I., Ricard, S., Diop, A., Boye, C.S.B., Landelouci, K., Daoust, B., Berthoux, L., Pépin, G. et al. 2021a. Biological investigation of Amaryllidaceae alkaloid extracts from the bulbs of *Pancratium trianthum* collected in the Senegalese Flora. *Molecules*, 26: 7382.

Ka, S., Merindol, N., Sow, A.A., Singh, A., Landelouci, K., Plourde, M.B., Pépin, G., Masi, M., Di Lecce, R., Evidente, A. et al. 2021b. Amaryllidaceae alkaloid cherylline inhibits the replication of dengue and Zika viruses. *Antimicrob. Agents Chemother.*, 65: e0039821.

Kerharo, J., and Adam, J.-G. 1974. *La pharmacopée sénégalaise traditionnelle: Plantes médicinales et toxiques*. Paris, France: Editions Vigot frères.

Kerharo, J., and Bouquet, A. 1950. *Plantes médicinales et toxiques de la Côte d'Ivoire-Haute-Volta: Mission d'étude de la pharmacopée indigène en AOF*. Paris, France: Editions Vigot.

Lamoral-Theys, D., Andolfi, A., Van Goietsenoven, G., Cimmino, A., Le Calvé, B., Wauthoz, N., Mégalizzi, V., Gras, T., Bruyère, C., and Dubois, J. 2009. Lycorine, the main phenanthridine Amaryllidaceae alkaloid, exhibits significant antitumor activity in cancer cells that display resistance to proapoptotic stimuli: An investigation of structure activity relationship and mechanistic insight. *Journal of Medicinal Chemistry*, 52: 6244–6256.

Lamoral-Theys, D., Decaestecker, C., Mathieu, V., Dubois, J., Kornienko, A., Kiss, R., Evidente, A., and Pottier, L. 2010. Lycorine and its derivatives for anticancer drug design. *Mini-Reviews in Medicinal Chemistry*, 10: 41–50.

Linnaeus, C. 1753. *Pancratium L. species plantarum* (Vol. 1, pp. 289–291). *Laurentii Salvii*, Stockholm.

Lopez, S., Bastida, J., Viladomat, F., and Codina, C. 2002. Acetylcholinesterase inhibitory activity of some Amaryllidaceae alkaloids and Narcissus extracts. *Life Sciences*, 71: 2521–2529.

Monteil, V. 1953. *Contribution à l'étude de la flore du Sahara occidental II*. Paris, France: Institut des Hautes Etudes Marocaines; Notes Documents VI; Éditions Larose.

Orch, H., Chaachouay, N., Douiri, E. M., Faiz, N., Zidane, L., & Douira, A. (2021). Use of medicinal plants in dermato-cosmetology: An ethnobotanical study among the population of Izarène. *Jordan Journal of Pharmaceutical Sciences*, *14*(3), 323–340.

Ozturk, M., Altay, V., Gucel, S., and Guvensen, A. 2014. Halophytes in the East Mediterranean-their medicinal and other economical values. In: Khan, M.A., Böer, B., Ozturk, M., Al Abdessalaam, T.Z., Clüsener-Godt, M., and Gul, B. (Eds.), *Sabkha ecosystems: Volume IV: Cash crop halophyte and biodiversity conservation tasks for vegetation science* (Vol. 47, pp. 247–272). Springer Netherlands: Springer Science+Business Media Dordrecht.

Peng, X., Sova, P., Green, R.R., Thomas, M.J., Korth, M.J., Proll, S., Xu, J., Cheng, Y., Yi, K., and Chen, L. 2014. Deep sequencing of HIV-infected cells: Insights into nascent transcription and host-directed therapy. *Journal of Virology*, 88: 8768–8782.

Prameela, R., Padal, S.B., and Rao, M.S. 2022. A new species of *Pancratium* Dill. ex L. (Amaryllidaceae) from Eastern Ghats of India. *Journal of Threatened Taxa*, 14(3): 20801–20804.

Schultes, E.R. 1978. *Atlas des Plantes Hallucinogènes du Monde*. Montréal, QC, Canada: Editions de l'Aurore.

Schultes, R.E. 1977. The botanical and chemical distribution of hallucinogens. *Journal of Psychedelic Drugs*, 9(3): 247–263.

Szlavik, L., Gyuris, A., Minarovits, J., Forgo, P., Molnar, J., and Hohmann, J. 2004. Alkaloids from *Leucojum vernum* and antiretroviral activity of Amaryllidaceae alkaloids. *Planta Medica*, 70: 871–873.

Tan, C.-X., Schrader, K.K., Mizuno, C.S., and Rimando, A.M. 2011. Activity of lycorine analogues against the fish bacterial pathogen *Flavobacterium columnare*. *Journal of Agricultural and Food Chemistry*, 59: 5977–5985.

USDA (United States Department of Agriculture Natural Resources Conservation Service). 2023. *Pancratium* L. https://plants.usda.gov/home/plantProfile?symbol=PANCR. Accesses on 25 July 2023.

Wang, P., Li, L.F., Wang, Q.Y., Shang, L.Q., Shi, P.Y., and Yin, Z. 2014. Anti-dengue-virus activity and structure-activity relationship studies of lycorine derivatives. *ChemMedChem*, 9: 1522–1533.

WFO. 2021. *Pancratium* L. http://www.worldfloraonline.org/-taxon/wfo-4000027860. Accessed on 01.10.2021.www.powo.science.kew.org. Accessed on 25 July 2023.www.zauberpilz.com/golden. Accessed on 25 June 2023.

Zou, G., Puig-Basagoiti, F., Zhang, B., Qing, M., Chen, L., Pankiewicz, K.W., Felczak, K., Yuan, Z., and Shi, P.-Y. 2009. A single-amino acid substitution in West Nile virus 2K peptide between NS4A and NS4B confers resistance to lycorine, a flavivirus inhibitor. *Virology*, 384: 242–252.

5 Dream herb (*Calea zacatechichi* Schltdl. Asteraceae)

Iqra Siddique and Alvina Gul

5.1 INTRODUCTION

The *Calea* genus, which falls under the Asteraceae family, encompasses approximately 125 species. Dream herb, scientifically known as *Calea zacatechichi* Schltdl, is a flowering plant belonging to the family Asteraceae and genus *Calea*, recognized for its potential medicinal properties. It is native to Mexico and is also found in other parts of Central America (Mata et al., 2022). It is commonly indicated as a 'dream herb', and derives its name from its reported ability to augment dream clarity and have significant effects on the central nervous system. This plant is utilized for medicinal purposes in various countries, but it was added to prohibited plants list in Poland in 2013, implying that its cultivation, sale, or use may be restricted in that particular country (Simonienko et al., 2013). Dream herbs have cultural importance and are recognized for their traditional usage in rituals and healing practices. The Chontal Indians give importance to this plant in their rituals to improve sensory perception and believe it helps in the reception of divine messages via dreams (Pérez, 2009). It is also used in Mexico, where it is employed for its ritualistic importance. While it has demonstrated the capability of inducting sensations of well-being and composure in healthy individuals. There is a lack of extensive research regarding its potential anxiolytic and antidepressant properties and its influence on sleep-waking levels. Historically, dream herbs have been employed for a diverse range of therapeutic purposes, for instance, anorexia, periodic fevers gastrointestinal problems, diarrhea, diabetes, skin problems, and bile-related issues. Dream herb became popular due to its association with dreams and is used to increase dream remembrance and promote lucid or vivid dreaming (Martinez-Mota et al., 2021a, 2021b; Mata et al., 2022). In Mexico, the concoction made from the entire plant is used alone or by combining it with different herbs, for the treatment of diabetes, colic, cough, and fever. Additionally, in the United States, a tincture derived from *Calea ternifolia* is available commercially and marketed for various medicinal uses, including diabetes treatment (Andrade-Cetto and Heinrich, 2005; Escandón-Rivera et al., 2017a). However, there are inadequate studies related to causing kidney toxicity of this herb, despite the sensitivity of the renal system to the detrimental effects of xenobiotics (Mossoba et al., 2016).

The human's ability to acquire knowledge from others and pass on this knowledge has been pivotal in shaping history. This is equally true in the realm of medicinal plant usage, where traditional understandings are typically transferred orally, often within families, groups, or communities. This transmission and preservation of knowledge have played a crucial role in the conservation of valuable knowledge regarding medicinal plants (Abbet et al., 2014; Saslis-Lagoudakis et al., 2014; Chaachouay et al., 2019). Despite the widespread utilization of various *Calea* species in folk medicine, limited studies have explored their efficacy. Given the prevalence of *Calea* species in traditional medicine and the documented biological activity of certain species, this chapter aims to contribute specific knowledge about *C. zacatechichi*, its botany, phytochemistry, and medicinal properties.

5.2 SCIENTIFIC CLASSIFICATION

This herb is known by the name of 'simonillo' in Central America. Scientific name commonly used to address this member of genus *Calea* is *C. zacatechichi* Schltdl. This name was given by Schlechtendal in 1835, on the basis of plant material collected from Colombia. In 1820, this plant was initially assigned the official botanical name *C. ternifolia*, but in 1835, it was reclassified as *C. zacatechichi* by the German botanist Diederich Franz Leonhard von Schlechtendal. Both names are used to refer to this plant species, but the latter one is mostly used. The term 'zacatechichi' in the name is derived from the Nahuatl language, in which 'zacatl' means 'grass' and 'chichic' means 'bitter'. Therefore, the term 'zacatechichi' translates to 'bitter grass', which refers to the plant's bitter taste (de Aguiar Amaral et al., 2017; Mata et al., 2022).

5.3 VERNACULAR NAMES

C. zacatechichi is a known dream herb. *C. ternifolia* and its associated species are commonly known by different common names in Mexico, including 'techichic', 'sacatechichi', 'sacachichie', 'zacatechi', 'zacachichi', and zacatechichi'. These names are used in different regions among different communities found in Mexico. Different names are often assigned to regional forms of this species, so a set

DOI: 10.1201/9781003460336-5

of synonyms like Amula, Jaral, Mala hierba, Prodigiosa, Hicxihuitl, and Xchim are used for referring *C. ternifolia* in different regions (de Aguiar Amaral et al., 2017; Mata et al., 2022).

5.4 BOTANICAL DESCRIPTION

Generally, the species included in the genus *Calea* exhibit a diverse range of growth forms, including perennial herbs, shrubs, subshrubs, and sometimes short trees or climbing vines. Some species may have tuberous type of roots and possess woody xylopodia. Leaves can be alternate, opposite, basal, or verticillate, having linear or ovate blade shape. They have a solitary inflorescence called 'capitula' and consist of pistillate ray florets, yellow, or whitish ligulate corollas and bisexual disc florets which have tubular corollas. *Calea* plants have black to brown colored obpyramidal cypselae or obconical fruit (dry and small fruits with one seed). Overall, *Calea* species display a diverse array of growth forms, leaf patterns, flower characteristics, and fruit attributes, contributing to the overall botanical diversity found within this genus (Kadereit et al., 2007).

Wussow et al. (1985) gave an entire description of the species, which can be summarized as: *C. zacatechichi* is characterized by shrubby growth, having 0.5 to 3 meters height, depending on environment. The stem can vary; it is either erect or lax and spreading, with varying degrees of hairs from glabrous (hairless) to pilose (covered in soft hairs) or tomentose (covered in dense, matted hairs). *C. zacatechichi* is a perennial herbaceous plant.

Leaves are arranged oppositely along the stem and are linear to ovate in shape. The leaf blades may vary in size and texture i.e., width (0.5–7 cm) and length (0.5–12 cm). Leaves are generally obtuse to acute or basally rounded. The leaves' apex is typically obtuse to acuminate. The leaf margins are coarsely crenate (have rounded teeth), serrate (have sharp, pointed teeth), or nearly entire (almost smooth). The leaves are trinerved, which means that they have three main veins extending from the base of the leaf to the tip. The leaf's upper surface can vary from scabrous (rough) to glabrous (hairless). Like the abaxial surface, they may also have resin dots. The underside of the leaves is covered in a tomentose (dense, matted hairs) to glabrescent (becoming hairless) texture. They may also possess resin dots, which are little translucent glands. The petioles' length ranges from 1 to 10 mm. They may vary from glabrous (hairless) to densely pubescent (covered in fine hairs). The petioles may contain resin dots.

C. zacatechichi produces solitary inflorescences known as capitula. These capitula consist of individual flowers arranged in a cluster. The capitula consists of two types of flowers, disc florets and ray florets. The disc florets are bisexual, having tubular corollas predominantly yellow in color, but they can also be white or purplish, albeit not frequently seen. *C. zacatechichi* has 4 to 12 sometimes up to 20-disc flowers in each flower head. The disc flowers have cream-colored corollas. The measurement of the corolla tube is about 2 to 3 mm in length and is often glandular-dotted and it flares at the base area. The throat of the corolla is approximately 0.5 mm in length and is dilated at its base. The lobes of the corolla are acute and measure 1 to 2 mm in length. The anthers, which are pollen-producing structures, protrude from the corolla tube. The ray florets are pistillate, characterized by ligulate-type corollas that are usually yellow-colored and sometimes whitish. Ray flowers in *C. zacatechichi* are mostly absent or occur in small numbers, typically 1 to 3, and are inconspicuous. When present, the ligules of the ray flowers are cream-colored to bright yellow-colored. They are shallow to deeply 2 to 3, and sometimes 4 lobed, and are approximately 2 mm in length. The length of the ligules does not exceed the disc florets. The ligules often exhibit resin dots on them. The flower heads are arranged in terminal or axillary clusters, forming corymbose or cymose clusters.

Primary peduncles are the main stalks that support clusters of flower heads. They range from 1 to 10 mm in length and can be tomentose (covered in dense, matted hairs) to sparsely pubescent (covered in scattered fine hairs). Secondary peduncles are short in length, ranging up to approximately 3 cm, and may be either absent or present. The flower heads can be either obscurely radiated (with inconspicuous ray florets) or discoid (having only tubular florets). They typically consist of 5 to 15 flowers. The involucres are cylindrical in shape, having a height of about 4 to 6 and sometimes up to 8 mm, and width of approximately 2 to 4 mm.

They have 4 to 6 seriate phyllaries, which are modified leaf-like structures enclosing the flower head. The outer phyllaries are elliptic to suborbicular or broadly ovate in shape, measuring 0.5 to 3.5 mm in width and 1 to 6 mm in length. They may be herbaceous or scarious (thin and membranous) throughout or with herbaceous tips. These phyllaries are often densely hirsute (covered in coarse hairs) to essentially glabrous (hairless) and usually exhibit resin dots. The inner phyllaries gradually transition from ovate to narrowly elliptic in shape, typically about 2.5–6 mm in length, and about 1–3 mm wide. They are scarious (thin and membranous) and have round to acute apices. The apices and margins are often seen to have cilia and membranous structures, sometimes appearing purplish. The receptacle of *C. zacatechichi* is conic in shape, measuring 0.5 to 1 mm in both length and width. The paleae, which are modified bracts surrounding the flower head, are oblong in shape. They have a length of 5 to 6 mm and 2 to 3 mm width. The paleae are scarious (thin and membranous), conduplicate (folded lengthwise), and have blunt apices with laciniate (fringed) edges. They have a slightly yellowish color.

The achenes or seeds of both ray and disc flowers in *C. zacatechichi* are cylindric to somewhat prismatic in shape, measuring 2 to 3.5 mm in length. They are either glabrous (smooth) or hirsute (covered in fine hairs). A light-colored carpopodium (structure at the base of the achene), is present. The pappus (feathery appendage attached to the seed), consists of 8 to 12 or sometimes up to 16 generally oblanceolate to sometimes oblong or lanceolate squamellae. These squamellae are stramineous (straw-colored), measuring 0.5 or 1 to 1.5 or 2 mm in length. They have rounded acute apices and may sometimes exhibit a purplish color at the base or throughout. The margins of the squamellae are minutely laciniate (finely divided). After successful pollination, *C. zacatechichi* develops obconical or obpyramidal fruits called cypselae. These cypselae are small, dry, and one-seeded fruits that may exhibit black or brown coloration. The texture of their surface ranges from glabrous (smooth) to densely pubescent (covered in fine hairs) or it may be glandular.

5.5 ORIGIN AND GEOGRAPHICAL DISTRIBUTION

C. zacatechichi is native to Mexico and Central America (Mata et al., 2022).

5.6 HABITAT AND ECOLOGY

Dream herb is an important medicinal herb naturally growing in Mexico and Costa Rica (Mata et al., 2022). This particular plant blooms for the majority of the year with its highest flowering activity occurring between June and January. It exhibits a broad range of adaptability to various climates, being present in both tropical and temperate regions. Its habitat ranges from deciduous or semideciduous tropical forests to temperate forests, including cloud, pine-oak, and pine forests. Furthermore, this plant demonstrates resilience to human-induced disruptions, surviving in proximity to roads and open spaces (Wussow et al., 1985).

5.7 ETHNOBOTANY

The initial record of documentation of *C. ternifolia* can be traced back to 16th century in Francisco Hernández's work, History of Plants from New Spain (Historia de las Plantas de la Nueva España). In this historical account, Hernández mentioned a plant named 'ahuapatli', believed to be utilized in fevers and have appetite-enhancing properties (Hernández and Ochoterena, 1942). Friar Francisco Ximénez documented a bitter herb known as 'çacachichic'. Interestingly, this herb exhibited a botanical description similar to 'ahuapatli' described by Hernández (Hernández, 1888). In 1801, Friar Juan Navarro published his book, *Natural History or American Garden* (Historia Natural o Jardín Americano), in which he mentioned plant 'ahuapatli' as a useful remedy for stomach ailments. The mentioned species was confirmed to be *C. zacatechichi* on the basis of the provided illustration in a book (Navarro, 1992). During the 19th century, a study titled 'Lecciones de Farmacología' (Pharmacology Lessons), described *C. zacatechichi* as a febrifuge and tonic. The manuscript highlighted the plant's medicinal properties in treating fevers and its potential to be used as a general tonic (Oliva, 1854). The tincture of *C. zacatechichi* was officially registered in Formulario de la Facultad Médica Mexicana (a form of Mexican Medical School) as an aperitif, known for its appetite-stimulating effects. It was recognized for its potential in treating conditions such as anorexia and hepatic illnesses (Parra, 1893). Between 1886 and 1895, the aerial parts of *C. ternifolia* underwent rigorous chemical, pharmacological, and clinical analysis. The findings of these studies showcased the plant's efficacy as an aperitif, febrifuge and in providing relief from diarrhea and stomach aches (Nacional, 1907). *C. zacatechichi* is documented under the local names 'prodigiosa' and 'ahuapahtli'. The plant's traditional usage involved preparing a decoction with sugar and egg, which was believed to aid in the treatment of digestive issues, bile problems, and flatulence (Redfield, 1928; Chaachouay et al., 2021a). The pioneering ethnobotanist named Thomas MacDougall was the first to document the oneirogenic ability of the leaves of *C. ternifolia*. The Chontal people of the Oaxaca state utilized the herb for sensory clarification purposes. According to their traditional practices, a small amount of crushed dry leaves would be infused in warm water and a cup of resulting tea would be consumed slowly. Subsequently, an individual who uses it would find a tranquil and secluded location to lie down and smoke a cigarette prepared from the dried plant leaves. It was believed that this ritual would induce a profound sense of well-being that could last for several days or more. This traditional method exemplifies the cultural significance and ceremonial use of the herb within the Chontal community (MacDougall, 1968).

Renowned ethnobotanist R. Gordon Wasson put forth a speculation that a specific flower depicted in the famous Xochipilli statue, known as the Aztec Prince of Flowers, present at The National Museum of Anthropology in Mexico, might belong to *C. ternifolia*. Based on his expertise and knowledge of psychedelic plants, Wasson hypothesized that the representation of the flower in the statue could be related with *C. ternifolia* due to its psychoactive properties, which he later clarified is not related to *C. ternifolia* (Wasson, 1973, 1982). The Chontal Curanderos (Healers) made a clear distinction between the medicinal use of the plant and its application in oneiromancy, which is the practice of using dreams for divination or spiritual purposes. The Chontal healers recognized that the plant, presumably *C. ternifolia*, possessed distinct properties that could be utilized both for medicinal purposes, such as treating ailments and promoting the well-being of human as well as to facilitate dream-like experiences and spiritual insights. This differentiation underscores the Chontal healers' comprehensive understanding of the plant's diverse applications in their traditional healing practices (Díaz, 1979; Chaachouay et al., 2021b). Toward the end of the 19th century, in the second and third editions of New Mexican Pharmacopea, *C. zacatechichi*'s galenical preparations were officially registered as digestive and aperitive agents. Interestingly, the botanical description provided in these editions aligns with the current understanding of *C. ternifolia* (Mexicana, 1896). The plant, believed to be *C. ternifolia*, is said to possess the ability to purify the senses and facilitate the reception of spiritual messages during the dream phase. As a part of traditional practice, a handful of dried leaves is recommended, which is known to induce sensations of drowsiness, serenity, nausea, and reduction in pulse rate and heartbeat. Additionally, it is suggested to place some of these leaves under the pillow before sleeping, which is believed to enhance the potential of dream experiences and spiritual insights (Toro and Thomas, 2007). In the latter half of the 20th century, multiple publications documented the diverse uses of this plant, believed to be *C. ternifolia*, for various ailments. In veterinary medicine, it is useful in treating stomach pain in rams (Martínez, 1991). Moreover, this plant was recognized for its potential to manage obesity (Alonso-Castro et al., 2015) as well as gynecological and respiratory conditions (Leonti et al., 2003) and diabetic conditions (de Aguiar Amaral et al., 2017). It was noted to be employed in the treatment of skin diseases, periodic fevers, bile illnesses, swollen scalp, 'muina' (tantrum), 'coraje' (anger), as well as serving as an appetizer and antidiarrheic agent (Lara Reimers et al., 2019; Ortega-Cala et al., 2019).

5.8 PHARMACOLOGY

A total of 257 compounds are extracted from various species of *Calea* genus. A remarkable diversity is observed in compounds among species. Among all compounds, 116 compounds are sesquiterpene lactones, 18 are derived from p-hydroxyacetophenone, 14 are phenolic compounds, 10 are chromenes, 8 are flavonoids, 7 are benzofurans, and 5 are chromanones. The majority of phytochemical studies focusing on *Calea* species have been conducted using plant material collected from Mexico (Escandón-Rivera et al., 2017a; Sonia et al., 2012). Some phytochemical analyses have also been done by using plants obtained from Guatemala (Bohlmann and Zdero, 1977). Honduras (Herz and Kumar, 1980), El Salvador (Köhler et al., 2002) or commercially bought (Sałaga et al., 2016). A comprehensive phytochemical investigation focusing on the plant parts (leaves and stems) of *C. zacatechichi* resulted in the isolation of a noteworthy series of 6 germacranolides. These compounds have exhibited significant antileishmanial activity, showcasing their potential as therapeutic agents against Leishmania, a parasitic disease (Wu et al., 2011). Only study on essential oils of the plant revealed that the constituents found in essential oil are caleochromene A, camphor, β-eudesmol, curcumene, caryophyllene oxide and spathulenol (Escandón-Rivera et al., 2017a). The phytochemical studies at different times led to the identification of several specialized metabolites from *C. ternifolia*. The isolated compounds include 37 sesquiterpene lactones (Bohlmann and Zdero, 1977; Fischer et al., 1984; Wu et al., 2011), including sesquiterpene calicein II cpmpound (Chávez-Soto, 1977), essential oils (Carvalho et al., 2014), 9 flavonoids, 6 flavones, 3 flavonols, 4 chromenes, chromanones, 2 simple coumarins, 2 triterpenoids and additionally, 2 phenolic acids, 4 acetylenes, and 2 common fatty acids were also identified (do Nascimento and de Oliveira, 2014; Escandón-Rivera et al., 2012; Escandón-Rivera et al., 2017a; Köhler et al., 2002; Lima et al., 2015, 2016; Mendes do Nascimento et al., 2007; Sałaga et al., 2016). A significant portion of constituents reported sesquiterpene characteristics (Ohguchi et al., 2009).

5.9 MEDICINAL EFFECTS

It has been traditionally used to address various ailments such as rheumatism, edema, respiratory tract disorders, cold fever, and painful limbs (Segura-Cobos et al., 2010). Originally recognized for its oneirogenic properties, *Calea* was initially employed to induce lucid dreams, enhance dreaming, and generate vivid dream experiences. However, the specific compounds responsible for the psychoactive effects, acting on the receptors of the central nervous system, have yet to be identified (Bachliński et al., 2013; Rodrigues and Carlini, 2006). *C. zacatechichi* exhibits plenty of pharmacological properties. Non-polar type extracts derived from *Calea ternifolia* exhibited a modest antiplasmodial effect against both *Plasmodium falciparum* strains (chloroquine-sensitive and chloroquine-resistant strains). The IC50 (half-maximal inhibitory concentration) values varied from 10 to 24.3 ppm, whereas the values for artemisinin were significantly lower at 0.0008 ppm and 0.004 ppm. The primary contributors to the observed activity were genkwanin, linoleic acid, and apigenin (Köhler et al., 2002). An aqueous extract of *C. zacatechichi*, administered at a dose rate of 100 and 10 mg kg^{-1}, exhibited anti-edema properties in rat paws induced by carrageenan. The extract effectively reduced neutrophil migration by 44% into the peritoneal cavity. This effect was comparative to positive control substances, indomethacin (10 mg kg^{-1}) and dexamethasone (1 mg kg^{-1}). *C. zacatechichi* exhibits compounds that have the power of anti-inflammation which may be linked to the lipoxygenase product and prostaglandins biosynthesis (Venegas-Flores et al., 2002). The methanol extract of *C. zacatechichi* (dose: 25–50 mg kg^{-1}), demonstrated dose-dependent antinociceptive action in the acetic acid writhing test. Additionally, the extract exhibited inhibitory effects on edema induced by carrageenan in rat paws at a dose rate of 150–300 mg kg^{-1}. The antinociceptive activity of a methanol extract (at 25 and 50 mg kg^{-1}) from *C. zacatechichi* and its fractions (30–50 mg kg^{-1}) were studied (Segura-Cobos et al., 2010). The hydroalcoholic extract of plant, at a concentration of 1 mg mL^{-1}, demonstrated a significant restrictive effect of 60% on the activity of crude enzymes in Sprague Dawley rat intestines (Ramírez et al., 2012). Germacranolides present in *C. zacatechichi* showed leishmanicidal activity (Wu et al., 2011). Extracts of *C. zacatechichi* demonstrated antinociceptive and antidiarrheal activity in mouse models having irritable bowel syndrome. A methanolic infuse was subjected to further extraction using dichloromethane to obtain a solid form of extract. The findings suggest that these extracts have the potential to be utilized as a therapeutic option for managing diarrhea and pain associated with bowel syndrome (Sałaga et al., 2015).

The antidepressant and anxiolytic activities were assessed in rodents using classical behavioral assays. The neuropharmacological activities of a commercially purchased aqueous extract of *C. zacatechichi* were investigated. The extract was evaluated using three convulsion seizure models, locomotor activity tests, elevated plus-maze, and forced swim and grip tests to assess antidepressant-like behavior. The researchers found that the aqueous extract did not alter the seizure thresholds in mice for phenyltetrazole, 6 Hz psychomotor, or maximal electroshock models. Furthermore, muscle strength, exploratory behavior, or anxiety levels remained unaffected by extract at the tested doses. At dose rates of 400 and 800 mg kg^{-1}, the immobility time was enhanced in mice. Based on their findings, the authors found that the oral administration of the aqueous extract of *C. ternifolia* did not induce significant changes in neurological events, anxiety levels, or exploratory behavior in mice (Sałaga et al., 2016). An aqueous extract obtained from above-ground parts of plant exhibited significant effects in attenuating postprandial hyperglycemia in mice suffering from hyperglycemia (induced by streptozotocin/nicotinamide). The extract given at doses of 100 and 316 mg kg^{-1} demonstrated improved glucose tolerance during oral sucrose and glucose tolerating tests, outperforming positive controls glibenclamide and acarbose, respectively. In addition, the extract displayed hypoglycemic effects in an experiment, with the 316 mg kg^{-1} dose exhibiting superior efficacy compared to glibenclamide. Furthermore, the extract exhibited inhibitory activity against yeast-α-glucosidase, with an IC50 value of 0.169 mg mL^{-1} as compared to 1.12 mg mL^{-1} for acarbose. These findings highlight the effective blood glucose level regulation by *C. ternifolia* in animal models, both during fasting and after meals (Escandón-Rivera et al., 2017b; Escandón-Rivera et al., 2012). In mice and rats, *C. zacatechichi* demonstrated potent anxiolytic and antidepressant effects comparable to well-established drugs within the dosage range of 0.5–50 mg kg^{-1}. Moreover, when administered at the rate of 100 mg kg^{-1}, *C. zacatechichi* induced mild sedation in rats, characterized by observable effects on episodes of Slow Wave Sleep over a 6-hour period, along with heightened fast frequencies in the hippocampus (specifically in the gamma-band range of 31–50 Hz) during paradoxical or REM sleep (Martinez-Mota et al., 2021a).

Oneirogenic properties refer to the ability of any substance to augment dream-related experiences. *C. zacatechichi* (dream herb), is renowned for its oneirogenic attributes and has been used for countless years to provoke and enhance dreams. Traditional indigenous cultures in Mexico have long recognized the remarkable dream-inducing abilities of *C. zacatechichi*. It is suggested as a prototypical dream-inducing plant. It is believed that the plant can grant access to the realm of dreams, enabling individuals to have meaningful, vivid, and even lucid dream experiences. It holds a prominent place in shamanic rituals, where it is believed to help in communication with ancestors and spirits via dream interactions. Scientific investigations have also explored the oneirogenic properties of this plant. It significantly possesses the ability to augment dream recall, making it easier for individuals to remember and document their dreams upon waking. Furthermore, it enhances the clarity of dreams, making them more immersive and realistic. The precise mechanisms behind *C. zacatechichi*'s oneirogenic properties are not yet fully known. However, it is believed that the plant's effects may be attributed to its unique chemical composition (i.e., compounds including sesquiterpenes and flavonoids are thought to interact with neurotransmitters and receptors in the brain associated with the dreaming process). An exceptional aspect of *C. zacatechichi*'s oneirogenic properties is its potential to induce lucid dreaming. This is a state in which the dreamer becomes aware that they are dreaming and can actively engage in and manipulate the content of the dream. It is important to note that the effects of *C. zacatechichi* can vary from person to person. Some individuals may experience profound dream enhancement and lucidity while others may have more subtle effects or no noticeable changes in their dream experiences. Using this plant as an oneirogenic herb should be approached with caution, and it is advisable to consult with a healthcare professional before use, especially if you have any pre-existing medical conditions or are taking medications (Mossoba et al., 2016; Toro and Thomas, 2007). Further research is still needed to fully understand the mechanisms behind its oneirogenic effects and to explore its potential therapeutic applications in the field of dream studies.

C. zacatechichi is recommended for short-term usage due to its limited toxicity against *A. salina*. Prolonged administration at a dose of 8.5 mg/kg resulted in unfavorable effects on the kidneys and liver of Winstar rats. However, in traditional medicine, it is commonly employed for extended durations, particularly in managing hyperglycemia (González-Yáñez et al., 2019). To comprehensively understand the plant's biological activity, toxicity, and mechanism of action, further research is imperative in identifying the specific chemical constituents responsible. The researchers tested the aqueous extract of the plant obtained from Loma Bonita in the state of Oaxaca using a modified Lorke procedure. No signs of toxic effects or mortality were observed at doses of 10 and 100 mg kg^{-1}. However, at raised doses of extract to 1000, 2600, and 5000 mg kg^{-1}, the animals exhibited symptoms such as loose stools, pain, and diarrhea soon after taking a dose. Despite these adverse effects, all animals survived the treatment period and maintained a healthy condition, characterized by bright and clear eyes, consistent fecal pellets, and intact mucous membranes (Martinez-Mota et al., 2021b). The lyophilized methanol extract of *C. zacatechichi* plant exhibited concentration-dependent cytotoxicity against human proximal tubule cells in vitro, with a lethal concentration 50 (LC50) value of 91.7 μg mL^{-1} as compared to 13.3 μg mL^{-1}

for cisplatin. Treatment with the extract led to the toxicity of mitochondria, as evidenced by elevated levels of reactive oxygen species (ROS) in cells, surpassing those observed in cisplatin-treated cells. Moreover, the extract caused faster mitochondrial injury as compared to cisplatin. Additionally, the extract of *C. ternifolia* (at concentrations of 111 and 333 $\mu g\ mL^{-1}$) caused elevated concentrations of nephrotoxicity-associated biomarkers, detected using human kidney toxicity kits (Mossoba et al., 2016). Collectively, the consumption of *C. zacatechichi* or its constituents may potentially result in adverse health effects, particularly when used over an extended period to manage hyperglycemia. Further comprehensive research is required to thoroughly investigate the toxicity of this plant.

The appropriate dose of *C. zacatechichi* is currently unknown due to insufficient scientific information. Factors such as age, health condition, and other individual circumstances should be taken into consideration when determining the appropriate dosage. There is a potential risk associated with the oral or inhalation use of *C. zacatechichi*, as its safety has not been thoroughly examined in clinical trials. Reported side effects include hallucinations, as well as nausea and vomiting. It is important to exercise caution when using *C. zacatechichi* and consult with a healthcare professional for appropriate guidance. It has the potential to slow down the central nervous system, so caution is advised for individuals taking sedatives or other central nervous system depressants, including alcohol. It may also cause allergic reactions in individuals sensitive to the Asteraceae family. The herb can affect breathing rate, so individuals with respiratory conditions such as asthma should use it with caution or avoid its usage altogether. Additionally, *C. zacatechichi* can cause low blood sugar levels, so people having diabetes should be cautious and monitor for signs of lower blood sugar. It may also lower blood pressure, so individuals with heart-related problems or using blood pressure medication should be cautious. Lastly, it should be avoided during psychiatric conditions or when taking psychiatric medication, as it can induce vivid images, disorienting effects, and hallucinations (Simonienko et al., 2013; US Food and Drug Administration, 2022).

5.10 CONCLUSION

The *Calea* genus offers potential pharmacophores for drug discovery, particularly chromenes and sesquiterpene lactones, with anti-inflammatory and anticancer properties. *C. zacatechichi* has been widely used in Mexican traditional medicine for different purposes. It possesses oneirogenic properties that have been valued by traditional cultures and are of interest to scientific researchers. Its ability to enhance dream recall, vividness and potentially induce lucid dreaming make it a fascinating herb for those interested in exploring the realms of dreams and consciousness. Pharmacological studies, both in vitro and in vivo, have explained the plant's diverse properties, including anti-inflammatory, antinociceptive, spasmolytic, antidepressive, antiprotozoal, antidiarrheic, antidiabetic, and anxiolytic effects. Some of these properties align with its traditional uses. Phytochemically, the plant produces bioactive sesquiterpene lactones, coumarins, chromenes, flavonoids acetylenes, and phenylpropanoids, with sesquiterpene lactones being particularly noteworthy. However, toxicological effects such as mitochondrial toxicity, cytotoxicity, and inhibition of eryptosis have also been reported in vitro. Before considering the application or recommendation of *C. zacatechichi* for disease prevention or treatment, further pharmacological toxicological studies are necessary.

LIST OF ABBREVIATIONS

C. ternifolia	*Calea ternifolia*
C. zacatechichi	*Calea zacatechichi*

ACKNOWLEDGMENTS

Our special thanks go to Prof. Dr. Munir Ozturk and Prof. Dr. Volkan Altay for their critical evaluation of this chapter.

REFERENCES

Abbet, C., Slacanin, I., Corradi, E., De Mieri, M., Hamburger, M., and Potterat, O. 2014. Comprehensive analysis of *Cirsium spinosissimum* Scop., a wild alpine food plant. *Food Chemistry*, 160: 165–170.

Alonso-Castro, A.J., Domínguez, F., Zapata-Morales, J.R., and Carranza-Álvarez, C. 2015. Plants used in the traditional medicine of Mesoamerica (Mexico and Central America) and the Caribbean for the treatment of obesity. *Journal of Ethnopharmacology*, 175: 335–345.

Andrade-Cetto, A., and Heinrich, M. 2005. Mexican plants with hypoglycaemic effect used in the treatment of diabetes. *Journal of Ethnopharmacology*, 99(3): 325–348.

Bachliński, R., Duszyńska, A., Kowalczuk, A., and Łozak, A. 2013. Zastosowanie metody GC-MS do charakterystyki *Calea zacatechichi* i Psychotria viridis-roślin kontrolowanych ustawą o przeciwdziałaniu narkomanii. *Problemy Kryminalistyki*, 279: 15–21.

Bohlmann, F., and Zdero, C. 1977. Neue germacrolide aus *Calea zacatechichi*. *Phytochemistry*, 16(7): 1065–1068.

Carvalho, C.C., Turatti, I.C.C., Lopes, N.P., Salvador, M.J., and Nascimento, A.M. 2014. Chemical composition and antimicrobial activity of essential oil from Brazilian plants *Acanthospermum australe, Calea fruticosa* and *Mikania glauca*. *African Journal of Pharmacy and Pharmacology*, 8(14): 392–398.

Chaachouay, N., Benkhnigue, O., Fadli, M., Ibaoui, H., & Zidane, L. 2019. Ethnobotanical and ethnopharmacological studies of medicinal and aromatic plants used in the treatment of metabolic diseases in the Moroccan Rif. *Heliyon*, *5*(10), e02191. https://doi.org/10.1080/10477845.2021.1942433

Chaachouay, N., Douira, A., & Zidane, L. 2021a. Herbal Medicine Used in the Treatment of Human Diseases in the Rif, Northern Morocco. *Arabian Journal for Science and Engineering*, 47, 131–153. https://doi.org/10.1007/s13369-021-05501-1

Chaachouay, N., Douira, A., & Zidane, L. 2021b. COVID-19, prevention and treatment with herbal medicine in the herbal markets of Salé Prefecture, North-Western Morocco. *European Journal of Integrative Medicine*, 42, 101285. https://doi.org/10.1016/j.eujim.2021.101285

Chávez-Soto, M. 1977. *Investigación química-etnobotánica de Calea zacatechichi Tesis de pregrado*. Mexico: Universidad Nacional Autónoma de Mexico.

de Aguiar Amaral, P., Costa, F.V., Antunes, A.R., Kautz, J., Citadini-Zanette, V., Lohézic-Le Dévéhat, F., Barlow, J., and DalBó, S. 2017. The genus *Calea* L.: A review of isolated compounds and biological activities. *Journal of Medicinal Plants Research*, 11(33): 518–537.

Díaz, J.L. 1979. Ethnopharmacology and taxonomy of Mexican psychodysleptic plants. *Journal of Psychedelic Drugs*, 11(1–2): 71–101.

do Nascimento, A.M., and de Oliveira, D.C. 2014. Three new chromanones from *Calea clausseniana*. *Helvetica Chimica Acta*, 97(1): 146–150.

Escandón-Rivera, S., González-Andrade, M., Bye, R., Linares, E., Navarrete, A., and Mata, R. 2017b. Correction to α-glucosidase inhibitors from *Brickellia cavanillesii*. *Journal of Natural Products*, 80(1): 233–233.

Escandón-Rivera, S., Pérez-Vásquez, A., Navarrete, A., Hernández, M., Linares, E., Bye, R., and Mata, R. 2017a. Anti-hyperglycemic activity of major compounds from *Calea ternifolia*. *Molecules*, 22(2): 289.

Escandón-Rivera, S., González-Andrade, M., Bye, R., Linares, E., Navarrete, A., and Mata, R. 2012. α-glucosidase inhibitors from *Brickellia cavanillesii*. *Journal of Natural Products*, 75(5): 968–974.

Fischer, N.H., Lee, I.-Y., Fronczek, F.R., Chiari, G., and Urbatsch, L.E. 1984. Three new furanone-type heliangolides from *Calea ternifolia* and the molecular structure of 8β-angeloyloxy-9α-hydroxycalyculatolide. *Journal of Natural Products*, 47(3): 419–425.

González-Yáñez, M.G.E., Rivas-Morales, C., Oranday-Cárdenas, M.A., Verde-Star, M.J., Núñez-González, M. A., Sanchez, E., and Leos-Rivas, C. 2019. Safety of aqueous extract of *Calea ternifolia* used in Mexican traditional medicine. *Evidence-Based Complementary and Alternative Medicine*, 2019: 7478152.

Hernández, F. 1888. Cuatro libros de la naturaleza y virtudes de las plantas y animales: de uso medicinal en la Nueva España. Oficina tip. de la Secretaría de Fomento.

Hernández, F., and Ochoterena, I. 1942. Historia de las plantas de Nueva España.

Herz, W., and Kumar, N. 1980. Sesquiterpene lactones of *Calea zacatechichi* and *C. urticifolia*. *Phytochemistry*, 19(4): 593–597.

Kadereit, J., Jeffrey, C., and Kubitzki, K. 2007. *VIII: Flowering plants*: Eudicots: Asterales.

Köhler, I., Jenett-Siems, K., Siems, K., Hernández, M.A., Ibarra, R.A., Berendsohn, W.G., Bienzle, U., and Eich, E. 2002. In vitro antiplasmodial investigation of medicinal plants from El Salvador. *Zeitschrift für Naturforschung C*, 57(3–4): 277–281.

Lara Reimers, E.A., Lara Reimers, D.J., Chaloupkova, P., Zepeda del Valle, J.M., Milella, L., and Russo, D. 2019. An ethnobotanical survey of medicinal plants used in Papantla, Veracruz, Mexico. *Plants*, 8(8): 246.

Leonti, M., Sticher, O., and Heinrich, M. 2003. Antiquity of medicinal plant usage in two Macro-Mayan ethnic groups (Mexico). *Journal of Ethnopharmacology*, 88(2–3): 119–124.

Lima, T.C., Santos, A.D.C., Costa, D., Souza, R.J., Barison, A., Steindel, M., and Biavatti, M.W. 2015. Chromenes from leaves of *Calea pinnatifida* and evaluation of their leishmanicidal activity. *Revista Brasileira de Farmacognosia*, 25: 7–10.

Lima, T.C., Souza, R.J., Santos, A.D., Moraes, M.H., Biondo, N.E., Barison, A., Steindel, M., and Biavatti, M.W. 2016. Evaluation of leishmanicidal and trypanocidal activities of phenolic compounds from *Calea uniflora* Less. *Natural Product Research*, 30(5): 551–557.

MacDougall, T. 1968. A composite with psychic properties. *Garden Journal*, 18(4): 105.

Martinez-Mota, L., Cruz-Tavera, A., Dorantes-Barrón, A.M., Arrieta-Báez, D., Ramírez-Salado, I., Cruz-Aguilar, M.A., Mayagoitia-Novales, L., Cassani, J., and Estrada-Reyes, R. 2021b. *Calea zacatechichi* Schltdl. (Compositae) produces anxiolytic- and antidepressant-like effects, and increases the hippocampal activity during REM sleep in rodents. *Journal of Ethnopharmacol*, 265: 113316.

Martinez-Mota, L., Cruz-Tavera, A., Dorantes-Barrón, A.M., Arrieta-Báez, D., Ramírez-Salado, I., Cruz-Aguilar, M.A., Mayagoitia-Novales, L., Cassani, J., and Estrada-Reyes, R. 2021a. *Calea zacatechichi* Schltdl. (Compositae) produces anxiolytic- and antidepressant-like effects, and increases the hippocampal activity during REM sleep in rodents. *Journal of Ethnopharmacology*, 265: 113316.

Martínez, M. 1991. Las plantas medicinales de México. *Las plantas medicinales de México*, 1991: 656–656.

Mata, R., Contreras-Rosales, A.J., Gutiérrez-González, J.A., Villaseñor, J.L., and Pérez-Vásquez, A. 2022. *Calea ternifolia* Kunth, the Mexican 'dream herb', a concise review. *Botany*, 100(2): 261–274.

Mendes do Nascimento, A., Costa, F.C., Thiemann, O.H., and Rodrigues de Oliveira, D.C. 2007. Chromanones with leishmanicidal activity from *Calea uniflora*. *Zeitschrift für Naturforschung C*, 62(5–6): 353–356.

Mexicana, S.F. 1896. *Nueva farmacopea mexicana de la Sociedad farmacéutica de México*. Oficina tipográfica de la Secretaría de Fomento.

Mossoba, M.E., Flynn, T.J., Vohra, S., Wiesenfeld, P., and Sprando, R.L. 2016. Evaluation of 'dream herb', *Calea zacatechichi*, for nephrotoxicity using human kidney proximal tubule cells. *Journal of Toxicology*, 2016: 9794570.

Nacional, I.M. 1907. Datos para la materia médica Mexicana. Vol. 4, Oficina tip. de la Secretaría de fomento.

Navarro, J.F. 1992. Historia natural. O, Jardin americano: manuscrito de 1801.

Ohguchi, K., Ito, M., Yokoyama, K., Iinuma, M., Itoh, T., Nozawa, Y., and Akao, Y. 2009. Effects of sesquiterpene lactones on melanogenesis in mouse B16 melanoma cells. *Biological and Pharmaceutical Bulletin*, 32(2): 308–310.

Oliva, L. 1854. Lecciones de farmacologia: dadas por el catedratico del ramo en la Universidad de Guadalajara. Vol. 2, D. Rodriguez.

Ortega-Cala, L.L., Monroy-Ortiz, C., Monroy-Martínez, R., Colín-Bahena, H., Flores-Franco, G., Luna-Cavazos, M., and Monroy-Ortiz, R. 2019. Plantas medicinales utilizadas para enfermedades del sistema digestivo en Tetela del Volcán, Estado de Morelos, México. *Boletín Latinoamericano y del Caribe de plantas medicinales y aromáticas*, 18(2): 106–129.

Parra, G. 1893. *Formulario de la facultad medica Mexicana formado por el doc tor Guillermo Parra*. Manero y Nava.

Pérez, P.M. 2009. Biblioteca Digital de la Medicina Tradicional Mexicana. *Revista Digital Universitaria*, 10(10): 1607–6079.

Ramírez, G., Zavala, M., Pérez, J., and Zamilpa, A. 2012. In vitro screening of medicinal plants used in Mexico as antidiabetics with glucosidase and lipase inhibitory activities. *Evidence-Based Complementary and Alternative Medicine*, 2012: 701261.

Redfield, R. 1928. Remedial plants of Tepoztlan: A Mexican folk herbal. *Journal of the Washington Academy of Sciences*, 18(8): 216–226.

Rodrigues, E., and Carlini, E.A. 2006. Plants with possible psychoactive effects used by the Krahô Indians, Brazil. *Brazilian Journal of Psychiatry*, 28: 277–282.

Sałaga, M., Fichna, J., Socała, K., Nieoczym, D., Pieróg, M., Zielińska, M., Kowalczuk, A., and Wlaź, P. 2016. Neuropharmacological characterization of the oneirogenic Mexican plant *Calea zacatechichi* aqueous extract in mice. *Metabolic Brain Disease*, 31: 631–641.

Sałaga, M., Kowalczuk, A., Zielinska, M., Błażewicz, A., and Fichna, J. 2015. *Calea zacatechichi* dichloromethane extract exhibits antidiarrheal and antinociceptive effects in mouse models mimicking irritable bowel syndrome. *Naunyn-schmiedeberg's Archives of Pharmacology*, 388: 1069–1077.

Saslis-Lagoudakis, C.H., Hawkins, J.A., Greenhill, S.J., Pendry, C.A., Watson, M.F., Tuladhar-Douglas, W., Baral, S.R., and Savolainen, V. 2014. The evolution of traditional knowledge: environment shapes medicinal plant use in Nepal. *Proceedings of the Royal Society B: Biological Sciences*, 281(1780): 20132768.

Segura-Cobos, D., Venegas-Flores, H., Baiza-Gutman, L.A., and Vázquez-Cruz, B. 2010. Antinociceptive and anti-inflammatory effects of the methanol extract of *Calea zacatechichi* leaves and its fractions. *Pharmacology*, 2: 1100–1110.

Simonienko, K., Waszkiewicz, N., and Szulc, A. 2013. Psychoactive plant species-actual list of plants prohibited in Poland. *Psychiatria Polska*, 47(3): 499–510.

Sonia, E.R., Martin, G.-A., Robert, B., Edelmira, L., Andrés, N., and Rachel, M. 2012. α-glucosidase inhibitors from *Brickellia cavanillesii. Journal of Natural Products*, 75(5): 968–974.

Toro, G., and Thomas, B. 2007. *Drugs of the dreaming: Oneirogens: Salvia divinorum and other dream-enhancing plants*. Simon and Schuster.

US Food and Drug Administration. 2022. FDA poisonous plant database. https://www.fda.gov/food/science-research-food/fda-poisonous-plant-database

Venegas-Flores, H., Segura-Cobos, D., and Vázquez-Cruz, B. 2002. Antiinflammatory activity of the aqueous extract of *Calea zacatechichi*. In: *Proceedings of the Western Pharmacology Society* (Vol. 45, pp. 110–111). Seattle, Wash.: The Society.

Wasson, R.G. 1973. The role of flowers'in nahuatl culture: A suggested interpretation. *Botanical Museum Leaflets, Harvard University*, 23(8): 305–324.

Wasson, R.G. 1982. Xochipilli, 'príncipe de las flores'. *Revista de la Universidad de Mexico*, 37: 10–18.

Wu, H., Fronczek, F.R., Burandt, C.L., and Zjawiony, J.K. 2011. Antileishmanial germacranolides from *Calea zacatechichi*. *Planta Medica*, 77(7): 749–753.

Wussow, J.R., Urbatsch, L.E., and Sullivan, G.A. 1985. *Calea* (Asteraceae) in Mexico, Central America, and Jamaica. *Systematic Botany*, 1985: 241–267.

6 *Tabernanthe iboga* Baill. Apocynaceae

Emmanuel Muhumuza and Godwin Anywar

6.1 INTRODUCTION

Tabernanthe iboga Baill., commonly known as iboga, is a perennial herb belonging to the Apocynaceae family. It traditionally holds great value in cultural practices and medicinal applications in Central African countries like Cameroon, Congo among others (Gallo et al., 2009; June & Dec, 2004). The plant has a long history of traditional use among Indigenous communities since, when taken in high doses, it causes hallucinogenic properties in initiation ceremonies, divine knowledge seeking, and settling social disputes by seeking divine advice (Marcelin et al., 2000; Pope, 1969; Winkelman, 2014). In moderate doses, it has been used for its healing properties to treat conjunctivitis, addictions, and an aphrodisiac and febrifuge (Forsyth et al., 2016). In very small quantities it is used as a tonic against cough, thirst, fatigue, and hunger (Forsyth et al., 2016; Paškulin et al., 2012) especially during hunting where long endurance is required. It is also used as a stimulant in small doses. The cortex of *T. iboga* has been traditionally and culturally utilized by tribes in the Congo basin in Central Africa as a psychoactive medium in initiation ceremonies of young men into adulthood through induction of trance and is considered to reveal one's purpose of life and his role in society (Pope, 1969; Chaachouay et al., 2019). Although the Tabernanthe iboga is not edible, its roots are the major source of ibogaine, which is an indole alkaloid and is naturally found in bark (Popik & Skolnick, 1999). The fruits and leaves have been known to contain compounds responsible for aphrodisiac and anesthetic properties against toothaches (Jenks, 2002; Winkelman, 2014).

Recent scientific research has revealed its pharmacological potential, particularly in the fields of addiction treatment and mental health. The precise neural mechanisms underlying these effects remain a subject of ongoing research. Studies using mammalian cell culture and animal models seemed to point to the fact that glial cell line-derived neurotrophic factor (GDNF) produced from glial cells' up-regulation contributed to reduced ethanol consumption ad libitum (Carnicella et al., 2010). Parallel research and studies have suggested the inhibition of numerous receptors in the central nervous system to block addiction. These include blockage of nicotinic acetylaldehyde receptors resulting in withdrawal from nicotine, and tobacco at large, and alcohol addiction (Arias et al., 2010; Glick et al., 2002).

Ibogaine overdose has been traditionally known to be fatal in Gabon but the ceremonies usually initiate young healthy individuals and are watched carefully throughout the ceremonies in case of deteriorating health (Marcelin et al., 2000). Increasing evidence has also revealed that under certain underlying conditions the drug can be cardiotoxic (Alper et al., 2012; Litjens & Brunt, 2016), neuro-toxic and even lethal (Litjens & Brunt, 2016; Ona et al., 2022). Most cases have pointed to its ability to cause sudden cardiac arrest in addiction withdraw subjects. This led to its categorization as a Schedule 4 drug in the United States hence becoming illegal to use, possess, sell, or move without a license (Alper et al., 2001).

6.2 SCIENTIFIC CLASSIFICATION

Tabernanthe iboga belongs to the family Apocynaceae.

6.3 VERNACULAR NAMES

Tabernanthe iboga, commonly known as iboga, is referred to by various vernacular names in different regions and languages. These include: Iboga (English), Bwiti, Kamba (Gabon), Bojogae, Mbobondo (Cameroon), Ngando, Eboka, Eboga, Mabonga (Congo), M'boyo (Brazzaville) (Gallo et al., 2009). The Bwiti tribe is specifically associated with the spiritual, shamanic, and religious practices that involve the use of iboga in Gabon and surrounding regions.

6.4 BOTANICAL DESCRIPTION

Tabernanthe iboga is an evergreen small equatorial shrub that typically grows from 2 to 10 meters. The plant has dark green, leathery leaves that are elliptical in shape and arranged oppositely along the stem. It produces small, tubular, yellowish-white flowers with a pleasant fragrance. It bears bright orange ovoid/elliptical-shaped fruits with a thick exocarp containing many seeds. It thrives in the humid understoreys of rainforests, riverine areas, and swamp forests of Central Africa, where it forms an essential part of the ecosystem (Marcelin et al., 2000; Pope, 1969).

DOI: 10.1201/9781003460336-6

6.5 ORIGIN AND GEOGRAPHIC DISTRIBUTION

Tabernanthe iboga is a perennial shrub native to the rainforests of West-Central Africa. It is primarily found in several countries within this region, including Gabon, Cameroon, Republic of the Congo, and the Democratic Republic of the Congo (DRC) (Pope, 1969):(June & Dec, 2004). The natural distribution of *T. iboga* is limited to these countries in West-Central Africa however, the plant has been cultivated in other parts of the world for its medicinal and psychoactive properties (Pope, 1969). The fact that this shrub is largely localized in a small geographical area, human activities like farming and overexploitation are threatening the ecological success of this plant (Nyongo Ndoua & Vaghar, 2018). Indeed, the usage and exploitation of this shrub by the native people and the adoption of Bwiti shamanic practices in surrounding communities (Fernandez, 2019; Samorini, 1995) have seen increased exploitation of the plant. These factors, coupled with its very slow growth, mean the plant population is under immense pressure.

6.6 ECOLOGY

T. iboga thrives best in forest understoreys, although it grows in riverine or swamp forests and relatively wet savannahs. These areas are typically rich, well-drained, loamy soils and can tolerate both acidic and alkaline conditions. The plant requires high humidity and a stable temperature range of 20–30°C for optimal growth hence has been found at elevations from 0 to 1500 meters above sea level (Marcelin et al., 2000; Samorini, 1995). Since it grows in understoreys, it flowers and fruits throughout the year regardless of the season (June & Dec, 2004; Marcelin et al., 2000). The ecological role of *T. iboga* within its native habitat is not extensively studied, but it likely serves as a food source for various animals and plays a role in maintaining the biodiversity of the rainforest ecosystem. (Cousins & Huffman, 2002) suggested that gorillas and chimpanzees used iboga fruits for food and their leaves for recreation purposes.

6.7 PHYTOCHEMISTRY

The pharmacologically active components of *Tabernanthe iboga* are primarily found in its root bark. Indole alkaloids are however also found in the stem bark, root, seeds, and leaves, with about 20 compounds so far discovered (Bouso et al., 2020; Jenks, 2002). A study that utilized plant metabolomics techniques to profile the compounds in the iboga plant found 5 phenolic compounds and 30 alkaloids, 21 of which were newly discovered from *T. iboga* root bark (Bading-Taika et al., 2018). The key alkaloids found in the plant include ibogaine, ibogamine, voacangine, ibogaline, tabernanthine, coronaridine (Iyer et al., 2021; Popik & Skolnick, 1999). Among these, ibogaine is the most widely studied and recognized for its psychoactive properties. Iboga alkaloids belong to the indolomonoterpenes class, and are characterized by an isoquinuclidine in their nuclei. Ibogaine is a selective antagonist of various neurotransmitter receptors, including N-methyl-D-aspartate (NMDA), serotonin (5-HT2A), and kappa-opioid receptors (Popik & Skolnick, 1999). Ibogine has been proposed as a drug cessation/anti-addiction treatment and has a wide range of activities in targeting opioids, cocaine, cocaine, methamphetamine, and alcohol. Additionally, *T. iboga* contains several other alkaloids, flavonoids, and terpenes, which contribute to its complex chemical profile (Bouso et al., 2020; Corkery, 2018).

6.8 ETHNOBOTANY

Tabernanthe iboga plays an important role in the cultural and spiritual practices of Indigenous communities in Central Africa. It is utilized in initiation ceremonies, healing rituals, and shamanism in these communities (Iyer et al., 2021; Pope, 1969). The plant is considered a sacred tool for accessing ancestral knowledge, growth, and promoting spiritual awakening. Its cultural significance and traditional uses highlight the deep-rooted connection between humans and nature (Gomes & Tofoli, 2022; Marcelin et al., 2000; Chaachouay et al., 2021a). Traditionally, the root barks are collected and air-dried, then pounded into a yellowish/brownish powder. The iboga root may be eaten whole, or crushed and mixed with other ingredients, sometimes other psychoactive compounds like marijuana (Jenks, 2002; June & Dec, 2004). Users believe the smoke generated represents the soul leaving the body and traveling to meet with the ancestors (Marcelin et al., 2000). In small doses, iboga is eaten for its stimulant effect to stay awake and alert while hunting, or to reduce fatigue, hunger, and thirst (Jenks, 2002; Winkelman, 2014).

The unique pharmacological profile of *Tabernanthe iboga*, particularly its active alkaloid ibogaine, has attracted attention as a potential therapeutic agent. Traditionally, the root bark is consumed wholly or it can be ground and rolled into sizable balls for chewing on (Pope, 1969). During ceremonies, it is commonly mixed with other plant ingredients like marijuana and decocted (Maciulaitis et al., 2008). The root bark is the major source of hallucinogenic compounds used for spiritual discovery and to connect to the ancestors and the spirit world where it is taken in large quantities (June & Dec, 2004; Winkelman, 2014).

Iboga extracts are also taken in smaller quantities for therapeutic purposes. The roots are considered to be aphrodisiac, febrifuge, and tonic (Paškulin et al., 2012; Pope, 1969). A decoction of the whole root is taken to combat tiredness, hunger, and thirst, especially while hunting and farming, to overcome physical stress (Maciulaitis et al., 2008). A root macerate in palm wine is used to soothe coughs and eye drops to treat conjunctivitis (Forsyth et al., 2016). The leaves of *T. iboga* are used by locals for their anesthetic and aphrodisiac properties (Gomes & Tofoli, 2022; Chaachouay et al., 2021b). The warmed leaves are rubbed on the gums to overcome toothache (Winkelman, 2014).

Ibogaine may be used in the treatment of substance addiction. It may help to reduce drug cravings and withdrawal symptoms, potentially providing a window of opportunity for individuals to break free from addiction (Gallo et al., 2009; Maciulaitis et al., 2008; Ona et al., 2023). Ibogaine has been shown to be effective in the treatment of addictions to a number of substances as well as alleviating withdrawal symptoms, specifically opioids, alcohol and nicotine (Gallo et al., 2009; Maciulaitis et al., 2008) and to alleviate withdrawal symptoms, reduce cravings, and facilitate psychological introspection, potentially aiding in the resolution of underlying psychological issues contributing to addiction (Souza et al., 2011). Ibogaine was first used as a pharmaceutical product in Western medicine under the brand name, Lambarene, as an extract of the *Tabernanthe manii* containing 8 mg of ibogaine (Ona et al., 2023). In 1905 ibogaine was also marketed as a treatment for depression and cardiac disorders under many brands. Iperton was also sold as a tonic for depression that contained ibogaine (Gallo et al., 2009; Ona et al., 2023).

6.9 HALLUCINOGENIC EFFECTS

Ibogaine is classified as a dissociative hallucinogen. This implies that it can induce profound alterations in perception, cognition, and consciousness. The effects of ibogaine can last for several hours usually between 4 and 24 hours, depending on the dose and individual sensitivity (Ona et al., 2022). The experience is often described as 'dream-like or visionary' (Gallo et al., 2009; June & Dec, 2004). The hallucinogenic effects of ibogaine can vary from person to person, but some common experiences include vivid visual imagery, enhanced introspection, and a sense of expanded awareness. Users also report experiencing altered time perception, synesthesia (mixing of sensory modalities), and the presence of spiritual or introspective insights (June & Dec, 2004; Chaachouay & Zidane, 2021). Ibogaine interacts with various receptors and neurotransmitters in the brain, producing a range of pharmacological effects. It acts as a non-selective serotonin reuptake inhibitor, a mild NMDA receptor antagonist, and a moderate κ-opioid receptor antagonist (Akendengue et al., 2005; Ona et al., 2023). These actions result in altered neuronal activity and modulation of neurotransmitter release, leading to profound psychological and physiological effects. Ibogaine induces visionary experiences, enhances introspection, and promotes neuroplasticity during radioligand binding studies (Deecher et al., 1992).

Due to its potent psychoactive effects and potential risks, ibogaine can have serious physiological effects on the cardiovascular system (Vlaanderen et al., 2014). In some instances, ibogaine has been associated with adverse events and fatalities (Forsyth et al., 2016; Vlaanderen et al., 2014). These hallucinogenic effects arise from ibogaine's interaction with multiple neural pathways and Central Nervous System, (CNS) receptor sites (Itzhak & Ali, 1998). It modulates glutamate signaling through NMDA receptor antagonism, affects serotonergic neurotransmission, and influences the opioid system. It is a widely accepted mechanism that ibogaine molecules bind to multiple receptors in the CNS, particularly the N-NMDA, opioid, serotonin, and nicotinic acetylcholine receptors (Corkery, 2018; Gallo et al., 2009; Itzhak & Ali, 1998). These interactions lead to alterations in neuronal activity and the release of various neurotransmitters. Ibogaine induces vivid visual imagery and hallucinations even in animal models (Cousins & Huffman, 2002; Palumbo & Winter, 1992; Wei et al., 1998). These hallucinations often include geometric patterns, complex scenes, and introspective experiences. Ibogaine can also induce profound changes in consciousness, including a dream-like state and an altered sense of self. These altered states are attributed to the modulation of neural circuits involved in perception, cognition, and self-awareness.

6.10 CONCLUSION

Tabernanthe iboga, and its main active compound ibogaine, exhibit potent hallucinogenic properties. These properties have attracted scientific interest and exploration for both potential therapeutic applications and a deeper understanding of consciousness. While ibogaine shows promise as a treatment for addiction and psychological exploration, its risks and legal status should be carefully considered. Further research is necessary to fully comprehend its mechanisms of action and unlock its therapeutic potential while ensuring the safety of individuals who may choose to explore its effects in alleviating addictions and substance withdrawal symptoms. There is also a great need to preserve the shrub, as overexploitation in its native habit as well as human activities like farming are a threat to its flourishment.

LIST OF ABBREVIATIONS

NMDA	N-methyl-D-aspartate
GDNF	Glial cell line-derived neurotrophic factor
hERG	human Ether-a-go-go Related Gene
CNS	central nervous system

REFERENCES

Akendengue, B., Lemamy, G. J., Bourobou, H. B., & Laurens, A. (2005). Bioactive natural compounds from medico-magic plants of Bantu area. *Studies in Natural Products Chemistry*, *32*, 803–820.

Alper, K. R., Beal, D., & Kaplan, C. D. (2001). A contemporary history of ibogaine in the United States and Europe. *The Alkaloids: Chemistry and Biology*, *56*, 249–281. https://doi.org/10.1016/S0099-9598(01)56018-6

Alper, K. R., Stajić, M., & Gill, J. R. (2012). Fatalities temporally associated with the ingestion of ibogaine. *Journal of Forensic Sciences*, *57*(2), 398–412.

Arias, H. R., Rosenberg, A., Targowska-Duda, K. M., Feuerbach, D., Yuan, X. J., Jozwiak, K., Moaddel, R., & Wainer, I. W. (2010). Interaction of ibogaine with human α3β4-nicotinic acetylcholine receptors in different conformational states. *The International Journal of Biochemistry & Cell Biology*, *42*(9), 1525–1535.

Bading-Taika, B., Akinyeke, T., Magana, A. A., Choi, J., Ouanesisouk, M., Torres, E. R. S., Lione, L. A., Maier, C. S., Bobe, G., & Raber, J. (2018). Phytochemical characterization of *Tabernanthe iboga* root bark and its effects on dysfunctional metabolism and cognitive performance in high-fat-fed C57BL/6J mice. *Journal of Food Bioactives: An Official Scientific Publication of the International Society of Nutraceuticals and Functional Foods (ISNFF)*, *3*, 111.

Bouso, J. C., Fornís, I., Vilamala, M. V., Loenen, B. D., Sainz-Cort, A., Jiménez-Garrido, D. F., Santos, R. G. D., Hallak, J. E., Alcázar-Córcoles, M. Á., & Jenks, C. W. (2020). An analytical study of iboga alkaloids contained in *Tabernanthe iboga*-derived products offered by ibogaine treatment providers. *Archives of Clinical Psychiatry (São Paulo)*, *47*, 51–54.

Chaachouay, N., Benkhnigue, O., Fadli, M., Ibaoui, H., & Zidane, L. (2019). Ethnobotanical and ethnopharmacological studies of medicinal and aromatic plants used in the treatment of metabolic diseases in the Moroccan Rif. *Heliyon*, *5*(10), e02191. https://doi.org/10.1080/10477845.2021.1942433

Chaachouay, N., & Zidane, L. (2021). The Symbolic Efficacy of Plants in Rituals and Socio-Religious Ceremonies in Morocco, Northwest of Africa. *Journal of Religious & Theological Information*, *21*(1–2), 34–53. https://doi.org/10.1080/10477845.2021.1942433

Chaachouay, N., Douira, A., & Zidane, L. (2021a). Herbal Medicine Used in the Treatment of Human Diseases in the Rif, Northern Morocco. *Arabian Journal for Science and Engineering*, 47, 131–153. https://doi.org/10.1007/s13369-021-05501-1

Chaachouay, N., Douira, A., & Zidane, L. (2021b). COVID-19, prevention and treatment with herbal medicine in the herbal markets of Salé Prefecture, North-Western Morocco. *European Journal of* ***Integrative Medicine***, 42, 101285. https://doi.org/10.1016/j.eujim.2021.101285

Carnicella, S., He, D. Y., Yowell, Q. V., Glick, S. D., & Ron, D. (2010). Noribogaine, but not 18-MC, exhibits similar actions as ibogaine on GDNF expression and ethanol self-administration. *Addiction Biology*, *15*(4), 424–433.

Corkery, J. M. (2018). Ibogaine as a treatment for substance misuse: Potential benefits and practical dangers. *Progress in Brain Research*, *242*, 217–257.

Cousins, D., & Huffman, M. A. (2002). Medicinal properties in the diet of gorillas: An ethno-pharmacological evaluation. *African Study Monographs*, *23*(2), 65–89.

Deecher, D. C., Teitler, M., Soderlund, D. M., Bornmann, W. G., Kuehne, M. E., & Glick, S. D. (1992). Mechanisms of action of ibogaine and harmaline congeners based on radioligand binding studies. *Brain Research*, *571*(2), 242–247.

Fernandez, J. W. (2019). *Bwiti: An ethnography of the religious imagination in Africa*. Princeton University Press.

Forsyth, B., Machado, L., Jowett, T., Jakobi, H., Garbe, K., Winter, H., & Glue, P. (2016). Effects of low dose ibogaine on subjective mood state and psychological performance. *Journal of Ethnopharmacology*, *189*, 10–13.

Gallo, C., Renzi, P., Loizzo, S., Loizzo, A., & Capasso, A. (2009). *Tabernanthe iboga*: A comprehensive review. *Pharmacologyonline*, *3*, 906–920.

Glick, S. D., Maisonneuve, I. M., Kitchen, B. A., & Fleck, M. W. (2002). Antagonism of α3β4 nicotinic receptors as a strategy to reduce opioid and stimulant self-administration. *European Journal of Pharmacology*, *438*(1–2), 99–105.

Gomes, B. R., & Tofoli, L. F. (2022). A sacred plant of neuronal effect: The use of ibogaine in addiction treatments in Brazil. *Anthropology of Consciousness*, *33*(2), 333–357.

Itzhak, Y., & Ali, S. F. (1998). Effect of Ibogaine on the various sites of the NMDA receptor complex and sigma binding sites in Rat Brain a. *Annals of the New York Academy of Sciences*, *844*(1), 245–251.

Iyer, R. N., Favela, D., Zhang, G., & Olson, D. E. (2021). The iboga enigma: The chemistry and neuropharmacology of iboga alkaloids and related analogs. *Natural Product Reports*, *38*(2), 307–329.

Jenks, C. (2002). Extraction studies of *Tabernanthe iboga* and Voacanga africana. *Natural Product Letters*, *16*(1), 71–76.

June, J. F. M. A. M., & Dec, J. A. S. O. N. (2004). Traditional use. *The Key Non-Timber Forest Products of Central Africa: State of the Knowledge*, *122*, 1–199.

Litjens, R. P., & Brunt, T. M. (2016). How toxic is ibogaine? *Clinical Toxicology*, *54*(4), 297–302.

Maciulaitis, R., Kontrimaviciute, V., Bressolle, F., & Briedis, V. (2008). Ibogaine, an anti-addictive drug: Pharmacology and time to go further in development. A narrative review. *Human and Experimental Toxicology*, *27*(3), 181.

Marcelin, T. M., Stella, A., Nouhou, N., & Paul, B. (2000). State of knowledge study On *Tabernanthe iboga* Baillon A report for the Central African Regional Program for the Environment.

Nyongo Ndoua, P. D., & Vaghar, K. (2018). Bwiti, iboga, trance and healing in Gabon. *Mental Health, Religion & Culture*, *21*(8), 755–762.

Ona, G., Kohek, M., & Bouso, J. C. (2023). The modern clinical use of hallucinogens. What can we learn from the natural product research? *Natural Product Research*, *37*(10), 1742–1745.

Ona, G., Rocha, J. M., Bouso, J. C., Hallak, J. E., Borràs, T., Colomina, M. T., & Dos Santos, R. G. (2022). The adverse events of ibogaine in humans: An updated systematic review of the literature (2015–2020). *Psychopharmacology*, *239*(6), 1977–1987.

Palumbo, P. A., & Winter, J. (1992). Stimulus effects of ibogaine in rats trained with yohimbine, DOM, or LSD. *Pharmacology Biochemistry and Behavior*, *43*(4), 1221–1226.

Paškulin, R., Jamnik, P., Danevčič, T., Koželj, G., Krstić-Milošević, D., Blagojević, D., & Štrukelj, B. (2012). Metabolic plasticity and the energy economizing effect of ibogaine, the principal alkaloid of *Tabernanthe iboga*. *Journal of Ethnopharmacology*, *143*(1), 319–324.

Pope, H. G. (1969). *Tabernanthe iboga*: An African narcotic plant of social importance. *Economic Botany*, *23*(2), 174–184.

Popik, P., & Skolnick, P. (1999). Pharmacology of ibogaine and ibogaine-related alkaloids. *Alkaloids-New York-Academic Press*, *52*, 197–232.

Samorini, G. (1995). The Bwiti religion and the psychoactive plant *Tabernanthe iboga* (Equatorial Africa). *Integration*, *5*, 105–114.

Souza, A., Mbatchi, B., & Herchuelz, A. (2011). Induction of insulin secretion by an aqueous extract of *Tabernanthe iboga* Baill.(Apocynaceae) in rat pancreatic islets of Langerhans. *Journal of Ethnopharmacology*, *133*(3), 1015–1020.

Vlaanderen, L., Martial, L., Franssen, E., Van Der Voort, P., Oosterwerff, E., & Somsen, G. (2014). Cardiac arrest after ibogaine ingestion. *Clinical Toxicology*, *52*(6), 642–643.

Wei, D., Maisonneuve, I., Kuehne, M., & Glick, S. (1998). Acute iboga alkaloid effects on extracellular serotonin (5-HT) levels in nucleus accumbens and striatum in rats. *Brain Research*, *800*(2), 260–268.

Winkelman, M. (2014). Psychedelics as medicines for substance abuse rehabilitation: Evaluating treatments with LSD, Peyote, Ibogaine and Ayahuasca. *Current Drug Abuse Reviews*, *7*(2), 101–116.

7 False Peyote (*Lophophora diffusa* (Croizat) Bravo Cactaceae)

Ehsan Ali, Muhammad Farooq Azhar, and Ume Laila

7.1 INTRODUCTION

Peyote, scientifically known as *Lophophora williamsii*, is a cactus species renowned for its primary alkaloid mescaline, which is a well-known psychedelic substance (Cassels and Sáez-Briones, 2018). It has long been used as a sacred sacrament in Native American religious ceremonies and continues to be practiced today. However, due to the psychedelic effects of mescaline, other cactus species have been mistakenly harvested and sold as peyote. When we refer to 'True peyote', we are typically talking about *L. williamsii* specifically. On the other hand, all other cacti species related to peyote are considered 'false peyote'. *Lophophora diffusa* and *Lophophora fricii* are commonly referred to as 'false peyote', but there are other species like *Ariocarpus retusus* that resemble *L. williamsii* in terms of appearance, alkaloid content, habitat, or genetic relationship (Chan et al., 2021). False peyotes thrive in identical environments and resemble peyote in appearance (Šnicer et al., 2009). In traditional situations, they have also been utilized as alternatives for peyote, albeit they often lack or have modest mescaline effects (Rätsch, 2005; Chaachouay et al., 2019). Many fake peyote species are likewise vulnerable in their native environments, suffering comparable difficulties as peyote except for ingestion.

The taxonomic categorization of the *Lophophora* genus has caused some consternation. For example, *L. williamsii* has been given several names, including *Ariocarpus williamsii, Echinocactus williamsii, Echinocactus lewisii,* and *Anhalonium lewinii.* Similarly, *Anhalonium williamsii* was formerly assigned to *L. diffusa* (Perrine, 2001). Because of this inconsistency, it was difficult to determine which name corresponded to which species. The present nomenclature is based on alkaloid content analyses. With their initial publications in 1967 and 1975, *L. diffusa* and *Lophophora fricii* Haberm were classified as the second and third species within the *Lophophora* genus, respectively. A fourth species, Lophophora alberto-vojtechii Bohata, Myák & Nice, was recently identified. It is crucial to clarify that Astrophytum Asterias (Zucc.) Lem (Terry, 2007) is a cactus that is not in the *Lophophora* genus and is known as 'false peyote'. (Terry, 2007).

L. diffusa, also known as false peyote, is a rare and endangered cactus species that faces threats from psychedelic tourism, medicinal use, and religious practices, leading to a decrease in its population size (Briseño-Sánchez et al., 2020). This perennial plant relies on insect pollinators due to its low production of pollen and nectar (Díaz-Segura et al., 2017; Briseño-Sánchez et al., 2020). Unfortunately, approximately half of its population has been lost, mostly as a result of land use changes (Díaz Segura et al., 2012). While *L. williamsii* and *L. diffusa* have some physical similarities, they may be recognized by variations in stem color, the presence of ribs and furrows, and the color of the inner perianth segments of the flower (Anderson, 1996). *L. williamsii* bears pink blooms and blue-green or reddish-green stems with ribs and furrows. L. diffusa, on the other hand, has yellowish-green stems with no ribs and frequently yields white blooms.

The chapter explores the captivating world of *L. diffusa*, commonly known as false peyote. It aims to provide a comprehensive investigation into this distinctive cactus species, covering its taxonomic classification, vernacular names, botanical characteristics, geographical distribution, ecology, phytochemistry, ethnobotany, as well as its pharmacological and hallucinogenic effects. By conducting an extensive review of scientific literature, ethnobotanical studies, phytochemical analyses, and research on pharmacology and hallucinogenic properties, this chapter offers a holistic understanding of false peyote. Its goal is to contribute to the growing body of knowledge surrounding this intriguing cactus species, shedding light on its biology, cultural significance, and potential applications.

7.2 SCIENTIFIC CLASSIFICATION

L. diffusa is a plant species belonging to the Cactaceae family and is classified as one of the species within the *Lophophora* genus. The detailed taxonomic classification is mentioned in Table 7.1.

7.3 VERNACULAR NAMES

L. diffusa, a globular cactus, is known by various common names including peyote (Briseño-Sánchez et al., 2020), false peyote (Saroya and Singh, 2020), and peyote queretano (Díaz-Segura et al., 2023).

7.4 BOTANICAL DESCRIPTION

L. diffusa exhibits globose and slightly flattened stems, which can grow individually or in clusters (Figure 7.1). The

DOI: 10.1201/9781003460336-7

TABLE 7.1
Taxonomic Classification of False Peyote (*Lophophora diffusa* (Croizat) Bravo)

Kingdom	Plantae
Phylum	Tracheophytes
Class	Angiosperms
Subclass	Eudicots
Order	Caryophyllales
Family	Cactaceae
Subfamily	Cactoideae
Genus	*Lophophora*
Species	*Lophophora diffusa*

Sources: Gómez-Hinostrosa et al., 2017.

FIGURE 7.1 False peyote (*Lophophora diffusa* (Croizat) Bravo). Source: Lin et al., 2023.

stems have a grayish-green epidermis, while the flowers are white with hints of pink, and the mature fruits range in color from light pink to purple (Bravo-Hollis, 1967; Scheinvar, 2004). The flowering of *L. diffusa* has been observed between March and September, with multiple flowering peaks occurring throughout this period (Anderson, 1996; Díaz-Segura et al., 2017; Briseño-Sánchez et al., 2020). The species is self-incompatible and relies on biotic pollination ((Mafham & Mafham, 1997; Díaz-Segura et al., 2017).

The flowers have a short lifespan, typically lasting one day but occasionally up to three days. Despite their small size, the flowers of *L. diffusa* produce less pollen and nectar when compared to other cactus species (Nassar and Ramírez, 2004; Flores-Martínez et al., 2013). They interact with a variety of insects, including beetles, orthopterans, and solitary bees, with the latter serving as the species' major pollinator (Díaz-Segura et al., 2017; Briseño-Sánchez et al., 2020). *L. diffusa* is a xenogamous plant that relies on biotic pollination to produce fruit and seeds (Briseño-Sánchez et al., 2020). After pollination, the tiny fruits develop in about two months. Each flower produces a limited number of ovules ($\bar{x} = 32.5 \pm 1.27$ SE), resulting in a small number of seeds per fruit ($\bar{x} = 27.1 \pm 1.1$ SE) (Briseño-Sánchez et al., 2020). The seeds possess an elaiosome and an impermeable coating that enables them to float, suggesting potential dispersal mechanisms involving ants or water during rainfall. Additionally, gravity plays a role in the close-range dispersal of seeds near the mother plant (Briseño-Sánchez et al., 2020).

7.5 ORIGIN AND GEOGRAPHIC DISTRIBUTION

L. diffusa is an endemic cactus found in Querétaro semidesert (Anderson, 1969). The rarity is due to its unique ecological niche and limited distribution (Díaz-Segura et al., 2017). It is currently listed as an endangered species on the NOM-059-SEMARNAT-2010 and as an endangered species on the Red List of the International Union for the Conservation of Nature and Natural Resources (IUCN). In addition, it is included in Annex II of the International Trade Convention on Endangered Species of Wild Animals and Plants (CITES) (Mafham & Mafham, 1997). Despite being locally abundant, *L. diffusa* faces threats from wild harvesting and habitat alteration, resulting in a population decline of approximately 50% (Aragane et al., 2020; Díaz Segura et al., 2012). This species is endemic to the Chihuahuan Desert, with populations confined to the arid zones of Querétaro, Mexico, at elevations ranging from 1300 to 1600 meters above sea level (Hernández-Magaña et al., 2012; Gómez-Hinostrosa et al., 2017).

7.6 ECOLOGY

L. diffusa thrives in a specific habitat found in a small portion of the southern Chihuahuan Desert. It can be found in stony hills with gentle slopes or alluvial deposits derived from limestone and shale. The soils in this habitat have a high pH level ranging from 7.6 to 7.8 and are characterized by low organic matter content (Ibarra-Laclette et al., 2015). Within this environment, *L. diffusa* coexists with various shrubs such as *Larrea tridentata* (DC.) Coville, *Fouquieria splendens* Engelm., *Acacia sororia* Standl., *Mimosa* L. sp., *Karwinskia* Zucc. sp., and *Bursera* Jacq. ex L. sp. However, it is worth noting that nurse-plant associations have only been documented with *L. tridentata* and *A. sororia* (Díaz Segura et al., 2012; Trujillo Hernández, 2002).

7.7 PHYTOCHEMISTRY

Tetrahydroisoquinolines (THIQs) and phenethylamines are two different groups of alkaloids found in *Lophophora*. From the *Lophophora* genus alone, more than 40 different alkaloids have been discovered. Heffter identified the main alkaloid as pellotine. Subsequently, the pharmacological effects of pellotine in *L. diffusa* were investigated and will be discussed in a later section. While pellotine is the major alkaloid in *L. diffusa*, abundantly present at 0.75–0.89%

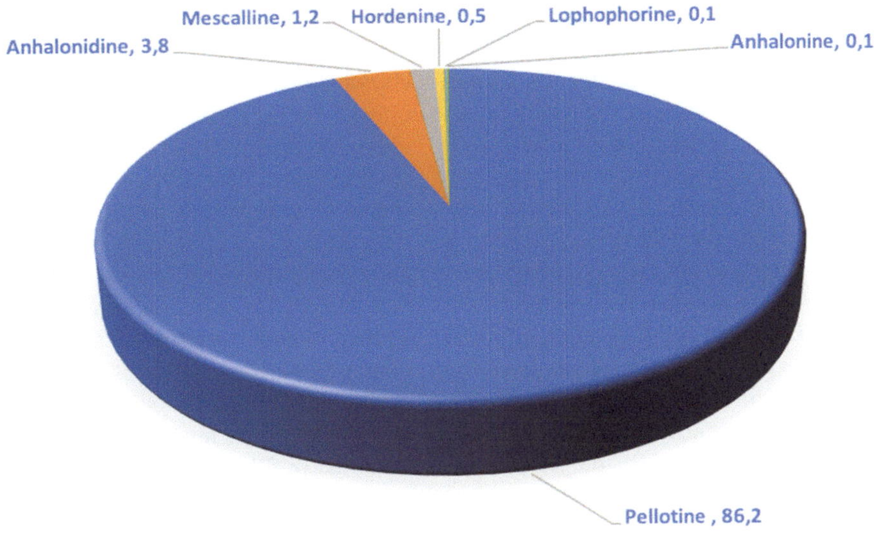

FIGURE 7.2 Percentage of total alkaloid content found in false peyote (*Lophophora diffusa* (Croizat) Bravo). (Source: Chan et al., 2021.)

of fresh weight, it is found in lower concentrations (0.14% fresh weight) in *L. williamsii* (Chan et al., 2021). The percentage of total alkaloid content found in *L. diffusa* mentioned in Figure 7.2 according to Chan et al. (2021). She has recorded the values of total alkaloid contents documented by different researchers and taken the mean value of total alkaloid content found in *L. diffusa*.

7.8 ETHNOBOTANY

There are not many ethnobotanical studies on how *L. diffusa* is used in different cultures, and we don't know much about how it's consumed. Additionally, the readily available information frequently offers conflicting experiences. Some individuals report feelings of clumsiness, confusion, and general discomfort after consuming *L. diffusa*, while others describe a sense of mental clarity and tranquility (Díaz, 1979; Chaachouay et al., 2021a). Indigenous populations have historically used *L. diffusa* for various purposes, including as a narcotic, for medicinal applications, and as an ornamental plant. It has been traditionally utilized as a religious sacrament in Native American ceremonies for many centuries. In the past, both American and Mexican communities used *L. diffusa* as a remedy for treating different ailments due to its morphological similarities with *L. williamsii* (Chan et al., 2021). Mexican Indigenous people, in particular, have employed *L. diffusa* as a medicinal plant for its therapeutic effects on respiratory, rheumatic, and venereal diseases (Ibarra-Laclette et al., 2015; Chaachouay et al., 2021b).

7.9 THE PHARMACOLOGICAL AND HALLUCINOGENIC EFFECTS

Pellotine is the primary and most abundant alkaloid in *Lophophora diffusa*, constituting approximately 0.75–0.89% of its fresh weight. Heffter studied the behavioral and hallucinogenic effects of pellotine in a variety of species, such as frogs, cats, rabbits, and dogs. Two distinct dosage ranges – 8–10 mg and 20–30 mg – were investigated in frog tests. After 30 to 40 minutes of the lower dosage range, there was an increase in reflex excitability and tetanic convulsions. The tetanic convulsions developed into total paralysis at the higher dosage level. At the 8–10 mg dose, the animals displayed a mild narcotic condition for the first 20 minutes, which was scarcely perceptible at the 20–30 mg level. A brief period of anesthesia was produced in rabbits at dosages of 50–70 mg/kg, and the animals were easily awakened but immediately returned to the anesthetic condition. Within 30 minutes of ingestion, tremors, ear twitching, and persistent chewing started to appear. These symptoms finally gave way to convulsions and tetanic seizures. The animals experienced convulsions and then died at dosages of 90–100 mg/kg. At dosages of 40 mg/kg, cats began to throw up and scream pitifully. At dosages of 60–90 mg/kg, increased salivation, dilated pupils, alertness, twitching of the ears, and tetanic convulsions were noted (Perrine, 2001). Dogs, on the other hand, displayed extraordinary tolerance to pellotine, with no hallucinogenic or behavioral effects observed even at doses of up to 50 mg/kg. In 1966, Brossi et al. conducted a study on the hallucinogenic effect of pellotine in mice and concluded that it had no effect, particularly when considering acute toxicity.

The pharmacological actions of pellotine were primarily studied in the early 20th century, with initial experiments conducted by Heffter. Heffter's reports on the sleep-inducing properties of pellotine in humans sparked a further investigation of the compound in clinical settings (Cassels and Sáez-Briones, 2018). Boehringer and Sohn briefly promoted pellotine as a sleep aid before the advent of less expensive synthetic barbiturates eclipsed it. The fact that pellotine had to be obtained from a natural source restricted its usage as a general sedative and was one of its key disadvantages at the time. Heffter, who discovered pellotine initially, also wrote about its effects on people. He was initially interested in figuring out what made peyote psychedelic but discovered that pellotine had a distinct sleep-inducing effect rather than any hallucinogenic effects (Ríos-Carrasco, 2017). In his self-experiment, dosages between 50 and 240 mg caused dizziness without any visual changes and slowed heart rate. Heffter was unable to stay awake and saw no aftereffects the next day. These findings prompted more research into the possible benefits of pellotine as a sleep aid. Jolly studied the use of pellotine as a possible painkiller and sleep aid in 1896, focusing mostly on patients with neuritis, multiple sclerosis, and tabes dorsalis (neurosyphilis). A modest analgesic effect, however distinct from morphine, was achieved by the oral or subcutaneous injection of 40–60 mg of pellotine. The effects of greater dosages were comparable to those of methylsulfonal (Trional) or chloral hydrate in the evening, when doses of 20–60 mg caused sleep. As adverse effects, vertigo, tinnitus, and a drop-in heart rate were noted. Jolly concluded that pellotine could be a beneficial sleeping aid, although some patients did not experience any effects. The same year, Pilcz administered pellotine beginning at a dosage of 20 mg to 58 patients from the Vienna Psychiatric Clinic. Pellotine showed no impact in 21% of patients; as a result, these instances were eliminated from the research. 49% of patients slept off within 90 minutes after treatment and slept well all night. 20 mg was adequate to provide equivalent results in 33% of subjects. When other hypnotics had failed, pellotine was discovered to work well with no obvious negative effects (Anderson, 1996). Hutchings administered doses of 15–30 mg of pellotine for up to 3 weeks to 11 mental patients to study their reactions to repeated dosages. No long-term side effects were observed, and no withdrawal symptoms were reported upon discontinuation of treatment. Patients experienced natural and calm sleep, waking up feeling refreshed. Additional daytime sleep of up to four hours was obtained when pellotine was administered during the day. Within the first hour after administration, minor side symptoms such as nausea, vomiting, vertigo, and dizziness appeared. When nausea and dizziness were reported, Hutchings observed a change in pulse rate (Chaachouay & Zidane, 2021). Even in the absence of sleep induction, pellotine seemed to have a soothing effect. When Langstein administered 10 mg of pellotine to a patient with tabes dorsalis, the patient instantly went into cyanosis and was on the verge of passing out. Heffter expressed doubts over the nature or composition of the administered alkaloid. Pellotine is an alkaloid produced by *L. diffusa* that has potent psychedelic effects akin to those of lysergic acid. In the Western world, these hallucinogenic effects may have spread considerably further (Ibarra-Laclette et al., 2015).

7.10 CONCLUSION

In the review presented in this chapter, we have provided an in-depth exploration of *Lophophora diffusa*, its taxonomic classification, vernacular names, botanical description, origin and geographic distribution, ecology, phytochemistry, ethnobotany, and the pharmacological and hallucinogenic effects. The available information in the scientific literature about the ethnobotanical and ethnomedicinal uses is often contradictory; therefore, it is recommended that phytochemical and pharmacological investigation should be carried out to validate the folk knowledge. It is concluded that pellotine is the main abundant alkaloid (86.2%) in *L. diffusa* followed by anhalonidine (3.8%) and mescaline (1.2%) while contains zero to trace amounts of hordenine, anhalinine, and lophophine by using available limited literature studies. Therefore, the study suggests that advanced analytical techniques must be carried out to trace the complete chemical profile of the plant which helps in the conservation of this unique threatened species.

CONFLICT OF INTERESTS

The authors declared that they have no conflict of interest concerning the authorship and/or publication of this chapter.

LIST OF ABBREVIATIONS

CITES Convention on International Trade in Endangered Species of Wild Fauna and Flora

IUCN International Union for Conservation of Nature and Natural Resources

REFERENCES

Anderson, E.F. (1969). The biogeography, ecology, and taxonomy of *Lophophora* (Cactaceae). *Brittonia*, 21, 299–310.

Anderson, E.F. (1996). *Peyote: The divine cactus*. University of Arizona Press.

Aragane, M., Sasaki, Y., Fukumori, N., Yoshizawa, M., Suzuki, Y., Kitagawa, S., Mori, K., Ogino, S., Yasuda, I., & Nagumo, S. (2020). Peyote identification on the basis of differences in morphology, mescaline content, and trnL/trnF sequence between *Lophophora williamsii* and *L. diffusa*. *Journal of Natural Medicines*, 65, 103–110. https://doi.org/10.1007/s11418-010-0469-7

Bravo-Hollis, H. (1967). *Las Cactaceas de Mexico*. Universidad Nacional Autónoma De México. Vol. 1 1–755

Briseño-Sánchez, M.I., Martínez-Peralta, C., & Mandujano, M.C. (2020). Population structure and reproductive biology of peyote (*Lophophora diffusa*, Cactaceae), a threatened species with pollen limitation1, 2. *The Journal of the Torrey Botanical Society*, 147(3), 243–257.

Cassels, B.K., & Sáez-Briones, P. (2018). Dark classics in chemical neuroscience: Mescaline. *ACS Chemical Neuroscience*, 9(10), 2448–2458.

Chaachouay, N., Benkhnigue, O., Fadli, M., Ibaoui, H., & Zidane, L. (2019). Ethnobotanical and ethnopharmacological studies of medicinal and aromatic plants used in the treatment of metabolic diseases in the Moroccan Rif. *Heliyon*, 5(10), e02191. https://doi.org/10.1080/10477845.2021.1942433

Chaachouay, N., & Zidane, L. (2021). The Symbolic Efficacy of Plants in Rituals and Socio-Religious Ceremonies in

Morocco, Northwest of Africa. *Journal of Religious & Theological Information*, 21(1–2), 34–53. https://doi.org/10.1080/10477845.2021.1942433

Chaachouay, N., Douira, A., & Zidane, L. (2021a). Herbal Medicine Used in the Treatment of Human Diseases in the Rif, Northern Morocco. *Arabian Journal for Science and Engineering*, 47, 131–153. https://doi.org/10.1007/s13369-021-05501-1

Chaachouay, N., Douira, A., & Zidane, L. (2021b). COVID-19, prevention and treatment with herbal medicine in the herbal markets of Salé Prefecture, North-Western Morocco. *European Journal of Integrative Medicine*, 42, 101285. https://doi.org/10.1016/j.eujim.2021.101285

Chan, C.B., Poulie, C.B., Wismann, S.S., Soelberg, J., & Kristensen, J.L. (2021). The alkaloids from *lophophora diffusa* and other 'false peyotes'. *Journal of Natural Products*, 84(8), 2398–2407.

Díaz, J.L. (1979). Ethnopharmacology and taxonomy of Mexican psychodysleptic plants. *Journal of Psychedelic Drugs*, 11(1–2), 71–101. doi:10.1080/02791072.1979.10472094

Díaz Segura, O., Jiménez Sierra, C.L., Matías-Palafox, M.L., & Vázquez Díaz, E. (2012). Evaluación del estado de conservación del peyote queretano *Lophophora diffusa* Croizat (Bravo), cactácea endémica del desierto Querétaro-Hidalguense, México. *Cactáceas y suculentas mexicanas*, 57(3), 68–85.

Díaz-Segura, O., Jiménez-Sierra, C.L., & Matías-Palafox, M.L. (2017). Some characteristics of the reproductive biology of the Querétaro peyote *Lophophora diffusa* (Croizat) Bravo, (Cactaceae). *Mexican Cactaceae and Succulents*, 62(4), 116–127.

Díaz-Segura, O., Sierra, C.J., Matias-Palafox, M., & Pérez, E.A. (2023). Is peyote *Lophophora diffusa* (Croizat) Bravo (Cactaceae) a species destined for extinction? A population dynamics approach. *Authorea*, 1–6. 10.22541/au.169155213.39207468/v1

Flores-Martínez, A., Manzanero, G.I., Golubov, J., & Mandujano, M.C. (2013). Biología floral de Mammillaria huitzilopochtli, una especie rara que habita acantilados. *Botanical Sciences*, 91(3), 349–356.

Gómez-Hinostrosa, C., Sánchez, E., Guadalupe Martínez, J., & Terry, M. (2017). *Lophophora diffusa*. *IUCN Red List of Threatened Species*, 2017, e.T40967A121501921. doi:10.2305/IUCN.UK.2017-3.RLTS.T40967A121501921.en. Retrieved 11 November 2021.

Hernández-Magaña, R., Hernández-Oria, J.G., & Chávez, R. (2012). Datos para la conservación florística en función de la amplitud geográfica de las especies en el Semidesierto Queretano, México. *Acta botánica mexicana*, 99, 105–140.

Mafham, K.P. & Mafham, R.P. (1997). *Cacti: The Illustrated Dictionary*. Timber Press. Vol 1 1–224.

Ibarra-Laclette, E., Zamudio-Hernández, F., Pérez-Torres, C.A., Albert, V. A., Ramírez-Chávez, E., Molina-Torres, J., ... Herrera-Estrella, L. (2015). De novo sequencing and analysis of *Lophophora* williamsii transcriptome, and searching for putative genes involved in mescaline biosynthesis. *BMC Genomics*, 16(1), 1–14.

Lin, J., Yang, S., Ji, J., Xiang, P., Wu, L., & Chen, H. (2023). Natural or artificial: An example of topographic spatial distribution analysis of mescaline in cactus plants by matrix-assisted laser desorption/ionization mass spectrometry imaging. *Frontiers in Plant Science*, 14, 1066595.

Nassar, J.M., & Ramírez, N. (2004). Reproductive biology of the melon cactus, *Melocactus curvispinus* (Cactaceae). *Plant Systematics and Evolution*, 248, 31–44.

Perrine, D.M. (2001). Visions of the night-western medicine meets peyote 1887–1899. *The Heffter Review of Psychedelic Research*, 2, 6–52.

Rätsch, C. (2005). *The encyclopedia of psychoactive plants: Ethnopharmacology and its applications*. Simon and Schuster.

Ríos-Carrasco, S (2017). *Ontogenia de los apéndices seminales de la familia Cactaceae. Tesis de Maestría*. México: Universidad Nacional Autónoma de México.

Saroya, A.S., & Singh, J. (2020). *Psychoactive medicinal plants and fungal neurotoxins*. Springer Singapore.

Scheinvar, L. (2004). *Flora cactológica del estado de Querétaro: diversidad y riqueza*. México, D.F.: Fondo de Cultura Económica. 1–390.

Šnicer, J., Bohata, J., & Myšák, V. (2009). The littlest *Lophophora*. *Cactus and Succulent Journal*, 81(6), 294–300.

Terry, M.K. (2007). A tale of two cacti: studies in Astrophytum asterias and *Lophophora* williamsii (Doctoral dissertation, Texas A&M University).

Trujillo Hernández, A. (2002). Physiological ecology of germination of cacti of the genus *Lophophora* (Unpublished master's thesis, National Autonomous University of Mexico).

8 San Pedro Cactus (*Echinopsis pachanoi* (Britton & Rose) H. Friedrich & G.D. Rowley Cactceae)

Talha Imtiaz, Ehsan Ali, and Muhammad Farooq Azhar

8.1 INTRODUCTION

If you are reading this chapter, you would probably have a special interest in the San Pedro cactus, scientifically known as *Echinopsis pachanoi* (formerly *Trichocereus pachanoi*), which stands as a remarkable botanical and cultural treasure of the Andes Mountains in South America (Anderson, 2001). This columnar cactus, revered for centuries by Indigenous peoples, has garnered increasing global attention due to its multifaceted significance. That's why Heaven (2012) calls it 'The cactus of Mistry'. Its scientific classification, ethnobotanical importance, and potent phytochemistry have made it a subject of both scientific inquiry and mysticism (Backeberg & Knuth 1936).

This chapter embarks on a journey to unveil the captivating story of the San Pedro Cactus. We will delve into its scientific classification, exploring the intricacies of its taxonomy. The vernacular names, often rich in cultural significance, will illuminate the diverse cultural contexts in which this cactus thrives. A comprehensive botanical description will paint a vivid picture of its physical attributes, and we will embark on a journey tracing its origin and the vast geographical range it calls home. The chapter will also unravel the ecological secrets that underlie its thriving presence in the Andean landscape.

One cannot discuss San Pedro cactus without delving into its rich phytochemistry, including the presence of mescaline which is a potent alkaloid that has intrigued both traditional healers and modern psychonauts (Carod-Artal & Vazquez-Cabrera, 2006). We will explore the ethnohistoric tapestry that links San Pedro cactus to Indigenous rituals and shamanic practices, where its hallucinogenic effects have played a pivotal role. Furthermore, we will examine the contemporary relevance of the cactus in the context of modern psychedelia and therapeutic research (Vamvakopoulou et al., 2022).

Through this exploration, we hope to illuminate the multifaceted nature of the San Pedro cactus, transcending its botanical existence to reveal its cultural, ecological, and pharmacological dimensions. This chapter is a tribute to the enduring legacy of a plant that has bridged the realms of science, spirituality, and human experience.

8.2 SCIENTIFIC CLASSIFICATION

The San Pedro cactus belongs to the plant family *Cactaceae*. Within this family, it is classified under the genus *Echinopsis*. The San Pedro cactus, scientifically known as *Trichocereus pachanoi*, derives its name from its historical and geographical associations (Yetman, 2007). The name 'San Pedro' can be traced back to a small northern coastal town, likely in Peru, where this remarkable cactus species was first identified and documented. The town of San Pedro, nestled in a region where the cactus thrived, became the namesake for this plant (Wallace & Gibson 2002). Moreover, the columnar form of the San Pedro cactus, which is distinctive and characteristic of this species, has been depicted on ancient Charvin ceramic vessels that date back to around 600 B.C. These depictions on ancient ceramics not only illustrate the enduring presence of the cactus in the region but also provide valuable historical evidence of its significance to Indigenous cultures (Evans, 1979).

8.2.1 Scientific Nomenclature

The scientific nomenclature reflects its evolutionary relationship and botanical characteristics, allowing for precise identification and study (Anderson, 2001).

Kingdom:	Plantae
Clade:	Tracheophytes
Clade:	Angiosperms
Clade:	Eudicots
Order:	Caryophyllales
Family:	Cactaceae
Subfamily:	Cactoideae
Genus:	Echinopsis
Species:	*Echinopsis pachanoi* (formerly *Trichocereus pachanoi*)

8.3 VERNACULAR NAMES

Beyond its scientific nomenclature, the San Pedro Cactus (*Trichocereus pachanoi*) (Figure 8.1) has acquired a diverse

DOI: 10.1201/9781003460336-8

array of vernacular names, reflecting its cultural significance across different regions and communities (Lin et al., 2023). These names offer a glimpse into the various roles and interpretations that this cactus has held throughout history.

8.3.1 San Pedro Cactus

As previously mentioned, the cactus takes its name from the small coastal town in northern Peru, where it was initially documented. The name 'San Pedro' resonates with the geographical origin of this remarkable plant (Heaven, 2012).

8.3.2 Peyote

The Indigenous Aztec people referred to the cactus as 'Peyotl'. This name signifies the cactus's presence and usage in their culture, where it was employed for its psychoactive properties in spiritual and shamanic rituals. It should be noted, however, that the more renowned peyote cactus (*Lophophora williamsii*) is a distinct species used by Native American tribes in the southwestern United States and Mexico (Muneta, 2020; Stewart, 1987).

8.3.3 Devil's Root

Upon the arrival of Christian explorers and colonizers in the Americas, the San Pedro cactus earned the ominous moniker 'devil's root'. This label reflected the European misunderstanding and mistrust of Indigenous practices involving entheogenic plants (Anderson, 1996).

8.3.4 Likuri (Mescal Buttons)

Among the Huichol tribe of Mexico today, the San Pedro cactus is known as 'Likuri'. The term 'mescal buttons' underscores the association with mescaline, a potent psychoactive compound found in the cactus. The Huichol people have preserved their sacred traditions involving the Likuri, emphasizing its spiritual significance in their culture (Evans, 1979).

8.3.5 Wachuma

The San Pedro cactus (*Echinopsis pachanoi*) is known as 'wachuma' or 'huachuma' in the Quechua languages of the Andean highlands of Peru, Bolivia, and Ecuador. Wachuma

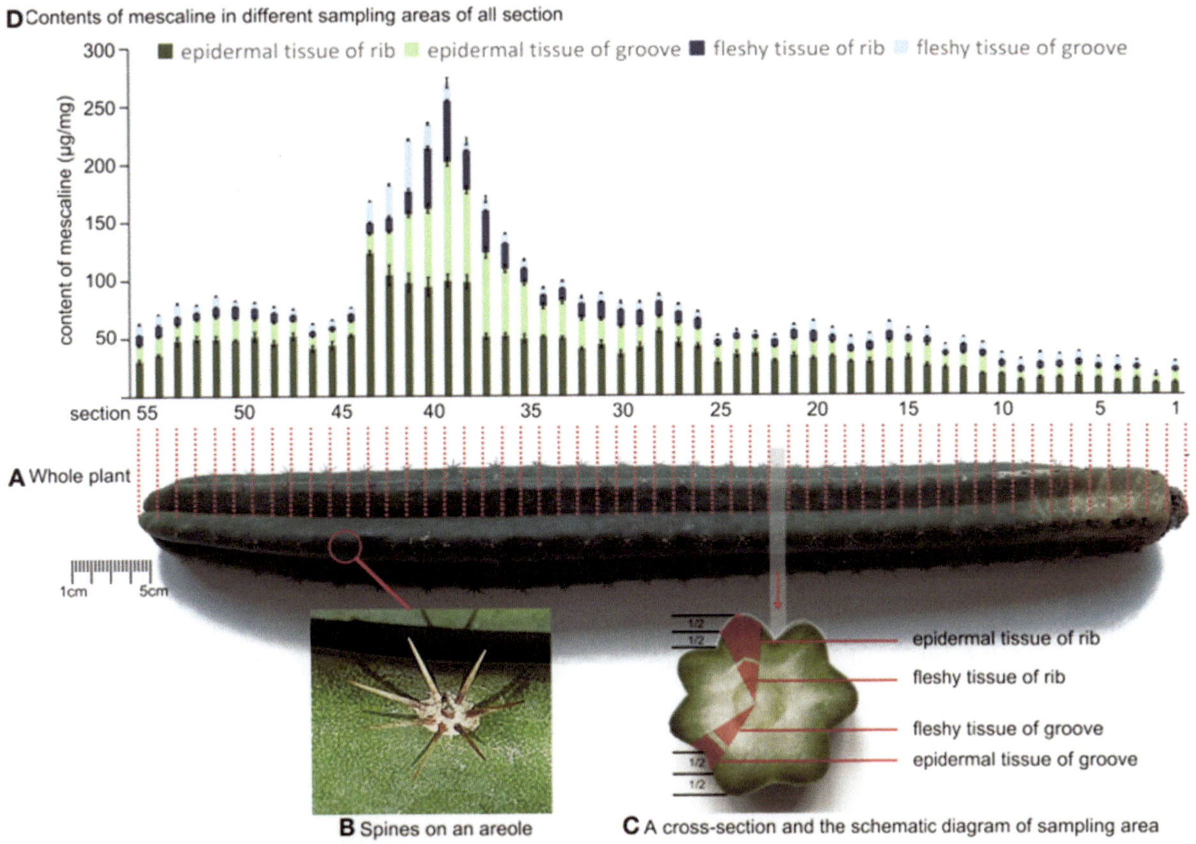

FIGURE 8.1 (A) Schematic diagram of the position of the longitudinal incision of *Trichocereus pachanoi*. (B) Spines on an areole of *Trichocereus pachanoi*. (C) A cross-section and the schematic diagram of the sampling area of *Trichocereus pachanoi*. (D) Mescaline content in different sampling areas of all sections of *Trichocereus pachanoi* (Lin et al., 2023).

translates to 'vision plant' or 'visionary cactus', referring to its long history of ritual use by Andean shamans and healers. In traditional ceremonies, wachuma is prepared from dried or processed slices of the cactus and ingested as a psychoactive vision-inducing sacrament. Shamans enter transcendent states to commune with spirits, diagnose illnesses, and gain mystical guidance from ancient plant teachers. The psychedelic effects come from mescaline and other alkaloids in the cactus. While the Spanish name San Pedro was given after Catholic colonization, the Indigenous Quechua name wachuma has persisted in the Andean regions where this sacred 'teacher plant' continues to be used in traditional rituals to this day (Pollan, 2022).

These vernacular names provide insight into the diverse cultural contexts in which the San Pedro cactus has played a role. From its Indigenous uses among the Aztecs to its encounter with Christian invaders and its continued importance among contemporary Indigenous communities, this cactus's names reflect the multifaceted history and significance of *Trichocereus pachanoi*.

8.4 BOTANICAL DESCRIPTION

The San Pedro cactus (*Trichocereus pachanoi*) boasts a distinct botanical profile characterized by its unique features, making it readily identifiable among cactus species. Its physical attributes and adaptations enable it to thrive in the diverse landscapes of its native regions, primarily the Andes Mountains of Peru and Ecuador.

8.4.1 Columnar Form

One of the most striking features of the San Pedro cactus is its columnar growth habit. It develops tall, upright stems that can reach impressive heights, often exceeding several meters. These cylindrical stems are vertically ribbed, and their striking blue-green coloration provides them with a distinct appearance (Albesiano, 2012).

8.4.2 Ribs and Areoles

The stems of the San Pedro cactus typically exhibit six to seven pronounced ribs that run vertically along their length. Along these ribs, you can find areoles – small, woolly structures that serve as the cactus's primary locations for spine growth. In some cultivated varieties, the cactus may appear mostly spineless or have reduced spines (Albesiano & Kiesling, 2012).

8.4.3 Spines

While the San Pedro cactus does possess spines, they are generally fewer in number and smaller compared to many other cactus species. The spines vary in color, ranging from yellow to brown. Despite their presence, the San Pedro cactus is less spiny with approximately seven to nine spines, and more accommodating to human interaction than some of its cactus relatives (Anderson, 2001).

8.4.4 Spacing and Growth

The areoles are evenly spaced along the ribs, typically around 2 cm (0.8 inches) apart. The cactus exhibits relatively rapid growth, with some individuals growing up to 30 cm (approximately 12 inches) per year, under optimal conditions (Niklas & Mauseth, 1981).

8.4.5 Branching

As the San Pedro cactus ages, it frequently sprouts several branches, particularly at the base of the main stem or in the vicinity of broken or damaged parts. This branching adds to its intimidating and unusual aspect (Anderson, 2001).

8.4.6 Flowers

The San Pedro cactus produces large, white, and highly fragrant flowers, which bloom at night. These blooms, measuring approximately 19 to 24 cm (7.5 to 9.4 inches) in length with a diameter of up to 20 cm (7.9 inches), add to the cactus's allure. Black hairs along the thick base of the flower lead to its distinctive appearance (Albesiano & Kiesling, 2012; Lin et al., 2023).

8.4.7 Fruits

After successful fertilization, the cactus yields oblong, dark green fruits, measuring about 3 cm (1.2 inches) in diameter and 5–6 cm (2.0–2.4 inches) in length. These fruits play a role in the cactus's reproduction and seed dispersal (Anderson, 2001).

The San Pedro cactus's botanical characteristics make it well-suited to the rugged Andean environment and have contributed to its enduring presence in the region. Its columnar growth, relatively low spines, and striking flowers have captivated the attention of botanists and enthusiasts alike, highlighting its unique place in the world of cacti (Britton & Rose, 1920).

8.5 ORIGIN AND GEOGRAPHIC DISTRIBUTION

The San Pedro cactus (*Trichocereus pachanoi*) finds its origins rooted in the breathtaking landscapes of the Andes Mountains of South America. Its geographic distribution spans across the high-altitude regions of Peru and Ecuador, where it has thrived for centuries, intertwining itself with the rich tapestry of Indigenous cultures and ecosystems (Engel et al., 2023).

8.5.1 Andean Heartland

The heartland of the San Pedro cactus lies within the Andes, a mountain range that stretches along the western edge of South America. Specifically, it is native to the central and northern regions of the Andes, where the cactus has adapted to the diverse microclimates, elevations, and ecosystems that characterize this mountainous terrain (Ogunbodede et al., 2010). Its core habitat is the arid slopes and valleys of the Peruvian Andes, particularly around the cities of Trujillo and Chiclayo (Rätsch, 2005).

8.5.2 Elevational Range

This remarkable cactus exhibits an impressive range in elevation, thriving at altitudes between approximately 1,000 meters (3,280 feet) and 3,000 meters (9,840 feet) above sea level. Such adaptability to varying elevations showcases its resilience in the face of the Andes' climatic fluctuations (Davis, 1983). The cactus thrives in the hot, dry climate of the coastal Andean valleys. It is well-adapted to drought conditions and intense sunlight. It is commonly found on west-facing slopes and hillsides, where humidity is very low (Rätsch, 2005).

8.5.3 Andean Flora

Within the Andes, the San Pedro cactus shares its habitat with a diverse array of flora and fauna, many of which are uniquely adapted to the harsh alpine conditions. Its ecological niche within this ecosystem has contributed to its distinct growth patterns and survival strategies (Ezcurra, 2006).

Understanding the origin and geographic distribution of the San Pedro cactus is crucial in appreciating its unique ecological adaptations and the cultural contexts in which it has played a pivotal role. Its ability to thrive in the challenging Andean environment and its enduring presence in Indigenous traditions make it a symbol of resilience and cultural heritage in this remarkable region.

8.6 HABITAT AND ECOLOGY

Echinopsis pachanoi is well-adapted to the dry, desert-like conditions of the Peruvian coastal valleys. Its succulent tissues allow it to store water and withstand prolonged droughts. The waxy cuticle and dense canopy of spines help reduce water loss through transpiration (Schlumpberger & Renner, 2012). The cactus has an extensive root system that spreads laterally to absorb water from light rains. Its roots also reach deep into the soil for groundwater. *Echinopsis pachanoi* employs crassulacean acid metabolism (CAM) photosynthesis, which minimizes water loss (Nobel, 2002). Recent research has isolated Actinobacteria from the rhizospheric soil of *Echinopsis pachanoi* that demonstrate potential as plant growth promoters, especially under salt stress conditions. Actinobacteria are gram-positive bacteria that are ubiquitous in soil and play important ecological roles (Ruiz Cruz, 2021). Stomata on the stems open at night to take in CO_2 and close during the hot daytime. Water loss is reduced because the openings are closed when temperatures are highest. The spines of *Echinopsis pachanoi* also help shade the stem and protect it from excessive evapotranspiration (Das & Panda, 1995). Reproduction occurs sexually through pollination by insects and birds. The flowers produce nectar and are visited by bees, butterflies, hummingbirds, and others. Being pollinated at night reduces water loss. Asexual reproduction also occurs readily from cuttings or fallen stems that take root (Glass-Coffin, 2010). *Echinopsis pachanoi* grows relatively slowly, maturing to columnar heights up to 6 meters over 10–15 years. It can live for over 100 years. The cactus is quite hardy and tolerant of a range of conditions once established (Nobel, 2002; Madsen 1989).

8.7 PHYTOCHEMISTRY

San Pedro cactus contains a number of alkaloids and other compounds that have psychoactive and medicinal effects. The most well-studied alkaloid is mescaline.

8.7.1 Mescaline

Trichocereus pachanoi (San Pedro cactus) contains a hallucinogenic alkaloid called mescaline (b-3,4,5-trimethoxy-phenethylamin). It is concentrated in the green skin and white inner pulp of the cactus, with levels ranging from 0.01 to 2.375% dry weight (Lin et al., 2023).

Mescaline is structurally related to the neurotransmitters dopamine and norepinephrine. It causes psychedelic effects by stimulating serotonin 5-HT2A receptors in the brain. The effects last about 10–12 hours and include visual hallucinations, euphoria, altered perception of space and time, and mystical experiences (Carvalho et al., 2023; Ley et al., 2023).

8.7.2 Other Alkaloids

In addition to mescaline, *Echinopsis pachanoi* contains other phenethylamine and tetrahydroisoquinoline alkaloids such as tyramine, hordenine, 3-methoxytyramine, anhalamine, anhalinine, and anhalonidine (Moreno-Pedraza et al., 2019). Some of these compounds inhibit monoamine oxidase (MAO) enzymes and enhance the effects of mescaline (Martins et al., 2023).

8.7.3 Other Compounds

The cactus also contains non-alkaloid compounds such as:

Trichocerine – a lignan that inhibits MAO
Pachanoid acid – enhances serotonin signaling
Flavonoids – antioxidant and anti-inflammatory effects
Betalains – antioxidant nitrogenous pigments
Sterols like ergosterol – antimicrobial effects

The combination of alkaloids, lignans, and other compounds contributes to the psychotropic potency and medicinal properties of *Echinopsis pachanoi*. The phytochemistry varies between different samples and growth conditions (Moreno-Pedraza et al., 2019). Further research is needed to fully elucidate the diverse metabolites produced by this cactus species.

In addition to mescaline, San Pedro contains other phenethylamine and tetrahydroisoquinoline alkaloids such as tyramine, hordenine, 3-methoxytyramine, anhalamine, anhalinine, anhalonidine, and anhalidine (Ogunbodede et al., 2010). Some of these compounds act as MAO inhibitors and enhance or prolong the effects of mescaline (Kapadia & Fayez, 1970). Anhalonidine also shows activity at 5-HT2A receptors (Pardanani et al., 1977). The alkaloid content varies between different San Pedro specimens (Ogunbodede et al., 2010). The combination of alkaloids, lignin, and other compounds likely contribute to the psychoactive potency and medicinal properties of San Pedro cactus. Further phytochemical analysis is needed to fully characterize its chemical makeup.

8.8 ETHNOBOTANY

The San Pedro cactus has a rich history of ethnobotanical usage across various Indigenous cultures in South America. Its psychoactive properties have made it a central component of shamanic rituals and healing practices for centuries.

8.8.1 Shamanic and Ritual Use

Among the Indigenous peoples of the Andes, including the Quechua, Aymara, and Mochica cultures, the San Pedro cactus is known as 'wachuma' or 'huachuma'. It plays a vital role in Andean shamanism and Indigenous medicine. Shamans ingest the psychoactive cactus during healing ceremonies and rituals to induce visionary states. They use these transcendent experiences to communicate with spirits, diagnose illnesses, and gain insight into their patients' troubles (Pollan, 2022). The cactus is viewed as a wise plant teacher that provides guidance. Shamans claim the plant enables them to journey to mystical realms and connects them to ancient Andean spirits. The mescaline in the cactus produces psychedelic effects that shamans harness for divination, soul retrieval, and invoking mystical experiences in patients (Heaven, 2012). To prepare San Pedro for shamanic use, the green skin and white inner pulp are cut into strips and dried. The dried cactus is then boiled for several hours to make a concentrated brew. Additives like chacruna leaves (*Psychotria viridis*) are sometimes added to enhance the visionary effects. The foul-tasting liquid is ritually ingested in ceremonial contexts (Glass-Coffin, 2010). In northern Peru, shamans known as curanderos hold San Pedro ceremonies called 'mesadas'. Participants drink the brew and engage in prayer, chanting, and introspection while the curandero guides the experience and performs healing rituals (Carod-Artal & Vazquez-Cabrera, 2006).

8.8.2 Modern Usage

Today, interest in traditional shamanic practices has led to a renewed popularity of the San Pedro cactus in Peru and worldwide. New Age seekers, psychonauts, and alternative healers use it as a psychedelic sacrament and tool for self-exploration. However, sustainability and ethical concerns have been raised about the booming demand for the cactus (Engel et al., 2023). Conservation efforts are needed to preserve San Pedro's natural habitats and traditional uses.

8.8.3 Medicinal Uses

In addition to its use in spiritual contexts, ethnomedical traditions have utilized San Pedro for treating various physical ailments. Andean healers use it to treat conditions like arthritis, inflammation, fever, and microbial infections (Heaven, 2012; Chaachouay et al., 2021a, 2021b). The cactus has demonstrated analgesic, anti-inflammatory, and antimicrobial properties in preliminary pharmacological studies, lending credence to some of its traditional uses (Ogunbodede et al., 2010). However, more research is still needed to fully validate its efficacy as medicine.

8.8.4 Cultural Symbolism

Beyond its visionary and medicinal significance, the San Pedro cactus holds essential symbolic meaning in Andean worldviews. It represents concepts like wisdom, ancestral connection, and the sacredness of nature. The cactus is invoked in cultural rituals like the Fiesta de la Cruz in Peru, where it carries agricultural and spiritual symbolism (Glass-Coffin, 2010; Chaachouay & Zidane, 2021). The endurance of such traditions highlights San Pedro's enduring cultural relevance in the Andes.

8.8.5 San Pedro Cactus in Wari Art

The integration of sacred plants and their symbolism in ancient art is a captivating facet of the Wari civilization, which thrived in the Andean region from approximately 600 to 1,000 C.E. Among the various elements woven into Wari textiles and iconography, the San Pedro cactus (*Trichocereus pachanoi*) emerges as a distinctive and culturally significant motif (Isbell, 2008).

In Wari art, the San Pedro cactus is often depicted in the form of its intricate and vibrant flower. This motif is especially notable on textiles, ceramics, and other artifacts. The cactus flower is rendered with meticulous detail, showcasing the Wari artists' skill and reverence for this sacred plant (Stone, 2023).

8.8.6 Moche Art

In the context of the Moche civilization, which thrived in north coastal Peru from approximately 200 to 900 AD, their artistic expressions were known for their naturalistic qualities and diverse range of subjects. However, intriguingly, their art didn't prominently feature imagery directly related to entheogenic plants, despite the culture's rich history of utilizing these substances for healing and spiritual experiences (Carod-Artal & Vazquez-Cabrera, 2006).

While the Moche art does not extensively depict entheogenic plants, there are noteworthy exceptions. Ceramic vessels reveal figures known as 'achumeras', and these figures are often portrayed holding segments of the *Echinopsis pachanoi*, commonly referred to as the San Pedro cactus. These depictions hint at a complex interplay of societal dynamics, gender roles, and the entheogenic plant's usage within Moche society (Engel et al., 2023; Scher, 2021).

8.9 HALLUCINOGENIC EFFECTS

The most profound effects of the San Pedro cactus stem from its potent psychoactive alkaloids, particularly mescaline. The hallucinogenic effects have been utilized for millennia in sacred shamanic contexts, and more recently for recreational psychedelic experiences (Engel et al., 2023).

8.9.1 Subjective Effects

The onset of San Pedro's effects occurs around 1–2 hours after ingesting it orally. The peak subjective effects last approximately 6–10 hours. Users describe visual and auditory hallucinations, euphoria, intense introspection, synesthesia, changed perceptions of space and time, and mystical states (Bohn et al., 2022). The experience has been described as more grounded in Figure 8.2.

Other common visual effects include geometric patterns, kaleidoscopic visions, enhanced colors and textures, moving images, and distortions of objects or the environment. Some users report encountering entities, mythical creatures, or ancestral spirits. Visual acuity may be enhanced. Open and closed-eye visuals are possible (Gonzalez, 2006).

8.9.2 Psychological Effects

In addition to sensory effects, San Pedro induces profound changes in thoughts, emotions, and consciousness. Users often gain fresh perspectives on relationships, personal issues, and life purpose. Some experience a dissolution of the ego or a temporary sense of unity with nature or spirit realms. However, the psychological effects are very dose-dependent and set & setting dependents (Bohn et al., 2022).

8.9.3 Neuropharmacology

The hallucinogenic effects are mediated primarily through mescaline's agonist activity at serotonin 5-HT2A receptors in the brain (Aragane et al., 2020). Mescaline has structural and functional similarities to serotonin and other phenethylamine psychedelics like psilocin. Binding to 5-HT2A sites triggers a cascade of neural changes leading to altered perception, cognition, and mood (Cassels & Sáez-Briones, 2018).

8.9.4 Safety Profile

Compared to other psychedelics, mescaline is relatively non-toxic, produces little physiological toxicity, and has a very wide therapeutic index. However, it can elevate heart rate and blood pressure. It is not generally considered addictive, but alcohol should be avoided. Traditional ceremonies emphasize mindfulness and introspection within a safe set and setting (Uthaug et al., 2022).

8.10 CONCLUSION

The San Pedro cactus, with its distinctive columnar form and psychoactive properties, stands as a captivating botanical and cultural treasure. This exploration has illuminated the multifaceted significance of *Echinopsis pachanoi*, from its taxonomic classification and vernacular names, to its unique adaptations, intricate phytochemistry, and rich ethnobotanical history. San Pedro's rich pharmacology begins with mescaline, which structurally resembles neurotransmitters like

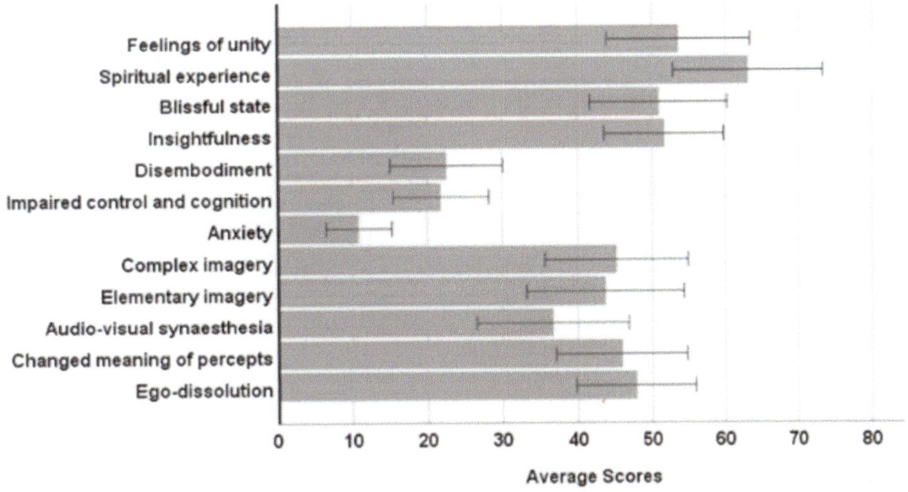

FIGURE 8.2 Altered states of consciousness as measured with the 11D-ASC and EDI associated with ceremonial San Pedro use. Horizontal bar chart of mean scores of the 11D-ASC subscales and EDI scale revealing significant differences from the sober state (Bohn et al., 2022).

dopamine. Mescaline primarily elicits psychedelic effects through agonist activity at serotonin 5-HT2A receptors in the brain. This triggers visual and auditory hallucinations, synesthesia, mystical states, and other subjective effects that last for 6–10 hours. In addition to mescaline, San Pedro contains various other alkaloids like tyramine, hordenine, anhalonidine, and anhalamine. Some of these compounds act as MAO inhibitors, enhancing and prolonging mescaline's actions. However, the San Pedro cactus is more than just its botanical and pharmacological attributes. Its intricate integration into Indigenous rituals, artwork, and worldviews reveals its cultural symbolism and visionary significance. From pre-Columbian societies like the Moche and Wari to present-day curanderos, San Pedro has maintained its sacred status as an ancient plant teacher across generations. Looking ahead, mescaline and San Pedro show promise in therapeutic applications for addiction, psychiatric disorders, and personal growth. However, these potential medicinal uses will require extensive clinical trials to validate safety and efficacy. Sustainable harvesting will also be crucial for preserving traditional and modern practices.

CONFLICT OF INTERESTS

The authors declared that they have no conflict of interests with respect to the authorship and publication of this chapter.

LIST OF ABBREVIATIONS

5-HT2A	Serotonin 2A receptor
ADHD	attention deficit hyperactivity disorder
CAM	crassulacean acid metabolism
CNS	central nervous system
EOL	encyclopedia of life
GC-MS	gas chromatography-mass spectrometry
HPLC	high-performance liquid chromatography
ITIS	integrated taxonomic information system
LD50	median lethal dose
MAO	monoamine oxidase

REFERENCES

Albesiano, S. (2012). A new taxonomic treatment of the genus *Trichocereus* (Cactaceae) in Chile. *Haseltonia, 2012*(18), 116–139.

Albesiano, S., & Kiesling, R. (2012). Identity and neotypification of Cereus macrogonus, the type species of the genus Trichocereus (Cactaceae). *Haseltonia, 2012*(17), 24–34.

Anderson, E. F. (1996). *Peyote: The divine cactus*. University of Arizona Press.

Anderson, E. F. (2001). *The cactus family*. Timber Press.

Aragane, M., Sasaki, Y., Fukumori, N., Yoshizawa, M., Suzuki, Y., Kitagawa, S., Mori, K., Ogino, S., Yasuda, I., & Nagumo, S. (2020). Peyote identification on the basis of differences in morphology, mescaline content, and trnL/trnF sequence between *Lophophora williamsii* and *L. diffusa*. *Journal of Natural Medicines*, 65, 103–110. https://doi.org/10.1007/s11418-010-0469-7

Backeberg, C., & Knuth, F. M. (1936). *Gymnocalycium proliferum Backbg*. Kaktus-ABC.

Bohn, A., Kiggen, M. H. H., Uthaug, M. V., van Oorsouw, K. I. M., Ramaekers, J. G., & van Schie, H. T. (2022). Altered states of consciousness during ceremonial San Pedro use. *The International Journal for the Psychology of Religion, 33*(4), 309–331. https://doi.org/10.1080/10508619.2022.2139502

Britton, N. L., & Rose, J. N. (1920). *The Cactaceae. Descriptions and illustrations of plants of the cactus family*, vol. 2. Carnegie Institute, 248 pp.

Carod-Artal, F. J., & Vazquez-Cabrera, C. B. (2006). Mescaline and the San Pedro cactus ritual: Archaeological and ethnographic evidence in northern Peru. *Revista de Neurologia, 42*(8), 489–498.

Carvalho, A., Barbosa, B. M., Flores, J. S., do Carmo Gonçalves, P., Diniz, R., Cordeiro, Y., Fernández, C. O., Cukierman, D. S., & Rey, N. A. (2023). New mescaline-related N-acylhydrazone and its unsubstituted benzoyl derivative: Promising metallophores for copper-associated deleterious effects relief in Alzheimer's disease. *Journal of Inorganic Biochemistry, 238*, 112033.

Cassels, B. K., & Sáez-Briones, P. (2018). Dark classics in chemical neuroscience: Mescaline. *ACS Chemical Neuroscience, 9*(10), 2448–2458.

Chaachouay, N., & Zidane, L. (2021). The Symbolic Efficacy of Plants in Rituals and Socio-Religious Ceremonies in Morocco, Northwest of Africa. *Journal of Religious & Theological Information, 21*(1–2), 34–53. https://doi.org/10.1080/10477845.2021.1942433

Chaachouay, N., Douira, A., & Zidane, L. (2021a). Herbal medicine used in the treatment of human diseases in the Rif, Northern Morocco. *Arabian Journal for Science and Engineering*, 47, 131–153. https://doi.org/10.1007/s13369-021-05501-1

Chaachouay, N., Douira, A., & Zidane, L. (2021b). COVID-19, prevention and treatment with herbal medicine in the herbal markets of Salé Prefecture, North-Western Morocco. *European Journal of Integrative Medicine*, 42, 101285. https://doi.org/10.1016/j.eujim.2021.101285

Das, P., & Panda, P. C. (1995). Protected cultivation of cacti and other succulents. *Advances in Horticulture, 12*(2), 819–851.

Davis, E. W. (1983). Sacred plants of the San Pedro cult. *Botanical Museum Leaflets, Harvard University, 29*(4), 367–386.

Engel, L., Barratt, M., Ferris, J., Puljevic, C., & Winstock, A. (2023). Mescaline, Peyote and San Pedro: Is sustainability important for cacti consumers? *Journal of Psychedelic Studies, 7*(2), 135–142 https://doi.org/10.1556/2054.2023.00252

Evans, P. (1979). Hallucinogenic Cacti. *The National Cactus and Succulent Journal, 34*(1), 14–16.

Ezcurra, E. (2006). *Global deserts outlook*. UNEP/Earthprint.

Glass-Coffin, B. (2010). Shamanism and San Pedro through time: Some notes on the archaeology, history, and continued use of an entheogen in northern Peru. *Anthropology of Consciousness, 21*(1), 58–82.

Gonzalez, F. J. (2006). The 2006 Bernard B. Brodie award lecture. Cyp2e1. *Drug Metabolism and Disposition: The Biological Fate of Chemicals, 35*(1), 1–8.

Heaven, R. (2012). *Cactus of mystery: The shamanic powers of the Peruvian San Pedro cactus*. Simon and Schuster.

Isbell, W. H. (2008). Wari and Tiwanaku: international identities in the central Andean Middle Horizon. In *The handbook of South American archaeology* (pp. 731–759). Springer.

Kapadia, G. J., & Fayez, M. B. E. (1970). Peyote constituents: Chemistry, biogenesis, and biological effects. *Journal of Pharmaceutical Sciences, 59*(12), 1699–1727.

Ley, L., Holze, F., Arikci, D., Becker, A. M., Straumann, I., Klaiber, A., Coviello, F., Dierbach, S., Thomann, J., & Duthaler, U. (2023). Comparative acute effects of mescaline, lysergic acid diethylamide, and psilocybin in a randomized, double-blind, placebo-controlled cross-over study in healthy participants. *Neuropsychopharmacology*, 1–9.

Lin, J., Yang, S., Ji, J., Xiang, P., Wu, L., & Chen, H. (2023). Natural or artificial: An example of topographic spatial distribution analysis of mescaline in cactus plants by matrix-assisted laser desorption/ionization mass spectrometry imaging. *Frontiers in Plant Science*, *14*, 1066595.

Madsen, J. (1989). *Cactaceae. Nordic Journal of Botany*, *9*(5), 45–70.

Martins, D., Gil-Martins, E., Cagide, F., da Fonseca, C., Benfeito, S., Fernandes, C., Chavarria, D., Remião, F., Silva, R., & Borges, F. (2023). Unraveling the in vitro toxicity profile of psychedelic 2C phenethylamines and their N-Benzylphenethylamine (NBOMe) analogues. *Pharmaceuticals*, *16*(8), 1158.

Moreno-Pedraza, A., Rosas-Román, I., Garcia-Rojas, N. S., Guillén-Alonso, H., Ovando-Vázquez, C., Díaz-Ramírez, D., Cuevas-Contreras, J., Vergara, F., Marsch-Martínez, N., & Molina-Torres, J. (2019). Elucidating the distribution of plant metabolites from native tissues with laser desorption low-temperature plasma mass spectrometry imaging. *Analytical Chemistry*, *91*(4), 2734–2743.

Muneta, J. D. (2020). Peyote crisis confronting modern indigenous peoples: The declining peyote population and a demand for conservation. *Am. Indian LJ*, *9*, 135.

Niklas, K. J., & Mauseth, J. D. (1981). Relationships among shoot apical meristem ontogenic features in *Trichocereus pachanoi* and Melocactus matanzanus (Cactaceae). *American Journal of Botany*, *68*(1), 101–106.

Nobel, P. S. (2002). *Cacti: biology and uses*. Univ of California Press.

Ogunbodede, O., McCombs, D., Trout, K., Daley, P., & Terry, M. (2010). New mescaline concentrations from 14 taxa/cultivars of Echinopsis spp.(Cactaceae)('San Pedro') and their relevance to shamanic practice. *Journal of Ethnopharmacology*, *131*(2), 356–362.

Pardanani, J. H., McLaughlin, J. L., Kondrat, R. W., & Cooks, R. G. (1977). Cactus alkaloids. XXXVI. Mescaline and related compounds from Trichocereus peruvianus. *Lloydia*, *40*(6), 585–590.

Pollan, M. (2022). *This is your mind on plants*. Penguin.

Rätsch, C. (2005). *The encyclopedia of psychoactive plants: Ethnopharmacology and its applications* (J. Baker, Trans.). Park Street Press.(Originally Published 1998, Enzyklopädie Der Psychoaktiven Pflanzen. Aarau, Switzerland: AT Verlag.).

Ruiz Cruz, L. G. (2021). *Actinobacterias rizosféricas de Echinopsis pachanoi 'San Pedro hembra' y su potencial como promotores de crecimiento en Solanum lycopersicum L. bajo estrés salino*.

Scher, S. (2021). The Achumera: gender, status, and the San Pedro cactus in moche ceramic motifs and iconography. *Andean foodways: Pre-Columbian, colonial, and contemporary food and culture*. In: Staller, J.E. (eds) Andean Foodways. The Latin American Studies Book Series. Springer, Cham. (pp. 237–256). https://doi.org/10.1007/978-3-030-51629-1_9

Schlumpberger, B. O., & Renner, S. S. (2012). Molecular phylogenetics of Echinopsis (Cactaceae): Polyphyly at all levels and convergent evolution of pollination modes and growth forms. *American Journal of Botany*, *99*(8), 1335–1349.

Stewart, O. C. (1987). *Peyote religion: A history* (Vol. 181). University of Oklahoma Press.

Stone, R. R. (2023). Representations of the san Pedro (Trichocereus spp.) cactus flower and the Anadenanthera colubrinaplant in Wari textiles and Tiwanaku art. *Boletín Del Museo Chileno de Arte Precolombino*, *28*(1), 221–246.

Uthaug, M. V., Davis, A. K., Haas, T. F., Davis, D., Dolan, S. B., Lancelotta, R., Timmermann, C., & Ramaekers, J. G. (2022). The epidemiology of mescaline use: Pattern of use, motivations for consumption, and perceived consequences, benefits, and acute and enduring subjective effects. *Journal of Psychopharmacology*, *36*(3), 309–320.

Vamvakopoulou, I. A., Narine, K. A. D., Campbell, I., Dyck, J. R. B., & Nutt, D. J. (2022). Mescaline: The forgotten psychedelic. *Neuropharmacology*, *22* 109294. https://doi.org/10.1016/j.neuropharm.2022.109294

Wallace, R. S., & Gibson, A. C. (2002). Evolution and systematics. *Cacti: Biology and uses* (pp. 1–21). University of California Press, Berkeley.

Yetman, D. (2007). *The great cacti: ethnobotany & biogeography*. University of Arizona Press.

9 Peyote (*Lophophora williamsii* (Lem. ex J.F. Cels) J.M. Coult.- Cactaceae)

Viktória L. Balázs, Lilla Nagy-Radványi, Ágnes Farkas, and Eszter Csikós

9.1 INTRODUCTION

Peyotl (peyotl) or mescaline cactus, or peyote (*Lophophora williamsii*) is a Central American spherical cactus with an exceptionally high alkaloid content (more than 60 types of alkaloids have been isolated from it so far). It has a prominent role in Central American Indian cultures, as a sacred plant. It is mainly consumed on ritual occasions, but it is also known as a universal medicinal plant, among other things, for its natural antibiotic content. One of its alkaloids, mescaline, has a hallucinogenic effect when ingested. Many varieties of the species are known. One of the best-known, strongly sprouting versions is *L. williamsii* var. *caespitosa*. Other versions: *L. williamsii* var. *echinata*, *L. williamsii* var. *fricii*, *L. williamsii* var. *koehresii*, *L. williamsii* var. *lewinii*, *L. williamsii* var. *lutea*.

The plant has been used by Indians for thousands of years on ritual occasions. Nowadays, both its cultivation and consumption are illegal in Mexico and the USA. Exceptions are the NAC (Native American Church) in the USA and some Indian tribes in Mexico, which are allowed by law to collect and consume the cactus for religious purposes. In Europe, it is commonly known as a hallucinogen. Among peyote's alkaloids, most people only know mescaline, the other 60 or so compounds are ignored. Its effect is unpredictable due to the many alkaloids. Unpleasant physical effects usually occur one hour after consumption (nausea, vomiting, difficulty breathing, muscle spasms, rapid heartbeat). Then begins the altered state of consciousness, which lasts from 6 to 12 hours. Hallucinations do not always occur. Navajo members of the NAC who consume it regularly report that they rarely experience visions, but when they do, they take it as a bad sign (Anderson 1969; WFO, Guzman-Chavez 2017; Dinis-Oliveira et al. 2019; Rucker et al. 2018; Chaachouay et al., 2019).

9.2 SCIENTIFIC CLASSIFICATION

Systematically, *Lophophora williamsii* belongs to the Cactaceae family and the Cactoideae subfamily, being a representative of the *Lophophora* genus. Members of this genus are generally found in sunny areas, but it is not uncommon to see them in the shade of shrubs (e.g., *Agave lecheguilla*, *Jatropha dioica*, *Euphorbia antisyphilitica*). The representatives of the genus have a common feature that they can be found on limestone, even on vertical limestone cliffs, and they show a varied morphology (WFO, Anderson 1969).

9.3 VERNACULAR NAMES

The Navaho Indians refer to the peyote cactus as: peyote azee ('azee'), 'azee' diyiní, 'azee' yit'aalii, 'azee' ch'íidii. The word 'azee' means medicine, and is commonly used by the Navaho for traditional medicine. The most frequently used name for the cactus is 'azee' yit'aalii, which translates to 'holy medicine'. Peyote is marketed illegally under various street names: *'half moon'*, *'britton'*, *'hikori'*. The street name for peyote combined with marijuana is *'bad seed'*. Other, less common names for peyote include mescal buttons and divine cactus (WFO, Guzman-Chavez 2017).

9.4 BOTANICAL DESCRIPTION

The peyotle's body has a flattened spherical shape, which continues under the surface in a thick taproot. In its habitat, it burrows into the soil. Its skin is bluish-gray, and in its areolae (thorn cushions) dirty white tufts of hairs (trichomes) grow instead of thorns. Only very young seedlings have small, insignificant thorns. Compared to other *Lophophora* species (*L. diffusa, L. fricii*), its ribs are clearly distinguishable, straight, or wavy. The number of ribs can vary depending on age and living conditions. Young plants are usually five-ribbed, which later divide. Eight ribs are common in adult plants, but there may be more (10–13) in very old plants. Their lines of response go into little depth. Peyote cacti are very slow growing, taking up to 30 years from the seedling to the blooming of flowers. The petals of its barely 1 cm large, narrow-petaled flower are pink with a darker central stripe (their color can vary from white to dark pink). Pistil and stamens are yellow. Its fruit is a red, soft-fleshed berry with relatively few small, black seeds. The fruit grows in the body of the plant, under the wool felt, and only emerges when ripe (Terry and Mauseth 2006; Olive and Triggle 2007; Hernandez and Gomez-Hinostrosa 2011).

DOI: 10.1201/9781003460336-9

9.5 ORIGIN AND GEOGRAPHIC DISTRIBUTION

It lives in the highlands and mountains from Central Mexico, the state of San Luis Potosí, to South Texas and New Mexico. It occurs primarily near the Rio Grande and in the Chihuahuan Desert. In addition, the peyote cactus can be found in Mexico in the state of Coahuila south of the border, west of Saltillo, and in Tamaulipas and the northeastern part of Zacatecas (Dinis-Oliveira et al. 2019; WFO 2023).

9.6 HABITAT AND ECOLOGY

The soil test results confirmed that *L. williamsii* can be observed in all cases on calcareous soil. It lives in limestone, rocky, or even sandy areas. Although it likes sunny areas, it can also be observed together with shrubs and bushes. Among the most common companion species is chaparral (*Larrea divaricata*), which is 75% companion plant to peyote. *L. williamsii* occurs in a wide range of vegetation types in the Chihuahuan Desert, as it occurs with many plants. It often forms a common association with *Echinocereus* species, *Prosopis juliflora*, and *Agave lechequilla*. It can also form associations with *Euphorbia antisyphilitica*, *Echinocactus horizonthalonius*, *Coryphantha* spp., and *Prosopis glandulosa* as well. Among the listed species, *Flourensia cernua*, *Agave lechuguilla*, *Larrea tridentata*, and *Prosopis glandulosa* are considered extremely important for the preservation of *L. williamsii* stands, as these different perennial plants act as nurse or facilitator species for *L. williamsii* (Anderson 1969; Schwertner-Charao et al. 2022).

9.7 ETHNOBOTANY

The use of the peyote cactus dates back to ancient times. Its use is related to the Native American religion, as it was used for its intoxicating and hallucinogenic effects. Its harvesting has an important tradition, since if it is not done properly, it can lead to the destruction of the plant (Chaachouay et al., 2021a). Since it is difficult to determine where the shoot ends and where the root part begins, unfortunately in many cases the photosynthesizing part of the plant cannot regenerate and dies. Populations of *L. williamsii* have declined significantly as a result of improper harvesting techniques and habitat destruction resulting from urban developments. Therefore, the plant is endangered in South Texas. The Government of Mexico classifies peyote as a species subject to special protection through the Official Mexican Standard (NOM-059-SEMARNAT-2010).

During ceremonial use, the upper two-thirds of the cactus body are cut off. These heads are then dried and disc-shaped slices are made. The discs are usually dried and chewed, or boiled in water to make a psychoactive tea. In order to achieve an enhanced effect, members of the Native American tribe fast before using peyote and consume it on an empty stomach. In this way, they achieve a higher degree of absorption of the active ingredients of the cactus in the gastrointestinal tract. The cactus has a bitter taste and causes nausea, so today's users consume it in powder form. Members of Indian tribes drink only tea during religious rituals, despite its bitter taste. In the case of Indian tribes, the consumption of peyote tea is a privilege, not all members of the tribe can enjoy it. In addition to religious ceremonies, the cactus is also used for its healing effect: Current Native American tribes use dried peyote cactus and/or tea for pain relief (toothache, labor pain), fever, and rheumatic complaints. (Schultes 1969; Amanda 2005; Salas et al. 2011; Díaz-Segura et al. 2017; Chaachouay et al., 2021b ; Ermakova et al. 2021).

9.8 PHARMACOLOGY

The most important alkaloid in peyote is mescaline, which was isolated by Arthur Heffner in 1897. The fresh plant contains approximately 0.4% of the active ingredient, in the dried plant this value can reach 3–6% (Spinella 2001; Ogunbodede et al. 2010). In addition to mescaline, several other alkaloids have also been identified, including N-methylmescaline, N-acetylmescaline, hordenine, pellotine, anhalonine, lobivine, tyramine, N-methyltyramine, anhalidine, anhalonidine, lophophorine, homopiperonylamine (Spiith 1919). It is probable that, in addition to mescaline, other alkaloids also play a role in the development of the hallucinogenic effect or strengthen the effect of mescaline. The distribution of alkaloids in the plant is uneven, they accumulate in different plant parts. Mescaline is present in the green and above-ground parts of the cactus, being the most concentrated in the buds. Hordenine can be found in the root, while lophorine can be detected both in the above-ground parts and in the root. Pelotine is the second most abundant alkaloid in *L. williamsii*. Anhalinine is also an alkaloid present in larger quantities (Lundstrom and Agurell 1968; Kapadia and Fayez 1970; Dinis-Oliveira et al. 2019).

The hallucinogenic effects of mescaline originate from the interference of neuronal serotonergic mechanisms. Although it shows an affinity for the 5-HT1A receptor, its effect can be mostly related to binding to the 5-HT2, 5-HT2A, and 5-HT2B receptors. Freedman et al. reported that low-dose mescaline results in reduced brain levels of 5-hydroxyindoleacetic acid (5-HIAA), on the other hand, when used at a higher dose, the level of 5-HIAA in the brain increased. Tilson and Sparber reported that mescaline enhanced serotonin release and/or reuptake. Since mescaline contains a phenylethylamine moiety (that is, a structural analog of amphetamine), it is not surprising that it is also associated with dopaminergic activity. Cross-tolerance of mescaline with other serotonergic agents, such as LSD and psilocybin, has been described in humans and other animals. Mescaline tolerance develops after a few days of consumption. Although its effects are similar to other traditional hallucinogens, mescaline is the least powerful drug of the group, but its effect can last longer than 10–12 hours (Freedman 1969; Freedman et al. 1970; Tilson and Sparber 1972; Trulson et al. 1983; Aghajanian and Marek 1999; Nichols 2004).

Mescaline is mainly used orally, but it can also be inhaled. Discs made from the cactus are chewed or made into tea. A mescaline tea usually contains 200–400 mg of mescaline sulfate or 178–356 mg of mescaline hydrochloride, which is found in an average of 3–6 cactus buds or 10–20 g of dried cactus buds. However, its concentration largely depends on the species, environmental factors (temperature, soil nutrient supply, water supply), the age of the cacti, and the harvested part. Mescaline is rapidly absorbed from the gastrointestinal tract. A large percentage of the dose of mescaline it is distributed in the kidney and liver, which leads to a delay in hallucinogenic effects. Mescaline has a low lipid solubility and therefore a low ability to cross the blood-brain barrier. Therefore, LSD is approximately 2000 times more potent as a hallucinogenic agent. Peak effects do not coincide with peak mescaline concentrations in the brain, suggesting that mescaline undergoes bioactivation (Nichols 2004; Palenicek et al. 2008; Carstairs and Cantrell 2010; Dasgupta 2017).

9.9 MEDICINAL EFFECTS

Along with mescaline, hordenine has significant pharmacological activity. The sympathomimetic effect of hordenine is based on its structural similarity to various neurotransmitters (dopamine and adrenaline). Thanks to this, it results in high blood pressure, and it also increases respiratory and cardiac functions (Frank et al. 1990). The main active component of peyote is mescaline; after its discovery, it was widely used in psychiatry until the 1970s. In 1936, Erich Guttman proposed its possible application in psychiatry. After self-experiments, mescaline isolated from cactus was used to treat manic depressives, schizophrenics, and morphine addicts. After conducting several clinical trials, the researchers came to the conclusion that mescaline is not suitable for alleviating the symptoms of schizophrenia (Guttmann 1936; Denber and Merlis 1955; Merlis 1957; Rucker et al. 2018).

9.10 CONCLUSION

Overall, it can be concluded that the more than 60 alkaloids of *Lophophora williamsii* exert their hallucinogenic effect, which manifests itself in the appearance of colorful visions, a state of stupor, and a slowing down of perception. The main alkaloid of the cactus is mescaline, which is the most responsible for the effect. The Indian tribes still use the dried form of the cactus in their religious ceremonies, as well as its tea.

LIST OF ABBREVIATIONS

5-HIAA	5-hydroxyindoleacetic acid
L. williamsii	*Lophophora williamsii*
LSD	lysergic acid diethylamide
NAC	Native American Church

REFERENCES

Aghajanian G.K., Marek G.J. (1999): Serotonin and hallucinogens. *Neuropsychopharmacology*. 21:6–23.

Amanda L. (2005): *Peyote*. Elsevier Inc. 379–381.

Anderson E.F. (1969): The biogeography, ecology, and taxonomy of *Lophophora* (Cactaceae). *Brittonia*. 21(4):299. doi: 10.2307/2805756

Carstairs S.D., Cantrell F.L. (2010): Peyote and mescaline exposures: A 12-year review of a statewide poison center database. *Clinical Toxicology (Phila)*. 48:350–353.

Chaachouay, N., Benkhnigue, O., Fadli, M., Ibaoui, H., & Zidane, L. (2019). Ethnobotanical and ethnopharmacological studies of medicinal and aromatic plants used in the treatment of metabolic diseases in the Moroccan Rif. *Heliyon*, 5(10), e02191. https://doi.org/10.1080/10477845.2021.1942433

Chaachouay, N., Douira, A., & Zidane, L. (2021a). Herbal Medicine Used in the Treatment of Human Diseases in the Rif, Northern Morocco. *Arabian Journal for Science and Engineering*, 47, 131–153. https://doi.org/10.1007/s13369-021-05501-1

Chaachouay, N., Douira, A., & Zidane, L. (2021b). COVID-19, prevention and treatment with herbal medicine in the herbal markets of Salé Prefecture, North-Western Morocco. *European Journal of Integrative Medicine*, 42, 101285. https://doi.org/10.1016/j.eujim.2021.101285

Dasgupta A. (2017): Chapter Five – Challenges in laboratory detection of unusual substance abuse: Issues with magic mushroom, peyote cactus, khat, and solvent abuse. In *Advances in clinical chemistry*, Makowski, G.S., Ed. Elsevier: 78:63–186.

Denber H., Merlis S. (1955): Studies on mescaline I. Action in schizophrenic patients. *Psychiatr*. 29:421–429. doi: 10.1007/BF01567467

Díaz-Segura O., Jimenez-Sierra C.L., Matías-Palafox M.L. (2017): Algunas características de la biología reproductiva del peyote queretano *Lophophora diffusa* (Croizat) Bravo, (Cactaceae). *Cactáceas y Suculentas Mexicanas* 62:116–127.

Dinis-Oliveira R.J., Lança Pereira C., Dias da Silva D. (2019): Pharmacokinetic and pharmacodynamic aspects of peyote and mescaline: Clinical and forensic repercussions. *Current Molecular Pharmacology*. 12:184–194.

Ermakova A., Whiting C.V., Trout K., Clubbe C., Terry M.K., Fowler N. (2021): Densities, plant sizes, and spatial distributions of six wild populations of *Lophophora williamsii* (Cactaceae) in Texas, U.S.A. *Journal of the Botanical Research Institute* 15:149–160. doi: 10.17348/jbrit.v15.i1.1057

Frank M., Weckman T.J., Wood T., Woods W.E., Tai C.L., Chang S.L., Ewing A., Blake J.W., Tobin T. (1990): Hordenine: Pharmacology, pharmacokinetics and behavioural effects in the horse. *Equine Veterinary Journal* 22:437–441.

Freedman D.X. (1969): The psychopharmacology of hallucinogenic agents. *Annual Review of Medicine* 20:409–418.

Freedman D.X., Gottlieb R., Lovell R.A. (1970): Psychotomimetic drugs and brain 5-hydroxytryptamine metabolism. *Biochemical Pharmacology*. 19:1181–1188.

Guttmann E. (1936): Artificial psychoses produced by mescaline. *British Journal of Psychiatry*. 82:203–221. doi: 10.1192/bjp.82.338.203

Guzman-Chavez M.G. (2017) Interculturalidad en torno al uso del peyote. Un patrimonio biocultural en condicion de ilegalidad. *Alteridades*. 27:95–106. ISSN 2448-850X.

Hernandez H.M., Gomez-Hinostrosa C. (2011): Mapping the cacti of Mexico. *Succulent plant research* (7). DH Books.

Kapadia G.J., Fayez M.B. (1970): Peyote constituents: Chemistry, biogenesis, and biological effects. *Journal of Pharmaceutical Sciences* 59:1699–1727.

Lundstrom J., Agurell S. (1968): Biosynthesis of mescaline and anhalamine in peyote. *IIa. Tetrahedron Letters* 9:4437–4440.

Merlis S. (1957): The effects of mescaline sulfate in chronic schizophrenia. *Journal of Nervous and Mental Disease*. 125:432.

Nichols D.E. (2004): Hallucinogens. *Pharmacology & Therapeutics* 101:131–181.

Ogunbodede O., McCombs D., Trout K., Daley P., Terry M. (2010): New mescaline concentrations from 14 taxa/cultivars of *Echinopsis* spp. (Cactaceae) ("San Pedro") and their relevance to shamanic practice. *Journal of Ethnopharmacology* 131:356–362.

Olive M.F., Triggle D.J. (2007): *Drugs the straight facts: Peyote and mescaline*. Chelsea House.

Palenicek T., Balikova M., Bubenikova-Valesova V., Horacek J. (2008): Mescaline effects on rat behavior and its time profile in serum and brain tissue after a single subcutaneous dose. *Psychopharmacology*. 196:51–62.

Rucker J.H., Iliff J., Nutt J. (2018): Psychiatry & the psychedelic drugs. Past, present & future. *Neuropharmacology*. 142:200–218.

Salas J.S., Muro-Perez G., Estrada-Castillon E., Garcia-Aranda M., Alba-Avila J.A. (2011): Registro de una nueva localidad de *Lophophora williamsii* (Cactaceae) a punto de extincion por saqueo en Coahuila. Mexico. *Journal of the Botanical Research*. 5:685–687.

Schultes R.E. (1969): Hallucinogens of plant origin. *Science*. 163:245–254.

Schwertner-Charao L., Delgado-Martínez R., Trevino-Carre J., Jimenez-Sierra C.L., Astudillo-Sanchez C.C., Osorio-Hernandez E. (2022): Interactions between facilitator species and *Lophophora williamsii* (Lem. ex Salm-Dyck) J.M.Coult. (Cactaceae) in a rosetophyllus desert scrub in Mexico. *Journal of Arid Environments*. 205:104824.

Spinella M. (2001): *The psychopharmacology of herbal medicine: Plant drugs that alter mind, brain, and behavior*. MIT Press.

Terry M., Mauseth J.D. (2006): Root-shoot anatomy and post-harvest vegetative clonal development in *Lophophora williamsii* (Cactaceae: Cacteae). *Implications for Conservation*. 22: 565–592.

Tilson H.A., Sparber S.B. (1972): Studies on the concurrent behavioral and neurochemical effects of psychoactive drugs using the pushpull cannula. *Journal of Pharmacology and Experimental Therapeutic*. 181:387–398.

Trulson M.E., Crisp T., Henderson L.J. (1983): Mescaline elicits behavioral effects in cats by an action at both serotonin and dopamine receptors. *European Journal of Pharmacology*. 96:151–154.

WFO (2023): *Lophophora williamsii* (Lem. ex Salm-Dyck) J.M.Coult. Published on the Internet. http://www.worldfloraonline.org/taxon/wfo-0001287424. Accessed on: 15 Sep 2023

10 Devil's Tobacco (*Lobelia tupa* L. Campanulaceae)

Lilla Nagy-Radványi, Eszter Csikós, Ágnes Farkas, and Viktória L. Balázs

10.1 INTRODUCTION

Lobelia species have been used in medicine for centuries, and their ethnobotanical application is extensive (Folquitto et al. 2019). This is primarily due to medicinally active piperidine alkaloids found in plants, such as lobeline (Felpin and Lebreton 2004; Kesting et al. 2009; Kuo et al. 2011). The best-known *L. inflata* L. is applied in the form of powder, tincture, syrup, or infusion due to its anti-asthmatic, emetic, and astringent impacts. In the past, the utilization of these medicinal plants was even more diverse, as an antiseptic (*L. urens* L.) and expectorant (*L. laxiflora* Humb.) or a preparation for syphilis (*L. siphilitica* L.) (Borio 1959).

Due to its hallucinogenic and anesthetic (e.g., toothache) effects, the Indigenous people of Chile, the Mapuche Indians, applied *L. tupa* L. with preference during various rituals and in healing (Mösbach 1992; Lammers 2000). The hallucinogenic property has not been proven, but the American ethnobotanist Richard Evans Schultes believed that several species of *Lobelia* are psychoactive (Lammers 2000). The main alkaloid of *L. tupa* is lobeline, which has significant pharmacological properties such as anti-asthmatic, respiratory stimulating, and expectorant effects, and is also used for smoking cessation (Felpin and Lebreton 2004). Thanks to this, nicotine addiction has also been treated with it in traditional medicine, as lobeline exerts a partial agonist effect on potential nicotinic acetylcholine receptor (nAChR) (Paz et al. 2015; Chaachouay et al. 2019). In addition, it is important to mention that several studies reported its properties similar to nicotine, its use resulted in tachycardia and high blood pressure in the human body (Olin et al. 1995).

10.2 SCIENTIFIC CLASSIFICATION

Lobelia tupa belonging to the family Campanulaceae, subfamily Lobelioideae, is one of the representatives of the genus *Lobelia*, which consists of approximately 415 species. Named after the Flemish physician and botanist Matthias de L'Obel, the genus includes many medicinal plants (Lammers 2011; WFO 2023). Among the three subgenera, *L. tupa* belongs to the subgenus *Tupa* (G. Don) Wimmer, where there are robust herbs and shrubs with large flowers (one-lipped corolla). The subgen. *Lobelia* (syn. subgen. *Lagotis* Wimmer) and subgen. *Mezleria* (Presl) Wimmer usually contains herbs of shorter stature (Matthews 1988).

10.3 VERNACULAR NAMES

The plant's Spanish name is 'Tabaco del Diablo' (Devil's tobacco), which can be attributed to the hallucinogenic effect when smoking its leaves. It is the sacred plant of the Mapuche Indians, with the original Indian name 'Tupa' or 'Trupa' (Mösbach 1992).

10.4 BOTANICAL DESCRIPTION

Perennial, evergreen semi-shrub up to three meters high. Its upright, square-shaped, hollow shoots contain white latex. Its simple, narrow lanceolate or ovate leaves are 10–23 (–30) cm long and 2.5–8.5 (–10) cm wide. The yellow-green leaf blade is white-tomentose on both sides, the edge finely serrated, and the apex is acuminate. Its simple cluster inflorescence consists of resupinate flowers, the shape of which is ideal for pollination by hummingbirds. The pedicels are up to 4.5 cm long and reddish-purple in color. The ovate or lanceolate bracts are usually shorter than the pedicels, but may be of equal length or slightly longer. The reddish-purple calyx made of five sepals is 8–10 (–12) cm in diameter, its lobes are lance-shaped, 4–7 mm long. The five-petalled corolla is long, curved upwards, and then reflexed, 3–5 cm long. Its color ranges from scarlet to crimson-red. The three middle lobes are 6–15 mm long, the two lateral lobes 12–20 mm long. The stamens (five pieces) fused into a tube are 22–26 mm long, reddish-purple. The pale-yellow ovary bears a broad ring of hairs below the two-lobed stigma. Its fruit is a subglobose, 14 mm long dry capsule (Matthews 1988).

10.5 ORIGIN AND GEOGRAPHIC DISTRIBUTION

The discovery of *L. tupa* is linked to the name of Feuillée, who found it in Chile at 37 degrees south latitude. In his *Species Plantarum* (second edition), Linnaeus refers to the work of Feuillée, despite this, the origin of the plant is listed in Peru. After the introduction of *L. tupa* to Britain (1824), the cultivation of the plant began from seeds collected in

DOI: 10.1201/9781003460336-10

Chile, and later its use in horticulture became common (Matthews 1988).

Lobelioideae are found in most regions of the world, with the exception of Central Asia, the Middle East, and the Arctic. Most species are concentrated in Africa, and the greatest genetic diversity can be observed in tropical areas (Lammers 2007). Among the species belonging to the genus *Lobelia* L., 4 are native to Chile: *L. tupa*, *L. polyphylla* Hook & Arn., *L. excelsa* Bonpl. and *L. bridgesii* Hook & Arn. The distribution area of *L. tupa* extends from the Coquimbo region to the Talca province (Riedemann and Aldunate 2001).

10.6 HABITAT AND ECOLOGY

In its native land, *L. tupa* prefers habitats with sandy soil and hilly areas covered with grass near the coast. During the flowering period, it requires a larger amount of water, but basically, soil with good drainage and a warm, sunny environment are the most ideal for it (Matthews 1988).

10.7 ETHNOBOTANY

There is an ancient tradition of using *Lobelia* species in folk medicine worldwide. In India, the species *L. inflata* ('Indian Tobacco'), *L. nicotianaefolia* Roth E (in Ayurveda), and *L. pyramidalis* Wall. were most often applied in the form of tincture or decoction to treat asthma and bronchitis (Felpin and Lebreton 2004; Joshi et al. 2011; Kurdekar et al. 2012). *Lobelia pyramidalis* also had an important role in the treatment of snakebites and pain relief according to tradition (Vigneshwaran et al. 2014). The dry leaves of *L. flaccida* Persl. served as an analgesic and antiepileptic in the Eastern Cape region of South Africa (Stolom et al. 2016). The medicinal use of *L. cardinalis* L., native to America, was also significant in cases of vomiting, typhus, and fever (Brown et al. 2016). In Chile, the leaves of *L. tupa* and *L. polyphylla* are primarily considered a narcotic, smoking them has a mind-altering effect (causes hallucinations) and is extremely harmful to health. *Lobelia tupa* contains toxic and caustic latex that can cause vomiting, intestinal irritation, and even delirium. Due to its corrosive properties, it is particularly dangerous if it gets in the eyes or on the skin, but it is useful for toothaches (Mösbach 1992; Villegas et al. 2014; Chaachouay et al. 2021a, 2021b).

10.8 PHARMACOLOGY

The following alkaloids were previously detected from the aerial parts *of L. tupa*: lobeline, norlobeline, lobelanidine, norlobelanidine, lobelanine, norlobelanine, 3-hydroxy-3-phenyl-propanoic allosedamine, 8-propyl-10-phenyl lobelionol, and 3-hydroxy-3-phenyl-propanoic norallosedamine (Folquitto et al. 2019).

In addition to lobelanidine and lobeline, Paz et al. (2015) isolated a new piperidine alkaloid called pentylsedinine 1-(6-(2-hydroxy-2-phenylethyl)-1-methyl-1,2,5,6-tetrahydropyridin-2-yl) pentan-2-ol from the fresh leaves of *L. tupa*, following extraction and purification. Its structure is very similar to that of sedinine and contains five carbon atoms in the side chain, which is not characteristic of *Lobelia* species (Felpin and Lebreton 2004). The compound has two alcohol groups on carbon atoms 8 and 10 and is characterized by a double bond in the central heterocyclic ring. During the tests, it was determined that the largest amount of lobelanidine is found in the leaves of the plant, followed by lobeline and finally pentylsedinine (Paz et al. 2015).

The three piperidine alkaloids isolated by Paz et al. (2015) are structurally very similar, yet functional differences were observed. Lobelanidine is a neutral antagonist of $\alpha3\beta2/\alpha3\beta4$ nAChR and $\alpha7$ nAChR, while lobeline and pentylsedinine act as nAChR partial agonists. Pentylsedinine, which is present in the plant in lower amounts compared to the other two alkaloids, has increased efficacy on the nAChR compared to lobeline. Knowing these facts, it can be assumed that pentylsedinine can induce an effect on neuronal receptors corresponding to ethnopharmacobotanical use.

10.9 MEDICINAL EFFECTS

Several pharmacologically active piperidine alkaloids have been isolated from *Lobelia* species over the years (e.g., lobeline, lobelanidine, lobelane), and several research groups have reported on the high phytosterol content of these plants (e.g., sitosterol, stigmasterol, campesterol). The latter compounds can be responsible for their analgesic, anti-inflammatory properties, and anticonvulsant effects. In addition, the high amount of secondary metabolic products (e.g., flavonoids, saponins, coumarins, triterpenes) also contributes significantly to its medicinal use (Folquitto et al. 2019).

Although the endemic plant of Chile, *L. tupa* is mainly known for its hallucinogenic effects (Mösbach 1992; Lammers 2000), the large amount of lobeline present in it did not cause such a reaction in human studies (Paz et al. 2015).

10.10 CONCLUSION

Based on earlier traditions, *Lobelia tupa* is currently used as a recreational drug, despite the fact that smoking the leaves has a harmful effect on health. In addition, very little data is available on the biological activity of this species, and this fact increases the risk of its use even more. In the future, it would definitely be important to reveal the exact mechanism of action of the piperidine alkaloids found in the plant, as well as its presumed psychoactive effect during in vivo experiments. Thanks to its spectacular flowers, the species is also popularly grown in gardens. In this case, too, caution is essential, since the latex in the stems becomes powdery after pruning and can cause serious symptoms when dispersed in the air.

LIST OF ABBREVIATIONS

L. inflata L.	*Lobelia inflata* L.
***L. laxiflora* Humb.**	*Lobelia laxiflora* Humb.
***L. siphilitica* L.**	*Lobelia. siphilitica* L.
L. tupa	*Lobelia tupa* L.
***L. urens* L.**	*Lobelia urens* L.
nAChR	nicotinic acetylcholine receptor

REFERENCES

Borio E.B.L. (1959): Lobelia langeana Dusén, Contribuição ao seu estudo farmacognóstico. *Curitiba*. 86 p. https://doi.org/10.1590/2175-78601980325418

Brown D.P., Rogers D.T., Pomerleau F., Siripurapu K.B., Kulshrestha M., Gerhardt G.A., Littleton J.M., (2016): Novel multifunctional pharmacology of lobinaline, the major alkaloid from *Lobelia cardinalis*. *Fitoterapia*. 111: 109–123. https://doi.org/10.1016/j.fitote.2016.04.013

Chaachouay, N., Benkhnigue, O., Fadli, M., Ibaoui, H., & Zidane, L. (2019). Ethnobotanical and ethnopharmacological studies of medicinal and aromatic plants used in the treatment of metabolic diseases in the Moroccan Rif. *Heliyon*, 5(10), e02191. https://doi.org/10.1080/10477845.2021.1942433

Chaachouay, N., Douira, A., & Zidane, L. (2021a). Herbal Medicine Used in the Treatment of Human Diseases in the Rif, Northern Morocco. *Arabian Journal for Science and Engineering*, 47, 131–153. https://doi.org/10.1007/s13369-021-05501-1

Chaachouay, N., Douira, A., & Zidane, L. (2021b). COVID-19, prevention and treatment with herbal medicine in the herbal markets of Salé Prefecture, North-Western Morocco. *European Journal of Integrative Medicine*, 42, 101285. https://doi.org/10.1016/j.eujim.2021.101285

Felpin F., Lebreton J. (2004): For a review of history, chemistry, and biology of *Lobelia* alkaloids. *Tetrahedron*. 60: 10127–10153. https://doi.org/10.1016/j.tet.2004.08.010

Folquitto D.G., Swiech J.N.D., Pereira C.B., Bobek V.B., Halila Possagno G.C., Farago P.V., Miguel M.D., Duarte J.L., Miguel O.G. (2019): Biological activity, phytochemistry and traditional uses of genus *Lobelia* (Campanulaceae): A systematic review. *Fitoterapia*. 134(2019): 23–28. https://doi.org/10.1016/j.fitote.2018.12.021

Joshi S., Mishra D., Bisht G., Khetwal K.S. (2011): Essential oil composition and antimicrobial activity of *Lobelia pyramidalis* Wall. *EXCLI Journal*. 10: 274–279. https://doi.org/10.17877/DE290R-5129

Kesting J.R., Tolderlund I.L., Pedersen A.F., Witt M., Jaroszewski J.W., Staerk D. (2009): Piperidine and tetrahydropyridine alkaloids from *Lobelia siphilitica* and *Hippobroma longiflora*. *Journal of Natural Products*. 72(2009): 312–315. https://doi.org/10.1021/np800743

Kuo P.C., Hwang T.L., Lin Y.T., Kuo Y.C., Leu Y.L. (2011): Chemical constituents from *Lobelia chinensis* and their anti-virus and anti-inflammatory bioactivities. *Archives of Pharmacal Research*. 34(2011): 715–722. https://doi.org/10.1007/s12272-011-0503-7

Kurdekar R.R., Hegde G.R., Hegde G., Hebbar S.S. (2012): Antimicrobial screening of medicinal plants against human pathogens- a comparative account of two different methods of extraction. *International Journal of Drug Development and Research*. 4(1): 24–31.

Lammers T.G. (2000): Revision of *Lobelia* sect. *Tupa* (Campanulaceace: Lobelioideae). *Journal of the Botanical Research Institute of Texas*. 19(1): 87–110.

Lammers T.G. (2007): World checklist and bibliography of Campanulaceae. Royal Botanic Gardens. *Kew*. 685 pp.

Lammers T.G. (2011): Revision of the infrageneric classification of *Lobelia* L. (Campanulaceae: Lobelioideae). *The Annals of the Missouri Botanical Garden*. 98(2011): 37–62. https://doi.org/10.3417/2007150

Matthews V. (1988): *Lobelia Tupa*: Campanulaceae. *Kew*. 5: 157–161.

Mösbach E.W. (1992): Botánica Indígena de Chile. Santiago, Chile. 140 pp.

Olin B. R., Hebel S. K., Gremp J. L., Hulbert M. K. (1995): Smoking deterrents. In: Olin B. R., Hebel S. K., Gremp J. L., Hulbertt M. K. *Drug facts and comparisons*. JB Lippincott. pp. 3087–3095.

Paz C., Becerra J., Silva M., Burgos V., Heydenreich M., Schmidt B. Tran T., Vetter I. (2015): (-)-Pentylsedinine, a new alkaloid from the leaves of lobelia tupa with agonist activity at nicotinic acetylcholine receptor. *Natural Product Communications*. 10(8): 1355–1357. https://doi.org/10.1177/1934578x1501000810

Riedemann P., Aldunate G. (2001): Flora nativa de valor ornamental: identificación y propagación, Chile, zona centro. Santiago, Chile.

Stolom S., Oyemitan I.A., Matewu R., Oyedeji O.O., Oluwafemi S.O., Nkeh-Chungag B.N., Songca S.P., Oyedeji A.O. (2016): Chemical and biological studies of *Lobelia flaccida* (C. Presl) A.DC leaf: a medicinal plant used by traditional healers in Eastern Cape, South Africa, *Tropical Journal of Pharmaceutical Research*. 15: 1715–1721. https://doi.org/10.4314/tjpr.v15i8.17

Vigneshwaran V., Somegowda M., Pramod S. (2014): Pharmacological evaluation of analgesic and antivenom potential from the leaves of folk medicinal plant *Lobelia nicotianaefolia*. *American Journal of Phytomedicine and Clinical Therapeutics*. 2: 1404–1415.

Villegas A., Espinoza J., Urzua A. (2014): Piperidine Alkaloids from *Lobelia polyphylla* Hook. & Arn. (Campanulaceae). *Boletín latinoamericano y del Caribe de plantas medicinales y aromáticas*. 13(2): 205–212.

WFO (The World Flora Online): https://www.worldfloraonline.org/taxon/wfo-0000815628 Retrieved 25 August 2023.

11 Marijuana (*Cannabis sativa* L. Cannabaceae)

Noureddine Chaachouay, Rahmatullah Qureshi, Ouafae Benkhnigue, Abdelhamid Azeroual, and Lahcen Zidane

11.1 INTRODUCTION

Throughout millennia, humans have harnessed the potential of plants for a wide range of uses. These have functioned as sources of medicine, nourishment, cosmetics, subsistence, building materials, and other purposes. Plants have played a vital role in the development of human civilization, providing crucial resources for constructing homes, making clothes, and producing other needed goods (Ansari et al., 2023; Chaachouay et al., 2020, 2020; Giannenas et al., 2020; Jamshidi-Kia et al., 2018). Currently, plants are of utmost importance due to their wide-ranging implementations in various sectors such as medicine, agriculture, food production, industry, and cosmetics (Ansari et al., 2023; Hassan, 2015; Srivastava, 2018).

Throughout countless millennia, humans have used plants for a multitude of purposes. These have served as supplies of sustenance, nourishment, beauty products, pharmaceuticals, construction materials, and several other purposes (Qureshi et al., 2016). Plants have been essential to the evolution of human culture throughout history because they provide resources for needs like food, clothing, and shelter (Ansari et al., 2023; Chaachouay et al., 2019a, 2019b; Giannenas et al., 2020; Jamshidi-Kia et al., 2018).

Certain plants contain psychoactive chemicals that may induce hallucinatory experiences. The drugs mentioned, including DMT, psilocybin, and mescaline, can influence an individual's perception, cognitive functions, and emotions, leading to hallucinations and other distinctive encounters (Alrashedy & Molina, 2016; Schultes, 1969). Indigenous cultures have traditionally used hallucinogenic plants for spiritual and medicinal reasons, while contemporary civilization has progressively embraced them for recreational ones. Furthermore, the effects of hallucinogenic plants might vary based on the situation and the user's objectives, including both advantageous and detrimental outcomes (Passos & Mironidou-Tzouveleki, 2016; Pierce & Peroutka, 1989). Under precise conditions and the right circumstances, hallucinogenic plants have the potential to induce profound spiritual insights, facilitate personal growth, and promote healing. Nonetheless, their use can be detrimental and result in adverse effects, including paranoia, anxiety, and additional mental health complications (Cunningham, 2008; Passos & Mironidou-Tzouveleki, 2016; Schultes, 1969).

Hemp or Marijuana (*Cannabis sativa* L.), which is endemic to the Himalayas, is said to have originated in Central Asia and is now globally acknowledged. This species has a great degree of adaptability to many climatic conditions, from low-lying plains to elevated regions reaching altitudes of up to 10,000 feet. Cannabis is cultivated in several nations worldwide, such as Iran, China, India, Pakistan, and Russia, in addition to its presence in the wild. Throughout the years, it has been used for sustenance, textiles, and oil extraction. Aside from its religious and recreational use, it is also employed as a kind of medicine (Piluzza et al., 2013). Cannabinoids, alkaloids, flavonoids, and terpenoids are among the active compounds present in *C. sativa* (Andre et al., 2016). Cannabidiol (CBD) is a very potent compound found in significant amounts. These terpene-phenolic compounds are primarily present in the trichomes of female flowers (Taura et al., 2007). Delta-9-tetrahydrocannabinol (D9-THC) is a considerable compound known for its psychoactive properties, which induce euphoria or pleasure in users (Atakan, 2012; Whiting et al., 2015). The early humans with sedentary living standards were the first to cultivate *C. sativa*. Initially, it was an agricultural crop grown in China, which then transferred to Africa, the Middle East, and Asia, where it was used as an analgesic and for the treatment of malaria, gout, poor memory, rheumatism, and recreational (Kalant, 1999). According to (UNODC, 2013), there are projected 182.5 million users of cannabis globally, and it is extensively cultivated, marketed, and used up in the form of a drug. The recent *C. sativa* varieties with low amounts of psychoactive D9-THC are essentially used for making fish nets, ropes, strings, paper, and textiles (Sabo et al., 2013).

The use of cannabis for medical purposes has been documented all over the world (I. Ahmad et al., 2011; K. S. Ahmad et al., 2012; S. Ahmad et al., 2006; Barkatullah & Ibrar, 2011; Haq et al., 2011; Hussain et al., 2010; Ibrar et al., 2007; Iqbal & Sher, 2011; Kalant, 2001). The dried leaves of *C. sativa* are smoked via a pipe called a Hukkah in Kashmir to treat depression, and its leaf extract is used to treat cholera, according to (Mir et al., 2021). According to (Jan et al., 2021), *C. sativa* leaf infusion can be used to treat pregnancy labor pain. According to (Jan et al., 2022), *C. sativa* could help develop novel medications for the treatment of female reproductive

DOI: 10.1201/9781003460336-11

health, which is being used in the region for the treatment of various gynecological issues. In several countries, dried female cannabis buds and flowers are used as medicines. In several nations, including those in the Middle East, North Africa, and Southwest Asia, the resin is scarce. Small stems or leaves (bhang, marijuana, kif, or dagga), resin (hash, hashish, polm, charas), female flower heads, or oil (alcoholic resin extract) are all common street forms of cannabis. All of these cannabis varieties range in purity from 1 to 60% THC, a psychoactive component, according to (Cunha-Oliveira et al., 2013).

One of these psychoactive plants that may have a variety of impacts on the body and mind of a person is hemp. The induction of hallucinations, a condition when a person has an altered view of reality and perceives sights, sounds, or feelings that are not there, is a commonly documented side effect of cannabis use (Desai et al., 2022; MCDOWELL, 2005; McGeeney, 2013). The psychoactive effects of marijuana are mostly caused by a hallucinogenic molecule known as delta-9-tetrahydrocannabinol (THC), which is the principal active constituent in cannabis. THC influences the brain by binding to specific receptors in the central nervous system, resulting in modified perceptions and emotions (Desai et al., 2022; McGeeney, 2013; Stephens & Banes, 2013).

The dried buds and flowers of female cannabis are consumed as a medicinal remedy in many countries. Its resin is limited in some countries like Southwest Asia, North Africa, and the Middle East. Cannabis is available in the form of small stems or leaves (bhang, marijuana, kif, or dagga), as a resin (hash, hashish, polm, charas), female flower heads, or oil (alcoholic resin extract). All these forms have varying purity levels with 1 to 60% THC psychoactive compound (Cunha-Oliveira et al., 2013). Afghanistan, Morocco, Lebanon, India, and Pakistan were the principal nations recognized as 'hashish' source nations in 2011. The main markets for cannabis in Morocco and resin in Afghanistan between 2009 and 2011 are reflected in the origins of 'hashish', which originated in both countries. Hashish was smuggled to Morocco from 17 nations, 11 in Central and Western Europe. Afghanistan has also been identified as a source of 'hashish' seizures by countries in the Middle East and Europe.

Researchers continue to explore this medicinal plant's potential benefits and risks and their possible use in medical treatments for conditions such as depression, anxiety, and addiction. The current chapter presents an overview of the results found about the botany, ecology, traditional uses, phytochemical, and hallucinogenic effects of *Cannabis sativa*.

11.2 SCIENTIFIC CLASSIFICATION

Cannabis sativa L. is an annual herbaceous plant belonging to the Cannabis *genus* and Cannabaceae family (ITIS, 2022).

Kingdom:	Plantae
Subkingdom:	Viridiplantae
Infrakingdom:	Streptophyta
Superdivision:	Embryophyta
Division:	Tracheophyta
Subdivision:	Spermatophytina
Class:	Magnoliopsida
Superorder:	Rosanae
Order:	Rosales
Family:	Cannabaceae
Genus:	*Cannabis*
Species:	*Cannabis sativa* L.

11.3 VERNACULAR NAMES

Arabic: Qinnab Lhendi, Hashish; **Bengali**: Bhâng, Charas, Ganja, Jia; **Burmese**: Segiyav; **Chinese**: Ta-ma-jen, Ye ma, Dàmáshū; **Czech**: Konopí seté; **Danois**: Hamp; **English**: Marijuana, Cannabis, Indian hemp, Mary jane, Hashish; **French**: Chanvre commun, Chanvre cultivé, Chanvre; **Finland**: Hamppu; **Italian**: Canapé, **German**: Gebauter hanf, Gewöhnlicher hanf; **Hindi**: Bhâng, Charas, Ganja, Sabji, Siddhi; **Japanese**: Mashinin, Asa; **Nepali**: Gajima, Cares, Ganja; **Norwegian**: Hamp; **Persian**: Shahdanah, Bang, Kinnah; **Polish**: Konopie siewne; **Portuguese**: Maconha, Cânhamo-comum; **Sweden**: Hampa; **Spanish**: Cáñamo común, Cañamón, Cáñamo; **Russian**: Konoplya; **Thai**: Porkanchaa; **Urdu**: Bhang; **Turkish**: Hint keneviri.

11.4 SYNONYMS

Cannabis sativa L. includes thirty-two synonyms, which are as follows (The Plant List, 2013):

(1) *Cannabis indica* Lam.; **(2)** *Cannabis sativa* var. *indica* (Lam.) Wehmer; **(3)** *Cannabis erratica* Siev.; **(4)** *Cannabis foetens* Gilib.; **(5)** *Cannabis chinensis* Delile; **(6)** *Cannabis gigantea* Crevost; **(7)** *Cannabis americana* Pharm. ex Wehmer; **(8)** *Cannabis indica* f. *afghanica* (Vavilov) Vavilov; **(9)** *Cannabis ruderalis* Janisch.; **(10)** *Cannabis lupulus* Scop.; **(11)** *Cannabis kafiristanica* (Vavilov) Chrtek; **(12)** *Cannabis intersita* Soják; **(13)** *Cannabis macrosperma* Stokes; **(14)-** *Cannabis indica* var. *kafiristanica* Vavilov; **(15)** *Cannabis sativa* f. *afghanica* Vavilov; **(16)** *Cannabis sativa* f. *chinensis* (Delile) A.DC.; **(17)** *Cannabis sativa* var. *gigantea* (Delile ex Vilm.) Alef.; **(18)** *Cannabis sativa* var. *ruderalis* Janisch.; **(19)** *Cannabis sativa* subsp. *indica* (Lam.) E.Small & Cronquist; **(20)** *Cannabis sativa* var. *indica* (Lam.) E. Small & Cronquist; **(21)** *Cannabis generalis* E.H.L.Krause; **(22)** *Cannabis sativa* var. *kafiristanica* (Vavilov) E.Small & Cronquist; **(23)** *Cannabis sativa* var. *kif* A.DC.; **(24)** *Cannabis sativa* subsp. *intersita* (Soják) Soják; **(25)** *Cannabis sativa* var. *monoica* Hol.; **(26)** *Cannabis sativa* var. *sativa*; **(27)** *Cannabis sativa* f. *pedemontana* A.DC.; **(28)** *Cannabis* *sativa* var. *macrosperma* (Stokes) Asch. & Graebn.; **(29)** *Cannabis sativa* var. *ruderalis* (Janisch.) S.Z.Liou; **(30)** *Cannabis sativa* var. *praecox* Serebr.; **(31)** *Cannabis sativa* var. *spontanea* Vavilov; **(32)** *Cannabis sativa* var. *vulgaris* Alef.

11.5 BOTANICAL DESCRIPTION

Cannabis is a perennial plant that may reach up to 12 feet. The stem is often characterized by its size, straightness, and strength, possessing a pungent, woody outer layer and a soft, spongy inner layer. It is highly branched, with variations in cannabinoid content, and the branches have a slightly angular shape. The stem is sparsely covered in hairs as it ages, while immature shoots have a denser hair coating. The stipules have a length of 4–6 mm and possess a papery texture. Leaves palmately compound, usually with 3–9 leaflets with white erect hairs; leaflets sessile, narrowly lanceolate, much narrowed at the base, palmi-nerved, margin serrate, apex accuminate-caudate, approximately 2–15 cm across; upper surface scabrid with stiff hairs topping the cystoliths; lower surface more or less densely pubescent, covered with sessile glands. Both male and female plants (Figure 11.1) separately grow possess small flowers (4–6 mm across) which are greenish-yellow and grow in clusters at the base of the leaves (Chandra, Lata, & ElSohly, 2017; Strzelczyk et al., 2022). The flowers have a strong, pungent odor and are used recreationally and medicinally. Female flowers are substantially bigger than males, and they are accompanied by leafy bracts (2–13 mm) coated in tiny glandular hairs. Female flowers are preferred because they have a higher concentration of resinous trichomes, which contain cannabinoids such as THC and CBD. The fruit is sub-globose, smooth, grown in large, and sometimes sub-spontaneous in various countries. The seeds (3–4 mm) are oval-shaped and can range in color from light gray to yellowish brown or black, with a hard outer shell (Chandra, Lata, Khan, et al., 2017; Farag & Kayser, 2017; Schilling et al., 2020).

11.6 HISTORY

Cannabis has been utilized for millennia. Hemp textiles from the late 8th century B.C. have been unearthed in Turkey and materials dating back nearly four millennia have been uncovered at an Egyptian archaeological site. Ancient Thebes used the herb to produce a beverage that had opium-like effects. The Scythians cultivated cannabis near the Volga River 3,000 years ago. They would scatter cannabis seeds and leaves upon heated stones in steam baths, resulting in the production of a mind-altering vapor (Schultes & Smith, 1976).

According to Chinese folklore, the plant has been used for 4,800 years. Before 1000 B.C., Indian physicians documented the therapeutic uses of cannabis. Names like 'heavenly guide' and 'soother of grief' indicate that the early Hindus valued its euphoric qualities. Cannabis was known by the Chinese names 'liberator of sin' and 'delight giver'. Around A.D. 160, the Greek physician Galen noted that eating hemp-based cakes regularly resulted in narcotic effects. The word 'assassin' perhaps derived from 'Hashashins', denoting organized assassins in Asia Minor around the third century who were compensated with hashish (Schultes & Smith, 1976).

FIGURE 11.1 Habit of *Cannabis sativa* L., (a) male plant, (b) female plant.

The Pilgrims brought hemp as a source of fiber to New England, and the Spanish and Portuguese brought it to their colonies in the New World. Artifacts related to cannabis use have been discovered in the frozen graves of the prehistoric Scythians, located in the Altai Mountains on the boundary between outer Mongolia and Russia. Covered with a felt or leather pad, the little tepee-shaped building stood above the copper center, a stool-like device with four legs. Nearby, carbonized cannabis kernels were discovered. The container with two handles was filled with cannabis fruits. The practice of inhaling cannabis vapors while in a steam bath was documented by the Greek naturalist Herodotus around 500 B.C. (Schultes & Smith, 1976).

11.7 ORIGIN AND GEOGRAPHIC DISTRIBUTION

Cannabis sativa L., commonly known as hemp or marijuana, is a plant species believed to have originated in Central Asia, possibly in the region of Northeastern Tibet to now Mongolia and southern Siberia (Brown, 1998; Chandra, Lata, Khan, et al., 2017), which was confirmed by a molecular study (Zhang et al., 2018) as well as a pollen study (McPartland et al., 2019). However, the native place is believed to be Central Asia to Xinjiang and Pakistan (GBIF, 2020). Throughout thousands of years, it has been cultivated for many purposes, including its use as a medical treatment, a fiber source, and a fundamental part of the diet. Over time, cannabis has been disseminated to many parts of the globe via the exchange and movement of human beings. The plant has a lengthy historical record of use across several civilizations, supported by archaeological findings indicating its therapeutic application in ancient China dating back to 4,000 years ago (Liu et al., 2017; Zhang et al., 2018). Currently, cannabis is extensively dispersed over the globe, with many strains and kinds being farmed for a multitude of reasons. Cannabis is now considerably grown worldwide (Figure 11.2), particularly in North and South America, Europe, Africa, Asia, and Australasia (McPartland et al., 2019).

The plant is mainly linked to nations like as Mexico, Jamaica, and the United States, where it has had a notable impact on cultural and societal aspects. Moreover, it is cultivated and used in many other countries, including Morocco, India, Afghanistan, and other European regions (Chaachouay et al., 2021, 2021b, 2021c; Kuddus et al., 2013). The cultivation of Cannabis in Morocco is mainly concentrated in the northern section of the nation, namely in the Rif Mountains. This area is especially suitable for producing cannabis because of its favorable conditions, including a temperate temperature, rich soil, and enough water resources (Figure 11.3). Cannabis growing has been a longstanding tradition in the Rif area for millennia, serving as a primary source of revenue for several small-scale farmers. In recent years, there has been an increasing effort by governmental and international bodies to suppress the production of cannabis because of its connection to the criminal drug market and its unlawful status. It is important to note that the legal quality of cannabis varies greatly across nations, with some countries allowing its usage for medicinal or recreational reasons while many others outlaw it entirely.

11.8 ECOLOGY

Cannabis sativa L. is a very adaptable plant capable of thriving in many settings and temperatures, ranging from tropical to temperate areas (Anwar et al., 2006). The plant exhibits

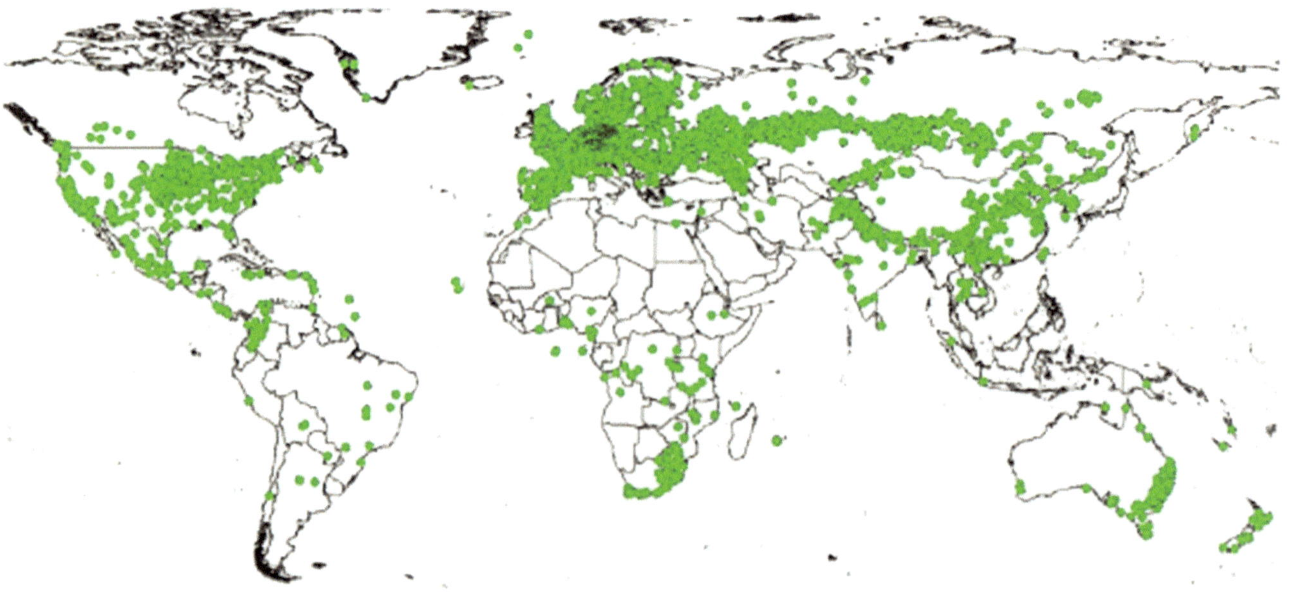

FIGURE 11.2 Global map of the distribution of hemp.

FIGURE 11.3 A cannabis field in Oulad Aissa (Ketama), Al Hoceima province, Morocco.

adaptability to many soil types and thrives in regions characterized by limited rainfall and nutrient availability. *Cannabis sativa* L. thrives in areas with abundant sunshine and well-drained soil, often in disrupted environments like roadsides, fields, and waste areas (Pavlovic et al., 2019). The plant is also shown to thrive near riverbanks, where it may avail itself of the water's moisture and nutrients. *C. sativa* is classified as an invasive plant in some areas, such as Australia and certain portions of the United States, where it has managed to spread beyond cultivation and develop self-sustaining populations (Yazici et al., 2020). These populations may adversely affect local ecosystems by displacing native plant species and modifying soil conditions. *C. sativa* is renowned for its phytoremediation capabilities, enabling it to absorb and eliminate pollutants from the soil effectively. The facility has been used to remediate soil that is polluted, namely in regions that have been contaminated with heavy metals and industrial waste. The ecological effects of *C. sativa* agriculture might differ according to the size and techniques used for growth (Haney & Kutscheid, 1975). The extensive cultivation of crops may have a substantial influence on water resources, soil quality, and biodiversity. Conversely, adopting small-scale, sustainable farming techniques can provide ecological advantages, such as enhancing soil health and promoting pollinator populations. The environmental dynamics of *C. sativa* are complex, exhibiting both beneficial and detrimental effects on nearby ecosystems, depending upon the specific circumstances of its cultivation and control.

11.9 PHYTOCHEMISTRY

The chemistry of *C. sativa* is complex. A multitude of chemical compounds have been found, some possessing narcotic properties while others do not. The plant has various compounds in several chemical classes, including steroids, flavonoids, mono and sesquiterpenes, hydrocarbons, amino acids, and nitrogenous substances (M. A. ElSohly & Slade, 2005). Due to broad interest by the researchers in this plant, the number of recognized compounds is increasing. In 1980, 423 compounds were reported (Turner et al., 1980), in 1995, 483; while, in 2008, the compounds reached over 525 (S. A. Ahmed et al., 2008; Radwan et al., 2008). The main constituents synthesized by a newly harvested plant are cannabidiolic acids, which function as the fundamental components for tetrahydrocannabinol and other associated substances like cannabinol, cannabidiol, tetrahydrocannabinol-carboxylic acid, stereoisomers of tetrahydrocannabinol, and cannabichromene. Additionally, it encompasses cannabinoids, terpenes, flavonoids, steroids, and lignans (Brenneisen, 2007; M. A. ElSohly et al., 2017; Pavlovic et al., 2019; Radwan et al., 2008; Rock & Parker, 2021). The most well-known cannabinoid found in cannabis is delta-9-tetrahydrocannabinol (THC), which is responsible for the plant's psychoactive effects. Other cannabinoids found in cannabis include cannabinol (CBN), cannabidiol (CBD), cannabitriol (CBT), and cannabigerol (CBG), among others (Brenneisen, 2007; dos Reis Rosa Franco et al., 2021; McGeeney, 2013; Radwan et al., 2021). The presence of terpenes, which are fragrant molecules, is responsible for the unique scent of cannabis. Tetrahydrocannabinol (THC) is a non-nitrogenous chemical molecule that is formed from terpenes. It includes an oily mixture consisting of many isomers. Until recently, our understanding of the effects of pure tetrahydrocannabinol on humans was limited (Schultes & Smith, 1976). Flavonoids, a kind of plant component known for their anti-oxidant and anti-inflammatory qualities, are often present in a wide range of plants (M. Ahmed et al., 2019; Orch et al., 2021; M. A. ElSohly, 2002; Radwan et al., 2021). The composition of cannabis might vary depending on the strain and the cultivation circumstances. Some cannabis strains may have high quantities of THC and low levels of CBD, while others have the opposite profile. Similarly, the terpene composition might vary amongst different themes, with some containing substantial amounts of myrcene while others exhibit elevated levels of limonene, pinene, or other terpenes (Table 11.1).

11.10 REGULATION AND COMMERCIAL PRODUCTION OF CANNABIS

The regulation of *C. sativa* varies significantly according to the nation and jurisdiction. In some jurisdictions, the act is strictly prohibited; however, in other regions, it is permissible for medicinal or recreational purposes (Gartner & Hall, 2020). Cannabis regulation encompasses legislation and rules about the production, distribution, possession, and use of the plant. Governments may establish regulations for the testing, labeling, and packaging of cannabis products to guarantee their safety and uniformity (Abuhasira et al., 2018; Pertwee, 2014).

As to our understanding, starting in 2021, the manufacturing and use of cannabis were prohibited in Morocco. Nevertheless, the nation has a long record of cultivating cannabis, namely in the Rif Mountains, and it is approximated that over 800,000 Moroccans are engaged in this sector (Elhamdaoui et al., 2019). In recent years, there has been a rising discussion in Morocco over cannabis legislation. Legalization and regulation, proponents claim, might provide economic gains, decrease criminal activity, and enhance public health results. In response, the Moroccan government established a panel in 2018 to investigate the possible consequences of legalizing. The Moroccan parliament approved in July 2021 to allow the cultivation of cannabis for therapeutic and industrial uses (Chaachouay et al., 2023; Zeroual & Danguir, 2023). Under the legislation, authorized farmers are permitted to cultivate cannabis for use in medicines, cosmetics, and other industrial goods. Additionally, a centralized national agency is established to regulate and supervise the business. Nevertheless, the use of cannabis for recreational purposes remains prohibited in Morocco, and the legislation imposes severe consequences for any illicit growing or distribution (Chouvy, 2022).

Cannabis policy in the United States is mainly determined at the state level, with several states permitting its usage for both medicinal and recreational purposes. Conversely, other jurisdictions have only authorized its therapeutic use or have not allowed it (Schauer, 2021). At the federal level, cannabis remains illegal and is classified as a Schedule I controlled substance under the Controlled Substances Act (Goundar et al., 2021). In 2018, Canada implemented the Cannabis Act, which legalized the recreational use of marijuana and created regulations for the production, distribution, sale, and possession of cannabis products (Kogan et al., 2019). The laws include restrictions on where and how marijuana can be sold and used, age limits, and limits on the amount of marijuana an individual can possess (Huntsman et al., 2021).

In some countries, like the Netherlands, marijuana is decriminalized, signifying that it is formally against the law, but the government has opted not to pursue legal action against people for the possession or use of minor quantities (Aguilar et al., 2018; Fischer, 2017). Other countries, including Uruguay and Portugal, have completely legalized and regulated marijuana (de Souza et al., 2022; Gartner & Hall, 2020). As an increasing number of countries and states reassess their legislation and policies concerning this contentious substance, global marijuana regulation changes and evolves.

11.11 ETHNOBOTANY

Hemp has been utilized historically by numerous cultures for medicinal, recreational, and spiritual intentions, among others (Shah, 2004). Due to the native plant of Pakistan, it has a rich traditional use by the local communities living in various areas. Since most of the parts of the country are occupied by wild hemp, cannabis is collected from the wild for preparing medicaments such as decoction, powder, juice, paste, or infusions. The people of Booni Valley, Chitral (Pakistan) use its leaves, seeds, and floral buds to treat jaundice, bronchitis, and malaria (S. A. Ahmed et al., 2008), while in the Ranyal hills of Shangla, the same is used as a sedative and narcotic (Ibrar et al., 2007). According to (Barkatullah & Ibrar, 2011) and (Haq et al., 2011), shoots of *C. sativa* are used as hypnotic, narcotic, and sedative from the Malakand and Battagram, respectively. The plant is reported as a diuretic and stimulant by Shuaib et al., (2014) from the Lower Dir. The leaves are given in stomach disorders, chicken pox, and measles in the Haramosh Valley (Abbas et al., 2019). Seeds are given to treat cancer and infertility in females. The powder of leaves is used as a narcotic agent (Munir et al., 2022). Paste comprises fresh leaves and turmeric, slightly warmed in rapeseed oil, and applied to treat pain and swellings (Qaseem et al., 2019). Zareef et al., (2023) reported the whole plant extract in treating cough, anorexia, arthritis, and migraine. Another study said that leaves are smoked as a narcotic, sedative, appetizer, and chronic pain. Leaves are ground and given in piles, whooping cough (Rehman et al., 2023; Ullah et al., 2023 Chaachouay et al., 2022a, 2022b, 2022c).

The plant comprises several chemicals, notably cannabinoids like THC and CBD, which are thought to possess numerous medicinal properties. Recently, there has been an increasing fascination with the potential industrial applications of cannabis, namely in the cultivation of hemp. Hemp is a variety of cannabis that has low levels of THC and is mostly farmed for its fiber and seeds. Hemp is a versatile plant that may be used in several industries, such as textile manufacturing, paper production, biofuel development, and

TABLE 11.1
Bioactive Compounds of *Cannabis sativa* L.

Cannabis Constituents and Chemical Class	References
Non-cannabinoid bioactive components (125 biomolecules): **Terpenes**: (120 biomolecules): Monoterpenes (61 compounds), Sesquiterpenes (51 biomolecules), Miscellaneous Terpenes (4 biomolecules), Diterpenes (2 biomolecules), and Triterpenes (2 biomolecules). **Flavonoids**: (34 biomolecules): Geranyl flavone, Apigenin, Canniflavin, Chrisoeriol, Luteolin, Kaempferol, Glycosides Iso-prenoid flavones, Cytisoside, Cytisoside glucoside, Isovitexin, Rutin, Naringin, Vitexin, Prenyl flavone, 6 – Prenylapigenin, Quercetin, Orientin; Phenols (42 biomolecules): Spiro-indans (16 compounds), Dihydrostilbenes (12 biomolecules), simple Phenols (7 biomolecules), and Dihydrophenanthrenes (7 biomolecules). **Steroids**: (11 biomolecules): β-sitosterol, sitosterol, stigmasterol, ergosterol types, and campesterol **Alkaloids** (2 biomolecules): Cannabisativine and Anhydrocannabisativine. **Lignans**: (Lignanamides and phenolic amides)	(Andre et al., 2016; Brenneisen, 2007; M. A. ElSohly, 2002; M. A. ElSohly et al., 2017; M. ElSohly & Gul, 2014; Gertsch et al., 2010; Haller et al., 2006; Hazekamp & Fischedick, 2012; Ibrahim et al., 2019; Lowe et al., 2021; Mudge et al., 2019; Pertwee, 2014; Radwan et al., 2008, 2021; Rock & Parker, 2021; Russo, 2011; Turner et al., 1973, 1980)
Cannabinoid bioactive components (125 biomolecules): Δ^9-THC Type (25 Cannabinoids); Δ^8-THC Type (5 Cannabinoids); Miscellaneous Types: (30 biomolecules); Cannabigerol (CBG) Type (16 Cannabinoids); Cannabinol (CBN) Type: (11 biomolecules); Cannabichromene (CBC) Type: (10 Cannabinoids); Cannabicyclol (CBL) Type: (9 biomolecules); Cannabitriol (CBT) Type: (9 biomolecules); Cannabinodiol (CBND) Type: (5 biomolecules); Cannabielsoin (CBE) Type: (3 biomolecules); Cannabidiol (CBD) Type: (2 Cannabinoids).	(Brenneisen, 2007; M. A. ElSohly, 2002; M. A. ElSohly et al., 2017; M. ElSohly & Gul, 2014; Gertsch et al., 2010; Haller et al., 2006; Ibrahim et al., 2019; Lowe et al., 2021; Mudge et al., 2019; Pertwee, 2014; Radwan et al., 2008, 2021; Rock & Parker, 2021; Turner et al., 1973, 1980)

building material fabrication. Hemp seeds are a very nourishing source of protein and other vital minerals.

11.11.1 Herbal Medicine

The therapeutic qualities of *C. sativa* have been comprehended since antiquity. Notably, it has a lengthy record of use in traditional medicine, and it has lately been included in Western pharmacopeias. It was highly esteemed, especially for its effectiveness in treating hysteria, and was included in the United States Pharmacopoeia in the 1930s. The advancement of contemporary research substantiates the notion that a chemical entity like cannabis has the inherent capacity to foster the creation of innovative pharmaceuticals (Schultes & Smith, 1976). Cannabis has historically been used in traditional medicine to alleviate a range of ailments, including pain, inflammation, and anxiety. In recent times, there has been an increasing fascination with the possible medicinal applications of cannabis, namely in the treatment of conditions such as nausea, neuralgia, cachexia, gout, multiple sclerosis, seizures, cancer, Alzheimer's, rheumatism, insanity, insomnia, and Crohn's disease (Benlarbi et al., 2023; Benkhnigue et al., 2022, 2023; Chaachouay & Zidane, 2021a, 2021b; Brunetti et al., 2020; Amin & Ali, 2019; Hazekamp, 2007; Liu et al., 2017; Shah, 2004).

The Food and Drug Administration (FDA) has recently approved three synthetic pharmaceuticals containing cannabidiol, nabilone, or dronabinol for treating various diseases. In 1985, these were approved for the treatment of nausea and vomiting brought on by emetic compounds that are resistant to cancer chemotherapy (Todaro, 2012). In 2018, the FDA approved CBD 2018, which is allowed to be given to children with various seizure disorders such as Dravet syndrome and Lennox-Gastaut syndrome (Lattanzi et al., 2020; Mammana et al., 2019). Hemp is a very commonly used drug in Pakistan, and approximately 3.6% (4 million) of the population utilize it throughout the country (UNODC, 2013). Most of the formulations include drink (bhang) and smoked as charas or hashish.

11.11.2 Textiles

Hemp has a long historical record of being used in the textile industry because of its robust and resilient fibers. Hemp fibers are considered one of the most durable natural fibers and possess exceptional resistance to damage, making them well-suited for textile manufacturing (Gedik & Avinc, 2018). Hemp fibers have been used in textile manufacture for millennia and were extensively employed in the manufacturing of clothes, rope, and other materials until the 20th century. Hemp fibers have versatile applications in textile items, including apparel, bags, and upholstery (Crini et al., 2020). Employing cannabis fibers in textiles offers both durability and sustainability while providing environmental benefits (Clarke, 2007; Vandepitte et al., 2020).

11.11.3 Paper Industry

Hemp fiber is very suitable for paper manufacture owing to its abundant cellulose content, which imparts strength and durability. Furthermore, hemp fibers may undergo processing without chlorine, a substance typically used in conventional paper manufacturing techniques that can harm the environment. (Placet et al., 2018; Robinson, 1996). Hemp fibers can manufacture a diverse range of paper goods, such as notebooks, stationery, and even top-notch printing paper. Utilizing cannabis fibers in paper manufacturing is both ecologically advantageous and sustainable (Crini et al., 2020; Gusovius et al., 2019). Currently, there is a surging inclination toward using hemp fibers in the manufacturing of paper, driven by the increasing number of individuals actively seeking environmentally sustainable alternatives to conventional paper production techniques (Fortenbery & Bennett, 2004; Horne, 2020).

11.11.4 Construction Materials

Hemp fiber is versatile and may produce many building materials, such as insulation, flooring, and bricks. Hemp insulation is a sustainable substitute for conventional insulating materials like fiberglass and foam, offering several benefits (Crini et al., 2020). Hemp insulation has exceptional resistance to mold and mildew, making it particularly suitable for settings characterized by high levels of moisture or humidity (Benfratello et al., 2013; Pochwala et al., 2020). Additionally, it is non-toxic and does not release harmful substances, therefore enhancing the quality of indoor air. Furthermore, hemp insulation has exceptional thermal insulation properties, contributing to mitigating energy expenses. Hemp flooring is a sustainable substitute for conventional materials like hardwood and carpet (Walker et al., 2014). Hemp flooring exhibits exceptional durability and robust resistance to abrasion, making it well-suited for places with heavy foot activity. Additionally, it is environmentally friendly, constructed from a sustainable source, and does not release toxic substances. Hemp bricks are an emerging innovation being investigated as an environmentally friendly substitute for conventional clay bricks (Fortenbery & Bennett, 2004; Horne, 2020).

11.11.5 Food

Hemp, particularly hemp seed, has a variety of culinary uses and is renowned for its exceptional nutritional content. Hemp seeds are a nutritious source of protein, fiber, and beneficial fats. They may be incorporated into various food items, such as hemp milk, hemp protein powder, hemp oil, hemp seeds, and hemp flour (Carus & Sarmento, 2016; Leonard et al., 2020). Hemp milk is a non-dairy substitute for cow's milk that provides a substantial amount of protein and beneficial lipids. Hemp protein powder is a botanical protein supplement that may be included in smoothies and other culinary preparations. Hemp oil is a rich source of essential fatty acids, including omega-3 and omega-6. It is often used in salad dressings and other culinary preparations (Callaway, 2002). Hemp seeds may be consumed in their raw or roasted form and are often manipulated into salads, granola, and many culinary preparations (Burton et al., 2022; Shen et al., 2021). Hemp flour, a gluten-free substitute, is a nutritious option abundant in protein and fiber. It may be used in various dishes, such as bread and baked goods. Overall, the ethnobotany of cannabis illustrates the historical use and cultural significance of this plant throughout many societies, and it remains a topic of continuous investigation and discovery.

11.12 HALLUCINOGENIC EFFECTS

Marijuana is renowned for its euphoric properties attributed to THC, a cannabinoid molecule that may induce diverse physiological and psychological effects. While some individuals may have little feelings of pleasure and relaxation after the use of cannabis, others may encounter more pronounced hallucinogenic effects (Desai et al., 2022; Keeler et al., 1971; Leuchtenberger & Leuchtenberger, 1971; MCDOWELL, 2005; McGeeney, 2013; Rinder, 1978; Sewell et al., 2009; Shrivastava et al., 2011). The hallucinatory effects of marijuana might differ according to the dosage and individual characteristics, such as tolerance and susceptibility to THC (Ibarra-Lecue et al., 2018; Van Went, 1978; Wilcox et al., 2002). Some of the potential hallucinogenic activities of *Cannabis sativa* L. include:

11.12.1 Altered Perceptions of Time and Space

Marijuana can modify the perceptions of time and space in some individuals. THC, the psychoactive component of marijuana, can impact the brain and induce alterations in perception. THC often distorts time experience (Barrett et al., 2018). Some people report that time is moving more slowly, while others may feel like it is passing quickly (Sewell et al., 2009; Shrivastava et al., 2011). These effects may be more noticeable during the 'elevated' period after the use of marijuana. THC may also alter spatial perception. Specific individuals may have a sensation of visual distortion or a heightened perception of their environment (Hall & Solowij, 1998). For instance, objects may exhibit magnification or reduction in size, while distances may look more substantial or abbreviated.

11.12.2 Changes in Sensory Perception

Marijuana may induce alterations in sensory perception in some individuals. The presence of the psychoactive component delta-9-tetrahydrocannabinol is responsible for this phenomenon. It can impact the brain and modify an individual's perception of their surroundings. THC is anticipated to enhance sensory perceptions. Consuming marijuana has been reported to enhance individuals' sensory perception, including taste, touch, smell, and sound (Hasin et al., 2016; Ibarra-Lecue et al., 2018). This may enhance the perception of food, music, and other sensory inputs. Furthermore, THC may induce sensory distortion, resulting in modified impressions of sensory stimuli. For instance, some individuals may encounter visual aberrations, such as alterations in color cognition or a distorted image of spatial dimensions. Some individuals may experience auditory distortions, such as seeing phantom sounds or undergoing alterations in sound fidelity (Hall & Solowij, 1998). It should be emphasized that visual and aural hallucinations resulting from cannabis use are often less frequent and less powerful compared to hallucinogenic substances like LSD or psilocybin (Pierce & Peroutka, 1989; Stephens & Banes, 2013).

11.12.3 Intense Emotions

Many individuals may have profound emotional responses after the use of *C. sativa*. The presence of the psychoactive component THC is responsible for this phenomenon since it can impact the brain and alter an individual's mood and feelings. THC may elicit a range of emotional reactions, such as pleasure, relaxation, anxiety, and paranoia (Schilling et al., 2020; Strajhar et al., 2016). Individuals may experience heightened relaxation and a sense of tranquility after cannabis use, while some individuals may also have feelings of anxiety, restlessness, or paranoia. THC can augment the user's existing emotional state, intensifying sensations of joy, sorrow, rage, or apprehension (Shrivastava et al., 2011).

It's important to remember that not all cannabis strains have the same amount of THC; in fact, some may contain none at all or just very little. Furthermore, depending on variables including dose, mode of ingestion, and individual sensitivity, each person's reaction to cannabis might differ significantly. While some individuals may benefit greatly from cannabis, others may suffer adverse side effects, including anxiety or paranoia (Hall & Solowij, 1998; Murray et al., 2007; Stephens & Banes, 2013). In general, cannabis has the potential to induce hallucinations, but the precise characteristics and strength of these hallucinations may significantly differ depending on many individual and contextual circumstances. Practicing responsible use of cannabis and being mindful of possible hazards and negative consequences is crucial.

11.13 PHARMACOLOGICAL PROPERTIES

Hemp is composed of a wide array of chemicals, including more than 125 cannabinoids. THC is the most well-recognized cannabinoid since it is responsible for the psychoactive properties of the plant (M. A. ElSohly et al., 2017; Seth & Sinha, 1991). Cannabidiol (CBD) is another kind of cannabinoid that has the potential to be medicinal but does not have any psychoactive effects. Cannabis' pharmacological qualities are determined by the particular cannabinoids it contains and how they interact with the body's endocannabinoid system (ECS). A complex network of receptors, enzymes, and endocannabinoids, the ECS aids in regulating several physiological functions, including hunger, pain, inflammation, and mood (Amin & Ali, 2019; Seth & Sinha, 1991). Some of the pharmacological properties of cannabis include the following (Table 11.2).

11.14 CLINICAL USES

Hemp has been employed historically for medicinal purposes, and in recent times, there has been a resurgence of curiosity regarding its potential therapeutic uses (Walsh et al., 2003). THC and CBD, the primary bioactive constituents of cannabis, have undergone extensive research to investigate their impact on many medical ailments, as shown in Table 11.3. It is crucial to remember that cannabis is not a panacea and should not be regarded as the primary therapy for any medical problem. Before using cannabis for therapeutic reasons, it is essential to get guidance from a healthcare practitioner.

11.15 CONCLUSION AND FUTURE PERSPECTIVES

C. sativa is a multifaceted and adaptable plant with a long history of traditional usage, particularly in the region where it originated. Thanks to its genetic adaptability and broad ecological range, it can thrive in many environmental circumstances and growing methods. Currently, this plant is cultivated in many regions around the globe for textile production, paper manufacturing, food consumption, and medicinal applications, owing to its acknowledged potential to provide therapeutic advantages. The plant contains more than 100 phytochemicals, including cannabinoids, terpenes, and flavonoids, which have different effects. The psychedelic and hallucinatory effects of the plant are mainly attributed to the presence of THC, which also has potential medicinal properties such as analgesic and anti-inflammatory effects. Nevertheless, the use of marijuana has inevitable psychological consequences, such as compromised driving abilities and addiction. Hence, individuals need to acquire knowledge about the prospective hazards and advantages associated with the utilization of marijuana in regions where it is lawfully allowed. The global legal status of marijuana and its derivatives varies, with several nations permitting its usage for medicinal or recreational purposes while others enforce stringent controls or outright bans. Consequently, the investigation of hemp continues to be a lively field of study, with continuous endeavors to comprehend its impacts and possible uses, as well as establish guidelines that strike a balance between public health and safety issues and the availability of potential medicinal advantages.

LIST OF ABBREVIATIONS

Cannabis sativa	*C. sativa*
CBC	Cannabichromen
CBD	Cannabidiol
CBE	Cannabielsoin
CBG	Cannabigerol
CBL	Cannabicyclol
CBN	Cannabinol

TABLE 11.2
Pharmacological Properties of *Cannabis sativa* L.

Pharmacological Properties	References
Analgesic	(Buxbaum, 1972; Formukong et al., 1988; H. M. Harris et al., 2019; L. S. Harris, 1976; Kraft et al., 2008; Liktor-Busa et al., 2021; Sofia et al., 1975; van de Donk et al., 2019)
Anti-anxiety	(Almeida et al., 2013; Blessing et al., 2015; Campos et al., 2013; Chaves et al., 2021; Crippa et al., 2011; de Oliveira Ribeiro et al., 2012; Fogaça et al., 2018; Gomes et al., 2011, 2012, 2013; Moreira et al., 2006; R de Mello Schier et al., 2014; de Schier et al., 2012)
Anticonvulsant	(Hill et al., 2012; Johnson et al., 1975; Lazarini-Lopes et al., 2020; Leite et al., 1982; Shirazi-Zand et al., 2013; Sofia et al., 1976; Thompson et al., 2016; Vilela et al., 2017; Zhu et al., 2014)
Anti-depressant	(Abdel-Salam et al., 2013; Ahn et al., 2021; Chaves et al., 2021; de Morais et al., 2018; R de Mello Schier et al., 2014; Réus et al., 2011; Sales et al., 2019; Silote et al., 2019)
Antifungal	(Khan & Javaid, 2020; Pal et al., 2013; Wanas et al., 2016)
Antihypertensive	(Girgih et al., 2014; Kosersky, 1978)
Anti-inflammatory	(Costa et al., 2004; Downer, 2020; Formukong et al., 1988; Gallily et al., 2018; Nallathambi et al., 2017; Rajan et al., 2016; Ryz et al., 2017)
Anti-melanogenesis	(Kim et al., 2020, 2021)
Anti-microbial	(Ali et al., 2012; Chakraborty et al., 2018; Essien, 2011; Karas et al., 2020; Martinenghi et al., 2020; Muscarà et al., 2021; Palmieri et al., 2021; Pasquali et al., 2020; Sarmadyan et al., 2014; Tahsin et al., 2021)
Anti-oxidant	(M. Ahmed et al., 2019; Cantele et al., 2020; Casares et al., 2020; di Giacomo et al., 2020; Girgih et al., 2011; Kitrytė et al., 2018; Kornpointner et al., 2021; Moccia et al., 2020; Muscarà et al., 2021; Nafis et al., 2019; Palmieri et al., 2021; Rajan et al., 2016; Smeriglio et al., 2020; Teh et al., 2016; Yan et al., 2015)
Antiproliferative	(Caffarel et al., 2008; Moccia et al., 2020; Omer et al., 2020; Sainz-Cort et al., 2020; Wong, 2017)
Antipyretic	(Bilir et al., 2018; Gercek et al., 2020; Morris et al., 2021)
Anti-tumor	(Baram et al., 2019; Grijo et al., 2019; Jaafari et al., 2007; Ligresti et al., 2006; Massi et al., 2004; McAllister et al., 2015; Todorova et al., 2021)
Anxiolytic	(Almeida et al., 2013; Crippa et al., 2011; Guimarães et al., 1994; Masataka, 2019; Moreira et al., 2006; R de Mello Schier et al., 2014; Schier et al., 2012)
Cytotoxic	(Nallathambi et al., 2018; Namdar et al., 2019; Sainz-Cort et al., 2020; Scott et al., 2015; Zengin et al., 2018)
Neurodegenerative disease	(Abdel-Salam et al., 2015; Amin & Ali, 2019; Cassano et al., 2020; di Giacomo et al., 2020; dos Reis Rosa Franco et al., 2021; Fernández-Ruiz et al., 2013; Iuvone et al., 2009; Mahdi et al., 2021; Milano et al., 2017; Rodriguez-Martin et al., 2019; Scotter et al., 2010; Watt & Karl, 2017)
Sedative	(de Souza Crippa et al., 2004; Nicholson et al., 2004; Pickens, 1981)

TABLE 11.3
Clinical Uses of *Cannabis sativa* L.

Medical conditions	References
Chronic pain: Cannabis treats chronic pain, including neuropathic pain, spasticity, and muscle pain.	(Haroutounian et al., 2016; Mücke et al., 2018; Romero-Sandoval et al., 2018)
Cancer: Cannabis has been used to help alleviate the symptoms associated with cancer treatment, including nausea, vomiting, and loss of appetite.	(Abrams & Guzman, 2015; Hall & MacPhee, 2002; Sasco & Besson, 2003; Weiss et al., 2022)
Multiple sclerosis: Cannabis has been shown to help reduce muscle stiffness and spasticity in people with multiple sclerosis.	(Clark et al., 2004; Fragoso et al., 2020; Ingram & Pearson, 2019; Zajicek et al., 2012)
Epilepsy: CBD, a non-psychoactive component of cannabis, effectively reduces the frequency and severity of seizures in some people with epilepsy.	(Huntsman et al., 2020; Karler & Turkanis, 1979; Reddy & Golub, 2016; Szaflarski & Martina Bebin, 2014)
Glaucoma: Cannabis has been used to help reduce intraocular pressure in people with glaucoma, which can help slow the progression of the disease.	(Lehrer & Rheinstein, 2022; Rafuse & Buys, 2019; Vallée et al., 2021; Weldy et al., 2020; West & Lockhart, 1978)
Anxiety and depression: Cannabis has been shown to have anti-anxiety and anti-depressant effects in some patients.	(Blessing et al., 2015; Gobbi et al., 2019; Kosiba et al., 2019; Turna et al., 2019; Vaysse et al., 2014)

CBND	Cannabinodiol
CBT	Cannabitriol
DMT	N, N-Dimethyltryptamine
ECS	Body's endocannabinoid system
THC	Tetrahydrocannabinol
Δ^8-THC	Delta-8-Tetrahydrocannabinol
Δ^9-THC	Delta-9-Tetrahydrocannabinol

REFERENCES

Abbas, Q., Hussain, A., Khan, S. W., Hussain, A., Shinwari, S., Hussain, A., Ullah, A., Zafar, M., & Ali, K. (2019). Floristic Diversity, Ethnobotany and Traditional Recipes of Medicinal Plants of Maruk Nallah, Haramosh Valley, District Gilgit, Gilgit Baltistan: Traditional recipes of Maruk Nallah, Haramosh Valley, District Gilgit. *Proceedings of the Pakistan Academy of Sciences: B. Life and Environmental Sciences*, *56*(3), 97–112.

Abdel-Salam, O. M., Abdel-Rahman, R. F., & Gaafar, A. E.-D. M. (2013). Behavioral and biochemical effects of Cannabis Sativa and their modulation by antidepressant drugs. *Revista latinoamericana de química*, *41*(1), 21–37.

Abdel-Salam, O. M., Youness, E. R., Khadrawy, Y. A., Mohammed, N. A., Abdel-Rahman, R. F., Omara, E. A., & Sleem, A. A. (2015). The effect of cannabis on oxidative stress and neurodegeneration induced by intrastriatal rotenone injection in rats. *Comparative Clinical Pathology*, *24*(2), 359–378.

Abrams, D. I., & Guzman, M. (2015). Cannabis in cancer care. *Clinical Pharmacology & Therapeutics*, *97*(6), 575–586.

Abuhasira, R., Shbiro, L., & Landschaft, Y. (2018). Medical use of cannabis and cannabinoids containing products–Regulations in Europe and North America. *European journal of internal medicine*, *49*, 2–6.

Aguilar, S., Gutiérrez, V., Sánchez, L., & Nougier, M. (2018). Medicinal cannabis policies and practices around the world. *Int Drug Policy Consort. Briefing paper*, 1–32.

Ahmad, I., Ibrar, M., & Ali, N. (2011). Ethnobotanical study of tehsil kabal, swat district, KPK, pakistan. *Journal of Botany*, *2011*.

Ahmad, K. S., Kayani, W. K., Hameed, M., Ahmad, F., & Nawaz, T. (2012). Floristic diversity and ethnobotany of senhsa, district Kotli, Azad Jammu & Kashmir (Pakistan). *Pak J Bot*, *44*(SI), 195–201.

Ahmad, S., Ali, A., Beg, H., & Dasti, A. A. (2006). Of Booni Valley, District Chitral Pakistan. *Journal of Weed Science Research*, *12*(3), 183.

Ahmed, M., Ji, M., Qin, P., Gu, Z., Liu, Y., Sikandar, A., Iqbal, M. F., & Javeed, A. (2019). Phytochemical screening, total phenolic and flavonoids contents and antioxidant activities of Citrullus colocynthis L. and Cannabis sativa L. *Appl. Ecol. Environ. Res*, *17*, 6961–6979.

Ahmed, S. A., Ross, S. A., Slade, D., Radwan, M. M., Zulfiqar, F., & ElSohly, M. A. (2008). Cannabinoid ester constituents from high-potency Cannabis sativa. *Journal of Natural Products*, *71*(4), 536–542.

Ahn, Y., Han, S. H., Kim, M. G., Hong, K.-B., Kim, W. J., Suh, H. J., & Jo, K. (2021). Anti-depressant effects of ethanol extract from Cannabis sativa (hemp) seed in chlorpromazine-induced Drosophila melanogaster depression model. *Pharmaceutical Biology*, *59*(1), 998–1007.

Ali, E. M., Almagboul, A. Z., Khogali, S. M., & Gergeir, U. M. (2012). *Antimicrobial activity of Cannabis sativa L.*

Almeida, V., Levin, R., Peres, F. F., Niigaki, S. T., Calzavara, M. B., Zuardi, A. W., Hallak, J. E., Crippa, J. A., & Abílio, V. C. (2013). Cannabidiol exhibits anxiolytic but not antipsychotic property evaluated in the social interaction test. *Progress in Neuro-Psychopharmacology and Biological Psychiatry*, *41*, 30–35.

Alrashedy, N. A., & Molina, J. (2016). The ethnobotany of psychoactive plant use: A phylogenetic perspective. *PeerJ*, *4*, e2546.

Amin, M. R., & Ali, D. W. (2019). Pharmacology of medical cannabis. *Recent Advances in Cannabinoid Physiology and Pathology*, 151–165.

Andre, C. M., Hausman, J.-F., & Guerriero, G. (2016). Cannabis sativa: The plant of the thousand and one molecules. *Frontiers in Plant Science*, *7*, 19.

Ansari, M. K. A., Iqbal, M., Chaachouay, N., Ansari, A. A., & Owens, G. (2023). The Concept and Status of Medicinal and Aromatic Plants: History, Pharmacognosy, Ecology, and Conservation. In *Plants as Medicine and Aromatics* (p. 129–144). CRC Press.

Anwar, F., Latif, S., & Ashraf, M. (2006). Analytical characterization of hemp (Cannabis sativa) seed oil from different agro-ecological zones of Pakistan. *Journal of the American Oil Chemists' Society*, *83*(4), 323–329.

Atakan, Z. (2012). Cannabis, a complex plant: Different compounds and different effects on individuals. *Therapeutic advances in psychopharmacology*, *2*(6), 241–254.

Baram, L., Peled, E., Berman, P., Yellin, B., Besser, E., Benami, M., Louria-Hayon, I., Lewitus, G. M., & Meiri, D. (2019). The heterogeneity and complexity of Cannabis extracts as antitumor agents. *Oncotarget*, *10*(41), 4091.

Barkatullah, B., & Ibrar, M. (2011). Plants profile of malakand pass hills, district malakand, Pakistan. *African Journal of Biotechnology*, *10*(73), 16521–16535.

Barrett, F. S., Schlienz, N. J., Lembeck, N., Waqas, M., & Vandrey, R. (2018). 'Hallucinations' following acute cannabis dosing: A case report and comparison to other hallucinogenic drugs. *Cannabis and Cannabinoid Research*, *3*(1), 85–93.

Benfratello, S., Capitano, C., Peri, G., Rizzo, G., Scaccianoce, G., & Sorrentino, G. (2013). Thermal and structural properties of a hemp–lime biocomposite. *Construction and Building Materials*, *48*, 745–754.

Benlarbi, F., Mimoune, N., Chaachouay, N., Souttou, K., Saidi, R., Mokhtar, M.R., Kaidi, R., & Benaissa, M. T. (2023). Ethnobotanical survey of the traditional antiparasitic use of medicinal plants in humans and animals in Laghouat (Southern Algeria). *Veterinary World*, *16*(2), 357–368. https://doi.org/10.14202/vetworld.2023.357-368

Benkhnigue, O., Chaachouay, N., Khamar, H., El Azzouzi, F., Douira, A, & Zidane, L. (2022). Ethnobotanical and ethnopharmacological study of medicinal plants used in the treatment of anemia in the region of Haouz-Rehamna (Morocco). *Journal of Pharmacy & Pharmacognosy Research*, *10*(2), 279–302. https://doi.org/10.56499/jppres21.1196_10.2.279

Benkhnigue, O., Chaachouay, N., & Zidane, L. (2023). A Floristical and Ecological Study of the Medicinal Flora Used by the Local Population of the Haouz-Rehamna Region (Middle Atlantic Morocco-4). *Acta Botanica Hungarica Akadémiai Kiadó*, *65*(3–4), 247–304. https://doi.org/10.1556/034.65.2023.3-4.3

Bilir, K. A., Anli, G., Ozkan, E., Gunduz, O., & Ulugol, A. (2018). Involvement of spinal cannabinoid receptors in the antipruritic effects of WIN 55,212–2, a cannabinoid receptor agonist. *Clinical and Experimental Dermatology*, *43*(5), 553–558.

Blessing, E. M., Steenkamp, M. M., Manzanares, J., & Marmar, C. R. (2015). Cannabidiol as a potential treatment for anxiety disorders. *Neurotherapeutics*, *12*(4), 825–836.

Brenneisen, R. (2007). Chemistry and analysis of phytocannabinoids and other Cannabis constituents. In *Marijuana and the Cannabinoids* (p. 17–49). Springer.

Brown, D. T. (1998). *Cannabis: The genus cannabis*. CRC Press.

Brunetti, P., Pichini, S., Pacifici, R., Busardò, F. P., & Del Rio, A. (2020). Herbal preparations of medical cannabis: A vademecum for prescribing doctors. *Medicina*, *56*(5), 237.

Burton, R. A., Andres, M., Cole, M., Cowley, J. M., & Augustin, M. A. (2022). Industrial hemp seed: From the field to value-added food ingredients. *Journal of Cannabis Research*, *4*(1), 1–13.

Buxbaum, D. M. (1972). Analgesic activity of Δ 9-tetrahydrocannabinol in the rat and mouse. *Psychopharmacologia*, *25*(3), 275–280.

Caffarel, M. M., Moreno-Bueno, G., Cerutti, C., Palacios, J., Guzman, M., Mechta-Grigoriou, F., & Sanchez, C. (2008). JunD is involved in the antiproliferative effect of Δ 9-tetrahydrocannabinol on human breast cancer cells. *Oncogene*, *27*(37), 5033–5044.

Callaway, J. C. (2002). Hemp as food at high latitudes. *Journal of Industrial Hemp*, *7*(1), 105–117.

Campos, A. C., Ortega, Z., Palazuelos, J., Fogaça, M. V., Aguiar, D. C., Diaz-Alonso, J., Ortega-Gutiérrez, S., Vázquez-Villa, H., Moreira, F. A., & Guzmán, M. (2013). The anxiolytic effect of cannabidiol on chronically stressed mice depends on hippocampal neurogenesis: Involvement of the endocannabinoid system. *International Journal of Neuropsychopharmacology*, *16*(6), 1407–1419.

Cantele, C., Bertolino, M., Bakro, F., Giordano, M., Jędryczka, M., & Cardenia, V. (2020). Antioxidant effects of hemp (Cannabis sativa L.) inflorescence extract in stripped linseed oil. *Antioxidants*, *9*(11), 1131.

Carus, M., & Sarmento, L. (2016). The European Hemp Industry: Cultivation, processing and applications for fibres, shivs, seeds and flowers. *European Industrial Hemp Association*, *5*, 1–9.

Casares, L., García, V., Garrido-Rodríguez, M., Millán, E., Collado, J. A., García-Martín, A., Peñarando, J., Calzado, M. A., de la Vega, L., & Muñoz, E. (2020). Cannabidiol induces antioxidant pathways in keratinocytes by targeting BACH1. *Redox biology*, *28*, 101321.

Cassano, T., Villani, R., Pace, L., Carbone, A., Bukke, V. N., Orkisz, S., Avolio, C., & Serviddio, G. (2020). From Cannabis sativa to cannabidiol: Promising therapeutic candidate for the treatment of neurodegenerative diseases. *Frontiers in Pharmacology*, *11*, 124.

Chaachouay, N., Benkhnigue, O., Fadli, M., Ibaoui, H., & Zidane, L. (2019a). Ethnobotanical and ethnopharmacological studies of medicinal and aromatic plants used in the treatment of metabolic diseases in the Moroccan Rif. *Heliyon*, *5*(10), e02191. https://doi.org/10.1080/10477845.2021.1942433

Chaachouay, N., Benkhnigue, O., Fadli, M., Ibaoui, H., El Ayadi, R., & Zidane, L. (2019b). Ethnobotanical and ethnopharmacological study of medicinal and aromatic plants used in the treatment of respiratory system disorders in the Moroccan Rif. *Ethnobotany Research and Applications*, *18*(22), 1–16.

Chaachouay, N., Douira, A., Hassikou, R., Brhadda, N., Dahmani, J., Belahbib, N., Ziri, R., & Zidane, L. (2020). Ethnobotanical and ethnopharmacological study of medicinal and aromatic plants used in the treatment of respiratory system disorders in the Moroccan Rif. PhDThesis, Département de Biologie - Université Ibn Tofail - Kénitra, Maroc, https://tel.archives-ouvertes.fr/tel-03376377

Chaachouay, N., Benkhnigue, O., & Zidane, L. (2020). Ethnobotanical study aimed at investigating the use of medicinal plants to treat nervous system diseases in the Rif of Morocco. *Journal of chiropractic medicine*, *19*(1), 70–81. https://doi.org/10.1016/j.jcm.2020.02.004

Chaachouay, N., & Zidane, L. (2021a). The Symbolic Efficacy of Plants in Rituals and Socio-Religious Ceremonies in Morocco, Northwest of Africa. *Journal of Religious & Theological Information*, *21*(1–2), 34–53. https://doi.org/10.1080/10477845.2021.1942433

Chaachouay, N., & Zidane, L. (2021b). Neurological Phytotherapy by Indigenous People of Rif, Morocco. In *Therapy Approaches in Neurological Disorders*. IntechOpen.

Chaachouay, N., Benkhnigue, O., Douira, A., & Zidane, L. (2021). Poisonous medicinal plants used in the popular pharmacopoeia of the Rif, northern Morocco. *Toxicon*, 189, 24–32. https://doi.org/10.1016/j.toxicon.2020.10.028

Chaachouay, N., Douira, A., & Zidane, L. (2021b). COVID-19, prevention and treatment with herbal medicine in the herbal markets of Salé Prefecture, North-Western Morocco. *European Journal of Integrative Medicine*, 42, 101285. https://doi.org/10.1016/j.eujim.2021.101285

Chaachouay, N., Douira, A., & Zidane, L. (2021c). Herbal Medicine Used in the Treatment of Human Diseases in the Rif, Northern Morocco. *Arabian Journal for Science and Engineering*, 47, 131–153. https://doi.org/10.1007/s13369-021-05501-1

Chaachouay, N., Benkhnigue, O., & Zidane, L. (2022a). Ethnobotanical and Ethnomedicinal study of medicinal and aromatic plants used against dermatological diseases by the people of Rif, Morocco. *Journal of Herbal Medicine*, 32, 100542. https://doi.org/10.1016/j.hermed.2022.100542

Chaachouay, N., Azeroual, A., Bencharki, B., Douira, A., & Zidane, L. (2022b). Ethnoveterinary medicines plants for animal therapy in the Rif, North of Morocco. *South African Journal of Botany*, *147*, 176–191. https://doi.org/10.1016/j.sajb.2021.12.037

Chaachouay, N., Azeroual, A., Douira, A., & Zidane, L. (2022c). Ethnoveterinary practices of medicinal plants among the Zemmour and Zayane tribes, Middle Atlas, Morocco. *South African Journal of Botany*, *151*, 826–840. https://doi.org/10.1016/j.sajb.2022.11.009

Chaachouay, N., Azeroual, A., Bencharki, B., Douira, A., & Zidane, L. (2023). Cannabis sativa L.: A Review on Traditional Uses, Botany, Phytochemistry, and Pharmacological Aspects. *Traditional and Integrative Medicine*.

Chakraborty, S., Afaq, N., Singh, N., & Majumdar, S. (2018). Antimicrobial activity of Cannabis sativa, Thuja orientalis and Psidium guajava leaf extracts against methicillin-resistant Staphylococcus aureus. *Journal of integrative medicine*, *16*(5), 350–357.

Chandra, S., Lata, H., & ElSohly, M. A. (2017). *Cannabis sativa L.-botany and biotechnology*. Springer.

Chandra, S., Lata, H., Khan, I. A., & ElSohly, M. A. (2017). Cannabis sativa L.: Botany and horticulture. *Cannabis sativa l.-botany and biotechnology*, 79–100.

Chaves, Y. C., Genaro, K., Crippa, J. A., da Cunha, J. M., & Zanoveli, J. M. (2021). Cannabidiol induces antidepressant and anxiolytic-like effects in experimental type-1 diabetic animals by multiple sites of action. *Metabolic Brain Disease*, *36*(4), 639–652.

Chouvy, P.-A. (2022). Moroccan hashish as an example of a cannabis terroir product. *GeoJournal*, 1–18.

Clark, A. J., Ware, M. A., Yazer, E., Murray, T. J., & Lynch, M. E. (2004). Patterns of cannabis use among patients with multiple sclerosis. *Neurology*, *62*(11), 2098–2100.

Clarke, R. C. (2007). Traditional cannabis cultivation in Darchula District, Nepal—Seed, resin and textiles. *Journal of Industrial Hemp*, *12*(2), 19–42.

Costa, B., Colleoni, M., Conti, S., Parolaro, D., Franke, C., Trovato, A. E., & Giagnoni, G. (2004). Oral anti-inflammatory activity of cannabidiol, a non-psychoactive constituent of cannabis, in acute carrageenan-induced inflammation in the rat paw. *Naunyn-Schmiedeberg's archives of pharmacology*, *369*(3), 294–299.

Crini, G., Lichtfouse, E., Chanet, G., & Morin-Crini, N. (2020). Applications of hemp in textiles, paper industry, insulation and building materials, horticulture, animal nutrition, food and beverages, nutraceuticals, cosmetics and hygiene, medicine, agrochemistry, energy production and environment: A review. *Environmental Chemistry Letters*, *18*(5), 1451–1476.

Crippa, J. A. S., Derenusson, G. N., Ferrari, T. B., Wichert-Ana, L., Duran, F. L., Martin-Santos, R., Simões, M. V., Bhattacharyya, S., Fusar-Poli, P., & Atakan, Z. (2011). Neural basis of anxiolytic effects of cannabidiol (CBD) in generalized social anxiety disorder: A preliminary report. *Journal of Psychopharmacology*, *25*(1), 121–130.

Cunha-Oliveira, T., Rego, A. C., Carvalho, F., & Oliveira, C. R. (2013). Medical toxicology of drugs of abuse. *Principles of Addiction-Comprehensive Addictive Behaviors and Disorders*, *1*, 159–175.

Cunningham, N. (2008). Hallucinogenic plants of abuse. *Emergency Medicine Australasia*, *20*(2), 167–174.

de Morais, H., Chaves, Y. C., Waltrick, A. P. F., Jesus, C. H. A., Genaro, K., Crippa, J. A., da Cunha, J. M., & Zanoveli, J. M. (2018). Sub-chronic treatment with cannabidiol but not with URB597 induced a mild antidepressant-like effect in diabetic rats. *Neuroscience Letters*, *682*, 62–68.

de Oliveira Ribeiro, N. P., Crippa, J. A., Nardi, A. E., & Zuardi, A. (2012). Cannabidiol, a Cannabis sativa constituent, as an anxiolytic drug. *Official Journal of the Brazilian Psychiatric Association*, *34*, 1.

Desai, S., Kulkarni, N., Gandhi, K., You, X. X., & Gerolemou, A. (2022). The Link Between Marijuana and Hallucinogen Use Among US Adolescents. *The Primary Care Companion for CNS Disorders*, *24*(5), 42561.

de Souza Crippa, J. A., Zuardi, A. W., Garrido, G. E., Wichert-Ana, L., Guarnieri, R., Ferrari, L., Azevedo-Marques, P. M., Hallak, J. E. C., & McGuire, P. K. (2004). Effects of cannabidiol (CBD) on regional cerebral blood flow. *Neuropsychopharmacology*, *29*(2), 417–426.

de Souza, M. R., Henriques, A. T., & Limberger, R. P. (2022). Medical cannabis regulation: An overview of models around the world with emphasis on the Brazilian scenario. *Journal of Cannabis Research*, *4*(1), 1–15.

di Giacomo, V., Chiavaroli, A., Recinella, L., Orlando, G., Cataldi, A., Rapino, M., Di Valerio, V., Ronci, M., Leone, S., & Brunetti, L. (2020). Antioxidant and neuroprotective effects induced by cannabidiol and cannabigerol in rat CTX-TNA2 astrocytes and isolated cortexes. *International Journal of Molecular Sciences*, *21*(10), 3575.

dos Reis Rosa Franco, G., Smid, S., & Viegas, C. (2021). Phytocannabinoids: General Aspects and Pharmacological Potential in Neurodegenerative Diseases. *Current Neuropharmacology*, *19*(4), 449–464.

Downer, E. J. (2020). Anti-inflammatory Potential of Terpenes Present in Cannabis sativa L. *ACS Chemical Neuroscience*, *11*(5), 659–662.

Elhamdaoui, O., Meiouet, M., Ajlal, A., Chergui, A., Zakariya, I., & Nejjari, R. (2019). Medicinal cannabis and challenges of legislation in Morocco. *International Journal of Pharmacognosy*, *6*, 323–328.

ElSohly, M. A. (2002). *Chemical constituents of cannabis*. Haworth Press, New York.

ElSohly, M. A., Radwan, M. M., Gul, W., Chandra, S., & Galal, A. (2017). Phytochemistry of Cannabis sativa L. *Phytocannabinoids*, 1–36.

ElSohly, M. A., & Slade, D. (2005). Chemical constituents of marijuana: The complex mixture of natural cannabinoids. *Life Sciences*, *78*(5), 539–548.

ElSohly, M., & Gul, W. (2014). Constituents of Cannabis sativa. *Handbook of Cannabis*, *3*, 1093.

Essien, E. E. (2011). Effect of extraction conditions on total polyphenol contents, antioxidant and antimicrobial activities of Cannabis sativa L. *Electronic Journal of Environmental, Agricultural and Food Chemistry (EJEAFChe)*, *11*(04), 300–307.

Farag, S., & Kayser, O. (2017). The cannabis plant: Botanical aspects. In *Handbook of cannabis and related pathologies* (p. 3–12). Elsevier.

Fernández-Ruiz, J., Sagredo, O., Pazos, M. R., García, C., Pertwee, R., Mechoulam, R., & Martínez-Orgado, J. (2013). Cannabidiol for neurodegenerative disorders: Important new clinical applications for this phytocannabinoid? *British Journal of Clinical Pharmacology*, *75*(2), 323–333.

Fischer, B. (2017). Legalisation of non-medical cannabis in Canada: Will supply regulations effectively serve public health? *The Lancet Public Health*, *2*(12), e536–e537.

Fogaça, M. V., Campos, A. C., Coelho, L. D., Duman, R. S., & Guimaraes, F. S. (2018). The anxiolytic effects of cannabidiol in chronically stressed mice are mediated by the endocannabinoid system: Role of neurogenesis and dendritic remodeling. *Neuropharmacology*, *135*, 22–33.

Formukong, E. A., Evans, A. T., & Evans, F. J. (1988). Analgesic and antiinflammatory activity of constituents of Cannabis sativa L. *Inflammation*, *12*(4), 361–371.

Fortenbery, T. R., & Bennett, M. (2004). Opportunities for commercial hemp production. *Applied Economic Perspectives and Policy*, *26*(1), 97–117.

Fragoso, Y. D., Carra, A., & Macias, M. A. (2020). Cannabis and multiple sclerosis. *Expert Review of Neurotherapeutics*, *20*(8), 849–854.

Gallily, R., Yekhtin, Z., & Hanuš, L. O. (2018). The anti-inflammatory properties of terpenoids from cannabis. *Cannabis and Cannabinoid Research*, *3*(1), 282–290.

Gartner, C., & Hall, W. (2020). Lessons from tobacco regulation for cannabis product regulation. *Legalizing Cannabis*, 233–258.

GBIF. (2020). *GBIF—the Global Biodiversity Information Facility—Is an international network and data infrastructure funded by the world's governments and aimed at providing anyone, anywhere, open access to data about all types of life on Earth.* https://www.gbif.org/species/3001509

Gedik, G., & Avinc, O. (2018). Bleaching of hemp (Cannabis sativa L.) fibers with peracetic acid for textiles industry purposes. *Fibers and Polymers*, *19*, 82–93.

Gercek, O. Z., Oflaz, B., Oguz, N., Demirci, K., Gunduz, O., & Ulugol, A. (2020). Role of Nitric Oxide in the Antipruritic Effect of WIN 55,212-2, a Cannabinoid Agonist. *Basic and Clinical Neuroscience*, *11*(4), 473.

Gertsch, J., Pertwee, R. G., & Di Marzo, V. (2010). Phytocannabinoids beyond the Cannabis plant–do they exist? *British Journal of Pharmacology*, *160*(3), 523–529.

Giannenas, I., Sidiropoulou, E., Bonos, E., Christaki, E., & Florou-Paneri, P. (2020). The history of herbs, medicinal and aromatic plants, and their extracts: Past, current situation and future perspectives. In *Feed additives* (p. 1–18). Elsevier.

Girgih, A. T., He, R., Malomo, S., Offengenden, M., Wu, J., & Aluko, R. E. (2014). Structural and functional characterization of hemp seed (Cannabis sativa L.) protein-derived antioxidant and antihypertensive peptides. *Journal of Functional Foods*, *6*, 384–394.

Girgih, A. T., Udenigwe, C. C., & Aluko, R. E. (2011). In vitro antioxidant properties of hemp seed (Cannabis sativa L.) protein hydrolysate fractions. *Journal of the American Oil Chemists' Society*, *88*(3), 381–389.

Gobbi, G., Atkin, T., Zytynski, T., Wang, S., Askari, S., Boruff, J., Ware, M., Marmorstein, N., Cipriani, A., Dendukuri, N., & Mayo, N. (2019). Association of Cannabis Use in Adolescence and Risk of Depression, Anxiety, and Suicidality in Young Adulthood: A Systematic Review and Meta-analysis. *JAMA Psychiatry*, *76*(4), 426–434. https://doi.org/10.1001/jamapsychiatry.2018.4500

Gomes, F. V., Alves, F. H., Guimarães, F. S., Correa, F. M., Resstel, L. B., & Crestani, C. C. (2013). Cannabidiol administration into the bed nucleus of the stria terminalis alters cardiovascular responses induced by acute restraint stress through 5-HT1A receptor. *European Neuropsychopharmacology*, *23*(9), 1096–1104.

Gomes, F. V., Reis, D. G., Alves, F. H., Corrêa, F. M., Guimaraes, F. S., & Resstel, L. B. (2012). Cannabidiol injected into the bed nucleus of the stria terminalis reduces the expression of contextual fear conditioning via 5-HT1A receptors. *Journal of Psychopharmacology*, *26*(1), 104–113.

Gomes, F. V., Resstel, L. B., & Guimarães, F. S. (2011). The anxiolytic-like effects of cannabidiol injected into the bed nucleus of the stria terminalis are mediated by 5-HT1A receptors. *Psychopharmacology*, *213*(2), 465–473.

Goundar, P., Macaulay, T., & Szafron, M. (2021). A comparative analysis of laws on recreational cannabis edibles between Canada and the United States of America. *International Journal of Drug Policy*, *94*, 103191.

Grijo, D. R., Bidoia, D. L., Nakamura, C. V., Osorio, I. V., & Cardozo-Filho, L. (2019). Analysis of the antitumor activity of bioactive compounds of Cannabis flowers extracted by green solvents. *The Journal of Supercritical Fluids*, *149*, 20–25.

Guimarães, F. S., De Aguiar, J. C., Mechoulam, R., & Breuer, A. (1994). Anxiolytic effect of cannabidiol derivatives in the elevated plus-maze. *General Pharmacology*, *25*(1), 161–164.

Gusovius, H.-J., Lühr, C., Hoffmann, T., Pecenka, R., & Idler, C. (2019). An alternative to field retting: Fibrous materials based on wet preserved hemp for the manufacture of composites. *Agriculture*, *9*(7), 140.

Hall, W., & MacPhee, D. (2002). Cannabis use and cancer. *Addiction*, *97*(3), 243–247.

Hall, W., & Solowij, N. (1998). Adverse effects of cannabis. *The Lancet*, *352*(9140), 1611–1616.

Haller, V. L., Cichewicz, D. L., & Welch, S. P. (2006). Non-cannabinoid CB1, non-cannabinoid CB2 antinociceptive effects of several novel compounds in the PPQ stretch test in mice. *European Journal of Pharmacology*, *546*(1–3), 60–68.

Haney, A., & Kutscheid, B. B. (1975). An ecological study of naturalized hemp (Cannabis sativa L.) in east-central Illinois. *American Midland Naturalist*, 1–24.

Haq, F., Ahmad, H., & Alam, M. (2011). Traditional uses of medicinal plants of Nandiar Khuwarr catchment (District Battagram), Pakistan. *Journal of Medicinal Plants Research*, *5*(1), 39–48.

Haroutounian, S., Ratz, Y., Ginosar, Y., Furmanov, K., Saifi, F., Meidan, R., & Davidson, E. (2016). The effect of medicinal cannabis on pain and quality-of-life outcomes in chronic pain. *The Clinical Journal of Pain*, *32*(12), 1036–1043.

Harris, H. M., Rousseau, M. A., Wanas, A. S., Radwan, M. M., Caldwell, S., Sufka, K. J., & ElSohly, M. A. (2019). Role of cannabinoids and terpenes in cannabis-mediated analgesia in rats. *Cannabis and Cannabinoid Research*, *4*(3), 177–182.

Harris, L. S. (1976). Analgesic and antitumor potential of the cannabinoids. In *The Therapeutic Potential Of Marijuana* (p. 299–312). Springer.

Hasin, D. S., Kerridge, B. T., Saha, T. D., Huang, B., Pickering, R., Smith, S. M., Jung, J., Zhang, H., & Grant, B. F. (2016). Prevalence and correlates of DSM-5 cannabis use disorder, 2012–2013: Findings from the National Epidemiologic Survey on Alcohol and Related Conditions–III. *American Journal of Psychiatry*, *173*(6), 588–599.

Hassan, H. M. A. (2015). A short history of the use of plants as medicines from ancient times. *Chimia*, *69*(10), 622–622.

Hazekamp, A. (2007). *Cannabis: Extracting the medicine*. Leiden University.

Hazekamp, A., & Fischedick, J. T. (2012). Cannabis-from cultivar to chemovar. *Drug Testing and Analysis*, *4*(7–8), 660–667.

Hill, A. J., Mercier, M. S., Hill, T. D. M., Glyn, S. E., Jones, N. A., Yamasaki, Y., Futamura, T., Duncan, M., Stott, C. G., & Stephens, G. J. (2012). Cannabidivarin is anticonvulsant in mouse and rat. *British Journal of Pharmacology*, *167*(8), 1629–1642.

Horne, M. R. (2020). Bast fibres: Hemp cultivation and production. *Handbook of Natural Fibres*, 163–196.

Huntsman, R. J., Kelly, L. E., Alcorn, J., Appendino, J. P., Bélanger, R. E., Crooks, B., Finkelstein, Y., Gilpin, A., Lewis, E., & Litalien, C. (2021). Improving the regulation of medical cannabis in Canada to better serve pediatric patients. *CMAJ*, *193*(41), E1596–E1599.

Huntsman, R. J., Tang-Wai, R., & Shackelford, A. E. (2020). Cannabis for Pediatric Epilepsy. *Journal of Clinical Neurophysiology*, *37*(1), 2. https://doi.org/10.1097/WNP.0000000000000641

Hussain, K., Nisar, M. F., Majeed, A., Nawaz, K., & Bhatti, K. H. (2010). Ethnomedicinal survey for important plants of Jalalpur Jattan, district Gujrat, Punjab, Pakistan. *Ethnobotanical Leaflets*, *2010*(7), 11.

Ibarra-Lecue, I., Mollinedo-Gajate, I., Meana, J. J., Callado, L. F., Diez-Alarcia, R., & Urigüen, L. (2018). Chronic cannabis promotes pro-hallucinogenic signaling of 5-HT2A receptors through Akt/mTOR pathway. *Neuropsychopharmacology*, *43*(10), 2028–2035.

Ibrahim, E. A., Wang, M., Radwan, M. M., Wanas, A. S., Majumdar, C. G., Avula, B., Wang, Y.-H., Khan, I. A., Chandra, S., & Lata, H. (2019). Analysis of terpenes in Cannabis sativa L. using GC/MS: Method development, validation, and application. *Planta Medica*, *85*(05), 431–438.

Ibrar, M., Hussain, F., & Sultan, A. (2007). Ethnobotanical studies on plant resources of Ranyal hills, District Shangla, Pakistan. *Pakistan Journal of Botany*, *39*(2), 329.

Ingram, G., & Pearson, O. R. (2019). Cannabis and multiple sclerosis. *Practical Neurology*, *19*(4), 310–315.

Iqbal, H., & Sher, Z. (2011). Medicinal plants from salt range pind dadan khan, district jhelum, punjab, pakistan. *Journal of Medicinal Plants Research*, *5*(11), 2157–2168.

ITIS. (2022). Integrated Taxonomic Information System (ITIS) online database, www.itis.gov, CC0 https://doi.org/10.5066/F7KH0KBK. http://itis-web-prod.snafu.cr.usgs.gov

Iuvone, T., Esposito, G., De Filippis, D., Scuderi, C., & Steardo, L. (2009). Cannabidiol: A promising drug for neurodegenerative disorders? *CNS Neuroscience & Therapeutics*, *15*(1), 65–75.

Jaafari, A., Mouse, H. A., Rakib, E. M., M'Barek, L. A., Tilaoui, M., Benbakhta, C., Boulli, A., Abbad, A., & Zyad, A. (2007). Chemical composition and antitumor activity of different wild varieties of Moroccan thyme. *Revista Brasileira de Farmacognosia*, *17*, 477–491.

Jamshidi-Kia, F., Lorigooini, Z., & Amini-Khoei, H. (2018). Medicinal plants: Past history and future perspective. *Journal of Herbmed Pharmacology*, 7(1), 1–7. doi: 10.15171/jhp.2018.01

Jan, M., Khare, R. K., & Mir, T. A. (2021). Medicinal plants used during pregnancy and childbirth in Baramulla District of Jammu and Kashmir, India. *Ethnobotany Research and Applications*, *22*, 1–19.

Jan, M., Mir, T. A., Jan, H. A., & Khare, R. K. (2022). Medicinal plants diversity and their uses for Gynecological Disorders of District Baramulla, Jammu and Kashmir, India. *Vegetos*, *35*(2), 438–452.

Johnson, D. D., McNeill, J. R., Crawford, R. D., & Wilcox, W. C. (1975). Epileptiform seizures in domestic fowl. V. The anticonvulsant activity of Δ9-tetrahydrocannabinol. *Canadian Journal of Physiology and Pharmacology*, *53*(6), 1007–1013.

Kalant, H. (Éd.). (1999). *The health effects of cannabis*. Centre for Addiction and Mental Health.

Kalant, H. (2001). Medicinal use of cannabis: History and current status. *Pain Research and Management*, *6*, 80–91.

Karas, J. A., Wong, L. J., Paulin, O. K., Mazeh, A. C., Hussein, M. H., Li, J., & Velkov, T. (2020). The antimicrobial activity of cannabinoids. *Antibiotics*, *9*(7), 406.

Karler, R., & Turkanis, S. A. (1979). Cannabis and Epilepsy. In G. G. Nahas & W. D. M. Paton (Éds.), *Marijuana Biological Effects* (p. 619–641). Pergamon. https://doi.org/10.1016/B978-0-08-023759-6.50052-4

Keeler, M. H., Ewing, J. A., & Rouse, B. A. (1971). Hallucinogenic effects of marijuana as currently used. *American Journal of Psychiatry*, *128*(2), 213–216.

Khan, I. H., & Javaid, A. (2020). Antifungal activity of leaf extract of Cannabis sativa against Aspergillus flavipes. *Pakistan Journal of Weed Science Research*, *26*(4), 447.

Kim, J. K., Heo, H.-Y., Park, S., Kim, H., Oh, J. J., Sohn, E.-H., Jung, S.-H., & Lee, K. (2021). Characterization of Phenethyl Cinnamamide Compounds from Hemp Seed and Determination of Their Melanogenesis Inhibitory Activity. *ACS omega*.

Kim, J. K., Kim, H. H., Kim, J., Lim, Y.-S., Jung, S.-H., & Lee, K. (2020). Inhibitory activity of ethyl acetate fraction derived from Cannabis sativaL. seed on melanogenesis in B16F10 cells. *Asian Journal of Complementary and Alternative Medicine*, 17.

Kitrytė, V., Bagdonaitė, D., & Venskutonis, P. R. (2018). Biorefining of industrial hemp (Cannabis sativa L.) threshing residues into cannabinoid and antioxidant fractions by supercritical carbon dioxide, pressurized liquid and enzyme-assisted extractions. *Food Chemistry*, *267*, 420–429.

Kogan, L. R., Hellyer, P. W., Silcox, S., & Schoenfeld-Tacher, R. (2019). Canadian dog owners' use and perceptions of cannabis products. *The Canadian Veterinary Journal*, *60*(7), 749.

Kornpointner, C., Martinez, A. S., Marinovic, S., Haselmair-Gosch, C., Jamnik, P., Schröder, K., Löfke, C., & Halbwirth, H. (2021). Chemical composition and antioxidant potential of Cannabis sativa L. roots. *Industrial Crops and Products*, *165*, 113422.

Kosersky, D. (1978). Antihypertensive effects of delta9-tetrahydrocannabinol. *Archives Internationales de Pharmacodynamie et de Therapie*, 76–81.

Kosiba, J. D., Maisto, S. A., & Ditre, J. W. (2019). Patient-reported use of medical cannabis for pain, anxiety, and depression symptoms: Systematic review and meta-analysis. *Social Science & Medicine*, *233*, 181–192. https://doi.org/10.1016/j.socscimed.2019.06.005

Kraft, B., Frickey, N. A., Kaufmann, R. M., Reif, M., Frey, R., Gustorff, B., & Kress, H. G. (2008). Lack of analgesia by oral standardized cannabis extract on acute inflammatory pain and hyperalgesia in volunteers. *The Journal of the American Society of Anesthesiologists*, *109*(1), 101–110.

Kuddus, M., Ginawi, I. A., & Al-Hazimi, A. (2013). Cannabis sativa: An ancient wild edible plant of India. *Emirates Journal of Food and Agriculture*, 736–745.

Lattanzi, S., Brigo, F., Trinka, E., Zaccara, G., Striano, P., Del Giovane, C., & Silvestrini, M. (2020). Adjunctive cannabidiol in patients with Dravet syndrome: A systematic review and meta-analysis of efficacy and safety. *CNS drugs*, *34*, 229–241.

Lazarini-Lopes, W., Do Val-da Silva, R. A., da Silva-Júnior, R. M., Leite, J. P., & Garcia-Cairasco, N. (2020). The anticonvulsant effects of cannabidiol in experimental models of epileptic seizures: From behavior and mechanisms to clinical insights. *Neuroscience & Biobehavioral Reviews*, *111*, 166–182.

Lehrer, S., & Rheinstein, P. H. (2022). Cannabis smoking and glaucoma in the UK Biobank cohort. *Journal Français d'Ophtalmologie*, *45*(4), 423–429. https://doi.org/10.1016/j.jfo.2021.12.012

Leite, R., Carlini, E. A., Lander, N., & Mechoulam, R. (1982). Anticonvulsant effects of the (–) and (+) isomers of cannabidiol and their dimethylheptyl homologs. *Pharmacology*, *24*(3), 141–146.

Leonard, W., Zhang, P., Ying, D., & Fang, Z. (2020). Hempseed in food industry: Nutritional value, health benefits, and industrial applications. *Comprehensive Reviews in Food Science and Food Safety*, *19*(1), 282–308.

Leuchtenberger, C., & Leuchtenberger, R. (1971). Morphological and cytochemical effects of marijuana cigarette smoke on epithelioid cells of lung explants from mice. *Nature*, *234*(5326), 227–229.

Ligresti, A., Moriello, A. S., Starowicz, K., Matias, I., Pisanti, S., De Petrocellis, L., Laezza, C., Portella, G., Bifulco, M., & Di Marzo, V. (2006). Antitumor activity of plant cannabinoids with emphasis on the effect of cannabidiol on human breast carcinoma. *Journal of Pharmacology and Experimental Therapeutics*, *318*(3), 1375–1387.

Liktor-Busa, E., Keresztes, A., LaVigne, J., Streicher, J. M., & Largent-Milnes, T. M. (2021). Analgesic Potential of Terpenes Derived from Cannabis sativa. *Pharmacological Reviews*, *73*(4), 98–126.

Liu, F.-H., Hu, H.-R., Du, G.-H., Deng, G., & Yang, Y. (2017). *Ethnobotanical research on origin, cultivation, distribution and utilization of hemp (Cannabis sativa L.) in China*.

Lowe, H., Steele, B., Bryant, J., Toyang, N., & Ngwa, W. (2021). Non-cannabinoid metabolites of cannabis sativa l. With therapeutic potential. *Plants*, *10*(2), 400.

Mahdi, O., Baharuldin, M. T., Nor, N. H. M., Chiroma, S. M., Jagadeesan, S., & Moklas, M. A. (2021). The Neuroprotective Properties, Functions, and Roles of Cannabis sativa in Selected Diseases Related to the Nervous System. *Central Nervous System Agents in Medicinal Chemistry (Formerly Current Medicinal Chemistry-Central Nervous System Agents)*, *21*(1), 20–38.

Mammana, S., Cavalli, E., Gugliandolo, A., Silvestro, S., Pollastro, F., Bramanti, P., & Mazzon, E. (2019). Could the combination of two non-psychotropic cannabinoids counteract neuroinflammation? Effectiveness of cannabidiol associated with cannabigerol. *Medicina*, *55*(11), 747.

Martinenghi, L. D., Jønsson, R., Lund, T., & Jenssen, H. (2020). Isolation, Purification, and antimicrobial characterization of cannabidiolic acid and cannabidiol from Cannabis sativa L. *Biomolecules*, *10*(6), 900.

Masataka, N. (2019). Anxiolytic effects of repeated cannabidiol treatment in teenagers with social anxiety disorders. *Frontiers in Psychology*, *10*, 2466.

Massi, P., Vaccani, A., Ceruti, S., Colombo, A., Abbracchio, M. P., & Parolaro, D. (2004). Antitumor effects of cannabidiol, a nonpsychoactive cannabinoid, on human glioma cell lines. *Journal of Pharmacology and Experimental Therapeutics*, *308*(3), 838–845.

McAllister, S. D., Soroceanu, L., & Desprez, P.-Y. (2015). The antitumor activity of plant-derived non-psychoactive cannabinoids. *Journal of Neuroimmune Pharmacology*, *10*(2), 255–267.

McDowell, D. (2005). Marijuana, hallucinogens, and club drugs. *Clinical Textbook of Addictive Disorders*, *3*, 175–176.

McGeeney, B. E. (2013). Cannabinoids and hallucinogens for headache. *Headache: The Journal of Head and Face Pain*, *53*(3), 447–458.

McPartland, J. M., Hegman, W., & Long, T. (2019). Cannabis in Asia: Its center of origin and early cultivation, based on a synthesis of subfossil pollen and archaeobotanical studies. *Vegetation History and Archaeobotany*, *28*, 691–702.

Milano, W., Tecce, M. F., & Capasso, A. (2017). Cannabinoids involvement in neurodegenerative diseases. *Curr Neurobiol*, *8*, 135–144.

Mir, T. A., Jan, M., Khare, R. K., & Dhyani, S. (2021). Ethnosurvey of traditional use of plants in Lolab Valley, Kashmir Himalaya. *Indian Forester*, *147*(3), 281–287.

Moccia, S., Siano, F., Russo, G. L., Volpe, M. G., La Cara, F., Pacifico, S., Piccolella, S., & Picariello, G. (2020). Antiproliferative and antioxidant effect of polar hemp extracts (Cannabis sativa L., Fedora cv.) in human colorectal cell lines. *International Journal of Food Sciences and Nutrition*, *71*(4), 410–423.

Moreira, F. A., Aguiar, D. C., & Guimarães, F. S. (2006). Anxiolytic-like effect of cannabidiol in the rat Vogel conflict test. *Progress in Neuro-Psychopharmacology and Biological Psychiatry*, *30*(8), 1466–1471.

Morris, E. M., Kitts-Morgan, S. E., Spangler, D. M., Gebert, J., Vanzant, E. S., McLeod, K. R., & Harmon, D. L. (2021). Feeding Cannabidiol (CBD)-Containing Treats Did Not Affect Canine Daily Voluntary Activity. *Frontiers in Veterinary Science*, *8*.

Mücke, M., Phillips, T., Radbruch, L., Petzke, F., & Häuser, W. (2018). Cannabis-based medicines for chronic neuropathic pain in adults. *Cochrane Database of Systematic Reviews*, *3*.

Mudge, E. M., Brown, P. N., & Murch, S. J. (2019). The Terroir of Cannabis: Terpene metabolomics as a tool to understand Cannabis sativa selections. *Planta Medica*, *85*(09/10), 781–796.

Munir, M., Sadia, S., Khan, A., Rahim, B. Z., Gagosh Nayyar, B., Ahmad, K. S., Khan, A. M., Fatima, I., & Qureshi, R. (2022). Ethnobotanical study of Mandi Ahmad Abad, District Okara, Pakistan. *Plos one*, *17*(4), e0265125.

Murray, R. M., Morrison, P. D., Henquet, C., & Forti, M. D. (2007). Cannabis, the mind and society: The hash realities. *Nature Reviews Neuroscience*, *8*(11), 885–895.

Muscarà, C., Smeriglio, A., Trombetta, D., Mandalari, G., La Camera, E., Occhiuto, C., Grassi, G., & Circosta, C. (2021). Antioxidant and antimicrobial activity of two standardized extracts from a new Chinese accession of non-psychotropic Cannabis sativa L. *Phytotherapy Research*, *35*(2), 1099–1112.

Nafis, A., Kasrati, A., Jamali, C. A., Mezrioui, N., Setzer, W., Abbad, A., & Hassani, L. (2019). Antioxidant activity and evidence for synergism of Cannabis sativa (L.) essential oil with antimicrobial standards. *Industrial Crops and Products*, *137*, 396–400.

Nallathambi, R., Mazuz, M., Ion, A., Selvaraj, G., Weininger, S., Fridlender, M., Nasser, A., Sagee, O., Kumari, P., & Nemichenizer, D. (2017). Anti-inflammatory activity in colon models is derived from δ9-tetrahydrocannabinolic acid that interacts with additional compounds in cannabis extracts. *Cannabis and Cannabinoid Research*, *2*(1), 167–182.

Nallathambi, R., Mazuz, M., Namdar, D., Shik, M., Namintzer, D., Vinayaka, A. C., Ion, A., Faigenboim, A., Nasser, A., & Laish, I. (2018). Identification of synergistic interaction between cannabis-derived compounds for cytotoxic activity in colorectal cancer cell lines and colon polyps that induces apoptosis-related cell death and distinct gene expression. *Cannabis and Cannabinoid Research*, *3*(1), 120–135.

Namdar, D., Voet, H., Ajjampura, V., Nadarajan, S., Mayzlish-Gati, E., Mazuz, M., Shalev, N., & Koltai, H. (2019). Terpenoids and phytocannabinoids co-produced in Cannabis sativa strains show specific interaction for cell cytotoxic activity. *Molecules*, *24*(17), 3031.

Nicholson, A. N., Turner, C., Stone, B. M., & Robson, P. J. (2004). Effect of Δ-9-tetrahydrocannabinol and cannabidiol on nocturnal sleep and early-morning behavior in young adults. *Journal of Clinical Psychopharmacology*, *24*(3), 305–313.

Omer, S., Boothe, D., Mansour, M., Dhanasekaran, M., & Pondugula, S. (2020). *Abstract PO-047: Anti-proliferative effect of cannabidiol (CBD) against B and T-cell lymphoma*. AACR.

Orch, H., Chaachouay, N., Douiri, E. M., Faiz, N., Zidane, L., & Douira, A. (2021). Use of medicinal plants in dermato-cosmetology: An ethnobotanical study among the population of Izarène. *Jordan Journal of Pharmaceutical Sciences*, *14*(3), 323–340.

Pal, G. K., Kumar, B., & Shahi, S. K. (2013). Antifungal activity of some common weed extracts against phytopathogenic fungi Alternaria spp. *International Journal of Universal Pharmacy and Life Sciences*, *3*(2), 6–14.

Palmieri, S., Maggio, F., Pellegrini, M., Ricci, A., Serio, A., Paparella, A., & Lo Sterzo, C. (2021). Effect of the Distillation Time on the Chemical Composition, Antioxidant Potential and Antimicrobial Activity of Essential Oils from Different Cannabis sativa L. Cultivars. *Molecules*, *26*(16), 4770.

Pasquali, F., Schinzari, M., Lucchi, A., Mandrioli, M., Toschi, T. G., De Cesare, A., & Manfreda, G. (2020). Preliminary data on the antimicrobial effect of Cannabis sativa L. variety Futura 75 against food-borne pathogens in vitro as well as against naturally occurring microbial populations on minced meat during storage. *Italian Journal of Food Safety*, *9*(2).

Passos, I. D., & Mironidou-Tzouveleki, M. (2016). Hallucinogenic plants in the Mediterranean countries. In *Neuropathology of drug addictions and substance misuse* (p. 761–772). Elsevier.

Pavlovic, R., Panseri, S., Giupponi, L., Leoni, V., Citti, C., Cattaneo, C., Cavaletto, M., & Giorgi, A. (2019). Phytochemical and ecological analysis of two varieties of hemp (Cannabis sativa L.) grown in a mountain environment of Italian Alps. *Frontiers in Plant Science*, *10*, 1265.

Pertwee, R. G. (2014). *Handbook of cannabis*. Oxford University Press, USA.

Pickens, J. T. (1981). Sedative activity of cannabis in relation to its Δ '-trans-tetrahydrocannabinol and cannabidiol content. *British Journal of Pharmacology*, *72*(4), 649–656.

Pierce, P. A., & Peroutka, S. J. (1989). Hallucinogenic drug interactions with neurotransmitter receptor binding sites in human cortex. *Psychopharmacology*, *97*, 118–122.

Piluzza, G., Delogu, G., Cabras, A., Marceddu, S., & Bullitta, S. (2013). Differentiation between fiber and drug types of hemp (Cannabis sativa L.) from a collection of wild and domesticated accessions. *Genetic Resources and Crop Evolution*, *60*, 2331–2342.

Placet, V., François, C., Day, A., Beaugrand, J., & Ouagne, P. (2018). Industrial hemp transformation for composite applications: Influence of processing parameters on the fibre properties. *Advances in Natural Fibre Composites: Raw Materials, Processing and Analysis*, 13–25.

Pochwala, S., Makiola, D., Anweiler, S., & Böhm, M. (2020). The heat conductivity properties of hemp–lime composite material used in single-family buildings. *Materials*, *13*(4), 1011.

Qaseem, M. F., Qureshi, R., Amjad, M. S., Ahmed, W., Masood, A., & Shaheen, H. (2019). Ethno-botanical evaluation of indigenous flora from the communities of Rajh Mehal and Goi union councils of district Kotli, Azad Jammu Kashmir Pakistan. *Applied Ecology & Environmental Research*, *17*(2).

Qureshi, R., Ghazanfar, S. A., Obied, H., Vasileva, V., & Tariq, M. A. (2016). Ethnobotany: A living science for alleviating human suffering. In *Evidence-Based Complementary and Alternative Medicine* (Vol. 2016). Hindawi.

R de Mello Schier, A., P de Oliveira Ribeiro, N., S Coutinho, D., Machado, S., Arias-Carrión, O., A Crippa, J., W Zuardi, A., E Nardi, A., & C Silva, A. (2014). Antidepressant-like and anxiolytic-like effects of cannabidiol: A chemical compound of Cannabis sativa. *CNS & Neurological Disorders-Drug Targets (Formerly Current Drug Targets-CNS & Neurological Disorders)*, *13*(6), 953–960.

Radwan, M. M., Chandra, S., Gul, S., & ElSohly, M. A. (2021). Cannabinoids, Phenolics, Terpenes and Alkaloids of Cannabis. *Molecules*, *26*(9), 2774.

Radwan, M. M., ElSohly, M. A., Slade, D., Ahmed, S. A., Wilson, L., El-Alfy, A. T., Khan, I. A., & Ross, S. A. (2008). Non-cannabinoid constituents from a high potency Cannabis sativa variety. *Phytochemistry*, *69*(14), 2627–2633.

Rafuse, P., & Buys, Y. M. (2019). Medical use of cannabis for glaucoma. *Canadian Journal of Ophthalmology*, *54*(1), 7–8. https://doi.org/10.1016/j.jcjo.2018.11.001

Rajan, T. S., Giacoppo, S., Iori, R., De Nicola, G. R., Grassi, G., Pollastro, F., Bramanti, P., & Mazzon, E. (2016). Anti-inflammatory and antioxidant effects of a combination of cannabidiol and moringin in LPS-stimulated macrophages. *Fitoterapia*, *112*, 104–115.

Reddy, D. S., & Golub, V. M. (2016). The Pharmacological Basis of Cannabis Therapy for Epilepsy. *Journal of Pharmacology and Experimental Therapeutics*, *357*(1), 45–55. https://doi.org/10.1124/jpet.115.230151

Rehman, S., Iqbal, Z., Qureshi, R., & Shah, G. M. (2023). Quantitative ethnobotanical study of medicinal plants used by the indigenous communities of Shawal Valley, District North Waziristan, Pakistan. *Ethnobotany Research and Applications*, *25*, 1–24.

Réus, G. Z., Stringari, R. B., Ribeiro, K. F., Luft, T., Abelaira, H. M., Fries, G. R., Aguiar, B. W., Kapczinski, F., Hallak, J. E., & Zuardi, A. W. (2011). Administration of cannabidiol and imipramine induces antidepressant-like effects in the forced swimming test and increases brain-derived neurotrophic factor levels in the rat amygdala. *Acta neuropsychiatrica*, *23*(5), 241–248.

Rinder, I. D. (1978). The effects of marijuana: A social psychological interpretation. *Psychiatry*, *41*(2), 202–206.

Robinson, R. (1996). *The great book of hemp: The complete guide to the environmental, commercial, and medicinal uses of the world's most extraordinary plant*. Inner Traditions/Bear & Co.

Rock, E. M., & Parker, L. A. (2021). Constituents of Cannabis Sativa. In *Cannabinoids and Neuropsychiatric Disorders* (p. 1–13). Springer.

Rodriguez-Martin, N. M., Toscano, R., Villanueva, A., Pedroche, J., Millan, F., Montserrat-de la Paz, S., & Millan-Linares, M. C. (2019). Neuroprotective protein hydrolysates from hemp (Cannabis sativa L.) seeds. *Food & Function*, *10*(10), 6732–6739.

Romero-Sandoval, E. A., Fincham, J. E., Kolano, A. L., Sharpe, B. N., & Alvarado-Vázquez, P. A. (2018). Cannabis for chronic pain: Challenges and considerations. *Pharmacotherapy: The Journal of Human Pharmacology and Drug Therapy*, *38*(6), 651–662.

Russo, E. B. (2011). Taming THC: Potential cannabis synergy and phytocannabinoid-terpenoid entourage effects. *British Journal of Pharmacology*, *163*(7), 1344–1364.

Ryz, N. R., Remillard, D. J., & Russo, E. B. (2017). Cannabis roots: A traditional therapy with future potential for treating inflammation and pain. *Cannabis and Cannabinoid Research*, *2*(1), 210–216.

Sabo, A., Horvat, O., Stilinovic, N., Berenji, J., & Vukmirovic, S. (2013). Industrial hemp decreases intestinal motility stronger than indian hemp in mice. *European Review for Medical and Pharmacological Sciences*, *17*(4), 486–490.

Sainz-Cort, A., Müller-Sánchez, C., & Espel, E. (2020). Anti-proliferative and cytotoxic effect of cannabidiol on human cancer cell lines in presence of serum. *BMC Research Notes*, *13*(1), 1–6.

Sales, A. J., Fogaça, M. V., Sartim, A. G., Pereira, V. S., Wegener, G., Guimarães, F. S., & Joca, S. R. (2019). Cannabidiol induces rapid and sustained antidepressant-like effects through increased BDNF signaling and synaptogenesis in the prefrontal cortex. *Molecular Neurobiology*, *56*(2), 1070–1081.

Sarmadyan, H., Solhi, H., Najarian-Araghi, N., & Ghaznavi-Rad, E. (2014). Determination of the Antimicrobial Effects of Hydro-Alcoholic Extract of Cannabis Sativa on Multiple Drug Resistant Bacteria Isolated from Nosocomial Infections. *Iranian Journal of Toxicology*, *7*(23), 967–972.

Sasco, A., & Besson, H. (2003). Cancer et cannabis. *Toxibase n*, *12*.

Schauer, G. L. (2021). Cannabis policy in the United States: Implications for public health. *JNCI Monographs*, *2021*(58), 39–52.

de Schier, A. R. M., de Ribeiro, N. P. O., de Silva, A. C. O., Hallak, J. E. C., Crippa, J. A. S., Nardi, A. E., & Zuardi, A. W. (2012). Cannabidiol, a Cannabis sativa constituent, as an anxiolytic drug. *Brazilian Journal of Psychiatry*, *34*, 104–110.

Schilling, S., Melzer, R., & McCabe, P. F. (2020). Cannabis sativa. *Current Biology*, *30*(1), R8–R9.

Schultes, R. E. (1969). Hallucinogens of Plant Origin: Interdisciplinary studies of plants sacred in primitive cultures yield results of academic and practical interest. *Science*, *163*(3864), 245–254.

Schultes, R. E., & Smith, E. W. (1976). *Hallucinogenic plants* (Vol. 35). Golden Press New York.

Scott, K. A., Dennis, J. L., Dalgleish, A. G., & Liu, W. M. (2015). Inhibiting heat shock proteins can potentiate the cytotoxic effect of cannabidiol in human glioma cells. *Anticancer Research*, *35*(11), 5827–5837.

Scotter, E. L., Abood, M. E., & Glass, M. (2010). The endocannabinoid system as a target for the treatment of neurodegenerative disease. *British Journal of Pharmacology*, *160*(3), 480–498.

Seth, R., & Sinha, S. (1991). Chemistry and pharmacology of cannabis. In M. J. Parnham, E. Graf, V. M. Kolb, R. Seth, S. Sinha, R. R. Ruffolo, J. P. Hieble, D. P. Brooks, G. Z. Feuerstein, A. J. Nichols, P. G. De Benedetti, & E. Jucker (Éds.), *Progress in Drug Research / Fortschritte der Arzneimittelforschung / Progrès des recherches pharmaceutiques* (p. 71–115). Birkhäuser. https://doi.org/10.1007/978-3-0348-7136-5_3

Sewell, R. A., Ranganathan, M., & D'Souza, D. C. (2009). Cannabinoids and psychosis. *International Review of Psychiatry*, *21*(2), 152–162.

Shah, N. C. (2004). Indigenous uses and ethnobotany of Cannabis sativa L.(Hemp) in Uttaranchal (India). *Journal of Industrial Hemp*, *9*(1), 69–77.

Shen, P., Gao, Z., Fang, B., Rao, J., & Chen, B. (2021). Ferreting out the secrets of industrial hemp protein as emerging functional food ingredients. *Trends in Food Science & Technology*, *112*, 1–15.

Shirazi-Zand, Z., Ahmad-Molaei, L., Motamedi, F., & Naderi, N. (2013). The role of potassium BK channels in anticonvulsant effect of cannabidiol in pentylenetetrazole and maximal electroshock models of seizure in mice. *Epilepsy & Behavior*, *28*(1), 1–7.

Shrivastava, A., Johnston, M., & Tsuang, M. (2011). Cannabis use and cognitive dysfunction. *Indian Journal of Psychiatry*, *53*(3), 187.

Shuaib, M., Khan, I., Sharifullah, Khan, R., Hashmatullah, Mubarik, S., & Naz, R. (2014). Ethnobotanical studies of spring flora of district Dir (Lower), Khyber Pakhtunkhwa, Pakistan. *Pakistan Journal of Weed Science Research*, *20*(1), 37–49.

Silote, G. P., Sartim, A., Sales, A., Eskelund, A., Guimarães, F. S., Wegener, G., & Joca, S. (2019). Emerging evidence for the antidepressant effect of cannabidiol and the underlying molecular mechanisms. *Journal of Chemical Neuroanatomy*, *98*, 104–116.

Smeriglio, A., Trombetta, D., Alloisio, S., Cornara, L., Denaro, M., Garbati, P., Grassi, G., & Circosta, C. (2020). Promising in vitro antioxidant, anti-acetylcholinesterase and neuroactive effects of essential oil from two non-psychotropic Cannabis sativa L. biotypes. *Phytotherapy Research*, *34*(9), 2287–2302.

Sofia, R. D., Solomon, T. A., & Barry III, H. (1976). Anticonvulsant activity of Δ9-tetrahydrocannabinol compared with three other drugs. *European Journal of Pharmacology*, *35*(1), 7–16.

Sofia, R. D., Vassar, H. B., & Knobloch, L. C. (1975). Comparative analgesic activity of various naturally occurring cannabinoids in mice and rats. *Psychopharmacologia*, *40*(4), 285–295.

Srivastava, A. K. (2018). Significance of medicinal plants in human life. In *Synthesis of Medicinal Agents from Plants* (p. 1–24). Elsevier.

Stephens, R. S., & Banes, K. E. (2013). *Cannabis and hallucinogens*.

Strajhar, P., Schmid, Y., Liakoni, E., Dolder, P. C., Rentsch, K. M., Kratschmar, D. V., Odermatt, A., & Liechti, M. E. (2016). Acute effects of lysergic acid diethylamide on circulating steroid levels in healthy subjects. *Journal of Neuroendocrinology*, *28*(3).

Strzelczyk, M., Lochynska, M., & Chudy, M. (2022). Systematics and botanical characteristics of industrial hemp Cannabis sativa L. *Journal of Natural Fibers*, *19*(13), 5804–5826.

Szaflarski, J. P., & Martina Bebin, E. (2014). Cannabis, cannabidiol, and epilepsy—From receptors to clinical response. *Epilepsy & Behavior*, *41*, 277–282. https://doi.org/10.1016/j.yebeh.2014.08.135

Tahsin, K. N., Watson, D., Rizkalla, A., Heinrichs, D., & Charpentier, P. (2021). Antimicrobial Studies of Cannabidiol as Biomaterials against superbug MRSA. *CMBES Proceedings*, *44*.

Taura, F., Sirikantaramas, S., Shoyama, Y., Yoshikai, K., Shoyama, Y., & Morimoto, S. (2007). Cannabidiolic-acid synthase, the chemotype-determining enzyme in the fiber-type Cannabis sativa. *FEBS Letters*, *581*(16), 2929–2934.

Teh, S.-S., Bekhit, A. E.-D. A., Carne, A., & Birch, J. (2016). Antioxidant and ACE-inhibitory activities of hemp (Cannabis sativa L.) protein hydrolysates produced by the proteases AFP, HT, Pro-G, actinidin and zingibain. *Food Chemistry*, *203*, 199–206.

The Plant List. (2013). The Plant List. Version 1.1. Published on the Internet; http://www.theplantlist.org (accessed 1st January).

Thompson, M. D., Ross, R. A., & Burnham, W. M. (2016). 9. The anticonvulsant effects of cannabidiol examined in animal seizure models. *Clinical Neurophysiology*, *127*(4), e164.

Todaro, B. (2012). Cannabinoids in the treatment of chemotherapy-induced nausea and vomiting. *Journal of the National Comprehensive Cancer Network: JNCCN*, *10*(4), 487–492. https://doi.org/10.6004/jnccn.2012.0048

Todorova, J., Lazarov, L. I., Petrova, M., Tzintzarov, A., & Ugrinova, I. (2021). The antitumor activity of cannabidiol on lung cancer cell lines A549 and H1299: The role of apoptosis. *Biotechnology & Biotechnological Equipment*, *35*(1), 873–879.

Turna, J., Simpson, W., Patterson, B., Lucas, P., & Van Ameringen, M. (2019). Cannabis use behaviors and prevalence of anxiety and depressive symptoms in a cohort of Canadian medicinal cannabis users. *Journal of Psychiatric Research*, *111*, 134–139. https://doi.org/10.1016/j.jpsychires.2019.01.024

Turner, C. E., Elsohly, M. A., & Boeren, E. G. (1980). Constituents of Cannabis sativa L. XVII. A review of the natural constituents. *Journal of Natural Products*, *43*(2), 169–234.

Turner, C. E., Hadley, K. W., Fetterman, P. S., Doorenbos, N. J., Quimby, M. W., & Waller, C. (1973). Constituents of Cannabis sativa L. IV: Stability of cannabinoids in stored plant material. *Journal of Pharmaceutical Sciences*, *62*(10), 1601–1605.

Ullah, H., Qureshi, R., Munazir, M., Bibi, Y., Saboor, A., Imran, M., & Maqsood, M. (2023). Quantitative ethnobotanical appraisal of Shawal Valley, South Waziristan, Khyber Pakhtunkhwa, Pakistan. *Ethnobotany Research and Applications*, *25*, 1–17.

UNODC. (2013). *United Nations Office on Drugs and Crime: Drug use in Pakistan 2013, National key findings*. Islamabad, Pakistan: United Nations Office on Drugs and Crime. Retrieved from http://www.unodc.org/documents/pakistan/Survey_Report_Final_2013.pdf

Vallée, A., Lecarpentier, Y., & Vallée, J.-N. (2021). Cannabidiol and the Canonical WNT/β-Catenin Pathway in Glaucoma. *International Journal of Molecular Sciences*, *22*(7), Article 7. https://doi.org/10.3390/ijms22073798

van de Donk, T., Niesters, M., Kowal, M. A., Olofsen, E., Dahan, A., & van Velzen, M. (2019). An experimental randomized study on the analgesic effects of pharmaceutical-grade cannabis in chronic pain patients with fibromyalgia. *Pain*, *160*(4), 860.

Van Went, G. F. (1978). Mutagenicity testing of 3 hallucinogens: LSD, psilocybin and‡9-THC using the micronucleus test. *Experientia*, *34*(3), 324–325. https://doi.org/10.1007/BF01923013

Vandepitte, K., Vasile, S., Vermeire, S., Vanderhoeven, M., Van der Borght, W., Latré, J., De Raeve, A., & Troch, V. (2020). Hemp (Cannabis sativa L.) for high-value textile applications: The effective long fiber yield and quality of different hemp varieties, processed using industrial flax equipment. *Industrial Crops and Products*, *158*, 112969.

Vaysse, B., Gignon, M., Zerkly, S., & Ganry, O. (2014). Alcohol, Tobacco, Cannabis, Anxiety and Depression Among second-year Medical Students. Identify in Order to Act. *Sante Publique*, *26*(5), 613–620.

Vilela, L. R., Lima, I. V., Kunsch, É. B., Pinto, H. P. P., de Miranda, A. S., Vieira, É. L. M., de Oliveira, A. C. P., Moraes, M. F. D., Teixeira, A. L., & Moreira, F. A. (2017). Anticonvulsant effect of cannabidiol in the pentylenetetrazole model: Pharmacological mechanisms, electroencephalographic profile, and brain cytokine levels. *Epilepsy & Behavior*, *75*, 29–35.

Walker, R., Pavia, S., & Mitchell, R. (2014). Mechanical properties and durability of hemp-lime concretes. *Construction and Building Materials*, *61*, 340–348.

Walsh, D., Nelson, K. A., & Mahmoud, F. (2003). Established and potential therapeutic applications of cannabinoids in oncology. *Supportive Care in Cancer*, *11*, 137–143.

Wanas, A. S., Radwan, M. M., Mehmedic, Z., Jacob, M., Khan, I. A., & Elsohly, M. A. (2016). Antifungal activity of the volatiles of high potency Cannabis sativa L. against Cryptococcus neoformans. *Records of Natural Products*, *10*(2), 214.

Watt, G., & Karl, T. (2017). In vivo evidence for therapeutic properties of cannabidiol (CBD) for Alzheimer's disease. *Frontiers in Pharmacology*, *8*, 20.

Weiss, M. C., Hibbs, J. E., Buckley, M. E., Danese, S. R., Leitenberger, A., Bollmann-Jenkins, M., Meske, S. W., Aliano-Ruiz, K. E., McHugh, T. W., & Larson, S. L. (2022). A Coala-T-Cannabis Survey Study of breast cancer patients' use of cannabis before, during, and after treatment. *Cancer*, *128*(1), 160–168.

Weldy, E. W., Stanley, J., Koduri, V. A., McCourt, E. A., Patnaik, J. L., Kahook, M. Y., & Seibold, L. K. (2020). Perceptions of Marijuana Use for Glaucoma from Patients, Cannabis Retailers, and Glaucoma Specialists. *Ophthalmology Glaucoma*, *3*(6), 453–459. https://doi.org/10.1016/j.ogla.2020.06.009

West, M. E., & Lockhart, A. B. (1978). *The treatment of glaucoma using a non-psychoactive preparation of cannabis sativa*. 16–25.

Whiting, P. F., Wolff, R. F., Deshpande, S., Di Nisio, M., Duffy, S., Hernandez, A. V., Keurentjes, J. C., Lang, S., Misso, K., Ryder, S., Schmidlkofer, S., Westwood, M., & Kleijnen, J. (2015). Cannabinoids for medical use: A systematic review and meta-analysis. *JAMA*, *313*(24), 2456–2473. https://doi.org/10.1001/jama.2015.6358

Wilcox, H. C., Wagner, F. A., & Anthony, J. C. (2002). Exposure opportunity as a mechanism linking youth marijuana use to hallucinogen use. *Drug and Alcohol Dependence*, *66*(2), 127–135.

Wong, J. C. M. (2017). *Contribution of intestinal lymphatic transport to the antiproliferative effect of delta-9-tetrahydrocannabinol* [PhD Thesis]. University of Nottingham.

Yan, X., Tang, J., dos Santos Passos, C., Nurisso, A., Simoes-Pires, C. A., Ji, M., Lou, H., & Fan, P. (2015). Characterization of lignanamides from hemp (Cannabis sativa L.) seed and their antioxidant and acetylcholinesterase inhibitory activities. *Journal of Agricultural and Food Chemistry*, *63*(49), 10611–10619.

Yazici, L., Yilmaz, G., Koçer, T., & Şakar, H. (2020). Investigation of some yield characteristics of hemp (Cannabis sativa L.) in Tokat Ecology. *Journal of International Environmental Application and Science*, *15*(2), 104–108.

Zajicek, J. P., Hobart, J. C., Slade, A., Barnes, D., Mattison, P. G., & Group, M. R. (2012). Multiple sclerosis and extract of cannabis: Results of the MUSEC trial. *Journal of Neurology, Neurosurgery & Psychiatry*, *83*(11), 1125–1132.

Zareef, H., Gul, M. T., Qureshi, R., Aati, H., & Munazir, M. (2023). Application of ethnobotanical indices to document the use of plants in traditional medicines in Rawalpindi district, Punjab-Pakistan. *Ethnobotany Research and Applications*, *25*, 1–29.

Zengin, G., Menghini, L., Di Sotto, A., Mancinelli, R., Sisto, F., Carradori, S., Cesa, S., Fraschetti, C., Filippi, A., & Angiolella, L. (2018). Chromatographic analyses, in vitro biological activities, and cytotoxicity of Cannabis sativa

L. essential oil: A multidisciplinary study. *Molecules*, *23*(12), 3266.

Zeroual, R., & Danguir, A. (2023). The Possibilities of Medical Cannabis Application in Dentistry in the Era of Cannabis Legislation in Morocco, One of the World's Largest Producers of Cannabis. *Open Access Library Journal*, *10*(3), 1–14.

Zhang, Q., Chen, X., Guo, H., Trindade, L. M., Salentijn, E. M., Guo, R., Guo, M., Xu, Y., & Yang, M. (2018). Latitudinal adaptation and genetic insights into the origins of Cannabis sativa L. *Frontiers in Plant Science*, *9*, 1876.

Zhu, H.-L., Wan, J.-B., Wang, Y.-T., Li, B.-C., Xiang, C., He, J., & Li, P. (2014). Medicinal compounds with antiepileptic/anticonvulsant activities. *Epilepsia*, *55*(1), 3–16.

12 Bushman's Tea (*Catha edulis* (Vahl) Endl. Celastraceae)

Priyatama Vijaysing Powar and Noureddine Chaachouay

12.1 INTRODUCTION

Khat, known as *Catha edulis* (Vahl) Endl., belongs to the Celastraceae family (The Plant List, 2013). It is a perennial tree that thrives in mountainous regions throughout East and Southern Africa, Afghanistan, Yemen, and Madagascar. *C. edulis.* is a botanical substance that has been historically used by many ethnicities in the Horn of Africa, Yemen, and the global diaspora communities associated with these traditions (Chaachouay et al., 2019a, 2019b; Al-Hebshi & Skaug, 2005; Brooke, 1960; Kiunga, 2017). *C. edulis* occupies a significant role in the socioeconomic customs of these civilizations. Yemen has the highest chewers among the countries closely related to *C. edulis*. Many houses in Yemen have dedicated rooms for *C. edulis* chewing. With the increasing demand for Khat in this area, economically profitable companies have emerged, supporting individuals closely tied to the plant (Gudata, 2020). *C. edulis* is often used in social contexts because it can enhance interpersonal contact and induce euphoria. Furthermore, *C. edulis* is used by laborers not just to augment their productivity but also to combat exhaustion. The Yemeni population attributes some medical properties to *C. edulis* and may use it to treat mild, self-limiting ailments like headaches and the common cold (Al-Motarreb et al., 2002). *C. edulis* is also known for its purported efficacy in treating depression and its alleged sexual effects. *C. edulis* was historically regulated for religious, ceremonial, and therapeutic use for decades. During the early 1990s, *C. edulis*, a plant native to the Horn of Africa, was introduced to Europe, North America, and Australia due to emigration from that region. Currently, the worldwide estimate of abusers exceeds 20 million, including both adult males and females, as well as college and middle school pupils (Balint et al., 2009).

The primary bioactive compounds found in *C. edulis* are phenylpropylamino alkaloids, namely cathine and cathinone, which are responsible for pharmacological and toxicological impacts. Of them, the β-keto analog of amphetamine, when viewed structurally, has the potential to cause dependency by exerting psychostimulatory effects on the neurological system (Getasetegn, 2016; Mohamed Abdoul-Latif et al., 2023; Murthy et al., 2016). Furthermore, extended exposure to *C. edulis* may result in addiction, psychosis, high blood pressure, heart-related issues, sexual dysfunction, liver damage, and other adverse effects (Al-Habori, 2005; Engidawork, 2017). Consequently, the misuse of *C. edulis* has become a significant public health issue. Nevertheless, apart from these negative consequences, there are further studies indicating possible therapeutic advantages of *C. edulis*, such as its antimicrobial properties, its ability to operate as an antidepressant, its potential use as a supplementary therapy for obesity, and its potential as a replacement for brain tissue (Al-Motarreb et al., 2002; Githua et al., 2022; N. Wabe & Mohammed, 2012; N. T. Wabe, 2011). These findings need thorough research. Therefore, this chapter provides a comprehensive and cohesive examination of the literature about the Khat plant's botany, ecology, traditional usage, and phytochemical makeup. Furthermore, it presents the toxicological and pharmacological characteristics of *C. edulis*.

12.2 SCIENTIFIC CLASSIFICATION

C. edulis, Khat or Qat, is a blossoming plant from the Horn of Africa and the Arabian Peninsula. This perennial shrub or little tree is classified under the botanical family Celastraceae (The Plant List, 2013).

Kingdom:	Plantae
Phylum:	Magnoliophyta
Class:	Magnoliopsida
Subclass:	Rosidae
Order:	Celastrales
Family:	Celastraceae
Genus:	*Catha*
Species:	*Catha edulis* (Vahl) Endl.

12.3 VERNACULAR NAMES

Arabic: Qat, Cat, Kat; **English**: Abyssinian Tea, Somali Tea, Miraa, Arabian Tea, Jimaa, and Kafta; **German**: Katstrauch; Khat; Arabischer Tee; **Spanish**: Té de Arabia; **Somali**: Jaad; **Amharic (Ethiopia)**: ጫት (Chat); **Somali**: Qaad, Jaad; **Oromo (Ethiopia)**: Jimaa; **Tigrinya (Eritrea and Ethiopia)**: ጫት (Chat); **Swahili (East Africa)**: Miraa, Veve.

DOI: 10.1201/9781003460336-12

12.4 BOTANICAL DESCRIPTION

C. edulis is a tree that may reach a height of over 15 meters. However, it can also be a smaller shrub. The top of the tree is erect and has few limbs that hang downward. The trunk is often straight and thin, accompanied by a small crown. The bark on juvenile branches is light gray, with powdered yellow pigments, a smooth texture, and the possibility of becoming pink (Kiunga, 2017). Over time, the bark transforms, becoming darker, rougher, and cracked. The bark of giant trees may exhibit a lattice-like structure. The shedding of bark exposes a powdered yellow tint (Figure 12.1(a)). Immature branches may have a rosy hue extending to the leaf stalk and central vein. A more intense shade of pink emerges on the sun-exposed side (Al-Hebshi & Skaug, 2005).

The size of leaves may range up to 11 × 5 cm; however, they are often smaller. This is seen in picture 611. They often dangle and sway while producing a flickering effect when exposed to the wind (Figure 12.1(b)). The leaves of the *Catha* plants are devoid of hair. They are characterized by their simplicity, consisting of a single blade that may include shallow perforations that do not entirely split the leaf into leaflets (Bally, 1945). The leaves on blooming branches are arranged opposingly and quickly shed the triangular stipules between the leaf bases. This results in a scar or ridge between the two petioles. The stipules are around 2 mm long and caducous, meaning they are readily separated and shed early (Pantelis et al., 1989). The petiole, which is somewhat flattened, has a grooved surface on its upper side and may have a pinkish coloration. The apex and base exhibit either a tapering or rounded shape. During the fall season, the leaves of deciduous trees have the potential to undergo a visually appealing transformation, adopting a vibrant golden hue. The upper surface of the blade is lustrous green, while the lower surface is a muted, paler shade of green. Vein characteristics are more readily seen when the leaf is placed against a powerful light source. At least ten pairs of lateral veins are observable, initially positioned at about a 45-degree angle to the midrib. These veins gradually bend and shift toward the apex, nearing the edge. A juvenile midrib may have a reddish hue. The margins of the object have a slightly undulating shape and are uniformly notched, with each tip terminating in a small gland (Krikorian, 1985). The serrations are oriented toward the apex.

The bisexual flowers, which are either creamy-white or green in color, have a very tiny width of 2 mm. Several planes traveling through the axis may divide flowers into two halves. They are found in two branching cymes, which are large, more or less flat-topped flower clusters, with the central one maturing first (Figure 12.1(c)). The receptacle is a protuberant structure made of flesh, an elevated and developed part of the receptacle (Pantelis et al., 1989). The object has a disc-like form and a border equally serrated with shallow teeth. The five little verdant sepals are arranged overlappingly and fused at the lower half. The sepals are fragmented into slender and asymmetrical halves. The corolla is adorned with five petals that are either creamy-white or green (Simmons et al., 2008). The petals are longer than the sepals and have an oblong to elliptic shape with coarse margins, including a slightly flattened tip. The shot reveals the noticeable green throat at the base of the petal. The five stamens, which alternate with the petals, start upright but soon bend backward. The basifixed anthers are connected by their base, and the longitudinal dispersion slots are seen here. The pistil, a component of the gynoecium, consists of the ovary, style, and stigma. The ovary is located inside the disc and includes three compartments. Each locule contains two ovules. The abbreviated style terminates with a trilobed stigma (N. T. Wabe, 2011).

The fruit lacks thorns and develops into a slender, three-lobed capsule. The elongated capsule is around 1 cm in length and has a roughly triangular shape. The capsule changes from green to red, ultimately transitioning to a reddish-brown hue before it opens up. The dehiscence

FIGURE 12.1 Plant parts of *C. edulis*: leaves (a), flowers (b), and young stem (c).

occurs by splitting up the chambers, or locules, of the capsule, dividing each of the three halves, and releasing the seeds while keeping the septa, or partitions, intact. The seeds are scarlet and slightly winged, each with a white aril. The funiculus is a stalk that connects an ovule or seed with the placenta, giving rise to it (Alles et al., 1961).

12.5 HISTORY AND GEOGRAPHIC DISTRIBUTION

The origins of Khat are often a subject of debate. It is widely believed that they originated in Ethiopia and then spread to the slopes of East Africa and Yemen. Some argue that Khat's origin may be traced back to Yemen when it traveled to Ethiopia and neighboring nations. In his book *First Footsteps in East Africa*, published in 1856, Sir Richard Burton elucidates that *C. edulis* was brought to Yemen from Ethiopia around the 15th century (Ye et al., 2021; Zyoud, 2015). Furthermore, compelling evidence indicates that this event may have occurred as early as the 13th century. The trees expanded from Ethiopia and Yemen to include Kenya, Somalia, Malawi, Uganda, Tanzania, Arabia, the Congo, present-day Zimbabwe, Zambia, and South Africa (Al-Thobhani et al., 2008; Ngari et al., 2019; Tembrock et al., 2017). The New Testament is said to contain the oldest documented instance of the medicinal usage of *C. edulis*. The ancient Egyptians revered the Khat plant as a 'divine sustenance' that had the power to awaken the divine nature inside people. The Egyptians used the herb for purposes beyond its energizing properties. In 1854, a Malay writer, Abdullah bin Abdul Kadir, remarked that chewing *C. edulis* was widespread in Al Hudaydah, Yemen. He stated that it was common to see people chewing leaves, likening it to how goats chew their food (Bally, 1945; Chaachouay et al., 2020a, 2020b; Ye et al., 2021). A kind of leaf, characterized by its broad shape and around two finger lengths, is commonly available for purchase. Unlike betel leaves, which need certain condiments to be consumed, these leaves are inserted into the mouth and chewed. Consequently, when individuals congregated, the residual fragments of these leaves would accumulate in their presence.

Khat cultivation has been practiced for millennia in the Horn of Africa and the Arabian Peninsula, primarily for its stimulating properties. In that region, chewing *C. edulis* predates the use of coffee and is employed in a comparable social setting. This plant's fresh leaves and tips are often eaten or sometimes dried and brewed into tea to induce a mood of exhilaration and stimulation. The foliage or the stem's tender portion may be masticated with chewing gum or roasted peanuts to facilitate the chewing process. In Uganda, the cultivation of this crop is mainly concentrated in the central region, particularly in Kasenge, Butambala District, Mabira Forest, and some areas of the western region. Kenya cultivates this crop in Meru County and Embu County (Tembrock et al., 2020). While Khat-chewing remains mainly limited to its native cultivation region in the Red Sea, it has gradually spread throughout southern Africa and tropical areas, where it thrives on rocky outcrops and woods. The shrub is now dispersed but remains indigenous to the KwaZulu-Natal, Eastern Cape, Western Cape, and Mpumalanga provinces of South Africa, as well as Eswatini and Mozambique. Recently, advancements in infrastructure, including better roads, off-road vehicles, and air transportation, have facilitated the worldwide dissemination of this perishable product (Al-Thobhani et al., 2008; Tembrock et al., 2017). Consequently, the plant has been discovered in various locations such as Australia, Amsterdam, Canada, England, the Levant, Italy, New Zealand, Wales, and the United States. Freshly packaged *C. edulis* leaves are available for sale in the African and Middle Eastern marketplaces of Dallas, Boston, New York City, and Los Angeles, the regions with the most significant demand in the United States (Hocking, 1963; Patel, 2019).

12.6 HABITAT AND ECOLOGY

C. edulis flourishes in temperate to subtropical temperatures, particularly in areas with moderate precipitation. The plant exhibits optimal development within the 1,500 to 2,000-meter height range. However, it may also adapt to lower elevations in some regions. *C. edulis* plants thrive on well-drained, loamy soils that harmonize moisture retention and drainage. These plants can endure a variety of soil conditions (Hirst, 1997; Ngarakana et al., 2022). However, their productivity is reduced in very sandy or heavy clay soils. *C. edulis*, a perennial shrub or small tree, has the potential to grow up to 20 meters tall. The required annual rainfall is between 800 and 1000 mm across 4–6 months. Growth is hampered by frost and heavy humidity (Ngari et al., 2019). *C. edulis* may be grown in various soil types, ranging from sandy loams to heavy clays, that are moderately acidic to alkaline, suitably deep and well-drained, and have a high organic matter concentration in the topsoil. It cannot tolerate salt. Yet, it often stays lower in size. The plant's leaves, which include psychotropic substances, are highly prized and characterized by their glossy, oval-shaped appearance and dense arrangement (Zahran et al., 2014). The agricultural and ecological ramifications of *C. edulis* are substantial. It is often planted on terraced slopes where it is cultivated for commercial purposes, such as in some districts in Ethiopia and Yemen (Hill, 1965). This agricultural technique is beneficial for efficient water control and preservation of soil quality. *C. edulis* cultivation requires significant manual work since it involves meticulous upkeep and hand-picking (Figure 12.2). The crop is usually harvested many times throughout the year (Khatib et al., 2013; Krikorian, 1985). Nevertheless, the vast production of *C. edulis* may have a significant environmental effect. Large-scale *C. edulis* cultivation in some locations has resulted in deforestation and soil degradation. At the same time, the need for irrigation water may significantly burden local water supplies, particularly in dry areas. *C. edulis* production in places with substantial farming practices may lead to a decline in biodiversity due to the clearance of other species to make way for its growth despite its ability to thrive in different environments alongside other plants (Tembrock et al., 2020; Orch et al., 2021; Chaachouay & Zidane, 2021).

12.7 ETHNOBOTANY

Khat consumption has historically been limited to the areas where *C. edulis* is cultivated since the necessary energizing effects can only be obtained from fresh leaves. The worldwide distribution of this perishable product has been enhanced in recent years due to the development of better road infrastructure, off-road motor vehicles, and air transport (Chaachouay et al., 2022a, 2022b, 2022c; Benkhnigue et al., 2023). Historically, *C. edulis* has been used as a substance for social interaction, and this practice is prevalent in Yemen, where *C. edulis* consumption is mainly seen among males. *C. edulis* consumption is mostly seen among those who are single and at social gatherings in many countries (Kiunga, 2017). Khat is used chiefly as a recreational substance in places where it is cultivated (Figure 12.3). However, it is also utilized by farmers and workers to alleviate physical exhaustion and by drivers and students to enhance focus and concentration (Kandari et al., 2014). Among the counter-culture factions of the affluent Kenyan populace, *C. edulis* is used as a remedy for the symptoms of a hangover or excessive alcohol consumption, akin to the utilization of the coca leaf in South America (Stevenson et al., 1996). In Yemen, chewing Khat is not done mainly by males. Women also have establishments where they engage in this activity and join their spouses in chewing Khat on weekends (Al-Motarreb et al., 2002; Hirst, 1997; Mwenda et al., 2003; Chaachouay et al., 2021a, 2021b, 2021c).

C. edulis cultivation is prevalent in several regions, where it has gained sufficient popularity for many adolescents to chew the plant before reaching puberty. *C. edulis* is used for its gentle, euphoric, and stimulating impacts and exhibits anorectic properties. Its use is often not restricted by religion. However, due to its stimulating properties, the Ethiopian Orthodox Tewahedo Church has prohibited Christians from using it. In Yemen, the cultivation of this particular crop is highly favored, to the extent that 40% of the nation's water resources are allocated for irrigation purposes (Githua et al., 2022; Gudata, 2020). Furthermore, the production of this crop experiences an annual growth rate of around 10% to 15%. The water supply of Sanaa is severely endangered by the cultivation of crops, to the extent that government authorities have suggested the possibility of moving significant segments of the city's population to the Red Sea shore (Al-Motarreb et al., 2002).

The use of Khat is considered to be prevalent among several million individuals. A significant portion of the users come from nations spanning from Sudan to Madagascar, as well as the southern region of the Arabian Peninsula, notably Yemen. A study conducted in Yemen revealed that 60% of the male and 35% of the female population were identified as chronic *C. edulis* users, engaging in daily chewing

FIGURE 12.2 A farmer's field in Yemen is dedicated to growing Khat, a local stimulant plant.

FIGURE 12.3 The harvesting process of the agricultural crop *C. edulis*.

for extended durations throughout their lives. In Yemen, chewing *C. edulis* is mainly limited to males, whereas females engage in this activity with less formality and frequency (Gezon, 2016).

The Khat plant has great importance in the social and cultural fabric of the communities where it has been used for many years. It is used to extend greetings and amuse visitors at joyous events like wedding ceremonies and during somber occasions like post-burial parties held at individuals' homes (Balint et al., 2009; Gebissa, 2010). There is a widespread belief that consuming it may alleviate exhaustion, enhance alertness, and diminish feelings of hunger. Additionally, it instills a sense of euphoria, enhances one's capacity to convey thoughts effectively, and boosts self-assurance (Kiunga et al., 2016). Khat cultivation and usage are prohibited in Saudi Arabia, and the restriction is rigorously enforced. The clergy also endorsed the prohibition of *C. edulis* based on the premise that the Qur'an prohibits anything damaging the physical well-being (Abdelwahab et al., 2015). According to reports, 61% of the people in Somalia confirmed their *C. edulis* use, with 18% being frequent users and 21% being occasional users (Lamina, 2010).

12.8 HALLUCINOGENIC EFFECTS

C. edulis consumption elicits moderate euphoria and exhilaration, comparable to the effects observed with potent coffee. Individuals become incredibly talkative when the plant is present. According to animal studies, *C. edulis* induces heightened levels of motor activity. When cathinone is administered orally, its effects manifest more swiftly than when amphetamine tablets are taken, approximately 15 minutes versus 30 minutes for amphetamine (Gambaro et al., 2012; Hassan et al., 2007; Karunamoorthi et al., 2010). *C. edulis* can elicit manic states and hyperactivity, exhibiting comparable effects to those induced by amphetamine. Users experience heightened vigilance, vitality, and sociability, accompanied by an enhanced mood and sometimes excessive activity (Pantelis et al., 1989). Chronic usage can result in mental health problems, possibly causing personality disorders and mental decline. Since *C. edulis* leaves have vasoconstrictor qualities, they may raise blood pressure, heart rate, and cause an acute myocardial infarction. After about 2 hours, tension, emotional instability, and irritability begin to appear, later leading to feelings of low mood and sluggishness (Al-Habori, 2005; Gebissa, 2010; Kotb El Sayed & Amin, 2015; Wabel, 2011). Nevertheless, prolonged use might result in heightened mental health complications, such as sleeplessness, loss of appetite, gastrointestinal difficulties, tooth discoloration, increased ulcer vulnerability, reduced libido, and depressive symptoms. *C. edulis* is a potent anorectic agent that induces hunger suppression (Kotb El Sayed & Amin, 2015). Notably, the use of *C. edulis* has cultural significance in several places where it is cultivated and eaten and possesses social and recreational importance.

12.9 REGULATION AND CONTROL STATUS

In 1965, the Fourteenth Report of the Expert Committee on Dependence-producing Drugs of the World Health Organization acknowledged the resolution of the Economic and Social Council regarding Khat (Organization, 1965). The resolution affirmed that the misuse of this substance is a localized issue and can be most effectively managed at the regional level. Due to this rationale, *C. edulis* was not included in the Single Convention on Narcotic Drugs (Faro et al., 2020). In 1980, the World Health Organization categorized *C. edulis* as a substance of misuse capable of inducing mild to severe psychological addiction (Organization, 2003). Khat is classified as a restricted drug in Canada and is included under Schedule IV of the Restricted Drugs and Substances Act. Any individual who acquires a substance without disclosing their authorization to do so within 30 days of obtaining another prescription from a medical professional is committing a serious crime. They may be sentenced to a maximum of 18 months if the substance in question is listed in Schedule IV. Alternatively, if the offense is considered less severe, they may be subject to a fine of up to one thousand dollars or imprisonment for up to six months for a first offense. For subsequent offenses, the penalty may be a fine of up to two thousand dollars or imprisonment for up to one year or both. The importation of Khat in Australia is regulated under the Customs (Prohibited Imports) (Douglas & Pedder, 2010) Regulations. Individual users, mainly immigrants from the Horn of Africa, can get several necessary permissions to import a maximum of five kilograms per month for personal consumption (Gebissa, 2010). 2003, there were 294 yearly licenses, and 202 individual *C. edulis* permits. Khat is not classified as a regulated drug in the United Kingdom. Due to these factors, namely Khat's limited durability and the U.K.'s strategic position, it acts as a primary conduit for Khat's air shipments to North America. Khat is consumed mainly by male members of the Somali and Yemeni communities, primarily located in London, Birmingham, Bristol, Cardiff, Manchester, and Sheffield. Presently, it is permissible by law. However, several factions within the Somali community are advocating for its prohibition. Cathine and cathinone are classified as Class C medicines in the U.K. (Thomas & Williams, 2013). The plant *C. edulis* is not subject to any regulation or control. Cathine is classified as a Schedule IV substance, whereas cathinone is classified as a Schedule I substance under the U.S. Controlled Substance Act. The 1993 DEA regulation classifying cathinone as a Schedule I substance explicitly acknowledged that it was effectively prohibiting the use of Khat as well (Bonson et al., 2019). Cathinone is the primary hallucinogenic compound found in the *C. edulis* plant, often known as Khat. Khat's youthful foliage is masticated to provide a stimulating impact. The implementation of this regulation leads to the categorization of any substance containing cathinone into Schedule I. Cathine is classified as a prohibited drug in Germany, making it unlawful to possess or sell the plant. Comparable degrees of regulation are present in the majority of other European nations. (Cathine and cathinone are subject to regulation in Hong Kong under Schedule 1 of the Dangerous Drugs Ordinance, Chapter 134.) Its lawful use is restricted to health practitioners and academic research endeavors alone (Bonson et al., 2019). Pharmacists may administer the drug with a prescription. Individuals who distribute the medicine without a prescription may face a fine of $10,000. The punishment for engaging in the illegal trade or production of the chemical is a fine of $5,000,000 (HKD) and a sentence of life imprisonment. Unauthorized possession of the drug for consumption is prohibited by law and may result in a fine of $1,000,000 (HKD) and a prison sentence of up to 7 years (Brooke, 1960). The Khat plant is classified as a Schedule III (Class C) substance in New Zealand. However, it is seldom found, although there have been unusual instances of its confiscation at airports. Pre-existing *C. edulis* trees that reached maturity before the plant's scheduling in 1998 do not need to be eradicated. However, it is against the law to collect the leaves or engage in any other kind of preparation for consumption. Khat is categorized as a narcotic substance in Norway, rendering its use, sale, and possession illegal. Most consumers consist of Somali immigrants and the illicit transportation of Khat occurs from the Netherlands and England. In Sweden, Khat is categorized as a narcotic substance, making it unlawful to use, trade, or possess. As per the authorities, the majority of users are Somali immigrants, and the majority of Khat is illicitly imported from the Netherlands and England. As of 17 November 2006, Somali Islamists in regions under their control prohibited the use and distribution of Khat (Klein, 2014). In Somalia, the Supreme Islamic Courts Council, which gained authority over a significant portion of the nation in 2006, prohibited using Khat during the holy month of Ramadan, leading to public demonstrations in Kismayo. In November 2006, Kenya banned all air travel to Somalia due to security apprehensions, which resulted in protests by Kenyan *C. edulis* cultivators. The Member of Parliament for Ntonyiri in Meru North District of Kenya said that the land in the area has been specifically designated for the production of Khat. He mentioned that a staggering 20 tons, valued at US$800,000, are exported to Somalia daily. Furthermore, he expressed concern that imposing a ban on flights may have a severely detrimental impact on the local economy. Following the unexpected triumph of the Provisional Government, supported by Ethiopian troops, in late December 2006, the sale of Khat has resumed in the streets of Mogadishu. However, Kenyan vendors have observed that the Khat demand has yet to reach the levels seen before the ban was imposed (Cochrane & O'Regan, 2016; Griffiths et al., 2010; Klein, 2014).

12.10 PHYTOCHEMISTRY

For over a century, the chemical composition of *C. edulis* has perplexed chemists and pharmacologists. Many chemical investigations are dedicated to identifying the active component of *C. edulis* and characterizing its alkaloid content. Initial efforts ascribed the stimulating effects of *C. edulis* to cathine, a component of the phenyl alkylamine type identified as beta-or pseudoephedrine. Subsequent research, however, presented preliminary proof of the existence of a different and more potent phenyl alkylamine compound in the plant's new leaves, which might potentially be a volatile precursor of cathine. The much-anticipated phenylalkylamine compound was ultimately separated, identified as (7)-a-amino propiophenone, and officially designated as (7)-Cathinone. Subsequent research revealed that cathinone is found in large amounts in the young leaves but is quickly transformed into cathine and, to a lesser degree, norephedrine in the mature leaves (Atlabachew et al., 2014; Getasetegn, 2016; N. T. Wabe, 2011). Norephedrine is another phenylalkylamine compound found in *C. edulis*. Both cathine and cathinone have a structural relationship with amphetamine. Szendrei's assessments thoroughly account for the advancements in comprehending *C. edulis* chemistry during that specific timeframe. Khat contains various chemicals, including alkaloids, terpenoids, flavonoids, sterols, glycosides, tannins, amino acids, vitamins, and minerals (Table 12.1). The phenylalkylamines and cathedulins are the primary alkaloids. The cathedulins are derived from a sesquiterpene skeleton that contains several hydroxyl groups (Alsanosy et al., 2020; Krizevski et al., 2007). They may be described as polyesters of euonyminol. A total of 62 distinct cathedulins were identified by characterizing fresh *C. edulis* leaves. The *C. edulis* phenylalkylamines consist of three compounds: Cathinone [S-(–)-cathinone], cathine [1S, 2S-(+)-norpseudoephedrine or (+)-norpseudoephedrine], and norephedrine [1R,2S-(–)-norephedrine]. These chemicals have a similar structure to amphetamine and noradrenaline. The plant only includes the (–)-enantiomer of cathinone. Therefore, the naturally existing S-(–)-cathinone has an identical absolute configuration to S-(+)-amphetamine. Cathinone is primarily present in juvenile foliage and sprouts. Cathinone undergoes a metabolic transformation throughout maturity into cathine, also known as (+)-or pseudoephedrine and (–)-norephedrine (Al-Meshal, 1988). The leaves contain [(+)-norpseudoephedrine] and (–)-norephedrine in a ratio of roughly 4 to 1 (Kiros, 2020; Mohamed Abdoul-Latif et al., 2023). The *C. edulis* leaves include additional phenylalkylamine alkaloids, including merucathinone, pseudomerucathine, and merucathine (Crombie et al., 1990; Guantai, 1982). These chemicals seem to have a reduced impact on the stimulating effects of *C. edulis*.

12.11 PHARMACOLOGICAL EFFECTS

Rats that are given *C. edulis* material, either in the form of an extract or as a whole, exhibit heightened locomotor activity and experience a decrease in weight gain (Kimani & Nyongesa, 2008). The deceleration of the growth rate was attributed to reduced food absorption rather than reduced food intake. Khat decreases food intake and maternal weight growth in pregnant rats while reducing the food efficiency index (Nigusse et al., 2019). Subsequent studies have verified the heightened locomotor activity. Furthermore, *C. edulis* extracts and (–)-cathinone induce predictable behavior patterns, self-administration, and appetite-suppressing effects in several animal species. In terms of quality, this behavior is comparable to the response elicited by amphetamine [S-(+)-amphetamine]. Both *C. edulis* extract and (–)-cathinone augment the inherent aggressive behavior of isolated rats. In addition, (–)-cathinone may induce conditioned place preference in rats when administered at a 1.6 mg/kg dosage (Al-Meshal, 1988). This leads to an increase in locomotor activity, indicating the drug's rewarding impact. A reduced dosage of cathinone (0.2 mg/kg), which did not increase locomotion, likewise did not exhibit conditioned site preference. Cathinone may also function as a discriminative stimulus in a food-reinforced operant task (Hyatt et al., 2020).

Locomotor sensitization and deficiencies in prepulse inhibition, caused by psychostimulants, are extensively investigated as animal behavioral models for amphetamine

TABLE 12.1
Checklist of Biochemical Components Derived from *C. edulis*

Chemical Category	Chemical Name	References
Amino acids	Alanine, Alpha-aminobutyric acid, Arginine, Asparagine acid, Choline, Glutamic acid, Glycine, Histidine, Isoleucine, Leucine, Ornithine, Phenylalanine, Proline, Serine, Threonine, Tryptophan, Tyrosine, Valine	Alles et al. (1961), Getasetegn (2016), Kiros (2020)
Flavonoids	Dihydromyricetin, Dihydromyricetin-3-O-rhamnoside, Kaempferol, Myricetin, Myricetin-3-O-b-D-galactoside, Myricetin-3-O-rhamnoside, Quercetin, Quercetin-3-O-b-D-galactoside	Fatima et al. (2017), Nigusse et al. (2019)
Phenylalkylarnine Alkaloids (Khatamines)	(+)-Cathine, (-)-Cathinone, 3,6-Dimethyl-2,5-diphenyl pyrazine, Merucathine, Merucathinone, (-)-Norephedrine, (-)-Norephedrine N-formyl, Cathedulin E1–E6, Cathedulin K1, 2, 6, 12, 15, 17, 19, 20, Cathedulin Y7–Y10, Cathidine A, B, D, Euonyminol, 1-Phenylpropane-1,2-dione, Pseudomerucathine, Cathedulin E1–E6	Al-Meshal (1988), Atlabachew et al. (2014), Baxter et al. (1979); Crombie et al. (1990), Krizevski et al. 2007)
Sterol and Triterpenes	Celastrol, Friedeline, Iguesterin, Pristimerin, beta-Sitosterol, beta-Sitosterol glycoside, Tingenin A, B	Getasetegn (2016), Murthy et al. (2016), Rizk et al. (1989)
Vitamins	Ascorbic acid, Niacin, Riboflavin, Thiamine	Al-Mehdar et al. (2012), Getasetegn (2016)
Volatiles	Fenchone, Linalool, Nerol, Ocimene, beta-Phellandrene, alpha-Pinene, beta-pinene, alpha–Terpineol, Terpinolene, alpha–Thujone, beta–thujone	Alsanosy et al. (2020), Int (2019), Mohamed Abdoul-Latif et al. (2023)

psychosis. Administering a standardized extract of *C. edulis* orally many times to rats at a dosage of 1 mg cathinone per kg body weight or giving them (–)-cathinone at a dose of 1.5 mg/kg resulted in a significant increase in locomotor sensitization and a progressive decrease in prepulse inhibition. The behavioral sensitization had a prolonged duration and continued to exist even after the treatments were stopped, similar to the sensitization generated by amphetamine (Feyissa & Kelly, 2008). Clozapine, an unconventional antipsychotic drug, successfully counteracted the behavioral sensitization and prepulse inhibition deficits caused by *C. edulis* extract or cathinone (Atlabachew et al., 2014). These findings perhaps corroborate the accounts of *C. edulis*-induced psychosis in people. A study of neurotransmitter levels revealed a substantial elevation in dopamine levels in the prefrontal cortex ($P < 0.05$). Additionally, there was a notable reduction in the concentration of serotonin in the nucleus accumbens ($P < 0.05$) and its metabolite 5-hydroxyindoleacetic acid in the prefrontal cortex ($P < 0.01$). No substantial alterations were seen in the anterior and posterior striatum areas (N. T. Wabe, 2011).

Cathinone has vasoconstrictor action in isolated perfused hearts obtained from guinea pigs. The observed impact is unlikely to be caused by an indirect mechanism involving the release of noradrenaline from sympathetic nerve terminals or by a direct interaction with alpha1-adrenoreceptors. (–)-Cathinone can enhance the contractions of the rat's right ventricle that are triggered by noradrenaline. The substance hinders the absorption of noradrenaline into ventricular slices by competitively blocking the noradrenaline transporter. The rise in blood pressure seen in humans may be attributed to the vasoconstrictor properties of cathinone. In animals, this phenomenon may be linked to the higher occurrence of myocardial infarction during *C. edulis* sessions, namely when Khat is effective and is connected with intense *C. edulis* chewing (Hassan et al., 2007; Stevenson et al., 1996). *C. edulis* extract administered orally to rabbits for 30 consecutive days increased urinary 17-hydroxycorticosteroids, serum-free fatty acids, and adrenal phosphorylase activity while decreasing adrenal cholesterol, glycogen, and ascorbic acid (Ahmed & El-Qirbi, 1993). The findings are evidence that *C. edulis* has an adrenocortical function-stimulating effect. This effect was also observed when cathinone and cathine (6.5 mg/kg) were administered orally (Engidawork, 2017; N. Wabe & Mohammed, 2012).

The administration of a *C. edulis* extract to male mice for six weeks resulted in a dose-dependent decrease in the fertility rate of female mice during the initial week following the *C. edulis* treatment (Asrade, 2013). Sperm count and motility decreased significantly in rodents treated with cathinone, while the number of abnormal sperm cells increased. In cathinone-treated animals, histopathological examination of the testes revealed interstitial tissue degeneration, cellular infiltration, and Sertoli and Leydig cell atrophy (Alles et al., 1961). Cathinone additionally induced a substantial reduction in the plasma testosterone concentrations of the rodents. While both enantiomers of cathinone exhibited detrimental effects on the male reproductive system, it was determined that (–)-cathinone was more toxic. Conversely, rabbits that were provided with *C. edulis* for three months exhibited a heightened rate of spermatogenesis and maintained healthy Leydig cells. Oral administration of crude *C. edulis* extract (equivalent to 250 g leaves and stems) to adult male olive baboons once per week for two months increased testosterone levels in the plasma and decreased prolactin and cortisol levels (Muema, 2015; Mwenda et al., 2006). The findings regarding testosterone contradict previous research conducted on rodents and humans (Githua et al., 2022). Histopathological examination of animal biopsies obtained one month following the final *C. edulis* administration revealed no discernible alterations in the testis, epididymis, liver, kidney, or pituitary gland. In contrast, tissue sections from rabbits administered 20% and 30% *C. edulis* showed progressive chronic inflammation accompanied by porto-portal fibrosis in response to cathinone treatment of the liver (N. T. Wabe, 2011). Variations in dosage and administration schedule accounted for the observed disparities. *C. edulis* administered to expectant guinea pigs inhibits placental blood flow and causes progeny to develop growth retardation.

12.12 TOXICITY OF KHAT

Several investigations worldwide have reported *C. edulis* chewing as a toxic activity on health. *C. edulis* consumption affects the circulatory, digestive, respiratory, endocrine, and genito-urinary systems. Furthermore, it impacts the neural system and has the potential to trigger paranoid psychosis and hypomanic disorder characterized by grandiose delusions. The impact on the neurological system is similar to that of amphetamine, with variations mostly in degree rather than in kind (Al-Meshal et al., 1991; N. T. Wabe, 2011). The primary harmful consequences include elevated blood pressure, rapid heart rate, sleeplessness, loss of appetite, difficulty with bowel movements, overall discomfort, irritability, severe headaches, and diminished sexual function in males. Incidents of mild depressed symptoms have been documented at the cessation of *C. edulis* use or the conclusion of a *C. edulis* session. Repeated use of large quantities might elicit psychotic responses (Al-Habori, 2005; Alsalahi et al., 2012). From a biochemical perspective, consuming *C. edulis* leaves reduced plasma cholesterol, glucose, and triglycerides in rabbits (AbdulMughni et al., 2018; Berihu et al., 2017; Nyongesa et al., 2017). Additionally, it led to an increase in plasma alkaline phosphatase and alanine aminotransferase levels, specifically in white rabbits. Histopathological symptoms of congestion of the central hepatic veins were found with acute hepatocellular injury and regeneration. Furthermore, the examination revealed the existence of kidney lesions characterized by fatty droplets in the upper cortical tubules, acute cellular edema, hyaline tubules, and acute tubular necrosis. The spleen remained unaffected, while the structure of the testes and cauda epididymis seemed normal. However, there was an observed increase in the rate of spermatogenesis (Belete et al., 2021; Engidawork, 2017). The quantity of *C. edulis* ingested by the rabbits cannot be assessed based on the provided information. The authors noted that the animals exhibited typical activity and behavior in general. The negative consequences of *C. edulis* may be categorized based on the affected system.

12.13 CONCLUSION

Khat is a plant of cultural importance in East African and Middle Eastern communities, known for its stimulating effects. *C. edulis* has extensive historical use in many cultures within these areas, often in social contexts, due to its stimulant properties that are somewhat comparable to, while less potent than, amphetamines. *C. edulis* leaf use is customary in Yemen, Ethiopia, Djibouti, and Somalia. Users often consume the leaves in a communal setting, similar to coffee or tea in different societies. The benefits include heightened vigilance, enhanced focus, and a sense of satisfaction.

Nevertheless, it also has some drawbacks, such as the possibility of developing a dependency and experiencing negative impacts on cardiovascular health. Moreover, the excessive or extended use of this substance might result in health complications such as dental ailments, gastrointestinal difficulties, and sleeplessness. Due to the growing global immigration from countries where *C. edulis* use is shared, the global prevalence of *C. edulis* usage is likewise increasing. Therefore, there is a potential for a worldwide rise in *C. edulis*-associated lesions. Tooth practitioners may need help understanding the different behaviors related to Khat use and its correlation with tooth illnesses. Therefore, it is essential to consider these behaviors when formulating differential diagnoses for oral lesions that display clinical symptoms seldom seen in typical instances, especially among persons from various ethnic and cultural backgrounds. We highly recommend including *C. edulis* and its potential adverse health effects in the curriculum of dental and medical colleges. Furthermore, it is crucial to develop and implement new governmental rules regarding the sale and importation of these goods to reduce the harmful effects of *C. edulis* on those who are addicted to this substance.

LIST OF ABBREVIATIONS

C. edulis	*Catha edulis (*Vahl) Endl.
DEA	Drug Enforcement Administration
HKD	Hong Kong dollar

REFERENCES

Abdelwahab, S. I., Alsanosy, R. M., Rahim, B. E., Mohan, S., Taha, S., Mohamed Elhassan, M., & El-Setouhy, M. (2015). Khat (*Catha edulis* Forsk.) dependence potential and pattern of use in Saudi Arabia. *BioMed Research International, 2015*. https://www.hindawi.com/journals/bmri/2015/604526/abs/

AbdulMughni, A., ElNahla, S., Hassan, S., & Ali, A. (2018). Teratogenic effects of Khat (*Catha edulis*) in New Zealand rabbit. *Journal of Advanced Veterinary and Animal Research*, *5*(1), 25. https://doi.org/10.5455/javar.2018.e242

Ahmed, M. B., & El-Qirbi, A. B. (1993). Biochemical effects of *Catha edulis*, cathine and cathinone on adrenocortical functions. *Journal of Ethnopharmacology*, *39*(3), 213–216. https://doi.org/10.1016/0378-8741(93)90039-8

Al-Habori, M. (2005). The potential adverse effects of habitual use of *Catha edulis* (khat). *Expert Opinion on Drug Safety*, *4*(6), 1145–1154. https://doi.org/10.1517/14740338.4.6.1145

Al-Hebshi, N., & Skaug, N. (2005). Khat (*Catha edulis*) – An updated review. *Addiction Biology*, *10*(4), 299–307. https://doi.org/10.1080/13556210500353020

Alles, G. A., Fairchild, M. D., Jensen, M., & Alles, A. (1961). Chemical Pharmacology of *Catha Edulis*. *Journal of Medicinal and Pharmaceutical Chemistry*, *3*(2), 323–352. https://doi.org/10.1021/jm50015a010

Al-Mehdar, A. A., El-Denshary, E. S., & Abdel-Wahhab, M. A. (2012). *Alpha Lipoic Acid and Alpha-Tocopherol Counteract the Oxidative Stress and Liver Damage in Rats Sub-Chronically Treated with Khat (Catha edulis) Extract. Global Journal of Pharmacology*, *6*(2), 94–105.

Al-Meshal, I. A. (1988). Effect of (−)-cathinone, an active principle of *Catha edulis* Forssk. (Khat) on plasma amino acid levels and other biochemical parameters in male wistar rats. *Phytotherapy Research*, *2*(2), 63–66. https://doi.org/10.1002/ptr.2650020203

Al-Meshal, I. A., Qureshi, S., Ageel, A. M., & Tariq, M. (1991). The toxicity of *Catha edulis* (khat) in mice. *Journal of Substance Abuse*, *3*(1), 107–115. https://doi.org/10.1016/S0899-3289(05)80011-2

Al-Motarreb, A., Baker, K., & Broadley, K. J. (2002). Khat: Pharmacological and medical aspects and its social use in Yemen. *Phytotherapy Research*, *16*(5), 403–413. https://doi.org/10.1002/ptr.1106

Alsalahi, A., Abdulla, M. A., Al-Mamary, M., Noordin, M. I., Abdelwahab, S. I., Alabsi, A. M., Shwter, A., & Alshawsh, M. A. (2012). Toxicological Features of *Catha edulis* (Khat) on Livers and Kidneys of Male and Female Sprague-Dawley Rats: A Subchronic Study. *Evidence-Based Complementary and Alternative Medicine*, *2012*, e829401. https://doi.org/10.1155/2012/829401

Alsanosy, R., Alhazmi, H. A., Sultana, S., Abdalla, A. N., Ibrahim, Y., Al Bratty, M., Banji, D., Khardali, I., & Khalid, A. (2020). Phytochemical screening and cytotoxic properties of ethanolic extract of young and mature khat leaves. *Journal of Chemistry*, *2020*, e7897435. https://doi.org/10.1155/2020/7897435

Al-Thobhani, M. A. H., Sathyanarayana, B. N., Simon, L., & Sondur, S. N. (2008). First comparative genotypic study on khat (*Catha edulis* Forsk). Genotypes from Yemen. *Middle Eastern and Russian Journal of Plant Science and Biotechnology*, *2*(1), 1–8.

Asrade, S. (2013). Evaluation of the reversibility and possible mechanisms of antifertility of *Catha edulis* F. (khat) extract following subacute administration in rodents. *African Journal of Pharmacy and Pharmacology*, *7*(40), 2693–2700. https://doi.org/10.5897/AJPP2013.3879

Atlabachew, M., Chandravanshi, B. S., & Redi, M. (2014). Selected Secondary Metabolites and Antioxidant Activity of Khat (*Catha edulis* Forsk) Chewing Leaves Extract. *International*

Journal of Food Properties, *17*(1), 45–64. https://doi.org/10.1080/10942912.2011.614367

Balint, E. E., Falkay, G., & Balint, G. A. (2009). Khat – a controversial plant. *Wiener Klinische Wochenschrift*, *121*(19–20), 604–614. https://doi.org/10.1007/s00508-009-1259-7

Bally, P. R. O. (1945). *Catha edulis. East African Medical Journal*, *22*(1), 2–3.

Baxter, R. L., Crombie, L., Simmonds, D. J., Whiting, D. A., Braenden, O. J., & Szendrei, K. (1979). Alkaloids of *Catha edulis* (khat). Part 1. Isolation and characterisation of eleven new alkaloids with sesquiterpene cores (cathedulins); identification of the quinone–methide root pigments. *Journal of the Chemical Society, Perkin Transactions*, *1*, 2965–2971. https://doi.org/10.1039/P19790002965

Belete, S., Asres, K., Bekuretsion, Y., Ashebir, R., Abebe, M. S., & Seyoum, G. (2021). Toxic Effect of Khat in Rat Embryos and Fetuses. *BioMed Research International*, *2021*, e9933389. https://doi.org/10.1155/2021/9933389

Benkhnigue, O., Chaachouay, N., & Zidane, L. (2023). A Floristical and Ecological Study of the Medicinal Flora Used by the Local Population of the Haouz-Rehamna Region (Middle Atlantic Morocco-4). *Acta Botanica Hungarica Akadémiai Kiadó*, *65*(3–4), 247–304. https://doi.org/10.1556/034.65.2023.3-4.3

Berihu, B. A., Asfeha, G. G., Welderufael, A. L., Debeb, Y. G., Zelelow, Y. B., & Beyene, H. A. (2017). Toxic effect of khat (*Catha edulis*) on memory: Systematic review and meta-analysis. *Journal of Neurosciences in Rural Practice*, *08*(1), 30–37. https://doi.org/10.4103/0976-3147.193524

Bonson, K. R., Dalton, T., & Chiapperino, D. (2019). Scheduling synthetic cathinone substances under the Controlled Substances Act. *Psychopharmacology*, *236*(3), 845–860. https://doi.org/10.1007/s00213-018-5129-8

Brooke, C. (1960). Khat (*Catha edulis*): Its production and trade in the Middle East. *The Geographical Journal*, *126*(1), 52–59.

Chaachouay, N., Benkhnigue, O., Fadli, M., Ibaoui, H., & Zidane, L. (2019a). Ethnobotanical and ethnopharmacological studies of medicinal and aromatic plants used in the treatment of metabolic diseases in the Moroccan Rif. *Heliyon*, *5*(10), e02191. https://doi.org/10.1080/10477845.2021.1942433

Chaachouay, N., Benkhnigue, O., Fadli, M., Ibaoui, H., El Ayadi, R., & Zidane, L. (2019b). Ethnobotanical and ethnopharmacological study of medicinal and aromatic plants used in the treatment of respiratory system disorders in the Moroccan Rif. *Ethnobotany Research and Applications*, *18*(22), 1–16.

Chaachouay, N., Douira, A., Hassikou, R., Brhadda, N., Dahmani, J., Belahbib, N., Ziri, R., & Zidane, L. (2020a). Ethnobotanical and ethnopharmacological study of medicinal and aromatic plants used in the treatment of respiratory system disorders in the Moroccan Rif. PhDThesis, Dèpartement de Biologie - Universitè Ibn Tofail - Kénitra, Maroc, https://tel.archives-ouvertes.fr/tel-03376377

Chaachouay, N., Benkhnigue, O., & Zidane, L. (2020b). Ethnobotanical study aimed at investigating the use of medicinal plants to treat nervous system diseases in the Rif of Morocco. *Journal of Chiropractic Medicine*, *19*(1), 70–81. https://doi.org/10.1016/j.jcm.2020.02.004

Chaachouay, N., & Zidane, L. (2021). The Symbolic Efficacy of Plants in Rituals and Socio-Religious Ceremonies in Morocco, Northwest of Africa. *Journal of Religious & Theological Information*, *21*(1–2), 34–53. https://doi.org/10.1080/10477845.2021.1942433

Chaachouay, N., Benkhnigue, O., Douira, A., & Zidane, L. (2021a). Poisonous medicinal plants used in the popular pharmacopoeia of the Rif, northern Morocco. *Toxicon*, 189, 24–32. https://doi.org/10.1016/j.toxicon.2020.10.028

Chaachouay, N., Douira, A., & Zidane, L. (2021b). COVID-19, prevention and treatment with herbal medicine in the herbal markets of Salé Prefecture, North-Western Morocco. *European Journal of Integrative Medicine*, 42, 101285. https://doi.org/10.1016/j.eujim.2021.101285

Chaachouay, N., Douira, A., & Zidane, L. (2021c). Herbal Medicine Used in the Treatment of Human Diseases in the Rif, Northern Morocco. *Arabian Journal for Science and Engineering*, 47, 131–153. https://doi.org/10.1007/s13369-021-05501-1

Chaachouay, N., Benkhnigue, O., & Zidane, L. (2022a). Ethnobotanical and Ethnomedicinal study of medicinal and aromatic plants used against dermatological diseases by the people of Rif, Morocco. *Journal of Herbal Medicine*, 32, 100542. https://doi.org/10.1016/j.hermed.2022.100542

Chaachouay, N., Azeroual, A., Bencharki, B., Douira, A., & Zidane, L. (2022b). Ethnoveterinary medicines plants for animal therapy in the Rif, North of Morocco. *South African Journal of Botany*, *147*, 176–191. https://doi.org/10.1016/j.sajb.2021.12.037

Chaachouay, N., Azeroual, A., Douira, A., & Zidane, L. (2022c). Ethnoveterinary practices of medicinal plants among the Zemmour and Zayane tribes, Middle Atlas, Morocco. *South African Journal of Botany*, *151*, 826–840. https://doi.org/10.1016/j.sajb.2022.11.009

Cochrane, L., & O'Regan, D. (2016). Legal harvest and illegal trade: Trends, challenges, and options in khat production in Ethiopia. *International Journal of Drug Policy*, *30*, 27–34. https://doi.org/10.1016/j.drugpo.2016.02.009

Crombie, L., Crombie, W. M. L., & Whiting, D. A. (1990). Chapter 4 Alkaloids of Khat (*Catha Edulis*). In A. Brossi (Éd.), *The Alkaloids: Chemistry and Pharmacology* (Vol. 39, p. 139–164). Academic Press. https://doi.org/10.1016/S0099-9598(08)60166-2

Douglas, H., & Pedder, M. (2010). Legal regulation of the drug khat in Australia. *J Law Med*, *18*(2), 284–301.

Engidawork, E. (2017). Pharmacological and Toxicological Effects of *Catha edulis* F. (Khat). *Phytotherapy Research*, *31*(7), 1019–1028. https://doi.org/10.1002/ptr.5832

Faro, A. F. L., Di Trana, A., La Maida, N., Tagliabracci, A., Giorgetti, R., & Busardò, F. P. (2020). Biomedical analysis of New Psychoactive Substances (NPS) of natural origin. *Journal of Pharmaceutical and Biomedical Analysis*, *179*, 112945.

Fatima, N., Rizwan, M., Hobani, Y. H., Marwan, A. E. S., Kumar, B. V., Sunosi, R. A., Abdulwahab, S. I., Areeshi, M. Y., Alvi, A., & Oriaby, M. E. (2017). Gas chromatography/Mass spectroscopy analysis of *Catha edulis* Forsk, A psycho stimulant revealed potent solvent dependent antimicrobial activity. *Journal of Pharmacognosy and Phytochemistry*, *6*(2), 197–204.

Feyissa, A. M., & Kelly, J. P. (2008). A review of the neuropharmacological properties of khat. *Progress in Neuro-Psychopharmacology and Biological Psychiatry*, *32*(5), 1147–1166. https://doi.org/10.1016/j.pnpbp.2007.12.033

Gambaro, V., Arnoldi, S., Colombo, M. L., Dell'Acqua, L., Guerrini, K., & Roda, G. (2012). Determination of the active principles of *Catha edulis*: Quali–quantitative analysis of cathinone, cathine, and phenylpropanolamine. *Forensic science international*, *217*(1–3), 87–92.

Gebissa, E. (2010). Khat in the Horn of Africa: Historical perspectives and current trends. *Journal of ethnopharmacology*, *132*(3), 607–614.

Getasetegn, M. (2016). Chemical composition of *Catha edulis* (khat): A review. *Phytochemistry Reviews*, *15*(5), 907–920. https://doi.org/10.1007/s11101-015-9435-z

Gezon, L. (2016). *Drug effects: Khat in biocultural and socioeconomic perspective* (Vol. 3). Routledge. https://books.google.com/books?hl=fr&lr=&id=0s5mDAAAQBAJ&oi=fnd&pg=PP1&dq=The+Khat+plant+has+great+importance+in+the+social+and+cultural+fabric+of+the+communities+where+it+has+been+used+for+many+years.+&ots=crRS2GqYdF&sig=4DKoDJIdxlzqEBXbNIsMce6ichI

Githua, K. K., Maitho, T. E., Nguta, J. M., & Okumu, M. O. (2022). Studies on the ethnopharmacology, antimicrobial activity, and toxicity of *Catha edulis* (Vahl.) Endl., in Sprague Dawley rats. *F1000Research*, *11*, 286.

Griffiths, P., Lopez, D., Sedefov, R., Gallegos, A., Hughes, B., Noor, A., & Royuela, L. (2010). Khat use and monitoring drug use in Europe: The current situation and issues for the future. *Journal of ethnopharmacology*, *132*(3), 578–583.

Guantai, A. N. (1982). *Catha edulis, (vahl) Forsk, (Miraa): Occurrence, active constituents and pharmacological activity*. [Thesis, University of Nairobi]. http://erepository.uonbi.ac.ke/handle/11295/56790

Gudata, Z. G. (2020). Khat culture and economic wellbeing: Comparison of a chewer and non-chewer families in Harar city. *Cogent Social Sciences*, *6*(1), 1848501. https://doi.org/10.1080/23311886.2020.1848501

Hassan, N., Gunaid, A. A., & Murray Lyon, I. M. (2007). Khat [*Catha edulis*]: Health aspects of khat chewing. *EMHJ-Eastern Mediterranean Health Journal*, *13*(3), 706–718. https://apps.who.int/iris/handle/10665/117302

Hill, B. G. (1965). Čat (*Catha edulis* Forsk). *Journal of Ethiopian studies*, *3*(2), 13–23.

Hirst, M. M. (1997). The utilization of *Catha Edulis* in the household economy of Xhosa farm inhabitants of the bolo reserve, Eastern cape. *Journal of Contemporary African Studies*, *15*(1), 119–142. https://doi.org/10.1080/02589009708729606

Hocking, G. M. (1963). *Catha edulis* ('Kath' or 'Kat'). *Quarterly Journal of Crude Drug Research*, *3*(1), 338–340. https://doi.org/10.3109/13880206309102224

Hyatt, W. S., Hirsh, C. E., Russell, L. N., Chitre, N. M., Murnane, K. S., Rice, K. C., & Fantegrossi, W. E. (2020). The synthetic cathinone 3,4-methylenedioxypyrovalerone increases impulsive action in rats. *Behavioural Pharmacology*, *31*(4), 309. https://doi.org/10.1097/FBP.0000000000000548

Int, C. (2019). *Composition of essential oil and antioxidant activity of Khat (Catha edulis Forsk), Ethiopia* [Preprint]. Arabixiv. https://doi.org/10.31221/osf.io/w7h8j

Kandari, L. S., Yadav, H. R., Thakur, A. K., & Kandari, T. (2014). Chat (*Catha edulis*): A socio economic crop in Harar Region, Eastern Ethiopia. *SpringerPlus*, *3*(1), 579. https://doi.org/10.1186/2193-1801-3-579

Karunamoorthi, K., Udeani, G., Babu, S. M., & Mossie, A. (2010). Psychopharmacosocial Aspects of *Catha edulis* Forsk (Fam. Celastraceae). *African Journal of Pharmaceutical Sciences and Pharmacy*, *1*(1), 112–129.

Khatib, M., Jarrar, Z., Bizrah, M., & Checinski, K. (2013). Khat: Social habit or cultural burden? A survey and review. *Journal of Ethnicity in Substance Abuse*, *12*(2), 140–153. https://doi.org/10.1080/15332640.2013.788908

Kimani, S. T., & Nyongesa, A. W. (2008). Effects of single daily khat (*Catha edulis*) extract on spatial learning and memory in CBA mice. *Behavioural Brain Research*, *195*(1), 192–197. https://doi.org/10.1016/j.bbr.2008.05.022

Kiros, T. (2020). Non-alkaloidal compounds from Khat (*Catha edulis*) leaves. *Biology, Medicine, & Natural Product Chemistry*, *9*(2), Article 2. https://doi.org/10.14421/biomedich.2020.92.81-89

Kiunga, J. K. (2017). *Ethnobotanical and taxonomic study of Catha Edulis plant grown in Embu and Meru counties, Kenya*. [PhD Thesis, University of Nairobi]. http://erepository.uonbi.ac.ke/bitstream/handle/11295/101262/Kiunga_Ethnobotanical%20and%20Taxonomic%20Study%20of%20Catha%20Edulis%20Plant%20Grown%20in%20Embu%20and%20Meru%20Counties,%20Kenya..pdf?sequence=1

Kiunga, J. K., Lukhoba, C. W., Dossaji, S. F., & Yenesew, A. (2016). *A survey of traditional medicinal uses of Catha edulis (Celastraceae) in Meru and Eembu counties of Kenya*. http://erepository.uonbi.ac.ke/handle/11295/107095

Klein, A. (2014). Framing the chew: Narratives of development, drugs and danger with regard to Khat (*Catha edulis*). In B. C. Labate & C. Cavnar (Éds.), *Prohibition, religious freedom, and human rights: Regulating traditional drug use* (p. 131–147). Springer Berlin Heidelberg. https://doi.org/10.1007/978-3-642-40957-8_7

Kotb El Sayed, M., & Amin, H.-K. (2015). *Catha edulis* chewing effects on treatment of paranoid schizophrenic patients. *Neuropsychiatric Disease and Treatment*, 1067. https://doi.org/10.2147/NDT.S81011

Krikorian, A. D. (1985). Growth mode and leaf arrangement in *Catha edulis* (Kat). *Economic Botany*, *39*(4), 514–521. https://doi.org/10.1007/BF02858759

Krizevski, R., Dudai, N., Bar, E., & Lewinsohn, E. (2007). Developmental patterns of phenylpropylamino alkaloids accumulation in khat (*Catha edulis*, Forsk.). *Journal of Ethnopharmacology*, *114*(3), 432–438. https://doi.org/10.1016/j.jep.2007.08.042

Lamina, S. (2010). Khat (*Catha edulis*): The herb with officio-legal, socio-cultural and economic uncertainty. *South African Journal of Science*, *106*(3), 1–4.

Mohamed Abdoul-Latif, F., Ainane, A., Houmed Aboubaker, I., Merito Ali, A., El Montassir, Z., Kciuk, M., Mohamed, J., & Ainane, T. (2023). Chemical composition of the essential oil of *Catha edulis* Forsk from Djibouti and its toxicological investigations in vivo and in vitro. *Processes*, *11*(5), Article 5. https://doi.org/10.3390/pr11051324

Muema, E. K. (2015). *Biochemical, hormonal and toxicological effects of catha edulis (khat) on pregnancy and fetal development in olive Baboons (papio anubis)* [Thesis]. http://erepository.uonbi.ac.ke/handle/11295/94691

Murthy, H. N., Joseph, K. S., Madiwal, A., Rajan, G. D., Badiger, M., Kolkar, L., Hiremath, R., & Shirugumbi, M. (2016). Chemical composition and fatty acid profile of Khat (*Catha edulis*) seed oil. *Journal of the American Oil Chemists' Society*, *93*(3), 405–409. https://doi.org/10.1007/s11746-015-2782-5

Mwenda, J. M., Arimi, M. M., Kyama, M. C., & Langat, D. K. (2003). Effects of khat (*Catha edulis*) consumption on reproductive functions: A review. *East African Medical Journal*, *80*(6), 318–323.

Mwenda, J. M., Owuor, R. A., Kyama, C. M., Wango, E. O., M'Arimi, M., & Langat, D. K. (2006). Khat (*Catha edulis*) up-regulates testosterone and decreases prolactin and cortisol levels in the baboon. *Journal of Ethnopharmacology*, *103*(3), 379–384. https://doi.org/10.1016/j.jep.2005.08.016

Ngarakana, E., Zimudzi, C., Kativu, S., & Stedje, B. (2022). *Range expansion of Catha edulis: Implications on plant communities in upland Zimbabwe*. https://www.intechopen.com/online-first/83144

Ngari, L. N., Njagi, S. N., Kamau, J. K., Lagat, R. C., Musyoki, A. M., Akoth, B. O., Kioko, F., Okwany, Z.-A., & Ngugi,

M. P. (2019). Phenotypic characterization of selected Kenyan Khat (*Catha edulis*) cultivars based on morphological traits. *African Journal of Biotechnology*, *18*(29), 865–874.

Nigusse, T., Zhang, L., Wang, R., Wang, X.-N., Li, J., & Liu, C. (2019). Flavonoids in a crude extract of *Catha edulis* inhibit rat intestinal contraction via blocking Ca^{2+} channels. *Neurogastroenterology & Motility*, *31*(7), e13602. https://doi.org/10.1111/nmo.13602

Nyongesa, A. W., Patel, N., Wango, E. O., & Onyango, D. W. (2017). High khat dose and long-term exposure impairs spermatogenesis: Experimental study using rabbit model. *Journal of Morphological Sciences*, *34*(3), 156–167. https://doi.org/10.4322/jms.111917

Orch, H., Chaachouay, N., Douiri, E. M., Faiz, N., Zidane, L., & Douira, A. (2021). Use of medicinal plants in dermato-cosmetology: An ethnobotanical study among the population of Izarène. *Jordan Journal of Pharmaceutical Sciences*, *14*(3), 323–340.

Organization, W. H. (1965). *WHO Expert Committee on Dependence-Producing Drugs [meeting held in Geneva from 19 to 24 July 1965]: Fourteenth report*. https://apps.who.int/iris/bitstream/handle/10665/39802/WHO_TRS_312.pdf?sequence=1

Organization, W. H. (2003). *WHO expert committee on drug dependence: Thirty-third report*. https://apps.who.int/iris/bitstream/handle/10665/42655/WHO_TRS_915.pdf

Pantelis, C., Hindler, C. G., & Taylor, J. C. (1989). Use and abuse of khat (*Catha edulis*): A review of the distribution, pharmacology, side effects and a description of psychosis attributed to khat chewing. *Psychological Medicine*, *19*(3), 657–668.

Patel, N. B. (2019). Khat (*Catha edulis* Forsk)–and now there are three. *Brain Research Bulletin*, *145*, 92–96.

Rizk, A. M. [ر. م. ا. عبد., Mobarak, Z. M., & El-Shihi, T. (1989). *Constituents of catha edulis (alkaloids, triterpenoids and related substances and saponins)*. http://qspace.qu.edu.qa/handle/10576/9983

Simmons, M. P., Cappa, J. J., Archer, R. H., Ford, A. J., Eichstedt, D., & Clevinger, C. C. (2008). Phylogeny of the Celastreae (Celastraceae) and the relationships of *Catha edulis* (qat) inferred from morphological characters and nuclear and plastid genes. *Molecular Phylogenetics and Evolution*, *48*(2), 745–757.

Stevenson, M., Fitzgerald, J., & Banwell, C. (1996). Chewing as a social act: Cultural displacement and *khat* consumption in the East African communities of Melbourne. *Drug and Alcohol Review*, *15*(1), 73–82. https://doi.org/10.1080/09595239600185691

Tembrock, L. R., Simmons, M. P., Richards, C. M., Reeves, P. A., Reilley, A., Curto, M. A., Meimberg, H., Ngugi, G., Demissew, S., Al Khulaidi, A. W., Al-Thobhani, M., Simpson, S., & Varisco, D. M. (2017). Phylogeography of the wild and cultivated stimulant plant qat (*Catha edulis*, Celastraceae) in areas of historical cultivation. *American Journal of Botany*, *104*(4), 538–549. https://doi.org/10.3732/ajb.1600437

Tembrock, L. R., Simmons, M. P., Richards, C. M., Reeves, P. A., Reilley, A., Curto, M. A., Meimberg, H., Ngugi, G., Demissew, S., Al-Khulaidi, A. W., Al-Thobhani, M., Simpson, S., & Varisco, D. M. (2020). Clonal diversity, cultivar traits, geographic dispersal, and the ethnotaxonomy of cultivated qat (*Catha edulis*, Celastraceae). *Economic Botany*, *74*(3), 273–291. https://doi.org/10.1007/s12231-020-09501-4

The Plant List. (2013). *The Plant List. Version 1.1*. Published on the Internet; http://www.theplantlist.org (accessed 1st January).

Thomas, S., & Williams, T. (2013). Khat (*Catha edulis*): A systematic review of evidence and literature pertaining to its harms to UK users and society. *Drug Science, Policy and Law*, *1*. https://doi.org/10.1177/2050324513498332

Wabe, N., & Mohammed, M. (2012). What science says about khat (*Catha edulis* Forsk)? Overview of chemistry, toxicology and pharmacology. *Journal of Experimental and Integrative Medicine*, *2*(1), 29. https://doi.org/10.5455/jeim.221211.rw.005

Wabe, N. T. (2011). Chemistry, pharmacology, and toxicology of khat (*catha edulis* forsk): A review. *Addiction & Health*, *3*(3–4), 137.

Wabel, N. T. (2011). Psychopharmacological aspects of *catha edulis* (khat) and consequences of long term use: A review. *Psychiatry and Behavioral Sciences*, *1*(4), 187.

Ye, S., Hu, J., Liu, Z., & Liang, M. (2021). Progress and research trends on *Catha edulis* (Vahl) Endl.(*Catha Edulis*): A review and bibliometric analysis. *Frontiers in Pharmacology*, *12*, 705376.

Zahran, M. A., Khedr, A., Dahmash, A., & El-Ameir, Y. A. (2014). Qat farms in Yemen: Ecology, dangerous impacts and future promise. *Egyptian Journal of Basic and Applied Sciences*, *1*(1), 1–8.

Zyoud, S. H. (2015). Bibliometric analysis on global *Catha edulis* (khat) research production during the period of 1952–2014. *Globalization and Health*, *11*(1), 39. https://doi.org/10.1186/s12992-015-0124-x

13 Chilean Holly (*Desfontainia spinosa* Ruiz & Pav. Columelliaceae)

Murad Muhammad, Nimrah Ameen, Gopal Dixit, Yasir Arafat, Abdullah Abdullah, and Abdul Wahab

13.1 INTRODUCTION

Desfontainia was once classified as belonging to the Linnaean clade Pentandria monogynia, but numerous authors later moved it to the *Solanaceae*. Later, it was transferred to the Genera Plantarum, *Gentianaceae*, and *Aquifoliaceae*. J. D. Hooker noted a connection to Diapensia, Bentham transferred it to the *Loganiaceae* tribe Fagreae, and Solereder (1895) positioned it at the extremity of the family. Like Fagraea, Gelsemieae, Retzieae, and Potalia, *Desfontainia* has a corolla, stamens, and aestivation. The seeds resemble those of the Potalieae, and the ovary is unilocular in the upper and quinquelocular in the lower section. The tribe *Desfontainieae* G. Don is moved from the *Gentianaceae* to the *Loganiaceae* by the author, who chooses to place the genus in a different tribe next to the Potalieae and Retzieae (Schultes & Smith, 1976). *Desfontainia spinosa* is a high Andean shrub used as a hallucinogen by Indians in both countries (Schultes, 1984).

Having opposite, simple, pinnately veined, petiolate, slightly connate, spinulose margins, coriaceous leaves with mucronate glands at the tips of teeth, *Columelliaceae* are shrubs or treelets. Flowers in terminal or axillary cymes, bisexual, actinomorphic, perianth hypogynous or semi-epigynous, sepals 5, fused, valvate or slightly imbricate in bud, persistent, stamens 2, 3, or 5, anthers basifixed, anthers basifixed, loculi-, and septicidal capsule opening with four valves, seeds numerous, minute, oblong, endosperm copious, embryo minute, straight, and without chlorophyll (Backlund, 2016).

Desfontainia spinosa is a spiny shrub indigenous to the Andes Mountains in South America. It is also known as 'chuchuhuasi' in the local Quechua language. The phytochemistry of this intriguing plant, which may also have medical applications, is not well understood. *Desfontainia* is a medicinal/hallucinogenic plant, usually used in natural dye and for ornamental purposes. Spinosa's historic usage as a narcotic and hallucinogenic by many Indigenous groups in South America may be another explanation for the absence of study on it. The plant's psychoactive qualities have led to its usage in traditional medicine for many conditions, such as rheumatism therapy, fever reduction, and pain alleviation (Houghton & Ming, 1985; Chaachouay et al., 2019).

Hallucinogens, commonly called psychedelics or psychomimetics, cause benign changes in mood, perception, and thinking. They have perceptual effects, depersonalization, distortions, illusions, and hallucinations and are typically considered nonaddictive (Akers, 2019). Common psychological effects include euphoria, heightened consciousness, and amplification of sensations. Since ancient times, hallucinogens have been used in cultural and religious rituals. According to estimates, between 0.62% and 2% of Americans take hallucinogens yearly. The U.S. Drug Enforcement Agency has categorized the majority of hallucinogens as Schedule I; however, there is increased interest in their possible therapeutic benefits (Tassadduq et al., 2022)

13.2 SCIENTIFIC CLASSIFICATION

Phylum:	Tracheophytes
Order:	Bruniales
Family:	Columelliaceae
Genus:	*Desfontainia*
Species:	*Desfontainia spinosa*

13.3 VERNACULAR NAMES

The common names of *Desfontainia spinose* are taique, Borrachero de Paramo ('intoxicator of the swamps'), Chapico ('chili water'), *Defontainia*, Intoxicator, Latuy, Michai, Muerdago, Trau-trau (Mapuche, 'unique'). *Desfontainia* is a Chilean plant famous for *Latua pubiflora*, Michay Blanco, and its glossy leaves. A white shrub that makes yellow dye is also known as 'Michay' in the Berberis genus. *Mapuche people use Desfontainia* as a poison and in psychedelics that, like Brugmansia, permanently impair brain function (Plowman et al., 1971).

13.4 BOTANICAL DESCRIPTION

Chilean holly (*Desfontainia spinosa*) is an evergreen shrub endemic to South America (Figure 13.1). It features orange, trumpet-shaped blooms and leaves that resemble common holly (Hessayon, 1994).

Desfontainia spinosa is an evergreen shrub that grows slowly to 2.5 m (8 ft) by 1.5 m (5 ft). It has glossy dark green

DOI: 10.1201/9781003460336-13

holly-like leaves and waxy crimson tubular blooms with yellow ends that reach 4 cm (1.6 in) in length. The fruit is a greenish-yellow berry around 1.5 cm (0.59 in) in diameter and contains approximately 44 gleaming, coffee-brown seeds. it is a calcifuge (needs a lime-free environment) and will survive in wetter circumstances in the natural than it is commonly given credit for in the horticultural literature, existing as it does in bogs and swamps. It is typically a terrestrial plant but may grow as an epiphyte (Ferguson, 2010).

Desfontainia spinosa is a lovely 1–6-foot-tall shrub with glossy dark green leaves that resemble Christmas holly and tubular red blooms with a yellow tip. The fruit is globose, white, or greenish-yellow, and has numerous glossy seeds. An evergreen shrub that can grow 10 feet tall or more in ideal locations, with branches coated in pale, glabrous, shiny bark. Its leaves are 1–212 in. long, 34–112 in. broad, oval, or ovate in the center, but armed at the margins with sharp triangular spines 18–13 in. long, shining dark green, glabrous; stem 13 in. long, shining dark green, and glabrous. Flowers are produced from July to late October on 13–12 in. tall stems with single flowers. Corolla funnel-shaped, red scarlet, 112 in. long, 12 in. broad at the mouth, with five rounded, yellow, shallow lobes; calyx green, with five oblong lobes 13 in. long, fringed with hairs; anthers five, with barely any stalk, connected at the base of the corolla lobes. The genus was named after the French botanist René Louiche Desfontaines. It is hardy to 5°C (23°F) and requires winter protection in locations with heavy frosts.

13.5 ORIGIN AND GEOGRAPHIC DISTRIBUTION

Desfontainia spinosa is a relatively recent species to cultivation, having been found by botanist Aimé Bonpland in the early nineteenth century and named after the French botanist Charles-François Desfontaines. In 1835, the plant was brought to cultivation for the first time in Europe, and it immediately became a popular decorative plant. *Desfontainia spinosa* is a South American evergreen shrub with holly-like leaves that grows slowly. It is only suitable for the warmer regions of the United Kingdom and will require protection from the cold winter winds. *D. spinosa* is a native of mountain slopes in America. The range starts from Costa Rica in the north and Tierra del Fuego's islands in the south, and Panama, Venezuela, Peru, Bolivia, and Ecuador (Dorling, 2008). *D. spinosa* is often growing in forests dominated by Nothofagus species present in southern beech, lenga (Nothofagus pumilio), and coihue (*Nothofagus dombeyi*) present in Argentina and Chile. It is also widespread from Columbia through southern Chile, Argentina, and Ecuador's upper Andean regions. It's also been discovered in Costa Rica. *D. spinosa* is a common garden plant in southern Chile, and it may be reproduced from seeds or cuttings. It requires damp soil since it develops in swampy places.

Native of the Andes from Colombia to the Straits of Magellan; described by Ruiz and Pavon in 1799 from plants gathered in Peru; brought by William Lobb from Chile in 1843 and again by Comber in 1925-6. This species grows in cool mountain cloud forests in Chile's northern range; further south, it descends to sea level and ranges through the coves and innumerable islands of the Pacific coast as far as the Magellan region, where the climate is exceptionally moist and equable. It is hardy to Zone 8 in the United Kingdom. It has leaves all year and flowers from August to October. The species is hermaphrodite, which possesses both male and female parts. It can grow in semi-shade (light forest). It likes damp soil. Light (sandy), medium (loamy),

FIGURE 13.1 *Desfontainia spinosa.*

and heavy (clay) soils are all appropriate. The suitable pH conditions include moderately acidic and neutral soils, which may grow in highly acidic soils (Govaerts et al. 1794).

13.6 HABITAT AND ECOLOGY

William Lobb brought *Desfontainia* to Europe in 1843. Royal Horticultural Society's Garden Merit awarded it. It will develop slowly (in 10–20 years) to around 2.5 m (8 ft 2 in) in height and width under cultivation, but in the wild, it can take the shape of a small tree and reach roughly 4 m (13 ft) in height and breadth.

It grows best in cultivation on or around Scotland's west coast and Northern Ireland, where it may reach enormous proportions. The examples are Stonefield, Argyll, Brodick, Isle of Arran, Inverewe, Wester Ross, and Rowallane, Co. Down. The final is a magnificent plant that stands 10 feet tall, 32 feet broad, and 118 feet in diameter. It grows and blossoms well in other sections of the nation, where the climate is not too dry and continental. It is the rock garden in Edinburgh, which is resilient in an exposed location. The walled garden at Nymans in Sussex, U.K., planted with Comber's seed, has two excellent examples in the south. When a blooming branch of one of these was shown at Vincent Square in 1955, it received an Award of Merit; the blossom hue differs somewhat from that of the earlier introduction, being vermilion that leans toward oriented red (Pattengale et al., 2010).

13.6.1 Epiphyte

Epiphytes, prevalent in tropical forests, account for 10% of world plant biodiversity and up to 25% of vascular plant species. They vary from holo-epiphytes, which grow on trees, and hemi-epiphytes, which spend only a portion of their lives as epiphytes until their aerial roots link to the earth (Spicer & Ortega, 2023). Vertical ecological gradients and individual host tree structures impact the distribution patterns of these plants. Studies are presently focusing on the influence of climbing plants and epiphytes and the feedback effects on ecosystem functioning. Increased liana abundance may improve forest dynamics; however, epiphyte dieback may reduce mist moisture efficiency (Spicer & Woods, 2022). The giant and extraordinarily long-lived conifer *Fitzroya cupressoides*, also known as the lahuán or alerce, served as the host tree for *D. spinosa*'s initial description as an observed growing (totally autotrophic) epiphyte in 2001. In addition to *Desfontainia*, two species, *Tepualia stipularis* and *Weinmannia trichosperma*, were also present in the communities of the epiphyte, have the most significant substrates, which were present in deep soils that were usually located 34 m up in the crowns of *Fitzroya*. A *Tepualia* specimen standing 4 m (13 ft) tall exhibited no signs of stress, demonstrating how these ordinarily terrestrial species survived in an epiphytic environment. There appeared to be two crowns on particular *Fitzroya* crowns because of the size of the epiphytic trees (Clement et al., 2001).

13.6.2 Pollinators

Like many plants with red flowers, *Desfontainia spinosa*'s pollination occurred by birds. The most southerly hummingbird species are the green-backed fire crown and sephaniodes (Martínez-Harms et al., 2010). Also involved is a species of bumble bee called *Bombus dahlbomii*. Due to their l-receptors, bee species have been reported to sense red flowers, particularly blueish-red ones, despite hardly being sensitive to wavelengths of red light at 600 nm. Although the majority of *Desfontainia* blooms are red (scarlet as opposed to deep pink), they nevertheless provide a contrast with the green to be apparent and hence pollen able when viewed by a bee with a green-sensitive part of their vision. Furthermore, bees could see the yellow flower lips at 590 nm on some *Desfontainia*. *Bombus dahlbomii*, a species with golden fur and the only endemic to southern Argentina's forests and in Chile, South America, is tragically threatened due to the spread of European *Bombus terrestris*. The loss of pollinating species threatens ecosystem services such as agricultural pollination. Many crops and most wild plants rely on animal pollination for sexual reproduction, supplying calories and micronutrients to humans. Approximately 70% of tropical crops have at least one variety for which animal pollination improves productivity (Zariman et al., 2022). European crops, on the other hand, rely on animal pollination for 84% of their yield. Honeybees continue to be the most economically valuable pollinators of agricultural monocultures worldwide. Still, their numbers are decreasing owing to pests, poor pesticide usage, aging beekeepers, and reduced market prices (Klein et al., 2007).

13.6.3 Seed dispersal

Although seed dispersal ecology has had little impact on conservation practice, it interacts with essential drivers of biodiversity loss in the twenty-first century, including habitat fragmentation, overharvesting, biological invasions, and climate change (Schiesari et al., 2019). Understanding these forces can aid in the identification of research gaps and the improvement of conservation management. Because of site-specific and species-specific variance, only a few generalizations directly apply to conservation management (van Wilgen et al., 2020).

The forest of *Fitzroya*, which were remains of Argentina and Chile, the marsupial which is *Dromiciops gliroides* with a dormouse appearance and 20 cm (7.9 in) length (including the tail), serves as the source seed dispersion for epiphytic and terrestrial populations. This little animal, which is both an insectivore and a frugivore, represents an evolutionary connection between the marsupials of South America and the fauna of Australia. The Spanish moniker is more commonly used to refer to it. The arboreal and nocturnal *Dromiciops* disperses the berry-bearing seeds and endemic plants that were found in its range in its feces, which were only, but two shrubs that cause human hallucinations: *Gaultheria insana*, and *Desfontainia spinosa*, which were known as *Pernettya furens* (*Ericaceae*) (Wang & Chen, 2008).

13.7 PHARMACOLOGY

Hallucinogens are psychoactive chemicals that can change perception, emotion, and cognition. Despite being listed by the DEA as a Schedule I narcotic, they can cause severe suffering and worsen psychosis (Castellanos et al., 2020). However, current studies show they may also provide benefits such as improved creative thinking or mental wellness. Plant-based hallucinogens have been utilized for centuries in diverse civilizations, and recent state initiatives have reevaluated their legal position (Carhart-Harris & Goodwin, 2017). Clinical research indicates significant therapeutic advantages, and the psychedelic drug industry is expected to be worth about $11 billion by 2027. On the other hand, increased recreational usage may have negative implications for the broader population (Tupper et al., 2015).

For thousands of years, hallucinogenic plants and fungi have been utilized for healing, divination, magic, and protection. The word 'hallucinogen' was used in the 1950s to describe chemicals that cause hallucinations (Metzner, 2014). These drugs have a dreamlike influence on perception, emotion, and mind, with no physical addiction and minor intellectual damage. Both the mental and physical states impact the consequences. Following a period of worry in the 1970s, hallucinogens were less popular, but a rebirth occurred in the 1990s, aided by the internet.

Serotoninergic psychedelics, often known as classical hallucinogens, are psychoactive medications that cause significant alterations in mood, perception, and cognition (Khan et al., 2021; Sajid et al., 2020). These medicines have a long history of ceremonial and religious use worldwide, emphasizing mystical experiences (Sellers et al., 2018). Until the late 1800s and early 1900s, the chemicals were primarily unknown. Still, they have since been explored for their potential therapeutic use in treating neurosis, obsessive-compulsive disorder, drug dependency, and existential anxiety and sadness (Kellner, 2022). Since the commencement of clinical research in the early 1990s, the number of human trials employing these medications has increased, with encouraging findings and no significant adverse responses documented (Petta et al., 2005).

Among the first substances that humans have used hallucinogenic plants. Natural hallucinogens may be found in cacti, spore-producing plants like mushrooms, and many other blooming plants (Hawksworth & Wiltshire, 2011). In the contemporary day, these drugs have been incorporated into innumerable songs and stories and even made passing references in artwork and children's books. They have long been connected to religion and medicine. Around half a million plant species are thought to exist in the world. More than 450 plant and mushroom species have misuse potential, around 120 of which are classified as psychoactive (Cunningham, 2008).

Shamanic healing practices and Western psychotherapy have employed plants or medicines with psychedelic properties for healing and knowledge-gaining. However, the fundamental worldviews and notions of reality are drastically different (Ali, Ketema, et al., 2021; Futterknecht, 2012). Four paradigms are examined: the traditional Western paradigm for psychedelic psychotherapy, syncretic folk religion ceremonies, healing and divination shamanic rituals, and hybrid shamanic therapeutic rituals. The shamanic and hybrid shamanic ceremonies' worldview, which holds that 'spirits' are equally as accurate as the physical organism and multiple realities can be explored in expanded states of consciousness, differs significantly from the conventional Western worldview (Metzner, 1998).

Drugs that cause hallucinations have been around for a while and are made from plants. They served as sacraments for religious rituals in prehistoric people. Recently, they have been used to encourage reality opening, perception improvement, spiritual growth, and personal development (Hensley-Dellefield, 2019). Due to reliable botanical verification, cutting-edge research methodologies, and the demand for active principles, most of our comprehensive chemical understanding of hallucinogenic plants has developed in the last ten years (Farnsworth, 1968).

It is well known that *Desfontainia spinosa* allegedly has narcotic characteristics. However, according to research, the plant contains 11 urs-12-ene triterpene compounds, including four well-known compounds and six unknown ones. Ursolic acid, mainly found in this plant's leaves and stems, has been thoroughly investigated for its anti-inflammatory and anti-cancer qualities. Ursolic acid acetate, pomolic acid, 23-hydroxytormentic acid, and nigaichigoside Fl, its ester glucoside, were the other four substances known to have been isolated from *Desfontainia spinosa*. The unique chemicals discovered were tormentic acid, 7-hydroxytormentic acid, 7-,23-dihydroxytormentic acid, and their corresponding ester glucosides (Gudoityte et al., 2021). It is significant to note that these substances are not likely to be responsible for the narcotic effects. Instead, it has been discovered that the triterpenes found in *Desfontainia spinosa* may have anti-inflammatory and anti-cancer effects. The discovery of these new chemicals expands the corpus of knowledge on the possible health advantages of this plant (Houghton & Lian, 1986b).

11-Deoxocucurbitacin I, a novel cytotoxic triterpene molecule isolated from *Desfontainia spinosa*, has been found by researchers. The lack of an 11-keto functionality made this substance chemically unique from other cucurbitacins (Li et al., 2017; Waheed et al., 2023). Field desorption mass spectrometry and a thorough study of the high field 1H and 13C NMR spectra were used to determine the molecule's structure. Numerous studies have been conducted on cytotoxic substances to examine their potential anti-cancer effects. Triterpenes, a group of chemical substances in various plants, have been recognized as a possible source of cytotoxic substances (Kellogg et al., 2016).

A new study on triterpenes as possible anti-cancer drugs, 11-deoxocucurbitacin I from *Desfontainia spinosa*, has been added to the increasing body of knowledge. In the cucurbitacin family, 11-deoxocucurbitacin-I is distinct due to the lack of an 11-keto functionality. The discovery of this substance sheds new light on the biosynthesis of cucurbitacins and may inspire the creation of fresh procedures for producing related substances (Rios et al., 2012). *Desfontainia spinosa*'s 11-deoxocucurbitacin I discovery is a significant development in natural product chemistry. This chemical is an exciting candidate for future investigation into its possible therapeutic applications due to its distinct structure and potential cytotoxic capabilities (Amonkar et al., 1985).

Secoxyloganin, dimethyl secologanoside, and sweroside, as well as loganin, loganetin, and loganic acid, were discovered to be present in *Desfontainia spinosa*. 7-O-(p-coumaroyl)-loganin, a new loganin derivative, was discovered. Desfontainic acid and desfontainoside, which were exclusively present in the stems, were also present in the leaves (Cao et al., 2019). These substances offer chemotaxonomic support for maintaining *Desfontainia* as a distinct tribe within the *Loganiaceae* family (Houghton & Lian, 1986a; Wahab et al., 2023).

Three novel derivatives of the parent chemical, 1-deoxocucurbitacin I, were found after fractionation of the active ethanol extract of *Desfontainia spinosa* based on cytotoxicity. Based on a thorough analysis of their high field 1H NMR, 13C NMR, high-resolution mass spectra (fast atom bombardment, field desorption, chemical ionization, and electron impact), and chemical interconversions, the structures of 23,24-dihydro-11-deoxocucurbitacin I, the glycosides, and the spinosides A and B were determined. In the KB cell culture experiment, spinosides A and B were cytotoxic (Reddy et al., 1988).

A plant species called *Desfontainia spinosa* has a long history of medical usage. Researchers discovered that the leaves of *Desfontainia spinosa* contain loganin and loganic acid, which are well-known iridoid glucosides, in a study aimed at determining the chemical components of this plant (Maurya et al., 2014). The isolation of loganetin, the aglucone of loganin from the plant is however what makes this discovery so intriguing. This is significant since there has never been information on loganetin occurring naturally. When loganin is hydrolyzed, which results in the loss of its glucose moiety, loganetin is created. It may have pharmacological effects, such as anti-inflammatory and anticancer qualities (Dinda et al., 2007).

Desfontainia spinosa's loganetin was shown to be a source of unique natural chemicals, which increases our knowledge of the chemical makeup of this plant. These findings may affect the development of novel medications or other uses. Additionally, they might serve as a starting point for additional research into *Desfontainia spinosa*'s and its constituents' potential medical benefits. Having opposite, simple, pinnately veined, petiolate, slightly connate, spinulose margins, coriaceous leaves with mucronate glands at the tips of teeth, *Columelliaceae* are shrubs or treelets. Flowers in terminal or axillary cymes, bisexual, actinomorphic, perianth hypogynous or semi-epigynous, sepals 5, fused, valvate or slightly imbricate in bud, persistent, stamens 2, 3, or 5, anthers basifixed, anthers basifixed, loculi- and septicidal capsule opening with four valves, seeds numerous, minute, oblong, endosperm copious, embryo minute, straight and without chlorophyll (Backlund, 2016).

13.8 ETHNOBOTANY

In 1895, American plant taxonomist John Harshberger created ethnobotany, a discipline of study that focuses on the links between humans and their environment. It rose to popularity throughout the early twentieth century, with John Schultes dubbed 'The Father of Modern Ethnobotany (Sheldrake, 2020). The discipline seeks to investigate the psychological aspects of human-plant connections and identify plant species with agricultural, industrial, or medicinal use (Muhammad et al., 2021). Ethnobotany can take several forms, such as analyzing gathered plant materials for physiologically active substances or recreating former societies' plant knowledge. These studies, however, may be exploited as a platform for exploitative colonialism and activism (McClatchey et al., 2009; Chaachouay et al., 2021a, 2021b).

The Tarahumara Indians of Mexico have utilized plants to modify their vision and establish spiritual connections. According to ethnobotanical studies, several plants used by the Tarahumara people, including several cactus and grasses, exhibit hallucinogenic effects (Wyndham, 2004). Early studies by academics described Tarahumara usage of the *peyote cactus* (*Lophophora williamsii*) for religious and therapeutic purposes. There is evidence of the psychotropic effects of several cactus species, including *Astrophytum asterias* and *Echinocactus platyacanthus* (Ventura-Aguilar et al., 2017). Additionally, recent fieldwork has discovered several fresh plants that the Tarahumara use for their hallucinogenic properties. *Coryphantha, Echinocereus, Mammillaria*, and *Scirpus* species are among them (Schultes, 2016).

Although the chemical elements that cause their psychoactive effects are still unknown, the Tarahumara's use of these plants serves as additional proof of this Indigenous people's extensive ethnobotanical knowledge and traditions. In addition to plants recognized to have hallucinogenic effects, additional plants may also cause hallucinations or be linked to plants that do (Rätsch, 2005). For instance, the Tarahumara employs the leaves of the *Mimosa tenuiflora* tree in an admixture with other hallucinogenic plants. At the same time, several Agave species are said to have psychoactive effects when consumed (Jay, 2019).

The Mapuche Indians of southern Chile use the leaves of this plant as a hallucinogenic stimulant and in folk medicine, although this use has recently come to light. Indians from the southern Andes of Colombia make the leaves into a tea, which medicine men consume when they want to dream and diagnose disease (Rojo et al., 2017). It is believed that the tea is vital because it makes the men 'crazy', which is why it is rarely utilized. Its common names include taique, chapico, micbai planco, trau-trau, and borrachero de paramo in Colombia. The peculiar genus *Desfontainia* has not yet been the subject of chemical investigations (Majeed et al., 2023; Mentioned, 1983).

Richard Evans Schultes has twice documented the use of *Desfontainia spinosa* as a hallucinogen in southern Andean Colombia. When they wish to check visions and diagnose disease, shamans in Colombia's Sibundoy Valley make tea. It is not used regularly due to its potency and partly because it cannot be grown and must be foraged in the wild in far-off páramos (Schultes & Hofmann, 1980).

The Sibundoy Valley's Camsá shamans were proficient in the use of the highly toxic hallucinogens, i.e., Iochroma and Brugmansia, and their use for analogous prediction purposes as indicated a plant that is equally dangerous, challenging to use, and producing a comparable bad experience and aftereffects. Carlos Mariani Ramirez identified that *Desfontainia spinosa* was used as a hypnotic by the people of Chile. The plant was also used as a yellow dye. It has been said that occasionally, a potently psychoactive kind of chicha has been made from the fruit of *D. spinosa*, baccate, which is supposed to be more toxic than plant foliage (Christian, 1998).

13.9 MEDICINAL EFFECTS

The holy, psychedelic, and mind-altering substances in many prehistoric societies are called Plants of the Gods. These plants are commonly utilized for medical and recreational purposes and in religious ceremonies to communicate with gods or cherished ancestors (Forte, 2012; Majeed et al., 2023). Hallucinogenic experiences, on the other hand, are not necessarily euphoric. Plants of the Gods: Their Sacred, Healing, and Hallucinogenic Powers is a thorough book that includes ethnographic, historical, and neuropharmacologic information on hallucinogenic plants and their psychological and social influence, particularly in primitive communities (Metzner, 2014; Sinha et al., 2023).

Desfontainia spinosa, has traditionally been used for its anti-inflammatory, analgesic, antispasmodic, digestive aid, sedative and anxiolytic effects, respiratory health, wound healing, antibacterial and antifungal properties, and immune system support. It has been used to relieve pain, decrease inflammation, calm muscular spasms, and enhance digestive health. More study, however, is required to prove its efficacy and potential advantages. The herb has also been used to cure respiratory problems, increase wound healing, and help the immune system (Srivastava et al., 2010).

Ingesting enough of the cytotoxic (toxic to cells) compounds called spinosides found in Chilean Holly can result in bewilderment and even death. Despite being poisonous, the plant is occasionally used by shamans or medicine men to cause altered states of perception, trigger visions, and identify illnesses (Jilek, 2005). While puffing on a leaf-filled cigar, a pot of powerful tea is made. Like Datura, another potent plant medicine, it is thought that this plant can drive one insane. The fruit of Chilean Holly is sometimes used to make chicha, an alcoholic beverage. Ayahuasca is a medical beverage containing DMT and is also used as an admixture (Ali, Muhammad et al., 2021; Schultes & Hofmann, 1980).

Medicine, Botany, and Pharmacology ancestors were born in Greece, the home of philosophical thinking and Western civilization. Greeks such as Hippocrates, Theophrastus, and Pedanius Dioscorides established the therapeutic usefulness of medicinal herbs (Chamorro-Cevallos et al., 2022). The Middle Ages, or 'Dark Ages', experienced a stall in scientific advancement, yet medicinal plant knowledge endured. Arabs, Persian chemists, and Avicena contributed to spreading knowledge. The growing population has increased interest in generating chemically synthesized chemicals, although medicinal plants' therapeutic potential remains. Chemical synthesis and biotechnology are the most common methods for producing biologically active compounds. The ceremonial use of psychoactive/hallucinogenic plant-based medicines like ayahuasca and psilocybin is becoming more popular in the United States and worldwide (Jamieson et al., 2021). Research conducted in 2016 by 15 plant medicine facilitators in the United States discovered that participants distinguish between ceremonial and recreational drug usage. Plant medicine is viewed as a natural therapeutic modality, although there are worries regarding interference with recovery and potential effects on recreational usage (Dorsen et al., 2019).

13.10 CONCLUSION

Desfontainia spinosa, famous as 'taique' or 'Taique del Sur', is a Columelliaceae family tiny evergreen shrub. It prefers well-drained soils and low temperatures and thrives in cool, temperate, and subalpine woods. The plant is endemic to South America, namely Chile and Argentina, where it may be found in the southern Andes and Patagonian areas. It is grown as an ornamental plant because of its beautiful blossoms and has been used in traditional medicine. *Desfontainia spinosa*'s conservation status may vary based on its location and habitat. Its phytochemical makeup, as well as possible pharmacological characteristics, have been investigated. *D. spinosa*, a medicinal plant, has traditionally been used for its anti-inflammatory, analgesic, antispasmodic, digestive aid, sedative and anxiolytic effects, respiratory health, wound healing, antibacterial and antifungal properties, and immune system support. It has been used to relieve pain, decrease inflammation, calm muscular spasms, and enhance digestive health. More study, however, is required to prove its efficacy and potential advantages. The herb has also been used to cure respiratory problems, increase wound healing, and help the immune system.

REFERENCES

Akers, J. (2019). Royal Perth Hospital. Dr Akers is a staff anaesthetist and pain specialist at RPH. His interests include acute and chronic pain. *Australasian Anaesthesia 2019, 82*(6), 202.

Ali, M., Ketema, S., Muhammad, M., Basit, A., Hayat, A., Khan, M. M., Khan, F., Yousaf, H., Ahmad, M., & Ahmad, M. (2021). Consumption of soybean products prevent cardiovascular disease and improve mental health: A systematic review. *NVEO-Natural Volatiles & Essential Oils Journal|NVEO, 8*(6), 3001–3013.

Ali, M., Muhammad, M., Basit, A., Ketema, S., Rasheed, A., Macadangdang Jr, R. R., Durrani, Y., Ahmad, M., Zeeshan, M., & Farooq, T. N. (2021). Exposure status of children to chemical pesticides in food: Systematic review. *NVEO-Natural Volatiles & Essential Oils Journal| NVEO, 8*(4), 14540–14553.

Amonkar, A. A., McCloud, T. O., Chang, C.-J., Saenz-Renauld, J. A., & Cassady, J. M. (1985). A cytotoxic deoxocucurbitacin from Desfontainia spinosa. *Phytochemistry, 24*(8), 1803–1805.

Backlund, A. (2016). Columelliaceae: Columelliaceae D. Don, Edinb. New Philos. J. 6: 46, 49 (1828), nom. cons., including Desfontainiaceae Endl. ex Pfeiff. (1873). *Flowering Plants. Eudicots: Aquifoliales, Boraginales, Bruniales, Dipsacales, Escalloniales, Garryales, Paracryphiales, Solanales (except Convolvulaceae), Icacinaceae, Metteniusaceae, Vahliaceae*, 141–144.

Cao, J., Yu, H., Wu, Y., & Wang, X. (2019). Occurrence and biological activities of phenylpropionyl iridoids. *Mini Reviews in Medicinal Chemistry, 19*(4), 292–309.

Carhart-Harris, R. L., & Goodwin, G. M. (2017). The therapeutic potential of psychedelic drugs: past, present, and future. *Neuropsychopharmacology, 42*(11), 2105–2113.

Castellanos, J. P., Woolley, C., Bruno, K. A., Zeidan, F., Halberstadt, A., & Furnish, T. (2020). Chronic pain and psychedelics: a review and proposed mechanism of action. *Regional Anesthesia & Pain Medicine, 45*(7), 486–494.

Chaachouay, N., Benkhnigue, O., Fadli, M., Ibaoui, H., & Zidane, L. (2019). Ethnobotanical and ethnopharmacological studies of medicinal and aromatic plants used in the treatment of metabolic diseases in the Moroccan Rif. *Heliyon, 5*(10), e02191. https://doi.org/10.1080/10477845.2021.1942433

Chaachouay, N., Douira, A., & Zidane, L. (2021a). Herbal Medicine Used in the Treatment of Human Diseases in the Rif, Northern Morocco. *Arabian Journal for Science and Engineering*, 47, 131–153. https://doi.org/10.1007/s13369-021-05501-1

Chaachouay, N., Douira, A., & Zidane, L. (2021b). COVID-19, prevention and treatment with herbal medicine in the herbal markets of Salé Prefecture, North-Western Morocco. *European Journal of Integrative Medicine*, 42, 101285. https://doi.org/10.1016/j.eujim.2021.101285

Chamorro-Cevallos, G., Mojica-Villegas, M. A., García-Martínez, Y., Pérez-Gutiérrez, S., Madrigal-Santillán, E., Vargas-Mendoza, N., Morales-González, J. A., & Cristóbal-Luna, J. M. (2022). A complete review of mexican plants with teratogenic effects. *Plants, 11*(13), 1675.

Christian, R. (1998). *The encyclopedia of psychoactive plants: Ethnopharmacology and its applications*. Rochester: Park Street Press.

Clement, J. P., Moffett, M. W., Shaw, D. C., Lara, A., Alarçon, D., & Larrain, O. L. (2001). Crown structure and biodiversity in Fltzroya cupressoides, the giant conifers of Alerce Andino National Park, Chile. *Selbyana, 22*(1), 76–88.

Cunningham, N. (2008). Hallucinogenic plants of abuse. *Emergency Medicine Australasia, 20*(2), 167–174.

Dinda, B., Debnath, S., & Harigaya, Y. (2007). Naturally occurring secoiridoids and bioactivity of naturally occurring iridoids and secoiridoids. A review, part 2. *Chemical and Pharmaceutical Bulletin, 55*(5), 689–728.

Dorling, K. (2008). RHS AZ Encyclopedia of garden plants. *United Kingdom, 1136*.

Dorsen, C., Palamar, J., & Shedlin, M. G. (2019). Ceremonial 'Plant Medicine' use and its relationship to recreational drug use: An exploratory study. *Addiction Research & Theory, 27*(2), 68–75.

Farnsworth, N. R. (1968). Hallucinogenic plants: Various chemical substances are known to be the active hallucinogenic principles in many plants. *Science, 162*(3858), 1086–1092.

Ferguson, N. (2010). *Right plant, right place: Over 1400 plants for every situation in the garden*. Simon and Schuster.

Forte, R. (2012). *Entheogens and the Future of Religion*. Simon and Schuster.

Futterknecht, V. (2012). *A quest for transpersonal ways of knowing in anthropology of religion and consciousness* [Ph. D. dissertation, Universität Wein].

Govaerts, R. Desfontainia Ruiz & Pav. (1794). Type species: Desfontainia spinosa Ruiz & Pav. Synonyms Homotypic Linkia Pers., Syn. Pl. 1: 219 (1805), nom. illeg. hom. vide Linkia Cav., Icon.[Cavanilles] iv. 61. t. 389.(1797)= Persoonia Sm.(Proteaceae).

Gudoityte, E., Arandarcikaite, O., Mazeikiene, I., Bendokas, V., & Liobikas, J. (2021). Ursolic and oleanolic acids: Plant metabolites with neuroprotective potential. *International Journal of Molecular Sciences, 22*(9), 4599.

Hawksworth, D. L., & Wiltshire, P. E. (2011). Forensic mycology: The use of fungi in criminal investigations. *Forensic Science International, 206*(1-3), 1–11.

Hensley-Dellefield, C. E. (2019). *Decolonizing Chinigchinich: Religious landscapes of Southern California*. San Diego State University.

Hessayon, D. G. (1994). *The flowering shrub expert*. Sterling Publishing Company, Inc.

Houghton, P. J., & Lian, L. M. (1986a). Iridoids, iridoid-triterpenoid congeners and lignans from Desfontainia spinosa. *Phytochemistry, 25*(8), 1907–1912.

Houghton, P. J., & Lian, L. M. (1986b). Triterpenoids from Desfontainia spinosa. *Phytochemistry, 25*(8), 1939–1944.

Houghton, P. J., & Ming, L. L. (1985). Iridoids from desfostainia spinosa. *Phytochemistry, 24*(8), 1841–1842.

Jamieson, C. S., Misa, J., Tang, Y., & Billingsley, J. M. (2021). Biosynthesis and synthetic biology of psychoactive natural products. *Chemical Society Reviews, 50*(12), 6950–7008.

Jay, M. (2019). *Mescaline: a global history of the first psychedelic*. Yale University Press.

Jilek, W. G. (2005). Transforming the shaman: Changing Western views of shamanism and altered states of consciousness. *Investigación en Salud, 7*(1), 8–15.

Kellner, M. (2022). Drug treatment of obsessive-compulsive disorder. *Dialogues in Clinica Neuroscience, 12*(2), 187–197.

Kellogg, J. J., Todd, D. A., Egan, J. M., Raja, H. A., Oberlies, N. H., Kvalheim, O. M., & Cech, N. B. (2016). Biochemometrics for natural products research: comparison of data analysis approaches and application to identification of bioactive compounds. *Journal of Natural Products, 79*(2), 376–386.

Khan, F., Ullah, W., Muhammad, M., Ullah, I., Basit, A., Ahmad, M., Uddin, B., Ullah, A., Zaib, A., & Ullah, I. (2021).

Surveillance of insect pollinators on cucumbers and their effect on its yield. *NVEO-Natural Volatiles & Essential Oils Journal|NVEO*, *8*(6), 5753–5762.

Klein, A.-M., Vaissière, B. E., Cane, J. H., Steffan-Dewenter, I., Cunningham, S. A., Kremen, C., & Tscharntke, T. (2007). Importance of pollinators in changing landscapes for world crops. *Proceedings of the Royal Society B: Biological Sciences*, *274*(1608), 303–313.

Rios, L. J., Andújar, I., Escandell, J. M., Giner, R. M., & Recio, M. C. (2012). Cucurbitacins as inducers of cell death and a rich source of potential anticancer compounds. *Current Pharmaceutical Design*, *18*(12), 1663–1676.

Li, P., Zhu, N., Hu, M., Wu, H., Yu, T., Wu, T., Zhang, D., Sun, Z., Yang, J., & Ma, G. (2017). New cucurbitane triterpenoids with cytotoxic activities from Hemsleya penxianensis. *Fitoterapia*, *120*, 158–163.

Majeed, M., Muhammad, M., Nawaz, S., Naz, T., Iqbal, M. M., Zahid, N., Hussain, M., Nawaz, A. N., Abbas, G., & Gulshan, A. B. (2023). Nanopriming for crop management for sustainable agriculture: An overview. *In book: Nanopriming Approach to Sustainable Agriculture edited by Abhishek Singh, et al., IGI Global*, 110–141. 10.4018/978-1-6684-7232-3.ch005

Martínez-Harms, J., Palacios, A., Márquez, N., Estay, P., Arroyo, M. T., & Mpodozis, J. (2010). Can red flowers be conspicuous to bees? Bombus dahlbomii and South American temperate forest flowers as a case in point. *Journal of Experimental Biology*, *213*(4), 564–571.

Maurya, A., Dwivedi, G. R., Darokar, M. P., & Srivastava, S. K. (2014). Preparative isolation of bioenhancer loganetin from sf Alstonia scholaris by fast centrifugal partition chromatography. *Separation Science and Technology*, *49*(5), 773–777.

McClatchey, W. C., Mahady, G. B., Bennett, B. C., Shiels, L., & Savo, V. (2009). Ethnobotany as a pharmacological research tool and recent developments in CNS-active natural products from ethnobotanical sources. *Pharmacology & Therapeutics*, *123*(2), 239–254.

Mentioned, P. P. (1983). Peruvian and Chilean Psychoactive Plants Mentioned in Ruiz's Relacibn (1777–1788). *Journal of Psychoactive Drugs*, *15*, 4.

Metzner, R. (1998). Hallucinogenic drugs and plants in psychotherapy and shamanism. *Journal of Psychoactive Drugs*, *30*(4), 333–341.

Metzner, R. (2014). *The ayahuasca experience: A sourcebook on the sacred vine of spirits*. Simon and Schuster.

Muhammad, M., Badshah, L., Shah, A. A., Shah, M. A., Abdullah, A., Bussmann, R. W., & Basit, A. (2021). Ethnobotanical profile of some useful plants and fungi of district Dir Upper, Tehsil Darora, Khyber Pakhtunkhwa, Pakistan. *Ethnobotany Research and Applications*, *21*, 1–15.

Pattengale, N. D., Alipour, M., Bininda-Emonds, O. R., Moret, B. M., & Stamatakis, A. (2010). How many bootstrap replicates are necessary? *Journal of Computational Biology*, *17*(3), 337–354.

Petta, C. A., Ferriani, R. A., Abrao, M. S., Hassan, D., Rosa e Silva, J. C., Podgaec, S., & Bahamondes, L. (2005). Randomized clinical trial of a levonorgestrel-releasing intrauterine system and a depot GnRH analogue for the treatment of chronic pelvic pain in women with endometriosis. *Human Reproduction*, *20*(7), 1993–1998.

Plowman, T., Gyllenhaal, L. O., & Lindgren, J. E. (1971). Latua pubiflora magic plant from Southern Chile. *Botanical Museum Leaflets, Harvard University*, *23*(2), 61–92.

Rätsch, C. (2005). *The encyclopedia of psychoactive plants: ethnopharmacology and its applications*. Simon and Schuster.

Reddy, K. S., Amonkar, A. J., McCloud, T. G., Chang, C.-J., & Cassady, J. M. (1988). Spinosides A and B. Two cytotoxic cucurbitacin glycosides from Desfontainia spinosa. *Phytochemistry*, *27*(12), 3781–3785.

Rojo, L., Cubillos-Roble, K., Troncoso-Fonseca, A., & Delatorre-Herrera, J. (2017). J. Kellogg, BM Schmidt, BL Graf, C. Jofré-Jimenez, H. Aguayo-. Kellogg, BM Schmidt, BL Graf, C. Jofré-Jimenez, H. Aguayo-Cumplido, C. Calfío. Ethnobotany: A Phytochemical Perspective, 183.

Sajid, M., Ahsanullah, B. A., Ullah, I., Shah, S., Haleema, B., Nawab, K., Ahmad, I., Ahmad, N., Khan, A., & Muhammad, M. (2020). Screening of citrus rootstocks for salinity tolerance based on initial growth attributes and leaves proline content. *Element. Educ. Online*, *19*, 3128–3140.

Schiesari, L., Matias, M. G., Prado, P. I., Leibold, M. A., Albert, C. H., Howeth, J. G., Leroux, S. J., Pardini, R., Siqueira, T., & Brancalion, P. H. (2019). Towards an applied metaecology. *Perspectives in Ecology and Conservation*, *17*(4), 172–181.

Schultes, R. E. (1984). Psychoactive plants in need of chemical and pharmacological study. *Proceedings: Plant Sciences*, *93*, 281–304.

Schultes, R. E. (2016). phytochemical gaps in Our Knowledge of Hallucinogens. *Progress in Phytochemistry*, *7(7)*, 301.

Schultes, R. E., & Hofmann, A. (1980). The botany and chemistry of hallucinogens. *(No Title)*.

Schultes, R. E., & Smith, E. W. (1976). *Hallucinogenic plants* (Vol. 35). Golden Press.

Sellers, E. M., Romach, M. K., & Leiderman, D. B. (2018). Studies with psychedelic drugs in human volunteers. *Neuropharmacology*, *142*, 116–134.

Sheldrake, M. (2020). The 'enigma'of Richard Schultes, Amazonian hallucinogenic plants, and the limits of ethnobotany. *Social Studies of Science*, *50*(3), 345–376.

Sinha, D., Odoh, U. E., Ganguly, S., Muhammad, M., Chatterjee, M., Chikeokwu, I., & Egbuna, C. (2023). Phytochemistry, history, and progress in drug discovery. In: Chukwuebuka Egbuna, Mithun Rudrapal, Habibu Tijjani (eds) *Phytochemistry, Computational Tools and Databases in Drug Discovery* (pp. 1–26). Elsevier.

Spicer, M. E., & Ortega, J. (2023). Source height and contact with terrestrial soil drive transplanted epiphyte performance. *Journal of Ecology*. *11*(11), 2388–2400.

Spicer, M. E., & Woods, C. L. (2022). A case for studying biotic interactions in epiphyte ecology and evolution. *Perspectives in Plant Ecology, Evolution and Systematics*, *54*, 125658.

Srivastava, J. K., Shankar, E., & Gupta, S. (2010). Chamomile: A herbal medicine of the past with a bright future. *Molecular Medicine Reports*, *3*(6), 895–901.

Tassadduq, S. S., Akhtar, S., Waheed, M., Bangash, N., Nayab, D.-E., Majeed, M., Abbasi, S., Muhammad, M., Alataway, A., & Dewidar, A. Z. (2022). Ecological distribution patterns of wild grasses and abiotic factors. *Sustainability*, *14*(18), 11117.

Tupper, K. W., Wood, E., Yensen, R., & Johnson, M. W. (2015). Psychedelic medicine: A re-emerging therapeutic paradigm. *CMAJ*, *187*(14), 1054–1059.

van Wilgen, N. J., van Wilgen, B. W., & Midgley, G. F. (2020). Biological invasions as a component of South Africa's global change research effort. In: van Wilgen, B., Measey, J., Richardson, D., Wilson, J., Zengeya, T. (eds) *Biological*

Invasions in South Africa. Invading Nature - Springer Series in Invasion Ecology, vol 14. Cham: Springer. 855. https://doi.org/10.1007/978-3-030-32394-3_29

Ventura-Aguilar, R. I., Bosquez-Molina, E., Bautista-Baños, S., & Rivera-Cabrera, F. (2017). Cactus stem (*Opuntia ficus-indica* Mill): anatomy, physiology and chemical composition with emphasis on its biofunctional properties. *Journal of the Science of Food and Agriculture*, *97*(15), 5065–5073.

Wahab, A., Muhammad, M., Munir, A., Abdi, G., Zaman, W., Ayaz, A., Khizar, C., & Reddy, S. P. P. (2023). Role of Arbuscular Mycorrhizal Fungi in Regulating Growth, Enhancing Productivity, and Potentially Influencing Ecosystems under Abiotic and Biotic Stresses. *Plants*, *12*(17), 3102.

Waheed, A., Li, C., Muhammad, M., Ahmad, M., Khan, K. A., Ghramh, H. A., Wang, Z., & Zhang, D. (2023). Sustainable potato growth under straw mulching practices. *Sustainability*, *15*(13), 10442.

Wang, B., & Chen, J. (2008). Tannin concentration enhances seed caching by scatter-hoarding rodents: an experiment using artificial 'seeds'. *Acta Oecologica*, *34*(3), 379–385.

Wyndham, F. S. (2004). *Learning ecology: ethnobotany in the Sierra Tarahumara, Mexico*. University of Georgia.

Zariman, N. A., Omar, N. A., & Huda, A. N. (2022). Plant attractants and rewards for pollinators: Their significant to successful crop pollination. *International Journal of Life Sciences and Biotechnology*, *5*(2), 270–293.

14 Christmas Vine (*Ipomoea corymbosa* (L.) Roth Convolvulaceae)

Eszter Csikós, Viktória L. Balázs, Ágnes Farkas, and Lilla Nagy-Radványi

14.1 INTRODUCTION

Ipomoea corymbosa (along with *I. violacea* and *I. tricolor*) was used in Central America for shamanic and ceremonial purposes and the Mazatec of Mexico also continued their usage (Schultes and Hofmann, 1980; Schultes and Hofmann, 1992; Carod-Artal, 2015). They are widely spread ornamental plants, investigated as potential pesticides (Kaur et al., 2020; Chen et al., 2021), and though commercial seeds are usually coated with emetic agents to discourage intoxication (Halpern, 2004), they are also naturally occurring sources of 'new psychoactive substances' according to its psychoactive ergot alkaloids (Lo Faro et al., 2020).

In Convolvulaceae species, the alkaloids always co-occur with symbiotic *Periglandula* fungus species (de Cárdenas, 1988; Beaulieu et al., 2015; Steiner and Leistner, 2018).

The Clavicipitaceae family of fungi are known to produce ergot alkaloids in several of its members. The recently discovered *Periglandula* genus of this family was found in the adaxial leaf surface of particular plants belonging to the Convolvulaceae family, associated with the secretory glands (Steiner et al., 2008, Steiner et al., 2011). These peltate glandular trichomes do not only secrete sesquiterpenes and palmitic acid but also seem to absorb ergot alkaloids from the *Periglandula* species (Steiner et al., 2015). The Convolvulaceae species proved to be hosts of *Periglandula* fungi are *Ipomoea asarifolia* (Markert et al., 2008), *Ipomoea corymbosa* (Steiner et al., 2011), *Argyreia nervosa*, *I. amnicola*, *I. argillicola*, *I. gracilis*, *I. hildebrandtii*, *I. leptophylla*, *I. muelleri*, *I. pes-caprae* (Beaulieu et al., 2015), and *Ipomoea violacea* (Steiner and Leistner, 2018).

14.2 SCIENTIFIC CLASSIFICATION

Ipomoea corymbosa (L.) Roth belongs to the Convolvulaceae (bindweeds or morning glories) family in the Solanales order, and is one of the *Ipomoea* genus, which consists of approximately 600 species (W.F.O.). The plant was previously classified in the now-dissolved plant genus *Turbina* under the name of *Turbina corymbosa* (L.) Raf. (WFO, 2024).

14.3 VERNACULAR NAMES

Ipomoea corymbosa is known as Christmas vine, Christmas pops, Christmas wreath or snakeplant; also as coatl xoxouhquij (meaning: green serpent) and coaxihuitl (meaning: snakeplant) in Aztec, as xtabentún (meaning: vines growing on stone) in Mayan, or as bejuco, corona de novia, flor de la virgin, hiedra, quiebracajete blanco or quiebraplatos in Spanish (Carod-Artal 2015).

The seeds are also referred to as ololiúqui (ololiuhqui/ ololiúqui; meaning: round thing) in Aztec, or as aguinaldo blanco, badoh or semilla de la Virgen in Spanish (Carod-Artal, 2015).

14.4 BOTANICAL DESCRIPTION

The stems are lianas, herbaceous at the tips, woody at the base, twining or glabrous. The leaves have slender petioles and 4–10 cm long, glabrous or rarely pubescent, cordate-ovate blades (Brazil Flora, 2014). The white flowers grow in axillary or terminal thyrses, have 8–12 mm long, oblong, glabrous, subequal sepals, and have trumpet-shaped, campanulate corollas with a 2.5–3 cm long, red, purple, or brown tube, and greenish radiating stripes (Woodson Jr. and Schery, 1980). The fruit is 1–1.5 cm long, ovoid-oblong indehiscent, the capsules contain one, rarely two brown, pubescent seeds (Brazilian Flora 2020 project, Flora Panama).

14.5 ORIGIN AND GEOGRAPHIC DISTRIBUTION

Christmas vine is native to Central and South America (from Mexico to Panama, West Indies, southern Florida, and South America) but it was also introduced as an ornamental plant to parts of North America, Australia, the Philippines, and Europe (Flora de Panama, 2024).

14.6 HABITAT AND ECOLOGY

It is a terrestrial, perennial, climbing vine, or liana, native to rainforests, common in disturbed, bushy, moister places near settlements or cultivations, usually below 1100 m (Brazilian Flora 2020 project, Flora Panama, Wood et al., 2020).

DOI: 10.1201/9781003460336-14

14.7 ETHNOBOTANY

The dried and ground seeds of *I. corymbosa* and *I. violacea* (badoh negro) have been consumed with water or alcoholic beverages for religious, ceremonial, and healing purposes since the Mayas and Aztecs. They were commonly used by Mixtec and Zapotec peoples in the now Mexican state of Oaxaca, had religious connotations in the 16th century, and they are used to this day by the local Mazatec healers (Schultes and Hofmann, 1980; Schultes and Hofmann, 1992; Carod-Artal, 2015, Chaachouay et al., 2019). It is mostly taken alone with the healer to induce sensory/perception alterations (visual illusions, synesthesia, euphoria, memory changes, and discrete somnolence), and trance states. The 'psychic void' of the first phase can be associated with vasovagal response and vertigo. The second phase occurs some hours later with intense serenity and sedation (Carod-Artal, 2015). The fully hallucinogenic dose was reported as 100–150 ground seeds dissolved in cold water (Hernández, 1959; Carod-Artal, 2015). The seeds were also used to treat gout (Hernandez et al., 1651; Fray Bernardino de Sahagún, 1985, Carod-Artal 2015, Steiner and Leistner, 2018; Chaachouay et al., 2021a, 2021b), syphilis, bone distortion and fracture, flatulence, and eye diseases, to alleviate pain, to control unnatural surge, to stimulate sexual interaction, and to increase body strength (Hernandez et al., 1651, Steiner and Leistner, 2018).

14.8 PHARMACOLOGY

Hofmann first isolated lysergic acid alkaloids from *I. corymbosa* seeds (Hofmann & Tscherter, 1960; Hofmann, 1961). Since then it has been reported to contain about 0.02–0.06% lysergic acid alkaloids (Dermarde and Youngken, 1966; Graziano et al., 2017), including chanoclavine, lysergic acid, lysergol, elymoclavine, agroclavine, lysergic acid amide (ergine, LSA), lysergic acid α-hydroxy-ethylamide, and ergonovine (ergometrine) alkaloids (Hofmann, 1961; Steiner et al., 2006; Ahimsa-Müller et al., 2007; Eich, 2008; Steiner et al., 2015; Steiner and Leistner, 2018).

These alkaloids have a variable affinity of the methylergoline core to the central nervous system receptors. The possible physiological effects occur by the cumulative (partial) agonism and/or antagonism at dopaminergic, adrenergic, and serotonergic receptors (Tasker and Wipf, 2021).

LSA has a similar structure to lysergic acid diethylamide (LSD) but has less potent psychedelic properties (Paulke et al., 2015). The hallucinogenic dose of LSA is 2–5 mg causing 4–8 hours of intoxication (Schultes and Hofmann, 1980).

14.9 MEDICINAL EFFECTS

Ergot alkaloids can cause gangrenous and convulsive ergotism (mental effects include hallucinations, mania, psychosis, headaches, nausea, and vomiting); but different naturally occurring or unnatural ergot alkaloids can be used as vasoconstrictor, treating acute migraines and cluster headaches, pituitary tumors, Parkinson's disease, age-related neurodegenerative disorders, such as dementia and Alzheimer's disease, vascular diseases, balance disorders, type 2 diabetes, hyperprolactinemia, and postpartum hemorrhaging (PPH) (Tasker and Wipf, 2021), which suggests that *I. corymbosa* can have medicinal purposes in the future but so far there is insufficient evidence for this plant.

14.10 CONCLUSION

I. corymbosa is a popular, widely planted ornamental, historically used as a common entheogen, today as a traditional remedy and a hallucinogen due to the seeds' ergot alkaloid content. However, more scientific evidence is needed with respect to the analytical possibilities, pharmacological effects, side effects, and possible medicinal properties of its components.

LIST OF ABBREVIATIONS

LSA d-lysergic acid amide, ergine
LSD lysergic acid diethylamide

REFERENCES

Ahimsa-Müller, M.A., Markert, A., Hellwig, S., Knoop, V., Steiner, U., Drewke, C., & Leistner, E. (2007). Clavicipitaceous fungi associated with ergoline alkaloid-containing convolvulaceae. *Journal of natural products*, 70(12), 1955–1960. https://doi.org/10.1021/np070315t

Beaulieu, W.T., Panaccione, D.G., Ryan, K.L., Kaonongbua, W., & Clay, K. (2015). Phylogenetic and chemotypic diversity of Periglandula species in eight new morning glory hosts (Convolvulaceae). *Mycologia*, 107(4), 667–678. https://doi.org/10.3852/14-239

Brazilian Flora 2020 project – Projeto Flora do Brasil 2020. http://floradobrasil.jbrj.gov.br

Brazil Flora, G. (2014). Brazilian Flora 2020 project - Projeto Flora do Brasil 2020. v393.147. Instituto de Pesquisas Jardim Botanico do Rio de Janeiro. Dataset/Checklist. https://doi.org/10.15468/1mtkaw

Carod-Artal F.J. (2015). Hallucinogenic drugs in pre-Columbian Mesoamerican cultures. *Neurologia (Barcelona, Spain)*, 30(1), 42–49. https://doi.org/10.1016/j.nrl.2011.07.003

Chaachouay, N., Benkhnigue, O., Fadli, M., Ibaoui, H., & Zidane, L. (2019). Ethnobotanical and ethnopharmacological studies of medicinal and aromatic plants used in the treatment of metabolic diseases in the Moroccan Rif. *Heliyon*, *5*(10), e02191. https://doi.org/10.1080/10477845.2021.1942433

Chaachouay, N., Douira, A., & Zidane, L. (2021a). Herbal Medicine Used in the Treatment of Human Diseases in the Rif, Northern Morocco. *Arabian Journal for Science and Engineering*, 47, 131–153. https://doi.org/10.1007/s13369-021-05501-1

Chaachouay, N., Douira, A., & Zidane, L. (2021b). COVID-19, prevention and treatment with herbal medicine in the herbal markets of Salé Prefecture, North-Western Morocco. *European Journal of Integrative Medicine*, 42, 101285. https://doi.org/10.1016/j.eujim.2021.101285

Chen, X.D., Kaur, N., Horton, D.R., Cooper, W.R., Qureshi, J.A., & Stelinski, L. L. (2021). Crude extracts and alkaloids derived from ipomoea-periglandula symbiotic association cause mortality of Asian Citrus Psyllid Diaphorina citri Kuwayama (Hemiptera: Psyllidae). *Insects*, 12(10), 929. https://doi.org/10.3390/insects12100929

Dermarde, A., & Youngken, H.W. (1966). The distribution of indole alkaloids among certain species and varieties of Ipomoea, Rivea and Convolvolus (Convolvulaceae). *Llodia*, 29:35–42.

Eich, E. (2008) *Solanaceae and Convolvulaceae: secondary Metabolites*. Berlin, Heidelberg: Springer

Flora de Panama http://www.tropicos.org/Project/PAC; Tropicos. org. Missouri Botanical Garden. 07 Dec 2024 https://tropicos.org

Woodson, Robert E. Jr. and Schery, Robert W. (1980). Annals of the Missouri Botanical Garden Missouri Botanical Garden Press, 67(4), 819–821.

Fray Bernardino de Sahagún. (1985) Bernardino de Sahagún. Historia general de las cosas de Nueva Espa¿na. Porrúa

Graziano, S., Orsolini, L., Rotolo, M.C., Tittarelli, R., Schifano, F., & Pichini, S. (2017). Herbal highs: Review on psychoactive effects and neuropharmacology. *Current Neuropharmacology*, 15(5), 750–761. https://doi.org/10.2174/1570159X14666161031144427

Halpern J.H. (2004). Hallucinogens and dissociative agents naturally growing in the United States. *Pharmacology & Therapeutics*, 102(2), 131–138. https://doi.org/10.1016/j.pharmthera.2004.03.003

Hernandez, F., Recchi, N.A., Lynceus, J.T. (1651). *Rerum Medicarum Novae Hispaniae Thesaurus*. 3rd ed. Typis Vitalis Mascardi, Rome. 145–146.

Hernández, F. (1959) Historia natural de Nueva Espa¿na. Obras completas. UNAM.

Hofmann A. (1961) Die Wirkstoffe der Mexikanischen Zauberdroge "Ololiuqui". *Planta Med*, 9, 354–367.

Hofmann, A., & Tscherter, H. (1960). Isolierung von Lysergsäure-Alkaloiden aus der mexikanischen Zauberdroge Ololiuqui (Rivea corymbosa (L.) Hall. f.). *Experientia*, 16, 414. https://doi.org/10.1007/BF02178840

de Cárdenas, J. (1988). Pr*oblemas y secretos maravillosos de las Indias*. Madrid: Alianza Editorial.

Kaur, N., Cooper, W.R., Duringer, J.M., Rashed, A., Badillo-Vargas, I.E., Esparza-Díaz, G., & Horton, D.R. (2020). Mortality of potato psyllid (Hemiptera: Triozidae) on host clippings inoculated with ergot alkaloids. *Journal of Economic Entomology*, 113(5), 2079–2085. https://doi.org/10.1093/jee/toaa144

Lo Faro, A.F., Di Trana, A., La Maida, N., Tagliabracci, A., Giorgetti, R., & Busardò, F.P. (2020). Biomedical analysis of New Psychoactive Substances (NPS) of natural origin. *Journal of Pharmaceutical and Biomedical Analysis*, 179, 112945. https://doi.org/10.1016/j.jpba.2019.112945

Markert, A., Steffan, N., Ploss, K., Hellwig, S., Steiner, U., Drewke, C., Li, S.M., Boland, W., & Leistner, E. (2008). Biosynthesis and accumulation of ergoline alkaloids in a mutualistic association between *Ipomoea asarifolia* (Convolvulaceae) and a clavicipitalean fungus. *Plant Physiology*, 147(1), 296–305. https://doi.org/10.1104/pp.108.116699

Paulke, A., Kremer, C., Wunder, C., Wurglics, M., Schubert-Zsilavecz, M., & Toennes, S.W. (2015). Studies on the alkaloid composition of the Hawaiian Baby Woodrose Argyreia nervosa, a common legal high. *Forensic Science International*, 249, 281–293. https://doi.org/10.1016/j.forsciint.2015.02.011

Schultes, R.E., & Hofmann, A. (1980). *The Botany and Chemistry of Hallucinogens*. Springfield, IL: Charles C. Thomas.

Schultes, R.E., & Hofmann A (1992). *Plants of the Gods: Their Sacred, Healing and Hallucinogenic Powers*. Rochester, VT: Healing Arts Press.

Steiner, U., & Leistner, E. (2018). Ergot alkaloids and their hallucinogenic potential in morning glories. *Planta Medica*, 84(11), 751–758. https://doi.org/10.1055/a-0577-8049

Steiner, U., Ahimsa-Müller, M.A., Markert, A., Kucht, S., Gross, J., Kauf, N., Kuzma, M., Zych, M., Lamshöft, M., Furmanowa, M., Knoop, V., Drewke, C., & Leistner, E. (2006). Molecular characterization of a seed transmitted clavicipitaceous fungus occurring on dicotyledoneous plants (Convolvulaceae). *Planta*, 224(3), 533–544. https://doi.org/10.1007/s00425-006-0241-0

Steiner, U., Hellwig, S., & Leistner, E. (2008). Specificity in the interaction between an epibiotic clavicipitalean fungus and its convolvulaceous host in a fungus/plant symbiotum. *Plant Signaling & Behavior*, 3(9), 704–706. https://doi.org/10.4161/psb.3.9.6432

Steiner, U., Kucht, S.H., Ahimsa-Müller, M.A., Grundmann, N., Li, S.M., Drewke, C., & Leistner, E. (2015). The key role of peltate glandular trichomes in symbiota comprising clavicipitaceous fungi of the genus periglandula and their host plants. *Toxins*, 7(4), 1355–1373. https://doi.org/10.3390/toxins7041355

Steiner, U., Leibner, S., Schardl, C.L., Leuchtmann, A., & Leistner, E. (2011). Periglandula, a new fungal genus within the Clavicipitaceae and its association with Convolvulaceae. *Mycologia*, 103(5), 1133–1145. https://doi.org/10.3852/11-031

Tasker, N.R., & Wipf, P. (2021). Biosynthesis, total synthesis, and biological profiles of Ergot alkaloids. The Alkaloids. *Chemistry and Biology*, 85, 1–112. https://doi.org/10.1016/bs.alkal.2020.08.001

W.F.O. (The World Flora Online). (2024). https://www.worldfloraonline.org/taxon/wfo-0001297447 (2023 September 11)

Wood, J.R.I., Muñoz-Rodríguez, P., Williams, B.R.M., & Scotland, R.W. (2020). A foundation monograph of Ipomoea (Convolvulaceae) in the New World. *PhytoKeys*, 143, 1–823. https://doi.org/10.3897/phytokeys.143.32821

15 Beach Moonflower (*Ipomoea violacea* L. Convolvulaceae)

Anywar Godwin, Ojelel Samuel, Tugume Patience, Ssenku Jamilu, and Kakudidi Esezah

15.1 INTRODUCTION

Hallucinogens are psychoactive substances linked to ancestral cultures, for magical, therapeutic, and religious practices (Burillo-Putze et al., 2013; Carod-Artal, 2015) and are still used in various forms of traditional medicine practices globally for divination. They are a group of drugs that induce trance states, alter a person's perception of reality, thoughts, and feelings. Hallucinogens can be synthesized while some of them come from compounds of natural origin that occur in plants and fungi (Fatur, 2021; Chaachouay & Zidane, 2021).

The family Convolvulaceae is one of the known sources of hallucinogens. It has a global occurrence, and is dominated by twining or climbing woody or herbaceous plants, with approximately 2000 species of 58 genera, of which *Ipomoea* (ca. 600 world species) is the largest genus in this family. Some members are wild while others are weeds or ornamental plants, medicines, foods such as *Ipomoea batatas* (sweet potato) (Austin, 2007; Mohanraj & Sivasankar, 2014).

Ipomoea species have several medicinal properties with several bioactive compounds of pharmacological importance. These compounds have antibacterial, antifungal, anti-inflammatory, antioxidant, antidiabetic, glycosidase inhibitory, wound healing, immunomodulatory, cardiovascular, anticancer, and anxiolytic effects (Mohanraj & Sivasankar, 2014; Singh, Navneet, & Kumar, 2020) although they are historical uses of members of Ipomoea in shamanism as sources of natural hallucinogens and they can be embryotoxic (Carod-Artal, 2015).

Ipomoea violacea possesses many phytochemicals. These include the naturally occurring tryptamine known as lysergic acid amide (LSA), which is closely related to lysergic acid diethylamide (LSD) (Hofmann, 1970; Srivastava & Rauniyar 2020a). The species is majorly of pantropical distribution but more so from South America (Alencar, Staples, & Budden, 2021). In this chapter we explore the scientific classification of the species, vernacular names, botanical description, its origin and geographic distribution, habitat and ecology, ethnobotany, pharmacology, and medicinal effects.

15.2 SCIENTIFIC CLASSIFICATION

Until an investigation of *Ipomoea violacea* distribution by Alencar, Staples, and Budden (2021), the species' long history with regard to its taxonomy and nomenclature had been complex and confusing. According to the Catalogue of Life (2023), the species belongs to: Kingdom: Plantae, Phylum: Tracheophyta, Class: Magnoliopsida, Order: Solanales Juss. ex Bercht. & J. Presl, Family: Convolvulaceae Juss., Subfamily: Convolvuloideae Burnett, Genus: Ipomoea L., species: *Ipomoea violacea* L.

The species has had taxonomic attention for many years (Der Marderosian, 1965; Wood, Muñoz-Rodríguez, Williams, & Scotland, 2020) and therefore with many synonyms some of which are ambiguous. The common synonyms include: *Calonyction comorensis* Bojer ex Desjardins, *Calonyction comospermum* Bojer, *Calonyction grandiflorum* (Jacq.) Choisy, *Calonyction jacquinii* G. Don, *Calonyction longiflorum* (R. Br.) Hassk., *Calonyction macranthum* (Schult.) Lem., *Calonyction muticum* Decne., *Calonyction noctolucum* G. Don, *Calonyction speciosum* var. laeve Choisy, *Calonyction tuba* (Schltdl.) Colla, *Convolvulus catharticus* Blanco, *Convolvulus grandiflorus* Desr., *Convolvulus muticus* (Decne.) Steud., *Convolvulus tuba* Schltdl., *Ipomoea browniana* F. Dietr., *Ipomoea comosperma* Drake, *Ipomoea glaberrima* Bojer ex Bouton, *Ipomoea grandiflora* (Jacq.) Hallier fil., *Ipomoea tuba* (Schltdl.) G. Don, *Pharbitis violacea* (L.) Bojer, *Tereietra violacea* (L.) Raf.

15.3 COMMON/VERNACULAR NAMES

Ipomea violacea is commonly known as the beach moonflower, sea moon flower, or morning glory (Meira, Silva, David, & David, 2012). The plant is traditionally known by different names among several cultures. These include Badoh Negro in Mexico, Bajucillo in Spain, La'aja Shnash or 'seeds of the Virgin' in Central America, Mantos de Cielo ('coat of heaven') in Ecuador, Ma:sung Pahk (Mixe, 'bones of the children'), Mehen Tu'xikin (Lacandon, 'little stink ear'), Pih Pu'ucte:sh (Mixe, 'flower of the broken plates') by the Mixe that live in Sierra Norte eastern highlands of

DOI: 10.1201/9781003460336-15

the state of Oaxaca, Quiebraplato (Mexico, 'breaker of plates') in Mexico, Tlitliltzin ('black divine') by theAztec, Xha'il (Mayan, 'that from the water') (Meira, Silva, David, & David, 2012), Kinarvel in Maharashtra state in India (Londhe et al., 2017) and as aboro aibá in Yoruba in Nigeria.

15.4 ETHNOBOTANY

The morning glory has traditionally been used in psychedelic and visionary practices across various cultures among the Chontal Indians of Oaxaca and the Indigenous people of Mexico, (Aztecs, and the Zapotec) (Ratsch 1998) who believed that seeds this plant of connected them with the sun Gods. The traditional use of *I. violacea* has persisted to the present time and majorly for spiritual and religious purposes. For instance, in southern Mexico, the seeds were majorly used as a narcotic in divination during religious rituals by Indians (Meira, Silva, David, & David, 2012, Chaachouay et al., 2019).

In India, the seeds are held in great veneration where they place offerings to the seeds in secret places so that the offerings cannot be found if a search is made. These seeds are placed among the idols of ancestors. However, its use is least known in many other parts of the world. Traditionally *I. violacea* has been used to treat different ailments. In Uttar-Pradesh, India, its roots are used as a diuretic, laxative, and expectorant for coughs while the leaves are used to manage headaches and indigestion (Srivastava & Rauniyar, 2020b). Apart from its use in herbal medicine, *I. violacea* is used for ornamental purposes (Sharma 2004). In Mexico, the plant was used in a magic potion and believed to possess an analgesic effect. Beach morning glory is employed in West Africa and in Bahia in spiritual baths (Verger 1995; Voeks 1997).

15.5 BOTANICAL DESCRIPTION

Ipomoea violacea is a perennial woody extensive climber with a large stout glabrous twiner that twines and climbs to treetops. Its stems and branches are terete longitudinally wrinkled and can grow up to 5 m (Vinod et al., 2021). The leaves are large, simple, alternate, heart-shaped, broadly ovate at the apex and cordate at the base; 6–15 cm long. The petioles are much longer at 3.5–11 cm; the leaf blade may be circular or ovate, 5–16 × 5–14 cm, with a base deeply cordate; lobes are round but may be angular in some instances, the apex is acuminate, mucronulate; it has seven or eight pairs of lateral veins.

The inflorescence may have one or few flowers with a peduncle which may be up to 7 cm. It has a thick pedicel of 1.5–3 cm in fruit. Flowers are nocturnal. Sepals are circular, equal or the outer two may be shorter, 1.5–2.5 cm, with a thin leathery apex which could be obtuse or emarginate, mucronulate, or enlarged in fruit and reflexed. The corolla varies from white to green with midpetaline bands salverform, 9–12 cm; a limb 8–10 cm in diameter. The flower has hairy stamens with filaments inserted near the base of the corolla tube. The pistil has a two-lobbed stigma with a glabrous ovary. The capsule is pale brown and may be ovoid to globose. Seeds are ovoid-rounded black in color with sizes ranging from 1–1.2 cm, densely short tomentose, and edges with ca. 3 mm long sericeous hairs. 2n = 30 (Humane 2014).

15.5.1 Origin And Geographic Distribution

Ipomoea violacea is native to North America and Central America, although their cultivation has spread throughout the world (https://www.iceers.org/ipomoea-violacea-basic-info/#). According to the Herberium catalogue of the Kew, the native range of *I. violacea* is Tropical & Subtropical coasts (The Herbarium Catalogue, 2023) including Aldabra, Andaman Is., Aruba, Assam, Bahamas, Bangladesh, Belize, Bismarck Archipelago, Borneo, Brazil Northeast, Brazil Southeast, Cambodia, Caroline Is., Cayman Is., China Southeast, Christmas I., Cocos (Keeling) Is., Colombia, Comoros, Cook Is., Cuba, Dominican Republic, Fiji, Florida, French Guiana, Ghana, Gilbert Is., Guinea, Guyana, Hainan, Haiti, India, Ivory Coast, Jamaica, Japan, Jawa, Kenya, Leeward Is., Line Is., Madagascar, Malaya, Maluku, Marianas, Marquesas, Marshall Is., Mauritius, Mexico Southeast, Mexico Southwest, Mozambique, Myanmar, Nauru, Netherlands Antilles, New Caledonia, New Guinea, Niue, Northern Territory, Ogasawara-shoto, Panamá, Philippines, Phoenix Is., Pitcairn Is., Puerto Rico, Queensland, Rodrigues, Samoa, Senegal, Seychelles, Society Is., Solomon Is., South China Sea, Sri Lanka, Sumatera, Suriname, Taiwan, Tanzania, Texas, Thailand, Tokelau-Manihiki, Tonga, Trinidad-Tobago, Tuamotu, Tubuai Is., Tuvalu, Vanuatu, Venezuela, Vietnam, Wake I., Wallis-Futuna Is., Western Australia, Windward Is (The Herbarium Catalogue, 2023).

15.5.2 Distribution

Ipomoea violacea is pantropical in distribution. Supplementary data for the distribution of *I. violacea* reveals 29 records of the species in African & Indian Ocean records; 146 records in the Americas (North, Middle, South, and Caribbean) records, and 226 records for Asian & Pacific records. Alencar, Staples, and Budden (2021) have postulated that *I. violacea* has become a globally distributed, pantropical littoral species today following historical routes and mechanisms:

15.6 HABITAT AND ECOLOGY

Ipomoea violacea is a perennial herbaceous vine that can grow to between ten and twenty feet long and climbs up to 3–4 m high on trees. The native range of this species is tropical and subtropical coasts of the wet tropical biome (Wood et al., 2020). It is a pantropical littoral plant species (Alencar et al., 2021) that grows near saltwater or brackish water bodies seldom far from sea beaches or coastal mangrove habitats, at an altitude of less than 100 m above sea level (Fang & Staples, 1995; Staples & Syahida-Emiza, 2015). It often grows over rocky outcrops or twines in littoral thickets or on top of coastal cliffs (Fosberg & Sachet 1977), hinterland vegetation, and along riverbanks near mangrove forests (Prospéri et al., 2005). *Ipomoea violacea* is rarely or never trails on sandy beaches or dunes. It also occurs as a cultivated or wild species in the tropics (Prospéri et al., 2005).

15.7 PHARMACOLOGY EFFECTS

Some of the pharmacology effects of *I. violacea* include: induction of nausea and vivid imagination (Nichols, 2004); induction of dizziness; hyperattention to sound; closed eye visuals, feeling of skin moving, fatigue, tactile enhancement, disconnection from self and childlike fantasies (Adovasio & Fry, 1976). This plant also possesses a hallucinogenic property due to the presence of ergoline alkaloids in the seeds which have been used by ancient people to attain a state of mind suitable for divination during religious ceremonies and magical rituals especially in Central America and Africa (Meira et al., 2012). Ergoline alkaloids are the main bioactive constituents responsible for the psychotomimetic effects of Convolvulaceae plants (Chen et al., 2018).

The ergot alkaloid can be found in the fresh leaves, stalks, and roots of the plant (Shinners, 1965). However, works by Hofmann (1963) showed that the psychotomimetic effects are unknown for ergometrine, which is used to a large extent in obstetrics as a utero tonic and hemostatic agent. Other alkaloids in *I. violacea* include ergometrine and elymoclavine and they all elicit an excitation syndrome in various animals that is caused by a central stimulation of the sympathicus (Yui & Takeo, 1958). Ergot alkaloids such as lysergic acid amides in the seeds of *I. violacea* may also have a pronounced narcotic component (Eich, 2008; Hofmann, 1961; Schultes & Hofmann, 1980) or even result in a hypnotic state in the case of ergot alkaloids present in seeds of *Argyreia nervosa* (Eich, 2008). Its seeds are also used in the creation of D-Lysergamide which is further used to make psychoactive drugs (Londhe et al., 2017).

According to Doughty et al. (2004), *I. violacea* acts in a similar manner to lysergic acid diethylamide (LSD) though at different sites in the central nervous system. This is because it contains ergot alkaloids such as lysergic acid amides in its seeds (Eich, 2008; Hofmann, 1961; Schultes & Hofmann, 1980). Lysergic acid diethylamide is a semisynthetic compound consisting of naturally occurring lysergic acid, which is found in all major ergots whose psychoactive effects begin within 20–60 minutes of ingesting the drug and typically last 8–12 hours (Liester, 2014).

Lysergic acid diethylamide is believed to exert its effects on the neurotransmitters and their receptors (Liester, 2014), as it is able to bind to 5-HT1A/1B/1D, 5-HT2A/2C, 5-HT5A, 5-HT6, and 5-HT7 receptors (Nichols, 2001). It also alters the patterns of expression of genes within cells even when a single dose is used. Some of the genes whose expression changes after treatment with LSD include c-fos, krox-20, NOR-1, arc, IKβ-α, sgk, and Ania3 (Nichols & Sanders-Bush, 2002).

Other specific pharmacology effects that can be brought about due to treatment with LSD include; somatic changes in which subjects may experience changes in heart rate and blood pressure, dilation of the pupils, sweating, hypersalivation, piloerection, nausea, diarrhea, vomiting, fatigue, increased muscular tension, tremors, headache, heaviness of the extremities, and sexual feelings can also arise (Liester, 2014; Chaachouay et al., 2021a, 2021b).

Perceptual changes due to LSD treatment which further cause visual changes such as blurring of vision, distortion of three-dimensional space, changes in faces and objects, colors changing or even becoming brighter, and areas of light or color may surround objects are also possible (Liester, 2014). Emotional responses can also result and these may include euphoria, depression, anxiety, panic, and irritation (Grof, 2009; Modell, 1968). Furthermore, administering LSD to psychiatric patients causes their negative moods to increase and may also cause suicidal ideation (Grof, 2009). Cognitive effects are also brought about by administering LSD and they include: impaired judgment; shortened concentration span; interposed thoughts; mind wandering; wavelike changes in thoughts; inability to control thoughts; and memory changes (Liester, 2014).

15.8 MEDICINAL EFFECTS

It can be used as a laxative to prevent constipation, as a diuretic in clinical pharmacology to treat extracellular fluid volume expansion, as an expectorant of purulent secretions, treatment for cough while its leaves are prescribed for headache & indigestion (Srivastava & Rauniyar, 2020b). It also contains lysergol (Khanuja et al., 2007), which enhances the killing activities of different antibiotics on bacteria and is a promising herbal bioenhancer (Singh et al., 2009).

As a bioenhancer, lysergol has properties like nontoxicity to humans and animals, being effective at low concentrations, easy to formulate, and also enhancing absorption and bioactivity of drugs or nutrients (Singh et al., 2009). Other medicinal effects of *I. violacea* include producing uterine contractions, stopping uterine bleeding, and reducing migraine headaches due to the presence of ergot alkaloids (Liester, 2014).

15.9 CONCLUSION

Ipomoea violacea is a versatile multi-purpose plant with a rich history of traditional use in various parts of the world for religio-cultural practices in shamanism, traditional medicine, and rituals. The plant contains several psychoactive metabolites of interest because of their pharmacological properties as potent hallucinogens. These metabolites also have several therapeutic properties attributed to them despite their potentially harmful effects.

BIBLIOGRAPHY

Alencar, J., Staples, G., & Budden, A. (2021). Distribution of *Ipomoea violacea* (Convolvulaceae): patterns, gaps and reports for its occurrence in Brazil and West Tropical Africa. *Rodriguésia*, 72. https://doi.org/10.1590/2175-7860202172034

Adovasio, J., & Fry, G.F. (1976). Prehistoric psychotropic drug use in northeastern Mexico and Trans-Pecos Texas. *Economic Botany*, 30(1), 94–96.

Austin, D.F. (2007). Water spinach (*Ipomoea aquatica*, Convolvulaceae) a food gone wild. *Ethnobotany Research & Applications*, 5, 123–146.

Austin, D. (1997). Convolvulaceae (Morning Glory Family) [on line].

Austin, D.F., & Huáman, Z. (1996). A synopsis of *Ipomoea* (Convolvulaceae) in the Americas. *Taxon*, 45, 3–38.

Bovell-Benjamin, A.C. (2007). Sweet potato: A review of its past, present, and future role in human nutrition. *Advances in Food and Nutrition Research*, 52, 1–59.

Burillo-Putze, G., López Briz, E., Climent Díaz, B., et al. (2013). Emergent drugs (III): hallucinogenic plants and mushrooms. *Anales del Sistema Sanitario de Navarra*, 36(3), 505–518. https://doi.org/10.4321/s1137-66272013000300015

Carod-Artal, F.J. (2015). Hallucinogenic drugs in pre-Columbian Mesoamerican cultures. *Neurologia (Barcelona, Spain)*, 30(1), 42–49. https://doi.org/10.1016/j.nrl.2011.07.003

Catalogue of Life. https://www.catalogueoflife.org/; accessed October 7, 2023

Der Marderosian, A.H. (1965). Nomenclatural history of the morning glory, *Ipomoea violacea* (L.). *Taxon*, 14(7),234–240.

Cao, S., Guza, R.C., Wisse, J.H., Miller, J.S., Evans, R., & Kingston, D.G. (2005). Ipomoeassins A– E, cytotoxic macrocyclic glycoresins from the leaves *of Ipomoea s quamosa* from the suriname rainforest. *Journal of Natural Products*, 68(4), 487–492.

Chaachouay, N., Benkhnigue, O., Fadli, M., Ibaoui, H., El Ayadi, R., & Zidane, L. (2019). Ethnobotanical and ethnopharmacological study of medicinal and aromatic plants used in the treatment of respiratory system disorders in the Moroccan Rif. *Ethnobotany Research and Applications*, *18*(22), 1–16.

Chaachouay, N., Douira, A., & Zidane, L. (2021a). Herbal Medicine Used in the Treatment of Human Diseases in the Rif, Northern Morocco. *Arabian Journal for Science and Engineering*, 47, 131–153. https://doi.org/10.1007/s13369-021-05501-1

Chaachouay, N., Douira, A., & Zidane, L. (2021b). COVID-19, prevention and treatment with herbal medicine in the herbal markets of Salé Prefecture, North-Western Morocco. *European Journal of Integrative Medicine*, 42, 101285. https://doi.org/10.1016/j.eujim.2021.101285

Chaachouay, N., & Zidane, L. (2021). The Symbolic Efficacy of Plants in Rituals and Socio-Religious Ceremonies in Morocco, Northwest of Africa. *Journal of Religious & Theological Information*, 21(1–2), 34–53. https://doi.org/10.1080/10477845.2021.1942433

Chen, G.T., Lu, Y., Yang, M., Li, J.L., & Fan, B.Y. (2018). Medicinal uses, pharmacology, and phytochemistry of Convolvulaceae plants with central nervous system efficacies: A systematic review. *Phytotherapy Research*, 32(5), 823–864.

Doughty, C., Walker, A., & Brenchley, J. (2004). Herbal mind altering substances: An unknown quantity? *Emergency Medicine Journal*, 21(2), 253–255.

Eich, E. (2008). *Solanaceae and Convolvulaceae: Secondary metabolites: Biosynthesis, chemotaxonomy, biological and economic significance (a handbook).* Springer Science & Business Media.

Fang, R.-C., & Staples, G. (1995). Convolvulaceae. In book: Flora of China vol. 16, Edition: 1Publisher. Science Press, Beijing, and Missouri Botanical Garden Press, 16: 271–325.

Fatur, K. (2021). Peculiar plants and fantastic fungi: An ethnobotanical study of the use of hallucinogenic plants and mushrooms in Slovenia. *Plos One*, 16(1), e0245022.

Fosberg, F.R. & Sachet, M.-H. (1977). Flora of micronesia 3, Convolvulaceae. *Smithsonian Contributions to Botany*, 36, 1–34.

Fields, F.H. (1969). Rivea corymbosa: Notes on some zapotecan customs. *Economic Botany*, 23(3), 206–209.

Grof, S. (2009). *LSD: Doorway to the numinous: The groundbreaking psychedelic research into realms of the human unconscious.* Simon and Schuster.

Hofmann, A. (1961). Die wirkstoffe der mexikanischen zauberdroge 'ololiuqui' 1. *Planta Medica*, 9(04), 354–367.

Hofmann, A. (1963). The active principles of the seeds of *Rivea corymbosa* and *Ipomoea violacea. Botanical Museum Leaflets, Harvard University*, 20(6), 194–212.

Hofmann, A. (1970). The discovery of LSD and subsequent investigations on naturally occurring hallucinogens. *Discoveries in Biological Psychiatry*, 8(3), 91–106.

Humane, P.T. (2014). Taxonomy and ethnobotany of convolvulaceae in bhandara district (ms). *International Journal of Researches in Biosciences, Agriculture & Technology*, II(3), 208–223

Khanuja, S., Arya, J., Srivastava, S., Shasany, A., Kumar, T. S., Darokar, M., & Kumar, S. (2007). Antibiotic pharmaceutical composition with lysergol as bio-enhancer and method of treatment. In: Google Patents.

Liester, B.M. (2014). A review of lysergic acid diethylamide (LSD) in the treatment of addictions: historical perspectives and future prospects. *Current Drug Abuse Reviews*, 7(3), 146–156.

Londhe, D., Neel, R., & Bhuktar, A. (2017). Ethno-medicinal uses of some species of genus Ipomoea L. from Maharashtra state. *Internat J Appl Res*, 3, 82–84.

Manitz, H. (1977). Was is it *Ipomoea violacea* L.? *Feddes Repertorium*, 88(4), 265–271.

Meira, M., Silva, E.P.D., David, J.M., & David, J.P. (2012). Review of the genus Ipomoea: traditional uses, chemistry and biological activities. *Revista Brasileira de Farmacognosia*, 22, 682–713.

Modell, W. (1968). *The Hallucinogens*. Wiley Online Library.

Mohanraj, R., & Sivasankar, S. (2014). Sweet potato (*Ipomoea batatas* [L.] Lam)—A valuable medicinal food: A review. *Journal of Medicinal Food*, 17(7), 733–741.

Nichols, C.D., & Sanders-Bush, E. (2002). A single dose of lysergic acid diethylamide influences gene expression patterns within the mammalian brain. *Neuropsychopharmacology*, 26(5), 634–642.

Nichols, D. (2004). Hallucinogens. *Pharmacol Therapeut*, 101, 131–181.

Nichols, D.E. (2001). LSD and its lysergamide cousins. *The Heffter Review of Psychedelic Research.* Santa Fe, New Mexico: Heffter Research Institute, 80–87.

Prospéri, M.J., Ramesh, B.R., Grard, P., Jayatissa, L.P., Aravajy, S., & Depommier, D. (2005). Mangroves v1. 0. A multimedia identification system of Mangroves species. French Institute of Pondicherry: India.

Ratsch, C. (1998). *The encyclopedia of psychoactive plants: Ethnopharmacology and its applications.* Rochester: Park Street Press.

Schultes, R.E., & Hofmann, A. (1980). The botany and chemistry of hallucinogens. (No Title).

Schultes, R.E. (1970). The plant kingdom and hallucinogens (part III). *Bulletin on Narcotics*, 22(1), 25–53.

Schultes, R.E. (1981). Iconography of new world plant hallucinogens. *Arnoldia*, 41(3), 80–125.

Shinners, L.H. (1965). Correct nomenclature of two Mexican narcotic plants. *Taxon*, 103–105.

Sharma, O.P. (2004). *Plant taxonomy*. New Delhi: Tata McGraw-Hill Publishing Company Limited

Singh, R., Devi, S., Patel, J., Patel, U., Bhavsar, S., & Thaker, A. (2009). Indian herbal bioenhancers: A review. *Pharmacognosy Reviews*, 3(5), 90.

Singh, R., Navneet, M.D., & Kumar, A. (2020). Ethnobotanical and pharmacological aspects of Ipomoea carnea Jacq. and Celosia cristata Linn. In Joshi P.C., Joshi N., Reshman Y, Mansotra D. K. (eds) *Current Status of Researches in Biosciences*. New Delhi: Today & Tomorrow's Printers and Publishers, 493–506.

Staples G.W. & Syahida-Emiza, S. (2015) Convolvulaceae. In R. Kiew, R. C. K. Chung, L. G. Saw & E. Soepadmo (Eds.), *Forest Research Institute Malaysia, Kepong*, 5, 55–198.

Srivastava, D., & Rauniyar, N. (2020a). *Medicinal plants of genus Ipomoea*. Beau Bassin, Mauritius: LAP Lambert Academic Publishing.

Srivastava, D. & Rauniyar N. 2020b. Medicinal plant of genus Ipomoea: A glimpse of potential Bioactive compounds of genus Ipomoea and its detail. Mauritius: Lambert Academic Publishing, 1–80.

The Herbarium Catalogue, Royal Botanic Gardens, Kew. 2023. Published on the Internet http://www.kew.org/herbcat. 07-12-2023

Yui, T., & Takeo, Y. (1958). Neuropharmacological studies on a new series of ergot alkaloids elymoclavine as a potent analeptic on reserpine-sedation. *The Japanese Journal of Pharmacology*, 7(2), 157–161.

Voeks, R.A. (1997). *Sacred leaves of Candomblé: African magic, medicine, and religion in Brazil*. Austin: University of Texas Press.

Verger, P (1995). *Ewé: O uso das plantas na sociedade Iorubá*. São Paulo: Editora Schwarcz.

Vinod, K., Asokan, P.K., Joshi, K.K., Narayanakumar, R., Zacharia, P.U., Varghese, M., … Gopalakrishnan, A. (2021). Glimpses of biodiversity in the Kadalundi-Vallikunnu Community Reserve, the first Community Reserve of Kerala.

Wood, J.R., Muñoz-Rodríguez, P., Williams, B.R., & Scotland, R.W. (2020). A foundation monograph of Ipomoea (Convolvulaceae) in the New World. *PhytoKeys*, 143, 1–823.

Zhao, G., Kan, J., Li, Z., & Chen, Z. (2005). Characterization and immunostimulatory activity of an (1→6)-aD-glucan from the root of *Ipomoea batatas*. *International Immunopharmacology*, 5(9), 1436–1445.

16 Coca (*Erythroxylum coca* Lam. Erythroxylaceae)

Shehar Bano, Alvina Gul, and Munir Ozturk

16.1 INTRODUCTION

Hallucinogenic plants have been used by humans for decades. These are referred to as hallucinogens by virtue of their potential to distort the five senses accompanied by hallucinations (experiences and feelings that are fanatical in nature) (Bouso et al., 2023). *E. coca*, commonly named as coca, is a shrub, commonly found growing in hot and humid (tropical) regions, with psychedelic effects and extends to about 2.4 meters in height. The branches of this plant are erect, oval-shaped leaves are frail, opaque, and the beautiful small flowers flourish and turn into red berries (Chaachouay et al., 2021; García-Cordero et al., 2021; Orch et al., 2021).

This plant grows well in hot and humid environments, most likely the forest glades, but the leaves most desirably are collected from dried-up regions (periphery of hills) (Alamgir and Alamgir, 2017). Constituents of coca leaves play an inherent part in the lives of the Andean people in terms of conventional medicine and cultural norms (Karch, 2017). It is also the source plant from which the alkaloid cocaine is obtained (Bravo et al., 2022).

The products formed from the constituents of the coca plant serve as a panacea for health-related issues specifically in South American regions. This species is recorded as an aboriginal plant of South America with multiple alkaloid constituents, the most notorious one is cocaine with psychotropic constituents (Castaldelli-Maia et al., 2023). Cocaine plays its part in reducing hunger and enabling tasks requiring consistent labor, a dependency-causing drug, known as 'blow' to addicts nowadays, and affects the central nervous system (Richards et al., 2019). It results in a boost in mental power, followed by a feeling of calmness, felicity, and elimination of the sense of exhaustion (Snyder, 2019). The factual reasons for such delirium are the chemical constituents present in coca – the true tranquilizers. A narcotic/tranquilizer causes a depressive effect, either mild or severe, on the CNS (central nervous system) (Cordovilla-Guardia et al., 2019). This chapter focuses on different ethnobotanical, ecological, phytochemical, and hallucinogenic effects of *E. coca*.

16.2 SCIENTIFIC CLASSIFICATION

Taxonomically coca plant belongs to the genus *Erythroxylum* (Table 16.1), which includes 230 taxa. A huge variety of the plants from this genus are tropical flowering (Islam, 2011).

16.3 VERNACULAR NAMES

Erythroxylum coca is commonly termed 'Coca'.

16.4 BOTANICAL DESCRIPTION

Timothy Plowman an ethno-botanist (Plowman, 1986) characterized the botanical diversity of coca plant. His analysis showed that all variants of coca were drawn from two intimately associated species: *Erythroxylum coca* and *E. novogranatense*. Each of these two species are divided into two diverse varieties, comprised of distinctive environmental and ethnobotanical traits (White et al., 2021). The four varieties of cocas are scientifically termed as: *Erythroxylum coca* var. *coca*, *E. coca* var. *ipadu*, *E. novogranatense* var. *novogranatense*, and *E. novogranatense* var. *truxillense* (Loyola et al., 2020). All these four varieties are lean shrubs with lively green leaves. Their tiny white flowers along the erect branches develop into yellow, red, or orange-colored fruits, eagerly consumed by birds in tropical regions (Nair and Nair, 2021). The elegant look of these shrubs make their common use as ornamental plants in various South American countries (De Luca, 2023). The coca shrubs are produced from seedlings, and can be garnered after 12 or 24 months, and at that time the leaves are collected from the whole plant. The most common coca plant variety (*Huánuco coca*), produces 46 yields annually, consistently for about 40 years. No other such plant shows such a leaf productivity and quality that remains so

TABLE 16.1
Taxonomic Ranking of Coca Plant (*Erythroxylum coca* Lam.)

Kingdom	Plantae
Division	Tracheophyta
Subdivision	Spermatophytina
Class	Magnoliopsida
Order	Malpighiales
Family	Erythroxylaceae
Genus	*Erythroxylum*
Species	*Erythroxylum coca* Lam.

DOI: 10.1201/9781003460336-16

abundant, even when it is being garnered repetitively with short intervals (Conzelman and White, 2016).

16.5 ORIGIN AND GEOGRAPHICAL DISTRIBUTION

E. coca is one of the indigenous plant species of South America, Mexico, and the West Indies. It was initially utilized by the inhabitants of the Andes Mountains and Puna (South America), termed Aymara (Kellogg et al., 2017). Dwellers of this region were the pioneers who gave this palangt its original name 'Khokha', which refers to the tree, and its species term 'coca' derived from it (Biondich and Joslin, 2016).

16.6 HABITAT AND ECOLOGY

The tropical alkaloid cocaine is derived from the coca plant of the Andes. It was first obtained in 1859 and is considered one of the most common, traditionally magical and celestial plants distributed in the Americas (Díaz, 2010). The indigenous region of this species is W. South America. It is a plant bushy in nature and grows mainly in the humid tropical biome. The plant propagates in hyper-humid regions, and the vegetation type is primarily rainforest (Arias-Gámez et al., 2023).

This plant is cultivated in the Amazonian zone of South America, particularly in Colombia, Chile, and Brazil. The maximum number of plants has been reported from the tropical Peruvian Andes, Ecuador, and Bolivia. Comparatively, a lower number of coca plants are found in Mexico and the West Indies, along with Indonesia (Biondich and Joslin, 2016). These plants grow and develop well on higher altitudes by virtue of their natural habitat being the Andes Mountains, where these plants burgeon in the wild habitat within the range of 300–2,000 meters. The most appropriate soil conditions for these plants include adequately watered, richly productive soil with a pH value ranging from 5.5 to 6.5, however, these can grow within a pH range of 4.3 to 8 (Mihai et al., 2023). Apart from the ideally best environmental conditions mentioned above for the growth and development of *E. coca*, it can also flourish well without any sort of complexities in its physiology in regions where rainfall ranges from 700 to 4,000 millimeters and temperatures lie around 14–27°C (Islam, 2011).

16.7 ETHNOBOTANY

Coca has played a central and culturally significant role in multiple South American communities over decades, where it is believed as a symbolic entity for cultural identification (Ablard, 2021). The utilization of *E. coca* continues up-till-date in various regions of the Andes, in Ecuador, Peru, Colombia, Bolivia, Argentina, and in the western Amazon Basin (Valdez and Taboada, 2016).

Before the arrival of the Spaniards, fabrics and coca leaves used to be the most precious commodities rather than gold and silver, hence they were frequently utilized for trading purposes or as reimbursement for labor, and sharing coca was also ritualized (Babidge, 2016). This kind of utilization of coca was only a part of the thorough system of 'barter exchange' through which merchandise and labor used to be traded within the communities instead of capital (Valdez et al., 2015; Chaachouay and Zidane, 2021).

Leafy parts of coca are chiefly evaluated by chewing, whereas there are some variations in the ways the leaves are used. The leaves are always made moisture-free prior to their use, which enables the swift release of the chemical components when chewed (Delfanti, 2023). Once moisture-free, one or a few leaves are chewed, and gradually dampened with saliva. The coca itself is never directly chewed, but rather the moistened chaw of coca leaves is slurped to obtain the fluid enriched with coca leaf constituents, and then gradually allowed to dribble down to the stomach (Pilcher, 2023). Conventional norms frequently escort this act of chewing coca leaves, which is usually practiced by men and women laborers in the meadows, excavations, marketplaces, study areas, and meeting auditoriums for its multiple work-time benefits (Valencia and Courtheyn, 2023).

One of the Bolivian campesino, when asked about the possible reasons for utilizing coca leaves, mentioned that 'chewing coca leaves enriches us with vigor, fervor, and motivation to work; being used by Latin Americans, expiates all the aches and fatigue, and satiates the hunger'. However, augmenting labor productivity rate is not the only lying cause of chewing coca leaves, it also exhibits esteem and admiration toward the goddess of crop fruitfulness and fertility (Conzelman and White, 2016). Various jargon is used particularly to refer to coca chewing: ('mambear' in Colombia), ('pijchear' in Peru and Bolivia), coquear, chacchar, acullicar, and mascara (Granda et al., 2014).

16.8 PHARMACOLOGY

The coca plant derives its name from the alkaloid cocaine, all of the grown coca plants are comprised of the chemical compound cocaine and vary substantially in the content of minute quantity alkaloids as well as other chemical components (De Luca, 2023; Dos Santos et al., 2021). The coca leaves contain protein with amino acid lysine in limited amounts, beta-carotene, vitamin E, a trivial quantity of vitamin D, zinc, magnesium, calcium, and iron (Granda et al., 2014). The biosynthesis of alkaloid cocaine in *E. coca* and the sequences of enzymes implied in the series of reactions are presented next (Figure 16.1). Currently available information on tropane alkaloid (cocaine) biosynthesis in *E. coca* is less than compared with other comprehended alkaloid-producing plant taxa (Kim et al., 2016). In *E. coca*, the chemical compounds (cocaine and cinnamoylcocaine) are stockpiled in the vacuoles and form complex compounds by adhering to HQAs (Hydroxycinnamoyl Quinate Esters). Chlorogenic acid is the chief HQA that forms a complex with alkaloid cocaine and cinnamoylcocaine. A BAHD acyltransferase enzyme (EcHQT) is categorized as the concluding enzymatic action of hydroxycinnamoyl quinate ester biosynthesis (Torre et al., 2013).

Coca is a well-known plant in therapeutic treatments, exclusively to safeguard and guarantee health (Hodge et al., 2023). In internal medicine, *E. coca* is taken as a potion and chewed. The most significant medicinal use of this plant is for GIT (gastrointestinal) disorders, chiefly dysentery, stomach-pain, diarrhea, stomach ulcers, indigestion, cramps and other situations (Stolberg, 2011). This plant is utilized to cure signs of altitude/motion sickness including nausea, light-headedness, spasms, and stern headaches. It is also generally used for treating toothache, rheumatoid arthritis, aftermaths of various ailments, and multiple other disorders, and is taken orally or applied as an ointment.

16.9 MEDICINAL EFFECTS

As pointed out above the drug cocaine has been extracted and purified from the plant *E. coca* and a few other related species, which have grown naturally in South America for thousands of years (Ozturk, 1976). The chewing of leaves with lime powder has been an ancient practice from the time of pre-Incans. This activity is similar to the 'Khat' leaf chewing in Yemen (*Catha edulis*) or 'Pawn' *Piper betle*, commonly known as Betel or Sireh, all used as a stimulant as well as to improve fatigue, reduce hunger and mood enhancement (Ozturk, 1976).

The narcotic drug cocaine (an alkaloid) is extracted from the leaves of two coca plant species (Wang et al., 2023). It is a bonanza for various territories, namely Colombia (Murillo-Sandoval et al., 2023), Ecuador, Peru, parts of Argentina and Bolivia (Briefings, 2023). It is an indirect adrenergic drug that crosses over the blood-brain barrier and hinders the passage of dopamine (Caccamise et al., 2021), serotonin, and norepinephrine, halting the consumption of these neurotransmitters at the presynaptic vessels, ultimately enhancing their amount, in addition to the stimulation of the receptors present in post-synaptic vessels of neurons (Faria, 2021). This results in the effects of cocaine inebriation, including conditions such as euphoria, cardiac infarction, sexual exhilaration, jolting, and day-dreaming states. Addiction is a supplementary issue as with all the other narcotics (Azizi, 2022). Some toxicologists mention that the toxic effects of alkaloid cocaine are due to their producing psychoactive drugs which can be categorized into five groups (Excitanica, Euphorica, Hypnotica, Inebriantia and Phantastica); the *E. coca* belongs to the group Euphorica (the euphoriants) (Díaz, 2010).

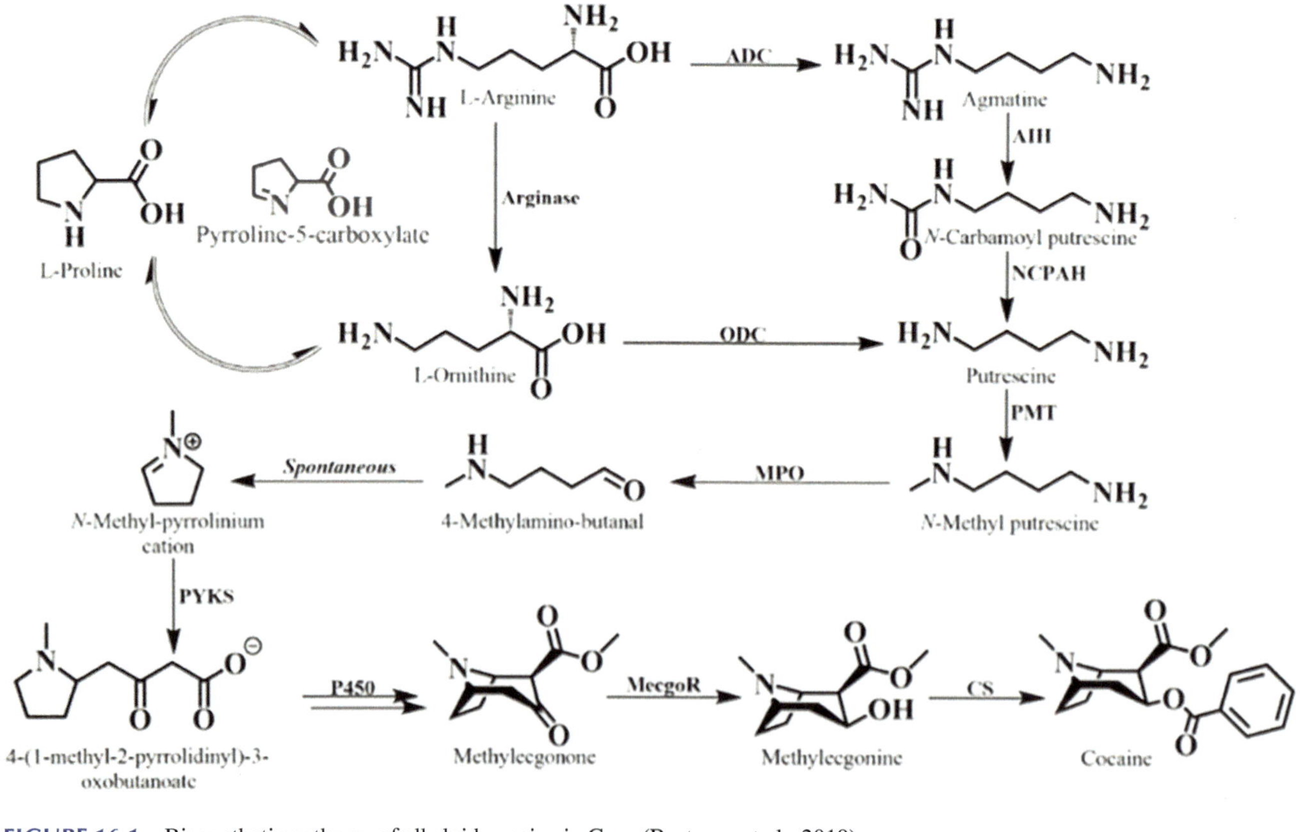

FIGURE 16.1 Biosynthetic pathway of alkaloid cocaine in Coca (Restrepo et al., 2019).

Coca leaves are popular in Europe because of their psychedelic features (Caroff and Berkowitz, 2022), and by virtue of the discovery of processes involved in the extraction of cocaine, utilization of cocaine became more widespread in Europe in the 19th century (Guerra-Doce et al., 2023). During this century, alkaloid cocaine was amalgamated into various pharmaceutical and user products, including Coca-Cola and cigarettes (Schwartz et al., 2023). The thin layers of coca leaves are open-dried in the sun, followed by packaging in sacks for the sake of transportation, however, to ensure the quality of cocaine, leaves must be kept dry. These leaves are comprised of about 0.3–1.5 percent of cocaine (Ozpınar and Camlıoğlu, 2022).

Cocaine intoxication acts as a magic bullet toward static mood, the aftermaths include strong elation, rapid heartbeat, higher blood pressure, and enhanced motor neuron functionality rate (Poyatos et al., 2023). Moreover, it causes increased anxiety, poor judgment, emotional disparity, frustration, and aggressiveness (Toyne, 2023). Extremely complex behavioral patterns, such as: touching one's nose or forehead repetitively, are observed in heavy cocaine addicts. A higher dosage of cocaine drug results in severe medical troubles including: death due to cardiac arrhythmia; extremely low body temperatures; dehydration; and local vasoconstriction, resulting in cardiac arrest or paralysis (Szöke et al., 2023). Repetitive nosebleeds and collapsed nasal septum are the symptoms observed in heavy intranasal cocaine addicts. The characteristic marks of persistent cocaine needle usage can be observed in the arms, legs, or neck of the injection users (He et al., 2023).

16.10 CONCLUSION

We can conclude that this plant originates from the Andes region, with botanical features like the shrubby nature of these plants. Taxonomically coca belongs to genus *Erythroxyllum*. The ideal habitat of coca is tropical regions, and phyto-chemically these are a source of alkaloid cocaine. However, the biosynthetic pathway of alkaloid cocaine still needs to be researched further in order to develop a thorough incite. The ecological and ethnobotanical attributes of *E. coca* have been pooled up here, as its leaves are chewed by the workers to keep themselves fatigue-free. The hallucinogenic effects of this plant have been discussed as these are said to result in fatal ailments with severe symptoms. In general, the constituents of coca leaves are being used to treat multiple agonies as well.

LIST OF ABBREVIATIONS

AHCT	anthocyanin O-hydroxycinnamoyl transferase
BEAT	benzoyl alcohol O-acetyltransferase
CNS	central nervous system
DAT	deacetyl vindoline 4-O-acetyltransferase
E. coca	*Erythroxyllum coca*
EcHQT	two of the *Erythroxyllum coca* sequences, EcBAHD3 and EcBAHD4 (later characterized as EcHQT)
HCBT	anthranilate N-hydroxycinnamoyl/benzoyl transferase
W. America	western America

ACKNOWLEDGMENTS

Our special thanks go to Prof. Dr. Volkan Altay for the critical evaluation of this chapter.

REFERENCES

Ablard, J.D. 2021. Framing the Latin American nutrition transition in a historical perspective, 1850 to the present. *História, Ciências, Saúde-Manguinhos*, 28: 233–253.

Alamgir, A.N.M., and Alamgir, A.N.M. 2017. Medicinal, non-medicinal, biopesticides, color-and dye-yielding plants; secondary metabolites and drug principles; significance of medicinal plants; use of medicinal plants in the systems of traditional and complementary and alternative medicines (CAMs). *In: Therapeutic Use of Medicinal Plants and Their Extracts: Volume 1.. Progress in Drug Research*, vol 73. Springer, Cham, pp. 61–104. https://doi.org/10.1007/978-3-319-63862-1_3

Arias-Gámez, J.M., Linares-Perea, E., Vicente-Orellana, J.A., and Galán-de-Mera, A. 2023. Biogeographical relationships and diversity in the Peruvian flora reported by Hipólito Ruiz and José Pavón: vegetation, uses and anthropology. *Biology*, 12(2): 294.

Azizi, S.A. 2022. Monoamines: Dopamine, norepinephrine, and serotonin, beyond modulation, 'Switches' that alter the state of target networks. *The Neuroscientist*, 28(2): 121–143.

Babidge, S. 2016. Contested value and an ethics of resources: Water, mining and indigenous people in the Atacama Desert, Chile. *The Australian Journal of Anthropology*, 27(1): 84–103.

Biondich, A.S., and Joslin, J.D. 2016. Coca: the history and medical significance of an ancient Andean tradition. *Emergency Medicine International*, 2016: 2016.

Bouso, J.C., Ona, G., Kohek, M., Dos Santos, R.G., Hallak, J.E., Alcázar-Córcoles, M.Á., and Obiols-Llandrich, J. 2023. Hallucinations and Hallucinogens: Psychopathology or Wisdom?. *Culture, Medicine, and Psychiatry*, 47(2): 576–604.

Bravo, R.R., Faria, A.C., Brito-da-Costa, A.M., Carmo, H., Mladěnka, P., Dias da Silva, D., … Oemonom, R. 2022. Cocaine: an updated overview on chemistry, detection, biokinetics, and pharmacotoxicological aspects including abuse pattern. *Toxins*, 14(4): 278.

Briefings, O.A. 2023. *Cocaine figures point to upward trends globally*. Emerald Expert Briefings, (oxan-es).

Caccamise, A., Van Newenhizen, E., and Mantsch, J.R. 2021. Neurochemical mechanisms and neurocircuitry underlying the contribution of stress to cocaine seeking. *Journal of Neurochemistry*, 157(5): 1697–1713.

Caroff, S.N., and Berkowitz, R.M. 2022. A magic bullet for the mind: Sigmund Freud and the quest for chemical transcendence. *Pharos*, 2022: 43.

Castaldelli-Maia, J.M., Wang, Y.P., Brunoni, A.R., Faro, A., Guimarães, R.A., Lucchetti, G., … Bensenor, I.M. 2023. Burden of disease due to amphetamines, cannabis, cocaine, and opioid use disorders in South America, 1990–2019: a systematic analysis of the Global Burden of Disease Study 2019. *The Lancet Psychiatry*, 10(2): 85–97.

Chaachouay, N., Douira, A., & Zidane, L. (2021). Herbal Medicine Used in the Treatment of Human Diseases in the Rif, Northern Morocco. *Arabian Journal for Science and Engineering*, 47, 131–153. https://doi.org/10.1007/s13369-021-05501-1

Chaachouay, N., & Zidane, L. (2021). The Symbolic Efficacy of Plants in Rituals and Socio-Religious Ceremonies in Morocco, Northwest of Africa. *Journal of Religious & Theological Information*, *21*(1–2), 34–53. https://doi.org/10.1080/10477845.2021.1942433

Conzelman, C.S., and White, D.M. 2016. *The botanical science and cultural value of Coca leaf in South America. Roadmaps to regulation: coca, cocaine, and derivatives*. The Beckley Foundation.

Cordovilla-Guardia, S., Lardelli-Claret, P., Vilar-López, R., López-Espuela, F., Guerrero-López, F., and Fernández-Mondéjar, E. 2019. The effect of central nervous system depressant, stimulant and hallucinogenic drugs on injury severity in patients admitted for trauma. *Gaceta Sanitaria*, 33: 4–9.

De Luca, V. 2023. Understanding how plants produce cocaine. *Proceedings of the National Academy of Sciences*, 120(1): e2218838120.

Delfanti, I. 2023. Coca cola and coca leaves: A case study on the use of FPIC. *Fourth World Journal*, 22(2): 40–54.

Díaz, J.L. 2010. Sacred plants and visionary consciousness. *Phenomenology and the Cognitive Sciences*, 9: 159–170.

Dos Santos, N.A., de Almeida, C.M., Gonçalves, F.F., Ortiz, R.S., Kuster, R.M., Saquetto, D., and Romão, W. 2021. Analysis of *Erythroxylum coca* Leaves by Imaging Mass Spectrometry (MALDI–FT–ICR IMS). *Journal of the American Society for Mass Spectrometry*, 32(4): 946–955.

Faria, M. 2021. 'Plants of the Gods' and their hallucinogenic powers in neuropharmacology – A review of two books. *Surgical Neurology International*, 2021: 12.

García-Cordero, J., Pino, A., Cuevas, C., Puertas-Martín, V., San Román, R., and de Pascual-Teresa, S. 2021. Neurocognitive effects of cocoa and red-berries consumption in healthy adults. *Nutrients*, 14(1): 1.

Granda, L., Rosero, M.G., Rosero, A. 2014. Medicinal plants in the Tropical Andean Region. Quinoa (*Chenopodium quinoa* Willd.) and Coca (*Erythroxylum* sp.), millenary treasures for medicinal treatment. *Folia*, 2014: 58.

Guerra-Doce, E., Rihuete-Herrada, C., Micó, R., Risch, R., Lull, V., and Niemeyer, H.M. 2023. Direct evidence of the use of multiple drugs in Bronze Age Menorca (Western Mediterranean) from human hair analysis. *Scientific Reports*, 13(1): 4782.

He, G., Huai, Z., Jiang, C., Huang, B., Tian, Z., Le, Q., … and Liu, X. 2023. Persistent increase of accumbens cocaine ensemble excitability induced by IRK downregulation after withdrawal mediates the incubation of cocaine craving. *Molecular Psychiatry*, 28(1): 448–462.

Hodge, A.T., Strayhan, R., and Sukpraprut-Braaten, S. 2023. Systematic review: Is there a medicinal use of cocaine in psychiatry?. *Current Treatment Options in Psychiatry*, 10(1): 50–58.

Islam, M. 2011. Tracing the evolutionary history of coca (*Erythroxylum*). Doctoral dissertation, University of Colorado at Boulder.

Karch, M.D.S.B. 2017. *A brief history of cocaine*. CRC Press.

Kellogg, J., Schmidt, B.M., Graf, B. L., Jofré-Jimenez, C., Aguayo-Cumplido, H., Calfío, C., … and Delatorre-Herrera, J. 2017. The Americas. In: Schmidt, B.M. and Klaser Cheng, D.M. (eds), *Ethnobotany: A phytochemical perspective*. John Wiley & Sons Ltd., pp. 183–226.

Kim, N., Estrada, O., Chavez, B., Stewart Jr, C., and D'Auria, J.C. 2016. Tropane and granatane alkaloid biosynthesis: A systematic analysis. *Molecules*, 21(11): 1510.

Loyola, D., Mendoza, R., Chiong, L., Rueda, M., Alvítez-Temoche, D., Gallo, W., and Mayta-Tovalino, F. 2020. Ethanol extract of *Schinus molle* L.(molle) and *Erythroxylum coca* Lam (coca): Antibacterial properties at different concentrations against *Streptococcus mutans*: An in vitro study. *Journal of International Society of Preventive & Community Dentistry*, 10(5): 579.

Mihai, R.A., Melo Heras, E.J., Terán Maza, V.A., Espinoza Caiza, I.A., Pinto Valdiviezo, E.A., and Catana, R.D. 2023. The panoramic view of ecuadorian soil nutrients (deficit/toxicity) from different climatic regions and their possible influence on the metabolism of important crops. *Toxics*, 11(2): 123.

Murillo-Sandoval, P.J., Kilbride, J., Tellman, E., Wrathall, D., Van Den Hoek, J., and Kennedy, R.E. 2023. The post-conflict expansion of coca farming and illicit cattle ranching in Colombia. *Scientific Reports*, 13(1): 1965.

Nair, K.P., and Nair, K.P. 2021. Cocoa (*Theobroma cacao* L.). In: Nair, K.P. (ed), *Tree crops*. Springer, Cham. https://doi.org/10.1007/978-3-030-62140-7_5

Orch, H., Chaachouay, N., Douiri, E. M., Faiz, N., Zidane, L., & Douira, A. (2021). Use of medicinal plants in dermato-cosmetology: An ethnobotanical study among the population of Izarène. *Jordan Journal of Pharmaceutical Sciences*, *14*(3), 323–340.

Ozpınar, H., and Camlıoğlu, D. 2022. Some examples of psychoactive plant species. In: Yalın, A.E. and Eraslan, M. (eds.), *Interdisciplinary medicine and health sciences concepts, researches and practice*. Livre de Lyon Press, pp. 193–208.

Ozturk, M. 1976. *Erythroxylum coca*. In: *'Economic Botany Course' Notes; Ege Univ*. Science Faculty Press.

Pilcher, J.M. 2023. *Food in world history*. Routledge Press.

Plowman, T. 1986. Coca chewing and the botanical origins of coca (*Erythroxylum* spp.) in South America. Coca and Cocaine. *Cultural Survival Report*, 23: 5–33.

Poyatos, L., Pérez-Mañá, C., Hladun, O., Núñez-Montero, M., de la Rosa, G., Martín, S., … Papaseit, E. 2023. Pharmacological effects of methylone and MDMA in humans. *Frontiers in Pharmacology*, 14: 328.

Restrepo, D.A., Saenz, E., Jara-Muñoz, O.A., Calixto-Botía, I.F., Rodríguez-Suárez, S., Zuleta, P., … and D'Auria, J.C. 2019. *Erythroxylum* in focus: an interdisciplinary review of an overlooked genus. *Molecules*, 24(20): 3788.

Richards, J.R., Gould, J.B., Laurin, E.G., and Albertson, T.E. 2019. Metoprolol treatment of dual cocaine and bupropion cardiovascular and central nervous system toxicity. *Clinical and Experimental Emergency Medicine*, 6(1): 84.

Schwartz, L.P., Toegel, F., Devine, J.K., Holtyn, A.F., Roma, P.G., and Hursh, S.R. 2023. Latent factor structure of behavioral economic heroin and cocaine demand curves. *Experimental and Clinical Psychopharmacology*, 31(2): 378.

Snyder, S. 2019. CNS stimulants and hallucinogens. In: Mulé, S.J., and Brill, H. (eds.), *Chemical and biological aspects of drug dependence*. CRC Press, pp. 55–64.

Stolberg, V.B. 2011. The use of coca: Prehistory, history, and ethnography. *Journal of Ethnicity in Substance Abuse*, 10(2): 126–146.

Szöke, É., Kéry, Á., and Lemberkovics, É. 2023. Biological activity of compounds of plant origin. In: Szöke, É., Kéry, Á., and Lemberkovics, É. (eds.), *From herbs to healing: Pharmacognosy-phytochemistry-phytotherapy-biotechnology*. Springer International Publishing, pp. 411–511.

Torre, J.C.P., Schmidt, G.W., Paetz, C., Reichelt, M., Schneider, B., Gershenzon, J., and D'Auria, J.C. 2013. The biosynthesis of hydroxycinnamoyl quinate esters and their role in the storage of cocaine in *Erythroxylum coca*. *Phytochemistry*, 91: 177–186.

Toyne, M. 2023. Transformation of coca to cocaine: An overview of traditional drug use and modern drug abuse. *The University of Western Ontario Journal of Anthropology*, 25(1): 94–115.

Valdez, L.M. and Taboada, J. 2016. Coca leaves in the context of the Central Andean Wari State. In: Patton, Margaret and Manion, Jessica (eds), Trading Spaces: The Archaeology of Interaction, Migration and Exchange. *Proceedings of the 46th Annual Chacmool Archaeology Conference*, pp. 136–151. Chacmool Archaeology Association, University of Calgary, Calgary, Alberta, CA.

Valdez, L.M., Taboada, J., and Valdez, J.E. 2015. Ancient use of coca leaves in the Peruvian central highlands. *Journal of Anthropological Research*, 71(2): 231–258.

Valencia, Ó.E., and Courtheyn, C. 2023. Peace through coca? Decolonial peacebuilding ecologies and rural development in the Territory of Conviviality and Peace of Lerma, Colombia. *Third World Quarterly*, 2023: 1–21.

Wang, Y.J., Tain, T., Yu, J.Y., Li, J., Xu, B., Chen, J., … and Huang, S.X. 2023. Genomic and structural basis for evolution of tropane alkaloid biosynthesis. *Proceedings of the National Academy of Sciences*, 120(17): e2302448120.

White, D.M., Huang, J.P., Jara-Muñoz, O.A., Madriñán, S., Ree, R.H., and Mason-Gamer, R.J. 2021. The origins of coca: museum genomics reveals multiple independent domestications from progenitor *Erythroxylum gracilipes*. *Systematic Biology*, 70(1): 1–13.

17 Tepezcohuite (*Senegalia tenuifolia* (L.) Britton & Rose [syn.: *Mimosa tenuiflora* (Willd.) Poir.] Fabaceae)

Mustafa Keskin, Volkan Altay, and Munir Ozturk

17.1 INTRODUCTION

Senegalia tenuifolia (syn.: *Mimosa tenuiflora*) is distributed naturally from Mexico to some of the States in Tropical America (www.powo.science.kew.org). In Northeastern Brazil; one of the characteristic countries in Latin American traditional medicine; the local people prepare a drink called 'Yurema or Jurema' from the root bark of this plant, a psychoactive drink that is popularly used in religious ceremonies. The drink is used by some Indigenous tribes of the country and is believed to be a miraculous drink (Lewis, 1987; De Albuquerque, 2001; Vepsalainen et al., 2005; De Souza et al., 2008; Majeed et al., 2021). It has also been reported that in Mexico the whole plant is used for hallucinogenic compounds (Evans and Hofmann, 2008; Majeed et al., 2021).

In this chapter, information has covered data on various features of *S. tenuifolia*, namely botanical aspects and vernacular names, geographical distribution and ecological features, traditional uses, and some pharmacological effects.

17.2 SCIENTIFIC CLASSIFICATION

The latest updated classification of this species according to the United States Department of Agriculture Natural Resources Conservation Service (USDA) (2023) is presented below:

Kingdom	Plantae
Subkingdom	Tracheobionta
Superdivision	Spermatophyta
Division	Magnoliophyta
Class	Magnoliopsida
Subclass	Rosidae
Order	Fabales
Family	Fabaceae Lindl.
Genus	*Senegalia* Raf.
Species	*Senegalia tenuifolia* (L.) Britton & Rose
Synonymous	*Acacia tenuifolia* (L.) Willd.; *Mimosa tenuifolia* L.

17.3 VERNACULAR NAMES

S. tenuifolia (syn.: *Mimosa tenuiflora*) is widely named 'Tepezcohuite' in English and is known by different local names in different Latin American countries. In Northeastern Brazil and Venezuela, locals call it 'Skin tree, Jurema-preta, Black jurema, Vinho de jurema, Calumbi'; and in Mexico as 'Tepescohuite' (Majeed et al., 2021).

17.4 BOTANICAL DESCRIPTION

The plants are found as shrubs or decumbent trees, armed with thorns. The stems, rachis, and petioles are puberulent, glabrous, strigulous, or tomentose branched. Leaves are up to 15 cm long, pinnae (8-)10–23 pairs, (1-)2–4.5 cm long; leaflets (40-)60–80 pairs per pinna, asymmetrically linear, 2–4 mm long and up to 0.5 mm broad, apex acute, base obliquely truncate, glabrous, short ciliate, main vein marginal; rachis (4-)6–11(–12) cm long, with several circular glands between the last pairs of pinnae, petioles 1.5–2 cm long, with one gland in the middle, stipules linear-triangular, up to 5 mm long, fleeting. The panicles of the capitula are axillary and terminal, (9-)15–20 cm long, main axis armed with stingers, tomentose, capitula 4–10 mm diameter, peduncles 0.5–1(–1.5) cm long, tomentose, basally with 1 linear bract 2–2.5(–4) mm long, strigose, floral bract 0.5 mm long, whitish flowers; calyx bell-shaped, ca 1.5 mm long, slightly five-lobed, stridulous; corolla bell-shaped, 2 mm long, five-lobed, stridulous; stamens glandular; ovary ca 1 mm long, strigose, stipe 1.5 mm long; nectary absent. The fruits are flat, 8–13 cm long, 0.9–3 cm wide, and 3 mm thick, straight, with triangular to rounded apex, base acute, dehiscent, valves light brown, somewhat reddish when ripe, glabrous, margins thicker than valves, stipe 7–10 mm long; ellipsoid seeds 9–10 mm long, 5–7 mm wide and 2–3 mm thick, light brown (Barros, 2011).

17.5 ORIGIN AND GEOGRAPHICAL DISTRIBUTION

The natural distribution of *S. tenuifolia* reveals that it is a scrambling shrub or tree distributed in the seasonally dry

DOI: 10.1201/9781003460336-17

tropical biome regions of Mexico extending to Tropical America. The natural distribution areas are outlined mainly as; Bolivia, Brazil, Colombia, Costa Rica, Cuba, Ecuador, El Salvador, French Guiana, Guyana, Honduras, Mexico, Panamá, Paraguay, Peru, Suriname, Trinidad-Tobago, and Venezuela (www.powo.science.kew.org).

Some species belonging to the Fabaceae family, such as *Mimosa* and *Acacia* are used as ornamental plants for landscaping purposes (Altay et al., 2010; Osma et al., 2010; Altay, 2012; Eskin et al., 2012), *S. tenuifolia* (syn.: *Mimosa tenuiflora*), as a tree species, appears to possess a potential to be used as an ornamental plant.

17.6 HABITAT AND ECOLOGY

The plant in general is distributed from Mexico and other areas in Tropical America, where its habitat preferences include northern Brazil, mainly northeastern swamps, seasonal semideciduous forests, dense ombrophilous forests, low areas, especially riparian forests, and edges. It is also found in other Brazilian phytogeographical domains (Barros, 2011). From an ecological perspective, it develops in the seasonally dry tropical biome regions (www.powo.science.kew.org).

17.7 ETHNOBOTANY

In Northeastern Brazil, stem bark, leaves, and flowers of the plant are generally used in Latin American traditional medicine to treat bronchitis, cough, burns, external ulcers, fever, headaches, hypertension, inflammations, menstrual colic, skin problems, and toothache (Albuquerque and de Andrade, 2002; de Albuquerque et al., 2007; De Souza et al., 2008; Cruz et al., 2016; Majeed et al., 2021; Chaachouay et al., 2021a, 2021b). In Mexico, the stem bark of the plant is used to treat skin burns, lesions, and inflammation (Camargo-Ricalde et al., 1994; Camargo-Ricalde, 2000; Rivera-Arce et al., 2007; Majeed et al., 2021).

17.8 PHARMACOLOGY

The studies undertaken on the determination of phytochemical compounds present in the flowers, leaves, root bark, stem bark, and seeds of this plant have revealed that these especially include; chalcones, flavones, flavonoids, phenols, phytoindole alkaloids, polysaccharides, saponins, steroids, terpenoids, tannins, and condensed tannins (Dominguez et al., 1989; Meckes-Lozoya et al., 1990a; Jiang et al., 1991, 1992; Anton et al., 1993; León et al., 2004; Vepsalainen et al., 2005; Bautista et al., 2011; Bezerra et al., 2011; Haddad et al., 2012; Oliveira et al., 2013; Gardner et al., 2014; Calvert and Sperry, 2015; Rizwan et al., 2022).

In many pharmacological experimental studies, different extracts obtained from the root and stem bark of this plant have good antimicrobial and antioxidant activities, in terms of potentially important biological activity (Silva et al., 2020; Majeed et al., 2021). Some studies have also proven its wound-healing, antimutagenic, and anti-inflammatory potentials (Rivera-Arce et al., 2007; Silva et al., 2013; Cruz et al., 2016; Silva et al., 2020; Orch et al., 2021).

It has been observed that different extracts obtained from different parts of the plant, especially the root and stem bark, show antimicrobial activity in many experimental studies conducted using different applied methods. It has been emphasized that this plant has the potential to successfully inhibit morphotypic strains such as *Candida albicans, Cryptococcus neoformans, Escherichia coli, Micrococcus lysodeikticus, Pseudomonas aeruginosa, Proteus mirabilis, Shigella sonnei, Staphylococcus aureus, Streptococcus pyogenes, Streptococcus pyogenes*, and *Staphylococcus* spp. (Meckes-Lozoya et al., 1990b; Gonçalves et al., 2005; Padilha et al., 2010; De Morais-Leite et al., 2015; Valencia-Gómez et al., 2016; De Souza-Araújo et al., 2018; Silva et al., 2020; Majeed et al., 2021).

Magalhaes et al. (2018), and Silva et al. (2020) have proved in their studies that ethyl acetate and ethanol extracts obtained from different parts of the plant have strong antioxidant activity against DPPH and ABTS radicals. da Silva-Leite et al. (2016) have reported that alcoholic extract prepared from the bark of the plant has the potential to heal acute inflammation in experimental wistar rat models.

In 2018, an experimental clinical study was conducted by Choi et al. They demonstrated the potential wound-healing effects of a herbal mixture consisting of *S. tenuifolia* (syn.: *M. tenuiflora*) leaves and *Alchemilla vulgaris* on human keratinocytes, umbilical vein endothelial cells, and mouse fibroblast. The histological results obtained by the same authors have shown that the herbal mixture promotes collagen synthesis, re-epithelialization, and renewal of the appendages of skin and hair follicles (Choi et al., 2018). Zippel et al. (2009) have shown that in a clinical study on wound-healing activity, ethanol extracts obtained from the bark of this plant significantly stimulate the mitochondrial activity and proliferation of dermal fibroblasts showing a small stimulation on human keratocytes.

17.9 MEDICINAL EFFECTS

In the Latin American traditional medical system, the whole plant and roots of this plant are popularly used as a psychoactive drink because of the presence of hallucinogenic compounds (Lewis, 1987; De Albuquerque, 2001; Vepsalainen et al., 2005; De Souza et al., 2008; Evans and Hofmann, 2008; Majeed et al., 2021). There is only one neuropharmacological activity study conducted on this plant. In this study, using acetylcholinesterase inhibitor therapy, the methanol extract obtained from this plant was shown to have potential anti-Alzheimer activity due to its rich compounds (Majeed et al., 2021).

17.10 CONCLUSION

S. tenuifolia (syn.: *M. tenuiflora*) is widely used in Latin American traditional folk medicine. Several pharmacological studies have been carried out on this plant. It is a natural genetic resource with good therapeutic potential, mainly because of its antimicrobial activity, as an antioxidant and anti-inflammatory, and possesses wound-healing characteristics. It is a promising potential candidate in the development of some innovative treatments. Detailed investigations are needed in order to make it available to humans at a global level, especially in the Latin Americas. There is a need to fully reveal its therapeutic potential in neuropharmacological activities and other potential diseases and to ensure its safe use in modern medicine.

LIST OF ABBREVIATIONS

S. tenuifolia	*Senegalia tenuifolia*
Syn.	Synonymous
USDA	United States Department of Agriculture Natural Resources Conservation Service

REFERENCES

Albuquerque, U.P., and de Andrade, L.D.H.C. 2002. Uso de recursos vegetais da caatinga: o caso do agreste do estado de Pernambuco (Nordeste do Brasil). *Interciência*, 27: 336–346.

Altay, V. 2012. Mustafa Kemal Üniversitesi Tayfur Ata Sökmen Kampüsü (Hatay)'nün süs bitkileri. *Karadeniz Fen Bilimleri Dergisi*, 2(6): 11–26.

Altay, V., Ozyıgıt, I.I., and Yarcı, C. 2010. Urban flora and ecological characteristics of the Kartal District (Istanbul): A contribution to urban ecology in Turkey. *Scientific Research and Essays*, 5(2): 183–200.

Anton, R., Jiang, Y., Weniger, B., Beck, J., and Rivier, L. 1993. Pharmacognosy of *Mimosa tenuiflora* (Willd.) poiret. *Journal of Ethnopharmacology*, 38: 145–152.

Barros, M.J.F. 2011. *Senegalia* Raf. (Leguminosae, Mimosoideae) do Domínio Atlântico, Brasil. Master Thesis. Instituto de Pesquisas Jardim Botânico do Rio de Janeiro Escola Nacional de Botânica Tropical Programa de Pós-graduação Stricto Sensu, Rio de Janerio.

Bautista, E., Calzada, F., Ortega, A., and Yépez-Mulia, L. 2011. Antiprotozoal activity of flavonoids isolated from *Mimosa tenuiflora* (Fabaceae-Mimosoideae). *Journal of the Mexican Chemical Society*, 55: 251–253.

Bezerra, D.A.C., Rodrigues, F.F.G., da Costa, J.G.M., Pereira, A.V., de Sousa, E.O., and Rodrigues, O.G. 2011. Abordagem fitoquímica, composiç˜ao bromatol´ogica e atividade antibacteriana de *Mimosa tenuiflora* (Wild) Poiret E Piptadenia stipulacea (Benth) Ducke. *Acta Scientiarum: Biological Sciences*, 33. https://doi.org/10.4025/actascibiolsci.v33i1.5366

Calvert, M.B., and Sperry, J. 2015. Bioinspired total synthesis and structural revision of yuremamine, an alkaloid from the entheogenic plant *Mimosa tenuiflora*. *Chemical Communications*, 51: 6202–6205.

Camargo-Ricalde, S., Grether, R., and Martínez-Bernal, A. 1994. Uso medicinal del 'tepescohuite', *Mimosa tenuiflora* (Leguminoseae) en Mexico. *Contacto*, 5: 29–34.

Camargo-Ricalde, S.L. 2000. Descripción, distribución, anatomía, composición química y usos de *Mimosa tenuiflora* (Fabaceae-Mimosoideae) en México. *Revista de Biología Tropical*, 48: 939–954.

Chaachouay, N., Douira, A., & Zidane, L. (2021a). Herbal Medicine Used in the Treatment of Human Diseases in the Rif, Northern Morocco. *Arabian Journal for Science and Engineering*, 47, 131–153. https://doi.org/10.1007/s13369-021-05501-1

Chaachouay, N., Douira, A., & Zidane, L. (2021b). COVID-19, prevention and treatment with herbal medicine in the herbal markets of Salé Prefecture, North-Western Morocco. *European Journal of Integrative Medicine*, 42, 101285. https://doi.org/10.1016/j.eujim.2021.101285

Choi, J., Park, Y.-G., Yun, M.-S., and Seol, J.-W. 2018. Effect of herbal mixture composed of *Alchemilla vulgaris* and *Mimosa* on wound healing process. *Biomedicine & Pharmacotherapy*, 106: 326–332.

Cruz, M.P., Andrade, C.M.F., Silva, K.O., de Souza, E.P., Yatsuda, R., Marques, L.M., David, J.P., David, J.M., Napimoga, M.H., and Clemente-Napimoga, J.T. 2016. Antinoceptive and anti-inflammatory activities of the ethanolic extract, fractions and flavones isolated from *Mimosa tenuiflora* (Willd.) poir (Leguminosae). *PLoS One*, 11: e0150839.

da Silva-Leite, K.E.S., Silva Queiroz, C., Da Costa Madeira, J., Marcos Gomes Soares, P., Pereira, M., and Assreuy, A.M. 2016. Polysaccharide extract of *Mimosa tenuiflora* stem barks stimulates acute inflammatory response via nitric oxide. *Acta Scientiarum: Biological Sciences*, 38: 473.

De Albuquerque, U.P. 2001. The use of medicinal plants by the cultural descendants of African people in Brazil. *Acta Farm Bonaer*, 20: 139–144.

de Albuquerque, U.P., de Medeiros, P.M., Luiz, A., de Almeida, S., Monteiro, M., de Freitas, E.M., Neto, L., and de Melo, J.G. 2007. Medicinal plants of the caatinga (semiarid) vegetation of NE Brazil: A quantitative approach. *Journal of Ethnopharmacology*, 114: 325–354.

De Morais-Leite, S.C., Medeiros, C., Soares, A.I., Maia, P.C.G.G.S., Magalhães, M.I.S., Freitas, F.O.R., Pessôa, H.D.L.F., Nogueira, T.B.D.S., Morais, A.M.B., et al. 2015. Antibacterial and hemolytic activities of *Mimosa tenuiflora* (Willd) Poir.(Mimosoidea). *African Journal of Microbiology Research*, 9: 2166–2171.

De Souza, R.O.S., de Albuquerque, U.P., Monteiro, J.M., and de Amorim, E.L.C. 2008. Jurema-Preta (*Mimosa tenuiflora* [Willd.] Poir.): A review of its traditional use, phytochemistry and pharmacology. *Braz. Arch. Biol. Technol.*, 51: 937–947.

De Souza-Araújo, E., Pimenta, A., Feijó, F., Castro, R., Fasciotti, M., Monteiro, T., and de Lima, K. 2018. Antibacterial and antifungal activities of pyroligneous acid from wood of *Eucalyptus urograndis* and *Mimosa tenuiflora*. *Journal of Applied Microbiology*, 124: 85–96.

Dominguez, X.A., Garcia, G.S., Williams, H.J., Ortiz, C., Scott, A.I., and Reibenspies, J.H. 1989. Kukulkanins A and B, new chalcones from Mimosa tenuefolia. *Journal of Natural Products*, 52: 864–867.

Eskin, B., Altay, V., Ozyıgıt, I.I., and Serın, M. 2012. Urban vascular flora and ecological characteristics of the Pendik District (Istanbul-Turkey). *African Journal of Agricultural Research*, 7(4): 629–646.

Evans, S., and Hofmann, A. 2008. *Plantas de los dioses - Orígenes Del Uso De Los Alucinógenos*. México City, México: Fondo de Cultura.

Gardner, D., Riet-Correa, F., Lemos, D., Welch, K., Pfister, J., and Panter, K. 2014. Teratogenic effects of *Mimosa tenuiflora* in a rat model and possible role of N-methyl-and N, N-dimethyltryptamine. *Journal of Agricultural and Food Chemistry*, 62: 7398–7401.

Gonçalves, A., Alves Filho, A., and Menezes, H. 2005. Estudo comparativo da atividade antimicrobiana de extratos de algumas árvores nativas. *Arquivos do Instituto Biológico*, 72: 353–358.

Haddad, H., Seyed Hoseini, E., Nikzad, H., and Hossein Aarabi, M. 2012. Pharmacological properties of medicinal herbs by focus on secondary metabolites. *Life Science Journal*, 9: 509–520.

Jiang, Y., Massiot, G., Lavaud, C., Teulon, J.-M., Guéchot, C., Haag-Berrurier, M., and Anton, R. 1991. Triterpenoid glycosides from the bark of *Mimosa tenuiflora*. *Phytochemistry*, 30: 2357–2360.

Jiang, Y., Weniger, B., Haag-Berrurier, M., Anton, R., Beck, J.P., and Italiano, L. 1992. Effects of saponins from *Mimosa tenuiflora* on lymphoma cells and lymphocytes. *Phytotherapy Research*, 6: 310–313.

León, L., Maldonado, E., Cruz, A., and Ortega, A. 2004. Tenuiflorins A-C: New 2-Phenoxychromones from the leaves of *Mimosa tenuiflora*. *Planta Medica*, 70: 536–539.

Lewis, G.P. 1987. *Legumes of Bahia*. London, UK: Royal Botanic Gardens.

Magalhaes, F.E.A., Batista, F.L.A., Serpa, O.F., Moura, L.F.W.G., Lima, M.d.C.L., da Silva, A.R.A., Guedes, M.I.F., Santos, S.A.A.R., de Oliveira, B.A., Nogueira, A.B., et al. 2018. Orofacial antinociceptive effect of *Mimosa tenuiflora* (Willd.) Poiret. *Biomedicine & Pharmacotherapy*, 97: 1575–1585.

Majeed, I., Rizwan, K., Ashar, A., Rasheed, T., Amarowicz, R., Kausar, H., Zia-Ul-Haq, M., and Marceanu, L.G. 2021. A comprehensive review of the ethnotraditional uses and biological and pharmacological potential of the genus *Mimosa*. *International Journal of Molecular Sciences*, 22: 7463.

Meckes-Lozoya, M., Lozoya, X., and González, J.L. 1990b. Propiedades farmacológicas in vitro de algunos extractos de *Mimosa tenuiflora* (tepescohuite). *Archivos de investigación médica (Mexicano)*, 21: 163–169.

Meckes-Lozoya, M., Lozoya, X., Marles, R.J., Soucy-Breau, C., Sen, A., and Arnason, J.T. 1990a. N, N-dimethyltryptamine alkaloid in *Mimosa tenuiflora* bark (tepescohuite). *Archives of Medical Research*, 21: 175.

Oliveira, L.M.B., Macedo, I.T.F., Vieira, L.S., Camurça-Vasconcelos, A.L.F., Tomé, A.R., Sampaio, R.A., Louvandini, H., and Bevilaqua, C.M.L. 2013. Effects of *Mimosa tenuiflora* on larval establishment of *Haemonchus contortus* in sheep. *Veterinary Parasitology*, 196: 341–346.

Orch, H., Chaachouay, N., Douiri, E. M., Faiz, N., Zidane, L., & Douira, A. (2021). Use of medicinal plants in dermato-cosmetology: An ethnobotanical study among the population of Izarène. *Jordan Journal of Pharmaceutical Sciences*, *14*(3), 323–340.

Osma, E., Ozyıgıt, I.I., Altay, V., and Serın, M. 2010. Urban vascular flora and ecological characteristics of Kadikoy district, Istanbul, Turkey. *Maejo International Journal of Science and Technology*, 4(1): 64–87.

Padilha, I.Q., Pereira, A.V., Rodrigues, O.G., Siqueira-Júnior, J.P., and Pereira, M.d.S.V. 2010. Antimicrobial activity of *Mimosa tenuiflora* (Willd.) Poir. from Northeast Brazil against clinical isolates of *Staphylococcus aureus*. *Revista Brasileira de Farmacognosia*, 20: 45–47.

Rivera-Arce, E., Antonio Chávez-Soto, M., Herrera-Arellano, A., Arzate, S., Agüero, J., Feria-Romero, I.A., Cruz-Guzmán, A., and Lozoya, X. 2007. Therapeutic effectiveness of a *Mimosa tenuiflora* cortex extract in venous leg ulceration treatment. *Journal of Ethnopharmacology*, 109: 523–528.

Rizwan, K., Majeed, I., Bilal, M., Rasheed, T., Shakeel, A., and Iqbal, S. 2022. Phytochemistry and diverse pharmacology of genus *Mimosa*: A review. *Biomolecules*, 12(1): 83.

Silva, S.A.D.N.M., Barros, A.B., Souza, J.M.T., Moura, A.F., de Araujo, A.R., Mendes, M.G.A., ... Marinho Filho, J.D.B. 2020. Phytochemical and biological prospection of *Mimosa* genus plants extracts from Brazilian northeast. *Phytochemistry Letters*, 39: 173–181.

Silva, V.A., Gonçalves, G.F., Pereira, M.S.V., Gomes, I.F., Freitas, A.F.R., Diniz, M.F.F.M., and Pess^oa, H.L.F. 2013. Assessment of mutagenic, antimutagenic and genotoxicity effects of *Mimosa tenuiflora*. *Brazilian Journal of Pharmacognosy*, 23: 329–334.

USDA (United States Department of Agriculture Natural Resources Conservation Service). 2023. *Senegalia* Raf. https://plants.usda.gov/home/plantProfile?symbol=SENEG. Accesses on 25 July 2023.

Valencia-Gómez, L.E., Martel-Estrada, S.A., Vargas-Requena, C., Rivera-Armenta, J.L., Alba-Baena, N., Rodríguez-González, C., and Olivas-Armendáriz, I. 2016. Chitosan/*Mimosa tenuiflora* films as potential cellular patch for skin regeneration. *International Journal of Biological Macromolecules*, 93: 1217–1225.

Vepsalainen, J.J., Auriola, S., Tukiainen, M., Ropponen, N., and Callaway, J. 2005. Isolation and characterization of yuremamine, a new phytoindole. *Planta Medica*, 71: 1053–1057.

www.powo.science.kew.org. Accesses on 30 July 2023.

Zippel, J., Deters, A., and Hensel, A. 2009. Arabinogalactans from *Mimosa tenuiflora* (Willd.) Poiret bark as active principles for woundhealing properties: Specific enhancement of dermal fibroblast activity and minor influence on HaCaT keratinocytes. *Journal of Ethnopharmacology*, 124: 391–396

18 Sensitive Plant (*Mimosa pudica* L. Fabaceae)

Mustafa Keskin, Volkan Altay, Alvina Gul, and Munir Ozturk

18.1 INTRODUCTION

The historical importance of the *Mimosa pudica* plant has been described by Reverend John Layfield after his visit to the West Indies in 1598 (Allen, 1969). Surprisingly, the leaves of this plant fold downwards when touched, overturning Reverend John Layfield's prejudices that plants were passive while facing their environment. This has led to some hypotheses worth testing experimentally in natural sciences (Allen, 1969).

M. pudica is distributed naturally in the wet tropical biome regions of Mexico to tropical America but is at the same time cultivated in many parts of South Asia, East Asia, and several Pacific islands (Joseph et al., 2013; Patro et al., 2015; www.powo.science.kew.org).

In this chapter, information has been gathered on various features of *M. pudica*, namely botanical aspects and vernacular names, geographical distribution and ecological features, traditional uses, and some pharmacological effects.

18.2 SCIENTIFIC CLASSIFICATION

The latest updated classification of this species according to the United States Department of Agriculture Natural Resources Conservation Service (USDA) (2023) is presented below:

Kingdom	Plantae
Subkingdom	Tracheobionta
Superdivision	Spermatophyta
Division	Magnoliophyta
Class	Magnoliopsida
Subclass	Rosidae
Order	Fabales
Family	Fabaceae
Genus	*Mimosa* L.
Species	*Mimosa pudica* L.

18.3 VERNACULAR NAMES

M. pudica, widely named as 'Sensitive plant' in English, is known by different local names in different countries. In North India, locals call it 'Chhui-mui'; in Nepal as 'Lajabati'; in Hawaii as 'Hila-hila'; in Brazil as 'sabiá or sansão-do-campo'; in Sri Lanka as 'Makahiya or Nidikumbha'; and in some countries as 'humble plant', 'shame plant', 'touch-me-not', 'sleeping grass', and 'prayer plant' (Niroula et al., 2009; Ahmad et al., 2012; Joseph et al., 2013; Silva et al., 2014; Hasan et al., 2016; Song et al., 2017; Sunday, 2021). In Turkey, people call it 'Küstüm otu'.

18.4 BOTANICAL DESCRIPTION

Prostrate or straggling annual or perennial herb up to c. 1 m high; stems sparsely armed with short prickles, more or less densely covered by rough hairs, sometimes woody near the base; leaves sensitive, double compound with (1-)2 pairs of pinnae; leaflets 10–20 pairs, 0.6–1.2-cm long, 0.3–0.4 cm broad, sessile, obliquely narrow or linear oblong; obliquely rounded at base, acute, nearly glabrous; yellowish green; rachis very short, making the leaf appearing palmately divided; flowers as ovoid-globose heads, c. 1.3 × 1 cm, in the upper leaf axils, lilac or pink; calyx very small, lobes 4, ovate oblong; stamens 4, much exerted; ovary sessile; ovules numerous. Pods in clusters, 1–2 cm long, densely prickly or bristly on the margins only (Ahmad et al., 2012; www.zimbabweflora.co.zw).

18.5 ORIGIN AND GEOGRAPHICAL DISTRIBUTION

The natural distribution of this plant reveals that it is a herb distributed in the wet tropical biome regions of Mexico and Tropical America. The natural distribution areas are outlined as Bolivia, Brazil, Colombia, Costa Rica, Cuba, Ecuador, El Salvador, Guatemala, Guyana, Haiti, Honduras, Jamaica, Mexico, Peru, and Venezuela (www.powo.science.kew.org).

M. pudica has been introduced globally in many countries outside its natural distribution area and has naturalized to a large extent (Bissanti, 2023). While some species belonging to the Fabaceae family, such as *Mimosa* and *Acacia*, are used as ornamental plants for landscaping purposes (Altay et al., 2010; Osma et al., 2010; Altay, 2012; Eskin et al., 2012), *M. pudica* is also used as an outdoor as well as indoor ornamental plant in many countries (Parker, 2017; Panyadee, 2022).

DOI: 10.1201/9781003460336-18

18.6 HABITAT AND ECOLOGY

The plant in general is widely distributed in Central America, South Asia, East Asia, and many islands in the Pacific region (Joseph et al., 2013; Patro et al., 2015), where its habitat preferences are well-drained soils, even scalped or eroded sub-soils and soils with low nutrient contents (Azmi et al., 2011; Ahmad et al., 2012). It is generally spread on farmlands, orchards, roadsides, waste areas, mowed areas, and pastures between sea level up to 1300 meters (Azmi et al., 2011; Li et al., 2011; Ahmad et al., 2012; Bissanti, 2023). From an ecological perspective, it develops in the wet tropical biome regions (Joseph et al., 2013; Patro et al., 2015). The morpho-anatomical features of this plant permit it to grow well alongside such habitats. It also grows as a single plant or in tangled thickets, does not compete with tall vegetation, and is frost-sensitive and intolerant to shades which have given these plants an important advantage in their ecological adaptations (Azmi et al., 2011; Bissanti, 2023).

18.7 ETHNOBOTANY

M. pudica has been used in the Asian continent, especially as an exotic ornamental plant. In Thailand, it has been evaluated in traditional medicine in the improvement of urological, nutritional and digestive disorders, treatment of musculoskeletal systems, and in the treatment of cardiovascular diseases. It has also been reported to be one of the preferred plants for the respiratory system, female genital problems, psychological disorders, neurological disorders, male genital problems, and eye-related disorders (Orch et al., 2021; Panyadee, 2022).

The roots of this plant have been potentially used in traditional folk medicine as antipyretic, against hemorrhoids, for headache, in leprosy, dysentery, epilepsy, diarrhea, constipation, as an emetic, in fevers, migraine, as a sedative, tonic, in tumors, insomnia, and urogenital infections (Chatterjee and Prakash, 2000; Joy et al., 2001; Vaidyaratanm, 2001; Merlin and Narsimhan, 2009; Tamilarasi and Ananthi, 2012; Patro et al., 2015; Sharma, 2016; Chaachouay et al., 2021a, 2021b; Kumar et al., 2022). In Cambodia, this plant is widely used as an antipyretic (Chassagne et al., 2016), has traditionally been used in the treatment of diabetes (Tunna et al., 2014). The leaves are used in the treatment of urinary tract infections and hemorrhoids (Sharma, 2016). Interestingly, seeds of *M. pudica* have also been utilized for the treatment of constipation, depression, piles, snake bites, smallpox, and ulcers (Muhammad et al., 2016a; Bukhari et al., 2022).

18.8 PHARMACOLOGY

The studies undertaken on the determination of phytochemical compounds present in this plant have revealed that these especially include; alkaloids, amino acids, crocetin dimethyl ester, fatty acids, flavonoids, flavonoid glycosides, glycoproteins, glycosides, nitrogen-containing compounds, phenolic compounds, phytosterols, sitosterols, tannins and terpenoids (Kaur et al., 2011; Tamilarasi and Ananthi 2012; Ahmad et al., 2012; Ranjan et al., 2013; Patro et al., 2015; Song et al., 2017; Sandhosh et al., 2022).

This plant is famous for the alkaloid mimosine which contains a type of 5-deoxyflavonol, one of the phytochemical compounds playing an important role as an anticancer treatment (Song et al., 2017). The seeds contain rich mucilage containing glucuronoxylan consisting of D-xylose and D-glucuronic acid (Saraswat and Pokharkar, 2012; Rezagholi et al., 2019). This mucilage is pharmacologically pH-sensitive, biocompatible, and has a non-toxic biomaterial feature, and many studies show that these seeds can be used commercially for pharmaceutical purposes at an industrial level (Ahuja et al., 2013; Muhammad et al., 2016b, 2018; Bukhari et al., 2022).

In numerous pharmacological experimental studies on the leaves and roots of this plant different extracts have been prepared which are reported to show important pharmacological activities namely; antiasthmatic, anticancer, antidepressant, antidiabetic, antifertility, antihepatotoxic, anti-implantation, anti-inflammatory, anticonvulsant, antimumps, antioxidant, antivenom, atherosclerosis, anxiety, aphrodisiac, diuretic, eczema, hepatoprotective, hyperglycemic, hypolipidemic, leprosy, psoriasis cure, sedative, swelling, and wound-healing (Molina et al., 1999; Mahanta and Mukherjee, 2001; Sivarajan and Balachandran, 2002; Valsala and Karpagaganapathy, 2002; Bum et al., 2004; Balunas and Kinghorn, 2006; Ganguly et al., 2007; Yuan, 2007; Ignacimuthu et al., 2008; Kokane et al., 2009; Rekha et al., 2009; Sutar et al., 2009; Paul et al., 2010; Goli et al., 2011; Ayissi et al., 2012; Chowdhury et al., 2012; Tamilarasi and Ananthi, 2012; Baghel et al., 2013; Malayan et al., 2013; Alasyam et al., 2014; Patel and Bhutani, 2014; Zaware et al., 2014; Sandhosh et al., 2022).

The whole plant shows antimicrobial activity and hypoglycemic effect (Chowdhury et al., 2012; Tamilarasi and Ananthi, 2012; Konsue et al., 2017; Yupparach and Konsue, 2017; Vijayalakshmi and Udayakumar, 2018; Ranjani et al., 2021).

18.9 MEDICINAL EFFECTS

Commonly, in the traditional medical system, the leaves and roots of *M. pudica* are used in the treatment of mental disorders such as headaches, as antidepressant, in epilepsy, insomnia, as a sedative, in psychological and some neurological disorders (Panyadee, 2022; Kumar et al., 2022). Some pharmacological studies conducted on this plant have revealed that it has sedative and antidepressant activity (Molina et al., 1999; Sivarajan and Balachandran,

2002; Ayissi et al., 2012; Zaware et al., 2014; Rahman and Debnath, 2015).

In 2015, Patro et al., in an applied clinical study conducted on experimental animals, showed that the ethyl acetate extract obtained from *M. pudica* has proven strong analgesic, antiepileptic, and motor activities. The same authors have reported that this plant may be good for the effective treatment of various motor or seizure disorders (Patro et al., 2015). These authors have also emphasized in their experimental studies that the plant has muscle relaxants and reduced locomotor activity (Patro et al., 2015). Moreover, it is thought that the significant reduction in pain may be due to the presence of analgesic principles acting through prostaglandin pathways (Sabina et al., 2009; Patro et al., 2015).

18.10 CONCLUSION

M. pudica is widely used in traditional folk medicine. Several pharmacological studies have been carried out on this plant. It is a natural genetic resource with good therapeutic potential, mainly as an antidepressant and antidiabetic as well as in antifertility, antimicrobial activities, as a sedative, and possesses wound-healing characteristics. Its chemical composition (in particular alkaloid mimosine and mucilages) offers promising potential in the development of some innovative treatments. Therefore, detailed investigations are needed in order to make it available to people at a global level, especially in the Americas and Asia. There is a need to fully reveal its therapeutic potential in neurodegenerative disorders and other potential diseases and to ensure its safe use in modern medicine.

LIST OF ABBREVIATIONS

M. pudica	*Mimosa pudica*
USDA	United States Department of Agriculture Natural Resources Conservation Service

REFERENCES

Ahmad, H., Sehgal, S., Mishra, A. and Gupta, R. 2012. *Mimosa pudica* L. (Laajvanti): an overview. *Pharmacognosy Reviews*, 6(12): 115–124.

Ahuja, M., Kumar, A., Yadav, P. et al. 2013. *Mimosa pudica* seed mucilage: isolation; characterization and evaluation as tablet disintegrant and binder. *International Journal of Biological Macromolecules*, 57: 105.

Alasyam, N., Sunil, M., Jayasree, T. et al. 2014. Evaluation of anticonvulsant activity of alcoholic extract of *Mimosa pudica* in swiss albino rats. *Journal of Chemical and Pharmaceutical Research*, 2014: 2014.

Allen, R.D. 1969. Mechanism of the seismonastic reaction in *Mimosa pudica*. *Plant Physiology*, 44: 1101–1107.

Altay, V. 2012. Mustafa Kemal Üniversitesi Tayfur Ata Sökmen Kampüsü (Hatay)'nün süs bitkileri. *Karadeniz Fen Bilimleri Dergisi*, 2(6): 11–26.

Altay, V., Ozyıgıt, I.I., and Yarcı, C. 2010. Urban flora and ecological characteristics of the Kartal District (Istanbul): A contribution to urban ecology in Turkey. *Scientific Research and Essays*, 5(2): 183–200.

Ayissi, M.R., Gartside, S., Ngo, B.E. et al. 2012. Effect of *Mimosa pudica* (Linn) extract on anxiety behaviour and GABAergic regulation of 5-HT neuronal activity in the mouse. *Journal of Psychopharmacology*, 26: 575–583.

Azmi, L., Singh, M.K. and Akhtar, A.K. 2011. Pharmacological and biological overview on *Mimosa pudica* Linn. *International Journal of Pharmaceutical and Life Sciences*, 2(11): 1226–1234.

Baghel, A., Rathore, D.S., and Gupta, V. 2013. Evaluation of diuretic activity of different extracts of *Mimosa pudica* Linn. *Pakistan Journal of Biological Sciences*, 16: 1223–1225.

Balunas, M.J., and Kinghorn, A.D. 2006. Drug discovery from medicinal plants. *Life Sciences*, 78: 431–441.

Bissanti, G. 2023. *Mimosa pudica*. https://antropocene.it/en/2023/04/05/mimosa-pudica-2. Accesses on 25.07.2023.

Bukhari, S.N.A., Ali, A., Hussain, M.A., Tayyab, M., Alotaibi, N.F., Elsherif, M.A., Junaid, K., and Ejaz, H. 2022. Extraction optimization of mucilage from seeds of *Mimosa pudica* by response surface methodology. *Polymers*, 2022, 14: 1904.

Bum, E.N., Dawack, D.L., Schmutz, M., Rakotonirina, A., Rakotonirina, S.V. Portet, C., Jeker, A., Olpe, H.R. and Herrling, P. 2004. Anticonvulsant activity of *Mimosa pudica* decoction. *Fitoterapia*, 75(3–4): 309–314.

Chaachouay, N., Douira, A., & Zidane, L. (2021a). Herbal Medicine Used in the Treatment of Human Diseases in the Rif, Northern Morocco. *Arabian Journal for Science and Engineering*, 47, 131–153. https://doi.org/10.1007/s13369-021-05501-1

Chaachouay, N., Douira, A., & Zidane, L. (2021b). COVID-19, prevention and treatment with herbal medicine in the herbal markets of Salé Prefecture, North-Western Morocco. *European Journal of Integrative Medicine*, 42, 101285. https://doi.org/10.1016/j.eujim.2021.101285

Chassagne, F., Hul, S., Deharo, E. and Bourdy, G. 2016. Natural remedies used by Bunong people in Mondulkiri province (Northeast Cambodia) with special reference to the treatment of 11 most common ailments. *Journal of Ethnopharmacology*, 191(4): 41–70.

Chatterjee, A., and Prakash, S.C. 2000. *The Treatise of Indian Medicinal Plants*. Publications and Information Directorate. Vol. 2. New Delhi: CSIR, pp. 65–66.

Chowdhury, S., Saha, D., and Paul S. 2012. In vitro cytotoxic activities of methanolic extract of *Mimosa pudica*. *Bulletin of Pharmaceutical Research Institute*, 2012: 42–45.

Eskin, B., Altay, V., Ozyıgıt, I.I., and Serın, M. 2012. Urban vascular flora and ecological characteristics of the Pendik District (Istanbul-Turkey). *African Journal of Agricultural Research*, 7(4): 629–646.

Ganguly, M., Devi, N., Mahanta, R. and Borthakur, M.K. 2007. Effect of *Mimosa pudica* root extract on vaginal estrous and serum hormones for screening of antifertility activity in albino mice. *Contraception*, 7(6): 482–485.

Goli, V., Bhaskar, K.V., Macharla, S.P., Bhaskar, J., Devi, P.S., and Ramchander, T. 2011. Effects of anti-inflammatory activity of *Mimosa pudica*. *Asian Journal of Pharmaceutical Research*, 1(3): 69–71.

Hasan, M.N., Azam, N.K., Ahmed, M.N. et al. 2016. A randomized ethnomedicinal survey of snakebite treatment in southwestern parts of Bangladesh. *Journal of Traditional & Complementary Medicine*, 6: 337–342.

Ignacimuthu, S., Ayyanar, M., and Sankarasivaraman, K. 2008. Ethnobotanical study of medicinal plants used by Paliyar tribals in theni district of Tamil Nadu, India. *Fitoterapia*, 79: 562–568.

Joseph, B., George, J., and Mohan J. 2013. Pharmacology and traditional uses of *Mimosa pudica*. *International Journal of Pharmaceutical Sciences & Drug Research*, 5: 41–44.

Joy, P.P., Thomas, J., Mathew, S., and Skaria, B.P. 2001. Medicinal Plants. *TropHort*, 2: 449–632.

Kaur, P., Kumar, N., Shivananda, T.N. et al. 2011. Phytochemical screening and antimicrobial activity of the plant extracts of *Mimosa pudica* L., against selected microbes. *Journal of Medicinal Plants Research*, 5: 5356–5359.

Kokane, D.D., More, R.Y., Kale, M.B., Nehete, M.N., Mehendale, P.C. and Gadgoli, C.H. 2009. Evaluation of wound healing activity of root of *Mimosa pudica*. *Journal of Ethnopharmacology*, 124: 311–315.

Konsue, A., Picheansoonthon, C., and Talubmook, C. 2017. Fasting blood glucose levels and hematological values in normal and streptozotocin-induced diabetic rats of *Mimosa pudica* L. *Pharmacognosy Journal*, 9(3): 315–322.

Kumar, D., Singh, H., Bhatt, U., Sharma, J., Sharma, S., and Soni, V. 2022. Physiological performance of *Mimosa pudica* L. under different light quality and photoperiods. *Physiologia*, 2: 132–153.

Li, Y.X., Zhu, J.X., Yang, H.X., and Yuan, K. 2011. Studies on antitumor activities of six glycosylflavones from *Mimosa pudica*. *2011 IEEE International Symposium on IT in Medicine and Education*, Vol. 1, pp. 186–189.

Mahanta, M. and Mukherjee, A.K. 2001. Neutralization of lethality, myotoxicity and toxic enzymes of *Naja kaouthia* venom by *Mimosa pudica* root extracts. *Journal of Ethnopharmacology*, 75: 55–60.

Malayan, J., Selvaraj, B., Warrier, A. et al. 2013. Anti-mumps virus activity by extracts of *Mimosa pudica*, a unique Indian medicinal plant. *Virus Disease*, 24: 166–173.

Merlin, F.F., and Narsimhan, D. 2009. Plant names and uses as indicators of knowledge patterns. *Indian Journal of Traditional Knowledge*, 8: 645–648.

Molina, M., Contreras, C.M., and Tellez-Alcantara, P. 1999. *Mimosa pudica* may possess antidepressant actions in the rat. *Phytomedicine*, 6(5): 319–323.

Muhammad, G., Hussain, M.A., Jantan, I., and Bukhari, S.N.A. 2016a. *Mimosa pudica* L., a high-value medicinal plant as a source of bioactives for pharmaceuticals. *Compr. Rev. Food Sci. Food Saf.*, 15: 303–315.

Muhammad, G., Hussain, M.A., Ashraf, M.U., Haseeb, M.T., Hussain, S.Z., and Hussain, I. 2016b. Polysaccharide based superabsorbent hydrogel from *Mimosa pudica*: Swelling-deswelling and drug release. *RSC Adv.*, 6: 23310–23317.

Muhammad, G., Ashraf, M.U., Naeem-ul-Hassan, M., and Bukhari, S.N.A. 2018. Appraisal of acute oral toxicity of glucuronoxylan hydrogel from *Mimosa pudica* seeds. *Brazilian Journal of Pharmaceutical Sciences*, 54: e17579.

Niroula, B., Parajuli, D., and Jha, S. 2009. Ecophysiology of *Mimosa pudica* L. at Biratnagar, Eastern Nepal. *Our Nature*, 7(1): 177–181.

Orch, H., Chaachouay, N., Douiri, E. M., Faiz, N., Zidane, L., & Douira, A. (2021). Use of medicinal plants in dermato-cosmetology: An ethnobotanical study among the population of Izarène. *Jordan Journal of Pharmaceutical Sciences*, *14*(3), 323–340.

Osma, E., Ozyıgıt, I.I., Altay, V., and Serın, M. 2010. Urban vascular flora and ecological characteristics of Kadikoy district, Istanbul, Turkey. *Maejo International Journal of Science and Technology*, 4(1): 64–87.

Panyadee, P. 2022. A review on the ethnobotany of exotic species in Thailand I: *Mimosa pudica* L. (Leguminosae). *Thai Journal of Botany*, 14(2): 79–94.

Parker, C. 2017. *Mimosa pudica* (sensitive plant). Available from: http://www.cabi.org/isc. Accessed on 11.04.2022.

Patel, N.K., and Bhutani, K.K. 2014. Suppressive effects of *Mimosa pudica* (L) constituents on the production of LPS induced proinflammatory mediators. *Excli Journal*, 13: 1011–1021.

Patro, G., Bhattamisra, S.K., and Mohanty, B.K. 2015. Analgesic, antiepileptic, and behavioral study of *Mimosa pudica* (Linn.) on experimental rodents. *International Journal of Nutrition Pharmacology Neurological Diseases*, 5(4): 144–150.

Paul, J., Khan, S., Mohammed, S. et al. 2010. Wound healing evaluation of chloroform and methanolic extracts of *Mimosa pudica* roots in rats. *Indian Journal of Physical Anthropology and Human Genetics*, 1: 223–227.

Rahman, A.H.M.M., and Debnath, A. 2015. Ethno-botanical study at the Village Pondit Para under Palash Upazila of Narsingdi District. Bangladesh. *International Journal of Advanced Research*, 3: 1037–1052.

Ranjan, R.K., Kumar, M.S., Seethalakshmi, I., and Rao, M.R. 2013. Phytochemical analysis of leaves and roots of *Mimosa pudica* collected from Kalingavaram, Tamil Nadu. *Journal of Chemical and Pharmaceutical Research*, 5(5): 53–55.

Ranjani, S., Faridha Begum, I., Santhoshini, J., Senthil Kumar, N., Ruckmani, K., and Hemalatha, S. 2021. *Mimosa pudica* floral nanoparticles: A potent antibiotic resistance breaker. Inorg. *Inorganic and Nano-Metal Chemistry*, 51(12): 1751–1758.

Rekha, R., Hemalatha, S., Akasakalai, K. et al. 2009. Hepatoprotective activity of *Mimosa pudica* leaves against carbontetrachloride induced toxicity. *Journal of Natural Products*, 2: 116–122.

Rezagholi, F., Hashemi, S.M.B., Gholamhosseinpour, A., Sherahi, M.H., Hesarinejad, M.A., and Ale, M.T. 2019. Characterizations and rheological study of the purified polysaccharide extracted from quince seeds. *Journal of the Science of Food and Agriculture*, 99: 143–151.

Sabina, E.P., Chandel, S., and Rasool, M.K. 2009. Evaluation of analgesic, antipyretic and ulcerogenic effect of Withaferin A. *International Journal of Integrative Biology*, 6: 52–56.

Sandhosh, K., Hoskeri, J., Pujari, N., and Priyanka Manjunatha, H. 2022. Antihypertensive and cardioprotective property of *Mimosa pudica* using zebra fish. *Pharmacog Res.*, 14(3): 333–337.

Saraswat, R., and Pokharkar, R. 2012. GCMS studies of *Mimosa pudica*. *Int. J. PharmTech Res.*, 4: 93–98.

Sharma, R. 2016. Medicinal plants diversity in Bhilai city district drug, Chhattisgarh, India. *International Journal of Pharmaceutical and Life Sciences*, 7(3): 4952–4966.

Silva, M.J.D., Carvalho, A.J.S., Rocha, C.Q. et al. 2014. Ethanolic extract of *Mimosa caesalpiniifolia* leaves: Chemical characterization and cytotoxic effect on human breast cancer MCF-7 cell line. *South African Journal of Botany*, 93: 64–69.

Sivarajan, V.V., and Balachandran, I. 2002. *Ayurvedic drugs and their plant sources*. New Delhi: Oxford and IBH Publishing Co. Pvt. Ltd., pp. 271–272.

Song, X.Y., Yao, G.D., Zhang, Y., and Song, S.J. 2017. Main chemical constituents and pharmacological bioactive of *Mimosa pudica* Linn. 亚洲传统医药, 12(4): 137–148.

Sunday, R.M. 2021. Toxicity profile of *Mimosa pudica* hydro-ethanolic seed extract. *Journal of Pharmaceutical & Allied Sciences*, 18(3): 3489–3496.

Sutar, N.G., Sutar, U.N. and Behera, B.C. 2009. Antidiabetic activity of *Mimosa pudica* Linn in albino rats. *Journal of Herbal Medicine*, 3(1): 123–126.

Tamilarasi, T., and Ananthi, T. 2012. Phytochemical analysis and anti-microbial activity of Mimosa pudica Linn. *Research Journal of Chemical Sciences*, 2(2):72–74.

Tunna, T.S., Ahmed, Q.U., Helal, A.B.M., and Zaidul, I.S.M. 2014. Weeds as alternative useful medicinal source: *Mimosa pudica* Linn. On diabetes mellitus and its complications. *Advanced Materials Research*, 995: 45–59.

USDA (United States Department of Agriculture Natural Resources Conservation Service). 2023. *Mimosa pudica* L. https://plants.usda.gov/home/plantProfile?symbol=MIPU8. Accessed on 25.07.2023.

Vaidyaratanm, P.S. 2001. *Indian Medicinal Plants Database*. 1st ed. Kottakkal II, Orient Longman: Arya Vidyashala, pp. 36–37.

Valsala, S., and Karpagaganapathy, P.R. 2002. Effect of *Mimosa pudica* root powder on oestrous cycle and ovulation in cycling female albino rat, Rattus norvegicus. *Phytotherapy Res.*,16: 190–192.

Vijayalakshmi, K. and Udayakumar, R. 2018. Antibacterial activity of leaf and root of *M. pudica* L. against selected human pathogenic microorganisms. *Biomedical and Pharmaceutical Sci.*, 1(2): 1–6.

www.powo.science.kew.org. Accesses on July 25, 2023.

www.zimbabweflora.co.zw. Accesses on July 25, 2023.

Yuan, J. 2007. Two new C-glycosylflavones from *Mimosa pudica*. *Chinese Chemical Letters*, 18: 1231–1234.

Yupparach, P., and Konsue. A. 2017. Hypoglycemic and hypolipidemic activities of ethanolic extract from *Mimosa pudica* L. in normal and streptozotocin-ınduced diabetic rats. *Pharmacog. J.*, 9(6): 834–837.

Zaware, B.B., Chaudhari, S.R., and Shinde, M.T. 2014. An overview of *Mimosa pudica* Linn. chemistry and pharmacological profile. *Research Journal of Pharmaceutical Biological and Chemical Sciences*, 5: 754–761.

19 Yopo (*Anadenanthera peregrina* (L.) Speg. Fabaceae)

Noureddine Chaachouay, Allal Douira, and Lahcen Zidane

19.1 INTRODUCTION

Plant-based medicines are utilized extensively across the world's traditional civilizations and are becoming increasingly well-liked as natural compounds for artificial chemicals (Chaachouay et al., 2019a, 2019b; Benkhnigue et al., 2022; Ahirwar, 2023). The usage of herbal medicine has skyrocketed in the previous several decades (Ekor, 2014). Because of its natural origin and few adverse effects, it has recently gained popularity in both developed and developing countries (Amadi & Orisakwe, 2018). *Anadenanthera peregrina* (L.) Speg., often called cohoba, jopo, Yopo, calcium tree, or parica, is a perennial tree belonging to the genus *Anadenanthera*. It is Indigenous to the Caribbean and South America. The tree may reach a height of 20 meters (66 feet) and has a rugged, horny bark (von Reis Altschul, 1964). The flowers of this plant form tiny, light yellow to white spherical clusters that resemble Acacia inflorescences, such as those seen on wattle trees. It is a psychoactive substance used in curative rites and rituals for millennia throughout northern South America and the Caribbean (Mota et al., 2017; Nascimento et al., 2009). Archaeological evidence indicates that *A. peregrina* beans have been used as hallucinogens for millennia (Athayde & Morellato, 2014). The first unequivocal evidence of use comes from smoking devices crafted from puma bone, discovered with *A. peregrina* beans, most likely belonging to the closely related species *A. colubrina* (Delices et al., 2023; Francisco et al., 2017). These artifacts were found in Inca Cueva, located in the Humahuaca Valley on the outskirts of the Puna region in Jujuy Province, Argentina.

Indigenous communities residing in the Orinoco basin in Colombia, Venezuela, and maybe the southern region of the Brazilian Amazon use Yopo snuff for spiritual healing (Childs, 2012; Llamazares & Sarasola, 2004). Yopo snuff was extensively used in ceremonial settings across the Caribbean region, including Puerto Rico and La Española, until the arrival of the Spanish (Guerra-Doce, 2015). Yopo snuff is often administered nasally to the user by another individual using bamboo tubes or sometimes self-administered by the user using bird bone tubes (Altschul, 1972). The hydroethanolic extract derived from the stem bark of *A. peregrina* was analyzed, revealing the presence of several chemical components. Furthermore, these compounds are responsible for several pharmacological activities such as antioxidant, anti-inflammatory, anesthetic, psychoactive, and toxic effects.

19.2 SCIENTIFIC CLASSIFICATION

Anadenanthera peregrina (L.) Speg., as a species, was initially recorded in 1737 in the Hortus Cliffortianus, based on a specimen found in Holland. However, Spanish explorers had already documented the use of the hallucinogenic snuff derived from the seeds and other components of this plant, commonly known as Yopo or Cohoba, as early as 1496 (Torres & Repke, 2014). Linnaeus included the species in his '*Species Plantarum*' and designated it *Mimosa peregrina*. The Western knowledge of Yopo's botanical identification remained elusive until June 1854, when British botanist Richard Spruce documented the components and production methods of Yopo. Spruce subsequently released *Notes of a Botanist*, a comprehensive account of his expeditions around South America, with detailed depictions of the flora and techniques used by the native populations to create mind-altering snuffs using *A. peregrina* and other botanical specimens. This species is one of only two widely recognized representatives of the genus *Anadenanthera* (Altschul, 1972). Previous classifications mainly focused on four subspecies or ecotypes. Following many categorization attempts across several genera, the majority of taxonomists now acknowledge the meticulous research conducted by Sirri von Reis Altschul in 1964. Altschul's revision of the genus taxonomy included just two members: *A. colubrina* and *A. peregrina*. Two latter varieties are acknowledged: *A. peregrina* var. *falcata* and *A. peregrina* var. *peregrina*.

Domain: Eukaryota
Kingdom: Plantae
Division: Magnoliophyta
Class: Magnoliopsida
Order: Fabales
Family: Fabaceae
Subfamily: Mimosoideae
Genus: *Anadenanthera*
Species: *Anadenanthera peregrina* (L.) Speg.

DOI: 10.1201/9781003460336-19

19.3 VERNACULAR NAMES

English: Cohoba, Hakudufha, Hataj, Kahhobba, Kahobba, Niopo, Parica, Yopa, Yopo, Yupa, Villca, Calcium tree; **French**: Bois l'écorce, Bois galle, Œuf de poule; **Spaish**: Candelón, Cojoba, Una de Gato, El Yopo;

19.4 BOTANICAL DESCRIPTION

Anadenanthera peregrina (L.) Speg. is a woody plant that may range in height from 3 to 27 meters, varying from a shrub to a towering tree. The trunk has a diameter of 20–40 cm at breast height and is often bent, twisted, and sometimes separated into many shafts at the base. Above the center of the sole stem, the branches are deformed and uneven, spreading out into an umbrella-like crown (Torres & Repke, 2014). The bark is gray to almost black and has many little lenticels. It lacks thorns, except for the lower trunk, which may produce conical thorns with wedge-shaped projections (Silva et al., 2019). These thorns may be particularly prominent while the tree is young, but the bark becomes rough and corky as it matures, forming tubercles and warty growths. In drier regions, the bark can get quite thick. The young twigs and leaves are covered in fine hairs, sometimes appearing bluish-green; adult leaves are smooth or almost smooth. The leaves, including the petioles, are 12–30 cm long, with the primary rachis being somewhat concave on the underside (Silva et al., 2019). The petioles have a slight darkening at their bases, with a flattish, oval, or oblong gland located 5–15 mm above. The length of the object is 5–5 mm. There are 1–4 glands that are smaller and similar in appearance. These glands are located either between or immediately below each of the pairs of ultimate pinna. There are often 10–30 or more pairs of pinnae, each with its own structure. The leaflets are typically 2–5 cm long and arranged in opposite or subopposite pairs. They are frequently imbricated, with 25–80 pairs. Each leaflet is 2–8 mm long (von Reis Altschul, 1964).

The leaves are 5–1.5 mm broad and have a linear or oblong shape. They are straight or curved like a sickle, with an oblique or truncated base and an acute to acuminate or apiculate tip. The mid-vein is off-center. The leaves have a membranous to leathery texture and are shiny (Figure 19.1), occasionally showing differences in color and texture on the upper and lower surfaces (Mota et al., 2017).

The heads, measuring 10–18 mm in diameter, consist of stamens and are greenish-white to creamy-yellow. They are placed in groups of 1–5 and may have a somewhat hairy or smooth surface while in the bud (Nascimento et al., 2009). The heads are found in the axils of the leaves and toward the tips of the branches, sometimes forming a racemose pattern. The peduncles are 1.75–4 cm long, covered in fine hairs, and are either thin or somewhat thicker. Each peduncle has a bell-shaped involucre with two teeth, covered in fine hairs, attached approximately three-quarters of the way up the peduncle (da Silva et al., 2023). The involucre may detach and glide loosely around the base of the peduncle—the calyx measures between 0.5 and 2.6 millimeters in length. The stamens are 2–3.5 mm long, while the anthers are 5–8 mm long. The anthers do not have glands while they are in the bud stage (Figure 19.2a). Bean: The object's length, including the stripe but not the peduncle, ranges from 5 to 35 cm. It is straight to falcate in shape, oblongish to elongated, and may be regularly or irregularly contracted around the seeds or not contracted at all. The object is more or less flattened with slightly thickened margins (Cortelete et al., 2021). At the base, it tapers to a point or is obtuse, while at the apex, it has a pointed or sharp tip, or if the tip is broken off, it is rounded. The surface of the object is scurfy to verrucose and dull in appearance. In dried specimens, it appears dark brown with rufous scales. The seeds, ranging from 8 to 16 in number, are

FIGURE 19.1 Aerial parts of *A. peregrina*.

FIGURE 19.2 Flowers (A) and seeds (B) of *A. peregrina*.

FIGURE 19.3 Geographical range of *A. peregrina*.

fragile and have a flat, round shape (Figure 19.2b). They have a dark chestnut brown to a black color and a lustrous appearance. The diameter of the seeds is between 10 and 20 mm (Athayde & Morellato, 2014).

19.5 ORIGIN AND GEOGRAPHIC DISTRIBUTION

A. peregrina is native to South America's tropical savannahs and semi-deciduous woods and several Caribbean islands. Within the Cerrado Biome of South America, this particular tree species is frequently observed in low-altitude tropical seasonal forests (Figure 19.3). It has been recorded in over 50% of surveyed locations in Sao Paulo state, Brazil, and in more than 70% of surveyed areas in the seasonally dry tropical forests of the central-west region of the Biome (Mangaravite et al., 2023). This tree is found in some regions of Colombia, namely in the Amazon basin. Venezuela is renowned for its growth in certain areas, particularly inside the Amazon jungle (Silva et al., 2019; Torres & Repke, 2014). In Argentina, it is located in the northern regions, which include areas characterized by a tropical or subtropical climate. *A. peregrina* can flourish in the nutrient-deficient soils of the Cerrado due to nitrogen-fixing rhizobia in the nodules of its roots (Viana et al., 2018; von Reis Altschul, 1964).

19.6 HABITAT AND ECOLOGY

A. peregrina is a flowering tree species Indigenous to South America, especially abundant in the tropical areas of Argentina, Brazil, Bolivia, and Paraguay. This species has remarkable ecological adaptability, allowing it to thrive in various settings. *A. peregrina* often occurs in open, disrupted habitats such as grasslands, forest clearings, and savannas, although it may also flourish in thickly forested surroundings (Athayde & Morellato, 2014). The plant's capacity to thrive in several soil types adds to its extensive distribution, enabling it to grow in alkaline and acidic soils. *A. peregrina* has a substantial ecological impact on nearby habitats (Silva et al., 2019). The tree can perform nitrogen fixation in the ground, which improves its fertility and provides advantages to nearby plant species. Moreover, the seeds of *A. peregrina* serve as a vital nourishment for many species, enhancing the biodiversity of the locations where it resides (Tótola & Borges, 2000). The ecological interactions associated with this species, ranging from nitrogen fixation to seed distribution, emphasize its significance in preserving the equilibrium and health of its habitats (Cortelete et al., 2021). *A. peregrina* has distinctive reproductive traits that are probably adaptations to its surroundings. The breeding system of this organism has been shown to be primarily xenogamous, a common reproductive strategy among many plants. This method promotes genetic variety, leading to increased chances of survival (Mangaravite et al., 2023). Nevertheless, when compelled to generate fruit by autogamy, *A. peregrina* progressively enhances its genetic self-compatibility, a tactic that promotes the existence of its offspring in a fragmented forest habitat (Francisco et al., 2017).

19.7 *A. PEREGRINA* CULTIVATION

The *A. peregrine* tree is Indigenous to the tropical regions of South America, including Brazil, Colombia, and Peru. *A. peregrine* tree thrives in arid areas such as savannas and open plains rather than rainforests. The height of *A. peregrina* trees may reach a maximum of 18 meters. They are distinguished by their gray-black bark covered in thorns and their green, pinnate leaves (Santos et al., 2020). Cultivating an *A. peregrina* tree outside the tropics is challenging, but it can be tried by maintaining it as a houseplant in a well-illuminated location. To initiate the seed germination process, it is necessary to immerse them in sterile water for 24 hours (Silva et al., 2019). Subsequently, plant the seeds in a blend of perlite and sand at a depth of roughly 1 cm in the soil. Ensure that the earth is adequately hydrated but not saturated (Gomes et al., 2012). Position the pot(s) in a well-lit area (avoiding direct sunlight) with a minimum temperature of 22°C. To inhibit mold growth, provide frequent watering and maintain proper ventilation. To foster the development of the saplings into young trees, it is necessary to transfer them to a bigger and deeper container with optimal drainage (Souza et al., 2020; Torres & Repke, 2014).

19.8 ETHNOBOTANY

19.8.1 Archaeological Evidences

A. peregrina has held cultural and historical importance among Indigenous communities, who have employed it for diverse purposes, such as shamanic and religious ceremonies. Archaeological evidence indicates that *A. peregrina* beans have been used as hallucinogens for millennia (Torres & Repke, 2014). The first unambiguous proof of use comes from pipes crafted from *Puma concolor* bone discovered with *A. peregrina* beans, perhaps from the related species *A. colubrina*, in Inca Cueva, a location situated in the Huma-huaca valley on the outskirts of the Puna of Jujuy Province, Argentina. The pipes were discovered to contain the hallucinogenic substance DMT, which is one of the chemicals present in *A. peregrina* beans. The radiocarbon analysis of the importance yielded a date of 2130 B.C., indicating that the use of *A. peregrine* as a hallucinogen dates back more than 4,000 years (Pagán-Jiménez & Carlson, 2014). Snuff trays and tubes like those often used for *A. peregrina* were discovered on the central Peruvian coast, with origins reaching back to 1200 B.C. This finding implies that insufflating *A. peregrina* beans is a relatively recent development (Torres, 1995). Studies indicate the presence of archaeological evidence showing the usage of insufflation in northern Chile within the time frame of A.D. 500–1000. Indigenous populations residing in the Orinoco basin in Colombia, Venezuela, and maybe the southern region of the Brazilian Amazon use Yopo snuff as a means of spiritual healing (Altschul, 1972).

19.8.2 Traditional Preparation

The preparation methods of *A. peregrina* differ across different tribes. The pods are first wet and then rolled into a paste, then slowly roasted over a moderate fire until dehydrated and toasted. Occasionally, the beans undergo fermentation before being processed into the paste. The solidified mixture may be preserved for future use after the toasting process. The Indigenous communities roast the beans and pulverize them into a paste, using a meticulously crafted block of hardwood designed explicitly for this task (Monteiro et al., 2006; Torres & Repke, 2014; Chaachouay et al., 2022a, 2022b, 2022c). The resultant grayish-green powder is often combined with about equal quantities of an alkaline substance, such as lime derived from snail shells or the ashes of plant matter. The ashes are derived from a diverse range of plant components, including the charred fruit of the monkey pot, the bark of various vines and trees, and even the roots of sedges (Pagán-Jiménez & Carlson, 2014). The inclusion of the ashes likely serves a functional purpose: to prevent the snuff from solidifying in the moist jungle environment.

19.8.3 Herbal Medicine

Yopo is a naturally occurring entheogenic substance from the Venezuelan and Brazilian Amazon and has been used for medicinal purposes since ancient times. The meaning is derived from the powdered seeds of the *A. peregrina* plant. It is combined with other ashes and plant seeds and then shaped into little plates or cakes of various colors (black, brown, or white). These plates are further crushed into a fine powder before being inhaled (Marinho et al., 2021; Chaachouay et al., 2021a, 2021b, 2021c). Yopo, in contrast to ayahuasca or Bufo Alvarius, transcends the experience of love and gives us the ability to examine both unconditionality and restrictions. It is as if we find ourselves in a room with two windows, each offering a unique perspective. *A. peregrina* is mainly believed for its hallucinogenic effects. However, some Indigenous cultures have also ascribed therapeutic benefits to this plant (Belhaj et al., 2021; Orch et al., 2021; Delices et al., 2023). Historically, Yopo has been used in medicinal rituals, where Indigenous communities integrate it into rites designed to treat both physical and psychological conditions. Shamans, or traditional healers who possess extensive knowledge about the qualities of plants, give Yopo in meticulously planned ceremonies (Torres & Repke, 2014; Benlarbi et al., 2023). They believe that the psychoactive substances in Yopo may cause altered states of consciousness that are beneficial for therapeutic purposes (Pagán-Jiménez & Carlson, 2014). Within the traditional medicine practiced by Indigenous communities, *A. peregrina* has been used as a diagnostic instrument to ascertain ailments and understand their underlying origins (Monteiro et al., 2006). The visionary encounters elicited by Yopo are said to provide shamans with a distinctive outlook that facilitates comprehension of the spiritual or energy irregularities that contribute to an individual's sickness (Torres & Repke, 2014). A tea derived from *A. peregrina* seeds is used to alleviate digestive ailments. When mixed with a chicha drink, the seeds may also relieve fever, depression, and other conditions. Honey is supplemented with seeds to enhance female fertility and act as a diuretic. However, the seeds are also abortifacients (Chaachouay et al., 2020a, 2020b; Marinho et al., 2021). The gum derived from the *A. peregrina* plant is used to treat coughs, while the snuff from this plant alleviates headaches, constipation, and flu symptoms. Yopo snuff is administered by inhaling hollow tubes made from bird bones or bamboo. The onset of the effects is almost instantaneous, characterized by muscle twitching, mild convulsions, and impaired motor coordination (Mota et al., 2017). This is subsequently accompanied by symptoms such as nausea, visual hallucinations, and disrupted sleep patterns. Objects often exhibit an unnatural magnification in size. According to an initial account, the Indians describe their residences as inverted, with men seemingly walking on their feet in the air.

19.8.4 Ceremonial Uses

A. peregrina is a botanical species traditionally used by Indigenous communities in South America and the Caribbean to conduct ceremonial rituals and engage in magical-religious practices. The plant has alkaloids in its seeds, leaves, and bark, which are said to possess hallucinogenic qualities (Pagán-Jiménez & Carlson, 2014). The plant has been used in several modalities, including nasal inhalation, smoking via pipes or cigars, and incorporation into fermented beverages. The herb was vital in the cultural advancement of Indigenous communities in South America (Torres & Repke, 2014; Chaachouay & Zidane, 2021). Starch grains that match those found in modern *A. peregrina* seeds were discovered in a coral milling base at a precolonial habitation site in Eastern Puerto Rico. This is the first time such grains have been found in the West Indies. The context of the discovery dates back to A.D. 1150–1250 (Altschul, 1972). A comprehensive analysis was conducted on historical, cultural, and archaeological information on *A. peregrina* in the West Indies and South America (Delices et al., 2023). The purpose was to provide credible sociocultural explanations for the discoveries.

19.8.5 Snuffing Rituals

Yopo snuff was extensively used in ceremonial settings across the Caribbean region, including Puerto Rico and La Española, until the arrival of the Spanish. Yopo snuff is often administered nasally to the user either by another individual using bamboo tubes or by the user using bird bone tubes (Altschul, 1972; Rodd, 2002). Blowing is a more efficient technique since it enables a more significant amount of powder to enter the nasal cavity and is considered less unpleasant. In some regions, the raw ground beans are inhaled or burned, resulting in a diminished impact and heightened bodily sensations (Schultes, 1967). Certain Indigenous groups use Yopo in conjunction with *Banisteriopsis caapi* to enhance and extend the hallucinogenic effects, resulting in a comparable experience to that of ayahuasca (Rodd & Sumabila, 2011).

19.9 HALLUCINOGENIC EFFECTS

The psychoactive effects of *A. peregrina* seeds are mainly due to potent chemicals such as DMT bufotenin, 5-HO-DMT, and 5-MeO-DMT. These chemicals are classified as tryptamines and are recognized for their significant hallucinogenic properties. Bufotenin, specifically, is linked to profound psychedelic encounters, and its presence in the seeds enhances the psychoactive properties of Yopo (Torres & Repke, 2014). DMT, a renowned hallucinogenic chemical, is acknowledged for its ability to induce altered states of consciousness marked by intense visual and aural hallucinations. The amalgamation of these mind-altering components in *A. peregrina* produces a potent concoction used for millennia in shamanic rites and ceremonies (Pagán-Jiménez & Carlson, 2014).

The Indigenous tribes, such as the Yanomami and Piaroa in Venezuela and the Yora in Peru, have included Yopo in their spiritual and shamanic rituals. The seeds are often processed into a powdered form known as snuff, which is then given during ceremonial rites (Bussmann, 2016). The hallucinatory effects caused by Yopo are considered essential in these rituals, as participants aim to establish a connection with the spiritual domain (Passos & Mironidou-Tzouveleki, 2016). The psychotropic substances are said to facilitate visionary experiences that enable contact with the divine, spirits, or ancestors. This link is a method of acquiring profound understanding, getting direction, and obtaining information otherwise beyond everyday awareness (Delices et al., 2023).

The divinatory and spiritual importance of Yopo in Indigenous cultures beyond simple psychedelic encounters (Chagnon et al., 1971). These rituals are firmly ingrained in the cultural essence of these tribes, influencing their perception of the universe, the interdependence of everything, and the significance of people in the broader spiritual realm (Moreira et al., 2015). The use of *A. peregrina* in shamanic rituals serves as a method for generating modified mental states and a significant cultural practice that promotes a heritage, a feeling of harmony, and the passing down of ancestral knowledge from one lineage to another.

19.10 PHYTOCHEMISTRY

An investigation of the hydroethanolic extract obtained from the stem bark of *A. peregrina* identified the existence of many compounds, including cardiac glycosides, organic acids, reducing sugars, hemolytic saponins, phenols, coumarins, condensed tannins, flavonoids, catechins, depsides, and depsidones generated from benzoquinones (Delices et al., 2023; Marinho et al., 2021). Other investigations have shown that *A. peregrina* includes many compounds, including 5-Methoxy-N, N-dimethyltryptamine, Isoorientin, Isovitexin, Leucopelargogidin, Orientin, Vitexin, Serotonin, Saponarentin, Viterine, Bufotenine, Bufotenin N-oxide, N, N-Dimethyltryptamine N-oxide, N, N-Dimethyltryptamin, Homoorientine, Leucoanthocyanin, Leucopelargonidol, 5-Methoxy-N-methyltryptamine, Catechol, O-Methylbufotenin, N-Methyltryptamine, N, N-Dimethyltryptamine-oxide, N, N-Dimethyltryptamine, Bufotenine, 5-methoxy-N, N-dimethyltryptamine, 5-Methoxy-N, N-dimethyltryptamine, 1,2,3,4-Tetrahydro-6-methoxy-2-methyl-beta-carboline, and 1,2,3,4-Tetrahydro-6-methoxy-2,9-dimethyl-beta-carboline (Marinho et al., 2022; Moreira et al., 2015; Mota et al., 2017).

19.11 PHARMACOLOGY

A. peregrina has historically been used by Indigenous cultures for several reasons, including medicinal and shamanic rituals. It has many pharmacologically active chemicals.

19.11.1 Anesthetic Effects

The plant's seeds contain bioactive chemicals, such as bufotenin and DMT, with local anesthetic qualities (Marinho et al., 2021). Indigenous tribes have historically used *A. peregrina* preparations as a means of local anesthetic for dental treatments and surgical interventions (Miller et al., 2019). When administered topically or by nasal insufflation, these substances can cause transient numbness and analgesia, which makes them practical for relieving pain and suffering in a specific area. Nevertheless, it is imperative to show prudence while using *A. peregrina* for its anesthetic capabilities since it has psychotropic attributes and is subject to regulatory constraints in different areas (Marinho et al., 2022, 2022).

19.11.2 Anti-Inflammatory Properties

A. peregrina has been associated with the possibility of possessing anti-inflammatory properties. According to some studies, anti-inflammatory properties in extracts derived from *A. peregrina* may be attributable to bioactive compounds within the plant (Marinho et al., 2022). These compounds have the potential to alleviate inflammation and its related symptoms. However, additional investigation is required to comprehend its anti-inflammatory properties' intricacies and scope comprehensively (Junior et al., 2021). The utilization of *A. peregrina* must be approached with utmost caution and in adherence to local regulations and cultural customs, given that its principal historical application occurs within the framework of Indigenous ceremonies and rituals (Passos & Mironidou-Tzouveleki, 2016; Torres & Repke, 2014).

19.11.3 Antioxidant Activity

Several studies have documented the antioxidant activity of *A. peregrina*. This characteristic is frequently ascribed to the botanical specimen's phenolic compounds and additional antioxidants (Marinho et al., 2021). Antioxidants are of the utmost importance in the body because they neutralize free radicals, which are known to cause oxidative stress and a multitude of other health complications. Although *A. peregrina* exhibits encouraging antioxidant properties, it is critical to acknowledge that its conventional application is predominantly centered around Indigenous rituals and ceremonies (Athayde & Morellato, 2014; Mota et al., 2017). Consequently, additional investigation is warranted to understand better the plant's distinct antioxidant mechanisms and potential health advantages.

19.11.4 Toxic Effects

A. peregrina can cause hazardous consequences if its seeds are consumed or utilized incorrectly. *A. peregrina* seeds contain psychoactive substances such as bufotenin and dimethyltryptamine, which may elicit altered states of consciousness and hallucinations when consumed by different methods (Delices et al., 2023). Nevertheless, these effects may be accompanied by unfavorable responses, such as nausea, vomiting, diarrhea, disorientation, confusion, and potentially more severe symptoms, contingent upon the dosage and individual susceptibility. It is essential to exercise utmost vigilance and adhere to local regulations and cultural customs while contemplating the use of

A. peregrina for its psychotropic attributes. (Francisco et al., 2017; Gaujac et al., 2012) To limit the danger of harmful responses and guarantee safe consumption, engaging in proper preparation, adhering to the correct dose, and seeking professional counsel are crucial. The need for cautious and well-informed use is emphasized by the toxicity linked to *A. peregrina* (Marinho et al., 2022). Those who are interested in its traditional or therapeutic benefits should consult with trained practitioners and be mindful of the possible hazards involved (Gomes et al., 2012). Moreover, psychoactive chemicals such as *A. peregrina* are subject to different legal limitations depending on the area, making it crucial to adhere to local legislation to prevent legal consequences. In addition, it is crucial to acknowledge that the pharmacological impacts of *A. peregrina* might differ based on the method of preparation and administration, and its utilization may be subject to regulatory limitations in some areas (Marinho et al., 2022).

19.12 MEDICINAL EFFECTS

Recent scientific studies increasingly recognize Yopo's potential for therapeutic purposes, aligning with old spiritual traditions. However, several bioactive substances in the seeds of *A. peregrina*, including bufotenin, 5-HO-DMT, and DMT, have been examined for their possible psychoactive and medicinal characteristics (Gaujac et al., 2012; Rodd, 2002; Rodd & Sumabila, 2011). Bufotenin, 5-HO-DMT, and DMT possess hallucinogenic qualities and can cause altered states of consciousness (Oroc, 2009). The primary emphasis of research on these chemicals has been their impact on serotonin receptors in the brain, which may lead to mood, perception, and cognition alterations (Sargin et al., 2019). Imperial College London initiated clinical studies in early 2021 to investigate the efficacy of DMT as a treatment for depression. The clinical research investigating the possible therapeutic uses of these substances, such as in the management of depression, anxiety, or post-traumatic stress disorder, are restricted and encounter regulatory obstacles (Andersen et al., 2021).

19.13 CONCLUSION

A. peregrina is a captivating plant that has undergone extensive botanical and phytochemical research, uncovering a plethora of bioactive chemicals. The historical and cultural significance of ayahuasca is seen in its traditional use by Indigenous communities in South America, primarily for its hallucinogenic effects. Nevertheless, using this botanical specimen entails potential hazards associated with its poisonous properties, such as emesis, gastrointestinal disturbances, and cardiovascular complications. Consequently, it is crucial to handle it with care. *A. peregrina* is a subject of ongoing fascination for researchers in botany, phytochemistry, and ethnobotany. Researching this plant's chemical components, biosynthesis, and genetic diversity is essential for gaining a deeper understanding of its therapeutic potential. Simultaneously, recording the Indigenous knowledge and cultural customs linked to this plant is crucial while promoting awareness of its possible hazards. Effective regulation is vital to mitigate the potential risks related to its use and encourage conscientious usage while safeguarding the cultural legacy of Indigenous people who hold it as holy.

LIST OF ABBREVIATIONS

A. peregrina	*Anadenanthera peregrina* (L.) Speg.
A.D.	Anno Domini
B.C.	Before Christ
DMT	Dimethyltryptamine

REFERENCES

Ahirwar, R. K. (2023). An Overview of Medicinal Plants: Drugs of Tomorrow. In D. Singh, A. K. Mishra, & A. K. Srivastava (Éds.), *Stress-responsive Factors and Molecular Farming in Medicinal Plants* (pp. 1–16). Springer Nature Singapore. https://doi.org/10.1007/978-981-99-4480-4_1

Altschul, S. R. (1972). *The Genus Anadenanthera in Amerindian cultures*. Harvard University. https://www.samorini.it/doc1/alt_aut/ad/altschul-the-genus-anadenanthera-in-amerindian-cultures.pdf

Amadi, C. N., & Orisakwe, O. E. (2018). Herb-induced liver injuries in developing nations: An update. *Toxics*, *6*(2), 24.

Andersen, K. A. A., Carhart-Harris, R., Nutt, D. J., & Erritzoe, D. (2021). Therapeutic effects of classic serotonergic psychedelics: A systematic review of modern-era clinical studies. *Acta Psychiatrica Scandinavica*, *143*(2), 101–118. https://doi.org/10.1111/acps.13249

Athayde, E. A., & Morellato, L. P. C. (2014). Anthropogenic edges, isolation and the flowering time and fruit set of *Anadenanthera peregrina*, a cerrado savanna tree. *International Journal of Biometeorology*, *58*(4), 443–454. https://doi.org/10.1007/s00484-013-0727-y

Belhaj, S., Chaachouay, N., & Zidane, L. (2021). Ethnobotanical and toxicology study of medicinal plants used for the treatment of diabetes in the High Atlas Central of Morocco. *Journal of Pharmacy & Pharmacognosy Research*, *9*(5), 619–662. https://doi.org/10.56499/jppres21.1014_9.5.619

Benlarbi, F., Mimoune, N., Chaachouay, N., Souttou, K., Saidi, R., Mokhtar, M.R., Kaidi, R, & Benaissa, M. T. (2023). Ethnobotanical survey of the traditional antiparasitic use of medicinal plants in humans and animals in Laghouat (Southern Algeria). *Veterinary World*, *16*(2), 357–368. https://doi.org/10.14202/vetworld.2023.357

Benkhnigue, O., Chaachouay, N., Khamar, H., El Azzouzi, F., Douira, A, & Zidane, L. (2022). Ethnobotanical and ethnopharmacological study of medicinal plants used in the treatment of anemia in the region of Haouz-Rehamna (Morocco). *Journal of Pharmacy & Pharmacognosy Research*, *10*(2), 279–302. https://doi.org/10.56499/jppres21.1196_10.2.279

Bussmann, R. W. (2016). Magic Plants. In U. P. Albuquerque & R. R. Nóbrega Alves (Éds.), *Introduction to Ethnobiology* (p. 163–169). Springer International Publishing. https://doi.org/10.1007/978-3-319-28155-1_24

Chaachouay, N., Benkhnigue, O., Fadli, M., Ibaoui, H., & Zidane, L. (2019a). Ethnobotanical and ethnopharmacological studies of medicinal and aromatic plants used in the treatment of metabolic diseases in the Moroccan Rif. *Heliyon*, *5*(10), e02191. https://doi.org/10.1080/10477845.2021.1942433

Chaachouay, N., Benkhnigue, O., Fadli, M., Ibaoui, H., El Ayadi, R., & Zidane, L. (2019b). Ethnobotanical and ethnopharmacological study of medicinal and aromatic plants used in the treatment of respiratory system disorders in the Moroccan Rif. *Ethnobotany Research and Applications*, *18*(22), 1–16.

Chaachouay, N., Douira, A., Hassikou, R., Brhadda, N., Dahmani, J., Belahbib, N., Ziri, R., & Zidane, L. (2020a). Ethnobotanical and ethnopharmacological study of medicinal and aromatic plants used in the treatment of respiratory system disorders in the Moroccan Rif. PhDThesis, Département de Biologie - Université Ibn Tofail - Kénitra, Maroc, https://tel.archives-ouvertes.fr/tel-03376377

Chaachouay, N., Benkhnigue, O., & Zidane, L. (2020b). Ethnobotanical study aimed at investigating the use of medicinal plants to treat nervous system diseases in the Rif of Morocco. *Journal of Chiropractic Medicine*, *19*(1), 70–81. https://doi.org/10.1016/j.jcm.2020.02.004

Chaachouay, N., & Zidane, L. (2021). The Symbolic Efficacy of Plants in Rituals and Socio-Religious Ceremonies in Morocco, Northwest of Africa. *Journal of Religious & Theological Information*, *21*(1–2), 34–53. https://doi.org/10.1080/10477845.2021.1942433

Chaachouay, N., Benkhnigue, O., Douira, A., & Zidane, L. (2021a). Poisonous medicinal plants used in the popular pharmacopoeia of the Rif, northern Morocco. *Toxicon*, 189, 24–32. https://doi.org/10.1016/j.toxicon.2020.10.028

Chaachouay, N., Douira, A., & Zidane, L. (2021b). COVID-19, prevention and treatment with herbal medicine in the herbal markets of Salé Prefecture, North-Western Morocco. *European Journal of Integrative Medicine*, 42, 101285. https://doi.org/10.1016/j.eujim.2021.101285

Chaachouay, N., Douira, A., & Zidane, L. (2021c). Herbal Medicine Used in the Treatment of Human Diseases in the Rif, Northern Morocco. *Arabian Journal for Science and Engineering*, 47, 131–153. https://doi.org/10.1007/s13369-021-05501-1

Chaachouay, N., Benkhnigue, O., & Zidane, L. (2022a). Ethnobotanical and Ethnomedicinal study of medicinal and aromatic plants used against dermatological diseases by the people of Rif, Morocco. *Journal of Herbal Medicine*, 32, 100542. https://doi.org/10.1016/j.hermed.2022.100542

Chaachouay, N., Azeroual, A., Bencharki, B., Douira, A., & Zidane, L. (2022b). Ethnoveterinary medicines plants for animal therapy in the Rif, North of Morocco. *South African Journal of Botany*, *147*, 176–191. https://doi.org/10.1016/j.sajb.2021.12.037

Chaachouay, N., Azeroual, A., Douira, A., & Zidane, L. (2022c). Ethnoveterinary practices of medicinal plants among the Zemmour and Zayane tribes, Middle Atlas, Morocco. *South African Journal of Botany*, *151*, 826–840. https://doi.org/10.1016/j.sajb.2022.11.009

Chagnon, N. A., Le Quesne, P., & Cook, J. M. (1971). Yanomamo hallucinogens: Anthropological, botanical, and chemical findings. *Current Anthropology*, *12*(1), 72–74. https://doi.org/10.1086/201170

Childs, D. (2012). *A Study of Religion, Culture, and Medicinal Plants of Three South American Indigenous Groups* [University of Delaware]. http://udspace.udel.edu/handle/19716/11328

Cortelete, M. A., Silva Júnior, A. L., Pereira, M. L. S., Miranda, F. D., & Caldeira, M. V. W. (2021). Molecular characterization as strategy for ex situ conservation of *Anadenanthera peregrina* (L.) Speg. *Scientia Forestalis*, *49*, e3443.

da Silva, R. C., Fonseca, A. C. C., Aguiar, B. A. C., de Almeida Sousa, H. G., Epifânio, M. L. F. G., Muroi, E. de A. R., Viana, R. H. O., & de Souza, P. B. (2023). Morphological parameters associated with the quality of seedlings of *Anadenanthera peregrina* (L.) speg. on different substrates. *Revista De Gestão Social E Ambiental*, *17*(9), 1–9.

Delices, M., Muller, J. de A. I., Arunachalam, K., & de Oliveira Martins, D. T. (2023). *Anadenanthera colubrina* (Vell) Brenan: Ethnobotanical, phytochemical, pharmacological and toxicological aspects. *Journal of Ethnopharmacology*, *300*, 115745.

Ekor, M. (2014). The growing use of herbal medicines: Issues relating to adverse reactions and challenges in monitoring safety. *Frontiers in Pharmacology*, *4*, 177.

Francisco, T. M., Lopes-Mattos, K. L. B., Picoli, E. A. D. T., Couto, D. R., Oliveira, J. A., Zanuncio, J. C., Serrão, J. E., De Oliveira Silva, I., & Boere, V. (2017). Feeding habits of marmosets: A case study of bark anatomy and chemical composition of *Anadenanthera peregrina* gum. *American Journal of Primatology*, *79*(3), e22615. https://doi.org/10.1002/ajp.22615

Gaujac, A., Navickiene, S., Collins, M., Brandt, S., & Andrade, J. (2012). Analytical techniques for the determination of tryptamines and β-carbolines in plant matrices and in psychoactive beverages consumed during religious ceremonies and neo-shamanic urban practices. *Drug Testing and Analysis*, *4*, 636–648. https://doi.org/10.1002/dta.1343

Gomes, M. P., Moreira Duarte, D., Silva Miranda, P. L., Carvalho Barreto, L., Matheus, M. T., & Garcia, Q. S. (2012). The effects of arsenic on the growth and nutritional status of *Anadenanthera peregrina*, a Brazilian savanna tree. *Journal of Plant Nutrition and Soil Science*, *175*(3), 466–473. https://doi.org/10.1002/jpln.201100195

Guerra-Doce, E. (2015). Psychoactive substances in prehistoric times: Examining the archaeological evidence. *Time and Mind*, *8*(1), 91–112. https://doi.org/10.1080/1751696X.2014.993244

Junior, O. C., Lima, N. M., Silva, M. G. A., Aguiar, V. B., Carli, G. P., Scherrer, E. C., Castro, S. B. R., Alves, C. C. S., Oliveira, M. A. L., & Carli, A. P. (2021). In vitro and in vivo evaluation of anti-inflammatory activity and free radical scavenging potential of leaves extract from *Anadenanthera colubrina*. *Natural Product Research*, *35*(22), 4819–4823. https://doi.org/10.1080/14786419.2020.1727472

Llamazares, A. M., & Sarasola, C. M. (2004). *Main Sacred Plants in South America - El lenguaje de los dioses. pp.259–285.as 'Principales plantas sagradas de Sudamérica'*. Editorial Biblos.

Mangaravite, É., Silveira, T. C. da, Vinson, C. C., Bueno, M. L., Silva, R. dos S., Carniello, M. A., Veldman, J. W., Garcia, M. G., & Oliveira, L. O. de. (2023). Unlocking the secret diversity of *Anadenanthera*: Insights from molecular genetics of four evolving species. *Botanical Journal of the Linnean Society*, *204*(1), 47–62.

Marinho, T. A., Nunes, P. S., Junior, R. de S. L., de Melo Reis, P. R., Machado, M. R. F., Junior, N. J. S., de Menezes Filho, A. C. P., & Ventura, M. V. A. (2022). Evaluation of angiogenic and embryotoxic activity of the extract of *Anadenanthera peregrina* (Angico-do-Cerrado). *Brazilian Journal of Science*, *1*(9), 20–29.

Marinho, T. A., Oliveira, M. G., Menezes-Filho, A. C. P., Castro, C. F. S., Oliveira, I. M. M., Borges, L. L., Melo-Reis, P. R., & Silva-Jr, N. J. (2021). Phytochemical characterization, and antioxidant and antibacterial activities of the hydroethanolic extract of *Anadenanthera peregrina* stem bark. *Brazilian*

Journal of Biology, *82*. https://www.scielo.br/j/bjb/a/XFGt5mmpbTrVBXtc6zZX9qg/?format=html&lang=en

Miller, M. J., Albarracin-Jordan, J., Moore, C., & Capriles, J. M. (2019). Chemical evidence for the use of multiple psychotropic plants in a 1,000-year-old ritual bundle from South America. *Proceedings of the National Academy of Sciences*, *116*(23), 11207–11212. https://doi.org/10.1073/pnas.1902174116

Monteiro, J. M., De Almeida, C. D. F. C. R., De Albuquerque, U. P., De Lucena, R. F. P., Florentino, A. T. N., & De Oliveira, R. L. C. (2006). Use and traditional management of *Anadenanthera colubrina* (Vell.) Brenan in the semi-arid region of northeastern Brazil. *Journal of Ethnobiology and Ethnomedicine*, *2*(1), 6. https://doi.org/10.1186/1746-4269-2-6

Moreira, L. A., Murta, M. M., Gatto, C. C., Fagg, C. W., & Santos, M. L. D. (2015). Concise synthesis of *N,N*-dimethyltryptamine and 5-methoxy- *N,N* -dimethyltryptamine starting with bufotenine from Brazilian *Anadenanthera* ssp. *Natural Product Communications*, *10*(4). https://doi.org/10.1177/1934578X1501000411

Mota, G. S., Sartori, C. J., Miranda, I., Quilho, T., Mori, F. A., & Pereira, H. (2017). Bark anatomy, chemical composition and ethanol-water extract composition of *Anadenanthera peregrina* and *Anadenanthera colubrina*. *PloS One*, *12*(12), e0189263.

Nascimento, G. de A., Pifano, D. S., de Lima, M. P., & Calegario, N. (2009). Floristic aspects and diversity of regenerated arboreal species under a stand of *Anadenanthera peregrina* Speg. *Cerne*, *15*(2), 187–195.

Orch, H., Chaachouay, N., Douiri, E. M., Faiz, N., Zidane, L., & Douira, A. (2021). Use of medicinal plants in dermato-cosmetology: An ethnobotanical study among the population of Izarène. *Jordan Journal of Pharmaceutical Sciences*, *14*(3), 323–340.

Oroc, J. (2009). *Tryptamine Palace: 5-MeO-DMT and the Sonoran Desert Toad*. Simon and Schuster.

Pagán-Jiménez, J. R., & Carlson, L. A. (2014). Recent archaeobotanical findings of the hallucinogenic snuff cojoba (*Anadenanthera peregrin* A (L.) Speg.) in precolonial Puerto Rico. *Latin American Antiquity*, *25*(1), 101–116.

Passos, I. D., & Mironidou-Tzouveleki, M. (2016). Hallucinogenic plants in the Mediterranean countries. In *Neuropathology of drug addictions and substance misuse* (pp. 761–772). Elsevier. Academic Press. https://doi.org/10.1016/B978-0-12-800212-4.00071-6

Rodd, R. (2002). Snuff Synergy: Preparation, Use and Pharmacology of *Yopo* and *Banisteriopsis Caapi* Among the Piaroa of Southern Venezuela. *Journal of Psychoactive Drugs*, *34*(3), 273–279. https://doi.org/10.1080/02791072.2011.566499

Rodd, R., & Sumabila, A. (2011). Yopo, ethnicity and social change: A comparative analysis of Piaroa and Cuiva Yopo use. *Journal of Psychoactive Drugs*, *43*(1), 36–45. https://doi.org/10.1080/02791072.2011.566499

Santos, C. C., Jorge, H. P. G., Dias, L. G. F., & Vieira, M. do C. (2020). Shading levels and substrates affect morphophysiological responses and quality of *Anadenanthera peregrina* (L.) Speg seedlings. *Floresta e Ambiente*, *27*. https://www.scielo.br/j/floram/a/qC8DdFVBzJSTMHyMhkM3L4S/?lang=en

Sargin, D., Jeoung, H.-S., Goodfellow, N. M., & Lambe, E. K. (2019). Serotonin regulation of the prefrontal cortex: Cognitive relevance and the impact of developmental perturbation. *ACS Chemical Neuroscience*, *10*(7), 3078–3093. https://doi.org/10.1021/acschemneuro.9b00073

Schultes, R. E. (1967). The Botanical Origins of. *Public Health Service Publication*, *1645*, 291.

Silva, G. O., Aguiar, B. A. C., Terra, D. C. V., Sousa, R. M., Fonseca, E. F., & de Souza, P. B. (2019). Substrate humidity and emergence performance of *Anadenanthera peregrina* (L) speg. *Magistra*, *30*, 336–341.

Souza, C. O., Silva, J. G. M., Arantes, M. D. C., Vidaurre, G. B., Dias Júnior, A. F., & Oliveira, M. P. (2020). Pyrolysis of *Anadenanthera peregrina* wood grown in different spacings from a forest plantation in Brazil aiming at the energy production. *Environment, Development and Sustainability*, *22*(6), 5153–5168. https://doi.org/10.1007/s10668-019-00418-0

Torres, C. M. (1995). Archaeological evidence for the antiquity of psychoactive plant use in the Central Andes. *Annuli dei Musei Civici Roverero*, *11*, 291–326.

Torres, C. M., & Repke, D. B. (2014). *Anadenanthera: Visionary Plant of Ancient South America*. Routledge. https://www.taylorfrancis.com/books/mono/10.4324/9781315864594/anadenanthera-constantino-torres-david-repke

Tótola, M. R., & Borges, A. C. (2000). Growth and nutritional status of Brazilian wood species Cedrella fissilis and *Anadenanthera peregrina* in bauxite spoil in response to arbuscular mycorrhizal inoculation and substrate amendment. *Brazilian Journal of Microbiology*, *31*, 257–265.

Viana, A. B. T., Torres, C., da Silva, L. F., & da Rocha, S. (2018). Plot size influence in the spatial distribution calculation of *Anadenanthera peregrina* (L.) Speng. and *Apuleia leiocarpa* JF Macbr in semideciduous seasonal forest. *Nativa: Pesquisas Agrárias e Ambientais*, *6*(2), 183–190.

von Reis Altschul, S. (1964). A taxonomic study of the genus *Anadenanthera*. *Contributions from the Gray Herbarium of Harvard University*, *193*, 3–65.

20 Vilca (*Anadenanthera colubrina* (Vell.) Brenan Fabaceae)

Ruvimbo Jessy Mapaya and Alfred Maroyi

20.1 INTRODUCTION

Plants or substances with substantial influence on perception of space and time and on emotional feelings are known as hallucinogens (Wink and Van Wyk, 2008). Hallucinogenic substances have also been termed psychotomimetics, psychotics, psychedelics, empathogens, and entheogens. Humans have been using hallucinogenic plants for thousands of years. The hallucinogens cause illusions that appear real to the person who is under their influence (Wink and Van Wyk, 2008; Chaachouay et al., 2019). A variety of plant species with hallucinogenic effects were and continue to be a major element of traditional and modern tribes' culture and religion. Hallucinogens later became popular in some mainstream culture and were used as illicit and often dangerous amusement. Their widespread usage as stimulants and adverse effects causes several societal issues as well as curiosity in the realm of scientific research. One of the species used as a hallucinogen is *Anadenanthera colubrina* (Vell.) Brenan. It is therefore, within this context that the current study was undertaken aimed at documenting the botany, habitat, and ecology, phytochemistry, and hallucinogenic effects of *A. colubrina*.

20.2 SCIENTIFIC CLASSIFICATION

Anadenanthera colubrina belongs to the Fabaceae family, subfamily Caesalpinioideae. The genus *Anadenanthera* Speg. is endemic to South America and the Caribbean Islands and comprises two species *A. colubrina* and *A. peregrina* (L.) Speg. (Altschul, 1964). The two species have few characters, which are morphologically distinct such as flowers with buds, fruits and leaves, and have also a partly overlapping geographical distribution patterns. The distinguishing morphological characters of the two species include pod texture (nitid, smooth to reticulated), a minute gland that may be present or absent on each anther and the peduncle that is surrounded by a small involucre (Altschul, 1964). Both species are well-known sources of the hallucinogenic snuffs commonly referred to as 'cebil', 'vilca', 'cohoba', 'niopo' and 'yopo' (Wink and Van Wyk, 2008; Van Wyk and Wink, 2015). The genus *Anadenanthera* is restricted in distribution to the southern hemisphere, from middle Peru through northern Argentina and northeastern Brazil (Altschul, 1964). It is important as a source of narcotics such as yopo and vilca. The synonyms of *A. colubrina* include *Acacia cebil* Griseb., *A. colubrina* (Vell.) Mart., *A. grata* Willd., *A. macrocarpa* (Benth.) Brenan, *Mimosa colubrina* Vell., *Piptadenia cebil* (Griseb.) Griseb., *P. colubrina* (Vell.) Benth., *P. grata* (Willd.) J.F.Macbr., *P. hassieriana* Chodat, *P. hassieriana* Chodat var. *fruticosa* Chodat & Hass, *P. macrocarpa* Benth., *P. macrocarpa* Benth. var. *cebil* (Griseb.) Chodat & Hass, *P. macrocarpa* Benth. var. *genuina* Chodat & Hass, *P. macrocarpa* Benth. var. *plurifoliata* Hoehne, *P. macrocarpa* Benth. var. *vestita* Chodat & Hass and *P. macrocarpa* Benth. (Altschul, 1964; Lorenzi, 2002; POWO, 2023). *Anadenanthera colubrina* has three varieties namely, *A. colubrina* var. *cebil* (Griseb.) Altschul, *A. colubrina* var. *glabra* O.G.Martínez & D.E.Prado and *A. colubrina* var. *colubrina* (Altschul, 1964; Cacharani et al., 2020). *Anadenanthera colubrina* var. *glabra* is distinguished from *A. colubrina* var. *cebil* and *A. colubrina* var. *glabra* by the trunk, which lacks conical projections or mamelons, the glabrous leaf rachis and the leaf venation pattern that forms a complex closed reticulum (Cacharani et al., 2020).

20.3 VERNACULAR NAMES

The vernacular names of *A. colubrina* include 'angico' or 'vilca' derived from 'wilka', which means 'sacred' in Quechua language (Monteiro et al. 2006b), 'cebil', 'curupau', 'curupay', and in Peru it is known as 'huilca', 'huilco', and 'wilco' (Monteiro et al., 2006a, 2006b). Other vernacular names reported by Delices et al. (2023) include 'angico gum', 'angico-branco', 'angico-do-caroço', 'angico-vermelho', 'angico-do-campo', 'angico-preto', 'arapiraca', 'yopo', 'huilca cohoba', and 'hatax'.

20.4 BOTANICAL DESCRIPTION

Anadenanthera colubrina is a deciduous tree with an open and sparse crown. It has a straight and cylindrical bole, which can be free of branches for up to seven meters and the tree can grow up to 30 meters in height and 30 centimeters in diameter at breast height (Altschul, 1964; Chudnoff, 1984). The bark is gray and smooth albeit it is striated with

DOI: 10.1201/9781003460336-20

longitudinal fissures at times (Altschul, 1964). Young twigs range in color from dark to light gray, with whitish or reddish warts. It consists of bipinnate compound leaves, specialized ant glands, and inflorescence with hermaphroditic flowers (Monteiro et al., 2006a). The species is characterized by having conspicuous thorns, which may be many. The involucre of *A. colubrina* can be glabrous or puberulous. It is carried directly below the receptacle and does not detach (Altschul, 1964). The shiny seeds of *A. colubrina* are thin and dark chestnut brown. One of the varieties *A. colubrina* var. *cebil* is a semi-deciduous shrub to tree and grows from 3 to 30 m high with long fruits (Altschul, 1964). Another variety *A. colubrina* var. *colubrina* is a shorter tree that grows up to 24 m high (Altschul, 1964). The tree has a smooth bark, is elongated, constricted, and darker than *A. colubrina* var. *cebil*. The leaflets are linear and nitidulous above and the mid-vein is noticeable. The heads are distinguished by the fact that they are whitish in the bud, with the bracts (Altschul, 1964). The paniculate heads have fewer flowers and are borne in paniculate patterns at the branch apices. The inflorescence is arranged distinctively with a stated flowery smell, and the relative showiness of the inflorescences may be attributed to the relatively enormous size of the petiole glands in this variety (Altschul, 1964).

20.5 ORIGIN AND GEOGRAPHIC DISTRIBUTION

The center of origin for the genus *Anadenanthera* may be Brazil, where *A. colubrina* and *A. peregrina* seem to be well represented as indicated by their geographical distributions (Altschul, 1964). *Anadenanthera colubrina* is postulated to have arisen in the Brazilian highlands (Tropical Plants Database, 2023). In Brazil, *A. colubrina* is regarded as a species of conservation concern and categorized as a 'high priority' species for *in situ* conservation (Monteiro et al., 2006b). *Anadenanthera colubrina* has been recorded in rainforests, lowland forests and edges of rivers at altitudes ranging from 400 m to 2100 m above sea level (Altschul, 1964). The species is a characteristic tree of secondary growth areas, common in well-drained soils, poor stony soils, fertile clay and basalt soils (Chudnoff, 1984; Lorenzi, 2002). Although *A. colubrina* is restricted to the southern hemisphere, it seems to be more suited to a wider range of environments. The tree has a wide distribution being found in various biomes in Argentina, Colombia, the semiarid region of Brazil, Bolivia, Dominican Republic, Ecuador, Paraguay, Puerto Rico, Peru, and Venezuela (Delgobo et al., 1998; Delices et al., 2023). The variety *A. colubrina* var. *colubrina* has an east coastal distribution in Argentina and Brazil (Altschul, 1964). It has been recorded along river edges, mountains, up to elevations of 700 m and xerophytic environments, and hence restricted distribution. The representatives of this variety are located in a narrow geographical area, preferring a maritime environment. The variety *A. colubrina* var. *cebil* grows on forests, sandy soils, arid hillsides, riverbanks, in mountains, high plateaus, and terraces up to 2000 m in elevation. At altitudes between 800 m to 2700 m, *A. colubrina* var. *cebil* emerges as a dominating tree in the xerophytic savanna-like formations of riparian valleys and on the western slopes, as far north as the Maraon Valley in Peru (Williams, 1945). A symbiotic relationship exists with some soil bacteria that fix atmospheric nitrogen, which can be utilized by the plant for defense (Huxley, 1992; Gross et al., 2002, Monteiro et al., 2006a).

20.6 PHYTOCHEMISTRY

Phytochemicals derived from *A. colubrina* extract have been studied extensively (Table 20.1). Flavonoids, alkaloids, phytosterols, tannins, and saponins have been reported for the species (Weber et al., 2011; Santos et al., 2013; Medeiros et al., 2020). The bark of *A. colubrina* has been shown to contain flavonoids, catequins, phenols, tannins, steroids, saponins, triterpenes, and xanthones (Santos et al., 2013). The phytochemical compounds isolated from the seeds of *A. colubrina* include N,N-dimethyltryptamine, N,N-dimethyl-5-methoxytryptamine, serotonin mimics such as bufotenine and traces of β-carboline alkaloids such as 2-methyltetrahydro-β-carboline (Wink and Van Wyk, 2008). The compounds bufotenine, N,N-dimethyltryptamine and N,N-dimethyl-5-methoxytryptamine are known to stimulate serotonin receptors, moderately hazardous, neurotoxin, and mind-altering (Wink and Van Wyk, 2008). The symptoms of bufotenine intoxication start with headache, salivation, and vomiting, and continues with trance-like conditions including dancing, singing, and shouting (Wink and Van Wyk, 2008). The N,N-dimethyltryptamine and closely related compounds are easily broken down in the digestive tract and hence are traditionally snorted or ingested with other plant additives that contain β-carbolines, especially harmaline and harmine, that block monoamine oxidase, the enzyme that breaks down tryptamines (Bussmann, 2016).

The plant also has pharmacological potential, for example, leaf and bark extracts have anti-inflammatory (Cardoso-Junior et al., 2021), antimicrobial (Araújo et al., 2014), antiproliferative, antinociceptive activities (Santos et al., 2013), and antifungal (Silva and Barbosa, 2000). An aqueous extract of the seeds of A. *colubrina* produced 100% mortality rate in the third stage mosquito larvae of *Aedes aegypti* (Farias et al., 2010).

20.7 ETHNOBOTANY

Anadenanthera colubrina is economically important and has a diverse variety of traditional medicinal uses. The seeds of *A. colubrina* are ground into a fine powder and used as a highly narcotic snuff (Uphof, 1959). When the

TABLE 20.1
Phytochemical Constituents Isolated from *Anadenanthera colubrina*

Phytochemical Compound	Plant Part	Reference
Alkaloids	Seeds	Medeiros et al. (2020); Medeiros et al. (2021)
Carbohydrates	Seeds	Pessoa et al. (2012); Barbosa et al. (2014)
Cyanogenic glycosides	Not specified	Delices et al. (2023)
Fatty acids	Not specified	Delices et al. (2023)
Flavonoids	Leaves, fruits and seeds	Pessoa et al. (2012); Araújo et al. (2014); Medeiros et al. (2020); Cardoso-Junior et al. (2021); Medeiros et al. (2021)
Phenolics	Leaves and stem bark	Sá et al. (2016); Costa et al. (2020); Cardoso-Junior et al. (2021)
Proteins	Seeds	Barbosa et al. (2014)
Saponins	Leaves and seeds	Pessoa et al. (2012); Medeiros et al. (2020); Cardoso-Junior et al. (2021); Medeiros et al. (2021)
Steroids	Leaves and fruits	Pessoa et al. (2012); Araújo et al. (2014)
Tannins	Leaves, fruits and seeds	Araújo et al. (2014); Sá et al. (2016); Pereira et al. (2018); Medeiros et al. (2020); Medeiros et al. (2021)
Triterpenes	Leaves and fruits	Pessoa et al. (2012); Araújo et al. (2014)

snuff is inhaled or blown in strongly, a temporary mild to deep intoxication, trance, visions, and spirit visitation may be experienced (Mors et al., 2000). The bark is astringent, bitter, depurative, and haemostatic (Uphof, 1959; Chaachouay et al., 2021a, 2021b). *Anadenanthera colubrina* is used to treat leucorrhea and gonorrhea (Mors et al., 2000). *Anadenanthera colubrina* can be used as a powder, tincture, infusion, or extract to treat lung diseases (Uphof, 1959). A gum obtained from the bark of *A. colubrina* is used to treat respiratory infections (Mors et al., 2000). Different parts of *A. colubrina* are used in the treatment of respiratory problems, flu symptoms, coughs, bronchitis, and infections, and inflammations; toothache, diarrhea, and wounds (Almeida et al., 2006; Monteiro et al., 2006a, 2006b; Agra et al., 2007; Agra et al., 2008). It is taken orally and is made by decoction, infusion, or juice derived following maceration of the plant's leaves or other parts (Agra et al., 2008; Orch et al., 2021). The bark and the seeds of *A. colubrina* are the most used plant parts (Delices et al., 2023) with the seeds used in traditional practices.

The plant is also used in shamanic rituals by the Indigenous people (Delices et al., 2023), as sources of energy (household cooking, charcoal made from wood) and the barks of the plants are used for wood construction – marine work, flooring, fence posts, tool handles, railroad crossties, cabinet making (Chudnoff, 1984; Lorenzi, 2002; Monteiro et al., 2006a; Chaachouay & Zidane, 2021; Delices et al., 2023) and leather tanning in industry. The bark is also used to make syrup (a locally prepared sugary drink) in northeastern Brazil, tinctures, decoction, and teas used for treating ailments (Monteiro et al., 2006b). The timber usually harvested from the wild and traded is durable and of high quality (Lorenzi, 2002). Cebil gum containing 80% arabin is also obtained from *A. colubrina* and mostly utilized as a source of mucilage (tropical.theferns.info/viewtropical.php?id=Anadenanthera+colubrina). It is also used as artifacts such as smoking pipes and pencil holders (Monteiro et al., 2006a, 2006b).

Anadenanthera colubrina is also used as an ornamental plant in general landscaping especially in parks and public areas (Lorenzi, 2002). The tree is used as a pioneer species in restoring natural forests and as a timber crop in plantations (Lorenzi, 2002). The flowers of *A. colubrina* are a source of pollen for the bees.

20.8 HALLUCINOGENIC EFFECTS

Hallucinogens are a broad class of psychoactive chemicals distinguished by their distinct perception-altering effects (Hallucinogens, 2022). From time immemorial, Indigenous peoples have used hallucinogenic plant, fungal, and animal alkaloids. These alkaloids and their synthetic derivatives, including other hallucinogenic chemicals are accessible on the illegal drug market. Many hallucinogenic plants are used for entertainment purposes around the world. Based on their chemical structure and mode of action, hallucinogens are categorized into classes and subclasses. Tryptamines, lysergamides, phenethylamines, arylcyclohexylamines, morphinans, 1,2-diarylethylamines, and others are among these subclasses (Hofmann, 1959; Hallucinogens, 2022).

Anadenanthera colubrina is used as an entheogen and, together with a related species *Anadenanthera peregrina* (L.) Speng., is known for its hallucinogenic effects. Several of the hallucinogenic tryptamines reported in variable amounts in *A. peregrina* have been detected in *A. colubrina*-related material (www.unodc.org/unodc/en/data-and-analysis/bulletin/bulletin_1969-01-01_4_page004.html). In *A. peregrina*, the seeds are used to make *yopo*, a hallucinogenic inhalant utilized by the *curandeiros* of the *Piaroa* tribe occupying southeastern Venezuela (Rodd, 2002). *Yopo* has been used as a hallucinogenic among the Indigenous people of Latin America as shown by archaeological evidence (Bßez et al., 2000).

20.9 CONCLUSION

The current review attempted to provide information available on the botany, habitat, and ecology, phytochemistry, and hallucinogenic effects of *A. colubrina*. A closely related species, *A. peregrina* is also regarded as an important source of traditional medicines in South America, particularly the Andes and Brazil, and is known to have hallucinogenic effects (Wink and Van Wyk, 2008). Such literature reports are supported by the findings from this review, which showed that *A. colubrina* is characterized by similar diverse medicinal and hallucinogenic effects. These reports of the ethnomedicinal uses of *A. colubrina* for similar ailments and diseases in different communities in Argentina, Bolivia, Brazil, Paraguay, and Peru calls for detailed ethnopharmacological research focusing on its effectiveness and safety. The hallucinogenic effects, pharmacological studies on the extracts and phytochemical compounds isolated from *A. colubrina* demonstrated various pharmacological activities such as anti-inflammatory, antinociceptive, antidiarrheal, wound healing, antibacterial, antifungal, anti-HIV, antitumour, antioxidant, antidiabetic, insecticide, antibiofilm, immunomodulatory, dermocosmetic, and allelopathic (Delices et al., 2023). Therefore, understanding of the ethnopharmacological properties of *A. colubrina* extracts and phytochemical compounds isolated from the species is important as these active ingredients act directly or indirectly to prevent or treat diseases and also maintain health. An important reason for evaluating the phytochemical properties of medicinal plants is therefore, not to identify the main phytochemical compounds but to understand the rationale behind the medicinal properties of the plant species from a scientific perspective. Hallucinogenic and medicinal plants such as *A. colubrina* are among the traditional medicines that have been the basis in the treatment and management of several diseases for centuries and drug discovery from such plant species due to improved phytochemical profiling and ethnopharmacological research has an important role in the development of current therapeutic systems.

The phytochemical compounds identified from various parts of *A. colubrina* include alkaloids, cyanogenic glycosides, fatty acids, flavonoids, phenolics, saponins, steroids, tannins, and triterpenes (Table 20.1). Some of these classes of secondary phytochemical constituents may be responsible for the documented hallucinogenic effects and biological activities of *A. colubrina*. Therefore, knowledge about the resultant hallucinogenic effects, and pharmacological and toxicological properties of these phytochemical constituents isolated from *A. colubrina* should be subjected to controlled clinical trials to determine if such phytochemical compounds are characterized by sufficient therapeutic properties and can be linked to a particular disease or indication associated with the species. Though various phytochemical compounds have been isolated from *A. colubrina*, purified and characterized, many of these phytochemical compounds are still to be studied in detail. Till now, limited hallucinogenic and ethnopharmacological research has been made in the pharmacokinetics studies related to the mechanism of action of the individual isolated phytochemical compound under in vivo conditions. Also, ethnopharmacological research should focus on toxicity, with preclinical and clinical trials aimed at accounting for the hallucinogenic effects, safety, effectiveness, and side effects of the phytochemical compounds in the formulation of *A. colubrina* as a source of traditional medicine.

LIST OF ABBREVIATION

POWO Plants of the World Online

REFERENCES

Agra, M.D.F., de Freitas, P.F., and Barbosa-Filho, J.M. 2007. Synopsis of the plants known as medicinal and poisonous in Northeast of Brazil. *Brazilian Journal of Pharmacognosy* 17(1): 114–140.

Agra, M.D.F., Silva, K.N., Basılio, I.J.L.D., De Freitas, P.F., and Barbosa-Filho, J.M. 2008. Survey of medicinal plants used in the region Northeast of Brazil. *Brazilian Journal of Pharmacognosy* 18 (3): 472–508.

Almeida, C.D.F.C.B.R., de Amorim, E.L.C., de Albuquerque, U.P., and Maia, M.B.S. 2006. Medicinal plants popularly used in the Xing-o region—A semi-arid location in Northeastern Brazil. *Journal of Ethnobiology and Ethnomedicine*, 2 (15): 1–7.

Altschul, S. Von R. 1964. A taxonomic study of the genus *Anadenanthera* Contributions from the Gray Herbarium of Harvard University 193: 3–65.

Araújo, D.R.C., Silva, L.C.N., Silva, A.G., Araújo, J.M., Macêdo, A.J., Correia, M.T.S., and Silva, M.V. 2014. Comparative analysis of anti-Staphylococcus aureus action of leaves and fruits of *Anadenanthera colubrina* var. cebil (Griseb.) Altschul. *African Journal of Microbiology Research* 8: 2690–2696.

Barbosa, O., Ortiz, C., Berenguer-Murcia, Á., Torres, R., Rodrigues, R.C., and Fernandez-Lafuente, R. 2014. Glutaraldehyde in bio-catalysts design: A useful crosslinker and a versatile tool in enzyme immobilization. *Royal Society of Chemistry Advances* 4: 1583–1600.

Bßez, H., Castro, M.M., Benavente, A.M., Kintz, P., Cirimele, V., Camargo, C., and Thomas, C. 2000. Drugs in prehistory: Chemical analysis of ancient human hair. *Forensic Science International* 108: 173–179.

Bussmann, R.W. 2016. Magic plants. In: Albuquerque, U.P., and Alves, R. (Editors), *Introduction to Ethnobiology*. Switzerland: Springer International Publishing.

Cacharani, D.A., Barrandeguy, M.E., García, M.V., Costas, M.L., Martínez, O.G. and Prado, D.E. 2020. Una nueva variedad de *Anadenanthera colubrina* (Leguminosae, Mimosoideae) de Argentina. *Boletin de la Sociedad Argentina de Botanica* 55: 403–410.

Cardoso-Junior, O., Lima, N.M., Silva, M.G.A., Aguiar, V.B., Carli, G.P., Scherrer, E.C., Castro, S.B.R., Alves, C.C.S., Oliveira, M.A.L., Carli, A.P. 2021. *In vitro* and *in vivo* evaluation of anti-inflammatory activity and free radical scavenging potential of leaves extract from *Anadenanthera colubrina. Natural Product Research* 35: 4819–4823.

Chaachouay, N., & Zidane, L. (2021). The Symbolic Efficacy of Plants in Rituals and Socio-Religious Ceremonies in Morocco, Northwest of Africa. *Journal of Religious & Theological Information, 21*(1–2), 34–53. https://doi.org/10.1080/10477845.2021.1942433

Chaachouay, N., Benkhnigue, O., Fadli, M., Ibaoui, H., & Zidane, L. (2019). Ethnobotanical and ethnopharmacological studies of medicinal and aromatic plants used in the treatment of metabolic diseases in the Moroccan Rif. *Heliyon, 5*(10), e02191. https://doi.org/10.1080/10477845.2021.1942433

Chaachouay, N., Douira, A., & Zidane, L. (2021a). Herbal Medicine Used in the Treatment of Human Diseases in the Rif, Northern Morocco. *Arabian Journal for Science and Engineering*, 47, 131–153. https://doi.org/10.1007/s13369-021-05501-1

Chaachouay, N., Douira, A., & Zidane, L. (2021b). COVID-19, prevention and treatment with herbal medicine in the herbal markets of Salé Prefecture, North-Western Morocco. *European journal of integrative medicine*, 42, 101285. https://doi.org/10.1016/j.eujim.2021.101285

Chudnoff, M. 1984. Tropical timbers of the World. Ag. Handbook No. 607. USDA Forest Service. Wisconsin.

Costa, O.J., Barbosa, R.S., Soares, I.M., Souza, E.E., Gellen, L.F.A., Lemos, J.P.P., Aguiar, R.W.S., Montel, A.L.B., Ascêncio, S.D. 2020. Inhibitory effects of *Anadenanthera colubrina* (Vell.) Brenan stem bark extract on α-glucosidase activity and oxidative stress. *Journal of Medicinal Plants Research* 14: 583–592.

Delgobo, C.L., Gorin, P.A.J., Jones, C., Iacomini, M. 1998. Gum heteropolysaccharide and free reducing monoandoligosaccharides of *Anadenanthera colubrina. Phytochemistry* 47: 1207–1214.

Delices, M., de AraujoIsaias Muller, J., Arunachalam, K., de Oliveira Martins, D.T. 2023. *Anadenanthera colubrina* (Vell) Brenan: Ethnobotanical, phytochemical, pharmacological and toxicological aspects. *Journal of Ethnopharmacology* 300: 115745.

Farias, D.F., Cavalheiro, M.G., Viana, M.P., Queiroz, V.A., Rocha Bezerra, L.C.B., Vasconcelos, I.M., de Morais, S.M., and Carvalho, A.F.U. 2010. Water extracts of Brazilian leguminous seeds as rich sources of larvicidal compounds against Aedes aegypti L. *Annals of the Brazilian Academy of Sciences* 82: 585–594.

Gross, E., Cordeiro, L., and Caetano, F.H. 2002. Nodule ultrastructure and initial growth of *Anadenanthera peregrina* (L.) Speg. var. *falcata* (Benth.) Altschul. plants infected with rhizobia. *Annals of Botany* 90: 175–183.

Hallucinogens 2022. *Encyclopedia of forensic sciences: Volume 1–4* (3rd Edition). Amsterdam, Netherlands. Elsevier 1–716.

Hofmann, A. 1959. Psychotomimetic drugs: chemical and pharmacological aspects. *Acta Physiologica et Pharmacologica Neerlandica* 8: 240–258.

United Nations. The plant kingdom and hallucinogens (part II) https://www.unodc.org/unodc/en/data-and-analysis/bulletin/bulletin_1969-01-01_4_page004.html (accessed on 01 August 2023).

Huxley, A. 1992. *The new RHS dictionary of gardening*. London: MacMillan Press. 1 (4): 173–186.

Lorenzi, H. 2002. *Brazilian trees. Volume 1* (4th Edition). Brazil: Instituto Plantarum De Estudos Da Flora.

Medeiros, J.G.F., Araujo-Neto, A.C., Silva, E.C., Rodrigues, R.M., Demartelaere, A.C.F., and Silva, J.V.B. 2021. Phytochemical profile and antifungal action of *Anadenanthera colubrina* extract on the quality of maize seeds. *Arquivos do Instituto Biológico* 88: 1–9.

Medeiros, J.G.F., Demartelaere, A.C.F., Silva, H.F., Silva, H.C., Nascimento, L.C. 2020. Phytochemical survey and antifungal activity of plant extracts in angico seeds (*Anadenanthera colubrina* Vell. Brenan). *Brazilian Journal* of *Development* 6: 53941–53953.

Monteiro J. M., Albuquerque U. P. D., Lins-Neto E. M., Araujo E. L. D., and Amorim E. L. C. D. 2006a. Use patterns and knowledge of medicinal species among two rural communities in Brazil's semi-arid northeastern region. *Journal of Ethnopharmacology* 105 (1–2): 173–186.

Monteiro, J.M., de Fátima, C., Rangel de Almeida, C.B., de Albuquerque, U.P., de Lucena, R.F. P., Florentino, A.T.N., and Rodrigo Leonardo, C., de Oliveira, R.L.C. 2006b. Use and traditional management of *Anadenanthera colubrina* (Vell.) Brenan in the semi-arid region of northeastern Brazil. *Journal of Ethnobiology and Ethnomedicine* 2: 6.

Mors, W.B., Rizzini, C.T., and Pereira, N.A. 2000. *Medicinal plants of Brazil*. Michigan: Reference Publication.

Orch, H., Chaachouay, N., Douiri, E. M., Faiz, N., Zidane, L., & Douira, A. (2021). Use of medicinal plants in dermatocosmetology: An ethnobotanical study among the population of Izarène. *Jordan Journal of Pharmaceutical Sciences, 14*(3), 323–340.

Pereira, A.V., Góis, M.B., Azevêdo, T.K.B., Ferraz, F.N., Vieira, S.L.V., Cavalcanti-Dantas, V.M., Pereira, J.V., Rodrigues, O.G., and Pereira, M.S.V. 2018. Effects of associations of tannins from *Anacardium occidentale* and *Anadenanthera colubrina* with cephalosporin against bovine *Staphylococcus aureus* isolates. *Arquivos do Instituto Biológico* 85: 1–8.

Pessoa, W.S., Estevão, L.R.D.M., Simões, R.S., Barros, M.E.G.D., Mendonça, F.D.S., Baratella-Evêncio, L., and Evêncio-Neto, J. 2012. Effects of angico extract (*Anadenanthera colubrina* var. *cebil*) in cutaneous wound healing in rats. *Acta Cirúrgica Brasileira* 27: 655–670

Plants of the World Online (POWO). 2023. Facilitated by the Royal Botanic Gardens, Kew. 2023. Available online: https://powo.science.kew.org (accessed on 23 June 2023).

Rodd, R. 2002. Snuff synergy: Preparation, use, and pharmacology of yopo and Bonisteriopsis Caopi among the Piaroa of southern Venezuela. *Journal of Psychoactive Drugs* 34: 273–279.

Sá, J.L.F., Siqueira, W.N., Silva, H.A.M.F., Santos, M.L.O., Santos, F.T.J., Silva, L.R.S., Cabral, D.L.V., Bezerra, I.C.F., Soares, L.A.L., and Melo, A.M.M.A. 2016. Evaluation of molluscicidal activity of *Anadenanthera colubrina* extracts on adult mollusc and embryos of the species *Biomphalaria glabrata* (Say, 1818). *Scientia plena* 12: 1–9.

Santos, J.S., Marinho, R.R., Ekundi-Valentim, E., Rodrigues, L., Yamamoto, M.H., Teixeira, S. A., Muscara, M.N., Costa, S.K., and Thomazzi, S.M. 2013. Beneficial effects of *Anadenanthera colubrina* (Vell.) Brenan extracton the inflammatory and nociceptive responses in rodent models. *Journal of Ethnopharmacology* 148: 218–222.

Silva, L.M.B. and Barbosa, D.C. 2000. Crescimento e sobrevivΩncia de *Anandenanthera macrocarpa* (Benth.) Brenan (Leguminosae), em uma ßrea de Caatinga, Alagoinha, P.E. *Acta Botanica Brasilica* 14: 251–261.

Tropical Plants Database, Ken Fern. tropical.theferns.info. tropical.theferns.info/viewtropical.php?id=*Anadenanthera+colubrina* (accessed 10 August 2023).

Uphof, J.C.T. 1959. *Dictionary of economic plants*. New York: Hafner Publishing Co.. 75, 400

Van Wyk, B.-E. and Wink, M. 2015. *Phytomedicines, herbal drugs and plant poisons*. Pretoria: Briza Publications.

Weber, C.R., Soares C.M.L., Lopes, A.B.D., Silva, T.S. Nascimento, M.S., and Ximenes, E.C.P.A. 2011. *Anadenanthera colubrina*: um estudo do potencial terapêutico. *Revista Brazileria De Farmácia* 92 (4): 235–244.

Williams, L. 1945. The phytogeography of Peru. In Verdoorn, F. (Editor) *Plants and plant science in Latin America. Chronica Botanica*. Waltham, MA: Frans Verdoorn.

Wink, M. and Van Wyk, B.-E. 2008. *Mind-altering and poisonous plants of the world: An illustrated scientific guide*. Pretoria: Briza Publications.

21 Canary Broom (*Cytisus canariensis* (L.) Kuntze Fabaceae)

Tamara Attard and Everaldo Attard

21.1 INTRODUCTION

Cytisus canariensis (L.) Kuntze, commonly known as Canary Island Bloom (Figure 21.1), is known for its characteristically yellow flowers. It is a coastal plant and an endemic of the Canary Islands. This archipelago is composed of seven islands, namely, Tenerife, Fuerteventura, Gran Canaria, Lanzarote, La Palma, La Gomera, and El Hierro, close to West Coast of Morocco (Chaachouay et al., 2019). Though being autonomous communities of Spain, the plant has been introduced to continental Europe and beyond. It is a member of the pea family, also known as legumes.

In spite of its long-standing history as a hallucinogen, it is currently being investigated as a viable alternative to tobacco or as a supplementary component to mitigate the inhalation of tobacco by individuals who smoke. When consumed in the form of cigarettes, *G. canariensis* exhibits effects that are comparable to, or somewhat less intense than, those of tobacco cigarettes. The observed aftertaste of electronic cigarettes is characterized by its sweetness, in contrast to the mildly bitter aftertaste commonly associated with traditional tobacco cigarettes (Fadiman, 1965).

FIGURE 21.1 *Genista canariensis* (L.) Kuntze.

21.2 SCIENTIFIC CLASSIFICATION

Genista canariensis L. was first described in 1753. It has 25 synonyms, four of which are homotypic (nomenclature) synonyms, and the most quoted one being *Cytisus canariensis* (L.) Kuntze. The other 21 are heterotypic (taxonomic) synonyms. Out of these binomial synonyms, eight are associated with *Genista* genera, seven with *Cytisus*, five with *Teline*, two with *Telinaria*, two with *Spartium*, and one with *Avornela* (Kew, 2023). The species epithet is mainly associated with *canariensis*, amongst others.

The classification of this plant is illustrated in Table 21.1. The binomial nomenclature is provided in Table 21.2.

21.3 VERNACULAR NAMES – COMMON NAMES

The vernacular names for *Genista canariensis L.* are mainly attributed to its association with the Canary Islands. In fact, the most commonly quoted vernacular names are Canary Island Bloom or Canary Island Broom. Other common names include cytisus, genista, kanarischer ginster, kytisos, spartion, spartium. There are several Spanish names attributed to *Genista canariensis*. These include Hierba de Ballesta, Retama de Gran Canaria, and Retama Loca (crazy broom). Vernacular names include Genêt des Canaries in French and Canarische Stechginster in German.

21.4 BOTANICAL DESCRIPTION

Genista canariensis L. is a resilient, erect evergreen shrub that reaches a height of 3 m and a width of 1.5 m. It is characterized by its hairy green stems. The foliage consists of

DOI: 10.1201/9781003460336-21

TABLE 21.1
The Classification of *Genista canariensis* L.

Kingdom	Plantae
Phylum	Tracheophyta
Class	Angiospermae
Sub-class	Dicotyledoneae
Order	Fabales
Family	Fabaceae
Subfamily	Faboideae
Genus	*Genista*
Species	*G. canariensis*
Binomial name	*Genista canariensis* L.

TABLE 21.2
Genera and Species Epithets Attributed to *Genista canariensis* L.

Genus	Species epithet and Authority/ies
Avornela	canariensis (L.) Raf.
Cytisus	canariensis (L.) Kuntze
	canariensis var. ramosissimus
	hillebrandtii (Christ) Briq.
	paniculatus Loisel
	philippii Lindl.
	discolour ramosissimus Poir.
	rhodopnoa (Webb & Berthel.) Schouw
Genista	candicans var. subspicata Webb ex Christ
	discolor Webb ex Lowe
	hillebrandii Christ
	paniculata (Loisel.) A.Braun & Asch.
	philippii Lindl.
	ramosissima (Poir.) Spach
	ramosissima var. discolor (Webb ex Lowe) Christ
	rhodopnoa Webb & Berthel.
Spartium	albicans Cav.
	terminale Salisb.
Telinaria	canariensis (L.) C.Presl
	ramosissima C.Presl
Teline	canariensis (L.) Webb & Berthel.
	canariensis var. lanuginosa G.Kunkel
	canariensis var. ramosissima (Poir.) G.Kunkel
	hillebrandii (Christ) G.Kunkel
	ramosissima (Poir.) Webb & Berthel.

Source: Kew, 2023.

elliptical blue-green leaflets, each measuring up to 10 mm in length, and exhibiting a dense covering of hair on their lower surfaces. The stipules are less than 2 mm and a petiole generally less than 6 mm. The leaflets are 5 to 10 mm and generally obovate to round in shape. The raceme inflorescence are 10 to 60 mm and may bear a maximum of 20 vibrant yellow blooms that resemble peas in shape, and have a delicate fragrance. The plant generally flowers during spring and the beginning of summer when nighttime temperatures are below 16°C. Flowering is maintained as long as these nighttime temperatures are cool. The fruit can be described as a legume pod measuring between 15 to 25 mm in length, which encloses many seeds of a dark brown color (Jespon eFlora, 2023).

21.5 ORIGIN AND GEOGRAPHIC DISTRIBUTION

Genista canariensis L. is native to the Canary Islands, hence the name. It was then introduced to Mexico, where the Yaqui Indians of northern Mexico used its seeds for their hallucinogenic effects (Schultes, 2016). It also spread to mainland Europe, the United States of America (California, Oregon, and Washington), Sri Lanka, and India.

21.6 ECOLOGY – ENVIRONMENT AND MICROORGANISMS

This plant has naturalized in various habitats typical of the Canary Islands. These include rocky slopes, volcanic terrain, and coastal areas.

G. canariensis is considered an exotic plant and is one of the major fire hazards in several areas where it occupies mainly formerly cultivated land, cover streams, fringe roads, or in sclerophyll forests. The following reasons substantiate this fire hazard fact for this plant. *G. canariensis* grows into a thick bush that facilitates the quick spread of intense fires. Another factor is the accumulation of highly combustible fuel within the dried vegetation from previous seasons in the lower strata. The density and height of this plant may supersede those of the indigenous scrub strata. On the other hand, the density and height may contribute to the increased risk of crown fires in eucalyptus due to the volatile component within the leady upper layer. In contrast, fires in lower native shrubs typically do not result in the ignition of the tree canopy during the initial phases. Within a span of two seasons, a substantial and sustainable repository of seeds is generated. The occurrence of a fire leads to the proliferation of compact growth of gorse and broom seedlings. This is facilitated by the heat-induced melting of the protective seed coverings, which are cuticularized, as well as the alkaline nature of bushfires. This plant exhibits a significantly higher rate of regeneration in comparison to indigenous shrubs. Within two years of a wildfire event, *G. canariensis* exhibits greater heights exceeding 2.5 m and significantly higher density compared to the indigenous scrub climax vegetation, which typically takes between 7 and 10 years to reach a similar level of growth. *G. canariensis* regenerations under *Eucalyptus obliqua* forest just over three and a half years after a bushfire (Cochrane, 1963).

The metabolites present in this plant, protect it from pathogenic microorganisms and help in the formation of root nodules (Sohn et al., 2021). In fact, isoflavones have been implicated in the promotion of spore germination, hyphal formation, and root colonization in symbiotic fungi (Abdel-Lateif et al., 2012).

21.7 PHYTOCHEMISTRY

Genista canariensis L. contains typical metabolites that are found in other *Genista* L. species. Numerous chemical components have been discovered within the *Genista* genus, including triterpene saponins, quinolizidine alkaloids, flavonoids, isoflavonoids, and chalcones (Benayache et al., 2018; Boutaghane et al., 2013; Harborne 1969). Among these compounds, lupinine-type quinolizidine alkaloids, flavonoids, and isoflavonoids are present in nearly all species, serving as chemotaxonomic indicators for the genus (Harborne, 1969).

In *G. canariensis* intact leaves, the main alkaloids include cytisine, N-methyl-Cytisine, and anagyrine, constituting 79% of the total alkaloidal content (Figure 21.2) (Schultes, 1970; Wink et al., 1983). However, in cell culture, the main alkaloid is lupinine, which constitutes 95% of the total alkaloidal content, alongside two other minor alkaloids. It was observed that in cell suspension, the biosynthetic route that converts lupinine and other lupin alkaloids to cytisine and other α-pyridone alkaloids is not expressed (Wink et al., 1983). As a result, these cell cultures accumulate lupinine alkaloids that are relatively less toxic to humans than the α-pyridone type. This implies that the ring A conformation renders the α-pyridone quinolizidine alkaloids more toxic than the lupinine quinolizidine alkaloids (Crews, 2014). Cytisine is chemically known as (1R,5S)-1,2,3,4,5,6-hexahydro-1,5-methano-8H-pyrido[1,2-a][1,5]diazocin-8-one. Several researchers identified cytisine in different species, hence given a number of synonyms such as ulexine (derived from *Ulex europaeus*), baptitoxine (found in *Baptisia* species), sophorine (found in *Sophora* species) and laburnin (derived from *Laburnum* spp.) (Cayman Chemical, 2023).

The main isoflavones present in *G. canariensis* are diadzein, genistein, formononetin and isoprunetin (Harborne and Baxter, 2001) (Figure 21.3). Several isoflavone aglycones are known to bind both ERα and ERβ (estrogen receptor isotypes) and often display biological activities quite similar to 17β-estradiol (estradiol). Fokialakis et al. (2019) suggest that 7-O-β-D-glucopyranosyl-genistein and 7,4′-di-O-β-D-glucopyranosyl-genistein exhibited higher

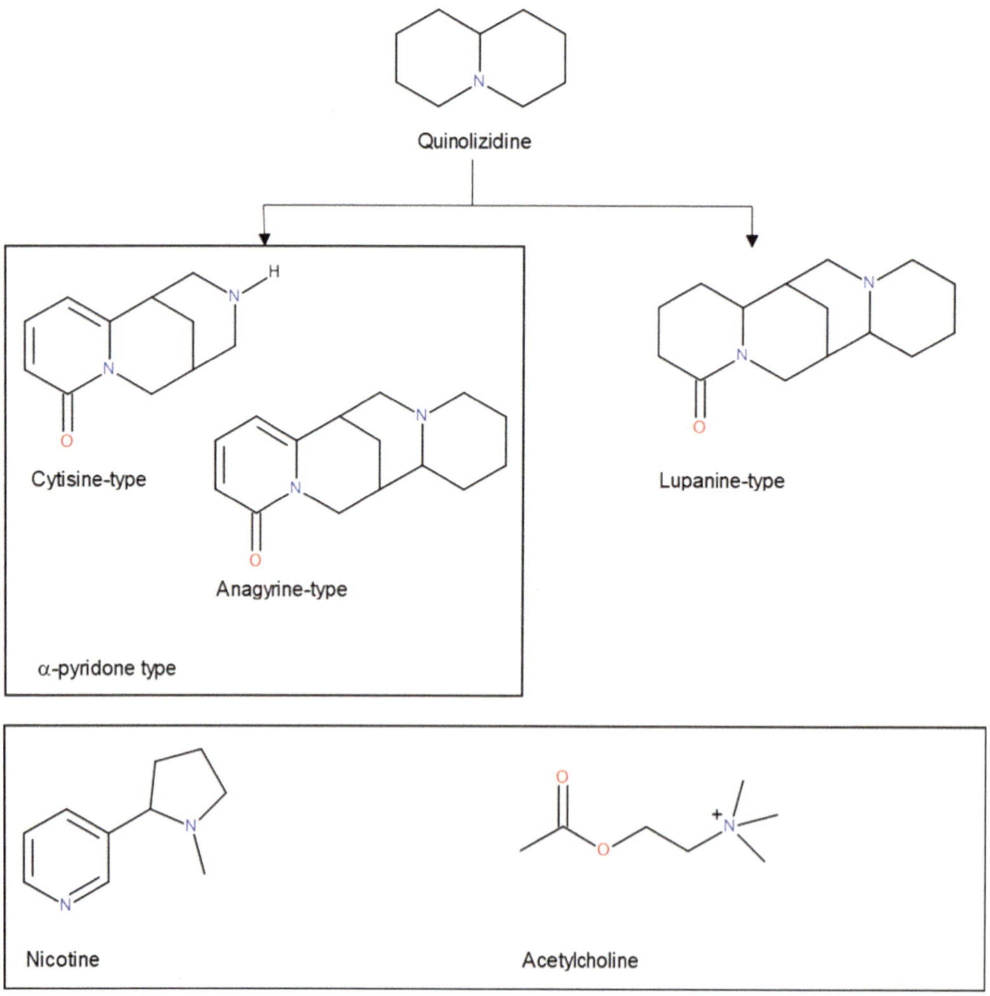

FIGURE 21.2 The main quinolizidine alkaloids in *Genista canariensis* compared to nicotine and acetylcholine.

estrogen-like and/or anti-inflammatory activity compared to the aglycone.

21.8 ETHNOBOTANY

The *Genista* L. genus has been utilized as a herbal cure for the treatment of various diseases since ancient times (Chaachouay et al., 2021a, 2021b). This was mentioned by Dioscorides (Wellmann, 1958), and later on by Gennadius (1914). The species belonging to the genus have been traditionally used in folk medicine in the Mediterranean region. They have been employed for the treatment of respiratory diseases (Bouchouka et al., 2012) and rheumatic disorders (Adams et al., 2009). Additionally, they have been recognized for their antalgic and anti-inflammatory properties (Agelet and Vall, 2003), antihyperglycemic properties (Rauter et al., 2009; Edwards et al., 2006), ability to alleviate ulcers (Guarrera and Lucia, 2007) and diuretic effects (Hamel et al., 2018). In addition to its medical applications, the genus is recognized for its production of a yellow pigment, as exemplified by species such as *G. tinctoria* and *G. pilosa* (Harborne and Baxter, 2001).

In spite that this plant originated from the Canary Islands, the use of *G. canariensis* as an entheogen has been ingrained into the traditions and magical rituals of the Yaqui medicine men in northern Mexico. The integration of non-native plant species into the magical or religious practices of Indigenous communities is a somewhat infrequent occurrence (Fadiman 1965). According to the Yaqui people, it is asserted that a shaman from their community consumed a different entheogenic substance, most probably peyote, and received a revelation during a visionary experience. This revelation purportedly conveyed the instruction to smoke the petals of the *Genista* plant. Nevertheless, it is conceivable that the awareness of Genista's potential as an entheogen was introduced alongside the plant from the Canary Islands (Ratsch, 1998).

The Genista blossoms were either consumed by inhalation or orally. The flowers are chopped, dried, and either smoked in pipes or rolled in the form of a cigarette. It was a common practice to use other herbs in combination. A single dosage consists of an amount of leaves equivalent to that found in one to three standard cigarettes (Fadiman, 1965). The dried flowers are also brewed in water and consumed as an aphrodisiac drink. This drink is claimed to calm the

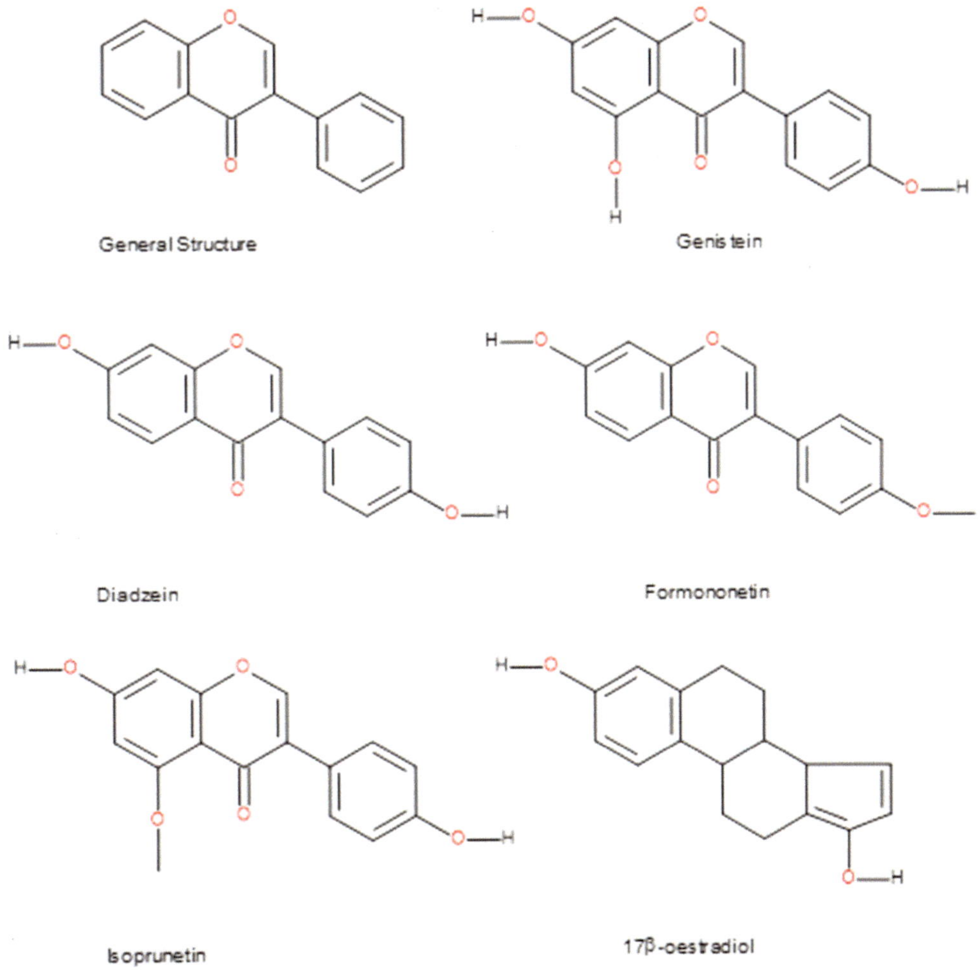

FIGURE 21.3 The major isoflavones in *G. canariensis* in comparison to the structure of 17β-oestradiol.

body and increase sexual excitement. In Mexico, the Yaqui people cautiously aged the flowers for ten days in an airtight glass jar, to prevent molding or fermentation. These were then dried and smoked. Apart from the flowers, the seeds were brewed by the Yaqui healers to prepare a beverage intended to aid in physical and spiritual healing. Besides its spiritual and aphrodisiac effects, *G. canariensis* was used for ailments of the kidney, heart, and bladder. It was also used for the treatment of rheumatism (Baker, 2010).

Since antiquity, it was well known that this plant is toxic to humans. In fact, this plant was used with caution for curative purposes. Toxicity is expressed as nausea, seizures and respiratory failure that may lead to death (Ott, 1993). On the other hand, it is believed that the leaves of the plant have no harmful side effects with potentially moderate psychedelic effects. It was reported that if one cigarette is smoked, the subject experiences a euphoric sensation for approximately two hours. However, if more cigarettes are consumed, then there is an increase in mental capacity and concentration that may last for up to five hours (Fadiman 1965).

21.9 HALLUCINOGENIC EFFECTS

The hallucinogenic effects of *G. canariensis* were challenged in *in vitro*, pre-clinical but also clinical studies. The effects are mainly attributed to the major alkaloid, cytisine. In an *in vitro* study conducted by Rao et al. (1997), rat striatal and hippocampal slices were incubated with radiolabeled dopamine and norepinephrine and placed in Krebs buffer. It was observed that both cytisine and nicotine increased the basal dopamine release within the striatum and basal norepinephrine release within the hippocampus. In a more recent study, it was found that cystisine has a neuronal protective effect via the downregulation of N-methyl-D-aspartate (NMDA) receptors (Li et al., 2013).

Several studies demonstrated the efficacy of cytisine in rat models. A study revealed its effects in the treatment of induced depression. It was observed that cytisine, a partial agonist of α4β2 nicotinic acetylcholine receptors (α4β2nAChRs) and a full agonist of α7 and α3β4 nicotinic acetylcholine receptors (α7nAChRs and α3β4nAChRs) (Scholze and Huck, 2020), relieved symptoms of chronic mild stress (Han et al., 2016). A mechanism was proposed for the biological activity of cytisine, which can upregulate α3β4nAChRs through the substrate-receptor interaction that leads to a chain of events within the cell; from the assembly of the pentameric α4β2 or α3β4 receptors in the endoplasmic reticulum, to the transfer of these receptors to the Golgi body via the COPII vesicles, toe the final incorporation into the plasma membrane (Gotti and Clementi, 2021). In another study on a mouse model, the combined administration of fluoxetine and cytisine demonstrated synergistic properties in relation to depression. Conversely, the antidepressant benefits of cytisine were impeded by serotonin depletion (Mineur et al., 2015). Igari et al. (2014) demonstrated that indicated that cytisine exhibited a mitigating effect on the psychological distress commonly experienced during nicotine withdrawal. In two independent rodent studies, it was shown that cytisine has the capacity to regulate chronic voluntary ethanol consumption, hence indicating its potential involvement in the control of alcohol addiction and reduction of alcohol intake (Sajja and Rahman, 2013; Sotomayor-Zárate et al., 2013). Cytisine was also challenged in a rat model of temporal lobe epilepsy to demonstrate its effects via the α7nAChRs (Zheng et al., 2021). The observed outcomes were linked to a reduction in glutamate concentrations, suppression of synaptic restructuring, and enhancement of cholinergic signaling in the hippocampus. This study not only indicates that cytisine has potential as an effective medication for epilepsy, but also give supporting evidence for the α7nAChRs as a new and promising target for the treatment of temporal lobe epilepsy.

Since antiquity, *Genista canariensis L.* was known to have minor psychedelic properties acquired through blossom smoking. However, this plant has been exploited in research for its effects on smoking cessation, given that its major constituent, cytisine, is biologically related to nicotine. As a matter of fact, early human studies dating back to the 1960s were undertaken, giving way to future scientific studies within this field. In research conducted by Fadiman (1965), subjects were exposed to the inhalation of a cigarette containing the aged and dried plant. Participants were instructed to retain their breath for a duration of 10 seconds after inhalation, aiming to optimize the absorption of the active components. This practice aimed to minimize the loss of the inhaled plant material upon exhalation. The experienced results encompass a state of relaxation, enhanced self-esteem, and increased sociability. When multiple cigarettes were consumed, the duration and intensity of the effects were increased. This effect was substantiated by De Smet (1985). These early studies described the effects of *G. canariensis* as a psychoactive plant rather than its effects on nicotine-smoking cessation. More recently, cytisine was challenged against nicotine for smoking cessation (Walker et al., 2014). The efficacy of cytisine in aiding smoking cessation was found to be superior to that of nicotine-replacement treatment. This is also substantiated by Zatoński and Zatoński (2015). However, it was observed that the use of cytisine was linked to a greater occurrence of self-reported side effects. Vinnikov et al. (2008) reported that cytisine should be contraindicated in cases of severe arrhythmias and ischemic heart disease, severe atherosclerosis, schizophrenia, tumor, in pregnancy and breastfeeding.

21.10 CONCLUSION

In spite of the limited existing literature, this review presented in this chapter demonstrates the two sides of *Genista canariensis*. On one hand, the negative implications as a potential drug of abuse, imply caution with its availability and use. On the other hand, several studies demonstrated its potential implications as a nicotine agonist and hence a possible replacement for nicotine for smoking cessation. In spite of all this and the fact that in some countries, the main alkaloid cytisine, is being used clinically, further studies are necessary to substantiate its clinical uses but also to render *G. canariensis* and cytisine more safe in the future.

LIST OF ABBREVIATIONS

NMDA	N-methyl-D-aspartate
$\alpha 4\beta 2$nAChRs	$\alpha 4\beta 2$ nicotinic acetylcholine receptors
$\alpha 7$nAChRs	$\alpha 7$ nicotinic acetylcholine receptors
$\alpha 3\beta 4$nAChRs	$\alpha 3\beta 4$ nicotinic acetylcholine receptors
ERα	estrogen receptor α isotype
ERβ	estrogen receptor β isotype

CONFLICT OF INTERESTS

The authors declared that they have no conflict of interests with respect to the authorship and/or publication of this chapter.

REFERENCES

Abdel-Lateif, K., Bogusz, D., and Hocher, V. 2012. The role of flavonoids in the establishment of plant roots endosymbioses with arbuscular mycorrhiza fungi, rhizobia and Frankia bacteria. *Plant Signaling & Behavior*, 7: 636–641.

Adams, M., Berset, C., Kessler, M., and Hamburger, M. 2009. Medicinal herbs for the treatment of rheumatic disorders—A survey of European herbals from the 16th and 17th century. *Journal of Ethnopharmacology*, 121(3): 343–359.

Agelet, A., and Vallès, J. 2003. Studies on pharmaceutical ethnobotany in the region of Pallars (Pyrenees, Catalonia, Iberian Peninsula). Part II. New or very rare uses of previously known medicinal plants. *Journal of Ethnopharmacology*, 84(2–3): 211–227.

Baker, G. 2010. Garden of Eden: The Shamanic use of psychoactive flora and fauna and the study of consciousness. *Australian Journal of Medical Herbalism*, 22(3): 107–108.

Benayache, F., D'Ambola, M., Cotugno, R., Chaouche, M., Benayache, S., Benayache, F., Braca, A. and De Tommasi, N. 2018. A new triterpene glucoside from *Genista numidica*. *Natural Product Communications*, 13(9): 1101–1104. https://doi.org/10.1177/1934578X1801300902

Bouchouka, E., Djilani, A., and Bekkouche, A. 2012. Antibacterial and antioxidant activities of three endemic plants from Algerian Sahara. *Acta Scientiarum Polonorum Technologia Alimentaria*, 11(1): 61–65.

Boutaghane, N., Voutquenne-Nazabadioko, L., Harakat, D., Simon, A., and Kabouche, Z. 2013. Triterpene saponins of *Genista ulicina* Spach. *Phytochemistry*, 93: 176–181.

Cayman Chemical. 2023. Cytisine. https://www.caymanchem.com/product/26437

Cochrane, G.R. 1963. Vegetation studies in forest-fire areas of the Mount Lofty Ranges, South Australia. *Ecology*, 44(1): 41–52.

Chaachouay, N., Benkhnigue, O., Fadli, M., Ibaoui, H., & Zidane, L. (2019). Ethnobotanical and ethnopharmacological studies of medicinal and aromatic plants used in the treatment of metabolic diseases in the Moroccan Rif. *Heliyon*, *5*(10), e02191. https://doi.org/10.1080/10477845.2021.1942433

Chaachouay, N., Douira, A., & Zidane, L. (2021a). Herbal Medicine Used in the Treatment of Human Diseases in the Rif, Northern Morocco. *Arabian Journal for Science and Engineering*, 47, 131–153. https://doi.org/10.1007/s13369-021-05501-1

Chaachouay, N., Douira, A., & Zidane, L. (2021b). COVID-19, prevention and treatment with herbal medicine in the herbal markets of Salé Prefecture, North-Western Morocco. *European Journal of Integrative Medicine*, 42, 101285. https://doi.org/10.1016/j.eujim.2021.101285

Crews C. 2014. Natural toxicants: Alkaloids. In: Y. Motarjemi (ed.), *Encyclopedia of Food Safety*. Academic Press, ISBN 9780123786135.

De Smet, P.A.G.M. 1985. Ethnopharmacological table on some reputedly psychoactive fumigatories among Middle and South American natives. *Pharmaceutisch Weekblad*, 7: 212–218.

Edwards, E. L., Rodrigues, J. A., Ferreira, J., Goodall, D. M., Rauter, A. P., Justino, J., and Thomas-Oates, J. 2006. Capillary electrophoresis-mass spectrometry characterisation of secondary metabolites from the antihyperglycaemic plant *Genista tenera*. *Electrophoresis*, 27(11): 2164–2170.

Fadiman, J. 1965. *Genista Canariensis*: A minor psychedelic. *Economic Botany*, 19: 383–384.

Fokialakis, N., Alexi, X., Aligiannis, N., Boulaka, A., Meligova, A.K., Lambrinidis, G., Kalpoutzakis, E., Pratsinis, H., Cheilari, A., Mitsiou, D.J. and Mitakou, S. 2019. Biological evaluation of isoflavonoids from *Genista halacsyi* using estrogen-target cells: Activities of glucosides compared to aglycones. *PLoS One*, 14(1): e0210247.

Gennadius, P. (1914). Lexikon phytologikon, Athens, p. 398 (in Greek).

Gotti, C., and Clementi, F. (2021). Cytisine and cytisine derivatives. More than smoking cessation aids. *Pharmacological Research*, 170: 105700.

Guarrera, P. M. and Lucia, L. M. 2007. Ethnobotanical remarks on central and southern Italy. *Journal of Ethnobiology and Ethnomedicine*, 3(1): 1–11.

Hamel, T., Zaafour, M. and Boumendjel, M. 2018. Ethnomedical knowledge and traditional uses of aromatic and medicinal plants of the Wetlands Complex of the Guerbes-Sanhadja Plain (Wilaya of Skikda in Northeastern Algeria). *Herbal Medicine*, 4(1): 1–9.

Han, J., Wang, D.S., Liu, S.B. and Zhao, M.G. 2016. Cytisine, a partial agonist of α4β2 nicotinic acetylcholine receptors, reduced unpredictable chronic mild stress-induced depression-like behaviors. *Biomolecules and Therapeutics* (Seoul), 24(3):291–297.

Harborne, J. B. 1969. Chemosystematics of the Leguminosae. Flavonoid and isoflavonoid patterns in the tribe Genisteae. *Phytochemistry*, 8(8): 1449–1456.

Harborne, J. B. and Baxter, H. (Eds.). 2001. *Chemical dictionary of economic plants*. John Wiley & Sons.

Igari, M., Alexander, J.C., Ji, Y., Qi, X., Papke, R.L. and Bruijnzeel, A.W. 2014. Varenicline and cytisine diminish the dysphoric-like state associated with spontaneous nicotine withdrawal in rats. *Neuropsychopharmacology*, 39(2): 455–465.

Jespon eFlora. 2023. *Genista canariensis* L. Fabaceae. *Index of California Plant Names*. University of California, Berkeley University and Jespon Herbaria. https://ucjeps.berkeley.edu/cgi-bin/get_cpn?26556

Kew. 2023. *The International Plant Names Index and World Checklist of Vascular Plants*. Published on the Internet at http://www.ipni.org and https://powo.science.kew.org/

Li, Y.J., Yang, Q., Zhang, K., Guo, Y.Y., Li, X.B., Yang, L., Zhao, M.G. and Wu, Y.M. 2013. Cytisine confers neuronal protection against excitotoxic injury by down-regulating GluN2B-containing NMDA receptors. *Neurotoxicology*, 34: 219–225.

Mineur, Y.S., Einstein, E.B., Bentham, M.P., Wigestrand, M. B., Blakeman, S., Newbold, S.A. and Picciotto, M.R. 2015. Expression of the 5-HT1A serotonin receptor in the hippocampus is required for social stress resilience and the antidepressant-like effects induced by the nicotinic partial agonist cytisine. *Neuropsychopharmacology*, 40(4): 938–946.

Ott, J. 1993. *Pharmacotheon: Entheogenic Drugs, Their Plant Sources and History*. Natural Products Company.

Rao, T.S., Correa, L.D. and Lloyd, G.K. 1997. Effects of lobeline and dimethylphenylpiperazinium iodide (DMPP) on N-methyl-D-aspartate (NMDA)-evoked acetylcholine release *in vitro*: evidence for a lack of involvement of classical neuronal nicotinic acetylcholine receptors. *Neuropharmacology*, 36(1): 39–50.

Ratsch, C. 1998. The *Encyclopedia of Psychoactive Plants: Ethnopharmacology and its Applications*. Park Street Press.

Rauter, A.P., Martins, A., Lopes, R., Ferreira, J., Serralheiro, L.M., Araújo, M.E., Borges, C., Justino, J., Silva, F.V., Goulart, M. and Thomas-Oates, J. 2009. Bioactivity studies and chemical profile of the antidiabetic plant *Genista tenera*. *Journal of Ethnopharmacology*, 122(2): 384–393.

Sajja, R.K. and Rahman, S. 2013. Cytisine modulates chronic voluntary ethanol consumption and ethanol-induced striatal upregulation of DeltaFosB in mice. *Alcohol*, 47(4): 299–307.

Scholze, P. and Huck, S. 2020. The α5 nicotinic acetylcholine receptor subunit differentially modulates α4β2 and α3β4 receptors. *Frontiers in Synaptic Neuroscience*, 12: 607959.

Schultes, R.E. 1970. The botanical and chemical distribution of hallucinogens. *Annual Review of Plant Physiology*, 21(1): 571–598.

Schultes, R.E. 2016. Phytochemical gaps in our knowledge of hallucinogens. *Progress in Phytochemistry*, 7(7): 301.

Sohn, S.I., Pandian, S., Oh, Y.J., Kang, H.J., Cho, W.S. and Cho, Y.S. 2021. Metabolic engineering of isoflavones: An updated overview. *Frontiers in Plant Science*, 12: 670103.

Sotomayor-Zárate, R., Gysling, K., Busto, U. E., Cassels, B. K., Tampier, L. and Quintanilla, M. E. 2013. Varenicline and cytisine: two nicotinic acetylcholine receptor ligands reduce ethanol intake in University of Chile bibulous rats. *Psychopharmacology*, 227: 287–298.

Vinnikov, D., Brimkulov, N. and Burjubaeva, A. 2008. A double-blind, randomised, placebo-controlled trial of cytisine for smoking cessation in medium-dependent workers. *Journal of Smoking Cessation*, 3(1): 57–62.

Walker, N., Howe, C., Glover, M., McRobbie, H., Barnes, J., Nosa, V., Parag, V., Bassett, B. and Bullen, C., 2014. Cytisine versus nicotine for smoking cessation. *New England Journal of Medicine*, 371(25): 2353–2362.

Wellmann M. 1958. Pedanii Dioscuridis Anazarbei, *De materia medica libri quinque*. Berlin, Weidmann, 1906–1914.

Wink, M., Witte, L., Hartmann, T., Theuring, C. and Volz, V. 1983. Accumulation of quinolizidine alkaloids in plants and cell suspension cultures: Genera *Lupinus, Cytisus, Baptisia, Genista, Laburnum*, and *Sophora*. *Planta Medica*, 48: 253–257.

Zatoński, W. and Zatoński, M. 2015. Cytisine versus nicotine for smoking cessation. *New England Journal of Medicine*, 372(11): 1072–1072.

Zheng, J.J., Zhang, T.Y., Liu, H.T., Huang, Z.X., Teng, J.M., Deng, J.X., Zhong, J.G., Qian, X., Sheng, X.W., Ding, J.Q. and He, S.Q. 2021. Cytisine exerts an anti-epileptic effect via α7nAChRs in a rat model of temporal lobe epilepsy. *Frontiers in Pharmacology*, 12: 706225.

22 Mescal Bean (*Sophora secundiflora* (Ortega) Lag. ex DC. Fabaceae)

Noureddine Chaachouay and Lahcen Zidane

22.1 INTRODUCTION

The vast array of plants, marine organisms, and animals in nature provides a rich supply of potential therapeutic compounds that may be explored for their medicinal properties (Malve, 2016). Throughout history, people have used botanical extracts to address various health conditions (Giannenas et al., 2020; Chaachouay et al., 2020a, 2020b; Orch et al., 2021). Early civilizations acquired knowledge about the biological properties of plants, which has been transmitted through generations till the present day (Jamshidi-Kia et al., 2018).

Sophora secundiflora (Ortega) Lag. ex DC., often known as Mescal Bean Tree or Texas Mountain Laurel, is a compact evergreen tree indigenous to the southwestern region of the United States, including Texas and some areas of Mexico (Forrester et al., 2020). The Mescal Bean Tree is renowned for its aesthetically pleasing, compact foliage and exquisite blossoms. The average height of this plant ranges from 10 to 30 feet (3 to 9 meters), while its width spans from 10 to 20 feet (3 to 6 meters). The leaves are complex, having a deep green color and a tough texture, consisting of several tiny leaflets (Eldahshan et al., 2019). The tree yields compact groups of conspicuous, aromatic, bluish-purple blossoms throughout spring, resembling miniature orchids and enticing pollinators such as bees and butterflies (Gilman & Watson, 1994; Chaachouay et al., 2019a, 2019b, Belhaj et al., 2021; Benkhnigue et al., 2022). *S. secundiflora* has excellent adaptation to its indigenous environment's dry and partially dry conditions. This plant has exceptional drought tolerance and flourishes when exposed to direct sunlight. The tree has a high tolerance toward various soil types, such as rocky or alkaline soils, as long as they possess good drainage properties (Merrill, 1977; Ruter & Ingram, 1990).

The Mescal Bean Tree has cultural importance among Indigenous populations in the southwestern region of the United States. The seeds of the tree have ceremonial and historical significance in some Native American communities despite their hazardous nature (Forrester et al., 2020; Merrill, 1977). Nevertheless, it is essential to be cautious while handling and consuming the seeds because of their dangerous nature. *S. secundiflora* is extensively grown as a decorative tree due to its appealing aesthetics and aromatic blossoms (Hussiny et al., 2020; Izaddoost et al., 1976). It is often used in landscaping to serve as a central element or as a component of indigenous and xeriscape gardens in areas characterized by arid temperatures. The tree's perennial leaves and flowers throughout the spring make it an attractive option for enhancing outdoor areas' aesthetic and olfactory aspects (Eldahshan et al., 2019; Krishna et al., 2012).

The Mescal Bean Tree is distinguished by the captivating aroma emitted by its blossoms. The flowers emit a strong, pleasant fragrance that evokes the smell of grape soda or synthetic grape essence. However, it's vital to remember that the seeds of this plant are very poisonous and should not be swallowed. They contain a chemical known as cytisine, which can be detrimental to people and animals if ingested (Aly et al., 2020; Hussiny et al., 2020). Although the Mescal Bean Tree seeds are poisonous, they play a crucial role in providing animal habitat and sustenance. The blooms lure pollinators, whilst certain avian species ingest the seeds (Aly et al., 2020). The plant also acts as a host for the larvae of the Genista moth, which has a particular symbiotic connection with the tree. This chapter explicitly examines the botanical characteristics, cultural applications, environmental interactions, chemical composition, and biological properties of *S. secundiflora*.

22.2 SCIENTIFIC CLASSIFICATION

Sophora secundiflora (Ortega) Lag. ex DC. is an evergreen, deciduous shrub that does not exceed 10 feet in height and, in certain regions, can develop into a 35-foot-tall tree (Bean, 1912). It belongs to the Fabaceae botanical family and *Sophora* genus (The Plant List, 2013).

Kingdom: Plantae
Subkingdom: Tracheobionta
Superdivision: Spermatophyta
Division: Magnoliophyta
Class: Magnoliopsida
Subclass: Rosidae
Order: Fabales
Family: Fabaceae
Genus: *Sophora*
Species: *Sophora secundiflora* (Ortega) Lag. ex DC.

DOI: 10.1201/9781003460336-22

22.3 VERNACULAR NAMES

English: Mescal Bean, Texas mountain laurel, Texas mescal bean; **Spanish**: Frijolillo, Frijolito, Pito, Paloverde, Coral-bean, Palo Fierro, Mescal Bean, Frijolito; **Nahuatl (Aztec language)**: Cacachilaniquilxochitl; **Navajo**: Tashkaah; **Hopi**: Moos; **Comanche**: Amash; **French**: Sophora du Texas; **German**: Texas-Mondnuss; **Portuguese**: Feijão-de-mescal; **Italian**: Fagiolo messicano; **Folk Names**: Big drunk bean, chilicote, colorín, colorines, coral bean tree, coral bean, frijolillo of Texas, k'awnk'odl, mescal bean, meskalbohne, mountain laurel, patiol, patol, red bean, red medicine, schnurbaum.

22.4 BOTANICAL DESCRIPTION

S. secundiflora is a perennial plant that often has many trunks and may range in height from a few feet to over 30 feet (Figure 22.1). However, when fully grown, its average size is generally between 10 and 15 feet (Eldahshan et al., 2019). The compound leaves are thick, dark green, and glossy. They consist of 7–9 lustrous and leathery leaflets that have rounded ends.

The leaflets measure up to 2 inches or more in length, with a more gradual taper toward the base than the tip. They are organized along an axis that ends with a single leaflet (Gilman & Watson, 1994). The flowers (Figure 22.2), which are blue lavender and arranged in clusters that hang and measure between 3 and 7 inches, are conspicuous and have a pleasant fragrance (Aly et al., 2020).

The fruit is a somewhat lignified pod containing vibrant red toxic seeds. *S. secundiflora* is versatile and may serve as a specimen, patio, or street tree. It is particularly well-suited for screen, bank cover, or espalier use (Forrester et al., 2020).

22.5 ORIGIN AND GEOGRAPHIC DISTRIBUTION

S. secundiflora is indigenous to the southwestern region of the United States and the northern part of Mexico. This plant species has a lengthy historical record of thriving in this area and has developed strong compatibility with the prevailing climatic and soil characteristics (Peña & Cassels, 1996). The geographical distribution of *S. secundiflora* is mainly limited to Texas, thus its popular designation as 'Texas mountain laurel'. It is present across a large portion of central and western Texas, including regions such as the Edwards Plateau, the Hill Country, and the Chihuahuan Desert section of the state (Forrester et al., 2020). This species also ranges into northern Mexico, namely in Coahuila, Nuevo León, and Tamaulipas. *S. secundiflora* often thrives in environments such as rugged inclines, calcareous hills, ravines, and unobstructed forests (Merrill, 1977; Peña et al., 2000). It is usually found in regions with soils that have good drainage. *S. secundiflora* is planted as attractive in places with a comparable climate. The plant's ability to withstand drought, pleasant-smelling blossoms, and appealing visual qualities contribute to its widespread use in landscaping (Ruter & Ingram, 1990).

FIGURE 22.1 Aerial parts of *S. secundiflora*.

FIGURE 22.2 The flowers of *S. secundiflora*.

22.6 HABITAT AND ECOLOGY

S. secundiflora thrives in the dry and semi-dry environments in the southwestern United States and northern Mexico. This little tree or perennial shrub flourishes in regions distinguished by permeable, bare soils, often seen on steep inclines, ravines, calcareous hills, and unobstructed forests (Furuya & Van Auken, 2009). The plant has evolved several adaptations to deal with these demanding circumstances, such as developing long roots that can reach water in more profound *S. secundiflora* is found as isolated specimens among shrubby vegetation in the Edwards Plateau and Rio Grande Plains of Texas (Aly et al., 2020; Forrester et al., 2020). These regions also frequently feature riparian deciduous forests that are primarily composed of *Quercus virginiana, Carya illinoensis, Celtis reticulata, Diospyros texana, Fraxinus berlandieriana, Platanus occidentalis,* and *Ulmus crassifolia* (Crowder, 1978).

The Texas mountain laurel fulfills a crucial ecological function by providing refuge and protection to various animal species Table 22.1. The thick vegetation offers shelter for miniature animals, insects, and avian species, increasing their likelihood of survival in these desert habitats (Ruter & Ingram, 1990). Although the seeds of this plant are deadly, some animals like deer and squirrels have acquired a partial resistance to the poisons and may ingest them in tiny amounts. Moreover, *S. secundiflora* has cultural importance among Indigenous populations in its original habitat, who traditionally used the seeds in ceremonial practices and as decorative items, enhancing the plant's artistic value and ecology in the area (Forrester et al., 2020; Krishna et al., 2012; Merrill, 1977; Schultes & Farnsworth, 1980).

TABLE 22.1
***S. secundiflora* Characteristics**

Duration	Perennial
Habit	Shrub, Tree
Leaf Retention	Evergreen
Leaf Arrangement	Alternate
Leaf Complexity	Pinnate
Leaf Pubescence	Glabrous
Leaf Margin	Entire
Leaf Texture	Leathery
Breeding System	Flowers Bisexual
Inflorescence	Raceme
Fruit Type	Legume
Size Notes	Typically reaching a maximum height of around 30 feet, but sometimes much less
Leaf	Green
Flower	Racemes 5–15 cm long
Fruit	The seedpods have a gray color and a surface texture that resembles felt. As they age, they become dark brown. The seeds are shiny crimson and measure between 2 and 12 centimeters

22.7 ETHNOBOTANY

S. secundiflora is a deciduous plant that produces flowers and belongs to the Fabaceae family. It is indigenous to the southwestern region of the United States and the northern part of Mexico. The plant is highly esteemed for its vibrant purple blossoms, delightful fragrance, and the vivid scarlet seeds it yields. The Indigenous populations in the southwestern United States and northern Mexico have a rich historical tradition of using several components of the *S. secundiflora* plant for medical, ceremonial, and practical objectives.

22.7.1 Traditional Medicine

S. secundiflora seeds are rich in alkaloids, including cytisine and various other compounds, that, when ingested in excess, can cause toxicity (Aly et al., 2022). Despite their toxicity, some Native American tribes utilized minute quantities for medicinal purposes. Native Americans have historically used *S. secundiflora* legume as an analgesic (Hussiny et al., 2020). Cytisine and other alkaloids present in the legumes possess analgesic properties. However, this application is hazardous due to the toxic character of these compounds. *S. secundiflora* legume has been employed in certain traditional practices to remedy respiratory ailments such as colds and coughing (Krishna et al., 2012; Chaachouay et al., 2021a, 2021b, 2021c; Benlarbi et al., 2023). The Indigenous Kickapoo people residing in northern Mexico use the seeds to remedy ear problems. A decoction is made by grinding one seed and some *Nicotiana spp.*, then it is applied by dripping it into the ear canal. An ear wash is made by extracting the crushed seeds in cold water (Abd-Alla et al., 2020). Ear drops may be prepared by boiling a bean together with *Juniperus sp.* It was formulated and administered as an infusion or decoction to alleviate respiratory distress. Toothaches have been treated topically with *S. secundiflora*. The seeds were either ingested or administered topically to the affected region to alleviate pain (Aly et al., 2020; Krishna et al., 2012).

22.7.2 Religion and Ritual

The archaeological record indicates that *S. secundiflora* has been used for a long time and has cultural value among Native Americans in the region that is now Texas and Mexico (Merrill, 1977; Chaachouay & Zidane, 2021). Nevertheless, the use of *S. secundiflora* extended much beyond the native range of the mountain laurel tree in the past. In the 19th century, Indian people experienced significant turmoil; during this time, several societies or revival movements emerged (Du Toit, 1977). One such movement was the mescalbean society, which gained prominence and expanded its influence. The specifics of these societies varied among different groups, but they usually restricted membership to males and typically ingested a beverage derived from *S. secundiflora* during certain rituals (Aly et al., 2020). It just invites partook of *S. secundiflora* in some societies. However, in a few tribes, such as the displaced Algonquian-speaking communities, there is no evidence of consumption, just documentation of the culture (Stewart, 1980). Membership was exclusive and sometimes required the new member to pay for admission into the group. The Texas mountain laurel fulfills a crucial ecological function by providing refuge and protection to various animal species (Wrede, 2010). The thick vegetation offers shelter for avian species, insects, and miniature animals, increasing their likelihood of survival in these desert habitats (Cook, 2012). Although the seeds of this plant are deadly, some animals like squirrels and deer have acquired a partial resistance to the poisons and may ingest them in tiny amounts (Janzen, 1971; Chaachouay et al., 2022a, 2022b, 2022c).

Moreover, the plant has cultural importance among Indigenous populations in its original habitat, who traditionally used the seeds in ceremonial practices and as decorative items, enhancing the plant's ecological and cultural value (Howard, 1957). Certain societies only affixed the mescal bean to artifacts connected to religious ceremonies. For example, they tied the seeds to various items or stored them in medicinal pouches, often possessed by male members of an *S. secundiflora* medicine society. The groups above included the Pawnee, Wichita, Iowa, Oto, Omaha, Ponca, Missouri, and Osage. The Wichita and Pawnee linked *S. secundiflora* to deer rituals (Forrester et al., 2020; Gilman & Watson, 1994).

Furthermore, there is evidence of the growing popularity of *S. secundiflora* consumption since many communities have related *S. secundiflora* to horses and triumph in raiding missions. The ceremonial use of *S. secundiflora* extended beyond the confines of communities that exclusively practiced *S. secundiflora* cultures (Merrill, 1977). Within several Coahuilteco-speaking communities, like the Hasinai Caddo and the Tonkawa, it was observed that both males and females took part in rituals based around the mescalbean plant. The Hasinai used the mescalbean in ways to induce hallucinations and before engaging in raids. The vivid crimson *S. secundiflora* was linked to military conflict and prominently shown in garments worn during combat (Schultes, 1969).

22.7.3 Clothing and Other Objects

Many groups incorporated *S. secundiflora* into their dress and accessories but did not establish a formal *S. secundiflora* society. The tribes mentioned include the Delaware, Shawnee, Tewa, Sac Fox, Western Apache, Kiowa-Apache, and Kickapoo (Merrill, 1977; Stewart, 1980). *S. secundiflora* was used to adorn garments, leggings, sacks, necklaces, skirts, bandoleers, warrior dolls, and medicine bags. In addition, these groups used the wood of this plant to make tools and implements (Boyd & Dering, 1996; Howard, 1957; Schultes & Farnsworth, 1980).

22.8 HALLUCINOGENIC EFFECTS

The seeds of *S. secundiflora* are believed to possess hallucinogenic properties due to their historical use in certain Native American rituals, which resulted in the consumption of the seeds leading to visionary experiences (Díaz, 2010; Schultes, 1969). Consuming *S. secundiflora* beans in isolation might induce hallucinations, however, accompanied by a plethora of adverse side effects, including prolonged episodes of intense vomiting, cognitive disarray, hypotension, stomach discomfort, restlessness, tachycardia, muscular tremors, followed by convulsions and death (Forrester et al., 2020; Krishna et al., 2012). In addition, the mescal bean plant's seeds, leaves, and blooms are toxic to cattle, sheep, and goats. Cattle are more susceptible to leaf poisoning, while goats and sheep show greater tolerance (Izaddoost et al., 1976). Animals that are impacted often have a complete recovery when they are provided with food of superior quality. *S. secundiflora* poisoning exhibits similarities to nicotine poisoning due to the action of cytisine chemicals in *S. secundiflora* on the same receptors targeted by nicotine (Schultes & Farnsworth, 1980). Cytisine, which is present in *S. secundiflora* beans, has been used for smoking cessation due to its impact on nicotine receptors (Aly et al., 2020; Schultes, 1984). Although they have a similar name, *S. secundiflora* does not contain mescaline, a hallucinogenic chemical in some cacti.

22.9 PHYTOCHEMISTRY

S. Secundiflora's phytochemical analysis has shown diverse chemicals in distinct plant components Table 22.2. The primary constituents identified include carbohydrates, tannins, flavonoids, terpenoids, and alkaloids, notably the poisonous alkaloid cytisine and several quinolizidine derivatives (Aly et al., 2020, 2022; Hussiny et al., 2020; Izaddoost et al., 1976). *S. Secundiflora* seeds contain several alkaloids, with sophorine being the most prevalent. Sophorine has hazardous qualities, and ingestion may result in symptoms such as nausea, vomiting, and other adverse effects (Aly et al., 2022; Chavez & Sullivan, 1984; Izaddoost et al., 1976).

TABLE 22.2
Phytochemical Constituents Isolated from *S. secundiflora*

Compounds	Chemical Class	References
Cytisine alkaloids	Baptitoxine, sophorine, ulexine, laburnine, cytitone, N-methylcytisine, and sparteine	Aly et al. (2020); Farias and Iturriaga (2022); Izaddoost et al. (1976); Makboul et al. (1987)
Quinolizidine alkaloids	epi-lupinine, Δ^5-dihydrolupanine, anagyrine, and thermopsine	Aly et al. (2022); Chavez and Sullivan (1984)
Other compounds	Carbohydrates, flavonoids, tannins, terpenoids	Aly et al. (2021); Hussiny et al. (2020); Shirataki et al. (1997, 1997)

Historically, Native American tribes have used it for ceremonial and therapeutic purposes; nonetheless, it has a high level of toxicity and should not be ingested. *S. secundiflora* also has quinolizidine alkaloids, which are alkaloids often present in plants belonging to the Fabaceae family (Aly et al., 2020; Chavez & Sullivan, 1984). These alkaloids exhibit various biological functions and are often linked to toxicity in some animals.

22.10 PHARMACOLOGY

S. secundiflora has been traditionally used by Indigenous communities for various functions, including medical and ceremonial applications. (Merrill, 1977) Nevertheless, it is crucial to highlight that the plant harbors poisonous substances and is not advisable for therapeutic or leisure purposes owing to its inherent health hazards. The primary pharmacological actions of *S. secundiflora* are mainly attributed to its alkaloid composition, namely sophorine and quinolizidine alkaloids. *S. secundiflora* contains alkaloids, particularly quinolizidine alkaloids, that can induce vasoconstriction, resulting in the constriction of blood vessels (Aly et al., 2020, 2022; Izaddoost et al., 1976; Krishna et al., 2012). This phenomenon has been linked to the plant's customary use in mitigating headaches and other illnesses, although it may also pose a risk and result in unfavorable cardiovascular consequences (Aly et al., 2021). The alkaloids in *S. secundiflora* may affect the respiratory system, resulting in symptoms such as difficulty breathing and respiratory discomfort. Some Indigenous communities have used *S. secundiflora* for its putative analgesic and anesthetic effects (Abd-Alla et al., 2020, 2020; Aly et al., 2022; Hussiny et al., 2020).

Nevertheless, these effects are probably subordinate to the plant's poisonous characteristics and are not advisable owing to the accompanying hazards. The main pharmacological effect linked to *S. secundiflora* is its toxic nature (Aly et al., 2019; Hussiny et al., 2020). Consuming any portion of the plant, especially the seeds, may lead to symptoms such as nausea, vomiting, diarrhea, seizures, and, in extreme instances, coma and fatality (Hussiny et al., 2020). It is essential to emphasize that the use of *S. secundiflora* for medical or recreational intentions is severely prohibited owing to its elevated toxicity. The likelihood of experiencing severe ill effects and possibly mortality is substantial (Izaddoost et al., 1976).

22.11 CONCLUSION

S. secundiflora is a plant with fascinating botanical traits, significant cultural uses, and intricate environmental relationships. The plant's vibrant purple blossoms and resilient leaves give it a unique appearance in the landscapes of the southwestern region. Throughout history, Native American tribes have used it for traditional medicinal purposes and spiritual rituals, highlighting its cultural importance. However, it is essential to approach it cautiously since its seeds contain deadly alkaloids, such as sophorine. The ecological significance of *S. secundiflora* is emphasized by its ability to improve the soil via nitrogen fixation and its function as a habitat and food source for animals. The potential of future studies on *S. secundiflora* is promising across several fields. Furthermore, it is imperative to sustain ethnobotanical investigations to investigate the cultural importance of *S. secundiflora* within Indigenous customs further while simultaneously honoring the traditions and legacy of Native American people. There is a need to broaden public awareness efforts to educate the general population about the plant's poisonous nature. This effectively decreases the likelihood of unintentional consumption and poisoning incidents. Hence, the comprehensive examination of *S. secundiflora* presents a promising opportunity for future investigations, integrating endeavors in cultural comprehension, chemical investigation, public health consciousness, and conservation.

LIST OF ABBREVIATION

S. secundiflora *Sophora secundiflora* (Ortega) Lag. ex DC.

REFERENCES

Abd-Alla, H. I., Ibrahim, M. T., Taie, H. A. A., Hasan, M. A., & Shalaby, N. M. (2020). Antioxidant and The Efficacy of Sophora Secundiflora and Methoxyisoflavones in the Immune Function of Pigeons Vaccinated Against Paramyxovirus Serotype-1. *Pharmacognosy Journal*, *12*(6). https://www.phcogj.com/Article/1243

Aly, S. H., Elissawy, A. M., Allam, A. E., Farag, S. M., Eldahshan, O. A., Elshanawany, M. A., & Singab, A. N. B. (2022). New Quinolizidine Alkaloid and Insecticidal Activity of *Sophora Secundiflora* and *Sophora Tomentosa* against *Culex Pipiens* (Diptera: Culicidae). *Natural Product Research*, *36*(11), 2722–2734. https://doi.org/10.1080/14786419.2021.1919108

Aly, S. H., Elissawy, A. M., Eldahshan, O. A., Elshanawany, M. A., Efferth, T., & Singab, A. N. B. (2019). The Pharmacology of the Genus Sophora (Fabaceae): An Updated Review. *Phytomedicine*, *64*, 153070.

Aly, S. H., Elissawy, A. M., Eldahshan, O. A., Elshanawany, M. A., & Singab, A. N. B. (2020). Variability of the Chemical Composition of the Essential Oils of Flowers and the Alkaloid Contents of Leaves of *Sophora Secundiflora* and *Sophora Tomentosa*. *Journal of Essential Oil Bearing Plants*, *23*(3), 442–452. https://doi.org/10.1080/0972060X.2020.1750489

Aly, S. H., Elissawy, A. M., Fayez, A. M., Eldahshan, O. A., Elshanawany, M. A., & Singab, A. N. B. (2021). Neuroprotective Effects of *Sophora Secundiflora*, *Sophora Tomentosa* Leaves and formononetin on Scopolamine-Induced Dementia. *Natural Product Research*, *35*(24), 5848–5852. https://doi.org/10.1080/14786419.2020.1795853

Bean, W. J. (1912). Some Gardens and Parks in S. Europe. *Bulletin of Miscellaneous Information (Royal Botanic Gardens, Kew)*, *1912*(6), 284–297.

Belhaj, S., Chaachouay, N., & Zidane, L. (2021). Ethnobotanical and toxicology study of medicinal plants used for the treatment of diabetes in the High Atlas Central of Morocco. *Journal of Pharmacy & Pharmacognosy Research*, *9*(5), 619–662. https://doi.org/10.56499/jppres21.1014_9.5.619

Benkhnigue, O., Chaachouay, N., Khamar, H., El Azzouzi, F., Douira, A, & Zidane, L. (2022). Ethnobotanical and ethnopharmacological study of medicinal plants used in the treatment of anemia in the region of Haouz-Rehamna (Morocco). *Journal of Pharmacy & Pharmacognosy Research*, *10*(2), 279–302. https://doi.org/10.56499/jppres21.1196_10.2.279

Benlarbi, F., Mimoune, N., Chaachouay, N., Souttou, K., Saidi, R., Mokhtar, M.R., Kaidi, R, & Benaissa, M. T. (2023). Ethnobotanical survey of the traditional antiparasitic use of medicinal plants in humans and animals in Laghouat (Southern Algeria). *Veterinary World*, 16(2), 357–368. https://doi.org/10.14202/vetworld.2023.357

Boyd, C. E., & Dering, J. P. (1996). Medicinal and Hallucinogenic Plants Identified in the Sediments and Pictographs of the Lower Pecos, Texas Archaic. *Antiquity*, *70*(268), 256–275.

Chaachouay, N., Benkhnigue, O., Fadli, M., Ibaoui, H., & Zidane, L. (2019a). Ethnobotanical and ethnopharmacological studies of medicinal and aromatic plants used in the treatment of metabolic diseases in the Moroccan Rif. *Heliyon*, *5*(10), e02191. https://doi.org/10.1080/10477845.2021.1942433

Chaachouay, N., Benkhnigue, O., Fadli, M., Ibaoui, H., El Ayadi, R., & Zidane, L. (2019b). Ethnobotanical and ethnopharmacological study of medicinal and aromatic plants used in the treatment of respiratory system disorders in the Moroccan Rif. *Ethnobotany Research and Applications*, *18*(22), 1–16.

Chaachouay, N., Douira, A., Hassikou, R., Brhadda, N., Dahmani, J., Belahbib, N., Ziri, R., & Zidane, L. (2020a). Ethnobotanical and ethnopharmacological study of medicinal and aromatic plants used in the treatment of respiratory system disorders in the Moroccan Rif. PhDThesis, Département de Biologie - Université Ibn Tofail - Kénitra, Maroc, https://tel.archives-ouvertes.fr/tel-03376377

Chaachouay, N., Benkhnigue, O., & Zidane, L. (2020b). Ethnobotanical study aimed at investigating the use of medicinal plants to treat nervous system diseases in the Rif of Morocco. *Journal of Chiropractic Medicine*, *19*(1), 70–81. https://doi.org/10.1016/j.jcm.2020.02.004

Chaachouay, N., & Zidane, L. (2021). The Symbolic Efficacy of Plants in Rituals and Socio-Religious Ceremonies in Morocco, Northwest of Africa. *Journal of Religious & Theological Information*, *21*(1–2), 34–53. https://doi.org/10.1080/10477845.2021.1942433

Chaachouay, N., Benkhnigue, O., Douira, A., & Zidane, L. (2021a). Poisonous medicinal plants used in the popular pharmacopoeia of the Rif, northern Morocco. *Toxicon*, 189, 24–32. https://doi.org/10.1016/j.toxicon.2020.10.028

Chaachouay, N., Douira, A., & Zidane, L. (2021b). COVID-19, prevention and treatment with herbal medicine in the herbal markets of Salé Prefecture, North-Western Morocco. *European Journal of Integrative Medicine*, 42, 101285. https://doi.org/10.1016/j.eujim.2021.101285

Chaachouay, N., Douira, A., & Zidane, L. (2021c). Herbal Medicine Used in the Treatment of Human Diseases in the Rif, Northern Morocco. *Arabian Journal for Science and Engineering*, 47, 131–153. https://doi.org/10.1007/s13369-021-05501-1

Chaachouay, N., Benkhnigue, O., & Zidane, L. (2022a). Ethnobotanical and Ethnomedicinal study of medicinal and aromatic plants used against dermatological diseases by the people of Rif, Morocco. *Journal of Herbal Medicine*, 32, 100542. https://doi.org/10.1016/j.hermed.2022.100542

Chaachouay, N., Azeroual, A., Bencharki, B., Douira, A., & Zidane, L. (2022b). Ethnoveterinary medicines plants for animal therapy in the Rif, North of Morocco. *South African*

Journal of Botany, *147*, 176–191. https://doi.org/10.1016/j.sajb.2021.12.037

Chaachouay, N., Azeroual, A., Douira, A., & Zidane, L. (2022c). Ethnoveterinary practices of medicinal plants among the Zemmour and Zayane tribes, Middle Atlas, Morocco. *South African Journal of Botany*, *151*, 826–840. https://doi.org/10.1016/j.sajb.2022.11.009

Chavez, P. I., & Sullivan, G. (1984). A Qualitative and Quantitative Comparison of the Quinolizidine Alkaloids of theFasciated of the Normal Stems of the Sophora Secundiflora. *Journal of the Natural Products*, *47*(4), 735–736. https://doi.org/10.1021/np50034a035

Cook, W. E. (2012). *Avian Desert Predators, Adaptations of Desert Organisms*. Springer Science & Business Media - 128 pages. https://www.amazon.fr/Avian-Desert-Predators-W-Cook/dp/3540592628

Crowder, C. A. (1978). *The Ecology of the Reproduction of the Sophora Leachiana Peck (Fabaceae)*. https://Ir.Library.Oregonstate.Edu/Concern/Graduate_Thesis_Or_Dissertations/R494vp238

Díaz, J. L. (2010). Sacred Plants of the Visionary Consciousness. *Phenomenology of the Cognitive Sciences*, *9*(2), 159–170. https://doi.org/10.1007/s11097-010-9157-z

Du Toit, B. M. (1977). *Drugs, Rituals of the Altered States of the Consciousness*. AA Balkema Rotterdam. https://Samorini.It/Doc1/Alt_Aut/Sz/Schultes-Botanical-Chemical-Distribution-Hallucinogens.Pdf

Eldahshan, O., Elissawy, A., Elshanawany, M., & Singab, A. N. (2019). Morphological of the Genetic Characteristics of the Sophora Secundiflora of the Sophora Tomentosa (Fabaceae) Cultivated in Egypt. *Taeckholmia*, *39*(1), 103–129.

Farias, A., & Iturriaga, P. (2022). A Systematic Review on Secondary Metabolites of the Genus Sophora: Chemical Diversity. *Journal of the Chilean Chemical Society*, *67*(3), 5571–5581.

Forrester, M. B., Layton, G. M., & Varney, S. M. (2020). Abuse of the Misuse of the *Sophora Secundiflora* in Texas. *Clinical Toxicology*, *58*(4), 302–303. https://doi.org/10.1080/15563650.2019.1648819

Furuya, M., & Van Auken, O. W. (2009). Gas Exchange Rates of the Sun of the Shade Leaves of the Sophora Secundiflora (Leguminosae, Texas Mountain Laurel). *The Texas Journal of the Science*, *61*(4), 243–259.

Giannenas, I., Sidiropoulou, E., Bonos, E., Christaki, E., & Florou-Paneri, P. (2020). The History of the Herbs, Medicinal of the Aromatic Plants, of the their Extracts: Past, Current Situation of the Future Perspectives. In *Feed Additives* (pp. 1–18). Elsevier.

Gilman, E. F., & Watson, D. G. (1994). *Sophora Secundiflora*. Texas Mountain Laurel) Florida Cooperative Extension Service, Fact Sheet ST .... https://hort.ifas.ufl.edu/Database/Documents/Pdf/Tree_Fact_Sheets/Sopseca.Pdf

Howard, J. H. (1957). The Mescal Bean Cult of the Central of the Southern Plains: An Ancestor of the Peyote Cult. *American Anthropologist*, *59*(1), 75–87.

Hussiny, S., Elissawy, A., Eldahshan, O., Elshanawany, M., & Singab, A.-N. (2020). Phytochemical Investigation Using GC/MS Analysis of the Evaluation of the Antimicrobial of the Cytotoxic Activities of theLipoidal Matter of the Leaves of the Sophora Secundiflora of the Sophora Tomentosa. *Archives of the Pharmaceutical Sciences Ain Shams University*, *4*(2), 207–214.

Izaddoost, M., Harris, B. G., & Gracy, R. W. (1976). Structure of the Toxicity of the Alkaloids of the Amino Acids of the Sophora Secundiflora. *Journal of the Pharmaceutical Sciences*, *65*(3), 352–354. https://doi.org/10.1002/jps.2600650309

Jamshidi-Kia, F., Lorigooini, Z., & Amini-Khoei, H. (2018). Medicinal Plants: Past History of the Future Perspective. *Journal of the Herbmed Pharmacology*, *7*(1), 1–7. 10.15171/jhp.2018.01

Janzen, D. H. (1971). Seed Predation by Animals. *Annual Review of the Ecology of the Systematics*, *2*(1), 465–492. https://doi.org/10.1146/annurev.es.02.110171.002341

Krishna, P. M., Knv, R., & Banji, D. (2012). A Review on Phytochemical, Ethnomedical of the Pharmacological Studies on Genus Sophora, Fabaceae. *Revista Brasileira De Farmacognosia*, *22*, 1145–1154.

Makboul, A. M., Abdel-Baky, A. M., & Bishay, D. W. (1987). Argentine, Lupinine, Cytisine and N-Ethylcytisine Alkaloids from Sophora Secundiflora Cultivated in Egypt. *Bulletin of the Pharmaceutical Sciences. Assiut*, *10*(2), 47–54.

Malve, H. (2016). Exploring the Ocean for New Drug Developments: Marine Pharmacology. *Journal of the Pharmacy & Bioallied Sciences*, *8*(2), 83–91. https://doi.org/10.4103/0975-7406.171700

Merrill, W. L. (1977). *An Investigation of the Ethnographic of the Archaeological Specimens of the Mescalbeans (Sophora Secundiflora) in American Museums* (Vol. 6). U of M Museum Anthro Archaeology. https://books.google.com/Books?Hl=Fr&Lr=&Id=4sbgdwaaqbaj&Oi=Fnd&Pg=PR5&Dq=ORIGIN+AND+GEOGRAPHIC+DISTRIBUTION++ofthe+Sophora+Secundiflora+&Ots=Bz1qmgsxkp&Sig=Tvlk7_Uhpcoleqgc_Ovusylyapi

Orch, H., Chaachouay, N., Douiri, E. M., Faiz, N., Zidane, L., & Douira, A. (2021). Use of medicinal plants in dermato-cosmetology: An ethnobotanical study among the population of Izarène. *Jordan Journal of Pharmaceutical Sciences*, *14*(3), 323–340.

Peña, R. C., & Cassels, B. K. (1996). Phylogenetic Relationships Among Chilean Sophora Species. *Biochemical Systematics of the Ecology*, *24*(7–8), 725–733.

Peña, R. C., Iturriaga, L., Montenegro, G., & Cassels, B. K. (2000). *Phylogenetic of the Biogeographic Aspects of the Sophora Sect. Edwardsia (Papilionaceae)*. https://Scholarspace.Manoa.Hawaii.Edu/Handle/10125/1611

Ruter, J. M., & Ingram, D. L. (1990). Effect of the Scarification Treatments on Germination of the Sophora Secundiflora Seeds. *Effect of the Scarification Treatments on Germination of the Sophora Secundiflora Seeds*, *40*, 423–426.

Schultes, R. E. (1969). Hallucinogens of the Plant Origin: Interdisciplinary Studies of the Plants Sacred In Primitive Cultures Yield Results of the Academic of the Practical Interest. *Science*, *163*(3864), 245–254.

Schultes, R. E. (1984). Psychoactive Plants in Need of the Chemical of the Pharmacological Study. *Proceedings/Indian Academy of the Sciences*, *93*(3), 281–304. https://doi.org/10.1007/bf03053083

Schultes, R. E., & Farnsworth, N. R. (1980). *Ethnomedical, Botanical of the Phytochemical Aspects of the Natural Hallucinogens*. Botanical Museum Leaflets (pp. 123–214). Harvard University.

Shirataki, Y., Yoshida, S., Sugita, Y., Yokoe, I., Komatsu, M., Ohyama, M., Tanaka, T., & Iinuma, M. (1997). Isoflavanones in Roots of the Sophora Secundiflora. *Phytochemistry*, *44*(4), 715–718.

Stewart, O. C. (1980). Peyotism of the Mescausm. *Plains Anthropologist*, *25*(90), 297–310. https://doi.org/10.1080/2052546.1980.11909000

The Plant List. (2013). *The Plant List. Version 1.1*. Published on the Internet; http://www.theplantlist.org (Accessed 1st January).

Wrede, J. (2010). *Trees, Shrubs, of the Vines of theTexas Hill Country: A Field Guide* (Vol. 39). Texas A&M University Press. https://Books.Google.Com/Books?Hl=Fr&Lr=&Id=Qqjkdwaaqbaj&Oi=Fnd&Pg=PR4&Dq=The+Texas+Mountain+Laurel+Fulfills+A+Crucial+Ecological+Function+By+Providing+Refuge+of the+Protection+To+Various+Animal+Species&Ots=Row-B9ozde&Sig=Mackxmuktpwy5ehbvphe_Moqnak

23 Coral Tree (*Erythrina americana* Mill. Fabaceae)

Anam Fatima, Muhammad Zafar, Mushtaq Ahmad, Wajiha Zaka Ansari, Salman Majeed, and Nida Zahra

23.1 INTRODUCTION

Medicinal plants are utilized for the therapeutic management of various illnesses. These plants lack distinctive physical features that set them apart from other plants, but they possess pharmacological properties and contain medicinal substances. Nevertheless, it is important to note that herbal medicine and herbalists may not have been scientifically validated as effective treatments for acute, severe illnesses, severe injuries, appendicitis, or heart attack, unlike conventional doctors who employ current diagnostic testing, surgical procedures, and pharmaceutical medications. Some significant drawbacks of herbal treatment include self-administration of herbs, excessive dosage, misidentification of wild-sourced herbs, potential interactions with prescription medications, and a lack of quality assurance (Alamgir, 2017; Chaachouay et al., 2021a, 2021b).

The term 'hallucinogens' is commonly used to refer to these substances. The word hallucinogen gained popularity in the 1950s and was first associated with chemicals that were believed to induce hallucinations. However, it is now understood that this effect is often not observed at lower doses (Chaachouay and Zidane, 2021; Passos and Mironidou-Tzouveleki, 2016). The primary impacts of these compounds are on cognition, emotional state, and mental processes. When consumed in smaller quantities, they exhibit negligible cognitive decline and do not cause any incapacitating consequences. Furthermore, they do not induce physical dependence. These effects are frequently described as having a dream-like quality and are distinguished by sensory distortions and occasional genuine hallucinations, typically visual or auditory. They involve an intensified emotional state, altered perception of time, mystical attributes, and a range of spiritual encounters and revelations, including ego dissolution, a sensation akin to a near-death experience, and a feeling of interconnectedness or profound comprehension of the universe. The significance of set and setting (psychological mood and physical surroundings) in the consumption of these substances is widely acknowledged (Fatur, 2021).

Erythrina is a genus of flowering trees that belongs to the Fabaceae family. It consists of approximately 115 species and is found in tropical and subtropical regions in both the northern and southern hemispheres. Several species within the genus, especially those found in the Amazon, generate phenolics and alkaloids that have various beneficial effects such as pain relief, reduction of inflammation, anti-malaria properties, ability to kill bacteria, muscle relaxation, and ability to kill fungi (Bernardino-Nicanor et al., 2016; San Miguel-Chavez et al., 2007 Erythrina is a notable genus comprising over 130 species of edible flowers, one of which being the colorin plant. Colorin is planted as ornamental trees and is extensively distributed in tropical and subtropical climates globally. The entire colorin plant, or certain components of it, are utilized in traditional folk medicine and as a food source. The phenols included in the methanolic extract of E. indica leaves have been found to contain antiulcer capabilities (Lopez-Agama et al., 2021).

Erythrina americana, commonly known as the American coral tree or Cherokee bean, is a medicinal plant with a rich history of traditional uses in herbal medicine. Native to the southeastern United States, Mexico, and Central America, this deciduous tree is characterized by striking red or coral-colored flowers and compound leaves (Fehling-Fraser and Ceccon, 2015). Various parts of *Erythrina americana*, including the bark, leaves, and seeds, have been employed in traditional medicinal practices for their potential therapeutic properties. The plant is known to contain alkaloids, isoflavonoids, flavonoids, terpenoids, triterpenes, phenolic compounds, and saponins, contributing to its pharmacological significance. Traditional uses include the treatment of inflammatory conditions, pain relief, and as a remedy for various ailments. As ongoing research continues to unveil the diverse phytochemical composition of *Erythrina americana*, it remains a subject of interest for its potential contributions to modern medicine and health applications (Oliva-Hernández et al., 2021).

23.2 SCIENTIFIC CLASSIFICATION

Kingdom: Plantae (Plants)

Subkingdom: Viridiplantae (Green plants)

Phylum: Tracheophyta (Vascular plants)

Subphylum: Spermatophytina (Seed plants)

Class: Magnoliopsida (Dicotyledons)

Subclass: Rosanae

DOI: 10.1201/9781003460336-23

Order: Fabales
Family: Fabaceae (Leguminosae) – Legume or pea family
Genus: *Erythrina* L.
Species: *Erythrina americana* Mill.

23.3 VERNACULAR NAMES

Erythrina americana is known by various vernacular names in different regions. Some of the common vernacular names for this tree include:

i. American Coral Tree
ii. Cherokee Bean
iii. Crybaby Tree
iv. Cockspur Coral Tree
v. Coral Bean
vi. Cardinal Spear
vii. Cherokee Bean
viii. Red Cardinal
ix. Dadap (in Indonesia)
x. Tiger's Claw
xi. Mother of Coral

23.4 BOTANICAL DESCRIPTION

Erythrina americana, commonly known as the American coral tree or Cherokee bean, is a deciduous tree belonging to the Fabaceae family (Figure 23.1). *Erythrina americana* typically grows as a small to medium-sized tree, reaching heights of 15 to 30 feet (4.5 to 9 meters). It has an open, spreading crown with a somewhat irregular shape. The leaves are compound and trifoliate, meaning each leaf is divided into three leaflets. The leaflets are lance-shaped and can be up to 6 inches (15 cm) long. The foliage is typically green, and the leaves are deciduous, meaning they are shed in response to seasonal changes. The bark is generally smooth and grayish when the tree is young. As the tree matures, the bark can become somewhat rough and fissured. The striking red or coral-colored flowers are one of the tree's most distinctive features. The flowers are borne in terminal clusters or racemes and are shaped like pea flowers. Flowering usually occurs in spring or early summer. The tree produces long, slender, brown pods that can reach lengths of 6 to 8 inches (15 to 20 cm). These pods contain seeds and persist on the tree for some time before eventually splitting open to release the seeds. The seeds are typically smooth, oval, and may have a shiny appearance. They are often red or brown in color (García-Mateos et al., 2002).

FIGURE 23.1 Botanical illustration of different parts of *Erythrina americana*.

23.5 ORIGIN AND GEOGRAPHIC DISTRIBUTION

Erythrina americana Mill., commonly known as the 'American coral tree' or 'Cherokee bean', boasts a diverse and extensive geographical heritage, originating from regions across multiple continents, including southern United States (Mexico), Central America, and various parts of South America. (García-Mateos et al., 2001). It is believed to have originated in the southern United States, particularly in the southeastern regions of Texas, Louisiana, and Florida (Neill, 1988). Its origin can be traced back to the pre-Columbian era when indigenous peoples of the Americas recognized its value for both practical and ceremonial purposes. Within these regions, it commonly thrives in tropical and subtropical settings, playing a role in enriching the biodiversity of the Amazon rainforest and its neighboring ecosystems (Fahmy et al., 2020; López-Agama et al., 2021). This coral tree made its way to France in the early 19th century after thriving in English cultivation. Currently, it is cultivated in various regions, including the southeastern United States, California, Cuba, Hawaii, and parts of Africa, such as Kenya and Natal. It is not known when it was introduced into California, however, the Evans and Reeves Nurseries (McClintock, 1993). It is an integral part of the flora in these countries and is appreciated or its aesthetic qualities and longstanding cultural significance. The vibrant red flowers and distinctive appearance of this plant have symbolized beauty and vitality within numerous indigenous cultures. Additionally, its durable wood was used for crafting tools and ceremonial items, while the plant's seeds had medicinal and ritualistic significance. Over time, *E. americana* has also gained recognition as an ornamental tree due to its stunning foliage and blossoms, thus extending the reach of its natural beauty well beyond its native habitat (García-Mateos et al., 2002; Mulík and Ozuna, 2020). Its adaptability to different climates and ecosystems contributes to its broad distribution, making it a notable and versatile member of the legume family throughout the America (Figure 23.2).

23.6 ECOLOGY

Erythrina americana, exhibits an interesting ecological profile that influences its interactions with the environment, wildlife, and other organisms. It predominantly thrives in areas characterized by warm and subtropical climates. Within this geographical range, it inhabits a diverse range of ecosystems, encompassing forests, wetlands, and

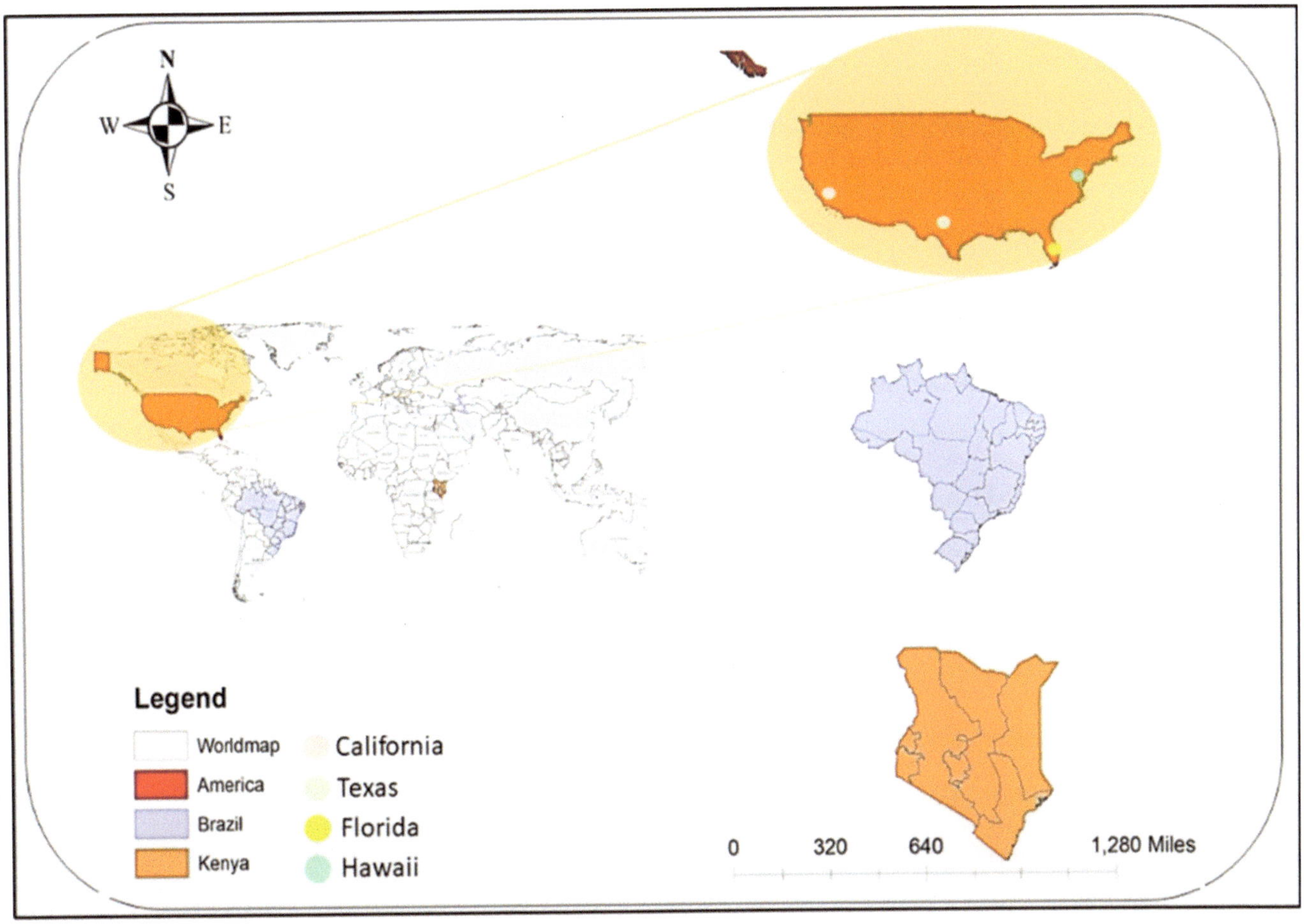

FIGURE 23.2 Geographical distribution of *Erythrina americana*.

open landscapes. It is distributed across lowland areas and moderately high elevations, typically ranging from 1,000 to 2,100 meters above sea level. It grows in seasonal tropical forest, perennial tropical forest, and xerophilous shrublands (Krukoff, 1979). This particular species holds a prominent position, being highly recognized and culturally significant to local communities, particularly among rural inhabitants. Its significance lies in its versatility, as it serves multiple purposes such as being employed in the construction of living fences, offering shade and organic material to crops, and finding extensive use as an ornamental tree (Dubeux et al., 2017). Like other members of the *Erythrina* genus, it possesses the remarkable capability of nitrogen fixation. It establishes symbiotic associations with nitrogen-fixing bacteria residing in its root nodules. This unique capacity to transform atmospheric nitrogen into a plant-accessible form contributes to soil enrichment with nitrogen, thereby fostering the well-being of neighboring plants and promoting overall ecosystem health (Fehling-Fraser and Ceccon, 2015). It is famous for its striking red or coral-colored flowers. These vibrant, tubular blossoms are rich in nectar, making it immensely appealing to pollinators, including bees, hummingbirds, and butterflies. Its flowers assume a pivotal role in the provision of nectar, serving as a vital resource for these insects and birds, thus playing a significant part in local pollination networks (San Miguel-Chávez et al., 2006). Upon reaching maturity, these pods have the propensity to burst open, dispersing the seeds in their immediate vicinity. Additionally, certain animals, such as birds and rodents, can contribute to the dispersal of these seeds by consuming them and subsequently depositing them in new locations (Parra-Gil et al. 2023). The seeds and foliage of *E. americana* serve as a food source for a diverse range of wildlife species. Seed-eating birds, with a focus on parrots and similar avian species, are known to consume the seeds. In addition, mammals such as squirrels may also rely on these seeds as a food source. Furthermore, the tree's branches and leaves provide valuable nesting sites and shelter for both birds and small animals (Sotelo et al., 2007).

23.7 PHYTOCHEMISTRY

The phytochemistry of *Erythrina americana* involves the identification and study of the chemical compounds present in various parts of the plant (Table 23.1). Different plant parts, such as leaves, flowers, bark, and seeds, may contain distinct phytochemicals (de Araújo-Júnior et al., 2012).

Erythrina alkaloids are well-known and are often present in various *Erythrina* species. These alkaloids can have diverse pharmacological activities. Examples include erysodine, erythratidine, erysopine, and β-erythroidine (Sanchez-Herrera et al., 2021). Isoflavonoids are a class of compounds with antioxidant properties. They are known for their potential health benefits. *Erythrina americana* may contain isoflavonoids such as genistein and daidzein (Kwesiga et al., 2020). Flavonoids are a large group of polyphenolic compounds with antioxidant and anti-inflammatory properties. Quercetin and kaempferol derivatives are examples of flavonoids found in *Erythrina americana* (López-Agama et al., 2021). Terpenoids are a diverse group of compounds that include essential oils and other bioactive substances. *Erythrina* species may contain terpenoids with potential biological activities. Triterpenes are a type of terpenoid with diverse biological activities. *Erythrina americana* may contain triterpenes with potential medicinal properties (Farag et al., 2016). Phenolic compounds are known for their antioxidant and anti-inflammatory effects. *Erythrina americana* may contain various phenolic acids and other phenolic compounds (Pinedo-Espinoza et al., 2020). Saponins are glycosides with soap-like properties. They have been studied for their potential medicinal benefits. *Erythrina americana* may contain saponins in certain parts of the plant (Son and Elshamy, 2021).

23.8 ETHNOBOTANY

Erythrina americana, or the American coral tree, has a rich ethnobotanical history, with various traditional uses by indigenous cultures in the regions where it is native. While specific uses may vary among different communities (Obakiro et al., 2021).

23.8.1 Anti-Inflammatory Properties

Infusions or decoctions made from different parts of *Erythrina americana*, such as the bark and leaves, have been traditionally used for their anti-inflammatory properties (Togola et al., 2008).

23.8.2 Analgesic Effects

Some indigenous communities have used parts of the plant to alleviate pain, and the analgesic properties are often attributed to the presence of alkaloids and other bioactive compounds (Jiménez-Cabrera et al., 2020).

23.8.3 Traditional Healing

In certain traditional healing practices, *Erythrina americana* may be employed for treating wounds, skin infections, and other external ailments. The plant's phytochemicals, including alkaloids and flavonoids, are believed to contribute to its healing properties (Magassouba et al., 2007).

23.8.4 Ceremonial and Ritual Uses

In some cultures, *Erythrina americana* has ceremonial significance. The vibrant red or coral-colored flowers are

TABLE 23.1
Active Phyto-Constituents in *Erythrina americana*

Class	Phyto-Constituent	Pharmacological Effects	References
Alkaloids	Erysodine	Analgesic, sedative, anticonvulsant, and muscle relaxant	Folkers and Koniuszy (1940)
	Erythroidine (Toxic)	muscle weakness, twitching, paralysis, respiratory failure	Sanchez-Herrera et al. (2021)
	Hypaphorine	Antioxidant, anti-inflammatory, anxiolytic, neuroprotective, antidepressant, anti-cancer properties	Rambo et al. (2019)
	Erysotrine	Neuro-pharmacological effects, muscle relaxant properties, insecticidal properties, cytotoxicity	Florian et al. (2021)
	Erythravine	Neuropharmacological effects, muscle relaxant properties, insecticidal properties, cytotoxicity, antimicrobial	Dos Reis et al. (2021)
Flavonoids	Lucenin	Antioxidant properties neuroprotective effects	Lopez-Agama et al. (2021)
	kaempferol	Antioxidant properties anti-Inflammatory effectsantibacterial and antifungal	Periferakis et al. (2022)
	Quercetin	antioxidant activities neuroprotective effectsanti-Inflammatory effects	Deviani et al. (2022)
Terpenoids	β-Amyrin	Anti-cancer agent, anti-retroviral, anti-malarial, and anti-inflammatory properties	Kaushal et al. (2020)
	Lupeol	Anti-cancer, anti-malarial, antimicrobial, anti-inflammatory potential	Alam and Akhtaruzzaman (2022)
	Erythrodiol	anti-inflammatory, Hepatoprotective, anti-cancer potential	Fofana et al. (2022)

visually striking, and the plant may be incorporated into rituals and ceremonies (Nyoni, 2021).

23.8.5 Toxicity in Indigenous Practices

While *Erythrina americana* has beneficial uses, some parts of the plant are known to contain toxic compounds. Indigenous communities may have methods to detoxify or prepare the plant for specific uses (Obakiro et al., 2021).

23.8.6 Ornamental and Shade

Beyond its medicinal and ritual uses, *Erythrina americana* is sometimes planted for its ornamental qualities. The tree attractive flowers and overall appearance make it a desirable choice for landscaping. The tree spreading canopy provides shade, and in some regions, it is intentionally planted for its shade-providing qualities (Kumar et al., 2011).

23.9 HALLUCINOGENIC EFFECTS

Erythrina americana, commonly known as the American coral tree, is a plant that belongs to the Fabaceae family. While it has been used traditionally in some cultures for various purposes, there is limited scientific information on its hallucinogenic effects. It's important to note that the term 'hallucinogenic' is often associated with substances that can alter perception, mood, and consciousness, typically leading to vivid sensory experiences that may not be based in reality. *Erythrina americana* can contain a variety of compounds, and their effects can vary widely based on factors such as dosage, individual sensitivity, and preparation methods. Additionally, using plants for their psychoactive properties without proper knowledge and guidance can pose health risks (Mulík and Ozuna, 2020).

Some species of the *Erythrina* genus, such as *Erythrina mulungu*, have been reported to have sedative and anxiolytic effects, but these are different from hallucinogenic effects. The alkaloids present in *Erythrina* species, such as erythrinine and erythratidine, are believed to contribute to these effects (Rodrigues et al., 2018). However, their specific mechanisms of action and potential psychoactive effects are not fully understood, and further research is needed to clarify their pharmacological properties. The psychoactive effects of *Erythrina mulungu* are primarily attributed to the presence of alkaloids, including erythravine, erythravine N-oxide, and β-erythroidine. These alkaloids are believed to interact with the central nervous system, producing calming and relaxing effects (Menezes et al., 2017).

Reports suggest that *Erythrina mulungu* is traditionally used to alleviate anxiety, promote relaxation, and improve sleep. It's important to note that while the plant is known for its calming effects, it is not commonly associated with strong hallucinations or intense psychedelic experiences (Silveira-Souto et al., 2014). The primary traditional uses are more aligned with anxiolytic and sedative properties. As with any substance with potential psychoactive effects, caution should be exercised. The effects of plants can vary, and individual responses may differ. Furthermore, the safety and dosage of *Erythrina mulungu* have not been extensively studied, so it's important to approach its use with care.

23.10 CONCLUSION

Erythrina americana, the American coral tree, presents an intriguing botanical profile, boasting compound leaves and vibrant coral-colored flowers. Despite its association with the *Erythrina* genus, known for certain psychoactive species, concrete evidence supporting hallucinogenic properties in *Erythrina americana* is currently lacking. While traditional uses hint at potential psychoactivity, rigorous scientific investigation is necessary to validate these claims. The plant's phytochemistry reveals the presence of alkaloids like erythraline and erythroidine, which are known in related species for their pharmacological activities. However, the specific effects of these compounds in *Erythrina americana* remain unclear. This book chapter underscores the need for further research to unravel the plant's psychoactive potential, encouraging a cautious approach in interpreting anecdotal reports and emphasizing the importance of evidence-based exploration in understanding the true nature of *Erythrina americana* botanical and psychoactive characteristics.

CONFLICT OF INTEREST

The authors declare no conflicts of interest.

DATA AVAILABILITY STATEMENT

The data that support the findings of this study are available from the corresponding author upon reasonable request.

REFERENCES

Alam, M. K., & Akhtaruzzaman, M. (2022). Coral Tree (Erythrina variegata L.). In *Handbook of Phytonutrients in Indigenous Fruits and Vegetables* (pp. 167–178). GB: CABI.

Alamgir, A. N. M. (2017). *Therapeutic use of medicinal plants and their extracts: volume 1*. Springer International Publishing AG.

Bernardino-Nicanor, A., Montañéz-Soto, JL, Vivar-Vera, MA, Juárez-Goiz, JM, Acosta-García, G., & González-Cruz, L. (2016). Effect of drying on the antioxidant capacity and concentration of phenolic compounds in different parts of the Erythrina Americana tree. *BioResources*. https://doi.org/10.15376/biores.11.4.9741-9755

Chaachouay, N., & Zidane, L. (2021). The Symbolic Efficacy of Plants in Rituals and Socio-Religious Ceremonies in Morocco, Northwest of Africa. *Journal of Religious & Theological Information*, 21(1–2), 34–53. https://doi.org/10.1080/10477845.2021.1942433

Chaachouay, N., Douira, A., & Zidane, L. (2021a). Herbal Medicine Used in the Treatment of Human Diseases in the Rif, Northern Morocco. *Arabian Journal for Science and Engineering*, 47, 131–153. https://doi.org/10.1007/s13369-021-05501-1

Chaachouay, N., Douira, A., & Zidane, L. (2021b). COVID-19, prevention and treatment with herbal medicine in the herbal markets of Salé Prefecture, North-Western Morocco. *European Journal of Integrative Medicine*, 42, 101285. https://doi.org/10.1016/j.eujim.2021.101285

de Araújo-Júnior, J. X., de Oliveira, M. S., Aquino, P. G., Alexandre-Moreira, M. S., & Sant'Ana, A. E. (2012). *A phytochemical and ethnopharmacological review of the genus Erythrina*. InTech.

Deviani, V., Hardianto, A., Farabi, K., & Herlina, T. (2022). Flavanones from Erythrina crista-galli twigs and their antioxidant properties determined through in silico and in vitro studies. *Molecules*, *27*(18), 6018.

Dos Reis, S. L., Gelfuso, E. A., Fachin, A. L., Pereira, A. M. S., & Beleboni, R. O. (2021). Pharmacological characterisation of anticonvulsant effects elicited by erythrartine. *Journal of Pharmacy and Pharmacology*, *73*(1), 93–97.

Dubeux, J. C. B., Muir, J. P., Apolinário, V. X. D. O., Nair, P. R., Lira, M. D. A., & Sollenberger, L. E. (2017). Tree legumes: an underexploited resource in warm-climate silvopastures. *Revista Brasileira de Zootecnia*, *46*(8), 689–703.

Fahmy, N. M., Al-Sayed, E., El-Shazly, M., & Nasser Singab, A. (2020). Alkaloids of genus Erythrina: An updated review. *Natural Product Research*, *34*(13), 1891–1912.

Farag, M. A., Mekky, H., & El-Masry, S. (2016). Metabolomics driven analysis of Erythrina lysistemon cell suspension culture in response to methyl jasmonate elicitation. *Journal of Advanced Research*, *7*(5), 681–689.

Fatur, K. (2021). Saps and Syrups, Tinctures and Teas: An Analysis of Medicinal Plant Usage in the Coastal Region of Primorska Province, Slovenia. *Economic Botany*, *75*(2), 112–125.

Fehling-Fraser, T. C., & Ceccon, E. (2015). Macropropagation of Erythrina americana in a greenhouse: a potential tool for seasonally dry tropical forest restoration. *Revista Chapingo serie ciencias forestales y del ambiente*, *21*(1), 5–16.

Florian, S. U., Sepulveda, B., Torres-Benítez, A., Simirgiotis, M. J., Fuentes-Estrada, M., Areche, C., & García-Beltrán, O. (2021). Erythrinoid and indol alkaloids isolated from the seeds of Erythrina rubrinervia Kunth: Chemotaxonomic significance. *Biochemical Systematics and Ecology*, *97*, 104295.

Fofana, S., Ouédraogo, M., Esposito, R. C., Ouedraogo, W. P., Delporte, C., Van Antwerpen, P., … Guissou, I. P. (2022). Systematic Review of Potential Anticancerous Activities of Erythrina senegalensis DC (Fabaceae). *Plants*, *11*(1), 19.

Folkers, K., & Koniuszy, F. (1940). Erythrina alkaloids. VII. Isolation and characterization of the new alkaloids, erythraline and erythratine. *Journal of the American Chemical Society*, *62*(2), 436–441.

García-Mateos, R., Peña-Valdivia, C. B., & Soto-Hernández, M. (2002). Phytotoxicity of crude alkaloid fractions from Erythrina americana. *Revista de la Sociedad Química de México*, *46*(1), 04–09.

García-Mateos, R., Soto-Hernández, M., & Vibrans, H. (2001). Erythrina Americana Miller ("Colorin," Fabaceae), a Versatile Resource from Mexico—A Review. *Economic Botany*, *55*, 391–400.

Herrera, S.G.S., Soto-Hernández, R.M., Garín-Aguilar, M.E., Vibrans-Lidermann, H., del Rosario García-Mateos, M., & Martínez-Vázquez, M. (2021). Alkaloids from the seed, calyx, and corolla of Erythrina americana Miller and Erythrina coralloides A. DC. *Agro Productivity*.

Jiménez-Cabrera, T., Bautista, M., Velázquez-González, C., Jaramillo-Morales, O. A., Guerrero-Solano, J. A., Urrutia-Hernández, T. A., & De la O-Arciniega, M. (2020).

Promising antioxidant activity of Erythrina genus: an alternative treatment for inflammatory pain?. *International Journal of Molecular Sciences*, *22*(1), 248.

Kaushal, A., Sharma, M., Navneet, N., & Sharma, M. (2020). Ethnomedicinal, phytochemical, therapeutic and pharmacological review of the genus Erythrina. *International Journal of Botany Studies*, *5*(6), 642–648.

Krukoff, B. A. (1979). Notes on the species of Erythrina. XII. *Annals of the Missouri Botanical Garden*, 422–445.

Kumar, A., Lingadurai, S., Shrivastava, T. P., Bhattacharya, S., & Haldar, P. K. (2011). Hypoglycemic activity of Erythrina variegata leaf in streptozotocin-induced diabetic rats. *Pharmaceutical Biology*, *49*(6), 577–582.

Kwesiga, G., Kelling, A., Kersting, S., Sperlich, E., Von Nickisch-Rosenegk, M., & Schmidt, B. (2020). Total Syntheses of Prenylated Isoflavones from Erythrina sacleuxii and Their Antibacterial Activity: 5-Deoxy-3′-praenylbiochanin A and Erysubin F. *Journal of Natural Products*, 83(11), 3445–3453. https://doi.org/10.1021/acs.jnatprod.0c00932

López-Agama, I., Ramos-García, M. D. L., Zamilpa, A., Bautista-Baños, S., & Ventura-Aguilar, R. I. (2021). Comparative analysis of the antioxidant compounds of raw edible flowers and ethanolic extracts of Cucurbita pepo, Tagetes erecta, and Erythrina Americana during storage. *Journal of Food Processing and Preservation*, *45* (10), e15842.

Magassouba, F. B., Diallo, A., Kouyaté, M., Mara, F., Mara, O., Bangoura, O., & Baldé, A. M. (2007). Ethnobotanical survey and antibacterial activity of some plants used in Guinean traditional medicine. *Journal of Ethnopharmacology*, *114*(1), 44–53.

McClintock, E. (1993). A botanical history of Erythrina introductions. *Westley, SB and.*

Menezes, K. A. S., Escobar, I. E. C., Fraiz, A. C. R., Martins, L. M. V., & Fernandes, P. I. (2017). Genetic variability and symbiotic efficiency of Erythrina velutina willd. Root nodule bacteria from the semi-arid region in Northeastern Brazil. *Revista Brasileira de Ciência do Solo*, *41*, e0160302.

Mulík, S., & Ozuna, C. (2020). Mexican edible flowers: Cultural background, traditional culinary uses, and potential health benefits. *International Journal of Gastronomy and Food Science*, *21*, 100235.

Neill, D. A. (1988). Experimental studies on species relationships in Erythrina (Leguminosae: Papilionoideae). *Annals of the Missouri Botanical Garden*, 886–969. 10.2307/2399377

Nyoni, L. (2021). *A comparative study of the relationship between the applications of Erythrina lysistemon (Umsinsi) within the traditional African and homeopathic medicinal systems* (Doctoral dissertation).

Obakiro, S. B., Kiprop, A., Kigondu, E., K'Owino, I., Odero, M. P., Manyim, S., & Bunalema, L. (2021). Traditional medicinal Uses, phytoconstituents, bioactivities, and toxicities of Erythrina abyssinica Lam. ex DC.(fabaceae): a systematic review. *Evidence-Based Complementary and Alternative Medicine*, *2021*(1), 5513484.

Oliva-Hernández, J., López-Herrera, M. A., & Castillo-Linares, E. B. (2021). Chemical composition and foliage production of Erythrina americana (Fabaceae) in living fences during two climatic seasons. *Revista de Biología Tropical*, *69*(1), 90–101.

Parra-Gil, P. D. J., Espinosa-Vásquez, G., Lucio-Cruz, C. Y., Romero-Nápoles, J., & Arce-Cervantes, O. (2023). Damage to the seeds of Erythrina americana Mill.,(Leguminosae: Faboideae: Erythrininae) by the bruchid Specularius impressithorax (Pic, 1932)(Coleoptera: Bruchidae) and its effect on germination. *Acta zoológica mexicana*, *39*.

Passos, I. D., & Mironidou-Tzouveleki, M. (2016). Hallucinogenic plants in the Mediterranean countries. In: Victor R. Preedy (ed) *Neuropathology of drug addictions and substance misuse* (pp. 761–772). Academic Press.

Periferakis, A., Periferakis, K., Badarau, I. A., Petran, E. M., Popa, D. C., Caruntu, A., & Costache, D. O. (2022). Kaempferol: antimicrobial properties, sources, clinical, and traditional applications. *International Journal of Molecular Sciences*, *23*(23), 15054.

Pinedo-Espinoza, J. M., Gutiérrez-Tlahque, J., Santiago-Saenz, Y. O., Aguirre-Mancilla, C. L., Reyes-Fuentes, M., & López-Palestina, C. U. (2020). Nutritional composition, bioactive compounds and 1antioxidant activity of wild edible flowers consumed in semiarid regions of Mexico. *Plant Foods for Human Nutrition*, *75*, 413–419.

Rambo, D. F., Biegelmeyer, R., Toson, N. S., Dresch, R. R., Moreno, P. R. H., & Henriques, A. T. (2019). The genus Erythrina L.: A review on its alkaloids, preclinical, and clinical studies. *Phytotherapy Research*, *33*(5), 1258–1276.

Rodrigues, D. R., Silva, A. F. D., Cavalcanti, M. I. P., Escobar, I. E. C., Fraiz, A. C. R., Ribeiro, P. R. D. A., ... Fernandes-Júnior, P. I. (2018). Phenotypic, genetic and symbiotic characterization of Erythrina velutina rhizobia from Caatinga dry forest. *Brazilian Journal of Microbiology*, *49*, 503–512.

San Miguel-Chavez, R., Soto-Hernandez, M., Ramos-Valdivia, A. C., & Kite, G. (2007). Alkaloid production in elicited cell suspension cultures of Erythrina americana Miller. *Phytochemistry Reviews*, *6*, 167–173.

San Miguel-Chávez, R., Soto-Hernández, M., Terrazas, T., & Kite, G. (2006). Morphology and alkaloidal profile of the seedlings of Erythrina americana MILL. and E. coralloides A. DC. *Feddes Repertorium: Zeitschrift für botanische Taxomonie und Geobotanik*, *117*(3-4), 232–239.

Silveira-Souto, M. L., São-Mateus, C. R., de Almeida-Souza, L. M., & Groppo, F. C. (2014). Effect of Erythrina mulungu on anxiety during extraction of third molars. *Medicina Oral, Patologia Oral y Cirugia Bucal*, *19*(5), e518.

Son, N. T., & Elshamy, A. I. (2021). Flavonoids and other nonalkaloidal constituents of genus Erythrina: phytochemical review. *Combinatorial Chemistry & High Throughput Screening*, *24*(1), 20–58.

Sotelo, A., López-García, S., & Basurto-Peña, F. (2007). Content of nutrient and antinutrient in edible flowers of wild plants in Mexico. *Plant Foods for Human Nutrition*, *62*, 133–138.

Togola, A., Austarheim, I., Theïs, A., Diallo, D., & Paulsen, B. S. (2008). Ethnopharmacological uses of Erythrina senegalensis: a comparison of three areas in Mali, and a link between traditional knowledge and modern biological science. *Journal of Ethnobiology and Ethnomedicine*, *4*(1), 6.

24 Snoutbean (*Rhynchosia pyramidalis* (Lam.) Urb. Fabaceae)

Syed Sabir Hussain Shah, Saba Latif, Rahmatullah Qureshi, Zia ur Rehman Mashwani, Mubashrah Munir, and Noureddine Chaachouay

24.1 INTRODUCTION

Rhynchosia species have played a significant role among the plants of the Fabaceae family that are utilized in traditional medicine. *Rhynchosia* genus contains some 300 hundred species found in both hemispheres of tropical and subtropical regions (Chadha, 1976; Willis 1966; Schultes and Hofmann 1980). This genus has both herbs and twining or erect shrub species. A few plants from this genus were employed in traditional medicine to cure a number of conditions, including boils, rheumatic pains, diabetes, abortifacients, wound healing, and skin infections. Previous studies on the phytochemistry of many *Rhynchosia* species revealed that the genus is only known for its prolific synthesis of C-glycosyl flavonoids (Rao et al. 1991). Some of the interesting biological properties such as antioxidant, anti-inflammatory, antimycobacterial, and antiproliferative are demonstrated by plant extracts and a few isolated compounds from the *Rhynchosia* genus (Kinjo et al. 2001; Coronado-Aceves et al. 2017; Rondo 2017).

Rhynchosia pyramidalis (Lam.) Urb., commonly known as snoutbean, is a vine that is widely distributed throughout tropical and subtropical regions of the world (Rammohan et al. 2020). It is commonly distributed from eastern Texas and Oklahoma through southern Arkansas, Louisiana, and southern Missouri to western Tennessee and Mississippi. To the best of our knowledge, no extensive, updated evaluation of the literature on ethnobotany, medicinal use, phytochemical, and pharmacology of *Rhynchosia pyramadalis* is known. The objective of this chapter is to provide an insight into the traditional usage, chemical components, and pharmacological activities of *R. pyramadalis*, for potential use in the future.

24.2 SCIENTIFIC CLASSIFICATION

Rhynchosia pyramidalis is a climbing shrub that is mainly part of the wet tropical biome. The scientific classification is shown in Table 24.1.

24.3 VERNACULAR NAMES

Arabic: Lubiat al'anf; **English**: Snoutbean, Praying Snout-Bean, Rosary snoutbean, virility-vine, rosary bean virility-vine; **America**: bird's eyes, rosary bean, and John Crow Bead; **Mexico**: Ojo de Chanate (eye of the thrush); **Bengali**: Rohita; **Italian**: Fagiolo nasuto; **German**: Schnabelbohne; **French**: Rhyncosie pyramidale; **Spanish**: Frijol pico de loro; ojo de pajarito (eye of the little bird), ojitos de picho (little eyes of the dove), ojo de cangrejo (crab's eye), ojo de culebra (eye of the snake); **Turkish**: Gagalı fasulye; **Portuguese**: Feijão-patola; **Dutch**: Snavelboon; **Persian**: Lubiyā-ye peykāni; **Russian**: Rinkhoziya piramidal'naya; **Chinese**: Jīn zhuī biǎn dòu; **Polish**: Rynchosia piramidalna; **Czech**: Rynchosia pyramidální; **Japanese**: Sunauto bīn; **Korean**: Bulikong (Cleversley, 2002).

24.4 SYNONYMS

According to the Kew list (https://powo.science.kew.org/taxon/urn:lsid:ipni.org:names:1006990-2#other-data), the species possess three synonyms viz., *Dolicholus phaseoloides* (Sw.) Kuntze, *Glycine corallina* Salisb. and *Glycine phaseoloides* Sw., while *Rhynchosia phaseoloides* var. *precatoria* Griseb. is listed as a heterotypic synonym. The WFO (2023) indicates that the plant has four synonyms as follows:

TABLE 24.1
Hierarchical Classification of *Rhynchosia pyramidalis*

Kingdom:	Plantae
Subkingdom:	*Tracheobionta*
Infrakingdom:	Streptophyta
Superdivision:	Tracheophytes
Subdivision:	Spermatophytina
Infradivision:	Angiospermae
Class:	Magnoliopsida
Clade:	Monophyletic
Superorder:	Rosanae
Order:	Fabales
Family:	Fabaceae
Genus:	*Rhynchosia* Lour.
Binomial name:	*Rhynchosia pyramidalis* (Lam.) Urb.

Source: https://www.gbif.org/species/5350702

DOI: 10.1201/9781003460336-24

FIGURE 24.1 Habit of *Rhynchosia pyramidalis*: (a) vegetative branch, (b) fruiting branch.

Dolicholus pyramidalis (Lam.) Britton & P. Wilson; *Dolicholus phaseoloides* (Sw.) Millsp

Dolichos pyramidalis Lam.; *Rhynchosia phaseoloides* var. *precatoria* Griseb.

24.5 BOTANICAL DESCRIPTION

Rhynchosia pyramidalis is a small to large climbing plant with twining stems that are mostly soft or herbaceous but sometimes become more or less hard or woody near their base. This perennial climbing vine attains up to 30 feet tall (Figure 24.1). It is a perennial twiner that intertwines around the associated host without any particular mechanism of attachments (Standley & Steyermark, 1976). Older stems are flattened and 3–5 cm broad; young stems are puberulent; most vegetative parts possess orange glands. Leaves are trifoliate measuring 2.5–8 cm long and 2.5–6.5 cm broad; petioles 2.5–5 cm long, swollen at base, hispidulous; rachis with a raised margin above; leaflets ovate to rhombic-ovate, apex acuminate, base obtuse to rounded, mostly 4–16 cm long, 3.5–11.5 cm wide, glabrous to sparsely pubescent especially on veins below, the lateral leaflets inequilateral, the veins palmate at base. Racemes apical, ca 15 cm long; 3–10 flowered; flowers dense, yellowish-green, short-pedicellate; calyx ca. 2 mm long, puberulent and glandular-dotted; standard obovate, clawed, ca 6 mm long, purple-striate, puberulent and glandular-dotted outside; wings narrow, puberulent outside near apex, fused to keel near the base, the keel petals fused in apical half, weakly enclosing stamens; stamens diadelphous; style-bearing antrorse trichomes near apex; stigma borne among the anthers. Legumes ca. 1.5 cm long and 1 cm wide, constricted between the seeds, puberulent; seeds 2, black and red. Flowering from January to March. The fruits develop soon, but seeds may persist throughout the rainy season (Croat, 1978; Secretariat, 2021).

24.6 ORIGIN AND GEOGRAPHIC DISTRIBUTION

The genus *Rhynchosia* Lour. was first established by Loureiro in 1790 (Hutchinson, 1964). The genus name is a Greek word '*Rhynchos*' that means beak of the keel, while the species name '*pyramidata*' is of Latin origin that means pyramid nature for broad leaves (Clarke et al., 2016).

The species native range stretches from Mexico to the Greater Antilles and west Bolivia (Figure 24.2). As a purported aphrodisiac and androgenic agent, this plant has recently gained popularity (Farnsworth et al., 1967). This species has a broad distribution in the Neotropics, where it thrives in Belize and Petén in Guatemala, which are parallel to the southern region of the Yucatán Peninsula Biotic Province (Arellano-Rodríguez et al., 2003). However, it does not flourish in the Mexican region (Schultes, 1969). This species can also be found in Belize's district. This plant species is native to the Antilles, Guatemala, Honduras, and the Mexican states of Oaxaca, Tabasco, and Veracruz (Martínez-Bernal et al., 2011). It is native to Mexico in Northern America and Hispaniola, Cuba, Jamaica, Belize, Guatemala, and Honduras in Southern America. This species is also found in the Caribbean (Cleversley, 2002).

24.7 HABITAT AND ECOLOGY

Rhynchosia pyramidalis is climbing vine which is indigenous to Mexico and Central America as well as the Caribbean, in thickets and woodland margins. According to Rätsch (1998), it is primarily found in fallow milpas (slash-and-burn gardens). It has been introduced to other regions of the earth, including the tropics and subtropics. It is a plant that can be found up to 1,500 meters above sea level in thickets that range from moist to dry and are typically in

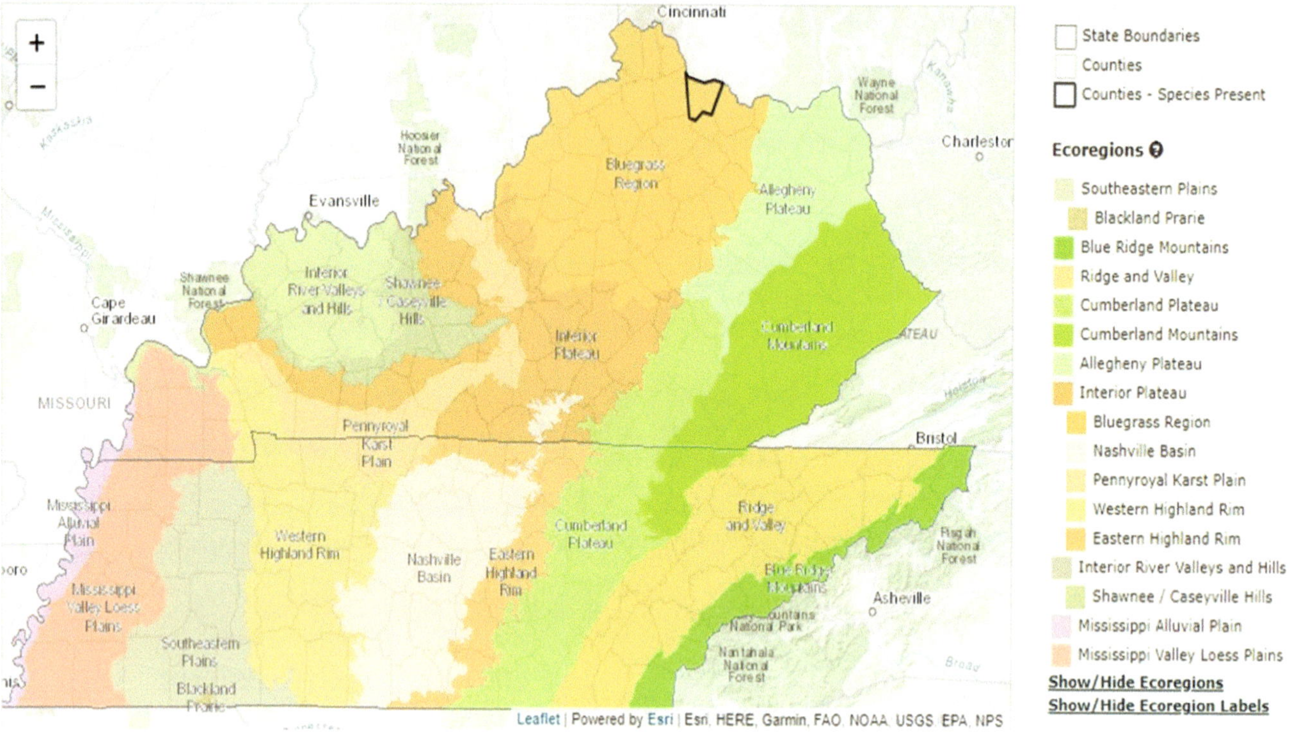

FIGURE 24.2 Distribution pattern of *Rhnchosia pyramidalis* (https://tennessee-kentucky.plantatlas.usf.edu/plant.aspx?id=2478).

secondary development. The moist tropical environment is where it primarily develops because it is a climber (Standley & Steyermark, 1976). The species coexists in harmony with a variety of soil bacteria, which are involved in the production of nodules in plant roots and fix atmospheric nitrogen. The plants utilize this nitrogen, but some of it can also be used by the surrounding plants that are also growing. This species, which is developed from seeds with a documented natural origin, is primarily found as a climbing shrub or vine. The species can be found in forests, woodlands, thickets, and disturbed regions like roadside ditches and agricultural fields. The species typically grows in moist, and well-drained soils. It prefers direct sunlight to light shade (Garau et al. 2009).

24.8 ETHNOBOTANY

The stunning seeds of this plant may have been employed in rituals by the Aztecs (Schultes & Hofmann, 1980). On the slopes of Popocatepetl (Mexico), in the community of San Pedro Nexapa, the red-black seeds known as piule (Santesson 1938), were or are utilized in rituals (Wasson & Wasson, 1957). This plant species is harvested from the wild for local use as a source of materials. It is used as a medicine (Maroyi, 2023). The seeds are carmine or scarlet with a black end or side, or almost completely black. The attractive seeds are used frequently for making necklaces and bracelets (Harper, 1932).

The leaves and stems are used by washerwomen to clean dirt from clothes. The species has been used to cover the ground and in projects of soil conservation (Standley & Steyermark, 1976). The seeds of this species have narcotic properties and were used by the peoples of southern Mexico (Schultes, 1969). It is used as a topical analgesic in different areas of Mexico. It is also used in tea or alcohol extraction as a male aphrodisiac in the Dominican Republic. It is considered to improve sexual desire and performance, although studies on animals in the labs have not confirmed this until now (Voogelbreinder 2009).

The Pima individuals of Mexico grind the seeds of this plant and sprinkle the powder in the eyes of patients who are under the curse of the 'evil eye' (Rätsch 1998). Approximately, 12 seeds are ingested with six pairs of *Psilocybe aztecorum* mushrooms per person in the Valleys of Mexico (Wasson & Wasson, 1957; Cleversley, 2002).

24.9 PHYTOCHEMISTRY

The phytochemistry of some species of *Rhynchosia* resulted in the isolation of 77 substances, including sterols, flavonoids, isoflavonoids, favan-3-ols, xanthones, and biphenyls. Some compounds such as prenylated isoflavonoids and C-glycosylflavonoids are abundant in the genus *Rhynchosia*. According to Rammohan et al. (2020), some unique bioactive chemicals such rhynchoviscin, cajanone, genistein, mangiferin, vitexin, and precatorin A are richly found in the genus. The biological activities such as antioxidant, anti-inflammatory, antimycobacterial and antiproliferative, anti-angiogenic activity, antityrosinase, and cytotoxic activity are exhibited by plant extracts and isolated compounds of *Rhynchosia* genus (Rammohan et al. 2020; Chaachouay et al., 2021a, 2021b).

Phytochemical analysis revealed that ethyl gallate and an unidentified alkaloid have been identified from seeds of *Rhynchosia pyramidalis*. The flavonoids cajanin, genistein, and 2′-OH-genistein are found in the leaf. The stems are known to contain inorganic acids, saponins, tannins, phenols, and 0.19% crude, unidentified alkaloids. In mice, a stem extract has CNS-depressant effects. Animals exposed to a seed extract exhibit some curare action (Voogelbreinder 2009).

24.10 HALLUCINOGENIC EFFECTS

Several *Rhynchosia* species were valued as narcotics by the ancient Mexicans. The poisonous seeds of *R. pyramidalis* are known to modern-day Oaxacan Indians by the name piule, which they use to describe the seeds as a hallucinogenic agent. Aztec murals depicting black *Rhynchosia* beans alongside mushrooms have been identified suggesting psychedelic use (Schultes 1937, 1965, 1969). *R. pyramidalis* has produced an unidentified alkaloid, according to Ristic and Thomas (1962). According to Rätsch (2005), the seeds of *R. pyramidalis* are sedative in nature.

Rhynchosia pyramidalis seeds are considered to be toxic and narcotic at the same time and cause insanity. Therefore, it is concluded that seeds have been powdered and given to unsuspecting victims for harmful purposes (Hofmann et al. 1992; Voogelbreinder 2009). According to Wasson and Wasson (1957), in the highlands of Mexico, 12 untreated seeds along with six pairs of *Psilocybe aztecorum* mushrooms are consumed per individual as an intoxicant. There are no recent reports of psychoactive effects (Cleversley, 2002; Chaachouay & Zidane, 2021). Some of the legumes contain toxic principles and are used for animal poisoning. Delgado-Salinas et al. (2021) reported vertebrate poisoning from *Rhynchosia pyramidalis*. According to them, such poisons may be used as candidates for pharmacological investigation and there is very little information available about health benefits derived from legumes.

24.11 PHARMACOLOGY

The traditional peoples of southern Mexico used the seeds of *R. pyramidalis* as narcotic effects (Allen & Allen, 1981). In Mexico, seeds are used as a topical analgesic, while in the Dominican Republic, it is used as an aphrodisiac and also as a tea or alcohol beverage. It possesses psychedelic, narcotic, and poisonous effects. It also alleviates coughs and impotence (Cleversley, 2002). Although *in vivo* activity of *Rhynchosia pyramidalis* did not establish androgenic and aphrodisiac effects in tests (Farnsworth et al., 1967).

24.12 MEDICINAL USES AND TOXICITY

Rhynchosia pyramidalis seeds are shown falling from the hands of the Aztec rain god Tlaloc in a Mexican Tepantitla mural that dates to about 300 and 400 A.D. This has led some people to hypothesize that these seeds might possess entheogenic qualities. The seeds are known as *'piule'*, a term that is also used to describe some kinds of morning glory and mushrooms that contain psilocybin. The seeds are regarded as lucky presents and are used as good luck charms, amulets, and necklaces (Schultes 1969; Rätsch 1998; Christian 1998; Chaachouay et al., 2019; Belhaj et al., 2021; Benkhnigue et al., 2022). The ceremonial use of the seeds can be found associated with psilocybe mushrooms. It's unclear whether this consumption results from the seeds' entheogenic effects or from the fact that their crimson and black coloring resembles bodiless eyes, a representation of transcendent, clairvoyant vision (Rätsch 1998). Rammohan et al. (2020) reported that *R. pyramidalis* has androgenic and aphrodisiac properties.

In some regions of Mexico, *R. pyramidalis* is used as a topical analgesic. As a male aphrodisiac, it is utilized in the Dominican Republic as a tea or alcohol extraction. Although laboratory tests on animals have not yet been validated, it is rumored that it increases sexual desire and performance (Voogelbreinder 2009). The Pima people of Mexico ground the seeds and sprinkled the powder in the eyes of people who were suffering from the 'evil eye' (Rätsch 1998).

24.13 CONCLUSION

Rhynchosia pyramidalis is native to Southern America in Hispaniola, Cuba, Honduras, and Guatemala and to Northern America in Mexico. The seeds of this species have been used as a narcotic agent and topical analgesic by the people of southern Mexico. The beautiful and attractive seeds are used commonly for making bracelets and necklaces. There is little known about biochemical and pharmacological properties for determining therapeutic potential compared to other species of *Rhynchosia* genus. Therefore, there is a need to undertake a detailed investigation of this species for exploring its potential and utilization.

CONFLICT OF INTERESTS

The authors declared that they have no conflict of interests with respect to the authorship and/or publication of this chapter.

REFERENCES

Allen, O.N., Allen, E.K. (1981). *The Leguminosae: A source book of characteristics, uses and nodulation*. University of Wisconsin.

Arellano-Rodríguez, J.A., Flores-Guido, J.S., Tun-Garrido, J., Cruz-Bojórquez, M.M. (2003). Nomenclatura, Forma de Vida, Uso, Manejo y Distribución de las Especies Vegetales de la Península de Yucatán. Etnoflora Yucatanense, Fascículo 20. Universidad Autónoma de Yucatán, Mérida.

Belhaj, S., Chaachouay, N., & Zidane, L. (2021). Ethnobotanical and toxicology study of medicinal plants used for the treatment of diabetes in the High Atlas Central of Morocco. *Journal of Pharmacy & Pharmacognosy Research*, 9(5), 619–662. https://doi.org/10.56499/jppres21.1014_9.5.619

Benkhnigue, O., Chaachouay, N., Khamar, H., El Azzouzi, F., Douira, A, & Zidane, L. (2022). Ethnobotanical and ethnopharmacological study of medicinal plants used in the treatment of anemia in the region of Haouz-Rehamna (Morocco). *Journal of Pharmacy & Pharmacognosy Research, 10*(2), 279–302. https://doi.org/10.56499/jppres21.1196_10.2.279

Chaachouay, N., Benkhnigue, O., Fadli, M., Ibaoui, H., & Zidane, L. (2019). Ethnobotanical and ethnopharmacological studies of medicinal and aromatic plants used in the treatment of metabolic diseases in the Moroccan Rif. *Heliyon, 5*(10), e02191. https://doi.org/10.1080/10477845.2021.1942433

Chaachouay, N., Douira, A., & Zidane, L. (2021b). Herbal Medicine Used in the Treatment of Human Diseases in the Rif, Northern Morocco. *Arabian Journal for Science and Engineering*, 47, 131–153. https://doi.org/10.1007/s13369-021-05501-1

Chaachouay, N., Benkhnigue, O., Douira, A., & Zidane, L. (2021a). Poisonous medicinal plants used in the popular pharmacopoeia of the Rif, northern Morocco. *Toxicon*, 189, 24–32. https://doi.org/10.1016/j.toxicon.2020.10.028

Chaachouay, N., & Zidane, L. (2021). The Symbolic Efficacy of Plants in Rituals and Socio-Religious Ceremonies in Morocco, Northwest of Africa. *Journal of Religious & Theological Information, 21*(1–2), 34–53. https://doi.org/10.1080/10477845.2021.1942433

Chadha, Y.R. (1976). *The wealth of India: a dictionary of Indian raw materials and industrial products*, vol 10. Publication Information Directorate, CSIR, New Delhi, pp 522–524.

Christian, R. (1998). *The encyclopedia of psychoactive plants: Ethnopharmacology and its applications*. Rochester: Park Street Press.

Clarke, H., Charters, M., & Moll, E. (2016). *The illustrated dictionary of southern African plant names*. Jacana Media.

Cleversley, K. (2002). *Rhynchosia pyramidalis*–Piule. https://entheology.com/plants/rynchosia-phaseoloides-piule

Coronado-Aceves, E.W., Gigliarelli, G., Garibay-Escobar, A., Zepeda, R.E.R., Curini, M., Cervantes, J.L., Espitia-Pinzón, C.I.I., Superchi, S., Vergura, S., & Marcotullio, M.C. (2017). New Isoflavonoids from the extract of Rhynchosia precatoria (Humb. & Bonpl. ex Willd.) DC. and their antimycobacterial activity. *Journal of Ethnopharmacology* 206, 92–100.

Croat, T. B. (1978). *The flora of Barro Colorado Island*. Stanford University Press.

Delgado-Salinas, A., Torres-Colín, L., Luna-Cavazos, M., & Bye, R. (2021). Diversity of useful Mexican legumes: Analyses of herbarium specimen records. *Diversity* 13, 267. https://doi.org/10.3390/d13060267

Farnsworth, N. R., Hess, M., Fisher, A. E., Martello, R. E., & Cammarato, L. V. (1967). Medicinal folklore evaluation I. Alleged androgenic and aphrodisiac action of pega palo (*Rhynchosia pyramidalis*). *Journal of Pharmaceutical Sciences*, 56(8), 967–970.

Garau, G., Yates, R. J., Deiana, P., & Howieson, J. G. (2009). Novel strains of nodulating Burkholderia have a role in nitrogen fixation with papilionoid herbaceous legumes adapted to acid, infertile soils. *Soil Biology and Biochemistry, 41*(1), 125–134.

Harper, R.M. 1932. Useful plants of Yucatan. *Bulletin of the Torrey Botanical Club*, 59(5): 279–288.

Hofmann, A., Ratsch, C., Schultes, R. (1992). *Plants of the Gods: Their Sacred, Healing, and Hallucinogenic Powers*. Rochester: Healing Arts Press, p. 340.

Hutchinson, J. (1964). *The genera of flowering plants Dicotylédones*, Vol. 1. Clarendon Press: Oxford University Press, London. p. 516.

Kinjo, J., Nagao, T., Tanaka, T., Nonaka, G.I., Okabe, H. (2001). Antiproliferative constituents in the plant 8. Seeds of Rhynchosia volubilis. *Biological and Pharmaceutical Bulletin* 24: 1443–1445.

Maroyi, A. (2023). Medicinal uses of the Fabaceae family in Zimbabwe: A review. *Plants, 12*(6), 1255.

Martínez-Bernal, A., Duno-de Stefano, R., & Lorena-Can, L. (2011). Los géneros Cajanus y Rhynchosia (Leguminosae, Papilionoideae, Phaseoleae, Cajaninae) en la península de Yucatán, México. *Revista mexicana de biodiversidad, 82*(4), 1098–1107.

Rammohan, A., Reddy, G. M., Bhaskar, B. V., Gunasekar, D., & Zyryanov, G. V. (2020). Phytochemistry and pharmacological activities of the genus Rhynchosia: a comprehensive review. *Planta, 251*, 1–15.

Rao, K.V., Sreeramulu, K., Gunasekar D. (1991). Chemical constituents of the genus Rhynchosia. *Indian Journal of Natural Products* 7: 3–12.

Rätsch, C. (1998). *The encyclopedia of psychoactive plants: Ethnopharmacology and its applications*. Park Street Press, p. 461.

Rätsch, C. (2005). *The encyclopedia of psychoactive plants: ethnopharmacology and its applications*. Simon and Schuster.

Ristic, S., & Thomas, A. (1962). Uber Die Inhaltsstoffe Von Catha Edulis. 1. *Archiv Der Pharmazie Und Berichte Der Deutschen Pharmazeutischen Gesselschaft, 295*(7), 524.

Rondo, M. (2017). Phytochemical and biological studies on some South African plants used in traditional medicine for skin hyperpigmentation; MS Dissertation submitted to university of the Western Cape, Africa, pp. 54–60.

Santesson, C.G. (1938). Noch eine mexikanische 'Piule'-Droge: Semina Rynchosiae phaseoloidis DC. [sic!]. *Etnologiska Studier* 6: 179–83.

Schultes, R. E. (1937). Peyote and plants used in the peyote ceremony. *Botanical Museum Leaflets, Harvard University, 4*(8), 129–152.

Schultes, R. E. (1965). Ein Halbes Jahrhundert Ethnobotanik Amerikanischer Halluzinogene1. *Planta Medica*, 13(02), 125–157.

Schultes, R. E. (1969). Hallucinogens of plant origin: Interdisciplinary studies of plants sacred in primitive cultures yield results of academic and practical interest. *Science*, 163(3864), 245–254.

Schultes, R.E., Hofmann, A. 1980. *Plants of the Gods: Origins of hallucinogenic use*. McGraw-Hill Co-Publication, p. 338.

Secretariat, G. B. I. F. (2021). GBIF backbone taxonomy. https://doi.org/10.15468/39omei

Standley, P.C. & Steyermark, J.A. (1976). Flora of Guatemala. http://www.archive.org

Voogelbreinder, Snu. (2009). Garden of Eden: The Shamanic use of psychoactive flora and fauna, and the study of consciousness. *Snu Voogelbreinder*, p. 296.

Wasson, R.G., and V.P. Wasson. (1957). *Mushrooms, Russia, and history*. Pantheon Books.

WFO (2023). World Flora Online. Published on the Internet; http://www.worldfloraonline.org. Accessed on: 28 Sep 2024.

Willis, J.C. (1966). *A dictionary of the flowering plants and ferns*. Cambridge University Press, p 970.

25 Agara (*Galbulimima belgraveana* (F. Muell.) Sprague Himantandraceae)

Daniyal Hussain, Ehsan Ali, and Muhammad Farooq Azhar

25.1 INTRODUCTION

Himantandraceae is a monogeneric family with only one genus, Galbulimima F.M. Bailey. Galbulimima is found in Eastern Asia, Australia, New Guinea, Indonesia (including the Moluccas, Celebes, and West Papua), and the Solomon Islands (Lan et al., 2018). Galbulimima was originally supposed to have one to three species, mostly found in northeastern Australia, New Guinea, and the Molucca Islands (Cronquist, 1981). Van Royen (Brophy et al., 2005), on the other hand, identified only one variable species, *Galbulimima belgraveana* (F. Muell.). Sprague. This species has been given various scientific names throughout the years, including *Eupomatia belgraveana*, *Himantandra belgraveana*, *Himantandra nitida*, and *Galbulimima nitida*, with the largely recognized scientific name presently being *Galbulimima belgraveana*.

Historically, there has been debate about the number of distinct species in the genus Galbulimima, with estimates ranging from four (*G. belgraveana, G. baccata, G. nitida*, and *G. parvifolia*) to just one (*Galbulimima belgraveana*) (Ritchie and Taylor, 1967). *G. baccata* is found in North Queensland, Australia, and *G. belgraveana* is found in Papua New Guinea (Endress, 1983).

G. belgraveana is a large, evergreen aromatic tree grown in Papua New Guinea, Indonesia, and Northern Australia (Jessup, 1985). This plant is well-known for its hallucinogenic effects and has been employed in native tribes' religious rites as well as herbal treatments to promote deep slumber and relieve abdominal problems (Thomas, 2005).

In this chapter, we will dive deep into the fascinating world of *G. belgraveana*, a plant species that is both unusual and intriguing, with important botanical and cultural significance. Throughout this chapter, we will look at the morphology, distribution, ecological relevance, and traditional usage of *G. belgraveana*. We also look at the scientific study that has been done on this species, which sheds insight into its phytochemical composition, therapeutic properties, hallucinogenic effects, and possible applications in a variety of sectors. Our goal with this in-depth look at *G. belgraveana* is to provide a full review of this unusual plant species, emphasizing its ecological significance, cultural significance, and prospective contributions to numerous scientific disciplines. Finally, we hope to inculcate a greater appreciation for nature's wonders and emphasize the necessity of maintaining our botanical legacy.

25.2 SCIENTIFIC CLASSIFICATION

The botanical name of a plant is *Galbulimima belgraveana* (F. Muell). Sprague belongs to the family Himantandraceae. The detain scientific classification of a plant is given in Table 25.1.

TABLE 25.1
Taxonomic Classification of (*Galbulimima belgraveana* (F. Muell.) Sprague Himantandraceae)

Kingdom	Plantae
Phylum	Tracheophytes
Class	Angiosperms
Sub Class	Magnoliids
Order	Magnoliales
Family	Himantandraceae
Genus	Galbulimima
Species	*Galbulimima belgraveana*

25.3 VERNACULAR NAMES

In English, *G. belgraveana* is known as Galbulimima, and it is also known as agara (Benjamin, 1999). Both *G. belgraveana* and *G. baccata* trees are known colloquially as Northern pigeonberry ash in English (Rinner et al., 2010). However, *G. belgraveana* has also been incorrectly described as white magnolia (Rinner et al., 2010), which can be misleading because this name is usually linked with taxonomically unrelated species such as *Magnolia denudata* (Lim, 2014).

25.4 BOTANICAL DESCRIPTION

G. belgraveana is a large scented evergreen tree (Womersley et al., 1978). It grows to incredible heights, reaching up to 35 meters in height and 60 cm in diameter. A buttress root may be present on occasion, reaching three meters in height, one meter in width, and varying in thickness from five to twenty centimeters. This tree's leaves are round or oval in

DOI: 10.1201/9781003460336-25

shape and covered in copper-colored scales, giving them a distinctive coppery-green appearance. Each leaf has a complicated network of veins. The tree produces large blossoms in white, cream, or brown after the flowering period, followed by small, meaty pink or red fruits. Its exterior bark is gray or grayish-brown in color, with a rough, scaly, and pimply surface (Figure 25.1). According to Benjamin (1999), this bark has a resinous fragrance and a harsh taste. *G. belgraveana* is distinguished by its unbuttressed growth habit, reaching heights of up to 90 feet, as described by Hofmann et al. (1992). This tree's bark is strongly scented, with a gray-brown color and a scaly texture that measures about half an inch thick. Its elliptic leaves are glossy, with a metallic green upper surface and a brown underside and are normally 6 inches long and 2 inches wide (Conran & Kubitzki, 1998).

25.5 ORIGIN AND GEOGRAPHIC DISTRIBUTION

G. belgraveana, first discovered in rainforests by Mueller in 1887 and mentioned by Rinner et al. (2010), has a diverse range that includes Papua New Guinea, Northern Australia, Indonesia, and Eastern Asia. This tree is notable for its scented evergreen leaves (Lan et al., 2018). Its range includes Papua New Guinea, Indonesia, the Solomon Islands, and Australia, with a strong presence in Irian Jaya (West Papua) and Papua New Guinea, stretching from the Vogelkop ('Bird's Head') region in the west to Milne Bay and New Britain (Womersley et al., 1978). The species is most commonly found in the highlands of Papua New Guinea's hilly rainforests. Additionally, *G. belgraveana* can be found in rainforests in the Solomon Islands and in the northeastern and southern regions of Queensland, Australia.

25.6 ECOLOGY

G. belgraveana flourishes in lush mountain rainforests, usually in the canopy of Nothofagus forests in Papua New Guinea and Australia. It is most usually found in the Papua New Guinea region at elevations ranging from 1,200 to 2,700 meters above sea level, while it has been observed growing as low as 5 meters above sea level (Womersley et al., 1978). This species is found in northeastern Queensland at elevations ranging from 400 to 1100 meters.

25.7 PHYTOCHEMISTRY

G. belgraveana phytochemistry has been widely examined, and it has been shown that the bark of both species is a rich source of alkaloids. A trans-decalin system and a piperidine ring are shared structural features of Galbulimima alkaloids. Based on their structural properties (Figure 25.2), these alkaloids can be categorized into four distinct classes, namely Classes I, II, III, and miscellaneous (Class IV) (Binns et al., 1965). Except for GB 18, all Galbulimima alkaloids in Class I have a -lactone linked to a trans-decalin core. An (S)-2-methylpiperidine ring is present in Class I alkaloids. The difference between GB 18, GB 20, and GB 21 is that they lack a 7-double bond and instead form a 7,20-annulation. The (S)-2-methylpiperidine ring distinguishes Class II Galbulimima alkaloids, which are linked to the decalin structure by an extra bond, resulting in a more complicated fused ring system. Galbulimima alkaloids in Class III differ from those in Class II in terms of the connectivity between the (S)-2-methylpiperidine unit and the trans-decalin system. All other Galbulimima alkaloids are classified as Class IV. Notably, Movassaghi synthetic research toward total synthesis of GB (Movassaghi et al.,

FIGURE 25.1 (a) Floral buds of *Galbulimima belgraveana* (F. Muell.) Sprague and (b). Fruit of *Galbulimima belgraveana* (F. Muell.) Sprague.

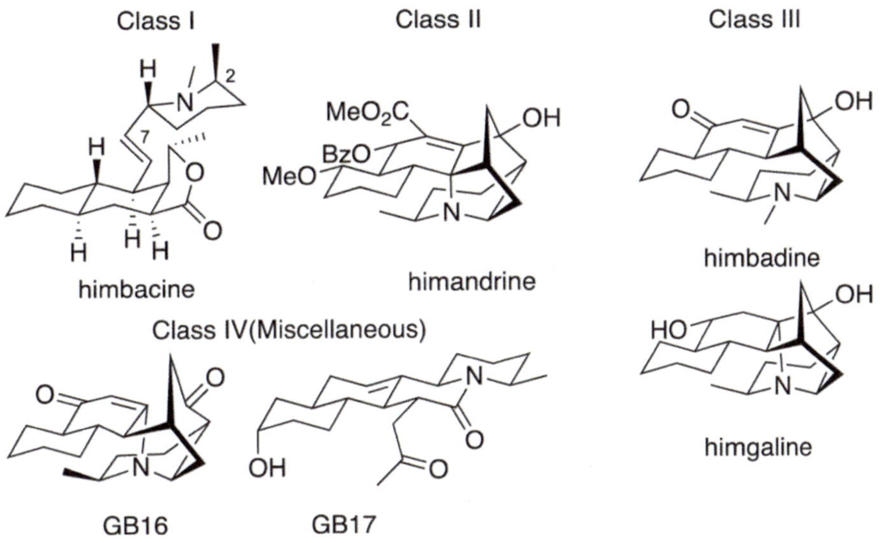

FIGURE 25.2 Classification of *G. belgraveana* alkaloids and representative structures.

2006) resulted in a correction of the stereochemical assignment at C-2, hence correcting the stereochemistry of all known Galbulimima alkaloids (Movassaghi et al., 2006).

Before this change, a total of 36 Galbulimima alkaloids had been identified, divided into four classes: Class I (himbacine type) (Knight et al., 2016); Class II (himandrine type) (Evans and Adams, 2007); Class III (himbadine and himgaline type) (Bradford et al., 2011); and Class IV (miscellaneous). Furthermore, Ritchie and Taylor (1971) and Lan et al. (2018) reported the isolation of new members in Classes III and IV, as well as two novel Class II individuals. They also added the first non-alkaloid member of Class I to the existing categorization.

25.8 ETHNOBOTANY

According to Thomas (2005), the bark of the rainforest tree *G. belgraveana* has been eaten for its hallucinogenic qualities in Papua New Guinea. *G. belgraveana* has been used historically in Papua New Guinea and has been claimed to contain psychedelic and psychotropic qualities (Thomas, 2005). Because of its propensity to bind preferentially to M2/M4 muscarinic receptors, the prototype Galbulimima alkaloid, himbacine, was shown to exhibit antispasmodic action (Gilani, 1992) and to be a powerful cardio-specific muscarinic antagonist (Zhang et al., 2005). As a result, it became a lead molecule in the hunt for novel medications to treat neurodegenerative disorders, leading to the development of vorapaxar, a first-in-class PAR-1 antagonistic agent (Knight et al., 2016). This is unquestionably a noteworthy success story from the postwar 'Australian phytochemical survey' (Price et al., 1992).

To generate vivid visions and intense dreams, the bark and leaves of *G. belgraveana* are cooked with the leaves of *Homalomena belgraveana* and the root of *Zingiber zerumbet* in Papua New Guinea. Psilocybe mushrooms are occasionally taken in addition to this combo (Schultes et al., 2001). Warriors also take the leaves and bark to gain vigor before combat. When painting the faces of male babies during their first initiation, the Nopoko tribe uses its leaves to store color. The Gimi tribe chews the bark to enter a trance and get an insight into perplexing situations or impending occurrences. Surprisingly, the effects of *Homalomena belgraveana* and *G. belgraveana* are stated to be the same, whether taken alone or in combination (Baker, 2010). For ritual purposes, indigenous tribes of Papua New Guinea boil the leaves and bark of *G. belgraveana* along with other hallucinogenic plants. Some people massage the bark on their legs to absorb the plant's alkaloids via their skin (Baker, 2010; Chaachouay et al., 2021a, 2021b; Belhaj et al., 2021).

People in the Okapa area of the Eastern Highlands chewed the bark of *G. belgraveana* with the leaves of an unnamed Homa/omena sp. *G. belgraveana* bark (agara) and Homalomena sp. leaves (ereriba) have been claimed to cause visions and a dream-like condition when chewed (Hamilton, 1960). Chewing agara with eririba can result in strong tremors and miosis. The visuals and severe tremors persist for approximately an hour, followed by calm, euphoria, and slumber (De Smet, 1985).

G. belgraveana has also been employed for divination and to induce trance-like states and visions among the Gimi people of the Eastern Highlands without the presence of Homalomena sp. leaves (Glick, 1963). Other cultures in the Eastern Highlands have employed the bark and leaves to make young males ferocious (Powell, 1976). The residents of Aseki in Morobe Province's southernmost region utilize waga, the bark of *G. belgraveana*, as an analgesic by chewing it, spitting it out into a bowl, combining salt with it, and then ingesting it again to treat pain (Woooley, 1991). The Oksapmin of West Sepik Province employ alusa, shredded *G. belgraveana* bark combined with wild ginger (*Zingiber* sp. [Zingiberaceae]), to treat sorcery and witchcraft-related disorders (e.g., fever, skin problems, and poisoning)

(Skingle, 1970). *G. belgraveana* has also been employed in rituals by the Bimin-Kiskusmin of the West Sepik Province (Poole, 1984).

25.9 HALLUCINOGENIC EFFECTS

Galbulimima bark has yielded 28 alkaloids. None of these alkaloids are psychoactive, according to direct pharmacological data (Thomas, 2005). Many active chemicals from *G. belgraveana* have been discovered to have antioxidant, dormant, and analgesic activities in recent decades (Thomas, 2005; Chaachouay & Zidane, 2021). Because of their high beneficial efficacy, Galbulimima alkaloids have garnered a lot of attention. Galbulimima alkaloids are found mostly in the bark of *G. belgraveana* and are made up of a piperidine ring and a trans-decalin system. Galbulimima alkaloids are classified into four groups (I-IV). Himbacine, a Galbulimima alkaloid of Class I, displays high antagonism to muscarinic receptors, which upregulates the production of acetylcholine and has a beneficial effect in the treatment of Alzheimer's disease (Takadoi et al., 2003).

28 alkaloids have been isolated from *G. belgraveana*, but a specific psychoactive principle has not yet been identified. The chemical makeup of *G. belgraveana* varies from tree to tree, even those growing together. The main alkaloids that have been isolated include himbacine, himgravine, himbosine, himandrine, and himbadine, which have antispasmodic and hypotensive effects. Some other alkaloids found in the plants also had depressant effects. The fruits contain an essential oil that smells similar to Juniperus, and traces of various alkaloids (Baker, 2010).

The indigenous peoples who consume *G. belgraveana* report seeing visions of men and animals they are supposed to kill and experiencing powerful dreams. One individual chewed 10 grams of dried, powdered *G. belgraveana* bark, swallowing it after ten minutes. After 30 minutes, reported effects included drowsiness, increased heart rate, concentration issues, and dizziness, followed by a very relaxed state, which eventually wore off with a mild euphoria after two hours (Schultes et al., 2001).

25.10 CONCLUSION

The comprehensive review offered in this chapter sheds light on the taxonomic classification, various names in vernacular language, unusual adaptations, intricate phytochemistry, and rich ethnobotanical history of *G. belgraveana*. Notably, *G. belgraveana* bark has yielded a total of 28 alkaloids. Recent research has revealed the presence of active chemicals within *G. belgraveana*, exhibiting antioxidative, dormancy induction, and analgesic characteristics. Because of their exceptional medicinal potential, these Galbulimima alkaloids have received a lot of attention. These beneficial alkaloids are predominantly found in the bark of *G. belgraveana* and are divided into four classes (I–IV) based on their structural makeup, which comprises a piperidine ring and a trans-decalin system. Himbacine, a Galbulimima alkaloid of Class I, has a significant antagonistic impact on muscarinic receptors. This characteristic has implications for enhancing acetylcholine expression, which might aid in the treatment of Alzheimer's disease.

However, the research on the phytochemistry, toxicology, pharmacology, and hallucinogenic properties of *G. belgraveana* is still scarce. As a result, the study advises that advanced analytical techniques be used to thoroughly trace the plant's chemical composition, a measure that could contribute to the conservation of this rare yet endangered species. Furthermore, while the plant appears to have potential for a variety of medical applications, thorough clinical trials will be required to establish its safety and efficacy. It will also be vital to ensure sustainable harvesting practices to preserve both traditional and contemporary applications of *G. belgraveana*.

CONFLICT OF INTERESTS

The authors have stated that they have no competing interests in the authorship and/or publishing of this chapter.

REFERENCES

Baker, G. (2010). Garden of Eden: the Shamanic use of psychoactive flora and fauna and the study of consciousness. *Australian Journal of Medical Herbalism*, *22*(3), 107–108.

Belhaj, S., Chaachouay, N., & Zidane, L. (2021). Ethnobotanical and toxicology study of medicinal plants used for the treatment of diabetes in the High Atlas Central of Morocco. *Journal of Pharmacy & Pharmacognosy Research*, *9*(5), 619–662. https://doi.org/10.56499/jppres21.1014_9.5.619

Benjamin, T. (1999). *Galbulimima belgraveana* (F. Muell.) *Sprague galbulimima*, agara. *Eleusis*, *2*, 82–88.

Binns, S. V., Dunstan, P. J., Guise, G. B., Holder, G. M., Hollis, A. F., McCredie, R. S.,... & Taylor, W. C. (1965). The chemical constituents of Galbulimima species. V. The isolation of further alkaloids. *Australian Journal of Chemistry*, *18*(4), 569–573.

Bradford, T. A., Willis, A. C., White, J. M., Herlt, A. J., Taylor, W. C., & Mander, L. N. (2011). The structures of four new himbacine-like Galbulimima alkaloids. *Tetrahedron Letters*, *52*(2), 188–191.

Brophy, J. J., Goldsack, R. J., & Forster, P. I. (2005). The leaf essential oil of *Galbulimima baccata* (Himantandraceae) from Queensland, Australia. *Journal of Essential Oil Research*, *17*(5), 536–538.

Chaachouay, N., Douira, A., & Zidane, L. (2021a). Herbal Medicine Used in the Treatment of Human Diseases in the Rif, Northern Morocco. *Arabian Journal for Science and Engineering*, 47, 131–153. https://doi.org/10.1007/s13369-021-05501-1

Chaachouay, N., Douira, A., & Zidane, L. (2021b). COVID-19, prevention and treatment with herbal medicine in the herbal markets of Salé Prefecture, North-Western Morocco. *European Journal of Integrative Medicine*, 42, 101285. https://doi.org/10.1016/j.eujim.2021.101285

Chaachouay, N., & Zidane, L. (2021). The Symbolic Efficacy of Plants in Rituals and Socio-Religious Ceremonies in Morocco, Northwest of Africa. *Journal of Religious &*

Theological Information, *21*(1–2), 34–53. https://doi.org/10.1080/10477845.2021.1942433

Conran, J. G., & Kubitzki, K. (1998). The families and genera of vascular plants. *Flowering plants: Monocotyledons; Lilianae (except Orchidaceae)*, *3*, 114–121.

Cronquist, A. (1981). *An integrated system of classification of flowering plants*. Columbia university press.

De Smet, P. A. (1985). *Ritual enemas and snuffs in the Americas*. CEDLA.

Endress, P. K. (1983). Dispersal and distribution in some small archaic relic families (Austrobaileyaceae, Eupomatiaceae, Himantandraceae, Idiospermoideae-Calycanthaceae). *Sonderh. Naturwiss. Vereins Hamburg*, *7*, 201–217.

Evans, D. A., & Adams, D. J. (2007). Total synthesis of (+)-Galbulimima alkaloid 13 and (+)-himgaline. *Journal of the American Chemical Society*, *129*(5), 1048–1049.

Gilani, A. U. H. (1992). Effects of alkaloids from Galbulimima baccata on guinea-pig ileum. *Fitoterapia*, *63*(3), 238–242.

Glick, L.B. (1963). Foundations oja Primitive MedicaI System: The Gimi ojthe New Guinea Highlands. Ph.D. Thesis. University of Pennsylvania.

Hamilton, L. (1960). An experiment to observe the effect of eating substances called ereriba leaves and agara bark. *Transactions of the Papua and New Guinea Scientific Society*, *1*, 16–18.

Hofmann, M.C., Narisawa, S., Hess, R.A., & Mill, N.J.L. (1992). Immortalization of germ cells and somatic testicular cells using the SV40 large T antigen. *Experimental Cell Research*, *201*(2), 417–435.

Jessup, L. W. (1985). *Flora of Australia* (pp. 170–187). Anacardiaceae, Australian Government Publishing Service, Camberra.

Knight, E., Robinson, E., Smoktunowicz, N., Chambers, R. C., Aliev, A. E., Inglis, G. G., … Caddick, S. (2016). Synthesis of novel and potent vorapaxar analogues. *Organic & Biomolecular Chemistry*, *14*(12), 3264–3274.

Lan, P., Herlt, A. J., Willis, A. C., Taylor, W. C., & Mander, L. N. (2018). Structures of new alkaloids from rain forest trees Galbulimima belgraveana and Galbulimima baccata in Papua New Guinea, Indonesia, and Northern Australia. *ACS Omega*, *3*(2), 1912–1921.

Lim, T. K. (2014). Magnolia grandiflora. In: Lim T. K. (ed) *Edible medicinal and non medicinal plants: Volume 8, flowers* (pp. 243–275). Springer Netherlands.

Movassaghi, M., Hunt, D. K., & Tjandra, M. (2006). Total synthesis and absolute stereochemical assignment of (+)-and (−)-galbulimima alkaloid 13. *Journal of the American Chemical Society*, *128*(25), 8126–8127.

Poole, F.J.P. (1984). Ritual rank, the self, and ancestraJ power: Jiturgy and substance in a Papua New Guinea society. In L. Lindstrom (Ed.), *Drugs in Western Pacific societies: Relations of substance* (pp. 149–196). University Press of America.

Powell, J.M., (1976). Ethnobotany. In K. Paijmans (Ed.), *New Guinea Vegeta/ion* (Vol. 1, pp. 06–183). Australian National University Press.

Price, J. R., Lamberton, J. A., & Culvenor, C. C. J. (1992). The Australian phytochemical survey: Historical aspects of the CSIRO search for new drugs in Australian plants. *Historical Records of Australian Science*, *9*(4), 335–356.

Rinner, U., Lentsch, C., &Aichinger, C. (2010). Syntheses of galbulimima alkaloids. *Synthesis*, *2010*(22), 3763–3784.

Ritchie, E., & Taylor, W. C. (1967). The galbulimima alkaloids. In: Manske R.H.F. (ed) *The Alkaloids: Chemistry and Physiology* (Vol. 9, pp. 529–543). Academic Press.

Ritchie, E., & Taylor, W. C. (1971). The galbulimima alkaloids. In *The alkaloids: Chemistry and physiology* (Vol. 13, pp. 227–271). Academic Press.

Schultes, R. E., Hofmann, A., & Rätsch, C. (2001). *Plants of the gods: Their sacred, healing, and hallucinogenic powers*. Healing Arts Press; 2nd edition. p. 208.

Skingle, D.C. (1970). Some medicinal herbs used by the natives of New Guinea Mankind, *7*, 223–225.

Takadoi, M., Yamaguchi, K., & Terashima, S. (2003). Synthetic studies on himbacine, a potent antagonist of the muscarinic M2 subtype receptor. Part 2: Synthesis and muscarinic M2 subtype antagonistic activity of the novel himbacine congeners modified at the C-3 position of lactone moiety. *Bioorganic & Medicinal Chemistry*, *11*(7), 1169–1186.

Thomas, B. (2005). Psychoactive properties of Galbulimima bark. *Journal of Psychoactive Drugs*, *37*(1), 109–111.

Womersley, J. S., Henty, E. E., & Conn, B. J. (1978). *Handbooks of the flora of Papua New Guinea*. Vol 1. Carlton South Australia: Melbourne University Press.

Woooley, E. (Ed.) (1991). *Medicinal plants of Papua New Guinea*. ParI 1: Morobe Province. Verlag lasef Margraf Scientific Books.

Zhang, H. M., Li, D. P., Chen, S. R., & Pan, H. L. (2005). M2, M3, and M4 receptor subtypes contribute to muscarinic potentiation of GABAergic inputs to spinal dorsal horn neurons. *Journal of Pharmacology and Experimental Therapeutics*, *313*(2), 697–704.

26 Queule (*Gomortega keule* (Molina) Baill. Gomortegaceae)

Saba Latif, Rahmatullah Qureshi, Syed Sabir Hussain Shah, and Noureddine Chaachouay

26.1 INTRODUCTION

Gomortega keule (Mol.) Baillon, commonly named queule, is an evergreen tree which is endemic to Chile (Hechenleitner et al., 2005). This is a monotypic species of a rare family, Gomortegaceae. It occurs in a coastal area and is restricted to the Maule and Biobío (Echeverría et al., 2006) and Ñuble (Anonymous, 2023). According to Schultes (1976), it is confined to nearly 100 miles^2 in Central Chile. According to the Royal Botanical Garden, Kew, it ranges to the Central & S. Central Chile and is a key species of the subtropical biome. This medium-sized evergreen tree is distributed in isolated ravines in the Cordillera de la Costa (Muñoz-Concha and Davey, 2011). It was found associated with roble beech (*Lophozonia obliqua*), forming a typical plant community that is part of the Central Chile Biodiversity Hotspot (González, 1998). It yields good quality fruits which are a good source of food and utilized for making jams, drinks, and syrups, in addition to the best quality timber for woodwork and structure (Lander et al., 2010). It is also used for flavoring, dye, and marbles.

According to IUCN, *C. keule* is an endangered species and is the sole representative in 25 biodiversity hotspots of Central Chile (Garcia-Gonzales et al., 2008). Due to habitat destruction coupled with climatic change, the tree is marked near extinction and under conservation (Muñoz-Concha and Davey, 2011). The species favorably grows in the Mediterranean climate (i.e. coastlines and warm temperatures) and is confined to 35°37′ to 37°41 in Chile in fragmented geographies (Delaveau et al., 2013). The plant is approximately 25 meters tall and has characteristic opposite and evergreen leaves. It has a drupe fruit that contains a solid stone in it.

The tree contains good quality timber which is used as a source of fuel and timber (José San Martín, 2003). Although it produces high-quality edible fruits unluckily, it is never planted on a large scale in Chile because it is very hard to propagate the tree by seed or cuttings. Additionally, it gives good quality jams and different sauces. (Hinojosa et al., 2006). Moreover, the plant contains highly oxygenated and rare compounds; rutin and chlorogenic acid. Other six important antioxidants are also present in it. Although this species is near to extinct, however, there is high genetic variability among geographical populations (Simirgiotis et al., 2013). In addition to using wood for construction and fuelwood, this tree produces wonderful fruits that are used fresh or processed to make jams, beverages, and syrups (Albert, 1924; Rodríguez et al., 1983; Ruiz, 1998; Chaachouay et al., 2019, 2020). According to Mariani (1965) and Schultes (1983), the fruits have an intoxicating and hallucinogenic character.

The fruit was used to make a sweet beverage by the Mapuche natives of Central Chile (Burdick, 2017). The machi shaman frequently employs intoxicating fruits (Mariani, 1965). Later, he revealed additional evidence of the Mapuche's use of fresh fruit as a hallucinogen. According to Schultes (1983), the fruits of *G. keule* have the same psychoactive qualities as those utilized by the Mapuche people; nevertheless, if consumed in excess, they can give people headaches.

26.2 SCIENTIFIC CLASSIFICATION

Gomortega keule (Mol.) Baillon is a large tree confined to Chile. The classification is as shown in Table 26.1.

26.3 VERNACULAR NAMES

English: Queule, *Gomortega keule*; **Spanish**: Queule, Hualhual; **Polish**: Queule lub *Gomortega keule*; **Danish**: Queule eller *Gomortega keule*; **Czech**: Queule nebo *Gomortega keule*.

26.4 SYNONYMS

According to The Plant List (2013) data, *Gomortega keule* (Molina) Baill. is a plant species that has three scientific synonyms include:

Gomortega nitida Ruiz & Pav.

Adenostemum nitidum (Ruiz & Pav.) Pers.

Lucuma keule Molina.

Keulia chilensis Molina.

DOI: 10.1201/9781003460336-26

TABLE 26.1
Taxonomic Hierarchy of *Lagochilus inebrians* Bunge

Kingdom:	Plantae
Subkingdom:	Viridaeplantae
Infrakingdom:	Streptophyta
Division:	Tracheophytes
Subdivision:	Spermatophytina
Infradivision:	Angiospermae
Class:	Magnoliopsida
Clade:	Magnoliid
Superorder:	Asterids
Order:	Laurales
Family:	Gomortegaceae
Genus:	*Gomortega*
Species:	*G. keule*
Binomial name:	*Gomortega keule* (Molina) Baill.

26.5 ETYMOLOGY

The name derived from a sort of blend word formed by combining 'Gom' and 'Ortega' for the genus *Gomortega* in recognition of the work of the Spanish botanist and physician Casimiro Gómez Ortega (1741–1818), who was the first professor at the Royal Botanic Garden of Madrid. The name was given by the Chilean naturalist Juan Ignacio Molina (1740–1829). Furthermore, the name of *Gomortega keule* was borrowed from the Mapuche people, indigenous to Chile and Argentina, who call this plant 'keule' or 'queule' (Simirgiotis et al., 2013).

26.6 BOTANICAL DESCRIPTION

Lucuma keule was the first time named by Molina (1782). Later on in 1869, this was renamed *Gomortega keale* by Baillon in 1924 and then *G. keule* by Johnston. After its confirmation as *G. keule* by Espinosa (1948), it is currently in use (Muñoz-Concha and Davey, 2011). This is an evergreen tree with a specific aroma. The tree stem has two foliar traces with unilacunar nodes. It possesses a 15- to 25-meter-tall trunk (Dohlen, 1999) which is 100 to 110 cm in diameter (Villegas et al., 2003). It is monopodic at its young stage that changes to sympodic in older trees (Cabello, 1987). The bark of the stem is longitudinal fissured which is gray in color. Leaves are evergreen, simple, petiolate, opposite, lack stipules (Kubitzki, 1993), approximately 5–11.5 cm long and 2–4.5 cm wide (Rodríguez et al., 1983), petiole ca. 0.8–1.5 cm long. The young stems are 4-angular (Hechenleitner et al., 2005). The adaxial surface is a thick cuticle (Rodríguez, 2005). The inflorescence assembled in terminal racemes, 6–14 in number. Flowers are bisexual green to creamy colored, 5–7 mm in diameter (Rodríguez et al., 1983), and aromatic (Del Fierro, 1998). Tepals 7 (5–9), approximately 3–3.5 mm long, spirally arranged becoming smaller in center (Brizicky, 1959). Stamens 7–13, about 1.5–3.5 mm long, spirally arranged, 1–3 exterior stamens are tepal-like, bearing two anthers at the top and sometimes 1 or 2 glands at their bases. The other 5–10 stamens (often 8) are typical in structure with a filament and valvate anther, usually bearing at the base one stipitate, ovate to reniform gland on each side; staminodes 1–4, ca. 1–1.5 mm long. The pollen is spherical or granular with 25–31 μm in diameter (Erdtman, 1952). The ovary is inferior, 2–3 locular with a single ovule; style 2 or 3 stigmatic branched (Brizicky 1959). The fruit is a yellow drupe that contains fleshy mesocarp which is juicy and edible (Figure 26.1d; Hechenleitner et al., 2005), approximately 3.5–7 cm long and 2–5 cm broad (Le Quesne and Stark, 2006) with an average of 28.2 g weight, having 71–78% weight (17.5 g) of pulp (Le Quesne and Stark, 2006). The endocarp is a very hard, stony, or woody nut that contains 1–2 seeds, ca. 21–24 mm long, 17–19 mm broad (Figure 26.1e; Aburto 2002). The seed is albuminous (Doweld, 2001) with a profuse endosperm that has oil droplets; the embryo is straight and symmetrical (Figure 26.1f; Heo et al., 2004).

26.7 HISTORY

G. keule was the first time reported in the Spanish chronicle of Jerónimo de Vivar in 1558 (Torrejón and Cisternas, 2003), describing its uses in syrup and beverages made up of the fruit. De Rosales (1877) mentioned that the fruit is tasty and edible, either eaten fresh or cooked, and described the existence of the endocarp (Baillon, 1869). The first Spanish-Mapuche dictionaries (Febrès, 1765) contained the term queul, asserting 'a yellow fruit'. Molina (1782) reported that the tree produces a delicious fruit; while, Ruiz and Pavon (1798) stated that the fruit is sweet, pleasing, and highly flavored.

26.8 ORIGIN AND GEOGRAPHIC DISTRIBUTION

G. keule, the only living member of Gomortegaceae, is restricted to the coastline of Central Chile. Its total population is estimated about thousand members. It is located at 35°47′S to, 37°41′S and the total occupied area is estimated at 300 km (Echeverría and Campos, 2019) as shown in Figure 26.2. There is a scattered distribution of trees up to 25 localities and each spot is not more than 4 hectares in measurement. The tree trunk is about 25 cm in diameter. There is the presence of a couple of trees in a circle that is 2–4 in diameter (Araus et al., 2003).

In recent decades, the population of Gomortega has drastically declined due to habitat destruction and the introduction of exotic plants in the area. On the other hand, fire has been playing a critical role in the decline of this species, ultimately trees are found in different patches and solely plants are also found (Jaksic et al., 2021). Currently, *G. keule* is confined to 145 ha in Chile and safeguarded. In

FIGURE 26.1 Habit of *Gomortega keule*. (a) Young trees; (b) flowers (scale bar =5 mm); (c) seedlings; (d) fruits, the halved one showing the endocarp; (e) open endocarp showing a seed (scale bar = 15 mm); (f) a zygotic embryo (scale bar =3 mm); (g) shoots circularly arranged around an old rotten stump.

other spots, it is threatened due to habitat destruction, fire, and rodents that are natural predators of the plant. Rodent *Rattus rattus* is a natural menace against the survival of this plant. This Chilean endemic species is considered to be about 100 million years old, going out on a taxonomic and evolutionary limb. While the Lauraceae, a related and significant ecological family, can be traced back to *G. keule*. The future of this species is questionable and it is

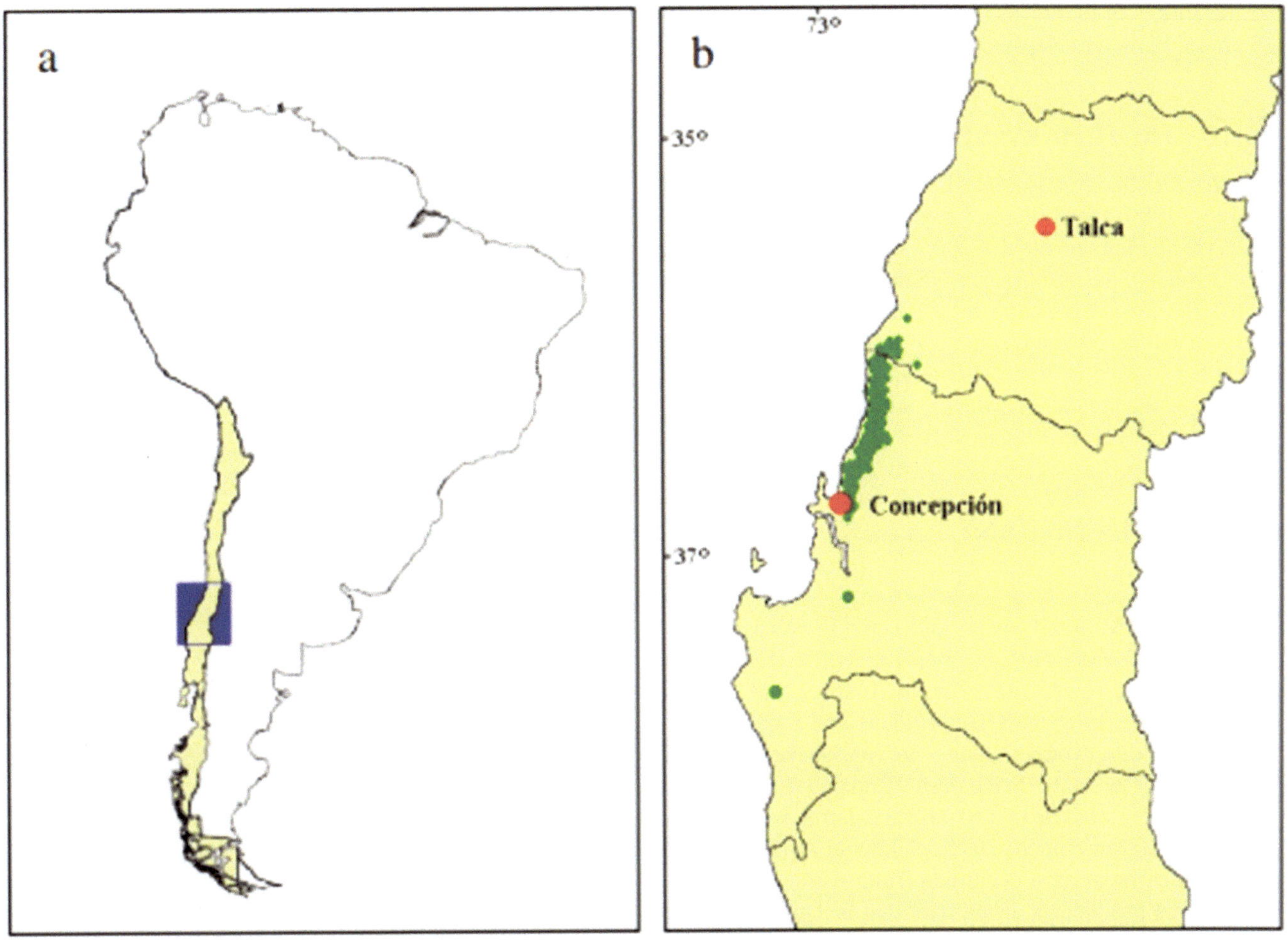

FIGURE 26.2 Natural distribution of *G. keule*: (a) Chile; (b) *G. keule* populations.

listed as an endangered species (Lander et al., 2010). The Maule and Bo Bo provinces of Chile's Cordillera de la Costa (coastal mountain range) are home to this medium-sized evergreen tree. With roble beech (*Lophozonia* (formerly *Nothofagus*) *obliqua*), *G. keule* forms a unique vegetation community that is part of the Central Chile Biodiversity Hotspot (González, 1998). Its fruits are a popular source of food, used to produce jams, alcoholic drinks, and syrups, with its high-quality timber favored for carpentry and construction.

Fires, logging for timber, recent land conversion to exotic conifer plantations, and other factors have all had a significant negative impact on the original native plant ecosystem near Gomortega over a long period (Muñoz-Concha and Davey, 2011). This rare tree only occasionally grows in small groups or as lone trees in the remnants of the natural forest. The seeds of this plant, which are currently distributed by rodents or gravity but may earlier have relied on larger, extinct creatures, regenerate poorly.

26.9 ECOLOGY

G. keule has an association with herbs, shrubs, trees, and epiphytes. There is the presence of four types of pteridophytes, specifically ferns, in coastal areas in Central Chile. This can originate in association with many species; it can also grow with pure strands (Le Quesne and Stark, 2006). Some reports suggested that *G. keule* favors growing in moist shady places. Additionally, it may favor north-exposed faces. Moreover, the tree favors growing under 55% canopy. It may grow best in the middle canopy but occasionally, it grows in the upper layer of the canopy (Gómez et al., 2022). The roots of the tree are devoid of roots hair because it has an abundance of mycorrhiza in it which help in the absorption of nutrients and mineral from the soil. The length of the root is not more than 25 cm in *G. keule*. The tree grows best in soil with 6% organic matter, thin and sloppy with sufficient granite. It requires 5 to 6 soil pH and phosphorus quantity should not exceed 5 ppm (Holland et al., 2014). The tree blooms in the March-April and syrphid flies are responsible for pollination and fruit set. Insects are inhabitants of the local forests of the Mediterranean. On the other hand, pine in the area doesn't favor the pollinators, hence it declines the fruit (Lander et al. 2009). As the seed of the tree is heavy, they are distributed in far-off places by barochory and rodents (Lander et al., 2010). *G. keule* has a special association with 15 kinds of lichens that grow on this plant. This indicates the hydrophilic nature of the species of this type. According to some reports, there is a presence of 14 species of lichens growing on the shoots of plants and these colonies may range up to 15 cm in diameter (Nepstad et al., 1996).

26.10 FRUIT HARVESTING

The fruit collection is regarded as a traditional activity (Munoz, 1986) traditionally known by the local communities (Hechenleitner et al., 2005). Fruits are abundantly found during autumn season (San Martín and Donoso, 1996) and have been used by them for years (Le Quesne and Stark 2006). Most of the people collect fruits from the forest from April to June. The collection of *G. keule* fruits is made by the local people during wood collection (Muñoz-Concha and Garrido-Werner, 2011).

26.11 ETHNOBOTANY

26.11.1 Food

According to Muñoz-Concha and Garrido-Werner (2011), 16% of Chile's communities consume fruit as food due to its sweet taste (Albert, 1924; Munoz, 1986; San Martín and Donoso, 1996; Ruiz, 1998; Chaachouay and Zidane, 2021). According to a woman, fruits are pre-soaked in water before being used to make jam (Rodríguez, 2004). De Rosales (1877) was the first to describe eating fruit that had been roasted over embers. Hechenleitner et al. (2005) documented the preparation of canned fruit.

The tree produces a drupe fruit that is a good source of beverages, syrups, and jams. Additionally, the fruit gives high-quality alcoholic drinks. Moreover, wood obtained from this species is durable and used to manufacture furniture and in construction (Barceló and Simonetti, 2021). As the fruit of the tree is tasty and juicy, it is consumed as either cooked or in flesh. In addition to this, the endocarp of the fruit is highly nutritious and consumed in any form (Cisternas et al., 2005). Seeds are dispersed by animals to far-off places. Some authors compare the *G. keule* fruit with the egg of a hen because it is yellow in color and contains a stone in it. Others suggested that although the endocarp of the fruit is not juicy it is sweet in taste (Arnold and Ruiz Pérez, 1998; Orch et al., 2021).

The previous figure describes the *G. keule* complete picture of fruit and fruit dispersal. The fruit falls on the ground after repining. It accumulated there for a long time due to the absence of wild animals, which may become extinct. The fruit contains a fleshy exocarp which is yellow in color. It has a pointed end and elongated stones. It has a fleshy juicy endocarp that is attached to the seed. When it ripens, releases seeds by pressure by rupture of the endocarp (Muñoz-Concha et al., 2020); see Figure 26.3. In addition to fruits, the leaves of the tree have a characteristic odor due to the presence of turpentine in it. It is sticky in appearance while rubbing. Its odor more resembles camphor in smell. Some authors reported the manufacture of soaps from the essential oil that is extracted from leaves through steam distillation (Vilegas and Lanças, 1995).

26.11.2 Medicinal Use

Leaves are aromatic in nature and contain a therapeutic oil (Albert, 1924). The species possesses balsamic and healing properties (Ruiz, 1998) and is used for medicinal purposes (Muñoz-Concha and Garrido-Werner, 2011; Chaachouay et al., 2021a, 2021b, Belhaj et al., 2021; Benkhnigue et al., 2022). The essential oil from the fruits showed significant antifungal activity against *Aspergillus niger.* It also demonstrated insecticidal activity against the beetle *Acanthoscelides obtectus* (Becerra et al., 2010).

26.11.3 Beverages

The fruit of this plant is used for the reparation of alcohol (Hechenleitner et al., 2005). The production of a very good beverage was reported by Spanish Jerónimo de Vivar in his chronicle (1558) from fruits of *G. keule* (Torrejón and Cisternas, 2003). The preparation of a particularly sweet alcoholic beverage using queule fruit was mentioned by Ovalle (1646). The flavor of hot drinks (known as mate) is improved by adding leaves of *G. keule.* The Spanish naturalist Hipolito Ruiz first recorded the scent of queule leaves in the 18th century (Ruiz, 1998).

26.11.4 Miscellaneous Uses

The fruits are used as forage mostly eaten on the forest floor (San Martín and Donoso, 1996). The bark of the tree is used to dye sheep wool in red hues. The harvested wood is used for construction, furniture (chairs and beds), and carpentry (doors and windows). The wood is prized for its high quality and vibrant red color. The tree is planted as ornamental in Chile for its lush green foliage and canopy. It is advised to plant it in public places and for home decor in the area (Vilegas and Lanças, 1995).

26.12 PHYTOCHEMISTRY

It is a medicinal plant that contains many chemical compounds. The leaves of the tree possess α- terpene, 1, 8 cineole, and β- pinene present. These secondary compounds are highly medicinal and have strong antimicrobial potential (Renner, 1999). Besides, the presence of camphene, 3-carene, limonene, and b-phellandrene are also reported from *G. keule.* The leaves of the tree are effective in controlling insects for the storage of grain for a long time (Oginuma and Tobe, 2006). Due to the presence of sitosterol, phytosterols, pentadecanoic acids, and lauric acid, the cortex and bark of *G. keule* were reported to possess strong healing properties such as improving skin health, lowering cholesterol, reducing hypertension and congestion.

26.13 HALLUCINOGENIC EFFECTS

Studies show that the tree possesses psychoactive properties, if it is consumed in excess, it may cause headaches in

FIGURE 26.3 *Gomortega keule*. (a) Fruit ripen and fall; (b) morphological variation of *G. Keule* fruit; (c) morphological variation of stones; (d) polar view of stone; (e) open stone showing endocarp (Muñoz-Concha et al., 2020).

humans. Other reports showed that the fruit is hot and bitter, while the majority said that the fruit is neither hot nor bitter; rather, it is soothing and sweet in taste (Muñoz-Concha and Garrido-Werner, 2011). The fruit may cause intoxication, nausea, vomiting, headaches, and stomachaches, especially after consuming more than three to four immature fruits. These adverse effects are consistent with those that have been reported in the literature as generating headaches (Ruiz, 1998) or vertigo (Rodríguez, 2004). According to some sources, fresh fruits exhibit euphoric effects that may be attributed to the presence of glycoalkaloids and the heating process would eliminate the sedative effect. The other authors (Mariani, 1965; Schultes, 1983) suggested that similar effects are brought on by the presence of essential oils.

26.14 CONCLUSION AND FUTURE PROSPECTS

G. keule is an endemic tree and an important part of the biodiversity hotspots of the subtropical biome. It yields nutritious pulpy fruit and medicinal essential oils, and provides good quality wood, natural landscape, and scenic beauty. The collected literature emphasizes the importance of the tree; however much of the outdated work is previously reported, and there is a gap of knowledge for the cultivation and nutritional compounds present in the fruit. A long term strategy must be adopted for *in situ* and *ex situ* conservation of this threatened and neglected species. For conserving this natural treasure, *in situ* conservation of the remnant vegetation be made by the government by declaring protected areas of Chile. Additionally, re-generated population must be protected from fire and herd. Last but not least, there should be formal research on medicinal importance, essential oil, agronomy, and conservation of *G. keule*.

BIBLIOGRAPHY

Aburto, A. (2002). Extracción, caracterización y germinación de semillas de queule (*Gomortega keule* (Mol.) Baillon). Memoria de Título. Facultad de Ciencias Forestales, Universidad de Concepción. Concepción, Chile. 40 p.

Albert, F. (1924). *Materias primas vegetalesy animales*. Santiago, Sociedad Imprentay Litografi´a Uni-verso. 35 p.

Anonymous. (2023). A checklist of the endemic plants of Chile. https://chileanendemics.rbge.org.uk/taxa/gomortega-keule-molina-baill. Date of access: 14.05.2023.

Araus, J. L., Bort, J., Steduto, P., Villegas, D., & Royo, C. (2003). Breeding cereals for Mediterranean conditions: ecophysiological clues for biotechnology application. *Annals of Applied Biology*, *142*(2), 129–141.

Arnold, J. E. M., & Ruiz Pérez, M. (1998). The role of non-timber forest products in conservation and development. In: Wollenberg, E. and Ingles, A. (eds.). *Incomes from the forest: methods for the development and conservation of forest products for local communities* (pp. 17–42). Cifor.

Baillon, H. (1869). Histoire des plantes. Tome 1. Librairie de L. Hachette et Cie. Paris, France. 488 p.

Barceló, M., & Simonetti, J. A. (2021). Rattus rattus, a potential threat to the endangered tree *Gomortega keule* in the

Maulino forest of Chile. *New Zealand Journal of Botany*, *59*(2), 285–290.

Becerra, J., Bittner, M., Hernandez, V., Brintrup, C., & Silva, M. (2010). Activity of essential oils of Canelo, Queule, Bailahuen y Culen against phythopatogenic fungi Boletin Latinoamericano Y Del Caribe. *De Plantas Medicinales Y Aromaticas*, *9*(3), 212–215.

Belhaj, S., Chaachouay, N., & Zidane, L. (2021). Ethnobotanical and toxicology study of medicinal plants used for the treatment of diabetes in the High Atlas Central of Morocco. *Journal of Pharmacy & Pharmacognosy Research*, *9*(5), 619–662. https://doi.org/10.56499/jppres21.1014_9.5.619

Benkhnigue, O., Chaachouay, N., Khamar, H., El Azzouzi, F., Douira, A, & Zidane, L. (2022). Ethnobotanical and ethnopharmacological study of medicinal plants used in the treatment of anemia in the region of Haouz-Rehamna (Morocco). *Journal of Pharmacy & Pharmacognosy Research*, *10*(2), 279–302. https://doi.org/10.56499/jppres21.1196_10.2.279

Brizicky, G. (1959). Variability in the floral parts of Gomortega. *Willdenowia*, *2*, 200–207.

Burdick, C. E. (2017). The remedies of the machi: visualizing Chilean medicinal botanicals in Alonso de Ovalle's Tabula geographica (1646). *Colonial Latin American Review*, *26*(3), 313–334.

Cabello, A. (1987). Proyecto de protección y recuperación de especies arbóreas y arbustivas amenazadas de extinción (II Parte y final). Documento Técnico N° 22, Corporación Nacional Forestal. Santiago, Chile, 8 p.

Cisternas, M., Atwater, B. F., Torrejón, F., Sawai, Y., Machuca, G., Lagos, M., … Husni, M. (2005). Predecessors of the giant 1960 Chile earthquake. *Nature*, *437*(7057), 404–407.

De Rosales, D. (1877). Historia General del Reyno de Chile. Libro Primero. Imprenta del Mercurio. Valparaíso, Chile. 506 p.

Del Fierro, P. (1998). Experiencia silvicultural del bosque nativo de Chile. Corporación Nacional Forestal. Publicaciones Lo Castillo S.A. Santiago, Chile. 420 p.

Delaveau, C., Fuentes-Arce, G., Ruiz, E., Hasbún, R., Uribe, M., & Valenzuela, S. (2013). Variabilidad genética mediante AFLP en tres relictos de *Gomortega keule* (Molina) Baillon: especie endémica chilena en peligro de extinción. *Gayana. Botánica*, *70*(2), 188–194.

Chaachouay, N., Benkhnigue, O., Fadli, M., Ibaoui, H., & Zidane, L. (2019). Ethnobotanical and ethnopharmacological studies of medicinal and aromatic plants used in the treatment of metabolic diseases in the Moroccan Rif. *Heliyon*, *5*(10), e02191. https://doi.org/10.1080/10477845.2021.1942433

Chaachouay, N., Douira, A., Hassikou, R., Brhadda, N., Dahmani, J., Belahbib, N., Ziri, R., & Zidane, L. (2020). Ethnobotanical and ethnopharmacological study of medicinal and aromatic plants used in the treatment of respiratory system disorders in the Moroccan Rif. PhD Thesis, Département de Biologie - Université Ibn Tofail - Kénitra, Maroc, https://tel.archives-ouvertes.fr/tel-03376377

Chaachouay, N., & Zidane, L. (2021). The Symbolic Efficacy of Plants in Rituals and Socio-Religious Ceremonies in Morocco, Northwest of Africa. *Journal of Religious & Theological Information*, *21*(1–2), 34–53. https://doi.org/10.1080/10477845.2021.1942433

Chaachouay, N., Douira, A., & Zidane, L. (2021a). Herbal Medicine Used in the Treatment of Human Diseases in the Rif, Northern Morocco. *Arabian Journal for Science and Engineering*, *47*, 131–153. https://doi.org/10.1007/s13369-021-05501-1

Chaachouay, N., Douira, A., & Zidane, L. (2021b). COVID-19, prevention and treatment with herbal medicine in the herbal markets of Salé Prefecture, North-Western Morocco. *European Journal of Integrative Medicine*, *42*, 101285. https://doi.org/10.1016/j.eujim.2021.101285

Dohlen, H. (1999). *Gomortega keule* Mol. Eine Tertiärpflanze mit schwierigem Dasein in heutiger Zeit. *Palmengarten*, *63*, 46–52.

Doweld, A. (2001). Carpology and phermatology of *Gomortega* (Gomortegaceae): systematic and evolutionary implications. *Acta Botanica Malacitana*, *26*, 19–37.

Echeverría, C., & Campos, S. (2019). *Gomortega keule*. The IUCN red list of threatened species 2019: e. T31357A2805379. doi: 10.2305/IUCN. UK. 2019-2. *RLTS. T31357A2805379. en.*

Echeverría, C., Coomes, D., Salas, J., Rey-Benayas, J. M., Lara, A., & Newton, A. (2006). Rapid deforestation and fragmentation of Chilean temperate forests. *Biological Conservation*, *130*(4), 481–494.

Erdtman, G. (1952). *Pollen morphology and plant taxonomy. Angiosperms*. Almqvist and Wiksells. 539 p.

Espinosa, M. (1948). Estudios botánicos. 2. Nomenclatura del queule. *Boletín del Museo Nacional de Historia Natural*, *24*, 52–79.

Febrès, A. (1765). Arte de la lengua general del reyno de Chile. Lima, Peru. 682 p.

Garcia-Gonzales, R., Carrasco, B., Penailillo, P., Letelier, L., Herrera, R., Lavandero, B., … Caligari, P. D. (2008). Genetic variability and structure of *Gomortega keule* (Molina) Baillon (Gomortegaceae) relict populations: geographical and genetic fragmentation and its implications for conservation. *Botany*, *86*(11), 1299–1310.

Gómez, P., Espinoza, S., Garrido, P., Martín, J. S., & Ormazábal, Y. (2022). Post-fire tree regeneration from seed of the endangered Nothofagus alessandrii Espinosa in the Maule region of central Chile. *Southern Forests: A Journal of Forest Science*, *84*(1), 75–82.

González, M. (1998) *Gomortega keule*. The IUCN Red List of Threatened Species 1998. http://www.iucnredlist.org/details/full/31357/0

Hechenleitner, P., Gardner, M.F., Thomas, P.I., Eche-Verría, C., Escobar, B., Brownless, P., & Martínez, C. (2005). *Plantas amenazadas del centro-sur de Chile*. Universidad Austral de Chile y Royal Botanic Garden of Edinburgh. 188 p.

Heo, K., Kimoto, Y., Riveros, M., & Tobe, H. (2004). Embryology of Gomortegaceae (Laurales): characteristics and character evolution. *Journal of Plant Research*, *117*, 221–228.

Hinojosa, L. F., Armesto, J. J., & Villagrán, C. (2006). Are Chilean coastal forests pre-Pleistocene relicts? Evidence from foliar physiognomy, palaeoclimate, and phytogeography. *Journal of Biogeography*, *33*(2), 331–341.

Holland, T. C., Bowen, P., Bogdanoff, C., & Hart, M. (2014). Arbuscular mycorrhizal fungal communities associated with Vitis vinifera vines under different frequencies of irrigation. *American Journal of Enology and Viticulture*, *65*(2), 222–229.

Jaksic, F. M., Castro, S. A., Jaksic, F. M., & Castro, S. A. (2021). Transport and Introduction. *Biological Invasions in the South American Anthropocene: Global Causes and Local Impacts* Springer, Cham. (pp. 49–67). https://doi.org/10.1007/978-3-030-56379-0_3

José San Martín, A. (2003). Floristic and vegetative structure of the Nothofagus alpina (Fagaceae) remanent forest of the Coastal Range, central Chile. *Revista Bosque*, *24*(1), 71–85.

Kubitzki, K. (1993). Gomortegaceae. In Kubitzki, K., J.G. Rohwer, & V. Bittrich (Eds.), *The families and genera of vascular plants*, Tome 2 (pp. 318–320). Springer Verlag.

Lander, T. A., Harris, S. A., & Boshier, D. H. (2009). Flower and fruit production and insect pollination of the endangered Chilean tree, *Gomortega keule* in native forest, exotic pine plantation and agricultural environments. *Revista Chilena de Historia Natural*, *82*(3), 403–412.

Lander, T.A., Boshier, D.H., & Harris, S.A. (2010). Fragmented but not isolated: contribution of single trees, small patches and long-distance pollen flow to genetic connectivity for *Gomortega keule*, an endangered Chilean tree. *Biological Conservation*, *143*, 2583–2590.

Le Quesne, C., & Stark, D. (2006). *Gomortega keule* (Mol.) Baillon. In Donoso, C. (Ed.), *Las especies arbóreas de los bosques templados de Chile y Argentina. Autoecología* (pp. 277–284). Marisa Cuneo Ediciones.

Mariani, C. (1965). *Temas de hipnosis*. Editorial Andrés Bello. 435 p.

Molina, G. (1782). *Saggio sulla storia naturale del Chili*. Stamperia di S. Tommaso d'Aquino. 367 p.

Munoz, R. (1986). El queule, una especie en peligro de extincio´ n. *Chile Forestal*, *134*, 16–18.

Muñoz-Concha, D., & Davey, M.R. (2011). *Gomortega keule*, the neglected and endangered Chilean fruit tree. *European Journal of Forest Research*, *130*, 677–693.

Muñoz-Concha, D., & Garrido-Werner, A. (2011). Ethnobotany of *Gomortega keule*, an endemic and endangered Chilean tree. *New Zealand Journal of Botany*, 49(4), 509–513.

Muñoz-Concha, D., Muñoz, K., & Loayza, A. P. (2020). Anachronic fruit traits and natural history suggest extinct megafauna herbivores as the dispersers of an endangered tree. *Plants*, *9*(11), 1492.

Nepstad, D. C., Uhl, C., Pereira, C. A., & Da Silva, J. M. C. (1996). A comparative study of tree establishment in abandoned pasture and mature forest of eastern Amazonia. *Oikos*, *76*(1), 25–39. https://doi.org/10.2307/3545745

Oginuma, K., & Tobe, H. (2006). Chromosome evolution in the Laurales based on analyses of original and published data. *Journal of Plant Research*, *119*, 309–320.

Orch, H., Chaachouay, N., Douiri, E. M., Faiz, N., Zidane, L., & Douira, A. (2021). Use of medicinal plants in dermato-cosmetology: An ethnobotanical study among the population of Izarène. *Jordan Journal of Pharmaceutical Sciences*, *14*(3), 323–340.

The Plant List (2013). Version 1.1. Published on the Internet; http://www.theplantlist.org/ (accessed 1st January).

Renner, S. S. (1999). Circumscription and phylogeny of the Laurales: Evidence from molecular and morphological data. *American Journal of Botany*, *86*(9), 1301–1315.

Rodríguez, V. (2004). *Estudio morfolo´gico y anato´mico de Gomortega keule (Mol.) Baillon y supotencial agrı´cola*. Universidad Cató licadel Maule. 51 p.

Rodríguez, R., Matthei O., & Quezada, M. (1983). *Flora arbórea de Chile*. Editorial Universidad de Concepción. 408 p.

Rodríguez, V. (2005). *Estudio morfológico y anatómico de Gomortega keule (Mol.) Baillon y su potencial agrícola.* Seminario de Título. Facultad de Ciencias Agrarias y Forestales, Universidad Católica del Maule. 51 p.

Ruiz, H, (1998). *The journals of Hipo´lito Ruiz, Spanish botanist in Peru and Chile, 1777–1788*. Timber Press. 357 p.

Ruiz, H., & Pavon, J. (1798). *Systema vegetabilium florae, peruvianae et chilensis*. Typis Gabrielis de Sancha. 455 p.

San Martín, J., & Donoso, C. (1996). Floristic structure and human impact on the Maulino Forest of Chile. In: Armesto, J.J., Villagrán, C., & Arroyo, M.K. (eds.). *Ecologı´a de los bosques nativos de Chile* (pp. 153–168). Editorial Universitaria.

Schultes, R.E. (1976). *Hallucinogenic plants* (pp. 150–151). Golden Press; Western Publishing Company, Inc.

Schultes, R.E. (1983). Peruvian and Chilean psychoactive plants mentioned in Ruiz's Relación (1777–1788). *Journal of Psychoactive Drugs*, *15*, 303–312.

Simirgiotis, M. J., Ramirez, J. E., Hirschmann, G. S., & Kennelly, E. J. (2013). Bioactive coumarins and HPLC-PDA-ESI-ToF-MS metabolic profiling of edible queule fruits (*Gomortega keule*), an endangered endemic Chilean species. *Food Research International*, *54*(1), 532–543.

Torrejón, F., & Cisternas, M. (2003). Early environmental impact in the Araucania territory inferred from Spanish chronicles and historiographical studies. *Bosque*, *24*, 45–55.

Vilegas, J. H., & Lanças, F. M. (1995). HRGC-MS identification of low-polarity compounds from *Gomortega keule*. *Journal of High Resolution Chromatography*, *18*(2), 129–130.

Villegas, P., Le Quesne, C., & Lusk, C.H. (2003). Structure and dynamics of a population of *Gomortega keule* (Mol.) Baillon in old-growth Valdivian rain forest in south-central Chile. *Gayana Botanica*, *60*, 107–113.

27 Catnip (*Nepeta cataria* L. Lamiaceae)

Belkıs Muca Yiğit, İlker Çinbilgel, Sefa Gözcü, Ahmet Zafer Tel, and Ahmad Ali

27.1 INTRODUCTION

Plants around the world serve many needs of people. They are utilized for various purposes such as food, pharmaceuticals, stuff, cosmetics, dyes, and ornamentation. Many plants, whose species have been identified and whose intended use is known, contribute to the country's economy. There are 405 plant families and 14,559 plant genera accepted today. The Lamiaceae family is one of the richest plant families in the world in terms of the number of species. Lamiaceae family is currently represented by 245 genera and 7,886 species (The Plant List, 2023). Many species belonging to the Lamiaceae family possess aromatic and medicinal properties. *Nepeta*, a genus of the Lamiaceae family, is an important and valuable genus because of its properties. (Herron, 2003; The Plant List, 2023).

Generally, *Nepeta* species are known catmint or catnip but 'catnep', 'catmint', 'catrup', 'catwort', 'nip', 'nep', and 'field balm' names are also used (Grognet, 1990; Herron, 2003; Watts, 2007). There are a lot of traditional uses for *Nepeta* species. It is known that species belonging to the genus *Nepeta* are used for healing purposes, as ornamental plants, as insecticide (Reichert, 2019) or for food purposes (making sauces, soups, or stews). There are different medicinal uses in different countries such as India, Pakistan, Nepal, Iran, China, and Türkiye (Baytop, 1999; Chin et al., 2005; Mall & Tripathi, 2015). For example, Chinese people use *Nepeta cataria* for dermatitis, chronic tonsillitis, dysentery, and measles diseases. *Nepeta cataria* is also traditionally employed in folk medicine for the treatment of painful swellings by English people. Indian people use *N. ciliaris* for antipyretic, upper respiratory tract diseases. Turkish people, like Indians, use *Nepeta* species for upper respiratory tract diseases. (Baytop, 1999; Chin et al., 2005; Ghendov et al., 2013; Salehi et al., 2018; Sharma et al. 2019; Sharma et al., 2021).

Nepeta cataria is known as 'catmint' or 'catnip', similar to the other *Nepeta* species. Similar to other genera within the Lamiaceae family, the *Nepeta* genus also possesses essential oil. *Nepeta* species have a rich essential oil consisting of monoterpenes, sesquiterpenes, flavonoids, and steroids. The essential oil of *Nepeta cataria* is produced in the leaves and flowers of the plant (Reichert, 2019). Essential oil and secondary metabolites (Nepetalactones, linalool, nerol, germacrene, nepeticin, rosmarinic acid, apigenin, stigmasterol, etc.) containing these compounds are used for the treatment of various diseases (weakness of the heart, scorpion bites, gum disease, liver diseases, etc.) (Tucker and Tucker, 1988; Zenasni et al., 2008; Ghendov et al., 2013; Sharma et al., 2021). *Nepeta cataria* has a lot of chemical components but the most important of these is nepetalactone. Catnip plants' essential oil has 70–99% nepetalactone. Despite nepetalactone being the main component of catnip, its most active constituent is a metabolic product known as nepetalic acid (Grognet, 1990). In addition, the thymol substance contained in catnip can be used as a fungicide when it is extracted, and the leaves of *Nepeta cataria* can be used to regulate menstruation periods (Chin et al., 2005).

It is known that some plants containing volatile oils contain certain chemicals in their oil composition with hallucinogenic effects. Among the members of the Lamiaceae family, there are species that possess such hallucinogenic effects. It is observed that the species *Nepeta cataria*, found within the *Nepeta* genus, also has these hallucinogenic effects due to certain chemicals it contains. The cause of these effects is especially related to nepetalactone (Jackson & Reed, 1969; Blackburn, 1976). It is known that *Nepeta cataria* is especially effective on cats (Tucker and Tucker, 1988). For this reason, this species is called 'catnip' or 'catmint'. Moreover, some research has shown that hallucinogenic effects of catnip have been used to treat people's psychological diseases recently. It is used for nervous disorders, headaches, insomnia, or hysteria (Grognet, 1990; Chin et al., 2005; Luo, 2019; Lungoci et al., 2022).

Although studies have increased in recent years, more studies are needed on the hallucinogenic effects of catnip on humans. The current chapter presents an overview of the findings found about the botany, ecology, traditional uses, phytochemical, and hallucinogenic effects of *Nepeta cataria* L.

27.2 SCIENTIFIC CLASSIFICATION

Nepeta cataria L. Sp. Pl.2: 570 (1753) is a perennial herbaceous plant belonging to the *Nepeta* genus and Lamiaceae (Labiatae) family (IPNI, 2023; ITIS, 2023; POWO, 2023).

Kingdom: Plantae
Subkingdom: Viridiplantae
Infrakingdom: Streptophyta
Superdivision: Embryophyta
Division: Tracheophyta

DOI: 10.1201/9781003460336-27

Subdivision: Spermatophytina
Class: Magnoliopsida
Superorder: Asteranae
Order: Lamiales
Family: Lamiaceae
Genus: *Nepeta*
Species: *Nepeta cataria* L.

27.3 VERNACULAR NAMES

The name '*Nepeta*' was first used by Rivinus in 1690, and this name was later used for the plants that were known as '*Mentha cataria*' (Ubera & Valdes, 1983). The term 'catmint' is coined from the potent allure it typically holds for most feline creatures (Ibrahim et al., 2017).

Arabic: Qatram, Fafdhang gabali, Hhash. El hirr. حشيشة الهر, قطرم, فافذنج جبلى Armenian: կատվախոտ, Azerbaijani: Pişik Nanesi, Bal Otu Chinese: jing jie, 荆芥, 猫薄荷 Danish: Katteurt Dutch: Kattenkruid Engilish: catmint, catnip, Nep. (Népète) French: Cataire, Herbe aux chats, Menthe de German: Katzenminze, Echtes katzenkraut, Nept. Italian: Cataria commune, Erba gattaja, Erba dei gatti, Erba Kurdish: Beybuna kitikan, Bizbizok, Belaziz, Qerengoz, Giyakunci Norwegian: Norveççe: Kattmynte Persian: Punesa Polish: Kocimiętka Russian: КОТОВНИК, Кошачья мята Spanish: Hierba gatera, Menta de gato Swedish: Kattmynta Turkish: Kedi nanesi, Pisik out, Nezle Otu (Britten & Holland, 1886; Baytop, 1994; Baytop, 1999; Watts, 2007; Formisano et al., 2011; Fırat, 2013; Luo, 2019)

27.4 SYNONYMS

Nepeta cataria L. includes 17 synonyms (Table 27.1) (The Plant List, 2023, Checklistbank, 2023; Bizimbitkiler, 2023; WFO, 2023; IPNI, 2023):

TABLE 27.1
Synonyms of *N. cataria*

No	Synonim Names	Reference
1	*Calamintha albiflora* Vaniot	Bull. Acad. Int. Géogr. Bot. 14: 181 (1904)
2	*Cataria tomentosa* Gilib.	Fl. Lit. Inch. 1: 78 1782
3	*Cataria vulgaris* Gaterau	Descr. Pl. Montauban: 105 (1789)
4	*Glechoma cataria* (L.) Kuntze	Revis. Gen. Pl. 2: 518 (1891)
5	*Glechoma macrura* (Ledeb. ex Spreng.) Kuntze	Revis. Gen. Pl. 2: 518 (1891)
6	*Nepeta americana* Vitman	Summa Pl. 3: 419 1789
7	*Nepeta bodinieri* Vaniot	Bull. Acad. Int. Géogr. Bot. 14: 172 (1904)
8	*Nepeta cataria* var. *citriodora* (Dumort.) Lej.	Rev. Fl. Spa: 114 (1824)
9	*Nepeta ceretana* Sennen	Treb. Inst. Catalana Hist. Nat. 15: 29 (1931)
10	*Nepeta citriodora* Dumort	Fl. Belg.: 47 (1827)
11	*Nepeta laurentii* Sennen	Bol. Soc. Ibér. Ci. Nat. 32: 20 (1933 publ. 1934)
12	*Nepeta macrura* Ledeb. ex Spreng.	Syst. Veg. 2: 729 (1825)
13	*Nepeta minor* Mill.	Gard. Dict. ed. 8: 2 (1768)
14	*Nepeta mollis* Salisb.	Prodr. Stirp. Chap. Allerton: 78 (1796)
15	*Nepeta ruderalis* Boiss.	Fl. Orient. 4: 643 (1879), [yasd. ad / nom. illeg.]
16	*Nepeta tomentosa* Vitman	Summa Pl. 3: 419 (1789)
17	*Nepeta vulgaris* Lam.	Fl. Franç. 2: 398 1779

27.5 BOTANICAL DESCRIPTION

Nepeta cataria is a perennial and semi-shrub plant belonging to the Cataria section. It has an erect stem, with a stem height ranging from 50 to 130 cm. The upper parts of the stem are branched. The leaves are ovate, with leaf dimensions of 2–6 × (1) 2–4 cm. Both leaf surfaces are slightly flattened and pilose, with the lower surface being grayish-green and more densely glandular and sessile. The leaves have serrated margins and are cordate or nearly truncate in shape. Leaf stalks can extend up to 2 cm. Both the upper and verticillate leaves have stalks. The inflorescence is a broad panicle. The verticillates are closely spaced above, with a few verticillates separate at the base. The flowers are numerous (Bedevian, 1936). The bracteoles are linear-oblong and notably shorter than the calyx tube. The calyx tube is nearly curved, slightly inclined or not, densely pilose, short-soft-haired, and densely glandular, with a size of (4–4.5) 5–6.5 mm. The calyx teeth are linear-lanceolate and measure (1) 1.5–2.75 mm. The corolla measures (5) 5.5–7.5 mm and is white, occasionally with blue-purple spots on the lower lip. It slightly exceeds the calyx teeth, and there are dense glandular structures on the outer parts of both the upper and lower lips. The seeds are broadly ellipsoid, oblong, approximately 1.5 × 1 mm in size, with a dull matte appearance. Tubercles are not entirely visible at the apex, and the surface is undulated with a flat areole. *Fl.: May-August.* Fr: *July-Sept.* (Linnaeus, 1753; George, 1933; Shishkin and Yuzepchuck, 1954; Davis, 1982; Rechinger, 1982).

Nepeta nuda is the closest species to *Nepeta cataria* but *N. cataria* is different from *N. nuda* with some features. Their differences relate to the calyx tube and petals (Pakistan Flora, 2023). The calyx tube of *N. cataria* is curved and its leaves are petiolate. *Nepeta cataria* is described from Europe. The type specimen of *N. cataria* was described by Linne (Figure 27.1).

27.6 ORIGIN AND GEOGRAPHIC DISTRIBUTION

Members of the Lamiaceae family can grow in various regions and altitudes, ranging from the Himalayas to Southwest Asia, Australia, Africa, and even the Americas,

FIGURE 27.1 *Nepeta cataria* specimen from the Linnaean Herbarium (http://linnaeus.nrm.se/botany/fbo/n/nepet/nepecat.html.en).

except for a few regions (Dirmenci, 2003). The origin of *Nepeta* genus is the western region of Asia on the world. The *Nepeta* genus has a wide distribution in Southwest Asia, India, China, Europe, North Africa, Saudi Arabia, and North and Central America. (Dirmenci, 2003; Lungoci et al., 2022). There are two regions with the highest diversity and abundance of *Nepeta* taxa: Southwestern Asia, notably Türkiye and Iran, and the Western Himalayas, encompassing the Hindu Kush (China Flora, 2023). Iran serves as one of the primary centers of origin, hosting 67 catmint species, with 53% of them being endemic (Dirmenci et al., 2004; Salehi et al., 2018).

A type sample of *Nepeta cataria* was described from Europe in London (Shishkin & Yuzepchuck, 1954). *N. cataria*, a highly renowned and extensively studied species within this genus, holds significant importance. It is indigenous to a wide range of habitats, spanning from coastal areas to alpine zones across Europe, and it thrives in arid roadside locations as well as mountainous regions of Asia, with altitudes reaching up to 1500 meters (Dirmenci, 2003; Salehi et al., 2018).

Nepeta cataria, which belongs to the Lamiaceae family, originates from Eastern and Southern Europe, but can also be found in the Middle East, China, and North America (Figure 27.2). The majority of *Nepeta* species are endemic, particularly in Southwestern Asia, with a focus on Türkiye and Iran (Dirmenci, 2003; Dirmenci et al., 2004; Ibrahim et al., 2017; Ibrahim et al., 2022). According to the literature records, the distribution of *Nepeta cataria* in the world can be listed as follows: Afghanistan, Alabama, Alaska, Albania, Austria, Belarus, Belgium, British Columbia, Bulgaria, California, Central European Russia, China, East European Russia, Finland, France, Georgia, Germany, Greece, Iran, Iraq, Italy, Japan, Kazakhstan, Kirgizstan, Korea, Lebanon-Syria, Morocco, North European Russia, Northwest European Russia, Norway, Pakistan, Pennsylvania, Poland, Portugal, Romania, Saskatchewan, Sicilia, South European Russia, Spain, Sweden, Switzerland, Tadzhikistan, Transcaucasus, Türkiye, Turkmenistan, Ukraine, Utah, Uzbekistan, West Himalaya (Miller, 1759; Lamarck, 1778; Kuntze, 1893; Shishkin & Yuzepchuck, 1954; Davis, 1982; Ubera & Valdes, 1983; Dirmenci, 2003; Luo, 2019, https://www.catalogueoflife.org/data/taxon/46XJB)

In ancient times, the catnip plant used to be cultivated in gardens as an ornamental plant and for making herbal tea. The commencement of large-scale catnip production can be attributed to Canadians (Gomes et al., 2020; Ibrahim et al.,

FIGURE 27.2 *Nepeta cataria* habitus. (Photo: Fatih Yayla.)

2022). Following the discovery of its effects on cats when in a dried state, an increase in literature pertaining to its chemical composition has been observed. Particularly, certain studies have suggested that, due to its hallucinogenic effects, the plant replaced marijuana after the 1960s. (Grognet, 1990; Setzer, 2016; Gomes et al., 2020; Ibrahim et al., 2022). Scientific records regarding the cultivation of *Nepeta* species date back to the 1700s. In *The Gardeners Dictionary*, published in 1759, Miller describes his cultivation in various regions, including England, Sicily, Portugal, Spain, the Alps, Italy, France (Paris), and North America. He notes that the growth and development characteristics of *Nepeta* species vary due to the diverse conditions in each region (Miller, 1759; Aćimović et al., 2021). Studies conducted from the 1700s to the present have revealed various developmental characteristics of the plant. For example, when looking at the records of the cultivation of the plant in North America, it was noted that the plant was not suitable for agriculture due to its low oil yield (Gomes et al., 2020; Aćimović et al., 2021). In a study conducted under growing conditions in Egypt, the plant is reported to exhibit a favorable oil content. The study also suggests that it could potentially be a new source of income for Egypt (Ibrahim et al., 2017). Furthermore, it is known that various species belonging to the Nepeta genus are cultivated in countries such as Ukraine, Poland, and Iran (Ibrahim et al., 2017; Salehi et al., 2018; Aćimović et al., 2021).

27.7 ECOLOGY

Catnip is a plant species with a long flowering period. It is possible to see the flower of the plant from May to August. For this reason, it can be preferred for the beekeeping industry (Shishkin & Yuzepchuck, 1954; Formisano et al., 2011; Chevallier, 2016; Salehi et al., 2018; Ibrahim et al., 2022)

Catnip can grow wild at various altitudes, from sea level to alpine zones. In the natural habitats of each country, *Nepeta cataria* can display distribution across varying altitudinal ranges and diverse ecological conditions (de Géographie Botanique, 1904). In North America, it is distributed in dry roadside areas and mountainous regions. *N. cataria* can usually be encountered up to an altitude of 1500 m. Its flowers are typically harvested during the summer and autumn seasons (Formisano et al., 2011; Chevallier, 2016; Salehi et al., 2018). In Türkiye, it has been recorded to spread in fallow fields, field edges, and wetlands at altitudes ranging from 1200 to 1500 meters. It is known to bloom in July and August (Davis, 1982; Dirmenci, 2003). According to U.S.S.R. flora records, it is indicated to be distributed in meadows, forest edges, forests, shrubs, riverbanks, mostly places invaded by wild herbs, fields, orchards, gardens, ditches, and slopes of mid-mountain and subalpine zone meadows (Moench, 1794). Its flowering period is noted as June to July, while fruiting occurs in July (Shishkin & Yuzepchuck, 1954). In the Iranian flora, while the altitude range is not specified, it is mentioned to be distributed in Iraq (900–1500 m), Afghanistan (1080–2500 m), Pakistan (1100–1800 m), and Persia (1020–2500 m) (Rechinger, 1982). In the Kashmir Himalaya region, catnip is distributed within an altitude range of 1600–2000 m. In India, it is noted that the plant flowers between May and August, with fruiting occurring between July and September. It is reported to prefer dry habitats and vacant areas (Hassan, 2010; Hassan et al., 2011; Salehi et al., 2018). In the Chinese flora, *N. cataria* is stated to use shrublands or areas around houses as its habitat, with a distribution range of 0–2500 m. In China, it flowers between July and September, with fruiting occurring between September and October. Furthermore, it is mentioned that the cultivation of the plant takes place in the regions of Gansu, Guizhou, Henan, Hubei, Shaanxi, Shandong, Shanxi, Sichuan, Xinjiang, and Yunnan (www.eFloras.org, 2023). According to Pakistan flora records, the plant is noted to flower in June and July (www.eFloras.org, 2023).

While numerous catmint varieties demonstrate excellent adaptation to drought conditions in arid or semi-arid fields, they are also recognized as species that exhibit resistance to winter weather. Catnip is generally classified as an anthropophilic and nitrophilic species (G.k.ür, 2021). *N. cataria* is a widely distributed plant and it is not an overly selective species when it comes to soil. However, it thrives better in slightly alkaline and well-drained soils. Catnip prefers sunny locations over lightly shaded areas and thrives within a temperature range of 7–19°C. It requires an annual rainfall of 0.4–1.3 meters and a soil pH level ranging from 4.9

to 7.5 (Pădure, 2003). Catnip can be propagated from seeds in countries where it is grown. For instance, in Egypt, seed planting is typically conducted between October and April. Subsequently, seedlings are transplanted into pots in greenhouse (Salehi et al., 2018; Ibrahim et al., 2022).

27.8 PHYTOCHEMISTRY

Nepeta cataria L. (Catnip) is a plant that produces a variety of biologically bioactive compounds, including terpenes, steroids, phenolic acids, phenylethanoid glycosides, and flavonoids (Süntar et al., 2018; Sharma et al., 2019). To date, phenolic acid derivatives have been isolated from the *Nepeta cataria* L., including caffeic acid (Modnicki et al., 2007; Mihaylova et al., 2013), caffeoyltartronic acid (Snook et al., 1993), *p*-coumaric acid (Mihaylova et al., 2013), ferulic acid, gallic acid, cinnamic acid, rosmarinic acid, vanillic acid (Modnicki et al., 2007; Mihaylova et al., 2013), and syringic acid (Modnicki et al., 2007; Mihaylova et al., 2013; Sharma et al., 2019) (Table 27.2). Moreover, phenylethanoid glycosides have been isolated from *N. cataria* including teucrioside, verbascoside, teupolioside (Prescott et al., 2011) (Table 27.2).

Terpenes are aromatic compounds found in many plants that contribute to catnip's characteristic aroma. Those identified/isolated as major mooterpenes include ajugol (Sharma et al., 2019), geranyl acetate (Baranauskiene et al., 2003), citronellyl acetate (Baranauskiene et al., 2003), citronellol (Baranauskiene et al., 2003), geraniol (Baranauskiene et al., 2003), linalool (Baranauskiene et al., 2003), *β*-pinene, ascaridole, 4*aα*,7*α*,7*aα*-nepetalactone, *α*-copaene, *β*-bourbonene (Safaei-Ghomi et al., 2009), 4a*α*,7*α*,7a*β*-nepetalactone (Adiguzel et al., 2009; Safaei-Ghomi et al., 2009), 4a*α*,7*β*,7a*α*-nepetalactone (Adiguzel et al., 2009; Safaei-Ghomi et al., 2009), 4*a*-dihydronepetalactone (Regnier et al., 1967a), 4*β*-dihydronepetalactone (Regnier et al., 1967b), 5,9-dehydronepetalactone (Sastry et al., 1972), *β*-caryophyllene (Adiguzel et al., 2009; Safaei-Ghomi et al., 2009), (*z*)-*β*-farnesene (Safaei-Ghomi et al., 2009), myrcene, limonene, 1,8-cineole, camphor, pulegone, piperitone, thymol, carvacrol (Adiguzel et al., 2009), nepetoidin A, nepetoidin B (Grayer et al., 2003), epideoxyloganic acid (Murai et al., 1984), nepetalactone, epinepetalactone (Regnier et al., 1967a), iridomyrmecin (Bates et al., 1958) (Table 27.2).

Flavonoids are secondary metabolites found in many plants with antioxidant and anti-inflammatory, diuretic, and antimicrobial properties (Heim et al., 2002; Xiao et al., 2005). When the phytochemistry of catnip is examined in detail, the presence of secondary metabolites containing acacetin (Jamzad et al., 2003; Mišić et al., 2015), apigenin (Akbar & Akbar, 2020; Mišić et al., 2015), apigenin-7-*O*-glucoside (Akbar & Akbar, 2020), apigenin-7-*O*-D-glucuronide (Modnicki et al., 2007; Mišić et al., 2015), 7-*O*-methylapigenin (Jamzad et al., 2003), eriodictyol (Proestos et al., 2006), luteolin (Jamzad et al., 2003; Mišić et al., 2015), luteolin-7-*O*-glucoside, luteolin-7-*O*-glucouronide (Modnicki et al., 2007), cirsimaritin (Süntar et al., 2018; Sharma et al., 2019), chrysin, quercetin, kaempferol, naringenin, pinocembrin, galangin, pinocembrin (Mišić et al., 2015), leucanthogenin (Modnicki et al., 2007), and isothymusin (Sharma et al., 2019) draws attention. In addition, *α*-amyrin, *β*-amyrin, ursolic acid, *β*-sitosterol, campesterol (Klimek & Modnicki, 2004), aesculin (Mišić et al., 2015), and *β*-sitosterol 3-*O*-*β*-D-glucopyranoside (Klimek & Modnicki, 2004) triterpenes, which are known to be responsible for many activities, have also been reported to be isolated from catnip (Table 27.2).

27.9 ETHNOBOTANY

Nepeta cataria (catnip) is mostly in its wild state, and limited breeding efforts have been made to enhance the plant's

TABLE 27.2
Bioactive Compounds of *Nepeta cataria* L.

Catnip Constituents and Chemical Class	References
Phenolics: caffeic acid, caffeoyltartronic acid, *p*-coumaric acid, ferulic acid, gallic acid, cinnamic acid, rosmarinic acid, vanillic acid, and syringic acid	Snook et al., 1993; Modnicki et al., 2007; Mihaylova et al., 2013; Sharma et al., 2019
Monoterpenes: ajugol, geranyl acetate, citronellyl acetate, citronellol, geraniol, linalool, *β*-pinene, ascaridole, 4*aα*,7*α*,7*aα*-nepetalactone, *α*-copaene, *β*-bourbonene, 4a*α*,7*α*,7a*β*-nepetalactone, 4a*α*,7*β*,7a*α*-nepetalactone, 4*a*-dihydronepetalactone, 4*β*-dihydronepetalactone, 5,9-dehydronepetalactone, *β*-caryophyllene, (*z*)-*β*-farnesene, myrcene, limonene, 1,8-cineole, camphor, pulegone, piperitone, thymol, carvacrol, nepetoidin A, nepetoidin B, epideoxyloganic acid, nepetalactone, epinepetalactone, iridomyrmecin	Bates et al., 1958; Regnier et al., 1967a; Regnier et al., 1967b; Sastry et al., 1972; Murai et al., 1984; Baranauskiene et al., 2003; Grayer et al., 2003; Adiguzel et al., 2009; Safaei-Ghomi et al., 2009; Sharma et al., 2019
Flavonoids: acacetin, apigenin, apigenin-7-*O*-glucoside, apigenin-7-*O*-D-glucuronide, 7-*O*-methylapigenin, eriodictyol, luteolin, luteolin-7-*O*-glucoside, luteolin-7-*O*-glucouronide, cirsimaritin, chrysin, quercetin, kaempferol, naringenin, pinocembrin, galangin, pinocembrin, and isothymusin, leucanthogenin	Jamzad et al., 2003; Proestos et al., 2006; Modnicki et al., 2007; Mišić et al., 2015; Süntar et al., 2018; Sharma et al., 2019; Akbar & Akbar, 2020
Triterpenes: *α*-amyrin, *β*-amyrin, ursolic acid, *β*-sitosterol, campesterol, aesculin, and *β*-sitosterol 3-*O*-*β*-D-glucopyranoside	Klimek & Modnicki, 2004; Mišić et al., 2015
Phenylethanoid glycosides: teucrioside, verbascoside, teupolioside	Prescott et al., 2011

horticultural characteristics. Catnip is predominantly cultivated for use in the pet toy industry to attract cats safely and for decorative purposes. However, recent research has uncovered that essential oils derived from catnip also function as a highly effective insect repellent.

Dried/fresh aromatic flowering tops and leaves are used as flavoring agents, primarily in sauces and preparing foods (Sharma et al., 2019). Together with other medicinal herbs, the plant is traditionally used to prepare various formulations to treat mastitis, diarrhea, chronic rhinitis, measles, chronic tonsillitis, eczema, and swollen sore throat (Zhang et al., 2015). The infusion obtained from the leaves is used as a sedative and to treat diarrhea, asthma, bronchitis, and cough (Formisano et al., 2011; Belhaj et al., 2021). Moreover, the herb is traditionally used in English folk medicine to treat painful swellings (Sharma et al., 2019). In another study, an infusion made from the plan*t N. cataria* L. was used to treat ovarian cysts and reduce menstrual pains in Serbian women (Zlatkovic̋, et al., 2014; Chaachouay et al., 2021a, 2021b).

27.10 HALLUCINOGENIC EFFECTS

In 1924, Lewis Lewin defined five classes of drugs. Hallucinogens were classified under the category 'Phantastica' (Emboden, 1979). Hallucinogens are substances within the drug category that are claimed to be physiologically safe, non-addictive, and have therapeutic effects. The effects of hallucinogens can vary depending on the dosage. They have the ability to alter the perception of the mind, mood, and various cognitive processes. When hallucinogens are used, the perception of physical space can change, and sensations from sounds, colors, or touched objects can manifest unusually (Nichols, 2004).

When psychoactive substances are used in controlled and specific dosages, hallucinatory effects do not occur. For this reason, some researchers suggest that terms like 'psyche-loosening', 'psychedelics', or 'psychotomimetics' may be more accurate than the word 'hallucinogen' (Emboden, 1979; Vollenweider, 2022). However, some researchers argue against the use of the term 'psychotomimetics' for hallucinogens since they rarely produce schizophrenic effects (Schuckit, 2006).

Plants rich in psychoactive substances have been used by humans for a very long time, especially in ritual activities where their significance is significant. The reason for the use of such plants in rituals is their ability to alter the perception of reality. They have been used by priests, shamans, or wizards to alter the perception of time and space. Plants containing psychoactive substances with the oldest history of use include *Ephedra* and poppy plants (Czigány, 1980; De Rios & Cardenas, 1980; Mutlu, 2018; Chaachouay & Zidane, 2021; Karaköz, 2022).

In the 1950s, serotonin (5-HT) and lysergic acid diethylamide (LSD-25) were discovered. During the same period, research related to the brain suggested the potential serotonergic effects of LSD and other hallucinogens (Nichols, 2004). Today, hallucinogenic substances have become a broad class that includes pharmacological substances such as cannabinoids, N-methyl-D-aspartate (NMDA) antagonists, anticholinergic agents, ecstasy (MDMA; 3,4-methylenedioxymethamphetamine), and many others (Nichols, 2004). According to the European Monitoring Center for Drug Addiction (EMCDDA) report in 2016, over 620 psychoactive substances were monitored, and in 2021, this number increased to 880. Among these substances, MDMA holds a significant presence (Palaska & Özkanlı, 2017; European Drug Report, 2022). When we examine the existing body of academic research, there is a limited presence of publications demonstrating the hallucinogenic effects of catnip (*Nepeta cataria*) on humans. Instead, most studies have primarily focused on its effects on cats, chick, and rodents. There are publications available that discuss the hallucinogenic properties of the plant itself or its extracts (Morris III, 1978; Sherry & Hunter, 1979; Grognet, 1990). The most well-known hallucinogenic effect of catnip is observed in cats. Humans, on the other hand, are known to use it as a tea for flavoring meals or for its calming properties. It is known that catnip contains substances like LSD or atropine, which can induce hallucinogenic effects. Furthermore, in the 1960s, the substitution of catnip for marijuana was linked to the plant chemicals in catnip being similar in nature to the cannabinoids found in marijuana (Schuckit, 2006). Similarly, the presence of PCP (Phencyclidine)-like compounds in the biochemical content of catnip enhances its hallucinogenic effects (Schuckit, 2006; Palaska & Özkanlı, 2017). Although the number of studies on the hallucinogenic effect and mechanism of *Nepeta cataria* is limited (Schultes, 1970), it is useful to know its different effects on the human body.

27.10.1 Central Nervous System Effects

Essential oils of related *Nepeta* species (*N. caessarea*) were shown to have strong analgesic and sedative effects. Since naloxone blocked the effect, an opioid mechanism is suspected (Soulimani et al., 2008). It was determined that nepelactone had an agonistic effect at a specific opioid receptor subtype (Aydin et al., 1998). The euphoric effects of inhaling dried catnip leaves have been reported to parallel those of cannabis (Soulimani et al., 2008).

27.10.2 Antispasmodic and Myorelaxant Effects

The bronchodilator and antispasmodic activities of *Nepeta cataria* essential oil were evaluated on guinea-pig trachea and rabbit jejunum. In the study, *Nepeta cataria* was reported to have spasmolytic and muscle relaxant activities mediated through dual inhibition of calcium channels and PDE (Gilani et al., 2009). This study explains the reason why *N. cataria* is used as a folk medicine in gastrointestinal and respiratory tract diseases (Formisano et al., 2011; Zhang et al., 2015).

27.10.3 Antimicrobial Effect

The effect of *N. cataria* extract on 44 *Staphylococcus aureus* strains was studied by Nostro et al. (2001). The study examined the effect of subminimum inhibitory concentrations on coagulase, deoxyribonuclease, thermonuclease, and lipase production, as well as in vitro adhesion. Inhibition of deoxyribonuclease, thermonuclease, and lipase was detected, and a reduction in adherence was shown (Nostro et al., 2001).

27.10.4 Repellent and Insecticide Effect

The essential oil from *N. cataria* was tested for its insecticidal effects on *Spodoptera littoralis* larvae, and it was discovered to be very toxic, with an LC_{50} value of 10.0 ml/m^3, or 50%, of fatal levels (Pavela, 2005). In another study, the major iridoid constituents and the essential oil of *N. cataria* were assessed on their capacity to repel *Anopheles gambiae*, *Culex quinquefasciatus*, *Rhipicephalus appendiculatus*, and *Dermanyssus gallinae* in which pathojen. The essential oil has been reported to moderate repellent activity (Birkett et al., 2011).

27.10.5 Analgesic and Anti-Inflammatory Effect

N. cataria is a plant with traditional usage in the treatment of inflammation. Calcineurin, a critical role in the regulation of T-cell-mediated inflammation, was reported to be suppressed by *N. cataria* extract. Secondary metabolites of teucrioside, verbascoside, teupolioside, which are phenylethanoid derivatives isolated from *N. cataria*, have been reported to have high analgesic and anti-inflammatory activity (Prescott et al., 2011).

27.10.6 Effects on Sexual Behavior

N. cataria has been utilized in the formulation of daily necessities such teas and infusions for a very long time in North American culture as a popular medication. The effect of *N. cataria* on sexual behavior and penile erection was evaluated in male rats fed a diet rich in *N. cataria* leaves. Sexual behavior apomorphine-induced penile erection and motor coordination were monitored 15 minutes after the end of the 4-hour period of feeding with catnip. At the end of this process, it was determined that penile erection increased in male rats fed catnip leaves (Bernardi et al., 2011).

27.11 CONCLUSION AND FUTURE PERSPECTIVES

In conclusion, *Nepeta cataria* L. is a plant species with hallucinogenic effects, the mechanism of which is not fully understood. The fact that the plant exhibits natural distribution under different ecological conditions and altitudes suggests its potential for easy cultivation. The production of the plant is primarily aimed at filling cat toys to induce relaxation in cats.

When examining the activities and effects of *Nepeta cataria*, we observe a range of properties, including antioxidant, antidiabetic, spasmolytic, sexual activity enhancement, anti-nociceptive, anti-inflammatory activity, antimicrobial activity, insecticidal activity, antiparasitic activity, allelopathic activity, as well as antidepressant and sedative activity. It is evident from research that studies related to its antidepressant activity have primarily focused on cats and mice. However, there is limited research on its calming effects in humans.

Future studies on *Nepeta cataria* should explore its potential use in the composition of tranquilizing medications for psychological treatments. It is particularly essential to compare the chemical content of this plant with that of other psychoactive plant species.

LIST OF ABBREVIATIONS

EMCDDA	The European Monitoring Center for Drug Addiction
IPNI	International Plant Name Index
ITIS	Integrated Taxonomic Information System
L.	Linneaus
LSD	Lysergic Acid Diethylamide
MDMA	3,4-methylenedioxymethamphetamine
PCP	Phencyclidine
Sp.	Species
WFO	World Flora Online

REFERENCES

Aćimović, M., Zeremski, T., Kiprovski, B., Brdar-Jokanović, M., Popović, V., Koren, A., & Sikora, V. (2021). *Nepeta cataria*–cultivation, chemical composition and biological activity. *Journal of Agronomy, Technology and Engineering Management (JATEM)*, *4*(4), 620–634.

Adiguzel, A., Ozer, H., Sokmen, M., Gulluce, M., Sokmen, A., Kilic, H., ... Baris, O. (2009). Antimicrobial and antioxidant activity of the essential oil and methanol extract of Nepeta cataria. *Polish Journal of Microbiology*, *58*(1), 69–76.

Akbar, S., & Akbar, S. (2020). *Nepeta cataria* L.(Lamiaceae) (Syns.: *N. bodinieri* Vaniot; *N. ceretana* Sennen; *N. citriodora* Dumort.; *N. laurentii* Sennen). *In: Handbook of 200 Medicinal Plants. Springer, Cham*, 1279–1285. https://doi.org/10.1007/978-3-030-16807-0_134

Aydin, S., Beis, R., Öztürk, Y., & Başer, K. H. C. (1998). Nepetalactone: A new opioid analgesic from *Nepeta caesarea* Boiss. *Journal of Pharmacy and Pharmacology*, *50*(7), 813–817.

Baranauskiene, R., Venskutonis, R. P., & Demyttenaere, J. C. (2003). Sensory and instrumental evaluation of catnip (*Nepeta cataria* L.) aroma. *Journal of Agricultural and Food Chemistry*, *51*(13), 3840–3848.

Bates, R. B., Eisenbraun, E., & McElvain, S. (1958). The configurations of the nepetalactones and related compounds. *Journal of the American Chemical Society*, *80*(13), 3420–3424.

Baytop, T. (1994). *Türkçe bitki adları sözlüğü* (Vol. 578, p. 227). Turk Dil Kurumu.

Baytop, T. (1999). *Türkiyede bitkiler ile tedavi* (p. 444). İstanbul Eczacılık Fakültesi Yayınları.

Bernardi, M. M., Kirsten, T. B., Lago, J. H. G., Giovani, T. M., & de Oliveira Massoco, C. (2011). *Nepeta cataria* L. var. *citriodora* (Becker) increases penile erection in rats. *Journal of Ethnopharmacology*, *137*(3), 1318–1322.

Bedevian, A. K. (1936). *Illustrated polyglottic dictionary of plant names*. Argus & Papazian Press. (ISBN: 978-625-405-323-8)

Belhaj, S., Chaachouay, N., & Zidane, L. (2021). Ethnobotanical and toxicology study of medicinal plants used for the treatment of diabetes in the High Atlas Central of Morocco. *Journal of Pharmacy & Pharmacognosy Research*, *9*(5), 619–662. https://doi.org/10.56499/jppres21.1014_9.5.619

Birkett, M. A., Hassanali, A., Hoglund, S., Pettersson, J., & Pickett, J. A. (2011). Repellent activity of catmint, *Nepeta cataria*, and iridoid nepetalactone isomers against Afro-tropical mosquitoes, ixodid ticks and red poultry mites. *Phytochemistry*, *72*(1), 109–114.

Blackburn, T. (1976). A query regarding the possible hallucinogenic effects of Ant ingestion in South-Central California. *The Journal of California Anthropology*, *3*(2), 78–81.

Britten, J. F.S., Holland, R. (1886) *A dictionary of English plant-names*. For The English Dialect Society, Trübner & CO.

Chaachouay, N., Douira, A., & Zidane, L. (2021a). Herbal Medicine Used in the Treatment of Human Diseases in the Rif, Northern Morocco. *Arabian Journal for Science and Engineering*, *47*, 131–153. https://doi.org/10.1007/s13369-021-05501-1

Chaachouay, N., Douira, A., & Zidane, L. (2021b). COVID-19, prevention and treatment with herbal medicine in the herbal markets of Salé Prefecture, North-Western Morocco. *European Journal of Integrative Medicine*, 42, 101285. https://doi.org/10.1016/j.eujim.2021.101285

Chaachouay, N., & Zidane, L. (2021). The Symbolic Efficacy of Plants in Rituals and Socio-Religious Ceremonies in Morocco, Northwest of Africa. *Journal of Religious & Theological Information*, *21*(1–2), 34–53. https://doi.org/10.1080/10477845.2021.1942433

Chevallier, A. (2016). *Encyclopedia of herbal medicine* (3rd ed.). Dorling Kindersley Limited.

Chin, K. L., Qi, Y., Berhane, M., & Simon, J. E. (2005). Biological characteristics, nutritional and medicinal value of catnip, Nepeta cataria. *Agricultural Research and Extension Center*, 70813, 225.

Czigány, L. G. (1980). The use of hallucinogens and the shamanistic tradition of the Finno-Ugrian people. *The Slavonic and East European Review*, 58(2), 212–217.

Davis, P. H., 1982. *Flora of Turkey and East Aegean Islands* (p. 7). University Press.

de Géographie Botanique, A. I. (1904). Bulletin de l'Académie Internationale de Géographie Botanique (Ancien Monde Des Plantes). Institut de Bibliographie V 14.

De Rios, M. D., & Cardenas, M. (1980). Plant hallucinogens, shamanism and Nazca ceramics. *Journal of Ethnopharmacology*, 2(3), 233–246.

Dirmenci, T. (2003) Taxonomic and chemical investigations on Nepeta L. (Lamiaceae) species growing in Turkey, Balikesir University, Institute of Science, Department of Biology Education, Ph. D. Thesis, p. 202.

Dirmenci, T., Yildiz, B., & Tümen, G. (2004). Threatened categories of four Nepeta L.(Lamiaceae) species endemic to the East Anatolia. *Turkish Journal of Botany*, 28(1), 221–226.

Emboden, W. (1979). *Narcotic plants* (pp.44–132). Macmillan.

European Drug Report (2022). Trends and developments. https://www.emcdda.europa.eu/publications/edr/trends-developments/2022_en Access on: 01.09.2023. ISBN 978-92-9497-764-9

Fırat, M., (2013) Ferhenga Naven Riweken Bi Kurdi, Sitav Yayınları, Van, p. 335 (ISBN: 978-605-63744-0-1).

Formisano, C., Rigano, D., & Senatore, F. (2011). Chemical constituents and biological activities of *Nepeta* species. *Chemistry & Biodiversity*, 8(10), 1783–1818.

George E. (1933) *Post flora of syria palestine and sinai A handbook of the flowering plants and ferns, Native and Naturalized from the Taurus to Ras Muhammad and from the Mediterranean Sea to the Syrian Desert* (Volume II, p. 363). American University of Beirut Publications The Faculty of Arts and Sciences.

Gilani, A. H., Shah, A. J., Zubair, A., Khalid, S., Kiani, J., Ahmed, A., … Ahmad, V. U. (2009). Chemical composition and mechanisms underlying the spasmolytic and bronchodilatory properties of the essential oil of *Nepeta cataria* L. *Journal of Ethnopharmacology*, *121*(3), 405–411.

Ghendov, V., Danila, D., & Ciocarlan, N. (2013). Habitat assessment of Nepeta L. species with anti-inflammatory potential in the flora of Republic of Moldova. In *Biotehnologii avansate–realizări şi perspective* (pp. 158–158). Chişinău, Moldova.

Gomes, N.E., Allen, K., Jaworski, K., Zorde, M., Lockhart, A., Besancon, T., Brown, T., Reichert, W., Wu, Q., Simon, J.E. (2020). In *Medicinal and aromatic plants of North America*. Á. Máthé (Ed.). Springer International Publishing.

Grayer, R. J., Eckert, M. R., Veitch, N. C., Kite, G. C., Marin, P. D., Kokubun, T., … Paton, A. J. (2003). The chemotaxonomic significance of two bioactive caffeic acid esters, nepetoidins A and B, in the Lamiaceae. *Phytochemistry*, *64*(2), 519–528.

Grognet, J. (1990). Catnip: Its uses and effects, past and present. *The Canadian Veterinary Journal*, 31(6), 455.

Gökçür, E. (2021) *Azerbaycan Bitki Adları Sözlüğü*. Palet Yayınları ISBN: 978-625-7675-38-3)

Hassan, T., 2010. Systematic studies on genus Nepeta L Lamiaceae in Kashmir Himalaya, Ph. D. Thesis, Post Graduate Department of Botany Faculty of Bıologıcal Scıence Unıversıty of Kashmır Srınagar.

Hassan, T., Dar, G. H., & Khuroo, A. A. (2011). Taxonomic status of genus Nepeta L.(lamiaceae) in kashmir himalaya, India.

Heim, K. E., Tagliaferro, A. R., & Bobilya, D. J. (2002). Flavonoid antioxidants: Chemistry, metabolism and structure-activity relationships. *The Journal of Nutritional Biochemistry*, *13*(10), 572–584.

Herron, S. (2003). Catnip, *Nepeta cataria*, a morphological comparison of mutant and wild type specimens to gain an ethnobotanical perspective. *Economic Botany*, 57(1), 135–142.

Ibrahim, M. E., El-Sawi, S. A., & Ibrahim, F. M. (2017). Nepeta cataria L, one of the promising aromatic plants in Egypt: seed germination, growth and essential oil production. *Journal of Materials and Environmental Science*, 8, 1990–1995.

Ibrahim, Z. S., Qayssar, R., Sardar, S., Senjawi, L. S. M., Sleman, E., Fazil, H., & Porwal, O. (2022). Catnip (Nepeta cataria L.): Recent advances in pharmacognosy, cultivation, chemical composition and biological activity. *Journal of Drug Delivery and Therapeutics*, 12(4-S), 254–263.

Jackson, B., & Reed, A. (1969). Catnip and the alteration of consciousness. *JAMA*, 207(7), 1349–1350.

Jamzad, Z., Grayer, R. J., Kite, G. C., Simmonds, M. S., Ingrouille, M., & Jalili, A. (2003). Leaf surface flavonoids in Iranian species of *Nepeta* (Lamiaceae) and some related genera. *Biochemical Systematics and Ecology*, 31(6), 587–600.

Karaköz, K.(2022) Eski Mezopotamya ve Anadolu Ritüellerinde Psikoaktif Bitkiler. Sosyal Bilimler Araştırma Dergisi (KOUSBAD) 15, Bahar, pp 31–68.

Klimek, B., & Modnicki, D. (2004). Terpenoids and sterols from *Nepeta cataria* L. var. citriodora (Lamiaceae). *Acta Poloniae Pharmaceutica - Drug Research*, 62(3), 231–235.

Kuntze, O. (1893). Revisio generum plantarum vascularium omnium atque cellularium multarum secundum Leges nomenclaturae internationales cum enumeratione plantarum exoticarum in itinere mundi collectarum: Pars I-[III]. A. Felix.

Lamarck, P. M. C. (1778). Flore française ou description succincte de toutes les plantes qui croissent naturellement en France, disposées selon une nouvelle méthode d'analyse, Imprimerie Royale, Paris Vol 3, 398.

Linnaeus, C. (1753). Species Plantarum Exhibentes Plantas Rite Cognitas ad Genera Relatas. Tomus II, 571–572.

Lungoci, C., Motrescu, I., Filipov, F., Jitareanu, C. D., Teliban, G. C., Ghitau, C. S., ... & Robu, T. (2022). The impact of salinity stress on antioxidant response and bioactive compounds of Nepeta cataria L. *Agronomy*, 12(3), 562.

Luo, J. (2019). Characterization of the complete chloroplast genome of *Nepeta cataria* L. of the herbal medicine from China and phylogenetic relationships. *Mitochondrial DNA Part B*, 4(1), 1452–1454.

Mall, T. P., & Tripathi, S. C. (2015). Nepeta herb-a nutrimental potent ethnomedicinal plant from north western tarai forests of up India. *World Journal of Pharmaceutical Research*, *4*(11), 123–1630.

Mihaylova, D., Georgieva, L., & Pavlov, A. (2013). *In vitro* antioxidant activity and phenolic composition of *Nepeta cataria* L. extracts. *International Journal of Agricultural and Technology*, *1*(4), 74–79.

Miller, P. (1759). *The gardeners dictionary*. P. Miller.

Mišić, D., Šiler, B., Gašić, U., Avramov, S., Živković, S., Nestorović Živković, J., ... Tešić, Ž. (2015). Simultaneous UHPLC/DAD/(+/−) HESI–MS/MS analysis of phenolic acids and nepetalactones in methanol extracts of *Nepeta* species: A possible application in chemotaxonomic studies. *Phytochemical Analysis*, *26*(1), 72–85.

Modnicki, D., Tokar, M., & Klimek, B. (2007). Flavonoids and phenolic acids of *Nepeta cataria* L. var. citriodora (Becker) Balb. (Lamiaceae). *Acta Poloniae Pharmaceutica*, *64*(3), 247–252.

Moench, C. (1794). Methodus plantas horti botanici et agri marburgensis. *In officina nova libraria academiae, Marburgi*, 2, 387–388.

Morris III, R. (1978). The effect of catnip, an hallucinogen, on the neurotransmitters serotonin, norepinephrine and dopamine (Doctoral dissertation).

Murai, F., Tagawa, M., Damtoft, S., Jensen, S. R., & Nielsen, B. J. (1984). (1R, 5R, 8S, 9S)-Deoxyloganic acid from *Nepeta cataria*. *Chemical and Pharmaceutical Bulletin*, *32*(7), 2809–2814.

Mutlu, E. (2018). Madde bağımlılığı ve insan evrimi. *Bağımlılık Dergisi*, 19(1), 17–22.

Nichols, D. E. (2004). Hallucinogens. *Pharmacology & Therapeutics*, 101(2), 131–181.

Nostro, A., Cannatelli, M. A., Crisafi, G., & Alonzo, V. (2001). The effect of Nepeta cataria extract on adherence and enzyme production of Staphylococcus aureus. *International Journal of Antimicrobial Agents*, *18*(6), 583–585.

Pădure, I. M. (2003). Chorological and ecological aspects of *Nepeta cataria* L.(Lamiaceae, Nepetoideae) In Romania. Scientifically Papers, 26.

Palaska, E., & Özkanlı, F. (2017). Yeni sentetik psikoaktif maddeler ve bonzai gerçeği. *Türk Farmakope Dergisi*, 2(2), 19–43.

Pavela, R. (2005). Insecticidal activity of some essential oils against larvae of Spodoptera littoralis. *Fitoterapia*, *76*(7–8), 691–696.

Prescott, T. A., Veitch, N. C., & Simmonds, M. S. (2011). Direct inhibition of calcineurin by caffeoyl phenylethanoid glycosides from Teucrium chamaedrys and Nepeta cataria. *Journal of Ethnopharmacology*, *137*(3), 1306–1310.

Proestos, C., Boziaris, I., Nychas, G.-J., & Komaitis, M. (2006). Analysis of flavonoids and phenolic acids in Greek aromatic plants: Investigation of their antioxidant capacity and antimicrobial activity. *Food Chemistry*, *95*(4), 664–671.

Regnier, F., Eisenbraun, E., & Waller, G. (1967a). Nepetalactone and epinepetalactone from *Nepeta cataria* L. *Phytochemistry*, *6*(9), 1271–1280.

Regnier, F., Waller, G., & Eisenbraun, E. (1967b). Studies on the composition of the essential oils of three *Nepeta* species. *Phytochemistry*, *6*(9), 1281–1289.

Reichert, W. (2019). The phytochemical investigation, breeding and arthropod repellent efficacy of Nepeta cataria (Doctoral dissertation, Rutgers The State University of New Jersey, School of Graduate Studies).

Rechinger, K. H. (1982). Nepeta L. In Rechinger, K. H. ed. *Flora Iranica. Lieferung* (Vol. 150, pp. 141–143). Akademische Druck und Verlagsanstalt.

Safaei-Ghomi, J., Djafari-Bidgoli, Z., & Batooli, H. (2009). Volatile constituents analysis of *Nepeta cataria* from central Iran. *Chemistry of Natural Compounds*, *45*, 913–915.

Salehi, B., Valussi, M., Jugran, A. K., Martorell, M., Ramírez-Alarcón, K., Stojanović-Radić, Z. Z., ... Sharifi-Rad, J. (2018). Nepeta species: From farm to food applications and phytotherapy. *Trends in Food Science & Technology*, 80, 104–122.

Sastry, S., Springstube, W., & Waller, G. (1972). Identification of 5, 9-dehydronepetalactone, a new monoterpene from *Nepeta cataria*. *Phytochemistry*, *11*(1), 453–455.

Schuckit, M. A. (2006). *Hallucinogens and related drugs. Drug and alcohol abuse: A clinical guide to diagnosis and treatment* In: Drug and Alcohol Abuse. Springer, Boston, MA. (pp. 210–235). https://doi.org/10.1007/0-387-25733-0_8

Setzer, W. N. (2016). Catnip essential oil: There is more to it than making your cat go crazy. *American Journal of Essential Oils and Natural Products*, 4(4), 12–15.

Sharma, A., Nayik, G., & Cannoo, D. S. (2019). Pharmacology and toxicology of *Nepeta cataria* (Catmint) species of genus Nepeta: A review. *Plant and human health, Volume 3: Pharmacology and therapeutic uses* Springer, Cham (pp. 285–299). https://doi.org/10.1007/978-3-030-04408-4_13

Sharma, A., Cooper, R., Bhardwaj, G., & Cannoo, D. S. (2021). The genus Nepeta: Traditional uses, phytochemicals and pharmacological properties. *Journal of Ethnopharmacology*, 268, 113679.

Sherry, C. J., & Hunter, P. S. (1979). The effect of an ethanol extract of catnip (*Nepeta cataria*) on the behavior of the young chick. *Experientia*, 35, 237–238.

Shishkin, B.K. & Yuzepchuck, S.V. (1954) Flora of the U.S.S.R, Vol XX, Moskova, pp. 234–236.

Schultes, R. E. (1970). The plant kingdom and hallucinogens (part III). *Bulletin on Narcotics*, 22(1), 25–53.

Snook, M. E., Blum, M. S., Whitman, D. W., Arrendale, R. F., Costello, C. E., & Harwood, J. S. (1993). Caffeoyltartronic acid from catnip (*Nepeta cataria*): A precursor for catechol in lubber grasshopper (*Romalea guttata*) defensive secretions. *Journal of Chemical Ecology*, *19*, 1957–1966.

Soulimani, R., Fleurentin, J., & Mortier, F. (2008). *Nepeta cataria. In Psychoactive Herbs in Veterinary Behavior Medicine*, 1–52.

Süntar, I., Nabavi, S. M., Barreca, D., Fischer, N., & Efferth, T. (2018). Pharmacological and chemical features of *Nepeta* L. genus: Its importance as a therapeutic agent. *Phytotherapy Research*, 32(2), 185–198.

Tucker, A. O., & Tucker, S. S. (1988). Catnip and the catnip response. *Economic Botany*, 42, 214–231.

Ubera, J.L., & Valdes, B. (1983), Revision Del Genero Nepeta (Labiatae) En La Peninsula Iberica E Islas Baleares, *Lagascalia*, 12 (1), 3–80.

Vollenweider, F. X. (2022). Brain mechanisms of hallucinogens and entactogens. *Dialogues in Clinical Neuroscience. 3*(4), 265–279.

Watts, D.C. (2007) *Dictionary of Plant Lore*. Elsevier Inc. (ISBN 13:978-0-12-374086-1)

Xiao, J., Jiang, X., & Chen, X. (2005). Antibacterial, anti-inflammatory and diuretic effect of flavonoids from *Marchantia convoluta. African Journal of Traditional, Complementary and Alternative Medicines*, *2*(3), 244–252.

Zenasni, L., Bouidida, H., Hancali, A., Boudhane, A., Amzal, H., Idrissi, A., … Benjouad, A. (2008). The essentials oils and antimicrobial activity of four Nepeta species from Morocco. *Journal of Medicinal Plants Research*, 2(5), 111–114.

Zhang, L., Wang, Y., Yang, D., Zhang, C., Zhang, N., Li, M., & Liu, Y. (2015). Platycodon grandiflorus–An Ethnopharmacological, phytochemical and pharmacological review. *Journal of Ethnopharmacology*, 164, 147–161.

Zlatković, B. K., Bogosavljević, S. S., Radivojević, A. R., & Pavlović, M. A. (2014). Traditional use of the native medicinal plant resource of Mt. Rtanj (Eastern Serbia): Ethnobotanical evaluation and comparison. *Journal of Ethnopharmacology*, 151(1), 704–713.

INTERNET REFERENCE

Bizimbitkiler (2023). *Nepeta cataria L.* https://bizimbitkiler.org.tr/yeni/demos/technical/, Accessed on 12 September 2023.

Checklistbank (2023). *Nepeta cataria* L. https://www.checklistbank.org/dataset/9923/taxon/46XJB, Accessed on 12 September 2023.

China Flora (2023). *Nepeta cataria* L. http://www.efloras.org/florataxon.aspx?flora_id=2&taxon_id=200019873 Accessed on 05 August 2023.

IPNI. (2023). *Nepeta cataria* L. International Plant Names Index. https://www.ipni.org/n/452333-1, Accessed on 17 September 2023.

ITIS. (2023). Integrated Taxonomic Information System (ITIS) on-line database, www.itis.gov, CC0, https://doi.org/10.5066/F7KH0KBK, Accessed on 17 September 2023.

The Plant List 2023. *Nepeta cataria* L. http://www.theplantlist.org, Accessed on 12 April 2023.

Pakistan Flora (2023). *Nepeta cataria* L. http://www.efloras.org/florataxon.aspx?flora_id=5&taxon_id=200019873, Accessed on 05 August 2023

POWO (2023). *Nepeta cataria* L.. https://powo.science.kew.org/taxon/urn:lsid:ipni.org:names:452333-1 Accessed on 17 September 2023.

WFO (2023). *Nepeta cataria* L. https://www.worldfloraonline.org/taxon/wfo-0000250846, Accessed on 17 September 2023.

28 Inebriating Mint (*Lagochilus inebrians* Bunge Lamiaceae)

Rahmatullah Qureshi, Ayesha Siddiqa, and Noureddine Chaachouay

28.1 INTRODUCTION

Lamiaceae, formerly known as Labiatae, a family of cosmopolitan in distribution, consists of 236 genera and 6900–7200 species, including herbs, shrubs, and trees of significant commercial and therapeutic value (Harley et al., 2004). The Lamiaceae family has a global distribution, particularly in the Mediterranean region and many of its members are grown for their attractive flowers and fragrant foliage (Orch et al., 2021). This family has great importance for human beings because, plants of this family have flavor, fragrance, or therapeutic benefits. Lamiaceae, a family with a cosmopolitan distribution, comprises 236 genera with 6900–7200 species, including economically and medicinally important herbs, horticultural shrubs, and trees (Harley et al., 2004; Chaachouay et al., 2020). Lagochilus is a significantly important genus that contains 46 species across the globe (www.theplantlist.org) and is distributed in Central, South-Central, and Eastern Asia (Mamadalieva et al., 2021). These species have been widely utilized for thousands of years, particularly in Asian nations, as herbal medicines to cure a variety of illnesses. This genus has a range of pharmacological properties, including hemostatic, antispasmodic, and anti-edemic qualities. These are also used as tranquilizers and to treat ailments like stomach pain, skin disorders, and coronary heart disease as well as to treat bleeding and chest pain. The most frequently reported effects are euphoria, dementia, sleeplessness, relaxation, and slight perceptual alterations. Allergies and skin conditions are also treated by Lagochilus (Dumaa et al., 2012; Qian et al., 2012; Sezik et al., 2004; Eisenman, Zaurov, and Struwe 2012). According to pharmacological research, diterpenoids are mostly capable of preventing bleeding (Jiao et al., 2015; Zainutdinov et al., 2002). Because many members of this genus are listed as endangered and difficult to collect, there have been few chemical investigations on this genus. However, some studies have been carried out to characterize the chemical composition of the genus and resulting in the separation and identification of several diterpenes, flavonoids, iridoids, triterpenes, and polysaccharides. One of the potentially effective hemostatic compounds is diterpene lagochiline and its derivatives (Qian et al., 2012; Jiao et al., 2015; Zainutdinov et al., 2002).

Lagochilus inebrians Bunge (*L. inebrians*) has been incorporated into the 8th edition of the Russian Pharmacopoeia and is used in traditional medicine. According to folk medicine, it is used to treat skin diseases, stop bleeding, and cause drowsiness in cases of nerve disorders. It has long served as a source of intoxication for the Tajik, Tartar, Turkoman, and Uzbek tribespeople. The decoction of *L. inebrians* is employed as a hemostatic agent by the local communities. In Central Asia, this species is still employed in this manner. A large quantity of plant material is harvested every year, particularly in the Uzbek provinces of Jizzax and Samarqand (Rahmonov et al., 2020).

Bozulbang (*L. inebrians*) is still used extensively as an astringent and coagulant. The diterpene-alcohol lagochilin, alkaloids, essential oils, tannins, resins, Ca, Mg, and 20 distinct microelements (Co, Ti, Au), vitamins C and K, and organic acids are present in all parts of the plant (Khojimatov et al., 2020). In public healthcare, drugs made from bozulbang are frequently used as blood coagulants, sedatives for nervous disorders, and for the treatment of various skin conditions including dermatitis, itching, etc. (Khojimatov et al., 2020; Belhaj et al., 2021; Benkhnigue et al., 2022).

To date, there is no review available on this plant that offers thorough insights into the significance of *L. inebrians* and its therapeutic use. Some work has been done that revealed the extremely promising potential of the investigated plant. This chapter made an effort to compile the most recent information on the plant. It also provides a thorough review that covers all of the necessary information about botanic characteristics, distribution, traditional uses, primary and secondary metabolites, and pharmacological activities. This appraisal may contribute to drug discovery for novel chemicals with exciting biological potential.

28.2 SCIENTIFIC CLASSIFICATION

The scientific classification of *L. inebrians* is shown in Table 28.1.

28.3 VERNACULAR NAMES

This species is commonly known by several vernacular names, depending on the region and culture. Some of the common vernacular names for this plant include: Na'nāmskr (Arabic); Inebriating mint, Intoxicating mint, Turkistan mint, Turkistan Skullcap, Bunge's Lagochilus, Tibetan

DOI: 10.1201/9781003460336-28

TABLE 28.1
Taxonomic Hierarchy of *Lagochilus inebrians* Bunge

Kingdom	Plantae
Subkingdom	Viridaeplantae
Infrakingdom	Streptophyta
Division	Tracheophytes
Subdivision	Spermatophytina
Infradivision	Angiospermae
Class	Magnoliopsida
Clade	Eudicots
Superorder	Asterids
Order	Lamiales
Family	Lamiaceae
Genus	Lagochilus
Species	*L. inebrians*
Binomial name	*Lagochilus inebrians* Bunge

Dragonhead (English); Môdodāŷaka pudina, Meyan-e Goli (Bengali); Nashīlā pudīnā (Hindi); Nasha aawar pudina (Urdu); Menta inebriante (Italian); Berauschende Minze (Miguez and Bollero, 2005); Menthe inebriante (French et al., 2021); Berusande mynta (Sweden); Menta embriagadora (Spanish); Turkestan Mint, Sarhoş edici nane (Turkish); Turkmenistanian mint (Turkmenistan); Turkistanian mint (Tajikistan); Sarxushbozulbang (Uzbek); Hortelã inebriante (Portuguese); Bedwelmende munt; Zunbesh (Dutch); Nanā'-e mast konande (Persian), Opyayushchaya myata (Russian); Táozuì bòhé (Chinese); Upojna mięta (Polish); Berusende mynte (Gondal et al., 2021); Opojná máta (Czech); Meitei shita minto (Japanese); Chwihan minteu (Korean). These names are often used by local people and herbalists who are familiar with the plant and its medicinal virtues.

28.4 ETYMOLOGY

The plant gets its name from the Greek words *lagos* and *cheilos*, which translate to 'hare' and 'lip/cheek', respectively. The term '*inebrians*' means intoxicating, resulting in the name 'intoxicating hare's lip'. This name is attributed to the appearance of the corolla's upper lip, which resembles a hare's lip.

28.5 BOTANICAL DESCRIPTION

Lagochilus inebrians is a deciduous sub-shrub up to 1 m tall (Figure 28.1). Stems numerous, woody at the base, and densely coated with white long-hairs and a shining crust. Leaves are opposite, broadly ovate, 3–5-lobed, cuneate at the base, and petioles about 2 cm long coated with scattered hairs and glands on both sides. The upper semi-whorl leaves contain 4–8 flowers in the axils. The calyx is five-partite, pubescent. Corolla is white or light pink with brown veins. The fruit is a nut with brown color. The flowers blossom from May through June that mostly ripen from August through September (LotusSoft, 2023).

Small-scale harvesting is done with specialized permits. Raw materials are collected when the plant is in bloom, which is often in June and July. The leafy flowering twigs are harvested from the soil surface. They are then spread out to dry in the sun in a thin layer. The dried raw materials are threshed, and the stems that don't have leaves are removed from the mixture. After a few years, the same thicket can be harvested again.

28.6 ORIGIN AND GEOGRAPHIC DISTRIBUTION

Lagochilus inebrians is endemic to western Pamir-Alay and adjacent areas including Kyzylkum, Karshi Steppe, the valley of the river Zeravshan and Fergana Valley. It is native to Central Asia, particularly the regions of Kazakhstan, Kyrgyzstan, Tadzhikistan, Uzbekistan, and Turkmenistan as shown in Figure 28.2 (Govaerts, 2003; Sezik et al., 2004; Tojibaev et al., 2020). *Lagochilus inebrians* is listed as threatened species in Uzbekistan's flora (Republic of Uzbekistan, 2009). It is a rare species that is included in the USSR's Red Book and occurs in Samarkand, Jizzakh, and Navoi only (Tojibaev et al., 2020). The plant is distributed in Kazakhstan regions such as Karatau, Karzhantau, Aksu-Zhabagly (Aimenova, Eshibaev, and Zaynutdinov 2015), Pamir-Alay Mountains of Tajikistan, Turkmenistan and Nuratau and Kyzylkum deserts, Tian-Shan of Uzbekistan (Shomurodov, Akhmedov, and Saribayeva, 2014; Akhmedov, Shomurodov, and Nomozova, 2016). It is also found in Kuldzhuktau and Kokchatau Mountains and their piedmonts. It has long been used for its intoxicating and sedative properties by Tajik, Tatar, Turkoman, and Uzbek tribesmen, ancient Central Asian tribes, and Shamans.

28.7 HABITAT AND ECOLOGY

The majority of the *Lagochilus* species are found in Iran, Central Asia, northern Pakistan, south-central Russia, Mongolia, and northwest China in arid slopes, valleys, and deserts. *L. inebrians* prefers soil that drains well and is light sandy or medium loamy in texture. It can survive soils with a pH range of weakly acidic to neutral to gently alkaline. The plant cannot grow in gloomy areas; it needs sunny plains. It can withstand drought because of its hard character and can thrive in moist or dry soil (Akhmedov, Shomurodov, and Nomozova, 2016).

28.8 ETHNOBOTANY

Lagochilus genus has numerous species that have been used to alleviate inflammation and hemorrhages in traditional medicine (Sezik et al., 2004; Eisenman, Zaurov, and Struwe, 2012). A decoction of herbs and roots from *Lagochilus* species is used as a styptic, as well as a treatment

FIGURE 28.1 Habit and habitat of *Lagochilus inebrians*.

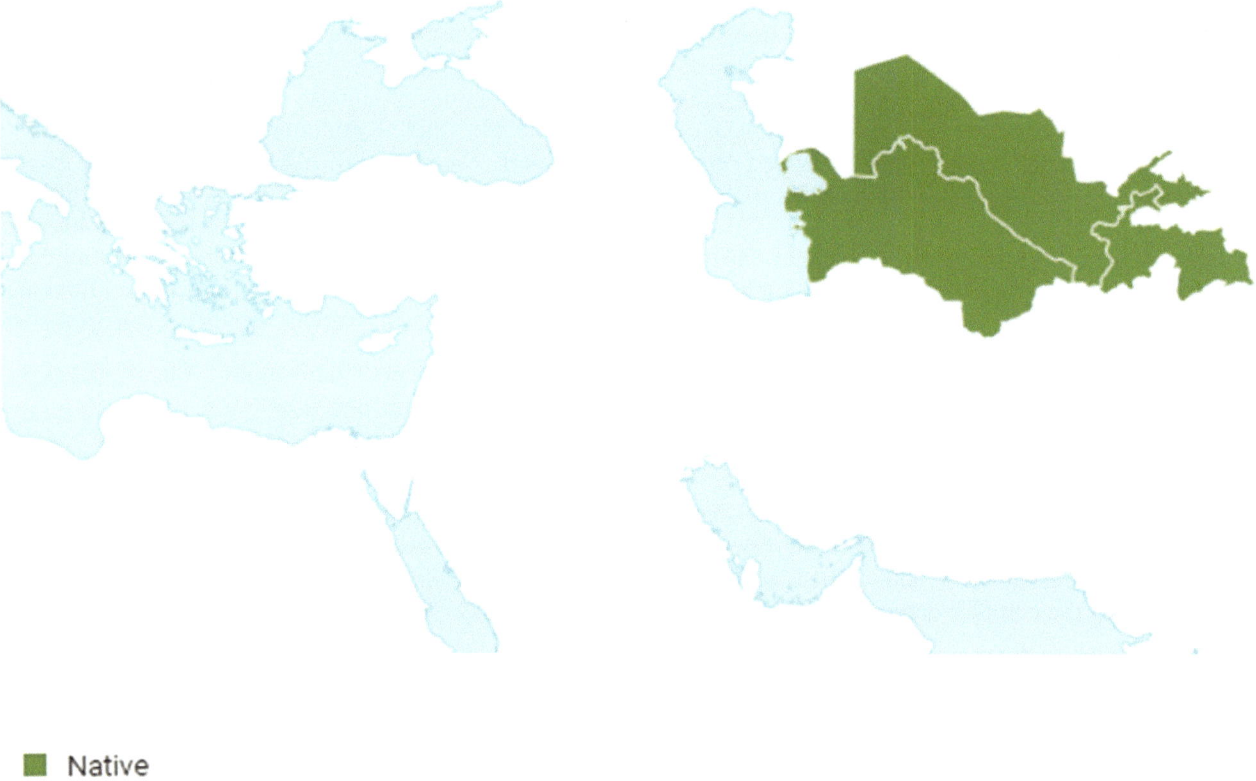

FIGURE 28.2 Global distribution of *Lagochilus inebrians*.

for stomachache, skin disorders, and agitation. The most frequently reported effects are exhilaration, relaxation, and slight perceptual adjustments. *L. inebrians* has been utilized for its euphoric and sedative effects in Central Asia during festivals (Pratov, Kholmatov, and Makhsumov, 2006). The dried leaves of *L. inebrians* are used to prepare infusions; honey and sugar are frequently added to this beverage to lessen its bitter flavor.

The herb is well-known for its effects on the mind and body and has long been utilized for its psychedelic characteristics. *L. inebrians* has been used ethnobotanically to treat a number of conditions in traditional Central Asian medicine, including fever, headaches, and stomachaches. However, as a psychedelic plant, it is most famously used. The flowers and leaves are made into a preparation called zaytseguba and employed for treating and preventing bleeding. In Central Asia, shamanic rituals and religious rites have long utilized the plant *L. inebrians*. It is frequently used in rituals to bring about altered states of consciousness since it is said to have spiritual and divinatory characteristics. In some areas, the herb is also smoked or made into tea for recreational purposes (Emboden, 1979; Chaachouay and Zidane, 2021).

The local populations use the decoction of *L. inebrians* as a hemostatic agent. This species is still used in this way in Central Asia. Every year, a significant amount of plant material is gathered, especially in the Uzbek regions of Jizzak and Samarqand. The plant name, *inebrians*, and its common names, inebriating mint, intoxicating mint, and Turkestan mint, derive from the plant's long history of use by local tribesmen as an intoxicant and sedative (Rahmonov et al., 2020).

28.9 HALLUCINOGENIC EFFECTS

For ages, the Tajik, Tartar, Turkoman, and Uzbek tribesmen have used *L. inebrians* as a source of hallucinogens. The leaves are collected in October and occasionally combined with stems, fruits, and flowers. Their aromatic scent is enhanced by drying and preservation. To minimize their incredibly harsh flavor, honey, and sugar are frequently added. This recipe is used to cure skin conditions, control hemorrhages, and used as a sedative drink for nerve illnesses (Schultes, Hofmann, and Rätsch, 2001). A diterpene crystalline substance known as lagochiline extracted from the plant has a stimulating effect on the central nervous system. It is thought that the interaction of these chemicals results in a special psychoactive effect that is characterized as a dreamy, relaxed state accompanied by minor visual and aural hallucinations (Pulatova and Khazanovich, 1962).

28.10 PHYTOCHEMISTRY AND PHARMACOLOGY

Various phytochemicals such as lagochilin (1), lagochilin and its mono, di-, tetraacetates (1–15), vulgarol and its acetate (37–38), 5-hydroxy-4′,7-dimethoxyflavone (48), β-sitosterol (121), nonacosane (139), hentriacontane (140), tritriacontane (141), 8-O-acetylharpagide (99), harpagide (100), stachydrine (127) have been identified from *L. inebrians* (Pulatova and Khazanovich, 1962; Pulatova, 1969; Islamov, Zainutdinov, and Aslanov, 1978; Zainutdinov and Islamov, 1986; Islamov, Zainutdinov, and Aslanov, 1981; Zainutdinova et al., 1994; Kotenko et al., 1994; Akramov, Sasmakov, et al., 2019; Akramov et al., 2020). The plant possesses essential oils such as linalool, β-ionone, trans-chrysanthenyl acetate, α-terpineol, trans-chrysanthenyl acetate, eugenol, trans-verbenol, bicyclo hept-3-en-2-one, pinocarvone, monoterpens linalool, trans-chrysanthenyl acetate, α-terpineol, and eugenol (Akramov, Sasmakov, et al., 2019). Stachydrine (127) is an alkaloid that has been associated with the plant ((Proskurnina and Utkin, 1960; Pulatova and Khazanovich, 1962). Stachydrine possesses potent neuroprotective, anti-fibrotic, and anti-inflammatory properties, according to numerous studies (Cheng et al., 2020).

The review of the literature indicated that several species of Lagochilus have been investigated for biological properties that include hemostatic, antibacterial, anti-inflammatory, anti-allergic, cytotoxic, enzyme inhibitory, antispasmodic, sedative, hypotensive, psychedelic, and others. Numerous biological processes have been connected to these compounds' essential oils, extracts, fractions, and separated components. Chinese and Central Asian traditional medicine have primarily used Lagochilus species as a folk remedy to treat hemostatic (Akopov, 1954; Akopow, 1971; Rakhmonberdiev et al., 1998; Zainutdinov et al., 2002; Jiao et al., 2015; Matchanov et al., 2017) and inflammatory conditions (Jiao et al., 2015). The reported biological activity of the *L. inebrians* is listed in the subsections below.

28.10.1 Hemostatic Properties

Various species of *Lagochilus* are frequently used as effective hemostatic agents in China and other Central Asian countries. According to Akopov (Akopov, 1954; Akopow, 1971) *L. inebrians* has a reputation for being therapeutically beneficial and is utilized as a powerful hemostatic agent in medicine. The mechanism of the hemostatic action of its various preparations was evaluated using a number of in vivo tests. The extracts have shown encouraging results when used to treat hemophilia.

28.10.2 Anti-Inflammatory Activity

In Chinese traditional medicine, certain *Lagochilus* species have been used to cure inflammation. It has been presumed that iridoid glucosides have anti-inflammatory properties (Jiao et al., 2015).

28.10.3 Antibacterial Activity

The antibacterial effect of the various components (i.e. extracts, essential oils, total alkaloids, and some individual compounds) from various *Lagochilus* species was evaluated by Akramov (Akramov, Sasmakov, et al., 2019). According to them, *L. inebrians* showed weak antibacterial effects on human pathogens.

28.10.4 Antioxidant Activity

The antioxidant activity of methanol extracts of *L. inebrans* was evaluated using a metal chelating (MC), ferric reducing power (FRAP), cupric reducing antioxidant capacity (CUPRAC),

2,2′-azino-bis (3-ethylbenzothiazoline-6-sulphonic acid) (ABTS), and 2,2-diphenyl-1-picrylhydrazyl (DPPH) assays. In these assays, the antioxidant activities were found as Trolox equivalents, whereas ethylene diaminetetra acetic acid (EDTA) was used for MC assay (Akramov et al., 2020). The DPPH radical activity had potent antioxidant activity collected from Djizzakh and the Surkhandarya region. Similar to DPPH, the best cupric and ferric reducing power (FRAP) activity was also observed. The antioxidant activity is probably due to the polyphenols present. The highest amount of total phenolic compounds was also found in the tested extract. The antioxidant potential of essential oil *in vitro* assays (Akramov, Bacher, et al., 2019) *L. inebrians* oil exhibited the best antioxidant effect. The activity of *L. inebrians* oil had 16-fold higher than that of *L. setulosus* in DPPH assay. Similarly, *L. inebrians* exhibited a significant reducing power effect with 2093.33 mgTE/g in FRAP assay. The observed activity might be due to its chemical profile. The main components *trans*-chrysanthenyl acetate, eugenol, and verbenol have already been reported as strong antioxidant compounds, and they could be considered as the main contributors to the observed activity (Akramov, Sasmakov, et al., 2019). On the other hand, only very rare work could be traced regarding the *in vivo* antioxidant activity of the plants belonging to the genus *Lagochilus*.

28.10.5 Enzyme Inhibition

The essential oil from *L. inebrians* contained acetylcholinesterase (AChE), which is one of the most efficient enzymes of the nervous system. Its inhibitors are used in treating Alzheimer's disease, myasthenia gravis, glaucoma, smooth muscle atony, and various ailments of autonomic nervous system functions (Akramov, Sasmakov, et al., 2019). The Methanolic extracts and compounds lagochilin (1), 5-hydroxy-4′,7-dimethoxyflavone (48) exhibited the strongest inhibitory effects on both AChE and BChE; the highest tyrosinase inhibitory effect. The same also exhibited stronger glucosidase inhibitory effects (Akramov, Bacher, et al., 2019; Akramov et al., 2020).

28.11 MEDICINAL USES AND TOXICITY

The aerial branches are said to be used in medicine as sedative, hypotensive, hallucinogenic, antispasmodic, and hemostatic effects (Emboden, 1979). The decoction of plant is internally used to treat allergies, and skin conditions. Some of the studies revealed that various parts of *L. inebrians* possess anti-hemorrhagic, anti-allergic, stomachic, tranquilizing, intoxicating, and sedative properties (Sezik et al., 2004; Eisenman, Zaurov, and Struwe, 2012; Chaachouay et al., 2021a, 2021b).

28.12 CONCLUSION

L. inebrians is endemic to western Pamir-Alay and adjoining areas. It is native to Central Asia, particularly the regions of Kazakhstan, Kyrgyzstan, Tadzhikistan, Uzbekistan, and Turkmenistan. Although the plant has long been used for its intoxicating and sedative properties by the natives, there is little knowledge in terms of therapeutic values since it has not been thoroughly investigated scientifically. Besides, the species possessed valuable phytochemical constituents, however, based on ethnobotanical use, there is a need to further explore new bioactive compounds. *L. inebrians* is a rare and endangered species and faces threats to its existence. The economic development, climate change, and over-exploitation by the natives, further exacerbated the survival of this species and therefore augments serious efforts to safeguard and protection of this species.

CONFLICT OF INTERESTS

The authors declared that they have no conflict of interest with respect to the authorship and/or publication of this chapter.

LIST OF ABBREVIATIONS

ABTS	2,2′-azino-bis (3-ethylbenzothiazoline-6-sulphonic acid)
AChE	Acetylcholinesterase
DPPH	2,2-diphenyl-1-picrylhydrazyl
EDTA	Ethylene diaminetetra acetic acid
FRAP	Ferric reducing power
MC	Metal chelating

REFERENCES

Aimenova, ZE, AA Eshibaev, and UN Zaynutdinov. 2015. Research of areas of distribution of valuable medicinal species of plants of Lagochilus Bunge (Lamiaceae) genus of South Kazakhstan territory. Paper read at *The Fourth European Conference on Biology and Medical Sciences.*

Akhmedov, AK, HF Shomurodov, and ZB Nomozova. 2016. The ontogenesis and ontogenetic structure of Lagochilus proskorjakovii Ikram (Lamiaceae) coenopopulations in Nuratau Mountain Range (Uzbekistan). *American Journal of Plant Sciences* 7 (6):928–936.

Akopov, IE. 1954. The effect of Lagochilus preparations upon clot retraction. *Byull. Eksperim. Biol. Med* 37:49–54.

Akopow, IE. 1971. Pharmacotherapy of hemophilia with the preparation of Lagochilus inebrians Bge. *Folia Haematologica (Leipzig, Germany: 1928)* 95 (1):72–83.

Akramov, D Kh, M Bacher, G Zengin, et al. 2019. Chemical composition and anticholinesterase activity of Lagochilus inebrians. *Chemistry of Natural Compounds* 55:575–577.

Akramov, Davlat Kh, Markus Bacher, Stefan Böhmdorfer, et al. 2020. Phytochemical analysis and biological evaluation of Lagochilus species from Uzbekistan. *Industrial Crops and Products* 154:112715.

Akramov, DK, SA Sasmakov, G Zengin, ON Ashirov, SS Azimova, and NZ Mamadalieva. 2019. Antimicrobial and antioxidant activities of the components of Lagochilus species from Uzbekistan. *Uzb. Biol. J* 3:3–7.

Belhaj, S., Chaachouay, N., & Zidane, L. (2021). Ethnobotanical and toxicology study of medicinal plants used for the treatment of diabetes in the High Atlas Central of Morocco. *Journal of Pharmacy & Pharmacognosy Research*, *9*(5), 619–662. https://doi.org/10.56499/jppres21.1014_9.5.619

Benkhnigue, O., Chaachouay, N., Khamar, H., El Azzouzi, F., Douira, A, & Zidane, L. (2022). Ethnobotanical and ethnopharmacological study of medicinal plants used in the treatment of anemia in the region of Haouz-Rehamna (Morocco). *Journal of Pharmacy & Pharmacognosy Research*, *10*(2), 279–302. https://doi.org/10.56499/jppres21.1196_10.2.279

Chaachouay, N., Douira, A., Hassikou, R., Brhadda, N., Dahmani, J., Belahbib, N., Ziri, R., & Zidane, L. (2020). Ethnobotanical and ethnopharmacological study of medicinal and aromatic plants used in the treatment of respiratory system disorders in the Moroccan Rif. PhDThesis, Dépertement de Biologie - Université Ibn Tofail - Kénitra, Maroc, https://tel.archives-ouvertes.fr/tel-0337637

Chaachouay, N., Douira, A., & Zidane, L. (2021a). Herbal Medicine Used in the Treatment of Human Diseases in the Rif, Northern Morocco. *Arabian Journal for Science and Engineering*, 47, 131–153. https://doi.org/10.1007/s13369-021-05501-1

Chaachouay, N., Douira, A., & Zidane, L. (2021b). COVID-19, prevention and treatment with herbal medicine in the herbal markets of Salé Prefecture, North-Western Morocco. *European Journal of Integrative Medicine*, 42, 101285. https://doi.org/10.1016/j.eujim.2021.101285

Chaachouay, N., & Zidane, L. (2021). The Symbolic Efficacy of Plants in Rituals and Socio-Religious Ceremonies in Morocco, Northwest of Africa. *Journal of Religious & Theological Information*, *21*(1–2), 34–53. https://doi.org/10.1080/10477845.2021.1942433

Cheng, Fang, Yanxi Zhou, Miao Wang, et al. 2020. A review of pharmacological and pharmacokinetic properties of stachydrine. *Pharmacological Research* 155:104755.

Dumaa, Mishig, P Xiang, Y Jinhai, Zhang Guolin, and L Yinggang. 2012. Chemical components of aerial parts of Lagochilus ilicifolius. *Chinese Journal of Applied and Environmental Biology* 18:924–927.

Eisenman, SW, DE Zaurov, and L Struwe. 2012. *Medicinal plants of central Asia: Uzbekistan and Kyrgyzstan*: Springer Science & Business Media.

Emboden, WA. 1979. *Narcotic Plants, revised and enlarged edition*. Studio Vista.

French, E, I Kaplan, A Iyer-Pascuzzi, CH Nakatsu, and L Enders. 2021. Emerging strategies for precision microbiome management in diverse agroecosystems. *Nature Plants* 7 (3):256–267.

Gondal, AH, H Zafar, H Yousaf, et al. 2021. Impacts of tillage technologies on soil, plant, environment and its management: A short communication. *Indian Journal of Pure and Applied Biosciences* 9 (3):76–83.

Govaerts, R 2003. World checklist of selected plant families database in ACCESS: 1-216203. *The board of trustees of the Royal Botanic Gardens, Kew*.

Harley, RM, S Atkins, AL Budantsev, et al. 2004. Labiatae. Flowering plants·dicotyledons: Lamiales (except In Flowering plants· Dicotyledons: Lamiales (except Acanthaceae including Avicenniaceae) (pp. 167–275). Berlin, Heidelberg: Springer Berlin Heidelberg.

Islamov, R, UN Zainutdinov, and KA Aslanov. 1978. Lagochilin 3-monoacetate from Lagochilus inebrians. *Chemistry of Natural Compounds* 14 (3):342–343.

Islamov, R, UN Zainutdinov, and KA Aslanov. 1981. Lagochilin diacetates from Lagochilus inebrians. *Chemistry of Natural Compounds* 17:50–53.

Jiao, Y, P-H Chen, A-Z Xiong, et al. 2015. Evaluation of hemostatic and anti-inflammatory activities of extracts from different Lagochilus species in experimental animals: Comparison of different extractives and sources. *Phytotherapy Research* 29 (1):22–29.

Khojimatov, OK, DT Khamraeva, AN Khujanov, and RW Bussmann. 2020. An overview of Ethnomedicinal plants of Uzbekistan. *Ethnobotany Research & Applications* 20 (08):1–19.

Kotenko, LD, M Ya Yakubova, NA Tselishcheva, MT Turakhozhaev, and TA Badalbaeva. 1994. Quantitative determination of the total iridoids in plants of the genus Lagochilus. *Chemistry of Natural Compounds* 30:669–672.

LotusSoft. 2023. *Intoxicating hare (lagohilus intoxicating; in Uzbek 'kukaparang'; in Tajik 'horak') Lagochilus inebrians Bunge Family Lamiaceae (Labiatae).* 2023 [cited 26.04.2023]. Available from https://med.com.ua/articles/99/1501.html.

Mamadalieva, NZ, DK Akramov, LA Wessjohann, et al. 2021. The Genus Lagochilus (Lamiaceae): A review of its diversity, ethnobotany, phytochemistry, and pharmacology. *Plants* 10 (1):132.

Matchanov, AD, DN Dalimov, UN Zainutdinov, NL Vypova, A Kh Islamov, and BM Bekpolatova. 2017. Preparation and physicochemical and biological properties of molecular associates of lagochilin and lagochirsine with glycyrrhizic acid and its monoammonium salt. *Chemistry of Natural Compounds* 53:665–669.

Miguez, FE, and GA Bollero. 2005. Review of corn yield response under winter cover cropping systems using meta-analytic methods. *Crop Science* 45 (6):2318–2329.

Orch, H., Chaachouay, N., Douiri, E. M., Faiz, N., Zidane, L., & Douira, A. (2021). Use of medicinal plants in dermato-cosmetology: An ethnobotanical study among the population of Izarène. *Jordan Journal of Pharmaceutical Sciences*, *14*(3), 323–340.

Pratov, UP, HK Kholmatov, and MM Makhsumov. 2006. *Natural medicaments*. Tashkent: Tashkent, Uzbekistan:208.

Proskurnina, NF, and LM Utkin. 1960. DL-stachydrine in Lagochilus. *Meditsinskaya Promyshlennost SSSR* 14:30–31.

Pulatova, TP. 1969. Alkaloid content of some plants of the family Labiatae. *Chemistry of Natural Compounds* 5 (1):55–55.

Pulatova, TP, and RL Khazanovich. 1962. On the alkaloid content of some *Lagochilus* species and on the nature of lagochiline. *Aptechnoe delo* 6:29–32.

Qian, JS, BF Zhang, W Wang, et al. 2012. Chemical constituents of *Lagochilus ilicifolius*. *Chin Tradit Herb Drugs* 43:869–72.

Rahmonov, O, DE Zaurov, BI Islamov, and SW Eisenman. 2020. Resources along the Silk Road in Central Asia: Lagochilus inebrians Bunge (Turkestan Mint) and Medicago sativa L.(Alfalfa). *Natural products of silk road plants*. (R Cooper, JJ Deakin, Eds). pp. 153–166.

Rakhmonberdiev, GR, AS Sidikov, RD Yusupov, UN Zainutdinov, and DS Kazantseva. 1998. Synthesis of a hemostatic drug from cellulose acetate. *Chemistry of Natural Compounds* 34:192–194.

Republic of Uzbekistan. 2009. *The Red Book. Vol. 1, Chinor ENK, Tashkent, 356 p.*

Schultes, RE, A Hofmann, and C Rätsch. 2001. *Plants of the gods: Their sacred, healing, and hallucinogenic powers*. (pp. 192). Healing Arts Press.

Sezik, E, E Yesilada, H Shadidoyatov, et al. 2004. Folk medicine in Uzbekistan: I. Toshkent, Djizzax, and Samarqand provinces. *Journal of Ethnopharmacology* 92 (2–3):197–207.

Shomurodov, HF, A Akhmedov, and SU Saribayeva. 2014. Distribution and the current state of Lagochilus acutilobus

(Lamiaceae) in connection with the oil and gas sector development in Uzbekistan. *Ecological Questions* 19:45–49.

Tojibaev, K, N Beshko, Avazbek Batoshov, D Azimova, ZY, and HST Deng. 2020. *Flora of the Dzhizak Province, Uzbekistan*. China Forestry Publishing House.

Zainutdinov, UN, and R Islamov. 1986. The active substances of *Lagochilus-inebrians*. *Khimiko-Farmatsevticheskii Zhurnal* 20 (5):583–584.

Zainutdinov, UN, R Islamov, DN Dalimov, TR Abdurakhmanov, OD Matchanov, and NL Vypova. 2002. Structure-activity relationship for hemostatic lagochilin diterpenoids. *Chemistry of Natural Compounds* 38:161–163.

Zainutdinova, UN, MP Pulatova, TA Badalbaeva, et al. 1994. Diterpene lactones and iridoid glycosides of plants of the genus *Lagochilus*. *Chemistry of Natural Compounds* 30:27–30.

29 Diviner's Sage (*Salvia divinorum* Epling & Játiva Lamiaceae)

Mai M. Younis, Ahmed M. Elissawy, Samir M. Osman, Gokhan Zengin, Omayma A. Eldahshan, and Abdel Nasser B. Singab

29.1 INTRODUCTION

The use of hallucinogenic plants for healing, mystical, or religious purposes has been an important part of human history for ages. The term hallucinogenic has been specified to numerous classes of agents that alter consciousness, and may produce loss of contact with reality, true hallucinations, delirium, and in some cases death. Cannabinoids, phenethylamines, lysergic acid derivatives, and other indole compounds are examples of hallucinogenic compounds (Prisinzano, 2013). They have been classified into deliriants, psychedelics, and dissociative hallucinogens (Feng et al., 2017).

Lamiaceae, sage or mint family, is one of the largest plant families that includes about 250 genera and 7000 species (Mesquita et al., 2019). The family includes a wide group of flowering plants and essential culinary herbs such as sage, rosemary, basil, mint, thyme, oregano, lavender, and marjoram. One species has unique medical significance, scientifically known as *Salvia divinorum* and simply called 'salvia', by people (Faudree, 2020). *S. divinorum* is an important member of the genus Salvia, one of the largest genera in the Lamiaceae family (Labiatae) (Murphy and Bola, 2013; Chaachouay et al., 2020).

Genus Salvia is composed of about 960 species of different herbaceous, woody, mint plants (Lopresti, 2017). Although *Salvia* species are widely distributed all over the world, they are abundantly located and utilized for several purposes in three main geographical areas, Europe around the Mediterranean, South-East Asia, Central and South America (Avula et al., 2022). Mexico is one of the greatest distribution areas of salvia with more than 300 species (Tlacomulco-Flores et al., 2020). Salvia is commonly used as a herbal tea worldwide. In addition, such plants act as significant medicinal plants as they are beneficial against cold, flu and stomach disorders (Hatipoglu et al., 2017). Salvia species produce various bioactive terpenes and terpenoids that support their effectiveness in folk medicines (Asadollahi et al., 2019). They have been reported for possessing significant biological activities such as antioxidant, antifungal and antibacterial properties (Pentea et al., 2015).

S. divinorum is a small perennial shrub native to Oaxaca, Mexico (Faudree, 2020). It is also known as diviner's sage, magic mint, Ska Maria Pastora, and Hierba Maria (Gonçalves et al., 2021). It is a potent psychoactive herb traditionally used by the Mazatec people for divination practices, spiritual rituals, and healing ceremonies. *S. divinorum* is consumed by masticating the fresh leaves or drinking their juice. As well, it may also be consumed via volatilization, inhalation, or smoking dried leaves. It induces very intense psychoactive and hallucinogenic effects, and promotes a pleasant state of introspection, increased sensitivity to auditory and luminous stimuli, and reduction of attention to surroundings (Carod-Artal, 2015). *S. divinorum* has also been medicinally used for centuries by natives of southern Mexico, where it is endemic, now it is studied by several pharmaceutical researchers (Faudree, 2020).

Phytochemical investigations have reported that salvinorin A (Sal-A) is the major active component of *S. divinorum* (Figure 29.1). It is a non-nitrogenous diterpene that is responsible for the potent hallucinogenic and pharmacological properties of the plant. Other pharmacologically inactive terpenoids have also been identified including salvinorins (B–J) and related compounds named salvidivins (A–D), salvinicins (A, B), and divinatorins (A–F) (Li et al., 2016; Mahendran et al., 2016; Cruz et al., 2017; Faro et al., 2020).

Sal-A has gained a great interest in the scientific community due to its unique properties and structure. It has been categorized as a dissociative hallucinogen as dextromethorphan, ketamine, and ibogaine (Feng et al., 2017). It has been known to be the first naturally occurring molecule that potently and selectively acts on κ opioid receptors (KOR). Its structure doesn't resemble other main classes of opioid receptor ligands (Prisinzano, 2013). It lacks the basic nitrogen atom existing in most known opioids therefore it interacts with KOR in a qualitatively different manner (Cruz et al., 2017).

Previous studies carried out on *S. divinorum* and its main bioactive ingredient Sal-A have shown its therapeutic potential in pain treatment, drug addiction, gastrointestinal, and neurological diseases (Gonçalves et al., 2021). The most recent research concerning Sal-A or *S. divinorum* has found that the reported therapeutic effectiveness is mostly related to their analgesic and anti-inflammatory activity.

Nowadays *S. divinorum* is broadly available through the internet and smart shops. The products are sold at different

DOI: 10.1201/9781003460336-29

FIGURE 29.1 Leaves and flowers of *Salvia divinorum* (https://reachoutrecovery.com/what-is-salvia-divinorum; https://www.inaturalist.org/taxa/281359-Salvia-divinorum).

'strengths', which correlate with the different concentrations of Sal-A (Hernández-Bello et al., 2015). It is mostly used through smoking the commercial 'enhanced leaves' which are fortified leaves preparations with additional Sal-A (Addy, 2016).

The psychoactive and dissociative activities of *S. divinorum* began to attract policymakers and media attention in the last decades. It was reported to serve as a legal high and cannabis alternative as well and it was classified as a Schedule I drug by the Florida Legislature in 2008 (Lin et al., 2014).

This chapter will provide a recent overview of the important hallucinogenic herb, *S. divinorum*, illustrating its botanical description, habitat, ecology, and ethnobotany. In addition, the plant pharmacology and therapeutic potential will be explained.

29.2 SCIENTIFIC CLASSIFICATION

S. divinorum was classified as a member of the Lamiaceae family in 1962 by Carl Epling and Carlos Ja´tiva (Phipps and Grundmann, 2017). It has been included within subgenus Calosphace, section Dusenostachys, and assumed to be an interspecific hybrid (Hao et al., 2015). Salvia subgenus Calosphace is the largest of four, it includes about 500 species native to North and South America (Jenks and Kim, 2013). Therefore, *S. divinorum* belongs to the Eukaryota Domain, Plantae Kingdom, Magnoliophyta Phylum, Magnoliopsida Class, Asteridae subclass, Lamiales Order, Lamiaceae Family, and therefore to the genus Salvia.

29.3 VERNACULAR NAMES

The name divinorum was specified by Epling and Játiva related to the divinatory use of the plant (Díaz, 2013). *S. divinorum* is known as diviner's sage, magic mint, hierba de Maria and, hojas de la Pastora (Gonçalves et al., 2021). It is also called mystic sage, purple sticky, lady salvia, or just salvia (Brito-da-Costa et al., 2021). Ska Maria pastora or ska pastora are also common names by the Mazatecs for *S. divinorum*, which means the 'leaves of Mary the shepherdess' or 'leaves of the shepherdess' (Mahendran et al., 2016).

29.4 BOTANICAL DESCRIPTION

S. Divinorum (Figure 29.1) is a small to medium-sized shrub that can grow up to 1–1.5 m in height (Kowalczuk et al., 2014). Its stem is translucent green and characterized by its crisp, hollow, and quadrangular structure. The stem is decumbent, enabling the vegetative reproduction of the plant. Two green, simple leaves arise from each node in opposite arrangement. The leaves are relatively large compared to other salvia species of length 10–25 cm and width 5–10 cm. They are elliptical, or oval in shape with serrated margins (Reisfield, 1993).

The flowers are about 3–4 cm long with purple calices and white corollas. They can grow on panicles of about 20–30 flowers. Flowering is rare and doesn't occur on a regular seasonal basis. The plant is observed to flower between October and June in Oaxaca. Flowering is induced by set durations of continual darkness greater than 12 hours. If the plant is exposed to light, flowering is stopped and returns to vegetative growth (Reisfield, 1993). It has been reported that *S. divinorum* seeds are sterile or rarely observed so its main propagation process occurs by vegetative reproduction of cuttings (Phipps and Grundmann, 2017). The diminished seed set, survival rate, and germination rate act as obstacles for *S. divinorum* growth from seeds (Reisfield, 1993;Brito-da-Costa et al., 2021). Consequently, the reproduction of the plant is based on cuttings. However, several researchers have achieved successful reproduction by growing new plants from seeds (Hernández-Alvarado et al., 2020).

Microscopical examination of *S. Divinorum* leaves has showed the presence of several kinds of trichomes on the leaf surface (Bagheri et al., 2016). Peltate glandular trichomes, containing the active component, salvinorin A, appear on the leaf abaxial surfaces, while capitate glandular trichomes appear on abaxial and adaxial surfaces (Chen et al., 2017;Dunkić et al., 2017). Other types of trichomes are also present as conical non-branched non-glandular trichomes. In addition, the leaf is characterized by the presence of frequent diacytic stomata and polygonal parenchymal cells (Kowalczuk et al., 2014).

29.5 ORIGIN AND GEOGRAPHIC DISTRIBUTION

S. divinorum is endemic to Sierra Mazateca, in Oaxaca region of Mexico in Central America, which is a vast woody mountainous region (Faudree, 2020). The plant grows mainly in the moist and shady areas in the Sierra Mazateca Mountains (Johnson et al., 2013). However, Now it is cultivated in different regions worldwide such as Hawaii, Canada, California and other regions in United States (Bischoff, 2018;Willard et al., 2020). However, in recent years it has become illegal in several US states after insertion in the drugs of concern list by the United States Drug Enforcement Administration (D.E.A.) (Faro et al., 2020).

29.6 HABITAT AND ECOLOGY

The native habitat of *S. divinorum* is the rainforests located in Oaxaca (Sierra Madre Mazateca), Southern Mexico. The state is surrounded in the west by the Pacific Ocean and in the north by the Sierra Mazateca Mountain range which is covered by tropical montane cloud forest (Reisfield, 1993;Ott, 1995). An ecosystem characterized by persistent cloud cover and high humidity. *S. divinorum* has been found in several sites between 500 and 1500 m altitude and mostly grows near water courses in full or partial shade (Ott, 1995). Its most common habitat is the nutrient-rich, moist, black soil underneath a gully, usually near a river or stream where bushes and small trees create an environment of high humidity and low light (Díaz, 2013). Although *S. divinorum* mainly spreads in regions of humid, semitropical climates (Bischoff, 2018), it has been successfully grown in moderate to warm climates in the presence of appropriate sunlight and humidity (Phipps and Grundmann, 2017).

29.7 ETHNOBOTANY

Salvia divinorum has been utilized as a legal hallucinogen for years. Since fourteenth century it has been used for shamanic purposes as a short-acting powerful hallucinogen inducing delirium and psychedelic effects (Sayin, 2016). It has been consumed for centuries as a ritual hallucinogen by the Indigenous ethnic groups of Oaxaca, possibly in pre-Columbian times (Skalicka-Woźniak and Gertsch, 2020). Mazatecs have mainly used *S. divinorum* during healing ceremonies, spiritual rituals, divination practices, and training of medical practitioners (Vázquez-León et al., 2021). They believe that *S. divinorum* is the reincarnation of the Virgin Mary and allows traveling to heaven and talking to the Saints and God about divination (Brito-da-Costa et al., 2021). Traditional use of *S. divinorum* has been mainly described as manifestations of divinities and good feeling states to promote internal peace and create energy balance. The consumption of *S. divinorum* leaves depends on the purpose, for ceremonies, illness treatment, or others. It has been accompanied by abstention from food, alcohol intake, and sexual activity (Hernández-Alvarado et al., 2020; Chaachouay et al., 2021a, 2021b).

Leaves of *S. divinorum* have been traditionally consumed by chewing, sucking, or grinding to produce a drinkable liquid. Additionally, short, potent hallucinations are elicited by inhalation of the vaporized Sal-A extract, or smoking the dried leaves (Feng et al., 2017). However, in recent years, the concern for *S. divinorum* has greatly increased among recreational users due to its awareness modification effects. The use of *S. divinorum* has widely spread to North America and Europe in a similar fashion to other natural drugs (González-Trujano et al., 2016).

Nowadays, the leaves of S. *divinorum* are smoked as a recreational or hallucinogenic drug worldwide. Due to its potent hallucinogenic property, *S. divinorum* is often smoked or co-consumed with alcohol (Vázquez-León et al., 2021). The use of *S. divinorum* is specifically high among the users of LSD, ecstasy, and phencyclidine (Salas-Wright et al., 2021). Recently, dried leaves of *S. divinorum* are sold on the internet as well as leaf extracts are available to consumers at different concentrations (Hernández-Alvarado et al., 2020). *S divinorum* is marketed on the internet and social media as a legal alternative to illicit drugs and marijuana (Diaz, 2016). Head shops and internet vendors now sell *S. divinorum* herb, dried leaves, leaf extracts of high Sal-A concentrations for smoking, also as tinctures for oral administration (Vázquez-León et al., 2021). It is sometimes known as 'the YouTube drug' because of the prevalence and popularity of online 'trip videos' (Faudree, 2020). Although it is legal to purchase *S. divinorum* from internet suppliers in some countries, there is great attention to the health hazards elicited by its abuse (Heal et al., 2018). Due to the common use of hallucinogenic plants as a recreational drug, several countries have begun to regulate either or both *S. divinorum* and Sal-A as controlled substances (Addy, 2016) and many states have passed legislation regulating its sale and use (Heal et al., 2018).

Although The Mazatec Indians have mainly used *S. divinorum* in spiritual rituals to create hallucinatory effects, they have also applied the plant in folklore medicine (Akaberi et al., 2015). *S. divinorum* has been used in semi-magical diseases (Bischoff, 2018), it has been specified for the recovery from a Mazatec illness called swollen belly or panzón de arrego. This illness has been believed to be caused by a brujo, a black or evil magic practitioner (Casselman et al., 2014).

The Mazatec Indians have used *S. divinorum* for therapeutic purposes to treat anemia, headaches, rheumatism, and others (Mendes et al., 2022). They have also used extracts of *S. divinorum* to regulate defecation and urination and cure eliminatory dysfunctions such as diarrhea (Hernández-Alvarado et al., 2020). Infusions of the plant's juices have been given as a palliative for people who are near death causing rapid recuperation (Casselman et al., 2014). In addition, they have used the plant for healing conditions, eczema, cystitis, candidiasis, menstrual cramps, insect bites as well as in alcohol addiction and depression (Martins and Brijesh, 2018;Coffeen and Pellicer, 2019). *Salvia divinorum* has been used externally to relieve pain, reduce swelling, invigorate, and cool blood (Wang et al., 2023). It has been reported that the traditional preparations of *S. divinorum* have been used in the treatment of diseases that are related to inflammation, rheumatism, pain, and gastrointestinal problems (Coffeen and Pellicer, 2019). It has also been used as an antidiabetic, antibacterial, and antioxidant agent (Loizzo et al., 2014). In addition, it has been added as an ingredient in Chinese herbal composition in the treatment of alopecia (Zhang et al., 2022).

29.8 PHARMACOLOGY

S. divinorum has been known to possess a wide range of pharmacological activities as analgesic, anti-inflammatory, and other activities. However, its hallucinogenic property is the most common pharmacological activity. The psychotropic effects elicited by *S. divinorum* is owed to its main active principle, a neoclerodane diterpene known as salvinorin A (Sal-A). (Aixalá et al., 2020;Brito-da-Costa et al., 2021;da Silva et al., 2021;Siegel et al., 2021). Sal-A is localized in the plant glandular trichomes (Yang et al., 2023) and represents about 0.18% of the dried plant (Farré et al., 2021).

Sal-A is well known as a dissociative atypical hallucinogen and considered as one of the most potent naturally occurring hallucinogens (Girn et al., 2023). It is more 20 times in potency than psilocybin and is equivalent in strength to LSD (Faudree, 2020). The active dose of Sal-A in humans is between 4.5 and 21 mg/kg (da Silva Maia et al., 2018).

The hallucinogenic mechanism of Sal-A differs from those of ketamine, mescaline, LSD and other classical hallucinating drugs (Vázquez-León et al., 2021). It has no affinity for the classical hallucinogenic drug receptors as a serotonin receptor or N-methyl=d-aspartate receptor (Walentiny et al., 2010). Sal-A acts as a selective agonist of kappa-opioid receptors (KOR) (Karila et al., 2019; Szczepaniak et al., 2022) in addition to allosteric modulation of cannabinoid 1 receptors (Coffeen and Pellicer, 2019; Silva et al., 2019). KOR is one of the opioid receptor subtypes, mu (MOR), delta (DOR), and kappa (KOR), that are differentially distributed in the brain and spinal cord and exerts unique pharmacological responses (Wang, 2019). The selectivity of Sal-A for KOR is greater than 5000 times than for mu and delta receptors (Clark and Abi-Dargham, 2019). KOR is a G protein-coupled receptor (GPCR) that influences pain perception, motor, and mood control (Hernández-Alvarado et al., 2020). In addition, its activation is known to elicit dysphoria, hallucination, and dissociation (Keasling and Zjawiony, 2016). The mechanism of the hallucinogenic properties of KOR agonists is unknown, but they are involved in perception disorders, such as depression and schizophrenia (Feng et al., 2017). Activation of KOR is involved in different neuropsychiatric symptoms, including hallucinations, social withdrawal, dysphoria, lack of motivation, and cognitive functions. It has been reported that healthy volunteers administered Sal-A experience dissociative mental states resembling positive symptoms of schizophrenia (Shekhar, 2019).

The selective interaction of Sal-A with KOR has been demonstrated in several assays. In radioligand binding assays, Sal-A has shown 18 nM affinity at KOR, and >10 000 nM affinity at DOR and MOR. Moreover, in functional assays, Sal-A has showed full agonism and EC_{50} of 7 nM at KOR, and >10 000 nM potency at DOR and MOR (Che et al., 2018; Chakraborty and Majumdar, 2020). Other studies have demonstrated that strong affinity of Sal-A for cloned KORs in human embryonic kidney-293 cells (Ki = 16 nM) and native KORs in guinea-pig brain (Roth et al., 2002). It was found that Sal-A nonbiased agonism toward KOR activates both the G protein signaling pathway with EC_{50} value 4.73 nM in cAMP assay and the β-arrestin recruitment pathway EC_{50} value 10.5 nM in tango assay (White et al., 2015). There are more than one binding site Sal-A in KOR, therefore, there may be multiple ways to form ligand–receptor complexes (Mahendran et al., 2016).

S. divinorum is consumed by smoking the dried leaves or chewing and swallowing the fresh leaves. It also can be crushed for water beverages or prepared as fresh leaves infusion yielding lower effects due to the poor solubility of Sal-A in water. Absorption of *S. divinorum* or its main active component Sal-A can occur through pulmonary route, oral mucosa or nasal mucosa (Brito-da-Costa et al., 2021). On masticating *S. divinorum* leaves, they must be chewed thoroughly, and the juice must be kept in the mouth for at least 10 min, before being expelled or swallowed, to permit the absorption of enough Sal-A through oral mucosa (Brito-da-Costa et al., 2021). The longer the leaves retained in mouth, the more potent the hallucinogenic effect (Griffin III, 2013). It has been reported that the percentage of absorption of Sal-A through the oral route in humans is around 85.8% (Brito-da-Costa et al., 2021). However, according to recreational users, smoking dried leaves or inhaling the vaporized compound are the most effective, rapid, desirable ways to elicit a psychotropic effect. (Karam et al., 2019). Previous studies have stated that smoking of Sal-A at a dose approximately 200–500 µg would be sufficient to induce extreme hallucinogenic and psychotomimetic properties (Gonçalves et al., 2021). In addition, it has been reported that *S. divinorum* use in the form of cigarettes can cause loss of body control, similar to phencyclidine and ketamine (Gürlek et al., 2017).

After a short time, consumers experience different sensations including depersonalization, dysphoria, and out-of-body experiences followed by proprioceptive sensations, colorful visions, and hallucinations that persist for an hour (Hernández-Alvarado et al., 2020). The hallucinatory effects of *S. divinorum* include loss of identity, spatiotemporal dislocation, convincing illusions, dissociative sensations, and entering another reality (Lin et al., 2014). The effects of Sal-A range from mild to profound, it causes sensations varying from pleasure and well-being to terror and fear (Margalho et al., 2016). Reports by sage abusers have documented the positive after-effects of Sal-A including improved mood and increased insight which last more than 24 hours (Heal et al., 2018).

The duration of action and the hallucinogenic potency of *S. divinorum* are closely related to its dose and route of administration. *S. divinorum* inhalation leads to rapid onset of intense hallucinogenic effects within 40 seconds (Gürlek et al., 2017). Its half-life time is about 8 minutes, quickly metabolized in GIT (Prisco et al., 2021) and can persist for 15–30 min (Phipps and Grundmann, 2017;Williams and Lundahl, 2019). Chewing the leaves leads to direct absorption from the oral cavity in half an hour to one hour, while excretion occurs through the urine and saliva (Gürlek et al., 2017). The oral doses are slightly absorbed through buccal mucosa and mostly degraded in GIT producing less pronounced psychoactive effects that can last up to 1 h (Phipps and Grundmann, 2017). It has been found that sublingual administration doesn't produce psychotropic activity in human (Zawilska and Wojcieszak, 2013).

The lack of a protonated amine group in Sal-A structure, commonly present in previously known opioids, offers a fast metabolism, swift elimination, and rapid loss of activity (Hernández-Alvarado et al., 2020). Since it lacks nitrogen atoms and basic amines, it is not ionized at cellular pH (Roach and Shenvi, 2018). Sal-A has several metabolic sites such as C-2 acetyl, C-4 ester, and lactone ring. Ester hydrolysis of the C-2 acetyl leads to salvinorin B (Sal-B), which is less active at KOR than Sal-A (White et al., 2014). The C-4 ester has been reported to be enzymatically stable, whereas the lactone ring has been found to be labile (Chakraborty and Majumdar, 2020). Sal-A is metabolized in humans by CYP450 (mainly CYP1A1, CYP2E1, CYP2D6, and CYP2C18 isoforms) while UGT2B7 and lipophilic metabolites are eliminated via biliary excretion and lipophilic ones through renal filtration (Zawilska and Wojcieszak, 2013).

The pharmacokinetics and psychological activities of *S divinorum* and Sal-A have been extensively evaluated. Several clinical assays have been conducted showing the dose-dependent responses of Sal-A and its psychotomimetic effects (Clark and Abi-Dargham, 2019). Participants have reported delusions, losing consciousness, and obvious changes in visual, auditory, and interoceptive sensory input (Addy et al., 2015). A double-blind, placebo-controlled, and randomized, clinical study has shown the dose-related subjective and cognitive effects of Sal-A at doses between 0.375 μg/kg to 21 μg/kg including memory impairment and strong dissociative effects with no adverse effects at one month follow-up (MacLean et al., 2013). In another study, 10 healthy volunteers, subjected to inhaled Sal-A (8, 10 mg), reported transient perceptual alterations and psychotomimetic effects as somesthetic and dissociative effects, elevated plasma levels of prolactin and cortisol, and reduced resting electroencephalogram spectral power. In addition, they have reported no change in vital signs, no cognitive deficit, no euphoria, low addictive potential, and no adverse effects (Ranganathan et al., 2012). Participants who smoked 1,017 or 100 μg Sal-A have reported laughing, hallucinogenic experiences, and no adverse effects. The effects are similar to use with 10% marijuana, 13% LSD, and 10% 3,4-methylenedioxy-methamphetamine (Addy, 2012). Other clinical studies conducted on healthy volunteers have shown that inhaled Sal-A has reached its Cmax (31 ng/mL) at 1–2 min, showing an area under the curve ranging between 370 to 487 ng/mL/min, Tmax at 2 min, with a rapid declining of the plasma concentration of the drug to 0 at 90 min post-inhalation (Johnson et al., 2016; Maqueda et al., 2016). Healthy volunteers vaporized Sal-A have explained a disconnection from external reality, modified interception, and auditory and visionary phenomena. The somatic sensations have been increased at low doses, but a complete loss of contact occurs at the highest dose (Maqueda et al., 2015). Rating scale and mysticism scale have been employed in double-blind placebo-controlled evaluations to assess the effects of inhaled Sal-A. The hallucinogenic effects of Sal-A have been found to seem like the intravenous administration of DMT and oral administration of psilocybin with no changes in heart rate and blood pressure (Hernández-Alvarado et al., 2020). Different human survey-based evaluations have been performed. The surveys have exhibited that most experiences are unique and pleasant, initiating well-being feelings and calmness. Participants have reported dreamlike experiences, depersonalization, synesthesia, and loss of awareness of their surroundings (Hernández-Alvarado et al., 2020).

Regarding oral administration, eight volunteers have experienced no psychoactive effects after sublingual Sal-A at doses up to 4 mg indicating the low bioavailability of Sal-A through this administration route (Brito-da-Costa et al., 2021). Immediate swallowing has elicited no hallucinogenic effects, and high doses (10 mg) have been subjected to extensive and rapid GIT enzymatic deactivation or a first-pass metabolism (Brito-da-Costa et al., 2021). Studies of Sal-A absorption via the intravenous administration are lacking in humans but have been performed in animal models (Feng et al., 2017).

Sal-A has been found to be an efficacious, potent and rapid KOR agonist in rodents and non-human primates with high KOR homology to human (Butelman and Kreek, 2015). Non-human primate studies have showed that Sal-A rapidly entered the brain, achieving the maximum concentration (3.3% of the total administered dose) in 40 seconds which could predict the duration of hallucinogenic property in humans (Roach and Shenvi, 2018). Rapid clearance from the brain has been reported with a half-life of 8–10 min (Zawilska and Wojcieszak, 2013). In addition, it has been demonstrated that Sal-B is the main metabolite of Sal-A in the blood of rhesus monkeys. Sal-B results from ester hydrolysis of C-2 acetoxy group by esterase (Margalho et al., 2013). Other *in vivo* studies conducted in rodents and non-human primates have shown that Sal-A causes locomotor-decreasing and sedative-like effects. These effects are similar to those of synthetic KOR agonists, and are sensitive to KOR receptor antagonists (Butelman and Kreek, 2015). In rodents and non-human primates, Sal-A has initiated a degree of impaired locomotion and sedation (Cichon et al., 2022).

KOR agonists have been indicated to inhibit dopamine accumulation in the midbrain and decrease dopamine levels in the mesocorticolimbic system. Their prolonged treatment could alter the function and density of dopamine D2 receptors (da Silva Maia et al., 2018;Snyder et al., 2018). Sal-A has been found to elicit a dose-dependent reduction in dopamine dialysates in the ventral, dorsal, and striatum of rodents at doses like those producing behavioral effects as place aversion and locomotor decreases (Butelman and Kreek, 2015; Serra et al., 2015;Mahendran et al., 2016). The effect of Sal-A in dopamine level decrease is similar to endogenous KOR receptor ligands and synthetic KOR agonists and may be attributed to the potentiation of dopamine re-uptake transporter (DAT) as a result of KOR/DAT heterodimers formation (Kivell et al., 2014a).

KOR receptor agonists have certain endocrine effects as they regulate hormonal processes. They stimulate cortisol, adrenocorticotropic, hypothalamo-pituitary-adrenocortical and prolactin release (Butelman and Kreek, 2015). Clinical studies have shown the elevated plasma concentration of prolactin and cortisol following Sal-A inhalation with certain individual variations (Maqueda et al., 2016). In another study, intravenous Sal-A administration in rhesus monkeys raised prolactin levels in a dose-dependent manner (Mahendran et al., 2016). Another study conducted on participants exposed to inhaled Sal-A also has shown that Sal-A generally increases prolactin, for a more prolonged and delayed time course than the drug itself (Johnson et al., 2016).

29.8.1 Adverse Effects and Toxicity

The toxicity of *S. divinorum* or Sal-A has been known to be relatively low and related to the doses of exposure during healing or ritual practices by the Mazateca group in Oaxaca, Mexico (Vázquez-León et al., 2021). In humans, acute or chronic toxicity of *S. divinorum* has been rarely reported compared to other hallucinogens such as amphetamine derivatives. No fatality case reports among *S. divinorum* users have been registered (Feng et al., 2017). No poisonings have been listed in domestic animals and no toxic effects have been observed in mice administered high salvia doses (Bischoff, 2018). *S. divinorum* is generally not abused by its users, supporting the belief that kappa-opioid agonists are not addictive or rewarding (Tejeda and Bonci, 2019). *S. divinorum* has shown low addictive potential in several animal studies (Feng et al., 2017). In addition, the recreational users of *S. divinorum* have not reported any adverse long-term effects, according to a survey (Karam et al., 2019).

The lack of somatic effects of *S. divinorum* suggests that is physiologically safe. However, like classic hallucinogens, *S. divinorum* may trigger psychiatric reactions in those with a predisposition for psychotic diseases (Herrmann et al., 2016). Sal-A often induces dysphoria, feelings of fear, depression, and sadness. In addition, it may cause slurred speech, a decreased heart rate, lack of coordination, and loss of consciousness (Farré et al., 2021). Confusion, yawning, mental slowness, tiredness, heaviness of the head, drowsiness, dizziness, fatigue, and insomnia are common adverse symptoms registered by *S. divinorum* consumers (Biggs et al., 2017; Badal et al., 2018). Other adverse effects of *S. divinorum* include psychosis, irritability, and Hallucinogen Persisting Perception Disorder (HPPD) (Schifano et al., 2021a; Schifano et al., 2021b). The activation of GPCRs results in signaling pathways involving β-arrestins and Gi/o proteins. β-arrestins recruitment is associated with many adverse effects including anxiety, sedation, dysphoria, and aversion (Hernández-Alvarado et al., 2020). Based on previous preclinical studies in primates and rodents, and in human laboratory studies, KOR activation has resulted in a variety of robust effects including sedative-like behaviors due to a reduction in locomotor activity, depressant-like behaviors, dysphoria, and aversion, and modulation in the effects of abused drugs (Butelman and Kreek, 2015). Several case studies have reported the psychiatric condition triggered by *S. divinorum* intoxication (Tejeda and Bonci, 2019). It has been found that adults administrating *S. divinorum* were more likely to show up experiencing depression and/or suicidality in the same year (Yang et al., 2022).

29.9 MEDICINAL EFFECTS

In addition to the potent hallucinogenic properties of *S. divinorum*, several studies have reported its therapeutic potential (Figure 29.2). The multiple therapeutic targets of *S. divinorum* have involved complex pharmacology and mechanism of action for this mystical plant (Eksi et al., 2020). *S. divinorum* or Sal-A is an attractive candidate for drug development due to its effect on κ-opioid receptors which are potential targets for treating depression, anxiety, obesity, and alcoholism and can regulate behavioral and cognitive alterations (Addy, 2016). Different psychotropic and therapeutic effects of Sal-A have been investigated, including antidepressant, psychedelic, and antinociceptive (Hernández-Alvarado et al., 2020) (Figure 29.3). Previous studies have reported the medicinal effectiveness of *S. divinorum* in the treatment of drug addiction, opioid-induced constipation, obesity, pain,

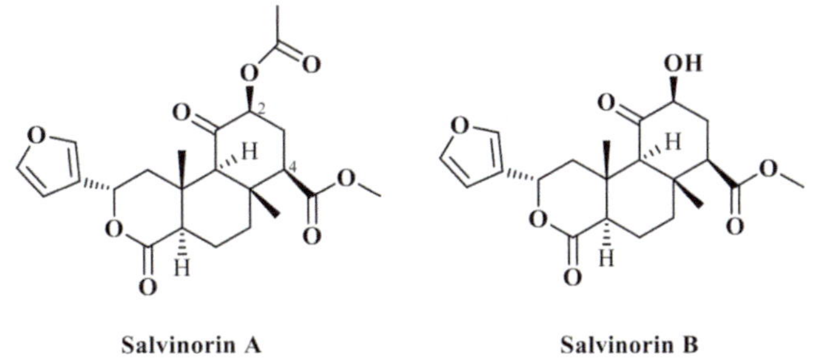

FIGURE 29.2 Chemical Structures of salvinorin A and its main metabolite, salvinorin B.

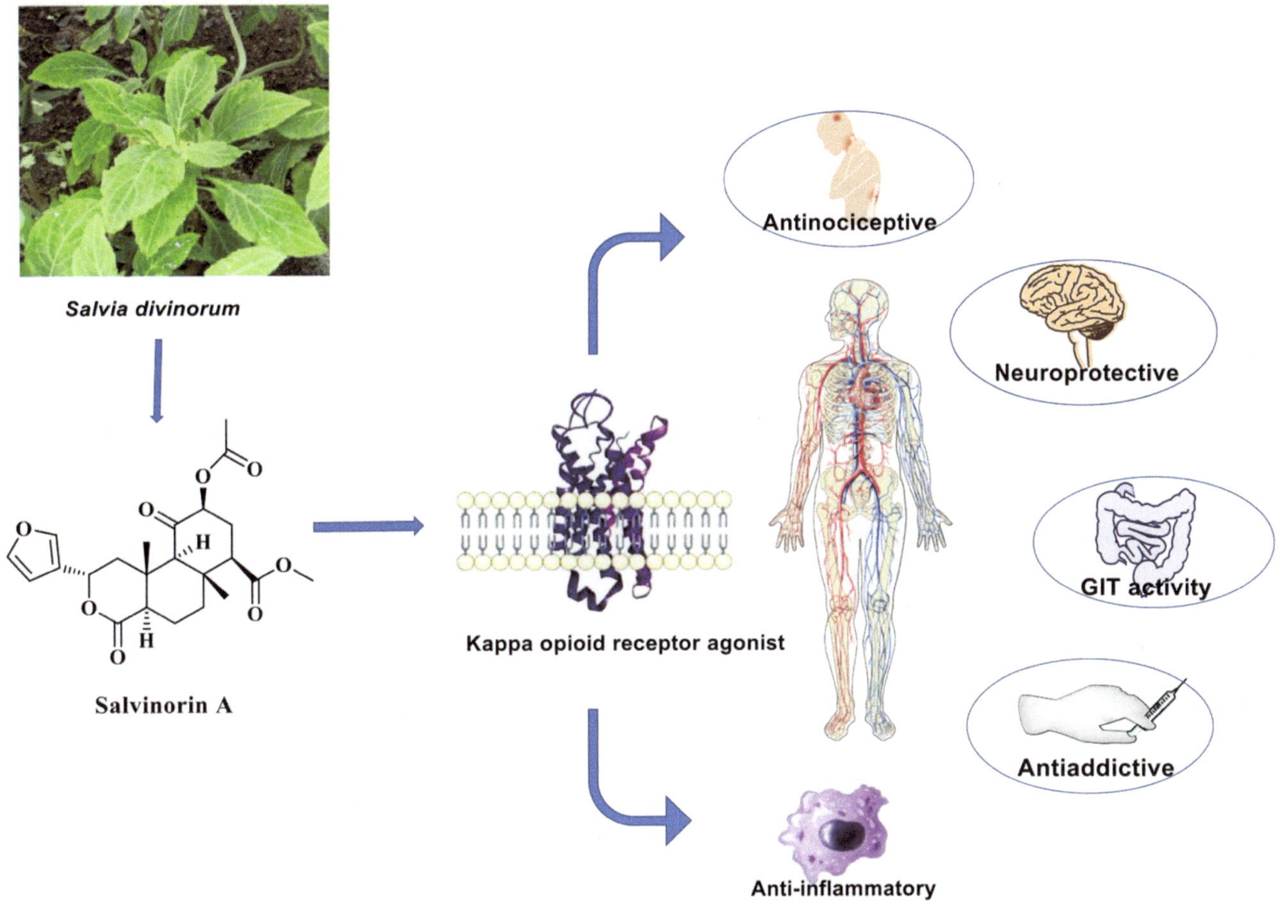

FIGURE 29.3 Summary of the medicinal effects of *Salvia divinorum* and its main bioactive component salvinorin A.

gastrointestinal, neurological, and inflammatory diseases (Gonçalves et al., 2021). Opioid agonists based on Sal-A have potential use in the treatment of cough, pain, diarrhea, mood disorders, and stimulant dependence. Thus, Sal-A and its analogs have been found to be excellent research tools as well as they offer a great insight into opioid receptor-mediated phenomena (Li et al., 2016).

29.9.1 Anti-Addiction Effect

Although *S. divinorum* is considered an infamous drug of abuse, its major component Sal-A has offered the opportunity for the evolution of novel treatments for addiction. The anti-addictive activity of Sal-A has been related to its unconventional structure and strong interaction with the KOR opioid receptor (Geck et al., 2021). KOR agonists have shown great anti-addictive activity as they have inhibited brain function alterations associated with chronic use of drugs such as cocaine (Mahendran et al., 2016). It has been reported that Sal-A could reduce dopamine levels, which play an important role in regulating the rewarding effects of abused drugs (Cruz et al., 2017). Studies carried out in rats and non-human models have demonstrated the effect of KOR agonists in attenuating the locomotive behavior related to cocaine use (da Silva Maia et al., 2018). Another study has shown the effectiveness of Sal-A to reduce cocaine-induced drug seeking activities in rats (Mahendran et al., 2016). Sal-A and its analogs have exerted anti-addictive effects in animal models by reducing dopamine levels in basal ganglia. Sal-A has also attenuated the neurochemical and behavioral alterations caused by acute and prolonged administration of psychostimulant drugs (dos Santos et al., 2014). Sal-A and its analogues have also reduced cocaine-prime induced cocaine seeking in rats without causing sedation or suppressing sucrose reinforcement in a similar manner to the traditional KOR agonists and spiradoline (Kivell et al., 2018). Sal-A has been considered a dose-dependent punisher of remifentanil and cocaine in rhesus monkeys, decreasing the self-administration of these drugs (Freeman et al., 2014). Therefore, it has been considered as a potential therapy for addiction and drug abuse. However, the rapid metabolism of Sal-A hinders its clinical development. Several Sal-A analogues with enhanced tolerability and pharmacokinetic profiles have been identified to have anti-addictive effects (Kivell et al., 2014b).

29.9.2 Antinociceptive Effect

Sal-A has received great attention as a putative antinociceptive compound due to its affinity for KOR and CB1 receptors (Cunningham et al., 2019; Turnaturi et al., 2019). Agonists of KOR have been reported to act as a safer alternative to classic opioids. Analgesic classical opioids have increased dopamine levels in rodent's nucleus accumbens inducing addictive properties like cocaine and other addictive drugs (Hernández-Alvarado et al., 2020). In addition, classic opioids acting on MOR have been reported to trigger several undesirable effects including respiratory depression (Madariaga-Mazón et al., 2017).

Several studies have indicated the analgesic and anti-inflammatory properties of *S. divinorum* and Sal-A in various models (Ujváry, 2014; Hesselink, 2018). The antinociceptive effect of Sal-A has been demonstrated by mice abdominal constriction, hot plate, and tail-flick tests. It has been found that μ and λ opioid receptor antagonists did not affect Sal-A antinociceptive response, indicating the pure and selective stimulation of Sal-A to KOR (Cruz et al., 2017). In another study, intraperitoneal and intrathecal administration of Sal-A elicited dose-dependent antinociception in hot plate and tail-flick tests in mice (Hernández-Alvarado et al., 2020). In addition, Sal-A has shown an antinociceptive effect in mice in hot plate, thermal, and chemo-nociceptive assays. intrathecal Sal-A administration increased the tail-flick latencies in mice (Coffeen and Pellicer, 2019). Sal-A (0.05–0.45 mg/kg) has also exerted an analgesic effect through the reduction of formalin pain scores in rat models (Mascarenhas et al., 2020). Sal-A injection produced a dose-dependent antinociception in mice that peaked at 10 minutes and rapidly returned to baseline levels (Mahendran et al., 2016). It has been reported that relief of neuropathic pain in rats' insular cortex is related to its interaction with kappa-opioid and cannabinoid receptors (Coffeen et al., 2018). Ethyl acetate extract of *S. divinorum* has been observed to exert sedative, antinociceptive effects in mice pain models (Tlacomulco-Flores et al., 2020). *S. divinorum* extract has exerted analgesic and antinociceptive potency in neuropathic and inflammatory pain models of rats (Simón-Arceo et al., 2017; Bodnar, 2020). Sal-A reduced neuronal hyperexcitability and glia hyperactivation, comparable to the anti-inflammatory and analgesic effects of autacoid palmitoylethanolamide (Hesselink, 2018).

Since oral Sal-A is not the most appropriate route of administration, intranasal delivery of Sal-A (which leads to immediate access of the drug to the brain) as well as inhaled Sal-A were suggested. Such new formulations could be of great benefit to treating neuropathic pain patients, due to the swift onset of action (Hesselink, 2018). Because of the rapid metabolism and certain side effects of Sal-A, similar analogs with non-addictive analgesic, anti-inflammatory antinociceptive effects have been developed (Paton et al., 2017).

29.9.3 Anti-Inflammatory Effect

(Rossi et al., 2016) has investigated the anti-inflammatory properties of Sal-A. Efficient inhibition of leukotriene (LT) synthesis has been demonstrated in isolated rat peritoneal macrophages as well as in carrageenan-induced pleurisy rats and zymosan-induced peritonitis mouse models. Sal-A has decreased inflammatory cells, exudate volume, and myeloperoxidase activity in rat models. In addition, inhibition of cell infiltration, vascular permeability, and myeloperoxidase activity has been reported in mouse models. In addition, Sal-A has reduced the inducible nitric oxide synthase expression in macrophages and TNF-α level and has restored the production of cytokine IL-10 (Coffeen and Pellicer, 2019). It has also inhibited LT synthesis in experimental models of inflammation (Howes, 2018). An *in vivo* study has revealed the effectiveness of Sal-A to inhibit airway hyperreactivity in an air pouch mouse model, through inhibition of leukotriene LT, IL-13, and mast cell degranulation. Therefore, Sal-A act as a promising drug candidate for allergic inflammatory diseases as asthma (Rossi et al., 2017).

29.9.4 Neuroprotective Effect

Sal-A has been reported to prompt a neuroprotective effect by inhibiting brain edema, neuronal death, and brain injury, as well as improving neurologic recovery in the cortex, striatum, and hippocampus (Xin et al., 2016). The rapid onset of action, crossing the blood–brain barrier (BBB), and selective KOR agonism of Sal-A make it a promising candidate for stroke treatment (Grothusen and Liu, 2019). Administration of Sal-A by early stroke patients might inhibit glutamate-induced hyperactivation and reduce brain injury. Thus, Sal-A could be useful in the treatment of post-traumatic stress and anxiety disorders that have been correlated to cortico-amygdala hyperactivation (Cichon et al., 2022). The rapid onset and short-acting neuroprotective effect of Sal-A have shown no undesirable effects as respiratory depression; therefore, it is suitable in the treatment of cardiac arrest, stroke, and asphyxia (Su et al., 2012).

The efficacy of Sal-A in the treatment of acute stroke has been exhibited in multiple animal models, including monkeys, piglets, and rodents (Cichon et al., 2022). Intranasal Sal-A has significantly decreased the Non-Human Primate Stroke Scale scores to normal in a 28-day observation period. It has improved neurological outcomes and decreased infarct volume in a rhesus monkey ischemic stroke model in autologous blood clot assay (Wu et al., 2021). Intravenous administration of Sal-A (20 μg/kg) has reduced apoptosis and damage in oxygen–glucose deprivation human brain microvascular endothelial cell model. It has also shown a reduction in cerebral edema and improvement of BBB permeability in the middle cerebral artery occlusion rat model. Protection of neuronal mitochondrial function has been verified through stabilization of mitochondrial membrane potential, reduction of oxidative stress, and preservation of functions and morphology of mitochondria in cerebrovascular endothelial cells (Dong et al., 2019). Intravenous Sal-A (10 and 20 μg/kg) has also improved edema and inflammation in a transient cerebral ischemic rat model. Reducing cell apoptosis and Autoregulating the cerebral pial artery have resulted in motor function improvement and ischemic injury prevention (Dong et al., 2018). Intraperitoneal Sal-A (2 and 10 μg/kg) has alleviated brain injury and cerebral vasospasm in subarachnoid hemorrhage (SAH) rat model via increasing expression of endothelial NO synthase and nitric oxide content and decreasing Endothelin-1 concentration and aquaporin-4 protein expression (Sun et al., 2018). It has also inhibited lateral ventricle expansion via activation of PI3K/Akt signaling pathway and decreasing apoptosis-related proteins in the hippocampus (Sun et al., 2019). Specific KOR activation by intranasal Sal-A has shown an improvement in neurological development, protection of the vascular integrity, and reduction in cerebral infarct volume in a middle cerebral artery occlusion mouse model. It also has also showed a decrease in caspase-3 activation reducing neuron apoptosis (Chen et al., 2016).

29.9.5 Antidepressant Effect

Several studies have demonstrated the variable effects of Sal-A to promote both pro- and antidepressive behaviors. The modulation of depressive behaviors by Sal-A may be influenced by drug administration, dose, or other procedural variables (Taylor and Manzella, 2016). The depressive effects of high doses of Sal-A (2.0 mg/kg) have been exhibited in behavioral assays as it has decreased rats' locomotor activity and motivational behavior (Cruz et al., 2017). Other study has showed the ability of Sal-A to increase immobility time of rats in FST and increase threshold in the ICSS model with a reduction in dopamine signaling suggesting pro-depressive activity (McHugh and Kelly, 2018).

However, several assays have revealed the antidepressant activity of Sal-A. Clinical studies have confirmed the therapeutic effect of *S. divinorum* in post-traumatic stress disorder and major depressive disorder (Siegel et al., 2021). Other clinical studies have demonstrated the effectiveness of oral extract of *S. divinorum* in attenuating depression (Abdelhalim and Hanrahan, 2021). Chewing *S. divinorum* leaves has been found to completely improve the symptoms of depression in a case study (Cruz et al., 2017). In addition, the antidepressant effect of *S. divinorum* has been verified in various animal models. It has promoted the antidepressive behaviors of mice in TST and rats in FST at doses of 0.001–1000 μg/kg. Sal-A has shown great enhancement of rats' attention and memory (Abdelhalim and Hanrahan, 2021). Animal studies have shown the anxiolytic activities of intraperitoneal Sal-A in the tail-suspension test in mice and the plus-maze test in rats at a dose of 1 μg/kg. Meanwhile, slightly higher doses of Sal-A have shown antidepressant activity in forced-swimming tests in mice. *S. divinorum* extracts have elicited sedative-like effects that could modify physiological sleep in rats, elicit sleep behavior disorders affecting the function of the REM stage due to indicating the role of SA as sleep disruptors (González-Trujano et al., 2016).

Thus, it was concluded that low doses of Sal-A may present anxiolytic effects and are used in the treatment of depression-like symptoms. While high doses might be involved in cognitive alterations in schizophrenia and other psychiatric disorders (Cruz et al., 2017).

29.9.6 Gastrointestinal Effect

The gastrointestinal effectiveness of *S. divinorum* has been attributed to the interaction of Sal-A with opioid and endocannabinoid receptors. KOR are functionally related to the gut hypermotility in the inflamed intestine (Badal et al., 2018). Therefore, KOR ligands have been reported to play an important role in the treatment of several GI disorders, such as irritable bowel syndrome, intestinal inflammation, and postoperative ileus. Additionally, CB receptor activation inhibits excitatory cholinergic neurotransmission (Coffeen and Pellicer, 2019).

The leaf extract of *S. divinorum* has been reported to inhibit the enteric cholinergic transmission in guinea-pig ileum, while Sal-A has significantly inhibited the inflamed gut motility (Calzada and Bautista, 2020). In another study carried out in mice, *S. divinorum* extract and Sal-A inhibited mice gut motility in physiological and inflammatory conditions (Hernández-Alvarado et al., 2020). In addition, the usefulness of *S. divinorum* in stopping diarrhea caused by protozoa has been reported. Sal-A has shown potent antiprotozoal activities with IC_{50} values of 49.0 μM against *Entamoeba histolytica* and 64.8 μM against *Giardia lamblia* (Calzada et al., 2015). (Coffeen and Pellicer, 2019) have reported the inhibitory effect of Sal-A on ileitis-induced hypermotility as well as on hypermotility of the mouse ileum, stomach, and colon; in addition, it has delayed GI transit. In addition, Sal-A has exhibited a potent anti-inflammatory effect in experimental colitis and an antinociceptive effect in mice with developed colitis, therefore it acts as an alternative treatment for inflammatory bowel syndrome (Szczepaniak et al., 2022).

29.9.7 Antioxidants Effect

Another important medicinal effect of *S. divinorum* is its antioxidant activity that inhibits degenerative diseases triggered by oxidative stress such as cancer, asthma, atherosclerosis, diabetes, dementia, cardiovascular, and inflammatory diseases. Ethanol, methanol, and aqueous extracts of *S. divinorum* have been found to scavenge free radicals with IC_{50} values of 60.9, 65.5, and 111.2 lg/mL (Jafri et al., 2023).

29.9.8 Antibacterial Effect

The traditional application of *S. divinorum* has been justified by several studies. Sal-A has attenuated acute lung injury by methicillin-resistant *Staphylococcus aureus* (MRSA) infection in mice. It has induced a reduction in pulmonary neutrophil infiltration and mRNA expression of pro-inflammatory cytokines such as IL-1β, IL-6, and TNF. While *in vitro* assays have shown the beneficial effects of Sal-A in apoptosis, oxidative stress, and inflammation in RAW264.7 cells (Zeng et al., 2021). However, other studies have reported very low antibacterial activity of the hydrolate of *S. divinorum* against *E. asburiae*, *E.coli*, *B. subtilis*, and *M. luteus* (Ürgeová et al., 2023).

29.9.9 Anticancer Effect

Sal-A and its analogs have been well known for their rapid penetration of BBB and antiproliferative activity in brain tumors. Analogues of Sal-A have been reported to exert antiproliferative effect against the MCF7 breast cancer cell line (Vasiljevik et al., 2014), Hep G2 and Caco 2 cell lines (Martinho et al., 2016). Quinone containing Sal-A analogs have shown significant antiproliferative efficiency in breast cancer cell lines in addition to the passive diffusion across the BBB *in vitro* model. Therefore, the neoclerodane core can act as a potent scaffold for the development of anticancer and antiproliferative drugs for the central nervous system (Hernández-Alvarado et al., 2020).

29.9.10 Anti-COV-2 Effect

It has been reported that Sal-A is effective in severe acute respiratory syndrome coronaviruses infections (SARS-COV-2) infection (Sharma et al., 2020). Molecular docking studies have shown the activity of Sal-A against SARS-COV-2 protease $3CL^{pro}$ with binding energy −32181 (Ullah et al., 2022).

29.9.10.1 Salvinorin A Analogues

The unique structure and distinctive biological characteristics of Sal-A have attracted the attention for exploring opioid pharmacology. It acts as an excellent template to develop new semisynthetic molecules used in human therapies and may be free from hallucinogenic properties (Pittaluga et al., 2013). The short action of Sal-A is coped by the analogs of Sal-A of extended half-life time (Cichon et al., 2022). The short duration of action of Sal-A is due to the rapid ester hydrolysis of the C2 ester group, yielding the inactive metabolite salvinorin B. However, Sal-A analogs and related drug candidates have been developed by replacing the C2 ester group with more stable functional groups such as ethers, amides, and carbamates. They are used to target different opioid receptors with extended half-life time (Roach and Shenvi, 2018; Keasling et al., 2019; Akins et al., 2022).

29.10 CONCLUSION

Until the mid-60s, the use of *S. divinorum* was confined to the Mazatec Indigenous people. They have used the mystic plant for centuries for spiritual and ritual purposes. However, its use as a recreational herb is a recent phenomenon. It has gained great popularity worldwide as a drug of abuse because of its potent hallucinogenic properties. Nowadays, smart shops and internet and social media vendors act as widespread sources of *S. divinorum* in the form of tinctures, dried leaves, or concentrated Sal-A leaf extract.

Sal-A, a neoclarodane diterpene, is the major bioactive component of *S. divinorum*. It acts as an important candidate in the scientific community because of the rapid onset of action, crossing BBB, and potent effectiveness. It represents a new class of KOR selective agonists due to the lack of the basic nitrogen atom. It interacts with KOR in qualitatively different manner. Sal-A induces multiple therapeutic activities in various conditions and diseases including addiction, pain, inflammation, depression, and stroke. In an effort to improve the pharmacokinetics and tolerability of this compound, kappa-opioid receptor agonists based on the structure of Sal-A have been synthesized. *S. divinorum* and the semisynthetic Sal-A analogues have been widely used for medicinal purposes because of their low dependence, safe physiological, and low toxicity profile.

Finally, both *S. divinorum* and Sal-A have attracted great concern as drugs of abuse due to easy availability and low cost. However, the therapeutic effectiveness of this mystic plant or the semisynthetic analogs of Sal-A should be further explored.

LIST OF ABBREVIATIONS

BBB	blood–brain barrier
DAT	dopamine re-uptake transporter
DOR	delta-opioid receptors
GPCR	G protein-coupled receptor

KOR	kappa-opioid receptors
LSD	lysergic acid diethylamide
LT	leukotriene
MOR	Mu-opioid receptors
S. divinorum	*Salvia divinorum*
Sal-A	salvinorin A
SAH	subarachnoid hemorrhage

REFERENCES

Abdelhalim, A. and J. Hanrahan (2021) Biologically active compounds from Lamiaceae family: Central nervous system effects. *Studies in Natural Products Chemistry* 68 255–315.

Addy, P. H. (2012) Acute and post-acute behavioral and psychological effects of salvinorin A in humans. *Psychopharmacology* 220 195–204.

Addy, P. H. (2016) Behavioral and psychological effects of *Salvia divinorum*: A focus on self-reported subjective acute behavioral effects and laboratory studies. *Neuropathology of Drug Addictions and Substance Misuse* 2 733–738.

Addy, P. H., A. Garcia-Romeu, M. Metzger and J. Wade (2015) The subjective experience of acute, experimentally-induced *Salvia divinorum* inebriation. *Journal of Psychopharmacology* 29 (4) 426–435.

Aixalá, M., G. Ona, Ò. Parés and J. C. Bouso (2020) Patterns of use, desired effects, and mental health status of a sample of natural psychoactive drug users. *Drugs: Education, Prevention and Policy* 27 (3) 191–198.

Akaberi, M., S. Mehri and M. Iranshahi (2015) Multiple pro-apoptotic targets of abietane diterpenoids from Salvia species. *Fitoterapia* 100 118–132.

Akins, N. S., N. Mishra, H. M. Harris, N. Dudhipala, S. J. Kim, A. W. Keasling, S. Majumdar, J. K. Zjawiony, J. J. Paris and N. M. Ashpole (2022) 6, 5-fused ring, C2-salvinorin ester, dual kappa and Mu opioid receptor agonists as analgesics devoid of anxiogenic effects. *ChemMedChem* 17 (7) e202100684.

Asadollahi, M., O. Firuzi, F. H. Jamebozorgi, M. Alizadeh and A. R. Jassbi (2019) Ethnopharmacological studies, chemical composition, antibacterial and cytotoxic activities of essential oils of eleven Salvia in Iran. *Journal of Herbal Medicine* 17 100250.

Avula, B., J.-Y. Bae, A. G. Chittiboyina, Y.-H. Wang, M. Wang, R. Srivedavyasasri, Z. Ali, J. Li, C. Wu and I. A. Khan (2022) Comparative analysis of five Salvia species using LC-DAD-QToF. *Journal of Pharmaceutical and Biomedical Analysis* 209 114520.

Badal, S., S. Turfus, R. Rajnarayanan, C. Wilson-Clarke and S. L. Sandiford (2018) Analysis of natural product regulation of opioid receptors in the treatment of human disease. *Pharmacology & Therapeutics* 184 51–80.

Bagheri, G., S. A. Mousavi and S. M. Mousavi (2016) Anatomy of the medicinal plants Salvia hydrangea, Salvia glutinosa and Salvia syriaca in Iran. *J Res Ecol* 4 (2) 216–220.

Biggs, J. M., J. A. Morgan, A. B. Lardieri, O. A. Kishk and W. Klein-Schwartz (2017) Abuse and misuse of selected dietary supplements among adolescents: a look at poison center data. *The Journal of Pediatric Pharmacology and Therapeutics* 22 (6) 385–393.

Bischoff, K. (2018) Toxicity of drugs of abuse. *n Veterinary toxicology (pp. 385-408). Academic Press.*

Bodnar, R. J. (2020) Endogenous opiates and behavior: 2017. *Peptides* 124 170223.

Brito-da-Costa, A. M., D. Dias-da-Silva, N. G. M. Gomes, R. J. Dinis-Oliveira and Á. Madureira-Carvalho (2021) Pharmacokinetics and pharmacodynamics of salvinorin a and *salvia divinorum*: Clinical and forensic aspects. *Pharmaceuticals* 14 (2) 116.

Butelman, E. R. and M. J. Kreek (2015) Salvinorin A, a kappa-opioid receptor agonist hallucinogen: pharmacology and potential template for novel pharmacotherapeutic agents in neuropsychiatric disorders. *Frontiers in Pharmacology* 6 190.

Calzada, F. and E. Bautista (2020) Plants used for the treatment of diarrhoea from Mexican flora with amoebicidal and giadicidal activity, and their phytochemical constituents. *Journal of Ethnopharmacology* 253 112676.

Calzada, F., E. Bautista, L. Yépez-Mulia, N. García-Hernandez and A. Ortega (2015) Antiamoebic and antigiardial activity of clerodane diterpenes from Mexican Salvia species used for the treatment of diarrhea. *Phytotherapy Research* 29 (10) 1600–1604.

Carod-Artal, F. J. (2015) Hallucinogenic drugs in pre-Columbian Mesoamerican cultures. *Neurología (English Edition)* 30 (1) 42–49.

Casselman, I., C. J. Nock, H. Wohlmuth, R. P. Weatherby and M. Heinrich (2014) From local to global—Fifty years of research on *Salvia divinorum*. *Journal of Ethnopharmacology* 151 (2) 768–783.

Chaachouay, N., Douira, A., Hassikou, R., Brhadda, N., Dahmani, J., Belahbib, N., Ziri, R., & Zidane, L. (2020). Ethnobotanical and ethnopharmacological study of medicinal and aromatic plants used in the treatment of respiratory system disorders in the Moroccan Rif. PhDThesis, Département de Biologie - Université Ibn Tofail - Kénitra, Maroc, https://tel.archives-ouvertes.fr/tel-03376377

Chaachouay, N., Douira, A., & Zidane, L. (2021a). Herbal Medicine Used in the Treatment of Human Diseases in the Rif, Northern Morocco. *Arabian Journal for Science and Engineering*, 47, 131–153. https://doi.org/10.1007/s13369-021-05501-1

Chaachouay, N., Douira, A., & Zidane, L. (2021b). COVID-19, prevention and treatment with herbal medicine in the herbal markets of Salé Prefecture, North-Western Morocco. *European Journal of Integrative Medicine*, 42, 101285. https://doi.org/10.1016/j.eujim.2021.101285

Chakraborty, S. and S. Majumdar (2020) Natural products for the treatment of pain: Chemistry and pharmacology of salvinorin A, mitragynine, and collybolide. *Biochemistry* 60 (18) 1381–1400.

Che, T., S. Majumdar, S. A. Zaidi, P. Ondachi, J. D. McCorvy, S. Wang, P. D. Mosier, R. Uprety, E. Vardy and B. E. Krumm (2018) Structure of the nanobody-stabilized active state of the kappa opioid receptor. *Cell* 172 (1–2) 55–67.

Chen, C., C. Xi, X. Liang, J. Ma, D. Su, T. Abel and R. Liu (2016) The role of kappa opioid receptor in brain ischemia. *Critical Care Medicine* 44 (12) e1219.

Chen, X., A. Berim, F. E. Dayan and D. R. Gang (2017) A (–)-kolavenyl diphosphate synthase catalyzes the first step of salvinorin A biosynthesis in *Salvia divinorum*. *Journal of Experimental Botany* 68 (5) 1109–1122.

Cichon, J., R. Liu and H. V. Le (2022) Therapeutic Potential of Salvinorin A and Its Analogues in Various Neurological Disorders. *Translational Perioperative and Pain Medicine* 9 (2) 452.

Clark, S. D. and A. Abi-Dargham (2019) The role of dynorphin and the kappa opioid receptor in the symptomatology of schizophrenia: a review of the evidence. *Biological Psychiatry* 86 (7) 502–511.

Coffeen, U., A. Canseco-Alba, K. Simón-Arceo, A. Almanza, F. Mercado, M. León-Olea and F. Pellicer (2018) Salvinorin A reduces neuropathic nociception in the insular cortex of the rat. *European Journal of Pain* 22 (2) 311–318.

Coffeen, U. and F. Pellicer (2019) *Salvia divinorum*: from recreational hallucinogenic use to analgesic and anti-inflammatory action. *Journal of Pain Research* 12 1069–1076.

Cruz, A., S. Domingos, E. Gallardo and A. Martinho (2017) A unique natural selective kappa-opioid receptor agonist, salvinorin A, and its roles in human therapeutics. *Phytochemistry* 137 9–14.

Cunningham, C. W., W. M. Elballa and S. U. Vold (2019) Bifunctional opioid receptor ligands as novel analgesics. *Neuropharmacology* 151 195–207.

da Silva, D. D., J. P. Silva, H. Carmo and F. Carvalho (2021) Neurotoxicity of psychoactive substances: A mechanistic overview. *Current Opinion in Toxicology* 28 76–83.

da Silva Maia, A. F., L. da Silva Neto, F. D. Garcia, R. B. Alves and Â. de Fátima (2018) Natural products in addiction treatment: A brief survey. *In Discovery and Development of Neuroprotective Agents from Natural Products Natural Product Drug Discovery* 415–434. https://doi.org/10.1016/B978-0-12-809593-5.00011-2

Díaz, J.-L. (2013) *Salvia divinorum*: A psychopharmacological riddle and a mind-body prospect. *Current Drug Abuse Reviews* 6 (1) 43–53.

Diaz, J. H. (2016) Poisoning by herbs and plants: rapid toxidromic classification and diagnosis. *Wilderness & Environmental Medicine* 27 (1) 136–152.

Dong, H., W. Zhou, J. Xin, H. Shi, X. Yao, Z. He and Z. Wang (2019) Salvinorin A moderates postischemic brain injury by preserving endothelial mitochondrial function via AMPK/Mfn2 activation. *Experimental Neurology* 322 113045.

Dong, H. P., W. Zhou, X. X. Ma, Z. Z. He and Z. H. Wang (2018) Salvinorin A preserves cerebral pial artery autoregulation after forebrain ischemia via the PI3K/AKT/cGMP pathway. *Brazilian Journal of Medical and Biological Research* 51 (5) e6714.

dos Santos, R. G., J. A. S. Crippa, J. P. Machado-de-Sousa and J. E. C. Hallak (2014) Salvinorin A and related compounds as therapeutic drugs for psychostimulant-related disorders. *Current Drug Abuse Reviews* 7 (2) 128–132.

Dunkić, V., D. Kremer, R. J. Grubešić, J. V. Rodríguez, D. Ballian, F. Bogunić, D. Stešević, I. Kosalec, N. Bezić and E. Stabentheiner (2017) Micromorphological and phytochemical traits of four Clinopodium L. species (Lamiaceae). *South African Journal of Botany* 111 232–241.

Eksi, G., S. Kurbanoglu and S. A. Erdem (2020) Analysis of diterpenes and diterpenoids *Recent advances in natural products analysis*, Elsevier. 313–345.

Faro, A. F. L., A. Di Trana, N. La Maida, A. Tagliabracci, R. Giorgetti and F. P. Busardò (2020) Biomedical analysis of New Psychoactive Substances (NPS) of natural origin. *Journal of Pharmaceutical and Biomedical Analysis* 179 112945.

Farré, M., E. Papaseit, F. Fonseca and M. Torrens (2021) Addiction of hallucinogens, dissociatives, designer drugs and 'Legal Highs': Update on potential therapeutic use. *Textbook of addiction treatment: International perspectives* Springer, Cham. 259–279.

Faudree, P. (2020) 'Making Medicine' with *Salvia divinorum*: Competing approaches and their implications. *Medical Anthropology* 39 (7) 582–596.

Feng, L.-Y., A. Battulga, E. Han, H. Chung and J.-H. Li (2017) New psychoactive substances of natural origin: A brief review. *Journal of Food and Drug Analysis* 25 (3) 461–471.

Freeman, K. B., J. E. Naylor, T. E. Prisinzano and W. L. Woolverton (2014) Assessment of the kappa opioid agonist, salvinorin A, as a punisher of drug self-administration in monkeys. *Psychopharmacology* 231 2751–2758.

Geck, M. S., D. Lecca, G. Marchese, L. Casu and M. Leonti (2021) Ethnomedicine and neuropsychopharmacology in Mesoamerica. *Journal of Ethnopharmacology* 278 114243.

Girn, M., F. E. Rosas, R. E. Daws, C. L. Gallen, A. Gazzaley and R. L. Carhart-Harris (2023) A complex systems perspective on psychedelic brain action. *Trends in Cognitive Sciences*. 27 (5) 433–445.

Gonçalves, J., Â. Luís, E. Gallardo and A. P. Duarte (2021) Psychoactive substances of natural origin: Toxicological aspects, therapeutic properties and analysis in biological samples. *Molecules* 26 (5) 1397.

González-Trujano, M. E., F. Brindis, E. López-Ruiz, I. Ramírez-Salado, A. Martínez and F. Pellicer (2016) Depressant effects of *Salvia divinorum* involve disruption of physiological sleep. *Phytotherapy Research* 30 (7) 1137–1145.

Griffin III, O. H. (2013) *Salvia divinorum*, Hallucinogens, and the determination of medical utility *Prohibition, religious freedom, and human rights: Regulating traditional drug use*, Springer. 149–164.

Grothusen, J. and R. Liu (2019) Salvinorin A as a potential medication for stroke. *Translational Perioperative and Pain Medicine* 6 115–119.

Gürlek, F., E. Tasdemir and T. Teker (2017) *Salvia divinorum*: A case report. *Journal of Clinical Toxicology* 7 340.

Hao, D. C., X. J. Gu, P. G. Xiao, D. C. Hao, X. J. Gu and P. G. Xiao (2015) Phytochemical and biological research of Salvia medicinal resources. *Medicinal plants. Woodhead Publishing*, 587–639.

Hatipoglu, S. D., B. Yalcinkaya, M. Akgoz, T. Ozturk, A. C. Goren and G. Topcu (2017) Screening of hallucinogenic compounds and genomic characterisation of 40 Anatolian Salvia species. *Phytochemical Analysis* 28 (6) 541–549.

Heal, D. J., J. Gosden and S. L. Smith (2018) Evaluating the abuse potential of psychedelic drugs as part of the safety pharmacology assessment for medical use in humans. *Neuropharmacology* 142 89–115.

Hernández-Alvarado, R. B., A. Madariaga-Mazón, A. Ortega and K. Martinez-Mayorga (2020) DARK classics in chemical neuroscience: salvinorin A. *ACS Chemical Neuroscience* 11 (23) 3979–3992.

Hernández-Bello, R., R. V. García-Rodríguez, K. García-Sosa, L. M. Peña-Rodríguez, M. Vázquez-Hernández, F. R. Ramos-Morales, O. Corcoran and A. Sánchez-Medina (2015) Salvinorin A content in legal high products of *Salvia divinorum* sold in Mexico. *Forensic Science International* 249 197–201.

Herrmann, E. S., P. S. Johnson, M. W. Johnson and R. Vandrey (2016) Novel drugs of abuse: Cannabinoids, stimulants, and hallucinogens. *Neuropathology of drug addictions and substance misuse*. Elsevier. 893–902

Hesselink, J. M. K. (2018) Salvinorin A, an herbal secondary metabolite from *Salvia Divinorum* as an anti-neuropathic pain scaffold. *Annals of Pain Medicine 1 (1)* 1006

Howes, M.-J. R. (2018) Phytochemicals as anti-inflammatory nutraceuticals and phytopharmaceuticals. *Immunity and inflammation in health and disease*. Elsevier. 363–388

Jafri, S. A. A., Z. M. Khalid, M. R. Khan, S. Ashraf, N. Ahmad, A. M. Karami, E. Rafique, M. Ouladsmane, N. M. S. Al

Suliman and S. Aslam (2023) Evaluation of some essential traditional medicinal plants for their potential free scavenging and antioxidant properties. *Journal of King Saud University-Science* 53 (1) 102562.

Jenks, A. A. and S.-C. Kim (2013) Medicinal plant complexes of Salvia subgenus Calosphace: An ethnobotanical study of new world sages. *Journal of ethnopharmacology* 146 (1) 214–224.

Johnson, L. A., R. L. Johnson and R.-B. Portier (2013) Current 'legal highs'. *The Journal of Emergency Medicine* 44 (6) 1108–1115.

Johnson, M. W., K. A. MacLean, M. J. Caspers, T. E. Prisinzano and R. R. Griffiths (2016) Time course of pharmacokinetic and hormonal effects of inhaled high-dose salvinorin A in humans. *Journal of Psychopharmacology* 30 (4) 323–329.

Karam, A., A. Said, C. Assaad, S. Hallit, G. Haddad, D. Hachem and F. Kazour (2019) Abuse and effects of *salvia divinorum* in a sample of patients hospitalized for substance dependence. *Community mental health journal* 55 702–708.

Karila, L., M. Marillier, B. Chaumette, J. Billieux, N. Franchitto and A. Benyamina (2019) New synthetic opioids: Part of a new addiction landscape. *Neuroscience & Biobehavioral Reviews* 106 133–140.

Keasling, A. W., P. Pandey, R. J. Doerksen, G. R. Pedrino, E. A. Costa, L. C. da Cunha, J. K. Zjawiony and J. O. Fajemiroye (2019) Salvindolin elicits opioid system-mediated antinociceptive and antidepressant-like activities. *Journal of Psychopharmacology* 33 (7) 865–881.

Keasling, A. W. and J. K. Zjawiony (2016) The Plant *Salvia divinorum* (Lamiaceae)—Chemistry and pharmacology. *Neuropathology of drug addictions and substance misuse*. Elsevier. 551–560.

Kivell, B., Z. Uzelac, S. Sundaramurthy, J. Rajamanickam, A. Ewald, V. Chefer, V. Jaligam, E. Bolan, B. Simonson and B. Annamalai (2014b) Salvinorin A regulates dopamine transporter function via a kappa opioid receptor and ERK1/2-dependent mechanism. *Neuropharmacology* 86 228–240.

Kivell, B. M., A. W. M. Ewald and T. E. Prisinzano (2014a) Salvinorin A analogs and other kappa-opioid receptor compounds as treatments for cocaine abuse. *Advances in Pharmacology* 69 481–511.

Kivell, B. M., K. F. Paton, N. Kumar, A. S. Morani, A. Culverhouse, A. Shepherd, S. A. Welsh, A. Biggerstaff, R. S. Crowley and T. E. Prisinzano (2018) Kappa opioid receptor agonist mesyl Sal B attenuates behavioral sensitization to cocaine with fewer aversive side-effects than salvinorin A in rodents. *Molecules* 23 (10) 2602.

Kowalczuk, A. P., V. Raman, A. M. Galal, I. A. Khan, D. J. Siebert and J. K. Zjawiony (2014) Vegetative anatomy and micromorphology of *Salvia divinorum* (Lamiaceae) from Mexico, combined with chromatographic analysis of salvinorin A. *Journal of Natural Medicines* 68 63–73.

Li, R., S. L. Morris-Natschke and K.-H. Lee (2016) Clerodane diterpenes: Sources, structures, and biological activities. *Natural Product Reports* 33 (10) 1166–1226.

Lin, P.-X., J.-H. Li, S.-H. Chen, H.-C. Chang and R. McKetin (2014) Quantitative determination of salvinorin A, a natural hallucinogen with abuse liability, in Internet-available *Salvia divinorum* and endemic species of Salvia in Taiwan. *Journal of Food and Drug Analysis* 22 (3) 370–378.

Loizzo, M. R., M. Abouali, P. Salehi, A. Sonboli, M. Kanani, F. Menichini and R. Tundis (2014) In vitro antioxidant and antiproliferative activities of nine Salvia species. *Natural Product Research* 28 (24) 2278–2285.

Lopresti, A. L. (2017) Salvia (Sage): A review of its potential cognitive-enhancing and protective effects. *Drugs in R&D* 17 (1) 53–64.

MacLean, K. A., M. W. Johnson, C. J. Reissig, T. E. Prisinzano and R. R. Griffiths (2013) Dose-related effects of salvinorin A in humans: dissociative, hallucinogenic, and memory effects. *Psychopharmacology* 226 381–392.

Madariaga-Mazón, A., A. F. Marmolejo-Valencia, Y. Li, L. Toll, R. A. Houghten and K. Martinez-Mayorga (2017) Mu-Opioid receptor biased ligands: A safer and painless discovery of analgesics? *Drug Discovery Today* 22 (11) 1719–1729.

Mahendran, R., H. A. Lim, J. Y. S. Tan, S. M. Chua and M. Winslow (2016) *Salvia divinorum*: An overview of the usage, misuse, and addiction processes. *Asia-Pacific Psychiatry* 8 (1) 23–31.

Maqueda, A. E., M. Valle, P. H. Addy, R. M. Antonijoan, M. Puntes, J. Coimbra, M. R. Ballester, M. Garrido, M. González and J. Claramunt (2015) Salvinorin-A induces intense dissociative effects, blocking external sensory perception and modulating interoception and sense of body ownership in humans. *International Journal of Neuropsychopharmacology* 18 (12) pyv065.

Maqueda, A. E., M. Valle, P. H. Addy, R. M. Antonijoan, M. Puntes, J. Coimbra, M. R. Ballester, M. Garrido, M. González and J. Claramunt (2016) Naltrexone but not ketanserin antagonizes the subjective, cardiovascular, and neuroendocrine effects of salvinorin-A in humans. *International Journal of Neuropsychopharmacology* 19 (7). pyw016.

Margalho, C., F. Corte-Real, M. López-Rivadulla and E. Gallardo (2016) *Salvia divinorum*: Toxicological aspects and analysis in human biological specimens. *Bioanalysis* 8 (13) 1415–1425.

Margalho, C., E. Gallardo, A. Castanheira, D. N. Vieira, M. López-Rivadulla and F. C. Real (2013) A validated procedure for detection and quantitation of salvinorin a in pericardial fluid, vitreous humor, whole blood and plasma using solid phase extraction and gas chromatography–mass spectrometry. *Journal of Chromatography A* 1304 203–210.

Martinho, A., S. M. Silva and E. Gallardo (2016) Cytotoxic effects of salvinorin A, a major constituent of *Salvia divinorum*. *Medicinal Chemistry* 12 (5) 432–440.

Martins, J. and S. Brijesh (2018) Phytochemistry and pharmacology of anti-depressant medicinal plants: A review. *Biomedicine & Pharmacotherapy* 104 343–365.

Mascarenhas, C. J., R. Liu and G. A. Barr (2020) Effects of plant-derived analgesic compounds sinomenine and salvinorin A in infant rats. *Journal of Integrative Medicine* 18 (2) 174–180.

McHugh, K. L. and J. P. Kelly (2018) Modulation of the central opioid system as an antidepressant target in rodent models. *Progress in Brain Research* 239 49–87.

Mendes, F. R., C. dos Santos Costa, V. D. Wiltenburg, G. Morales-Lima, J. A. B. Fernandes and R. Filev (2022) Classic and non-classic psychedelics for substance use disorder: a review of their historic, past and current research. *Addiction Neuroscience*. 3 100025.

Mesquita, L. S. S. D., T. R. S. A. Luz, J. W. C. D. Mesquita, D. F. Coutinho, F. M. M. D. Amaral, M. N. D. S. Ribeiro and S. Malik (2019) Exploring the anticancer properties of essential oils from family Lamiaceae. *Food Reviews International* 35 (2) 105–131.

Murphy, T. M. and G. Bola (2013) DNA identification of *Salvia divinorum* samples. *Forensic Science International: Genetics* 7 (1) 189–193.

Ott, J. (1995) Ethnopharmacognosy and human pharmacology of *Salvia divinorum* and salvinorin A. *Curare* 18 103–129.

Paton, K. F., N. Kumar, R. S. Crowley, J. L. Harper, T. E. Prisinzano and B. M. Kivell (2017) The analgesic and anti-inflammatory effects of Salvinorin A analogue β-tetrahydropyran Salvinorin B in mice. *European Journal of Pain* 21 (6) 1039–1050.

Pentea, M., M. Butu, I. Samfira, R. T. Cristina and M. Butnariu (2015) Extraction and analytical study of salvinorin from leaves of *Salvia divinorum. Digest Journal of Nanomaterials and Biostructures* 10 (1) 291–297.

Phipps, S. M. and O. Grundmann (2017) Pharmacology and structure-activity relationship of natural Products with psychoactive effects from *Salvia divinorum*, Mitragyna speciosa, and ayahuasca. *Studies in Natural Products Chemistry* 53 1–44.

Pittaluga, A., G. Olivero, S. Di Prisco, E. Merega, A. Bisio, G. Romussi, M. Grilli and M. Marchi (2013) Effects of the neoclerodane Hardwickiic acid on the presynaptic opioid receptors which modulate noradrenaline and dopamine release in mouse central nervous system. *Neurochemistry International* 62 (4) 354–359.

Prisco, L., A. Sarwal, M. Ganau and F. Rubulotta (2021) Toxicology of psychoactive substances. *Critical care clinics* 37 (3) 517–541.

Prisinzano, T. E. (2013) Neoclerodanes as Atypical opioid receptor ligands: 2012 David W. Robertson award for excellence in medicinal chemistry. *Journal of Medicinal Chemistry* 56 (9) 3435–3443.

Ranganathan, M., A. Schnakenberg, P. D. Skosnik, B. M. Cohen, B. Pittman, R. A. Sewell and D. C. D'Souza (2012) Dose-related behavioral, subjective, endocrine, and psychophysiological effects of the κ opioid agonist Salvinorin A in humans. *Biological Psychiatry* 72 (10) 871–879.

Reisfield, A. S. (1993) The botany of *Salvia divinorum* (Labiatae). *Sida, Contributions to Botany* 15 (3) 349–366.

Roach, J. J. and R. A. Shenvi (2018) A review of salvinorin analogs and their kappa-opioid receptor activity. *Bioorganic & Medicinal Chemistry Letters* 28 (9) 1436–1445.

Rossi, A., E. Caiazzo, R. Bilancia, M. A. Riemma, E. Pagano, C. Cicala, A. Ialenti, J. K. Zjawiony, A. A. Izzo and R. Capasso (2017) Salvinorin A inhibits airway hyperreactivity induced by ovalbumin sensitization. *Frontiers in Pharmacology* 7 525.

Rossi, A., S. Pace, F. Tedesco, E. Pagano, G. Guerra, F. Troisi, M. Werner, F. Roviezzo, J. K. Zjawiony and O. Werz (2016) The hallucinogenic diterpene salvinorin A inhibits leukotriene synthesis in experimental models of inflammation. *Pharmacological Research* 106 64–71.

Roth, B. L., K. Baner, R. Westkaemper, D. Siebert, K. C. Rice, S. Steinberg, P. Ernsberger and R. B. Rothman (2002) Salvinorin A: a potent naturally occurring nonnitrogenous κ opioid selective agonist. *Proceedings of the National Academy of Sciences* 99 (18) 11934–11939.

Salas-Wright, C. P., J. C. Hodges, A. H. Hai, A. Alsolami and M. G. Vaughn (2021) Toward a typology of hallucinogen users in the United States. *Drug and Alcohol Dependence* 229 109139.

Sayin, H. U. (2016) Psychoactive plants used during religious rituals. *Neuropathology of drug addictions and substance misuse*. Elsevier. 17–28

Schifano, F., S. Chiappini, J. M. Corkery, N. Scherbaum and A. Guirguis (2021a)a The e-psychonaut drugs' psychopharmacology. *Current Opinion in Pharmacology* 57 165–174.

Schifano, F., S. Chiappini, A. Miuli, J. M. Corkery, N. Scherbaum, F. Napoletano, D. Arillotta, C. Zangani, V. Catalani and A. Vento (2021b) New psychoactive substances (NPS) and serotonin syndrome onset: A systematic review. *Experimental Neurology* 339 113638.

Serra, V., L. Fattore, M. Scherma, R. Collu, M. S. Spano, W. Fratta and P. Fadda (2015) Behavioural and neurochemical assessment of salvinorin A abuse potential in the rat. *Psychopharmacology* 232 91–100.

Sharma, N., M. Muthamilarasan, A. Prasad and M. Prasad (2020) Genomics approaches to synthesize plant-based biomolecules for therapeutic applications to combat SARS-CoV-2. *Genomics* 112 (6) 4322–4331.

Shekhar, A. (2019) Role of kappa opioid receptors in symptoms of schizophrenia: what is the neurobiology? *Biological psychiatry* 86 (7) 494–496.

Siegel, A. N., S. Meshkat, K. Benitah, O. Lipsitz, H. Gill, L. M. W. Lui, K. M. Teopiz, R. S. McIntyre and J. D. Rosenblat (2021) Registered clinical studies investigating psychedelic drugs for psychiatric disorders. *Journal of Psychiatric Research* 139 71–81.

Silva, A. R., C. Grosso, C. Delerue-Matos and J. M. Rocha (2019) Comprehensive review on the interaction between natural compounds and brain receptors: Benefits and toxicity. *European Journal of Medicinal Chemistry* 174 87–115.

Simón-Arceo, K., M. E. González-Trujano, U. Coffeen, R. Fernández-Mas, F. Mercado, A. Almanza, B. Contreras, O. Jaimes and F. Pellicer (2017) Neuropathic and inflammatory antinociceptive effects and electrocortical changes produced by *Salvia divinorum* in rats. *Journal of Ethnopharmacology* 206 115–124.

Skalicka-Woźniak, K. and J. Gertsch (2020) Antipsychotic natural products. *Annual reports in medicinal chemistry*. Elsevier. 55 481–515.

Snyder, L. M., M. C. Chiang, E. Loeza-Alcocer, Y. Omori, J. Hachisuka, T. D. Sheahan, J. R. Gale, P. C. Adelman, E. I. Sypek and S. A. Fulton (2018) Kappa opioid receptor distribution and function in primary afferents. *Neuron* 99 (6) 1274–1288.

Su, D., J. Riley, W. M. Armstead and R. Liu (2012) Salvinorin A pretreatment preserves cerebrovascular autoregulation after brain hypoxic/ischemic injury via extracellular signal-regulated kinase/mitogen-activated protein kinase in piglets. *Anesthesia and Analgesia* 114 (1) 200.

Sun, J., X. Yang, Y. Zhang, W. Zhang, J. Lu, Q. Hu, R. Liu, C. Zhou and C. Chen (2019) Salvinorin A attenuates early brain injury through PI3K/Akt pathway after subarachnoid hemorrhage in rat. *Brain Research* 1719 64–70.

Sun, J., Y. Zhang, J. Lu, W. Zhang, J. Yan, L. Yang, C. Zhou, R. Liu and C. Chen (2018) Salvinorin A ameliorates cerebral vasospasm through activation of endothelial nitric oxide synthase in a rat model of subarachnoid hemorrhage. *Microcirculation* 25 (3) e12442.

Szczepaniak, A., W. Machelak, J. Fichna and M. Zielińska (2022) The role of kappa opioid receptors in immune system–An overview. *European Journal of Pharmacology* 933 175214.

Taylor, T.G. and F. Manzella (2016) Kappa opioids, salvinorin A and major depressive disorder. *Current Neuropharmacology* 14 (2) 165–176.

Tejeda, H. A. and A. Bonci (2019) Dynorphin/kappa-opioid receptor control of dopamine dynamics: Implications for negative affective states and psychiatric disorders. *Brain Research* 1713 91–101.

Tlacomulco-Flores, L. L., M. Déciga-Campos, M. E. González-Trujano, A. I. Carballo-Villalobos and F. Pellicer (2020) Antinociceptive effects of *Salvia divinorum* and bioactive salvinorins in experimental pain models in mice. *Journal of Ethnopharmacology* 248 112276.

Turnaturi, R., S. Chiechio, L. Salerno, A. Rescifina, V. Pittalà, G. Cantarella, E. Tomarchio, C. Parenti and L. Pasquinucci (2019) Progress in the development of more effective and safer analgesics for pain management. *European Journal of Medicinal Chemistry* 183 111701.

Ujváry, I. (2014) Psychoactive natural products: overview of recent developments. *Annali dell'Istituto superiore di sanita* 50 12–27.

Ullah, S., B. Munir, A. G. Al-Sehemi, S. Muhammad, I.-U. Haq, A. Aziz, B. Ahmed and A. Ghaffar (2022) Identification of phytochemical inhibitors of SARS-CoV-2 protease 3CLpro from selected medicinal plants as per molecular docking, bond energies and amino acid binding energies. *Saudi Journal of Biological Sciences* 29 (6) 103274.

Ürgeová, E., Ľ. Uváčková, M. Vaneková and T. Maliar (2023) Antibacterial potential of microwave-assisted extraction prepared hydrolates from different salvia species. *Plants* 12 (6) 1325.

Vasiljevik, T., C. E. Groer, K. Lehner, H. Navarro and T. E. Prisinzano (2014) Studies toward the development of antiproliferative neoclerodanes from salvinorin A. *Journal of Natural Products* 77 (8) 1817–1824.

Vázquez-León, P., U. Arenas-Martínez, D. Fregoso-Aguilar, E. Juan and A. Miranda-Páez (2021) *Salvia divinorum* increases alcohol intake and tonic immobility whilst decreasing food intake in Wistar rats. *Acta Neurobiologiae Experimentalis* 81 (1) 34–42.

Walentiny, D. M., R. E. Vann, J. A. Warner, L. S. King, H. H. Seltzman, H. A. Navarro, C. E. Twine, B. F. Thomas, A. F. Gilliam and B. P. Gilmour (2010) Kappa opioid mediation of cannabinoid effects of the potent hallucinogen, salvinorin A, in rodents. *Psychopharmacology* 210 275–284.

Wang, C., J. Zhou, S. Wang, Y. Liu, K. Long, T. Sun, W. Zhi, Y. Yang, H. Zhang and Y. Zhao (2023) Guanxining injection alleviates fibrosis in heart failure mice and regulates SLC7A11/GPX4 axis. *Journal of Ethnopharmacology* 310 116367.

Wang, S. (2019) Historical review: Opiate addiction and opioid receptors. *Cell Transplantation* 28 (3) 233–238.

White, K. L., J. E. Robinson, H. Zhu, J. F. DiBerto, P. R. Polepally, J. K. Zjawiony, D. E. Nichols, C. J. Malanga and B. L. Roth (2015) The G protein–biased κ-opioid receptor agonist RB-64 is analgesic with a unique spectrum of activities in vivo. *Journal of Pharmacology and Experimental Therapeutics* 352 (1) 98–109.

White, K. L., A. P. Scopton, M.-L. Rives, R. V. Bikbulatov, P. R. Polepally, P. J. Brown, T. Kenakin, J. A. Javitch, J. K. Zjawiony and B. L. Roth (2014) Identification of novel functionally selective κ-opioid receptor scaffolds. *Molecular Pharmacology* 85 (1) 83–90.

Willard, M. A. B., J. E. Hurd, R. W. Smith and V. L. McGuffin (2020) Statistical comparison of mass spectra of salvinorins in *Salvia divinorum* and related Salvia species. *Forensic Chemistry* 17 100192.

Williams, J. F. and L. H. Lundahl (2019) Focus on adolescent use of club drugs and 'other' substances. *Pediatric Clinics* 66 (6) 1121–1134.

Wu, L., D. Wu, J. Chen, C. Chen, T. Yao, X. He, Y. Ma, X. Zhi, R. Liu and X. Ji (2021) Intranasal salvinorin A improves neurological outcome in rhesus monkey ischemic stroke model using autologous blood clot. *Journal of Cerebral Blood Flow & Metabolism* 41 (4) 723–730.

Xin, J., Y. Zhang, Z. He and Z. Wang (2016) Highly selective non-opioid kappa opioid receptor (KOR) agonist salvinorin A protects against forebrain ischemia-induced brain injury in rats. *Brain Research* 1637 168–176.

Yang, K. H., B. H. Han and J. J. Palamar (2022) Past-year hallucinogen use in relation to psychological distress, depression, and suicidality among US adults. *Addictive Behaviors* 132 107343.

Yang, M.-J., S.-H. Luo, K. Guo, Y. Liu and S.-H. Li (2023) Chemical investigation of Buddleja officinalis leaves and localization of defensive triterpenoids to its glandular trichomes. *Fitoterapia* 164 105379.

Zawilska, J. B. and J. Wojcieszak (2013) *Salvia divinorum*: from Mazatec medicinal and hallucinogenic plant to emerging recreational drug. *Human psychopharmacology: clinical and experimental* 28 (5) 403–412.

Zeng, S., D. Chen, G. Liu, Y.-X. Wu, Z.-Q. Gao, Y. Su, J.-N. Yuan, L. Liu, J.-C. Shan and Q.-F. Pang (2021) Salvinorin A protects against methicillin resistant staphylococcus aureus-induced acute lung injury via Nrf2 pathway. *International Immunopharmacology* 90 107221.

Zhang, Z., X. Xu and X. Zhang (2022) Exploring the effects of Chinese herbal ingredients on the signaling pathway of alopecia and the screening of effective Chinese herbal compounds. *Journal of Ethnopharmacology* 294 115320.

30 Sinicuichi (*Heimia salicifolia* (Kunth) Link Lythraceae)

Rahmatullah Qureshi, Bushra Jabeen, and Noureddine Chaachouay

30.1 INTRODUCTION

Heimia salicifolia (Kunth) Link is a flowering plant belonging to the Lythraceae family. The genus name *Heimia* is given in honor of the German scientist E.L. Heim, and the species name/epithet *salicifolia* refers to the close resemblance of the leaves to the willow tree, called *Salix* (Malone & Rother, 1994). The species possesses several common names such as Abre-o-sol (sun opener), Herva de la vida (herb of life), Hierba de San Francisco (herb of St. Francis), Yerba de las ánimas (herb of the spirits), Shrubby yellow crest, and Sinicuichi or Sinicuiche (Martinez, 1959; Hoehne, 1978; Dominguez et al., 1975).

H. salicifolia is a spreading deciduous, densely branched, blooming shrub. From the far southern corner of Texas through Mexico and Central America to Argentina, it is native to brushlands and the Rio Grande basin. The plant is popularly known by its common names including shrubby yellowcrest, sinicuichi, sun opener, willow-leaf heimia, and sini (Chamorro-Cevallos, et al., 2022, Heim, 2023). The history of medical use is traced back to an Aztec herbal book that describes a variety of local plants, their therapeutic benefits, and techniques for treating various conditions. Libellus de Medicinalibus Indorum Herbis, originally composed in the Nahuatl language of the Aztecs was then translated into Latin.

The leaves have been used as traditional medicine, particularly in Mexico (Martinez, 1959). The local practitioners used leaves for a wide range of disorders, including dysentery, uterine inflammation, bronchitis, and other chest problems. According to Martínez (1959), Biomster et al. (1964), Douglas et al. (1964), Oliver-Bever (1972), Lema (1986), and Malone (1986), the plant possesses various medicinal properties such as diuretic, laxative, antisyphilitic, emetic, and tonic. These purported medical and spiritual benefits appear to be the reason for the continued use in current times. Many stories specifically mention the hallucinogenic and psychoactive effects of leaf decoction.

The Aztecs traditionally prepared sinicuichi tea from *H. salicifolia* leaves by drying them until wilted, and later on, placed in a cup or bowl of cool water, and left in the sun for at least 24 hours. They had no idea that this had permitted fermentation to take place. Instead, practitioners thought that the fermentation process – a crucial step in making the tea or elixir of the sun-transferred the sun's wisdom and spirit into the beverage (Bechelli, n.d.). A historical topic involving this plant is the symbolism of the sun and its energy. It has long been believed that fermenting tea before drinking it lessens the negative effects of doing so. The term *Sun Opener* refers to the strong mood booster and visual stimulation. *H. salicifolia* has been employed in shamanic rites since the time of the Aztecs, and tribes of Central and South America still use sinicuichi as a medicinal drink.

Consumers of the tea have described moderately pleasing effects such as euphoria, and auditory delirium (Diaz, 1979; Schultes, 1976). However, in larger quantities, there have also been reports of visual abnormalities and problems with time perception (Rother, 1989). J.B. Calderon originally described the sinicuichi tea's features in his 1897 publication, explaining its unique physiological action, a pleasant intoxication that causes objects to seem yellow and causes the sound of bells and human voices to be audible far away (Calderon, 1897). Victor Reko expanded on the psychedelic and hallucinogenic properties of *H. salicifolia* three decades later (Reko, 1926).

A variety of alkaloids have been identified from *H. salicifolia* including the three most abundant ones such as cryogenine, lyfoline, and lythrine (Rother, 1989). They are all members of the biphenyl quinolizidine lactone class of alkaloids, which constitutes the majority of physiological activity in the *Heimia* genus. There are 24 known alkaloids from the plant including nesodine, heimidine, sinicuichine, dehydrodecodine, lythridine (Kitajima et al., 2018; Kitajima et al., 2019). The principal source of the effects of traditional *H. salicifolia* brew, including anticholinergic, anti-inflammatory, sedative, tranquilizing, and spasmolytic activities, is thought to be vertine (cryogenine), the alkaloid displaying the highest concentration in the plant material (Kitajima et al., 2018).

30.2 SCIENTIFIC CLASSIFICATION

H. salicifolia is a deciduous shrub native to the brushlands and the Rio Grande basin of S.E. Texas, Mexico, and Central America to Argentina. The scientific classification is shown in Table 30.1.

DOI: 10.1201/9781003460336-30

TABLE 30.1
Taxonomic Hierarchy of *H. salicifolia*

Kingdom:	Plantae
Subkingdom:	*Tracheobionta*
Infrakingdom:	Streptophyta
Superdivision:	Tracheophytes
Subdivision:	Spermatophytina
Infradivision:	Angiospermae
Class:	Magnoliopsida
Clade:	Monophyletic
Superorder:	Rosanne
Order:	Myrtales
Family:	Lythraceae
Genus:	*Heimia*
Species:	*H. salicifolia*
Binomial name:	*Heimia salicifolia* (Kunth) Link

Source: https://plants.usda.gov/home/plantProfile?symbol=HESA

30.3 VERNACULAR NAMES

Numerous common names are also used to refer to it, including the Arabic Fettaht Echams, Sinicuichi, Sinichuiche, Sini, Sun Opener, Shrubby Yellowcrest, Hachinal, and Willow-leaf Heimia in English. In Spanish, it is known as Escoba colorada, Jarilla, Hierba jonequil, Sinicuilchi, Flores de San Francisco, Jara mansa, Escoba de arroyo, Trinchinole, Huanoli, Escoba de río, Garañona, Granadilla, Sinicuil, Trenchinole, Granadina, Jara, Quiebra arado, Jara amarilla, Sinicuiche, Escobilla, Hierba huachinol, Hachinal, Hauchinal, Hanchinol, Retama. While Heimia à feuilles de saule, Sini, and Sinicuichi are frequently used in French, Abre-o-sol (sun opener), Detta anche Sinicuiche, and Herva da Vida are common names in Italian (Malone and Rother, 1994).

30.4 SYNONYMS

The Plants of the World Online is an extensive online database of all known plant species and has information on the plant species *Heimia salicifolia* (Kunth) Link. According to it (Quattrocchi, 2012), 13 synonyms exist in the database as follows:

Chrysolyga salicifolia Willd. ex Steud.
*Decodon salicifolius (*Kunth) Kuntze
Ginoria flava Moc. & Sessé ex DC.
Ginoria syphilitica Moc. & Sessé ex DC.
Heimia grandiflora Hook.
Heimia linariifolia G.Don
Heimia salicifolia var. grandiflora Lindl.
Heimia syphilitica DC.
Lythrum flavum Spreng.
Nesaea linariifolia Steud.
Nesaea salicifolia Kunth
Nesaea syphilitica (DC.) Steud.
Ludwigia bertonii H.Lév.

The International Plant Names Index and World Checklist of Vascular Plants, 2023. Published on the internet at http://www.ipni.org and https://powo.science.kew.org/

30.5 BOTANICAL DESCRIPTION

A dwarf spreading perennial semi-woody shrub approximately 2–3 meters tall (Figure 30.1). Stems are woody, twisted, and somewhat ribbed or angled, usually reddish-brown in color. Leaves are decussate, narrow-oblong to linear-lanceolate, and slender, measuring 2–10 cm long and 2–10 mm in broad, simple, alternate, glossy, dark green color (García-Mendoza, 2006). Flowers are solitary, axillary, campanulate, yellow, 1–2 cm in diameter. Calyx campanulate. Petals lobes 5–7, ca. 2 cm across. Stamens 10–13. Style narrow. Fruits are dry, ribbed, spherical, four-celled capsules (5 mm) filled with tiny seeds (Link and Otto, 1828).

30.6 ORIGIN AND GEOGRAPHIC DISTRIBUTION

This shrub is most common in Mexico's highlands, although it can also be found in Baja California. It can be found all over South America, all the way down to Argentina (Martinez, 1959). As a native to Mexico and Central America, *H. salicifolia* is distributed to South America (Uruguay to Argentina), Texas, Paraguay, Bolivia, South to Southeast and West-Central Brazil, El Salvador, and Jamaica (Figure 30.2). It may be found in many environments, including desert areas, grasslands, and disturbed areas (Morton, 1975). It was also introduced to Azores and Queensland (Schäfer, 2021). The species was successfully cultivated in Baja, California, and south of Argentina (Graham, 1997).

30.7 HABITAT AND ECOLOGY

H. salicifolia is a perennial shrub that is adapted to dry and semiarid climates. It prefers hot, sunny, moist places. It is found in grasslands, open forests, scrublands, and disturbed environments. It is also frequently found along highways, railway lines, etc. (Graham, 1997).

H. salicifolia may thrive in locations with poor soil quality and can withstand severe temperatures and little soil moisture in its natural habitat (Martínez, 1959). It may thrive in highly salinized soils and can tolerate dry spells. In certain places, its natural populations have decreased due to habitat loss and overharvesting. It may be preserved by conservation initiatives and commercial production of the plant (Ortega and López, 2011).

30.8 ETHNOBOTANY

30.8.1 Medicinal Use

The traditional usage of *H. salicifolia* dates back many years in Mexico and other regions of South and Central

FIGURE 30.1 Habit and habitat of *H. salicifolia*.

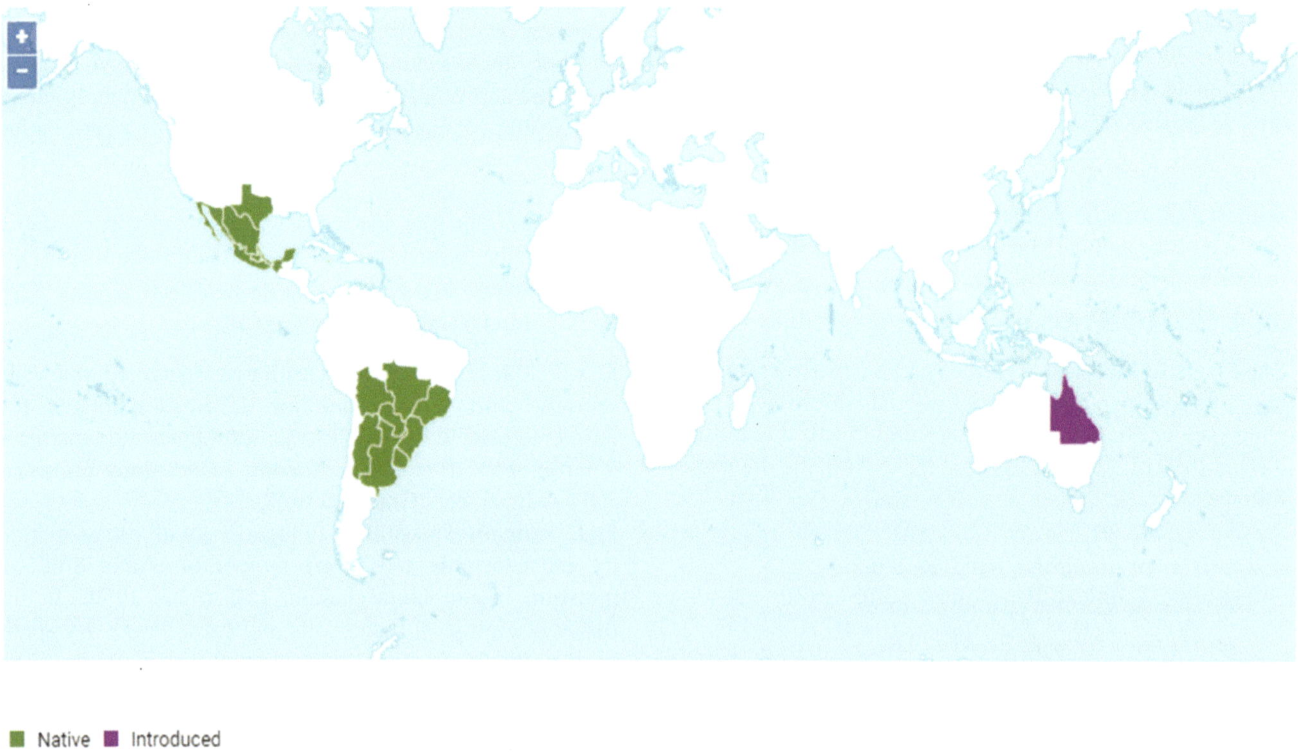

FIGURE 30.2 Global distribution of *H. salicifolia* (https://powo.science.kew.org/taxon/urn:lsid:ipni.org:names:96613-3).

America. Both its therapeutic benefits and its hallucinogenic effects have been utilized.

The plant has a wide reputation in the Americas as a folk medicine for treating a range of complaints. It is often harvested from the wild for local use and is also sold in local markets. A tea brewed from the leaves is said to aid with digestion in Mexican folk medicine. The herbage is also used to treat rabies and to ward off the 'evil eye'. Syphilis is commonly treated with the plant (Malone and Rother, 1994; Belhaj et al., 2021; Benkhnigue et al., 2022). The wilted leaves are crushed in water, and the mixture is then allowed to ferment for three days in a sunny location, producing an intoxicating and euphoric drink. Larger doses of this might cause hallucinations and a vision that is often yellow and clouded.

According to Standley (1926), Emboden (1979), and Mors et al. (2000), the leaves have astringent, emetic, diuretic, febrifuge, hemostatic, hallucinogenic, laxative, sedative, sudorific, tonic, and vulnerary properties. They also possess antisyphilitic and antispasmodic properties (Standley, 1926). An infusion helps maintain blood pressure and lower anxiety (Emboden, 1979). It is said to have a sedative effect that reduces anxiety, slows down the central nervous system's activity, and encourages sleep (Malone and Rother, 1994). It is meant to create a serene, tranquil environment (Malone and Rother, 1994). To treat wounds and skin diseases, a paste made from the leaves and stems is utilized (Malone and Rother, 1994). The aerial parts of the plant are said to be helpful in preparing postpartum baths and speed up wound healing. They also have a widespread folk reputation for having antisyphilitic, sudorific, antipyretic, laxative, and diuretic properties (Malone and Rother, 1994; Chaachouay et al., 2020).

A decoction of the plant is used to treat the symptoms of poison ivy (Rhus toxicodendron) (Standley, 1926). It is used as a postpartum bath, after performing intense physical labor, and to cure myalgia in aged people (Malone and Rother, 1994; Chaachouay et al., 2021a, 2021b). According to Standley (1926), consuming the plant's juice or decoction internally is claimed to provide a mild and enjoyable intoxication that causes all objects to appear yellow. The leaves are claimed to contain 14% of resin, which is the active component, and 9% of nesine, a bitter principle (Standley, 1926).

30.8.2 Other Uses

Leaves are applied to the body to repel flies and other insects (Malone and Rother, 1994).

30.9 PHYTOCHEMISTRY

Leaves contain 9% nesine, a bitter ingredient, and 14% resin (Standley, 1926). The decoction of leaves results in a light and enjoyable intoxication during which all objects appear to be yellow (Standley, 1926). The sinicuichi also includes pigments, gum, glucose, starch, fat, traces of volatile oil, and a variety of common minerals. Hollick (1936) first time reported the presence of an alkaloid. Later on, Webb (1948) verified the presence of an alkaloid. Hegnauer and Herfst (1958) also recorded polyphenolic compounds and acidic mucilage. The presence of alkaloids in the aerial parts of *H. salicilia* was reported by Biomster et al. (1964).

Various alkaloids are identified from *H. salicifolia* including the three most common such as cryogenine, lyfoline, and lythrine (Rother, 1990). These are all members of the biphenyl quinolizidine lactone class, constituting the most physiologically active elements in the *Heimia* genus (Watson and Malone, 1974). There are 24 known alkaloids in the class of minor alkaloids including nesodine, heimidine, sinicuichine, dehydrodecodine, lythridine, and others (Kitajima et al., 2018; Kitajima et al., 2019). This group of chemicals forms in plants far more frequently than the biosynthetically easier phenyl quinolizidinyl esters group.

Cryogenine ($C_{26}H_{29}NO_5$), the most prevalent alkaloid in *H. salicifolia*, has long been supposed to cause psychedelic and hallucinogenic effects (Calderon, 1897; Schultes, 1969; Rother, 1989). Despite the fact that none of the individual alkaloids from the genus *Heimia* have been identified as psychotomimetic or psychodysleptic (Malone & Rother, 1994). In addition to being a diuretic, cryogenine medication in animal experiments decreased spontaneous motor activity, muscular relaxation, hypothermia, blepharoptosis, and ataxia (Singh, 2011). According to Malone and Rother (1994), cryogenine also had effects on rats that were anticholinergic, anti-inflammatory, antispasmodic, hyperglycemic, hypotensive, sedative, tranquilizer, and vasodilator. The majority of the anti-inflammatory activity of cryogenine may be explained by the fact that it is 2.48 times more effective than aspirin as a prostaglandin synthetase inhibitor (Lema et al., 1986). Lyfoline, being the 2nd common alkaloid, had no beneficial effects.

The third most abundant alkaloid in the plant is lythrine ($C_{26}H_{29}NO_5$), which is a diastereomer of the allegedly psychotropic cryogenine (Figure 30.3). According to Weisbach, 1965; Byrne and Malone, 1981; Guzmán-Hernández et al. (2018), lythrine is the most effective vasorelaxant alkaloid of *H. salicifolia*. It is also said to be a diuretic and has anti-inflammatory properties. Lythrine showed a concentration-dependent vasorelaxation effect, as shown in a phenylephrine driven model of vasoconstriction employing isolated and perfused rat mesenteric vascular bed (MVB) preparations with intact endothelium.

The isolated alkaloids have been experimentally demonstrated to qualitatively and semi-qualitatively produce the activity of the whole alkaloid extracts of *H. salicifolia*, particularly vasodilation (Schultes, 1969; DeKorne, 2011; Guzmán-Hernández et al., 2018). Since isolated lythrine may have therapeutic benefits that have been documented for millennia and effects that are similar to those of the whole plant folk decoction of sinicuichi, it is important to emphasize that more therapeutic and medicinal research is needed. Additionally, specific applications of beneficial substances and associated constituents can be carried out

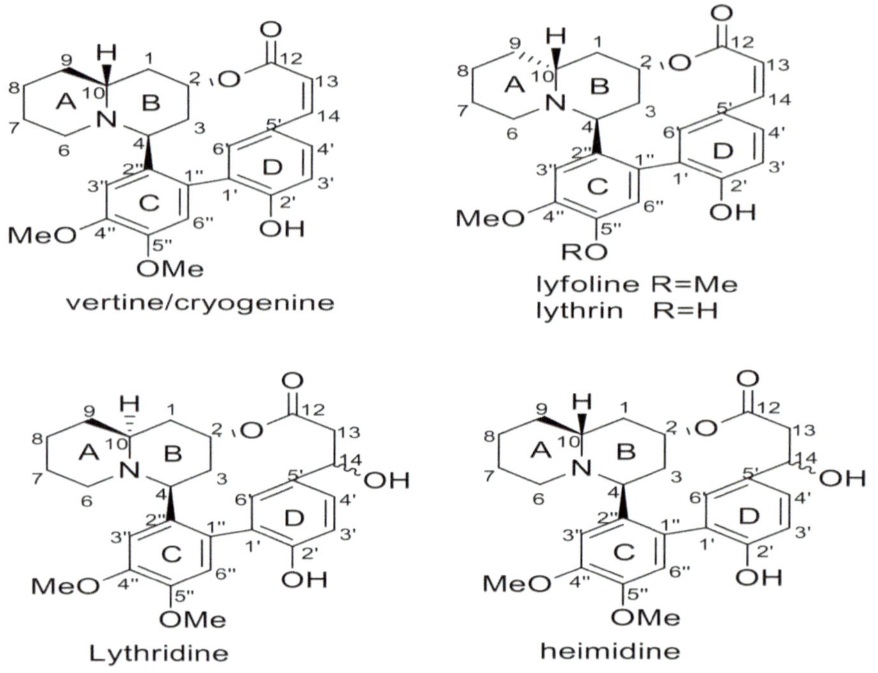

FIGURE 30.3 Structures of some important alkaloids.

with additional research and understanding of lythrine and the medicinal alkaloids and components of *H. salicifolia*. This suggests that lythrine may have psychological benefits, particularly in terms of reducing anxiety, in addition to its diuretic, anti-inflammatory, sedative, anti-hypertensive, and vasorelaxant properties.

The main alkaloids of *Heimia* are vertine/cryogenine, lyfoline, lythrine, lythridin, and nesodine, listed in order of prevalence (Appel et al. 1965). It has been established that the roots and seeds of H. salicifolia are devoid of alkaloids (Rother, 1989). Rother (1989) has provided a summary of the alkaloid concentrations of *H. salicifolia* as reported in several research (Rother, 1989). Some important structures are given below in Figure 30.3.

Additionally, Kitajima et al. (2019) reported the isolation of six novel biphenyl quinolizidine lactone alkaloids from cultivated sinicuichi. Chemical reactions and spectroscopic investigations were used to identify their structures. Other natural products found in *Heimia salicifolia* include flavonoids like quercetin and kaempferol, a sterol known as stigmasterol, as well as triterpenes like ursolic acid (Kitajima et al. 2019).

30.10 HALLUCINOGENIC EFFECTS

Sinicuichi (*H. salicifolia*) is an aural captivating hallucinogen native to central Mexico. The slightly wilted leaves are crushed and soaked in water. This juice is subjected to the sun for fermentation and used as a mild intoxicant beverage. It may cause an auditory hallucination, with a distorted voice that seems to come from a distance. It is also believed that drinking too much can be quite hazardous (Schultes, 1976). The researchers recently discovered six alkaloids belonging to the quinolizidine class from this plant. Although none of the psychedelic effects brought on by the consumption of the entire plant have yet been matched by any of the alkaloids discovered so far, one, cryogenine or vertine, appears to be the most active (Schultes, 1976).

The psychoactive properties of *H. salicifolia* are well-known, and these effects are frequently characterized as heightening sensory experience or generating a trance-like condition. The plant is frequently referred to as sinicuichi or sun opener because of its alleged capacity to improve sensitivity to sunlight. It has a long history of use in shamanic ceremonies (Kitajima et al., 2019). Alkaloids such lythrine, lythridine, and vertine are chiefly responsible for *H. salicifolia's* hallucinogenic properties. The serotonin and dopamine systems in the brain are thought to interact with these alkaloids, which may help explain why the plant has hallucinatory effects. Compared to other hallucinogenic plants like ayahuasca or Psilocybin mushrooms, *H. salicifolia* often has softer, less potent psychedelic effects. Users frequently describe slight auditory alterations, relaxed or euphoric feelings, and mild visual distortions. The dosage, mode of administration, and individual susceptibility to the plant can all affect the results, which can then vary. It is significant to highlight that it has not been thoroughly investigated in Western medicine, nor have clinical investigations adequately characterized its psychotropic effects. As with the usage of any psychoactive drug, it should only be used under the supervision of a qualified professional (Quattrocchi, 2012; Chaachouay & Zidane, 2021).

30.11 PHARMACOLOGY

Aerial parts have been used as an antisyphilitic, anti-inflammatory, and antispasmodic, sudorific, antipyretic, ataractic, hallucinogenic, laxative, diuretic, postpartum medicine, prostaglandin synthetase inhibitors, and to hasten wound healing. Diuretic, vulnerary, and purgative roots and stems. A drink made from leaves has an unpleasant taste and auditory hallucinations (Quattrocchi, 2012).

The plant contains a variety of alkaloids, including lythrine, lythridine, vertine, cryogenine, tsoiadinine, and others. These alkaloids are regarded to be the root of the plant's medicinal and hallucinogenic effects. Studies have demonstrated that lythrine, the most common alkaloid in *H. salicifolia*, relaxes smooth muscle tissue. Lythridine, a different alkaloid present in the plant, has been demonstrated to have anti-inflammatory properties. Vertine, a psychoactive alkaloid, is believed to contribute to the plant's hallucinatory effects.

Two more alkaloids found in *H. salicifolia* are cryogenine, which has been demonstrated to have antispasmodic properties, and tsoiadinine, which has been suggested to have anxiolytic properties. The plant also contains additional, less-studied alkaloids such heimidine. More research is needed to fully understand *H. salicifolia's* phytochemistry because the plants exact chemical composition phytochemistry of *H. salicifolia* because the exact chemical composition of the plant may vary depending on growing conditions and other factors.

30.12 MEDICINAL PROPERTIES

Using the observational (Hippocratic) techniques developed by Robichaud et al. (1964), preliminary pharmacologic/toxicological evaluation of finely powdered aerial portions of *H. salicifolia* and alkaloid-bearing extracts was carried out in unanesthetized rats. Vertine (2.0–8.6 mg/g of dry-weight aerial parts), lyfoline (1.8–6.6 mg/g), lythrine (0.55–0.66 mg/g), and nesodine (0.09–0.55 mg/g) were the first four main alkaloids to be identified. Vertine and lyfoline were the two most prevalent alkaloids, and whereas lyfoline lacked these effects, vertine consistently exhibited dose-related hypothermia, drowsiness, and passivity without significantly affecting skeletal muscle. Lythrine appeared to be mostly inactive of the two less abundant alkaloids, whereas nesodine shown some weaker vertine-like action. Due to vertine's relative abundance in plant material and the fact that it seems to both qualitatively and semiquantitatively contribute to the majority of the activity recorded in folk medicine, the following study has tended to concentrate on this alkaloid (Biomster et al., 1964).

The majority of biological investigations so far have described and explained the plant's possible diuretic, antispasmodic, anti-inflammatory, ataractic, and various other potential properties (Trottier and Malone, 1969). Some are as follows:

30.12.1 Ataractic Potential

Vertine consistently induced a passive response to head-tap and body-grip trials in unanesthetized rats. At levels that do not impair skeletal muscular function, this reaction is suggestive of real ataractic (tranquilizing) potential. The dog, cat, rabbit, and guinea pig all displayed passivity. Vertine has been demonstrated to selectively reduce conditioned behavior responses in both discrete- and continuous-avoidance/escape scenarios at levels that cause mild hypothermia in rats. Major tranquilizers with this property include reserpine and chlorpromazine, which both cause dose-related hypothermia. Vertine, however, obviously lacked the autonomic nervous system modifications unique to chlorpromazine and reserpine at ataractic dosages. Vertine showed no morphine-like analgesic potential in rats in similar experiments, but it did have the ability to lessen joint pain brought on by inflammation (García-Mendoza, 2006).

Vertine also showed antipyretic efficacy in rats with peptone-induced illness at anti-inflammatory oral dosages. Vertine demonstrated little anticonvulsant ability against either pentylenetetrazol- or electroshock-induced seizures when administered in near-lethal intraperitoneal dosages to rats. The findings of these various studies, combined with the finding that norepinephrine and serotonin levels in the brain and heart were unaffected by near-lethal doses, suggest that the central nervous system depression caused by vertine is selective by nature and pharmacologically distinct from that caused by the currently available commercial neuroleptics. However, the potential therapeutic use of this profile of activity can only be predicted by simian and human investigations (García-Mendoza, 2006).

30.12.2 Anti-Inflammatory Potential

Vertine demonstrated dose-dependent anti-inflammatory efficacy against carrageenan-induced pedal edema and against adjuvant-induced polyarthritis in rats at oral dosages that did not affect core body temperature. Vertine was administered orally to normal rats for two and a half days at a dose of 100 mg/kg/day for anti-inflammatory purposes, and histologic investigations revealed no signs of chronic toxicity (29 tissues were examined per rat) or adrenocortical activation. Vertine, 100 mg/kg/day orally, reduced adjuvant-induced adrenal hypertrophy and thymus involution but did not influence splenic enlargement, according to additional research in rats with adjuvant polyarthritis. Contrarily, hydrocortisone (10 mg/kg/day orally) significantly inhibited thymus and adrenal development while sparing animals of splenic hypertrophy (Watson and Malone, 1977).

Vertine has also been reported to shield rats against adjuvant-induced alterations in hematology, sedimentation rate, and serum turbidity in chronic oral anti-inflammatory doses (Kaplan and Malone, 1966). These results show that the glucocorticoid-like action of vertine is minimal to nonexistent. Vertine lost all of its action in the first phase (+ I h) of edema production and half of its estimated activity in the second phase (+3 and +5 h) when it was tested in rats with adrenalectomies against carrageenan-induced pedal edema. It is typically claimed that one of the finest pharmacologic indications for therapeutic value is activity against the second phase of edema. This decrease in activity in rats that had their adrenal glands removed would appear to suggest that in order to ensure the complete functioning of the vermis, the pituitary-adrenal axis may need to remain intact (Trottier Jr. and Malone, 1971).

30.12.3 Antispasmodic Potential

Early cardiovascular disease studies included intact and vagotomized dogs that received vertine intravenously until they passed away. In order to gather data that could identify mechanisms of action, transient physiologically significant challenges were made during the cumulative dosages in both types of experiments (Robichaud et al., 1964).

Vertine (0.10 mg/kg) displayed dose-related bradycardia in the vagotomized preparations but had no impact on the electrical vagal stimulation, bilateral carotid occlusion, or epinephrine-induced pressor responses. At 100 mg/kg, death by apparent collapse of respiration was seen.

Bradycardia did not occur in the preparations using intact dogs. Instead, at 0.128 mg/kg, the carotid occlusion pressor reaction started to gradually decrease along with a dose-response decrease in blood pressure. There was a T wave inversion alteration in the electrocardiogram at 0.128 mg/kg as well. The QRS spike decreased during the course of the consecutive dosages. There seemed to be corresponding compensatory increases in heart rate and breathing rate/depth when blood pressure dropped. The cholinergic and histamine challenges were also lessened, although these modifications appeared to be related more to the decline in resting blood pressure than to the actual inhibition of acetylcholine and histamine receptors. Although the results were not as striking as in tests employing reserpine and cocaine, the improvement in the epinephrine precursor response also appeared to be related to the decrease in resting blood pressure. Vertine dosed at 65.5 mg produced death (Kaplan et al., 1967).

The vertine-induced low blood pressure in the intact dog preparations indicated dilatation of the blood vessels in the peripheral region by some non-autonomic mechanism; this mechanism may be a direct musculotropic effect of vascular smooth muscle or it may be a centrally mediated effect because in the unaffected animals the decreased blood pressure was well correlated with reductions in the carotid occlusion pressor reflex that exists (Kaplan et al., 1967).

Wiebelhaus reported that decinine, an alkaloid from *Decodon verticillutus* with a similar molecular makeup, showed a diuretic effect in rats in a 1973 German patent. In a recent study, Wiebelhaus and colleagues provided evidence that both lythrine and deinine have the ability to antagonize antidiuretic hormones and result in a hydro-diuresis as opposed to the saluresis that is linked with the majority of the diuretics now used in medicine. These researchers have focused on evaluating decinine because of its enhanced efficacy in this aspect. They even chemically change lythrine to decinine for this reason. Before this discovery, ethanol was the only substance known to block the production of ADH from the pituitary, which affects the function of antidiuretic hormone and results in a rather specific hydro-diuresis (Wiebelhaus et al., 1974).

30.13 CONCLUSION

H. salicifolia is native to Central and South America and has a long history of traditional use for both its therapeutic and euphoric effects. In addition to being used as an aphrodisiac sometimes, the herb has been used to cure a wide range of conditions, such as anxiety, depression, sleeplessness, and gastrointestinal issues. Alkaloids such lythrine, lythridine, and vertine are chiefly responsible for their hallucinogenic properties. The serotonin and dopamine systems in the brain are thought to interact with these alkaloids, which may help explain why the plant has hallucinatory effects. However, compared to other hallucinogenic plants, *H. salicifolia* often has softer, less potent psychedelic effects. Though *H. salicifolia* has a long history of traditional use, little is known about its possible therapeutic advantages because it hasn't been thoroughly researched in Western medicine. When employing *H. salicifolia*, caution should be taken, and it should only be done under the supervision of a qualified professional.

REFERENCES

Appel, H.G., Rother, A. and Schwarting, A. E. 1965. Alkaloids of *Heimia salicifolia*. Isolation of nesodine and lyfoline and their correlation with other Lythraceae alkaloids. *Lloydia*, 2(28), 84–89.

Bechelli, J. (n.d.). Via Oneira: Three Dreaming Herbs of Mexico. Unpublished manuscript.

Belhaj, S., Chaachouay, N., & Zidane, L. (2021). Ethnobotanical and toxicology study of medicinal plants used for the treatment of diabetes in the High Atlas Central of Morocco. *Journal of Pharmacy & Pharmacognosy Research*, *9*(5), 619–662. https://doi.org/10.56499/jppres21.1014_9.5.619

Benkhnigue, O., Chaachouay, N., Khamar, H., El Azzouzi, F., Douira, A, & Zidane, L. (2022). Ethnobotanical and ethnopharmacological study of medicinal plants used in the treatment of anemia in the region of Haouz-Rehamna (Morocco). *Journal of Pharmacy & Pharmacognosy Research*, *10*(2), 279–302. https://doi.org/10.56499/jppres21.1196_10.2.279

Biomster, R.N., Sehwarting, A.E. and Bobbitt, J.M. 1964. Alkaloids of Heimia salicifolia. 1: A preliminary report, Lloydia, 27, 15–24.

Byrne, J. A. and Malone, M. H. 1981. Protective effects of cryogenine, lythrine, and certain derivatives in croton oil induced inflammation. *Proceedings of the Western Pharmacology Society*, 24, 183–187.

Calderon, J.B. 1897. Study on the shrub called sinicuiche. *Anales Mediconational Institute (Mexico)* 2, 36–42.

Chaachouay, N., Douira, A., Hassikou, R., Brhadda, N., Dahmani, J., Belahbib, N., Ziri, R., & Zidane, L. (2020). Ethnobotanical and ethnopharmacological study of medicinal and aromatic plants used in the treatment of respiratory system disorders in the Moroccan Rif. PhDThesis, Département de Biologie - Université Ibn Tofail - Kénitra, Maroc, https://tel.archives-ouvertes.fr/tel-03376377

Chaachouay, N., & Zidane, L. (2021). The Symbolic Efficacy of Plants in Rituals and Socio-Religious Ceremonies in Morocco, Northwest of Africa. *Journal of Religious & Theological Information*, *21*(1–2), 34–53. https://doi.org/10.1080/10477845.2021.1942433

Chaachouay, N., Douira, A., & Zidane, L. (2021a). Herbal Medicine Used in the Treatment of Human Diseases in the Rif, Northern Morocco. *Arabian Journal for Science and Engineering*, 47, 131–153. https://doi.org/10.1007/s13369-021-05501-1

Chaachouay, N., Douira, A., & Zidane, L. (2021b). COVID-19, prevention and treatment with herbal medicine in the herbal markets of Salé Prefecture, North-Western Morocco. *European Journal of Integrative Medicine*, 42, 101285. https://doi.org/10.1016/j.eujim.2021.101285

Chamorro-Cevallos, G., Mojica-Villegas M.A., García-Martínez Y., Pérez-Gutiérrez S., Madrigal-Santillán E., Vargas-Mendoza N., Morales-González J. A. and Cristóbal-Luna J.M. 2022. A complete review of Mexican plants with teratogenic effects. *Plants*, 11(13), 1675.

DeKorne, J. 2011. *Psychedelic shamanism: The cultivation, preparation, and shamanic use of psychotropic plants*. Berkeley, CA: North Atlantic Books.

Diaz, J.L. (1979). Ethnopharmacology and taxonomy of Mexican psychodysleptic plants. *Journal of Psychedelic Drugs*, 11. 1(1-2), 71–101.

Dominguez, X.A., Marroquín, J., Quintero, B.S., and Vargas, S.B. 1975. Two new quinolizidine alkaloids from *Heimia salicifolia*. *Phytochemistry* 14(8), 1883–1884.

Douglas, B., Kirkpatrick, J.L., Raffauf, R.F., Ribeiro, O., and Weisbach, J.A., 1964. Problems in chemotaxonomy. II. The major alkaloids of the genus *Heimia*. *Lloydia*, 27, 25–31.

Emboden, W. 1979. *Narcotic plants*. Studio Vista. London.

García-Mendoza, A.J. 2006. The genus Heimia (Lythraceae) in Mexico. *Harvard Papers in Botany*, 11(2), 285–294.

Graham, S. 1997. *Type Species*: *Heimia Salicifolia*. Archive. Kent University. http://web.archive.org/web/19970624061507/http://simon.kent.edu/Biology/Research/Shirley_Graham/Genera/heimia.html

Guzmán-Hernández, E.A., Segura-Cobos, D., Amato, D., Avila-Acevedo, J.G., and Vazquez-Cruz, B. 2018. Evaluation of antihypertensive and vasorelaxant effects of *Heimia salicifolia* (family: Lythraceae). *African Journal of Pharmacy and Pharmacology*, 12(3), 41–51. doi: 10.5897/ajpp2017.4860

Hegnauer, R. and Herfst, A. (1958) Heimia salicifolia Link & Otto. *Pharmaceutisch weekblad*, 93(19), 849–865.

Heim, E.L. (2023) *Heimia salicifolia*. Missouri Botanical Garden, 4344 Shaw Blvd., St. Louis, MO 63110(314) 577–5100. https://www.missouribotanicalgarden.org/PlantFinder/PlantFinderDetails.aspx?taxonid=295557. Retrieved 2023 July 21.

Hoehne, F.C. 1978. *Plants and toxic and medicinal plant substances*. Graphicars.

Hollick, C. A. (1936). *The tertiary floras of Alaska* (Vol. 182). US Government Printing Office. https://books.google.com/books?hl=fr&lr=&id=bmvnAAAAMAAJ&oi=fnd&pg=PP7&dq=Alas+(1880)+plants&ots=dCjC5msF5B&sig=x1Iraq0ys6GHFtftfhVI7N1yAS0

Kaplan, H.R. and Malone, M.H. (1966) A pharmacologic study of nesodine, cryogenine and other alkaloids of *Heimia saiicifolia*. *Lloydia*, 29, 348–359.

Kaplan, H.R., Wolke, R.E., and Malone, M.H. 1967. Antiinflammatory evaluation of cryogenine. *Journal of Pharmaceutical Sciences*, 56, I385–1392.

Kitajima, M., Yamaguchi, Y., Yanagisawa, T., Kogure, N., Ogata, J., Kikura-Hanajiri, R., and Takayama, H. 2019. Biphenyl quinolizidine lactone alkaloids from 'sinicuichi' (Heimia salicifolia). *Tetrahedron*, 75(27), 3733–3739. doi:10.1016/j.tet.2019.05.045

Kitajima, M., Yanagisawa, T., Tsukahara, M., Yamaguchi, Y., Kogure, N., Kikura- Hanajiri, R., and Takayama, H. 2018. Biphenyl ether and biphenyl quinolizidine lactone alkaloids from *Heimia salicifolia*. *Tetrahedron*, 74(4), 441–452. doi:10.1016/j.tet.2017.12.012

Lema, W.J., Blakenship J.W., and Malone M.H. 1986. Prostaglandin synthetase inhibition by alkaloids of *Heimia salicifolia*. *Journal of Ethnopharmacology*, 15, 161–167

Link, H.F., and Otto, F.S. 1828. Icons of selected plants from the Royal Botanic Garden of Berlin with descriptions and methods of cultivation 10. Royal Botanic Garden of Berlin.

Malone, M.H. 1986. Cryogenine. *Drugs of the Future*, 11, 95–96.

Malone, M.H., and Rother, A. 1994. Heimia salicifolia: a phytochemical and phytopharmacologic review. *Journal of Ethnopharmacology*, 42(3), 135.

Martínez, M. 1959. *The medicinal plants of Mexico* (4th ed.). Mexico: Botas, Boots Editions, 293–295.

Mors, W.B., Rizzini, C.T., and Pereira, N.A. 2000. *Medicinal Plants of Brazil*. Reference Publications.

Morton, J. F. 1975. Mexican narcotic plants: A review of an aspect of the ethnobotany of the New World. *Journal of Ethnopharmacology*, 1(2), 111–193.

Oliver-Bever, B. 1972. Drug plants in ancient and modern Mexico. *Q J Crude Drug Res*, 12, 1957–1972.

Ortega, A., and López, M.G. 2011. *Heimia salicifolia* (DC.) Link & Otto, a medicinal plant with potential anxiolytic and antidepressant properties. *Journal of Ethnopharmacology*, 134(3), 768–774.

Quattrocchi, U. 2012. *CRC world dictionary of Medicinal and poisonous plants: Common Names, scientific names, eponyms, synonyms, and Etymology*. CRC Press.

Reko, V. 1926. Sinicuichi. *The Yucatan Medical Journal*, 14, 22–27.

Robichaud, R.X. Malone, M.H., and Schwarting, A.E. 1964. Pharmacodynamics of cryogenine, an alkaloid from *Heimia salicifolia* link and Otto. Part I. *Archives Internationaies de Pharmocodynamie et de ThPrapie*, 150, 220–232.

Rother A. 1989. *Heimia salicifolia*: In vitro culture and the production of Phenyl- and Biphenylquinolizidines. In Bajaj Y.P.S. (eds.) *Medicinal and aromatic plants II. Biotechnology in agriculture and forestry* (Vol. 7). Springer.

Rother, A. 1990. Alkaloids of *Heimia montana*. *Phytochemistry*, 29(5), 1683–1686. doi:10.1016/0031-9422(90)80146-8

Schäfer, H. 2021. *Flora of the Azores a field guide* (pp. 1–445). Margraf Publishers GmbH.

Schultes, R.E. 1969. The new world Indians and their hallucinogenic plants. *7th Annual Lecture Honoring Dr. Laura L. Barnes*.

Schultes, R.E. 1976. *Hallucinogenic plants*. Golden Press; Western Publishing Company, Inc.

Singh, A. 2011. Herbalism, phytochemistry and ethnopharmacology. https://www.academia.edu/17381789/Herbalism_Phytochemistry_and_Ethnopharmycology Retrieved 2019 April 3

Standley, P.C. 1926. *Trees and shrubs of Mexico*. United States National Herbarium Vol 23, Smithsonian Institution.

Trottier, R.W., Jr. and Malone, M.H. 1969. Comparative in vitro evaluation of cryogenine, cyproheptadine and diphenhydramine as antagonists of furtrethonium, histamine and serotonin. *Journal of Pharmaceutical Sciences*, 58, 1250–1253.

Trottier, R.W., Jr. and Malone, M.H. 1971. Re-evaluation of thermal edema in the rat as a model for studying antiinflammatory activity. *Proceedings of the Western Pharmacology Society*, 14, 50–54.

Watson, W.C. and Malone, M.H. 1974. Effects of cryogenine on developing and established adjuvant polyarthritis in the rat. *The Pharmacologist*, 16, 291 (Abstr. 571).

Watson, W.C. and Malone, M.H. 1977. Evaluation of immunosuppressive potential of cryogenine using developing and established adjuvant arthritis in rats. *Journal of Pharmaceutical Sciences*, 66, 1304–1308.

Webb, L. J. (1948). Guide to the medicinal and poisonous plants of Queensland. No. 232, 202? https://www.cabidigitallibrary.org/doi/full/10.5555/19480602234

Weisbach, J.A. 1965. U.S. Patent No. 3184446. Washington, DC: U.S. Patent and Trademark Office. *Heimia* Alkaloids.

Wiebelhaus, V.D., Sosnowski, G., Maass, A.R. and Brennan, F.T. 1974. A water diuretic. How and why. In: L.G. Wesson and G.M. Fanetli (Eds.), *Recent advances in renal physiology and pharmueology* (pp. 331–347). University Park Press.

31 Ayahuasca (*Banisteriopsis caapi* (Spruce ex Griseb.) Morton Malpighiaceae)

Noureddine Chaachouay, Abdelkarim El Djiri, Abderrahmaine Katiri, and Lahcen Zidane

31.1 INTRODUCTION

Since antiquity, the utilization of plants has been an integral component of traditional cultures across the globe (Merlin, 2003; Chaachouay et al., 2019, 2019; Orch et al., 2021). The medicinal compounds found in these plants include analgesic, antibacterial, anti-inflammatory, hallucinogenic, and anti-painful properties (Lisa et al., 2020). Some of these plants are still used in religion, ceremonies, practices, and psychoactive and medicinal properties, either in their unprocessed state or as processed extracts or supplements (Ratsch, 2005; Chaachouay & Zidane, 2021). *Banisteriopsis caapi* (Spruce ex Griseb.) Morton is the primary constituent of ayahuasca, a hallucinogenic herbal infusion used in the Amazon region for ceremonial and therapeutic purposes, as well as by persons worldwide who are curious about its effects (De Oliveira et al., 2023; Morales-García et al., 2017). Ayahuasca is the Quechuan term for *B. caapi*, a Malpighiaceae family forest liana indigenous to the Orinoco and Amazon River basins (Baer & Snell, 1974).

Additionally, the tea produced by steeping the stems of *B. caapi* in water, either alone or in conjunction with other plants, is referred to by this name. Ayahuasca tea holds significant importance within the ceremonies of passage and ancient shamanic rituals of the indigenous peoples residing in the northwest region of South America (Santos et al., 2020). Ayahuasca has emerged as a pivotal sacrament within the traditions of Brazilian syncretic religious communities in recent times. The widespread adoption of ayahuasca has been facilitated by the proliferation of so-called ayahuasca religions in North America and Europe (Rodd, 2008; Smith, 1979).

The hallucinogenic effect of *B. caapi* is attributed to the presence of beta-carboline alkaloids in its bark. Over nine alkaloids have been extracted from *B. caapi*. A typical composition of ayahuasca medication consists of a combination of beta-carboline and tryptamine alkaloids (Morales-García et al., 2017). Nevertheless, several indigenous Amazonian societies exclusively concoct a ceremonial beverage using the ayahuasca vine. The beta-carbolines, including harmine, harmaline, and tetrahydroharmine, are derived from the ayahuasca vine, scientifically known as *B. caapi*. Harmine and harmaline have hallucinogenic effects at close to toxic concentrations, while at lower doses, they often induce a state of calmness and purging (Callaway et al., 2005; B. W. L. Santos et al., 2020; Stanković et al., 2015). Ayahuasca contains substantial amounts of tetrahydroharmine. It may be accountable for some of its more significant impacts in contrast to similar plants like *Peganum harmala* L.

The proportion of harmala alkaloids in ayahuasca exhibits significant variation across different regions within the Amazon basin. The diverse effects described by shamans from various 'types' of ayahuasca, although all being categorized as *B. caapi* botanically, might likely be attributed to the varying quantities in which they are present (Morales-García et al., 2017). Harmala alkaloids are short-term monoamine oxidase inhibitors. They work by temporarily lowering the amounts of monoamine oxidase in the body, which is responsible for quickly breaking down tryptamines (Chaachouay et al., 2022, 2022, 2022; De Oliveira et al., 2023). This allows tryptamines to become active when taken orally. Co-administering selective serotonin reuptake inhibitors (SSRIs) like Prozac and other antidepressants with ayahuasca or other monoamine oxidase inhibitors might potentially lead to critical medical problems or even death. The primary constituents of ayahuasca possess a shared indole structure that, via various ways, modulates specific processes of the central nervous system (Tupper, 2008). The crucial determinant is the molecular resemblance of these molecules to the neurotransmitter serotonin. Ayahuasca contains harmala alkaloids, namely harmine, and tetrahydroharmine, which can temporarily block the brain enzyme monoamine oxidase activity (Pic-Taylor et al., 2015). This enables dimethyltryptamine to exhibit its effects when consumed via the mouth.

Additionally, it promotes the buildup of biogenic amines, such as 5-HT, which are typically broken down by monoamine oxidase enzymes. Dimethyltryptamine is an endogenous biological compound that is synthesized and released by the pineal gland in the human body. It is present in several plant species globally (Oroc, 2009). When consumed in its unadulterated state or ingested orally in ayahuasca, it has

DOI: 10.1201/9781003460336-31

the potential to induce very potent hallucinogenic experiences (Morales-García et al., 2017). Characterizing the ayahuasca experience just as an oral dimethyltryptamine encounter facilitated by a beta-carboline inhibitor is an inaccurate representation (Santos et al., 2020). The proportion of harmala alkaloids in ayahuasca exhibits significant variation across different regions within the Amazon basin. The diverse effects described by shamans from various 'types' of ayahuasca, although all being botanically classed as *B. caapi*, might likely be attributed to the varying quantities in which they are present (Stanković et al., 2015). This chapter aims to thoroughly describe the cultural usage, botanical traits, habitat, and geographic distribution, chemical makeup, therapeutic applications, and biological impacts of *B caapi*.

31.2 SCIENTIFIC CLASSIFICATION

Banisteriopsis caapi (Spruce ex Griseb.) Morton is classified as a species within the Malpighiaceae (Barbados-cherry family) genus *Bogisteriopsis*, which comprises approximately 62 to 65 species (The Plant List, 2013). The description of *Banisteriopsis caapi* (Spruce ex Griseb.) Morton, which Richard Spruce initially provided, failed to satisfy the criteria necessary for publication validity. Subsequently, August Heinrich Rudolf Grisebach published a validation of the description. In 1931, Julius Sterling Morton revised and reclassified the facility (Souza, 2023).

Kingdom: Plantae
Subkingdom: Viridiplantae
Infrakingdom: Streptophyta
Superdivision: Embryophyta
Division: Tracheophyta
Subdivision: Spermatophyta
Order: Malpighiales
Family: Malpighiaceae
Genus: *Banisteriopsis*
Species: *Banisteriopsis caapi* (Spruce ex Griseb.) Morton

31.3 VERNACULAR NAMES

B. caapi is an indigenous plant of South America, specifically found in the Amazon rainforest area. It is referred to by numerous vernacular names in the languages spoken in the areas where it is found. Below are some vernacular names for *B. caapi* in different languages:

English: Caapi, ayahuasca, Soul vine, Yage, Yagé; **German**: ayahuasca; **French**: ayahuasca, Liane d'Ayahuasca; **Spanish**: ayahuasca; Caapi, Cipo, Mariri, Pildé, Yagugue, Yage; **Italian**: ayahuasca, Capii, Yagè; **Portuguese**: Cipó, Caapi, Yagê, aysahuasca, Mariri, Jagube, Santo Daime; **Dutch**: Ayahuascawijnstok; **Russian**: Лоза аяваски (Loza ayavaski); **Quechua (Peru)**: Ayawaska; **Shipibo-Conibo (Peru)**: Oni, Uní; **Yagua (Peru/Colombia)**: Tahuari, Natem; **Matsés (Peru)**: Kawa; **Kichwa (Amazon)**: Natema, Natem, Waska; **Huni Kuin (Brazil)**: Nixi Pae, Uni

31.4 BOTANICAL DESCRIPTION

B. caapi is a woody vine with stems covered in fine hairs that may become smooth with age (Figure 31.1). The lamina of the leaf measure (4.8-) 8.2–15.9(-20.5) cm in length and (2.5-)3.5–7.7(-11.5) cm in width (Figure 31.2). It has an ovate shape, with a base that is either obtuse or truncate and an apex that is short- to long-acuminate (B. W. L. Santos et al., 2020; Souza, 2023). The upper surface of the lamina is smooth, while the lower surface is sparsely covered in fine hairs or smooth. There are 2–5 pairs of sessile glands below the lamina's margin and another pair near the midrib at the base. The petiole, which connects the lamina to the stem, measures 9–25 mm long and is sparsely covered in fine hairs or smooth. It may or may not have glands near the apex (Schultes et al., 1969). The stipules, small leaf-like structures at the base of the petiole, are triangular and measure 0.5–1 mm in length (de Frias et al., 2012). The inflorescence is a cluster of four-flowered umbels that arise from the axils of the leaves. It is covered with a sparse tomentose to velutinous layer. The bracts and bracteoles, 1–1.8 mm long, fall off before or during flowering and sometimes soon after (de Carvalho et al., 2023; Smith, 1979).

The pedicel is 7–11 mm long and attached directly to the stem. It is covered with fine, closely pressed, or hairy silky fibers. The sepals are covered with fine hairs on the lower surface and have an excellent layer of hair on the upper surface. None of the sepals have any glands except for the two lateral sepals, which have two large glands measuring 0.5–2.2 mm in length. The petals start as pale pink and turn yellow as they age (Rodd, 2008). They are smooth and fringed and have a 1–1.5 mm length for the four lateral petals and 2.5–3 mm for the posterior petals. The limb of the lateral petals is 5–8.5 mm long and 4–6 mm wide, while the limb of the posterior petal is 5–7 mm long and 2.5–4.5 mm wide (Smith, 1979). The posterior petal is constricted at the apex and has a broad, obovate shape. The basal fimbriae of the petals have gland-tipped ends. The filaments are 2–4 mm long, with the posterior three bent inward between the posterior styles. The anthers have sparsely hairy to nearly hairless locules measuring 0.3–1.2 mm in length (Furquim, 2021; Rodd & Sumabila, 2011).

The connectives are 0.2–1.6 mm long. The five longest connectives are glandular and located opposite the antero-lateral sepals. These connectives extend beyond the locules by 0.5–1 mm. The anterior style is 2.8–3.2 mm long and straight, while the posterior styles are 3–f long, divergent, and lyrate at the base (De Oliveira et al., 2023). The stigmas are capitate. The samara is smooth to hairless, with the wing's upper surface measuring 18–42 mm in length and 8–22 mm in width. It has a rounded tooth at the inner base.

FIGURE 31.1 Leaf of *B. caapi*.

FIGURE 31.2 The twisted stem of *B. caap*.

The nut has pronounced ribs on the side perpendicular to the central area, sometimes with a little spiky growth. The inside part of the nut is covered in hair (Smith, 1979).

31.5 ORIGIN AND GEOGRAPHIC DISTRIBUTION

B. caapi is a plant native to the Amazon jungle in South America (Peru, Colombia, Brazil, Argentina, Bolivia, Ecuador, and Venezuela). It is believed that this species originated in the western and northwestern regions of the Amazon basin. Indigenous communities in these areas have used it for millennia for its spiritual, shamanic, and therapeutic properties (De Oliveira et al., 2023; B. W. L. Santos et al., 2020). The historical and cultural importance of the site is intricately linked to the indigenous civilizations, and the practice of using it in ancient ceremonies has been transmitted over centuries (Rodd, 2008). *B. caapi* is found in several nations in the Amazon basin, such as Brazil, Peru, Colombia, Ecuador, Bolivia, and Venezuela (Thevenin & Sambuichi, 2020, 2020). *B. caapi* thrives in the abundant rainforest habitats of these nations. The species flourishes in the humid and tropical environment of the Amazon, often climbing big trees and creating dense thickets in the understory. The distribution of this plant spans a large and ecologically varied area, which remains significant in indigenous communities' spiritual and healing traditions (Callaway, 2005; Morales-García et al., 2017). Additionally, it attracts global attention due to its distinctive psychoactive and therapeutic characteristics.

31.6 HABITAT AND ECOLOGY

B. caapi flourishes in a distinct environment found only in the Amazon rainforest, enhancing its distinctive ecology. This species is often seen in the lower layer of fully developed tropical rainforests, where it displays a climbing growth pattern (Souza, 2023). It often meanders around towering trees and other plants and uses them as structural support to get sunlight (do Carmo Bastos et al., 2011; Furquim, 2021; Gragson, 1997). By engaging in this climbing behavior, the organism can reach the uppermost layer of the jungle and thrive in the dimly lit and densely populated surroundings (Coe & Gaoue, 2023). *B. caapi* has developed to succeed in the tropical and humid environment of the Amazon basin, with ample precipitation year-round (Thevenin & Sambuichi, 2020). *B. caapi* is an integral part of the complex ecosystem of the rainforest. The ecosystem offers a home and serves as a food supply for diverse mammals, birds, and insects. In addition, the stems and foliage of *B. caapi* contain alkaloids, including harmaline and harmine, which serve as chemical deterrents against herbivores (Tupper, 2008). Certain studies suggest that these alkaloids further contribute to the vine's customary use in *B. caapi* preparations, where they synergistically combine with other botanical components to elicit hallucinogenic effects (De Oliveira et al., 2023).

31.7 ETHNOBOTANY

Within a cultural framework, the ayahuasca beverage is historically recognized as a 'master plan' that offers a structure for exploring the mythical, dramatic, and psychological aspects of human experience. Master plants have addressed physical, emotional, and spiritual imbalances. They also enable individuals to establish a connection with profound internal resonances, diagnose ailments, and grant direct entry into the spiritual realm and the vast reserves of wisdom that would otherwise remain inaccessible. In addition, master plants facilitate the development of existential intelligence (De Oliveira et al., 2023; Benkhnigue et al., 2022; Belhaj et al., 2021; Rodd, 2002).

31.7.1 Phytotherapy

The ayahuasca beverage is primarily made from the bark of the lianas *B. caapi* or *B. inebrians* and other ingredients derived from other species, including *P. viridis*, often known as chacruna. Indigenous populations have traditionally used these components for various reasons. In most syncretic churches, the ayahuasca beverage is usually made as follows: shards of *B. caapi* bark are gathered and cleansed with water, then crushed with a wooden hammer and meticulously put in a cauldron, alternating with cleaned leaves of *P. viridis* (Callaway et al., 2005; Politi et al., 2021). Subsequently, water is introduced until the plant material is fully submerged, and the concoction is then subjected to boiling and condensation for a minimum duration of 8 hours, producing multiple liters (Serrano-Dueñas et al., 2001). Ayahuasca is widely recognized in several indigenous cultures as a potent remedy for both physical and mental well-being, fostering a state of balance and unity with other individuals. Ayahuasca is a psycho-integrative plant since it effectively combines the aspects of mind, body, spirit, and emotions within a secure religious context that is socially accepted. Ayahuasca, when taken appropriately, may be regarded as an effective therapeutic instrument with confirmed positive benefits. In several Latin American nations such as Peru, Brazil, Ecuador, Colombia, Argentina, Chile, and Mexico, ayahuasca has been used for the treatment of addictions (Rodd, 2002; B. W. L. Santos et al., 2020).

Nevertheless, it is essential to note that its usage may also pose specific hazards. During the medicinal use of ayahuasca, the shaman, rather than the patient, is the one who consumes the substances (Schmid et al., 2010). Occasionally, the patient has previously experimented with traditional medicine. In the event of failure, they will seek the shaman's assistance, attributing the sickness to potential magical origins (Chaachouay et al. 2021a, 2021b, 2021c). During this therapeutic process, the shaman people will consume the beverage and interpret the hallucinations experienced while under the effect of the substance (Mabit, 2007). They analyze these images to identify the ailment's underlying source and combat it metaphorically. The shaman will chant to the patient of the battle, liberating him from evil forces. There are also medical applications for the beverages. The shaman and the patient will consume the beverage within some cultural contexts (Chaachouay et al., 2020, 2020; de Frias et al., 2012; Mabit et al., 1996).

31.7.2 Ayahuasca Ceremonies

Plant-derived psychoactive substances have been integral to South American traditional religions for countless years. Some communities have historically used ayahuasca for religious and spiritual intentions (Trichter, 2006). However, it has gained global popularity among those who want to expand their consciousness, recover from previous traumas, or just go on an ayahuasca expedition (Baer & Snell, 1974). Ayahuasca rituals often take place throughout the nighttime and conclude after the effects of ayahuasca have subsided (Callon et al., 2021). Following the shaman's preparation and blessing of the room, participants are administered ayahuasca, which may be divided into many doses. Following ingestion of ayahuasca, the majority of individuals typically experience its effects within a timeframe of 20 to 60 minutes (Ambrosini et al., 2023). The substance's results vary according to the dosage, and the duration of the experience might range from 2 to 6 hours (Gonzalez et al., 2021). People who use ayahuasca may encounter symptoms such as emesis, diarrhea, sensations of intense happiness, vivid visual and aural hallucinations, transformative psychedelic effects, apprehension, and paranoia (Williams et al., 2023).

31.8 HALLUCINOGENIC EFFECTS

In western cultures, hallucinogens are psychoactive chemicals that may cause alterations in perception and emotions without causing memory loss or loss of consciousness. The use of these hallucinogenic plants and mushrooms is prevalent among several cultures in South America (Bennett, 1992; Carod-Artal, 2015). Ayahuasca is a psychoactive beverage prepared by combining the stem and bark of the tropical *B. caapi* with many additional botanical components (Rodd, 2008; Trevisani et al., 2021). Numerous Amazon rainforest indigenous communities consume beverages derived from this plant, which they refer to by various names, including ayahuasca, caapi, dapa, datem, pinde, yaje, yage, and natem. Numerous tribe cultures have utilized it for centuries, which is significant in contemporary societies and religions (Morales-García et al., 2017). Ayahuasca, initially developed by Indigenous South Americans residing in the Amazon basin, is being used in many regions around the globe (Pic-Taylor et al., 2015). When preparing ayahuasca, it is frequently augmented with additional plant species that impart hallucinogenic attributes to the beverage. *B. caapi* in isolation inside ayahuasca has restricted hallucinogenic effects, but it is sometimes used in this manner (Santos et al., 2022; Serrano-Dueñas et al., 2001). However, the beverage produced probably consists of ingredients other than *B. caapi*. When these plant above species are used in the concoction of the beverage, they introduce their chemical components and amplify the psychoactive effects of the drink (Chaachouay et al., 2022, 2022, 2022; Politi et al., 2021). Certain users report visual and sensory perceptions, while others claim the drink has curative properties.

31.9 PHYTOCHEMISTRY

B. caapi serves as the fundamental component of ayahuasca, a psychoactive herbal infusion used in the Amazon region for ceremonial and therapeutic purposes, as well as by persons globally who are intrigued by its effects. Multiple studies indicate harmaline, harmine, and tetrahydroharmine (Figure 31.3), the primary alkaloids found in *B. caapi* (Morales-García et al., 2017). The first alkaloid isolated from *B. caapi* was identified as harmine, previously reported in *P. harmala*. Additional indole alkaloids, including β-carbolines called harmaline and tetrahydroharmine, were also detected in *B. caapi* (Callaway et al., 2005; Estrella-Parra et al., 2019; Santos et al., 2022). The ayahuasca beverage also includes DMT derived from *P. viridis* leaves, which was first identified in *Mimosa tenuiflora* and is classed as a simple tryptamine chemical. The chemical analysis of ayahuasca beverage has been motivated by a focused scientific curiosity in studying natural resources, conducting therapeutic research, and discovering illicit substances included in its components (Santos et al., 2020). *B. caapi* comprises tetra-hydronorharmine, harmine, and harmaline as the primary active constituents, along with smaller amounts of harmol, harmalol, and norharmane (Callaway, 2005; Rodd, 2002; Stanković et al., 2015). Furthermore, the chemical compositions of β-carboline alkaloids, namely banistenoside A and banistenoside B, as well as procyanidins, procyanidin B2, and epicatechin, have been documented in *B. caapi* (B. W. L. Santos et al., 2020). The biomarkers DMT, harmine, and harmaline have been identified in six different ayahuasca beverages. These beverages are prepared using various combinations of *B. caapi*, *P. viridis*, *Mimosa hostilis*, *P. harmala*, *Dyplopteris cabrerana*, and *Tetrapterys mucronate* (de Frias et al., 2012; Gonçalves et al., 2020).

31.10 PHARMACOLOGY

Ayahuasca, Daime, Yajé, and vegetal are among the several appellations attributed to *B. caapi*, an Amazonian liana belonging to the Malpighiaceae family. It creates diverse water infusions and decoctions from this vine or in conjunction with other botanicals (Rodd, 2002). The global use of this psychoactive herbal infusion is undergoing unparalleled growth and is the subject of escalating biological investigation (Morales-García et al., 2017). This mixture is an exceptional component of the native pharmacopeias of the Americas, which include a wealth of psychoactive plants capable of inducing altered states of consciousness (Bennett, 1992). The indigenous societies in the New World held these plants in high regard and considered them

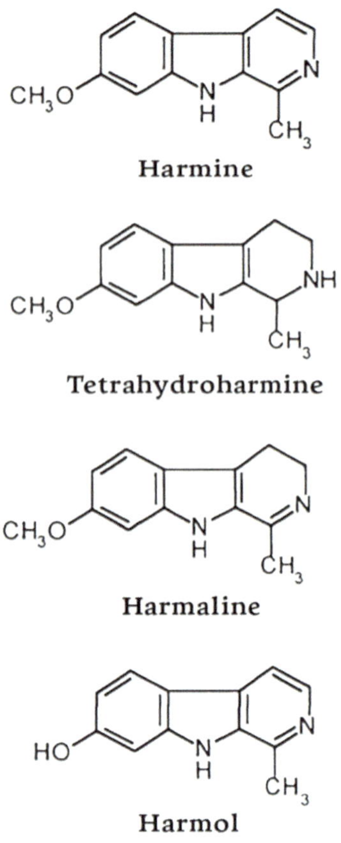

FIGURE 31.3 Chemical structures of the main alkaloids found in *B. caapi*.

integral to their worldview. They used these plants for medicinal purposes, religious rituals, and rites of passage (Rist & Dahdouh-Guebas, 2006).

B. caapi comprises the alkaloids harmine, tetrahydroharmine (THH), and minimal quantities of harmaline. These compounds possess a shared tricyclic β-carboline structure. Due to this rationale, they are often denoted as 'beta-carbolines' and 'harmala alkaloids' since harmine was first extracted from an unrelated plant, *P. harmala* (Callaway et al., 2005; Morales-García et al., 2017; Stanković et al., 2015). These beta-carbolines possess diverse pharmacological effects. Humans can temporarily inhibit the activity of subtype A of the monoamine oxidase (MAO) enzyme. The leaves of *B. caapi* contain significant quantities of the psychedelic N,N-dimethyltryptamine (DMT), which is considered one of the 'potentially dangerous' extraterrestrial amines (Domínguez-Clavé et al., 2016; Oroc, 2009).

Ayahuasca has physiological effects that mimic the action of the sympathetic nervous system, leading to an increase in the turnover of norepinephrine and producing dilation of the pupils (Hamill et al., 2019). Additionally, it elevates the blood concentrations of the stress chemicals cortisol and prolactin. Nevertheless, in contrast to the notable cardiovascular effects documented for pure DMT in trials using intravenous delivery, we found only modest elevations in systolic and diastolic blood pressure after ayahuasca ingestion and almost no alterations in heart rate (Greene, 2022; McKenna, 2004).

DMT has a range of putative molecular targets at the receptor level. It affects serotonin transmission in the brain because it has a structure similar to the natural neurotransmitter serotonin and firmly attaches to specific serotonin receptors (Carbonaro & Gatch, 2016). DMT has agonistic action at the 5-HT2A and 5-HT1A receptor sites. The 5-HT1A receptor primarily functions as a pre-synaptic receptor linked to inhibitory effects (Brito-da-Costa et al., 2020). The raphe nuclei of the brain stem contain significant amounts of it, and its activation leads to a decrease in serotonergic tone (Hornung, 2003).

DMT also binds to and affects additional biochemical sites unrelated to serotonin. Fontanilla and colleagues found a high affinity in the micromolar range for the sigma-1 receptor located within the cell (Su et al., 2009). The SR1 protein is linked to the endoplasmic reticulum and regulates the function of other proteins, facilitating the development of brain plasticity by increasing the production of dendritic spines (Szabo et al., 2014). DMT induces biochemical and behavioral changes in animals via activating sigma-1 receptors. It inhibits sodium channels and causes increased movement in mice (Frecska et al., 2013). The sigma-1 knockout mice do not exhibit these behavioral consequences. DMT acts as an agonist at the trace amine-associated receptor and is processed via the vesicle monoamine and serotonin transporters (Carbonaro & Gatch, 2016). The antagonistic action of DMT at the 5-HT2A receptor sites may potentially provide antidepressant and anxiolytic effects. This phenomenon has been shown in animals using the selective agonist DOI (Duan et al., 2023). This hypothesis is substantiated by the favorable outcomes of recent treatment experiments using diverse psychedelics that have the common characteristic of activating this receptor.

The pharmacological properties of beta-carbolines are closely linked to their therapeutic efficacy in treating depression and anxiety. Inhibiting MAO is a recognized treatment strategy for treating these illnesses (Domínguez-Clavé et al., 2016). The three primary beta-carbolines, harmol, harmalol, and tetrahydroharmol, possess characteristics that suppress the activity of MAO. Furthermore, THH functions as a serotonin reuptake inhibitor. The primary pharmacological mechanism of many antidepressants now used in clinical practice is the inhibition of the serotonin transporter (Pathak et al., 2016). The ingestion of ayahuasca might lead to elevated levels of monoamines in the synapse, which may contribute to the antidepressant and antianxiety effects of *B. caapi* formulations (Palhano-Fontes et al., 2014). Harmine exhibits potent and specific inhibition of DYRK1A, which stands for dual specificity tyrosine-(Y)-phosphorylation regulated kinase 1A. The up-regulation of this kinase, which impacts the development and maturation of neurites, is seen in individuals with Down Syndrome due to an extra copy of chromosome 21 (Göckler et al., 2009).

An investigation showed that the β-carboline harmane elicited an antidepressant response in the forced swim test (B. W. L. Santos et al., 2020). This action seems to be caused by an inverse-agonist mechanism targeting the benzodiazepine receptor. Subsequent research demonstrated that harmane, norharmane, and harmine also had similar antidepressant properties. Similarly, a study found that harmine, when given in a single dose, had antidepressant effects in both the forced swimming and open-field tests. Harmine, unlike the antidepressant imipramine, elevated BDNF levels, specifically in the hippocampus (Stanković et al., 2015). In a subsequent study conducted by the same research team, the authors evaluated the effects of harmine on rats subjected to the Chronic Mild Stress (CMS) technique, which serves as an animal model for depression. Remarkably, administration of harmine effectively alleviated anhedonia, reversed adrenal gland hypertrophy, and restored blood levels of ACTH and BDNF proteins to normal. In a subsequent investigation, they further showcased that prolonged administration of all tested quantities of harmine resulted in a reduction in the duration of immobility in rats during the forced swimming test (Cruz, 2023). The research examined the beneficial impacts of imipramine and harmine on oxidative stress indicators, which are believed to play a role in depression. This study found that harmine caused an increase in the activities of superoxide dismutase and catalase while also reducing lipid and protein oxidation.

Ayahuasca's effects on substance use disorders and depression-anxiety have been investigated in human subjects, mirroring the findings of animal studies to date (Mabit, 2007). Research documented a cessation of drug or alcohol abuse/dependence in a sample of individuals belonging to an ayahuasca community, with an average membership duration of 6.5 years (Domínguez-Clavé et al., 2016). In a separate study including a series of cases, the researchers observed a noteworthy decrease in cocaine use after a treatment session involving the use of ayahuasca among individuals from a First Nations community in Canada who had not before encountered ayahuasca (Argento et al., 2019). Additionally, they documented enhancements in mindfulness, empowerment, hopefulness, viewpoint on quality of life, and sense of purpose in life. Two case-control studies had comparable outcomes on the impact of drug use. An analysis revealed that a 15-person long-term sample of individuals who used ayahuasca had remission of alcohol, depressive, and anxiety disorders (Argento et al., 2019).

Additionally, they exhibited changes in behavior, attitude toward others, and view on life. These findings were compared to a group of 15 individuals who had no previous history of ayahuasca consumption and were matched to the sample group. Another found that symptoms related to anxiety and sadness decreased following the first use of ayahuasca among a group of Santo Daime members (R. G. Santos et al., 2007). In addition, they documented behavioral alterations, including heightened assertiveness, liveliness, and happiness among individuals belonging to two cohorts of ayahuasca consumers. Recently, two open-label studies examined the impact of a solitary administration of ayahuasca on individuals with mental depression who were hospitalized (Barbosa et al., 2009). An investigation documented substantial and statistically significant decreases, ranging from 82%, in scores related to depression throughout the period from the first assessment to 1, 7, and 21 days after the treatment. In addition, the administration of ayahuasca did not induce bouts of mania or hypomania, as assessed by the Young Mania Rating Scale (Osório et al., 2015).

31.11 CLINICAL USES

The therapeutic applications of ayahuasca have been extensively documented. In 1979, ayahuasca and harmaline were administered to 30 psycho-patients (Estrella-Parra et al., 2019). Ten patients showed favorable improvements, in contrast to the outcomes seen with other intense psychotherapy procedures. Australian individuals have a cathartic impact and a unique psychic phenomenon generated by ayahuasca and DMT, leading to heightened psycho-spiritual awareness (Oroc, 2009). Within the religious framework in Colombia, a group of 40 individuals residing in Bogota city partook in ayahuasca use, resulting in the attainment of inner tranquility and enhanced cognitive clarity, enabling them to address intricate personal predicaments effectively (Weiskopf, 2005). In addition, the use of ayahuasca in Brazilian teens resulted in a decrease in anxiety symptoms, a reduction in image dysmorphia, and an improvement in attention deficit disorders, as compared to teenagers who did not take ayahuasca (McKenna, 2004).

Domínguez-Clavé et al., (2016) documented that ayahuasca enhances mindfulness-related abilities in a group of 25 participants. Shanon, via an interview conducted with around 200 ayahuasca users, concluded that the effects of this beverage should be linked to the human psyche, particularly in the realm of cognitive psychology, rather than being attributed only to the brain and society (Shanon, 2010). The heightened mental state brought forth by the psychoactive elixir implies an altered state of awareness that warrants further investigation. In addition, Schmid documented that a group of 15 individuals who used ayahuasca for one year and were afflicted with various ailments such as cancer, asthma, depression, alcohol addiction, hepatitis C, migraine, fibromyalgia, and influenza exhibited an improved outlook on their life and diseases (Schmid et al., 2010).

According to Winkelman's assessment, ayahuasca improves the individuals' self-awareness, allows them to reach deeper depths of their being, and facilitates personal development in a group of 11 individuals from Europe and North America (Winkelman, 2014). Kuypers conducted a study with 26 Caucasian individuals and found that ayahuasca improves divergent creative thinking, reduces fear, promotes forgiveness toward others and oneself, and encourages self-acceptance (Kuypers et al., 2016). These findings suggest that ayahuasca increases mental flexibility and can be beneficial in supporting psychotherapeutic interventions for clinical trial initiatives.

Regarding the use of ayahuasca as an antiaddictive substance, a study conducted on 81 ayahuasca users from Santo Daime and 'União de Vegetal' groups in North America revealed a decrease in alcohol consumption, adoption of a more nutritious diet, enhanced mood, and more self-acceptance (dos Santos et al., 2021). Additionally, the participants had heightened feelings of affection and empathy within their interpersonal connections. Grob documented that 15 individuals belonging to the União de Vegetal syncretic religion practiced thoroughly refraining from alcohol and other substances as a result of using ayahuasca. The participants saw decreased anger, resentment, and violence while gaining improved self-control and a sense of responsibility toward their families (Metzner, 2005). In Brazil, teens who take ayahuasca as part of the União de Vegetal religious group consume a lower amount of alcohol (32.5%) compared to non-users of ayahuasca (65.1%), as well as other psychoactive substances such as cannabis, amphetamines, cocaine, sedatives, tranquilizers, nicotine, and other narcotics (Barbosa et al., 2018). A further instance of ayahuasca serving as a purifying substance occurred with a Brazilian patient who restored her well-being after a period of taking alcohol, cocaine, and nicotine for a minimum of two years, resulting in a transformation of her outlook on life (Da Silveira et al., 2005).

31.12 CONCLUSION

Because of its pharmacological, commercial, and spiritual significance, the traditional ayahuasca beverage garners significant attention from many groups worldwide, including believers, government officials, and scientists. Due to its enduring cultural importance, this practice has been traditionally observed throughout the Amazon area up until the present day. The usage of *B. caapi* in ayahuasca has become widespread globally, even in places that are highly technologically advanced and have made significant medical advancements. Nevertheless, the sensible use of this treatment remains a formidable task for contemporary scientific research. Further investigation is necessary for clinical populations, using bigger sample sizes, comparable control groups, randomized experimental designs, and evaluators unaware of the treatment conditions to validate its effectiveness. Finally, future research should prioritize the implementation of appropriate settings and the involvement of doctors with specialized training to guarantee the safety of users.

LIST OF ABBREVIATIONS

5-HT 5-hydroxytryptamine
ACTH adrenocorticotropic hormone
B caapi *Banisteriopsis caapi* (Spruce ex Griseb.) Morton
BDNF brain-derived neurotrophic factor
CMS chronic mild stress
DMT *N,N*-dimethyltryptamine
MAO monoamine oxidase
THH tetrahydroharmine

REFERENCES

Ambrosini, F., Pariset, L., & Biolcati, R. (2023). Ayahuasca ceremonies: Set and setting features across Italy and Colombia. *Drugs, Habits and Social Policy*, *24*(4), 351–371.

Argento, E., Capler, R., Thomas, G., Lucas, P., & Tupper, K. W. (2019). Exploring ayahuasca-assisted therapy for addiction: A qualitative analysis of preliminary findings among an Indigenous community in Canada. *Drug and Alcohol Review*, *38*(7), 781–789. https://doi.org/10.1111/dar.12985

Baer, G., & Snell, W. W. (1974). An ayahuasca ceremony among the Matsigenka (Eastern Peru). *Zeitschrift für Ethnologie*, *H. 1/2*, 63–80.

Barbosa, P. C. R., Cazorla, I. M., Giglio, J. S., & Strassman, R. (2009). A six-month prospective evaluation of personality traits, psychiatric symptoms and quality of life in Ayahuasca-Naïve subjects. *Journal of Psychoactive Drugs*, *41*(3), 205–212. https://doi.org/10.1080/02791072.2009.10400530

Barbosa, P. C. R., Tófoli, L. F., Bogenschutz, M. P., Hoy, R., Berro, L. F., Marinho, E. A. V., Areco, K. N., & Winkelman, M. J. (2018). Assessment of alcohol and tobacco use disorders among religious users of Ayahuasca. *Frontiers in Psychiatry*, *9*. https://www.frontiersin.org/articles/10.3389/fpsyt.2018.00136

Belhaj, S., Chaachouay, N., & Zidane, L. (2021). Ethnobotanical and toxicology study of medicinal plants used for the treatment of diabetes in the High Atlas Central of Morocco. *Journal of Pharmacy & Pharmacognosy Research*, *9*(5), 619–662. https://doi.org/10.56499/jppres21.1014_9.5.619

Benkhnigue, O., Chaachouay, N., Khamar, H., El Azzouzi, F., Douira, A, & Zidane, L. (2022). Ethnobotanical and ethnopharmacological study of medicinal plants used in the treatment of anemia in the region of Haouz-Rehamna (Morocco). *Journal of Pharmacy & Pharmacognosy Research*, *10*(2), 279–302. https://doi.org/10.56499/jppres21.1196_10.2.279

Bennett, B. C. (1992). Hallucinogenic plants of the Shuar and related indigenous groups in Amazonian Ecuador and Peru. *Brittonia*, *44*(4), 483–493. https://doi.org/10.2307/2807199

Brito-da-Costa, A. M., Dias-da-Silva, D., Gomes, N. G. M., Dinis-Oliveira, R. J., & Madureira-Carvalho, Á. (2020). Toxicokinetics and toxicodynamics of Ayahuasca alkaloids N,N-dimethyltryptamine (DMT), harmine, harmaline and tetrahydroharmine: clinical and forensic impact. *Pharmaceuticals*, *13*(11), Article 11. https://doi.org/10.3390/ph13110334

Callaway, J. C. (2005). Various alkaloid profiles in decoctions of banisteriopsis caapzl. *Journal of Psychoactive Drugs*, *37*, 2.

Callaway, J. C., Brito, G. S., & Neves, E. S. (2005). Phytochemical analyses of Banisteriopsis caapi and Psychotria viridis. *Journal of Psychoactive Drugs*, *37*(2), 145.

Callon, C., Williams, M., & Lafrance, A. (2021). 'Meeting the Medicine Halfway': Ayahuasca ceremony leaders' perspectives on preparation and integration practices for participants. *Journal of Humanistic Psychology*. https://doi.org/10.1177/00221678211043300

Carbonaro, T. M., & Gatch, M. B. (2016). Neuropharmacology of N,N-dimethyltryptamine. *Brain Research Bulletin*, *126*, 74–88. https://doi.org/10.1016/j.brainresbull.2016.04.016

Carod-Artal, F. J. (2015). Hallucinogenic drugs in pre-Columbian Mesoamerican cultures. *Neurología (English Edition)*, *30*(1), 42–49. https://doi.org/10.1016/j.nrleng.2011.07.010

Chaachouay, N., Azeroual, A., Bencharki, B., Douira, A., & Zidane, L. (2022). Ethnoveterinary medicines plants for animal therapy in the Rif, North of Morocco. *South African Journal of Botany*, *147*, 176–191. https://doi.org/10.1016/j.sajb.2021.12.037

Chaachouay, N., Azeroual, A., Douira, A., & Zidane, L. (2022). Ethnoveterinary practices of medicinal plants among the Zemmour and Zayane tribes, Middle Atlas, Morocco. *South African Journal of Botany*, *151*, 826–840. https://doi.org/10.1016/j.sajb.2022.11.009

Chaachouay, N., Benkhnigue, O., Douira, A., & Zidane, L. (2021a). Poisonous medicinal plants used in the popular pharmacopoeia of the Rif, northern Morocco. *Toxicon*, 189, 24–32. https://doi.org/10.1016/j.toxicon.2020.10.028

Chaachouay, N., Benkhnigue, O., Fadli, M., Ibaoui, H., El Ayadi, R., & Zidane, L. (2019). Ethnobotanical and ethnopharmacological study of medicinal and aromatic plants used in the treatment of respiratory system disorders in the Moroccan Rif. *Ethnobotany Research and Applications*, *18*(22), 1–16.

Chaachouay, N., Benkhnigue, O., Fadli, M., Ibaoui, H., & Zidane, L. (2019). Ethnobotanical and ethnopharmacological studies of medicinal and aromatic plants used in the treatment of metabolic diseases in the Moroccan Rif. *Heliyon*, *5*(10), e02191. https://doi.org/10.1080/10477845.2021.1942433

Chaachouay, N., Benkhnigue, O., & Zidane, L. (2020). Ethnobotanical study aimed at investigating the use of medicinal plants to treat nervous system diseases in the Rif of Morocco. *Journal of Chiropractic Medicine*, *19*(1), 70–81. https://doi.org/10.1016/j.jcm.2020.02.004

Chaachouay, N., Benkhnigue, O., & Zidane, L. (2022). Ethnobotanical and Ethnomedicinal study of medicinal and aromatic plants used against dermatological diseases by the people of Rif, Morocco. *Journal of Herbal Medicine*, 32, 100542. https://doi.org/10.1016/j.hermed.2022.100542

Chaachouay, N., Douira, A., Hassikou, R., Brhadda, N., Dahmani, J., Belahbib, N., Ziri, R., & Zidane, L. (2020). Ethnobotanical and ethnopharmacological study of medicinal and aromatic plants used in the treatment of respiratory system disorders in the Moroccan Rif. PhDThesis, Département de Biologie - Université Ibn Tofail - Kénitra, Maroc, https://tel.archives-ouvertes.fr/tel-03376377

Chaachouay, N., Douira, A., & Zidane, L. (2021b). COVID-19, prevention and treatment with herbal medicine in the herbal markets of Salé Prefecture, North-Western Morocco. *European Journal of Integrative Medicine*, 42, 101285. https://doi.org/10.1016/j.eujim.2021.101285

Chaachouay, N., Douira, A., & Zidane, L. (2021c). Herbal Medicine Used in the Treatment of Human Diseases in the Rif, Northern Morocco. *Arabian Journal for Science and Engineering*, 47, 131–153. https://doi.org/10.1007/s13369-021-05501-1

Chaachouay, N., & Zidane, L. (2021). The Symbolic Efficacy of Plants in Rituals and Socio-Religious Ceremonies in Morocco, Northwest of Africa. *Journal of Religious & Theological Information*, *21*(1–2), 34–53. https://doi.org/10.1080/10477845.2021.1942433

Coe, M. A., & Gaoue, O. G. (2023). Increased clonal growth in heavily harvested ecosystems failed to rescue ayahuasca lianas from decline in the Peruvian Amazon rainforest. *Journal of Applied Ecology*, *60*(10), 2105–2117. https://doi.org/10.1111/1365-2664.14488

Cruz, R. V. L. da. (2023). *Effects of N,N-Dimethyltryptamine on the phenotype and diathesis of a mice model of depression* [doctoralThesis, Universidade Federal do Rio Grande do Norte]. https://repositorio.ufrn.br/handle/123456789/56775

Da Silveira, D. X., Grob, C. S., de Rios, M. D., Lopez, E., Alonso, L. K., Tacla, C., & Doering-Silveira, E. (2005). Ayahuasca in adolescence: A preliminary psychiatric assessment. *Journal of Psychoactive Drugs*, *37*(2), 129–133. https://doi.org/10.1080/02791072.2005.10399792

de Carvalho, A. B., da Silva, C. J., Zortéa, K. É. M., Fernandes, J. M., Alves, A. O., Corrêa, M. A., & Rossi, A. A. B. (2023). Morfologia floral e polínica e aspectos reprodutivos de Banisteriopsis caapi (Spruce ex Griseb.) CV Morton nativos e cultivados no Município de Alta Floresta, Mato Grosso. *Delos: Desarrollo Local Sostenible*, *16*(45), 1632–1655.

de Frias, U. A., Costa, M. C. M., Takahashi, J. A., & Oki, Y. (2012). Banisteriopsis species: A source of bioactive of potential medical application. *International Journal of Biotechnology for Wellness Industries*, *1*, 163–171.

De Oliveira, R. C., Behrens, C. S. B., Nagamine-Pinheiro, N., Fagg, C. W., E Silva, M. S., Martins-Silva, T., & Sonsin-Oliveira, J. (2023). Ethnobotany and Wood Anatomy of Banisteriopsis caapi Ethnotaxa and Diplopterys cf. Pubipetala, Components of Ayahuasca in Brazilian Rituals. *Economic Botany*, *77*(1), 18–47. https://doi.org/10.1007/s12231-023-09567-w

do Carmo Bastos, M. de N., de Sousa, J. dos S., & Alexandrino, V. H. D. (2011). *The Taxonomic Study of the Malpighiaceae Juss. Family of the Restingas of Algodoal/Maiandeua, Maracanã, Pará, Brazil*. https://agris.fao.org/search/en/providers/122436/records/64747eaa2d3f560f80b05256

Domínguez-Clavé, E., Soler, J., Elices, M., Pascual, J. C., Álvarez, E., de la Fuente Revenga, M., Friedlander, P., Feilding, A., & Riba, J. (2016). Ayahuasca: Pharmacology, neuroscience and therapeutic potential. *Brain Research Bulletin*, *126*, 89–101. https://doi.org/10.1016/j.brainresbull.2016.03.002

dos Santos, R. G., Bouso, J. C., Rocha, J. M., Rossi, G. N., & Hallak, J. E. (2021). The use of classic hallucinogens/psychedelics in a therapeutic context: Healthcare policy opportunities and challenges. *Risk Management and Healthcare Policy*, *14*, 901–910. https://doi.org/10.2147/RMHP.S300656

Duan, W., Cao, D., Wang, S., & Cheng, J. (2023). Serotonin 2A receptor (5-HT2AR) agonists: Psychedelics and non-hallucinogenic analogues as emerging antidepressants. *Chemical Reviews*. https://doi.org/10.1021/acs.chemrev.3c00375

Estrella-Parra, E. A., Almanza-Pérez, J. C., & Alarcón-Aguilar, F. J. (2019). Ayahuasca: Uses, phytochemical and biological activities. *Natural Products and Bioprospecting*, *9*(4), 251–265. https://doi.org/10.1007/s13659-019-0210-5

Frecska, E., Szabo, A., Winkelman, M. J., Luna, L. E., & McKenna, D. J. (2013). A possibly sigma-1 receptor mediated role of dimethyltryptamine in tissue protection, regeneration, and immunity. *Journal of Neural Transmission*, *120*(9), 1295–1303. https://doi.org/10.1007/s00702-013-1024-y

Furquim, M. A. (2021). *Variedades de Banisteriopsis caapi (Spruce, 1873). Anatomia, cultivo e aspectos anatômicos de importância etnofarmacológica do cipó mariri*. https://repositorio.ufpb.br/jspui/handle/123456789/20823

Göckler, N., Jofre, G., Papadopoulos, C., Soppa, U., Tejedor, F. J., & Becker, W. (2009). Harmine specifically inhibits protein kinase DYRK1A and interferes with neurite formation. *The FEBS Journal*, *276*(21), 6324–6337. https://doi.org/10.1111/j.1742-4658.2009.07346.x

Gonçalves, J., Luís, Â., Gradillas, A., García, A., Restolho, J., Fernández, N., Domingues, F., Gallardo, E., & Duarte, A. P. (2020). Ayahuasca beverages: Phytochemical analysis and biological properties. *Antibiotics*, *9*(11), Article 11. https://doi.org/10.3390/antibiotics9110731

Gonzalez, D., Cantillo, J., Perez, I., Carvalho, M., Aronovich, A., Farre, M., Feilding, A., Obiols, J. E., & Bouso, J. C. (2021). The Shipibo ceremonial use of ayahuasca to promote well-being: An observational study. *Frontiers in Pharmacology*, 1059.

Gragson, T. L. (1997). The use of underground plant organs and its relation to habitat selection among the Pumé Indians of Venezuela. *Economic Botany*, *51*(4), 377–384. https://doi.org/10.1007/BF02861048

Greene, S. L. (2022). Chapter18—Tryptamines. In P. Dargan & D. Wood (Éds.), *Novel Psychoactive Substances* (2nd Edition) (p. 495–532). Academic Press. https://doi.org/10.1016/B978-0-12-818788-3.00014-0

Hamill, J., Hallak, J., Dursun, S. M., & Baker, G. (2019). Ayahuasca: Psychological and physiologic effects, pharmacology and potential uses in addiction and mental illness. *Current Neuropharmacology*, *17*(2), 108–128. https://doi.org/10.2174/1570159X16666180125095902

Hornung, J.-P. (2003). The human raphe nuclei and the serotonergic system. *Journal of Chemical Neuroanatomy*, *26*(4), 331–343. https://doi.org/10.1016/j.jchemneu.2003.10.002

Kuypers, K. P. C., Riba, J., de la Fuente Revenga, M., Barker, S., Theunissen, E. L., & Ramaekers, J. G. (2016). Ayahuasca enhances creative divergent thinking while decreasing conventional convergent thinking. *Psychopharmacology*, *233*(18), 3395–3403. https://doi.org/10.1007/s00213-016-4377-8

Lisa, S. R., Islam, M. K., & Qais, N. (2020). Plants and plant constituents with analgesic and anti-inflammatory activities: A systematic review. *Dhaka University Journal of Pharmaceutical Sciences*, *19*(2), 207–224.

Mabit, J. (2007). Ayahuasca in the treatment of addictions. *Psychedelic Medicine: New Evidence for Hallucinogenic Substances as Treatments*, *2*, 87–105.

Mabit, J., Giove, R., & Vega, J. (1996). Takiwasi: The use of Amazonian shamanism to rehabilitate drug addicts. *Yearbook of Cross-cultural Medicine and Psychotherapy*, *6*, 257–286.

McKenna, D. J. (2004). Clinical investigations of the therapeutic potential of ayahuasca: Rationale and regulatory challenges. *Pharmacology & Therapeutics*, *102*(2), 111–129. https://doi.org/10.1016/j.pharmthera.2004.03.002

Merlin, M. D. (2003). Cover article: Archaeological evidence for the tradition of psychoactive plant use in the old world. *Economic Botany*, *57*(3), 295–323. https://doi.org/10.1663/0013-0001(2003)057[0295:AEFTTO]2.0.CO;2

Metzner, Ralph (2005). *Sacred Vine of Spirits: Ayahuasca*. Inner Traditions / Bear & Co.

Morales-García, J. A., De la Fuente Revenga, M., Alonso-Gil, S., Rodríguez-Franco, M. I., Feilding, A., Perez-Castillo, A., & Riba, J. (2017). The alkaloids of Banisteriopsis caapi, the plant source of the Amazonian hallucinogen Ayahuasca, stimulate adult neurogenesis in vitro. *Scientific Reports*, *7*(1), 5309.

Orch, H., Chaachouay, N., Douiri, E. M., Faiz, N., Zidane, L., & Douira, A. (2021). Use of medicinal plants in dermato-cosmetology: An ethnobotanical study among the population of Izarène. *Jordan Journal of Pharmaceutical Sciences*, *14*(3), 323–340.

Oroc, J. (2009). *Tryptamine Palace: 5-MeO-DMT and the Sonoran Desert Toad*. Simon and Schuster.

Osório, F. de L., Sanches, R. F., Macedo, L. R., dos Santos, R. G., Maia-de-Oliveira, J. P., Wichert-Ana, L., de Araujo, D. B., Riba, J., Crippa, J. A., & Hallak, J. E. (2015). Antidepressant effects of a single dose of ayahuasca in patients with recurrent depression: A preliminary report. *Brazilian Journal of Psychiatry*, *37*, 13–20. https://doi.org/10.1590/1516-4446-2014-1496

Palhano-Fontes, F., Alchieri, J. C., Oliveira, J. P. M., Soares, B. L., Hallak, J. E. C., Galvao-Coelho, N., & de Araujo, D. B. (2014). The Therapeutic Potentials of Ayahuasca in the Treatment of Depression. In B. C. Labate & C. Cavnar (Éds.), *The Therapeutic Use of Ayahuasca* (p. 23–39). Springer. https://doi.org/10.1007/978-3-642-40426-9_2

Pathak, A. K., Srivastava, A. K., Singour, P., & Gouda, P. (2016). Synthetic and natural monoamine oxidase inhibitors as potential lead compounds for effective therapeutics. *Central Nervous System Agents in Medicinal ChemistryChemistry - Central Nervous System Agents*, *16*(2), 81–97.

Pic-Taylor, A., da Motta, L. G., de Morais, J. A., Junior, W. M., Santos, A. de F. A., Campos, L. A., Mortari, M. R., von Zuben, M. V., & Caldas, E. D. (2015). Behavioural and neurotoxic effects of ayahuasca infusion (Banisteriopsis caapi and Psychotria viridis) in female Wistar rat. *Behavioural Processes*, *118*, 102–110.

Politi, M., Friso, F., Saucedo, G., & Torres, J. (2021). Traditional use of *Banisteriopsis caapi* alone and its application in a context of drug addiction therapy. *Journal of Psychoactive Drugs*, *53*(1), 76–84. https://doi.org/10.1080/02791072.2020.1820641

Ratsch, C. (2005). *The Encyclopedia of Psychoactive Plants: Ethnopharmacology and its Applications*. Simon and Schuster. https://books.google.com/books?hl=fr&lr=&id=8V0oDwAAQBAJ&oi=fnd&pg=PT7&dq=Some+of+these+plants+are+still+used+in+religion,+ceremonies,+practices,+and+psychoactive+and+medicinal+properties,+either+in+their+unprocessed+state+or+as+processed+extracts+or+supplements&ots=Upd0WukzWZ&sig=SSR0zXItH8u6qYfqtjlHzLBVfOQ

Rist, S., & Dahdouh-Guebas, F. (2006). Ethnosciences—A step towards the integration of scientific and indigenous forms of knowledge in the management of natural resources for the future. *Environment, Development and Sustainability*, *8*(4), 467–493. https://doi.org/10.1007/s10668-006-9050-7

Rodd, R. (2002). Snuff synergy: Preparation, use and pharmacology of Yopo and Banisteriopsis Caapi among the. *Journal of Psychoactive Drugs*, *34*, 3.

Rodd, R. (2008). Reassessing the cultural and psychopharmacological significance of *Banisteriopsis caapi*: Preparation, classification and use among the Piaroa of Southern Venezuela. *Journal of Psychoactive Drugs*, *40*(3), 301–307. https://doi.org/10.1080/02791072.2008.10400645

Rodd, R., & Sumabila, A. (2011). Yopo, ethnicity and social change: A comparative analysis of Piaroa and Cuiva Yopo use. *Journal of Psychoactive Drugs*, *43*(1), 36–45. https://doi.org/10.1080/02791072.2011.566499

Santos, R. G., Landeira-Fernandez, J., Strassman, R. J., Motta, V., & Cruz, A. P. M. (2007). Effects of ayahuasca on psychometric measures of anxiety, panic-like and hopelessness in Santo Daime members. *Journal of Ethnopharmacology*, *112*(3), 507–513. https://doi.org/10.1016/j.jep.2007.04.012

Santos, B. W. L., Moreira, D. C., Borges, T. K. dos S., & Caldas, E. D. (2022). Components of *Banisteriopsis caapi*, a plant used in the preparation of the psychoactive ayahuasca, induce anti-inflammatory effects in microglial cells. *Molecules*, *27*(8), Article 8. https://doi.org/10.3390/molecules27082500

Santos, B. W. L., Oliveira, R. C. de, Sonsin-Oliveira, J., Fagg, C. W., Barbosa, J. B. F., & Caldas, E. D. (2020). Biodiversity of β-carboline profile of *Banisteriopsis caapi* and ayahuasca, a plant and a brew with neuropharmacological potential. *Plants*, *9*(7), 870.

Schmid, J. T., Jungaberle, H., & Verres, R. (2010). Subjective theories about (self-)treatment with ayahuasca. *Anthropology of Consciousness*, *21*(2), 188–204. https://doi.org/10.1111/j.1556-3537.2010.01028.x

Schultes, R. E., Holmstedt, B., & Lindgren, J.-E. (1969). De plantis toxicariis e mundo novo tropicale commentationes III: Phytochemical examination of Spruce's original collection of Banisteriopsis caapi. *Botanical Museum Leaflets, Harvard University*, *22*(4), 121–132.

Serrano-Dueñas, M., Cardozo-Pelaez, F., & Sánchez-Ramos, J. R. (2001). Effects of Banisteriopsis caapi extract on Parkinson's disease. *The Scientific Review of Alternative Medicine*, *5*(3), 127–132.

Shanon, B. (2010). The epistemics of ayahuasca visions. *Phenomenology and the Cognitive Sciences*, *9*(2), 263–280. https://doi.org/10.1007/s11097-010-9161-3

Smith, E. W. (1979). BANISTERIOPSIS Caapi. *Ethnopharmacologic Search for Psychoactive Drugs*, *1645*, 47.

Souza, R. M. de. (2023). *Taxonomia e morfologia das espécies filogeneticamente relacionadas a Banisteriopsis caapi (Spruce ex Griseb.) CV Morton (Malpighiaceae), componente do Chá Ayahuasca.* http://www.realp.unb.br/jspui/handle/10482/46266

Stanković, D., Mehmeti, E., Svorc, L., & Kalcher, K. (2015). New electrochemical method for the determination of β-carboline alkaloids, harmalol and harmine, in human urine samples and in Banisteriopsis caapi. *Microchemical Journal*, *118*, 95–100. https://doi.org/10.1016/j.microc.2014.08.007

Su, T.-P., Hayashi, T., & Vaupel, D. B. (2009). When the endogenous hallucinogenic trace amine N,N-dimethyltryptamine meets the sigma-1 receptor. *Science Signaling*, *2*(61), pe12–pe12. https://doi.org/10.1126/scisignal.261pe12

Szabo, A., Kovacs, A., Frecska, E., & Rajnavolgyi, E. (2014). Psychedelic N,N-dimethyltryptamine and 5-methoxy-N,N-dimethyltryptamine modulate innate and adaptive inflammatory responses through the sigma-1 receptor of human monocyte-derived dendritic cells. *PLOS ONE*, *9*(8), e106533. https://doi.org/10.1371/journal.pone.0106533

The Plant List. (2013). *The Plant List. Version 1.1. Published on the Internet*; http://www.theplantlist.org/ (accessed 1st January).

Thevenin, J. M. R., & Sambuichi, R. H. R. (2020). Phytogeography and floristics of the arbor component in União do Vegetal territories intended for the cultivation of Banisteriopsis caapi and Psychotria viridis in Rondônia. *RAEGA-O Espaço Geográfico em Análise*, *49*, 42–63.

Trevisani, A. C., Trevisani, J. P. C., Haverroth, M., Melo, A. D. C., Santos, I. C. D., Barbosa, L. N., & Gonçalves, D. D. (2021). A systematic literature review of the potential therapeutic effects of ayahuasca tea prepared from the use of *Banisteriopsis Caapi* and Psychotria Viridis. *Journal of Agricultural Studies*, *9*(1), 112. https://doi.org/10.5296/jas.v9i1.17910

Trichter, S. M. (2006). *Changes in spirituality among ayahuasca ceremony novice participants* [PhD Thesis, Argosy University/San Francisco Bay Area]. https://search.proquest.com/openview/bee8bd7f87dc06c14b41e0696165f621/1?pq-origsite=gscholar&cbl=18750&diss=y

Tupper, K. W. (2008). The globalization of ayahuasca: Harm reduction or benefit maximization? *International Journal of Drug Policy*, *19*(4), 297–303.

Weiskopf, J. (2005). *Yajé: The New Purgatory: Encounters with Ayahuasca*. Villegas Asociados.

Williams, M., Kingston Miller, A., Loizaga-Velder, A., Files, N., & Lafrance, A. (2023). 'Getting to the Root': Ayahuasca ceremony leaders' perspectives on eating disorders. *Journal of Psychoactive Drugs*, *55*(4), 425–433. https://doi.org/10.1080/02791072.2022.2113484

Winkelman, M. J. (2014). Therapeutic Applications of Ayahuasca and Other Sacred Medicines. In B. C. Labate & C. Cavnar (Éds.), *The Therapeutic Use of Ayahuasca* (p. 1–21). Springer. https://doi.org/10.1007/978-3-642-40426-9_1

32 Muiratinga (*Maquira sclerophylla* (Ducke) C.C. Berg Moraceae)

Noureddine Chaachouay and Lahcen Zidane

32.1 INTRODUCTION

Maquira sclerophylla (Ducke) C.C.Berg is a member of the Moraceae family often found in the tropical Amazonian Forest and is well-known for its excellent wood and latex-producing characteristics (Berg, 1969). *M. sclerophylla* is a tall, perennial tree that may grow to a remarkable height of 50 meters. The white wood of this plant is used in building. At the same time, Indigenous people employ pulverized seeds and barks to create a powdered substance utilized in religious rituals (Nebel et al., 2001; Chaachouay & Zidane, 2021). The tree has a strong and sturdy trunk that is protected by a thick layer of black bark. Its dense upper part comprises shiny and tough leaves, offering shade and enhancing its impressive look. *M. sclerophylla* is notable for its capacity to produce latex, a characteristic that has led to its economic importance in the area (Piva et al., 2018). The latex derived from this tree has historically been used by indigenous tribes for several applications, notably as a rubber resource (Shrestha et al., 1992). *M. sclerophylla* is not only economically significant but also serves a crucial ecological function in maintaining the equilibrium of the Amazon rainforest by serving as a home and food source for a diverse array of fauna. *M. sclerophylla*, while ecologically and economically important, is endangered due to deforestation, habitat loss, and unsustainable exploitation of its lumber and latex (Dahms, 1993; Ferraz & Kato, 1990; Piva et al., 2018). Conservation initiatives are being implemented to safeguard and responsibly oversee this important tree species to guarantee its long-term existence in the Amazon rainforest.

M. sclerophylla has yet to be well studied in terms of its phytochemistry, in contrast to several other plant species in the Amazon rainforest. However, preliminary studies and examinations have identified several types of phytochemical compounds present in different parts of the Muiratinga tree (Dahms, 1993). Preliminary pharmacological investigations have shown intraperitoneal injections of *M. sclerophylla* extracts in mice elicited indications of central nervous system excitation, followed by protracted depression (De Carvalho et al., 1997). The extracts induced a temporary decrease in blood pressure followed by a prolonged increase in rats under anesthesia. Atropine effectively prevented the occurrence of low blood pressure, whereas yohimbine, pithing, and reserpine therapy reduced high blood pressure but did not wholly prevent it (De Carvalho & Lapa, 1990). This chapter explicitly examines the botanical characteristics, ethnobotanical usage, habitat and ecological preferences, phytochemical composition, and hallucinogenic properties of *M. sclerophylla*.

32.2 SCIENTIFIC CLASSIFICATION

Maquira sclerophylla (Ducke) C.C.Berg, commonly known as 'Muiratinga', is a tree species native to the Amazon rainforest in South America. It belongs to the family Moraceae and *Maquira* genus.

Domain: Biota
Reign: Plantae
Sub-kingdom: Viridaeplantae
Infra-Kingdom: Streptophyta
Class: Equisetopsida
Clade: Tracheophyta
Clade: Spermatophyta
Subclass: Magnoliidae
Super-Order: Rosanae
Order: Rosales
Family: Moraceae
Tribe: Olmedieae
Genre: *Maquira*
Species: *Maquira sclerophylla* (Ducke) C.C.Berg

32.3 VERNACULAR NAMES

English: Muiratinga; **Portuguese**: Muiratinga, Pau-tanino, (Brazil) Pau-de-índio, Rapé-de-índio

32.4 BOTANICAL DESCRIPTION

M. sclerophylla is a tall tree, reaching a height of up to 25 meters. It may have separate male and female flowers (dioecious) or male and female flowers on the same tree (monoecious) (Figure 32.1). The branchlets are covered with fine hairs. The stipules are 0.5–1(–1.5) cm long and covered in fine hairs. The petiole is 0.8–2.5 cm long (Dahms, 1993). The blade is leathery, elliptic to narrowly elliptic, measuring 13–30(–38) × 5–15 cm. The apex is pointed, and the base is either blunt or sharp. The upper surface of the blade

DOI: 10.1201/9781003460336-32

is primarily smooth, while the lower surface is often slightly rough. The veins are visible on the lower surface, with 13–30 pairs of secondary veins and a tertiary venation pattern that resembles a ladder (Berg, 1969; MIRANDA & Ferraz, 1999). The staminate inflorescences may occur in groups of up to four, with a diameter ranging from 0.3 to 1.2 cm. The peduncle, the stalk supporting the inflorescence, is between 0.5 and 1.3 cm long. The number of stamens is typically three or less—the number 4. The pistillate inflorescences are solitary or accompanied by 1 or 2 staminate ones. They have a diameter of around 0.5 cm and are either subsessile or have a peduncle up to 0.8 cm long. The inflorescences include 2–4 basally connate flowers, sometimes only one. The stigmas are 2.5–3.5 mm long and have an acute shape. The infructescence has a diameter of up to 3(−4) cm (De Carvalho & Lapa, 1990; Piva et al., 2018).

32.5 ORIGIN AND GEOGRAPHIC DISTRIBUTION

M. sclerophylla is indigenous to the Amazon Basin, including regions of Brazil, Colombia, Peru, and adjacent countries (Figure 32.2). The species naturally inhabits the Amazon rainforest, which is recognized as one of the most biologically varied areas on the planet (Berg, 1969; MIRANDA & Ferraz, 1999). *M. sclerophylla* is located in

FIGURE 32.1 Giant tree of *M. sclerophylla* (height of up to 25 meters).

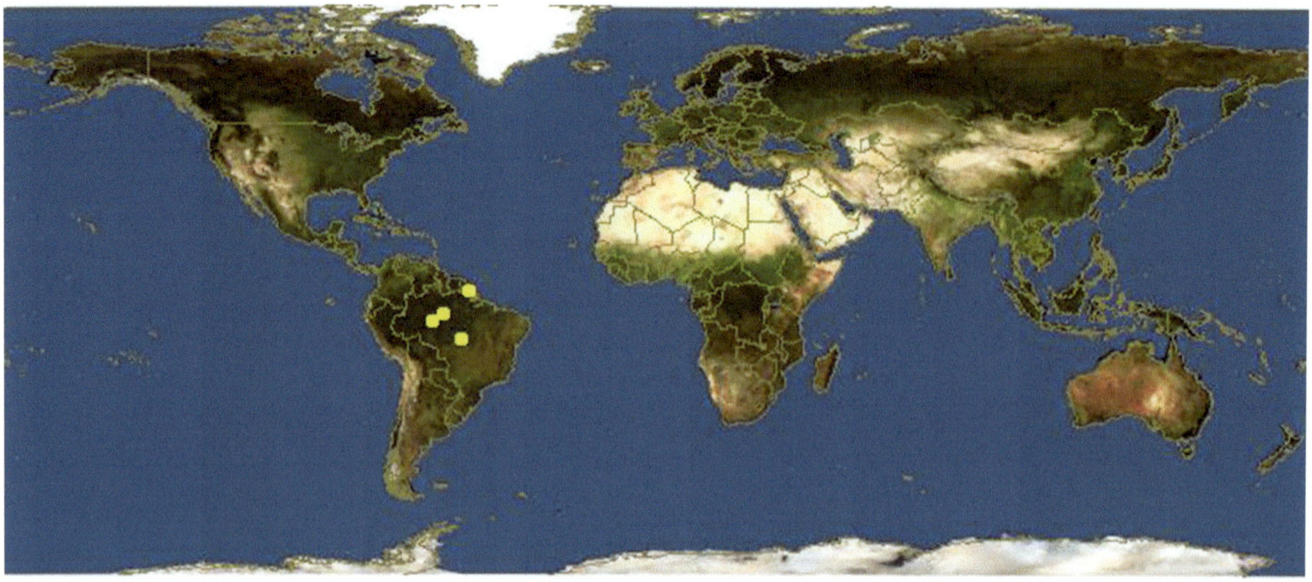

FIGURE 32.2 Geographical distribution of *M. sclerophylla*.

the Brazilian Amazon, namely in the states of Amazonas, Acre, Amapá, Pará, Roraima, and other states within the Amazon region (Carvalho, 2002; Dahms, 1993). In Colombia, it is also present in some areas of the Colombian Amazon, particularly in the southern regions that share borders with Brazil and Peru. Muiratinga is found in several locations of the Peruvian Amazon, such as Loreto, Ucayali, and Madre de Dios. Muiratinga may also be found in Venezuela, Ecuador, and Bolivia, which are adjacent nations located within the Amazon Basin, but it is less prevalent in these areas (FERRAZ & KATO, 1990; Piva et al., 2018; RAMOS, 1999).

32.6 HABITAT AND ECOLOGY

M. sclerophylla is a tree species indigenous to the vast Amazon rainforest in South America. The origin of this can be traced to the central area of the Amazon Basin, which includes parts of Brazil, Colombia, Peru, and other nations (Berg, 1969; Nebel et al., 2001). This species flourishes in the tropical rainforest habitat, distinguished by its elevated humidity, ample precipitation, and luxuriant vegetation. Muiratinga enhances the remarkable biodiversity of the Amazon by providing a suitable environment and nourishment for a diverse range of animal species (MIRANDA & Ferraz, 1999). Muiratinga is widely distributed over the Brazilian Amazon, encompassing Amazonas, Acre, Amapá, Pará, and Roraima. It is also found in some areas of the Colombian Amazon, namely in the southern border zones next to Brazil and Peru. *M. sclerophylla* is recorded in several parts of the Peruvian Amazon, particularly in places such as Loreto, Ucayali, and Madre de Dios in Peru (Carvalho, 2002). Although less often, it may manifest in some areas of adjacent nations within the Amazon Basin, such as Venezuela, Ecuador, and Bolivia. This tree species fulfills a crucial ecological function in the Amazon rainforest, adding to the complex life network inside this extraordinary environment. *M. sclerophylla* may be grown using either seed or cuttings. Seeds should be planted in a light soil mixture with good drainage and kept consistently wet—plant cuttings from semi-hardwood or hardwood in a well-draining soil mix. Maintain the soil's moisture and provide enough exposure to indirect light (Piva et al., 2018).

32.7 ETHNOBOTANY

32.7.1 Traditional Medicinal Uses

M. sclerophylla plays a prominent part in the traditional medical practices of indigenous and local groups in the Amazon rainforest (De Carvalho & Lapa, 1990). The several components of the Muiratinga tree, including the bark, leaves, and roots, are used to prepare cures for a diverse array of diseases (Schultes, 1972). Its traditional use includes the management of gastrointestinal issues, pyrexia, dermatological ailments, heart tonics, projectile poisons, emetics, and diuretics (Shrestha et al., 1992; Belhaj et al., 2021; Benkhnigue et al., 2022). The understanding of Muiratinga's therapeutic characteristics has been transmitted throughout many generations, emphasizing the rich ethnobotanical legacy of the indigenous tribes in the Amazon region (De Carvalho et al., 1997).

32.7.2 Practical Uses

In addition to its medical properties, Muiratinga is a multifunctional resource for these communities. The tree's resilient timber is much esteemed for its use in construction, erecting dwellings, fabricating boats, and fashioning diverse implements (Piva et al., 2018). In addition, the bark of the Muiratinga tree contains saponins, which are used in fishing (Sarquis et al., 2019). Historically, Indigenous people have pulverized the bark and added it to water to temporarily incapacitate fish, facilitating their capture (Bean, 1974; Chaachouay et al., 2019a, 2019b). This exemplifies the versatile usefulness of Muiratinga in the everyday existence of these Indigenous communities in the Amazon region (De Carvalho & Lapa, 1990).

32.7.3 Cultural and Spiritual Significance

Muiratinga has a unique and significant position in some indigenous communities' spiritual and cultural customs (Diniz & da Silva Cabral, n.d.; Yajahuanca et al., 2015). The tree is regarded as holy and often used in rituals and celebrations (Jayne, 2015). The tree may be strategically positioned close to settlements or used as a gift to spiritual entities, representing the profound connection between nature and spirituality throughout indigenous belief systems (Trigger & Mulcock, 2005). The significance of Muiratinga in these cultural rituals highlights its value beyond its practical functions, as it contributes to the conservation of indigenous traditions and spiritual legacy (Jayne, 2015).

32.8 HALLUCINOGENIC EFFECTS

M. sclerophylia is not often acknowledged for its hallucinogenic properties. There are few studies, evidence, or historical documents indicating that *M. sclerophylia* exhibits psychedelic qualities. As stated in prior comments, this tree species is mainly recognized for its conventional medicinal, practical, and cultural use among indigenous and local groups in the Amazon rainforest. *M. sclerophylia* snuff is claimed to induce euphoria and hallucinations by stimulating the central nervous system (Schultes, 1972; Yazbek et al., 2018; Chaachouay et al., 2020a, 2020b). Nevertheless, no scientific or pharmacological investigations have been undertaken on people about this drug. It is noteworthy that hallucinogenic effects are typically associated with particular plant species, such as specific types of psychedelic mushrooms (e.g., *Psilocybe* species) or plants like *Lophophora williamsii* and *Banisteriopsis caapi*, which contain psychoactive substances like psilocybin, mescaline, and DMT (Brito-da-Costa et al., 2020). When ingested in specific manners and quantities, these chemicals can elicit modified states of consciousness and hallucinations (Schultes, 1979). The technique for producing the traditional snuff derived from *M. sclerophylia* is said to be known just by old individuals belonging to tribes residing in the tree's native areas. Although *M. sclerophylia* is known for its many traditional applications, there is currently no generally acknowledged or well-documented scientific proof indicating its hallucinogenic effects or its use for such intentions in indigenous societies (Piva et al., 2018). Practicing prudence and doing a comprehensive study is crucial while investigating the possible impacts of any plant, particularly hallucinogenic drugs, due to the associated hazards and legal consequences.

32.9 PHYTOCHEMISTRY

M. sclerophylla has yet to be well investigated regarding its phytochemistry compared to several other plant species in the Amazon jungle. Nevertheless, initial research and investigations have revealed several categories of phytochemical substances found in different components of the Muiratinga tree (De Carvalho et al., 1997; Piva et al., 2018; Schultes, 2016). Available research indicates that the plant has a range of chemical components, including alkaloids, flavonoids, and tannins. The tree's bark has been discovered to contain tannins, saponins, and alkaloids. The tree's latex comprises isoprenoids, terpenoids, and steroids (Moretti et al., 2006; Schultes, 1984; Shrestha et al., 1992). The tree's leaves include flavonoids, tannins, and saponins. These chemical components in *M. sclerophylla* imply that the plant could possess promising therapeutic characteristics (De Carvalho & Lapa, 1990).

32.10 PHARMACOLOGY

The ethnopharmacological data on Amazonian *M. sclerophylla* shows that indigenous people use the plant in religious rituals and celebrations. Several additional snuffs are purportedly used for the same purpose (De Carvalho & Lapa, 1990; Chaachouay et al., 2021a, 2021b, 2021c). The indigenous people have used many components of this tree in traditional medicine due to its possible therapeutic capabilities. Nevertheless, the pharmacological properties of *M. sclerophylla* have yet to be well investigated compared to other plant species (Dahms, 1993).

A study demonstrates that the intraperitoneal injection of either the crude extract or its hydrosoluble fraction in mice and rats resulted in effects that resemble the stimulation of the central nervous system caused by amphetamines (De Carvalho & Lapa, 1990). These effects include initial symptoms such as rapid breathing, heightened excitability, and tremors, followed by impaired motor coordination, reduced exploratory behavior, unsteady movements, and muscle relaxation. The observed effects exhibited a gradual and proportional increase with dosage but diminished rapidly, often within 30 minutes. Regrettably, these indications did not manifest following the ingestion of medication at a dosage of 1 gram per kg. When injected, the crude extract consistently modified blood pressure, heart rate, and breathing (Moretti et al., 2006; Schultes, 1984).

Nonetheless, definitive outcomes were achieved only after dividing the unrefined extract and isolating the active compounds in a water-soluble fraction (Hosain et al., 2019). Hypertension manifestation was more prominent than the decrease in blood pressure caused by purified hydro soluble extract (Shrestha et al., 1992; Chaachouay et al., 2022a, 2022b, 2022c). The origin of this phenomenon is partly connected to the activation of alpha-sympathomimetic receptors since it was diminished but not completely inhibited following alpha-blockade. Chronic administration of reserpine or blocking of ganglionic activity with hexamethonium affected the time it took for hypertension to peak, suggesting that transmitter release may have a function (Piva et al., 2018; Shrestha et al., 1992). Still, it is not the primary mechanism by which purified hydrosoluble extract generates hypertension. Pithing effectively suppressed the hypertensive reaction.

Dogs exhibited a sensitivity to purified hydrosoluble extract that was 5–10 times greater than that of rats. However, the observed effects were qualitatively similar, with dogs experiencing rapid mortality at dosages above 5 mg/kg (Elisabetsky & Shanley, 1994). In guinea pigs, the pressure effects of purified hydrosoluble extract were nearly the same as those found in dogs and rats. Death followed the development of arrhythmias, ventricular extrasystoles, and fibrillation once hypertension was established. The water-soluble portion of *M. sclerophylla* has been purified and partly characterized (De Carvalho et al., 1997). The primary active component in the heart has a molecular structure with a cyclopentanoperhydrophenanthrene nucleus (Repke et al., 1991). However, when analyzed using high-pressure liquid chromatography, this molecule did not exhibit the same retention time as commonly found cardenolides like digoxin, digitoxin, and ouabain (Masso, 2020).

32.11 CLINICAL USES

M. sclerophylla snuff is reported to stimulate the central nervous system, resulting in euphoria and hallucinations. Nevertheless, no laboratory or pharmacological investigations have been undertaken on people about this drug (Piva et al., 2018). A study conducted on mice and rats yielded amphetamine-like effects. Subsequently, symptoms of 'motor incoordination, decreased exploratory activity, ataxia, and muscle relaxation' persisted for around 30 minutes. The chemical does not seem to have the ability to have an effect when taken orally (Moretti et al., 2006). Subsequent experiments were conducted on dogs, guinea pigs, and rats, whereby the animals were given a refined plant extract via injection (De Carvalho et al., 1997). This resulted in a two-phase alteration in carotid blood pressure. The duration of hypertension extended for 30 minutes. Guinea pigs and canines had a greater susceptibility than rats, resulting in mortality due to heart attack. The effects resemble those of medicines similar to digitalis (De Carvalho & Lapa, 1990).

32.12 CONCLUSION

M. sclerophylla is an indigenous tree species found in the Amazon rainforest. It has pharmacological qualities that have been traditionally used in indigenous medicine. Although minimal scientific study validates its many medical assertions, it is believed that *M. sclerophylla* contains anti-inflammatory, analgesic, antioxidant, wound healing, and antibacterial qualities, primarily based on anecdotal data and a few early investigations. Nevertheless, a more extensive study is required to substantiate these assertions and understand the processes behind its bioactive components. Creating safety protocols and precise dosage parameters for medical use is essential. Before taking *M. sclerophylla* as an alternative or supplementary therapy for particular health conditions, exercising care and consulting a medical practitioner for guidance is crucial. The potential advantages of herbal medicine provide captivating opportunities for future research and discovery in the sector.

LIST OF ABBREVIATION

M. sclerophylla *Maquira sclerophylla* (Ducke) C.C.Berg

REFERENCES

Bean, L. J. (1974). *Mukat's people: The Cahuilla Indians of southern California*. Univ of California Press. https://books.google.com/books?hl=fr&lr=&id=kWBFfooGTdcC&oi=fnd&pg=PA23&dq=indigenous+people+have+pulverized+the+bark+and+added+it+to+water+to+temporarily+incapacitate+fish,+facilitating+their+capture&ots=yOzTUgA1bl&sig=m50ZNLAsYoGHeehLNNYHBGnyYeY

Belhaj, S., Chaachouay, N., & Zidane, L. (2021). Ethnobotanical and toxicology study of medicinal plants used for the treatment of diabetes in the High Atlas Central of Morocco. *Journal of Pharmacy & Pharmacognosy Research*, *9*(5), 619–662. https://doi.org/10.56499/jppres21.1014_9.5.619

Benkhnigue, O., Chaachouay, N., Khamar, H., El Azzouzi, F., Douira, A, & Zidane, L. (2022). Ethnobotanical and ethnopharmacological study of medicinal plants used in the treatment of anemia in the region of Haouz-Rehamna (Morocco). *Journal of Pharmacy & Pharmacognosy Research*, *10*(2), 279–302. https://doi.org/10.56499/jppres21.1196_10.2.279

Berg, C. C. (1969). New Taxa and Combinations in the Neotropical Olmedieae. *Acta Botanica Neerlandica*, *18*(3), 462–465. https://doi.org/10.1111/j.1438-8677.1969.tb00611.x

Brito-da-Costa, A. M., Dias-da-Silva, D., Gomes, N. G. M., Dinis-Oliveira, R. J., & Madureira-Carvalho, Á. (2020). Toxicokinetics and Toxicodynamics of Ayahuasca Alkaloids N,N-Dimethyltryptamine (DMT), Harmine, Harmaline and Tetrahydroharmine: Clinical and Forensic Impact. *Pharmaceuticals*, *13*(11), Article 11. https://doi.org/10.3390/ph13110334

Carvalho, J. O. P. (2002). Changes in the spatial distribution of tree species in a terra firme rain forest in Brazilian Amazonia after logging. *Revista de Ciências Agrárias Amazonian Journal of Agricultural and Environmental Sciences*, *37*, 53–70.

Chaachouay, N., Benkhnigue, O., Fadli, M., Ibaoui, H., & Zidane, L. (2019a). Ethnobotanical and ethnopharmacological studies of medicinal and aromatic plants used in the treatment of metabolic diseases in the Moroccan Rif. *Heliyon*, *5*(10), e02191. https://doi.org/10.1080/10477845.2021.1942433

Chaachouay, N., Benkhnigue, O., Fadli, M., Ibaoui, H., El Ayadi, R., & Zidane, L. (2019b). Ethnobotanical and ethnopharmacological study of medicinal and aromatic plants used in the treatment of respiratory system disorders in the Moroccan Rif. *Ethnobotany Research and Applications*, *18*(22), 1–16.

Chaachouay, N., Douira, A., Hassikou, R., Brhadda, N., Dahmani, J., Belahbib, N., Ziri, R., & Zidane, L. (2020a).

Ethnobotanical and ethnopharmacological study of medicinal and aromatic plants used in the treatment of respiratory system disorders in the Moroccan Rif. PhDThesis, Département de Biologie - Université Ibn Tofail - Kénitra, Maroc, https://tel.archives-ouvertes.fr/tel-03376377

Chaachouay, N., Benkhnigue, O., & Zidane, L. (2020b). Ethnobotanical study aimed at investigating the use of medicinal plants to treat nervous system diseases in the Rif of Morocco. *Journal of chiropractic medicine*, *19*(1), 70–81. https://doi.org/10.1016/j.jcm.2020.02.004

Chaachouay, N., & Zidane, L. (2021). The Symbolic Efficacy of Plants in Rituals and Socio-Religious Ceremonies in Morocco, Northwest of Africa. *Journal of Religious & Theological Information*, *21*(1–2), 34–53. https://doi.org/10.1080/10477845.2021.1942433

Chaachouay, N., Benkhnigue, O., Douira, A., & Zidane, L. (2021a). Poisonous medicinal plants used in the popular pharmacopoeia of the Rif, northern Morocco. *Toxicon*, 189, 24–32. https://doi.org/10.1016/j.toxicon.2020.10.028

Chaachouay, N., Douira, A., & Zidane, L. (2021b). COVID-19, prevention and treatment with herbal medicine in the herbal markets of Salé Prefecture, North-Western Morocco. *European Journal of Integrative Medicine*, 42, 101285. https://doi.org/10.1016/j.eujim.2021.101285

Chaachouay, N., Douira, A., & Zidane, L. (2021c). Herbal Medicine Used in the Treatment of Human Diseases in the Rif, Northern Morocco. *Arabian Journal for Science and Engineering*, 47, 131–153. https://doi.org/10.1007/s13369-021-05501-1

Chaachouay, N., Benkhnigue, O., & Zidane, L. (2022a). Ethnobotanical and Ethnomedicinal study of medicinal and aromatic plants used against dermatological diseases by the people of Rif, Morocco. *Journal of Herbal Medicine*, 32, 100542. https://doi.org/10.1016/j.hermed.2022.100542

Chaachouay, N., Azeroual, A., Bencharki, B., Douira, A., & Zidane, L. (2022b). Ethnoveterinary medicines plants for animal therapy in the Rif, North of Morocco. *South African Journal of Botany*, *147*, 176–191. https://doi.org/10.1016/j.sajb.2021.12.037

Chaachouay, N., Azeroual, A., Douira, A., & Zidane, L. (2022c). Ethnoveterinary practices of medicinal plants among the Zemmour and Zayane tribes, Middle Atlas, Morocco. *South African Journal of Botany*, *151*, 826–840. https://doi.org/10.1016/j.sajb.2022.11.009

Dahms, K. G. (1993). [New knowledge on imported wood. Pt. 4. Latin America. Muiratinga (Maquira sclerophylla CC Berg., M. Coriacea CC Berg., Syn. Oldmediaperebea sclerophylla Ducke, family of Moraceae)].[German]. *Holz Zentralblatt*.

De Carvalho, J. E., & Lapa, A. J. (1990). Pharmacology of an Indian-snuff obtained from Amazonian Maquira sclerophylla. *Journal of ethnopharmacology*, *30*(1), 43–54.

De Carvalho, J. E., Souccar, C., Tersariol, I. L., Torres, L. B., & Lapa, A. J. (1997). Pharmacological properties and identification of cardiotonic principles from the Indian Snuff, Maquira sclerophylla, Ducke. *Phytotherapy Research*, *11*(2), 136–141. https://doi.org/10.1002/(SICI)1099-1573(199703)11:2<136::AID-PTR53>3.0.CO;2-S

Diniz, C. S. G., & da Silva Cabral, C. (n.d.). *We need to 'ikarar the kutipados': Intercultural understanding and health care in the Peruvian Amazon.* Consulté 5 janvier 2024, à l'adresse https://www.scielosp.org/pdf/csc/v20n9/en_1413-8123-csc-20-09-2837.pdf

Elisabetsky, E., & Shanley, P. (1994). Ethnopharmacology in the Brazilian Amazon. *Pharmacology & Therapeutics*, *64*(2), 201–214.

Ferraz, I., & Kato, A. (1990). Germinação de sementes de Maquira sclerophylla (Ducke) CC Berg.-Moraceae. *Congresso Florestal Brasileiro*, *6*, 644–648.

Hosain, N. A., Ghosh, R., Bryant, D. L., Arivett, B. A., Farone, A. L., & Kline, P. C. (2019). Isolation, structure elucidation, and immunostimulatory activity of polysaccharide fractions from Boswellia carterii frankincense resin. *International Journal of Biological Macromolecules*, *133*, 76–85.

Jayne, L. (2015). *« The River That Suffers From the Tide: » A Discovery Memoir of Self Among Others in the Brazilian Amazon* [PhD Thesis, Union Institute and University]. https://search.proquest.com/openview/817fc3ae6a82c5b7f7d8875c4c742515/1?pq-origsite=gscholar&cbl=18750

Masso, Z. F. (2020). *Assessing cardiotonic steroids involvement in hypertensive rat models with Helicobacter pylori infections* [PhD Thesis, University of Pretoria]. https://repository.up.ac.za/handle/2263/75537

Miranda, P. R., & Ferraz, I. D. K. (1999). Efeito da temperatura na germinação de sementes e morfologia da plântula de Maquira sclerophylla (Ducke) CC Berg. *Brazilian Journal of Botany*, 22, 303–307.

Moretti, C., Gaillard, Y., Grenand, P., Bévalot, F., & Prévosto, J.-M. (2006). Identification of 5-hydroxy-tryptamine (bufotenine) in takini (Brosimum acutifolium Huber subsp. Acutifolium CC Berg, Moraceae), a shamanic potion used in the Guiana Plateau. *Journal of Ethnopharmacology*, *106*(2), 198–202.

Nebel, G., Dragsted, J., Simonsen, T. R., & Vanclay, J. K. (2001). The Amazon flood plain forest tree *Maquira coriacea* (Karsten) CC Berg: Aspects of ecology and management. *Forest Ecology and Management*, *150*(1–2), 103–113.

Piva, L. R. O., Jardine, K. J., Cobello, L. O., Gimenez, B. O., Durgante, F. M., Higuchi, N., & Chambers, J. Q. (2018). Demonstration of a strict molecular oxygen requirement of yellow latex oxidation in the central Amazon canopy tree muiratinga (Maquira sclerophylla (Ducke) CC Berg). *Volume 10, Número 5*, Pags. 1316–1326. https://repositorio.inpa.gov.br/handle/1/15684

Ramos, M. (1999). Miranda, PRM; Ferraz, IDK Efeito da temperatura na germinação de sementes e morfologia da plântula de Maquira sclerophylla (Ducke) CC Berg. *Revista Brasileira de Botânica, São Paulo*, *22*(2), 303–307.

Repke, K. R. H., Weiland, Jür., & Menke, K.-H. (1991). Chemical models for the chemical nature of endogenous digitalis. *Journal of Enzyme Inhibition*, *5*(1), 25–32. https://doi.org/10.3109/14756369109069057

Sarquis, R. Do S. F. R., Sarquis, Í. R., Sarquis, I. R., Fernandes, C. P., Da Silva, G. A., de Silva, R. B. L., Jardim, M. A. G., Sánchez-Ort, B. L., & Carvalho, J. C. T. (2019). Research Article The Use of Medicinal Plants in the Riverside Community of the Mazagão River in the Brazilian Amazon, Amapá, Brazil: Ethnobotanical and Ethnopharmacological Studies. *History*, *4*, 5.

Schultes, R. E. (1972). From witch doctor to modern medicine: Searching the American tropics for potentially new medicinal plants. *Arnoldia*, *32*(5), 198–219.

Schultes, R. E. (1979). Evolution of the identification of the myristicaceous hallucinogens of South America. *Journal of Ethnopharmacology*, *1*(3), 211–239.

Schultes, R. E. (1984). Psychoactive plants in need of chemical and pharmacological study. *Proceedings / Indian Academy of Sciences*, *93*(3), 281–304. https://doi.org/10.1007/BF03053083

Schultes, R. E. (2016). Phy to chemical gaps in our knowledge of hallucinogens. *Progress in Phytochemistry*, *7*(7), 301.

Shrestha, T., Kopp, B., & Bisset, N. G. (1992). The Moraceae-based dart poisons of South America. Cardiac glycosides of Maquira and Naucleopsis species. *Journal of Ethnopharmacology*, *37*(2), 129–143.

Trigger, D., & Mulcock, J. (2005). Forests as spiritually significant places: Nature, culture and 'belonging' in Australia. *The Australian Journal of Anthropology*, *16*(3), 306–320. https://doi.org/10.1111/j.1835-9310.2005.tb00313.x

Yajahuanca, R. A., Diniz, C. S. G., & Cabral, C. da S. (2015). We need to 'ikarar the kutipados': Intercultural understanding and health care in the Peruvian Amazon. *Ciência & Saúde Coletiva*, *20*, 2837–2846.

Yazbek, P., Sauini, T., Cassas, F., Negri, G., & Rodrigues, E. (2018). Plants Used for Central Nervous System Disorders by Brazilian Indians Priscila Yazbek, Thamara Sauini, Fernando Cassas, Gi. In *Ethnobotany* (pp. 62–74). CRC Press. https://api.taylorfrancis.com/content/chapters/edit/download?identifierName=doi&identifierValue=10.1201/9780429453137-9&type=chapterpdf

33 Epená (*Virola theiodora* (Spruce ex Benth.) Warb. Myristicaceae)

Muhammad Younas, Ayesha Siddiqa, Rahmatullah Qureshi, Sabith Rehman, and Noureddine Chaachouay

33.1 INTRODUCTION

Virola is a genus of nutmeg family (Myristicaceae). The genus contains 60 species distributed in the lowlands and cloud rainforests of Central and South America spanning Mexico and Brazil's southern regions (Franklin, 2007; González-Rodríguez et al., 2021). It holds significant ethnobotanical value for the customary societies found in these regions. The indigenous people use the plant for different purposes. The cambial area of a thin tree produces a reddish exudate that is either snuffed or swallowed (Miller, 1999). *Virola* species are used to cure a number of illnesses, such as rheumatoid arthritis, chronic bronchitis, gastroenteritis, halitosis, ulcers, and intestinal worms (González-Rodríguez et al., 2021). The indigenous societies in South America frequently use *Virola* for their hallucinogenic qualities, which are well-known in addition to its non-medical benefits, in order to deepen their spiritual experiences (McKenna et al., 1984).

Virola theiodora (Spruce ex Benth.) Warb., an evergreen tree is native Central and South American usually inhabits in moist tropical woodlands. Owing to attractive foliage and tiny white flowers, it is utilized as an ornamental plant in gardens and parks. *Virola theiodora* contains psychoactive compounds, including N, N-Dimethyltryptamine (DMT) which is a stimulating tryptamine. Because of the intense hallucinogenic effects, it is popularly known as the soul atom (McKenna and Riba, 2018). The species also contains 5-Methoxy-N, N-Dimethyltryptamine (5-MeO-DMT), which is a psychoactive substance that has demonstrated its therapeutic potential and shown to be an effective, fast-acting medication for the treatment of post-traumatic stress disorder (PTSD), anxiety, and depression (Forner, 2022). In addition to being used as an anti-bleeding agent to treat ulcerating sores and wounds, hemorrhoids, and colic, reddish exudate of this species is also used to cure skin diseases, relieve tooth pain, and calm colic. Furthermore, it is used in many folk medical conditions with seed oils, such as asthma, rheumatism, joint tumors, intestinal worms, skin conditions, erysipelas, hemorrhoids, and foul breath (Santamaría-Aguilar, Aguilar, and Lagomarsino, 2019; Chaachouay et al., 2021a, 2021b). The wood is used in building and carpentry, while its seeds and arils are utilized to flavor beverages, such as chocolate drinks as shown in Figure 33.1 (Santamaría-Aguilar, Aguilar, and Lagomarsino, 2019). This chapter shed light on the *Virola theiodora*. It offers a comprehensive evaluation of botanical characteristics, geographical distribution, traditional utilization, metabolites, and pharmacological applications. This assessment may provide to work on novel drugs for fascinating biologically active new compounds.

33.2 SCIENTIFIC CLASSIFICATION

According to the Integrated Taxonomic Information System (ITIS, 2022), the scientific classification of *Virola theiodora* (Spruce) Warb., shown in Figure 33.2 and Table 33.1 as follows:

33.3 SYNONYMS

The Plant List (TPL) is a database that provides information on plant species' accepted names and synonyms. According to TPL (2013), *Virola theiodora* (Spruce ex Benth.) Warb., has three synonyms, which means that three different scientific names have been used to refer to this same plant species. These are as follows:

1. *Virola rufula* Warb.;
2. *Palala theiodora* (Spruce ex Benth.) Kuntze;
3. *Myristica theiodora* Spruce ex Benth.

33.4 BOTANICAL DESCRIPTION

It is a large-sized evergreen tree typically reaching up to 40 meters in height. The stem is a straight cylindrical trunk with a rounded canopy of up to 1.5 feet in diameter; when cut, it may exude a resinous substance. Shoots are monopodial and orthotropic, while lateral branches are plagiotropic, with a distichous phyllotaxy. The bark is smooth and grayish to brown-mottled having gray patches throughout on younger trees, becoming rougher and darker with age. Leaves are simple, alternate, shiny, and glossy. The shape of leaves is elliptic with entire margins, spirally arranged along the branches. Flowers are four-petaled, yellow, and small; inconspicuous, unisexual, arranged in clusters. The staminate flowers are

DOI: 10.1201/9781003460336-33

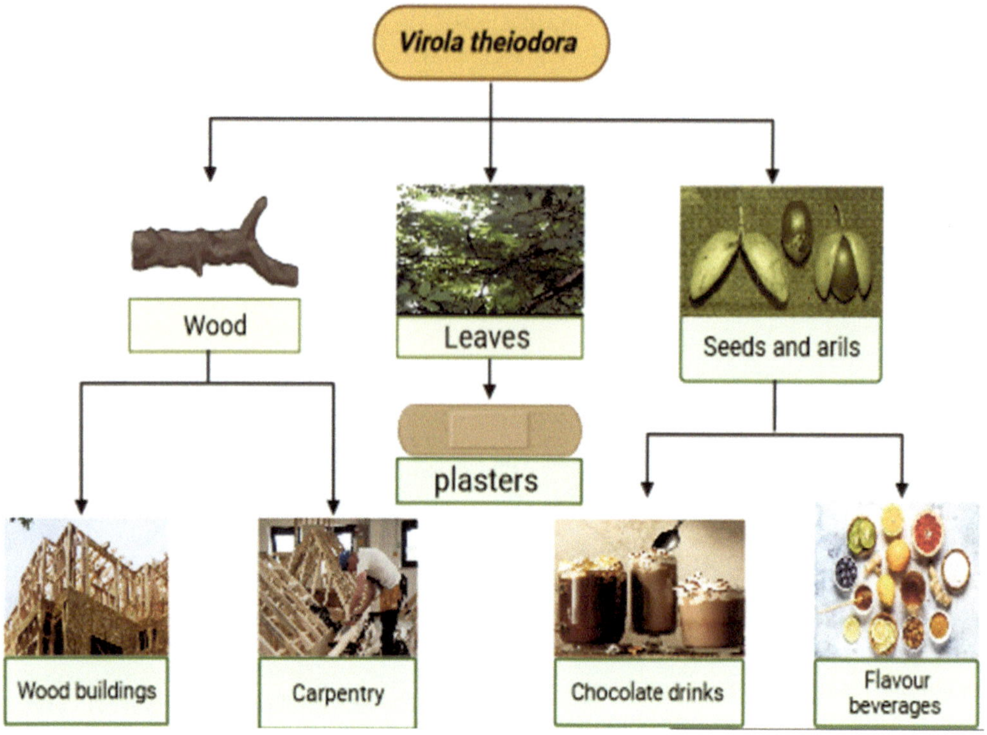

FIGURE 33.1 Applications of wood, leaves seeds, and aril of *Virola theiodora*.

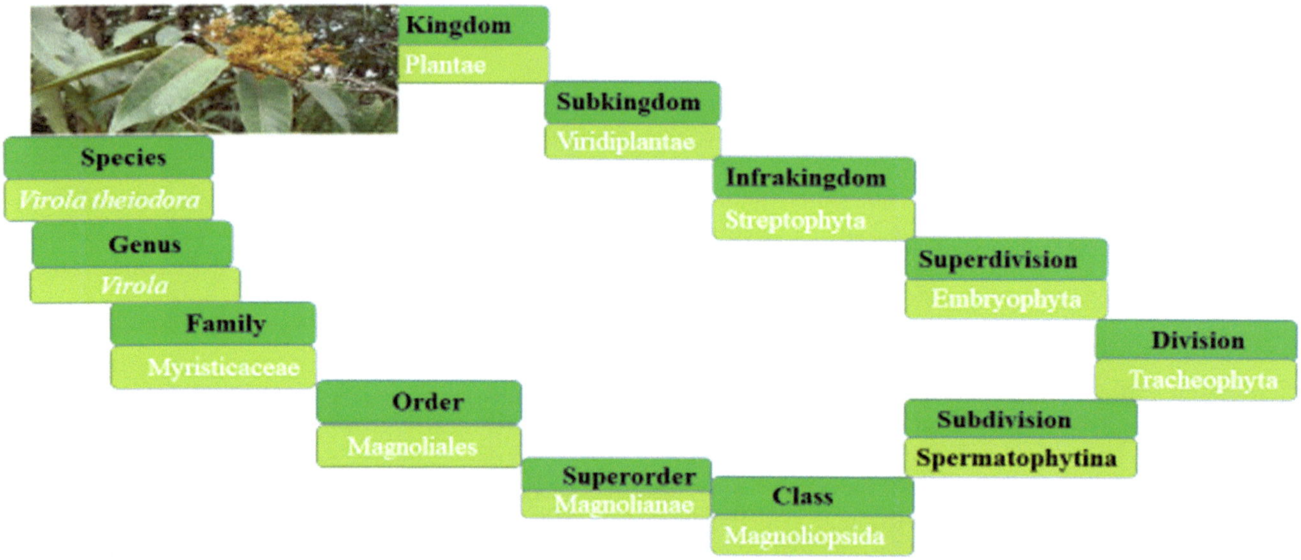

FIGURE 33.2 Taxonomic hierarchy of *Virola theiodora*.

numerous, paniculate, and typically up to 15 cm shorter than the leaves; while pistillate flowers, on the other hand, are smaller (Miller Jr, 1999). Fruits are fleshy, ovoid often brightly colored, ranging from vibrant red to orange. The seed capsule splits into two valves as the fruit ripens, exposing a single dark-brown seed, encased in a colorful laciniate aril that is packed with proteins, lipids, and sugars (Santamaría-Aguilar, Aguilar, and Lagomarsino, 2019). The morphological sketch is provided under Figure 33.3.

33.5 ORIGIN AND GEOGRAPHIC DISTRIBUTION

Virola theiodora is widely distributed throughout South America (Figure 33.4) and grows from Honduras to Brazil (Santamaría-Aguilar, Aguilar, and Lagomarsino, 2019). The species is native to the Amazon rainforest from southern Colombia, Venezuela to northern Brazil, where it thrives in humid and tropical climates. It is often found in lowland rainforests and riverine habitats. It is a tree that grows

TABLE 33.1
The Hierarchal Classification of *Virola theiodora* (Spruce) Warb

Kingdom:	Plantae
Subkingdom:	Viridaeplantae
Infrakingdom:	Streptophyta
Superdivision:	Embryophyta
Division:	Tracheophytes
Subdivision:	Spermatophytina
Infradivision:	Angiospermae
Class	Magnoliopsida
Clade	Magnoliid
Superorder	Magnolianae
Order	Magnoliales
Family	Myristicaceae
Genus	*Virola* Aubl.
Species	*V. theiodora*
Binomial name	*Virola theiodora* (Spruce) Warb.

mostly in the tropical wet biome. It has societal applications in addition to being a poison and medication. *Virola theiodora* plays a crucial role in the ecology of the Amazon rainforest by contributing to the overall biodiversity and providing habitat and sustenance for various organisms.

33.6 ETHNOBOTANY

Virola theiodora has a great ethnobotanical significance, especially among native communities in the Amazon rainforest area of South America. A significant number of the *Virola* species are taken advantage of for their medicinal properties by local occupants to treat or potentially oversee normal human infirmities, like psychological sickness, tainted injuries, skin diseases, colic, and vitiligo (Castro et al., 2004; Chaachouay et al., 2020; Belhaj et al., 2021; Benkhnigue et al., 2022). It is one of the numerous species of *Virola* that is utilized in the Amazon region as a psychedelic snuff. The plant contrasts in its pattern of utilization by the native people of this region. A ruddy gum like exudate is extracted from the cambial area of the bark of this slender tree and either snuffed or ingested orally. The significant

FIGURE 33.3 *Virola theiodora*: (a) leaves; (b) floral branch; (c) fruiting twig, and (d) red bark.

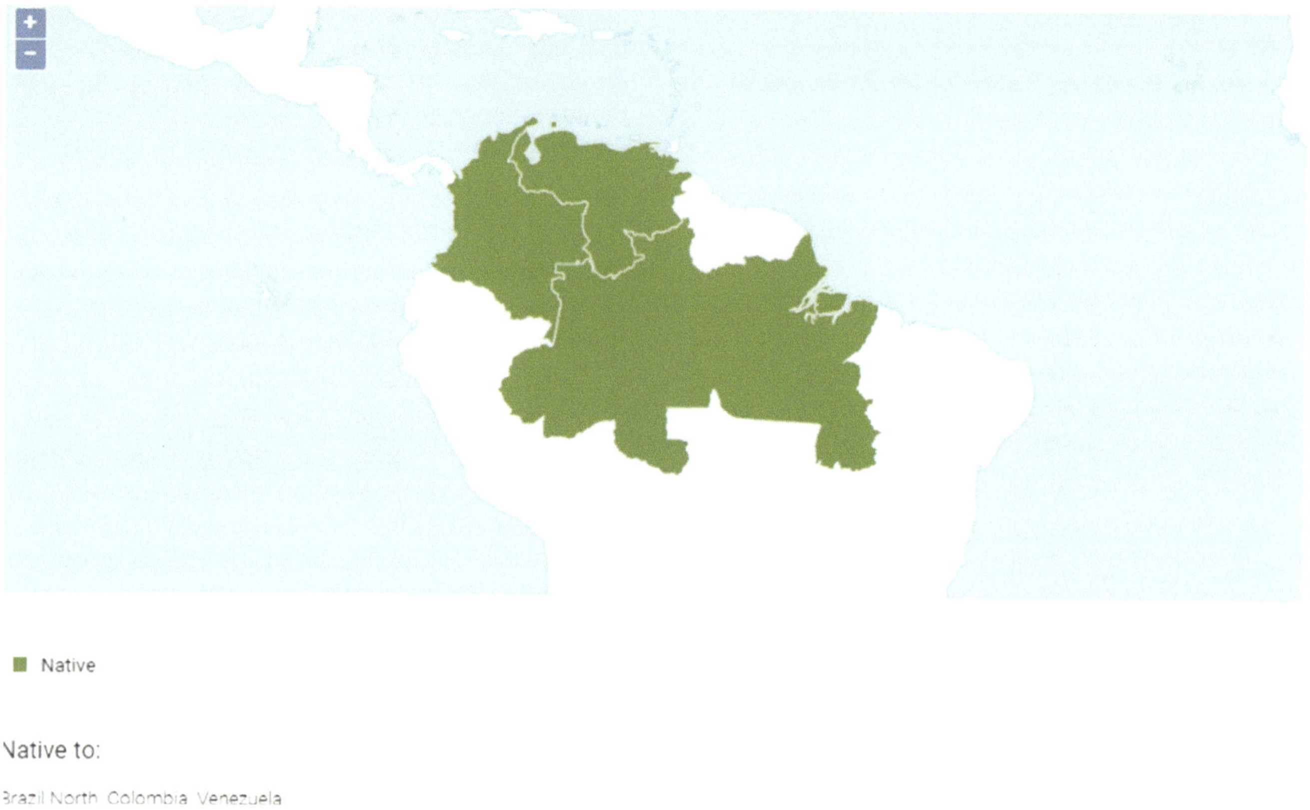

FIGURE 33.4 Distribution map of *Virola thoedora*.

alkaloid constituent of this snuff appears to be 5-MeODMT with small amounts of DM (Miller Jr, 1999). The hydroethanolic concentrate of the stem bark of *Virula theodora* is utilized for the treatment of stomach agony, heartburn, and gastric ulcers. This concentrate lessens gastric ulceration by upgrading gastro-protection through its antioxidant properties and capacity to significantly decrease gastric emission and causticity in mouse and rate samples of acute (acidified ethanol, piroxicam, and restraint-in-water stress) and chronic (acetic acid) gastric ulcers at doses of 100, 300 and 900 mg/kg p.o. (per os, orally) (de Almeida et al., 2019). Furthermore, *Virola theiodora* has shown sedating impacts and creates locomotor central nervous system wretchedness, being utilized as a hallucinogenic drug in light of its alkaloids (Macre and Towers, 1984).

33.7 HALLUCINOGENIC EFFECTS

Virola theiodora is a native tree of the Amazon rainforest in South America. In the midst of native cultures, it has a prolonged history of conventional use for its psychoactive properties, particularly in shamanic and healing rituals. The Brazilian botanist Ducke was a pioneer in making a conclusive relationship between the psychedelic snuff utilized by the locals in 1938. The most popular use of species is for hallucinogenic properties, which are frequently integrated into indigenous social acts of South America. A reddish gum-like exudate extracted from the cambial area of this slender tree was either inhaled or taken orally. The tryptamine chemicals particularly N, N-Dimethyltryptamine (DMT) and 5-Methoxy-N, N-Dimethyltryptamine (5-MeO-DMT), are mainly responsible for psychoactive properties (Schultes, 1954; Benington, Morin, and Clark Jr, 1965; Agurell et al., 1968; Schultes, 1969; Schultes, 1979; Prance, 1970; Soares Maia and Rodrigues, 1974; Barman et al., 2021). The wellspring of inebriating snuffs utilized by specific South Native American clans, has been displayed to contain the hallucinogen 5-methoxy-N, N-dimethyltryptamine, as well as various other indoles. One Indian snuff ended up being strangely high in alkaloid content. Significant contrasts in the alkaloid composition of various parts of the plant were encountered, N, N-dimethyltryptamine being the significant amount in the leaves and 5-methoxy-N,N-dimethyltryptamine in the bark (Agurell, 1969). The utilization of snuff is limited to shamans (Miller Jr, 1999). *Virola theiodora* is reported to produce a variety of hallucinogenic effects, when prepared and used in shamanic or ceremonial conditions, including:

33.7.1 Auditory Hallucinations

Auditory hallucinations are commonly used in some tribes. People report hearing voices, music, or other sounds that appear to have supernatural or spiritual origins (Lancelotta and Davis, 2021; Chaachouay & Zidane, 2021).

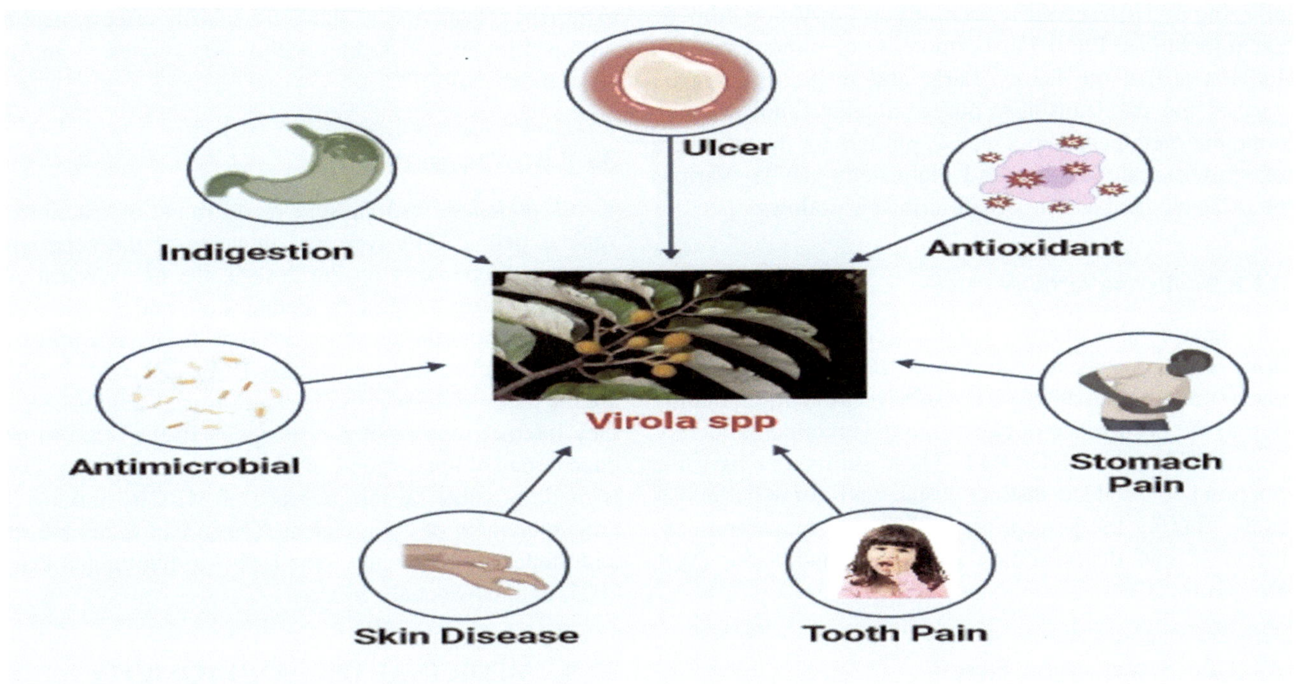

FIGURE 33.5 Pharmacological potential of *Virola theiodora*.

33.7.2 Visual Hallucinations

Consumers usually describe having powerful visual hallucinations. These include fractals, geometric patterns, and vibrant, detailed images that resemble the surrounding landscape (Perry et al., 2015; Schultes, 1979).

33.7.3 Altered Sense of Time and Space

The effects of *Virola theiodora* can alter as people perceive space and time. Users could experience a timeless or unreal zone where the significance of conventional ideas of time and space is lost (Miller Jr, 1999).

33.7.4 Ego Dissolution

Some people state that throughout the event, their feelings of self or ego vanished. This may cause a loss of sense of self and emotions of equanimity with everything in existence (Shepard, 2005).

33.7.5 Profound Spiritual and Mystical Experiences

An intense recognition of unity with nature, the universe, or Superior spiritual worlds are described by several users. An overwhelming sensation of wonder, veneration, and oneness frequently characterizes these encounters (Schultes, 1995).

33.7.6 Insights and Epiphanies

Virola theiodora users occasionally report experiencing life-altering epiphanies. Their personal experiences, spirituality, or understanding of reality may be related to these revelations (Lancelotta and Davis, 2020; Labate, 2014).

33.7.7 Emotional Intensity

The emotional impacts of *Virola theiodora* can arouse deep, complex feelings that range from happy, loving, and amazed feelings to fearful and anxious ones. These emotional states could vary significantly between individuals and situations (Shepard, 2005; Ermakova et al., 2022)

33.8 PHARMACOLOGY

Phytotherapy and the utilization of medicinal plants are conventionally important for the well-known information on various populations, clients, and practitioners. The World Health Organization assessed that 80% of the emerging country population actually relies on traditional medicines (González-Rodríguez et al., 2021). The therapeutic plants are mostly by low-income communities for their essential healthcare (Gottlieb, 1979). The ethnopharmacology of *Virola* indicates that these are utilized for the treatment of stomach torment gastric ulcers, skin conditions, tooth treatment, antioxidant, acid reflux, and antimicrobial activity (Figure 33.5). as well as hemorrhoids (González-Rodríguez et al., 2021; Santamaría-Aguilar, Aguilar, and Lagomarsino, 2019).

The Waika Indians of northwest Brazil and adjoining Venezuela set up their hallucinogenic epená or nyakwana-snuff primarily from the resin of *Virola theiodora*. Subsequent fieldwork uncovered a fascinating technique for utilizing myristicaceous resin orally as a psychotomimetic agent (Schultes, 1969). Furthermore, some minor groups of Indians utilize the leaves, bark, and resin to get ready washes that are cleansing to purge tainted wounds or accelerate their recuperating as well as plasters for the treatment of skin issues (Schultes and Holmstedt, .1971). Various pharmacological actions are described as follows:

33.8.1 Psychoactivity

The most notable pharmacological action of *Virola theiodora* is its capacity to elicit powerful psychedelic experiences and altered states of consciousness. This is mainly due to the occurrence of two strong tryptamine hallucinogens DMT and 5-MeO-DMT. These substances have the potential to produce distinct visual and aural hallucinations, altering the perception of time and space and cognition of ego disintegration (McKenna and Riba, 2018; Hesselink, 2019).

33.8.2 Serotonergic Effects

In the brain, DMT and 5-MeO-DMT predominantly interact with serotonin receptors to produce their effects. They bind to these receptors, especially the 5-HT2A receptor, and share structural similarities with serotonin. The significant shifts in perception, emotion, and consciousness related to *Virola theiodora* use are believed to be the result of this activation (Shen et al., 2010; Reckweg et al., 2022).

33.8.3 Spiritual and Shamanic Use

In the Amazon rainforest, *Virola theiodora* has a long history of usage in traditional Indigenous rituals and shamanic practices. It is often used to facilitate visions, spiritual journeys, and communication with the spiritual realm. The pharmacological effects of the plant are deeply intertwined with the cultural and spiritual beliefs of Indigenous communities (Labate, 2014).

33.8.4 Monoamine Oxidase Inhibition (MAOI)

Harmine and harmaline are the beta-carboline alkaloids predominantly found in the *Virola theiodora*. These substances have the ability to inhibit monoamine oxidase (MAOI). The monoamine oxidase enzyme, which breaks down neurotransmitters including serotonin and DMT in the body, is inhibited by MAOIs. DMT can be consumed orally and still be active because the presence of MAOIs in *Virola theiodora* delays the rate at which it degrades in the digestive tract (Lowe et al., 2022).

33.8.5 Short Duration

Compared to other psychoactive compounds, the effects of DMT and 5-MeO-DMT obtained from *Virola theiodora* are rather transient. Typically, the powerful psychedelic experience lasts for a short time with effects peaking within minutes and fading within 30 minutes to an hour (Ermakova et al., 2022).

33.8.6 Variability

Virola theiodora can cause a broad range of reactions in various people. The type and intensity of the experience may vary depending on the dosage, the user's sensitivity, set and setting (the user's mental state and physical surroundings), and the presence of other drugs. The powerful tryptamine chemicals, especially DMT and 5-MeO-DMT, are the focus of the plant's pharmacological actions because they interact with serotonin receptors in the brain to produce strong hallucinogenic effects. The oral action of DMT is made possible by the presence of MAOIs, making it a crucial element of both ancient Amazonian shamanic rites and modern psychonautic practices (Lancelotta and Davis, 2021; Ermakova et al., 2022).

33.9 MEDICINAL USES AND TOXICITY

Virola species are taken advantage of for their medicinal properties by local inhabitants to treat as well as manage human diseases, such as psychological sickness, infected wounds, skin infections, colic, and vitiligo (Castro et al., 2004). Recent investigations have uncovered the use amongst certain Indians of eastern Colombia and northwestern Brazil of several species of *Virola* in preparing violently toxic snuff which is employed by the medicine men in witchcraft divination and the diagnosis of illness (Schultes, 1954). The presence of 5-MeO-DMT in the plant extract does have toxicity traits. Studies with monkeys, mice, sheep, and rats revealed remarkable ataxia, tremor, mydriasis, shivering, and convulsion after being administered 5-MeO-DMT. Apart from the drug-induced discriminative stimulus control, 5-MeO-DMT also provokes a variety of other behavioral effects in animal models; these include head shaking, forepaw treading, straub tail, flat-body posture, and hind limb abduction. Indigenous peoples have a long history of use of *Virola theiodora* as a traditional remedy in the Amazon rainforest and other regions of South America. The varieties of uses are given next:

33.9.1 Psychoactive Properties

Two alkaloids such as 5-MeO-DMT and DMT are very commonly found in *Virola theiodora*. Due to their hallucinatory effects, these substances are employed in traditional shamanic ceremonies. They contain psychoactive effects and are mostly prepared for smoking or snuffing, used for healing and spiritual purposes (Lancelotta and Davis, 2021).

33.9.2 Pain Relief

Local people commonly used *Virola theiodora* for its potential analgesic (pain relieving) properties. The bark of the tree and resin have been applied topically or consumed orally to treat a variety of pains, including toothaches, headaches, and muscular discomfort, respectively (Schultes and Farnsworth, 1980).

33.9.3 Anti-inflammatory Properties

Virola theiodora extracts or formulations have been used in traditional medicine to treat inflammation and skin problems (Neuwinger, 1998).

33.9.4 Respiratory Disorders

Virola theiodora has been utilized in several traditional medical practices to treat respiratory disorders such as cold and cough, and asthma. It is said to have properties that might facilitate better breathing and reduce congestion (Schultes, 1969).

33.9.5 Antimicrobial Uses

Certain Indigenous cultures have used *Virola theiodora* preparations for their potential antimicrobial properties. The resin has been applied topically on wounds to prevent infection (Reckweg et al., 2022).

33.9.6 Aphrodisiac and Ritual Uses

Traditional Amazonian rites and ceremonies have employed *Virola theiodora* as an aphrodisiac agent, a vision-inducer, and for spiritual purposes. The psychoactive compounds play a significant role in such kind of practices (Schultes, 1979). The main psychoactive compounds (DMT and 5-MeO-DMT) can cause intense hallucinations and altered states of consciousness. These experiences can be overwhelming and psychologically challenging, potentially leading to worse psychological reactions, especially in individuals with underlying mental health conditions (Schultes and Farnsworth, 1980). Although DMT and 5-MeO-DMT are not considered physically harmful in the same way as some other compounds, they can have physical effects such as alterations in blood pressure and heartbeat rate. These effects can be problematic for those with pre-existing cardiovascular conditions (Gable, 2007; Tittarelli et al., 2015; Malcolm and Thomas, 2022). *Virola theiodora* experiences can be so powerful and unexpected that they can cause serious psychological discomfort, such as anxiety, paranoia, and in some cases even flashbacks. Any drug containing DMT or 5-MeO-DMT can interact with pharmaceuticals, particularly those that influence serotonin levels. *Virola theiodora* can interact dangerously with some drugs, including antidepressants, to cause serotonin syndrome (Lancelotta and Davis, 2020).

33.10 PHYTOCHEMISTRY

The psychotropic properties of *Virola theiodora* are mostly related to the presence of several phytochemical substances, especially tryptamines. An overview of some of the major phytochemicals present in *Virola theiodora* is given below:

33.10.1 N, N-Dimethyltryptamine (DMT)

The N, N-Dimethyltryptamine (DMT) is one of the most well-known and potent psychoactive compounds present in *Virola theiodora*. It is a powerful hallucinogenic tryptamine that can induce intense visual and auditory hallucinations and altered states of consciousness (López-Arnau et al., 2022).

33.10.2 5-Methoxy-N, N-Dimethyltryptamine (5-MeO-DMT)

Another hallucinogenic tryptamine discovered in *Virola theiodora* is 5-MeO-DMT. It is well-known for its dominant psychedelic effects, which include ego dissolution and intense spiritual encounters (López-Arnau et al., 2022).

33.10.3 Bufotenin (5-Hydroxy-N, N-Dimethyltryptamine)

A tryptamine molecule called buffotenin is associated with DMT and 5-MeO-DMT. It can have psychoactive effects and is also present in *Virola theiodora*, despite it typically being accounted as being less strong than DMT (Barman et al., 2021).

33.10.4 Other Tryptamines

Other tryptamine chemicals may also be present in *Virola theiodora*, some of which may participate in the psychedelic effects. The accurate composition of these tryptamines can differ among various plant samples (Schultes and Winkelman, 1995).

33.10.5 Beta-carbolines

Virola theiodora has been found to contain beta-carboline alkaloids, such as harmine and harmaline. These substances have the potential to function as monoamine oxidase inhibitors (MAOIs), which can enhance the effects of DMT and other tryptamines by preventing the body's natural enzymes from breaking them down (Schultes and Winkelman, 1995).

33.10.6 Alkaloids and Flavonoids

In addition to tryptamines, *Virola theiodora* may also include a wide range of additional alkaloids and flavonoids. These substances may exhibit their own pharmacological effects or engage in intricate interactions with the psychoactive tryptamines (Macre and Towers, 1984; Mesquita et al., 2020; Bourscheid et al., 2022).

Phytochemical makeup of *Virola theiodora* can change according to the location, plant age, and environmental circumstances. The concentration of these substances can also vary across the plant parts, the bark being the most frequently component utilized for its psychedelic properties. The main psychoactive substances found in the plant, including N, N-Dimethyltryptamine (DMT) and 5-Methoxy-N,N-Dimethyltryptamine (5-MeO-DMT), as well as other alkaloids and chemicals purported to be responsible for the therapeutic effects.

33.11 CONCLUSION

Virola theiodora is an evergreen tree native to southern Colombia, Venezuela, and northern Brazil and an important floristic component of the tropical wet biome. The plant is used as a hallucinogenic snuff in the Amazon region. Its bark is known for its psychedelic properties and is used to treat variety of ailments. Besides medicinal use, the wood is used in building and carpentry, while its seeds and arils are utilized to flavor beverages, such as chocolate drinks. Its ethnobotanical significance in indigenous communities and ongoing pharmacological research could further highlight the multifaceted importance of this species.

LIST OF ABBREVIATIONS

PTSD	post-traumatic stress disorder
5-MeO-DMT	5-methoxy-N, N-dimethyltryptamine
DMT	N, N-dimethyltryptamine

REFERENCES

Agurell, S, Holmstedt, B, Lindgren, J E, & Schultes, R E. 1968. Identification of two new beta-carboline alkaloids in South American hallucinogenic plants. *Biochemical Pharmacology* 17(12): 2487–2488.

Agurell, S. 1969. American Plants of Ethnopharmacologic Interest. *Acta Chem. Scand* 23: 3–13.

Barman, R, Bora, P K, Saikia, J, Kemprai, P, Saikia, S P, Haldar, S, & Banik, D. 2021. Nutmegs and wild nutmegs: An update on ethnomedicines, phytochemicals, pharmacology, and toxicity of the Myristicaceae species. *Phytotherapy Research 35*(9): 4632–4659.

Belhaj, S., Chaachouay, N., & Zidane, L. (2021). Ethnobotanical and toxicology study of medicinal plants used for the treatment of diabetes in the High Atlas Central of Morocco. *Journal of Pharmacy & Pharmacognosy Research*, *9*(5), 619–662. https://doi.org/10.56499/jppres21.1014_9.5.619

Benington, F, Morin, R D, and Clark Jr., L C 1965. 5-methoxy-N, N-dimethyltryptamine, a possible endogenous psychotoxin. *The Alabama Journal of Medical Sciences* 2 (4):397–403.

Benkhnigue, O., Chaachouay, N., Khamar, H., El Azzouzi, F., Douira, A, & Zidane, L. (2022). Ethnobotanical and ethnopharmacological study of medicinal plants used in the treatment of anemia in the region of Haouz-Rehamna (Morocco). *Journal of Pharmacy & Pharmacognosy Research*, *10*(2), 279–302. https://doi.org/10.56499/jppres21.1196_10.2.279

Bourscheid, Guimarães T, Cardoso C L, Santana H N M, et al. 2022. Use of snuff and its main constituents for religious purposes in an alternative community with shamanic practices in the south of Brazil. *Journal of Psychedelic Studies* 6 (2):154–165.

Castro, Ernesto, Suarez, E C L, Siengalewicz, P, Gutmann, R, Czermak, G, and Brueggeller, P. 2004. 3, 6-Dihydroxy-2-(11-phenylundecanoyl) cyclohex-2-en-1-one from *Virola* venosa bark. *Acta Crystallographica Section C: Crystal Structure Communications* 60 (7):o467–o469.

Chaachouay, N., Douira, A., & Zidane, L. (2021a). Herbal Medicine Used in the Treatment of Human Diseases in the Rif, Northern Morocco. *Arabian Journal for Science and Engineering*, 47, 131–153. https://doi.org/10.1007/s13369-021-05501-1

Chaachouay, N., Douira, A., & Zidane, L. (2021b). COVID-19, prevention and treatment with herbal medicine in the herbal markets of Salé Prefecture, North-Western Morocco. *European Journal of Integrative Medicine*, 42, 101285. https://doi.org/10.1016/j.eujim.2021.101285

Chaachouay, N., & Zidane, L. (2021). The Symbolic Efficacy of Plants in Rituals and Socio-Religious Ceremonies in Morocco, Northwest of Africa. *Journal of Religious & Theological Information*, *21*(1–2), 34–53. https://doi.org/10.1080/10477845.2021.1942433

Chaachouay, N., Douira, A., Hassikou, R., Brhadda, N., Dahmani, J., Belahbib, N., Ziri, R., & Zidane, L. (2020). Ethnobotanical and ethnopharmacological study of medicinal and aromatic plants used in the treatment of respiratory system disorders in the Moroccan Rif. PhDThesis, Département de Biologie - Université Ibn Tofail - Kénitra, Maroc, https://tel.archives-ouvertes.fr/tel-03376377

de Almeida, Botelho V G, Arunachalam, K, Balogun, O S, et al. 2019. Chemical characterization and evaluation of gastric antiulcer properties of the hydroethanolic extract of the stem bark of *Virola* elongata (Benth.) Warb. *Journal of ethnopharmacology* 231:113–124.

Ermakova, Anna O, Dunbar, F, Rucker, J, and Johnson, W M. 2022. A narrative synthesis of research with 5-MeO-DMT. *Journal of Psychopharmacology* 36 (3):273–294.

Forner. 2022. 5-Methoxy-N, N-dimethyltryptamines pharmacokinetics and viable routes of synthetization, NTNU. 1–18.

Franklin, Francesca T. 2007. *Diversity and Distribution of the Myristicaceae Family of Plants in Terra Firme and Seasonally Flooded Forests in Madre de Dios, Peru* (pp. 1–50). Texas Christian University.

Soares Maia, G J, and Rodrigues, William A. 1974. *Virola theiodora* como alucinógena e tóxica. *Acta Amazonica* 4 (1): 21–23.

Gable, Robert S. 2007. Risk assessment of ritual use of oral dimethyltryptamine (DMT) and harmala alkaloids. *Addiction* 102 (1): 24–34.

González-Rodríguez, María, Ruiz-Fernández, C, Francisco, V, et al. 2021. Pharmacological extracts and molecules from *Virola* species: traditional uses, phytochemistry, and biological activity. *Molecules* 26 (4):792.

Gottlieb, Otto R. 1979. Chemical studies on medicinal Myristicaceae from Amazonia. *Journal of Ethnopharmacology* 1 (4):309–323.

Hesselink, Keppel M J. 2019. Transformative Psychopharmacology: the Case of 5-methoxy-n,n-dimethyltryptamine. *International Journal of Psychotherapy Practice and Research* 1 (3):9–15.

Integrated Taxonomic Information System (ITIS). 2022. https://www.gbif.org/species/3152844.

Labate, B C. 2014. Legal, ethical, and political dimensions of ayahuasca consumption in Brazil. *Erowid.org* 2: 397–457.

Lancelotta, Rafael, and Davis, K A. 2021. Therapeutic potential of fast-acting synthetic tryptamines. In *Handbook of medical hallucinogens* (pp. 215–232). New York: The Guilford Press.

Lancelotta, Rafael L, and Davis, K A. 2020. Use of benefit enhancement strategies among 5-methoxy-N, N-dimethyltryptamine (5-MeO-DMT) users: Associations with mystical, challenging, and enduring effects. *Journal of Psychoactive Drugs* 52 (3):273–281.

López-Arnau, R, Camarasa, J, Carbó, M L, Nadal-Gratacós, N, Puigseslloses, P, Espinosa-Velasco, M, & Pubill, D. 2022. 3, 4-Methylenedioxy methamphetamine, synthetic cathinones and psychedelics: From recreational to novel psychotherapeutic drugs. *Frontiers in Psychiatry 13*, 990405.

Lowe, Henry, Toyang, N, Steele, B, et al. 2022. Psychedelics: alternative and potential therapeutic options for treating mood and anxiety disorders. *Molecules* 27 (8):2520.

Macre, Donald W, and Towers, N G H. 1984. An ethnopharmacological examination of*Virola* elongata bark: A South American arrow poison. *Journal of Ethnopharmacology* 12 (1):75–92.

Malcolm, Benjamin, and Thomas, K. 2022. Serotonin toxicity of serotonergic psychedelics. *Psychopharmacology* 239 (6):1881–1891.

McKenna, D, and Riba, J. 2018. New world tryptamine hallucinogens and the neuroscience of ayahuasca. In: Halberstadt, A.L., Vollenweider, F.X., Nichols, D.E. (eds) *Behavioral Neurobiology of Psychedelic Drugs. Current Topics in Behavioral Neurosciences*, vol 36. 283–311. Berlin, Heidelberg: Springer. https://doi.org/10.1007/7854_2016_472

McKenna, D J, Towers, G N, and Abbott, F. 1984. Monoamine oxidase inhibitors in South American hallucinogenic plants: tryptamine and β-carboline constituents of ayahuasca. *Journal of Ethnopharmacology* 10(2): 195–223.

Mesquita, Rochelly da S, de Oliveira, André C, Sá, Ingrity S C, et al. 2020. Essential oils from leaves of *Virola calophylla, Virola multinervia*, and *Virola pavonis* (Myristicaceae): Chemical composition and larvicidal activity against *Aedes aegypti. Journal of Essential Oil Bearing Plants* 23 (3):453–463.

Miller Jr, Terry R. 1999. The use of the genus *Virola* as a hallucinogen in South America. *Ethnobotanical Leaflets* 1999 (2):4.

Neuwinger, H. D. 1998. Alkaloids in arrow poisons. In: Margaret, F. R., Michael, W (eds) *Alkaloids: Biochemistry, Ecology, and Medicinal Applications* (pp. 45–84). Springer.

Perry, E, Mosimann, U P, & Collerton, D. 2015. The treatment of visual hallucinations at present and in the future. *The Neuroscience of Visual Hallucinations*, 321–341.

Prance, Ghillean T. 1970. Notes on the use of plant hallucinogens in Amazonian Brazil. *Economic Botany* 24 (1):62–68.

Reckweg, Johannes T, Uthaug, Malin V, Szabo, A, et al. 2022. The clinical pharmacology and potential therapeutic applications of 5-methoxy-N, N-dimethyltryptamine (5-MeO-DMT). *Journal of Neurochemistry* 162 (1):128–146.

Santamaría-Aguilar, D, Aguilar, R, & Lagomarsino, L P. 2019. A taxonomic synopsis of *Virola* (Myristicaceae) in Mesoamerica, including six new species. *PhytoKeys* 134: 1.

Schultes, R, & Holmstedt, B. 1971. nMiscellaneous notes on Myristicaceous plants of South America. *Lloydia* 34(1): 61–78.

Schultes, R E. 1954. A new narcotic snuff from the northwest Amazon. *Botanical Museum Leaflets, Harvard University* 16 (9):241–260.

Schultes, R E. 1969. Hallucinogens of plant origin: Interdisciplinary studies of plants sacred in primitive cultures yield results of academic and practical interest. *Science* 163 (3864):245–254.

Schultes, R E. 1979. Evolution of the identification of the myristicaceous hallucinogens of South America. *Journal of Ethnopharmacology* 1 (3):211–239.

Schultes, Evans R, and Farnsworth, R N. 1980. *Ethnomedical, Botanical and Phytochemical Aspects of Natural Hallucinogens* (pp. 123–214). Botanical Museum Leaflets, Harvard University.

Schultes, R E, and Winkelman, M. 1995. The principal American hallucinogenic plants and their bioactive and therapeutic properties. In Michael Winkelman and Walter Andritzky (eds.), *Sacred Plants, Consciousness and Healing Cross-Cultural and Interdisciplinary In book: Sacred Plants, Consciousness and Healing Cross-Cultural and Interdisciplinary Perspectives Yearbook of Cross-cultural Medicine and Psychotherapy Volume 5.Publisher: Berlin: Verlag* (pp. 205–240).

Shen, Hong-Wu, Jiang, X, Winter, C J, and Yu, A. 2010. Psychedelic 5-methoxy-N, N-dimethyltryptamine: metabolism, pharmacokinetics, drug interactions, and pharmacological actions. *Current Drug Metabolism* 11 (8):659–666.

Shepard, Glenn. 2005. Psychoactive botanicals in ritual, religion and shamanism. *Ethnopharmacology: Encyclopedia of life support systems (EOLSS)* 2:128–182.

The Plant List, TPL. 2013. Version 1.1. Published on the Internet; http://www.theplantlist.org/ (accessed 1st January).

Tittarelli, Roberta, Mannocchi, G, Pantano, F, and Romolo, S F. 2015. Recreational use, analysis and toxicity of tryptamines. *Current Neuropharmacology* 13 (1):26–46.

34 Virola (*Virola calophylla* (Spruce) Warb. Myristicaceae)

Rahmatullah Qureshi, Ayesha Siddiqa, Muhammad Ishaq, Mehmooda Munazir, Muhammad Maqsood, Sehrish Sadia, and Noureddine Chaachouay

34.1 INTRODUCTION

Myristicaceae, commonly known as the nutmeg family, has a broad distribution in tropical regions of the world. The family consists of about 20 genera with nearly 500 species (DE WILDE 2011). Members of Myristicaceae are recognized for their ecological and economic significance (Kühn and Kubitzki, 1993; Chaachouay et al., 2020, 2020; Orch et al., 2021). In various growing regions, some Myristicaceae plants have been utilized medicinally. *Virola* is a genus of this family that consists of nearly 70 species of trees and shrubs native to tropical rainforests in Central and South America (Franklin, 2007). The species of this genus are usually recognized for their cultural, ecological, and ethnobotanical significance. The hallucinogenic effects of such species are well known, which are frequently used in indigenous cultural practices in South America. According to Schultes (1954, 1969a, 1969b, 1976; Schultes and Holmstedt, 1968; Prance, 1970; Soares Maia and Rodrigues; 1974; Bennett and Alarcón, 1970), the hallucinogen is typically extracted from the inner bark's exudate. Some of the species such as *Virola calophylla, V. calophyloidea, V. carinata* and *V. elongata* were utilized by indigenous Amerindians in Central and South America as hallucinogenic snuff, smoke, vapor, beverages, and oral pellets in ceremonial practices (Schultes, 1969a, 1969b; Schultes, 1976; McKenna et al., 1984a, 1984b; Martinez et al., 1995). *Virola theiodora* and *Virola elongata* contain psychoactive compounds, including N, N-Dimethyltryptamine, sometimes known as DMT, is a hallucinogenic tryptamine. Due to its potent psychedelic effects, some call it the spirit molecule (McKenna and Riba, 2018). Seeds of these plants have been used in traditional shamanic rituals and as a source of hallucinogenic snuffs by indigenous peoples in the Amazon rainforest (Bourscheid et al., 2022). Beyond their ethnobotanical importance, Virola species are ecologically valuable for providing food to wildlife and contributing to the biodiversity of tropical rainforests (DeWalt et al., 2003; Prance, 1970; Schultes, 1954, 1969a, 1969b; Maia and Rodrigues, 1974).

Virola calophylla is a tree species native to S. Tropical America. It is typically found in the Amazon rainforest, a wet tropical ecosystem, but it is also distributed in other parts of the world, including Brazil, Colombia, Ecuador, and Peru in South America. It flourishes in the region's humid tropical temperatures. This species can be found all over the Amazon Basin in its natural habitat, frequently growing in lowland rainforests. This species is found along riverbanks with well-drained soils (Franco-Moraes et al., 2019). The plant is characterized by glossy leaves, small inconspicuous flowers, and fleshy fruits (Santamaría-Aguilar et al., 2019).

Virola calophylla has been utilized for centuries by the indigenous people in the Amazon region. The bark and seeds are particularly valued for their traditional uses. These usages frequently require for making psychedelic snuffs and potions and have a strong cultural and spiritual foundation. The plant is known to have chemicals, which includes psychoactive compounds. The most notable compounds found in *Virola calophylla* are N, N-dimethyltryptamine (DMT) and 5-methoxy-N, N-dimethyltryptamine (5-MeO-DMT). These tryptamines are known for their hallucinogenic and psychoactive properties (Mesquita et al., 2020).

Due to its medicinal and cultural uses and phytochemical composition, *Virola calophylla* has great value. There is no detailed review exists about this plant that provides in-depth explanations into the significance and its therapeutic use. This chapter entails the most recent information on the plant. Additionally, it offers a comprehensive evaluation that has all the necessary details regarding the botanic properties, distribution, conventional applications, primary and secondary metabolites, and pharmacological functions. This enumeration may aid in the development of new drugs for interesting biologically active novel compounds.

34.2 SCIENTIFIC CLASSIFICATION

The scientific classification of *Virola calophylla* (Spruce) Warb. is shown in Table 34.1 (Itis, 2022).

34.3 SYNONYMS

According to The Plant List (2013), *Virola calophylla* (Spruce) Warb. has five synonyms as follows:

(1) *Virola lepidota* A.C. Sm.;
(2) *Myristica calophylla* Spruce;

DOI: 10.1201/9781003460336-34

TABLE 34.1
The Scientific Classification of *Virola calophylla* (Spruce) Warb

Kingdom:	Plantae
Subkingdom:	Viridaeplantae
Infrakingdom:	Streptophyta
Superdivision:	Embryophyta
Division:	Tracheophytes
Subdivision:	Spermatophytina
Infradivision:	Angiospermae
Class	Magnoliopsida
Clade	Magnoliid
Superorder	Magnolianae
Order	Magnoliales
Family	Myristicaceae
Genus	*Virola*
Species	*V. calophylla*
Binomial name	*Virola calophylla* (Spruce) Warb.

(3) *Virola incolor* Warb.;
(4) *Palala calophylla* (Spruce) Kuntze;
(5) *Otoba incolor* H.Karst.

34.4 BOTANICAL DESCRIPTION

A medium to big size evergreen tree with 20 to 30 meters or taller trunk (Figure 34.1). The bark is smooth and varies from grayish to brownish colored (Santamaría-Aguilar and Lagomarsino 2022). Leaves ca. 20–60 × 10–16 cm across, simple, alternating, aggregated at the tips of branches, oblong to elliptic in shape, base deeply cordate to truncate or obtuse, margin entire, glossy, and leathery textured. The plant is dioecious; staminate inflorescence ca. 6 to 30 cm long. Flowers are small, unisexual arranged in spikes or clusters, and greenish to yellow in color. Perianth of staminate flowers is ca. 1–2.1 mm, filament column, 0.2–0.6 mm long, constricted at apex, anther 0.4–0.5 mm long. Fruit is a drupe with fleshy fruit and a hard, woody seed inside. Fruits ca. 2.5–4 cm across, spherical, base truncate. The fruit color might change from green to yellow or reddish when it is ripe. The fruit's seed is located in its fleshy exterior (Santamaría-Aguilar et al., 2019).

34.5 HISTORY

Virola calophylla, popularly known as the Ivory Nut Tree or Laurel Nut Tree, is a native of Central and South America, where its history is closely linked with those of the indigenous people. These traditional communities in the Amazon rainforest have used it for millennia especially seeds (Schultes, 2007). These seeds have economic, spiritual, and cultural value in developing the lives of many indigenous people. Because of their distinctive look, these are believed to be appropriate for carving. Native American carvers have fixed the seeds in a sophisticated manner depicting features of animals, flora, and legendary characters. These artifacts have cultural and spiritual significance in addition to being beautiful (Duke, 2008). In traditional shamanic rites, such as ayahuasca ceremonies, indigenous shamans have employed mixtures produced from the seeds. These rituals are thought to aid in communication with the spirit world, healing, and spiritual understanding. Native American groups have made money from the carved ivory nuts. Tourists and collectors frequently buy them as handicrafts, supporting the livelihoods of these villages (Mesquita et al., 2020).

34.6 ORIGIN AND GEOGRAPHIC DISTRIBUTION

Virola calophylla originated in the tropical rainforests of Central and South America. The Amazon rainforest is believed to be the birthplace of this tree. Its native range is thought to be in the Amazon Basin and surrounding areas (Figure 34.1). Several counties in Central and South America are included in the geographical distribution (Figure 34.2). Although it can also be found in higher-elevation forests and cloud forests, the tree is most prevalent in low-elevation rainforests (De Oliveira Brito et al., 2023). It can be found throughout Brazil's tropical regions, including the Amazonian rainforest. It occurs specifically in the states of Amazonas, Acre, and Rondônia. Additionally, this species can be found in Colombia's Amazon forest, specifically in the departments of Amazonas, Caquetá, and Putumayo (Uscategui, 1959). It has been found in Peru's Amazon rainforest, specifically in the areas of Loreto, Ucayali, and Madre de Dios. It is found throughout Ecuador's Amazonian provinces, including Sucumbos, Napo, and Pastaza (Janovec and García, 2004).

34.7 HABITAT AND ECOLOGY

The plant grows at heights of about 125 meters in dense and dry woodlands, although it can be found in higher-elevation forests and cloud forests (Williams, 1936). It prefers regions with a lot of rain and high humidity, which are found in the Amazon Basin. This species can be found frequently in terra firme woods and floodplain forests when the soil is not subject to seasonal floods (Parolin et al., 2013). It is a shade-tolerant tree that can survive in locations with minimal sunlight. It tolerates some floods but prefers well-drained soils. This species frequently flourishes on nutrient-rich soils. The extensively weathered soils of the Amazonian rainforests are renowned for being nutrient-poor on the surface but rich in nutrients deeper in the soil profile. The habitat consists of cloud forests, higher-elevation woods, and low-elevation rainforests (Franco-Moraes et al., 2019).

Virola callophylla plays a significant role in the ecosystem of the rainforest. This species like many canopy trees in tropical rainforests, performs a number of

FIGURE 34.1 *Virola calophylla*: (A) leaves with inflorescence; (B) bark and leaves of the mature tree; (C) nearly mature florescence; (D) close-up of inflorescence with ants; (E) abaxial leaf pubescence (scale bar = 2 mm); (F) indehiscent and dehiscent mature fruit with lacinate aril (Steeves, 2011).

essential ecological functions, including shading understory plants, preserving a constant microclimate, and minimizing temperature and light changes on the forest floor (Janovec and García, 2004). Numerous bird, insect, and other canopy-dwelling species can find habitat and nesting locations in the tree's canopy and branches. A variety of creatures, including bats, birds, monkeys, and other mammals eat the fruit of the tree, which helps spread the seed and promotes the tree's reproductive success (Howe, 2016).

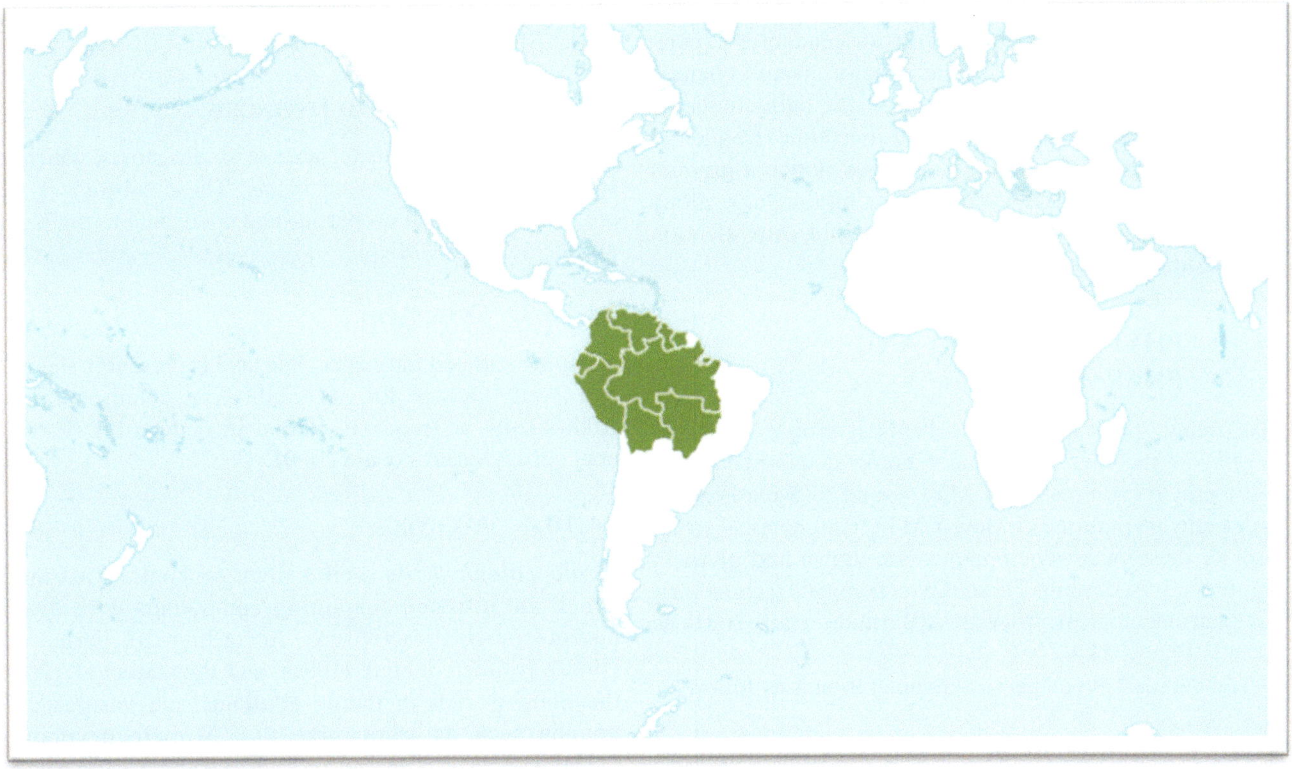

FIGURE 34.2 Geographical distribution of *Virola calophylla*.

34.8 ETHNOBOTANY

Virola callophylla has significant ethnobotanical importance, particularly among indigenous communities in the Amazon rainforest region of South America. Various parts such as the bark, leaves, and resin, have been used by Indigenous peoples for their purported medicinal properties (Schultes and Holmstedt, 1968; Belhaj et al., 2021; Chaachouay et al., 2021a, 2021b). Infusions, decoctions, or extracts from these plant parts have been employed to treat a range of ailments, including digestive issues, skin conditions, and respiratory problems (McKenna et al., 1984a). Mouth sores and thrush are the two medical conditions for which *Virola calophylla* is most frequently used. The affected area is treated directly with an exudate from the stems. It has also been used to treat malaria, cutaneous mycosis, and inflammation ((DeFilipps, Maina, and Crepin, 2004)).

The seeds of *Virola callophylla* have been used to prepare a traditional psychoactive snuff called 'yopo' or 'parica' by indigenous groups like the Yekwana and Yanomami. The seeds are ground into a powder and then mixed with other plant substances to create the snuff. Yopo is used in shamanic and spiritual rituals to induce altered states of consciousness and for divinatory purposes (Torres, 1995; Chaachouay and Zidane, 2021).

In the oral traditions of indigenous people, the tree may have mythological or metaphorical meaning, reflecting their strong contact with the natural environment. While the primary use of *Virola callophylla* is related to its medicinal and cultural significance, its wood and resin have been used for practical purposes (Roumy et al., 2007). Although it is not usually a principal timber species, the wood may have been utilized for construction, while the resin can be used for materials for waterproofing and sealing. The red sap is frequently applied with cotton on wounds to treat. The wood is employed in light construction, general carpentry, millwork, veneer and plywood, particle and fiberboard, furniture parts, boxes, and crates (Chudnoff, 1984). The pulp of the seeds is used as a substitute for soap (Williams, 1936). Candles and soap are also made by employing seed oil (Longwood, 1962; Williams, 1936; Chudnoff, 1984; Fern, 2018)

34.9 HALLUCINOGENIC EFFECTS

Virola callophylla is known for its hallucinogenic properties, primarily attributed to the psychoactive compounds present in its seeds. The primary hallucinogenic compound is 5-methoxy-N, N-dimethyltryptamine (5-MeO-DMT) that can induce significant alterations in consciousness (Miles et al., 1987; Agurell, 1969). When using this plant, some people report experiencing great emotional arousal and bliss. Users may experience a variety of emotional states, including fast emotional fluctuations (Schultes and Smith, 1976). *Virola callophylla* can provide physical feelings like tingling, vibrations, and a sense of energy moving through the body in addition to the mental and emotional effects. Many people claim to have had highly mystical or spiritual experiences when its seeds are utilized in shamanic or ritualistic contexts. These encounters could result in transcendental experiences, spiritual encounters, or a sense of cosmic oneness (Perry, 2002). Although the length of the hallucinogenic effects can vary usually peak after a short time, frequently between 15 and 30 minutes. Users may notice a gradual return to normal consciousness following the peak, along with some residual effects and a sense of introspection (Lancelotta and Davis, 2021).

34.10 PHYTOCHEMISTRY AND PHARMACOLOGY

Like many species, the phytochemical profile of *Virola calophylla* includes three major compounds such as N, N-dimethyltryptamine (DMT) and 5-methoxy-N, N-dimethyltryptamine (5-MeO-DMT). The chemical structures of these major coumpounds are drawn and given in Figure 34.3. According These phytochemicals trigger various pharmacological properties (Barman et al., 2021) as given in Figure 34.4.

The detailed phytochemical enumeration is as follows:

34.10.1 Myristic Acid

Myristic acid is found in the seed, which is a saturated fatty acid. It makes up a sizable portion of ucuuba oil and contributes to its distinctive qualities (Pardauil et al., 2017).

34.10.2 Terpenes and Terpenoids

The fragrant chemicals terpenes and terpenoids, which are frequently present in plant resins. These substances may also have biological properties and contribute to the plant's aroma (Schultes, 1969b).

34.10.3 Oleic Acid

Monounsaturated fatty acid oleic acid is present in the plant and well known for its ability to retain moisture. Additionally, it frequently found in ucuuba oil (Pardauil et al., 2017; Saraiva et al., 2009).

34.10.4 Alkaloids

Virola calophylla is well known to contain alkaloids, which are nitrogen-containing compounds with diverse pharmacological activities. According to Farnsworth (1968), Holmstedt et al. (1980), and Barman et al. (2021), the plant is rich in indole alkaloids and various compounds such as tryptamine (T), N′-methyltryptamine (MMT), 2-methyl-1,2,3,4-tetrahydro-β-carboline (MTHC). Leaves possess 0.15% N′-N′ -dimethyltryptamine (DMT), 0.006% N′-methyltryptamine (NMT or

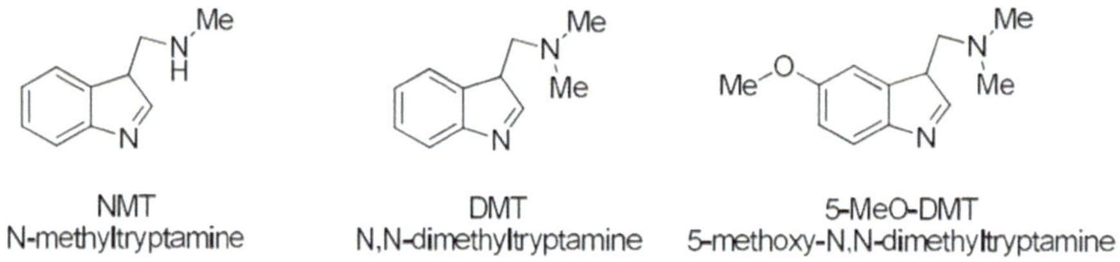

FIGURE 34.3 Chemical structures of three main compounds isolated from *Virola caryophylla*.

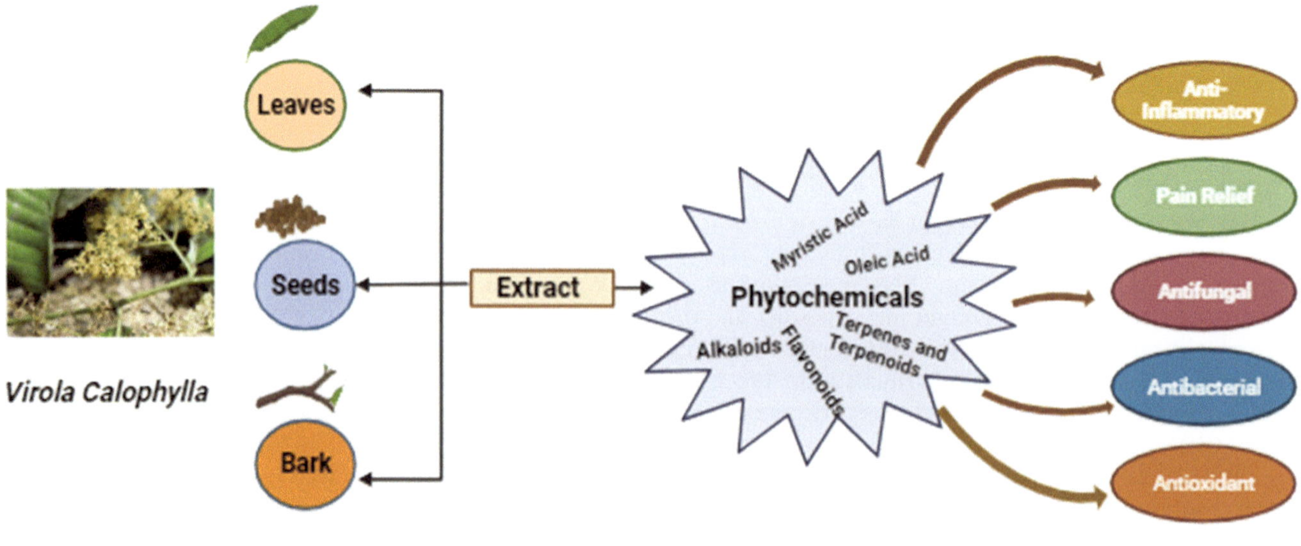

FIGURE 34.4 Phytochemicals present in different parts of *Virola calophylla* and their pharmacological activities.

NMT). The bark has 5-methoxy N-N-dimethyl tryptamine (5-MeO-DMT.

34.10.5 Flavonoids and Phenolic Compounds

Secondary metabolites are discovered from plants including flavonoids and phenolic compounds. Their antioxidant qualities and potential health advantages are well established. No detailed investigation has been made of *Virola calophylla* and it needs to be fully described (Martínez and Cuca Suarez, 2021; Martínez et al., 2021).

34.11 PHARMACOLOGICAL PROPERTIES

Myristic acid, oleic acid, and palmitic acid are only a few of the many fatty acids found in the oil derived from *Virola calophylla* seeds. When applied topically to the skin, these fatty acids have a variety of health advantages, including anti-inflammatory and moisturizing effects (Figure 34.3) (Schultes, 1969a).

34.11.1 Anti-Inflammatory Properties

Virola calophylla has been traditionally used for its anti-inflammatory properties, and it can be used topically to reduce skin inflammation (Vásquez-Ocmín et al., 2018).

34.11.2 Analgesic Activity

Virola calophylla has been utilized for pain alleviation by indigenous people, especially as topical creams or ointments. Traditional Amazonian medicine has utilized the plant for a number of functions, including as an analgesic, anti-inflammatory, and antipyretic agent (Roumy et al., 2007).

34.12 ANTIMICROBIAL ACTIVITY

Virola calophylla extracts possess antibacterial activity, according to some investigations. This could make it useful for treating infections in traditional medicine (Rodrigues et al., 2014).

34.13 MEDICINAL USES AND TOXICITY

Virola callophylla has significant ethnobotanical importance, particularly among indigenous communities in the Amazon rainforest region of South America. Various parts of Virola callophylla, such as the bark, leaves, and resin, have been used by Indigenous peoples for their purported medicinal properties (Schultes and Holmstedt, 1968). Bark, root, leaves, and fruit are psychotomimetic in nature and used in Colombia for skin infections and also inhaled as snuff (Davis and Yost, 1983; Holmstedt et al., 1980). Infusions, decoctions, or extracts from these plant parts have been employed to treat a range of ailments, including digestive issues, skin conditions, and respiratory problems (McKenna et al., 1984a). The seeds of *Virola callophylla* have been used to prepare a traditional psychoactive snuff called 'yopo' or 'parica' by indigenous groups like the Yekwana and Yanomami. The seeds are ground into a powder and then mixed with other plant substances to create the snuff. Yopo is used in shamanic and spiritual rituals to induce altered states of consciousness and for divinatory purposes (Torres, 1995; McKenna et al., 1984a; Martinez et al., 1995; Schultes, 1969a, 1969b, 1976; Davis and Yost, 1983).

Mouth sores and thrush are treated using an exudate from the stems (DeFilipps et al., 2004). The reddish bark is used for preparing a hallucinogenic snuff. The resin is boiled, dried, powdered, and mixed with the powdered leaves of a *Justicia* species, and the bark-ash of *Theobroma subincanum*, or *Elizabetha princeps*, acts nausea, hallucinations, and finally a deep sleep (DeFilipps, Maina, and Crepin, 2004).

34.14 CONCLUSION

Virola calophylla is a native to South America is a valuable tree both its historical and contemporary uses. The plant is used in traditional medicines and possesses analgesic, anti-inflammatory, antioxidant, and antimicrobial activities. Its bark is used for making hallucinogenic snuff by the Native Americans used in religious and spiritual ceremonies. Infusions, decoctions, and extracts made from various plant components have been used to treat a wide range of illnesses, including digestive issues, skin eruption, and respiratory issues. Some of the compounds have been isolated from the plant; however, these studies are preliminary in nature and still to determine the plant's therapeutic potential. Therefore, there is a need to undertake a detailed investigation of this species to explore its potential and utilization.

LIST OF ABBREVIATIONS

5-MeO-DMT	5-methoxy-N, N-dimethyltryptamine
DMT	N, N-dimethyltryptamine
MMT or NMT	N′-methyltryptamine
MTHC	2-methyl-1,2,3,4-tetrahydro-β-carboline
T	tryptamine

ACKNOWLEDGMENT

The chemical structures are drawn by Dr. Bushra Jabeen.

REFERENCES

Agurell, Stig. 1969. American plants of ethnopharmacologic interest. *Acta Chem. Scand* 23:3–13.

Barman, Rubi, Pranjit Kumar Bora, Jadumoni Saikia, et al. 2021. Nutmegs and wild nutmegs: An update on ethnomedicines, phytochemicals, pharmacology, and toxicity of the Myristicaceae species. *Phytotherapy Research* 35 (9):4632–4659.

Belhaj, S., Chaachouay, N., & Zidane, L. (2021). Ethnobotanical and toxicology study of medicinal plants used for the treatment of diabetes in the High Atlas Central of Morocco. *Journal of Pharmacy & Pharmacognosy Research*, *9*(5), 619–662. https://doi.org/10.56499/jppres21.1014_9.5.619

Bourscheid, Thais Guimarães, Leonardo Corrêa Cardoso, Marcelo Henrique Nascimento Santana, et al. 2022. Use of snuff and its main constituents for religious purposes in an alternative community with shamanic practices in the south of Brazil. *Journal of Psychedelic Studies* 6 (2):154–165.

Chaachouay, N., & Zidane, L. (2021). The Symbolic Efficacy of Plants in Rituals and Socio-Religious Ceremonies in Morocco, Northwest of Africa. *Journal of Religious & Theological Information*, *21*(1–2), 34–53. https://doi.org/10.1080/10477845.2021.1942433

Chaachouay, N., Douira, A., Hassikou, R., Brhadda, N., Dahmani, J., Belahbib, N., Ziri, R., & Zidane, L. (2020). Ethnobotanical and ethnopharmacological study of medicinal and aromatic plants used in the treatment of respiratory system disorders in the Moroccan Rif. PhDThesis, Département de Biologie - Université Ibn Tofail - Kénitra, Maroc, https://tel.archives-ouvertes.fr/tel-03376377

Chaachouay, N., Benkhnigue, O., & Zidane, L. (2020). Ethnobotanical study aimed at investigating the use of medicinal plants to treat nervous system diseases in the Rif of Morocco. *Journal of Chiropractic Medicine*, *19*(1), 70–81. https://doi.org/10.1016/j.jcm.2020.02.004

Chaachouay, N., Douira, A., & Zidane, L. (2021a). Herbal Medicine Used in the Treatment of Human Diseases in the Rif, Northern Morocco. *Arabian Journal for Science and Engineering*, 47, 131–153. https://doi.org/10.1007/s13369-021-05501-1

Chaachouay, N., Douira, A., & Zidane, L. (2021b). COVID-19, prevention and treatment with herbal medicine in the herbal markets of Salé Prefecture, North-Western Morocco. *European Journal of Integrative Medicine*, 42, 101285. https://doi.org/10.1016/j.eujim.2021.101285

Chudnoff, Martin. 1984. *Tropical timbers of the world*. US Department of Agriculture, Forest Service.

Davis, E Wade, and James A Yost. 1983. The ethnomedicine of the Waorani of Amazonian Ecuador. *Journal of Ethnopharmacology* 9 (2–3):273–297.

De Oliveira Brito, Wellison Rafael, Cristian De Sales Dambros, Domingos Cardoso, Veridiana Vizoni Scudeller, and Charles E Zartman. 2023. Divergent patterns of intraspecific trait variation among floral and vegetative characters in the hyperdominant dioecious Neotropical tree Virola sebifera (Myristicaceae). *Botanical Journal of the Linnean Society* 202 (2):233–248.

de Wilde, Willem JJO. 2011. T equally HE 20 genera Family divided and Myristicaceae over ca 500 the species tropics, is worldwide. a well-defined with middle-sized family, with the number of genera is about six genera (ca 350 species) in Southeast Asia, including New Guinea. The largest genus is Myristica, which is the most speciose in New Guinea and contains many endemics. The next most speciose genus is Horsfieldia. The genus Paramyristica is monotypic to New Guinea. *Ecology of Indonesian Papua Part One*.

DeFilipps, Robert A, Shirley L Maina, and Juliette Crepin. 2004. Medicinal Plants of the Guianas (Guyana, Surinam, French Guiana). *Medicinal Plants of the Guianas (Guyana, Surinam, French Guiana). Smithsonian Libraries and Archives*.

DeWalt, Saara J, Satya K Maliakal, and Julie S Denslow. 2003. Changes in vegetation structure and composition along a tropical forest chronosequence: implications for wildlife. *Forest Ecology and Management* 182 (1–3):139–151.

Duke, James A. 2008. *Duke's handbook of medicinal plants of Latin America*. CRC Press.

Fern, K. 2018. Tropical plants database, Ken Fern. tropical. the-ferns. info.

Franco-Moraes, Juliano, Armindo FMB Baniwa, Flávia RC Costa, Helena P Lima, Charles R Clement, and Glenn H Shepard Jr. 2019. Historical landscape domestication in ancestral forests with nutrient-poor soils in northwestern Amazonia. *Forest Ecology and Management* 446:317–330.

Franklin, Tiana Francesca. 2007. *Diversity and distribution of the Myristicaceae family of plants in terra firme and seasonally flooded forests in Madre de Dios, Peru*. Texas Christian University.

Holmstedt, Bo, JE Lindgren, T Plowman, L Rivier, RE Schultes, and O Tovar. 1980. Indole alkaloids in Amazonian Myristicaceae: Field and laboratory research. *Botanical Museum Leaflets, Harvard University* 28 (3):215–234.

Howe, Henry F. 2016. Making dispersal syndromes and networks useful in tropical conservation and restoration. *Global Ecology and Conservation* 6:152–178.

Janovec, J.P., and R. García. 2004. Tropical Forests | Myristicaceae. In *Encyclopedia of Forest Sciences*, edited by J. Burley. Elsevier.

Kühn, U, and K Kubitzki. 1993. Myristicaceae. In *Flowering Plants Dicotyledons: Magnoliid, Hamamelid and Caryophyllid Families*. Vol 2, Springer Science & Business Media.

Lancelotta, Rafael, and Alan K Davis. 2021. Therapeutic Potential of Fast-Acting Synthetic Tryptamines. *Handbook of Medical Hallucinogens*, New York: The Guilford Press. 215. 215–232.

Longwood, Franklin R. 1962. The Caribbean.

Maia, JG, and William A Rodrigues. 1974. Virola theiodora como alucinógena e tóxica. *Acta Amazonica* 4:21–23.

Martínez, Cabrera Xiomara Alejandra, and Luis Enrique Cuca Suarez. 2021. Metabólitos secundários isolados de espécies da família Myristicaceae que produzem inibição enzimática. *Revista Colombiana de Ciencias Químico-Farmacéuticas* 50 (2):352–385.

Martinez, V, C Juan, and E Luis, S Cuca. 1995. *Chemical studies of Myristicaceae species of the Colombian Amazon*. ACS Publications.

McKenna, Dennis J, GH Neil Towers, and F Abbott. 1984a. Monoamine oxidase inhibitors in South American hallucinogenic plants: Tryptamine and β-carboline constituents of ayahuasca. *Journal of Ethnopharmacology* 10 (2):195–223.

McKenna, Dennis J., G.H. Neil Towers, and FS Abbott. 1984b. Monoamine oxidase inhibitors in South American hallucinogenic plants Part 2: Constituents of orally-active Myristicaceous hallucinogens. *Journal of Ethnopharmacology* 12 (2):179–211.

McKenna, Dennis, and Jordi Riba. 2018. New world tryptamine hallucinogens and the neuroscience of ayahuasca. In: Halberstadt, A.L., Vollenweider, F.X., Nichols, D.E. (eds) *Behavioral Neurobiology of Psychedelic Drugs. Current Topics in Behavioral Neurosciences*, Vol 36. Berlin, Heidelberg: Springer. 283–311.

Mesquita, Rochelly da S, André C de Oliveira, Ingrity SC Sá, et al. 2020. Essential oils from leaves of *Virola calophylla*, *Virola multinervia*, and *Virola pavonis* (Myristicaceae): Chemical composition and larvicidal activity against *Aedes aegypti*. *Journal of Essential Oil Bearing Plants* 23 (3):453–463.

Miles, D. Howard, Ana Maria Ly, Shirley A. Randle, Paul A. Hedin, and Marcus L. Burks. 1987. Alkaloidal insect antifeedants from Virola calophylla Warb. *Journal of Agricultural and Food Chemistry* 35 (5):794–797.

Orch, H., Chaachouay, N., Douiri, E. M., Faiz, N., Zidane, L., & Douira, A. (2021). Use of medicinal plants in dermato-cosmetology: An ethnobotanical study among the population of Izarène. *Jordan Journal of Pharmaceutical Sciences 14*(3), 323–340.

Pardauil, Juliana JR, Fábio Alberto de Molfetta, Miguel Braga, et al. 2017. Characterization, thermal properties and phase transitions of amazonian vegetable oils. *Journal of Thermal Analysis and Calorimetry* 127:1221–1229.

Parolin, Pia, Florian Wittmann, and Leandro V Ferreira. 2013. Fruit and seed dispersal in Amazonian floodplain trees–A review. *Ecotropica* 19 (1/2):15–32.

Perry, E. K. 2002. Plants of the Gods. *Neurochemistry of consciousness* (Vol. 36, pp. 205–225). Amsterdam: John Benjamins Publishing Company.

Prance, Ghillean T. 1970. Notes on the use of plant hallucinogens in Amazonian Brazil. *Economic Botany* 24 (1):62–68.

Rodrigues, Katiane, Daniela F Ramos, Lillian L Carrion, et al. 2014. Antifungal activity of brazilian amazon plants extracts against some species of Candida spp. *International Journal of Phytopharmacology* 5 (6):445–453.

Roumy, Vincent, Gaston Garcia-Pizango, A-L Gutierrez-Choquevilca, et al. 2007. Amazonian plants from Peru used by Quechua and Mestizo to treat malaria with evaluation of their activity. *Journal of Ethnopharmacology* 112 (3):482–489.

Santamaría-Aguilar, Daniel, Reinaldo Aguilar, and Laura P Lagomarsino. 2019. A taxonomic synopsis of *Virola* (Myristicaceae) in Mesoamerica, including six new species. *PhytoKeys* 134:1.

Santamaría-Aguilar, Daniel, and Laura P Lagomarsino. 2022. New species of Virola (Myristicaceae) from South America. *PhytoKeys* 197:81.

Saraiva, Sergio A, Elaine C Cabral, Marcos N Eberlin, and Rodrigo R Catharino. 2009. Amazonian vegetable oils and fats: fast typification and quality control via triacylglycerol (TAG) profiles from dry matrix-assisted laser desorption/ionization time-of-flight (MALDI– TOF) mass spectrometry fingerprinting. *Journal of Agricultural and Food Chemistry* 57 (10):4030–4034.

Schultes, Richard Evans. 1954. A new narcotic snuff from the northwest Amazon. *Botanical Museum Leaflets, Harvard University* 16 (9):241–260.

Schultes, Richard Evans. 1969a. De Plantis Toxicariis E Mundo Novo Tropicale Commentationes V: Virola as an orally administered hallucinogen. *Botanical Museum Leaflets, Harvard University* 22 (6):229–240.

Schultes, Richard Evans. 1969b. Hallucinogens of plant origin: Interdisciplinary studies of plants sacred in primitive cultures yield results of academic and practical interest. *Science* 163 (3864):245–254.

Schultes, Richard Evans. 1976. Indole alkaloids in plant hallucinogens. *Journal of Psychedelic Drugs* 8 (1):7–25.

Schultes, Richard Evans. 2007. Amazonian ethnobotany and the search for new drugs. Paper read at Ciba Foundation Symposium 185-Ethnobotany and the Search for New Drugs: Ethnobotany and the Search for New Drugs: Ciba Foundation Symposium 185.

Schultes, Richard Evans, and Bo Holmstedt. 1968. De plantis toxicariis e mundo novo tropicale commentationes II: The vegetal ingredients of the myristicaceous snuffs of the northwest Amazon. *Rhodora* 70 (781):113–160.

Schultes, Richard Evans, and Elmer W Smith. 1976. *Hallucinogenic plants*. Vol. 35. Golden Press.

Steeves, R. A. D. 2011. *An intrageneric and intraspecific study of morphological and genetic variation in the Neotropical Compsoneura and Virola (Myristicaceae).* University of Guelph.

Torres, Constantino Manuel. 1995. Archaeological evidence for the antiquity of psychoactive plant use in the Central Andes. *Annuli dei Musei Civici Roverero* 11:291–326.

Uscategui, M. N. 1959. The present distribution of narcotics and stimulants amongst the Indian tribes of Colombia. *Botanical Museum Leaflets, Harvard University* 18 (6):273–304.

Vásquez-Ocmín, Pedro, Sandrine Cojean, Elsa Rengifo, et al. 2018. Antiprotozoal activity of medicinal plants used by Iquitos-Nauta road communities in Loreto (Peru). *Journal of Ethnopharmacology* 210:372–385.

Williams, Llewelyn. 1936. Woods of northeastern Peru. *Botanical series, v. 15.*

35 Red Ucuuba (*Virola sebifera* Aubl. Myristicaceae)

Noureddine Chaachouay, Allal Douira, and Lahcen Zidane

35.1 INTRODUCTION

Humans utilize plants for practical purposes, such as sustenance, attire, medicinal applications, and symbolic meanings, such as artistic expression, mythology, and literary works (Ratsch, 2005; Chaachouay et al., 2019, 2019). Agriculture, which ensures a consistent food supply, is the foundation of civilization. Ethnobotany is the discipline that examines the use of plants by Indigenous communities, while economic botany is concerned with the analysis of contemporary cultivated plants (Gerique, 2006; Chaachouay et al., 2020, 2020; Orch et al., 2021). Plants have been used in medicine since ancient times, serving as a source of several pharmaceuticals (Rout et al. 2009). Additionally, they serve as the raw material for various industrial goods such as paper, lumber, and different chemicals (Demirbaş, 2001).

One of the Amazonian people's most often employed species is *Virola sebifera* Aubl., which belongs to the Myristicaceae family and is a neotropical tree (Ribeiro et al., 2021). The tree is extracted from its natural habitat to obtain its lumber, oil, and medicinal properties. The Indigenous population in the Amazon region cultivates it for its oil and therapeutic properties (González-Rodríguez et al., 2021). This tree is very decorative and has a graceful form. It has been used in reforestation initiatives and is highly recommended for landscaping, particularly street planting (Steeves, 2011).

V. sebifera is a mostly deciduous tree characterized by a wide, pyramid-shaped canopy and a maximum height of 40 meters. The tall and erect main stem, often reinforced by buttresses, may be free of branches for up to 23 meters and have a diameter between 30 and 40 centimeters (De Oliveira Brito et al., 2023). *V. sebifera* mainly inhabits the Amazon Rainforest in South America. The species is indigenous to many countries, including Brazil, Peru, Colombia, Venezuela, and Ecuador, all within the Amazon Basin. *V. sebifera* mainly occupies the lush and varied ecosystem of the Amazon Rainforest in South America (Howe, 1981). The species thrives in the lowland tropical rainforests, known for their high annual rainfall, increased temperatures, and significant humidity. *V. sebifera* has a notable degree of adaptability to the humid conditions in the rainforest and can endure intermittent flooding throughout the rainy season (Tonelli et al., 2014).

V. sebifera has significant ethnobotanical importance since it has been used by Indigenous communities and local populations in the Amazon Rainforest for many reasons over an extended period. The sap is used for the management of ulcers and dermatological conditions (Santamaría-Aguilar et al., 2019). The bark has astringent, emetic, and stimulant characteristics. In Venezuela, ritual dances were organized to treat fevers among Yanomam shamans and other groups living close to the source of the Orinoco River. During these rites, individuals would either inhale or ingest the hallucinogenic inner bark of *V. sebifera* (Corothie and Nakano, 1969). Furthermore, they would concoct a decoction to ward off destructive creatures. The seed yields an oil that was once used as an illuminant. This oil finds application in the manufacturing of candles and soap. The seeds readily ignite, producing a dazzling blaze (Da Rocha Filho et al., 1992). Indigenous people attach them to a thin, sturdy piece of wood as lighting tools for nighttime outings (Enders et al., 2005).

V. sebifera has a variety of chemical constituents, primarily found in its seeds, that are used to extract ucuuba butter. The fully developed seeds contain glycerides of fatty acids, namely laurodimyristin and trimyristin. Indigenous communities have traditionally used the plant for many purposes, including medicinal and ceremonial practices (González-Rodríguez et al., 2021). While the scientific research on the pharmacology of *V. sebifera* is limited, several potential pharmacological properties and uses of this medicinal plant have been examined (Ribeiro et al., 2021). This chapter aims to thoroughly explore the *V. sebifera* plant, including its botanical features, habitat and ecology, chemical constituents, and pharmacological properties.

35.2 SCIENTIFIC CLASSIFICATION

Virola sebifera Aubl., also known as Red ucuuba and Ucuúba-do-Cerrado, is a tree belonging to the Myristicaceae family (The Plant List, 2013). It is native to Central and South America.

Domain: Biota

Reign: Plantae

Sub-kingdom: Viridaeplantae

Infra-Kingdom: Streptophyta

DOI: 10.1201/9781003460336-35

Class: Equisetopsida
Clade: Tracheophyta
Clade: Spermatophyta
Subclass: Magnoliidae
Superorder: Magnolianae
Order: Magnoliales
Family: Myristicaceae
Genre: *Virola*
Species: *Virola sebifera* Aubl.

35.3 VERNACULAR NAMES

Arabic: شجرة الأوكوبا الحمراء (Shajarat al-ukuba al-hamra); **Chinese**: 红色乌库巴树 (Hóngsè wūkùbā shù); **Dutch**: Rode ucuuba; **English**: Red ucuuba; **French**: Ucuuba rouge; **German**: Rotes Ucuuba; **Hindi**: लाल उकुबा पेड़ (Lāla ukubā ped); **Italian**: Ucuuba rossa; **Japanese**: 赤いウクウバの木 (Akai ukūba no ki); **Portuguese**: Ucuuba-vermelha; **Russian**: Красное укууба дерево (Krasnoye ukūba derevo); **Spanish**: Ucuuba roja.

Amazonian local names

Moulomba (Guyana); Sangre; Cumala Blanca (Peru); Camaticaro (Venezuela); Palo De Sangre; Colorado; Ucuuba (Brazil); Sebo (Colombia); Nuanamo (Colombia); Birma; Shempo (Ecuador); Cumala Colorada (Peru); Cumala (Peru); Pintri (Surinam); Cajuea (Trinidad & Tobago); Otivo (Venezuela); Virola (Venezuela); Cuajo (Venezuela); Sangrino (Venezuela); Chaliviande (Ecuador); Dalli (Guyana); Banak; Bicuiba Branca (Brazil); Tapsava; Virola (Brazil); Ucuhuba (Brazil); Yayamadou (Guyana); Baboen (Surinam); Chalviande (Ecuador); Tzimbo (Ecuador); Gabón (Bolivia); Sangrino (Venezuela); Cuajo (Venezuela); Virola (Venezuela); Cumala (Peru); Banak; Veuhuba; Venhuba; Babun; Sangre De Toro (Colombia); Bagabani; Sangre; Camaticaro; Baboen (Surinam); Dalli (Guyana)

35.4 BOTANICAL DESCRIPTION

V. sebifera is a semi-deciduous tree with a broad, pyramid-shaped canopy and a maximum height of 40 meters. The vertical, upright trunk, often supported by buttresses, may remain devoid of branches for a distance of up to 23 meters and have a diameter ranging from 30 to 40 centimeters. *V. sebifera* is distinguished by its abundant presence of pediculate and ferruginous trichomes, which densely cover almost all aspects of the plant (dos Santos, 2014; Teixeira and Rodrigues, 2006). The leaf blades have a limited number of lateral veins (9–17 per side), with well-defined spacing between them (3–4 (–5) veins per 5 cm, 1.3–2.7 cm apart) (Figure 35.1). The flowers have a perianth that may vary in color from brown, cream, orange, yellow-brown, to yellowish. Some of these flowers also emit a pleasant fragrance with a lemon or sweet scent. The staminate inflorescences are characterized by their extensive branching (Rambo et al., 2021; Teles and de Paula, 1980). The staminate flowers have a short filament column measuring 0.1–0.4 mm, while the anthers are more than twice as long as the column, measuring 0.6–0.8 mm and ending in an apiculate apex. The small fruits are round to slightly round and are covered by a dense layer of easily detachable trichomes (De Oliveira Brito et al., 2023). The pericarp of the fruits is thin, measuring 0.3–0.6 (–0.9) mm (Figure 35.2). Additionally, the aril surrounding the seed is light and has narrow, laciniate bands. Infructescence often includes a substantial number of fruits (Holmstedt et al., 1980).

35.5 ORIGIN AND GEOGRAPHIC DISTRIBUTION

V. sebifera is a tropical tree species that is native to the Amazon Rainforest and other regions in South America, including Colombia, Costa Rica, Ecuador, French Guiana, Guyana, Honduras, Bolivia, Brazil North, Brazil Northeast, Brazil Southeast, Brazil West-Central, Nicaragua, Panamá, Peru, Suriname, Venezuela (Oliveira-Filho and Ratter, 1995). *V. sebifera* is mainly found in the Amazon Rainforest in South America (Figure 35.3). The species is naturally found in many nations, including Brazil, Peru, Colombia, Venezuela, and Ecuador, all located within the Amazon Basin. It is notably prevalent in Brazil in states such as Amazonas, Acre, Pará, and Roraima (González-Rodríguez et al., 2021;

FIGURE 35.1 A branch of *V. sebifera* composed of flowers and leaves.

FIGURE 35.2 The fruits of *V. sebifera*.

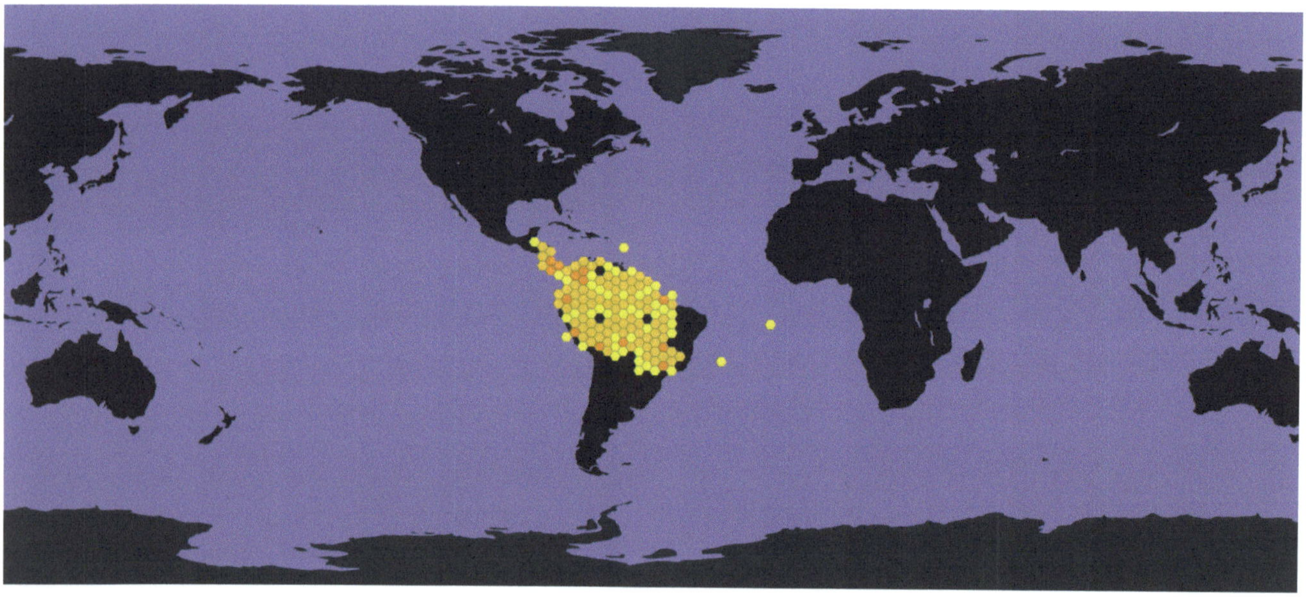

FIGURE 35.3 Geographical distribution areas of *V. sebifera*.

Santamaría-Aguilar et al., 2019). The tropical tree flourishes in the humid and varied environments of the Amazon, serving a crucial function in the local ecology. *V. sebifera* is renowned for its fruit, which is used by Indigenous tribes and local people for several reasons, such as the production of ucuuba butter (Lenza and Oliveira, 2006; Svenning et al., 2006).

35.6 HABITAT AND ECOLOGY

V. sebifera primarily inhabits the rich and diversified environment of the Amazon Rainforest in South America. It flourishes in the lowland tropical rainforests, with abundant yearly precipitation, elevated temperatures, and substantial humidity levels. *V. sebifera* has a high level of adaptation to the moist environment of the rainforest and can withstand periodic inundation during the rainy season (Steeves, 2011). They often thrive in the forest's lower and upper levels, contributing significantly to the intricate vertical composition of the Amazon environment (Wei et al., 2013). The abundant and diverse range of plant and animal species in the Amazon Rainforest plays a crucial role in sustaining *V. sebifera*, as their fruits and seeds serve as a vital food source for a wide array of animals, including bats, birds, and rodents (Howe, 1981). *V. sebifera* is essential to the complex ecosystem of the Amazon Rainforest (González-Rodríguez et al., 2021). The tree yields fruits encased in durable shells that protect its seeds. Animals ingest these fruits, facilitating seed dissemination and enhancing the tree's reproductive and genetic variety. *V. sebifera* trees also play a role in ecosystem services, including capturing carbon and cycling nutrients, which are essential for the overall well-being and operation of the rainforest environment (Howe and Smallwood, 1982; Svenning et al., 2006). In addition, *V. sebifera* tree has cultural and economic importance for Indigenous tribes and local inhabitants. They use its fruit for traditional medicinal purposes and extract ucuuba butter, which is employed in cosmetics and serves as a source of revenue for some communities.

35.7 ETHNOBOTANY

V. sebifera has considerable ethnobotanical significance due to its long-standing use by Indigenous people and local populations in the Amazon Rainforest for several purposes. The sap is used for the treatment of ulcers and skin problems (González-Rodríguez et al., 2021). The bark has astringent, emetic, and stimulating properties. An infusion manages dyspepsia, intestinal colic, erysipelas, and inflammations. The bark is simmered with water and used to provoke emesis. The pulverized bark is immersed in water and ingested as an antidiarrheal. The bark juice is topically administered as an analgesic for cracked heels. A decoction of the bark is used as a topical solution for cleansing bruises and sores (Corothie and Nakano, 1969). One might gargle or consume a decoction from the bitter crimson sap extracted from the bark to cure thrush and oral sores. Upon contact with air, the sap transforms, becoming resinous (Howe, 1981). This resin is then used to alleviate toothache and cauterize aphthae. *V. sebifera*, a homeopathic remedy, is made from the fresh, crimson sap obtained from the damaged bark of the tree. It is mainly used for treating conditions such as abscesses, phlegmon, paronychia, furuncle, anal fissures, infections of the parotid gland, bacterially infected tonsillitis, and other similar disorders. The tree's bark contains many tannins and the hallucinogens dimethyltryptamine and 5-MeO-DMT (Corothie and Nakano, 1969; Chaachouay and Zidane, 2021). Indigenous communities use it for the treatment of skin ailments. The mature seeds contain glycerides of fatty acids, particularly laurodimyristin and trimyristin. The Indigenous populations of South America use a substance extracted from the bark, known as exudate, to create concoctions that possess hallucinogenic qualities (Kawanishi, Uhara, and Hashimoto, 1985; Wei et al., 2013). In Venezuela, particularly Yanomam shamans and other communities residing near the origins of the Orinoco River would arrange ceremonial dances to treat fevers. During these rituals, they would either smoke or consume the hallucinogenic inner bark of *V. sebifera*. In addition, they would prepare a decoction with the purpose of repelling malevolent entities (Rambo et al., 2021; Chaachouay et al., 2021, 2021b, 2021c, Belhaj et al., 2021, Benkhnigue et al., 2022).

The seed produces an oil once used as a light source. Oil is used in the production of candles and soap (Da Rocha Filho et al., 1992). The seeds combust easily, creating a brilliant flame. Indigenous people affix them onto a slender plank of rigid timber and use them as illuminating devices for nocturnal excursions. When trees are cut or harmed, they release a crimson, diluted latex resembling the commercially used kino. The heartwood has a consistent light reddish, grayish, or brown hue, eventually transitioning into pale golden to pale rose-brown sapwood. The wood has a straight grain pattern, a medium texture, and a medium to high shine. It needs a distinguishing aroma and flavor (Pimenta et al., 2020). The object has a modest weight, limited durability, and is very vulnerable to termite and pinhole borer infestation. This material exhibits effortless compatibility with standard tools and achieves a satisfactory finish (Ribeiro et al., 2021). It demonstrates a propensity for easy nail and screw insertion without the risk of breaking. Furthermore, it can be effectively stained, lacquered, and polished, yielding desirable outcomes. It also exhibits satisfactory turning capabilities and effortlessly adheres when glued. It is used in constructing and interior finishing buildings, toys, light boxes, etc., and is well-suited for manufacturing toothpicks (Enders et al., 2005; Chaachouay et al., 2022, 2022, 2022C).

35.8 PHYTOCHEMISTRY

V. sebifera. has a range of chemical components, mainly concentrated in its seeds, which are used to extract ucuuba butter. The tree's bark has a high concentration of tannins and the hallucinogens dimethyltryptamine (DMT) and 5-MeO-DMT (Danelutte et al., 2000; González-Rodríguez et al., 2021; Rambo et al., 2021). The mature seeds contain glycerides of fatty acids, particularly laurodimyristin and trimyristin. The bark has alkaloids ranging from 0.065% to 0.25%, mostly DMT and 5-MeO-DMT. The extract derived from the bark possesses the most elevated levels of alkaloids (Pimenta et al., 2020; Ribeiro et al., 2021; Tonelli et al., 2014). The fruit of *V. sebifera* contains butyrolactones ((2R, 3S)-3-(3,4-dimethoxybenzyl)-2-(3,4-methylenedioxybenzyl)-butyrolactone in its seeds, and (2R, 3R)-3 -(3,4-dimethoxybenzyl)-2-(3,4-methylenedioxybenzyl)-butyrolactone, (2R, 3R)-2,3-di-(3, 4-dimethoxybenzyl)-butyrolactone and (2R, 3R)-2,3-di-(3,4-methylenedioxybenzyl)-butyrolactone in its pericarp), as well as naphthalene tetralines and neolignans (2,3-Dimethyl-4-piperonyl-4-veratrylbutan-1-ol), compounds of ω-phenylundecanoyl (2,6-dihydroxybenzene, 2,6-dihydroxy-4-methoxybenzene and 3-hydroxycyclohexan-2,6-dione), tetralone neolignans (including 2,4-dihydroxy-6,7-methylenedioxy-2,3-dimethyl-4-veratryltetralin-1-one), podophyllotoxinanalogues36,ahydroxy-otobainlignan,dibenzylbutane lignans (rac-(8a,8′P)-4,4′-dihydroxy-3 diacetate, 3′-dimethoxylignan-9,9′-diyl and rac-(8a,8′P)-4-hydroxy-3-

methoxy-3′,4′-methylenedioxylignan-9,9′-diyl diacetate). Studies have extracted 3-indolcarboxaldehyde from *V. sebifera* wood and diarylpropanes: 1-(4'-hydroxy-2'-methoxyphenyl-3-(3'-hydroxy-4'-methoxyphenyl)-propane, Virolanol B and Virolanol C (Carlos et al., 2018; Da Rocha Filho et al., 1992; Denny et al., 2008; Enders et al., 2005; Harrowven, Newman, and Knight, 1998; Kato et al., 1985; Lopes et al., 1983, 1984; Rezende and Kato, 2002; Ribeiro et al., 2021).

35.9 PHARMACOLOGY

V. sebifera, is an indigenous tree found in the tropical rainforests of South America, namely in the Amazon Basin. Indigenous tribes have historically used the herb for several reasons, including medical and ceremonial applications. Although scientific study on the pharmacology of *V. sebifera* is limited, several possible pharmacological qualities and applications of this medicinal plant have been investigated. The samples exhibited different levels of antioxidant properties (González-Rodríguez et al., 2021). The ethanol extract showed the highest antioxidant activity, accompanied by a notable abundance of phenolic chemicals and flavonoids, renowned for their antioxidant qualities (Rambo et al., 2021). Ribeiro et al. revealed that the extracts exhibited diverse antioxidant properties (Ribeiro et al., 2021). The ethanol extract showed superior antioxidant activity, possibly due to its elevated phenolic compounds and total flavonoid levels (Tonelli et al., 2014). The DPPH radical scavenging technique was used to evaluate the antioxidant activity.

An investigation assessed the antibacterial efficacy of the extracts against microorganisms, including *Escherichia coli*, *Staphylococcus epidermidis*, *Staphylococcus aureus*, *Salmonella typhimurium*, and the fungus *Candida albicans*. The extracts exhibited favorable outcomes against *S. epidermidis* and *S. aureus*, while some also hindered the growth of *S. typhimurium* and *E. coli*. Ribeiro et al. (2021) tested several strains of bacteria (*S. aureus*, *S. epidermidis*, *S. typhimurium*, *E. coli*) and the fungus *Candida albicans* using the extracts. The study showed favorable outcomes against *S. aureus* and *S. epidermidis*, and some extracts also showed inhibition of *S. Typhimurium* and *E. coli*. The study reviews the pharmacological characteristics of *V. sebifera*, highlighting its noteworthy antioxidant, antiproliferative, and antibacterial activities (Denny et al., 2008). These features make it a relevant species in ethnopharmacology and a potential subject for future pharmaceutical studies.

35.10 CONCLUSION

V. sebifera embodies a seamless fusion of customary practices, ecological soundness, and contemporary uses. *V. sebifera* has been highly esteemed by Indigenous tribes due to its medicinal and cultural importance, establishing a profound bond between this plant and their way of life. The seeds contain high levels of beneficial fatty acids, which provide excellent nutritional and aesthetic benefits that align with modern civilization's demands. Nevertheless, it is crucial to use *V. sebifera* responsibly to guarantee its sustained accessibility in the future. Using sustainable harvesting procedures and implementing conservation initiatives to save the ecosystem is vital. By achieving a harmonious equilibrium between conventional knowledge and contemporary innovation, we may effectively use the inherent capabilities of *V. sebifera* without endangering its survival or the fragile environments in which it flourishes. This method positively affects Indigenous communities and significantly contributes to the broader purposes of biodiversity conservation and global sustainability. *V. sebifera* exemplifies using Indigenous knowledge and natural resources for the collective advantage in a society where the convergence of culture, tradition, and science has great significance.

LIST OF ABBREVIATIONS

5-MeO-DMT	5-Methoxy-N, N-dimethyltryptamine
DMT	Dimethyltryptamine
V. sebifera	*Virola sebifera* Aubl.

REFERENCES

Belhaj, S., Chaachouay, N., & Zidane, L. (2021). Ethnobotanical and toxicology study of medicinal plants used for the treatment of diabetes in the High Atlas Central of Morocco. *Journal of Pharmacy & Pharmacognosy Research*, *9*(5), 619–662. https://doi.org/10.56499/jppres21.1014_9.5.619

Benkhnigue, O., Chaachouay, N., Khamar, H., El Azzouzi, F., Douira, A, & Zidane, L. (2022). Ethnobotanical and ethnopharmacological study of medicinal plants used in the treatment of anemia in the region of Haouz-Rehamna (Morocco). *Journal of Pharmacy & Pharmacognosy Research*, *10*(2), 279–302. https://doi.org/10.56499/jppres21.1196_10.2.279

Chaachouay, N., Benkhnigue, O., Fadli, M., Ibaoui, H., & Zidane, L. (2019). Ethnobotanical and ethnopharmacological studies of medicinal and aromatic plants used in the treatment of metabolic diseases in the Moroccan Rif. *Heliyon*, *5*(10), e02191. https://doi.org/10.1080/10477845.2021.1942433

Chaachouay, N., Benkhnigue, O., Fadli, M., Ibaoui, H., El Ayadi, R., & Zidane, L. (2019). Ethnobotanical and ethnopharmacological study of medicinal and aromatic plants used in the treatment of respiratory system disorders in the Moroccan Rif. *Ethnobotany Research and Applications*, *18*(22), 1–16.

Chaachouay, N., Douira, A., Hassikou, R., Brhadda, N., Dahmani, J., Belahbib, N., Ziri, R., & Zidane, L. (2020). Ethnobotanical and ethnopharmacological study of medicinal and aromatic plants used in the treatment of respiratory system disorders in the Moroccan Rif. PhDThesis, Départment de Biologie - Université Ibn Tofail - Kénitra, Maroc, https://tel.archives-ouvertes.fr/tel-03376377

Chaachouay, N., Benkhnigue, O., & Zidane, L. (2020). Ethnobotanical study aimed at investigating the use of medicinal plants to treat nervous system diseases in the Rif of Morocco. *Journal of Chiropractic Medicine*, *19*(1), 70–81. https://doi.org/10.1016/j.jcm.2020.02.004

Chaachouay, N., & Zidane, L. (2021). The Symbolic Efficacy of Plants in Rituals and Socio-Religious Ceremonies in

Morocco, Northwest of Africa. *Journal of Religious & Theological Information*, *21*(1–2), 34–53. https://doi.org/10.1080/10477845.2021.1942433

Chaachouay, N., Benkhnigue, O., Douira, A., & Zidane, L. (2021). Poisonous medicinal plants used in the popular pharmacopoeia of the Rif, northern Morocco. *Toxicon*, *189*, 24–32. https://doi.org/10.1016/j.toxicon.2020.10.028

Chaachouay, N., Douira, A., & Zidane, L. (2021b). COVID-19, prevention and treatment with herbal medicine in the herbal markets of Salé Prefecture, North-Western Morocco. *European Journal of Integrative Medicine*, *42*, 101285. https://doi.org/10.1016/j.eujim.2021.101285

Chaachouay, N., Douira, A., & Zidane, L. (2021c). Herbal Medicine Used in the Treatment of Human Diseases in the Rif, Northern Morocco. *Arabian Journal for Science and Engineering*, *47*, 131–153. https://doi.org/10.1007/s13369-021-05501-1

Chaachouay, N., Benkhnigue, O., & Zidane, L. (2022). Ethnobotanical and Ethnomedicinal study of medicinal and aromatic plants used against dermatological diseases by the people of Rif, Morocco. *Journal of Herbal Medicine*, *32*, 100542. https://doi.org/10.1016/j.hermed.2022.100542

Chaachouay, N., Azeroual, A., Bencharki, B., Douira, A., & Zidane, L. (2022). Ethnoveterinary medicines plants for animal therapy in the Rif, North of Morocco. *South African Journal of Botany*, *147*, 176–191. https://doi.org/10.1016/j.sajb.2021.12.037

Chaachouay, N., Azeroual, A., Douira, A., & Zidane, L. (2022). Ethnoveterinary practices of medicinal plants among the Zemmour and Zayane tribes, Middle Atlas, Morocco. *South African Journal of Botany*, *151*, 826–840. https://doi.org/10.1016/j.sajb.2022.11.009

Carlos, Amanda Aparecida, Karla da Silva Malaquias, Rafael Camargo Consolmagno, André Lucio Franceschini Sarria, João Batista Fernandes, et Odair Correa Bueno. 2018. Virola Sebifera Aubl. (Myristicaceae) Leaf Chemical Composition and Implications on Leaf-Cutter Ant Foraging Choice. *Arthropod-Plant Interactions* *12*(3):361–368. https://doi.org/10.1007/s11829-017-9582-9

Corothie, E., et T. Nakano. 1969. Constituents of the bark of *Virola sebifera. Planta Medica 17*(2):184–188. https://doi.org/10.1055/s-0028-1099844

Da Rocha Filho, G. N., M. H. S. Bentes, D. Brodzki, et G. Djéga-Mariadassou. 1992. Catalytic Conversion ofHevea Brasiliensis and *Virola sebifera* Oils to Hydrocarbon Fuels. *Journal of the American Oil Chemists Society 69*(3): 266–271. https://doi.org/10.1007/BF02635899

Danelutte, Ana Paula, Alberto José Cavalheiro, et Massuo Jorge Kato. 2000. Lignoids in Seedlings of Virola Sebifera. *Phytochemical Analysis 11*(6):383–386. https://doi.org/10.1002/1099-1565(200011/12)11:6<383::AID-PCA543>3.0.CO;2-M

De Oliveira Brito, Wellison Rafael, Cristian De Sales Dambros, Domingos Cardoso, Veridiana Vizoni Scudeller, et Charles E. Zartman. 2023. Divergent patterns of intraspecific trait variation among floral and vegetative characters in the hyperdominant dioecious Neotropical tree *Virola sebifera* (Myristicaceae). *Botanical Journal of the Linnean Society 202*(2): 233–248. https://doi.org/10.1093/botlinnean/boac069

Demirbaş, Ayhan. 2001. Biomass Resource Facilities and Biomass Conversion Processing for Fuels and Chemicals. *Energy Conversion and Management 42*(11):1357–1378. https://doi.org/10.1016/S0196-8904(00)00137-0

Denny, Carina, Marcela Engelbrecht Zacharias, Ana Lúcia Tasca Gois Ruiz, Maria Do Carmo E. Do Amaral, Volker Bittrich, Luciana Konecny Kohn, Ilza Maria de Oliveira Sousa, Rodney Alexandre Ferreira Rodrigues, João Ernesto de Carvalho, et Mary Ann Foglio. 2008. Antiproliferative Properties of Polyketides Isolated from Virola Sebifera Leaves. *Phytotherapy Research 22*(1):127–130. https://doi.org/10.1002/ptr.2251

Enders, Dieter, Mile Milovanovic, Elena Voloshina, Gerhard Raabe, et Jörg Fleischhauer. 2005. First Asymmetric Synthesis and Determination of the Absolute Configuration of a Lignan Isolated from Virola Sebifera. *European Journal of Organic Chemistry 2005*(10):1984–1990. https://doi.org/10.1002/ejoc.200400885

Gerique, Andrés. 2006. *An Introduction to ethnoecology and ethnobotany. Theory and Methods - Integrative assessment and planning methods for sustainable agroforestry in humid and semiarid regions.* Advanced Scientific Training (ed.), 20p. Loja, Ecuador.

González-Rodríguez, María, Clara Ruiz-Fernández, Vera Francisco, Djedjiga Ait Eldjoudi, Yousof Ramadan Farrag AbdElHafez, Alfonso Cordero-Barreal, Jesus Pino, Francisca Lago, Manuel Campos-Toimil, Glaucimeire Rocha Carvalho, Thiago Melo Costa Pereira, et Oreste Gualillo. 2021. Pharmacological Extracts and Molecules from Virola Species: Traditional Uses, Phytochemistry, and Biological Activity. *Molecules 26*(4):792. https://doi.org/10.3390/molecules26040792

Harrowven, David C., Nicola A. Newman, et Carly A. Knight. 1998. On the identity of a neo-lignan from the fruits of Virola sebifera. *Tetrahedron Letters 39*(37):6757–6760. https://doi.org/10.1016/S0040-4039(98)01419-1

Holmstedt, Bo, J. E. Lindgren, T. Plowman, L. Rivier, R. E. Schultes, et and O. Tovar. 1980. Indole Alkaloids in Amazonian Myristicaceae: Field and Laboratory Research. *Botanical Museum Leaflets, Harvard University 28*(3):215–234.

Howe, H. F., et J. Smallwood. 1982. Ecology of Seed Dispersal. *Annual Review of Ecology and Systematics 13*(1):201–228. https://doi.org/10.1146/annurev.es.13.110182.001221

Howe, Henry F. 1981. « Dispersal of a Neotropical Nutmeg (*Virola sebifera*) by Birds. *The Auk 98*(1):88–98. https://doi.org/10.1093/auk/98.1.88

Kato, Massuo J., Lucia M. X. Lopes, Hipólito F. Paulino Fo, Massayoshi Yoshida, et Otto R. Gottlieb. 1985. Acylresorcinols from *Virola sebifera* and *Virola elongata*. *Phytochemistry 24*(3):533–536. https://doi.org/10.1016/S0031-9422(00)80762-1

Kawanishi, K., Y. Uhara, et Y. Hashimoto. 1985. « Alkaloids from the Hallucinogenic Plant *Virola sebifera*. *Phytochemistry 24*(6):1373–1375. https://doi.org/10.1016/S0031-9422(00)81137-1

Lenza, Eddie, et Paulo Eugênio Oliveira. 2006. Biologia reprodutiva e fenologia de *Virola sebifera* Aubl. (Myristicaceae) em mata mesofítica de Uberlândia, MG, Brasil. *Brazilian Journal of Botany 29*:443–541. https://doi.org/10.1590/S0100-84042006000300011

Lopes, Lucia M. X., Massayoshi Yoshida, et Otto R. Gottlieb. 1983. Dibenzylbutyrolactone lignans from Virola sebifera. *Phytochemistry 22*(6):1516–1518. https://doi.org/10.1016/S0031-9422(00)84055-8

Lopes, Lucia M. X., Massayoshi Yoshida, et Otto R. Gottlieb. 1984. Further lignoids from Virola sebifera. *Phytochemistry 23*(11):2647–2652. https://doi.org/10.1016/S0031-9422(00)84118-7

Oliveira-Filho, A. T., et J. A. Ratter. 1995. A Study of the Origin of Central Brazilian Forests by the Analysis of Plant Species Distribution Patterns. *Edinburgh Journal of Botany* *52*(2):141–194. https://doi.org/10.1017/S0960428600000949

Orch, H., Chaachouay, N., Douiri, E. M., Faiz, N., Zidane, L., & Douira, A. (2021). Use of medicinal plants in dermatocosmetology: An ethnobotanical study among the population of Izarène. *Jordan Journal of Pharmaceutical Sciences*, *14*(3), 323–340.

Pimenta, Raphael, Rayele Silava, Rachel Fernandes, Mirelle Araujo, Ilsamar Soares, Juliana Silva, Guilherme Nascimento, et Elisandra Scapin. 2020. Chemical Assessment and Antioxidant, Antimicrobial, and Cytotoxicity Activity of Leafs Extracts of *Virola sebifera*, an Amazonian Medicinal Plant. *Frontiers in Pharmacology*. *10*(1):e578101523068. https://doi.org/10.33448/rsd-v10i15.23068

Rambo, Magale Karine Diel, Elisandra Scapin, Claudiane Lima Ribeiro, Maria Angélica Melo Rodrigues, et Ana Clara Alcantara Rodrigues. 2021. « Chemical Characterization and Cytotoxicity of Extracts of Leaves of *Virola sebifera* of the cerrado tocantinense ». *Revista Ibero-Americana de Ciências Ambientais* *12*(8):116–126.

Ratsch, Christian. 2005. *The encyclopedia of psychoactive plants: ethnopharmacology and its applications*. Simon and Schuster.

Rezende, Kênnia R., et Massuo J. Kato. 2002. Dibenzylbutane and aryltetralone lignans from seeds of Virola sebifera. *Phytochemistry* *61*(4):427–432. https://doi.org/10.1016/S0031-9422(02)00242-X

Ribeiro, Claudiane Lima, Rayele Moraes Silva, Rachel de Moura Nunes Fernandes, Mirelle Ribeiro Araújo, Ilsamar Mendes Soares, Juliana Fonseca Moreira da Silva, Guilherme Nobre Lima do Nascimento, Raphael Sanzio Pimenta, et Elisandra Scapin. 2021. Chemical Assessment, Antioxidant and Antimicrobial of Leafs Extracts of Virola Sebifera, an Amazonian Medicinal Plant. *Research, Society and Development* *10*(15):e578101523068. https://doi.org/10.33448/rsd-v10i15.23068

Rout, Soumya, K. Choudary, D. Kar, Das Lopamudra, et Avijeet Jain. 2009. Plants in traditional medicinal system—Future source on new drugs. *International Journal of Pharmacy and Pharmaceutical Sciences* *1*(1):1–23.

Santamaría-Aguilar, Daniel, Reinaldo Aguilar, et Laura P. Lagomarsino. 2019. A taxonomic synopsis of *Virola* (Myristicaceae) in Mesoamerica, including six new species. *PhytoKeys* *134*:1–82. https://doi.org/10.3897/phytokeys.134.37979

dos Santos, Jaqueline Lustosa. 2014. « Análise da estrutura de uma floresta primária na Reserva Extrativista-Rio Jutaí, Alto Solimões, Amazonas, Brasil. Universidade Federal Do Amazonas Pró-Reitoria De Pesquisa E Pós-Graduação Departamento De Apoio A Pesquisa Programa Institucional De Iniciação Científica. 1–46.

Steeves, Royce Allan David. 2011. *An Intrageneric and Intraspecific Study of Morphological and Genetic Variation in the Neotropical Compsoneura and Virola (Myristicaceae)*. University of Guelph.

Svenning, Jens-Christian, Bettina M. J. Engelbrecht, David A. Kinner, Thomas A. Kursar, Robert F. Stallard, et S. Joseph Wright. 2006. The Relative Roles of Environment, History and Local Dispersal in Controlling the Distributions of Common Tree and Shrub Species in a Tropical Forest Landscape, Panama. *Journal of Tropical Ecology* *22*(5):575–586. https://doi.org/10.1017/S0266467406003348

Teixeira, Aloysio de Pádua, et Ricardo Ribeiro Rodrigues. 2006. Floristic and Phytosociological Analysis of the Shrub/Tree Stratum in a Gallery Forest in Cristais Paulista, São Paulo State, Brazil. *Acta Botanica Brasilica* *20*:803–813. https://doi.org/10.1590/S0102-33062006000400006

Teles, A. A., et J. E. de Paula. 1980. Wood Anatomy of *Virola sebifera* and *Pseudobombax tomentosum*. *Brasil Florestal* *10*(42):25–35.

The Plant List. 2013. The Plant List. Version 1.1. Published on the Internet; http://www.theplantlist.org/ (accessed 1st January). »

Tonelli, Fernanda, Joao Maximo, Grazielle Maia, L. Soares, D. Silva, C. Carollo, et Ângela Sartori. 2014. Bioautography as a Search Tool to Identify the Allelopathic Compounds in *Virola sebifera*. *Allelopathy Journal* *33*:277–288.

Wei, Na, Christopher W. Dick, Andrew J. Lowe, et Michael G. Gardner. 2013. Polymorphic Microsatellite Loci for *Virola sebifera* (Myristicaceae) Derived from Shotgun 454 Pyrosequencing. *Applications in Plant Sciences* *1*(4):1200295. https://doi.org/10.3732/apps.1200295

36 Fragrant Nutmeg (*Myristica fragrans* Houtt. Myristicaceae)

Abdulrahman Mahmoud Dogara, Harmand A. Hama, Aisha Abdullahi Mahmud, Bawale Sani Halliru, Isah Labaran, and Hussaini Danlami

36.1 BACKGROUND

The utilization of medicinal plants as therapeutic agents for human ailments has been a practice that spans several centuries. This is mostly due to the presence of bioactive constituents within these plants, which possess significant therapeutic properties. *Fragrant myristica*. Houtt. (Myristicaceae), often known as nutmeg or mace, is an evergreen aromatic tree used as a spice and medicinally as a carminative, digestive, and expectorant in the Indian system of medicine. Nutmeg is widely used as a spice and in alternative medicine due to its aphrodisiac properties (S Ahmad et al., 2003). *Myristica fragrans* Houtt. (Myristicaceae), also known as nutmeg or mace and known as pala in Indonesia, luk jan in Thailand, and nikuzuku in Japan (J Chung et al., 2006). Endemic to Indonesia's Spice Island or Moluccas (M Olaleye et al., 2006; Parimalaa and Amerjothyb, 2013). The Moluccas, often known as the Spice Islands, are the origin of the nutmeg tree, or *Myristica fragrans* Houtt. It is now grown throughout the West Indies. It yields nutmeg and mace, two spices (M Jukić et al., 2006). Many different types of baked items, sweets, puddings, meats, sausages, dishes, veggies, and beverages use it as a flavoring (M Olaleye et al., 2006). *Fragrant myristica* hypolipidemic, antithrombotic, antiplatelet aggregating, antifungal, aphrodisiac, anxiogenic, anti-ulcerogenic, anticancer, and anti-inflammatory properties have been studied in recent years. The potential therapeutic applications of *Myristica fragrans* have garnered significant attention as a novel approach for the treatment of several ailments (Prakash and Gupta, 2013; Chaachouay and Zidane, 2021). The assessment of the physiological effects exhibited by plants and their derived compounds holds significant importance within the realm of natural products. Considering the widespread utilization of nutmeg in culinary, beverage, and medicinal applications, we aim to update the following reviews (Asgarpanah and Kazemivash, 2012; Ashokkumar et al., 2022; Warsito, 2021) and furnish current and pertinent insights pertaining to the fragrant nutmeg.

ANTIOXIDANT

The peroxide value is commonly employed as a means of quantifying the first oxidation process, hence serving as an indicator of primary oxidation. (G Singh et al., 2005). Numerous antioxidant assays have been developed and documented in scholarly literature, utilizing different approaches to determine the existence of antioxidant activity in samples obtained from natural sources (JA Kazlauskaite et al., 2023). Various techniques such as DPPH, FRAP, and ABTS are commonly utilized in this study (Table 36.1). The aforementioned approaches exhibit simplicity, expediency, and sensitivity in assessing antioxidant activity (JR Assa et al., 2014; M Jukić et al., 2006). The pursuit of antioxidants derived from natural sources has garnered considerable attention, prompting researchers to diligently explore chemical compounds that have the potential to serve as substitutes for synthetic antioxidants (MD Abdulrahman et al., 2022). Acetone extract has demonstrated the highest antioxidant activity of all extracts. At a concentration of 1 mg/mL extract, the DPPH scavenging activity was 63.04%, the chelating activity was 64.11%, and the bleaching of -carotene was inhibited by 74.36% (AD Gupta et al., 2013). The seed extract has the highest efficacy in scavenging free radicals, as demonstrated by the DPPH and FRAP tests, with values of 154.55 (IC_{50}=µg/ml) and 82.33 (mg GAE/g extract) respectively. The DPPH and FRAP values of mace extract are recorded as 201.97 and 56.31, respectively. In contrast, the flesh exhibits DPPH and FRAP values of 1372.91 and 13.45, respectively (JR Assa et al., 2014). Linoleic acid: The absorbance of a linoleic acid emulsion without an addition rose quickly, and at the $P < 0.05$ level, there was a difference between the antioxidants and the blank. Volatile oil is an effective radical scavenger of DPPH, according to the results (I Kapoor et al., 2013). The extracts of aril, seed kernel, and shell, obtained using an 80% methanol solution, demonstrated the highest total phenolic content. Among these extracts, the shell extract exhibited the most significant primary antioxidant activity, as evidenced by its superior ferric-reducing antioxidant power (FRAP) activity (EC_{50} 9.7 µg/mL), β-carotene-bleaching activity (EC_{50} 21.5 µg/mL), and 1,1-diphenyl-2-picrylhydrazyl (DPPH) scavenging activity (EC_{50} 160.9 µg/mL). On the other hand, the pericarp extract demonstrated the highest secondary antioxidant activity, acting as a metal chelator (EC_{50} 75.6 µg/mL) (Sulaiman and Ooi, 2012). The assessment of antioxidant activity was conducted using spectrophotometric analysis utilizing the DPPH test. After incubation, the nutmeg essential

DOI: 10.1201/9781003460336-36

TABLE 36.1
Profile of the Documented Studies

S/N	Activity	Methods	Plant parts	Solvents	Concentrations	Reference
1	Antioxidant	DPPH, Chelating activity on Fe2+, β-carotene bleaching activity	Seeds	Acetone, ethanol, methanol, butanol, aqueous	0.025–2 mg/mL, 0.20–1.2 mg/mL,1 mg/mL	Gupta et al., (2013)
		In vivo	Seeds	Hot aqueous	100–500 mg/kg	Olaleye et al., (2006)
		DPPH, FRAP	Flesh, seed, mace			Assa et al., (2014)
		he extracts of	Fleshy pericarp (husk), aril (mace), seed kernel (endosperm), shell (seed-coat)	80% methanol		Sulaiman and Ooi (2012)
			Bark	n-hexane		(Ginting, 2021)
				Aqueous	50 mg/kg	(Akinboro et al., 2011)
		DPPH, ABTS, FRAP	Essential oil		1%	(Kazlauskaite et al., 2023)
	Anti-inflammatory	In vivo	Mace	Methanol	1.5 g/kg, 0.9 g/kg, 0.5 g/kg, 0.19 g/kg	(Ozaki et al., 1989)
			Wood, nutmeg and mace	Ethanol		SC BATM (2016)
			seeds		0%, 2%, 4%, 8%, 12%	Azis Ikhsanudin and Rais (2021)
2	Antibacterial	Disk diffusion	Seeds	Aceton, ethanol, methanol, butanol, aqueous	10 mg/mL	Gupta et al. (2013)
		Ager well	Seeds	Essential oil, acetone		Singh et al., (2005)
		Ager well	Dihydroguaiaretic acid			Bhamarapravati et al., (2006)
			Macelignan		0.2, 2, 10 μg/mL	Yanti et al., (2008)
			Fleshy pericarp (husk), aril (mace), seed kernel (endosperm), shell (seed-coat)	80% methanol		Sulaiman and Ooi (2012)
		Ager well	Essential oil		500, 250, 125, 62.5, 31.25, 15.62, 7.81, 3.90, 1.95, 0.97 mg/mL	Lima et al. (2012)
		Tetrazolium microplate	Essential oil			Piaru, Mahmud, and Perumal (2012b)
		Twofold serial microdilutio, MIC	Flesh mace, seed	Ethyl acetate, ethanol	0.075 g/mL to 40 mg/mL and 1.25 mg/mL to 640 mg/mL	Shafiei et al. (2012)
		MIC	Seed	Aqueous, methanol, ethyl acetate, hexane	25%, 50%, 75%, and 100%	Panggabean et al. (2019)
		Disc diffusion	Mace	Hexane, Chloroform, methanoli, Ethanolic		Pooja et al. (2012)
		Disc, ager well	Seed, mace	Aqueous, ethanol	50%, 70%, and 95%	Singh et al. (2016)
3	Antiparasitic	In vivo	Seed		100, 200, and 300 mg/kg	Najafzadeh et al. (2014)
4	Antidiabetic	In vivo		Ethanol	500 mg/kg	Ram et al. (1996)
		In vivo				Lestari et al. (2012)

(*Conitnued*)

TABLE 36.1 (Continued)

S/N	Activity	Methods	Plant parts	Solvents	Concentrations	Reference
		In vivo	Ethanol	Fruit	150, 450 mg/kg	Arulmozhi et al. (2007)
5	Hematological indices	In vivo	Ethanol	Seed	100, 250, and 500 mg/kg	Bamidele et al. (2011)
		In vivo	50% ethanolic	Seeds	500 mg/kg	Ahmad et al. (2003)
6	Behavioral action	In vivo		n-hexane	10–100 mg/kg	Sonavane et al. (2001)
7	Anti-papillomagenesis	In vivo	Mace	Acetone	150 μg in 100 μL	Jannu et al. (1991)
8	Hepatoprotective	In vivo	Seed kernel	n 50% ethanol	10 mg/kg body weight/day	Sharma and Kumar (2007)
		MTT assay	Rhizomes		1–50 μM	Hwang, (2010)
		In vivo	Seed	Methanol, n-hexane	500 and 1,000 mg/kg	Anaduaka et al. (2022)
9	Anticancer	MTT assay	Mace		25 to 125 μg/mL	Rengasamy et al. (2018)
		MTT assay	Bark	Methanol	20, 50, and 100 μg/mL	Sa-nguanmoo and Poovorawan, 2007)
		MTT assay	Bark	n-hexane	25, 50, 100, and 200 ppm	Ginting (2021)
		SRB assay	Seed	Ethanol	100 μg/mL	Prakash and Gupta (2013)
		In vivo	Leaves	Aqueous	4, 20, 50, 100, 200, 500, and 1000 μg/well	Akinboro et al. (2014)
			Leaves	Methanol		Akinboro et al. (2012)
10	Toxicity	MTT assay	Mace		100 μg/mL to the lowest 3.125 μg/mL	Pillai et al. (2012)
		In vivo	Seed	Methanol, n-hexane	500 and 1,000 mg/kg	Anaduaka et al. (2022)

oil exhibited notable antioxidant activity, as evidenced by its EC_{50} value of 1.35 mg/mL (V Nikolic et al., 2021). The aglycone fraction exhibits scavenging activity, as seen by the IC_{50}-values of 3.5 g/L and 22 g/L for the aglycones and nutmeg oil, respectively (M Jukić et al., 2006). The efficacy of the essential oils was evaluated using the β-carotene/linoleic acid assay, yielding an IC_{50} value of 976 μg/mL(RK Lima et al., 2012). The antioxidant activity was determined and quantified using the IC_{50} value of 99.76 ppm (B Ginting, 2021). The extract alone, as well as in combination, suppressed cell division, and induced chromosomal aberrations that were insignificantly different from the negative control ($P \leq 0.05$) (A Akinboro et al., 2011). The antioxidant activity of mace essential oil was assessed through the utilization of the free radical-scavenging test (at a concentration of 21.95 μg/mL), as well as the β-carotene in linoleic acid and percent inhibition in linoleic acid (at a level of 67.9%) systems (A Ali et al., 2021). The inhibition percentage of the essential oil was (DPPH; 8.69, ABTS; 92.14 and FRAP; 176.05 μg TE/g dw, respectively (JA Kazlauskaite et al., 2023). In the histological examinations, it was demonstrated that the extent of the degenerative characteristics in the organs was inversely related to the extract concentration. The lymphoid follicles of the spleen were decreased and deteriorated. Vascular congestion, bile duct enlargement, and testicular germinal epithelial cells hepatic degeneration and necrosis, desquamation of tubular epithelial cells, and congested renal blood vessels heart muscle fiber deterioration, myocardial necrosis, and vascular injury in the kidney groups receiving treatments in comparison to those who are not(M Olaleye et al., 2006). The results of the study indicate that nutmeg and mace exhibited significant antioxidant activity, whilst wood shown a moderate level of activity. The IC_{50} values for nutmeg, mace, and wood were determined to be 21.164, 28.897, and 71.830 μg/ml, respectively (SC BATM, 2016). The study demonstrated notable radical-scavenging and metal-reducing characteristics of the substance, as well as a considerable inhibition of butyrylcholinesterase at a concentration of 4.78 mg GALAE/g (A Trifan et al., 2023). The extracts lacking additional egg white exhibit a significant antioxidant activity in the scavenging of the stable free radical ABTS (89.980μg/mL) and a moderate antioxidant activity in the scavenging of the stable free radical DPPH (105.669µg/mL) (I Antasionasti et al., 2021). The DPPH assay demonstrated robust antioxidant activity, as seen by the IC_{50} value of 12.67 ppm (B Ginting et al., 2020). The antioxidant activity of the methanol extract was evaluated using the DPPH assay at concentrations of 25 ppm, 50 ppm, and 100 ppm. The results indicated a significant level of antioxidant activity, with an IC_{50} value of 36.31 ppm. Additionally, Vitamin C demonstrated antioxidant activity with an IC_{50} value of 3.657 ppm (N Nurmilasari et al., 2017). Antioxidant power is thought to be highly associated with total phenolic content (Y Velioglu et al., 1998). The number and location of hydroxyl groups, which are essential for scavenging free radicals, chelating transitional metals, and inhibiting lipoxygenase, determine the antioxidant properties of phenolic antioxidants, which are produced primarily through secondary metabolism in plants (Y Velioglu et al., 1998). As a result of the thorough drying, there may be a significant reduction in antioxidant activity, which may be brought on by the interaction of free or oxidized phenolics with cell polysaccharides (K-T Chung et al., 1998). Phenolic groups are known to exert a significant influence on antioxidant activity. Therefore, the antioxidative activity shown in all antioxidant assays investigated can be attributed to the existence of phenolic substances, namely methyl eugenol, anisole (p-pentyl), eugenol, methyl eugenol, trans-isoeugenol, and trans-methyl eugenol, present in the essential oil and mace extract (JR Assa et al., 2014; G Singh et al., 2005). The iron (II) chelating ability of nutmeg essential oil can be ascribed to its inherent chelating agent, predominantly phenolics, including methyl eugenol and trans-isoeugenol (I Kapoor et al., 2013). The antioxidative activity of both eugenol and isoeugenol can be attributed to the presence of the hydroxyl group in the phenolic ring (M Jukić et al., 2006). The inherent antioxidants found in the plant extract possess the potential to be chemically synthesized and utilized as nutraceuticals, which have the capacity to mitigate oxidative harm within the human body. The antioxidants included in the plant extract can be synthesized into nutraceuticals that can lessen the risk of oxidative injury to the body.

36.2 ANTI-INFLAMMATION

Nonsteroidal anti-inflammatory medications (NSAIDs) and corticosteroids are the predominant pharmacological agents employed for the mitigation of acute and chronic inflammation. The utilization of these drugs in combination is common due to their subtly different effects on granulation tissue. Nonsteroidal anti-inflammatory drugs (NSAIDs) are known to exert their inhibitory effects on the cyclooxygenase enzymes, specifically COX-1 and COX-2 (Azis Ikhsanudin and Rais, 2021). These enzymes play a crucial role in the regulation of various cellular processes involved in the inflammatory response. Steroids have the potential to exert a substantial inhibitory effect on inflammation through their interaction with receptors responsible for regulating phospholipase A2, COX-2, iNOS, and interleukin. Regrettably, it is necessary to exercise caution when consuming two categories of medications due to their propensity to induce harmful effects such as gastrointestinal bleeding, myocardial infarction, and kidney diseases (Azis Ikhsanudin and Rais, 2021). The key alternative option for mitigating the long-term risks associated with the use of anti-inflammatory medicines is the development of a pharmaceutical product derived from plants. Considering these conditions, medicinal plants have notably reexamined cellular pathways to offer potent active molecules as therapeutic phytochemicals. The potencies of the methanol extract (1.5 g/kg), ether fraction (0.9 g/kg), n-hexane fraction (0.5 g/kg), Fr-II (0.19 g/kg), and Fr-VI (0.17 g/kg) were almost equivalent to those of indomethacin (10 mg/kg) and displayed a long-lasting anti-inflammatory effect(Y Ozaki et al., 1989). The extract had inhibitory effects on the expression of interleukin-6 (IL-6) and tumor necrosis factor α (TNFα), which are inflammation-related cytokines mediated by free fatty acids (FFAs), in the cells(W Zhao et al., 2020). The ethanolic extract of seeds and the quercetin compound had the lowest cytotoxic action and were safe substances at concentrations of 50 and 10 g/mL, respectively, when tested on RAW 264.7 cells. The generation of NO, TNF-, IL-6, and IL-1 by LPS-stimulated-RAW 264.7 was suppressed by *M. fragrans*. The strongest TNF-, IL-6, IL-1, and nitrite-associated NO inhibitory action was found in 50 µg/mL of seed ethanolic extract. The anti-inflammatory activity of the ethanolic extract of wood was shown to be more potent compared to nutmeg and mace, with IC_{50} values of 40.26, 65.42, and 75.40 µg/ml, respectively (SC BATM, 2016). The findings indicated that all formulae exhibited a visually non-transparent physical appearance, characterized by a brownish-yellow hue, a soft texture, and an aromatic scent (Azis Ikhsanudin and Rais, 2021). The secretion of interleukin-6 (IL-6) serves as a crucial inflammatory stimulus, leading to the activation of matrix metalloproteinases, macrophages, and the generation of neutrophils. Additionally, IL-6 has a role in autoimmune responses observed in chronic arthritis, osteoporosis, and psoriasis (I Matulyte et al., 2020).

36.3 ANTIMICROBIAL

All throughout their lives, humans are exposed to potentially dangerous microorganisms, which can cause a variety of diseases and have a significant negative impact on human health. Humans have developed defense mechanisms to deal with these issues, but frequently, this first line of protection fails, and different chemotherapeutic medications are employed to treat infections brought on by microorganisms. In the past, a variety of chemical agents have been discovered or created to treat and cure these infections, but their widespread and indiscriminate use has resulted in the emergence of numerous drug-resistant strains, which pose a serious problem globally because the drugs currently in use are no longer able to control them. As a result, finding alternatives to synthetic antibiotics and other medications is urgently needed. All the examined bacteria and fungus were susceptible to the antibacterial effects of the acetone extract. The most effective antibacterial and antifungal activity was demonstrated by acetone extract with *Staphylococcus aureus* (13.8 mm) and *Aspergillus niger* (14.4 mm), respectively (AD Gupta et al., 2013). With MIC values between 125 and 250 μg/ml, the ethanolic extract of mace was effective against H. pylori ATCC 43504 and six clinical strains. The MIC of the aqueous extract against the *H. pylori* ATCC reference strain and six clinical strains were 500 μg/ml and 0.5 g/ml, respectively (N Suthisamphat et al., 2020). MIC values between 2.1 and 213 μg/mL, were similarly susceptible to macelignan's (Figure 36.1) preferred action. These included *Streptococcus sobrinus, Streptococcus salivarius, Streptococcus sanguis, Lactobacillus acidophilus*, and *Lactobacillus casei* (J Chung et al., 2006). The extract exhibited a statistically significant positive clear zone in the agar diffusion test at a significance level of $P<0.05$ across all ten clinical cultures. The MIC and MBC were determined to be 100 μg/mL and 125 μg/mL, respectively. The results of the experiment demonstrated that the extract exhibited a significant inhibitory effect on the development of *Staphylococcus aureus* at a concentration of 3000 ppm, as evidenced by the presence of a 47 mm inhibition zone. The essential oil exhibits the greatest level of action against *Escherichia coli* at a concentration of 22 m (G Singh et al., 2005). The aril and seed-kernel extracts exhibited inhibitory effects on food-borne bacteria, specifically *Staphylococcus aureus* and *Bacillus cereus*, with a minimum inhibition concentration of 50 μg/mL (Yanti et al., 2008). The aril and seed-kernel extracts exhibited inhibitory effects on food-borne bacteria,

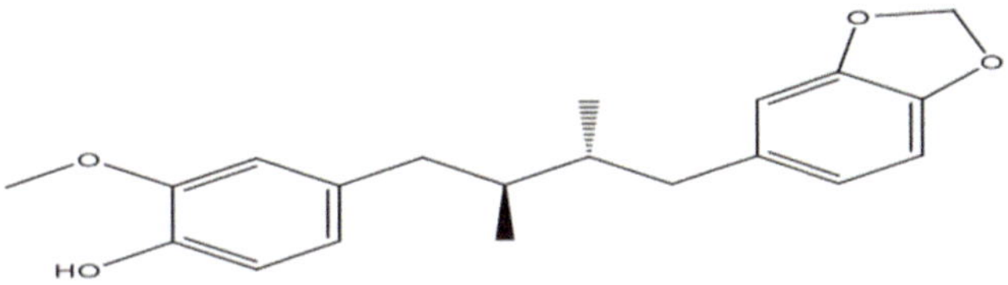

Macelignan

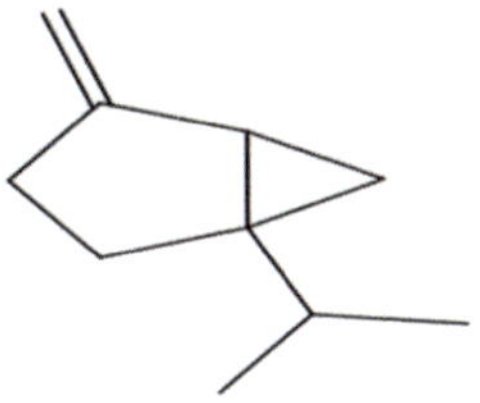

Sabinene

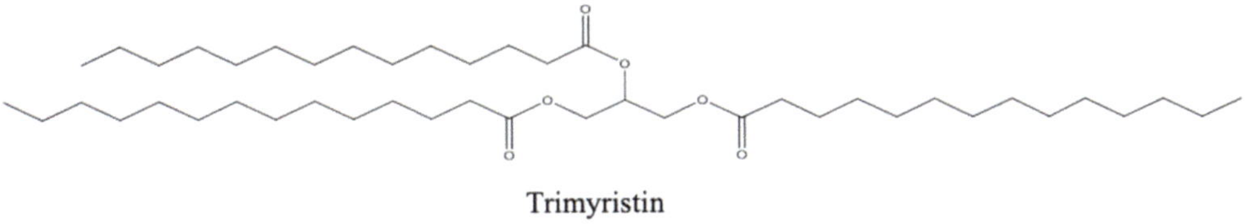

Trimyristin

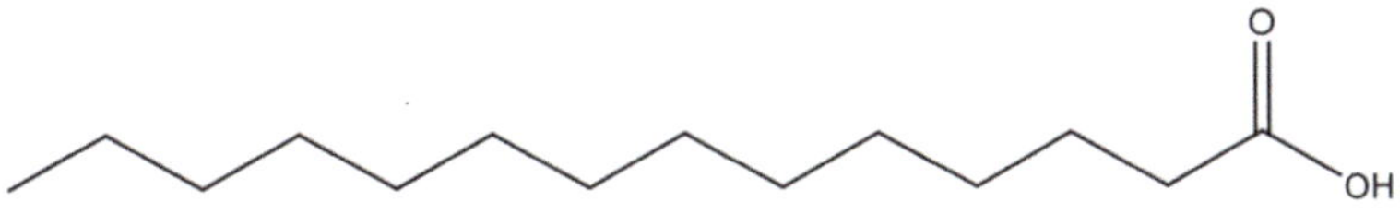

Myristic acid

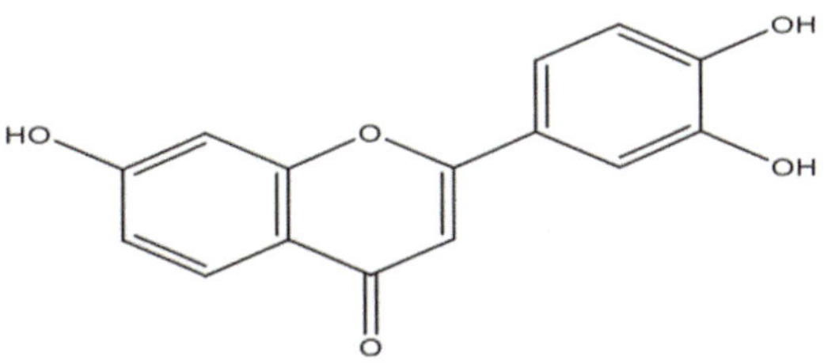

FIGURE 36.1 Some of the compounds identified responsible for biological activity.

specifically *Staphylococcus aureus* (ATCC12600) and *Bacillus cereus* (ATCC10876), with a minimum inhibition concentration of 50 µg/mL (Sulaiman and Ooi, 2012). Nutmeg essential oil was tested for antibacterial activity using disc diffusion. The investigation examined its effects on Gram-positive bacteria *Staphylococcus aureus*, *Bacillus cereus*, *Bacillus luteus*, and *Listeria monocytogenes*. It was also tested against Gram-negative bacteria such as *Escherichia coli*, *Klebsiella pneumoniae*, *Pseudomonas aeruginosa*, and Proteus vulgaris. Finally, nutmeg essential oil was tested against *Candida albicans*. The essential oil was antifungal and antibacterial. The essential oil had 28 mm anticandidal action, equivalent to nystatin (17 mm). The inhibitory zone diameter of 23 mm showed that Bacillus luteus was the most sensitive Gram-positive bacterium tested. The Gram-negative bacteria tested were less sensitive(V Nikolic et al., 2021). The essential oil exhibited the greatest level of action against the bacterial species *Aeromonas hydrophila* and *Salmonella choleraesuis*, as indicated by MIC values ranging from 3.90 to 7.81 µg/mL. Additionally, the MIC values for other tested species were within the range of 7.81 to 15.62 µg/mL(RK Lima et al., 2012). The crude essential oil (EO) at a concentration of 0.1% demonstrated inhibitory effects on the radial growth of various fungal species. Specifically, it reduced the growth of Colletotrichum gloeosporoides by 98%, Colletotrichum musa by 97%, *Fusarium oxysporum* by 75%, *Fusarium semitectum* by 78%, Aspergillus niger by 71%, and A. glaucus by 60%. The growth inhibition exhibited a significant rise, rising from 85% to 100%, when the concentration reached 0.3% (VMM Valente et al., 2011). The MIC values for nutmeg oil was seen to be between 0.031 and 1 mg/mL. Notably, a very notable level of efficacy was observed against Shigella dysenteriae, with a MIC value of 0.031 mg/mL(Piaru, Mahmud, and Perumal, 2012b). The ethyl acetate extract obtained from the flesh demonstrated the highest level of inhibitory activity against both Gram-positive and Gram-negative bacteria compared to all other extracts that were tested. The study observed a range of MIC values, varying from 0.625 to 1.25 mg/mL. The statistical analysis revealed a significant P-value of 0.017. Furthermore, the present excerpt exhibited the most pronounced bactericidal properties, as evidenced by the average MBC values spanning from 0.625 mg/mL to 20 mg/mL (Z Shafiei et al., 2012). The microorganisms *Aspergillus niger* and *Streptococcus mutans* exhibited the highest sensitivity, as indicated by their MIC values of 1.01 and 1.72 mg/mL, respectively (A Ali et al., 2021). The crude extract exhibited antibacterial properties, as evidenced by MIC ranging from 32 to 1024 µg/mL against a majority of the 29 Gram-negative bacterial strains that were tested. The fraction of methanol extract had inhibitory effects on the growth of all tested bacterial strains (100%, 29/29). Additionally, the molecule 3′,4′,7-trihydroxyflavone (Figure 36.1) exhibited inhibitory effects on all tested strains (12/12), with MIC values ranging from 32 to 1024 µg/mL and 4 to 128 µg/mL, respectively (JK Dzotam et al., 2018). The results of the experiments conducted to evaluate the antifungal efficacy of nutmeg oleoresin at different doses revealed that a concentration of 10,000 ppm exhibited significant growth inhibition and demonstrated evident efficacy against five different species of fungi (I Rodianawati et al., 2015). The MIC was observed at concentrations of 0.1% and 0.25% of nutmeg seed extract against *Staphylococcus aureus* and *Escherichia coli*, respectively (K Panggabean et al., 2019). The proliferation of vegetative cells of B. cereus was strongly suppressed by macelignan, with a MIC of 4 µg/mL. The vegetative cells of Bacillus cereus were effectively eradicated at a MBC of 8 µg/mL of macelignan (Y Rukayadi et al., 2009). The range of MIC of essential oils spans from 0.313% to 10% (DP Wibowo et al., 2018). The findings indicate that the extracts exhibit significant efficacy against both fungal strains. The efficacy of methanol and hexane extracts has been observed to be highest among the examined organisms. The MIC for the methanol extract was determined to be 0.237 mg/ml against *Candida albicans* and 0.232 mg/ml for *Aspergillus niger*(V Pooja et al., 2012). All toxigenic isolates and non-toxigenic isolates exhibited sensitivity to nutmeg oil, with a percentage of 100% and 93.33% respectively. Additionally, 85% of toxigenic isolates demonstrated sensitivity to nutmeg oil. The range of minimal inhibitory concentration was shown to be between 2.5µl/ml and 10µl/ml (Justin and Antony, 2016). The growth of *S. aureus* and *E. coli* was seen to be inhibited by the application of methanol extract derived from nutmeg fruit, resulting in inhibition rates of 3.81 and 81%, respectively. S. aureus exhibits a lower MIC value (8%) compared to *E. coli* (10%) (N Nurhasanah, 2014). With the ager well method, 95% of seed ethanolic extract gives the best inhibition of 30 mm against *E.coli* (S Singh et al., 2016). The presence and concentration of several chemical compounds in an extract can be used to explain its antimicrobial action. Nutmegs contain a variety of chemicals, many of which have important antibacterial properties. The primary antibacterial components identified from nutmeg are reported to be trimyristin and myristic acid (Figure 36.1) (AD Gupta et al., 2013). Antifungal efficacy has been attributed to three lignans isolated from nutmeg's methanolic extract: erythro-austrobailignan-6, meso-dihydroguaiaretic acid, and nectandrin-B(JY Cho et al., 2007). The majority of the antibacterial properties found in essential oils made from culinary herbs and spices come from phenolic chemicals (G Singh et al., 2005), but other components are thought to contribute little to antibacterial effects (G Singh et al., 2005). According to research, myristicin's bactericidal effect is related to the unsaturated side chain connected to the aromatic ring (RK Lima et al., 2012). The antibacterial properties of different parts of *Myristica fragrans* demonstrate their noteworthy impact, suggesting their potential medical application in combating microbial infections. Because of the synergistic effect of the chemicals, which increases the contact area and extract access to the microbial strain, it is reasonable to assume that the rise in extract concentration has antifungal action.

36.4 ANTIPARASITIC

Parasitic infections, which can cause morbidity and even death in their victims, are a major public health concern (MD Abdulrahman et al., 2022). Although chemical treatments are effective for parasite treatment, they have drawbacks such as drug resistance, drug residues, and negative side effects. Alternative treatments must be researched (MD Abdulrahman et al., 2022). In the in vitro study targeting *T. gondii*, the oil extract exhibited a noteworthy inhibitory effect, as evidenced by an EC_{50} value of 24.45 µg/mL (S Pillai et al., 2012). The findings indicated that the mean number of brain cysts was somewhat decreased in the group treated with the extract, with a reduction of 0.8% (TI FARAG et al., 2019). Group 2 had dramatically raised blood TNF- levels, however treatment with 300 mg/kg nutmeg resulted in a dose-dependent, but significant, reduction in high TNF- levels. The levels of TNF- were not noticeably reduced by the flunixin. Ankle joint pathology was significantly reduced in nutmeg-treated rats (H Najafzadeh et al., 2014). The use of flunixin to treat pannus was successful, but the drug had no effect on other types of lesions(H Najafzadeh et al., 2014).

36.5 ANTIDIABETIC

At present, the therapeutic options for diabetes encompass insulin as well as a range of oral antidiabetic medicines, including sulfonylureas, metformin, α-glucosidase inhibitors, and the thiazolidinedione class of glitazones (D Arulmozhi et al., 2007). Each of these oral medicines has the potential to induce several significant adverse effects (D Arulmozhi et al., 2007). Therefore, it is crucial to develop improved therapeutic drugs that have fewer adverse effects for the pharmacotherapy for diabetes. Plants have served as valuable sources of pharmacologically active compounds, with numerous medications being developed either directly or indirectly from botanical origins. After 60 days, the experimental group's levels of lipoprotein lipids were significantly lower than those of the control group of rabbits: total cholesterol was 573 mg/dl compared to 209 mg/dl, low-density lipoprotein (LDL) cholesterol was 493 mg/dl compared to 131 mg/dl, and triglycerides were 108 mg/dl compared to 67 mg/dl ($P< 0.001$) (A Ram et al., 1996). The experimental findings indicated that each administered dose had hypoglycemic action, with statistical significance observed at P-values of 0.01 and 0.05, respectively. The investigation additionally showed that an escalation in dosage resulted in a corresponding augmentation in hypoglycemic activity (K Lestari et al., 2012). The administration of the extract at doses of 150 and 450 mg/kg resulted in the improvement of metabolic abnormalities induced by chlorpromazine. This was demonstrated by a significant decrease in glucose and triglyceride (TG) levels. The maximum reduction observed was 41% for glucose and 53% for TG at the 450 mg dose, with a P-value of less than 0.01(D Arulmozhi et al., 2007). The findings indicated that the blood glucose level exhibited a reduction of 20% following a two-day treatment period, a 30% reduction after four days of treatment, and a 40% reduction after six days of therapy with nutmeg extracts. There was no significant alteration observed in the blood triglyceride level during the initial two days of treatment. However, after four days of treatment, a reduction of 25% in the blood triglyceride level was noted (K Lestari et al., 2019). Furthermore, after six days of treatment, an additional 25% decrease in the blood triglyceride level was observed (K Lestari et al., 2019). The most potent inhibition of -amylase was likewise seen with the aqueous phenolic (IC_{50}=360.18 mg/L) (H Hasbullah et al., 2023). According to preliminary research on the plant extract's ability to block alpha-glucosidase and alpha-amylase, the species has been proven to be a promising antidiabetic source in both in vitro and in vivo studies. The plant extract's preventative and therapeutic actions may be effective for diabetes-related disorders. Clinical trials are essential because this plant has been researched both in vitro and in vivo.

36.6 ANTICHOLINESTERASE

The fraction demonstrated inactivity against acetylcholinesterase (AChE), but it exhibited inhibitory effects on butyrylcholinesterase (BChE) with an IC_{50} value of 68.16 µg/mL. This value was compared to the reference medication, donepezil, which had an IC_{50} value of 1.97 µg/mL (A Rastegari et al., 2022). The administration of the extract results in the downregulation of the expression of genes associated with lipid production, namely fatty acid synthase and sterol regulatory element-binding protein 1, leading to a reduction in cellular lipid content (W Zhao et al., 2020). BChE was suppressed by the extract at 4130 µg/mL (D Gomes da Rocha Voris et al., 2018). The study revealed that none of the fractions exhibited any anti-acetylcholinesterase (AChE) activity. However, they did exhibit activity against butyrylcholinesterase (BChE), with IC_{50} values of 361.8, 215.0, and 145.8 µg/mL, respectively (S Omidpanah et al., 2022). These values were compared to the IC_{50} value of 1.97 µg/mL exhibited by the reference medication, donepezil (S Omidpanah et al., 2022). Compound 8, specifically [(7S)-8-(4-hydroxy-3-methoxyphenyl)-7-hydroxypropyl]benzene-2,4-diol, exhibited the highest level of efficacy with an IC_{50} value of 35.1 µM. Following this, compounds 2, denoted as [(8R,8S)-7-(3,4-methylenedioxyphenyl)-8,8-dimethyl-7-(3,4-dihydroxyphenyl)-butane], shown subsequent activity. Compound 10 (malabaricone B) and compound 11 (malabaricone C) had IC_{50} values of 42.1 and 44.0 µM, respectively(TD Cuong et al., 2014).

36.7 ANTICONVULSANT ACTIVITIES

Epilepsy is a prevalent and significant neurological condition. Approximately 50 million individuals worldwide are affected by epilepsy, with approximately 5% of the overall population experiencing at least one seizure throughout their lifetime, excluding febrile seizures (Bell and Sander, 2001). The antiepileptic medications (AEDs) currently on the market are synthetic compounds that have been found to produce significant side effects, teratogenicity, and withdrawal symptoms (A Wahab et al., 2009). The application of nutmeg oil demonstrated a rapid initiation of action and a brief duration of anticonvulsant efficacy (A Wahab et al., 2009). The compound exhibited notable anticonvulsant action in hind limb tonic extension caused by electroshock (A Wahab et al., 2009). The compound demonstrated anticonvulsant activity that was dependent on the dosage, specifically in relation to pentylenetetrazole-induced tonic seizures (A Wahab et al., 2009). The administration of the substance effectively postponed the initiation of hind limb tonic extensor jerks that were caused by the presence of strychnine (A Wahab et al., 2009).

36.8 HEMATOLOGICAL INDICES

The findings of the study demonstrated statistically significant reductions ($P < 0.05$) in the levels of red blood cells (RBC), packed cell volume (PCV), hemoglobin concentration (HbC), and platelet count, particularly when administered at higher doses. A statistically significant increase ($P < 0.05$) was seen in the total count of white blood cells (WBC) (O Bamidele et al., 2011). The application of nutmeg and clove extracts was seen to elicit the mounting behavior in male mice, as well as substantially enhance their mating performance (S Ahmad et al., 2003). The testis, epididymis, seminal vesicle, ventral prostate, and vas deferens all had considerable weight reductions because of the oil, according to the results. At the highest dose level examined, both sperm motility and density were severely decreased, which caused a 100% drop in male fertility (M Agarwal et al., 2009).

36.9 OTHER DISEASES

Histopathology: All of the tested doses of nutmeg seed ethanol extract increased blood urea nitrogen levels. However, there was no discernible change in creatinine levels between the two groups. There was no alteration in kidney morphology, according to the histopathology results. nutmeg seeds extracted with ethanol (MS Bachri, 2013). Behavioral action: The extract enhanced occupancy in the closed arm while decreasing occupancy in the open arm. In both arms, fewer entries were submitted. The 5-HT3 receptor antagonist ondansetron (0.5 mg/kg) and the sedative diazepam (1 mg/kg) both decreased the amount of time spent in the closed arm. The extract was able to counteract the negative effects of diazepam and ondansetron on the amount of time spent on the closed arm (G Sonavane et al., 2001). Anti-papillomagenesis: The incidence of skin papilloma was observed to decrease by 50%, while the average number of tumors per mouse with tumors was found to be 1.75. The observed decrease in papilloma was statistically significant ($P < 0.05$) (L Jannu et al., 1991). Antiallergic: The extracts derived from the three components (wood, nutmeg, and mace) exhibited significant antiallergic activity, as indicated by their IC_{50} values of 13.29, 20.90, and 12.95 µg/ml, respectively (SC BATM, 2016). The treated group exhibited a significant decrease in eosinophil count in the blood, as well as reduced levels of IgE, AEC, and the number of inflammatory cells in both bronchoalveolar lavage fluid (BALF) and lung tissues (SR Menthe et al., 2020). Analgesic activity: In female mice, but not in male mice, alkaloid extract at a dose of 1 g/kg considerably decreased the number of writhing reactions; alkaloid extract at a level of 0.5 g/kg had no impact in either sex (AA-S Hayfaa et al., 2013). 5.1 g/kg was the LD_{50}. Animals given a dose of 4 g/kg or more displayed signs of anomalous behavior such as hypoactivity, erratic walking, and fainting; the abnormal behavior persisted for several hours after the alkaloids' administration (AA-S Hayfaa et al., 2013). Anti-ulcer: The administration of *Myristica fragrans* demonstrated a statistically significant reduction ($P < 0.05$) in stomach lesions by 41.68% in an ethanol-induced ulcer model at a dosage of 200 mg/kg, in comparison to the efficacy of sucralfate (60.41%) (A Sattar et al., 2019).

36.10 HEPATOPROTECTIVE

Myristicin was found to have an incredibly potent hepatoprotective effect (T Morita et al., 2003). Myristicin considerably reduced the effects of LPS/D-GalN on the hepatic DNA fragmentation and serum TNF- levels in mice (T Morita et al., 2003). The extract was administered, it dramatically increased hepatic glutathione (GSH) and lowered testicular lipid peroxidation (LPO), but had no discernible effect on acid phosphatase (ACP) or alkaline phosphatase (ALP) activity(Sharma and Kumar, 2007). The LPO level and ACP activity were significantly increased as a result of irradiation, while the GSH content and ALP activity were lowered (Sharma and Kumar, 2007). A drop in LPO level and ACP activity and an increase in GSH and ALP activity show that MF pretreatment effectively prevented radiation-induced metabolic change(Sharma and Kumar, 2007). After UVB exposure, macelignan enhanced the viability of HaCaT cells and, in a concentration-dependent manner, reduced MMP-9 production and COX-2 expression (J-K Hwang, 2010). The lipid profile of the irradiated spice was characterized by a dose-dependent decrease in the triacylglycerol concentration and a concurrent rise in free fatty acids. This suggested that acylglycerols broke down while being exposed to radiation, releasing free fatty acids(Z Niyas et al., 2003). The results demonstrated a notable level of neuroprotection, with an observed percentage of 86.28% at a concentration of 100 µg/mL, against neurotoxicity generated by H_2O_2. Additionally, the substance exhibited metals such as Zn^{2+}, Fe^{2+}, and Cu^{2+} modest ability to chelate (A Rastegari et al., 2022). The administration of n-hexane extract at a dosage of 1000 mg/kg resulted in histological alterations that are indicative of hepatotoxicity (EG Anaduaka et al., 2022). These findings suggest that myristicin's capacity to stop macrophages from producing tumor necrosis factor alpha may be responsible for its hepatoprotective effects, at least in part. Although myristicin has been demonstrated to protect the liver, more analysis is still required to ascertain how (T Morita et al., 2003).

36.11 ANTICANCER

Uncontrolled cell division characterizes the disease known as cancer. Around 14 million new instances of cancer were diagnosed globally in 2012, while 8.2 million people died from cancer (MD Abdulrahman et al., 2022). The development and extension of the contemporary healthcare system have benefited from the use of medicinal herbs. The only path forward, as their acceptability and acknowledgment spread around the world, is through medicinal herbs. In recent years, there has been a lot of interest in discovering naturally occurring chemo preventive drugs that can suppress or reverse multistage carcinogenesis. Numerous phenolic compounds, particularly those found in food and medicine plants, have been discovered to have anti-carcinogenic and anti-mutagenic characteristics (Prakash and Gupta, 2013). Highest cytotoxic activity was displayed by the OVCAR-5 (Ovary) cell line (97%) Colon 502713 and PC-5 (Prostrate) cell lines had an 86% growth suppression. Hep-2 (Liver) demonstrated 72% growth inhibition, SF-295 (CNS) and Colo205 (73%). A-549 (Lung) displayed no appreciable activity (50%) at all(Prakash and Gupta, 2013). The mace extract displayed cytotoxicity and anticancer activity against KB cell lines, as well as suppressing cancer cell development; thus, a growth inhibitory impact was observed in extract-treated cell lines. When compared to untreated KB cells, the apoptotic potential of mace extract was accompanied by decreased gene expression of Bcl-2. The mace extract exhibited cytotoxic activity and promoted apoptosis in KB cell lines via regulation of its target genes Bcl-2, implying the possibility of mace as a candidate for oral cancer chemoprevention (G Rengasamy et al., 2018). The methanol extract had strong inhibitory effects on Jurkat cell proliferation and triggered apoptosis, as evidenced by annexin V staining, at doses of 50 and 100 μg/mL. A decrease in the expression of SIRT1 mRNA was detected in Jurkat cells, even at a concentration of 10 μg/mL of methanol extract (Sa-Nguanmoo and Poovorawan, 2007). The anticancer activity against MCF-7 cell lines was evaluated, and the IC_{50} value was determined to be 10.75 ppm (B Ginting, 2021). The cytotoxicity of nutmeg oil against the Vero cell line was found to be modest, with an IC_{50} value of 24.83 μg/mL (Piaru, Mahmud, and Perumal, 2012b). The results indicated that the mutagenicity of AFMF was below 2.0. Additionally, the administration of the extract at doses of 500, 1,000, 2,000, and 4,000 mg/kg did not substantially increase the occurrence of micronucleated polychromatic and normochromatic erythrocytes compared to the negative control ($P \geq 0.05$) (A Akinboro et al., 2014). The serum concentrations of magnesium (Mg), zinc (Zn), iron (Fe), manganese (Mn), calcium (Ca), and selenium (Se) exhibited a substantial decrease in the methylcholanthrene tumor models compared to both the control and mace groups ($P < 0.001$). The finding aligns with the observed tumor incidence. The anticancer action of mace may be attributed to the presence of trace elements, either directly or indirectly (C Zhang et al., 2012). In the strain of Salmonella typhimurium known as TA100, the mutagenicity ratio was found to be greater than 2 at concentrations of 200, 500, and 1,000 μg per well. The process of cell division in the root tips of *A. cepa* and the bone marrow of mice was found to be significantly inhibited at doses of 2,000 and 4,000 mg/kg. However, the presence of chromosomal aberrations and micronucleated polychromatic erythrocytes did not show a dose-dependent relationship and were not significantly different from the negative control at a significance level of $P \geq 0.05$(A Akinboro et al., 2012). The extract decreased LDH enzymatic activity, and in HT29 human colon cancer cells, it substantially inhibited cell proliferation and the overall Warburg effect. Both lactate generation and LDH activity were reduced in extract-treated cells in normoxic and hypoxic conditions, even though extract had no effect on the expression of LDH-A (E-Y Kim et al., 2016).

36.12 TOXICITY

The utilization of medicinal plant extracts is significant in the management of many ailments owing to its distinctive phytochemical contents, cost-effectiveness, safety, and affordability. There is a prevailing belief among individuals that these extracts possess natural properties and are likely to exhibit superior efficacy compared to synthetic pharmaceuticals. Hence, it is imperative to conduct an inquiry into the symptoms of toxicity that may arise from the extended utilization of medicinal plant extracts to ascertain their safety. While some plant extracts have negative or no effects, others are toxic (RK Lima et al., 2012; N Chaachouay et al., 2021a, 2021b). Nutmeg is used as a spice and for medical purposes, which suggests that some of its ingredients may be responsible for the biological activities that have been noted. Some of these active principles might simultaneously have some negative effects (M Olaleye et al., 2006). The cytotoxic activity of nutmeg essential oil against Vero cells was found to be significantly low, as indicated by an IC_{50} value of 24.83 μg/mL(S Pillai et al., 2012). The mace essential oil demonstrated toxicity, as seen by its LC_{50} value of 31.05 μg/mL, indicating cytotoxicity against nauplii of *Artemia salina*. Additionally, a minimum hemolytic activity of 5.21% was observed at a dosage of 0.5 mg/mL (A Ali et al., 2021). No apparent indications of toxicity or mortality were observed in the acute toxicity test within 24 hours of administering extracts at doses up to 5000 mg/kg body weight (EG Anaduaka et al., 2022). The LD_{50} of alkaloids isolated from the seeds of *M. fragrans* was calculated to be 5.1 μg/kg. This relates to somewhat hazardous and practically nontoxic grades (AA-S Hayfaa et al., 2013). The essential oil's LC_{50} in the brine shrimp toxicity test was 5434.78 23.2 µg/mL, indicating that it is not harmful (LC_{50} > 1000 µg/mL) (Piaru, Mahmud, and Ismail, 2012a). The current investigation demonstrates that the acetone mace extract derived from *M. fragrans* does not induce observable in vivo toxicity in an animal model (R Gayathri et al., 2018).

36.13 CONCLUSION

Numerous medicinal plants and the substances generated from them have been shown to have pharmacological and neurotherapeutic effects. To organize the various existing and future studies aiming more efficiently at treating a wide range of human ailments, we need a complete understanding of the pharmacological effects of medicinal plants. The literature on the pharmacological and therapeutic effects of *M. fragrans* was reviewed in this study. According to the review, *M. fragrans*' anticancer, antibacterial, and antidiabetic activities have been investigated using both in vitro and in vivo animal models. The preclinical findings must be confirmed by additional well-planned clinical trials because the safety and efficacy in humans have not yet been thoroughly studied. The importance of developing a safe dosage standard cannot be overstated. To identify the chemical components of *M. fragrans*, ascertain their biological activity, ascertain whether they are safe and efficient for human use, and eventually obtain Food and Drug Administration approval, more study involving biologists, pharmacologists, and medical professionals is required.

LIST OF ABBREVIATIONS

EO	Essential oil, DPPH: 2,2-Diphenyl-1-picrylhydrazyl
MTT	3-(4,5-Dimethylthiazol-2-yl)-2,5-Diphenyltetrazolium Bromide
MBC	minimum bactericidal concentration
b.wt	Body weight
fw	Fresh weight
FRAP	Ferric-Reducing Antioxidant Power Assay
ABTS	2,2'-azino-bis(3-ethylbenzothiazoline-6-sulfonic acid)
IZD	Inhibition Zone Diameter

REFERENCES

MD Abdulrahman, AM Zakariya, HA Hama, SW Hamad, SS Al-Rawi, SW Bradosty, & AH Ibrahim. 2022. Ethnopharmacology, biological evaluation, and chemical composition of Ziziphus spina-christi (L.) Desf.: A review. *Advances in Pharmacological and Pharmaceutical Sciences* 2022. 2022(1): 4495688. https://doi.org/10.1155/2022/4495688

M Agarwal, A Chauhan, & T Gehani. 2009. Effects of *Myristica fragrans* oil on the reproductive organs of male albino rats. *Pharmaceutical Biology* 47(6): 509–515.

S Ahmad, A Latif, & IA Qasmi. 2003. Aphrodisiac activity of 50% ethanolic extracts of *Myristica fragrans* Houtt. (nutmeg) and Syzygium aromaticum (L) Merr. & Perry. (clove) in male mice: A comparative study. *BMC Complementary and Alternative Medicine* 3(1): 1–5.

A Akinboro, KB Mohamed, MZ Asmawi, AS Othman, TH Ying, & SM Maidin. 2012. Mutagenic and antimutagenic assessment of methanol leaf extract of *Myristica fragrans* (Houtt.) using in vitro and in vivo genetic assays. *Drug and Chemical Toxicology* 35(4): 412–422.

A Akinboro, KB Mohamed, MZ Asmawi, SF Sulaiman, & OA Sofiman. 2011. Antioxidants in aqueous extract of *Myristica fragrans* (Houtt.) suppress mitosis and cyclophosphamide-induced chromosomal aberrations in Allium cepa L. cells. *Journal of Zhejiang University Science B* 12: 915–922.

A Akinboro, KB Mohamed, MZ Asmawi, & TA Yekeen. 2014. Antimutagenic effects of aqueous fraction of *Myristica fragrans* (Houtt.) leaves on Salmonella typhimurium and Mus musculus. *Acta Biochimica Polonica* 61(4).

A Ali, M Yasir, MI Jilani, S Shoaib, M Latif, A Ahmad, S Naz, F Aslam, M Iqbal, & A Nazir. 2021. Chemical composition and in vitro evaluation of cytotoxicity, antioxidant and antimicrobial activities of essential oil extracted from *Myristica fragrans* houtt. *Polish Journal of Environmental Studies* 30(2).

EG Anaduaka, IU Okagu, NO Uchendu, LUS Ezeanyika, & BC Nwanguma. 2022. Hepato-renal toxicity of *Myristica fragrans* Houtt. (Myristicaceae) seed extracts in rats. *Journal of King Saud University-Science* 34(1): 101694.

I Antasionasti, OS Datu, US Lestari, SS Abdullah, & I Jayanto. 2021. Correlation Analysis of Antioxidant Activities with Tannin, Total Flavonoid, and Total Phenolic Contents of Nutmeg (*Myristica fragrans* Houtt) Fruit Precipitated by Egg white. *Borneo Journal of Pharmacy* 4(4): 301–310.

D Arulmozhi, R Kurian, A Veeranjaneyulu, & S Bodhankar. 2007. Antidiabetic and antihyperlipidemic effects of Myristica fragrans. in animal models. *Pharmaceutical Biology* 45(1): 64–68.

J Asgarpanah, & N Kazemivash. 2012. Phytochemistry and pharmacologic properties of *Myristica fragrans* Hoyutt.: A review. *African Journal of Biotechnology* 11(65): 12787–12793.

K Ashokkumar, J Simal-Gandara, M Murugan, MK Dhanya, & A Pandian. 2022. Nutmeg (*Myristica fragrans* Houtt.) essential oil: A review on its composition, biological, and pharmacological activities. *Phytotherapy Research* 36(7): 2839–2851.

JR Assa, SB Widjanarko, J Kusnadi, & S Berhimpon. 2014. Antioxidant potential of flesh, seed and mace of nutmeg (*Myristica fragrans* Houtt). *Int J Chem Tech Res* 6(4): 2460–2468.

LL Azis Ikhsanudin, & DD Rais. 2021. Anti-inflammatory activity of Indonesian nutmeg seeds (*Myristica fragrans* Houtt): A topical gel formulation. *International Journal of Public Health* 10(3): 689–695.

MS Bachri. 2013. Sub chronic effect of ethanol extract of nutmeg (*Myristica fragrans* Houtt) seed in rat kidney. *IN Proceeding of International Safety Management of Central Cytotoxic Reconstitution* Yogyakarta, Indonesia. 165–199.

O Bamidele, A Akinnuga, I Alagbonsi, O Ojo, J Olorunfemi, & M Akuyoma. 2011. Effects of ethanolic extract of *Myristica fragrans* Houtt. (nutmeg) on some heamatological indices in albino rats. *International Journal of Medicine and Medical Sciences* 3(6): 215–218.

SC BATM. 2016. Bioactivities of Ethanolic Extracts of Three Parts (Wood, Nutmeg and Mace) from *Myristica fragrans* Houtt. *J Med Assoc Thai* 99(4): S124–S130.

G Bell, & J Sander. 2001. CPD—Education and self-assessment The epidemiology of epilepsy: the size of the problem. *Seizure* 10(4): 306–316.

S Bhamarapravati, S Juthapruth, W Mahachai, & G Mahady. 2006. Antibacterial activity of *Boesenbergia rotunda* (L.) Mansf. and *Myristica fragrans* Houtt. against Helicobacter pylori. *Songklanakarin Journal of Science and Technology* 28(1): 157–163.

N Chaachouay, & L Zidane. 2021. The Symbolic Efficacy of Plants in Rituals and Socio-Religious Ceremonies in Morocco, Northwest of Africa. *Journal of Religious & Theological Information*, *21*(1–2), 34–53. https://doi.org/10.1080/10477845.2021.1942433

N Chaachouay, A Douira, & L Zidane. 2021a. Herbal Medicine Used in the Treatment of Human Diseases in the Rif, Northern Morocco. *Arabian Journal for Science and Engineering*, 47, 131–153. https://doi.org/10.1007/s13369-021-05501-1

N Chaachouay, A Douira, & L Zidane. 2021b. COVID-19, prevention and treatment with herbal medicine in the herbal markets of Salé Prefecture, North-Western Morocco. *European Journal of Integrative Medicine*, 42, 101285. https://doi.org/10.1016/j.eujim.2021.101285

JY Cho, GJ Choi, SW Son, KS Jang, HK Lim, SO Lee, ND Sung, KY Cho, & JC Kim. 2007. Isolation and antifungal activity of lignans from *Myristica fragrans* against various plant pathogenic fungi. *Pest Management Science: Formerly Pesticide Science* 63(9): 935–940.

J Chung, J Choo, M Lee, & J Hwang. 2006. Anticariogenic activity of macelignan isolated from *Myristica fragrans* (nutmeg) against Streptococcus mutans. *Phytomedicine* 13(4): 261–266.

K-T Chung, TY Wong, C-I Wei, Y-W Huang, & Y Lin. 1998. Tannins and human health: a review. *Critical Reviews in Food Science and Nutrition* 38(6): 421–464.

TD Cuong, TM Hung, HY Han, H Sik Roh, J-H Seok, JK Lee, JY Jeong, JS Choi, JA Kim, & BS Min. 2014. Potent acetylcholinesterase inhibitory compounds from Myristica fragrans. *Natural Product Communications* 9(4): 1934578X1400900418.

JK Dzotam, IK Simo, G Bitchagno, I Celik, LP Sandjo, P Tane, & V Kuete. 2018. In vitro antibacterial and antibiotic modifying activity of crude extract, fractions and 3′, 4′, 7-trihydroxyflavone from *Myristica fragrans* Houtt against MDR Gram-negative enteric bacteria. *BMC Complementary and Alternative Medicine* 18: 1–9.

TI Farag, MA Salama, SH Yahia, & RA Elfeqy. 2019. Therapeutic efficacy of thymus vulgaris and myristica fragrance houtt (nutmeg) ethanolic extractagainst toxoplasmosis in murine model. *Journal of the Egyptian Society of Parasitology* 49(1): 73–79.

R Gayathri, A Venkataraman, V Vishnupriya, & M Jainu. 2018. Acute toxicity studies of acetone mace extract of *Myristica fragrans* houtt on rats. *Drug Invention Today* 10(8). p1508.

B Ginting. 2021. Evaluation of antioxidant and anticancer activity of *Myristica fragrans* Houtt. bark. *Pharmacognosy Journal* 13(3): 780–786. DOI: 10.5530/pj.2021.13.99

B Ginting, N Saidi, & P Simanjuntak. 2020. Lignan compound isolated from n-Hexane extract *myristica fragrans* Houtt root as antioxidant and antitumor activities against MCF–7 cell lines data. *Data in Brief* 31: 105997.

D Gomes da Rocha Voris, L dos Santos Dias, J Alencar Lima, K dos Santos Cople Lima, JB Pereira Lima, & AL dos Santos Lima. 2018. Evaluation of larvicidal, adulticidal, and anticholinesterase activities of essential oils of *Illicium verum* Hook. f., *Pimenta dioica* (L.) Merr., and *Myristica fragrans* Houtt. against Zika virus vectors. *Environmental Science and Pollution Research* 25: 22541–22551.

AD Gupta, VK Bansal, V Babu, & N Maithil. 2013. Chemistry, antioxidant and antimicrobial potential of nutmeg (*Myristica fragrans* Houtt). *Journal of Genetic Engineering and Biotechnology* 11(1): 25–31.

H Hasbullah, DN Faridah, D Indrasti, FNA Dewi, & N Andarwulan. 2023. In vitro antioxidant and antidiabetic activity of mace from *Myristica Fragrans* houtt. *The Annals of the University Dunarea de Jos of Galati. Fascicle VI-Food Technology* 47(1): 95–109.

AA-S Hayfaa, AMA-S Sahar, & MA-S Awatif. 2013. Evaluation of analgesic activity and toxicity of alkaloids in *Myristica fragrans* seeds in mice. *Journal of Pain Research* 6:611–615.

J-K Hwang. 2010. Effects of macelignan isolated from *Myristica fragrans* Houtt. on UVB-induced matrix metalloproteinase–9 and cyclooxygenase–2 in HaCaT cells. *Journal of Dermatological Science* 57(2): 114–122.

L Jannu, S Hussain, & A Rao. 1991. Chemopreventive action of mace (Myristica fragrans, Houtt) on DMBA-induced papillomagenesis in the skin of mice. *Cancer Letters* 56(1): 59–63.

M Jukić, O Politeo, & M Miloš. 2006. Chemical composition and antioxidant effect of free volatile aglycones from nutmeg (*Myristica fragrans* Houtt.) compared to its essential oil. *Croatica chemica ACTA* 79(2): 209–214.

S Justin, & B Antony. 2016. Antibacterial activity of the essential oils of Syzygium aromaticum (L.) Merr. Perry (Clove), *Myristica fragrans* Houtt. (Nutmeg) and Zingiber officinale Roscoe (Ginger) against clinical isolates of Clostridium difficile: an in vitro study. *International Journal of Contemporary Medical Research* 3: 1085–1089.

I Kapoor, B Singh, G Singh, CS De Heluani, M De Lampasona, & CA Catalan. 2013. Chemical composition and antioxidant activity of essential oil and oleoresins of nutmeg (*Myristica fragrans* Houtt.) fruits. *International Journal of Food Properties* 16(5): 1059–1070.

JA Kazlauskaite, I Matulyte, M Marksa, R Lelesius, A Pavilonis, & J Bernatoniene. 2023. Application of Antiviral, Antioxidant and Antibacterial *Glycyrrhiza glabra* L., *Trifolium pratense* L. Extracts and *Myristica fragrans* Houtt. Essential Oil in Microcapsules. *Pharmaceutics* 15(2): 464.

E-Y Kim, H-J Choi, M-J Park, Y-S Jung, S-O Lee, K-J Kim, J-H Choi, T-W Chung, & K-T Ha. 2016. *Myristica fragrans* suppresses tumor growth and metabolism by inhibiting lactate dehydrogenase A. *The American Journal of Chinese Medicine* 44(05): 1063–1079.

K Lestari, A Diantini, MI Barliana, TH Achmad, A Subarnas, R Abdulah, R Lesmana, & JK Hwang. 2019. Potential natural dual agonist PPARα/γ-induced antidiabetic and antidyslipidemic properties of safrole-free nutmeg seed (*Myristica fragrans* Houtt) extract. *The Natural Products Journal* 9(3): 248–253.

K Lestari, J Hwang, SH Kariadi, A Wijaya, T Ahmad, A Subarnas, & M Muchtaridi. 2012. Screening for PPAR [gamma] agonist from *Myristica fragrans* Houtt seeds for the treatment of Type 2 diabetes by in vitro and in vivo. *Medical and Health Science Journal* 12: 7–16.

RK Lima, MDG Cardoso, MA Andrade, PL Guimarães, LR Batista, & DL Nelson. 2012. Bactericidal and antioxidant activity of essential oils from *Myristica fragrans* Houtt and Salvia microphylla HBK. *Journal of the American Oil Chemists' Society* 89(3): 523–528.

I Matulyte, A Jekabsone, L Jankauskaite, P Zavistanaviciute, V Sakiene, E Bartkiene, M Ruzauskas, DM Kopustinskiene,

A Santini, & J Bernatoniene. 2020. The essential oil and hydrolats from *Myristica fragrans* seeds with magnesium aluminometasilicate as excipient: antioxidant, antibacterial, and anti-inflammatory activity. *Foods* 9(1): 37.

SR Menthe, G Vedantam, & S Patil. 2020. Effect of Jatiphala (*Myristica fragrans* houtt.) in experimentally induced allergic asthma in male wistar Rats. *International Journal of Herbal Medicine* 8(3): 90–96.

T Morita, K Jinno, H Kawagishi, Y Arimoto, H Suganuma, T Inakuma, & K Sugiyama. 2003. Hepatoprotective effect of myristicin from nutmeg (Myristica fragrans) on lipopolysaccharide/d-galactosamine-induced liver injury. *Journal of Agricultural and Food Chemistry* 51(6): 1560–1565.

H Najafzadeh, S Esmailzadeh, M Ghorbanpour, A Farajzadeh, & A Asadirad. 2014. Comparing the effect of *Myristica fragrans* and flunixin on adjuvant-induced arthritis in rat. *Zahedan Journal of Research in Medical Sciences* 16(2): 33–36.

V Nikolic, L Nikolic, A Dinic, I Gajic, M Urosevic, L Stanojevic, J Stanojevic, & B Danilovic. 2021. Chemical composition, antioxidant and antimicrobial activity of nutmeg (*Myristica fragrans* Houtt.) seed essential oil. *Journal of Essential Oil Bearing Plants* 24(2): 218–227.

Z Niyas, PS Variyar, AS Gholap, & A Sharma. 2003. Effect of γ-irradiation on the lipid profile of nutmeg (*Myristica fragrans* Houtt.). *Journal of Agricultural and Food Chemistry* 51(22): 6502–6504.

N Nurhasanah. 2014. Antimicrobial activity of nutmeg (*Myristica fragrans* Houtt) fruit methanol extract against growth *Staphylococus aureus* and *Escherichia coli*. *Jurnal Bioedukasi* 2(1): 277–286.

N Nurmilasari, B Ginting, & H Helwati. 2017. Isolation of antioxidant compounds of methanol extract of nutmeg leaves (*Myristica fragrans* Houtt). *Jurnal Natural* 17(1): 49–57.

M Olaleye, C Akinmoladun, & A Akindahunsi. 2006. Antioxidant properties of *Myristica fragrans* (Houtt) and its effect on selected organs of albino rats. *African Journal of Biotechnology* 5(15).

S Omidpanah, Y Vahedi-Mazdabadi, A Manayi, A Rastegari, R Hariri, E Mortazavi-Ardestani, M Eftekhari, M Khanavi, T Akbarzadeh, & M Saeedi. 2022. Phytochemical investigation and anticholinesterase activity of ethyl acetate fraction of *Myristica fragrans* Houtt. seeds. *Natural Product Research* 36(2): 610–616.

Y Ozaki, S Soedigdo, YR Wattimena, & AG Suganda. 1989. Antiinflammatory effect of mace, aril of *Myristica fragrans* Houtt., and its active principles. *The Japanese Journal of Pharmacology* 49(2): 155–163.

K Panggabean, H Rusmarilin, & D Suryanto. (2019). *The utilization of nutmeg seed (Myristica fragrans Houtt) extract as an antimicrobial on tempeh sausage.* Paper presented at the *IOP Conference Series: Earth and Environmental Science.*

N Parimalaa, & S Amerjothyb. 2013. Histological and histochemical investigations of *Myristica fragrans* Houtt. (Myristicaceae). *Journal of Pharmacognosy and Phytochemistry* 1(5): 106–111.

SP Piaru, R Mahmud, & S Ismail. 2012a. Studies on the phytochemical properties and brine shrimp toxicity of essential oil extracted from *Myristica fragrans* Houtt. (Nutmeg). *Journal of Essential Oil Bearing Plants* 15(1): 53–57.

SP Piaru, R Mahmud, & S Perumal. 2012b. Determination of antibacterial activity of essential oil of *Myristica fragrans* Houtt. using tetrazolium microplate assay and its cytotoxic activity against vero cell line. *International Journal of Pharmacology* 8(6): e6.

S Pillai, R Mahmud, WC Lee, & S Perumal. 2012. Anti-parasitic activity of *Myristica fragrans* Houtt. essential oil against Toxoplasma gondii parasite. *APCBEE Procedia* 2: 92–96.

V Pooja, H Sanwal, A Goyal, S Bhatnagar, & A Srivastava. 2012. Activity of *Myristica fragrans* and its effect against filamentous and non-filamentous fungus. *International Journal of Pharmacy and Pharmaceutical Sciences* 4(1): 538–540.

E Prakash, & DK Gupta. 2013. Cytotoxic activity of ethanolic extract of *Myristica fragrans* (Houtt) against seven human cancer cell lines. *Journal of Food and Nutrition Science* 1(1): 1–3.

A Ram, P Lauria, R Gupta, & V Sharma. 1996. Hypolipidaemic effect of *Myristica fragrans* fruit extract in rabbits. *Journal of Ethnopharmacology* 55(1): 49–53.

A Rastegari, A Manayi, M Rezakazemi, M Eftekhari, M Khanavi, T Akbarzadeh, & M Saeedi. 2022. Phytochemical analysis and anticholinesterase activity of aril of *Myristica fragrans* Houtt. *BMC Chemistry* 16(1): 106.

G Rengasamy, A Venkataraman, VP Veeraraghavan, & M Jainu. 2018. Cytotoxic and apoptotic potential of *Myristica fragrans* Houtt. (mace) extract on human oral epidermal carcinoma KB cell lines. *Brazilian Journal of Pharmaceutical Sciences* 54.

I Rodianawati, P Hastuti, & MN Cahyanto. 2015. Nutmeg's (*Myristica fragrans* Houtt) oleoresin: effect of heating to chemical compositions and antifungal properties. *Procedia Food Science* 3: 244–254.

Y Rukayadi, K Lee, S Han, S Kim, & J-K Hwang. 2009. Antibacterial and sporicidal activity of macelignan isolated from nutmeg (*Myristica fragrans* Houtt.) against Bacillus cereus. *Food Science and Biotechnology* 18(5): 1301–1304.

P Sa-Nguanmoo, & Y Poovorawan. 2007. *Myristica fragrans* Houtt. methanolic extract induces apoptosis in a human leukemia cell line through SIRT1 mRNA downregulation. *J Med Assoc Thai* 90(11): 2422–2428.

A Sattar, A Abdo, MN Mushtaq, I Anjum, & A Anjum. 2019. Evaluation of gastro-protective activity of *Myristica fragrans* on ethanol-induced ulcer in albino rats. *Anais da Academia Brasileira de Ciências* 91.

Z Shafiei, NN Shuhairi, N Md Fazly Shah Yap, C-A Harry Sibungkil, & J Latip. 2012. Antibacterial activity of *Myristica fragrans* against oral pathogens. *Evidence-Based Complementary and Alternative Medicine* 2012.

M Sharma, & M Kumar. 2007. Radioprotection of Swiss albino mice by *Myristica fragrans* houtt. *Journal of Radiation Research* 48(2): 135–141.

G Singh, P Marimuthu, CSd Heluani, & C Catalan. 2005. Antimicrobial and antioxidant potentials of essential oil and acetone extract of *Myristica fragrans* Houtt. (aril part). *Journal of Food Science* 70(2): M141–M148.

S Singh, S Morya, P Saini, M Singh, & S Rout. 2016. Studies on antioxidant and antibacterial activity of nutmeg seed and its Mace. *Advances in Life Sciences* 5(8): 3207–3210.

G Sonavane, V Sarveiya, V Kasture, & SB Kasture. 2001. Behavioural actions of *Myristica fragrans* seeds. *Indian Journal of Pharmacology* 33(6): 417–424.

SF Sulaiman, & KL Ooi. 2012. Antioxidant and anti food-borne bacterial activities of extracts from leaf and different fruit parts of *Myristica fragrans* Houtt. *Food Control* 25(2): 533–536.

N Suthisamphat, B Dechayont, P Phuaklee, O Prajuabjinda, R-K Vilaichone, A Itharat, K Mokmued, & N Prommee. 2020.

Anti-Helicobacter pylori, anti-inflammatory, cytotoxic, and antioxidant activities of mace extracts from Myristica fragrans. *Evidence-Based Complementary and Alternative Medicine*. 2020(1): 7576818.

A Trifan, G Zengin, I Korona-Glowniak, K Skalicka-Woźniak, & SV Luca. 2023. Essential oils and sustainability: In vitro bioactivity screening of *Myristica fragrans* Houtt. Post-distillation by-products. *Plants* 12(9): 1741.

VMM Valente, GN Jham, OD Dhingra, & I Ghiviriga. 2011. Composition and antifungal activity of the Brazilian *Myristica fragrans* Houtt essential oil. *Journal of Food Safety* 31(2): 197–202.

Y Velioglu, G Mazza, L Gao, & B Oomah. 1998. Antioxidant activity and total phenolics in selected fruits, vegetables, and grain products. *Journal of Agricultural and Food Chemistry* 46(10): 4113–4117.

A Wahab, RU Haq, A Ahmed, RA Khan, & M Raza. 2009. Anticonvulsant activities of nutmeg oil of *Myristica fragrans*. *Phytotherapy Research: An International Journal Devoted to Pharmacological and Toxicological Evaluation of Natural Product Derivatives* 23(2): 153–158.

MF Warsito. 2021. A Review on Chemical Composition, Bioactivity, and Toxicity of *Myristica fragrans* Houtt. Essential Oil. *Indonesian Journal of Pharmacy/Majalah Farmasi Indonesia* 32(3).

DP Wibowo, Y Febriana, H Riasari, & DL Auilifa. 2018. Essential oil composition, antioxidant and antibacterial activities of nutmeg (*Myristica fragrans* Houtt) from Garut West Java. *Indonesian Journal of Pharmaceutical Science and Technology* 5(3): 82–87.

Yanti, Y Rukayadi, KH Kim, & JK Hwang. 2008. In vitro anti-biofilm activity of macelignan isolated from *Myristica fragrans* Houtt. against oral primary colonizer bacteria. *Phytotherapy Research: An International Journal Devoted to Pharmacological and Toxicological Evaluation of Natural Product Derivatives* 22(3): 308–312.

C Zhang, X Qi, Y Shi, Y Sun, S Li, X Gao, & H Yu. 2012. Estimation of trace elements in mace (*Myristica fragrans* Houtt) and their effect on uterine cervix cancer induced by methylcholanthrene. *Biological Trace Element Research* 149: 431–434.

W Zhao, F Song, D Hu, H Chen, Q Zhai, W Lu, J Zhao, H Zhang, W Chen, & Z Gu. 2020. The protective effect of *Myristica fragrans* Houtt. extracts against obesity and inflammation by regulating free fatty acids metabolism in nonalcoholic fatty liver disease. *Nutrients* 12(9): 2507.

37 Wild Rue (*Peganum harmala* L. Nitrariaceae)

Fathia Benlarbi and Noureddine Chaachouay

37.1 INTRODUCTION

For many years, medicinal plants have been employed to treat illnesses in humans and animals. *Peganum harmala* L., a glabrous perennial plant from the Nitrariaceae family, has a long history as a traditional herbal remedy for multiple disorders (Singhai & Patil., 2021). It can be found naturally in the Middle East, Central Asia, and North Africa, and has also been introduced to America and Australia. Known by different names such as 'Harmel', 'Espand', 'Turkish rue', 'African rue', and 'Mexican rue', this plant thrives in semi-arid conditions, steppe areas, and sandy soils. Its origins lie in the eastern Mediterranean region (Moloudizargari et al., 2013). *P. harmala*, or 'Harmel', is widely utilized in traditional medicine across various countries, with Algeria, including the High Plateaus, southern and northern Sahara, and the mountains of the Central Desert, being a significant user (Maire, 1933; Chopra et al., 1960; Ozenda, 1991; Rezzagui et al., 2020). Extensive research has been conducted on *P. harmala* for its potential as a traditional medicine and pharmacological applications.

P. harmala is widely used in traditional medicine for various purposes. These include the treatment of ulcers, diabetes, cardiovascular and gastrointestinal issues, nervous and endocrine disorders, neoplasms, tumors, and pain relief. It is also utilized for respiratory illnesses, as an antipyretic, and for skin and hair health. Additionally, it is used to address rheumatism, arthritis, inflammation, and colic disorders and as an insect-repellent and carminative. *P. harmala* has effectively managed conditions like menstrual disorders, urinary problems, paralysis, asthma, and epilepsy. It has been employed in treating diabetes, hypertension, wound healing, skin inflammation, hemorrhoids, cough, diarrhea, and subcutaneous tumors, as well as for Parkinson's disease, jaundice, lumbago, and as a central nervous system stimulant. Recent studies have explored its antileishmanial activities, while various authors worldwide have reported on the chemical diversity and biologically active compounds found in different parts of *P. harmala*. Over 308 compounds have been identified, including carbohydrates, flavonoids, anthraquinones, alkaloids, amino acids, vitamins, triterpenoids, fatty acids, phenylpropanoids, proteins, volatile oils, carotenes, sterols, and other trace elements. These bioactive substances contribute to its diverse pharmacological properties, such as anticancer, antiparasitic, antiviral, antimicrobial, antioxidant, neuroprotective, and pesticidal effects, which have been extensively documented through research.

The psychoactive properties of *P. harmala*'s alkaloid content has been established through research, and several studies indicate that *P. harmala* and its active alkaloids have diverse effects on both the central and peripheral nervous systems. These effects include pain relief, hallucinations, increased excitement, and an antidepressant impact (Moloudizargari et al., 2013). The seeds of *P. harmala* contain β-carbolines such as harmine, harmalol, and harman, with harmine being the active hallucinogenic alkaloid. Additionally, the seeds possess a red pigment used for dyeing wool and carpets and are used as a spice. Furthermore, in traditional medicine, they are regarded as valuable aphrodisiacs. This chapter aims to show the botanical features, geographical distribution, botany, ecology, and biological activities of *P. harmala*, along with traditional and modern medicinal uses. It highlights the phytochemical profile and hallucinogenic effects *of P. harmala*.

37.2 SCIENTIFIC CLASSIFICATION

Peganum harmala L. is an herbaceous perennial plant belonging to the *Peganum* genus and Nitrariaceae family (ITIS, 2022).

Kingdom: Plantae
Subkingdom: Viridiplantae
Infrakingdom: Streptophyta
Superdivision: Embryophyta
Division: Tracheophyta
Subdivision: Spermatophytina
Class: Magnoliopsida
Superorder: Rosanae
Order: Sapindales
Family: Nitrariaceae
Genus: *Peganum* L.
Species: *Peganum harmala* L.

37.3 VERNACULAR NAMES

Arabic: Harmel, hre-milan, sadab-sami, legherma, Álqat al-dib, sadab-bari, huraymilan, harjal; **English**: Syrian Rue, African Rue, Turkish Rue, Wild rue, Harmal peganum,

DOI: 10.1201/9781003460336-37

Harmal shrub, Harmel, Mexican Rue, Isband, Peganum, Ozallaik, Esfand; **German**: Harmalkraute, Steppenraute; **Spanish**: Alharma, Gamarza; **Swedish**: Harmelbuske; **French**: Rue sauvage, Péganum de Syrie, Rue de Syrie; **Armenian**: Spand sovorakan; **Azerbaijan**: Üzərrik; **Turkish**: Ilezik, uzarih, üzerlik; **Persian**: Asbatan, ispand, sifand, fašars īn, sipand, isfanj, isirik; **Chinese**: Luo tuo hao, Luo tuo peng; **Indian**: Hurmul.

37.4 SYNONYMS

Below is a compilation of five alternative terms used to refer to the botanical species *P. harmala* L., as documented in The Plant List (2013): **(1)** *Harmala multifida* All.**; (2)** *Harmala peganum* Crantz; **(3)** *Harmala syrica* Bubani**; (4)** *Peganum harmalum* (L.) St.–Lag.; **(5)** *Peganum dauricum* L. These various names serve as synonyms for *P. harmala* and are employed interchangeably within botanical literature and discussions.

37.5 BOTANICAL DESCRIPTION

P. harmala is a perennial herbaceous plant that grows between 30 and 100 cm tall and has thick rhizomes (Figure 37.1). It lacks hair and has a dense, bushy appearance. It emits a strong, pungent scent similar to rue, and its bitter sap deters animals from approaching it (Quezel et Santa., 1963; Khan et al., 2017; Rachid, 2021). During the winter, the erect and vigorous stems disappear. The plant's leaves are alternate and divided into narrow strips, retaining their green color for a portion of the dry season. The flowers are solitary, with five large elliptic petals measuring 25–30 mm and a yellowish-white-green hue. The plant produces tiny white blossoms in the branch axils, and its fruit is globose with numerous flattened seeds (Weckesser, 2013; Rachid, 2021). The fruit, which resembles tiny spherical capsules, has three chambers with a 6–10 mm diameter. It stands erect on the stem, featuring a depression at the top. Each capsule contains over 50 tiny triangular seeds that are small, angular, and dark brown, measuring approximately 3–4 mm in diameter and 2 mm in length (Moloudizargari et al., 2013; Khan et al., 2017).

37.6 ORIGIN AND GEOGRAPHIC DISTRIBUTION

P. harmala is extensively studied and considered one of the most researched species. Its distribution ranges from Mediterranean countries like Algeria, Morocco, Palestine, Italy, Cyprus, Spain, Greece, Turkey, and Jordan to Central Asia. This kind of plant, which can withstand dryness, is found in Central Asia in places like Tibet, Kashgaria, Tsaidam, and Mongolia (Amartuvshin et al., 2006; Passos & Mironidou-Tzouveleki., 2016; Sharifi-Rad et al., 2021). According to Khan et al. (2017), *P harmala* is indigenous to the eastern Mediterranean area and broadly distributed across the Middle East, India, China, and Mongolia. This species is extensively spread in North Africa, the Middle East, Turkey, Iran, and many more nations, including Pakistan, India, Kazakhstan, Mexico, and South America. Which is endemic to arid and semi-arid environments (Seilkhan et al., 2019; Zhao et al., 2011; Sharifi-Rad et al., 2021). In Algeria, *P. harmala* is expected in the High Plateaux, the northern and southern Sahara, and the Central Desert mountains. (Maire, 1933; Chopra et al., 1960 et Ozenda, 1991; Rezzagui et al., 2020).

37.7 HABITAT AND ECOLOGY

P. harmala can be found in countries such as Iran, Pakistan, India, Afghanistan, Turkey, Egypt, Morocco, and Spain (Chaachouay et al., 2020a, 2020b). It thrives in arid and semi-arid regions with dry, rocky soils and is well-suited to hot, dry climates. It can endure high temperatures and is commonly observed in desert and steppe ecosystems. The plant species thrive in areas with a Mediterranean climate, distinguished by scorching, arid summers and gentle, damp winters. It tolerates various soil conditions, including sandy, loamy, and rocky soils (Narantsetseg, 2014). It is frequently found in areas with well-drained soils, such as sandy or gravelly soils. The plant is adaptable to different elevations, ranging from lowlands to high mountain regions, and can be found at elevations up to 2,000 meters above sea level

FIGURE 37.1 Various parts of *Peganum harmala* L. (A) Flowers, (B) leaves, (C) fruit, (D) seeds.

(El-Bakatoushi et al., 2011). *P. harmala* is often encountered in open, disturbed habitats such as wastelands, roadsides, abandoned fields, and pastures. It can also grow on rocky slopes, hillsides, and shrubby areas. The plant is classified as a ruderal species, thriving in areas with disrupted vegetation. Due to its vigorous growth and abundant seed production, *P. harmala* can become invasive in certain regions. It can outcompete native plants and establish dense stands, particularly in arid and semi-arid ecosystems (Naderi et al., 2011).

P. harmala is a perennial shrub that typically grows to a height of 1 to 1.5 meters. It has a spreading growth habit with multiple stems from a woody base. The plant features deeply divided leaves and small, fragrant, yellow flowers. *P. harmala* has adapted well to hot and dry climates, developing mechanisms to withstand prolonged periods of drought and tolerate arid conditions (Davoodi et al., 2015). It contains various alkaloids, including harmine, harmaline, and tetrahydroharmine, which serve as defensive compounds against herbivores and pests. The plant produces numerous tiny seeds in seed capsules, characterized by a hard outer coat that keeps them viable for an extended period (Nedjimi, 2020; Grebennikov, 2017). This adaptation facilitates the colonization of disturbed habitats by the plant and contributes to its potential invasiveness in certain regions. In areas where *P. harmala* has become invasive, it can outcompete native vegetation and form dense stands, impacting the diversity and composition of plant communities, particularly in arid and semi-arid ecosystems (Abbott et al., 2007).

37.8 PHYTOCHEMISTRY

Phytochemical analysis refers to the study of bioactive compounds present in plants. In the case of *P. harmala*, several studies have been conducted to identify and characterize its phytochemical constituents (Table 37.1). The information provided outlines the findings from various studies regarding compounds in *P. harmala*. Khan et al. (2017) identified alkaloids, flavonoids, and anthraquinones as common substances in *P. harmala*. Alkaloids are nitrogenous compounds known for their diverse biological activities and are often found in medicinal plants. Flavonoids are a group of plant pigments with antioxidant and anti-inflammatory properties, while anthraquinones are organic compounds with potential therapeutic effects. Farouk et al. (2009) and Khan et al. (2011) conducted separate studies on *P. harmala* and reported the presence of several compounds in addition to those mentioned above. These compounds include anthraquinones, flavonoids, tannins, alkaloids, flavones, saponins, chalcones, terpenoids, cardiac glycosides, and phlobatannins. Tannins are astringent compounds commonly found in plants and can have antimicrobial and antioxidant properties. Flavones are a type of flavonoid, while saponins are glycosides with diverse biological activities. Chalcones are another class of compounds known for their various pharmacological properties. Terpenoids, or terpenes, are a large group of natural products with a wide range of biological functions. Cardiac glycosides are compounds that can affect the heart muscle, and phlobatannins are a group of phenolic compounds. Shaheen et al. (2022)

TABLE 37.1
Bioactive Compound Present in Different Plant Parts of *P. harmala*

Plant parts	Bioactive compounds	References
Flowers	Harmine; peganine	(Iranshahy et al., 2019)
	Lysine; asparagine; proline; 4 hydroxyisoleucine; phosphorylcholine; sucrose; vasicine; choline	(Li et al., 2018)
Stem	Asparagine; asparagine; malic acid; lysine; 4-hydroxyisoleucine; succinic acid; vasicine; sucrose; choline; vasicinone; proline; phosphorylcholine	(Li et al., 2018)
Fruits	Acetic acid; asparagine; α-Glucose; choline; lysine; 4-hydroxyisoleucine; sucrose; choline; proline; vasicine; phosphorylcholine	(Li et al., 2018)
	Harmine; peganine; harmaline	(Iranshahy et al., 2019)
Roots	Harmine; vasicine; vasicinone; harmaline	(Ayoob et al., 2017)
	Asparagine; acetic acid; β-glucose; formic acid; threonine; lysine; sucrose; proline; 4-hydroxyisoleucine; phosphorylcholine	Li et al., 2018
	Harmol	Zheng et al., 2009
Seeds	Harmol; harmine; harmalol; norharmane; harmane; harmaline	(Tascon et al., 2015)
	Tetrahydroharmine; harmine; harmalol; diglycoside vasicine; ruine; 8-hydroxy-harmine; vasicinone; harmaline; harmol; vasicine;	(Liu et al., 2013)
	Anthraquinoneglucoside; stearic acid; quercetin-3-O-gentiobioside; kaempferol3-methyoxyl-5-Orutinoside; quercetin-3-O-rutinoside; isorhamnetin-7-O-glucoside; isorhamnetin-3-O-rutinoside; 9,14-dihydroxyoctadecanoicacid; 3,5-dimethyxyl-7-O-glucoside; kaempferol-kaempferol-3-methyoxyl-7-O-glucoside;	(Li, 2005)
Leaves	Alanine; acetic acid; asparagine; valine; proline; betaine; maleic acid; isoleucine; choline; lysine; sucrose; 4-hydroxyisoleucine; phosphorylcholine vasicine	(Li et al., 2018)

conducted a screening of *P. harmala* leaves and reported the presence of several compounds, including terpenes, cardiac glycosides, flavonoids, carbohydrates, saponins, alkaloids, steroids, tannins, and phenolic compounds. Steroids are a class of lipids with various biological activities, and phenolic compounds are a diverse group of secondary metabolites known for their antioxidant properties.

Zhu et al. (2022) studied various compounds found in *P. harmala*. From this plant, they identified more than 308 compounds, which consist of 97 alkaloids, 24 flavonoids, 18 carbohydrates, 17 amino acids, ten triterpenoids, three anthraquinones, two phenylpropanoids, 26 fatty acids, 99 volatile oils, three sterols, one carotene, one protein, one vitamin, and six other trace elements. Among these substances, the predominant content is β-carboline alkaloids (βCs), with alkaloids comprising up to 10% of the seeds' composition, followed by the roots. At the same time, leaves contain the most minor proportion. The primary substances in *P. harmala* include harmaline, harmine, harmane, and harmalol.

According to the authors, the main bioactive substances found in *P. harmala* are alkaloids, including harman, harmine, harmalol, and harmaline (Tascon et al., 2015; Li, 2017; Rachida et al., 2021). The richest sources of these alkaloids are the seeds and roots, while the stems and leaves have lower levels, and the flowers do not contain them (Khan et al., 2017). It is important to note that the alkaloids in *P. harmala* can interact dangerously with antihistamines, antidepressants, decongestants, expectorants, and certain stimulants (Dastagir et al., 2014; Ataee et al., 2018). Lamchouri et al. (2002) also demonstrated that harmaline, the main active compound in *P. harmala* seeds, and its derivatives can.

37.9 ETHNOBOTANY

37.9.1 Traditional Medicine Uses

P. harmala is a plant used for various traditional medicinal purposes in different cultures throughout history. *P. harmala* exhibits a range of therapeutic effects, such as its ability to protect the liver, induce abortion, act as a pesticide, and address conditions like candidiasis, inflammation, and cholinesterase inhibition. Additionally, it demonstrates antibacterial, antimicrobial, anticancer, angiogenic, antiparasitic, antioxidant, and cytotoxic properties (Miraj, 2016; Majid, 2018; Chaachouay et al., 2021a, 2021b, 2021c; Orch et al., 2021; El Khomsi et al., 2022). In traditional medicine, *P. harmala* is commonly employed to treat ulcers, diabetes, cardiovascular issues, gastrointestinal disorders, nervous system ailments, endocrine imbalances, neoplasms, tumors, pain, respiratory illnesses, fever, skin and hair conditions, rheumatism, arthritis, inflammation (Lamchouri, 1999; Wang et al., 2016; Chaachouay et al., 2019a, 2019b; Belhaj et al., 2021).

Wang et al. (2016) highlighted the traditional use of *P. harmala* in Chinese and Uygur medicine for cancer treatment, explicitly identifying its potential as an anti-lung cancer drug by targeting Osteoarthritis (OA). In 2023, Benlarbi et al. reported on the antileishmanial activities of *P. harmala* seeds. Additionally, *P. harmala* is recognized in traditional Iranian medicine for its diverse properties, including analgesic, intoxicating, abortive, disinfectant, anthelmintic, insect-repellent, carminative effects, as well as its effectiveness in treating colic, asthma, paralysis, gastrointestinal and urinary problems, epilepsy, and menstrual disorders (Shahrajabian et al., 2021; Benkhnigue et al., 2022; Chaachouay et al., 2022a, 2022b, 2022c).

The medicinal properties of *P. harmala* have gained significant recognition due to their extensive use in various treatments. These include diabetes, asthma, arthritis, hypertension, wound healing, skin inflammation, hemorrhoids, and cough (Sharifi-Rad et al., 2021; Zhu et al., 2022). Additionally, *P. harmala* has been utilized for managing conditions like diarrhea, pain, abortion, rheumatism, fever, subcutaneous tumors, and Parkinson's disease (Moloudizargari et al., 2013). Moreover, it has been employed for treating jaundice, lumbago and as a central nervous system stimulant (Sharifi-Rad et al., 2021; Chaachouay et al., 2022b). Traditional herbal medicine in Algeria and Azerbaijan recognizes Syrian rue as beneficial for addressing insomnia and depression (Bakiri et al., 2016; Mamedov et al., 2018).

37.9.2 As a Charm Against the Evil Eye

P. harmala, a plant known for its various superstitious applications, is utilized in different cultures and regions to protect against supernatural beings and the malevolent influence of the evil eye. For instance, the plant is set on fire in Morocco to ward of ghosts and goblins, serving as a protective measure (Chaachouay & Zidane, 2021). Similarly, in China and other Arab countries in Africa and Asia, *P. harmala* is employed to safeguard individuals or environments from the harmful gaze of the evil eye (Sharifi-Rad et al., 2021). In Turkey, dried capsules from *P. harmala* are hung in households as a means of protection against the influence of the evil eye. In Iran, a blend of dried capsules of *P. harmala* and other ingredients is burned, releasing fragrant smoke that acts as an air and mental purifier. This practice is primarily used as a charm against the evil eye and as a purification method (Frison et al., 2008). Additionally, Khan et al. (2017) reported that dried capsules of *P. harmala* are burned after childbirth in Northwestern India and West Pakistan. The seeds are placed on burning charcoal, and the resulting smoke permeates the rooms for several days, particularly in households where a baby has been born. The burning of *P. harmala* seeds is also incorporated into marriage ceremonies. Throughout history, this plant has gained significant recognition in traditional medicine, being revered as a remedy for various ailments.

37.9.3 As a Natural Colorant

P. harmala is a plant traditionally employed for various purposes, as El Khomsi et al. mentioned in 2022. While it possesses specific compounds that can serve as natural colorants, its utilization is limited and limited. However, the seeds of *P. harmala* are precious as a source of natural colorants, according to Adeel et al. (2022). These seeds offer an excellent alternative to carcinogenic yellow synthetic colorants. They contain a red pigment that has been utilized for dyeing wool and carpets, and they are also used as a spice and sexual stimulant in Central Asian countries, as reported by Mamedov et al. (2018).

37.10 HALLUCINOGENIC EFFECT

P. harmala has traditionally been used in folk medicine in some Levant areas as a sedative and abortifacient. Similar to ayahuasca, it is also employed as a hallucinogen (Aronson, 2016). *P. harmala* is typically ingested in crushed grains (3–4 g) or combined with other plants containing DMT. When taken in small doses (25–50 mg), it provides mild brain stimulation that leads to mild depression, resulting in drowsiness or a dream-like state lasting 1 to 2 hours. Higher doses (300–750 mg) induce hallucinations. *P. harmala*, particularly harmaline, can be combined with plants containing DMT (N, N-dimethyltryptamine) to create the beverage 'ayahuasca'. Harmine, a constituent of *P. harmala*, acts as a central nervous system stimulant and a reversible inhibitor of MAO (MAOI), which can contribute to hallucinogenic effects. The recommended amount of *P. harmala* in preparing the ayahuasca drink is 5–10 g. Therefore, due to its β-carboline content, *P. harmala* can serve as a critical ingredient in the herbal recipe for ayahuasca (Passos & Mironidou-Tzouveleki, 2016). Ayahuasca induces a psychedelic and visionary state of mind, often used for medical and religious purposes. Shamans may employ it to gain insights and make diagnoses. During South American shamanic ceremonies, individuals gather to consume ayahuasca and collectively enter a trance-like state through chanting. When taken orally, the hallucinogenic effects of ayahuasca become noticeable after 30–60 minutes, peak at 60–120 minutes, and subside by 240 minutes (Aronson, 2016).

37.11 TOXICITY

P. harmala contains several bioactive compounds, some of which can be toxic if misused or in large amounts (Chaachouay et al., 2020a). This plant's beta-carboline alkaloids (harmine, harmaline, and tetrahydroharmine) are responsible for toxic effects. The toxic effects of this plant are attributed to its beta-carboline alkaloids, namely harmine, harmaline, and tetrahydroharmine. These compounds can stimulate the central nervous system (CNS) by inhibiting the breakdown of amine neurotransmitters or directly interacting with specific receptors. Plant preparations containing these beta-carboline alkaloids can have harmful consequences, such as locomotor issues, ataxia, nausea, vomiting, confusion, agitation, and visual and auditory hallucinations (Aronson, 2016). Harmaline and other beta-carbolines have demonstrated diverse effects on the CNS. Moderate doses of these compounds can generate hallucinogenic effects resembling those caused by other psychedelic substances. However, taking higher doses or using them for an extended period can result in more severe and potentially hazardous outcomes. According to clinical studies on acute intoxications caused by Harmel, it was observed that Harmalin, when administered at a dose of 4 mg/kg, induces psycho-mimetic effects in humans (Frison et al., 2008). The toxicity of *P. harmala* can manifest in different ways depending on the dosage and individual factors. Some of the potential toxic effects include:

- **Central Nervous System Effects**: *P. harmala* can cause central nervous system stimulation, leading to restlessness, tremors, increased heart rate, and high blood pressure. In severe cases, it may result in seizures or convulsions.
- **Serotonin Syndrome**: *P. harmala* contains harmaline, a monoamine oxidase inhibitor (MAOI). When used alongside specific medications or substances that impact serotonin levels, such as antidepressants or certain recreational drugs, there is an elevated chance of developing serotonin syndrome. Serotonin syndrome is a potentially perilous condition characterized by restlessness, elevated body temperature, fast heart rate, dilated pupils, and muscular stiffness.
- **Gastrointestinal Disturbances**: Ingesting *P. harmala* can cause gastrointestinal issues such as vomiting, nausea, and diarrhea.
- **Allergic Reactions**: Some individuals may be allergic to *P. harmala*. Allergenic responses can vary from minor skin rashes to a grave form of allergic reaction known as anaphylaxis, which can be life-threatening.
- **Drug Interactions**: As mentioned earlier, *P. harmala* contains harmaline, a potent MAOI.

It can interact with various medications and substances, including certain antidepressants, stimulants, and other psychoactive drugs. These interactions can be dangerous and may result in adverse effects or potentiation of the substances involved. It is worth noting that *P. harmala* has been used safely for various purposes throughout history when used in appropriate doses and with caution. However, due to its potential toxicity and interactions with other substances, it is crucial to exercise caution and consult a healthcare professional before using *P. harmala* for any purpose, particularly if you are taking other medications or have preexisting medical conditions.

37.12 PHARMACOLOGICAL PROPERTIES

The pharmacological properties of *P. harmala* are primarily attributed to the presence of several bioactive alkaloids, including harmine, harmaline, and tetrahydroharmine. *P. harmala* contains numerous compounds, with over 308 identified, including 97 alkaloids, as Zahu et al. (2022) mentioned. The primary biologically active substances in *P. harmala* are β-carboline alkaloids (βCs), which include harman, harmine, harmalol, and harmaline, as noted by Tascon et al. (2015), Li (2017), and Rachida et al. (2021). In the present, we know this plant possesses various pharmacological actions extending beyond its hallucinogenic properties. These actions include combating viruses, cancer, convulsions, and fever. The β-carboline family of compounds is responsible for a broad range of pharmacological effects, as highlighted by Singhai and Patil (2021) and Moghadam et al. (2010). Specifically, they have shown effectiveness against specific inflammatory dermatoses such as impetigo, pityriasis alba, cutaneous leishmaniasis, and others. The alkaloid extract derived from *P. harmala* seeds is believed to possess antibacterial, anticancer, antifungal, antipruritic, and potentially antiprotozoal effects, as Khan et al. (2017) reported. Table 37.2 provides several examples of the pharmacological properties of *P. harmala*.

37.13 CLINICAL USES

It is important to note that the clinical uses of *P. harmala* have not been extensively studied, and its safety and efficacy for medical purposes have yet to be well established. Some potential clinical uses have been explored or suggested, but further research is needed to validate these claims. Thus, caution should be exercised when using *P. harmala*, and it should only be used under the supervision of qualified healthcare professionals. Additionally, *P. harmala* may interact with certain medications and should not be used by individuals with certain medical conditions, such as psychiatric disorders or liver disease. Always consult a healthcare provider before using *P. harmala* or any other herbal remedy for clinical purposes. The main findings are mentioned in (Table 27.3).

37.14 CONCLUSION

P. harmala is a hairless plant that grows perennially as an herb in arid and semi-arid regions of North Africa, the Middle East, and Central Asia. Its unpalatable taste deters animals from grazing on it. The plant contains a wide array of compounds, with over 308 identified so far, including alkaloids, flavonoids, carbohydrates, amino acids, triterpenoids, anthraquinones, phenylpropanoids, vitamins, volatile oils, sterols, fatty acids, carotenes, proteins, and trace elements. These alkaloids have psychoactive properties, and both in vitro and in vivo studies have shown diverse effects on the central and peripheral nervous systems, such as hallucination, excitation, analgesia, and antidepressant effects. The unique properties of *P. harmala* include its anticancer, neuroprotective, antiviral, antioxidant, antimicrobial, pesticide, and antiparasitic activities. The alkaloids in *P. harmala* possess significant health benefits and exhibit various biological activities, making them potentially useful for pharmaceutical and clinical purposes. However, it is essential to exercise caution when using *P. harmala* extracts for therapeutic purposes due to potential cytotoxic effects, such as hepatic nephritic toxicities, visual and auditory hallucinations, confusion, and agitation, especially when consumed in high doses over prolonged periods. Based on the available data, scientists and medical professionals need to exercise caution when considering the utilization of this particular plant for therapeutic reasons until rigorous scientific investigations can confirm its safety and quality. With the progress of biotechnology, advancements in this field could allow for the extraction of bioactive compounds from the plant, thereby providing safe and effective therapeutic options.

TABLE 37.2
Pharmacological Properties of *P. harmala*

Pharmacological properties	References
Anticancer	(Lamchouriet al., 2000; Daoud et al., 2014; Bournine et al., 2017)
Antiviral	(Kiani et al., 2007; Moradi et al., 2017; Edziri et al., 2018; Chen et al., 2018; Fahmy et al., 2022)
Anti-inflammatory	(Kumar et al., 2015; Liu et al., 2017a; Edziri et al., 2018)
Anti-bacteria	(Akbary et al., 2015; Shaheen & Issa, 2020; Sharifi-Rad et al., 2021)
Anti-proliferative	(Li et al., 2020)
Antioxidant	(Hayet et al., 2010; Soliman et al., 2013; Sharifi-Rad et al., 2021)
Antihyperglycemic	(Naresh et al., 2012)
Antidepressant	(Bakim et al., 2012; Moloudizargari et al., 2013; Sassoui et al., 2015)
Herbicidal	(Sodaeizadeh et al., 2009; Sodaeizadeh et al., 2010)
Anti-tuberculosis	(Davoodi et al., 2015; Jahanpour et al., 2015)
Anti-diabetes	(Tahraoui et al., 2007; Komeili et al., 2016)
Antileishmania	(Soosaraei et al., 2017; Lala, et al., 2004; Mirzaie et al., 2007; Khaliq et al., 2009)
Alzheimer	(Liu et al., 2017b; Filali et al., 2015)
Insecticidal	(Jbilou et al. 2008; Salari et al., 2012; Ismahane et al., 2016)
Immune-modulatory	(Farzin and Mansouri, 2006; Moloudizargari et al., 2013)

TABLE 37.3
Clinical Uses of *P. harmala*

Clinical uses	References
Knee Osteoarthritis: the efficacy of *P. harmala* oil in pain manage ment, stiffness, tenderness	(Abolhassanzadeh et al., 2015; Sharifi-Rad et al., 2021).
Dermatoses: Multiple skin conditions were assessed in Egypt through a clinical trial that investigated the antibacterial and antiprotozoal effects of *P. harmala*.	(El-Rifaie, 1980; Sharifi-Rad et al., 2021).
Gastroesophageal Cancer: In patients diagnosed with gastroesophageal cancer, Spinal-Z's therapeutic effects and potential adverse events were assessed in three separate studies. Spinal-Z contains *P. harmala* and Dracocephalum kotschyi Boiss as its primary constituents.	(Yazdanbod et al., 2001; Mashak et al., 2017; Panahi et al., 2018)
Benign Prostatic Hyperplasia: Protective against prostatic hyperplasia	(Shirani-Boroujeni et al., 2017; Sharifi-Rad et al., 2021).

LIST OF ABBREVIATIONS

β-Cs	β-carboline alkaloids
AChEIs	Acetylcholinesterase inhibitors
BChE	Butyrylcholinesterase
MAO-A	monoamine oxidase-A
MAO-B	monoamine oxidase-B
MAOI	monoamine oxidase inhibitor
CNS	Central Nervous System
DMT	N, N-dimethyltryptamine
OA	Osteoarthritis

REFERENCES

Abbott, L. B., Lepak, D., & Daniel, D. L. (2007). Vegetative and reproductive phenology of African rue (*Peganum harmala*) in the northern Chihuahuan Desert. *The Southwestern Naturalist*, *52*(2), 209–218.

Abolhassanzadeh, Z., Aflaki, E., Yousefi, G., & Mohagheghzadeh, A. (2015). Randomized clinical trial of peganum oil for knee osteoarthritis. *Journal of Evidence-Based Complementary & Alternative Medicine*, *20*(2), 126–131.

Adeel, S., Anjum, F., Zuber, M., Hussaan, M., Amin, N., & Ozomay, M. (2022). Sustainable extraction of colourant from harmal seeds (*Peganum harmala*) for dyeing of bio-mordanted wool fabric. *Sustainability*, *14*(19), 12226.

Akbary, P., Fereidouni, M. S., & Akhlaghi, M. (2015). In vitro antibacterial activity of *Peganum harmala* (L) extract to some fish pathogenic bacteria. *Sustainable Aquaculture and Health Management Journal*, *1*(1), 7–16.

Amartuvshin, N., Dariimaa, S., & Tserenbaljid, G. (2006). Taxonomy of the genus Peganum L. (Peganaceae VAN TIEGHEM) in Mongolia. *Mongolian Journal of Biological Sciences*, *4*(2), 9–13.

Aronson, J. K. (2016). *Meyler's side effects of drugs (Sixteenth Edition): The international encyclopedia of adverse drug reactions and interactions*. Elsevier. 611–612.

Seilkhan, A. S., Kudrina, N. O., Kulmanov, T. E., Kurmanbayeva, M. S., Inelova, Z. A., & Shalgimbayeva, S. M. (2019). Anatomical and morphological structure of *Peganum harmala* of Almaty region and its therapeutic properties. *Pakistan Journal of Botany*, *51*(2), 649–655.

Ataee, Z., Dadpour, B., Najari, F., Rahimpour, M., & Najari, D. (2018). Acute Poisoning With *Peganum Harmala*, Esfand: A Rare Case Report. *Int. J. Med. Toxicol. Forensic Med*, *8*, 119–121.

Ayoob, I., Hazari, Y. M., Lone, S. H., Khuroo, M. A., Fazili, K. M., & Bhat, K. A. (2017). Phytochemical and cytotoxic evaluation of *peganum harmala*: structure activity relationship studies of harmine. *ChemistrySelect*, *2*(10), 2965–2968.

Bakiri, N., Bezzi, M., Khelifi, L., & Khelifi-Slaoui, M. (2016). Enquête ethnobotanique d'une plante médicinale *Peganum harmala* L. dans la région de M'sila. *Revue Agriculture*, *1*, 38–42.

Bakim, B., Sertcelik, S., & Tankaya, O. (2012). A case of serotonin syndrome with. antidepressant treatment and concomitant use of the herbal remedy (*peganum harmala*). *Klinik Psikofarmakoloji Bülteni-Bulletin of Clinical Psychopharmacology*, *22*(4), 359–361.

Belhaj, S., Chaachouay, N., & Zidane, L. (2021). Ethnobotanical and toxicology study of medicinal plants used for the treatment of diabetes in the High Atlas Central of Morocco. *Journal of Pharmacy & Pharmacognosy Research*, 9(5), 619–662.

Benlarbi, F., Mimoune, N., Chaachouay, N., Souttou, K., Saidi, R., Mokhtar, M. R., … Benaissa, M. H. (2023). Ethnobotanical survey of the traditional antiparasitic use of medicinal plants in humans and animals in Laghouat (Southern Algeria). *Veterinary World*, *16*(2), 357–368.

Benkhnigue, O., Chaachouay, N., Khamar, H., El Azzouzi, F., Douira, A., & Zidane, L. (2022). Ethnobotanical and ethnopharmacological study of medicinal plants used in the treatment of anemia in the region of Haouz-Rehamna (Morocco). *Journal of Pharmacy & Pharmacognosy Research*, 10(2), 279–302.

Bournine, L., Bensalem, S., Fatmi, S., Bedjou, F., Mathieu, V., Iguer-Ouada, M., … Duez, P. (2017). Evaluation of the cytotoxic and cytostatic activities of alkaloid extracts from different parts of *Peganum harmala* L.(Zygophyllaceae). *European Journal of Integrative Medicine*, *9*, 91–96.

Chaachouay, N., Benkhnigue, O., Fadli, M., Ibaoui, H., & Zidane, L. (2019a). Ethnobotanical and ethnopharmacological studies of medicinal and aromatic plants used in the treatment of metabolic diseases in the Moroccan Rif. *Heliyon*, *5*(10), e02191. https://doi.org/10.1080/10477845.2021.1942433

Chaachouay, N., Benkhnigue, O., Fadli, M., Ibaoui, H., El Ayadi, R., & Zidane, L. (2019b). Ethnobotanical and ethnopharmacological study of medicinal and aromatic plants used in the treatment of respiratory system disorders in the Moroccan Rif. *Ethnobotany Research and Applications*, *18*(22), 1–16.

Chaachouay, N., Douira, A., Hassikou, R., Brhadda, N., Dahmani, J., Belahbib, N., Ziri, R., & Zidane, L. (2020a). Ethnobotanical and ethnopharmacological study of medicinal and aromatic plants used in the treatment of respiratory system disorders in the Moroccan Rif. PhDThesis, Département de Biologie - Université Ibn Tofail - Kénitra, Maroc, https://tel.archives-ouvertes.fr/tel-03376377

Chaachouay, N., Benkhnigue, O., & Zidane, L. (2020b). Ethnobotanical study aimed at investigating the use of medicinal plants to treat nervous system diseases in the Rif of Morocco. *Journal of chiropractic medicine*, *19*(1), 70–81. https://doi.org/10.1016/j.jcm.2020.02.004

Chaachouay, N., & Zidane, L. (2021). The Symbolic Efficacy of Plants in Rituals and Socio-Religious Ceremonies in Morocco, Northwest of Africa. *Journal of Religious & Theological Information*, *21*(1–2), 34–53. https://doi.org/10.1080/10477845.2021.1942433

Chaachouay, N., Benkhnigue, O., Douira, A., & Zidane, L. (2021a). Poisonous medicinal plants used in the popular pharmacopoeia of the Rif, northern Morocco. *Toxicon*, 189, 24–32. https://doi.org/10.1016/j.toxicon.2020.10.028

Chaachouay, N., Douira, A., & Zidane, L. (2021b). COVID-19, prevention and treatment with herbal medicine in the herbal markets of Salé Prefecture, North-Western Morocco. *European Journal of Integrative Medicine*, 42, 101285. https://doi.org/10.1016/j.eujim.2021.101285

Chaachouay, N., Douira, A., & Zidane, L. (2021c). Herbal Medicine Used in the Treatment of Human Diseases in the Rif, Northern Morocco. *Arabian Journal for Science and Engineering*, 47, 131–153. https://doi.org/10.1007/s13369-021-05501-1

Chaachouay, N., Benkhnigue, O., & Zidane, L. (2022a). Ethnobotanical and Ethnomedicinal study of medicinal and aromatic plants used against dermatological diseases by the people of Rif, Morocco. *Journal of Herbal Medicine*, 32, 100542. https://doi.org/10.1016/j.hermed.2022.100542

Chaachouay, N., Azeroual, A., Bencharki, B., Douira, A., & Zidane, L. (2022b). Ethnoveterinary medicines plants for animal therapy in the Rif, North of Morocco. *South African Journal of Botany*, *147*, 176–191. https://doi.org/10.1016/j.sajb.2021.12.037

Chaachouay, N., Azeroual, A., Douira, A., & Zidane, L. (2022c). Ethnoveterinary practices of medicinal plants among the Zemmour and Zayane tribes, Middle Atlas, Morocco. *South African Journal of Botany*, *151*, 826–840. https://doi.org/10.1016/j.sajb.2022.11.009

Chopra C., Abrol, B.K. and Handa K.L. (1960). Les plantes médicinales des régions arides. Recherche sur les zones arides. Ed UNESCO, Rome, 97.

Chen, D., Tian, X., Zou, X., Xu, S., Wang, H., Zheng, N., & Wu, Z. (2018). Harmine, a small molecule derived from natural sources, inhibits enterovirus 71 replication by targeting NF-κB pathway. *International Immunopharmacology*, *60*, 111–120.

Daoud, A., Song, J., Xiao, F., & Shang, J. (2014). B-9–3, a novel β-carboline derivative exhibits anti-cancer activity via induction of apoptosis and inhibition of cell migration in vitro. *European Journal of Pharmacology*, *724*, 219–230.

Dastagir, G, Hussain, F, & Rehman, IU. (2014). Essential oil composition of some plants of family Zygophyllaceae and Euphorbiaceae. *Pakistan Journal of Botany*, *46*(6), 2043–2049.

Davoodi, H., Ghaemi, E., Mazandarani, M., Shakeri, F., Javid, S. N., & Klishadi, M. (2015). Anti-mycobacterial and anti-inflammatory activity of *Peganum harmala*. *Journal of Chemical and Pharmaceutical Research*, *7*(4), 1611–1616.

Edziri, H., Marzouk, B., Mabrouk, H., Garreb, M., Douki, W., Mahjoub, A., ... Mastouri, M. (2018). Phytochemical screening, butyrylcholinesterase inhibitory activity and anti-inflammatory effect of some Tunisian medicinal plants. *South African Journal of Botany*, *114*, 84–88.

El-Bakatoushi, R., Hegazy, A. K., Saad, H., & Fawzy, M. (2011). Genetic diversity in coastal and inland desert populations of *Peganum harmala* L.(Peganaceae). *African Journal of Biotechnology*, 10(71), 15883–15890.

El Khomsi, M., Dandani, Y., Chaachouay, N., & Hmouni, D. (2022). Ethnobotanical study of plants used for medicinal, cosmetic, and food purposes in the region of Moulay Yacoub, Northeast of Morocco. *Journal of Pharmacy & Pharmacognosy Research*, 10(1), 13–29.

El-Rifaie, M. E. S. (1980). *Peganum harmala*: its use in certain dermatoses. *International Journal of Dermatology*, *19*(4), 221–222.

Fahmy, S. A., Mahdy, N. K., Al Mulla, H., ElMeshad, A. N., Issa, M. Y., & Azzazy, H. M. E. S. (2022). PLGA/PEG nanoparticles loaded with cyclodextrin-*Peganum harmala* alkaloid complex and ascorbic acid with promising antimicrobial activities. *Pharmaceutics*, *14*(1), 142.

Farouk, L., Laroubi, A., Ouachrif, A., Aboufatima, R., Benharref, A., & Chait, A. (2009). Antinociceptive activity of various extracts of *Peganum harmala* L. and possible mechanism of action. *Iranian Journal of pharmacology and Therapeutics* *8*, 29–35.

Farzin, D., & Mansouri, N. (2006). Antidepressant-like effect of harmane and other β-carbolines in the mouse forced swim test. *European Neuropsychopharmacology*, 16, 324–328.

Filali, I., Bouajila, J., Znati, M., Bousejra-El Garah, F., & Ben Jannet, H. (2015). Synthesis of new isoxazoline derivatives from harmine and evaluation of their anti-Alzheimer, anti-cancer and anti-inflammatory activities. *Journal of Enzyme Inhibition and Medicinal Chemistry*, *30*(3), 371–376.

Frison, G., Favretto, D., Zancanaro, F., Fazzin, G., Ferrara, S.D. (2008). A case of β-carboline alkaloid intoxication following ingestion of *Peganum harmala* seed extract. *Forensic Science International*, *179*, 37–43.

Grebennikov, K.A. (2017). Distribution and ecological features of *Peganum harmala* L. in the environs of Lake Baskunchak (Astrakhan oblast). *Arid Ecosystems*, 7(1), 49–51.

Hayet, E., Maha, M., Mata, M., Mighri, Z., Laurent, G., & Mahjoub, A. (2010). Biological activities of *Peganum harmala* leaves. *African Journal of Biotechnology*, *9*(48), 8199–8205.

Iranshahy, M., Bazzaz, S. F., Haririzadeh, G., Abootorabi, B. Z., Mohamadi, A. M., & Khashyarmanesh, Z. (2019). Chemical composition and antibacterial properties of *Peganum harmala* L. *Avicenna Journal of Phytomedicine*, *9*(6), 530.

ITIS (2022). Integrated Taxonomic Information System (ITIS) on-line database, www.itis.gov, CC0 https://doi.org/10.5066/F7KH0KBK; http://itis-web-prod.snafu.cr.usgs.gov

Ismahane, L., Sghir, M. M., Ibrahim, M., Khadija, B., & Laid, O. M. (2016). Ovicidal, larvicidal and adulticidal activities of essential oils from *Peganum harmala* L.(Zygophyllacae) against date moth Ectomyelois ceratoniae Zeller (Lepidoptera: Pyralidae). *International Journal of Biosciences*, *8*(5), 146–52.

Jahanpour, S., Ghazisaidi, K., Davoodi, H., Mazandarani, M., Samet, M., Jahanpour, N., & Ghaemi, E. A. (2015). Antimicrobial effects of folk medicinal plants from the North of Iran against Mycobacterium tuberculosis. *Archives of Pediatric Infectious Diseases*, *3*(1 TB), e18098. DOI: 10.5812/pedinfect.18098

Jbilou, R., Amri, H., Bouayad, N., Ghailani, N., Ennabili, A., & Sayah, F. (2008). Insecticidal effects of extracts of seven plant species on larval development, α-amylase activity and offspring production of *Tribolium castaneum* (Herbst) (Insecta: Coleoptera: Tenebrionidae). *Bioresource Technology*, *99*(5), 959–964.

Khaliq, T., Misra, P., Gupta, S., Reddy, K. P., Kant, R., Maulik, P. R., … Narender, T. (2009). Peganine hydrochloride dihydrate an orally active antileishmanial agent. *Bioorganic & Medicinal Chemistry Letters*, *19*(9), 2585–2586.

Khan, N. A., Raina, A., Wagay, N. A., & Tantray, Y. R. (2017). Distribution, status, pharmacological, and traditional importance of *Peganum harmala* L. *International Journal of Advanced Research in Science, Engineering*, *6*(8), 1887–1893.

Khan, A. M., Qureshi, R. A., Ullah, F., Gilani, S. A., Nosheen, A., Sahreen, S., … Murad, W. (2011). Phytochemical analysis of selected medicinal plants of Margalla Hills and surroundings. *Journal of Medicinal Plants Research*, *5*(25), 6017–6023.

Kiani, S. J., Shamsi Shahrabadi, M., Ataei, A., & Sajjadi, N. (2007). *Peganum harmala* seed extract can prevent HSV-1 replication in vitro. مجله ویروس شناسی ایران, *1*(4), 11–16.

Komeili, G., Hashemi, M., & Bameri-Niafar, M. (2016). Evaluation of antidiabetic and antihyperlipidemic effects of *Peganum harmala* seeds in diabetic rats. *Cholesterol*, *2016*. https://doi.org/10.1155/2016/7389864

Kumar, M. P., Joshi, S. D., Kulkarni, V. H., & Savant, C. (2015). Phytochemical screening and evaluation of analgesic, anti-inflammatory activities of *Peganum harmala* Linn., seeds in rodents. *Journal of Applied Pharmaceutical Science*, *5*(5), 052–055.

Lala, S., Pramanick, S., Mukhopadhyay, S., Bandyopadhyay, S., & Basu, M. K. (2004). Harmine: evaluation of its antileishmanial properties in various vesicular delivery systems. *Journal of Drug Targeting*, *12*(3), 165–175.

Lamchouri, F., Settaf, A., Cherrah, Y., Hassar, M., Zemzami, M., Atif, N., … Lyoussi, B. (2000). In vitro cell-toxicity of *Peganum harmala* alkaloids on cancerous cell- lines. *Fitoterapia*, *71*(1), 50–54.

Lamchouri, F., Settaf, A., Cherrah, Y., Zemzami, M., Lyoussi, B., Zaid, A., … Hassar, M. (1999). Antitumour principles from *Peganum harmala* seeds. *Therapie*, 54(6), 753–758.

Lamchouri, F., Settaf, A., Cherrah, Y., El Hamidi, M., Tligui, N. S., Lyoussi, B., & Hassar, M. (2002, March). Experimental toxicity of *Peganum harmala* seeds. In *Annales Pharmaceutiques Françaises*, *60*(2), 123–129.

Li, H., Wang, Z., Wang, Y., Xu, J., & He, X. (2020). Triterpenoids with anti-proliferative effects from the seeds of *Peganum harmala* L. *Phytochemistry*, *174*, 112342. https://doi.org/10.1016/j.phytochem.2020.112342

Li, S., Cheng, X., & Wang, C. (2017). A review on traditional uses, phytochemistry, pharmacology, pharmacokinetics and toxicology of the genus Peganum. *Journal of Ethnopharmacology*, *203*, 127–162.

Li, Y., He, Q., Du, S., Guo, S., Geng, Z., & Deng, Z. (2018). Study of methanol extracts from different parts of *Peganum harmala* L. Using 1 H-NMR plant metabolomics. *Journal of Analytical methods Chemistry*, *2018*. 1–9. https://doi.org/10.1155/2018/6532789

Li, Y. K. (2005). *Study on the Chemical Constituents of Above Ground of Peganum harmala L.* Northwest Agriculture Forestry University, Shanxi.

Liu, L., Zhao, T., Cheng, X. M., Wang, C. H., & Wang, Z. T. (2013). Characterization and determination of trace alkaloids in seeds extracts from *Peganum harmala* Linn. using LC-ESI-MS and HPLC. *Acta Chromatographica*, *25*(2), 221–240.

Liu, W., Zhu, Y., Wang, Y., Qi, S., Wang, Y., Ma, C., … Wang, C. (2017a). Anti-amnesic effect of extract and alkaloid fraction from aerial parts of *Peganum harmala* on scopolamine-induced memory deficits in mice. *Journal of Ethnopharmacology*, *204*, 95–106.

Liu, X., Li, M., Tan, S., Wang, C., Fan, S., & Huang, C. (2017b). Harmine is an inflammatory. inhibitor through the suppression of NF-κB signaling. *Biochemical and Biophysical Research Communications*, *489*(3), 332–338.

Maire, R. (1933). Études sur la flore et la végétation du Sahara central. Imprimerie 'La Typo-Litho'. La Typo-lit., 1, 272.

Majid, A. (2018). A review study of the chemical constituents and therapeutic effects of *Peganum harmala* L. *Global Journal of Pure and Applied Chemistry Research*, *6*(2), 12–9.

Mamedov, N. A., Pasdaran, A., & Mamadalieva, N. Z. M. (2018). Pharmacological studies of Syrian rue (*Peganum harmala* L., Zygophyllaceae). *International Journal of Secondary Metabolite*, *5*(1), 1–6.

Mashak, B., Hoseinzadeh, M., Ehsanpour, A., Ghanbaran, A. R., & Vakili, M. (2017). Evaluation of treatment response and side effects of spinal-Z in patients with metastatic gastroesophageal adenocarcinoma: a double-blind randomized controlled trial. *Jundishapur Journal of Chronic Disease Care*, *6*(3), e57870. doi: 10.5812/jjcdc.57870

Moloudizargari, M., Mikaili, P., Aghajanshakeri, S., Asghari, M., & Shayegh, J. (2013). Pharmacological and therapeutic effects of *Peganum harmala* and its main alkaloids. *Pharmacognosy Reviews*, *7*(14), 199.

Moradi, M. T., Karimi, A., Fotouhi, F., Kheiri, S., & Torabi, A. (2017). In vitro and in vivo effects of *Peganum harmala* L. seeds extract against influenza A virus. *Avicenna Journal of Phytomedicine*, *7*(6), 519.

Miraj S. (2016). A review study of therapeutic effects of *Peganum harmala* L. *Der Pharmacia Lettre*, *8*(13):161–6.

Mirzaie, M., Nosratabadi, S. J., Derakhshanfar, A., & Sharifi, I. (2007). Antileishmanial activity of *Peganum harmala* extract on the in vitro growth of Leishmania major promastigotes in comparison to a trivalent antimony drug. *Veterinarski arhiv*, *77*(4), 365–375.

Moghadam, M. S., Maleki, S., Darabpour, E., Motamedi, H., & Nejad, S. M. S. (2010). Antibacterial activity of eight Iranian plant extracts against methicillin and cefixime restistant Staphylococcous aureus strains. *Asian Pacific Journal of Tropical Medicine*, *3*(4), 262–265.

Naderi, G., Hemami, M. R., & Mohammadi, S. (2011). Investigation of habitat preferences of *Iranian jerboa* (*Allactaga firouzi* Womochel 1978).

Narantsetseg, A. (2014). Seed germination and abnormality of cotyledon of *Peganum harmala* populations in Mongolia. *African Journal of Plant Science*, 8(6), 254–259.

Naresh, G., Jaiswal, N., Sukanya, P., Srivastava, A. K., Tamrakar, A. K., & Narender, T. (2012). Glucose uptake stimulatory effect of 4-hydroxypipecolic acid by increased GLUT 4 translocation in skeletal muscle cells. *Bioorganic & Medicinal Chemistry Letters*, *22*(17), 5648–5651.

Nedjimi, B. (2020). Germination characteristics of *Peganum harmala* L. (Nitrariaceae) subjected to heavy metals: Implications for the use in polluted dryland restoration. *International Journal of Environmental Science and Technology*, 17(4), 2113–2122.

Orch, H., Chaachouay, N., Douiri, E. M., Faiz, N., Zidane, L., & Douira, A. (2021). Use of medicinal plants in dermato-cosmetology: An ethnobotanical study among the population of Izarène. *Jordan Journal of Pharmaceutical Sciences*, 323–340.

Ozenda P. (1991). Flore et végétation du Sahara 3éme édition, augmentée. Ed CNRS, Paris, 662.

Panahi, Y., Saadat, A., Seifi, M., Rajaee, M., Butler, A. E., & Sahebkar, A. (2018). Effects of Spinal-Z in patients with gastroesophageal cancer. *Journal of Pharmacopuncture*, *21*(1), 26.

Passos, I. D., & Mironidou-Tzouveleki, M. (2016). Hallucinogenic plants in the Mediterranean countries. *In Neuropathology of drug addictions and substance misuse*. Academic Press, 761–772.

Quezel, P., Santa, S. (1963). Nouvelle flore de l'Algérie et des régions désertiques méridionales. *CNRS*, 2, 59.

Rachida, A., Loubna, F., Mehdi, A. L., Sara, O., Laadraoui, J., Abderrazak, E. A., Zineb, E. G., Kenza, B., Fatim, Z. M., Zahra, S. and Abderrahman, C. (2021). The Antinociceptive Effects of Central and Peripheral Administration of *Peganum harmala* L. Extracts in Animal Models. *Biological Sciences*, *21*(2), 267.279.

Rachid, D. (2021). Intoxication by Harmel. In *Medical Toxicology*. IntechOpen. DOI: 10.5772/intechopen.92936

Rezzagui, A., Merghem, M., Derafa, I., & Dahamna, S. (2020). Acute and sub-acute toxic effects of Algerian *Peganum harmala* L. crud extract. *Journal of Drug Delivery and Therapeutics*, *10*(2), 115–121.

Salari, E., Ahmadi, K., Dehyaghobi, R. Z., Purhematy, A., & Takalloozadeh, H. M. (2012). Toxic and repellent effect of harmal (*Peganum harmala* L.) acetonic extract on several aphids and *Tribolium castaneum* (Herbst). *Chilean Journal of Agricultural Research*, *72*(1), 147.

Sassoui, D., Seridi, R., Azin, K., & Usai, M. (2015). Evaluation of phytochemical constituents by GC-MS and antidepressant activity of *Peganum harmala* L. seeds extract. *Asian Pacific Journal of Tropical Disease*, *5*(12), 971–974.

Shaheen, G., Ashfaq, A., Shamim, T., Asif, H. M., Ali, A., Rehman, S., & Sumreen, L. (2022). Antioxidant, Antimicrobial, Phytochemical and FTIR Analysis of Peganum harmala (Fruit) Ethanolic Extract from Cholistan Desert, Pakistan. *Dose-Response*, 20(3), 15593258221126832. https://doi.org/10.1177/15593258221126832

Shaheen, H. A., & Issa, M. Y. (2020). In vitro and in vivo activity of *Peganum harmala* L. alkaloids against phytopathogenic bacteria. *Scientia Horticulturae*, *264*, 108940. doi.org/10.1016/j.scienta.2019.108940.

Shahrajabian, M. H., Sun, W., & Cheng, Q. (2021). Improving health benefits with Considering traditional and modern health benefits of *Peganum harmala*. *Clinical Phytoscience*, *7*(1), 1–9.

Sharifi-Rad, J., Quispe, C., Herrera-Bravo, J., Semwal, P., Painuli, S., Özçelik, B., ... Cho, W. C. (2021). Peganum spp.: A comprehensive review on bioactivities and health-enhancing effects and their potential for the formulation of functional foods and pharmaceutical drugs. *Oxidative Medicine and Cellular Longevity*, *2021*, Article ID 5900422, 20, https://doi.org/10.1155/2021/5900422

Shirani-Boroujeni, M., Heidari-Soureshjani, S., & Hafshejani, Z. K. (2017). Impact of oral capsule of *Peganum harmala* on alleviating urinary symptoms in men with benign prostatic hyperplasia; a randomized clinical trial. *Journal of Renal Injury Prevention*, *6*(2), 127.

Singhai, A., & Patil, U. K. (2021). Amelioration of oxidative and inflammatory changes by *Peganum harmala* seeds in experimental arthritis. *Clinical Phytoscience*, *7*, 1–11.

Sodaeizadeh, H., Rafieiolhossaini, M., & Van Damme, P. (2010). Herbicidal activity of a medicinal plant, *Peganum harmala* L., and decomposition dynamics of its phytotoxins in the soil. *Industrial Crops and Products*, *31*(2), 385–394.

Sodaeizadeh, H., Rafieiolhossaini, M., Havlík, J., & Van Damme, P. (2009). Allelopathic activity of different plant parts of *Peganum harmala* L. and identification of their growth inhibitors substances. *Plant Growth Regulation*, *59*, 227–236.

Soliman, A. M., Abu-El-Zahab, H. S., & Alswiai, G. A. (2013). Efficacy evaluation of the. protein isolated from *Peganum harmala* seeds as an antioxidant in liver of rats. *Asian Pacific Journal of Tropical Medicine*, *6*(4), 285–295.

Soosaraei, M., Fakhar, M., Teshnizi, S. H., Hezarjaribi, H. Z., & Banimostafavi, E. S. (2017). Medicinal plants with promising antileishmanial activity in Iran: a systematic review and meta-analysis. *Annals of Medicine and Surgery*, *21*, 63–80.

Tahraoui, A., El-Hilaly, J., Israili, Z. H., & Lyoussi, B. (2007). Ethnopharmacological survey of plants used in the traditional treatment of hypertension and diabetes in south-eastern Morocco (Errachidia province). *Journal of Ethnopharmacology*, *110*(1), 105–117.

Tascon, M., Benavente, F., Sanz-Nebot, V. M., & Gagliardi, L. G. (2015). Fast determination of harmala alkaloids in edible algae by capillary electrophoresis mass spectrometry. *Analytical and Bioanalytical Chemistry*, *407*, 3637–3645.

The Plant List. (2013). The Plant List. Version 1.1. Published on the Internet; http://www.theplantlist.org/ (accessed 1st January).

Wang, C., Zhang, Z., Wang, Y., & He, X. (2016). Cytotoxic constituents and mechanism from *Peganum harmala*. *Chemistry & Biodiversity*, *13*(7), 961–968.

Weckesser, W. (2013). First record of *Peganum harmala* (Zygophyllaceae) in Val Verde County, Texas, and subsequent eradication treatment. *Phytoneuron*, *71*, 1–5.

Yazdanbod, A., Sadeghifard, N., & Niknam, M. (2001). Therapeutic effects, adverse events and spinal-Z pharmaceutical indicators in the treatment of cancer of the upper gastrointestinal tract. *Ardebil University of Medical Sciences*, *1*.

Zhao, T., Wang, Z. T., Branford-White, C. J., Xu, H., & Wang, C. H. (2011). Classification and differentiation of the genus Peganum indigenous to China based on chloroplast trnL-F and psbA-trnH sequences and seed coat morphology. *Plant Biology*, *13*(6), 940–947.

Zheng, X. Y., Zhang, Z. J., Chou, G. X., Wu, T., Cheng, X. M., Wang, C. H., & Wang, Z. T. (2009). Acetylcholinesterase inhibitive activity-guided isolation of two new alkaloids from seeds of Peganum nigellastrum Bunge by an in vitro TLC-bioautographic assay. *Archives of Pharmacal Research*, *32*(9), 1245–1251.

Zhu, Z., Zhao, S., & Wang, C. (2022). Antibacterial, antifungal, antiviral, and antiparasitic activities of *Peganum harmala* and its ingredients: A review. *Molecules*, *27*(13), 4161.

38 Opium Poppy (*Papaver somniferum* L. Papaveraceae)

Abdelhamid Azeroual, Noureddine Chaachouay, Ouafae Benkhnigue, Rahmatullah Qureshi, Bouchaib Bencharki, Khadija Temmante, and Aziz Baidani

38.1 INTRODUCTION

Opium poppy (*Papaver somniferum* L.) is one of the oldest well-known medicinal plants used for medicinal and therapeutic value and has been used for thousands of years. The milky sap of the poppy, known as opium, emerged from the Fertile Crescent between the Tigris and Euphrates rivers and was unquestionably therapeutic to ancient civilizations. Egyptians and people from Mesopotamia understood its ability to reduce pain, promote sleep, and cure a wide range of illnesses. Its benefits were praised by Greek doctors such as Hippocrates and Dioscorides, who opened the door for its widespread usage throughout Europe and Asia (Gümüşçü et al., 2008; Carod-Artal, 2013; Chaachouay et al., 2022a, 2022b, 2022c; Verma et al., 2016; Tavlı, 2022). It is also used for non-medicinal purposes and in culinary as a food ingredient in cooking (seed). In oriental cuisine, seeds are a binding agent to prepare curry or other spice blends. The plant is used to obtain seed oil, the oil cake as fodder for animals, and a narcotic drug (Lim, 2013; Van Wyk and Wink, 2004; Evans, 2009; Bernath, 1998; Kapoor, 1995; Duke, 1983).

Presently, the opium poppy holds significant agricultural value as it finds application in the food and pharmaceutical industries. Intensive poppy production has led to intensive cultivation to obtain highly productive varieties of food or industrial importance (Tetenyi, 1997; Matyasova et al., 2011). Opium, which is abundant in pharmaceutical alkaloids including morphine, codeine, papaverine, and others, is present in its capsules and latex (Prajapati et al., 2002; Frick et al., 2005; Fejer, 2007; Németh-Zámbori et al., 2011; Stranska et al., 2013).

38.2 SCIENTIFIC CLASSIFICATION

Papaver somniferum L. (2n = 22) belongs to the Ranunculales order, the family *Papaveraceae*, and the genus *Papaver* (Christenhusz et al., 2017; ITIS, 2023).

Kingdom: Plantae
Subkingdom: Viridiplantae
Infrakingdom: Streptophyta
Superdivision: Embryophyta
Division: Tracheophyta
Subdivision: Spermatophytina
Class: Magnoliopsida
Superorder: Ranunculanae
Order: Ranunculales
Family: Papaveraceae
Genus: *Papaver*
Species: *Papaver somniferum* L.

Danert (1958) proposed an intraspecific classification of *P. somniferum*, with three subspecies: *songaricum* Basil., Danert *somniferum*, and *setigerum* (DC.) Corb. (1958) described a total of 52 botanical varieties.

38.3 SYNONYMS OF *PAPAVER SOMNIFERUM* L. (WFO PLANT LIST, 2023)

Papaver amoenum Lindl., *Papaver album* Mill., *Papaver somniferum var. nigrum* DC., *Papaver indehiscens* Dumort., *Papaver amplexicaule* Stokes, *Papaver album-nigrum* Crantz, *Papaver paeoniflorum* (Alef.) Corrêa, *Papaver sylvestre* Godr., *Papaver opiiferum* Forssk., *Papaver officinale* C.C.Gmel., *Papaver somniferum var. album* (Mill.) DC.

38.4 VERNACULAR NAMES

English: Opium poppy, White poppy, Opium; **Arabic**: Abunom, Afyiun; **Bengali**: Pasto, Post; **French**: Pavot des jardins, Pavot somnifère, Pavot à opium; **Greek**: Agria; **German**: Schlafmohn; **Gujrati**: Aphina; **Hindi**: Afin, Khashkash, Afyun; **Italian**: Papavero da oppio, Papavero domestico; **Persian**: Afiun, Khashkhash, Koknar; **Portuguese**: papoila-dormideira; **Russian**: Мак наркотический (Mak narkoticheskiy); **Tamil**: Abini; **Spanish**: Adormidera, Planta del opio; **Urdu**: Khashkash safaid (Masihuddin et al., 2018; Yoganarasimhan, 2000; Chatterjee and Pakrashi, 1994).

DOI: 10.1201/9781003460336-38

38.5 BOTANICAL DESCRIPTION

The *Papaver* genus belongs to the cosmopolitan Papaveraceae family, which includes some 45 genera and 775 species (Christenhusz et al., 2017), and has colonized the majority of ecosystems, from the tropics to the Alps (Egan et al. 2011). According to POWO (2023), there are 132 recognized species of Papaver. These plants are upright and may be annual, biennial, or perennial. They have a milky-white, yellowish, or orange juice (Figure 38.1).

The opium poppy is a deep-rooted annual herbaceous plant. It is an erect plant (up to 2 m long) with a strong, hollow, fistulous stem 0.5 to 1.5 cm thick, with few or no branches (Cano Ortiz et al., 2022). The plant's leaves are characterized by their simplicity, opposite arrangement, oblong shape, colossal size, lack of hair, lobed structure, glaucous green color, wax coating that imparts a glossy appearance, and irregularly serrated or scalloped edges. The flowers, single or very double, are attractive, bisexual, large, some 12–13 cm wide (5 to 6 cm in diameter), varied with a dark spot at the base. Flower ranges from white to pink, purple, and red (Shasany et al., 2007).

The center of the flower contains a large globose or ovoid capsule, globous, up to 4 cm in diameter, pedunculated with a flat top, enclosing numerous tiny white or gray-brown to black, kidney-shaped seeds. The flowering and fruiting period extends from April to August (Gamble, 1967; Mukerji, 1953; Hooker, 1973; Chauhan, 1999; Collett, 1971). The fruit is generally a porous capsule containing numerous tiny seeds of various shapes (kidney-shaped, longitudinally striated, honeycombed, or pitted). The white-seeded somniferous poppy, *P. somniferum* L. subsp. *album*, officially banned from sale, is used to produce opium. The blue-seeded variety, *P. somniferum* L. subsp. *nigrum*, also known as blue poppy, is cultivated for its seeds, which are used in pastries and to make poppy oil.

P. somniferum is generally grown for its latex, produced in the unripe seed capsule. Latex is a milky sap that contains alkaloids, including codeine, papaverine, thebaine, and morphine (EFSA, 2019), which have powerful narcotic properties. These alkaloids are extracted from the sap and used to manufacture analgesics and other medicines. The oil derived from the seed is used for culinary and pharmacological applications and in producing soaps, paints, and varnishes (Krist et al., 2005). The seeds, often devoid of alkaloids, are used as birdseed and as ingredients in confectionary and baking items (Rahimi et al., 2011).

Poppy seed flour, a protein-rich by-product of oil extraction, is the primary dietary item obtained from poppies. 'Poppy seed meal' and 'poppy seed cake' denote the solvent and mechanical extraction results, respectively. Overall, *P. somniferum* has a complex chemical profile and significant medical uses. However, its cultivation and service are highly regulated in many countries due to its abuse and addiction potential.

FIGURE 38.1 The aerial parts of *P. Somniferum*.

38.6 ORIGIN AND GEOGRAPHIC DISTRIBUTION

The Papaver genus, belonging to the Papaveraceae family, consists of around 80 herbs that are either annual, biennial, or perennial. These plants are primarily distributed in Central and Southwest Asia, Central and Southern Europe, and North Africa. *P. somniferum* is native to the eastern Mediterranean and Asia Minor (Akinci & Bayram, 2003; Chaachouay et al., 2021a, 2021b, 2021c; Shasany et al., 2007), but is now widely cultivated worldwide as a food and ornamental plant, as well as for its medicinal and psychoactive properties. Because of its potent narcotic characteristics, opium poppy cultivation is extensively practiced in nations such as Afghanistan, Pakistan, Myanmar, Mexico, and Laos. The United Nations has authorized the legal cultivation of poppy seeds in many countries. Turkey and the Czech Republic dominate the global production of poppy seeds, contributing over 65% of the total output (White and Raymer, 1985; Tetenyi, 1997; Bieniek, 2007; Rahimi et al., 2011).

Many regional cultivars, chemotypes, and varieties adapted to different climates and applications have emerged due to the poppy's extensive history of domestication and selection. From Bombay to Moscow in the north to Tanzania in the southern hemisphere, poppy cultivation encompasses a vast geographical area (Krist et al., 2005). Today, *P. somniferum* can be found in many parts of the world, including Asia, Europe, Africa, and America. It is commonly cultivated in Afghanistan, Pakistan, India, and other parts of Asia for the production of opium, which is used to manufacture a variety of medicines, including analgesics and cough suppressants. The plant is also cultivated as an ornamental flower in many parts of the world. However, the cultivation and use of opium poppies is highly regulated in many countries due to its abuse and addiction potential.

38.7 ECOLOGY

The opium poppy is a very hardy plant that can grow in different climates, tropical and temperate regions, throughout 6 to 7 months. The extreme cold coupled with climatic conditions, particularly humidity result in lowering opium yield. *P. somniferum* is primarily pollinated by bees and other insects attracted to its nectar and pollen. The plant reproduces sexually through seeds. The plant typically produces large, showy flowers that can be white, pink, red, or purple. After pollination, the flowers set seed pods containing the plant's highly prized opium. The plant is often grown alongside other crops, such as wheat, barley, and cereals (Kumar, 2022).

Opium poppy has several adaptations that help it survive in its environment. The plant has a deep taproot that allows it to access nutrients and water from more profound into the soil. Plants also produce a thick, waxy cuticle on their leaves that helps prevent water loss. *P. somniferum* is considered an invasive species in some areas where it has been introduced, as it can displace native plant species. The plant is also highly susceptible to fungal diseases, which can decimate crops. Additionally, the production and sale of opium are highly regulated, as it is a controlled substance in many parts of the world. Research on the carbon emissions resulting from the production of opium poppy is quite scarce. An example of such a study is the research conducted by McAlister et al. (2016) on the cultivation of legal opium poppies on Australian farms. This study examined the whole environmental life cycle, starting with poppy cultivation and ending with the manufacturing of 100 mg of packaged morphine intended for intravenous injection. They determined that the carbon footprint of 100 mg of morphine was 204 grams of CO_2. Furthermore, they found that for every 1 kilogram of morphine, the carbon footprint would be 2040 kg of CO_2. The research revealed that about 90% of the carbon footprint associated with morphine prepared for medicinal purposes may be attributed to the latter phases of manufacture, namely the processes of sterilization and packing.

Mansfield (2019) carried a study between 2011 and 2017 in Afghanistan and reported that the use of herbicides and pesticides made it possible to grow poppies over larger areas, while in terms of irrigation, solar energy was gradually replacing diesel-powered pumps and generators. In a short time, poppy cultivation helped diversify crops in dry areas. At the same time, in the long term, soil salinization, falling water tables, and lower yields led to farmers migrating to other regions (SIGAR, 2018; Orch et al., 2021).

38.8 PHYTOCHEMISTRY

All varieties of *P. somniferum* contain several classes of alkaloids and non-alkaloid compounds.

This plant is mainly cultivated for its psychoactive alkaloids, which have important medicinal properties. The phytochemistry of *P. somniferum* is distinguished by the presence of various classes of alkaloids (80 in number), notably morphine, codeine, thebaine, noscapine, and papaverine; and other non-alkaloid compounds like phenolic acids and flavonoids (Vadhel et al., 2023; Sariyar, 2002). Its seeds contain 35–55% oil, 25–35% protein, abundant vitamins E and F, pantothenic acid, biotin, niacin, thiamine, and mineral elements (Ca, K, Mg, and Fe). In addition, the triglyceride composition of *P. somniferum* seed oil was analyzed, enabling us to identify the main triglyceride components as linoleic, oleic, and palmitic acids (around 70% of oil) (Krist et al., 2005). Poppy seed oil is appropriate for use in food to promote cardiovascular health due to its high concentration of unsaturated fatty acids (Krist et al., 2005; Dilek et al., 2018; Butnariu et al., 2022).

In addition to these alkaloids, *P. somniferum* also contains several phenolic acids, such as caffeic acid and syringic acid, as well as flavonoids, such as kaempferol and quercetin (Butnariu et al., 2022; EFSA, 2019). These compounds are believed to have antioxidant properties and may contribute to the plant's medicinal effects. Overall, the phytochemistry of the opium poppy is complex and diverse, with numerous alkaloids and other compounds with important medicinal properties. However, it is essential to note that the use of this plant and its derivatives is heavily regulated due to the potential for abuse and addiction. Opium poppy, namely its alkaloids, has become an essential component of medicine (Tavlı Ö. F., 2022).

38.9 ETHNOBOTANY

Throughout the history, *P. somniferum* has been valued for its medicinal properties. The plant contains several alkaloids, including morphine, codeine, and thebaine, which have powerful pain-relieving properties. These alkaloids are used to make various pharmaceutical products, including painkillers, cough suppressants, and sedatives. In addition to its medicinal uses, *P. somniferum* has a long history of recreational and ceremonial use. The plant's psychoactive properties have been known for thousands of years, and it has been used in various cultural contexts (Butnariu et al., 2022; Chaachouay et al., 2020a, 2020b; Benlarbi et al., 2023). The opium poppy has been used for ages to treat ailments and control pain globally. Its ability to promote sleep and reduce anxiety was acknowledged by the ancient Sumerians and Egyptians. In India, therapies based on poppies were used to treat anxiety, diarrhea, and coughs. The herb was used in Chinese medicine to treat a wide range of conditions, such as pain, diarrhea, and sleeplessness. In some cultures, opium was used as a pain reliever and sedative, while in others, it was used for religious and spiritual purposes. The poppy was used in many cultures for ritual purposes beyond simple healing. Poppy sap was used as a sacrifice to gods connected to death and rebirth in Aztec rites. Poppy seeds were associated with Demeter, the fertility and agricultural goddess in ancient Greece. Poppies were used to memorialize the dead and to signify remembering in certain European traditions.

Butnariu et al. (2022) have compiled a list of the traditional medical uses of opium. In Turkey, *P. somniferum* seeds are used to treat gastrointestinal, nervous, and respiratory diseases; in Italy, they are used as a tranquilizer for toothache; and in Pakistan, they are used as a narcotic, analgesic, or sedative to increase excitement and physical vigor. Despite its many uses, the cultivation and use of *P. somniferum* is highly regulated in many countries due to its potential for abuse and addiction. The plant is listed as a controlled substance under international drug control conventions, and the production, sale, and use of opium and its derivatives are strictly controlled. However, the plant continues to be cultivated and used illegally in many parts of the world.

38.10 HALLUCINOGENIC EFFECTS

The opioids derived from the opium poppy (opium and opium alkaloids, such as morphine) are a source of severe drug abuse worldwide, creating a dependency on these drugs (Lal, 2022). After taking opium, the effects are generally very rapid, especially in the inhaled form, and persist for a few hours (Chaachouay et al., 2019a, 2019b; Mohanty and Panda, 2018). The user quickly experiences a relaxing, anxiolytic, and sedative effect, providing an intense sense of well-being. Higher doses may even cause hallucinations. *P. somniferum* is a plant that produces opium, a powerful narcotic substance that has been used for its pain-relieving and psychoactive effects for thousands of years. The psychoactive compounds in opium can produce a range of effects on the body and mind, including:

- **Pain relief**: Opium has been used for centuries as a natural painkiller. The alkaloids in opium bind to receptors in the spinal cord and brain, blocking the perception of pain (Laux-Biehlmann et al., 2013).
- **Euphoria**: Opium can produce intense pleasure and well-being, sometimes called a 'rush' or 'high'. These effects are caused by the release of dopamine, a neurotransmitter that plays a crucial role in reward and motivation (Nyman & Hall, 1976; Thakur, 1983).
- **Sedation**: Opium can cause drowsiness and relaxation, making it useful as a sleep aid or to reduce anxiety and tension (Heinrich et al., 2017).
- **Hallucinations**: In high doses, opium can cause vivid and sometimes disturbing hallucinations, particularly when smoked or injected. Respiratory depression: One of the most dangerous effects of opium is its ability to slow down breathing, which can be fatal in high doses (Lal, 2022).

38.11 PHARMACOLOGICAL PROPERTIES

P. somniferum is a plant that has several pharmacological properties, primarily attributed to its alkaloids (Masihuddin et al., 2018). The plant has a long history as a traditional medicine. It has been used for several therapeutic applications including wound healing, infection control, anti-inflammatory effects, fever reduction, sleep regulation, anxiety alleviation, depression management, and other uses (Table 38.1). The plant contains many crucial alkaloids, including codeine, morphine, noscapine, thebaine, and papaverine (Dittbrenner et al., 2012; Sharopov et al., 2018; Stranska et al., 2013).

38.11.1 Analgesic Activity

The opium poppy has pain-relieving solid properties due to the presence of potent alkaloids. These alkaloids affect the central nervous system by attaching to specific receptors, namely the mu-opioid receptors, which significantly alleviate pain (Masihuddin et al., 2018). Morphine, a primary alkaloid in opium, is extensively used in medicine due to its potent analgesic qualities. It is particularly beneficial in the management of intense pain related to illnesses like cancer or post-operative recovery. The analgesic benefits are attained by modulating pain perception (Gómez-Serranillos et al., 1998). Although opium poppy derivatives are efficient, their use requires cautious deliberation owing to the possibility of adverse effects such as drowsiness and respiratory depression (Hijazi et al., 2017). Although *P. somniferum* has historically been used for pain treatment, its usage is heavily controlled owing to apprehensions about misuse and addiction. Healthcare practitioners prudently prescribe and oversee the use of drugs derived from opium, ensuring a delicate equilibrium between the advantages of pain alleviation and the hazards linked to their propensity for addiction.

38.11.2 Anticancer Properties

P. somniferum has fascinating anticancer effects due to its wide range of alkaloids, such as morphine, codeine, thebaine, papaverine, and noscapine. These alkaloids have different ways of affecting cancer cells, demonstrating their diverse potential to stop the development of tumors (Mondal et al., 2019; Thawabteh et al., 2019). Morphine, known for its opioid analgesic properties, not only relieves pain and generates a feeling of pleasure but also hinders the production of genes essential for cell growth and survival. Codeine, a relatively weak opioid painkiller, hinders the activity of enzymes that are crucial for the process of DNA synthesis and repair. Thebaine, a precursor of

TABLE 38.1
Some Pharmacological Properties of *P. somniferum*

Alkaloids	Chemical Formula	Structural Formula	Effect on the Human Body	Pharmaceutical Use
Morphine (Vadhel et al., 2023)	$C_{17}H_{19}NO_3$	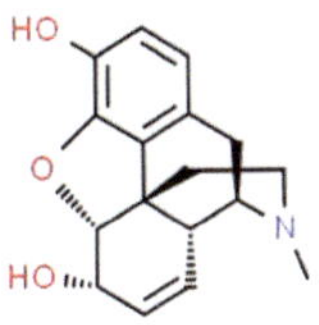	Consuming morphine may induce severe hypotension because it momentarily lowers systemic arterial pressure due to a decrease in vascular resistance	Analgesia, general anesthesia, antitussive, antidiarrheal, alleviating myocardial infarction pain
Codeine (Vadhel et al., 2023)	$C_{18}H_{21}NO_3$	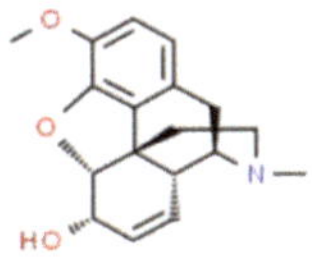	It alters bodily sensations and modulates the brain's pain response	Helps to relieve mild to moderate pain
Thebaine (Vadhel et al., 2023)	$C_{19}H_{21}NO_3$	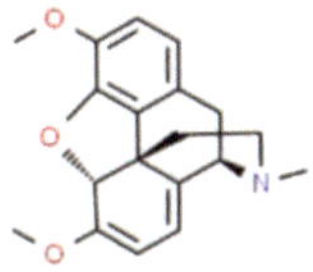	The primary substance used to create oxycodone, oxymorphone, buprenorphine, naloxone, and other similar semi-synthetic opioids	They were pharmaceutically utilized in the production process of oxycodone, oxymorphone, buprenorphine, and naloxone
Narcotine (Noscapine) (Vadhel et al., 2023)	$C_{22}H_{24}NO_7$	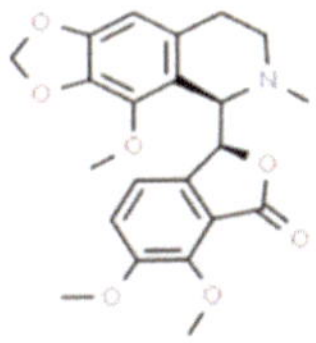	It exhibits antitussive properties and also demonstrates antimitotic activity	Mild analgesic, antitussive (cough-suppressing) effects, potential antineoplastic activities (anticancer activity)
Papaverine (Vadhel et al., 2023)	$C_{20}H_{21}NO_4$	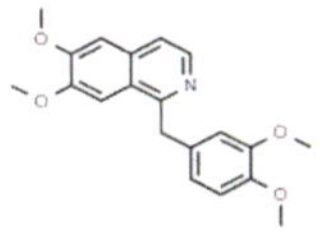	Relaxes various smooth muscles	Vasodilator

morphine and codeine, has comparable effects but possesses more potency, leading to apoptosis by activating caspase. Papaverine, a vasodilator, induces the relaxation of smooth muscles in blood vessels, promoting increased blood circulation. Additionally, it interferes with the development of microtubules, which are essential for cell division. Noscapine, an anticancer medication, hinders the development of tumors by disrupting cell division and death (Güler et al., 2016). Additionally, it regulates genes linked to inflammation and oxidative stress. The results emphasize the possible medical uses of *P. somniferum* alkaloids in cancer treatment.

38.11.3 Antimicrobial Activity

Several investigations have shown that *P. somniferum* and its alkaloids have antibacterial properties against various bacteria and fungi. Alkaloids such as morphine, codeine, thebaine, papaverine, noscapine, and sanguinarine have been investigated for their potential impact on microorganisms (Jafaar et al., 2021). Studies have shown that these alkaloids exhibit antibacterial activity against bacteria, including *E. coli*, *Pseudomonas aeruginosa*, *Vibrio cholerae*, *Enterococcus faecalis*, *Streptococcus pyogenes*, and *Bacillus subtilis*. The antibacterial characteristics are often ascribed to certain alkaloids, namely morphine and codeine (Ismaili et al., 2017; Kačániová et al., 2010). The effectiveness of this action seems to rely on the concentration level since more significant concentrations show more pronounced effects.

Nevertheless, there are notable deficiencies in research, necessitating more investigations to validate the effectiveness, safety, and possible uses of opium poppy alkaloids in antimicrobial treatment. Exercise caution while considering the use of opium poppy and its derivatives since they have the potential for addiction, abuse, and adverse effects (Dilek et al., 2018). Therefore, it is necessary to implement appropriate regulations and monitoring for therapeutic purposes. Furthermore, it is essential to have a comprehensive viewpoint while examining the antimicrobial characteristics of the plant, taking into account many aspects, such as the existence of several compounds, the combined effects they may have, and any possible adverse reactions.

38.11.4 Antidiarrheal Properties

P. somniferum is a plant renowned for its many medical applications, notably its antidiarrheal qualities. According to some reports, the alkaloids included in this plant can decelerate intestinal motility and alleviate stomach discomfort often linked to diarrheal (Gupta et al., 1993). Nevertheless, this plant also harbors opium, which may induce addiction and detrimental consequences when ingested excessively. Hence, getting medical advice before using any natural therapies for diarrhea (Agrawal et al., 2023). Additional plants with documented antidiarrheal properties include *P. somniferum*, a shrub indigenous to India and Pakistan, as well as *P. glaucum* and *P. decaisnei*, wild varieties of opium poppy found in Northern Iraq (Masihuddin et al., 2018). These plants possess several alkaloids that might exhibit antibacterial and anti-inflammatory activities. Nevertheless, more investigation is required to validate their safety and effectiveness in treating diarrhea.

38.11.5 Antitussive Properties

P. somniferum, is widely recognized for its medicinal properties, specifically its potential as an antitussive or cough suppressant (Tavlı, 2022). This botanical specimen contains various alkaloids, such as morphine and codeine, renowned for their antitussive and analgesic attributes (they alleviate pain). It subdues the cough reflex; morphine, a primary alkaloid, is a highly effective opioid analgesic via its influence on the central nervous system. Another opium poppy alkaloid, codeine, is employed for its analgesic and cough-suppressant properties (Lim, 2013). Nevertheless, it is imperative to acknowledge that opium-derived compounds, despite their efficacy in congestion relief, possess intrinsic hazards, including respiratory depression and dependence (Sharopov et al., 2018). As a result, healthcare professionals generally take care when it comes to their utilization. Anxiety-inducing cough syrups, which are pharmaceutical products derived from the opium poppy, have historically been utilized to treat cough. However, their controlled application is indicative of apprehensions regarding dependence and abuse (Mani & Dhawan, 2011). When it comes to medical practice, synthetic alternatives that possess similar antitussive properties are frequently preferred.

38.11.6 Spasmolytic Properties

P. somniferum, sometimes referred to as the opium poppy, has significant spasmolytic effects, which have attracted attention in the area of pharmacology (Mani & Dhawan, 2011; Sharopov et al., 2018). The plant's latex, rich in alkaloids, is generally known as opium. It includes substances like morphine and codeine, which have muscle-relaxing solid properties (Lim, 2013). The medical applications of these spasmolytic characteristics mainly focus on pain management and treating gastrointestinal disorders (Masihuddin et al., 2018). The opium poppy's capacity to relieve muscular spasms is ascribed to its influence on the central nervous system, where the active chemicals function as agonists on specific receptors, regulating the release of neurotransmitters and finally resulting in the relaxing of smooth muscles (Labanca et al., 2018). Although *P. somniferum* has medicinal promise, its usage is strictly controlled owing to the risk of misuse and addiction linked to its opioid alkaloids.

38.11.7 Sedative and Hypnotic Properties

P. somniferum, also referred to as the opium poppy or the bread seed poppy, is a botanical species used for ages due to its sedative and hypnotic properties (Masihuddin et al., 2018). The sedative and hypnotic effects of *P. somniferum* mainly stem from the activity of morphine, which is the predominant and potent alkaloid found in the plant (Agrawal et al., 2023). Morphine attaches to opioid receptors in the brain and spinal cord, imitating the effects of endorphins, which are the body's pain-relieving substances (Lim, 2013). Morphine furthermore influences other neurotransmitters, including dopamine, serotonin, and norepinephrine, and regulates mood, emotion, and reward. Morphine can provide sedation, relaxation, tiredness, and sleep while decreasing anxiety, tension, and terror (Mani & Dhawan, 2011). Nevertheless, morphine is associated with detrimental consequences, including respiratory depression, constipation, nausea, vomiting, itching, and addiction.

38.12 CLINICAL USES AND ADVERSE EFFECTS

P. somniferum has several applications in medicine and various other domains. One of its primary uses is as the origin of opium, a drug with analgesic, soothing, and euphoric properties. Opium is derived from the latex of the immature capsules of the plant. Opium consists of a variety of alkaloids, including papaverine, morphine, thebaine, and codeine, which include analgesic, antitussive, antispasmodic, and vasodilatory effects. These alkaloids treat several ailments, including cough, asthma, diarrhea, dysentery, menstrual cramps, and erectile dysfunction (Chaachouay et al., 2021c; Belhaj et al., 2021; Benkhnigue et al., 2022; Cano Ortiz et al., 2022).

P. somniferum has other phytochemicals, including phenolic substances (anthocyanins, flavonols), essential oils (camphor), and indole derivatives (nudicaulins), which exhibit antioxidant, anti-inflammatory, antibacterial, anticancer, and antidiabetic activities. These chemicals are accountable for many of the biological functions shown by *P. somniferum*. It has been used in traditional medicine for several therapeutic applications. Seeds are used to alleviate symptoms such as fever, thirst, gastric discomfort, and inflammation. Roots are used to produce a variety of cooling medications. The powder is used for the treatment of hepatic disorders. The root is often used as a tonic due to its stimulating effects. Plant roots also create an infusion as a medicinal remedy for alleviating fever.

Notwithstanding this, the utilization of *P. somniferum* is accompanied by certain risks and adverse reactions. Withdrawal symptoms, overdose symptoms (e.g., respiratory depression), constipation (caused by decreased bowel movements), nausea (augmented gastric secretion), vomiting (resulting from increased gastric motility), dry mouth (resulting from decreased saliva production), blurred vision (caused by decreased aqueous humor production), drowsiness (resulting from sedative effects), confusion (caused by central nervous system depression), and hallucinations (as a consequence of increased gastric motility) are all potential adverse effects of opium. Depending on their dose and route of administration, additional alkaloids may induce adverse effects. Morphine has been associated with a range of adverse effects, including but not limited to renal depression, constipation, nausea, vomiting, lethargy, confusion, hallucinations, and memory impairment (Chaachouay & Zidane, 2021). Nicotine has the potential to induce respiratory depression, nausea, vomiting, constipation, lethargy, and bewilderment. Typhoid, tremors, and convulsions are all potential side effects of thebaine. Paracetamol may cause flushing, headache, and hypotension. Consequently, prudent usage and medical oversight are recommended when administering *P. somniferum*.

38.13 CONCLUSION

Opium poppy is a complex and versatile plant that has been used by human beings for thousands of years for food, medicine, and traditional purposes. The phytochemistry reveals that the plant contains over 80 opioids which exhibit psychoactive, therapeutic, and hallucinogenic effects, mainly due to the presence of morphine, codeine, thebaine, noscapine, and papaverine. The production or harvesting of opium is restricted in many parts of the world. Still, a few countries currently authorize its cultivation for research and strictly scientific purposes to produce morphine, for example, for its efficacy in pain management. Opium and opioids are potent analgesics with psychoactive effects, and their non-therapeutic use can entail risks for consumers. Hence, there is a need to inform consumers about the risks and dangers of opium consumption and its medium- and short-term effects. All these products are physiologically addictive, resulting in both increased tolerance (i.e., the need to increase doses to feel the same impact) and signs of physical withdrawal when doses are stopped or reduced. The production of poppy seeds and legal medical cultivation coexist with the criminal opium trade. Balancing between abusive use and the demand for pain alleviation is a great challenge for combating drug abuse.

REFERENCES

Agrawal, N., Kumar, S., & Gupta, N. (2023). Pharmacological benefits of *papaver somniferum* l.: nutritional and medicinal aspects of opium poppy. In *Pharmacological Benefits of Natural Agents* (pp. 298–317). IGI Global. https://www.igi-global.com/chapter/pharmacological-benefits-of-papaver-somniferum-l/327315

Akinci, Z. & Bayram, I. (2003). Effects of poppy seed meal on egg production and hatching results in quail (*Coturnix coturnix japonica*). *Res. Vet. Sci.*, 75 (2): 141–147.

Belhaj, S., Chaachouay, N., & Zidane, L. (2021). Ethnobotanical and toxicology study of medicinal plants used for the treatment of diabetes in the High Atlas Central of Morocco. *Journal of Pharmacy & Pharmacognosy Research, 9*(5), 619–662. https://doi.org/10.56499/jppres21.1014_9.5.619

Benlarbi, F., Mimoune, N., Chaachouay, N., Souttou, K., Saidi, R., Mokhtar, M.R., Kaidi, R, & Benaissa, M. T. (2023). Ethnobotanical survey of the traditional antiparasitic use of medicinal plants in humans and animals in Laghouat (Southern Algeria). *Veterinary World, 16*(2), 357–368. https://doi.org/10.14202/vetworld.2023.357-368

Benkhnigue, O., Chaachouay, N., Khamar, H., El Azzouzi, F., Douira, A, & Zidane, L. (2022). Ethnobotanical and ethnopharmacological study of medicinal plants used in the treatment of anemia in the region of Haouz-Rehamna (Morocco). *Journal of Pharmacy & Pharmacognosy Research, 10*(2), 279–302. https://doi.org/10.56499/jppres21.1196_10.2.279

Bernath, J. (1998). Cultivation of poppy under tropical conditions. In J. Bernath (Ed.), *Poppy: The genus Papaver* (pp. 237–248). Harwood Academic Publishers.

Bieniek, A. 2007. Neolithic plant husbandry in the Kujawy region of central Poland. In S. Colledge & J. Conolly (dir.), *The origins and spread of domestic plants in southwest Asia and Europe* (pp. 327–342). Left Coast Press; UCL Institute of Archaeology publications.

Butnariu, M., Quispe, C., Herrera-Bravo, J., Pentea, M., Sarac, I., Küşümler, A.S., Özçelik, B., Painuli, S., Semwal, P., Imran, M., Gondal, T.A., Emamzadeh-Yazdi, S., Lapava, N., Yousaf, Z., Kumar, M., Eid, A.H., Al-Dhaheri, Y., Suleria, H.A.R., Del Mar Contreras, M., Sharifi-Rad, J., & Cho, W.C. (2022). Papaver plants: Current insights on phytochemical and nutritional composition along with biotechnological applications. *Oxid Med Cell Longev*, 3: 2041769. DOI: 10.1155/2022/2041769

Cano Ortiz, A., Piñar Fuentes, J. C., & Cano, E. (2022). Some medicinal plants of interest for their content in alkaloids I. *Biomedical Journal Scientific & Technical Research*, 42(3): 33702–33705. DOI: 10.26717/BJSTR.2022.42.006765

Carod-Artal, F. J. (2013). Psychoactive plants in ancient Greece. *Neurosci. Hist.*, 1(1): 28–38.

Chaachouay, N., Benkhnigue, O., Fadli, M., Ibaoui, H., & Zidane, L. (2019a). Ethnobotanical and ethnopharmacological studies of medicinal and aromatic plants used in the treatment of metabolic diseases in the Moroccan Rif. *Heliyon, 5*(10), e02191. https://doi.org/10.1080/10477845.2021.1942433

Chaachouay, N., Benkhnigue, O., Fadli, M., Ibaoui, H., El Ayadi, R., & Zidane, L. (2019b). Ethnobotanical and ethnopharmacological study of medicinal and aromatic plants used in the treatment of respiratory system disorders in the Moroccan Rif. *Ethnobotany Research and Applications, 18*(22), 1–16.

Chaachouay, N., Douira, A., Hassikou, R., Brhadda, N., Dahmani, J., Belahbib, N., Ziri, R., & Zidane, L. (2020a). Ethnobotanical and ethnopharmacological study of medicinal and aromatic plants used in the treatment of respiratory system disorders in the Moroccan Rif. PhDThesis, Dέpartement de Biologie - Université Ibn Tofail - Kénitra, Maroc, https://tel.archives-ouvertes.fr/tel-03376377

Chaachouay, N., Benkhnigue, O., & Zidane, L. (2020b). Ethnobotanical study aimed at investigating the use of medicinal plants to treat nervous system diseases in the Rif of Morocco. *Journal of Chiropractic Medicine, 19*(1), 70–81. https://doi.org/10.1016/j.jcm.2020.02.004

Chaachouay, N., & Zidane, L. (2021). The Symbolic Efficacy of Plants in Rituals and Socio-Religious Ceremonies in Morocco, Northwest of Africa. *Journal of Religious &*

Theological Information, *21*(1–2), 34–53. https://doi.org/10.1080/10477845.2021.1942433

Chaachouay, N., Benkhnigue, O., Douira, A., & Zidane, L. (2021a). Poisonous medicinal plants used in the popular pharmacopoeia of the Rif, northern Morocco. *Toxicon*, 189, 24–32. https://doi.org/10.1016/j.toxicon.2020.10.028

Chaachouay, N., Douira, A., & Zidane, L. (2021b). COVID-19, prevention and treatment with herbal medicine in the herbal markets of Salé Prefecture, North-Western Morocco. *European Journal of Integrative Medicine*, 42, 101285. https://doi.org/10.1016/j.eujim.2021.101285

Chaachouay, N., Douira, A., & Zidane, L. (2021c). Herbal Medicine Used in the Treatment of Human Diseases in the Rif, Northern Morocco. *Arabian Journal for Science and Engineering*, 47, 131–153. https://doi.org/10.1007/s13369-021-05501-1

Chaachouay, N., Benkhnigue, O., & Zidane, L. (2022a). Ethnobotanical and Ethnomedicinal study of medicinal and aromatic plants used against dermatological diseases by the people of Rif, Morocco. *Journal of Herbal Medicine*, 32, 100542. https://doi.org/10.1016/j.hermed.2022.100542

Chaachouay, N., Azeroual, A., Bencharki, B., Douira, A., & Zidane, L. (2022b). Ethnoveterinary medicines plants for animal therapy in the Rif, North of Morocco. *South African Journal of Botany*, *147*, 176–191. https://doi.org/10.1016/j.sajb.2021.12.037

Chaachouay, N., Azeroual, A., Douira, A., & Zidane, L. (2022c). Ethnoveterinary practices of medicinal plants among the Zemmour and Zayane tribes, Middle Atlas, Morocco. *South African Journal of Botany*, *151*, 826–840. https://doi.org/10.1016/j.sajb.2022.11.009

Chatterjee, A., & Pakrashi, S. C. (1994). *The Treatise on Indian Medicinal Plants*. Publications and Information Director, CSIR, New Delhi. vol. 1. p. 9.

Chauhan, N. S. (1999). *Medicinal and Aromatic Plants of Himachal Pradesh*, Indus Publishing Company. p. 597.

Christenhusz, M. J. M., Fay, M. F., & Chase, M. W. (2017). *Plants of the world, An illustrated encyclopedia of vascular plants*. Kew Publishing, Royal Botanic Gardens, 816 p.

Collett, H. (1971). *Flora Simlensis, Flowering Plants of Simla*, M/s Bishen Singh Mahendra Pal Singh, p. 405.

Danert, S. (1958). Zur Systematik von *P. somniferum*. *Kulturpflanze* 6, 61–88.

Dilek, M., Gültepe, A., & Öztaşan, N. (2018). Determination of essential oil composition and investigation of, antimicrobial properties of poppy (*Papaver Somniferum* L.) flower. *AKU Journal of Science and Engineering*, 18(3): 786–795.

Dittbrenner, A., Mock, H.-P., Börner, A., & Lohwasser, U. (2012). Variability of alkaloid content in *Papaver somniferum* L. *Journal of Applied Botany and Food Quality*, 82(2): 103–107.

Duke, J. (1983). *Papaver somniferum*. In *Handbook of energy crops*. Unpublished. https://hort.purdue.edu/newcrop/duke_energy/Papaver_somniferum.html

EFSA (European Food Safety Authority) (2019). EFSA, Opium alkaloids in poppy seeds. 2019. https://www.efsa.europa.eu/en/press/news/180516 (accessed on 30 October 2023).

Egan, P. A., Pendry, C. A., & Shrestha, S. (2011). Papaveraceae. In M. F. Watson et al. (Eds.), *Flora of Nepal* (Volume 3), Royal Botanic Garden Edinburgh, pp. 78–95.

Evans, V. C. (2009) *Pharmacognosy*. 16th ed., Elsevier.

Fejer, J. 2007. New trends and perspectives in poppy breeding (*Papaver somniferum* L.). In *8th Conference and Technical Seminar on Seed and Seedling. Czech University of Life Sciences, Prague*. Osivo a sadba: VII Odbory a Vedecky Seminar, 124–130.

Frick, S., Kramell, R., Schmidt, J., Fist, A. J., & Kutchan, T. M. 2005. Comparative qualitative and quantitative determination of alkaloids in narcotic and condiment *Papaver somniferum* cultivars. *J. Nat. Prod.*, 68, 666–673.

Gamble, J. S. (1967). *Flora of the Presidency of Madras*, reprinted edition, Botanical Survey of India, vol. 1. p. 157.

Gómez-Serranillos, M. P., Palomino, O. M., Carretero, E., & Villar, A. (1998). Analytical study and analgesic activity of oripavine from*Papaver somniferum* L. *Phytotherapy Research*, 12(5): 346–349. DOI: 10.1002/(SICI)1099-1573(199808)12:5<346::AID-PTR307>3.0.CO;2-A

Güler, D. A., Aydin, A., Koyuncu, M., & Parmaksiz, I. (2016). Anticancer activity of *papaver somniferum. Journal of the Turkish Chemical Society Section A: Chemistry*, 3(3): 349–366.

Gümüşçü, A., Arslan, N., & Sarihan, E. O. (2008). Evaluation of selected poppy (*Papaver somniferum* L.) lines by their morphine and other alkaloids contents. *European Food Research and Technology*, 226: 1213–1220.

Gupta, S., Yadava, J. N. S., & Tandon, J. S. (1993). Antisecretory (Antidiarrhoeal) activity of Indian medicinal plants against *Escherichia Coli* enterotoxin-induced secretion in rabbit and guinea pig ileal loop models. *International Journal of Pharmacognosy*, 31(3): 198–204. DOI: 10.3109/13880209309082942

Hijazi, M. A., El-Mallah, A., Aboul-Ela, M., & Ellakany, A. (2017). Evaluation of analgesic activity of *Papaver libanoticum* extract in mice: Involvement of opioids receptors. *Evidence-Based Complementary and Alternative Medicine, 2017*. https://www.hindawi.com/journals/ecam/2017/8935085/abs/

Heinrich, M., Barnes Joanne, Prieto-Garcia J., Gibbons S. & Williamson, E. M. (2017). *Fundamentals of Pharmacognosy and Phytotherapy*, Elsevier, 3rd Edition, 352 pp.

Hooker JD (1973). Flora of British India, Bishen Singh Mahendra Pal Singh, DehraDun and M/S Periodical Experts, Delhi. vol. IV. p. 35.

Ismaili, A., Sohrabi, S. M., Azadpour, M., Heydari, R., & Rashidipour, M. (2017). Evaluation of the antimicrobial activity of alkaloid extracts of four Papaver species. *Herbal Medicines Journal (Herb Med J)*, 146–152.

ITIS (Integrated Taxonomic Information System). 2023. *Papaver somniferum* L., Available online: https://www.itis.gov/servlet/SingleRpt/SingleRpt?search_topic=TSN&search_value=18894#null (accessed on 13 October 2023).

Jafaar, H. J., Isbilen, O., Volkan, E., & Sariyar, G. (2021). Alkaloid profiling and antimicrobial activities of *Papaver glaucum* and *P. decaisnei*. *BMC Research Notes*, 14(1): 348. DOI: 10.1186/s13104-021-05762-x

Kačániová, M., Nóžková, J., Fatrcová-Šramková, K., Kropková, Z., & Kubincová, J. (2010). Antioxidant, antimicrobial actiyity and heavy metals content in pollen of *Papaver somniferum* L. *Ecological Chemistry and Engineering A*, 17(1): 97–106.

Kapoor, L. D. (1995). *Opium poppy – Botany, chemistry, and pharmacology*. The Haworth Press.

Krist, S., Stuebiger, G., Unterweger, H., Bandion, F., & Buchbauer, G. (2005). Analysis of volatile compounds and triglycerides of seed oils extracted from different poppy varieties (*Papaver somniferum* L.). *J Agric Food Chem*. 53(21): 8310–8316. DOI: 10.1021/jf0580869

Kumar Vasanth (2022). *Handbook on Opium: History and Basis of Opioids in Therapeutics*, Academic Press Inc, 244 pages (ISBN: 978–0-323–90903–7).

Labanca, F., Ovesnà, J., & Milella, L. (2018). *Papaver somniferum* L. taxonomy, uses and new insight in poppy alkaloid pathways. *Phytochemistry Reviews*, 17(4): 853–871. DOI: 10.1007/s11101-018-9563-3

Lal, R. K. (2022). The opium poppy (*Papaver somniferum* L.): Historical perspectives recapitulate and induced mutation towards latex less, low alkaloids in capsule husk mutant: A review. *Journal of Medicinal Plants Studies*, 10(3): 19–29.

Laux-Biehlmann, A., Mouheiche, J., Vérièpe, J., & Goumon, Y. (2013). Endogenous morphine and its metabolites in mammals: history, synthesis, localization and perspectives. *Neuroscience* 233: 95–117. DOI: 10.1016/j.neuroscience.2012.12.013

Lim, T. (2013). *Papaver somniferum*. In T. Lim (Ed.), *Edible medicinal and non-medicinal plants* (pp. 201–217). Springer Science Business Media. DOI: 10.1007/978-94-007-5653-3_12.

Mani, D., & Dhawan, S. S. (2011). Scientific basis of therapeutic uses of opium poppy (*Papaver somniferum*) in Ayurveda. *International Symposium on Papaver 1036*: 175–180. https://www.actahort.org/books/1036/1036_20.htm

Mansfield, D. 2019. On the frontiers of development: Illicit poppy and the transformation of the deserts of Southwest Afghanistan. *Journal of Illicit Economies and Development*, 1(3): 330–345. DOI: 10.31389/jied.46

Masihuddin, M., Jafri, M., Siddiqui, A., & Chaudhary, S. (2018). Traditional uses, phytochemistry and pharmacological activities of *Papaver somniferum* with special reference of unani medicine: An updated review. *Journal of Drug Delivery & Therapeutics*, 8(5-s): 110–114. DOI: 10.22270/jddt.v8i5-s.2069

Matyasova, E., Novak, J., Stranska, I., Hejtmankova, A., Skalicky, M., Hejtmankova, K., & Hejnak, V. 2011. Production of morphine and variability of significant characters of *Papaver somniferum* L. *Plant Soil Environ*, 57, 423–428. DOI: 10.17221/222/2011-PSE

McAlister, S., Ou, Y., Neff, E., et al. (2016). The Environmental footprint of morphine: A life cycle assessment from opium poppy farming to the packaged drug. *BMJ Open*, 6: e013302. DOI: 10.1136/bmjopen-2016-013302

Mondal, A., Gandhi, A., Fimognari, C., Atanasov, A. G., & Bishayee, A. (2019). Alkaloids for cancer prevention and therapy: Current progress and future perspectives. *European Journal of Pharmacology*, 858, 172472.

Mukerji, B. (1953). *The Indian Pharmaceutical Codex, Indigenous Drugs, Council of Scientific and Industrial Research*, New Delhi. vol. I. pp. 6–8.

Németh-Zámbori, É., Jászberényi, C., Rajhárt, P., & Bernáth, J. 2011. Evaluation of alkaloid profiles in hybrid generations of different poppy (*Papaver somniferum* L.) genotypes. *Ind. Crops Prod.*, 33, 690–696.

Nyman, U., & Hall, O.. (1976). Some varieties of *Papaver somniferum* L. with changed morphine alkaloid. *Hereditas*: 84: 69.

Orch, H., Chaachouay, N., Douiri, E. M., Faiz, N., Zidane, L., & Douira, A. (2021). Use of medicinal plants in dermato-cosmetology: An ethnobotanical study among the population of Izarène. *Jordan Journal of Pharmaceutical Sciences*, *14*(3), 323–340.

Plants of the World Online [POWO]. 2023. Facilitated by the Royal Botanic Gardens, Kew. Available online: http://www.plantsoftheworldonline.org (accessed on 13 October 2023).

Prajapati, S., Bajpai, S., Singh, D., Luthra, R., Gupta, M. M., & Kumar, S. 2002. Alkaloid profiles of the Indian land races of the opium poppy *Papaver somniferum* L. *Genet. Resour. Crop Evol.*, 49: 183–188.

Mohanty, R. B., & Panda, T. (2018). The journey of opium poppy: from ancient Mesopotamia to golden triangle. *Everyman's Science*, LIII(3) (August–September '18): 165–167

Rahimi, A., Kiralan, M., Arslan, N., Bayrak, A., & Dogramac, S. (2011). Variation in fatty acid composition of registered poppy (*Papaver somniferum* L.) seed in Turkey. *Akademik Gida*, 9 (3): 22–25.

Sariyar, G. (2002). Biodiversity in the alkaloids of Turkish *Papaver* species, *Pure and Applied Chemistry*, 74(4): 557–574.

Sharopov, F., Valiev, A., Gulmurodov, I., Sobeh, M., Satyal, P., & Wink, M. (2018). Alkaloid content, antioxidant and cytotoxic activities of various parts of *Papaver somniferum*. *Pharmaceutical Chemistry Journal*, 52(5): 459–463. DOI: 10.1007/s11094-018-1839-9

Shasany, A.K., Shukla, A.K., & Khanuja, S.P.S. (2007). Medicinal and aromatic plants. In Chittaranjan Kole (Ed.), *Genome Mapping and Molecular Breeding in Plants*, Volume 6. Technical crops. Springer-Verlag, pp. 175–196.

SIGAR (2018). Counternarcotics: Lessons from the U.S. Experience in Afghanistan, Arlington, 16 pp.

Stranska, I., Skalicky, M., Novak, J., Matyasova, E., & Hejnak, V. (2013). Analysis of selected poppy (*Papaver somniferum* L.) cultivars: pharmaceutically important alkaloids. *Ind. Crops Prod.*, 41: 120–126.

Tavlı, Ö.F. (2022). *Papaver somniferum* L. (pp 479–494). In F. T. Gürağaç Dereli, M. Ilhan, & T. Belwal (Eds.), *Novel drug targets with traditional herbal medicines*. Springer. DOI: 10.1007/978-3-031-07753-1_31

Tetenyi, P. (1997). Opium poppy (*Papaver somniferum*); botany and horticulture. In J. Janick (Ed.), *Horticultural reviews*, 19 (pp. 373–408). John Wiley & Sons.

Thakur, R.S. (1983). Opium: Chemistry and uses. In A. Husain, & J. R. Sharma (Eds.), *The opium poppy* (pp. 117–133). CIMAP.

Thawabteh, A., Juma, S., Bader, M., Karaman, D., Scrano, L., Bufo, S. A., & Karaman, R. (2019). The biological activity of natural alkaloids against herbivores, cancerous cells and pathogens. *Toxins*, 11(11): 656.

Vadhel, A., Bashir, S., Mir, A.H., Girdhar, M., Kumar, D., Kumar A., Mohan, A., Malik, T., & Mohan A. (2023). Opium alkaloids, biosynthesis, pharmacology and association with cancer occurrence. *Open Biol*, 13(5): 220355. DOI: 10.1098/rsob.220355

Van Wyk, B.E., & Wink, M. (2004). *Medicinal Plants of the World*. Briza Publications.

Verma, N., Jena, S.N., Shukla, S., & Yadav, K. (2016). Genetic diversity, population structure and Marker trait associations for alkaloid content and licit opium yield in india-wide collection of poppy (*Papaver somniferum* L.). plant *Gene*, 7: 26–41.

WFO Plant List (2023). https://wfoplantlist.org/plant-list/taxon/wfo-0000480963-2023-06?page=1&hide_syns=true (accessed on 04 October 2023).

White, P. T., and Raymer, S. (1985). The poppy. *Natl. Geograph.* February 143–189.

Yoganarasimhan, S.N. (2000). Medicinal plants of India (Tamil Nadu), S.N. Yoganarasimhan, Bangalore. vol. II. p. 24.

39 Kava (*Piper methysticum* G. Forst. Piperaceae)

Frank Kalungi and Godwin Anywar

39.1 INTRODUCTION

Kava (*Piper methysticum* G.Forst.) is a plant that is widely distributed across all regions of the Pacific Islands except New Zealand, New Caledonia, and most of the Solomon Islands (Singh, 1992). It is thought to have originated from a wild progenitor *Piper Wichmanii* C. DC which grows in Fiji, New Guinea, Solomon Islands, and Vanuatu (Prasad, 2018). Kava species consist of non-reproducing cultivars that were ultimately bred from *P. wichmanii* (Lebot et al., 1997).

Kava has been used for social, recreational, and medicinal purposes. For instance, a wide range of kava products have been developed for the treatment of anxiety since the 1990s (Sarris et al., 2011). The roots and rhizomes of kava are used to prepare a traditional beverage by cold extraction (Lebot & Legendre, 2016) which is used to relieve pain and reduce stress by natives of the South Pacific Islands (Xiang et al., 2016).

Kava has been widely consumed in the Pacific Islands and also popularized in many western counties, such as the United States and Australia, where it is being used for its anxiolytic effects (Food et al., 2016). Recent advances in phytochemistry, extraction processes, and pharmacology have led to an increase in kava bioprospecting by different countries. Germany for example approved kava-kava for anxiety and nervous disorders such as restlessness and stress in 1990. This was followed by other countries such as the United Kingdom, Switzerland and Austria (Bilia et al., 2002). In addition to being used as a drug and traditional beverage, kava has been used as a dietary supplement or as a recreational drink in kava bars (Bian et al., 2020).

Besides its neurological activities such as restlessness, kava has anticancer activities. It has preventive effects on lung cancer, urothelial cell carcinoma, prostate cancer, and colon cancer through suppressing cell proliferation, induction of apoptosis, and up-regulation of cancer suppressor genes (Bian et al., 2020). However, kava products have been reported to have hepatotoxic effects and also potential pharmacokinetic interactions with pharmaceutical drugs via modulation of the CYP450 system or the P-glycoprotein pump. Thus numerous studies are underway to evaluate its efficacy and safety (Sarris et al., 2011). In this chapter, we give an over view of the ethnobotany, phytochemical composition, and hallucinogenic effects of *Piper methysticum* which may guide the further applications of this plant.

39.2 SCIENTIFIC CLASSIFICATION

Kava is a member of the Piperaceae family, which contains more than 5,000 species of plants. The binomial name for kava is *Piper methysticum* G.Forst. *Piper methysticum's* most likely ancestor is believed to be *Piper wichmanii* C. DC (Lebot et al., 1997) because *P. methysticum* is sterile while *Piper wichmanii* is fertile and has long inflorescences and wide roots (Applequist & Lebot, 2006). These two taxa are morphologically similar and there are no significant morphological differences between them than those between different cultivars of *Piper methysticum* (Lebot et al., 1997). *Piper wichmanni*is now recognized as a variety of *P. methysticum* and was validated as *Piper. methysticum var. Methysticum* whereas the wild varieties are known as *Piper methysticum var. Wichmannii* (Applequist & Lebot, 2006). *P. wichimanii* is does not produce kavalactones that produce the pharmacological effects associated with kava (Prasad, 2018). Previously, kava was classified basing on morphological characteristics such as flowers, leaves and roots however the recent advances in genome sequencing and other molecular genotyping tools may help to accelerate its classification (Jiang et al., 2009).

39.3 VERNACULAR NAMES

The vernacular names of kava reflect its cultural significance and historical use in different parts of the world. Both kava and the kawa are commonly used to refer to the *Piper methysticum* and its beverages, however other names exist (Singh, 1992). In Oceania where the plant is widely grown for its medicinal and cultural purposes, it is known in various names in different languages. The Proto-Polynesian reconstructed term for kava is kawa which refers to sour or bitter (Lynch, 2002). In Fiji, kava terms were based on the root 'gona' and Proto-Fijan may have had the form kawa when they are referring to both *Piper sp.* and fish poison plant (Lynch, 2002). In some Polynesian areas such as Hawaii, Tahiiti, Marquesas, the initial k of the word was often removed to become awa or awa (Singh, 1992). In Tonga, it is called kava, or kava-kava in Vanuatu, and waild in Papua New Guinea. In addition to its Pacific Island names, kava is also known by various native names in different parts of the world. It is also known as Chicha

DOI: 10.1201/9781003460336-39

or Chica in some parts of South America and as Sakau or Yangona in some African countries.

39.4 BOTANICAL DESCRIPTION

Kava is a fairly succulent branched perennial shrub. It grows up to 3 meters tall, has a stout stem and large leaves that can reach up to 30 m long. It grows at altitudes of 150 m–300 m above the sea level (Singh, 1992). Kava is characterized by heart-shaped leaves, yellowish-green flowers, and a thick woody stem. The internodes have a chicken leg-like appearance can be short and thick, or long and thin in (Tasi, 2021). It has large, fresh brown roots that can reach 10 cm in diameter and 2 m in length. The roots/rhizomes are widely harvested for their high composition of kavalactones (Ferreira et al., 2021).

The leaves have an alternate arrangement on the stem, and they are large in size, green, and heart shaped. Kava has both inconspicuous male and female flowers which are small in size and are arranged in spikes at the end of the branches. Kava does not reproduce by sexual means and thus it is usually propagated vegetatively from stem cuttings or root sections (Prasad, 2018).

39.5 ORIGIN AND GEOGRAPHIC DISTRIBUTION

Kava is a plant widely distributed in all regions of the Pacific Islands with the exception of New Caledonia, New Zealand, and most of the parts of the Solomon islands (Singh, 1992). Kava is believed to have originated from outside Southeast Asia and New Guinea. However, Vanuatu farmers were the first to select and multiply it vegetatively (Lebot et al., 1997). Kava is believed to have been first domesticated less than 3000 years ago in Vanuatu, East Malanesia (Bilia et al., 2002).

Captain James Cook was accompanied by Daniel Scholander (Swedish Botanist) and Sydney Parkinson (artist) on his voyage from 1986 to 1971 and they were the first white men to see and record the Kava Plant (Singh, 1992). Recent archaeological evidence about the arrival of humans in eastern Melanesia and the Vanuatu origin of kava suggest that it is less than 3000 years old. In addition, kava was transferred from Vanuatu to Fiji and Polynesia and westward into New Guinea and central Micronesia (Lebot et al., 1997; Lewin, 1886). The plant was then integrated into the social, cultural, religious, and medicinal aspects of the Indigenous communities in the Pacific Islands (Bian et al., 2020).

39.6 ECOLOGY

Kava is a tropical plant that requires warm temperatures, high humidity, and plenty of rainfall to grow. It is typically grown in well-draining, loamy soil that is rich in organic matter and the ideal conditions for growing include 21–35°C and 70–100% relative humidity (Tasi, 2021). After cultivation, the plant can be harvested when it is about 2–2.5 m tall although it can grow up to 6 m during warm humid conditions with a lot of sunlight (Singh, 1992).

Kava grows on well at altitudes of 150–300 m above sea level. It requires an annual rainfall of over 2200 m for low altitudes (Prasad, 2018). The pH of the soil is also important for optimal growth of the plant with a PH range of 5.0 to 6.5 being optimal (Davis and Brown, 1999). The plant is also susceptible to moisture stress and measures such as sustainable irrigation, mulching plus shading and wind breaks can be taken to prevent moisture loss (Prasad, 2018).

Kava plays an important role in natural habitats across the humid tropics and particularly in lowland rainforests. *Piper* spp. act as a vital component of the understory flora (Dyer & Palmer, 2004). Over recent years, there has been a growing interest in understanding herbivory relationships, ant mutualism, and seed dispersal in the *Piper* genus because of its large species diversity, easily recognizable swollen nodes, and the aromatic smell of crushed leaves and broken stems (Dyer & Palmer, 2004).

Biotic factors that affect the growth and reproduction of kava include pests, microbes, weeds, and animals. A number of weeds interfere with kava cultivation and some may harbor severe diseases and infectious vectors. For example, *Commelina diffusa* and *Glycine sp.* harbor aphids that host cucumber mosaic cucumorvirus (cmv) that causes awa die back disease leading to wilt collapse, yellowing, ring spots, wrinkling, necrosis, and deformation in leaves (Nelson, 2005). In addition, a number of insects such as *Aleurodicus dispersus*, *Aleurotrachelus sp.* and mites such as *Tetranchyus cinnarbarinus* and *Polyphagotarsonemus latus* pose potential risks to kava if their numbers are not controlled. Similarly, a growing number of fungi such as *Fusarium* sp., and nematodes such as *Meloidogyne sp.* that cause node rot and root-knot disease respectively also pose a potential risk to kava cultivation (Nelson, 2005). However, management practices such as the use of approved pesticides and insecticides, irrigation during drought, wind breaks, intercropping, and mulching are being put in place to mitigate the threats posed by abiotic and biotic factors.

39.7 PHYTOCHEMISTRY

Plants produce a variety of chemical compounds for specific reasons such as protection against pests and diseases. In kava, the major bioactive compounds that contribute to its social and pharmacological importance are kavalactones which are characterized by arylethylene-alpha-pyrone skeleton (Li & Ramzan, 2010). Over 18 lactones have been characterized but only six of them are potent (Cairney et al., 2002). The kavalactones have been divided into major and minor kavalactones. The major kavalactones include kavain,7,8-Dihydrokavain, yangonin, methysticin,7,8-hidhydromethysticin and desmeoxyyangonin (Bilia et al., 2002).

The bioactivity of kava products is affected by different factors such as the type of cultivar, storage conditions, processing methods, extraction methods, age of the plant, and time of harvesting. In order to improve the quality of kava products, different extraction and detection methods for isolation of kava phytochemicals have been developed (Bilia et al., 2004; de Jager et al., 2004).

Kavalactones are majorly present in the rhizome but they are also present in other parts of the plant and their variability is strongly affected by geographical location of the cultivation (Bilia et al., 2004). The composition of kava products is largely affected by extraction methods, preparation methods, storage conditions, cultivar type, time of harvesting and processing conditions, among others (Côté et al., 2004; Teschke et al., 2011). Traditional preparations of kava vary but most commonly, the pulverized root is soaked in a bowl of water, and then filtered to produce the brew, and then it is drunk from a cup (Cairney et al., 2002).

39.8 ETHNOBOTANY

Humans have relied on plants for a long period of time for different purposes such as medical, social, religious, and cultural aspects. Kava been widely cultivated for its cultural significance and medicinal properties such as anxiolytic and ant-inflammatory (Bian et al., 2020). Traditional forms of kava have been used in Pacific cultures in important events such as marriages, funerals, welcoming guests and celebrating child births in addition to taking a cup of the beverage before a meal (Bian et al., 2020). The beverage was related to drinking of wine in western countries and virgin boys and girls traditionally prepared it by chewing the fresh rhizome and then spitting it in a bowl (Bilia et al., 2002).

The plant was also used for magico-religious purposes such as pouring libations of beverage or kava root to the gods (Bilia et al., 2002). Kava was also reported to induce sociability, calmness and in large doses, it would induce sleep or provoke nausea (Lebot et al., 1997). There are many varieties of kava, which are mainly characterized by differences in morphology. These include noble, medicinal, two-day, and wichmanni (Bian et al., 2020).

Traditionally, the noble cultivar has been used for regular consumption because of its high concentration of kavalactones. It is the only permitted cultivar for export as it is associated with a balanced and favorable composition of kavalactones, resulting in a more desirable and safer experience compared to other varieties (Bian et al., 2020; Chaachouay et al., 2021a, 2021b). The kava beverage is majorly prepared from the peeled rhizome of the noble variety although other varieties are used either alone or mixed with the noble variety. In addition, some kava parts such as the stem or the peel can be mixed with the rhizome and then used to prepare the beverage (Food et al., 2016). Kava and its related practices have also been linked to social and cosmological configurations that concern gender, rank, generation, and ancestry in some communities whose use as a social narcotic date back to the pre-European past (Taylor, 2010). In Vanuatu, it was believed that kava is a woman and thus females were not allowed to take the plant because a woman cannot take a woman (Taylor, 2010).

39.9 PHARMACOLOGY

Understanding the pharmacology of kava is important in determining its efficacy and safety. Kavalactones are the major compounds that are responsible for the pharmacological activity of the plant (Food et al., 2016). Kavapyrones enhanced the binding to GABA-A receptors in a concentration-dependent manner. Regional differences in the modulation of [3H] muscimol binding to GABA-A receptor complexes were observed using membrane fractions from various brain regions including the hippocampus, amygdala, medulla oblongata, frontal cortex, and cerebellum in rats (Jussofie et al., 1994). This study suggested that kavapyrones may induce their sedative effects through binding to the GABA-A receptors.

Kava may also exert its effects through other mechanisms. Kavalactones have been reported to cause differential inhibition of human cytochrome P450. Additionally, there was significant inhibition of CYP2C9 by desmethoxyyangonin, methysticin and dihydromethysticin; 2C19 by dihydromethysticin; of 2D6 by methysticin; and of 3A4 by desmethoxyyangonin, methysticin, and dihydromethysticin (Mathews et al., 2002). The administration of *Piper methystine* to F-344 rats showed a significant increase in hepatic glutathionine, cytosolic superoxide dismutase (Cu/ZnSOD), tumor necrosis factor α mRNA expression, and cytochrome P450 (CYP) 2E1 and 1A2 hence suggesting adaptation to oxidative stress by the rats and possible drug-drug interactions (Lim et al., 2007). Kava extracts showed analgesic effects via non-opiate pathways as naloxone was completely ineffective in reversing the antinociceptive activities induced by kava extracts (Jamieson & Duffield, 1990).

Safety concerns related to kava include reports of hepatotoxicity, primarily associated with certain commercial preparations and the use of aerial parts of the plant (Teschke et al., 2011). These cases highlight the importance of quality control and the selection of reputable sources for kava products. Therefore, caution should be exercised when combining kava with other medications due to potential interactions, particularly in individuals with pre-existing liver conditions. There is a potential for harmful herb-drug interactions or toxicity when different herbs (Anywar et al., 2020; Anywar et al., 2021; Anywar et al., 2022; Kaggwa et al., 2022).

39.10 MEDICINAL EFFECTS

Kava has been used for centuries for its potential medicinal properties and extensive research has been conducted to explore its therapeutic effects. While further studies are needed, several medicinal effects of kava have been reported as shown below.

39.10.1 Hallucinogenic Effects

Hallucinogens, also known as pyschomimetics or psychedelics, are a class of compounds that produce changes in a person's feelings, perception, and thoughts. They have been characterized with a lot of psychological effects such as illusions, distortions, hallucinations and changes to perception of time and space (Chaachouay & Zidane, 2021; Waters, 2021). They are categorized as classic hallucinogens and dissociative drugs. Hallucinogens can be obtained naturally from plants or synthesized artificially. They include ibogaine, ketamine, lysergic acid diethylamide (LSD) and 3,4-methylenedioxymethamphetamine (MDMA), psilocybin (Begola & Schillerstrom, 2019).

For generations, societies have used these compounds for spiritual, religious, and recreational purposes. Surprisingly, the difference between a lethal and non-lethal hallucinogen cannot be determined based on the blood levels because the mind-altering effects which may be considered harmful are the benefits the user is hoping to get (Singh, 1992; Waters, 2021).

Over the past years, some people have started to use kava-kava for its hallucinogenic effects. However, little research has been done to prove kava's hallucinogenic effects. People who drink the kava beverage are expected to get intoxicating effects. They experienced effects like those of marijuana and other hallucinogens (Singh, 1992). When kava is taken in large quantities or concentrated forms, it can induce intense or vivid hallucinogens.

In a study conducted on a patient after taking the drink, a reduced ability to focus on nearby objects, pupil diameter increase, perturbations in the oculomotor balance were noted and there were no changes in the stereoacuity or ocular refractive error (Garner & Klinger, 1985). Another study conducted on individuals who were intoxicated with 205 g of kava powder each in a group session showed that they had saccadic slowing and reduced accuracy when performing a visual task that was evidenced as the complexity of the task increased. The researchers concluded that kava intoxication is associated with specific movement coordination and visual abnormalities and that these abnormalities suggest malfunctions in the cerebellar and gamma-aminobutyric acid (Cairney et al., 2003). Research studies examining cognitive performance in individuals intoxicated with kava have shown improved accuracy and speed in tasks assessing visual attention and short-term memory retrieval, such as the Sperling partial report and Stenberg item recognition tasks. These findings suggest beneficial effects on cognitive function (Thompson et al., 2004). Therefore, research studies have found little or no hallucinogenic effects from the kava plant.

39.10.2 Anxiolytic and Anticancer Activity

Six kavalactones namely kavain, methysticin, yangonin, dihyrokavain, dihydromethysticin, and desmethoxyyangonin were shown to inhibit NE-mediated intracellular calcium influx in H1299 lung cells, potentially through antagonizing β-AR signaling hence suggesting kava's anxiolytic and anticancer activity (Botello et al., 2020). Another study showed that kava extract induced anxiolytic-like behavioral changes in assays conducted in mice (Garrett et al., 2003).

Kava caused a significant increase in acetylcholinesterase levels in the striatum of mice after the first week. This was followed by a decrease in the enzyme activity in three brain areas of the cortex, hippocampus, and striatum four weeks later, implying that this cholinergic system may play a central role in the anxiolytic action of kava (Noor, 2010).

Furthermore, kava was shown to inhibit lung tumorigenesis in A/J mice by suppressing cell proliferation, suggesting its anticancer activity (Johnson et al., 2011). Kava mechanistically inhibited proliferation and enhanced apoptosis in lung tumors, shown by a reduction in proliferating cell nuclear antigen, increase in caspase-3, and cleavage of poly(ADP-ribose) polymerase (PARP) in A/J mice (Johnson et al., 2008).

39.10.3 Anti-inflammatory and Antioxidant Properties

Kava has anti-inflammatory and antioxidant properties as shown by flavokavain A which inhibited IκB kinase (IKK), PRAK (p38-regulated/activated kinase), MAPKAP-K3 (MAPK-activated protein kinase 3), DYRK1A (dual-specificity tyrosine-phosphorylated and regulated kinase 1A) and Aurora B (Folmer et al., 2006). In addition, inducible transcription factor nuclear factor κB (NF-κB) was inhibited by flavokavain A, flavokavain B, kavain, and dihydrokavain (Folmer et al., 2006). Flavokawain A also decreased ratios of pro- and anti-inflammatory cytokines (IL-6/IL-10; TNF-α/IL-10) (Yang et al., 2020). However, mast cells were activated by single or combined kavalactones including methysticin, dihydromethysticin, and kavain suggesting that these water-extractable ingredients have proinflammatory responses in mast cells (Shimoda et al., 2012).

Kava active ingredients such as yangonin, methysticin, dihyrokawain have antioxidant activities (Wu et al., 2002). Flavokawain A exhibited its antioxidant activity by suppressing lipopolysaccharide induced reactive oxygen species production and also blocked NFκB (p65) pathway activation leading to the significant suppression of iNOS, COX-2, TNF-α, and IL-1β protein expressions (Yang et al., 2020).

39.10.4 Hepatotoxicity

In 2002, kava was withdrawn from European and UK markets due to concerns of its hepatotoxicity risks (Sarris et al., 2011). Flavokawains A and B in combination in mice have induce hepatotoxicity in C57BL/6 mice (Narayanapillai et al., 2014). In addition, kava overdose or prolonged treatment increased the probability of toxic liver injury in some patients (Teschke et al., 2008). Therefore, the consumption of kava should be done with caution following the safety guidelines provided by different authorities. Even when herbs have been used traditionally for long periods of time, it is important to test for their safety before being widely promoted. It is also important to document how they are processed and consumed locally because such methods could actually play an important role in herbal detoxication (Anywar, 2020; Anywar et al., 2020).

39.11 CONCLUSION

Kava has been used for centuries for religious, cultural, social, and medicinal purposes. It may thus act as an important herb as the world is returning to the use of natural products to treat various diseases. However, the commercialization of kava products has been hampered by many factors such as pests and diseases, poor cultivation methods, associated liver toxicity, and other side effects. In addition, there are few phytochemicals in kava that have been studied for their pharmacological and other biological activities. Therefore, in order to develop kava into a sustainable natural product, various strategies have to be adopted such developing new methods for isolation, detection and quantification of kava phytochemicals, intercropping of kava with other crops, sensitization of the public about its benefits, increasing scientific research in the exploration of the bioactivities of kava, standardization of its products and finally developing measures to control post-harvest losses.

LIST OF ABBREVIATIONS

CYP450	Cytochrome P450.
HPLC	High performance liquid chromatography
LC-MS	Liquid chromatography-mass spectrometry
LC-UV	Liquid chromatography-ultra violent
USA	United States of America

REFERENCES

Anywar, G (2020). Historical use of toxic plants. *Poisonous Plants and Phytochemicals in Drug Discovery*. Wiley Online Books. https://doi.org/10.1002/9781119650034.ch1

Anywar, G., Kakudidi, E., Byamukama, R., Mukonzo, J., Schubert, A., & Oryem-Origa, H. (2020). Indigenous traditional knowledge of medicinal plants used by herbalists in treating opportunistic infections among people living with HIV/AIDS in Uganda. *Journal of Ethnopharmacology*, 246, 112205. doi: https://doi.org/10.1016/j.jep.2019.112205

Anywar, G. U., Kakudidi, E., Oryem-Origa, H., Schubert, A., & Jassoy, C. (2022). Cytotoxicity of medicinal plant species used by traditional healers in treating people suffering from HIV/AIDS in Uganda. *Frontiers in Toxicology*, *2*(4):832780. https://doi.org/10.3389/ftox.2022.832780

Anywar, G. U., Kakudidi, E. K., Byamukama, R., Mukonzo, J. K., Schubert, A., Oryem-Origa, H., & Jassoy, C. (2021). A review of the toxicity and phytochemistry of medicinal plant species used by herbalists in treating people living with HIV/AIDS in Uganda. *Frontiers in Pharmacology*, *12*:615147. doi: 10.3389/fphar.2021.615147

Applequist, W. L., & Lebot, V. (2006). Validation of Piper methysticum var. wichmannii (Piperaceae). *Novon: A Journal for Botanical Nomenclature*, *16*(1), 3–4.

Begola, M. J., & Schillerstrom, J. E. (2019). Hallucinogens and their therapeutic use: A literature review. *Journal of Psychiatric Practice®*, *25*(5), 334–346.

Bian, T., Corral, P., Wang, Y., Botello, J., Kingston, R., Daniels, T.,… Lu, J. (2020). Kava as a clinical nutrient: promises and challenges. *Nutrients*, *12*(10), 3044.

Bilia, A. R., Gallori, S., & Vincieri, F. F. (2002). Kava-kava and anxiety: growing knowledge about the efficacy and safety. *Life Sciences*, *70*(22), 2581–2597.

Bilia, A. R., Scalise, L., Bergonzi, M. C., & Vincieri, F. F. (2004). Analysis of kavalactones from Piper methysticum (kava-kava). *Journal of Chromatography B*, *812*(1–2), 203–214.

Botello, J. F., Corral, P., Bian, T., & Xing, C. (2020). Kava and its kavalactones inhibit norepinephrine-induced intracellular calcium influx in lung cancer cells. *Planta Medica*, *86*(01), 26–31.

Chaachouay, N., Douira, A., & Zidane, L. (2021a). Herbal Medicine Used in the Treatment of Human Diseases in the Rif, Northern Morocco. *Arabian Journal for Science and Engineering*, 47, 131–153. https://doi.org/10.1007/s13369-021-05501-1

Chaachouay, N., Douira, A., & Zidane, L. (2021b). COVID-19, prevention and treatment with herbal medicine in the herbal markets of Salé Prefecture, North-Western Morocco. *European Journal of Integrative Medicine*, 42, 101285. https://doi.org/10.1016/j.eujim.2021.101285

Chaachouay, N., & Zidane, L. (2021). The Symbolic Efficacy of Plants in Rituals and Socio-Religious Ceremonies in Morocco, Northwest of Africa. *Journal of Religious & Theological Information*, *21*(1–2), 34–53. https://doi.org/10.1080/10477845.2021.1942433

Cairney, S., Maruff, P., & Clough, A. R. (2002). The neurobehavioural effects of kava. *Australian & New Zealand Journal of Psychiatry*, *36*(5), 657–662.

Cairney, S., Maruff, P., Clough, A. R., Collie, A., Currie, J., & Currie, B. J. (2003). Saccade and cognitive impairment associated with kava intoxication. *Human Psychopharmacology: Clinical and Experimental*, *18*(7), 525–533.

Côté, C. S., Kor, C., Cohen, J., & Auclair, K. (2004). Composition and biological activity of traditional and commercial kava extracts. *Biochemical and biophysical research communications*, *322*(1), 147–152.

Davis, R. I., & Brown, J. F. (1999). Kava (*Piper methysticum*) in the South Pacific: its importance, methods of cultivation, cultivars, diseases and pests.

de Jager, L., Perfetti, G., & Diachenko, G. (2004). LC-UV and LC-MS analysis of food and drink products containing kava. *Food Additives and Contaminants*, *21*(10), 921–934.

Dyer, L. A., & Palmer, A. D. (2004). *Piper: a model genus for studies of phytochemistry, ecology, and evolution* (Vol. 25): Springer.

Ferreira, J. V., Pierotte, I. C., Pianetti, G. A., & César, I. C. (2021). Simultaneous quantitation of kavalactones in kava dry extracts: comparison of multi-standards and single standard validation approaches. *Phytochemical Analysis*, *32*(5), 740–748.

Folmer, F., Blasius, R., Morceau, F., Tabudravu, J., Dicato, M., Jaspars, M., & Diederich, M. (2006). Inhibition of TNFα-induced activation of nuclear factor κB by kava (Piper

methysticum) derivatives. *Biochemical Pharmacology, 71*(8), 1206–1218.
Food, Nations, A. O. o. t. U., & Organization, W. H. (2016). Kava: A review of the safety of traditional and recreational beverage consumption. *Food and Agriculture Organization of the United Nations and World Health Organization, Rome, Italy.*
Garner, L. F., & Klinger, J. D. (1985). Some visual effects caused by the beverage kava. *Journal of Ethnopharmacology, 13*(3), 307–311.
Garrett, K. M., Basmadjian, G., Khan, I. A., Schaneberg, B. T., & Seale, T. W. (2003). Extracts of kava (*Piper methysticum*) induce acute anxiolytic-like behavioral changes in mice. *Psychopharmacology, 170*, 33–41.
Jamieson, D., & Duffield, P. (1990). The antinociceptive actions of kava components in mice. *Clinical and Experimental Pharmacology and Physiology, 17*(7), 495–507.
Jiang, S., Jianhua, X., & Li, X. (2009). A study on the RAPD and SCAR molecular markers of piper species. *Journal of Agriculture and Rural Development in the Tropics and Subtropics (JARTS), 110*(2), 127–135.
Johnson, T. E., Hermanson, D., Wang, L., Kassie, F., Upadhyaya, P., O'Sullivan, M. G.,... Xing, C. (2011). Lung tumorigenesis suppressing effects of a commercial kava extract and its selected compounds in A/J mice. *The American Journal of Chinese Medicine, 39*(04), 727–742.
Johnson, T. E., Kassie, F., O'Sullivan, M. G., Negia, M., Hanson, T. E., Upadhyaya, P., ... Xing, C. (2008). Chemopreventive Effect of Kava on 4-(Methylnitrosamino)-1-(3-pyridyl)-1-butanone plus Benzo [a] pyrene–Induced Lung Tumorigenesis in A/J Mice. *Cancer Prevention Research, 1*(6), 430–438.
Jussofie, A., Schmiz, A., & Hiemke, C. (1994). Kavapyrone enriched extract from Piper methysticum as modulator of the GABA binding site in different regions of rat brain. *Psychopharmacology, 116*, 469–474.
Kaggwa, B., Kyeyune, H., Munanura, E. I., Anywar, G., Lutoti, S., Aber, J., Bagoloire, L. K., Weisheit, A., Tolo, C. U., Kamba, P. F., & Ogwang, P. E. (2022). Safety and efficacy of medicinal plants used to manufacture herbal products with regulatory approval in Uganda: A cross-sectional study. *Evidence-Based Complementary and Alternative Medicine, 2022*, 1304839. https://doi.org/10.1155/2022/1304839
Lebot, V., & Legendre, L. (2016). Comparison of kava (Piper methysticum Forst.) varieties by UV absorbance of acetonic extracts and high-performance thin-layer chromatography. *Journal of Food Composition and Analysis, 48*, 25–33.
Lebot, V., Merlin, M., & Lindstrom, L. (1997). *Kava: The Pacific elixir: The definitive guide to its ethnobotany, history, and chemistry*: Inner Traditions/Bear & Co.
Lewin, L. (1886). *Ueber Piper methysticum (Kawa): Untersuchungen von L. Lewin. Mit 1 lithographirten Tafel*: Aug. Hirschwald.
Li, X., & Ramzan, I. (2010). Role of ethanol in kava hepatotoxicity. *Phytotherapy Research: An International Journal Devoted to Pharmacological and Toxicological Evaluation of Natural Product Derivatives, 24*(4), 475–480.
Lim, S. T., Dragull, K., Tang, C.-S., Bittenbender, H. C., Efird, J. T., & Nerurkar, P. V. (2007). Effects of kava alkaloid, pipermethystine, and kavalactones on oxidative stress and cytochrome P450 in F-344 rats. *Toxicological Sciences, 97*(1), 214–221.
Lynch, J. (2002). Potent roots and the origin of kava. *Oceanic Linguistics, 41*(2), 493–513.
Mathews, J. M., Etheridge, A. S., & Black, S. R. (2002). Inhibition of human cytochrome P450 activities by kava extract and kavalactones. *Drug Metabolism and Disposition, 30*(11), 1153–1157.
Narayanapillai, S. C., Leitzman, P., O'Sullivan, M. G., & Xing, C. (2014). Flavokawains a and B in kava, not dihydromethysticin, potentiate acetaminophen-induced hepatotoxicity in C57BL/6 mice. *Chemical Research in Toxicology, 27*(10), 1871–1876.
Nelson, S. (2005). Integrated pest management for'Awa (*Piper methysticum*).
Noor, N. A. (2010). Anxiolytic action and safety of Kava: Effect on rat brain acetylcholinesterase activity and some serum biochemical parameters. *African Journal of Pharmacy and Pharmacology, 4*(11), 823–828.
Prasad, A. (2018). Kava (Piper methysticum)-An important source of income for the rural farmers in Fiji Islands. *Advances in Crop Science and Technology, 6*(325), 2.
Sarris, J., LaPorte, E., & Schweitzer, I. (2011). Kava: a comprehensive review of efficacy, safety, and psychopharmacology. *Australian & New Zealand Journal of Psychiatry, 45*(1), 27–35.
Shimoda, L. M., Park, C., Stokes, A. J., Gomes, H. H., & Turner, H. (2012). Pacific island'Awa (Kava) extracts, but not isolated kavalactones, promote proinflammatory responses in model mast cells. *Phytotherapy Research, 26*(12), 1934–1941.
Singh, Y. N. (1992). Kava: an overview. *Journal of Ethnopharmacology, 37*(1), 13–45.
Tasi, A. (2021, August 30). Kava guides. *Botany of the Kava plant.* Retrieved 29/03/2023, 2023, from https://kavaguides.com/kava-plant/
Taylor, J. P. (2010). Janus and the siren's call: kava and the articulation of gender and modernity in Vanuatu. *Journal of the Royal Anthropological Institute, 16*(2), 279–296.
Teschke, R., Sarris, J., & Lebot, V. (2011). Kava hepatotoxicity solution: A six-point plan for new kava standardization. *Phytomedicine, 18*(2–3), 96–103.
Teschke, R., Schwarzenboeck, A., & Hennermann, K.-H. (2008). Kava hepatotoxicity: a clinical survey and critical analysis of 26 suspected cases. *European Journal of Gastroenterology & Hepatology, 20*(12), 1182–1193.
Thompson, R., Ruch, W., & Hasenöhrl, R. U. (2004). Enhanced cognitive performance and cheerful mood by standardized extracts of Piper methysticum (Kava-kava). *Human Psychopharmacology: Clinical and Experimental, 19*(4), 243–250.
Waters, K. (2021). Pharmacologic similarities and differences among hallucinogens. *The Journal of Clinical Pharmacology, 61*, S100–S113.
Wu, D., Yu, L., Nair, M., De Witt, D., & Ramsewak, R. (2002). Cyclooxygenase enzyme inhibitory compounds with antioxidant activities from Piper methysticum (kava kava) roots. *Phytomedicine, 9*(1), 41–47.
Xiang, C.-P., Shi, Y.-N., Liu, F.-F., Li, H.-Z., Zhang, Y.-J., Yang, C.-R., & Xu, M. (2016). A survey of the chemical compounds of Piper spp.(Piperaceae) and their biological activities. *Natural Product Communications, 11*(9), 1403–1408 https://doi.org/10.1177/1934578X1601100
Yang, H.-L., Yang, T.-Y., Gowrisankar, Y. V., Liao, C.-H., Liao, J.-W., Huang, P.-J., & Hseu, Y.-C. (2020). Suppression of LPS-induced inflammation by chalcone flavokawain a through activation of Nrf2/ARE-mediated antioxidant genes and inhibition of ROS/NFκB signaling pathways in primary splenocytes. *Oxidative Medicine and Cellular Longevity, 2020, 2020*(1), 3476212. https://doi.org/10.1155/2020/3476212

40 Belladonna (*Atropa belladonna* L. Solanaceae)

Noureddine Chaachouay, Abdelhamid Azeroual, Mohd Kafeel Ahmad Ansari, Rahmatullah Qureshi, Bouchaib Bencharki, and Lahcen Zidane

40.1 INTRODUCTION

Medicinal plants have therapeutic properties and may be used to treat various illnesses. These plants have been used for therapeutic reasons for millennia and continue to be extensively employed in traditional and contemporary medicine (Ansari et al., 2023; Chaachouay et al., 2019a, 2019b). An example of such a plant is *Atropa belladonna* L., the deadly nightshade. This plant is classified within the family Solanaceae, which encompasses other plants, including tomatoes, potatoes, and peppers (Butcher, 1947; Chaachouay et al., 2021a, 2021b, 2021c; Lee, 2007). The plant has a lignified stem that has the potential to reach a height of 2 meters. The leaves have a dark green coloration and possess an ovate shape. They are arranged alternately on the stem and have a length ranging from 10 to 20 cm, as well as a width ranging from 5 to 10 cm. This plant is indigenous to North Africa, Western Asia, and Europe. It is distinguished by its unique bell-shaped blooms and glossy black berries (Akbar, 2020; Ali, 2000; Chaachouay et al., 2020a, 2020b).

Although deadly nightshade has poisonous qualities, it has a lengthy medicinal history owing to its therapeutic advantages. The plant harbors many alkaloids, including scopolamine, hyoscyamine, and atropine, which have anti-inflammatory, analgesic, and antispasmodic properties (Al-Ashaal et al., 2013; Liu et al., 2014; Owais et al., 2014). Atropine is used in medicine to induce pupil dilation and address bradycardia, a decreased heart rate. Scopolamine is used to alleviate symptoms of motion sickness and nausea, while hyoscyamine is prescribed for the management of gastrointestinal diseases, such as irritable bowel syndrome (Al-Ashaal et al., 2013).

Deadly Nightshade has historically been used in herbal therapy to address a range of conditions, such as asthma, epilepsy, and menstrual disturbances (Almubayedh and Ahmad, 2018; Kwakye et al., 2018). Nevertheless, because of its poisonous characteristics, it is essential to exercise caution and employ it only under the guidance of a skilled healthcare professional. Deadly Nightshade has been used for its hallucinogenic properties alongside its medical applications. When consumed in small amounts, the plant may induce a state of pleasure and enhance sensory awareness. However, consuming higher quantities of the plant can lead to delirium and hallucinations (Almubayedh and Ahmad, 2018; Berdai et al., 2012).

Many documented results discuss the toxicity cases, clinical uses, and adverse effects of *A. belladonna*. However, there has not been a comprehensive and up-to-date review that examines the comparative effectiveness of *A. belladonna* in treating different diseases and the documented adverse effects of its toxicity (Almubayedh and Ahmad, 2018; Çaksen et al., 2003; Lee, 2007; Owais et al., 2014). This chapter highlights the taxonomy, habitat, ethnobotany, pharmacology, and medicinal effects of *A. belladonna*.

40.2 SCIENTIFIC CLASSIFICATION

Atropa belladonna L., often referred to as deadly nightshade or belladonna, is a profoundly toxic plant that is native to the Eastern Hemisphere. It is a member of the Solanaceae family, which encompasses several other plants, such as eggplants, tomatoes, potatoes, tobacco, chili peppers, and jimsonweed. The scientific categorization of it is as follows (ITIS, 2022):

Kingdom: Plantae
Subkingdom: Viridiplantae
Infrakingdom: Streptophyta
Superdivision: Embryophyta
Division: Tracheophyta
Subdivision: Spermatophytina
Class: Magnoliopsida
Superorder: Asteranae
Order: Solanales
Family: Solanaceae
Genus: *Atropa*
Species: *Atropa belladonna* L.

DOI: 10.1201/9781003460336-40

40.3 VERNACULAR NAMES

Arabic: Louffah, Yabruj-us-sanam, Badenjan moumit, Nabat sett Alhassan,; **English**: Deadly nightshade, Belladonna, Sleeping nightshade, Devil's cherries, Poison black cherry, Dwale; **Bengali**: Yebruj, Belladona; **Hindi**: Sagangur, Suchi, Lacchmana lac-chmani; **Urdu**: Yabrujussanam, Angurshefa, Luffah; **Italian**: Belladonna comune; **German**: Schwarze Tollkirsche, Gewöhnliche Tollkirsche; **French**: Belle-dame, Bouton-noir, Belladonne commune; **Sweden**: Vanlig belladonna, Belladonna; **Spanish**: Solano furioso, Belladona común; **Turkish**: Güzelavratotu; **Portuguese**: Erva-midriática, Erva-moura-furiosa, Beladona-comum; **Dutch**: Wolfskers, Dodelijke nachtschade, **Persian**: Luffana, Inab-es-salib; **Russian**: Атропа белладонна; **Chinese**: 颠茄; **Polish**: Pokrzyk wilcza jagoda, Pokrzyk; **Danish**: Dødsøvehjelm; **Czech**: Rulík zlomocný; **Japanese**: アトロパ・ベラドンナ; **Korean**: 아트로파 벨라돈나.

40.4 SYNONYMS

Based on data from The Plant List (2013), this plant exhibits nine synonyms and alternative scientific names that pertain to the same plant species. The synonyms for *Atropa belladonna* L. include **(1)** *Atropa cordata* Pascher; **(2)** *Atropa borealis* Kreyer ex Pascher; **(3)** *Atropa lutescens* Jacquem. ex C.B.Clarke; **(4)** *Atropa digitaloides* Pascher; **(5)** *Atropa mediterranea* Kreyer ex Pascher; **(6)** *Atropa lethalis* Salisb.; **(7)** *Belladonna trichotoma* Scop.; **(8)** *Boberella belladonna* (L.) E.H.L.Krause; and **(9)** *Belladonna baccifera* Lam. These synonyms are significant as they prevent ambiguity when several appellations pertain to the same botanical species. Moreover, they might be used in scientific investigations, where it is crucial to identify and categorize different plant species precisely.

40.5 BOTANICAL DESCRIPTION

A. belladonna is a perennial herbaceous plant classified under the Solanaceae family. The plant may reach a maximum height of 1.5 meters. It has a stem that branches out and has a purplish-brown hue. Additionally, the stem is adorned with tiny hairs (Ali, 2000; Butcher, 1947; Lee, 2007). The leaves exhibit a substantial size, are uncomplicated in structure, and are arranged in a staggered pattern, measuring around 10–16 cm in length and 4–8 cm in breadth. Their color is a shade of dark green, and their surface is uniformly smooth (Figure 40.1). The flowers have a bell-shaped morphology and possess a deep purple hue. Their length ranges from around 2.5 to 3 centimeters, while their width varies between 2.5 and 3.5 centimeters. The plant yields fruit in the shape of a lustrous, ebony, round, or elliptical berry, measuring about 1–2 cm in diameter, housing countless minuscule seeds (Kurioka et al., 1992; Lacković, 2017). Ingesting the berries might be harmful due to their toxicity. The seeds possess an elliptical shape, measuring roughly 2 mm long, and exhibit a rigid and sleek surface. Every berry has the potential to hold a maximum of 50 seeds. The roots have a robust and succulent nature, capable of attaining a length of up to 50 cm (Liu et al., 2014). Their exterior displays a pale-yellow hue, while a white shade characterizes their interior. The plant has a strong and discernible aroma. (Campbell, 2007).

40.6 ORIGIN AND GEOGRAPHIC DISTRIBUTION

A. belladonna, originally native to temperate southern, Eastern Central and Europe, the Caucasus, Turkey, North Africa, and Iran, has been extensively imported and grown in regions beyond its natural habitat (Butcher, 1947; Genova et al., 1997). The plant was first documented in Flora of Skåne in 1870, where it was reportedly grown in apothecary gardens near Malmö in southern Sweden (Hylander, 1971). Nevertheless, because of its extensive cultivation and utilization in traditional medicine, it has been imported and acclimatized in several other global locations, such as North America, Australia, and some parts of Asia (Figure 40.2). The species is indigenous to Britain and is limited to calcareous soils. It may be seen in many habitats, including hedgerows, disturbed areas, open woodland, and field margins (Berdai et al., 2012; Stuart, 2004). *A. belladonna* is mainly distributed in the eastern United States and southeastern Canada in North America, where it is recognized as an invasive species. Additionally, it has been documented in the western region of the United States, but its occurrence is less frequent in that area. *A. belladonna* is mainly distributed in the southern regions of Australia, where it has become established in some locations. It is distributed in some regions of China, Japan, and Korea in Asia and is also used in traditional herbal treatment (Bekkouche et al., 2001; Rudnyk-Ivashchenko and Yaruta, 2016).

40.7 HABITAT AND ECOLOGY

A. belladonna exhibits versatility in its ability to thrive in many environments, such as forests, hedgerows, and meadows. However, it prefers fertile, damp soils found in somewhat shaded regions. It may also be seen in disrupted environments such as road edges and abandoned land (Bekiesch et al., 2021; Lacković, 2017). Deadly Nightshade flourishes in excessively humid habitats with abundant sunshine throughout its gestation phase. Upon reaching full maturity as a sporophyte, the plant thrives in soil that is regularly wet and slightly shaded. However, it is also capable of growing in regions that are shaded or unshaded (Kutlymuratova, 2022; Lee, 2007; Stateva and Desheva, 2018). *A. belladonna* is versatile in its ability to thrive in several soil types and pH levels, ranging from acidic to alkaline and from sandy to rocky or loamy. However, it is more often found in soil with a pH of 6 or below, which is classified as acidic. Recent research has shown that the

FIGURE 40.1 Morphological characteristics of *A. belladonna* plant.

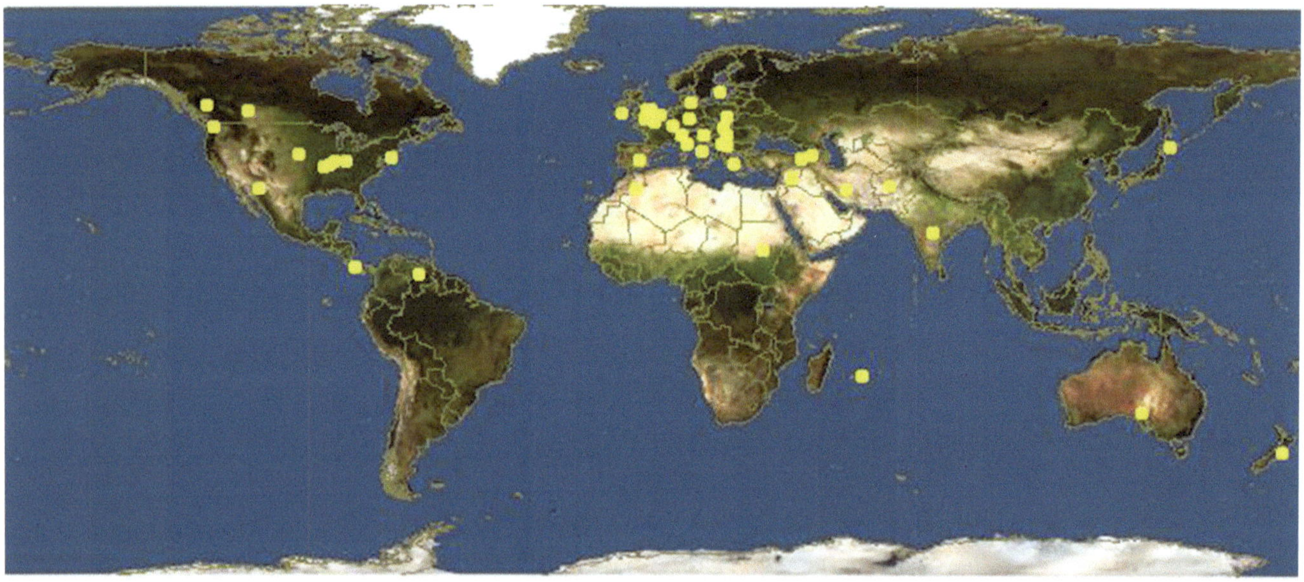

FIGURE 40.2 Geographic distribution map of *A. belladonna*.

plant exhibits the most outstanding production of alkaloids, used in medicinal applications when cultivated in well-lit, chalky soil with optimal sunlight exposure (Ali, 2000; Arquinet, 2019). Deadly Nightshade adheres to an annual growth cycle, blossoming from June to September. *A. belladonna* is a vital plant for several insects, especially bees and butterflies since they are drawn to its nectar. The plant's berries serve as a vital nourishment for avian species, including blackbirds and thrushes, that possess the ability to withstand the deadly alkaloids included in them. The plant is sometimes a natural pesticide for managing pests like aphids and spider mites (Arquinet, 2019; Butcher, 1947; Genova et al., 1997).

40.8 ETHNOBOTANY

The ethnobotany of *A. belladonna* pertains to the cultural and traditional use of the plant by various civilizations across time. Researchers and historians have extensively investigated the ethnobotany of deadly nightshade. The plant has been used for several reasons, ranging from the treatment of diseases to the deliberate poisoning of adversaries.

40.8.1 Traditional Medicine

The therapeutic characteristics of *A. belladonna* have made it a staple in traditional medicine for generations. The plant has been used for several applications, such as treating respiratory and gastrointestinal ailments, eye diseases, and inducing drowsiness (Nedelcheva and Draganov, 2014). The leaf is used externally to alleviate neurologic discomfort and ingested to address asthma and whooping cough (Berdai et al., 2012; Orch et al., 2021; Chaachouay et al., 2022a, 2022b, 2022c; Benkhnigue et al., 2023; Benlarbi et al., 2023). The plant contains alkaloids, such as atropine and hyoscyamine, which have various effects on the body. These alkaloids function as pain relievers, relax the muscles of the respiratory system, decrease the production of stomach acid, raise the tone of the muscles in the gastrointestinal tract, widen the pupils, and have a soothing impact on the central nervous system (Chaachouay & Zidane, 2021; Ivancheva and Stantcheva, 2000; Kwakye et al., 2018; Rajput, 2013). However, it is essential to emphasize that the plant is very poisonous and has been traditionally used as a lethal substance for arrows and assassinations. Although the plant is toxic, it is used in contemporary medicine as a means of obtaining atropine due to its antispasmodic and antiarrhythmic properties (Chaachouay et al., 2021a Belhaj et al., 2021; Toporcer et al., 2006). It is crucial to use deadly nightshade only under the supervision of a skilled healthcare professional to mitigate any potential health risks.

40.8.2 Cosmetic Uses

A. belladonna is a profoundly poisonous plant that includes tropane alkaloids, including atropine, hyoscyamine, and scopolamine (Bajaj and Simola, 1991; Kwakye et al., 2018). These chemicals possess a diverse range of medicinal and pharmacological applications, but they may also pose a significant risk if consumed or absorbed excessively via the skin. Although *A. belladonna* has been historically used in cosmetics, particularly in Victorian-era beauty products, to enlarge the pupils and create the illusion of more enormous, more captivating eyes, its contemporary usage in cosmetics is largely discouraged owing to possible health issues (Berdai et al., 2012; Gál et al., 2009; Toporcer et al., 2006). *A. belladonna* has been prohibited in cosmetics by the European Union, and its use in cosmetics has not been authorized by the U.S. Food and Drug Administration (FDA) (Cornell et al., 2019; Janetos and Thyparampil, 2019). Additionally, it is essential to mention that several nations have stringent restrictions regarding using botanical components in cosmetics. Therefore, any incorporation of *A. belladonna* would be required to adhere to these regulations. It is generally advised against utilizing any harmful ingredient in cosmetics due to the potential damage it might cause to the skin and overall well-being. Employing secure and controlled components in cosmetic items is vital to prevent unfavorable responses.

40.8.3 Poisonous Uses

A. belladonna is renowned for its toxic characteristics, and its alkaloids have been used as lethal substances throughout history. This plant has very poisonous features that were well-recognized in ancient times. It was used for poisoning arrows and as a method of carrying out assassinations (Lee, 2007; Shah et al., 2013). It is believed that the Roman Emperor Augustus was poisoned with *A. belladonna* by his wife, Livia (Cilliers and Retief, 2000). While a temporary ceasefire was in place, Macbeth, who held the position of a subordinate to King Duncan I of Scotland, purportedly used *A. belladonna* to lethally poison the soldiers of Harold Harefoot, the invading monarch of England. The English troops could not engage in combat and were compelled to withdraw to their naval vessels (Cutler, 1992). Furthermore, medical historians hypothesize that *A. belladonna*, in conjunction with laudanum, could have been used to administer poison to Solomon Northup, a free African-American man who was abducted and sold into slavery in 1841. The plant's foliage, roots, and fruits contain the poisonous alkaloids atropine, hyoscyamine, and scopolamine. The presence of these alkaloids might result in various symptoms, including pupil dilation, dryness of the mouth, impaired swallowing, cognitive disorientation, hallucinations, and even a state of coma. Consuming the plant in extreme instances might result in fatality (Berdai et al., 2012; Lee, 2007). It is crucial to acknowledge that *A. belladonna* has a high level of toxicity and should only be ingested or used with adequate information and supervision (Almubayedh and Ahmad, 2018). Even consuming modest quantities of the plant may pose a risk, especially for youngsters and dogs susceptible to its poisonous properties.

40.8.4 Mythology

Atropos, in Greek mythology, was one of the triumvirate of Fates responsible for determining the fate of individuals. Atropos was the agent responsible for severing the thread of life, representing the inescapable nature of mortality. The name Atropa and Belladonna, meaning 'beautiful woman' in Italian, may have been influenced by this mythical character (Cikla et al., 2011; Rajput, 2013). According to Norse mythology, the deity Odin acquired his intelligence and supernatural abilities via a nine-day and nine-night ordeal of suspending himself from the branches of the World Tree, known as Yggdrasil (Malik et al., 2021). According to some tale variations, Odin consumed a mixture of *A. belladonna* and other toxic herbs to enter a near-death state and acquire his wisdom (Attrell, 2013; Rätsch, 2005).

40.8.5 Folklore and Witchcraft

A. belladonna has been used in folklore and witchcraft for centuries. It was formerly thought to have magical abilities and was utilized in different ceremonies and rites (Kwakye et al., 2018; Lee, 2007). *A. belladonna* was often linked to enchanting and esoteric abilities in legend. It was thought to be able to evoke visions and facilitate communion with the spiritual realm. The plant was often used in divination rituals since it was thought to enhance the ability to forecast future events and interpret dreams. *A. belladonna* was used to safeguard against evil entities and deter maleficent spells and enchantments. The plant was often hung over entrances or worn as an amulet to save humans from danger (Berdai et al., 2012; Campbell, 2007; Harper, 1977; Shah et al., 2013).

A. belladonna was a commonly used component in several potions and spells within witchcraft. It was said to possess the ability to elicit feelings of love, inflict damage upon adversaries, and produce glimpses of the future (Kizi, 2022). The plant was often used in offensive actions aimed at subjugating or manipulating others, as well as in incantations specifically crafted to allure prosperity and achievement. A prominent use of *A. belladonna* in witchcraft was its purported capacity to metamorphose humans into animals and give them the power of flight (Malik et al., 2021). This idea was often linked to the notion of shapeshifting when humans were believed to possess the ability to assume the physical appearance of an animal at their own will. It is crucial to acknowledge that many of these beliefs and practices associated with *A. belladonna* are based on superstition and folklore (Ostling, 2016). Due to its hazardous properties, the use of this plant in contemporary witchcraft is strongly forbidden and should be addressed with care and reverence (Chaachouay et al., 2021b).

40.9. HALLUCINOGENIC EFFECTS

Many alkaloids found in *A. belladonna* have the potential to produce hallucinatory effects when consumed in high doses (Cikla et al., 2011; Kwakye et al., 2018). Deadly Nightshade may induce psychedelic effects, characterized by visual distortions such as impaired vision, dilated pupils, and heightened (Lacković, 2017). Individuals may also have auditory hallucinations, cognitive impairment, disorientation, and a feeling of dissociation from reality. Deadly Nightshade, apart from its hallucinogenic properties, may also induce many additional symptoms, such as xerostomia, dysphagia, tachycardia, hyperthermia, and urine retention. In extreme instances, it might potentially result in a state of unconsciousness or fatality (Çaksen et al., 2003).

40.10. PHYTOCHEMISTRY

The phytochemistry of *A. belladonna* has been extensively studied. The plant is known to contain a variety of alkaloids, including hygrine, hygroline A, hygroline B, tropinone, tropine 1, pseudotropine, 3a-acetoxytropane, alkaloid c, alkaloid d, cuscohygrine, 3a-tigloyloxytropane, 3a-phenylacetoxytropane, apoatropine, aposcopolamine, hyoscyamine, 6-hydroxyapoatropine, scopolamine, 6-hydroxyhyoscyamine, 6-hydroxyhyoscyamine (Al-Ashaal et al., 2013; Kizi, 2022). These alkaloids are responsible for the plant's toxic effects and medicinal properties (Ineloffer et al., 2019; Schwarzenberger et al., 2012). Furthermore, *A. belladonna* harbors a diverse array of flavonoids, which possess promising therapeutic properties, including enhancing cardiovascular well-being, mitigating the likelihood of certain malignancies, and alleviating inflammation. The plant species contains several flavonoids, such as quercetin, kaempferol, isorhamnetin, luteolin, apigenin, rutin, myricetin, diosmetin, hesperidin, naringenin, taxifolin, fisetin, chrysin, genistein, quercitrin, kaempferide, and acacetin (Al-Ashaal et al., 2018; Malik et al., 2021; Schwarzenberger et al., 2012).

Tannins are hydrophilic chemicals in plants that possess astringent characteristics and are often used to produce leather and ink. While *A. belladonna* does not contain a substantial quantity of tannins, trace levels of these compounds may be found in the plant (Imaseki, 1957). Although *A. belladonna* may include tannins, they do not serve as the main poisonous element in the plant. Coumarins are a group of plant-derived compounds known as phytochemicals, which may be found in various plants, including *A. belladonna*. Scopoletin, commonly called scopoline or coumarin lactone, is a coumarin compound that may be found in *A. belladonna* (Godara et al., 2014; Schwarzenberger et al., 2012). Scopoletin has been extracted from the leaves, roots, and berries of the plant. Scopoletin has shown many biological actions, including antioxidant, anti-inflammatory, and anticoagulant effects. Animal investigations have shown that it has analgesic and anticonvulsant properties.

Deadly Nightshade includes many steroidal alkaloids, a class of chemical substances characterized by a steroidal nucleus and an amino group. The plant contains many steroidal alkaloids, including Solanine, Solanidine, and Belladonnine. In addition to its primary components, *A. belladonna* includes many secondary chemicals, including fatty acids, amino acids, and carbohydrates (Hedayati et al., 2021; Heltmann, 1979). Nevertheless, the plant's pharmacological activities or toxicities are only sometimes investigated for these substances (Al-Ashaal et al., 2013; Kwakye et al., 2018). *A. belladonna* is a multifaceted plant containing various chemical components that have different effects on the body. The tropane alkaloids are the main active components of the substance, and they are responsible for its poisonous and pharmacological effects (Lee, 2007). Although several additional chemicals in *A. belladonna* may possess potential medicinal advantages, the plant is mainly recognized for its toxicity (Avci et al., 2022). It is essential to use this only under the supervision of a competent healthcare practitioner.

40.11 PHARMACOLOGY

A. belladonna is a toxic plant containing multiple alkaloids with diverse pharmacological effects on the body. These alkaloids are employed in medicine due to their pharmacological effects, namely in the fields of ophthalmology, anesthesia, and antispasmodics (Akbar, 2020; Grynkiewicz and Gadzikowska, 2008; Schmeller and Wink, 1998; Shah et al., 2013).

40.11.1 Analgesic Properties

The pain-relieving properties of *A. belladonna* have been documented in traditional medicine for ages. The plant has been used externally to alleviate pain and inflammation and internally to address diverse ailments such as dysmenorrhea, cephalalgia, and odontalgia. Nevertheless, more research is required to provide a more comprehensive and consistent body of scientific information about the analgesic properties of *A. belladonna* (Owais et al., 2014). Research has shown that the alkaloids found in the plant may possess analgesic properties by obstructing the transmission of pain signals in the neurological system (Al-Ashaal et al., 2013). Nevertheless, other inquiries have shown that the plant's pain-relieving qualities may be more robust and consistent. In addition, *A. belladonna* is not advised for pain management because of its inherent toxicity and negative consequences (Ramoutsaki et al., 2002). High quantities of the plant's alkaloids may induce a range of adverse effects, including but not limited to impaired vision, urine retention, constipation, dry mouth, delirium, and hallucinations.

40.11.2 Antispasmodic Properties

The main reason for the antispasmodic properties of *A. belladonna* is the presence of alkaloids like atropine and hyoscyamine. These alkaloids work on the neurological system to block the function of acetylcholine, a neurotransmitter that triggers muscular spasms. *A. belladonna* may diminish the occurrence and severity of muscular spasms, especially in the gastrointestinal tract and urinary system, by inhibiting the effects of acetylcholine (Sepehri et al., 2022). *A. belladonna* is often used for the treatment of illnesses such as irritable bowel syndrome (IBS), spastic colon, and bladder spasms (Rauf et al., 2021). Additionally, it has the potential to alleviate muscular spasms linked to multiple sclerosis, Parkinson's disease, and other neurological disorders (Rajput, 2013). Nevertheless, the use of *A. belladonna* should only be done under the supervision of a proficient healthcare practitioner since it has the potential to be poisonous and perhaps fatal when consumed in significant quantities.

40.11.3 Anticholinergic Properties

A. belladonna contains alkaloids such as atropine, hyoscyamine, and scopolamine, which enable it to provide potent anticholinergic effects (Rajput, 2013). These alkaloids inhibit the function of acetylcholine, which is essential for modulating the activity of the parasympathetic nervous system. *A. belladonna* can induce a range of anticholinergic effects, such as dilated pupils (mydriasis), decreased secretion of saliva, tears, and sweat, reduced gastrointestinal motility, elevated heart rate, and possible central nervous system effects like sedation, confusion, and hallucinations, especially at higher doses (Berdai et al., 2012). The medicinal use and potential harm of *A. belladonna* stem from its anticholinergic properties, manifesting when ingested excessively.

40.11.4 Anti-Inflammatory Properties

Despite the scarcity of studies, several studies indicate that *A. belladonna* may possess anti-inflammatory properties. In research published in the *Pakistan Journal of Pharmaceutical Sciences* in 2014, it was discovered that an extract of *A. belladonna* showed noteworthy anti-inflammatory effects in a rat model of carrageenan-induced inflammation (Owais et al., 2014). A separate study, conducted in 2020 and published in the *Journal of Plant Cell Reports*, examined the anti-inflammatory properties of silver nanoparticles derived from Indian *A. belladonna* (Rajput et al., 2020). Additional study is required to identify and thoroughly understand the potential therapeutic uses of *A. belladonna* in treating inflammation (Akbar, 2020). It is crucial to acknowledge that *A. belladonna* is a very toxic plant and should only be used under the supervision of a competent healthcare professional.

40.11.5 Antioxidant Properties

Antioxidants are compounds that can counteract detrimental free radicals inside the body and safeguard cells from oxidative damage. One of the primary antioxidant components discovered in the poisonous nightshade is atropine (Munir et al., 2014; S. Rajput et al., 2020). This alkaloid has been used for millennia as a medicinal remedy for various ailments, such as muscular spasms and asthma. *A. belladonna* contains additional antioxidant metabolites, including flavonoids like quercetin, kaempferol, and myricetin, as well as phenolic substances like chlorogenic and caffeic acid (Öz et al., 2021).

40.11.6 Anti-Microbial Properties

It has been observed that *A. belladonna* exhibits antibacterial action against various microorganisms (Azeem et al., 2021; Sultana and Naveed, 2014). The plant contains several alkaloids, including atropine and hyoscyamine, that have been extensively investigated for their antibacterial characteristics. Various investigations have examined the antibacterial properties of extracts and constituents derived from *A. belladonna*. For example, a 2014 study published in the *African Journal of Traditional, Complementary, and*

Alternative Medicines discovered the ethanol extract of *A. belladonna* (Munir et al., 2014). The antibacterial activity of the substance was notable against a range of bacterial species, such as *Staphylococcus aureus* and *Escherichia coli*. A further investigation revealed that the ethyl acetate extract derived from *A. belladonna* had potent antibacterial properties against several bacterial strains, including methicillin-resistant *Staphylococcus aureus* (MRSA) (Öz et al., 2021). Additionally, a research investigation discovered that the alkaloids isolated from *A. belladonna* showed noteworthy antibacterial efficacy against various bacterial and fungal strains (Avci et al., 2022).

40.12 MEDICINAL USES AND TOXICITY

The presence of atropine and hyoscyamine alkaloids in *A. belladonna* contributes to its many applications in contemporary medicine. Atropine is an anticholinergic medication that inhibits the activity of acetylcholine, a neurotransmitter responsible for smooth muscle contraction, gland secretion, and the functioning of the central and peripheral nervous systems (Ramoutsaki et al., 2002; Srivastava et al., 2013). Atropine is a remedy for poisoning caused by organophosphate nerve agents and pesticides. Additionally, it addresses bradycardia, irritable bowel syndrome, and dilated pupils seen during eye assessments (Anderson, 2012; Taylor, 2006). Hyoscyamine, an alkaloid present in *A. belladonna*, is used as an antispasmodic medication for treating gastrointestinal diseases, including irritable bowel syndrome, colic, and peptic ulcers. Additionally, it is used for the treatment of bladder spasms and urine incontinence, as well as for alleviating symptoms associated with Parkinson's disease (Grynkiewicz and Gadzikowska, 2008; Maurya et al., 2020; Shah et al., 2013). Furthermore, *A. belladonna* serves as a reservoir of scopolamine, an alkaloid possessing anticholinergic characteristics, which is used to avert the occurrence of nausea and vomiting linked to motion sickness and anesthesia (Rajput, 2013). Scopolamine is furthermore used for the treatment of irritable bowel syndrome and the alleviation of symptoms associated with Parkinson's disease (Poirier et al., 2016; Veeranjaneyulu et al., 2012).

A. belladonna is a very poisonous plant. Every component of the plant, such as the leaves, berries, and roots, contains tropane alkaloids, which are robust and possibly lethal poisons (Almubayedh and Ahmad, 2018; Çaksen et al., 2003). The toxicity of *A. belladonna* may be influenced by the concentration of tropane alkaloids in the plant and the quantity ingested (Kwakye et al., 2018). Consuming even small amounts of the plant may result in symptoms such as xerostomia, visual impairment, vertigo, hallucinations, cognitive disarray, restlessness, and extreme mental disorientation. In extreme situations, it may lead to respiratory failure, convulsions, coma, and death (Avci et al., 2022; Berdai et al., 2012; Butcher, 1947; Lacković, 2017; Lee, 2007). Extreme caution should be used while handling *A. belladonna*, and consumption should be avoided entirely. It is crucial to acknowledge that *A. belladonna* is a poisonous plant, and a skilled healthcare professional should supervise its use to mitigate any possible health problems.

40.13 CONCLUSION

A. belladonna is a profoundly poisonous plant that includes a range of alkaloids, such as hyoscyamine, scopolamine, and atropine. With a lengthy record of use in both medical and toxicological contexts, this substance has also been employed for recreational purposes due to its hallucinogenic qualities. The plant has several conventional therapeutic applications, such as inducing sedation, providing anesthesia, and alleviating pain. Nevertheless, its elevated toxicity renders it hazardous without enough expertise and oversight. Even consuming minuscule quantities of lethal nightshade may lead to symptoms such as xerostomia, visual impairment, hallucinations, and sometimes even unconsciousness or fatality. It should be emphasized that the plant's toxicity might fluctuate based on seasonal variations, growing circumstances, and other influencing variables. Given its poisonous properties, the use of *A. belladonna* should be restricted to the supervision of a qualified healthcare practitioner. Furthermore, it is imperative to avoid consuming the plant in its raw state or any other form without appropriate preparation since doing so may result in significant health hazards. *A. belladonna* is an intriguing botanical specimen with a profound historical background and intricate characteristics. Nevertheless, it is essential to handle it prudently and acknowledge its inherent risks.

LIST OF ABBREVIATIONS

FDA	Food and Drug Administration
IBS	Irritable bowel syndrome
MRSA	Methicillin-resistant *Staphylococcus aureus*

REFERENCES

Akbar, S. (2020). *Handbook of 200 medicinal plants: A comprehensive review of their traditional medical uses and scientific justifications*. Springer Cham. 2156 https://doi.org/10.1007/978-3-030-16807-0

Al-Ashaal, H. A., Aboutabl, M. E., Maklad, Y. A., & El-Beih, A. A. (2013). Tropane alkaloids of Atropa belladonna L.: In vitro production and pharmacological profile. *Egyptian Pharmaceutical Journal*, 12(2), 130.

Al-Ashaal, H. A. A.-H., Abuelnaga, A., & El-Beih, A. A. (2018). Anti food poisoning pathogenic bacteria in correlation to coumarins and flavonoids of Atropa belladonna field plant and in vitro cultures. *Journal of Pharmaceutical Sciences and Research*, 10(2), 235–239.

Ali, R. M. (2000). Role of putrescine in salt tolerance of Atropa belladonna plant. *Plant Science*, 152(2), 173–179.

Almubayedh, H., & Ahmad, R. (2018). Clinical uses and toxicity of Atropa belladonna; An evidence based comprehensive retrospective review (2003–2017). *Bioscience Biotechnology Research Communications*, 11. 41–48.

Anderson, P. D. (2012). Emergency management of chemical weapons injuries. *Journal of Pharmacy Practice*, 25(1), 61–68.

Ansari, M. K. A., Iqbal, M., Chaachouay, N., Ansari, A. A., & Owens, G. (2023). *The concept and status of medicinal and aromatic plants: History, pharmacognosy, ecology, and conservation. in plants as medicine and aromatics* (pp. 129–144). CRC Press.

Arquinet, M. (2019). Écologie, utilisations et risques sanitaires associés à la belladone: Atropa belladonna L. [PhD Thesis].

Attrell, D. (2013). Dionysian Semiotics: Myco-Dendrolatry and Other Shamanic Motifs in the Myths and Rituals of the Phrygian Mother [Master Thesis, University of Waterloo]. https://uwspace.uwaterloo.ca/handle/10012/7857

Avci, M. O., Muzoglu, N., Yilmaz, A. E., & Yarman, B. S. (2022). Antibacterial, cytotoxicity and biodegradability studies of polycaprolactone nanofibers holding green synthesized Ag nanoparticles using atropa belladonna extract. *Journal of Biomaterials Science, Polymer Edition*, 33(9), 1157–1180.

Azeem, M., Nengroo, Z. R., Ahamad Khan, M. O., & Ganie, A. S. (2021). Fatty acid composition, antibacterial and antioxidant potential of Atropa belladonna and Thymus linearis seeds grown in Kashmir. S.N. *Applied Sciences*, 3, 1–8.

Bajaj, Y. P. S., & Simola, L. K. (1991). *Atropa belladonna L.: In vitro culture, regeneration of plants, cryopreservation, and the production of tropane alkaloids. Medicinal and aromatic plants III*. Springer.

Belhaj, S., Chaachouay, N., & Zidane, L. (2021). Ethnobotanical and toxicology study of medicinal plants used for the treatment of diabetes in the High Atlas Central of Morocco. *Journal of Pharmacy & Pharmacognosy Research*, *9*(5), 619–662. https://doi.org/10.56499/jppres21.1014_9.5.619

Bekiesch, P., Oberhofer, M., Sykora, C., Urban, E., & Zotchev, S. B. (2021). Piperazic acid containing peptides produced by an endophytic *Streptomyces* sp. Isolated from the medicinal plant Atropa belladonna. *Natural Product Research*, 35(7), 1090–1096.

Bekkouche, K., Daali, Y., Cherkaoui, S., Veuthey, J.-L., & Christen, P. (2001). Calystegine distribution in some solanaceous species. *Phytochemistry*, 58(3), 455–462.

Benlarbi, F., Mimoune, N., Chaachouay, N., Souttou, K., Saidi, R., Mokhtar, M.R., Kaidi, R, & Benaissa, M. T. (2023). Ethnobotanical survey of the traditional antiparasitic use of medicinal plants in humans and animals in Laghouat (Southern Algeria). *Veterinary World*, *16*(2), 357–368. https://doi.org/10.14202/vetworld.2023.357-368

Benkhnigue, O., Chaachouay, N., & Zidane, L. (2023). A Floristical and Ecological Study of the Medicinal Flora Used by the Local Population of the Haouz-Rehamna Region (Middle Atlantic Morocco-4). *Acta Botanica Hungarica Akadémiai Kiadó*, *65*(3-4), 247–304. https://doi.org/10.1556/034.65.2023.3-4.3

Berdai, M. A., Labib, S., Chetouani, K., & Harandou, M. (2012). Case report-Atropa Belladonna intoxication:A case report. *Pan African Medical Journal*, 11(1), 1–4.

Butcher, R. W. (1947). Atropa Belladonna L. *Journal of Ecology*, 34(2), 345–353.

Çaksen, H., Odabaş, D., Akbayram, S., Cesur, Y., Arslan, Ş., Üner, A., & Öner, A. F. (2003). Deadly nightshade (Atropa belladonna) intoxication: An analysis of 49 children. *Human & Experimental Toxicology*, 22(12), 665–668.

Campbell, E. A. (2007). Don't say it with nightshades: Sentimental botany and the natural history of Atropa Belladonna. *Victorian Literature and Culture*, 35(2), 607–615.

Chaachouay, N., Benkhnigue, O., Fadli, M., Ibaoui, H., & Zidane, L. (2019a). Ethnobotanical and ethnopharmacological studies of medicinal and aromatic plants used in the treatment of metabolic diseases in the Moroccan Rif. *Heliyon*, *5*(10), e02191. https://doi.org/10.1080/10477845.2021.1942433

Chaachouay, N., Benkhnigue, O., Fadli, M., Ibaoui, H., El Ayadi, R., & Zidane, L. (2019b). Ethnobotanical and ethnopharmacological study of medicinal and aromatic plants used in the treatment of respiratory system disorders in the Moroccan Rif. *Ethnobotany Research and Applications*, *18*(22), 1–16.

Chaachouay, N., Douira, A., Hassikou, R., Brhadda, N., Dahmani, J., Belahbib, N., Ziri, R., & Zidane, L. (2020a). Ethnobotanical and ethnopharmacological study of medicinal and aromatic plants used in the treatment of respiratory system disorders in the Moroccan Rif. PhDThesis, Département de Biologie - Université Ibn Tofail - Kénitra, Maroc, https://tel.archives-ouvertes.fr/tel-03376377

Chaachouay, N., Benkhnigue, O., & Zidane, L. (2020b). Ethnobotanical study aimed at investigating the use of medicinal plants to treat nervous system diseases in the Rif of Morocco. *Journal of Chiropractic Medicine*, *19*(1), 70–81. https://doi.org/10.1016/j.jcm.2020.02.004

Chaachouay, N., & Zidane, L. (2021). The Symbolic Efficacy of Plants in Rituals and Socio-Religious Ceremonies in Morocco, Northwest of Africa. *Journal of Religious & Theological Information*, *21*(1–2), 34–53. https://doi.org/10.1080/10477845.2021.1942433

Chaachouay, N., Benkhnigue, O., Douira, A., & Zidane, L. (2021a). Poisonous medicinal plants used in the popular pharmacopoeia of the Rif, northern Morocco. *Toxicon*, 189, 24–32. https://doi.org/10.1016/j.toxicon.2020.10.028

Chaachouay, N., Douira, A., & Zidane, L. (2021b). COVID-19, prevention and treatment with herbal medicine in the herbal markets of Salé Prefecture, North-Western Morocco. *European Journal of Integrative Medicine*, 42, 101285. https://doi.org/10.1016/j.eujim.2021.101285

Chaachouay, N., Douira, A., & Zidane, L. (2021c). Herbal Medicine Used in the Treatment of Human Diseases in the Rif, Northern Morocco. *Arabian Journal for Science and Engineering*, 47, 131–153. https://doi.org/10.1007/s13369-021-05501-1

Chaachouay, N., Benkhnigue, O., & Zidane, L. (2022a). Ethnobotanical and Ethnomedicinal study of medicinal and aromatic plants used against dermatological diseases by the people of Rif, Morocco. *Journal of Herbal Medicine*, 32, 100542. https://doi.org/10.1016/j.hermed.2022.100542

Chaachouay, N., Azeroual, A., Bencharki, B., Douira, A., & Zidane, L. (2022b). Ethnoveterinary medicines plants for animal therapy in the Rif, North of Morocco. *South African Journal of Botany*, *147*, 176–191. https://doi.org/10.1016/j.sajb.2021.12.037

Chaachouay, N., Azeroual, A., Douira, A., & Zidane, L. (2022c). Ethnoveterinary practices of medicinal plants among the Zemmour and Zayane tribes, Middle Atlas, Morocco. *South African Journal of Botany*, *151*, 826–840. https://doi.org/10.1016/j.sajb.2022.11.009

Cikla, U., Turkmen, S., Karaca, Y., Ayaz, F. A., Turedi, S., & Gunduz, A. (2011). An Atropa belladonna L. poisoning with acute subdural hematoma. *Human & Experimental Toxicology*, 30(12), 1998–2001. https://doi.org/10.1177/0960327111407225

Cilliers, L., & Retief, F. P. (2000). Poisons, poisoning and the drug trade in ancient Rome. *Akroterion*, 45(1), 88–100. https://doi.org/10.10520/EJC20508

Cornell, E. M., Janetos, T. M., & Xu, S. (2019). Time for a makeover-cosmetics regulation in the United States. *Journal of Cosmetic Dermatology*, 18(6), 2041–2047. https://doi.org/10.1111/jocd.12886

Cutler, H. G. (1992). An historical perspective of ancient poisons. In H. N. Nigg & D. Seigler (Éds.), *Phytochemical Resources for Medicine and Agriculture* (pp. 1–13). Springer US. https://doi.org/10.1007/978-1-4899-2584-8_1

Gál, P., Toporcer, T., Grendel, T., Vidová, Z., Smetana Jr., K., Dvořánková, B., Gál, T., Mozeš, Š., Lenhardt, L., Longauer, F., Sabol, M., Sabo, J., & Bačkor, M. (2009). Effect of *Atropa belladonna* L. on skin wound healing: Biomechanical and histological study in rats and in vitro study in keratinocytes, 3T3 fibroblasts, and human umbilical vein endothelial cells. *Wound Repair and Regeneration*, 17(3), 378–386. https://doi.org/10.1111/j.1524-475X.2009.00475.x

Genova, E., Komitska, G., & Beeva, Y. (1997). Study on the germination of *Atropa belladonna* L seeds. *Bulgarian Journal of Plant Physiology*, 23(1–2), 61–66.

Godara, R., Katoch, M., Katoch, R., Yadav, A., Parveen, S., Vij, B., Khajuria, V., Singh, G., & Singh, N. K. (2014). In vitro acaricidal activity of *Atropa belladonna* and its components, Scopolamine and Atropine, against Rhipicephalus (Boophilus) microplus. *The Scientific World Journal*, 2014(1), 713170.

Grynkiewicz, G., & Gadzikowska, M. (2008). Tropane alkaloids as medicinally useful natural products and their synthetic derivatives as new drugs. *Pharmacological Reports*, 60(4), 439.

Harper, C. (1977). The witches' flying-ointment. *Folklore*, 88(1), 105–106. https://doi.org/10.1080/0015587X.1977.9716057

Hedayati, A., Hemmaty, S., Nourozi, E., & Amirsadeghi, A. (2021). Effect of yeast extract on h6h gene expression and tropane alkaloids production in *Atropa belladonna* L. hairy roots. *Russian Journal of Plant Physiology*, 68, 102–109.

Heltmann, H. (1979). Morphological and phytochemical studies in *Atropa* species [*Atropa belladonna*, *A.* baetica, *A. acuminata*, *A. komarovii*, *A. pallidiflora*]. Herba Hungarica (Hungary). Herba Hungarica 18(3): 101-110

Hylander, N. (1971). Prima loca plantarum vascularium Sueciae. Första litteraturuppgift för Sveriges vildväxande kärlväxter jämte uppgifter om första svenska fynd. Förvildade eller i senare tid inkomna växter. *Svensk Botanisk Tidskrift*, 64, 1–332.

Imaseki, I. (1957). Phytochemical investigation on cultivation of medicinal plants. XIII. On the alkaloid biogenesis in datura. (3). *Pharmaceutical Bulletin*, 5(5), 447–451.

Ineloffer, M., Heidari, M., Naghdi Badi, H., Tolyat Abulhassani, S. M., Makarian, H., & Ameryan, M. R. (2019). Phytochemical and morphological responses of *Atropa* (*Atropa belladonna* L.) to PGPR under greenhouse conditions. *Journal of Medicinal Plants*, 18(72), 228–240.

ITIS. (2022). Integrated Taxonomic Information System (ITIS) online database, www.itis.gov, CC0. https://doi.org/10.5066/F7KH0KBK. http://itis-web-prod.snafu.cr.usgs.gov

Ivancheva, S., & Stantcheva, B. (2000). Ethnobotanical inventory of medicinal plants in Bulgaria. *Journal of Ethnopharmacology*, 69(2), 165–172. https://doi.org/10.1016/S0378-8741(99)00129-4

Janetos, T. M., & Thyparampil, P. (2019). Ocular cosmetics: Public safety problem or harmless products? *Ophthalmic Plastic & Reconstructive Surgery*, 35(3), 294. https://doi.org/10.1097/IOP.0000000000001337

Kizi, K. S. A. (2022). *Atropa belladonna*, deadly nightshade and its peculiarities. *International Journal of Medical Science and Public Health Research*, 3(04), Article 04. https://doi.org/10.37547/ijmsphr/Volume03Issue04-04

Kurioka, Y., Suzuki, Y., Kamada, H., & Harada, H. (1992). Promotion of flowering and morphological alterations in *Atropa belladonna* transformed with a CaMV 35S-rolC chimeric gene of the Ri plasmid. *Plant Cell Reports*, 12, 1–6.

Kutlymuratova, G. A. (2022). Ecological features and agrotechnology of growing *Atropa belladonna* L. and *Echinops ritro* L in the conditions of karakalpakstan. *Journal of Pharmaceutical Negative Results*, 13(8), 2400–2404.

Kwakye, G. F., Jiménez, J., Jiménez, J. A., & Aschner, M. (2018). *Atropa belladonna* neurotoxicity: Implications to neurological disorders. *Food and Chemical Toxicology*, 116, 346–353.

Lacković, Z. (2017). « Bunanje »:XX century abuse of *Atropa belladonna* halucinogenic berries in continental Croatia. *Psychiatria Danubina*, 29(3), 379–382.

Lee, M. R. (2007). Solanaceae IV: *Atropa belladonna*, deadly Nightshade. *Journal-Royal College of Physicians of Edinburgh*, 37(1), 77.

Liu, C.-H., Zhang, J., Kang, S., Liu, T.-S., & Zhao, J. (2014). Pharmacognostical study of *Atropa belladonna*. Zhongguo Zhong yao za zhi= Zhongguo Zhongyao Zazhi= *China Journal of Chinese Materia Medica*, 39(9), 1589–1592.

Malik, M., Hussain, S., Sajjad, N., Nazir, S., Gondal, M., & Malik, J. (2021). Physicochemical analysis, Qualitative and quantitative investigation of atropine in medicinal plant Atropa belladonna. *European Journal Of Pharmaceutical And Medical Research*, 8(5), 149–157.

Maurya, V. K., Kumar, S., Kabir, R., Shrivastava, G., Shanker, K., Nayak, D., Khurana, A., Manchanda, R. K., Gadugu, S., & Kar, S. K. (2020). Dark classics in chemical neuroscience:An evidence-based systematic review of belladonna. *ACS Chemical Neuroscience*, 11(23), 3937–3954.

Munir, N., Iqbal, A. S., Altaf, I., Bashir, R., Sharif, N., Saleem, F., & Naz, S. (2014). Evaluation of antioxidant and antimicrobial potential of two endangered plant species *Atropa belladonna* and *Matricaria chamomilla*. *African Journal of Traditional, Complementary and Alternative Medicines*, 11(5), 111–117.

Nedelcheva, A., & Draganov, S. (2014). Bulgarian medical ethnobotany: The power of plants in pragmatic and poetic frames. In A. Pieroni & C. L. Quave (Éds.), *Ethnobotany and biocultural diversities in the balkans: Perspectives on sustainable rural development and reconciliation* (pp. 45–65). Springer. https://doi.org/10.1007/978-1-4939-1492-0_4

Orch, H., Chaachouay, N., Douiri, E. M., Faiz, N., Zidane, L., & Douira, A. (2021). Use of medicinal plants in dermato-cosmetology: An ethnobotanical study among the population of Izarène. *Jordan Journal of Pharmaceutical Sciences*, *14*(3), 323–340.

Ostling, M. (2016). Babyfat and *Belladonna*: Witches' ointment and the contestation of reality. *Magic, Ritual, and Witchcraft*, 11(1), 30–72. https://doi.org/10.1353/mrw.2016.0008

Owais, F., Anwar, S., Muhammad, S., Ishtiaque, S., & Mohiuddin, O. (2014). Analgesic, anti-inflammatory and neuropharmacological effects of Atropa belladonna. *Pakistan Journal of Pharmaceutical Sciences*, 27(6). 2183-7.

Öz, M., Fidan, M. S., Baltaci, C., Ücüncü, O., & Karatas, S. M. (2021). Determination of antimicrobial and antioxidant activities and chemical components of volatile oils of *Atropa belladonna* L. growing in Turkey. *Journal of Essential Oil Bearing Plants*, 24(5), 1072–1086.

Poirier, A.-A., Aubé, B., Côté, M., Morin, N., Di Paolo, T., & Soulet, D. (2016). Gastrointestinal dysfunctions in Parkinson's disease: Symptoms and treatments. *Parkinson's Disease*, 2016, 1–23. http://dx.doi.org/10.1155/2016/6762528

Rajput, H. (2013). *Effects of Atropa belladonna as an anticholinergic*. Natural Products Chemistry & Research.

Rajput, S., Kumar, D., & Agrawal, V. (2020). Green synthesis of silver nanoparticles using Indian Belladonna extract and their potential antioxidant, anti-inflammatory, anticancer and larvicidal activities. *Plant Cell Reports*, 39, 921–939.

Ramoutsaki, I. A., Askitopoulou, H., & Konsolaki, E. (2002). Pain relief and sedation in Roman Byzantine texts: *Mandragoras officinarum*, *Hyoscyamos niger* and *Atropa belladonna*. *International Congress Series*, 1242, 43–50.

Rätsch, C. (2005). *The Encyclopedia of Psychoactive Plants: Ethnopharmacology and Its Applications*. Simon and Schuster.

Rauf, A., Akram, M., Semwal, P., Mujawah, A. A., Muhammad, N., Riaz, Z., Munir, N., Piotrovsky, D., Vdovina, I., & Bouyahya, A. (2021). Antispasmodic potential of medicinal plants: A comprehensive review. *Oxidative Medicine and Cellular Longevity*, 2021(1), 4889719.

Rudnyk-Ivashchenko, O., & Yaruta, O. (2016). Specificity of heavy metals distribution in atropa belladonna plants. *Agricultural Science and Practice*, 3(3), 66–72.

Schmeller, T., & Wink, M. (1998). Utilization of alkaloids in modern medicine. *Alkaloids: Biochemistry, Ecology, and Medicinal Applications*, In: Roberts, M.F., Wink, M. (eds) Alkaloids. Springer, Boston, MA. 435–459. https://doi.org/10.1007/978-1-4757-2905-4_18

Schwarzenberger, M., Stintzing, F., Meyer, U., & Lindequist, U. (2012). Biochemical, microbiological and phytochemical studies on aqueous-fermented extracts from Atropa belladonna L. Part 2–Phytochemistry. *Die Pharmazie-An International Journal of Pharmaceutical Sciences*, 67(5), 460–466.

Sepehri, E., Hosseini, B., & Hedayati, A. (2022). The effect of iron oxide nanoparticles on the production of tropane alkaloids, h6h gene expression and antioxidant enzyme activity in *Atropa belladonna* hairy roots. *Russian Journal of Plant Physiology*, 69(6), 1–11.

Shah, V. V., Shah, N. D., & Patrekar, P. V. (2013). Medicinal plants from Solanaceae family. *Research Journal of Pharmacy and Technology*, 6(2), 143–151.

Srivastava, V., Kaur, R., Chattopadhyay, S. K., & Banerjee, S. (2013). Production of industrially important cosmaceutical and pharmaceutical derivatives of betuligenol by Atropa belladonna hairy root mediated biotransformation. *Industrial Crops and Products*, 44, 171–175.

Stateva, S., & Desheva, G. (2018). Adaptation of the species *Atropa belladonna* L. grown in vitro to the environment. *UARD Jubilee International Scientific Conference*.

Stuart, D. C. (2004). *Dangerous garden: The quest for plants to change our* lives. Harvard University Press.

Sultana, T., & Naveed, S. (2014). In-vitro anti-microbial activity of methanolic extract of Atropa belladonna using the agar well diffusion method. *World Journal of Pharmaceutical Sciences*, 2(7) 654–656.

Taylor, P. (2006). Anticholinesterase agents. *Goodman and Gilman's the Pharmacological Basis of Therapeutics*, 11, 201–216.

The Plant List. (2013). The Plant List. Version 1.1. Published on the Internet; http://www.theplantlist.org/ (accessed 1st January).

Toporcer, T., Grendel, T., Vidinský, B., Gál, P., Sabo, J., & Hudák, R. (2006). Mechanical properties of skin wounds after Atropa belladonna application in rats. *Journal of Metals, Materials and Minerals*, 16(1), 25–29.

Veeranjaneyulu, P., Rao, T. B., Mantha, S., & Vaidyanathan, G. (2012). A Sensitive Method for the Estimation of Scopolamine in Human Plasma Using ACQUITY UPLC and Xevo TQ-S. Waters Corporation, Bangalore, India, 1–7.

41 Black Henbane (*Hyoscyamus niger* L. Solanaceae)

Ouafae Benkhnigue, Noureddine Chaachouay, Jamaa Habchaoui, Khadija Temmante, Abdelhamid Azeroual, and Lahcen Zidane

41.1 INTRODUCTION

For thousands of years and throughout history, plants have played an essential role in human life, regardless of culture or civilization. Thanks to intuition and experimentation, man has succeeded in classifying plants according to their uses: food, medicinal, veterinary, hallucinogenic, toxic, ornamental, ritual, religious-magical, and recreational (Ansari et al., 2023; Belhaj et al., 2021; Benkhnigue et al., 2022, 2023; Chaachouay et al., 2022; Benlarbi et al., 2023; Fatur, 2021; Fatur, 2019; Passos & Mironidou-Tzouveleki, 2016; Bellakhdar, 1997; Sueur et al., 1999). Hallucinogenic plants are plants rich in hallucinogenic substances. These substances are defined as 'chemical compounds which, in non-toxic doses, cause changes in perception, thought and mood, but rarely mental confusion, memory disorders or disorientation in relation to people, space and time' (Hoffer & Osmond, 1967). The therapeutic, initiatory, or religious use of hallucinogenic substances is a constant in the history of civilizations. Historical research reports their use in Mesopotamia, India, Persia, Egypt, Africa, China, Japan, Europe, and pre-Columbian America, among others (Sueur et al., 1999). The use of hallucinogenic plants and mushrooms, thought to date back thousands of years, is an important part of the human experience (Adovasio & Fry, 1976).

The Solanaceae family, which belongs to the large Angiosperms group, is one of the most economically and culturally important families, as it includes numerous species that have been used in many areas including food, medicine, recreation, ornamentation, as well as witchcraft, and crime for poisonous plants (Carruthers, 2015; Hawkes, 1999). It is also one of the few botanical families to contain a high number of psychoactive and hallucinogenic plants, some species of which contain tropane alkaloids (hyoscyamine, atropine, and scopolamine) with well-known psychotropic powers (Adamse et al., 2014; Alrashedy & Molina, 2016). From a taxonomic point of view and according to databases such as POWO (https://powo.science.kew.org/taxon/urn:lsid:ipni.org:names:30000631-2), and GBIF (https://www.gbif.org/fr/species/7717) consulted on August 28, 2023). The Solanaceae family comprises around 100 genera and 3982 species worldwide, with great morphological and ecological diversity. *Hyoscyamus L.*, a member of this family, is one of the most important genera in medical terms, due to its high percentage of hyoscyamine and scopolamine. (Satıl et al., 2015). *H. niger* L., commonly known as black henbane, is one of 46 species in the genus *Hyoscyamus* L. It is an annual or biennial plant that thrives in many temperate climates. It is native to Europe, North Africa, and Central and Western Asia, but has been introduced to other parts of the world such as North America, South Australia, and other regions (POWO, 2023). This plant contains several alkaloids such as scopolamine, hyoscyamine, and atropine, which are used for their pharmacological properties (e.g., analgesic, anti-inflammatory, antipyretic, anticonvulsant, spasmolytic, antidiarrheal, antisecretory, bronchodilator, bladder relaxant, hypotensive, cardio-depressant, vasodilator, antitumor, antioxidant and sedative) (Yadav et al., 2016; Begum et al., 2010; Kintzios et al., 2002; Matés et al., 1999; Sabahi et al., 1987). In traditional medicine, *H. niger* is widely used for a variety of ailments: abdominal pain, lung infections, tumors, and those associated with urinary tract diseases, particularly kidney stones (Hajipoor et al., 2015a; Aparna et al., 2015; El Bazaoui et al., 2012; Matsuda et al., 1991; Duke et al., 1987). This plant is also used against tooth and ear pain and sinusitis (Batool et al., 2023; Karaköse, 2022; Sargın et al., 2013). In addition to its medicinal uses, black henbane has been used for its psychoactive effects and in the treatment of mental disorders, epilepsy, paralysis, convulsions, neuralgia, spasmodic cough, and asthma. (Fatur, 2019; Patel et al., 2018). Despite the ethnobotanical, phytochemical, and pharmacological richness of *H. niger*, it remains a toxic and even lethal plant in high doses due to its high content of tropane alkaloids, particularly scopolamine. (Alizadeh et al., 2014; Collak et al., 2016). A bibliographical analysis of published works (Google, Google Scholar, Pubmed, and Science Direct), concerning ethnobotanical, pharmacological, and toxicological evidence for *H. niger*, shows that this plant is already cited in other studies of the same type. However, no up-to-date and exhaustive work has been documented on the two axes of our study: on the one hand, taxonomy, morphological description, ecology, and geographical repair; on the other hand, the comparative impact of *H. niger* in the

DOI: 10.1201/9781003460336-41

treatment of various diseases and the side effects of toxicity. With this in mind, this chapter highlights the taxonomy, morphological description, ecology, chorology, biogeographical distribution, ethnobotany, pharmacology, and toxicity of *H. niger*. We hope that the information contained in this chapter will enrich our knowledge of this species.

41.2 SCIENTIFIC CLASSIFICATION

According to the classification published on the INPN (https://inpn.mnhn.fr/espece/cd_nom/103185/tab/taxo?lg=fr) databases consulted on August 28, 2023, *Hyoscyamus niger* L. is an annual herbaceous plant belonging to the *Hyoscyamus* L. genus and Solanaceae family. The scientific name used is the one accepted in the WFO (http://www.worldfloraonline.org/) and IPNI (https://www.ipni.org) databases consulted on August 28, 2023).

Domain: Eukaryota (eukaryotes)
Reign (Kingdom): Plantae Haeckel, 1866
Subkingdom: Viridaeplantae
Infra-Reign (InfraKingdom): Streptophyta John, Williamson & Guiry, 2011
Superdivision: Embryophyta
Class: Equisetopsida C. Agardh, 1825
Clade: Tracheophyta Sinnott ex Cavalier-Smith, 1998
Clade: Spermatophyta
Subclass: Magnoliidae Novák ex Takht., 1967
Super-Order: Asteranae Takht., 1967
Order: Solanales Juss. ex-Bercht. & J. Presl, 1820
Family: Solanaceae Juss., 1789
Subfamily: Solanoideae Kostel., 1834
Tribe: Hyoscyameae Endl., 1839
Subtribe: Hyoscyaminae Kitt., 1840
Genus: *Hyoscyamus* L., 1753
Species: *Hyoscyamus niger* L., 1753

41.3 VERNACULAR NAMES

The common name of *H. niger* differs from country to country. In this study, we have tried to cite the different names published in the WFO (http://www.worldfloraonline.org/), IPNI (https://www.ipni.org/), GBIF (https://www.gbif.org/fr/species/2928978), Tela Botanica (https://www.tela-botanica.org/bdtfx-nn-35153-ethnobotanique), USDA databases and Hocking (1947); Aparna et al. (2015); Marionneau, 2018; Fatur (2021); Paulsen, (2010): **Arabic**: Ssikrane, Kinkite, Bou Rendjouf, Guenquitt, Bû-narjûf, Bû-renjûf, Sikerane**; Berber**: Guenkite, Waye-lilou, Taylilût**; Catalan**: Jusquíam negre**; Chinese**: Tianxianzi 天仙; **Danish**: Almindelig buldies, Buldies, Bulmeurt, Bulmeurtslægten; **Deutschland**: Altsitzerkraut, Schwarzes Bilsenkraut; **English**: Henbane, Black Henbane, Cassilago, Common Henbane, Hog's-bean; Stinking-nightshade, Jupiter–bean, Syfonica, Cassilata, Caballinus, Henbell, Jusqiam, Bazrolbang, Mashe atar, Ajavainekhorasani, Devil's eye, Stinking-nightshade, Insane root, Poison tobacco, Benele, Chenilsenkeraut; **Finnish**: (Piha) Hullukaali; **French**: Jusquiame noire, Herbe aux dents, Herbe aux engelures, Main du diable ou Herbe des teigneux; **India**: Gya-lantang, Lantang, Parasika Yavani; **Irish**: Gafann, Bainne cích na n-éan, Deodha, Crann gafainn, Beann mhear, Caoch na gcearc, Deoda gafann; **Italian**: Giusquiamo nero, Disturbo; **German**: Bilsenkraut, Schwarzes Bilsenkraut; **Greek**: Υοςχυάμον, Υοσκύαμος ο μέλας; **Korean**: Purpurbolmört; **Ndl**: Bilzekruid**; Norwegion**; Bulmeurt; **Pinyin**: Tian xian zi; **Portuguese**: Meimendro-negro; **Rhaeto-Romance**: Flur da santeri; **Sanskrit**: Prasikayavani; **Slovenian**: Črni zobnik; **Spanish**: Beleño negro, Chupa mieles, Hierba loca, Črni zobnik; **Transcribed Korean**: saripul; **Turkish**: Kurtluca, kulak otu, Banotu; **Slovenian**: Črni zobnik; **Swedish**: Bolmört, (Piha) Hullukaali.

41.4 SYNONYMS

According to the information published based on IPNI (https://www.ipni.org/?q=Hyoscyamus%20niger) database consulted on August 28, 2023, *Hyoscyamus niger* L. has 27 synonyms, two of which are homotypic, and we symbolize them with (≡) in front of the name.

Homotypic Synonyms

(1) ≡ *Hyoscarpus niger* (L.) Dulac in Fl. Hautes-Pyrénées: 444 (1867)
(2) ≡ *Hyoscyamus lethalis* Salisb. in Prodr. Stirp. Chap. Allerton: 131 (1796)

Heterotypic Synonyms

(1) = *Hyoscyamus agrestis* Kit. ex Schult. in Oestr. Fl., ed. 2, 1: 383 (1814)
(2) = *Hyoscyamus agrestis* var. *integrifolius* Wallr. in Sched. Crit.: 83 (1822)
(3) = *Hyoscyamus agrestis* var. *sinuatus* Wallr. in Sched. Crit.: 83 (1822), not validly publ.
(4) = *Hyoscyamus auriculatus* Ten. in Fl. Napol. 3: 222 (1829)
(5) = *Hyoscyamus bohemicus* F.W.Schmidt in Fl. Boëm. 3: 31 (1794)
(6) = *Hyoscyamus bohemicus* var. *pallidus* (Waldst. & Kit. ex Willd.) Pojark. in V.L.Komarov (ed.), Fl. URSS 22: 95 (1955)
(7) = *Hyoscyamus niger* proles agrestis (Kit. ex Schult.) Rouy in G.Rouy & J.Foucaud, Fl. France 10: 370 (1908)
(8) = *Hyoscyamus niger* var. *agrestis* (Kit. ex Schult.) Nyman in Consp. Fl. Eur.: 524 (1881), nom. superfl.
(9) = *Hyoscyamus niger* subsp. *agrestis* (Kit. ex Schult.) Hultén in Kongl. Svenska Vetensk. Acad. Handl., ser. 4, 13(1): 286 (1971)

(10) = *Hyoscyamus niger* var. *annuus* Sims in Bot. Mag. 59: t. 2394 (1823)
(11) = *Hyoscyamus niger* subsp. *auriculatus* (Ten.) Arcang. in Comp. Fl. Ital.: 495 (1882)
(12) = *Hyoscyamus niger* var. *biennis* Wimm. & Grab. in Fl. Siles. 1: 201 (1827)
(13) = *Hyoscyamus niger* var. *chinensis* Makino in J. Jap. Bot. 3: 3 (1926)
(14) = *Hyoscyamus niger* var. *intermedius* Iverus in Bot. Not. 1875: 81 (1875)
(15) = *Hyoscyamus niger* var. *pallidus* (Waldst. & Kit. ex Willd.) W.D.J.Koch in Syn. Fl. Germ. Helv.: 509 (1837)
(16) = *Hyoscyamus niger* subsp. *pallidus* (Waldst. & Kit. ex Willd.) Juel, Cedergr. & Örtendahl in Index Seminum (UPS, Upsaliensi) 1917: 18 (1918)
(17) = *Hyoscyamus niger* var. *pygmaeus* Wallr. ex Klett & Richt. in Fl. Leipzig: 211 (1830)
(18) = *Hyoscyamus officinarum* Crantz in Inst. Rei Herb. 2: 325 (1766)
(19) = *Hyoscyamus pallidus* Waldst. & Kit. ex Willd. in Enum. Pl. 1: 227 (1809)
(20) = *Hyoscyamus persicus* Boiss. & Buhse in Nouv. Mém. Soc. Imp. Naturalistes Moscou 12: 158 (1860)
(21) = *Hyoscyamus pictus* Roth in Nov. Pl. Sp.: 119 (1821)
(22) = *Hyoscyamus syspirensis* K.Koch in Linnaea 22: 736 (1849)
(23) = *Hyoscyamus verviensis* Lej. in Fl. Spa 1: 116 (1811)
(24) = *Hyoscyamus vulgaris* Neck. in Delic. Gallo-Belg. 1: 122 (1768)
(25) = *Hyoscyamus vulgaris* L. in Fl. Angl.: 12 (1754).

41.5 BOTANICAL DESCRIPTION

Morphologically, *Hyoscyamus niger* is an annual or biennial herbaceous plant, depending on variety, 20–80 cm high, hairy-glandular, slimy, and very leafy. Roots are sometimes stubby, fleshy, and 2 to 3 cm in diameter. Stem simple or twining. Basal leaves are petiolate and sometimes grouped in a rosette, ovate-lanceolate, or oblong, approx. 30 × 10 cm, coarsely toothed or with pinnate or separate lobes, acute apex (Figure 41.1a). Stem leaves alternate and simple, broadly cuneiform at the base, lobed or entire, 4–10×2–6 cm, pubescent with sticky hairs; lower ones petiolate; upper ones sessile-amplexicaul, ovate-oblong, sinuate-lobed-pinnatifid, with unequal triangular lobes with large acute teeth (Figure 41.1b). Inflorescence foliate in a helical uniparous cyme (Figure 41.1c). Flowers subsessile or on 3–5 mm pedicels. Calyx accrescent green, campanulate-tubular, 1–1.5 cm; lobes unequal and acute with five teeth. Fruiting calyx dilated at base, 2–2.5×1–1.5 cm; lobes generally erect, with acuminate teeth. Corolla is narrowly infundibuliform, almost regular yellowish veined with purple, throat purple-black, 2–3 cm long, hairy on the outside, more or less bilabiate. Stamens exserted, inserted at the base of the corolla (Figure 41.1d). Ovoid-rounded capsules, approx. 1.5×1.2 cm, with pyxidial dehiscence. Seeds yellow-brown, discoid, about 1 mm in diameter (Fennane et al., 2007; Lauber & Wagner, 2007; Guinochet & de Vilmorin, 1973; Quézel et al., 1963) et WFO (2023).

Flowering and fruiting: Flowering generally takes place from May to August and fruiting from July to October (Fennane et al., 2007).

41.6 ORIGIN, GEOGRAPHIC DISTRIBUTION, AND CHOROLOGY

Native to the Mediterranean, Balkan, and European regions, *Hyoscyamus niger* has since spread throughout Eurasia, including Europe (France and Switzerland), Russia, Asia, northern Africa, and North America, where it normally thrives as a weed in uncultivated areas (Info flora, 2023; Daunay et al., 2007; Hocking, 1947; Passos & Mironidou-Tzouveleki, 2016). According to information published in the POWO (https://powo.science.kew.org) database consulted on September 11, 2023, *H. niger* L. it is an annual or biennial species, the native range of this species is temperate Eurasia, North-West Africa, Afghanistan, Albania, Algeria, Altay, Amur, Austria, Baltic States, Belarus, Belgium, Bulgaria, Buryatiya, Central European Rus, China North-Central, China South-Central, China Southeast, Chita, Corse, Czechoslovakia, Denmark, East Aegean Is., East European Russia, East Himalaya, Finland, France, Germany, Greece, Hungary, Inner Mongolia, Iran, Iraq, Irkutsk, Italy, Japan, Kazakhstan, Kirgizstan, Korea, Krasnoyarsk, Krym, Lebanon-Syria, Manchuria, Mongolia, Morocco, Nepal, Netherlands, North Caucasus, North European Russi, Northwest European R, Norway, Pakistan, Poland, Portugal, Primorye, Qinghai, Romania, Sakhalin, Sardegna, Sicilia, South European Russi, Spain, Sweden, Switzerland, Tadzhikistan, Transcaucasus, Tunisia, Turkey, Turkey-in-Europe, Turkmenistan, Tuva, Ukraine, Uzbekistan, West Himalaya, West Siberia, Xinjiang, Yugoslavia and Introduced and grows primarily in the temperate biome (Colorado, Great Britain, Illinois, Ireland, Maryland, Myanmar, New Mexico, New South Wales, New York, New Zealand North, New Zealand South, South Australia, Vermont, Victoria, Wisconsin, Yakutskiya) (Figure 41.2).

41.7 ECOLOGY

The various species belonging to the Solanaceae family are mainly cosmopolitan, except in Arctic regions. They are generally distributed in tropical, subtropical, and temperate regions, but are strongly centered in America in the southern United States, where a temperate climate prevails (https://www.futura-sciences.com/planete/definitions/botanique-solanacee-18655/); (GBIF, 2023; POWO, 2023); and (https://plantes-botanique.org/famille_solanaceae) consulted on September 11, 2023. *Hyoscyamus niger* is an invasive, toxic weed introduced to North America, the United States, and across Canada. This spontaneous

FIGURE 41.1 Photos of *Hyoscyamus niger* L. (**A**) basal leaves; (**B**) leafy stem with flowers; (**C**) inflorescence and (**D**) flower. (Source: https://www.bioimages.org.uk/html/r165523.htm?6.)

therophyte occupies a variety of habitats and generally grows in open, disturbed areas (Fettig, 2016). Henbane is a monocarpic plant with annual and biennial forms in its native Eurasian range (Correns 1904; Schläppi 2011; Selleck, 1964). The life cycle of introduced populations in North America is currently unknown. Early research into the life cycle duration of henbane suggests that it is determined by simple allele dominance (Correns, 1904). In India, *H. niger* grows in Western Himalayas from Kashmir to Kumaon at an altitude of 1600–3500 m; frequent in Spiti near villages and roadsides; sporadic elsewhere in Lahul and Kinnaur regions of Himachal Pradesh. Cultivated in some parts of Kashmir, Punjab, U.P., Rajasthan, Maharashtra, and Nilgiri (Sharma et al., 2000).

In Saskatchewan, *H. niger* survives only in wet depressions or disturbed areas and has invaded croplands, pastures, and disturbed areas, particularly in east-central Saskatchewan. This species is unlikely to seriously affect agricultural production in this area. Black henbane thrives during cycles of abundant moisture and disappears during cycles of drought (Selleck, 1964).

In Greece, *H. niger* occupies rocky terrain and roadsides where an arid climate prevails (Passos & Mironidou-Tzouveleki, 2016). In Switzerland, *H. niger* grows in fallow or plowed land, often near the sea or close to farm buildings. It has naturalized in temperate regions and is protected in Vaud and Bern. (Info flora, 2023). Throughout France and Corsica, *H. niger* thrives on rubble, paths, and wasteland (Guinochet & de Vilmorin, 1973).

In Morocco, as a weed of cultivation, *H. niger* grows particularly in nitrophilous areas and rubble on the plains and in the low and medium mountains. It prefers a semi-arid, subhumid to humid climate (Fennane et al., 2007). In Algeria, *H. niger* is a fairly rare species, colonizing nitrophilous zones and especially mountainous areas throughout the country (Quézel et al., 1963). This variability in the ecological conditions of *H. niger's* habitats could be at the origin of its life cycle.

41.8 CONSERVATION STATUS: ASSESSMENT, PROTECTION, AND THREAT STATUS

Due to habitat destruction; inadequate path maintenance and herbicide use; *H. niger* is considered a threatened species in Switzerland and is listed as Vulnerable and facing a high risk of extinction in the wild according to IUCN assessment criteria (2012a, 2012b) and the 2016 Red List (Info flora, 2023) (Figure 41.2). The regional status according to the 2019 Red List showed that *H. niger* is threatened in several biogeographical regions: Jura (CR(PE)), Plateau (CR), Alpine North Slope (CR), Alpine North Slope (CR(PE)), Central Eastern Alps (EN), Central Western Alps (Vu) (https://www.infoflora.ch/fr/flore/hyoscyamus-niger.html) consulted on September 11, 2023.

In France, *H. niger* is listed on the Red List of the vascular flora of mainland France under the assessment category **LC** (Least Concern). It is also listed on the Red List in other regions of France (https://inpn.mnhn.fr/espece/cd_nom/103185/tab/statut#ancreStatutEspece) consulted on 11/09/2023. Similarly, in Morocco, Fennane (2021) indicates in his book 'Livre rouge de la flore vasculaire du Maroc' that *H. niger* is also listed under the **LC** category.

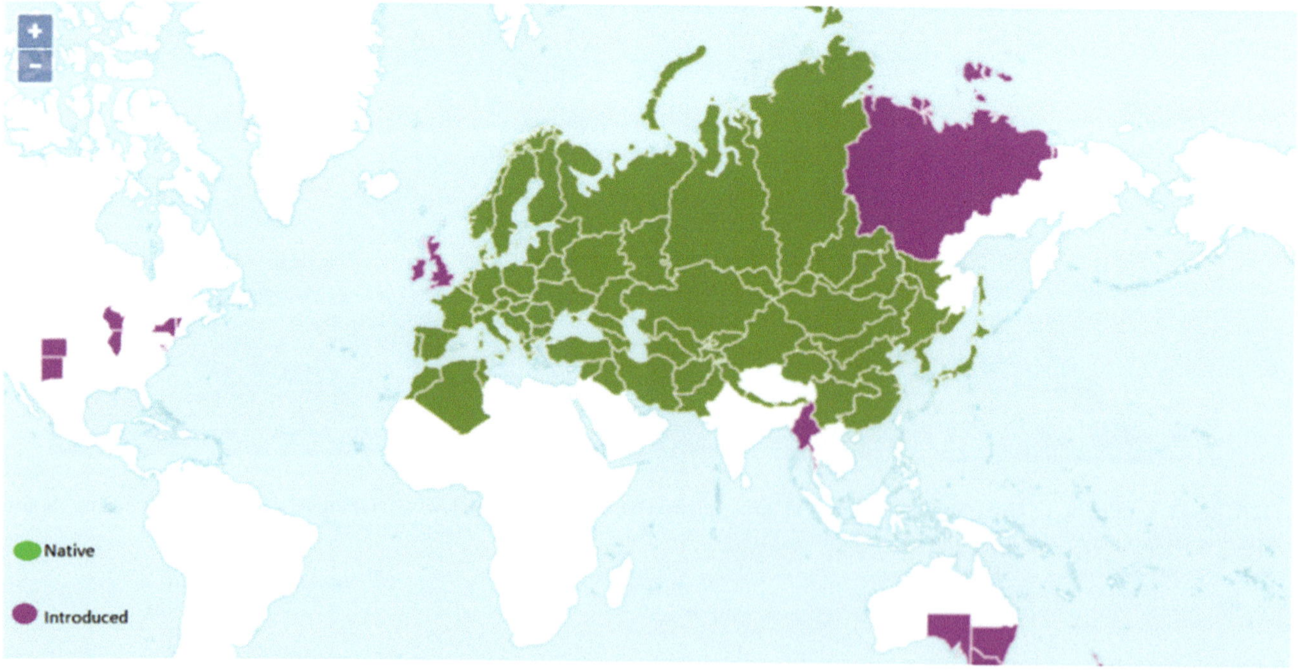

FIGURE 41.2 General ranges of *Hyoscyamus niger* L. Kew Backbone Distributions.The International Plant Names Index and World Checklist of Vascular Plants 2024. (Published at http://www.ipni.org and https://powo.science.kew.org/ © Copyright 2023 World Checklist of Vascular Plants. http://creativecommons.org/licenses/by/3.0.)

According to IUCN (2012a), a taxon is said to be of Least Concern when it has been assessed against the criteria and does not meet the criteria for Critically Endangered, Endangered, Vulnerable, or Near Threatened Figure 41.3. This category includes widespread and abundant taxa where the risk of extinction is low.

41.9 PHYTOCHEMISTRY

The importance of *Hyoscyamus niger* is due to its compounds rich in tropic alkaloids, such as hyoscyamine, atropine, and scopolamine, of which hyoscyamine sulfate and scopolamine hydrobromide are drugs with anticholinergic and antispasmodic properties respectively. The plant also has analgesic and sedative properties (Alaghemand et al., 2013; Bahmanzadegan et al., 2009; Supria, 1998). The roots, leaves, and seeds of *H. niger* contain three main alkaloids known for their psychoactive properties: scopolamine (over 50% of total alkaloid content), hyoscyamine, and atropine (Passos & Mironidou-Tzouveleki, 2016; Garnier et al., 1961; Paris & Moyse, 1976–1981). Scopolamine is the strongest psychoactive, with eight times more activity on the central nervous system than atropine (Ketchum et al., 1973) with more intense hallucinogenic properties. This explains its widespread historical use in Europe since ancient Greece as a narcotic, anesthetic, soporific, and analgesic, against insomnia and in the treatment of respiratory ailments (Carter, 1996; Daunay et al., 2007; Gorun et al., 2011; Hocking, 1947; Kala & Ratajc, 2012; Passos & Mironidou-Tzouveleki, 2016; Piomelli & Pollio, 1994; Schultes, 1977).

Leaves and flowers contain 0.05–0.15% total alkaloids (Arroo et al., 2007; Gaillard & Pepin, 1999; Garnier et al., 1961; Paris & Moyse, 1976–1981). Leaves also contain many salts: 18–23% ash (relative to dry matter); traces of essential oil; resin; choline; sugars; succinic acid; traces of scopoletin; rutoside; a volatile nitrogen base, tetramethyldiaminobutane (or tetramethyl putrescine) in fairly large quantities; a bitter glucoside: hyoscypicrin. (Garnier et al., 1961; Paris & Moyse, 1976–1981). Leaves contain 0.4±0.0 chlorogenic acid, 0.4±0.0 Quercetin-3O-glucoside-rhamnoside-rhamnoside (QGRR), and 9.2±0.5 mg/g rutin on a dry weight basis. Epicalice contains chlorogenic acid 1.1±0.1, quercetin-3O-glucoside-rhamnoside-rhamnoside (QGRR), and rutin 3.5±0.4 mg/g dry weight (Jassbi et al., 2014).

H. niger seeds contain several withanolide steroids (Ma et al., 1999) and steroid glycosides (hyoscyamoside G, E, F), of which hyoscyamoside E has cytotoxicity against human lung cancer cells (Zhang et al., 2013). Lunga et al. (2006, 2007) reported that the seeds were rich in two furostanol saponin compounds and four spirostanol saponin compounds. In addition, Ma et al. (2002) found four lignan-amides and ten other non-alkaloid compounds isolated from *H. niger* seeds. Also, one lignan, hyoscyamal, and three other compounds, balanophonine, pongamoside C, and pongamoside D, were isolated from *H. niger* seeds (Begum et al., 2009). In addition, coumarinolignans, hyosgerin, venkatasin, cleomiscosin A methyl ether, cleomiscosin A, cleomiscosin B, cleomiscosin A-9'-acetate and cleomiscosin B-9'-acetate, have also been isolated from *H. niger* seeds (Sajeli et al., 2006; Begum et al., 2010). Seeds contain 15–25% oil (Garnier et al., 1961; Paris & Moyse, 1976–1981). The plant also has Non-alkaloidal compounds including hyosmin, Canabisin D, Canabisin G, Grossamide, Hyosciamide, Hyoscyamal Balanophonin, Cleomiscosin A, Cleomiscosin B, Hyosgerin, Rutin, Atroposide and Riboflavin (Zhang et al., 2012).

41.10 ETHNOBOTANY

41.10.1 Traditional Medicine

In phytotherapy, almost all parts of *H. niger* (leaves, roots, and seeds) were used. But it's the leaves that are mainly used in traditional medicine, as they have a parasympatholytic action, resulting in an antispasmodic, analgesic, and sedative effect on the central nervous system. (Sijelmassi, 2011; Bellakhdar, 1997; Passos & Mironidou-Tzouveleki, 2016; The Herbal Resource, 2023). A bibliographical analysis of published works (Google, Google Scholar, Pubmed, and Science Direct) concerning ethnobotanical evidence for *H. niger* has shown that this plant enjoys a wide reputation throughout the world, both as a medicine and as a deadly poison, and by way of example, we can cite its uses in several countries:

In Turkey, Sargın et al. (2013) reported in their work that fumigation of *H. niger* seeds locally called 'kurtluca, kulak otu' is indicated against earache and sinusitis; In North-East Turkey, Karaköse (2022) cited in his ethnobotanical study that raw seeds, chewed in the mouth, are used against toothache. In the traditional Indian medical system, Ayurveda, powdered seeds of *H. niger* (Gourie-Devi et al., 1991) is one of the four plant-ingredients used for the treatment of Parkinson's disease (Nagashayana et al., 2000). Also, Batool et al. (2023) cited in their study 'Medicinal plants traditionally used in healing practices by indigenous communities in the Trans-Himalayan region of Ladakh, India' that smoke or poultice from the seeds or flowers of

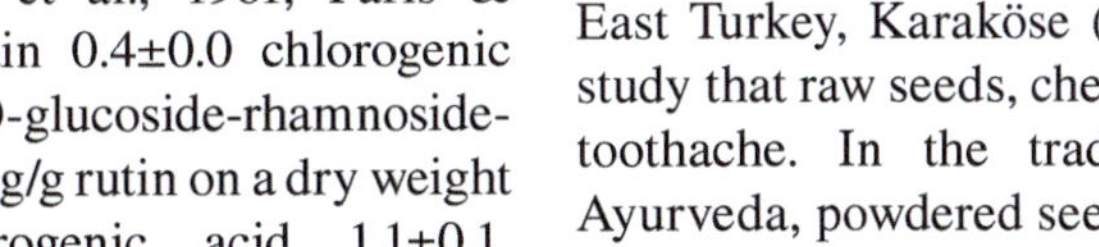

FIGURE 41.3 National status of *H. niger* in Switzerland according to the IUCN assessment criteria (2012a, 2012b) and the 2016 Red List (https://www.infoflora.ch/fr/flore/hyoscyamus-niger.html).

H. niger locally called 'Gya-lantang, Lantang' are administered orally or topically, against toothache and tooth cavity pain. This use is confirmed by other similar studies (Adhikari, 2017; Angmo et al., 2012; Dorjey et al., 2012; Rinchen & Pant, 2014). *H. niger* is also used for the treatment of mental disorders, epilepsy, paralysis, convulsions, neuralgia, spasmodic cough, and asthma (Patel et al., 2018). In Pakistan, seeds are used in traditional medicine for treating a variety of ailments including asthma, cough, colic, diarrhea, and genito-urinary complaints such as irritable bladder (Duke et al., 1987; Usmanghani et al., 1998). In traditional Chinese medicine, the seeds of *Hyoscyamus niger* are used as an antispasmodic, sedative, and analgesic agent, as well as for the treatment of stomach cramps, heavy coughs, neuralgia, and manic psychosis (Ma et al., 2002). In the Norwegian tradition, the leaves of *H. niger* were mixed with fat and made into an ointment. This was put by women, in poultice, in places of the body to have thin skin and the active ingredients were absorbed and caused hallucinations (Paulsen, 2010). In Morocco, Jusquiames are widely used in traditional medicine, both externally and internally, for their strong sedative and anesthetic properties. Their neurogenic properties mean that they are sometimes used in oniro-analysis to lift censorship and obtain a psychic arousal effect with delusions which are then interpreted by traditional psychotherapists. Taken in moderate quantities, the seeds are reputed to provoke erotic dreams and arouse sexual desire. In the Souss region, women mix black henbane seeds with their food to put on weight (Bellakhdar, 1997). In the Al-Haouz Rehamna region and according to the results of our field surveys that were conducted during the period from 2012 to 2017, we have documented some of the most cited uses of *H. niger* by the local population of this region:

- A poultice of seed powder is reputed to be effective against wounds and skin problems. The seeds, fumigated rectally, are used against intestinal glasses known locally as 'Toutiya: توتية'; at nose and ear level, this fumigation is indicated against nose and ear parasites.
- A recipe based on a pinch of *H. niger* seeds, combined with steamed dried fruits of *Ficus carica* L. and *Phoenix dactylifera* L., is used by women to gain weight.
- The seeds, taken in moderate quantities and cooked with *Allium cepa* L. bulbs and rock salt, are fed to cows to facilitate conception. The powdered seeds, a pinch in a bowl of *Hordeum vulgare* L. seeds « Tchicha » once a day, are administered orally against cancer.
- A vaginal poultice made from fresh, chopped leaves is reputed to be effective against cystitis and female sterility, provided the treatment is carried out in the traditional bath 'Hammam'.
- Nursing mothers apply one or two drops of sap from fresh *H. niger* leaves to their nipples to calm down irritated babies for sleep.

41.10.2 Industrial and Agricultural Applications

In industry, the seeds are used to synthesize biodiesel (Chen et al., 2023). The methanol extract of *H. niger* could be used as a natural antioxidant in the food industry (Matés et al., 1999).

In agronomy, Hu et al. (2022) have demonstrated the use of *H. niger* as a biopesticide against tobacco mosaic virus (TMV) and as a plant immune elicitor. The antifungal activity of steroidal saponins, particularly against agricultural pathogens, has been known for a long time (Dimoglo et al., 1985; Imai et al.; 1967; Wolters, 1966). According to the results of the study conducted by Behravan et al. (2017), henbane flower extract is recommended as an effective insecticide of natural origin against *Anopheles* spp. mosquito larvae.

41.10.3 Witchcraft, Folklore, and Religious Applications

In addition to its medicinal uses, *H. niger* is often claimed to have been a plant used by the witches of Europe (Schultes et al., 2001). It has been suggested that it was used in love potions, witches' brews, and the famous hallucinogenic and psychoactive flying ointments (Carruthers, 2015; Aulakh & Mukerjee, 1984; Disel et al., 2015; Ostling, 2016). In warfare, *Hyoscyamus niger* is used by berserker warriors instead of the famous psychoactive mushroom Amanita muscaria L. to achieve a special trance-like state that helps them in battle (Fatur, 2019).

41.11 HALLUCINOGENIC EFFECTS

Hallucinogenic plants contain psychoactive compounds known to affect the human brain, altering perception, mood, and consciousness. These effects are often referred to as dreamlike and are characterized by sensory distortions and sometimes outright hallucinations (often visual or auditory), exaggerated emotional states, time distortion, mystical qualities, and various spiritual experiences and revelations such as the dissolution of the ego, sense of near-death experience, and a sense of universal time connection, mystical qualities, and various spiritual experiences and revelations such as ego dissolution, sense of near-death experience, and a sense of universal connection or understanding (Nichols, 2004; Sanz et al., 2018; Frecska, 2007). These plants have been used for centuries for medicinal, spiritual, magico-religious, and recreational purposes (Alonzo & Fu, 2023; Passos & Mironidou-Tzouveleki, 2016, Sueur et al., 1999). *H. niger* is well-known for its traditional use as a hallucinogen (Li et al. 2011), it's very rich in alkaloids (hyoscyamine, atropine, and scopolamine) known for their delusional effects, can produce complex and convincing hallucinations (Maior et al., 2017).

First isolated in the 1800s, scopolamine has been found in all parts of the plant (Passos & Mironidou-Tzouveleki, 2016). This explains its widespread historical use in Europe since ancient Greece as a narcotic, anesthetic, analgesic, remedy for insomnia, in the treatment of respiratory ailments, in the famous soporific sponge, and many others (Carter, 1996; Daunay et al., 2007; Gorun et al., 2011; Hocking, 1947; Kala & Ratajc, 2012; Passos et Mironidou-Tzouveleki, 2016; Piomelli & Pollio, 1994; Schultes, 1977). *H. niger* has also found use in Tibetan medicine, traditional Chinese medicine, and Ayurveda for similar purposes (Patočka & Jelínková, 2018). In Western biomedicine, pure scopolamine is still used to treat motion sickness (Grynkiewicz & Gadzikowska, 2008). In Slovenia, the results of a recent study by Fatur (2021) showed that the majority of people using these substances were in their twenties and that curiosity and the desire to experience interesting effects, overcome personal difficulties, and achieve individual and spiritual fulfillment were at the root of their consumption. These results also showed that the spread of the Internet has increased the use of these hallucinogenic substances.

41.12 PHARMACOLOGICAL EFFECTS

Modern pharmacological and biological experiments on evidence of *H. niger* demonstrated that this plant possesses several therapeutic properties (e.g., antiparkinsonian, antioxidant, analgesic, anti-inflammatory, antipyretic, anticholinergic, anticonvulsant, spasmolytic, antidiarrheal, antisecretory, bronchodilator, Cardiovascular, vasodilator, antitumor, antimicrobial).

- **Antiparkinsonian effects**: Sengupta et al. (2011) report that the methanolic extract of *Hyoscyamus niger* seeds has antiparkinsonian activity via its ability to inhibit the increase in OH^- generated in mitochondria.
- **Anticholinergic properties**: *H. niger* is very rich in tropane alkaloids such as hyoscine and hyoscyamine, which have been identified as anticholinergic (Fatur & Kreft, 2020; Gilani et al., 2008; Kemmerer, 2007; Li et al., 2006; Brown & Taylor, 2001; Asano et al., 1996).
- **Antispasmodic**: Hyoscyamine and hyoscine, previously identified as the anticholinergic constituents of *H. niger*, are considered responsible for their antispasmodic effect (Brown & Taylor, 2001). Plants for medicinal use in hyperactive disorders of the intestine and respiratory tract generally have a blocking action on Ca^{2+} channels in combination with other spasmolytic mechanisms (Gilani et al., 2005a, b; Gilani & Rahman, 2005).
- **Antioxidant effect**: The methanol extract of *Hyoscyamus niger* could be used as natural antioxidants in food industries (Matés et al., 1999). The results of several biological and pharmacological tests confirmed the antioxidant activity of the methanolic extract of the seeds and aerial parts of *H. niger* extracts (Koçpinar, 2020; Hajipoor et al., 2015b; Ismeel, 2011; Jassbi et al., 2014; Souri et al., 2004).
- **Antimicrobial effect**: Results of several biological studies have shown the antimicrobial activity of methanolic or protein extracts of *H. niger* seeds against several urinary tract pathogens (Dulger & Dulger, 2015; Mateen et al., 2015). Similarly, results from Dulger G and Dulger B (2015) showed strong antimicrobial activity against *Enterococcus faecalis*, *Klebsiella pneumoniae*, and *Candida albicans* and *moderate activity* against *Escherichia coli*, *Pseudomonas aeruginosa*, *Proteus mirabilis*. The aqueous extract of *H. niger* seeds has anti-clostridial (*Clostridium petfringens*) activity (Afzal et al., 1992). Also, Lunga et al. (2007) demonstrated the antifungal activity of *H. niger* seeds against six *Candida* species (*C. albicans*, *C. tropicalis*, *C. guilliermondii*, *C. krusei*, *C. glabrata*, and *C. parapsilosis*) and greater activity against two *Cryptococcus species* (*C. neoformans* and *C. laurentii*).
- **Antifungal effect**: In addition to its alkaloid constituents, *H. niger* is rich in several steroidal glycosides that exhibit antifungal activity against human pathogenic fungi, yeasts as well as dermatophytes, and filamentous species (Lunga et al., 2007; Lacaille-Dubois & Wagner, 1996).
- **Insecticidal effect**: Methanol extract of *H. niger* aerial parts had positive effects on destroying the *Anopheles* spp. larvae (Behravan et al., 2017). The results of Küpeli Akkol et al. (2020) confirmed the folkloric use of *H. niger* seeds against eyeworms. These authors suggest that the alkaloid content (mainly hyoscyamine and scopolamine) may be responsible for its insecticidal activity against *Lucilia sericata*, responsible for myiasis.
- **Effect on gastrointestinal, respiratory, and urinary smooth muscles**: The crude extract of *H. niger* seeds has antidiarrheal and antisecretory activity against diarrhea, and antispasmodic, anticholinergic, and Ca^{2+} antagonist activity (Gilani et al., 2008). In addition, Khan and Gilani (2010) have shown that the crude extract of *H. niger* seeds has a positive effect on respiration.
- **Cardiovascular effects**: The results of pharmacological studies (Khan and Gilani, 2008; Vallabi & Elango, 2016) concerning the cardioprotective activity of *H. niger* crude powder in rats, showed that *H. niger* crude extract exhibited a cardio-depressant and vasodilator effect on the rate and strength of spontaneous atrial contractions. In addition to these effects, Vallabi and Elango (2016) confirmed the

cardioprotective effect of *H. niger* in isoproterenol-induced myocardial injury.

- **Anti-inflammatory, analgesic, and antipyretic effects**: The results of several biological studies (Begum et al., 2010; Khan & Gilani, 2010; Ghosian et al., 2012) carried out on experimental animal models at different doses have confirmed the analgesic, anti-inflammatory, and antipyretic activities of the methanolic extract of *H. niger* seeds.
- **Anti-Parkinsonian**: The protective effect of methanolic extract of *H. niger* seeds against parkinsonism in mice could be attributed to its inhibitory effects on the increase of OH^- generated in mitochondria (Sengupta et al., 2011). In addition, Khatri and Juvekar (2015) demonstrated the neuroprotective effect of methanolic extract of *H. niger* seeds.
- **Anticonvulsant effect**: The results of Kiasalari et al. (2011) and Reza et al. (2009) show that the alcoholic or methanolic extract of *H. niger* seeds has an anticonvulsant effect.
- **Antidepressant effect**: Many diseases in our lives occur due to damage to cell structure, which likely destroys tissue due to increased oxidative stress (Mollica et al., 2018). The methanolic extract of *H. niger* could be used as a natural antioxidant in food industries (Hajipoor et al., 2015b; Matés et al., 1999). Also, Patil et al. (2013) showed in the results of their study 'Antidepressant-like property of *H. niger* L. in a mouse model of depression' the antidepressant effect of ethanolic extract of *H. niger*.
- **Cytotoxic effect**: The alkaloid extract contained in *H. niger* showed cytotoxic activity against the cancer cell line (Ismeel, 2011). Grossamide and cannabisins D and G isolated from *H. niger* seeds showed moderate cytotoxic effects on human prostate cancer cells (Ma et al., 2002).

41.13 MEDICINAL USES

Tropane alkaloids such as hyoscyamine and scopolamine are among the oldest drugs in medicine, the latter being a more valuable drug due to their higher pharmacological activities and fewer side effects. These affect the parasympathetic nervous system and are used in ophthalmology, anesthesia, and the treatment of cardiac and gastrointestinal diseases (Oksman-Caldentey, 2007; Tyler et al., 1988; Chaachouay et al., 2019b).

The compound scopolamine is today used in travel sickness plaster. This is placed behind the ear and will stabilize the balance system in the brain; thus a lesser degree of travel sickness will be the case (Paulsen, 2010). Gilani et al. (2008) have demonstrated the therapeutic properties of *H. niger* e.g. spasmolytic, antidiarrheal, antisecretory, bronchodilator, bladder relaxant, and anticholinergic. This study explains the medicinal use of this plant in the treatment of gastrointestinal and respiratory disorders and bladder hyperactivity.

41.14 TOXICITY

Hyoscyamus niger is one of the common plants most widely used by people interested in traditional medicines (85%) (Cirak et al., 2004). However, despite its widespread use, it can cause poisoning (Oztekin-Mat, 1994). Ingestion of 15 to 20 seeds can lead to death (Sijelmassi, 2011). Toxicity is due to the tropane alkaloids hyoscyamine, atropine, and scopolamine. These alkaloids antagonize muscarine receptors in the central and peripheral nervous systems in place of acetylcholine, earning them the title of anticholinergic substances (Sayin, 2016). Their effects are particularly pronounced in the body's smooth muscles and secretory glands, as well as in the brain (Moulton & Fryer, 2011; Simone & Margarucci, 2008).

Symptoms of intoxication: at low doses, digestive disorders with nausea, vomiting, dry mouth, and mucous membranes in general with difficulty swallowing and burning sensation; long-lasting mydriasis, anxiety, dizziness, daze, and atropine delirium. Subsequently, tachycardia and respiratory distress set in. At higher doses, these symptoms are accompanied by increased drunkenness, confusion, agitation, blurred vision, auditory hallucinations, polyuria, profuse sweating, and sometimes skin rashes with urticaria. Significant nervous disorders follow, accompanied by convulsions and delirium. Subsequently, coma accompanied by cardiovascular depression and death follow in severe cases (Alizadeh et al., 2014; Verstraete, 2010; Sijelmassi, 2011; Charnot, 1945). Repeated doses of over 3 mg per day cause lasting behavioral and mental changes (Maior et al., 2017). *H. niger* is therefore a highly dangerous plant, whose use in family medicine is prohibited and must be entrusted to qualified personnel.

41.15 CONCLUSION

Hyoscyamus niger is an annual or biennial plant native to Europe, North Africa, and Central and Western Asia. It has long been of medical importance as a sedative, psychoactive, analgesic, and sleep aid. It contains various alkaloids, notably hyoscyamine, atropine, and over 50% scopolamine, whose hallucinogenic properties are more intense. This explains its historical use in various civilizations (Arab-Muslim, European, Sumerian, Chinese, Greek, etc.) against a variety of illnesses. *H. niger* was often used in witches' potions and other toxic preparations, causing visual, auditory, tactile, olfactory, and gustatory hallucinations, as well as the sensation of flying. It is also widely used for its pharmacological effects as a narcotic, anesthetic, analgesic, in the treatment of respiratory ailments, etc. Pharmacological studies have confirmed the plant's influence on the central nervous system, hence its use as a hallucinogen and psychoactive agent. Other toxicological studies have confirmed the toxicity of this poisonous plant, whose use, in any case, should be entrusted exclusively to qualified personnel and should not be used for domestic or family purposes. In addition to its medicinal and psychoactive uses, *H. niger* deserves a great deal of attention as it is used in agronomy as a biopesticide, insecticide, and fungicide, as a natural antioxidant in the food industry and also to synthesize biodiesel to protect the environment. It is within this framework that the present chapter aims to establish a more exhaustive and up-to-date study, both taxonomically, ecologically, and botanically, as well as ethnobotanically, pharmacologically, phytochemically and toxicologically, of *H. niger*, to make its uses rational and more reasonable.

LIST OF ABBREVIATIONS

AChE	Acetylcholinesterase
EN	Endangered
EX	Extinct
CR (PE)	Critically endangered, probably extinct
CR	Critically endangered
DD	Data deficient
GBIF	Global Biodiversity Information Facility
INPN	National Inventory of Natural Heritage
IPNI	International Plant Names Index
IUCN	International Union for Conservation of Nature
LC	Least concern
NA	Not applicable
NE	Not evaluated
NT	Near threatened
RE	Regionally extinct
USDA	Agricultural Research Service, National Plant Germplasm System
VU	Vulnerable
WFO	World flora online

REFERENCES

Adamse, P., Egmond, V. H. P., Van Egmond, H. P., Noordam, M.Y., Mulder, P. P. J., De Nijs, M. (2014). Tropane alkaloids in food: poisoning incidents. *Quality Assurance and Safety of Crops & Foods*, 6 (1), 15–24. https://doi.org/10.3920/QAS2013.0314.

Adhikari, B. S. (2017). Sowa-rigpa: a healthcare practice in the trans-Himalayan region of Ladakh, India. *SDRP J. Plant Sci*, 2, 621–630. https://doi.org/10.25177/jps.2.1.3

Adovasio, J. M., & Fry, G. F. (1976). Prehistoric psychotropic drug use in northeastern Mexico and Trans-Pecos Texas. *Economic Botany*, 30(1), 94–96.

Afzal, H., Anjum, A. A., Ashfaque, M., Akhtar, M. S., & Gill, S. A. (1992). Evaluation of anticlostridial efficacy of Indigenous medicinal plant drugs: Rasoot, Ajwain Khurasani, Neem and Bakain. *Pakistan Journal of Agricultural Sciences*, 29(4), 371–375.

Alaghemand, A., Ghorbanpour, M., Asli, D. E., & Moghaddasian, B. (2013). Calcium fertilization effects on hyoscyamine and scopolamine accumulation in henbane (*Hyoscyamus niger* L.) under hydroponic culture. *European Journal of Experimental Biology*, 3(3), 228–32.

Alizadeh, A., Moshiri, M., Alizadeh, J., & Balali-Mood, M. (2014). Black henbane and its toxicity–a descriptive review. *Avicenna Journal of Phytomedicine*, 4(5), 297–311.

Alonzo, M., & Fu, S. (2023). Hallucinogens. *Encyclopedia of Forensic Sciences, Third Edition*, 3, 107–121. https://doi.org/10.1016/B978-0-12-823677-2.00090-8

Alrashedy, N. A., Molina, J. (2016). The ethnobotany of psychoactive plant use: a phylogenetic perspective. *Peer J.*, 4, e2546. https://doi.org/10.7717/peerj.2546.

Angmo, K., Adhikari, B. S., Rawat, G. S. (2012). Changing aspects of traditional healthcare system in western Ladakh, India. *J. Ethnopharmacol*, 143, 621–630. https://doi.org/10.1016/j.jep.2012.07.017

Ansari, M. K. A., Iqbal, M., Chaachouay, N., Ansari, A. A., & Owens, G. (2023). The Concept and Status of Medicinal and Aromatic Plants: History, Pharmacognosy, Ecology, and Conservation. In: Ansari M.K., Unal B.T., Ozturk M., Owens G (eds) *Plants as Medicine and Aromatics* (p. 129–144). CRC Press.

Arroo, R. R. J., Woolley, J. G., & Oksman-Caldentey, K. M. (2007). Tropane Alkaloid Containing Plants: Henbane, Belladonna, Datura, and Duboisia. In: Transgenic Crops VI, Section II: Medicinal Crops (EC Pua, M Davey, eds.), *Biotechnology in Agriculture and Forestry*, 61, 2–20.

Aparna, K., Joshi, A. J., & Vyas, M. (2015). Phyto-chemical and pharmacological profiles of Hyoscyamus niger Linn (Parasika Yavani)-A review. *Pharma Science Monitor*, 6(1), 153–159.

Asano, N., Kato, A., Yokoyama, Y., Miyauchi, M., Yamamoto, M., Kizu, H., & Matsui, K. (1996). Calystegin N1, a novel nortropane alkaloid with a bridgehead amino group from Hyoscyamus niger: Structure determination and glycosidase inhibitory activities. *Carbohydrate Research*, 284(2), 169–178.

Aulakh, G. S., & Mukerjee, T. (1984). Plants associated with witchcraft and the evil eye. *Ancient Science of Life*, 4(1), 58.

Bahmanzadegan, J. A., Sefidkon, F., & Sonboli, A. (2009). Determination of hyoscyamine and scopolamine in four Hyoscyamus species from Iran. *Iranian Journal of Pharmaceutical Research*, 8 (1), 65–70.

Batool, Z., Singh, K., & Gairola, S. (2023). Medicinal plants are traditionally used in the health care practices by the indigenous communities of the Trans-Himalayan region of Ladakh, India. *Journal of Ethnopharmacology*, 116837. https://doi.org/10.1016/j.jep.2023.116837

Begum, A. S., Verma, S., Sahai, M., Schneider, K., & Süssmuth, R. (2009). Hyoscyamal, a new tetrahydrofurano lignan from *Hyoscyamus niger* Linn. *Natural Product Research*, 23(7), 595–600.

Begum, S., Saxena, B., Goyal, M., Ranjan, R., Joshi, V. B., Rao, C. V.,... & Sahai, M. (2010). Study of anti-inflammatory, analgesic, and antipyretic activities of seeds of *Hyoscyamus niger* and isolation of a new coumarinolignan. *Fitoterapia*, *81*(3), 178–184. https://doi.org/10.1016/j.fitote.2009.08.024.

Behravan, M., Vaezi Kakhki, M. R. and Baharshahi, A. (2017). Comparing the larvicidal effect of methanol extract of the aerial parts of henbane (*Hyoscyamus niger* L.) and oleander (*Nerium oleander* L.) plants on *Anopheles* spp larvae (Diptera: Culicidae) in vitro. *Zahedan Journal of Researches in Medical Sciences*, 19(3), e8631, 1–6. Doi: 10.5812/zjrms.8631.

Belhaj, S., Chaachouay, N., & Zidane, L. (2021). Ethnobotanical and toxicology study of medicinal plants used for the treatment of diabetes in the High Atlas Central of Morocco. *Journal of Pharmacy & Pharmacognosy Research*, *9*(5), 619–662. https://doi.org/10.56499/jppres21.1014_9.5.619

Bellakhdar, J. (1997) *La pharmacopée marocaine traditionnelle: Médecine arabe ancienne et savoirs populaires*. Le Fennec, Paris. 764 p.

Benlarbi, F., Mimoune, N., Chaachouay, N., Souttou, K., Saidi, R., Mokhtar, M.R., Kaidi, R, & Benaissa, M. T. (2023). Ethnobotanical survey of the traditional antiparasitic use of medicinal plants in humans and animals in Laghouat (Southern Algeria). *Veterinary World*, *16*(2), 357–368. https://doi.org/10.14202/vetworld.2023.357-368

Benkhnigue, O., Chaachouay, N., Khamar, H., El Azzouzi, F., Douira, A., Zidane, L. (2022). Ethnobotanical and ethnopharmacological study of medicinal plants used in the treatment of anemia in the region of Haouz-Rehamna (Morocco). *J. Pharm Pharmacogn Res*, 10(2), 279–302.

Benkhnigue, O., Chaachouay, N., & Zidane, L. (2023). A Floristical and Ecological Study of the Medicinal Flora Used by the Local Population of the Haouz-Rehamna Region (Middle Atlantic Morocco-4). *Acta Botanica Hungarica Akadémiai Kiadó*, *65*(3–4), 247–304. https://doi.org/10.1556/034.65.2023.3-4.3

Brown, J. H., & Taylor, P. (2001). Muscarinic receptor agonists and antagonists. *Goodman & Gilman's the pharmacological basis of therapeutics*, 11, 183–200.

Carruthers, D. M. J. (2015). Lines of flight of the deadly Nightshade: An inquiry into the properties of the magical plant, its literature, and history. Mosaic: *Interdiscip. Study Lit.* 48 (2), 119–132. https://doi.org/10.1353/mos.2015.0025.

Carter, A.J. (1996). Narcosis and nightshade. BMJ (Clin. Res. Ed.) 313 (7072), 1630–1632.http://www.ncbi.nlm.nih.gov/pubmed/8991015%0Ahttp://www.pubmedcentral.nih.gov/articlerender.fcgi?artid=PMC2359130.

Chaachouay, N., Benkhnigue, O., Fadli, M., Ibaoui, H., & Zidane, L. (2019b). Ethnobotanical and ethnopharmacological studies of medicinal and aromatic plants used in the treatment of metabolic diseases in the Moroccan Rif. *Heliyon*, *5*(10), e02191. https://doi.org/10.1080/10477845.2021.1942433

Chaachouay, N., Benkhnigue, O., Fadli, M., Ibaoui, H., El Ayadi, R., & Zidane, L. (2019a). Ethnobotanical and ethnopharmacological study of medicinal and aromatic plants used in the treatment of respiratory system disorders in the Moroccan Rif. *Ethnobotany Research and Applications*, *18*(22), 1–16.

Chaachouay, N., Azeroual, A., Douira, A., & Zidane, L. (2022). Ethnoveterinary practices of medicinal plants among the Zemmour and Zayane tribes, Middle Atlas, Morocco. *South African Journal of Botany*, 151, 826–840.

Charnot, A. (1945). La toxicologie au Maroc. Mémoire de la société des sciences naturelles du Maroc. Rabat Inst. Sci. Chérif. I, 945p.

Chen, Q., Wang, A., Quan, W., & Gong, W. (2023). Efficient synthesis of biodiesel from Hyoscyamus niger L. seed oil by base catalysis. *Fuel Processing Technology*, 241, 107630. https://doi.org/10.1016/j.fuproc.2022.10763.

Cirak, C., Kevseroglu, K., & Sağlam, B. (2004). Physical and physiological dormancy in black henbane (*Hyoscyamus niger* L.) seeds. *Journal of Plant Biology*, 47, 391–395. DOI: 10.1007/BF03030556

Collak, Z., Kara, D., Karaman, E., Karakoc, F., & Collak, A. (2016). *Hyoscyamus niger* intoxication: A case report. *British Journal of Medicine and Medical Research*, 11(6), 1–3.

Correns, C. E. (1904). Ein typisch spaltender Bastard zwischen einer einjährigen und einer zweijährigen Sippe des *Hyoscyamus niger*. *Berichte der Deutschen Botanischen Gesellschaft*, 22, 517–524.

Daunay, M. C., Laterrot, H., & Janick, J. (2007). The iconography of the Solanaceae from antiquity to the XVIIth century: A rich source of information on genetic diversity and uses. In *VI International Solanaceae Conference: Genomics Meets Biodiversity 745* (pp. 59–88). https://doi.org/10.17660/ActaHortic.2007.745.3

Díaz, J. L. (2010). Sacred plants and visionary consciousness. Phenomenol. *Cognitive Sci*, 9(2), 159–170. https://doi.org/10.1007/s11097-010-9157-z.

Dimoglo, A. S., Choban, I. N., Bersuker, I. B., Kintia, P. K., & Balashova, N. N. (1985). Structural traits of the antioxidant and fungicidal activity of steroid glycosides. *Bioorganicheskaia khimiia*, 11(3), 408–413.

Disel, N. R., Yilmaz, M., Kekec, Z., & Karanlik, M. (2015). Poisoned after dinner: dolma with *Datura stramonium*. *Turkish Journal of Emergency Medicine*, 15(1), 51–55. https://doi.org/10.5505/1304.7361.2015.70894.

Dorjey, K., Tamchos, S., Kumar, S. (2012). Ethnobotanical observations in the Trans-Himalayan region of Ladakh. *Journal of Plant Development Sciences*, 4, 459–464.

Duke, J. A., Bogenschutz-Godwin, M. J., duCellier, J., & Duke, P. K. (1987). *Handbook of Medicinal Herbs*, CRC, Press. Inc., Boca, Raton, Florida, 102–104.

Dulger, G., & Dulger, B. (2015). Antimicrobial activity of the seeds of *Hyoscyamus niger* L. (Henbane) on microorganisms isolated from urinary tract infections. *J Med Plants Stud*, 3(5), 92–95.

El Bazaoui, A., Bellimam, M. A., Lançar, I. T., Soulaymani, A. (2012). Gas-liquid chromatography-mass spectrometry investigation of tropane alkaloids in *Hyoscyamus albus* L. from Morocco. *Z Naturforsch C Biosci*, 67c, 461–465.

Fatur, K. (2021). Peculiar plants and fantastic fungi: An ethnobotanical study of the use of hallucinogenic plants and mushrooms in Slovenia. *Plos one*, 16(1), e0245022. https://doi.org/10.1371/journal.pone.0245022

Fatur, K., & Kreft, S. (2020). Common anticholinergic solanaceaous plants of temperate Europe. A review of intoxications from the literature (1966–2018). *Toxicon*, 177, 52–88.

Fatur, K. (2019). Sagas of the Solanaceae: Speculative ethnobotanical perspectives on the Norse berserkers. *Journal of Ethnopharmacologie*, 244, 112151.

Fettig, C. E. (2016). Ecology and plant defense of two invasive plants, *Hyoscyamus niger* and *Verbascum thapsus* (Doctoral dissertation, Colorado State University), 100p.

Fennane, M. (2021). *Livre rouge de la flore vasculaire du Maroc*. Institut scientifique, Rabat, Maroc. 750p.

Fennane M., Ibn Tattou M., El Oualidi J., Ouyahya, A. (2007) Flore pratique du Maroc. Vol. 2. Série botanique N 38. Rabat: Trav. Inst Sci. Université Mohamed 5. 636p.

Frecska, E. (2007). *Therapeutic guidelines: Dangers and contraindications in therapeutic applications of hallucinogens. Psychedelic medicine: new evidence for hallucinogenic substances as treatments*. Westport: Praeger Publishers, 69–95.

Gaillard, Y., & Pepin, G. (1999). Poisoning by plant material: review of human cases and analytical determination of main toxins by high-performance liquid chromatography–(tandem) mass spectrometry. *Journal of Chromatography B: Biomedical Sciences and Applications*, 733(1–2), 181–229. https://doi.org/10.1016/S0378-4347(99)00181-4

GBIF. (2023). https://www.gbif.org/fr/species/7717

Garnier, G., Bézanger-Beauquesne, L., & Debranx, G. (1961). *Ressources médicinales de la flore française, tome 2, Vigot frères*. Paris. 1511 p.

Ghosian, M. H., Moradi M. & Yaghout Poor, E. (2012). Assessment of Hyoscyamus niger seeds alcoholic extract effects on acute and chronic pain in male NMRI rats. *Journal of Basic and Clinical Pathophysiology*, 1(1), 29–36.

Gilani, A. H., Khan, A. U., Subhan, F., & Khan, M. (2005a). Antispasmodic and bronchodilator activities of St John's wort are putatively mediated through dual inhibition of calcium influx and phosphodiesterase. *Fundamental & Clinical Pharmacology*, 19(6), 695–705.

Gilani, A. H., Shah, A. J., Ghayur, M. N., & Majeed, K. (2005b). Pharmacological basis for the use of turmeric in gastrointestinal and respiratory disorders. *Life Sciences*, 76(26), 3089–3105.

Gilani, A. H., Khan, A. U., Raoof, M., Ghayur, M. N., Siddiqui, B. S., Vohra, W., & Begum, S. (2008). Gastrointestinal, selective airways, and urinary bladder relaxant effects of *Hyoscyamus niger* are mediated through dual blockade of muscarinic receptors and $Ca2^{+}$ channels. *Fundamental & Clinical Pharmacology*, 22(1), 87–99. https://doi.org/10.1111/j.1472-8206.2007.00561.x

Gilani, A. H., Rahman, A. U. (2005). Trends in ethnopharmacology. *J. Ethnopharmacol*, 100, 43–49.

Gorun, G., Curcă, G. C., Hostiuc, S., & Buda, O. (2011). 'Legal highs' in Romania: Historical and present facts. *Romanian Journal of Legal Medicine*, 19, 73–76. https://doi.org/10.4323/rjlm.2011.73

Gourie-Devi, M., Ramu, M. G., & Venkataram, B. S. (1991). Treatment of Parkinson's disease in 'Ayurveda' (ancient Indian system of medicine): discussion paper. *Journal of the Royal Society of Medicine*, 84(8), 491–492.

Guinochet, M., & De Vilmorin, R. (1973). Flore de France: Fascicule 2. Editions du Centre National de la Recherche Scientifique. Paris, France. Pages: 568 à 570.

Grynkiewicz, G., Gadzikowska, M. (2008). Tropane alkaloids as medicinally useful natural products and their synthetic derivatives as new drugs. *Pharmacological Reports*, 60(4), 439–463. http://www.embase.com/search/results?subaction=viewrecord&from=export&id=L352378025%5Cnhttp://www.if-pan.krakow.pl/pjp/pdf/2008/4_439.pdf

Hajipoor, K., Sani, A. L., Mohammadi, A. (2015a). In vitro antioxidant potential and phenolic profile of *Hyoscyamus niger*. *International Journal of Biology, Pharmacy and Allied Sciences*, 4(7), 4882–4890.

Hajipoor, K., Sani, A. M., & Mohammad, A. (2015b). In vitro antioxidant activity and phenolic profile of Hyoscyamus niger. *IJBPAS*, 4(7), 4882–4890.

Hawkes, J. G. (1999). The economic importance of the family Solanaceae. Solanaceae IV, 1–8. In *Solanaceae IV: Advances in Biology and Utilization*; Nee, M., Symon, D.E., Lester, R.N., Jessop, J.P., Eds.; Royal Botanic Gardens: Kew, UK, 1999; pp. 1–8.

Hocking, G. M. (1947). Henbane—Healing herb of Hercules and of Apollo. *Economic Botany*, 1(3), 306. https://doi.org/10.1007/BF02858575.

Hoffer A., Osmond H. (1967). The hallucinogens. New York and London: Academic Press. 626p.

Hu, Z., Bo, X., Sun, G., Zhao, L., Shi, C., Huang, L., & Tian, X. (2022). Identification of vanillic acid and its new amide derivative from *Hyoscyamus niger* and their modes of action in controlling tobacco mosaic virus. *Industrial Crops and Products*, 189, 115853. https://doi.org/10.1016/j.indcrop.2022.115853.

Imai, S., Murata, T., Fujioka, S., Murata, E., & Goto, M. (1967). Crude drugs and oriental crude drug preparations by bioassay. XXI. Search for biologically active plant ingredients using antimicrobial tests. 3. Antifungal principles of Dioscorea species. *Takeda Kenkyusho Nenpo*, 26, 66–75.

Info flora. 2023. Centre national de données et d'informations sur la flore de Suisse. Fiche espèce (*Hyoscyamus niger* L.). https://www.infoflora.ch/fr/flore/hyoscyamus-niger.html (consulted on August 28, 2023).

Ismeel, A. O. (2011). Cytogenetic and cytotoxic studies on the effect of phytoinvestigated active compounds of *Hyoscyamus niger* (in vivo and ex vivo). PhD thesis, Al-Nahrain University-College of Science, Iraq.

IUCN (2012a). Catégories et Critères de la Liste rouge de l'UICN. Version 3.1. Deuxième édition. Gland, Suisse et Cambridge, Royaume-Uni, UICN, vi + 32 p.

IUCN (2012b). Lignes directrices pour l'application des Critères de la Liste rouge de l'UICN aux niveaux régional et national. Version 4.0. Gland, Suisse et Cambridge, Royaume-Uni, UICN. iv + 44 p.

Jassbi, A. R., Miri, R., Masroorbabanari, M., Asadollahi, M., Attarroshan, M., & Baldwin, I. T. (2014). HPLC-DAD-ESIMS analyses of *Hyoscyamus niger* and *H. reticulatus* for their antioxidant constituents. *Austin Chromatography*, 1(5), 1022.

Kala, C. P., & Ratajc, P. (2012). High altitude biodiversity of the Alps and the Himalayas: ethnobotany, plant distribution and conservation perspective. *Biodiversity and Conservation*, 21, 1115–1126. https://doi.org/10.1007/s10531-012-0246-x.

Karaköse, M. (2022). An ethnobotanical study of medicinal plants in Güce district, north-eastern Turkey. *Plant Diversity*, 44(6), 577–597.

Kemmerer, D. A. (2007). Anticholinergic syndrome. *Journal of Emergency Nursing*, 33(1), 76–78.

Ketchum, J. S., Sidell, F. R., Crowell Jr, E. B., Aghajanian, G. K., & Hayes Jr, A. H. (1973). Atropine, Scopolamine, and Ditran: Comparative pharmacology and antagonists in man. *Psychopharmacologia*, 28(2), 121–145.

Khan, A. U., & Gilani, A. H. (2010). In vivo studies on the bronchodilatory and analgesic activities of *Hyoscyamus niger* and *Aspalathus linearis*. *Latin American Journal of Pharmacy*, 29(5), 777–782.

Khan, A. U., Gilani, A. H. (2008). Cardiovascular inhibitory effects of *Hyoscyamus niger*. *Methods and Findings in Experimental and Clinical Pharmacology* 30(4):295–300. Doi: 10.1358/mf.2008.30.4.1186075.

Khatri, D. K., & Juvekar, A. R. (2015). The propensity of *Hyoscyamus niger* seeds methanolic extract to allay stereotaxically rotenone-induced Parkinson's disease symptoms in rats. *Oriental Pharmacy and Experimental Medicine*, 15, 327–339.

Kiasalari, Z., Khalili, M., Heidari, H., & Azizi, Y. (2011). Anticonvulsant effect of alcoholic *Hyoscyamus niger* L seed extract on PTZ model of kindling in male mice. *Razi Journal of Medical Sciences*, 18(85), 27–33.

Kintzios, S., Barberaki, M., Tourgielis, P., Aivalakis, G., & Volioti, A. (2002). Preliminary evaluation of somaclonal variation for the in vitro production of new toxic proteins from *Viscum album L. Journal of Herbs, Spices & Medicinal Plants*, 9(2–3), 217–221.

Koçpinar, E. F. (2020). Antioxidant behavior of Hyoscyamus niger having a narcotic effect on heavy metal reduction and radical scavenging. *Journal of Physical Chemistry and Functional Materials*, 3(2), 68–71.

Küpeli Akkol, E., Ilhan, M., Kozan, E., Gürağaç Dereli, F. T., Sak, M., & Sobarzo-Sánchez, E. (2020). Insecticidal activity of *Hyoscyamus niger* l. on Lucilia sericata causing myiasis. *Plants*, 9(5), 655, 1–10. Doi:10.3390/plants9050655

Lauber, K., Wagner, G. (2007). Flora Helvetica. Flore illustrée de Suisse. 3eme Edition. Edition Haupt. Berne- Stuttgart-Vienne. Suisse. 1631p.

Li, R., Reed, D. W., Liu, E., Nowak, J., Pelcher, L. E., Page, J. E., & Covello, P. S. (2006). Functional genomic analysis of alkaloid biosynthesis in *Hyoscyamus niger* reveals a cytochrome P450 involved in littorine rearrangement. *Chemistry & Biology*, 13(5), 513–520.

Lacaille-Dubois, M. A., & Wagner, H. (1996). A review of the biological and pharmacological activities of saponins. *Phytomedicine*, 2(4), 363–386.

Li, J., Shi, J., Yu, X., Sun, J. K., Men, Q., & Kang, T. (2011). Chemical and pharmacological research on Hyoscyamus niger. *Chinese Herbal Medicines*, 3(2), 117–126.

Lunga, I., Chintea, P., Shvets, S., Favel, A. and Pizza, C. (2007). Steroidal glycosides from the seeds of *Hyoscyamus niger* and their antifungal activity. *Chemistry Journal of Moldova*, 2 (1), 108–113.

Lunga, I., Kintia, P., Şveţ, S., Bassarello, C., Pizza, C., & Piacente, S. (2006). Three spirostanol glycosides from the seeds of *Hyoscyamus Niger* L. *Chemistry Journal of Moldova*, 1(1), 100–103. https://doi.org/10.19261/cjm.2006.01(1).02

Ma, C. Y., Liu, W. K., & Che, C. T. (2002). Lignanamides and Nonalkaloidal Components of Hyoscyamus n iger Seeds. *Journal of Natural Products*, 65(2), 206–209.

Ma, C. Y., Williams, I. D., & Che, C. T. (1999). Withanolides from Hyoscyamus niger seeds. *Journal of Natural Products*, 62(10), 1445–1447.

Maior, M.C., Olah, N.-K., Burtescu, R.F., Dumitru, D.-V., Carpa, R., Dobrotă, C. (2017). Biochemical analysis of *Datura stramonium* extract. *Studia Universitatis Babeş-Bolyai Biologia*, 62(2), 5–19. https://doi.org/10.24193/subbbiol.2017.2.01

Marionneau, C. (2018). État des lieux de la consommation des Plantes et champignons hallucinogènes en France. Thèse de Doctorat en pharmacie, Université d'Angers, France. 111p.

Mateen, A., Tanveer, Z., Janardhan, K., & Gupta, V. C. (2015). Screening and purification of antibacterial proteins and peptides from some of the medicinal plant's seeds. *International Journal of Pharma and Bio Sciences*, 6(4), 774–781.

Matsuda, J., Okabe, S., Hashimoto, T., & Yamada, Y. (1991). Molecular cloning of hyoscyamine 6 beta-hydroxylase, a 2-oxoglutarate-dependent dioxygenase, from cultured roots of *Hyoscyamus niger*. *Journal of Biological Chemistry*, 266(15), 9460–9464.

Matés, J. M., Pérez-Gómez, C., & De Castro, I. N. (1999). Antioxidant enzymes and human diseases. *Clinical Biochemistry*, 32(8), 595–603.

Mollica, A., Zengin, G., Stefanucci, A., Ferrante, C., Menghini, L., Orlando, G., … Onaolapo, O. J. (2018). Nutraceutical potential of Corylus avellana daily supplements for obesity and related dysmetabolism. *Journal of Functional Foods*, 47, 562–574.

Moulton, B. C., & Fryer, A. D. (2011). Muscarinic receptor antagonists, from folklore to pharmacology; find drugs that work in asthma and COPD. *British Journal of Pharmacology*, 163(1), 44–52. https://doi.org/10.1111/j.1476-5381.2010.01190.x.

Nagashayana, N., Sankarankutty, P., Nampoothiri, M. R. V., Mohan, P. K., & Mohanakumar, K. P. (2000). Association of L-DOPA with recovery following Ayurveda medication in Parkinson's disease. *Journal of the Neurological Sciences*, 176(2), 124–127.

Nichols, D. E. (2004). Hallucinogens. *Pharmacol Therapeut*, 101, 131–181.

Oksman-Caldentey, K. M. (2007). Tropane and nicotine alkaloid biosynthesis-novel approaches towards biotechnological production of plant-derived pharmaceuticals. *Current Pharmaceutical Biotechnology*, 8(4), 203–210.

Ostling, M. (2016). Babyfat and Belladonna: Witches' ointment and the contestation of reality. *Magic, Ritual, and Witchcraft*, 11(1), 30–72. https://doi.org/10.1353/mrw.2016.0008

Oztekin-Mat, A. (1994). Plant poisoning cases in Turkey. *In Annales Pharmaceutiques Francaises*, 52(5), 260–265.

Quézel, P., Santa, S., Schotter, O., & Emberger, L. (1963). Nouvelle flore de l'Algérie et des régions désertiques méridionales. Tome 2. CNRS, Ed. Paris, 494p.

Passos, I. D., & Mironidou-Tzouveleki, M. (2016). Chapter 71-Hallucinogenic plants in the Mediterranean countries. Editor(s): Victor R. Preedy, *In Neuropathology of drug addictions and substance misuse*, Academic Press, 761–772. https://doi.org/10.1016/B978-0-12-800212-4.00071-6

Patel, S. S., Raguwanshi, R., Masood, M., Acharya, A., Jain, S. K. (2018). Medicinal plants with acetylcholinesterase inhibitory activity'. *Reviews in the Neurosciences*, 29(5): 491–529.

Patil, A. D., Patil, A. Y., & Raje, A. A. (2013). Antidepressant-like property of *Hyoscyamus niger* Linn. in a mouse model of depression. *Innovations in Pharmaceuticals and Pharmacotherapy*, 1(2), 60–69.

Patočka, J., Jelínková, R. (2018). Atropine and atropine-like substances are used in warfare. *Military Medical Science Letters*, 86(2), 58–69. https://doi.org/10.31482/mmsl.2017.010.

Paris, R. R. & Moyse, H. (1976–1981). Matière médicale, 3 tomes, Masson, Paris. 420, 518 et 509p.

Paulsen, B. S. (2010). Highlights through the history of plant medicine. Bioactive compounds in plants-benefits and risks for man and animals. Proceedings from a symposium held at The Norwegian Academy of Science and Letters. Aksel Bernhoft editor. Pp, 18–29.

Piomelli, D., & Pollio, A. (1994). In Upupa o strige. A study in Renaissance Psychotropic Plant Ointments. *History and Philosophy of the Life Sciences*, 241–273. http://www.ncbi.nlm.nih.gov/pubmed/7724723

POWO. (2023): https://powo.science.kew.org/ (consulted on September 11, 2023).

Reza, H. M., Mohammad, H., Golnaz, E., & Gholamreza, S. (2009). Effect of methanolic extract of Hyoscyamus niger L. on the seizure induced by picritoxin in mice. *Pakistan Journal of Pharmaceutical Sciences*, 22(3), 308–312.

Rinchen, T., & Pant, S. (2014). Ethnopharmacological uses of plants among inhabitants surrounding Suru and Zanskar valleys of cold desert, Ladakh. *International Journal of Pharmacy and Biological Sciences*, 5(1), 486–494.

Sabahi, M., Mansouri, S. H., Ramezanian, M., & Gholam-Hoseinian, A. (1987). Screening of plants from the southeast of Iran for antimicrobial activity. *International Journal of Crude Drug Research*, 25(2), 72–76.

Sajeli, B., Sahai, M., Suessmuth, R., Asai, T., Hara, N., & Fujimoto, Y. (2006). Hyosgerin, a new optically active coumarinolignan, from the seeds of *Hyoscyamus niger*. *Chemical and Pharmaceutical Bulletin*, 54(4), 538–541.

Sanz, C., Zamberlan, F., Erowid, E., Erowid, F., & Tagliazucchi, E. (2018). The experience elicited by hallucinogens presents the highest similarity to dreaming within a large database of psychoactive substance reports. *Frontiers in Neuroscience*, 12(7), 1–19. https://doi.org/10.3389/fnins.2018.00007

Sargın, S. A., Akçicek, E., & Selvi, S. (2013). An ethnobotanical study of medicinal plants used by the local people of Alaşehir (Manisa) in Turkey. *Journal of Ethnopharmacologir*, 150(3), 860–874.

Satıl, F., Aslan, M., Erdoğan, E., Polat, R., & Selvi, S. (2015). Comparative anatomical studies on some species of

Hyoscyamus L. (Solanaceae) growing in Turkey. *Bangladesh Journal of Botany*, 44(1), 37–43.

Sayin, H. U. (2016). Psychoactive plants are used during religious rituals. In *Neuropathology of Drug Addiction and Substance Misuse*, Vol. 3. Elsevier, London, pp. 17–28. Academic Press.

Schläppi, M. (2011). Genetic and physiological analysis of biennialism in *Hyoscyamus niger*. *Plant Biology*, *13*(3), 534–540.

Schultes, R. E., Hofmann, A., & Rätsch, C. (2001). *Plants of the Gods: Their Sacred, Healing, and Hallucinogenic Powers*. Healing Arts Press, Rochester Vt Retrieved from. https://www.worldcat.org/title/plants-of-the-gods-their-sacred-healing-andhallucinogenic-powers/oclc/47666585.

Schultes, R. E. (1977). The botanical and clinical distribution of hallucinogens. *Journal of Psychedelic Drugs*, 9(3), 247–263. https://doi.org/10.1080/02791072.1977.10472055

Selleck, G. W. (1964). Ecology of black henbane in Saskatchewan. *Weeds*, 12(2), 148–150.

Sengupta, T., Vinayagam, J., Nagashayana, N., Gowda, B., Jaisankar, P., & Mohanakumar, K. P. (2011). Antiparkinsonian effects of aqueous methanolic extract of Hyoscyamus niger seeds result from its monoamine oxidase inhibitory and hydroxyl radical scavenging potency. *Neurochemical Research*, 36, 177–186.

Sharma, P. C., Yelne, M. B., Dennis, T. J., Joshi, A., & Billore, K. V. (2000). *Database on medicinal plants used in Ayurveda*. Vol 7. New Delhi: Central Council for Research in Ayurveda & Siddha; page 319.

Sijelmassi, A. (2011). *Les plantes médicinales du Maroc*. 8^{éme} Ed. Le Fennec, Casablanca. 285p.

Simone, D., Margarucci, V. D. F. (2008). Tropane alkaloids: an overview. *Pharmacologyonline* 89, 70–89.

Souri, E., Amin, G., Dehmobed-Sharifabadi, A., Nazifi, A., & Farsam, H. (2004). Antioxidative activity of sixty plants from Iran. *Iranian Journal of Pharmaceutical Research*, 3(1), 55–59.

Sueur, C., Benezech, A., Deniau, D., Lebeau, B., & Ziskind, C. (1999). Les substances hallucinogènes et leurs usages thérapeutiques. Partie 1 Revue de la littérature. *Toxibase*, 1–28.

Supria, K. B. (1998). *Handbook of Medicinal Plants*. Poiter Publishers, India. 607p.

Tyler, V. E., Brady, L. R., & Riobbers, J. K. (1988). Pharmacognosy. 9th ed. Lea and Febiger, Philadelphia. Wood, AS, Reinhart, BS Rajaratham G. and Summers, JD 1971. A Comparison of Blood Constituents of Dwarf Versus Non dwarf Birds. *Poultry Sci*, 50(6), 804–807.

The Herbal Resource (2023). Henbane, side effects, and health benefits. https://www.herbal-supplement-resource.com/henbane.html (consulted on August 28, 2023).

USDA: Agricultural Research Service, National Plant Germplasm System. (2023). *Germplasm Resources Information Network (GRIN Taxonomy)*. National Germplasm Resources Laboratory, Beltsville, Maryland. https://npgsweb.ars-grin.gov/gringlobal/taxon/taxonomydetail?id=19532. Accessed 18 August 2023.

Usmanghani, K., Saeed, A., & Alam, M. T. (1998). Indusyunic Medicine, Traditional Medicine of Herbal, Animal and Mineral Origin in Pakistan, BCC &T. Press, University of Karachi, Pakistan (1997). 1 () GO Adegoke, LJM Rao and NB Shankaracharya. *Flavour and Fragrance Journal*, 13, 349–352.

Vallabi, D. E., & Elango, V. (2016). Preliminary studies on the cardioprotective effect of Hyoscyamus niger Linn in male albino rats. *Journal of Chemical and Pharmaceutical Research*, 8(7), 860–864.

Verstraete, F. (2010). Management and regulation of certain bioactive compounds present as inherent toxins in plants intended for feed and food. In Bioactive compounds in plants-benefits and risks for man and animals. *The Norwegian Academy of Science and Letters*, Aksel Bernhoft editor. Pp, 159–191.

Wolters, B. (1966). On the antimicrobial activity of plant steroids and triterpenes. *Planta medica*, 14(4), 392–399.

WFO. (2023). https://www.worldfloraonline.org/search?query=Hyoscyamus+niger+L. (consulted on August 28, 2023).

Yadav, R., Rathi, M., Pednekar, A., & Rewachandani, Y. (2016). A detailed review of the Solanaceae family. *European Journal of Pharmaceutical and Medical Research*, 3(1), 369–378.

Zhang, W. N., Luo, J. G., & Kong, L. Y. (2012). Phytotoxicity of lignanamides isolated from the seeds of Hyoscyamus niger. *Journal of Agricultural and Food Chemistry*, 60(7), 1682–1687.

Zhang, W., Zhang, W., Luo, J., & Kong, L. (2013). A new steroidal glycoside from the seeds of *Hyoscyamus niger*. *Natural Product Research*, 27(21), 1971–1974.

42 Mandrake (*Mandragora officinarum* L. Solanaceae)

Jamaa Habchaoui, Noureddine Chaachouay, and Ouafae Benkhnigue

42.1 INTRODUCTION

Mandragora officinarum L. is an exceptionally old plant that humans have recognized for a long time. However, it is still one of the most misunderstood plants on the planet (Berry & Jackson, 1976). This peculiar plant has been associated with both the medieval and Renaissance eras. This factory is renowned for producing packages with magical, aphrodisiac, healing, psychedelic, and poisonous properties (Chaachouay et al., 2021, 2021b, 2021c). The produce of the factory, also known as love apples, was thought to enhance fertility. The efficacy of mandrake as both an aphrodisiac and a general tonic is referenced in Genesis 30:14–17 and Song of Solomon 7:12–14; however, several writers cast doubt upon its effectiveness (Monadi et al., 2021). 'Mandragora' is derived from two Greek words meaning 'harmful to livestock'. The fruit is referred to as the 'Satan's apple' by the Arabs. Ancient Greece and Rome highly prized it for its intoxicating and narcotic properties, and it was used as a surgical anesthetic (Fakir & Özçelik, 2009; Whitmont, 1958). Mandrake has been extensively used as a therapeutic botanical from ancient biblical eras and recorded chronicles in Western society. Derived medications have always had significant importance in both medical and magical rituals. Specifically, the angiosperm family Solanaceae is particularly vital for human well-being.

The plant has several medicinal characteristics; it is known for its narcotic and anesthetic properties (Chidiac et al., 2012). It can be used in hallucination, mania, and delirium as a cure and to relieve some joint pain; it is a varied phytochemistry composed of alkaloids, non-alkaloids, and certain sugars (Waniakowa, 2007). Many species belonging to the family Solanaceae possess aromatic and medicinal properties (Shah et al., 2013). It is an important and valuable genus because of its properties.

Since the seventeenth century, another significant species of *M. officinarum* has been used under the name 'mandrake factory.' The species Mandragora (mandrake factory – *M. officinarum*) has long been linked to magical properties and held beliefs in ancient Greek and Roman communities (Riddle, 2010). Ancient Greeks and Romans thought that the mandrake plant had the power to assist infertile women in conceiving a child. In the era of Pliny (23–79 C.E.), fragments of *M. officinarum* roots were administered to patients before surgery due to their ability to decrease brain sensitivity and calm the nervous system (Dafni et al., 2021; Waniakowa, 2007). The chapter summarizes the history, constituents' phytochemical and toxicological effects, and pharmacological benefits of *M. officinarum*. Worldwide, there have been significant advances in medical technology.

42.2 SCIENTIFIC CLASSIFICATION

In his book *Species Plantarum*, published in 1753, Carl Linnaeus identified *M. officinarum* as the only species belonging to the rubric *Mandragora* (Linnaeus, 1753). In the subsequent editions (1762 and 1764) of his book, *M. officinarum* (Linné, 1753) is described as an enduring herbaceous plant that belongs to the *Mandragora* genus and Solanaceae family, according to APG III, 2009 (Table 42.1). The taxonomic groupings proposed by Linnaeus for the species *M. officinarum sensu lato* are widely recognized and accepted. The Flora Europaea, Kew's Medicinal Plant Names Services, and Kew's Plants of the worldonline.org all include *M. officinarum* in their listings (POWO, 2023).

42.3 VERNACULAR NAMES

M. officinarum has various vernacular names in different regions and languages. It is somewhat unexpected that these languages (German, Dutch, English, French) do not originate from the region where the plant is native (Dafni et al., 2021). Additionally, it is noteworthy that the plant has been deemed extinct in at least one other language (Serbo-Croatian), while it continues to grow naturally in Greece, Arabic countries in North Africa, Spain, and Turkey (Bellakhdar, 1997). These nine languages account for 74% of all names, whereas the other 33 languages comprise 26% of words. It's worth noting that the plant has a rich history and is often associated with various myths and folklore, contributing to its diverse range of names across cultures. Additionally, '*officinarum*' in the scientific name refers to its historical use as a medicinal plant, as '*officina*' traditionally referred to an apothecary or pharmacy.

Amazigh: Tryala, Yabruh; Arabic: الطلسم, At-Tilasm, Bid Elghol; **Albanian**: Mandragora; **Armenian**: Մանդրագորա (Mandragora); **Bengali**: ম্যানড্রেগোরা (Myāndregōrā); **Bulgarian**: Мандрагора (Mandragora); **Czech**: Mandragora lékařská; **Chinese (Mandarin)**: 曼陀罗 (Màntuóluó); **Croatian**: Mandragora; **Dutch**:

DOI: 10.1201/9781003460336-42

TABLE 42.1
Scientific Hierarchy of *M. officinarum*

Kingdom:	Plantae
Subkingdom:	Viridiplantae
Infrakingdom:	Streptophyta
Superdivision:	Embryophyta
Division:	Tracheophyta
Subdivision:	Spermatophytina
Class:	Magnoliopsida
Superorder:	Asteranae
Order:	Solanales
Family:	Solanaceae
Genus:	*Mandragora*
Species:	*Mandragora officinarum* L.

Alruin; **English**: Mandrake; **French**: Mandragore; **Greek**: Μανδραγόρα (Mandragóra); **Georgian**: მანდრაგორა (Mandragora); **German**: Alraune; **Hindi**: मैंड्रेगोरा (Maindregōrā); **Hungarian**: Mandragóra; **Italian**: Mandragora; **Spanish**: Mandrágora; **Portuguese**: Mandrágora; **Russian**: Мандрагора (Mandragora); **Turkish**: Mandragora; **Japanese**: マンドラゴラ (Mandoragora); **Korean**: 만드라고라 (Mandeuragola); **Persian**: مندرگورا (Mandragora); **Romanian**: Mandragora; **Polish**: Mandragora lekarska; **Slovak**: Mandragora lekárska; **Serbian**: Мандрагора (Mandragora); **Slovenian**: Mandragora; **Ukrainian**: Мандрагора (Mandragora); **Tamil**: மான்டிராகோரா (Mān̲r̲irākōrā); **Telugu**: మాండ్రాగోరా (Māṇḍrāgōrā); **Marathi**: मॅन्ड्रेगोरा (Mandregōrā).

42.4 SYNONYMS

According to world flora online, *M. officinarum* includes 13 synonyms, indicating alternative names that have been used to refer to *M. officinarum* in various botanical contexts: **(1)** *Mandragora acaulis* Gaertn.; **(2)** *Mandragora autumnalis* Bertol.; **(3)** *Mandragora foemina* Garsault; **(4)** *Mandragora haussknechtii* Heldr.; **(5)** *Mandragora hispanica* Vierh.; **(6)** *Mandragora hybrida* Hausskn. & Heldr. ex Heldr.; **(7)** *Mandragora mas* Garsault; **(8)** *Mandragora microcarpa* Bertol.; **(9)** *Mandragora neglecta* G.Don ex Loudon; **(10)** *Mandragora officinalis* Moris; **(11)** *Mandragora praecox* Sweet; **(12)** *Mandragora turcomanica* Mizg.; and **(13)** *Mandragora vernalis* Bertol (WFO, 2023).

42.5 BOTANICAL DESCRIPTION

M. officinarum is a perennial herb with stout, erect, often bifid, sometimes anthropomorphic, belonging to the area Mediterranean (Waniakowa, 2007). It is one of the plants known as 'magic' with such magnitude that no other plant could compete. It has a fleshy taproot; it is acaulescent or very short, up to 60 cm long (Fakir & Özçelik, 2009; Ramoutsaki, Askitopoulou, et al., 2002a). It is bi or trifurcated roots in human form. *M. officinarum* is characterized by: Leaves are densely rosette, ovate to ovate-lanceolate, petiolate, simple, entire, undulate, and sparsely villous on veins at least when young, maximum of 45 cm long (Zemanek & Zemanek, 2008). Pedicels are generally shorter than leaves (Figure 42.1).

Generally, in spring, *M. officinarum* flowers. Their flowers are axillary, solitary, and greenish-white to pale blue and violet, at least 2.5 cm (Whitmont, 1958). Calyx is campanulate, five-lobed, slightly accrescent, much shorter than berry, with narrowly triangular lobes. In the reproductive system of *M. officinarum*, embrocate is formed by five filtered stamens inserted in the lower half of the corolla villus under the anther filaments and generous (Berry & Jackson, 1976). At the base of the ovary is surrounded by a glandular disc, and its stigma is capitalized. It has spherical and yellow berries that become unilocular by obliteration of the septum; the fruits of *M. officinarum* take the form of shiny berries, 3–4 cm long and of yellow or red color (Figure 42.2).

42.6 ORIGIN AND GEOGRAPHIC DISTRIBUTION

Mandrake is a member of the Solanaceae family and a perennial plant of Mediterranean origin. It is appropriate that warm regions of the Middle East may be exceptionally rare in other parts of the world. *M. officinarum* is distributed in the Western Mediterranean region from Turkey to Spain and Morocco (Figure 42.3). Subsequently, the plant species was identified as *M. officinarum* by the botanical records of Turkey and Europaea (Volis et al., 2018). The local people call it 'Adam otu', meaning 'first prophet' in Turkish. This species is uncommon in Turkey due to the need to conserve its natural environment. The plant species *M. officinarum* was gathered and seen in the area during research done between 2005 and 2008 in the province of Sütçüler, which is situated in the Western Mediterranean region and falls within the administrative boundaries of Isparta (Chaachouay et al., 2020b, 2020a; Rudgley, 1994; Volis et al., 2015).

Generally, mandrake originates in the Mediterranean basin 1: North Africa: Algeria, Morocco, Tunisia; Southern Europe (Italy, Greece, former Yugoslavia, Spain, Portugal) and the Near East (Palestine, Israel, Jordan, Lebanon, Syria, Turkey, Cyprus). It is not found in mainland France or Corsica. It grows in open woods, olive groves, fallow, roadsides, and ruins. This plant has become very rare, even in its original area. Populations are dispersed in the area, and some are even vulnerable, such as those in northern Italy (Dafni et al., 2021; Volis et al., 2015).

42.7 HABITAT AND ECOLOGY

M. officinarum, commonly known as mandrake, is a perennial herbaceous plant indigenous to the Mediterranean

FIGURE 42.1 The stage of *M. officinarum* plant wilting (A) and its subterranean parts (B).

FIGURE 42.2 The flowers and fruit of *M. officinarum*.

region, including Southern Europe, North Africa, and Western Asia (Dafni et al., 2021). Thriving in well-drained, sandy, or loamy soils, mandrake often populates disturbed habitats like cultivated fields and forest edges, demonstrating a preference for partial shade. Its distinctive features include a forked taproot, resembling a human figure, contributing to its historical mystique (Volis et al., 2015).

The plant was brought to England around the 12th century and to Germany during the 16th century (Van Arsdall, 2009). Cultivating mandrake from seeds is quite easy, and successful cultivation has allowed the testing of mandrake-derived medications in regions where mandrake is not naturally found (Al-Qura'n, 2009). *M. officinarum* exhibits sexual dimorphism, with distinct male and female individuals. Male mandrakes bloom in March, displaying greenish-white blossoms, and bear fruit in July (Volis et al., 2015). On the other hand, female mandrakes produce purple or blue flowers, which emerge in August or September. In Morocco, *M. officinarum* is distributed throughout locations exhibiting a Mediterranean climate, which is distinguished by scorching and arid summers and temperate and rainy winters (Chaachouay et al., 2019b, 2019a; Fakir & Özçelik, 2009). The plant often thrives on substrates characterized by rocks or sand, exhibiting a preference for locations that get moderate shade. It is documented that it exists in many places in Morocco, encompassing the northern and central areas of the nation. It may be found in the Rif Mountains, the Middle Atlas, and other appropriate environments.

42.8 ETHNOBOTANY

42.8.1 Herbal Medicine

M. officinarum is a broadly utilized traditional medicinal plant in Persia, Europe, and Africa. Various ailments have been treated with the root of this herb for an extended

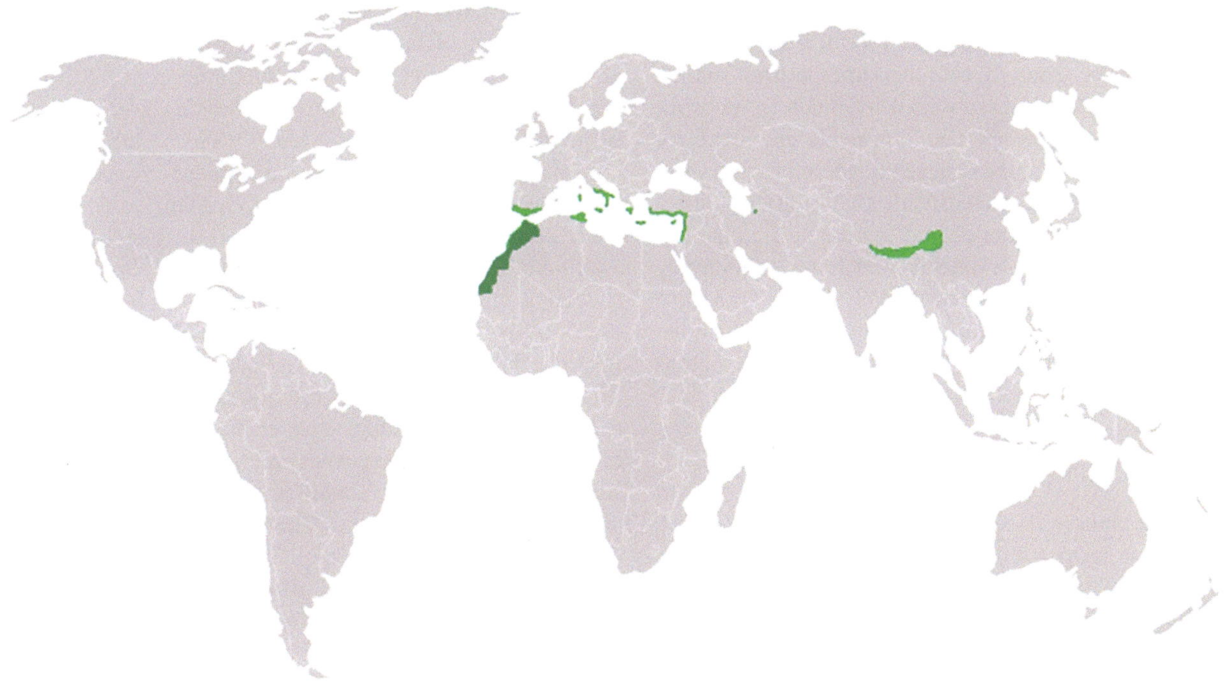

FIGURE 42.3 Geographical distribution of *M. officinarum*.

period. Small amounts of research have been conducted on the phytotherapeutic potential of the diverse components (roots, leaves, flowers, fruits, and seeds) of the plant *M. officinarum*. Nevertheless, the scope of research conducted to substantiate its therapeutic attributes (Berry & Jackson, 1976). Anyway, traditional medicine makes extensive use of the root for the following reasons: it can be used to balance the hormonal composition of the blood, stimulate worship among infertile women, demonstrate narcosis to aid in pregnancy for these women, and is regarded as a potent stimulant (Benítez et al., 2023). After being stripped and immersed in vinegar, the root is reputed to have medicinal properties for erysipelas, rheumatism, inducing sleep, and blending with wine or vinegar to create love concoctions (Ramoutsaki, Askitopoulou, et al., 2002a).

In Cyprus, roots were used as a therapeutic measure for respiratory and dermatological ailments. In Turkey, the infusion of the root is regarded as an aphrodisiac. The flowers and roots of *M. officinarum* are administered to rams, sheep, and goats to enhance their sexual drive. The mandrake root is also suggested for rectal inflammations (Chidiac et al., 2012). It is ideal to use the root while it is fresh, and it should be carefully washed and sliced before boiling it in diluted wine (Monadi et al., 2021). Afterward, it may be used as a plaster. Hippocrates proposed that the underlying cause of quartan fevers should be halted and pacified (Randolph, 1905).

M. officinarum plays a prominent role in Morocco's ethnobotanical traditions. Indigenous cultures have extensively used this plant for its alleged therapeutic and magical qualities (Benkhnigue et al., 2022, 2023; Habchaoui et al., 2023; Chaachouay et al., 2022c, 2022a, 2022b; Orch et al., 2021). The roots of *M. officinarum* are said to have strong therapeutic properties, and local herbalists often use them in traditional treatments for a wide range of conditions, including digestive problems and pain management (Chaachouay & Zidane, 2021). Furthermore, the plant has profound cultural origins, notably in traditional beliefs and ceremonies. *M. officinarum* has religious significance in some Moroccan communities, where its use goes beyond its therapeutic properties to include spiritual rituals and rites of passage.

42.8.2 Folklore and Magical Traditions

The *M. officinarum*, namely its root, has had a significant position in mythology, imbued with mystical beliefs and rites that have endured throughout the years (Dafni et al., 2021; Raeside, 1966). An extensively held superstition about mandrake is its purportedly lethal cry upon being removed from the soil. According to legend, the mandrake's vocalization was a warning of calamity and even mortality. Also, anyone who heard this cry was said to be susceptible to unfortunate circumstances, severe illnesses, or even mortality (Chaachouay & Zidane, 2021; Van den Berg & Dircksen, 2008; Waniakowa, 2007). This concept enhanced the aura around the plant and was often used in folklore and enchanting narratives.

Due to its bifid shape sometimes resembling the human silhouette, the mandrake root was interpreted as a manifestation of supernatural forces. Some believed that spirits or demons inhabited the plant (Randolph, 1905). The resemblance to a human figure likely contributed to the association of mandrake with magical rituals and esoteric practices. In numerous magical ways, mandrake was utilized on account of its legendary status (Harrison, 1956). There was speculation that the plant could grant good fortune, increase fecundity, or even grant access to supernatural realms (Monadi et al., 2021). Preparations were made to incorporate mandrake into potions and ointments intended for magical rituals; however, the efficacy of these practices has yet to be conclusively demonstrated. Occasionally, desiccated mandrake was fashioned into amulets or talismans (Benítez et al., 2023). These objects, which were frequently worn as jewelry or around the neck, were believed to have the ability to ward off evil entities and attract good fortune. This application strengthens the correlation between mandrake and magical rituals that seek to manipulate fate.

42.9 HALLUCINOGENIC EFFECTS

The mandrake plant is also familiar and known for its magic and aphrodisiac properties. Its fruit is known as an apple. Furthermore, the root that was crushed of the mandrake plant possessed hallucinations, which may cause death-like trances and sleep (Passos & Mironidou-Tzouveleki, 2016). The heart has the potential to cause insanity and can be utilized in flying potions. Another worthy blessing of mandrake is that it can be employed as a surgical anesthetic. For many years, the mandrake has analgesic and hypnotic properties (Faria, 2021). Other medicinal properties of mandrake were noted and found to have a small amount of mandrake (Gonçalves et al., 2021). An infusion can alter the numbness of the tongue, dry mouth, confusion in vision, restlessness, and exaggerated sensations (Frasca et al., 1997). Based on these findings, it was claimed that the plant could be isolated for its alkaloid constituents. It can be potentially used as an anesthetic and thus can be superseded atropine as a mydriatic.

The plant is often used for the treatment of motion sickness. The plant's root contains the alkaloid hyoscine, which serves as a cathartic, emetic, psychedelic, and narcotic. Historically, people used the plant for its pain-relieving and sleep-inducing properties (Dafni et al., 2021; Waniakowa, 2007). Additionally, they applied the freshly grated root extract topically to address ulcers, rheumatic discomfort, and crow's foot tumors (Ramoutsaki, Askitopoulou, et al., 2002b). Internally, the root extract may be used to treat depression, convulsions, and insanity, while topically, the leaf paste can be used to cure ulcers (Ramón et al., 2006). In addition, there are significant worries over various side effects, including somnolence, xerostomia, cardiac complications, visual impairments, hyperthermia, urinary difficulties, hallucinations, and potential overdose.

42.10 PHYTOCHEMISTRY

M. officinarum is very abundant in alkaloids, including hyoscyamine, scopolamine, and atropine (Table 42.2). It includes trace amounts of N-oxide alkaloids such as hyoscine and hyoscyamine in its leaves, roots, fruits, and stems (Schlesinger et al., 2019). In addition, its origins, stems, leaves, and fruits contain a minimal quantity of N-oxide alkaloids such as hyoscine and hyoscyamine (Belhaj et al., 2021; Monadi et al., 2021; Suleiman et al., 2010). The alkaloid components of *M. officinarum* were analyzed using thin-layer chromatography (TLC) and gas-liquid chromatography (GLC). The plant species' fresh and dried roots were used to extract hydroxytropane esters, namely 3-beta-tigloyloxytropane and 3-6-ditigloyloxytropane. Additionally, hyoscyamine, scopolamine, apo atropine, cuscohygrine, cysteines A3, A5, B1, B2, B3, B4, C1, and till esters of hydroxyproline were isolated (Piatti et al., 2023). In addition, the presence of belladonnas was detected in dried roots, but fresh seeds did not exhibit these chemicals (Mou et al., 2019; Schlesinger et al., 2019). The presence of tropane alkaloids in several sections of *M. officinarum*, such as the roots, leaves, and fruits, highlights their importance throughout the plant. These compounds have historical significance owing to their therapeutic characteristics; nevertheless, their powerful effects and toxicity need prudence (Hanuš et al., 2005).

M. officinarum contains steroids and triterpenoids, which are lipid-like molecules. These substances contribute to the plant's chemical complexity and may have important therapeutic implications (Al-Maharik et al., 2022). The existence of these particular chemicals highlights the complex nature of mandrake's chemical makeup, but the precise specifics may vary. Coumarins play a vital role in the pharmacological properties of mandrake within the field of aromatic chemicals. These compounds demonstrate various biological actions, suggesting potential anti-inflammatory and anticoagulant effects (Don-Lawson et al., 2022). The analysis of mandrake's coumarins, including herniarin, umbelliferone, angelicin, scopoletin, scopoline, and chlorogenic acid, highlights their impact on several physiological systems. This reveals possibilities for further investigation and possible use in therapy.

M. officinarum contains sesquiterpene lactones, compounds consisting of a lactone ring and a sesquiterpene moiety. These substances have attracted interest in pharmacological research because of their investigated potential for anti-inflammatory and immunomodulatory actions (Ramoutsaki, Askitopoulou, et al., 2002a). The investigation of sesquiterpene lactones adds dimension to the plant's complex biochemistry (Hanuš et al., 2006). *M. officinarum* has antioxidant capabilities within the range of phenolic substances, such as flavonoids and tannins. These chemicals have a protective function, aiding the plant in countering oxidative stress. Flavonoids, renowned for their many biological functions and possible health advantages, also enhance mandrake's already abundant chemical composition (Al-Maharik et al., 2022; Hanuš et al., 2005). Moreover, some plants, such as *M. officinarum*, generate volatile oils that enhance their unique fragrance (Fleisher & Fleisher, 1992). These oils include a diverse range of chemicals, including terpenes and aromatic compounds. The main volatile components found in *M. officinarum* berries include ethyl butyrate (22%), hexanol (9%), and hexyl acetate (7%),

TABLE 42.2
Secondary Metabolites Derived from *M. officinarum*

Plant Parts	*Different Alkaloids* of *M. officinarum*	References
Dried roots	*Cuscohygrine, apoatropine, 3a tigloyloxytropane, 3,6-ditigloyloxytropane, α-and β – belladonnine*	Benítez et al. (2023); Monadi et al. (2021)
Roots	Calystegines A3 (48 µg/g dm), B1 (35 µg/g dm) and B2 (39 µg/g dm), *Scopoletin (β – methyl-esculetin), sitosterol, sugars (fructose, glucose, rhamnose, sucrose), Cuscohygrine, apoatropine, 3α – tigloyloxytropan, 3,6 – ditigloyloxytropane*	Fleisher and Fleisher, (1992); Hanuš et al. (2005); Mou et al. (2019); Suleiman et al. (2010)
Leaves	Calystegine A3 (87µg/g dm), B1 (44 µg/g dm) and B2 (80 µg/g dm)	Don-Lawson et al. (2022); Piatti et al. (2023); Suleiman et al. (2010)
Young leaves	Calystegine A3 (200 µg/g dm), B1 (105 µg/g dm) and B2(138 µg/g dm)	Berry and Jackson (1976); Chioma and Oforibika (2022)
Flowers	Calystegine A3 (76 µg/g dm), B1 (40 µg/g dm) and B2 (69µg/g dm)	Monadi et al. (2021); Piatti et al. (2023); Schlesinger et al. (2019)
Fruits	Calystegine A3 (20 µg/g dm), B1 (17 µg/g dm), and B2 (30µg/g dm), Scopoletin (β-methyl-esculetin) (Volatiles (ethyl butyrate (22%), hexanol (9%), hexyl acetate (7%) and sulfur-containing compounds (>7%) among 55 compounds)	Al-Maharik et al. (2022); Hanuš et al. (2006)
Whole plant	Whitanolides (mandragorolide B, larnaxolide A, whitanolide B, withanicandrin, mandragorolide A, datura lactone, salpichrolide C) Fatty compounds, coumarins (scopetlin, scopolin), sterols (β-sitosteryl glucoside, β-sitosteryl glucoside-6-octadecanoate), 3-senecioyloxytropane, Isomeric N-oxides of (-)-hyoscyamine and isomeric N-oxides of (-)-hyoscine	Al-Maharik et al. (2022); Berry and Jackson (1976); Fleisher and Fleisher (1992); Hanuš et al. (2005); Monadi et al. (2021); Mou et al. (2019); Suleiman et al. (2010)

as well as sulfur compounds and γ-lactones. The aromatic richness of *M. officinarum* contributes to its total chemical profile by adding an olfactory component (Fleisher & Fleisher, 1992).

It is crucial to acknowledge that the phytochemical makeup of plants may vary depending on variables such as environmental circumstances, plant maturity, and the specific plant component being examined. Furthermore, the existence of these substances only automatically guarantees their safety and effectiveness for medical purposes, particularly considering the poisonous nature of some alkaloids found in *M. officinarum.* Before contemplating any medical use of plants containing potentially hazardous substances, it is crucial to exercise prudence and seek guidance from healthcare experts.

42.11 PHARMACOLOGY

M. officinarum is a botanical species that has been traditionally linked to several therapeutic and esoteric attributes. The tropane alkaloids included in it, such as scopolamine and atropine, may be poisonous when consumed in high quantities (Schlesinger et al., 2019; Suleiman et al., 2010). Traditionally, mandrake has been associated with pain alleviation, anesthetic, and aphrodisiac qualities in medicine. Furthermore, it has been linked to mystical ceremonies and traditional beliefs, often portrayed as having human-like characteristics (Al-Maharik et al., 2022; Monadi et al., 2021; Ramoutsaki, Askitopoulou, et al., 2002a). This section will address the biological activity and historical applications related to *M. officinarum.*

42.11.1 Aphrodisiac Activity

M. officinarum, sometimes called mandrake, had a rich historical background in several societies due to its purported aphrodisiac qualities. Nevertheless, it is crucial to acknowledge that the use of mandrake as an aphrodisiac is mostly rooted in historical ideas and tradition rather than substantiated by scientific facts (BİLDİK et al., 2011; Suleiman et al., 2010).

The mandrake's alleged aphrodisiac properties were first documented in ancient Greece, Rome, and Iran (Dafni et al., 2021; Ramoutsaki, Askitopoulou, et al., 2002a). The plant's roots possess these characteristics, and nowadays, nations such as Spain, Turkey, Armenia, Iraq, Lebanon, and Morocco continue to acknowledge its use for this purpose. The plant has a bifurcated root resembling the human figure, hence giving rise to superstitions and stories around its alleged abilities (BİLDİK et al., 2011). According to tradition, the mandrake root was thought to have the ability to augment fertility and stimulate sexual desire. Scientifically speaking, the study scores for more of the aphrodisiac properties of *M. officinarum*, and the existing data is mostly based on personal accounts rather than rigorous studies. The plant has various alkaloids, including hyoscyamine, scopolamine, and atropine, which possess psychotropic properties (Hanuš et al., 2005; Piatti et al., 2023). Nevertheless, the levels of these compounds may vary, and their impact on human physiology is intricate and needs to be better understood.

42.11.2 Analgesic and Anesthetic Properties

Mandrake has an extensive history of use in traditional medicine, particularly in ancient Greek, Roman, and Middle Eastern civilizations, due to its analgesic characteristics, which indicate its potential to alleviate pain (Benítez et al., 2023). Historical records suggest that mandrake extracts or ointments were used topically to relieve pain for conditions like rheumatism and neuralgia (Chidiac et al., 2012). The soothing properties of *M. officinarum* are ascribed to the tropane alkaloids, namely scopolamine and atropine. At high doses, these chemicals may cause sedation or even anesthesia. In ancient times, mandrake was used as an anesthetic during surgical procedures (Randolph, 1905). However, the use of mandrake as an anesthetic is mostly obsolete due to its low efficacy and the availability of safer alternatives.

Mandrake root is significant as a medicinal due to its analgesic and anesthetic properties. Throughout the Middle Ages and Renaissance, analgesic and anesthetic qualities were frequently mentioned (Chidiac et al., 2012; Don-Lawson et al., 2022). Mandrake has been shrouded in stories and superstitions since its roots often resemble human forms. According to mythology, removing the plant from the ground was thought to result in a loud and perhaps deadly scream or induce insanity (Benítez et al., 2023; Dafni et al., 2021). These beliefs led to the plant's correlation with mystical and ceremonial behaviors. The ethnomedical field studies conducted in the Middle East and Northern Africa continue to document these characteristics: the intravenous administration of scopolamine butyl-bromide (20 mg) reduces the duration of labor without any adverse effects on the mother or the newborn. This correlation is linked to the historical use of mandrake as an abortifacient (Berry & Jackson, 1976; Ramoutsaki, Askitopoulou, et al., 2002a). Nevertheless, it is important to note that atropine and scopolamine do not possess analgesic properties. Consuming solanaceous alkaloids when somewhat drunk might indeed cause significant discomfort (Frasca et al., 1997; Schlesinger et al., 2019). The inhibition of antinociceptive effects of acetylcholine may be achieved by blocking muscarinic and nicotinic receptors. By contrast, the use of medications containing atropine or scopolamine to induce analgesia is different.

42.11.3 Antispasmodic Effects

M. officinarum, often called mandrake, is a botanical species that has been used for several therapeutic applications since antiquity. One of its characteristics is that it has tropane alkaloids, including atropine, scopolamine, and hyoscyamine, which exhibit antispasmodic activities on smooth muscles (Monadi et al., 2021). Antispasmodic drugs inhibit or diminish spasms or contractions of the smooth muscles, namely those found in the digestive, respiratory, urinary, and reproductive systems (Dafni et al., 2021; Mou et al., 2019). They can alleviate symptoms such as stomach discomfort, cramps, diarrhea, asthma, cough, urine incontinence, and menstruation pain (Ramoutsaki, Askitopoulou, et al., 2002a).

M. officinarum has been used as a multifaceted antispasmodic agent. For instance, Aristotle classified it as a sedative, but Plato characterized it as a potent analgesic (Khalsa, 2007). Historically, the juice extracted from the finely grated root was topically used to alleviate rheumatic aches (Dafni et al., 2021). Additionally, it was used for obtaining relief and repose in persistent agony, as well as in cases of sadness, seizures, rheumatic pains, and scrofulous tumors (Al-Qura'n, 2009). Nevertheless, *M. officinarum* also exhibits several detrimental effects, including delirium, hallucinations, xerostomia, impaired eyesight, tachycardia, and urine retention (Benítez et al., 2023; Randolph, 1905). Hence, it is important to exercise prudence and use it under the careful guidance of medical professionals. Pregnant or nursing women, children, and individuals with glaucoma, heart illness, or prostate issues are advised against using this product.

42.11.4 Anticholinergic Effects

M. officinarum has alkaloids such as atropine and scopolamine, recognized for their anticholinergic properties. These chemicals inhibit the activity of acetylcholine, a neurotransmitter involved in several physiological processes, including the control of the urinary system. Anticholinergic effects may cause a reduction in bladder contraction, leading to urine retention (Monadi et al., 2021; Piccillo et al., 2006). An experiment done on Wistar albino rats exposed to *M. officinarum* leaf extract revealed that prolonged use of this extract heightens the likelihood of subacute renal problems (Ramoutsaki, Askitopoulou, et al., 2002a). The research showed that levels of urea and creatinine in the blood rose depending on the dosage and duration of treatment. The rise in urea was statistically significant when delivered at a dosage of 5.0 mL/kgBW for 14 days. This tendency was seen across all dosage levels over the 28-day dosing period (BİLDİK et al., 2011; Chidiac et al., 2012; Shim et al., 2022). The histology analysis revealed that the kidney tissue architecture remained normal after administering different dosages over 14 days. Abnormalities such as damaged renal tubules, missing glomeruli, and blocked bowman's capsular gaps were seen in all dosages within the 28-day group.

42.11.5 Antimicrobial Effects

Although *M. officinarum* has been mentioned in folklore and historical writings for its traditional therapeutic purposes, there needs to be more scientific material about its antibacterial qualities (Jodallah, 2013). The primary research focus on this plant has been its alkaloid content, including hyoscyamine, scopolamine, and atropine. These alkaloids have been studied to determine their possible pharmacological effects. One of the purported advantages of it is its ability to inhibit the growth of certain microbes (Al-Maharik et al., 2022). *M. officinarum* has been shown to have antioxidant properties and decrease the activity of enzymes such as tyrosinase, α-amylase, α-glucosidase, and acetylcholinesterase. Additionally, it has been found to contain immunomodulatory and antibacterial effects.

Mou et al conducted a comprehensive analysis that outlined the phytochemical composition and therapeutic characteristics of *M. officinarum* (Benítez et al., 2023; Monadi et al., 2021; Mou et al., 2019). The authors also highlighted the plant's historical usage as an antibacterial agent in traditional medicine (Passos & Mironidou-Tzouveleki, 2016). They referenced research conducted by (Al-Maharik et al., 2022) which documented the antimicrobial properties of the methanolic extract derived from *M. officinarum* against *Staphylococcus aureus*, *Bacillus subtilis*, *Escherichia coli*, and *Pseudomonas aeruginosa*. In addition, they referenced research conducted by Al-Maharik et al., which demonstrated the fungicidal properties of the methanolic extract derived from *M. officinarum* against *Candida albicans*, *Aspergillus niger*, and *Fusarium oxysporum*. Table 42.3 provides a concise overview of the pharmacological effects of *M. officinarum*.

42.12 MEDICINAL USES

Throughout history, mandrake has been linked to several medical purposes; however, the validity of these uses is mostly rooted in traditional beliefs rather than empirical proof (Chaachouay et al., 2019b; Dafni et al., 2021; Randolph, 1905). The practical aspects of the plant were explored in several disciplines, such as Persian medicine, traditional use, phytochemistry, pharmacology, and the toxicity profile. Consequently, the investigation needs to consider the medical use attributed to mandrake throughout the ancient Mesopotamian and Egyptian eras (Berry & Jackson, 1976; Waniakowa, 2007). These applications include the management of injured feet, induction of vomiting, alleviation of toothache, facilitation of childbirth, resolution of anal issues like hemorrhoids, therapy of rigid hips, and utilization as a protective charm against snake bites (Benítez et al., 2023; Don-Lawson et al., 2022; Tsiligianni et al., 2009).

Modern medicine employs artificially created variations of mandrake alkaloids, such as atropine and scopolamine, in well-regulated and uniform formats for targeted medicinal applications, such as pre-anesthetic medication for the management of certain ailments (Berry & Jackson, 1976; Ramoutsaki, Askitopoulou, et al., 2002a; Randolph, 1905). There are 587 records of usage, which are connected to 88 therapeutic applications. The dataset consists of 135 documents related to skin disorders, 48 recordings related to psychological intestinal diseases, and 45 records about neurological illnesses (Chidiac et al., 2012; Fakir & Özçelik, 2009). The predominant therapeutic applications were the treatment of dermatitis (68 URs), other dermatological conditions (55 URs), sleep induction (46 URs), analgesia (42 URs), and management of gastrointestinal

42.13 TOXICITY

M. officinarum is a plant with a rich historical background, both revered for its magical qualities and valued for its medical properties. It is crucial to acknowledge that many components of the mandrake plant include potentially harmful substances, including tropane alkaloids like atropine, scopolamine, and hyoscyamine (Piatti et al., 2023; Raeside, 1966). Significant ingestion of these alkaloids may result in dramatic effects on the central nervous system, including symptoms such as hallucinations, delirium, and paralysis (Monadi et al., 2021). The roots of the mandrake plant are renowned for their severe toxicity, and historical records indicate that the plant was sometimes used as a deadly poison (Frasca et al., 1997).

The residents of Delos Island in Greece intentionally avoided using *M. officinarum* due to their perception of the herb as hazardous, despite its acceptance by many populations (Berry & Jackson, 1976; Harrison, 1956). In traditional Jordanian folk medicine, it was believed that the whole plant of *M. officinarum* had poisonous properties and was associated with the development of disorders such as madness, convulsions, and hysteria (Tsiligianni et al., 2009). The Indigenous community particularly categorized immature fruits as toxic while considering the mature ones suitable for ingestion. Examination of the ethanolic (95%)

TABLE 42.3
Pharmacological Actions of *M. officinarum*

Activity	Species	Plant Part	Solvent	Test Model	Tested Living System/ Organ/Cell	Results
Antimicrobial	*M. officinarum*	Fruit (unripe)	Ac, acetone 95% EtOH	Agar diffusion (extract content 50 μg/disk)	*B. subtilis*	Nil
					E. coli	Nil
					C. albicans	Nil

Source: Jodallah, 2013.

extract of *M. officinarum* fruits demonstrated the absence of any harmful effects on the brine shrimp in the brine shrimp bioassay, even at concentrations as high as 10 mg/mL, after 24 hours. In addition, the extract did not show any mutagenic activity when tested at a dose of 5 mg/mL using a plate incorporation assay (Helbling et al., 2000). Although *M. officinarum* has historical importance, instances of systemic poisoning highlight its very toxic properties, resulting in severe toxicological symptoms such as vomiting, diarrhea, and bradycardia, frequently leading to fatality (Nikolaou et al., 2012).

The primary cause of the harmful effects of *M. officinarum* is its tropane alkaloids, which possess anticholinergic characteristics. These chemicals disrupt the regular operation of the neurotransmitter acetylcholine, impacting several physiological systems in the body (Suleiman et al., 2010). Consuming mandrake or its derivatives might result in negative effects such as impaired eyesight, parched oral cavity, elevated heart rhythm, and urinary blockage (Ramoutsaki, Askitopoulou, et al., 2002a). Consuming mandrake in severe instances may lead to convulsions, respiratory insufficiency, and even fatal outcomes (Passos & Mironidou-Tzouveleki, 2016). An extensive study on *M. officinarum* leaf extract has demonstrated a heightened susceptibility to subacute renal problems when used over a longer period. Renal studies on mice demonstrated intact tissue structure after 14 days of treatment (Ramoutsaki, Dimitriou, et al., 2002b). However, anomalies were seen in the 28-day group, including damaged renal tubules, missing glomeruli, and blocked bowman capsular gaps.

42.14 CONCLUSION

M. officinarum is a perennial herbaceous plant distinguished by a cluster of leaves and a dense taproot. This plant flourishes in Mediterranean conditions and has a long-standing connection with mystical and ceremonial traditions. Comprehending the plant's botanical characteristics and ecological needs is essential for the success of cultivation and conservation endeavors. This plant has a substantial historical background of customary use, often originating from folklore and mythology. It has been used for many reasons, including medical uses, aphrodisiac qualities, and ceremonial acts. The plant's cultural importance enhances its traditional uses with added complexity. *M. officinarum* has a diverse assortment of phytochemicals, which include alkaloids, flavonoids, and sterols. These chemicals have a role in the plant's pharmacological activities and might have implications for medical and therapeutic applications. Mandrake has a wide range of pharmacological activities, including analgesic and anti-inflammatory effects and possible neuroprotective benefits. Exploration in this field has the potential to provide the foundation for the creation of medicines or herbal therapies that may be used in contemporary medical practices. Mandrake has historically been linked to its psychedelic qualities, which are likely attributed to the presence of tropane alkaloids. Gaining comprehension of the processes and possible hazards linked to these characteristics is crucial from both scientific and social standpoints. However, it is important to exercise care, particularly regarding its hallucinogenic properties, since this feature might have legal and health consequences. Future studies on *M. officinarum* should prioritize the investigation of its pharmacological mechanisms, validate traditional usage via clinical studies, and explore sustainable farming approaches. Moreover, the integration of botany, ethnobotany, chemistry, and medicine may provide a comprehensive comprehension of this captivating plant. Thorough examination and understanding of possible dangers are essential from both a scientific and social perspective when dealing with any plant that has psychoactive characteristics.

LIST OF ABBREVIATIONS

GLC	Gas–liquid chromatography
M. officinarum	*Mandragora officinarum* L.
TLC	Thin-layer chromatography
URs	Use records

REFERENCES

Al-Maharik, N., Jaradat, N., Bassalat, N., Hawash, M., & Zaid, H. (2022). Isolation, identification and pharmacological effects of Mandragora autumnalis fruit flavonoids fraction. *Molecules*, *27*(3), 1046.

Al-Qura'n, S. (2009). Ethnopharmacological survey of wild medicinal plants in Showbak, Jordan. *Journal of Ethnopharmacology*, *123*(1), 45–50.

Belhaj, S., Chaachouay, N., & Zidane, L. (2021). Ethnobotanical and toxicology study of medicinal plants used for the treatment of diabetes in the High Atlas Central of Morocco. *Journal of Pharmacy & Pharmacognosy Research*, *9*(5), 619–662. https://doi.org/10.56499/jppres21.1014_9.5.619

Bellakhdar, J. (1997). *La pharmacopée marocaine traditionnelle*.

Benítez, G., Leonti, M., Böck, B., Vulfsons, S., & Dafni, A. (2023). The rise and fall of mandrake in medicine. *Journal of Ethnopharmacology*, *303*, 115874.

Benkhnigue, O., Chaachouay, N., Khamar, H., El Azzouzi, F., Douira, A, & Zidane, L. (2022). Ethnobotanical and ethnopharmacological study of medicinal plants used in the treatment of anemia in the region of Haouz-Rehamna (Morocco). *Journal of Pharmacy & Pharmacognosy Research*, *10*(2), 279–302. https://doi.org/10.56499/jppres21.1196_10.2.279

Benkhnigue, O., Chaachouay, N., & Zidane, L. (2023). A Floristical and Ecological Study of the Medicinal Flora Used by the Local Population of the Haouz-Rehamna Region (Middle Atlantic Morocco-4). *Acta Botanica Hungarica Akadémiai Kiadó*, *65*(3–4), 247–304. https://doi.org/10.1556/034.65.2023.3-4.3

Berry, M. I., & Jackson, B. P. (1976). European mandrake (Mandragora officinarum and M. autumnalis). *Planta medica*, *30*(07), 281–290.

Bildik, F., Kahveci, O., Aygencel, G., Keleş, A., Demircan, A., Aksel, G., & Temizkan, E. (2011). Two herbal poisoning cases from Turkey; Datura for haemorrhoidal pain relief,

Mandragora for aphrodisiac effect. *Journal of Emergency Medicine Case Reports*, *2*(2), 17–20.

Chaachouay, N., Benkhnigue, O., Fadli, M., Ibaoui, H., & Zidane, L. (2019b). Ethnobotanical and ethnopharmacological studies of medicinal and aromatic plants used in the treatment of metabolic diseases in the Moroccan Rif. *Heliyon*, *5*(10), e02191. https://doi.org/10.1080/10477845.2021.1942433

Chaachouay, N., Benkhnigue, O., Fadli, M., Ibaoui, H., El Ayadi, R., & Zidane, L. (2019a). Ethnobotanical and ethnopharmacological study of medicinal and aromatic plants used in the treatment of respiratory system disorders in the Moroccan Rif. *Ethnobotany Research and Applications*, *18*(22), 1–16.

Chaachouay, N., Douira, A., Hassikou, R., Brhadda, N., Dahmani, J., Belahbib, N., Ziri, R., & Zidane, L. (2020b). Ethnobotanical and ethnopharmacological study of medicinal and aromatic plants used in the treatment of respiratory system disorders in the Moroccan Rif. PhDThesis, Département de Biologie - Université Ibn Tofail - Kénitra, Maroc, https://tel.archives-ouvertes.fr/tel-03376377

Chaachouay, N., Benkhnigue, O., & Zidane, L. (2020a). Ethnobotanical study aimed at investigating the use of medicinal plants to treat nervous system diseases in the Rif of Morocco. *Journal of Chiropractic Medicine*, *19*(1), 70–81. https://doi.org/10.1016/j.jcm.2020.02.004

Chaachouay, N., & Zidane, L. (2021). The Symbolic Efficacy of Plants in Rituals and Socio-Religious Ceremonies in Morocco, Northwest of Africa. *Journal of Religious & Theological Information*, *21*(1–2), 34–53. https://doi.org/10.1080/10477845.2021.1942433

Chaachouay, N., Benkhnigue, O., Douira, A., & Zidane, L. (2021). Poisonous medicinal plants used in the popular pharmacopoeia of the Rif, northern Morocco. *Toxicon*, 189, 24–32. https://doi.org/10.1016/j.toxicon.2020.10.028

Chaachouay, N., Douira, A., & Zidane, L. (2021b). COVID-19, prevention and treatment with herbal medicine in the herbal markets of Salé Prefecture, North-Western Morocco. *European Journal of Integrative Medicine*, 42, 101285. https://doi.org/10.1016/j.eujim.2021.101285

Chaachouay, N., Douira, A., & Zidane, L. (2021c). Herbal Medicine Used in the Treatment of Human Diseases in the Rif, Northern Morocco. *Arabian Journal for Science and Engineering*, 47, 131–153. https://doi.org/10.1007/s13369-021-05501-1

Chaachouay, N., Benkhnigue, O., & Zidane, L. (2022c). Ethnobotanical and Ethnomedicinal study of medicinal and aromatic plants used against dermatological diseases by the people of Rif, Morocco. *Journal of Herbal Medicine*, 32, 100542. https://doi.org/10.1016/j.hermed.2022.100542

Chaachouay, N., Azeroual, A., Bencharki, B., Douira, A., & Zidane, L. (2022a). Ethnoveterinary medicines plants for animal therapy in the Rif, North of Morocco. *South African Journal of Botany*, *147*, 176–191. https://doi.org/10.1016/j.sajb.2021.12.037

Chaachouay, N., Azeroual, A., Douira, A., & Zidane, L. (2022b). Ethnoveterinary practices of medicinal plants among the Zemmour and Zayane tribes, Middle Atlas, Morocco. *South African Journal of Botany*, *151*, 826–840. https://doi.org/10.1016/j.sajb.2022.11.009

Chidiac, E. J., Kaddoum, R. N., & Fuleihan, S. F. (2012). Mandragora: Anesthetic of the ancients. *Anesthesia & Analgesia*, *115*(6), 1437–1441.

Chioma, Don-Lawson & Oforibika, G. A. (2022). Renal Effects of Mandragora officinarum Leaf Extracts on Wistar Albino Rats. *Scholars International Journal of Chemistry and Material Sciences*, *5*(9), 151–155.

Dafni, A., Blanché, C., Khatib, S. A., Petanidou, T., Aytaç, B., Pacini, E., Kohazurova, E., Geva-Kleinberger, A., Shahvar, S., Dajic, Z., Klug, H. W., & Benítez, G. (2021). In search of traces of the mandrake myth: The historical, and ethnobotanical roots of its vernacular names. *Journal of Ethnobiology and Ethnomedicine*, *17*(1), 68. https://doi.org/10.1186/s13002-021-00494-5

Don-Lawson, C., Oforibika, G. A., & Okah, R. (2022). Hepatic Impact of Mandragora officinarum Leaf Extract on Wistar Albino Rats. *American Journal of Analytical Chemistry*, *13*(12), 531–537.

Fakir, H., & Özçelik, H. (2009). *Mandragora officinarum* L.(Solanaceae): A new record for the flora of Turkey. *African Journal of Biotechnology*, *8*(15), 3560.

Faria, M. (2021). 'Plants of the Gods' and their hallucinogenic powers in neuropharmacology—A review of two books. *Surgical Neurology International*, *12*. https://www.ncbi.nlm.nih.gov/pmc/articles/PMC8326085/

Fleisher, Z., & Fleisher, A. (1992). The odoriferous principles of mandrake, Mandragora officinarum L. Aromatic plants of the Holy Land and the Sinai. Part IX. *Journal of Essential Oil Research*, *4*(2), 187–188.

Frasca, T., Brett, A. S., & Yoo, S. D. (1997). Mandrake toxicity: A case of mistaken identity. *Archives of Internal Medicine*, *157*(17), 2007–2009.

Gonçalves, J., Luís, Â., Gallardo, E., & Duarte, A. P. (2021). Psychoactive substances of natural origin: Toxicological aspects, therapeutic properties and analysis in biological samples. *Molecules*, *26*(5), 1397.

Habchaoui, J., Saad, I., El Khomsi, M., Fadli, M., & Bourkhiss, B. (2023). Ethnobotanical Study of Medicinal Herbs in the Saïss Urban Commune (Region of Fez/Morocco). *Tropical Journal of Natural Product Research*, *7*(11), 5034–5039.

Hanuš, L. O., Dembitsky, V. M., & Moussaieff, A. (2006). Comparative study of volatile compounds in the fresh fruits of Mandragora autumnalis. *Acta Chromatographica*, *17*, 151–160.

Hanuš, L. O., Řezanka, T., Spížek, J., & Dembitsky, V. M. (2005). Substances isolated from Mandragora species. *Phytochemistry*, *66*(20), 2408–2417.

Harrison, R. K. (1956). The mandrake and the ancient world. *The Evangelical Quarterly*, *28*(2), 87–92.

Helbling, A., Brander, K. A., Pichler, W. J., & Müller, U. B. (2000). Anaphylactic shock after subcutaneous injection of mandragora D3, a homeopathic drug. *Journal of Allergy and Clinical Immunology*, *106*(5), 989–990.

Jodallah, N. B. E. (2013). *Antioxidant and antimicrobial activity of Mandragora autumnalis Bertol extracts* [PhD Thesis]. https://repository.najah.edu/handle/20.500.11888/8438

Khalsa, K. P. S. (2007). Low Dose Herbs. *Journal of Herbal Pharmacotherapy*, *7*(1), 87–98. https://doi.org/10.1080/J157v07n01_08

Linnaeus, C. (1753). *Species plantarum*, vol. 2. Stockholm: Laurentius Salvius.

Monadi, T., Azadbakht, M., Ahmadi, A., & Chabra, A. (2021). A comprehensive review on the ethnopharmacology, phytochemistry, pharmacology, and toxicology of the mandragora genus; from folk medicine to modern medicine. *Current Pharmaceutical Design*, *27*(34), 3609–3637.

Mou, K. M., Parvin, M. N., & Dash, P. R. (2019). Phytochemistry and medicinal properties of Mandragora officinarum: A review. *International Journal of Pharmacognosy and Pharmaceutical Research*, *1*(1), 05–09. https://doi.org/10.33545/26647168.2019.v1.i1a.20

Nikolaou, P., Papoutsis, I., Stefanidou, M., Dona, A., Maravelias, C., Spiliopoulou, C., & Athanaselis, S. (2012). Accidental poisoning after ingestion of 'aphrodisiac' berries: Diagnosis by analytical toxicology. *The Journal of Emergency Medicine*, *42*(6), 662–665.

Orch, H., Chaachouay, N., Douiri, E. M., Faiz, N., Zidane, L., & Douira, A. (2021). Use of medicinal plants in dermato-cosmetology: An ethnobotanical study among the population of Izarène. *Jordan Journal of Pharmaceutical Sciences*, *14*(3), 323–340.

Passos, I. D., & Mironidou-Tzouveleki, M. (2016). Hallucinogenic plants in the Mediterranean countries. In *Neuropathology of drug addictions and substance misuse* (p. 761–772). Elsevier.

Piatti, D., Marconi, R., Angeloni, S., Caprioli, G., Maggi, F., Ribatti, E., Cattaneo, V., & Sagratini, G. (2023). Simultaneous quantification of hyoscyamine and scopolamine using HPLC-DAD in four Solanaceae: *Hyoscyamus niger, Datura stramonium, Atropa belladonna* and *Mandragora officinarum*. *Natural Product Research*, 1–6. https://doi.org/10.1080/14786419.2023.2269595

Piccillo, G. A., Miele, L., Mondati, E., Moro, P. A., Musco, A., Forgione, A., Gasbarrini, G., & Grieco, A. (2006). Anticholinergic syndrome due to 'Devil's herb': When risks come from the ancient time: DEVIL'S HERB. *International Journal of Clinical Practice*, *60*(4), 492–494. https://doi.org/10.1111/j.1368-5031.2006.00864.x

POWO. (2023). *Plants of the World Online, Mandragora officinarum L.* [jeu de données]. https://wcvp.science.kew.org/.

Raeside, J. R. (1966). A proving of Mandragora officinarum. *British Homeopathic Journal*, *55*(02), 68–75.

Ramón, M. F., Ballesteros, S., Martinez-Arrieta, R., & Bandrés, F. (2006). *Intentional Abuse of Plants. New Research on Street Drugs*. New York: Nova Science Publishers, Inc, 145–166.

Ramoutsaki, I. A., Askitopoulou, H., & Konsolaki, E. (2002a). Pain relief and sedation in Roman Byzantine texts: Mandragoras officinarum, Hyoscyamos niger and Atropa belladonna. *International Congress Series*, *1242*, 43–50.

Ramoutsaki, I. A., Dimitriou, H., & Kalmanti, M. (2002b). Management of childhood diseases in the Byzantine period: I – Analgesia. *Pediatrics International*, *44*(3), 335–337. https://doi.org/10.1046/j.1442-200X.2002.01548.x

Randolph, C. B. (1905). The Mandragora of the Ancients in Folk-lore and Medicine. *Proceedings of the American Academy of Arts and Sciences*, *40*(12), 487–537. https://www.jstor.org/stable/20021986

Riddle, J. M. (2010). Mandrake, the Love Apple, and the World's Religions. In J. M. Riddle, *Goddesses, Elixirs, and Witches* (p. 55–77). Palgrave Macmillan US. https://doi.org/10.1057/9780230105515_4

Rudgley, R. (1994). *Essential substances: A cultural history of intoxicants in society*. Kodansha International. https://www.cabdirect.org/cabdirect/abstract/19950307436

Schlesinger, D., Rikanati, R. D., Volis, S., Faigenboim, A., Vendramin, V., Cattonaro, F., Hooper, M., Oren, E., Taylor, M., & Sitrit, Y. (2019). Alkaloid chemodiversity in Mandragora spp. Is associated with loss-of-functionality of MoH6H, a hyoscyamine 6β-hydroxylase gene. *Plant Science*, *283*, 301–310.

Shah, V. V., Shah, N. D., & Patrekar, P. V. (2013). Medicinal plants from Solanaceae family. *Research Journal of Pharmacy and Technology*, *6*(2), 143–151.

Shim, K. H., Kang, M. J., Sharma, N., & An, S. S. A. (2022). Beauty of the beast: Anticholinergic tropane alkaloids in therapeutics. *Natural Products and Bioprospecting*, *12*(1), 33. https://doi.org/10.1007/s13659-022-00357-w

Suleiman, R. K., Zarga, M. A., & Sabri, S. S. (2010). New withanolides from Mandragora officinarum: First report of withanolides from the genus Mandragora. *Fitoterapia*, *81*(7), 864–868.

Tsiligianni, I. G., Vasilopoulos, T. K., Papadokostakis, P. K., Arseni, G. K., Eleni, A., & Lionis, C. D. (2009). A two cases clinical report of mandragora poisoning in primary care in Crete, Greece: Two case report. *Cases Journal*, *2*(1), 9331. https://doi.org/10.1186/1757-1626-2-9331

Van Arsdall, A. (2009). Exploring What was Understood by 'Mandragora' in Anglo-Saxon England. *Old Names-New Growth: Proceedings of the 2nd ASPNS Conference, University of Graz, Austria, 6-10 June 2007, and Related Essays*, 57–74. https://books.google.com/books?hl=fr&lr=&id=C47PiEbmQL4C&oi=fnd&pg=PA57&dq=+Mandragora+officinarum+in+FOLKLORE+AND+MAGICAL+TRADITIONS&ots=nyzt1ORBum&sig=t4-rHV17D-h879Ps9HGXgxz_b00

Van den Berg, M., & Dircksen, M. (2008). Mandrake from antiquity to Harry Potter. *Akroterion*, *53*(1), 67–79.

Volis, S., Fogel, K., Tu, T., Sun, H., & Zaretsky, M. (2018). Evolutionary history and biogeography of Mandragora L.(Solanaceae). *Molecular Phylogenetics and Evolution*, *129*, 85–95.

Volis, S., Tu, T., Deng, T., Zaretsky, M., Fogel, K., & Sun, H. (2015). Phylogeographic study of Mandragora L. reveals a case of ancient human assisted migration. *Israel Journal of Plant Sciences*, *62*(3), 176–186.

Waniakowa, J. (2007). « Mandragora » and' Belladonna': The names of two magic plants. *Studia Linguistica Universitatis Iagellonicae Cracoviensis*, *124*. https://ruj.uj.edu.pl/xmlui/bitstream/handle/item/81068/waniakowa_mandragora_and_belladonna_2007.odt?sequence=2&isAllowed=y

WFO. (2023). *World Flora Online, SPECIES - Mandragora officinarum L.* [jeu de données]. https://wfoplantlist.org/plant-list/

Whitmont, E. C. (1958). Mandragora. *British Homeopathic Journal*, *47*(04), 264–269. https://doi.org/10.1016/S0007-0785(58)80006-X

Zemanek, A., & Zemanek, B. (2008). The morphology of illustrated plants. In *Drawn after Nature* (pp. 78–85). KNNV Publishing. https://brill.com/downloadpdf/book/9789004278004/B9789004278004-s015.pdf

43 Cultivated Tobacco (*Nicotiana tabacum* L. Cannabaceae)

Bouchaib Bencharki, Noureddine Chaachouay, Abdelhamid Azeroual, and Aziz Baidani

43.1 INTRODUCTION

Humans have used plants for various purposes, including food, clothing, housing, gardening, medicine, smoking, snuffing, etc. Plants and plant products are smoked or snuffed for pleasure, narcotic, or psychological benefits in many regions of the globe (Hassan-Abdallah et al., 2013; Chaachouay & Zidane, 2021; Orch et al., 2021). Various plants' leaves, seeds, or latex as fumitories are prevalent in many places worldwide. The plants that produce fumitories are used in traditional medicine and as medicinal herbs (Udoye & Okeke, 2021). *Nicotiana tabacum* L. is a significant species of fumitory plants. It originated in the Andes area of South America, notably in what is now Peru and Ecuador. Archaeological evidence demonstrates that Indigenous peoples have extensively used it for thousands of years, playing a pivotal part in many Native American societies (Kishore, 2014).

The plant is a tall, perennial herb with a strong stem with glandular hairs. The leaves are broad, hairy, green, and shaped like an oblong-lanceolate. The flowers are pink or scarlet, compact, and arranged in many-flowered inflorescences, often in panicles (Belliard et al., 1978). The fruits are capsules that measure 1.5–1.8 cm in length. This plant is indigenous to tropical and subtropical regions of America and is now farmed globally as a profitable agricultural crop. It reaches a vertical measurement of 1 to 2 meters (Dewey & Xie, 2013). Tobacco refers to the desiccated and treated leaves of *N. tabacum*, a member of the Solanaceae family of flowering plants. Plant products are mostly ingested by smoking, chewing, snuffing, or dipping. The leaves produce cigarettes, cigars, snuff, pipe, and chewing tobacco (Kenton et al., 1993).

Medicinal herbalists have historically utilized tobacco as a sedative, but due to its potent addictive properties, it is administered with great caution. This medicinal plant has diverse phytochemicals and chemical compounds that have biological effects on living beings. Certain phytochemicals have beneficial effects, while others have harmful effects (Rawat et al., 2013). Nicotine, an alkaloid present in tobacco, is the most renowned phytochemical. The primary biological effects of *N. tabacum* are mostly attributed to its nicotine content, which is recognized for its ability to activate nicotine receptors (Demole & Berthet, 1972). This activation subsequently triggers the release of many substances, including acetylcholine, norepinephrine, dopamine, serotonin, growth hormone, and vasopressin (Xu et al., 2020). This plant is purported to manage a wide range of human ailments. *If well harnessed, N. tabacum* has many pharmacological potentials that benefit people and animals. Nevertheless, its pharmacological qualities have received little attention. Indeed, there is a need for more data about the biological properties of *N. tabacum*. This chapter aims to analyze the taxonomy, morphological description, phytochemical ingredients, pharmacological characteristics, toxicity, clinical applications, and ethnomedicinal uses of *N. tabacum*.

43.2 SCIENTIFIC CLASSIFICATION

Nicotiana tabacum L. belongs to the order Solanales, the family Solanaceae, and the genus *Nicotiana* (ITIS, 2022).

Kingdom: Plantae
Superdivision: Embryophyta
Division: Tracheophyta
Class: Magnoliopsida
Order: Solanales
Family: Solanaceae
Genus: *Nicotiana*
Species: *Nicotiana tabacum* L.

43.3 VERNACULAR NAMES

N. tabacum has several vernacular names depending on where it is grown: Tobacco: تبغ (**Arabic**); Tambākū: तम्बाकू (**Hindi**); Tamāk: তামাক (**Bengali**); Dambae: 담배 (**Korean**); Yan cao (**Chinese**); Tabako: タバコ (**Japanese**); Virginian tobacco, Tobacco (**English**); Turkish (**Tütün**); Kapnós: Καπνός (**Greek**); Tumbaku (**Swahili**); Tabac, Tabac commun, Tabac de Virginie, Herbe du Grand Prieur, Herbe de la Reine (**French**); Virginischer Tabak (**Germany**); Duma, Poga (**India**); Tabacco (**Italy**); Табак: Tabak (**Russian**); Tabaco comum (**Portuguese**); Tabaco (**Spanish**).

DOI: 10.1201/9781003460336-43

43.4 BOTANICAL DESCRIPTION

N. tabacum is a plant of great non-food agricultural value and the main feedstock for tobacco products worldwide. Tobacco is an annual herbaceous plant that can reach 2 to 3 meters (Knapp et al., 2004). It prefers rich, cool, well-drained soils, sunny to semi-shady exposure, and a Mediterranean climate. Its roots are tap-rooted, long, and fibrous (Kishore, 2014). The stem is rough, slimy, and covered with fine, woolly down branches, especially at the top. The leaves are numerous and large but fragile (30 to 60 cm long and 10 to 20 cm wide). They are oval or spearhead-shaped, light-green with glandular hairs, oblong-lanceolate, alternate, and strongly veined. The leaves give off a pungent (Figure 43.1), narcotic odor due to nicotine (Sanjeev & Kapgate, 2021). Flowering takes place from June to November. Loose panicles of small, dark pink flowers with white centers, hermaphroditic and axillary. The flowers, grouped in loose panicles, are tubular with five downy lobes and have a light-green calyx and pink petals. The corolla tube is greenish-white, and the lobes are tinged with pink (Loughrin et al., 1990; Sakai & Shimamoto, 1965).

In North America and Europe, the plant flowers between July and September. The ovoid capsule-shaped fruits, 10 to 15 mm long, contain numerous small brown oleaginous seeds. True tobacco is sometimes confused with other *Nicotiana* species. It is propagated from seed (Wu et al., 2013). In the tropics, all that's needed is to scatter the seeds over the soil, usually fertilized with ashes. In Central Europe, seeds should be sown in porous, sandy soil in greenhouses or window boxes between mid- and late March (Lisuma et al., 2019; Liu et al., 2021). The seeds should be lightly pressed into the ground and germinate in ten to twenty days if kept at 18 to 20°C. The young plants should then be transplanted to larger pots or beds. Tobacco needs plenty of sun, fertilizer, and water. It grows best in sheltered locations (Xing et al., 2021).

43.5 ORIGIN AND GEOGRAPHIC DISTRIBUTION

Tobacco originated in the Andes area of South America, notably in what is now Peru and Ecuador. Archaeological evidence demonstrates that indigenous peoples have extensively used it for thousands of years, playing a pivotal part in many Native American societies (Gray et al., 1974; Kawashima et al., 1976). Although tobacco was formerly limited to the Americas, its cultivation and commerce-enabled its widespread distribution worldwide (Figure 43.2). Currently, it is grown in several nations around the globe, mostly in locations with favorable conditions for its cultivation. In ancient Mesoamerica, tobacco was the object of religious veneration and was considered a plant of the gods (Arslan & Okumus, 2006). Tobacco played an important role in the pre-Columbian culture of the Maya. In addition to its many social meanings, tobacco was also regarded by the Indians as a universal antidote for all types of animal bites and poisoning (Lewis & Nicholson, 2007). When real tobacco was first introduced to Europe from America, it was used as a medicinal plant and listed in early modern herbariums as 'Indian beinwell'.

Tobacco is grown across multiple continents, with significant production areas in North America (particularly in states like Virginia, Kentucky, and North Carolina), South America (including countries like Brazil that contribute significantly beyond its native territories), Africa (with pivotal roles played by countries like Morocco, Zimbabwe,

FIGURE 43.1 Fresh (A) and dried (B) *N. tabacum* leaves and their root systems and stems (C).

FIGURE 43.2 How to cultivate *N. tabacum* plants in striped patterns in Malawi.

Tanzania, and Malawi), Asia (where China and India are major producers), and even certain European nations like Turkey (L. Cui et al., 2023a; L.-L. Cui et al., 2023b). Tobacco flourishes in temperate regions characterized by distinct growth seasons and well-drained soils, making it versatile and adaptable to many environmental circumstances (Arslan & Okumus, 2006). Tobacco production and commerce have had huge worldwide economic implications, acting as a crucial cash crop for several nations and making large contributions to their economy via local consumption and international exports (Hong et al., 2022). Real tobacco is mainly grown for the manufacture of nicotine-containing products such as cigarettes, cigars, and smokeless tobacco products.

43.6 ECOLOGY

N. tabacum can tolerate different soil types: light (sandy), medium (loamy), and heavy (clay) soils, with pH tolerance to both acidic and basic soils (Xing et al., 2021). It has been grown in various habitats, including forests, plains, mountains, wetlands, savannahs, dry valleys, and, apparently, even on a volcano (Figure 43.3). *N. tabacum* is extensively grown for its foliage, which produces tobacco items such as cigarettes, cigars, snuff, and chewing tobacco (Wu et al., 2013). It is an autogamous plant capable of interbreeding with other *Nicotiana* species, resulting in hybrids exhibiting distinct traits. *N. tabacum* grows best in well-drained, nutrient-rich soil in full sun but has been grown in the shade in Java and Connecticut (Upadhyay et al., 2009). The species is not very hardy in Britain, but plants can be grown as biennials in areas where winter temperatures don't fall below –5°C. This plant has a moderate tolerance to frost but cannot withstand severe cold or heat (Hovell & Hughes, 2009; Liu et al., 2021). The plant is vulnerable to various pests and diseases, including aphids, caterpillars, nematodes, fungi, and viruses. *N. tabacum* is a cultivated plant hybridizing several wild *Nicotiana* species, including *N. tomentosiformis*, *N. otophora*, and *N. sylvestris* (Cooper & Harrison, 1973). Humans have tamed and selectively produced cannabis for millennia, leading to the development of several types and cultivars exhibiting distinct traits, including variations in leaf size, shape, color, taste, and nicotine concentration (Figure 43.4).

43.7 PHYTOCHEMISTRY

Tobacco has a wide range of phytochemicals, which are chemical substances with biological impacts on living organisms (Table 43.1). Some of these phytochemicals are advantageous, while others are detrimental (Kaushik et al., 2010; Rawat et al., 2013). Nicotine, an alkaloid found in tobacco, is the most well-known phytochemical. It functions as both a stimulant and a neurotoxin. Nicotine is the primary causative agent behind the addictive and euphoric properties of tobacco, as well as a significant contributor to its detrimental health consequences. Nicotine has damaging effects on the cardiovascular, respiratory, neurological, and endocrine systems, hence contributing to the development of diverse ailments like cancer, heart disease, stroke, and diabetes (Ameya et al., 2017; Sharma et al., 2016).

In 1828, Ludwig Reimann and Wilhelm Heinrich Posselt, chemists from the University of Heidelberg, extracted the potent and volatile alkaloid nicotine from the tobacco plant (Chambers, 2002). It was then shown that nicotine had powerful properties that allowed it to be

FIGURE 43.3 A specialized nursery focused on growing and nurturing the Tobago plant.

FIGURE 43.4 Batek variety, a native tobacco; Mindanao – Philippines.

absorbed through the skin, making it a very efficient pesticide with insecticidal capabilities, which continues to be used globally. In addition to nicotine, tobacco also includes additional alkaloids, including cotinine, anabasine, nornicotine, and anatabine (Dewey & Xie, 2013). These alkaloids have structures and effects comparable to nicotine but at lower doses. Additionally, they may impede the metabolic processes and removal of nicotine, hence increasing the toxicity of tobacco (Saitoh et al., 1985; Wei et al., 2005). Flavonoids, a significant class of phytochemicals found in tobacco, are polyphenolic substances known for their antioxidant, anti-inflammatory, and anti-cancer characteristics. Flavonoids can regulate the function of many enzymes, receptors, and communication pathways. Additionally, they may shield cells against oxidative stress and inflammation. Quercetin, kaempferol, rutin, and naringenin are among the

TABLE 43.1
The Major Chemical Compounds Found in the Different Parts of *N. tabacum*

Plant Parts	Chemicals Compounds
Leaf	pyridine alkaloids, the principal one being a liquid alkaloid, nicotine. Other alkaloids present include nicotine, nicotimine, anabaine anatalline and nornicotine. It also contains a high percentage of organic acids. Leaves also have glucosides, tahacinin, tahacilin, and isoquercitrin, 1-quinic, chlorogenic, caffeic, and oxalic acids. They also contain terpenic and carcinogenic substances. Anatabine and (+) nornicotine have been isolated from roots (Kaushik et al., 2010; Kishore, 2014; Rawat et al., 2013; Weber et al., 2019).
Flowers	Quercetin-3,3'- dimethyl ether, Butanoic acid, 3-methyl, Hexanoic acid, 5-methyl, Solanone, pentanoic acid, 3-methyl, quercetin-3-Me ether. Three new gibberellins- nicotiana α, β, and γ, and gibberellins A and A3 (He et al., 2016; Loughrin et al., 1990; Popova et al., 2020; C. Yang et al., 2021).
Seed	cycloartanol, cycloartenol 2-4-daturadiol and solavetivone. Cholesterol, coolest-7- enol, 2-4-methylenecholesterol, campesterol, stigmasterol, sitosterol, citrostadienol, cycloeucalenol, β-amyrin, 2-8- isofucosterol, lanosterol, 31- norlanosterol, granisterol, lanost-8-enol, obtusifoliol, lupeol, 3-1-norcycloartenol, cycloartanol and 2-4- methylenecycloartanol have also been reported in seed oil (Frega et al., 1991; Maestri & Guzmán, 1993; Popova et al., 2020; Stanisavljević et al., 2007).
Roots	Quercetin-3,3-dimethyl ether, 7-(4-hydroxyphenyl)-6H-[1, 3]dioxolo[4,5-g]chromen-6-one, 7-(2-hydroxy-3,4-dimethoxyphenyl)-6H-[1,3]dioxolo[4,5-g]chromen-6-one, 7-hydroxy-6,3′,4′,5′-tetramethoxy-isoflavone (1) and 6-hydroxy-7,3′,4′,5′-tetramethoxy-isoflavone, quercetin-3-Me ether (Al-Lahham et al., 2020; Chen et al., 2012; Kishore, 2014; Lei et al., 2015; Wei et al., 2005).

flavonoids present in tobacco (Fu et al., 2016; Ru et al., 2012; Xiao et al., 2022).

Tobacco furthermore includes triterpenoids, which are chemical substances produced from squalene. Triterpenoids possess anti-microbial, anti-tumor, and anti-inflammatory properties and can regulate the expression of several genes involved in cell development, differentiation, and death (Gossart et al., 2023; He et al., 2016; Sallaud et al., 2012). Three triterpenoids present in tobacco include solanesol, alpha-tocopherol, and beta-sitosterol (Putri et al., 2023). Additional phytochemicals found in tobacco include phenols, saponins, malic acid, and citric acid (Rawat et al., 2013; Sanjeev & Kapgate, 2021). Phenols are aromatic chemicals with antioxidant and antibacterial effects (Sahai & Shuler, 1984; Zou et al., 2021). Saponins are a glycoside with detergent properties and may destroy red blood cells (Kesselmeier, 1980). Malic and citric acids are organic acids crucial in the Krebs cycle and energy creation (Grünweller et al., 1990). Tobacco's phytochemical content and activity are contingent upon several variables, including the plant's species, variety, cultivation, harvesting, curing, and processing. Various modalities of tobacco use, including smoking, chewing, or snuffing, impact the exposure and assimilation of the phytochemicals (Fu et al., 2016). Hence, the impact of tobacco on human health is intricate and diverse, contingent upon the quantity, duration, and frequency of tobacco use, as well as the individual's genetic and environmental circumstances.

43.8 TOXICITY AND SIDE EFFECTS

Tobacco is a plant that has a well-documented level of toxicity as a result of its elevated nicotine concentration. Nicotine, an alkaloid found in tobacco, functions as a powerful neurostimulator, impacting the central nervous system and causing addiction in those who use it (Kishore, 2014). When tobacco is taken, especially by smoking, it not only produces nicotine but also a multitude of detrimental chemicals. Inhaling tobacco smoke brings harmful substances such as tar, carbon monoxide, and many cancer-causing agents into the body (Ali Esmail Al-Snafi, 2022). Extended exposure to these compounds greatly heightens the likelihood of getting severe health ailments, such as cardiovascular illnesses, respiratory disorders, including chronic obstructive pulmonary disease, and many types of cancer (Bussmann et al., 2011; Maodzeka et al., 2017). Indications of acute poisoning include vertigo, excessive salivation, emesis, diarrhea, tremors in the hands, and sensations of frailty in the legs; very large doses may swiftly result in convulsions, loss of consciousness, cardiac arrest, and respiratory insufficiency (Schep et al., 2009). Instances of poisoning mostly arise from youngsters ingesting cigarettes and individuals coming into contact with pesticides containing nicotine, particularly during tobacco harvesting. Nicotine patches pose a risk to youngsters as well. The highly addictive properties of nicotine compound the difficulty of quitting smoking, hence adding to the long-lasting health hazards linked to *N. tabacum* (Latifi et al., 2022).

Secondhand smoking is an additional issue associated with the toxicity of *N. tabacum*. People who are exposed to secondhand smoke sometimes referred to as passive smoke, might suffer from negative health consequences, especially individuals who do not smoke, children, and pregnant women (Sabanayagam & Shankar, 2011). Inhaling smoke emitted by combusting tobacco products heightens the likelihood of contracting respiratory infections, experiencing sudden infant death syndrome in babies, and encountering developmental problems in youngsters (Panagiotakos et al., 2004). Furthermore, the presence of carcinogenic elements in secondhand smoke is a health hazard since it can induce cancer in those who do not smoke (Raupach et al., 2006). To effectively tackle the toxicity of *N. tabacum*, a comprehensive strategy is necessary, which includes public health campaigns, smoking cessation programs, and regulatory measures. Public awareness campaigns are vital in teaching individuals about the health hazards linked to tobacco use to prevent people from starting and promoting quitting (Hagstad et al., 2014). Smoking cessation programs assist those attempting to stop smoking by offering tools, counseling, and nicotine replacement treatments to manage the symptoms of withdrawal effectively (Naeem, 2015). In addition, regulatory interventions, such as levying tobacco taxes, imposing advertising constraints, and enforcing age restrictions on tobacco product purchases, are adopted to mitigate tobacco usage and safeguard public health. To effectively treat the harmful effects of *N. tabacum*, a comprehensive approach is required that targets both individual behavior and wider social causes that contribute to tobacco use.

43.9 REGULATORY FRAMEWORKS

The toxicity of tobacco smoke inhaled by smokers affects almost every organ in the human body, causing numerous illnesses and affecting smokers' health (Steinberg et al., 2007). Tobacco consumption increases the risk of cardiovascular disease, cancer, and respiratory illness. However, you don't have to smoke a pack of cigarettes a day to affect your health (Prabhu et al., 2014; Saffer & Chaloupka, 2000). Indeed, a Norwegian survey showed that people who smoked 1 to 4 cigarettes a day for an average of 18.7 years significantly increased their risk of developing lung cancer and dying of cardiovascular problems (Singh & Kathiresan, 2015). Consumers are generally unaware of the composition of cigarettes (including the additives and flavors added by companies) or the toxic products they absorb when smoking. As a result, they often underestimate the risk of cancer and are unaware that heart disease and heart attack are among the leading causes of death due to smoking (Saffer & Chaloupka, 2000).

Fifty percent of all regular smokers will die as a result of smoking, half in middle age (35–69 years) and the other half in old age (>70 years). According to available data, in 1998 in Canada, tobacco use was responsible for 47,581 deaths annually, while in Quebec, 13,296 deaths were attributable to smoking, representing a quarter of all deaths. Premature death awaits those who smoke during their working lives (McDermott et al., 2008). On average, North American smokers die 14 years earlier than non-smokers. The WHO Framework Convention on Tobacco Control includes provisions setting international standards on all important aspects of tobacco control (advertising, promotion and sponsorship, price and tax, sales, passive smoking, smoke-free environment, treatment of tobacco dependence, and illicit trade) (Degenhardt et al., 2008). It also provides for cooperation and support for developing countries and the transfer of knowledge and skills between Parties (Shibuya et al., 2003). Finally, the Convention recognizes the industry's responsibility for the tobacco epidemic and commits to monitoring its attempts to counter public health measures. Like the 159 other countries that have ratified the Convention, Canada, supported by Quebec, is committed to implementing the measures included in the treaty (Yach, 2014). These measures are based on best practices in tobacco control.

43.10 ETHNOBOTANY

Tobacco has a long history of usage as a sedative by medical herbalists, but since it is a higher additive medication, it is used with extreme care. Plant components include, but are not limited to, leaves, stems, stem bark, flowers, fruits, seeds, and roots (Groark, 2010; Chaachouay et al., 2019a, 2019b). Whether fresh, dried, or crushed, tobacco leaves are the primary component of traditional medicine. They have been used in treating human illnesses for several decades. For example, the leaves are used topically to treat conditions such as rheumatic swelling, skin ailments, and scorpion stings. Damp tobacco leaves may alleviate the discomfort caused by a sting (Anderson, 1993). Additionally, they serve as an effective remedy for uncomfortable hemorrhoids. A leaf poultice has been employed to relieve headaches, while green tobacco juice has been used as a remedy for eye irritation. Tobacco leaves are used with the stem and leaves of *Rhabdadenia biflora* to treat wounds caused by sting rays (Kishore, 2014; Steinberg et al., 2007). A homeopathic medication is derived from desiccated leaves and used to manage nausea and motion sickness.

In addition, *N. tabacum* leaves are used to treat skin ailments, localized infections, bronchitis, asthma, and inflammation. A salve produced by boiling the leaves in lard has been used to treat chronic sores and agonizing growths (Charlton, 2004). Furthermore, the infusion of leaves is used topically for sprains and bruises; the leaf extract is ingested to cause vomiting and narcosis; and the plant is utilized for treating ulcerated abscesses, fistulas, ulcers, invertebrate polyps, and several other illnesses (Udoye & Okeke, 2021). The leaves also possess antispasmodic, discutient, diuretic, emetic, expectorant, irritating, soothing, and sialagogue properties. Moist tobacco leaves are typically used to treat rheumatic inflammation, dermatological conditions, insect bites, and alleviate uncomfortable hemorrhoids. *N. tabacum* has been traditionally used to treat asthma, worms, wounds, and dysmenorrhea (Maroyi, 2013). The infusion of leaves is used to achieve muscular relaxation and alleviate discomfort. Tobacco leaves are said to be able to treat a range of conditions, including stomach discomfort, constipation, urinary tract blockage, tooth pain, gastrointestinal diseases, and dermatitis (Benkhnigue et al., 2022, 2023; Chaachouay et al., 2020a, 2020b).

Additional therapeutic applications of *N. tabacum* include using a compress made from fresh leaves on boils and infected wounds and combining crushed leaves with palm oil to prevent hair loss. Moreover, a mixture of dried leaves and *Securinega virosa* is employed topically to eradicate worms in sores (Bidarigh fard et al., 2019). The leaf extract is used to treat various ailments such as general discomfort, coldness, and snakebite and to manage pulmonary illnesses, strangulated hernia, orchitis, and skin conditions. Warmed leaves are used topically to cure hydrocele, while the oil derived from the leaves is used to treat arthralgia, gout, and lumbago (Sharma et al., 2021). Additional ethnomedicinal applications of *N. tabacum* include utilizing dried leaves as an insecticide, treating nausea and motion sickness, alleviating scorpion bites, managing distemper in veterinary medicine, and applying leaves externally to address mycosis, headaches, wounds, ringworms, ulcers, bruises, sores, and stomatitis (Berlowitz et al., 2020). The leaves greatly treat convulsions, dizziness, and other nervous system disorders. They are also used to alleviate sadness, fatigue, and certain types of pain.

43.11 HALLUCINOGENIC EFFECTS

Tobacco is one of the most widely used psychoactive pleasure agents in the world. Tobacco, derived from the leaves of *N. tabacum*, has long been used in the Americas (Berlowitz et al., 2020), where the plant was known to be cultivated in pre-Columbian Mexico and Peru (Kishore, 2014; Prabhu et al., 2014). Tobacco products consist of desiccated tobacco leaves, which are used to exploit the euphoric properties of the alkaloid nicotine (Bennett, 1992). Cigarettes are acidic due to the pH level of smoking tobacco, which ranges from 7.8 to 8.2. Chewing tobacco, the oldest form of tobacco consumption becomes basic after adding flavorings, sweeteners, or taste stabilizers. *N. tabacum* is the main source of nicotine. Its stems and leaves are used commercially (Carod-Artal, 2015). The plant contains between 0.5 and 9% nicotine, the main toxin. Nicotine from tobacco usage does not often cause hallucinations. Nevertheless, an overabundance of nicotine use or heightened susceptibility to the chemical might result in symptoms such as vertigo, faintness, and nausea (Groark, 2010). Occasionally, elevated levels of nicotine might lead to worsened symptoms, such as cognitive disorientation and perceptual distortions (Berlowitz et al., 2020; Belhaj et al., 2021; Chaachouay et al., 2021a, 2021b, 2021c). However, it is important to note that these consequences are not the intended or typical results of tobacco use. Nicotine attaches to certain acetylcholine receptors in various parts of the body, which are often referred to as nicotine receptors. This elicits initial excitation, followed by subsequent suppression, at receptors distributed throughout the nervous system (Siegel, 1976). Small dosages enhance the release of catecholamines and stimulate the sympathetic nervous system. Increased dosages elicit parasympathetic stimulation.

43.12 PHARMACOLOGICAL ACTIVITIES

N. tabacum is a widely cultivated plant highly esteemed for its leaves, mostly used for smoking, chewing, and snuffing (Ali Esmail Al-Snafi, 2022; Chaachouay et al., 2022a, 2022b, 2022c; Charlton, 2004). The primary mechanism by which this medicinal plant produces its pharmacological effects is the action of nicotine, a naturally occurring alkaloid present in the plant. Nicotine affects both the central and peripheral nervous systems, leading to various physiological and psychological consequences (Loughrin et al., 1990).

43.12.1 Antifungal Activity

N. tabacum extracts are used to eliminate fungal infections. Studies were conducted to isolate, identify, and manage the fungus responsible for fruit rot in *Lycopersicon esculentum*. The identified fungi include *Penicillium digitatum*, *Rhizopus spp*, and *Aspergillus viridae* (Ponstein et al., 1994). The researchers used botanical extracts derived from *Azadirachta indica* and *N. tabacum*. When the sections were applied in vitro to control fungi, it was shown that *N. tabacum* had a fungi-toxic effect (C. Yang et al., 2021). This action completely inhibited the development of mycelia at a concentration of 60% on *Penicillium digitatum* and *Aspergillus viridae*. The study examined the antifungal properties of various plant extracts from *N. tabacum*, Calotropis *procera*, and *Azadirachta indica* against specific phytopathogenic fungi: *Rhizoctonia solani* AG2-2 (MB), *Rhizoctonia solani* AG2-2 (IV), *Corynespora cassiicola*, *Rhizoctonia solani* AG2-1, and *Trichoderma harzianum* (Sela-Buurlage et al., 1993). The findings indicated that the leaf extracts' organic solvents, including *N. tabacum*, effectively suppressed mycelia development in phytopathogens, demonstrating potent antifungal properties (Huang et al., 2010). An investigation was conducted to assess the antifungal properties of *N. tabacum* leaf extracts against certain dermatophytes. The findings revealed that the crude extracts exhibited acidity and antifungal properties against all the tested microorganisms, namely *Trichophyton rubrum*, *Microsporum canis*, and *Epidermophyton floccosum*.

43.12.2 Antibacterial Effect

N. tabacum has been extensively researched for its possible antibacterial effects, specifically about the alkaloid nicotine found in the plant. Nevertheless, it is crucial to acknowledge that little data indicates antibacterial properties. Extracts derived from *N. tabacum* have been shown to have strong antibacterial activity (Bakht et al., 2012; Rawat et al., 2013). From this standpoint, research has examined the antibacterial properties of *N. tabacum* leaves. The findings revealed that a 60% methanol extract was antibacterial against six of nine bacterial isolates when administered at 25 mg/l (Tariq et al., 2021). Furthermore, the nicotine extracted from the leaves of *N. tabacum* was combined with zinc and examined for its ability to inhibit the growth of ten distinct types of bacteria, including both Gram-negative and Gram-positive species (Al-Lahham et al., 2020). The findings indicated combining nicotine and zinc complexes is more effective against bacterial strains than zinc metal salts employed for complexation and nicotine in isolation (Akinpelu & Obuotor, 2000; Prommaban et al., 2022). The investigation examined the antibacterial efficacy of the *N. tabacum* extract against three bacterial species: *Staphylococcus aureus*, *Pseudomonas aeruginosa*, and *Escherichia coli*. The results indicated that tobacco hindered the development of the bacteria, with *S. aureus* exhibiting the largest inhibition zone (26 mm) at the greatest concentration of the plant extract (Malik et al., 2015; Sharma et al., 2016). Comparatively, the lowest dose of plant extract had a little inhibitory effect with a diameter of 11 mm, while *E. coli* demonstrated the maximum inhibition at 20 mm and the lowest inhibition at 13 mm (Ameya et al., 2017). An investigation was conducted to assess the antibacterial properties and analyze the phytochemical composition of *N. tabacum* extract in various solvents against nine pathogenic bacteria.

43.12.3 Antiviral Properties

Studies have revealed the antiviral activity of several compounds derived from *N. tabacum*. The research examined the antiviral properties of flavones derived from the leaves of *N. tabacum*. The leaves of *N. tabacum* yielded three newly discovered flavones, named taba-flavones A (1–3), along with three previously identified flavones (4–6). Compounds 1–6 were assessed for their efficacy against tobacco *mosaic virus* (TMV) infection (Shen et al., 2016). Compound 1 demonstrated a modest level of anti-TMV activity, as shown by an inhibition rate of 29.2%, similar to that of the positive control (Yuan et al., 2015). The other drugs exhibited anti-TMV activity, with inhibition rates ranging from 14.2% to 20.8%. In addition, the antiviral properties of sesquiterpenes extracted from the leaves of *N. tabacum* have been investigated. This study identified three previously unreported sesquiterpenes, tabasequiterpenes A-C (1–3), with two new chemical structures (Shan-Zhai et al., 2016). Additionally, three known sesquiterpenes (3–5) were also isolated from the leaves of *N. tabacum* (S.-Z. Shang et al., 2015). The compounds 1–6 were assessed for their efficacy in inhibiting the TMV. Compound 2 had significant anti-TMV action, with an inhibition rate of 35.2%, surpassing that of the positive control (S.-Z. Shang et al., 2015). The other compounds had promising anti-TMV actions, with inhibition rates ranging from 20.5% to 28.6% (G.-Y. Yang et al., 2022). Furthermore, the experiment examined other plants, such as *N. tabacum*, as potential reservoirs of natural antiviral substances and oral immunizations against the COVID-19 Coronavirus (Mamedov et al., 2021). The assessment considers plant species, such as N. tabacum, viable candidates for antiviral medicines against the novel Coronavirus.

43.12.4 Antioxidant Activities

The antioxidant activities of flavonoids and polysaccharides derived from the leaves of *N. tabacum* were assessed using DPPH, ABTS, and reducing power assays. Tobacco leaves flavonoids shown similar capacities to ascorbic acid in scavenging superoxide anion, DPPH, and ABTS radicals at high doses (600 µg/mL) (Zou et al., 2021). Moreover, flavonoids were seen to have a notable influence on the antioxidant capacity, comparable to that of ascorbic acid, and were much greater than polysaccharides (Al-Lahham et al., 2020). In addition, the antioxidant activity of the *N. tabacum* stem extracts were assessed by measuring the levels of superoxide dismutase, catalase, glutathione, glutathione S transferase, and lipid peroxidase. The methanolic extract exhibited significant antioxidant action (Putri et al., 2023). The antioxidant properties of polyphenols derived from tobacco leaf were verified using several tests, including DPPH free radical scavenging (with an IC50 value of 5.02 µg/mL), hydroxyl radical scavenging (with an IC50 value of 49.6 µg/mL), superoxide anion radical scavenging (with an IC50 value of 44.0 µg/mL), and suppression of lipid peroxidation (with an IC50 value of 132 µg/mL) (Sharma et al., 2016). The antioxidant properties of polyphenolic flavonoids in the ethanolic extract of *N. tabacum* stem were investigated using DPPH free radical scavenging, enzymatic biochemical test, and nonenzymatic biochemical assay (measuring GSH level and lipid peroxidation) (Gratão et al., 2008; Ru et al., 2012). The extract exhibited antioxidant activity, with flavonoid and phenolic content measuring 12.5±0.1322 mg QE/g and 1133.25±0.02 mg QE/g in the section. Additionally, the ethanolic extract demonstrated the greatest levels of SOD, GST, and GSH, as well as MDA content (Prommaban et al., 2022).

43.12.5 Antiparasitic Effect

The efficacy of a crude aqueous methanol extract from *N. tabacum* leaves in combating oxfendazole-resistant *Haemonchus contortus* in sheep was assessed using a hatch assay, adult motility test, and fecal egg count reduction test (Ali Esmail Al-Snafi, 2022). The extract exhibited nematocidal activity that varied based on the dosage and duration, as seen by LC50 values of 0.566 and 1.91 mg/mL in the egg hatch assay and adult motility test, respectively (Nouri et al., 2016). The anthelmintic activity of extracts from *N. tabacum* leaves namely the crude aqueous and methanol extracts, was investigated utilizing both in vitro and in vivo models (Forouzan et al., 2012; Iqbal et al., 2006). Both extracts induced paralysis and death of worms, which was seen 6 hours after treatment in vitro. In research conducted in vivo, sheep spontaneously infected with mixed gastrointestinal nematodes were given escalating dosages (1.0–3.0 g/kg) of both extracts (Rakhshandehroo et al., 2015). The study examined the ovicidal and adulticidal activities of extracts from *N. tabacum* leaves, *Allium sativum* bulbs, soft soap, and their combinations against *Tetranychus urticae*. According to the findings, the tobacco leaf extract, mild soap, and the combo of garlic extract and soap were shown to be the most harmful to adult females. The acaricidal efficacy of acetone and aqueous extracts derived from the leaves of *N. tabacum* and deltamethrin were evaluated against *Rhipicephalus microplus larvae* using the larval packet test (Hamad & Iqbal, 2013). The larvicidal efficacy of leaf extracts from *N. tabacum* was examined against the larvae of *Anopheles* and *Culex mosquitoes*. The application of tobacco leaf extract resulted in a complete death rate of 100% when administered at doses of 80% and 100%. An investigation assessed the efficacy of a nano-emulsion derived from tobacco extract against *Aedes aegypti larvae*. The bioassay conducted on the larvicidal nano-emulsion demonstrated that the reduction in particle size directly corresponded to a drop in LC50 values. The LC50 values were minimized when the average particle size was 631 nm. The larvicidal action of tobacco is ascribed to its nicotine and other harmful constituents (Weber et al., 2019).

43.12.6 Nervous System Activity

N. tabacum extracts have diverse effects on the central nervous system. The primary chemical ingredient in *N. tabacum* is nicotine alkaloid, which exhibits many pharmacological effects, such as central nervous system activation (Ali Esmail Al-Snafi, 2022). The results of nicotine arise from the combination of activities at smooth muscle, ganglionic sites, and motor end plates. The central nervous system is first stimulated at low dosages, causing tremors. However, at large levels, it transitions to depression and, ultimately, at toxic concentrations, may lead to convulsions. Nicotine prominently stimulates breathing and triggers vomiting via both central and peripheral mechanisms (Balfour & Fagerström, 1996). The main locations where nicotine exerts its effects in the central nervous system are the pre-junctional sites, which stimulate the release of other neurotransmitters. The stimulating and pleasurable effects of nicotine occur due to the release of excitatory amino acids, dopamine, and other biogenic amines from different regions in the central nervous system (Rawat et al., 2013). The pharmacological effects of *N. tabacum* mostly arise from its nicotine content, which activates nicotine receptors and triggers the release of many chemicals, including dopamine, acetylcholine, serotonin, vasopressin, growth hormone, and norepinephrine. *N. tabacum* has significant effects on the peripheral nervous system (PNS). Nicotine mostly induces a temporary increase in activity followed by a longer-lasting decrease in the functioning of all independent ganglia (Jassbi et al., 2017). Nicotine has a biphasic effect on the adrenal medulla, wherein low levels stimulate the release of catecholamines, while high doses inhibit their firing in response to splanchnic nerve stimulation (Gray et al., 1974; Udoye & Okeke, 2021). Studies have shown that nicotine, similar to acetylcholine, can activate several sensory receptors.

43.12.7 Analgesic Properties

The analgesic effects of the methanolic leaf extract (administered orally at doses of 100, 200, and 300 mg/kg bw) of *N. tabacum* were assessed in mice utilizing tail immersion, hot plate, and acetic acid-induced abdominal constrictions or writhing tests (Ali Esmail Al-Snafi, 2022). Extracts from *N. tabacum* had a notable analgesic effect, as shown by a dose-dependent increase in the pain reaction time or latency period in mice. The methanolic extract derived from the leaves of *N. tabacum* showed a notable reduction in the average total count of abdominal constrictions or writhes in a manner that was dependent on the dosage (C. Yang et al., 2021). Additionally, the extract exhibited an improved percentage of protection against the abdominal constriction reflex, particularly at a dosage of 300 mg/kg (Iqbal et al., 2006). The methanolic leaf extract of *N. tabacum* demonstrated a notable analgesic effect, which may be ascribed to both central and peripheral nociceptive pathways (Kishore, 2014; Udoye & Okeke, 2021).

43.12.8 Antidiabetic Effect

The hypoglycemic effects of the hydroethanolic leaf extract of *N. tabacum* were assessed by an oral glucose tolerance test and observing normoglycemic rats (Emordi et al., 2015). The single-dose trial showed that administering 40 and 80 mg/kg bw of the extract resulted in a substantial ($P < 0.05$) reduction in blood glucose levels at 2 h compared to the control group (Kazeem et al., 2014). The reduction was 27.35% and 28.37% respectively (Oguntibeju, 2020). Similarly, after 6 h, the extract caused a decrease of 75.40% compared to the control group. The oral glucose tolerance test also showed that the extract resulted in a noteworthy decrease ($P < 0.05$) in blood glucose levels (Chittasupho et al., 2023).

43.12.9 Antinociceptive Activity

An antinociceptive quality refers to the ability of a chemical or medicine to inhibit the perception of a painful or harmful stimulation by the sensory nerve system (Rawat et al., 2013; Udoye & Okeke, 2021). *N. tabacum* has an active compound called nicotine, which has antinociceptive properties. Nevertheless, the investigation assessed the analgesic effects of the methanolic leaf extract of *N. tabacum* in albino Wister mice utilizing hot plate, tail immersion, and acetic acid-induced aberrant contractions or writhing models (Kishore, 2014). The findings demonstrated that *N. tabacum* extracts had a substantial amount of antinociceptive activity by substantially augmenting the pain reaction time or latent period in mice in a dose-dependent fashion (Ali Esmail Al-Snafi, 2022). The methanolic leaf extract of *N. tabacum* exhibited notable antinociceptive activity, perhaps exerting its effects via both central and peripheral nociceptive processes (Ezeja & Omeh, 2010).

43.12.10 Antiarthritic Effect

Studies have shown that the components derived from *N. tabacum* had antiarthritic properties. Studies have investigated the antiarthritic efficacy of many medicinal plants, including *N. tabacum* (Choudhary et al., 2015; Obinna et al., 2022). These studies have focused on 485 plant species from 100 different families. The findings demonstrated that the leaf extract of *N. tabacum* had an antiarthritic action, hence providing a rationale for its conventional use in treating arthritis (Udoye & Okeke, 2021).

43.12.11 Other Activities

Administration of aqueous extract of *N. tabacum* at doses of 20 and 30 mg/kg for 21 days in rats resulted in a notable reduction in sperm motility and concentration and a drop in spermatogenic cells seen in the histological sections of the testis (Gambo et al., 2013). A clinical experiment investigated the effects of a mouthwash made from *N. tabacum* leaves on mild recurrent aphthous (Yusuf et al., 2023). The trial was randomized, double-blinded, and placebo-controlled. The treatment group saw a substantial drop in ulcer pain scores by 79.2% and 93.8% on days 3 and 5, respectively (Chien et al., 2018). Additionally, ulcer size was reduced by 69.1% and 92.2% at the same time points. These reductions were considerably larger than those in the control group ($P < 0.01$) (Kusumawardani et al., 2020). There were no observed instances of mild or serious side effects (S. Shang et al., 2019). An isolated compound called tabsesquiterpene A, which belongs to the sesquiterpene class, was extracted from the leaves of *N. tabacum*. Its cytotoxic activity was evaluated against five different cell lines using the MTT technique (Zhou et al., 2015). The compound displayed moderate cytotoxicity against PC3, MCF7, and SHSY5Y cells, with IC50 values of 4.5, 8.8, and 6.9μM, respectively (Al-Lahham et al., 2020). Alzheimer's disease is a degenerative condition that gradually impairs memory and other crucial cognitive abilities. Alzheimer's disease is the primary cause of dementia, characterized by memory loss and impaired judgment. Extracts from the *N. tabacum* plant have anti-Alzheimer's properties and enhance memory function (Kishore, 2014; López-Arrieta & Sanz, 2001). Furthermore, the research indicated that extracts derived from tobacco plants had anti-Alzheimer's properties since these extracts improved memory function (Zhao et al., 2012).

43.13 CLINICAL USES

It is crucial to specify that the medicinal applications of *N. tabacum* are restricted, and the plant is mostly recognized for its connection with health hazards, especially when consumed via tobacco smoking (Maodzeka et al., 2017). Tobacco use is a prominent contributor to avoidable mortality and is associated with a multitude of health complications, such as cardiovascular ailments, respiratory conditions, and diverse forms of cancer (Maodzeka et al., 2017). Certain chemicals included in tobacco, such as nicotine, have been extensively researched for their possible medicinal uses (Charlton, 2004). Research has examined the impact of nicotine on cognitive function and its potential use in neurodegenerative disorders such as Alzheimer's disease. Nevertheless, some chemicals present in tobacco have been examined for their potential in medicinal uses (Dewey & Xie, 2013; Rawat et al., 2013; Udoye & Okeke, 2021). It is essential to acknowledge that these prospective uses are now being read, and the hazards linked to tobacco use should be thoroughly evaluated. Some study fields have investigated the components of *N. tabacum*, namely nicotine. Tobacco plants have been used in biotechnology to produce certain proteins and vaccinations (Mason & Arntzen, 1995; Tiwari et al., 2009). Scientists have devised techniques to use genetically engineered tobacco plants to synthesize proteins that may be employed in the creation of vaccinations (Ma et al., 2005). Furthermore, scientists have investigated the advancement of medications that imitate or engage with nicotine receptors for medicinal intentions (Ahmad et al., 2012). These medications are now under investigation for a range of illnesses, such as neurodegenerative disorders and mental problems.

43.14 CONCLUSION

N. tabacum L. is a perennial herbaceous plant species. It is exclusively cultivated and widely grown among all plants in the *Nicotiana* genus. Its leaves are commercially cultivated in several nations for tobacco production. Various plant components, including leaves, roots, stems, flowers, fruits, and seeds, have been associated with these functions. However, the leaf is considered the most crucial portion. *N. tabacum* has a range of bioactive components. However, nicotine, a pyridine alkaloid, is all plant sections' primary active chemical molecule. Nicotine is the main ingredient responsible for most of the plant's pharmacological actions and medical applications. Other prominent phytochemical components in *N. tabacum* include flavonoids, phenols, glycosides, steroids, terpenoids, saponins, tannins, cardiac glycosides, organic acids, essential oils, etc. It displays several pharmacological effects that may be attributed to its component phytochemicals. Moreover, numerous medicinal uses of *N. tabacum* have also been demonstrated. Nevertheless, the main component analysis was suggested for further investigation. Furthermore, it is necessary to conduct an additional investigation to evaluate the safety of *N. tabacum* extracts, specifically in terms of their pharmacological uses, cytotoxicity, and irritating potential.

REFERENCES

Ahmad, P., Ashraf, M., Younis, M., Hu, X., Kumar, A., Akram, N. A., & Al-Qurainy, F. (2012). Role of transgenic plants in agriculture and biopharming. *Biotechnology Advances*, *30*(3), 524–540. https://doi.org/10.1016/j.biotechadv.2011.09.006

Akinpelu, D. A., & Obuotor, E. M. (2000). Antibacterial activity of Nicotiana tabacum leaves. *Fitoterapia*, *71*(2), 199–200. https://doi.org/10.1016/S0367-326X(99)00148-3

Ali Esmail Al-Snafi. (2022). Pharmacological and toxicological effects of Nicotiana tabacum. *World Journal of Advanced Pharmaceutical and Medical Research*, *3*(1), 006–018. https://doi.org/10.53346/wjapmr.2022.3.1.0034

Al-Lahham, S., Sbieh, R., Jaradat, N., Almasri, M., Mosa, A., Hamayel, A., & Hammad, F. (2020). Antioxidant, antimicrobial and cytotoxic properties of four different extracts derived from the roots of Nicotiana tabacum L. *European Journal of Integrative Medicine*, *33*, 101039. https://doi.org/10.1016/j.eujim.2019.101039

Ameya, G., Manilal, A., & Merdekios, B. (2017). In vitro antibacterial activity and phytochemical analysis of Nicotiana tabacum L. extracted in different organic solvents. *The Open Microbiology Journal*, *11*, 352.

Anderson, E. F. (1993). Plants and people of the Golden Triangle. Ethnobotany of the hill tribes of northern Thailand. *Plants and People of the Golden Triangle. Ethnobotany of the Hill Tribes of Northern Thailand.* https://www.cabdirect.org/cabdirect/abstract/19940300925

Arslan, B., & Okumus, A. (2006). Genetic and geographic polymorphism of cultivated tobaccos (Nicotiana tabacum) in Turkey. *Russian Journal of Genetics*, *42*(6), 667–671. https://doi.org/10.1134/S1022795406060123

Bakht, J., Azra, A., & Shafi, M. (2012). *Antimicrobial activity of nicotiana tabacum using different solvents extracts.* http://142.54.178.187:9060/xmlui/handle/123456789/14439

Balfour, D. J. K., & Fagerström, K. O. (1996). Pharmacology of nicotine and its therapeutic use in smoking cessation and neurodegenerative disorders. *Pharmacology & Therapeutics*, *72*(1), 51–81. https://doi.org/10.1016/S0163-7258(96)00099-X

Belhaj, S., Chaachouay, N., & Zidane, L. (2021). Ethnobotanical and toxicology study of medicinal plants used for the treatment of diabetes in the High Atlas Central of Morocco. *Journal of Pharmacy & Pharmacognosy Research*, *9*(5), 619–662. https://doi.org/10.56499/jppres21.1014_9.5.619

Belliard, G., Pelletier, G., Vedel, F., & Quetier, F. (1978). Morphological characteristics and chloroplast DNA distribution in different cytoplasmic parasexual hybrids of Nicotiana tabacum. *Molecular and General Genetics MGG*, *165*(3), 231–237. https://doi.org/10.1007/BF00332521

Benkhnigue, O., Chaachouay, N., Khamar, H., El Azzouzi, F., Douira, A, & Zidane, L. (2022). Ethnobotanical and ethnopharmacological study of medicinal plants used in the treatment of anemia in the region of Haouz-Rehamna (Morocco). *Journal of Pharmacy & Pharmacognosy Research*, *10*(2), 279–302. https://doi.org/10.56499/jppres21.1196_10.2.279

Benkhnigue, O., Chaachouay, N., & Zidane, L. (2023). A Floristical and Ecological Study of the Medicinal Flora Used by the Local Population of the Haouz-Rehamna Region (Middle Atlantic Morocco-4). *Acta Botanica Hungarica Akadémiai Kiadó*, *65*(3–4), 247–304. https://doi.org/10.1556/034.65.2023.3-4.3

Bennett, B. C. (1992). Hallucinogenic plants of the Shuar and related indigenous groups in Amazonian Ecuador and Peru. *Brittonia*, *44*(4), 483–493. https://doi.org/10.2307/2807199

Berlowitz, I., Torres, E. G., Walt, H., Wolf, U., Maake, C., & Martin-Soelch, C. (2020). 'Tobacco Is the Chief Medicinal Plant in My Work': Therapeutic Uses of Tobacco in Peruvian Amazonian Medicine Exemplified by the Work of a Maestro Tabaquero. *Frontiers in Pharmacology*, *11*. https://www.frontiersin.org/articles/10.3389/fphar.2020.594591

Bidarigh Fard, A., Dehghan Nayeri, F., & Habibi Anbuhi, M. (2019). Transient expression of etanercept therapeutic protein in tobacco (Nicotiana tabacum L.). *International Journal of Biological Macromolecules*, *130*, 483–490. https://doi.org/10.1016/j.ijbiomac.2019.02.153

Bussmann, R. W., Malca, G., Glenn, A., Sharon, D., Nilsen, B., Parris, B., Dubose, D., Ruiz, D., Saleda, J., Martinez, M., Carillo, L., Walker, K., Kuhlman, A., & Townesmith, A. (2011). Toxicity of medicinal plants used in traditional medicine in Northern Peru. *Journal of Ethnopharmacology*, *137*(1), 121–140. https://doi.org/10.1016/j.jep.2011.04.071

Carod-Artal, F. J. (2015). Hallucinogenic drugs in pre-Columbian Mesoamerican cultures. *Neurología (English Edition)*, *30*(1), 42–49. https://doi.org/10.1016/j.nrleng.2011.07.010

Chaachouay, N., Benkhnigue, O., Fadli, M., Ibaoui, H., & Zidane, L. (2019a). Ethnobotanical and ethnopharmacological studies of medicinal and aromatic plants used in the treatment of metabolic diseases in the Moroccan Rif. *Heliyon*, *5*(10), e02191. https://doi.org/10.1080/10477845.2021.1942433

Chaachouay, N., Benkhnigue, O., Fadli, M., Ibaoui, H., El Ayadi, R., & Zidane, L. (2019b). Ethnobotanical and ethnopharmacological study of medicinal and aromatic plants used in the treatment of respiratory system disorders in the Moroccan Rif. *Ethnobotany Research and Applications*, *18*(22), 1–16.

Chaachouay, N., Douira, A., Hassikou, R., Brhadda, N., Dahmani, J., Belahbib, N., Ziri, R., & Zidane, L. (2020a). Ethnobotanical and ethnopharmacological study of medicinal and aromatic plants used in the treatment of respiratory system disorders in the Moroccan Rif. PhDThesis, Département de Biologie - Université Ibn Tofail - Kénitra, Maroc https://tel.archives-ouvertes.fr/tel-03376377

Chaachouay, N., Benkhnigue, O., & Zidane, L. (2020b). Ethnobotanical study aimed at investigating the use of medicinal plants to treat nervous system diseases in the Rif of Morocco. *Journal of Chiropractic Medicine*, *19*(1), 70–81. https://doi.org/10.1016/j.jcm.2020.02.004

Chaachouay, N., & Zidane, L. (2021). The Symbolic Efficacy of Plants in Rituals and Socio-Religious Ceremonies in Morocco, Northwest of Africa. *Journal of Religious & Theological Information*, *21*(1–2), 34–53. https://doi.org/10.1080/10477845.2021.1942433

Chaachouay, N., Benkhnigue, O., Douira, A., & Zidane, L. (2021a). Poisonous medicinal plants used in the popular pharmacopoeia of the Rif, northern Morocco. *Toxicon*, 189, 24–32. https://doi.org/10.1016/j.toxicon.2020.10.028

Chaachouay, N., Douira, A., & Zidane, L. (2021b). COVID-19, prevention and treatment with herbal medicine in the herbal markets of Salé Prefecture, North-Western Morocco. *European Journal of Integrative Medicine*, 42, 101285. https://doi.org/10.1016/j.eujim.2021.101285

Chaachouay, N., Douira, A., & Zidane, L. (2021c). Herbal Medicine Used in the Treatment of Human Diseases in the Rif, Northern Morocco. *Arabian Journal for Science and Engineering*, 47, 131–153. https://doi.org/10.1007/s13369-021-05501-1

Chaachouay, N., Benkhnigue, O., & Zidane, L. (2022a). Ethnobotanical and Ethnomedicinal study of medicinal and aromatic plants used against dermatological diseases by the people of Rif, Morocco. *Journal of Herbal Medicine*, 32, 100542. https://doi.org/10.1016/j.hermed.2022.100542

Chaachouay, N., Azeroual, A., Bencharki, B., Douira, A., & Zidane, L. (2022b). Ethnoveterinary medicines plants for animal therapy in the Rif, North of Morocco. *South African Journal of Botany*, *147*, 176–191. https://doi.org/10.1016/j.sajb.2021.12.037

Chaachouay, N., Azeroual, A., Douira, A., & Zidane, L. (2022c). Ethnoveterinary practices of medicinal plants among the Zemmour and Zayane tribes, Middle Atlas, Morocco. *South African Journal of Botany*, *151*, 826–840. https://doi.org/10.1016/j.sajb.2022.11.009

Chambers, K. L. (2002). *Novel analytical methods for the investigation of nicotine metabolism*. Sheffield Hallam University (United Kingdom). https://search.proquest.com/openview/c76af6d46b4d54440c578218e7bccc2e/1?pq-origsite=gscholar&cbl=2026366&diss=y

Charlton, A. (2004). Medicinal uses of Tobacco in History. *Journal of the Royal Society of Medicine*, *97*(6), 292–296. https://doi.org/10.1177/014107680409700614

Chen, Z., Tan, J., Yang, G., Miao, M., Chen, Y., & Li, T. (2012). Isoflavones from the roots and stems of Nicotiana Tabacum and their anti-tobacco mosaic virus activities. *Phytochemistry Letters*, *5*(2), 233–235. https://doi.org/10.1016/j.phytol.2012.01.001

Chien, S.-K., Chen, L.-C., Huang, H.-C., Chen, L.-C., Hsiao, J.-W., Cheng, M.-J., & Chen, J.-J. (2018). A New Flavone and Cytotoxic Constituents of Nicotiana tabacum. *Chemistry of Natural Compounds*, *54*(6), 1044–1047. https://doi.org/10.1007/s10600-018-2551-4

Chittasupho, C., Chaobankrang, K., Sarawungkad, A., Samee, W., Singh, S., Hemsuwimon, K., Okonogi, S., Kheawfu, K., Kiattisin, K., & Chaiyana, W. (2023). Antioxidant, Anti-Inflammatory and Attenuating Intracellular Reactive Oxygen Species Activities of Nicotiana tabacum var. Virginia Leaf Extract Phytosomes and Shape Memory Gel Formulation. *Gels*, *9*(2), Article 2. https://doi.org/10.3390/gels9020078

Choudhary, M., Kumar, V., Malhotra, H., & Singh, S. (2015). Medicinal plants with potential anti-arthritic activity. *Journal of Intercultural Ethnopharmacology*, *4*(2), 147–179. https://doi.org/10.5455/jice.20150313021918

Cooper, J. I., & Harrison, B. D. (1973). The role of weed hosts and the distribution and activity of vector nematodes in the ecology of tobacco rattle virus. *Annals of Applied Biology*, *73*(1), 53–66. https://doi.org/10.1111/j.1744-7348.1973.tb01309.x

Cui, L., Chen, H., Yuan, Y., Zhu, F., Nie, J., Han, S., Hou, H., Hu, Q., & Chen, Z. (2023b). Tracing the geographical origin of tobacco at two spatial scales by stable isotope and element analyses with chemometrics. *Food Chemistry: X*, *18*, 100716.

Cui, L.-L., Chen, H., Chen, Z.-P., Yuan, Y.-W., Han, S.-L., Fu, Y.-N., Hou, H.-W., & Hu, Q.-Y. (2023a). Geographical origin classification of tobacco by stable isotope and multi-elemental analysis in combination with chemometric methods. *Microchemical Journal*, *193*, 109163.

Degenhardt, L., Chiu, W.-T., Sampson, N., Kessler, R. C., Anthony, J. C., Angermeyer, M., Bruffaerts, R., De Girolamo, G., Gureje, O., & Huang, Y. (2008). Toward a global view of alcohol, tobacco, cannabis, and cocaine use: Findings from the WHO World Mental Health Surveys. *PLoS Medicine*, *5*(7), e141.

Demole, E., & Berthet, D. (1972). A Chemical Study of *Burley* Tobacco Flavour (*Nicotiana tabacum* L.). I. Volatile to medium-volatile constituents (b.p. ≦ 84°/0.001 Torr). *Helvetica Chimica Acta*, *55*(6), 1866–1882. https://doi.org/10.1002/hlca.19720550603

Dewey, R. E., & Xie, J. (2013). Molecular genetics of alkaloid biosynthesis in Nicotiana tabacum. *Phytochemistry*, *94*, 10–27.

Emordi, J. E., Agbaje, E. O., Oreagba, I. A., Iribhogbe, O. I., & Ota, D. A. (2015). Evaluation of hypoglycemic activities of hydroethanolic leaf extract of Nicotiana tabacum (Solanaceae). *International Journal of Herbs and Pharmacological Research*, *4*(2), Article 2.

Ezeja, M. I., & Omeh, Y. S. (2010). Anti-nociceptive activities of the methanolic leaf extract of Nicotiana tabacum (Linn). *Continental Journal of Pharmacology and Toxicology Research*, *3*, 5–10.

Forouzan, S., Bahmani, M., Parsaei, P., Mohsenzadegan, A., Gholami-Ahangaran, M., Sadeghi, E., Saki, K., & Delirrad, M. (2012). Anti-Parasitic Activites of Zingiber officinale Methanolic Extract on Limnatis nilotica. *Global Veterinaria*, *9*, 144–148. https://doi.org/10.5829/idosi.gv.2012.9.2.63134

Frega, N., Bocci, F., Conte, L. S., & Testa, F. (1991). Chemical composition of tobacco seeds (*Nicotiana tabacum* L.). *Journal of the American Oil Chemists' Society*, *68*(1), 29–33. https://doi.org/10.1007/BF02660305

Fu, B., Ji, X., Zhao, M., He, F., Wang, X., Wang, Y., Liu, P., & Niu, L. (2016). The influence of light quality on the accumulation of flavonoids in tobacco (Nicotiana tabacum L.) leaves. *Journal of Photochemistry and Photobiology B: Biology*, *162*, 544–549.

Gambo, I. M., Galam, N. Z., Adamu, G., Bello, N., Rabiu, A. M., & Odeh, S. O. (2013). *The Effect of Aqueous Leave Extract of Nicotiana Tabacum (Tobacco) On Some Reproductive Parameters and Micro-Anatomical Architecture Of The Testis In Male Albino Wistar Rats*.

Gossart, N., Berhin, A., Sergeant, K., Alam, I., André, C., Hausman, J.-F., Boutry, M., & Hachez, C. (2023). Engineering Nicotiana tabacum trichomes for triterpenic acid production. *Plant Science*, *328*, 111573.

Gratão, P. L., Pompeu, G. B., Capaldi, F. R., Vitorello, V. A., Lea, P. J., & Azevedo, R. A. (2008). Antioxidant response of Nicotiana tabacum cv. Bright Yellow 2 cells to cadmium and nickel stress. *Plant Cell, Tissue and Organ Culture*, *94*(1), 73–83. https://doi.org/10.1007/s11240-008-9389-6

Gray, J. C., Kung, S. D., Wildman, S. G., & Sheen, S. J. (1974). Origin of Nicotiana tabacum L. detected by polypeptide composition of Fraction I protein. *Nature*, *252*(5480), 226–227.

Groark, K. P. (2010). The Angel in the Gourd:Ritual, Therapeutic, and Protective Uses of Tobacco (Nicotiana Tabacum) Among the Tzeltal and Tzotzil Maya of Chiapas, Mexico. *Journal of Ethnobiology*, *30*(1), 5–30. https://doi.org/10.2993/0278-0771-30.1.5

Grünweller, S., Schröder, E., & Kesselmeier, J. (1990). Biological activities of furostanol saponins from Nicotiana tabacum. *Phytochemistry*, *29*(8), 2485–2490.

Hagstad, S., Bjerg, A., Ekerljung, L., Backman, H., Lindberg, A., Rönmark, E., & Lundbäck, B. (2014). Passive Smoking Exposure Is Associated With Increased Risk of COPD in Never Smokers. *Chest*, *145*(6), 1298–1304. https://doi.org/10.1378/chest.13-1349

Hamad, K. K., & Iqbal, Z. (2013). Antinematicidal Activity of Nicotiana tabacum L. Leaf Extracts to Control

Benzimidazole- Resistant Haemonchus contortus in Sheep. *Pak Vet J.*

Hassan-Abdallah, A., Merito, A., Hassan, S., Aboubaker, D., Djama, M., Asfaw, Z., & Kelbessa, E. (2013). Medicinal plants and their uses by the people in the Region of Randa, Djibouti. *Journal of Ethnopharmacology*, *148*(2), 701–713.

He, X.-F., Hou, X.-D., Ren, X., Guo, K., Li, X.-Z., Yan, Z.-Q., Du, Y.-M., Zhang, Z.-F., & Qin, B. (2016). Two new cembranic diterpenoids from the flowers of Nicotiana tabacum L. *Phytochemistry Letters*, *15*, 238–244.

Hong, L., Wang, W., Su, Y., Zhang, G., Su, Y., Zhang, C., Chen, J., Zhe, W., Liu, Z., & Cui, J. (2022). Identifying the geographical origin of tobacco leaf by strontium and lead isotopic with mineral elemental fingerprint. *International Journal of Chemical Engineering*, *2022*. https://www.hindawi.com/journals/ijce/2022/5949770/

Hovell, M. F., & Hughes, S. C. (2009). The behavioral ecology of secondhand smoke exposure:A pathway to complete tobacco control. *Nicotine & Tobacco Research*, *11*(11), 1254–1264.

Huang, H., Qi, S.-D., Qi, F., Wu, C.-A., Yang, G.-D., & Zheng, C.-C. (2010). NtKTI1, a Kunitz trypsin inhibitor with antifungal activity from Nicotiana tabacum, plays an important role in tobacco's defense response. *The FEBS Journal*, *277*(19), 4076–4088. https://doi.org/10.1111/j.1742-4658.2010.07803.x

Iqbal, Z., Lateef, M., Jabbar, A., Ghayur, M. N., & Gilani, A. H. (2006). In vitro and In vivo anthelmintic activity of Nicotiana tabacum L. leaves against gastrointestinal nematodes of sheep. *Phytotherapy Research*, *20*(1), 46–48. https://doi.org/10.1002/ptr.1800

ITIS. (2022). Integrated Taxonomic Information System (ITIS) online database, www.itis.gov, CC0 https://doi.org/10.5066/F7KH0KBK. http://itis-web-prod.snafu.cr.usgs.gov

Jassbi, A. R., Zare, S., Asadollahi, M., & Schuman, M. C. (2017). Ecological Roles and Biological Activities of Specialized Metabolites from the Genus Nicotiana. *Chemical Reviews*, *117*(19), 12227–12280. https://doi.org/10.1021/acs.chemrev.7b00001

Kaushik, S., Sharma, P., Jain, A., & Sikarwar, M. S. (2010). Preliminary phytochemical screening and HPTLC fingerprinting of Nicotiana tabacum leaf. *Journal of Research in Pharmacy*, *3*(5), 144.

Kawashima, N., Tanabe, Y., & Iwai, S. (1976). Origin of Nicotiana tabacum detected by primary structure of fraction I protein. *Biochimica et Biophysica Acta (BBA)-Protein Structure*, *427*(1), 70–77.

Kazeem, M. I., Ogungbe, S. M., Saibu, G. M., & Aboyade, O. M. (2014). In vitro study on the hypoglycemic potential of Nicotiana tabacum leaf extracts. *Bangladesh Journal of Pharmacology*, *9*(2), Article 2.

Kenton, A., Parokonny, A. S., Gleba, Y. Y., & Bennett, M. D. (1993). Characterization of the Nicotiana tabacum L. genome by molecular cytogenetics. *Molecular and General Genetics MGG*, *240*(2), 159–169. https://doi.org/10.1007/BF00277053

Kesselmeier, J. (1980). Development of chloro-etioplasts containing prolamellar bodies and steroidal saponins in suspension cultures of Nicotiana tabacum. *Protoplasma*, *104*(3–4), 295–306. https://doi.org/10.1007/BF01279774

Kishore, K. (2014). Monograph of tobacco (Nicotiana tabacum). *Indian Journal of Drugs*, *2*(1), 5–23.

Knapp, S., Chase, M. W., & Clarkson, J. J. (2004). Nomenclatural changes and a new sectional classification in *Nicotiana* (Solanaceae). *TAXON*, *53*(1), 73–82. https://doi.org/10.2307/4135490

Kusumawardani, B., Febi, Q. N., Rosidah, M., Azis, D. A., Puspitasari, E., & Nugraha, A. S. (2020). Cytotoxic Potential of Flavonoid from Nicotiana tabacum Leaves on MCF-7 Human Breast Cancer Cells. *Indonesian Journal of Cancer Chemoprevention*, *11*(2), Article 2. https://doi.org/10.14499/indonesianjcanchemoprev11iss2pp97-10

Latifi, R., Parasitology Department, Qazvin University of Medical Sciences, Qazvin, Iran, Azadmehr, A., Immunology Department, Babol University of Medical Sciences, Babol, Iran & Immunology Department, Qazvin University of Medical Sciences, Qazvin, Iran, Mosalla, S., Parasitology Department, Qazvin University of Medical Sciences, Qazvin, Iran, Saleki, K., Student Research Committee, Babol University of Medical Sciences, Babol, Iran & USERN Office, Babol University of Medical Sciences, Babol, Iran, Hajiaghaee, R., & Medicinal Plants Research Center, Institute of Medicinal Plants, ACECR, Karaj, Iran. (2022). Scolicidal effects of the *Nicotiana tabacum* L. extract at various concentrations and exposure times. *Journal of Medicinal Plants*, *21*(82), 111–118. https://doi.org/10.52547/jmp.21.82.111

Lei, C., Xu, W., Wu, J., Wang, S., Sun, J.-Q., Chen, Z., & Yang, G. (2015). Coumarins from the Roots and Stems of Nicotiana tabacum and their Anti-Tobacco Mosaic Virus Activity. *Chemistry of Natural Compounds*, *51*(1), 43–46. https://doi.org/10.1007/s10600-015-1199-6

Lewis, R. S., & Nicholson, J. S. (2007). Aspects of the evolution of Nicotiana tabacum L. and the status of the United States Nicotiana Germplasm Collection. *Genetic Resources and Crop Evolution*, *54*(4), 727–740. https://doi.org/10.1007/s10722-006-0024-2

Lisuma, J. B., Mbega, E. R., & Ndakidemi, P. A. (2019). Dynamics of nicotine across the soil–tobacco plant interface is dependent on agro-ecology, nitrogen source, and rooting depth. *Rhizosphere*, *12*, 100175.

Liu, T., Gu, Y., Zhou, Z., Liu, Z., Yin, H., Qin, C., Yi, T., & Tao, J. (2021). Ecological strategies of biological and chemical control agents on wildfire disease of tobacco (Nicotiana tabacum L.). *BMC Microbiology*, *21*(1), 184. https://doi.org/10.1186/s12866-021-02237-8

López-Arrieta, J., & Sanz, F. J. F. (2001). Nicotine for Alzheimer's disease. *Cochrane Database of Systematic Reviews*, *2*. https://doi.org/10.1002/14651858.CD001749

Loughrin, J. H., Hamilton-Kemp, T. R., Andersen, R. A., & Hildebrand, D. F. (1990). Headspace compounds from flowers of Nicotiana tabacum and related species. *Journal of Agricultural and Food Chemistry*, *38*(2), 455–460. https://doi.org/10.1021/jf00092a027

Ma, J. K.-C., Barros, E., Bock, R., Christou, P., Dale, P. J., Dix, P. J., Fischer, R., Irwin, J., Mahoney, R., Pezzotti, M., Schillberg, S., Sparrow, P., Stoger, E., & Twyman, R. M. (2005). Molecular farming for new drugs and vaccines. *EMBO Reports*, *6*(7), 593–599. https://doi.org/10.1038/sj.embor.7400470

Maestri, D. M., & Guzmán, C. A. (1993). Chemical composition of tobacco seeds (*Nicotiana tabacum* L) from Argentina. *Journal of the Science of Food and Agriculture*, *61*(2), 227–230. https://doi.org/10.1002/jsfa.2740610214

Malik, R., Bokhari, T., Siddiqui, M., Younis, U., Hussain, M., & Khan, I. (2015). Antimicrobial activity of Nerium oleander L. and Nicotiana tabacum L.:A comparative study. *Pakistan Journal of Botany*, *47*, 1587–1592.

Mamedov, T., Gurbuzaslan, I., Yuksel, D., Ilgin, M., Mammadova, G., Ozkul, A., & Hasanova, G. (2021). Soluble Human Angiotensin- Converting Enzyme 2 as a Potential

Therapeutic Tool for COVID-19 is Produced at High Levels In Nicotiana benthamiana Plant With Potent Anti-SARS-CoV-2 Activity. *Frontiers in Plant Science*, *12*. https://www.frontiersin.org/articles/10.3389/fpls.2021.742875

Maodzeka, A., Hussain, N., Wei, L., Zvobgo, G., Mapodzeke, J. M., Adil, M. F., Jabeen, S., Wang, F., Jiang, L., & Shamsi, I. H. (2017). Elucidating the physiological and biochemical responses of different tobacco (Nicotiana tabacum) genotypes to lead toxicity. *Environmental Toxicology and Chemistry*, *36*(1), 175–181. https://doi.org/10.1002/etc.3522

Maroyi, A. (2013). Traditional use of medicinal plants in south-central Zimbabwe:Review and perspectives. *Journal of Ethnobiology and Ethnomedicine*, *9*(1), 1–18.

Mason, H. S., & Arntzen, C. J. (1995). Transgenic plants as vaccine production systems. *Trends in Biotechnology*, *13*(9), 388–392. https://doi.org/10.1016/S0167-7799(00)88986-6

McDermott, R. J., Vamos, C. A., & Klein, K. (2008). Tobacco Consumption and Control in Canada, Germany, and the United States. *Umwelt Und Gesundheit Online*, *1*, 111–116.

Naeem, Z. (2015). Second-hand smoke – ignored implications. *International Journal of Health Sciences*, *9*(2), V–VI.

Nouri, F., Nourollahi-Fard, S. R., Foroodi, H. R., & Sharifi, H. (2016). In vitro anthelmintic effect of Tobacco (Nicotiana tabacum) extract on parasitic nematode, Marshallagia marshalli. *Journal of Parasitic Diseases*, *40*(3), 643–647. https://doi.org/10.1007/s12639-014-0550-3

Obinna, C. N., Damilola, E. O., & Oziegbe, O. (2022). Use of Nicotiana Tabacum, Jatropha Curcas, and Ficus Exasperata for Treatment of Pus-Producing Bacteria. *IOP Conference Series: Earth and Environmental Science*, *1054*(1), 012010. https://doi.org/10.1088/1755-1315/1054/1/012010

Oguntibeju, O. O. (2020). Antidiabetic and biological activities of nicotiana tabacum: Possible mechanisms. *Asian Journal of Pharmaceutical and Clinical Research*, 1–4. https://doi.org/10.22159/ajpcr.2020.v13i3.35546

Orch, H., Chaachouay, N., Douiri, E. M., Faiz, N., Zidane, L., & Douira, A. (2021). Use of medicinal plants in dermato-cosmetology: An ethnobotanical study among the population of Izarène. *Jordan Journal of Pharmaceutical Sciences*, *14*(3), 323–340.

Panagiotakos, D. B., Pitsavos, C., Chrysohoou, C., Skoumas, J., Masoura, C., Toutouzas, P., & Stefanadis, C. (2004). Effect of exposure to secondhand smoke on markers of inflammation:The ATTICA study. *The American Journal of Medicine*, *116*(3), 145–150. https://doi.org/10.1016/j.amjmed.2003.07.019

Ponstein, A. S., Bres-Vloemans, S. A., Sela-Buurlage, M. B., van den Elzen, P. J. M., Melchers, L. S., & Cornelissen, B. J. C. (1994). A Novel Pathogen- and Wound-Inducible Tobacco (Nicotiana tabacum) Protein with Antifungal Activity. *Plant Physiology*, *104*(1), 109–118. https://doi.org/10.1104/pp.104.1.109

Popova, V., Ivanova, T., Stoyanova, A., Nikolova, V., Hristeva, T., & Zheljazkov, V. D. (2020). GC-MS composition and olfactory profile of concretes from the flowers of four Nicotiana species. *Molecules*, *25*(11), 2617.

Prabhu, A., Obi, K. O., & Rubenstein, J. H. (2014). The Synergistic Effects of Alcohol and Tobacco Consumption on the Risk of Esophageal Squamous Cell Carcinoma:A Meta-Analysis. *Official Journal of the American College of Gastroenterology | ACG*, *109*(6), 822. https://doi.org/10.1038/ajg.2014.71

Prommaban, A., Kheawfu, K., Chittasupho, C., Sirilun, S., Hemsuwimon, K., & Chaiyana, W. (2022). Phytochemical, Antioxidant, Antihyaluronidase, Antityrosinase, and Antimicrobial Properties of *Nicotiana tabacum* L. Leaf Extracts. *Evidence-Based Complementary and Alternative Medicine*, *2022*, e5761764. https://doi.org/10.1155/2022/5761764

Putri, D. A., Solihah, R., Oktavia, R., & Fatmawati, S. (2023). Phytochemical and antioxidant activity of Nicotiana tabacum extracts. *AIP Conference Proceedings*, *2642*(1). https://pubs.aip.org/aip/acp/article-abstract/2642/1/060002/2872884

Rakhshandehroo, E., Ahmadi, A., Ghane, M., Foroodi, E., & Rastegari, F. (2015). *The in vitro efficacy of Nicotiana tabacum (Tobacco) and Zataria multiflora extracts on the viability of equid cyathostome (Nematoda:Cyathostominae) infective larvae*.

Raupach, T., Schäfer, K., Konstantinides, S., & Andreas, S. (2006). Secondhand smoke as an acute threat for the cardiovascular system:A change in paradigm. *European Heart Journal*, *27*(4), 386–392. https://doi.org/10.1093/eurheartj/ehi601

Rawat, A., Mali, R. R., Saini, A. K., Chauhan, P. K., Singh, V., & Sharma, P. (2013). Phytochemical properties and pharmcological activities of Nicotiana tabacum:A review. *Indian J Pharm Biol Res*, *1*, 74–82.

Ru, Q.-M., Wang, L.-J., Li, W.-M., Wang, J.-L., & Ding, Y.-T. (2012). In vitro antioxidant properties of flavonoids and polysaccharides extract from tobacco (Nicotiana tabacum L.) leaves. *Molecules*, *17*(9), 11281–11291.

Sabanayagam, C., & Shankar, A. (2011). The association between active smoking, smokeless tobacco, second-hand smoke exposure and insufficient sleep. *Sleep Medicine*, *12*(1), 7–11. https://doi.org/10.1016/j.sleep.2010.09.002

Saffer, H., & Chaloupka, F. (2000). The effect of tobacco advertising bans on tobacco consumption. *Journal of Health Economics*, *19*(6), 1117–1137. https://doi.org/10.1016/S0167-6296(00)00054-0

Sahai, O. P., & Shuler, M. L. (1984). Environmental parameters influencing phenolics production by batch cultures of *Nicotiana tabacum. Biotechnology and Bioengineering*, *26*(2), 111–120. https://doi.org/10.1002/bit.260260202

Saitoh, F., Noma, M., & Kawashima, N. (1985). The alkaloid contents of sixty Nicotiana species. *Phytochemistry*, *24*(3), 477–480.

Sakai, K.-I., & Shimamoto, Y. (1965). Developmental instability in leaves and flowers of Nicotiana tabacum. *Genetics*, *51*(5), 801.

Sallaud, C., Giacalone, C., Töpfer, R., Goepfert, S., Bakaher, N., Rösti, S., & Tissier, A. (2012). Characterization of two genes for the biosynthesis of the labdane diterpene *Z* -abienol in tobacco (*Nicotiana tabacum)* glandular trichomes. *The Plant Journal*, *72*(1), 1–17. https://doi.org/10.1111/j.1365-313X.2012.05068.x

Sanjeev, A., & Kapgate, S. M. (2021). Tobacco (Nicotiana tabacum)-A systemic review. *Population*, *51*, 7.

Schep, L. J., Slaughter, R. J., & Beasley, D. M. G. (2009). Nicotinic plant poisoning. *Clinical Toxicology*, *47*(8), 771–781. https://doi.org/10.1080/15563650903252186

Sela-Buurlage, M. B., Ponstein, A. S., Bres-Vloemans, S. A., Melchers, L. S., van den Elzen, P. J. M., & Cornelissen, B. J. C. (1993). Only Specific Tobacco (Nicotiana tabacum) Chitinases and β-1,3-Glucanases Exhibit Antifungal Activity. *Plant Physiology*, *101*(3), 857–863. https://doi.org/10.1104/pp.101.3.857

Shang, S., Shi, J., Tang, J., Jiang, J., Zhao, W., Zheng, X., Lei, P., Han, J., Wang, C., Yuan, D., Yang, G., Chen, Y.,

& Miao,M. (2019). New isolates from leaves of Nicotiana tabacum and their biological activities. *Natural Product Research*, *33*(11), 1577–1583. https://doi.org/10.1080/14786419.2018.1425840

Shang, S.-Z., Xu, W.-X., Li, L., Tang, J.-G., Zhao, W., Lei, P., Miao, M.-M., Sun, H.-D., Pu, J.-X., Chen, Y.-K., & Yang, G.-Y. (2015). Antiviral isocoumarins from the roots and stems of Nicotiana tabacum. *Phytochemistry Letters*, *11*, 53–56. https://doi.org/10.1016/j.phytol.2014.11.008

Shan-Zhai, S., Wei, Z., Jian-Guo, T., Jian-Xin, P., Dong-Lai, Z., Liu, Y., Han-Dong, S., Guang-Yu, Y., & Yong-Kuan, C. (2016). 14-Noreudesmane sesquiterpenes from leaves of Nicotiana tabacum and their antiviral activity. *Phytochemistry Letters*, *17*, 173–176. https://doi.org/10.1016/j.phytol.2016.07.019

Sharma, Y., Dua, D., Nagar, A., & Srivastava, N. S. (2016). Antibacterial activity, phytochemical screening and antioxidant activity of stem of Nicotiana tabacum. *International Journal of Pharmaceutical Sciences and Research*, *7*(3), 1156.

Sharma, Y., Kaur, A., Bhardwaj, R., Srivastava, N., Lal, M., Madan, S., & Bala, K. (2021). Preclinical assessment of stem of Nicotiana tabacum on excision wound model. *Bioorganic Chemistry*, *109*, 104731. https://doi.org/10.1016/j.bioorg.2021.104731

Shen, Q.-P., Xu, X.-M., Li, L., Zhao, W., Xiang, N.-J., Yang, G.-Y., Chen, Y.-K., Miao, M.-M., Liu, C.-B., & Liu, Z.-H. (2016). Sesquiterpenes from the leaves of Nicotiana tabacum and their anti-tobacco mosaic virus activity. *Chinese Chemical Letters*, *27*(5), 753–756. https://doi.org/10.1016/j.cclet.2016.01.048

Shibuya, K., Ciecierski, C., Guindon, E., Bettcher, D. W., Evans, D. B., & Murray, C. J. L. (2003). WHO Framework Convention on Tobacco Control:Development of an evidence based global public health treaty. *BMJ*, *327*(7407), 154–157. https://doi.org/10.1136/bmj.327.7407.154

Siegel, R. K. (1976). Herbal Intoxication:Psychoactive Effects From Herbal Cigarettes, Tea, and Capsules. *JAMA*, *236*(5), 473–476. https://doi.org/10.1001/jama.1976.03270050029024

Singh, C. R., & Kathiresan, K. (2015). Effect of cigarette smoking on human health and promising remedy by mangroves. *Asian Pacific Journal of Tropical Biomedicine*, *5*(2), 162–167. https://doi.org/10.1016/S2221-1691(15)30337-3

Stanisavljević, I., Lakićević, S., Veličković, D., Lazić, M., & Veljković, V. (2007). The extraction of oil from tobacco (Nicotiana tabacum L.) seeds. *Chemical Industry and Chemical Engineering Quarterly/CICEQ*, *13*(1), 41–50.

Steinberg, J. F., Cristine Delnevo, Douglas M. Ziedonis, Michael B. (2007). Health Effects of Tobacco, Nicotine, and Exposure to Tobacco Smoke Pollution. In *Handbook of the Medical Consequences of Alcohol and Drug Abuse* (2e éd.). Routledge.

Tariq, M., Ahmad, Z., Shah, S. A., Gul, Z., & Khan, S. A. (2021). Phytochemical Analysis and Antibacterial Activity of Nicotiana tabacum and Nicotiana rustica. *RADS Journal of Biological Research & Applied Sciences*, *12*(1), Article 1. https://doi.org/10.37962/jbas.v12i1.317

Tiwari, S., Verma, P. C., Singh, P. K., & Tuli, R. (2009). Plants as bioreactors for the production of vaccine antigens. *Biotechnology Advances*, *27*(4), 449–467. https://doi.org/10.1016/j.biotechadv.2009.03.006

Udoye, B. O., & Okeke, S. I. (2021). Phytochemical Properties, Pharmacological Activities and Ethnomedicinal Uses of Nicotiana tabacum L. *A Comprehensive Review*, *1*, 1–11.

Upadhyay, K. M. R. H., Chamola, O. B., & Dueby, P. (2009). *Integratd pest and disease management*. APH Publishing. https://books.google.com/books?hl=fr&lr=&id=9CUD07Fcp_kC&oi=fnd&pg=PR5&dq=Nicotiana+species,+resulting+in+hybrids+exhibiting+distinct+traits.+N.+tabacum+grows+best+in+well-drained,+nutrient-rich+soil+in+full+sun+but+has+been+grown+in+the+shade+in+Java+and+Connecticut.+&ots=0CF4y0tWfw&sig=-Cmlimayw5oT5m19E_MmW99HqFs

Weber, S. S., Kaminski, K. P., Perret, J.-L., Leroy, P., Mazurov, A., Peitsch, M. C., Ivanov, N. V., & Hoeng, J. (2019). Antiparasitic properties of leaf extracts derived from selected Nicotiana species and Nicotiana tabacum varieties. *Food and Chemical Toxicology*, *132*, 110660.

Wei, X., Sumithran, S. P., Deaciuc, A. G., Burton, H. R., Bush, L. P., Dwoskin, L. P., & Crooks, P. A. (2005). Identification and synthesis of novel alkaloids from the root system of Nicotiana tabacum:Affinity for neuronal nicotinic acetylcholine receptors. *Life Sciences*, *78*(5), 495–505.

Wu, W., Tang, X.-P., Yang, C., Liu, H.-B., & Guo, N.-J. (2013). Investigation of ecological factors controlling quality of flue-cured tobacco (Nicotiana tabacum L.) using classification methods. *Ecological Informatics*, *16*, 53–61.

Xiao, Q., Zhu, Y., Cui, G., Zhang, X., Hu, R., Deng, Z., Lei, L., Wu, L., & Mei, L. (2022). A comparative study of flavonoids and carotenoids revealed metabolite responses for various flower colorations between Nicotiana tabacum L. and Nicotiana rustica L. *Frontiers in Plant Science*, *13*, 828042.

Xing, L., Yang, J., Jia, Y., Hu, X., Liu, Y., Xu, H., Yin, H., Li, J., & Yi, Z. (2021). Effects of ecological environment and host genotype on the phyllosphere bacterial communities of cigar tobacco (*Nicotiana tabacum* L.). *Ecology and Evolution*, *11*(16), 10892–10903. https://doi.org/10.1002/ece3.7861

Xu, K., Liu, Q., Ni, L., Du, Y.-M., & Zhang, Z.-F. (2020). An update on chemical constituents from Nicotiana tabacum L. And their bioactivities from 2017 to 2020. *Medicine Research*, *4*(4), 200019.

Yach, D. (2014). The origins, development, effects, and future of the WHO Framework Convention on Tobacco Control:A personal perspective. *The Lancet*, *383*(9930), 1771–1779. https://doi.org/10.1016/S0140-6736(13)62155-8

Yang, C., Xie, S.-N., Ni, L., Du, Y.-M., Liu, S., Li, M.-Y., & Xu, K. (2021). Chemical Constituents from *Nicotiana tabacum* L. and Their Antifungal Activity. *Natural Product Communications*, *16*(11), 1934578X2110595. https://doi.org/10.1177/1934578X211059578

Yang, G.-Y., Dai, J.-M., Li, Z.-J., Wang, J., Yang, F.-X., Liu, X., Li, J., Gao, Q., Li, X.-M., Li, Y.-K., Wang, W.-G., Zhou, M., & Hu, Q.-F. (2022). Isoindolin-1-ones from the stems of Nicotiana tabacum and their antiviral activities. *Archives of Pharmacal Research*, *45*(8), 572–583. https://doi.org/10.1007/s12272-022-01399-x

Yuan, L., Huang, W., Zhang, C., Xiang, N., Liu, C., Yang, G., Chen, Y., Miao, M., & Ma, Y. (2015). Antiviral flavones from the leaves of Nicotiana tabacum. *Phytochemistry Letters*, *12*, 75–78. https://doi.org/10.1016/j.phytol.2015.02.003

Yusuf, B., Halimat, L., Okeme, U., Ganiyu, A., Salawu, M., Moronfolu, I., Olakunle, O., Abdulfatah, L. B., Aiwinnilomo, M., Ibrahim, R., Abdulazeez, A., & Oloyede, H. (2023). Cow Urine Extract of Nicotiana tabacum Leaf Portends Hepatotoxicity in Male Wistar Rats. *International Journal of Biochemistry Research & Review*, 34–43. https://doi.org/10.9734/ijbcrr/2023/v32i1795

Zhao, T., Zeng, Y., & Kermode, A. R. (2012). A plant cell-based system that predicts aβ42 misfolding:Potential as a drug discovery tool for Alzheimer's disease. *Molecular Genetics and Metabolism*, *107*(3), 571–579. https://doi.org/10.1016/j.ymgme.2012.08.010

Zhou, M., Zhou, K., Lou, J., Wang, Y.-D., Dong, W., Li, G.-P., Jian, Z.-Y., Du, G., Yang, H.-Y., Li, X.-M., & Hu, Q.-F. (2015). New biphenyl derivatives from the leaves of Nicotiana tabacum and their cytotoxic activity. *Phytochemistry Letters*, *14*, 226–229. https://doi.org/10.1016/j.phytol.2015.10.025

Zou, X., Bk, A., Rauf, A., Saeed, M., Al-Awthan, Y. S. A., Al-Duais, M., Bahattab, O., Hamayoon Khan, M., & Suleria, H. A. R. (2021). Screening of Polyphenols in Tobacco (*Nicotiana tabacum*) and Determination of Their Antioxidant Activity in Different Tobacco Varieties. *ACS Omega*, *6*(39), 25361–25371. https://doi.org/10.1021/acsomega.1c03275

44 Having Hairy Flowers (*Latua pubiflora* (Griseb.) Baill. Solanaceae)

Ume Laila, Ehsan Ali, and Muhammad Farooq Azhar

44.1 INTRODUCTION

The Solanaceae, commonly recognized as the nightshade family, is famous for its narcotic, toxic, therapeutic, and various significant food plants. Several genera, including *Datura, Mandragora, Zfyoscyamus*, and *Atropa* possess alkaloids with physiological effects on humans. Native peoples from all over the world discovered these plants and their characteristics separately. Various species of *Ztafura* have been frequently utilized as drugs and poisons by the Hindus, ancient Greeks, and American Indians in both the old and modern eras of the world (Plowman et al., 1971). Many solanaceous species include belladonna alkaloids, such as hyoseyamine, scopolamine, and tropane. These chemicals have a strong influence on mental health, causing delirium, vomiting, hallucinations, and trance-like states that are frequently similar to psychosis. As a result, solanaceous substances have been commonly used in indigenous societies for witchcraft and shamanism. These species, like other hallucinogenic plants, possess greater therapeutic potential and have frequently been used for exclusively medicinal purposes due to presence of secondary metabolites (Sánchez-Montoya et al., 2017; Chaachouay & Zidane, 2021).

Latua, being an individual endemic species in Southern Chile, is one of the most unusual and intriguing Solanaceae genera with outstanding narcotic and poisonous qualities. Because of its tremendous beauty and toxicity, *Latua* piqued the curiosity of botanists and pharmacologists quickly after its findings in the middle of the nineteenth century an interest that resulted in a number of papers on the plant's qualities and effects. *Latua pubiflora* (Griseb) Phil. a natural plant, grows abundantly in Southern Chile and Cordillera de-la Costa extending from Valdivia to Llanquihue areas (38°S to 43°S at 500–900 m above sea level). *Latua pubiflora* is the individual member of the genus *Latua*, and it possesses a significant mystic role in Mapuche native culture due to its purposed spiritual, therapeutic, and sedative properties (Olivares, 1995). In reality, the common name for this species is 'Latúe' (Mapuche) or 'Palo brujo' (Spanish), which literally means 'magical stick'. This species of Solonaceae family, chemotaxonomically and has been frequently utilized in Asian and European countries (Hanus et al., 2005; Beyer et al., 2009). These highlighted species consist of atropine and scopolamine as major alkaloids that play a key role in various pharmacological properties viz; pain relief, psychoactive effects, and sedation (Ramoutsaki et al., 2002; Beyer et al., 2009; Soni et al., 2012).

Anthropological study in the areas of Chile has revealed that few Solanaceae species, including *L. pubiflora*, have been used for smoking purposes in ancient ceremonies (Echeverria and Niemeyer, 2013; Echeverria et al., 2014; Carrasco et al., 2015). *L. pubiflora* has been ingested in various forms (e.g., through cigarette smoking and infusions), which might produce drowsiness, delirium, mouth dryness, mild fever, convulsions, and pupillary dilatation (Olivares, 1995). Mapuche medicine men employ *L. pubiflora* to induce drowsiness, to achieve hysteria or mystical experience.

Silva and Mancinelli (1959) performed the experiment and analyzed the extracts of leaves and fruits of *L. pubiflora* and identified the various alkaloids including tropane, atropine, and scopolamine. Recently, Muñoz and Casale (2003) demonstrated the results of gas chromatography indicating the presence of 3-α-cinnamoiloxytropane and apoatropine in *L. pubiflora*. However, the herbal extract of this *L. pubiflora* contained various secondary metabolites including alkaloids and flavonoids (Echeverria and Niemeyer, 2012); in this study, the authors analyzed two various partitions for behavioral and pharmacological assays to check the concentration of alkaloids and flavonoids.

44.2 SCIENTIFIC CLASSIFICATION

L. pubiflora is a plant species belonging to the family Solanaceae and is classified as a single species in the genus *Latua*. The genus contains only one species; however, there is the distinction between two forms based on the presence or absence of the pronounced thorns; due to this reason, the Mapuche people of India usually make the difference between male and female plants. The taxonomic classification for *Latua pubiflora* is described in Table 44.1.

44.3 VERNACULAR NAMES

L. pubiflora is synonymously recognized as *L. pubiflora* (Griseb) Phil., *Lycioplesium pubiflorum*, and *L. venenata* Phil. The most commonly encountered vernacular name in literature is mainly árbol de-los brujos, which literally

DOI: 10.1201/9781003460336-44

means 'tree of a sorcerer'. According to the Chilean Machi the vernacular names observed consist of *palo de-bruja*, which means 'witch tree' as well as *pal-mato*, which means 'tree stick that kills', and later recognized by the word *Latua* in reference to the Mapudungun.

44.4 BOTANICAL DESCRIPTION

It is recognized as a small tree or a shrub and grows up to 4–10 meters in height. It possesses various main trunks with a diameter of 3–25 cm. It possesses a spiny stem and greenish leaves. Its branches are mainly smooth and covered with spines. Spines are usually erect and arise from leaf axils in the form of modified branches. Spines are rigid and extend up to 2 cm, mainly having small leaf-like structures at the base or a pair of small cataphylls assembled on dwarf shoots or dispersed on the long shoots. The petiole of this unique plant is mainly short (2 mm), glabrescent, and usually pilose. Stipules are not present, while the peduncle arises from the axil of the spines or from the base of the leaf. It blooms (with attractive magenta flowers mainly pollinated by hummingbirds) in early spring or late winter. It is considered a semi-deciduous shrub, at bloom stage, many of its leaves are assembled in significantly reduced size, making the flowers dominant in the center. It has beautiful magenta-colored hairy flowers (Figure 44.1) and its tomato-type pale yellow fruits are up to 2 cm which are much attractive to use it as an ornamental plant (Plowman et al., 1971). Seeds are 2 mm in diameter, uniform or irregular in shape most commonly flattened ventrally. The seed coat is thick and seeds are known as albuminous.

44.5 ORIGIN AND DISTRIBUTION

Latua was first time described by Rudolph A. Philippi (1858). During pre-spanish period, this plant was used for shamanic uses in Southern Chile. However, its use has not been observed during the colonial period. The term Latúe is a Mapudrungun word meaning 'that which causes death'. There are various rumors regarding the properties of *Latua*, especially in Chile. Most people believe that it has lethal properties, so they avoid talking about the tree – which makes it more difficult for researchers to study this plant. It has been studied that *Latua* shrub was used as fish poison in earlier times by Chilean fishers. They used to blend the bark of the canelo plant (considered a shamanic tree) with the juice of *Latua* shrub using it as fish poison (Drimys winter frost; seen incense) (Plowman et al., 1971). This plant juice, having an unpleasant taste, is still used to poison food as in cases where there is a matter of taking revenge on people having more materialistic and selfish nature that do not take care of sentiments. *Latua pubiflora* is a native species in Chile (Hoffmann, 1994); however, it has been found even in extremely delimited areas such as in the mountain ranges of Cordillera Pelada. The shrub has been found more frequently on the mountain ridges of Cordillera de San Juan la Costa in central Mapuche. *Latua* has also been reportedly native to the island of Chiloé, but it is found very rarely in this region (Plowman et al., 1971). One of the unique and rarest psychotropic plants is *Latua* pubiflora. Until today, this plant has never been cultivated in any part of the land, so it has not been dispersed by humans.

TABLE 44.1
Taxonomic Classification of *Latua pubiflora* (Griseb.) Baill.-Solanaceae

Kingdom	Plantae
Phylum	Tracheophytes
Class	Angiosperms
Subclass	Eudicots
Order	Solanales
Family	Solanaceae
Subfamily	Solanideae
Genus	*Latua*
Species	*Latua Pubiflora*

Source: (Baillon, 1888)

44.6 HABITAT AND ECOLOGY

The plant is a tall shrub that usually grows in clear areas and secondary forests. Deforestation activities have played a role in the spread and excessive growth of *Latua* in various fields and pastures. The plant is highly adapted to open conditions, acting as a weed on roadsides and open grounds. Moreover, *Latua* has the ability to grow and spread via its underground parts and adventitious branches, therefore, making it difficult to be destroyed easily through cutting techniques. It has been observed that *Latua* species may grow up to 10 meters during the growing season when they make their association with *Laurelia, Chusquea*, and *Eucryphia*. *Latua* undergoes flowering at the start of October during the rainy season and produces fruit in the months of February and March, especially in the northern areas such as Osomo and Valdivia.

In southern areas, relatively fewer seasonal changes occur; it flowers in the month of March (recognized by the name of Llanquihue) and July (Chiloé'). More than one flowering season has also been observed per year. This plant is pollinated by hummingbirds due to the presence of their tubular, reddish-colored pendulous flowers. Mariani Ramires (1965) stated that this plant is avoided by animals and birds, which is why it is now considered a folk tale.

44.7 ETHNOBOTANY

For more than 3000 years, native people have employed plant extracts rich in alkaloids to cure a variety of illnesses,

FIGURE 44.1 *Latua. pubiflora.* (Source: Baillon, 1888.)

including fever, snakebites, and insanity. Due to their vast structural variety and varied mechanisms of action, several plant alkaloids are still classified as primary plant poisons despite having major benefits for both the human and pharmaceutical industries.

Toxic alkaloids can be ingested, vaporized, or directly contacted by humans and other animals. This exposure results in a specific mechanism involving receptors, transporters, enzymes, and genetic material in particular cells or tissues, which may have hepatotoxic effects and result in musculoskeletal deformities. Numerous *Latua* plant alkaloids are highlighted, including tropane, indolizidine, pyrrolizidine, and piperidine, which negatively affect humans and animals causing nausea, gastrointestinal perturbation, itching, psychosis, itching, paralysis, arrhythmias, mild teratogenicity, and even sudden death (Chaachouay et al., 2021a, 2021b; Benkhnigue et al., 2023).

44.8 PHYTOCHEMISTRY

Sánchez-Montoya et al. (2017) performed the experiment regarding UHPLC mass spectrometry for the extract of *L. pubiflora* to assess the occurrence of secondary metabolites. According to their findings, the concentrations per g dry biomass of *L. pubiflora* assessed by using HPLC were 1260±15.20 µg for atropine, 640±18.85 µg for scopoletine, and 340±6.30 µg for scopolamine. The retention time (11.7 min) was found higher for atropine than that of scopolamine. The whole plant consists of various alkaloids including atropine, tropane, and scopolamine. The amount of atropine (21%) is higher than scopolamine (5.7%) (Sánchez-Montoya et al., 2017). The leaves contain high contents of atropine and less content of scopolamine as per findings of Bodendorf and Kummer (1962). Previously, spectrometry analyses of aerial parts (leaves) of plants showed high contents of alkaloids. However, the stem and fruits were found with a very minute quantity of alkaloids. Until now, various other analyses have suggested that the stem consists of high amounts of alkaloids but leaves and fruits have low amounts of secondary metabolites (Plowman et al., 1971).

44.9 MEDICINAL USE

Latua was taken into account as a love potion and was used as a component in love drinks (Bodendorf and Kummer 1962). Mapuche believe that a small concentration of plant aids physical strength. Thereby, their children take very small amounts of *Latua* to become healthy and strong. Mapuche shamans recommend a minute intake of the leaves and fruits to mitigate pain (Nakashima Degarrod, n.d.).

In Chile, tayu is recognized as the bark of the bush and is utilized in folk medicine as a tea brewed for curing contusions and bruises (Schultes and Raffauf, 1970). Shamans suggested *Latua*. Fabiana imbricata, and palqui as tea brewed for curative drug additive. A *Latua* can be implemented to treat arthritis, coughing, rheumatism, and pain.

The Huilliche (a Mapuche tribe) believe that *L. pubiflora* is inhabited by a spirit and is thus sacred. It is thought that the plant can bring physical strength, healing, death, and good luck due to having shamanic powers. These powers, however, are only accompanied by the female plant. Such kind of female plant usually produces fruits and has thorns, which are lacking in the male plant. Only the female plant is used by the Huilliche. It is thought that the plant is jealous and possessive. For this reason, the people keep the rituals associated with it secret. In addition, they show the plant a deep respect, for they believe that they will otherwise be severely punished. Several conceptions about the plant exhibit syncretism with Christianity. The plant is known as sacred for some people. It is accompanied by Jesus: the thorns are said to have been used to crown Christ before the crucifixion.

44.10 HALLUCINOGENIC EFFECTS

Latúe maily causes numerous hallucinogenic effects, including acute delirium, visual hallucinations, and mainly induces pupillary dilation, confusion, mouth dryness, and headache. The effects remained for 2–3 days; after-effects can be longer for several weeks just like *Brugmansia*. The leaves of *Latua* are used to make a tea that induces cramps and hallucinations. Thereby, it has been also shown that *Latua* causes permanent 'imbecility' (Murillo 1889). Nevertheless, the shamans are usually not affected by such types of plants; in other words, it helps to learn about the hidden potential of the things.

Smoking the *L. pubiflora* dried leaves badly affects arousal and metal relaxation, similar to the influence of various species *Brugmansia* genus.

44.11 CONCLUSION

Latua pubiflora, being a unique member of the family Solanaceae, is a native shrub possessing a spiny stem and greenish leaves. It blooms in early spring or late winter with magenta flowers. *L. pubiflora* is primarily endemic to Southern Chile. However, it has been extended to the mountains of Chiloé and cordilleras. Its beautiful hairy flowers and tomato-type pale yellow fruits are fascinating, hence its use as an ornamental plant. The *L. pubiflora* is highly poisonous, causing delirium, an increase in sleeping time, and even induces madness. Its toxicity is due to the number of tropane alkaloids, including atropine as well as scopolamine compounds present in it. The concentration of atropine is higher in the leaves of *L. pubiflora*. It has psychotropic characteristics and antioxidant potential, indicating the uses of this effective plant in medicines, mystical, and religious activities carried out by Indigenous Chilean peoples.

CONFLICT OF INTERESTS

The authors declared that they have no conflict of interest with respect to the authorship and publication of this chapter.

LIST OF ABBREVIATION

UHPLC-MS Ultra-high-performance liquid chromatography-mass spectrometry

REFERENCES

Baillon, H.E. (1888). Histoire des plantes 9:225–491.

Benkhnigue, O., Chaachouay, N., & Zidane, L. (2023). A Floristical and Ecological Study of the Medicinal Flora Used by the Local Population of the Haouz-Rehamna Region (Middle Atlantic Morocco-4). *Acta Botanica Hungarica Akadémiai Kiadó*, *65*(3–4), 247–304. https://doi.org/10.1556/034.65.2023.3-4.3

Beyer, J., Drummer O. H., Maurer H. H. (2009). Analysis of toxic alkaloids in body samples. *Forensic Science International*, 185, 1–9.

Bodendorf, K., and H. Kummer. (1962). Über die Alkaloide in Latua venenosa. *Pharmazeutische Zentralhalle Deutschlands* 101:620–22.

Carrasco, C., Echeverria, J., Ballester, B., Niemeyer, H. M. (2015). Of pipes and substances: fumatorial customs during the formative period in the desert coast of Atacama (North of Chile). *Latin American antiquity*. 26, 143–161.

Chaachouay, N., Douira, A., & Zidane, L. (2021a). Herbal Medicine Used in the Treatment of Human Diseases in the Rif, Northern Morocco. *Arabian Journal for Science and Engineering*, 47, 131–153. https://doi.org/10.1007/s13369-021-05501-1

Chaachouay, N., Douira, A., & Zidane, L. (2021b). COVID-19, prevention and treatment with herbal medicine in the herbal markets of Salé Prefecture, North-Western Morocco. *European Journal of Integrative Medicine*, 42, 101285. https://doi.org/10.1016/j.eujim.2021.101285

Chaachouay, N., & Zidane, L. (2021). The Symbolic Efficacy of Plants in Rituals and Socio-Religious Ceremonies in Morocco, Northwest of Africa. *Journal of Religious & Theological Information*, *21*(1–2), 34–53. https://doi.org/10.1080/10477845.2021.1942433

Echeverria, J., Niemeyer H. M. (2013). Nicotine in the hair of mummies from San Pedro de Atacama (Northern Chile). *Journal of Archaeological Science*. 40, 3561–3568.

Echeverria, J., Niemeyer, H. M. (2012). Alkaloids from the native flora of Chile: a review. *Bol. Latinoam. Caribe Plant. Med. Arom*. 11, 291–305.

Echeverria, J., Planella M. T., Niemeyer, H. M. (2014). Nicotine in residues of smoking pipes and other artifacts of the smoking complex from an early ceramic period archaeological site in central Chile. *Journal of Archaeological Science*, 44, 55–60.

Hanus, L. O., Řezanka T., Spížek J., Dembitsky V. M. (2005). Substances isolated from Mandragora species. *Phytochemistry*, 66, 2408–2417.

Hoffmann, J. Adriana E. (1994). *Flora silvestre de Chile: Zona araucana*. Santiago: Ediciones Fundación Claudio Gay.

Mariani Ramires, C. (1965). *Temas de Hipnosis. Editorial Andrès Bello, Bibliotèca de Estudiòs Mèdicos*, Santiago, p.336.

Muñoz, O., Casale, E. (2003). Tropane alkaloids from Latua pubiflora. *Z. Naturforsch*. 58, 626–628.

Murillo, A. (1889). *Plantes Médicinales de Chile*. Paris: Imprimerie de Lagny.

Nakashima Degarrod, Lydia. n.d. Contemporary uses of the Latua pubiflora among the Huilliche of Chile. Unpublished manuscript, written ca. (1988).

Olivares, J. C. (1995). *El Umbral Roto*. Santiago, Fondo Matta: Museo Chileno de Arte Precolombino.

Philippi, Rudolph A. (1858). Latua Ph., ein neues Genus der Solanaceen. *Botanische Zeitung* 33 (Aug.): 241–42.

Plowman, T., Gyllenhaal, L. O., & Lindgren, J. E. (1971). Latua pubiflora magic plant from Southern Chile. *Botanical Museum Leaflets, Harvard University*, *23*(2), 61–92.

Ramoutsaki, I. A., Askitopoulou, H., Konsolaki, E. (2002). Pain relief and sedation in Roman Byzantine texts: *mandragoras officinarum*, *Hyoscyamos niger* and *Atropa belladonna*. *International Congress Series*. 1242, 43–50.

Sánchez-Montoya, E. L., Reyes, M. A., Pardo, J., Nuñez-Alarcón, J., Ortiz, J. G., Jorge, J. C., & Simirgiotis,

M. J. (2017). High resolution UHPLC-MS metabolomics and sedative-anxiolytic effects of Latua pubiflora: A mystic plant used by Mapuche Amerindians. *Frontiers in Pharmacology*, *8*, 494.

Schultes, R.E. and Raffauf, R.F. (1970). *The Healing Forest.* Portland, (SAR): Dioscorides Press.

Silva, M., and Mancinelli, P. (1959). Atropina en Latua pubiflora (Griseb.) Phil. *Boletín de la Sociedad Chilena de Química* 9:49–50.

Soni, P., Siddiqui A. A., Dwivedi J., Soni V. (2012). Pharmacological properties of Datura stramonium L. as a potential medicinal tree: an overview. *Asian Pacific Journal of Tropical Biomedicine*. 2, 1002–1008.

45 Pricklyburr (*Datura innoxia* Miller Solanaceae)

Volkan Altay, Mustafa Keskin, Alvina Gul, and Munir Ozturk

45.1 INTRODUCTION

Various hallucinogenic species have played an essential role in the culture and religion of earlier tribes as well as advanced civilizations, both in the past and present (Widelski and Kukula-Koch, 2017; Calina et al., 2021). Substances that are psychoactive, or psychotropic, have an impact on the nervous system, which in turn affects mental processes and behavior (Rätsch, 2005). Psychoactive substances include hallucinogens that encompass a range of effects, which can alter one's perception of reality, to sedatives or narcotics that cause sleep, to antidepressants, anxiolytics or calmatives, and stimulants that can stimulate the mind (Van Wyk and Wink, 2014). Therefore, their current usage is of considerable interest from several perspectives, including medicinal, ethnobotanical, social, and legal (Fatur, 2021; Orch et al., 2021).

The Solanaceae family is amongst the most extensive and widely distributed plant families containing around 2,500 species and 100 genera. These species have historically held significant cultural importance in various societies and have played a critical role in human nutrition, medicine, and religion (Ghatak et al., 2017; Fatur, 2019; Rosario et al., 2019). Numerous hallucinogenic and potentially dangerous tropane alkaloids, including atropine, scopolamine, and hyoscyamine, are derived from plants such as *Datura* spp., *Atropa belladonna*, *Brugmansia* spp., and *Hyoscyamus niger* (Bock and Norris, 2015). Each part of the *Datura* taxa contains atropine and scopolamine, which are known to be toxic (Romera-Torres et al., 2018; Kerchner et al., 2015; Jakabová et al., 2012).

The widespread occurrence of *Datura* taxa across the globe is driven by their successful reproductive strategies and ornamental cultivation, which has led to an increase in their toxicological significance. This effect is not limited to tropical regions but is also observed in temperate regions worldwide (Kerchner and Farkas, 2020). The interest in the plants of this genus is due to their attractive large leaves and flowers.

In this chapter, information has been pooled covering data on various features of *Datura innoxia*, namely botanical aspects, geographical distribution, and ecological features, traditional uses, and some pharmacological effects.

45.2 SCIENTIFIC CLASSIFICATION

The latest updated classification of this species according to USDA (2023) is presented below:

Kingdom	Plantae
Subkingdom	Tracheobionta
Superdivision	Spermatophyta
Division	Magnoliophyta
Class	Magnoliopsida
Subclass	Asteridae
Order	Solanales
Family	Solanaceae
Genus	*Datura* L.
Species	*Datura innoxia* Miller

45.3 VERNACULAR NAMES

Datura genus is commonly known as thornapple, stinkweed, locoweed, augushka, ditch weed, devil's snare, devil's seed, devil's trumpet, Korean morning glory, Jamestown weed, beelzebub's twinkie, madhatter, crazy tea, and 'Tatula' in Turkiye (Ozturk, 1980). In some parts of the world people also name it as 'Angel's trumpet' however, this name is more used for the *Brugmansia* species. *D. innoxia* is widely named as 'Pricklyburr' in English and is known by different local names in different countries. In America, locals call it 'Angel's trumpet, desert thornapple, downy thorn-apple, hoary thorn-apple, Indian-apple, pricklyburr, recurved thorn-apple, sacred datura, and thorn-apple'; in Mexico as 'Toloache and Nahuatl toloatzin', and in Spain as 'Toloache, Tártago and Yerba del diablo' (Valdés, 2012; Benítez et al., 2018; McFarland et al., 2022).

45.4 BOTANICAL DESCRIPTION

The genus name '*Datura*' comes from the Hindi name of *Datura stramonium* (dhatura) which is generally known as Jimsonweed in English. *D. innoxia* is up to 130 cm tall, branched, dense pubescent-villous; leaves 7–18 × 4.5–12 cm, broadly ovate, repand to sinuate-dentate, base oblique to cuneate; petiole shorter than lamina. It has bisexual flowers,

DOI: 10.1201/9781003460336-45

the pedicel 10–25 mm long, reflexed in fruit, tomentose with brownish hairs; calyx 7.5–10.5 cm long, 1/2–2/3rd the corolla length, 5-lobed, slightly inflated toward the base; basal portion persistent and strongly reflexed in fruit; lobes unequal, triangular, or ovate-lanceolate, acuminate, up to 22 mm long; corolla tube 15–17 cm long; limb 6–8.5 cm broad, 10 (-12) lobed; lobes broad triangular, acute or acuminate; anthers 9–11 mm long, white; capsule globose, 4–5 cm broad, dense spiny, nodding; spines acicular, 5–10 mm long; seeds 4.5–5 mm long, reniform, compressed, brown, minutely reticulate, rugose-lineate on the back (Figure 45.1) (Zhang et al., 1994). The flowers are followed by downy, spherical fruit, which is covered with stiff spines; as such, people also call it downy thorn apple.

45.5 ORIGIN AND GEOGRAPHICAL DISTRIBUTION

The natural distribution of *D. innoxia* reveals that it is a herb distributed in the subtropical biome regions of Arizona to Texas and Mexico. The natural distribution areas are outlined mainly as; Arizona, Texas, and Mexico (www.powo.science.kew.org). The places where this plant has been introduced are: Afghanistan, Algeria, Angola, Argentina, Australia, Bangladesh, Benin, Bolivia, Brazil, Bulgaria, Chile, China, Colombia, Costa Rica, Cuba, Cyprus, Czechoslovakia, Egypt, Ethiopia, France, Great Britain, Greece, Haiti, Honduras, India, Iran, Italy, Jamaica, Kenya, Korea, Kuwait, Lebanon, Libya, Madagascar, Malawi, Mali, Mauritania, Mauritius, Morocco, Mozambique, Namibia, Netherlands, Niger, Nigeria, Oman, Pakistan, Palestine, Paraguay, Peru, Portugal, Romania, Saudi Arabia, Senegal, Somalia, Spain, Sudan, Syria, Tanzania, Tunisia, Turkiye, Ukraine, Uruguay, Uzbekistan, Venezuela, Vietnam, Yemen, Yugoslavia, Zaïre, and Zimbabwe (www.powo.science.kew.org).

Some of the species are used as ornamental plants for landscaping purposes (Altay, 2012). *D. innoxia*, as a herb species, appears to possess the potential to be used as an ornamental plant in Asia, Africa, Europe, and Australia (www.powo.science.kew.org).

45.6 HABITAT AND ECOLOGY

As with other ruderal taxa of the Solanaceae family, species of the *Datura* genus generally develop in characteristically degraded and nitrogen-rich soils in terms of their habitat preferences. The ecological tolerance helps these plants to develop in urban areas where anthropogenic activities are intense, especially alongside the roads, construction sites, and residential areas (Bye et al., 1991; Osma et al., 2010; Altay and Ozturk, 2012; Eskin et al., 2012; Ozturk et al., 2012; Altay et al., 2010, 2012, 2015, 2020; Altay and Karahan, 2017; Benítez et al., 2018).

The distribution of species belonging to the *Datura* genus is well known in the American continent (especially in Central and South America), where they have a natural distribution (Luna-Cavazos et al., 2009; Benítez et al., 2018). Since most species of the *Datura* genus originate from Mexico, the Balsas River Basin in Mexico is considered the center of gene diversity for the *Datura* genus (Bye and Sosa, 2013; Benítez et al., 2018).

Among species belonging to the genus *Datura*, *D. innoxia* has been documented to be widely distributed in Central America (Benítez et al., 2018). Furthermore, it is classified as naturalized in many European countries and is widely distributed (Moore, 1972; Gallego, 2012). Considering the habitats where it is distributed in Europe, it is seen that the ecological demands of this species are generally nitrophilous areas, ditches, fallow areas, river and stream edges, sandy and clayey soils exposed to some degree of anthropogenic influence (Sanz-Elorza et al., 2001; Gallego, 2012; Benítez et al., 2018).

45.7 ETHNOBOTANY

Datura taxa were named as devil's trumpet due to the flower shape and toxic nature of these plants. The plants of this genus appear as one of the most interesting ones with hallucinogenic characteristics (Chaachouay et al., 2021a, 2021b). In some religions and old cultures, these plants were regarded as a symbol of power by dangerous ones. According to Hindu religious belief, Lord Shiva, after drinking the poison, the *Datura* plant emerged from his chest. All throughout history, we come across stories about *Datura* plants being used to make poisons, witches' brews or love potions. The *Datura* root powder is used by rain priests to pray for good rains, whereas some of the shamans among the tribal priests in some areas of Texas as well as North Mexico use these plants to transcend the borders of life and death and communicate with spirits. The native American tribal people from central and southern coastal regions of California and surroundings known locally as 'Chumash', datura has been personified as an old woman named Momoy. These people used these plants to enter a hallucinogenic trance, so as to talk to spiritual beings and diagnose serious illnesses. There are reports that the *Datura* extract is mixed with comestibles, mainly as cookies. This is, therefore, also known as 'roadside poison', as accidental poisoning can occur with seeds (Benkhnigue et al., 2023). The plants of some *Datura* species also produce addictive narcotic nectar to entice hawk moths to their flowers. The browsing animals generally do not eat these plants because of their bad taste and foul odor; the leaves are bitter with a nauseating taste (Ozturk, 1980).

As far as *D. innoxia* is concerned all parts are toxic if consumed in high quantities. Throughout history, it has been popularly used as a hallucinogenic plant by many societies in both the Old World and the New World, and its use continues today. For this reason, although its economic use is limited, many alkaloids found in the plant have been in demand and curiosity (Bonde, 1998). It has been reported

FIGURE 45.1 Some morphological features of *Datura innoxia*: (A) general view of the plant; (B) leaves and flower; (C) bracts; and (D) fruit.

that *D. innoxia* was widely used as a narcotic, anesthetic, and poultice to treat wounds and bruises, especially traditionally by the Zuni people. In addition, it is noteworthy that *D. innoxia* is used for hallucinatory effects, divination, magic, or religious purposes among Southwest Native American tribes (Yetman and Van Devender, 2002; UNODC, 2023).

It has been reported that *D. innoxia* was traditionally used in some foods and beverages or in so-called love affairs by mixing it with chocolate, especially by some

Indian women, in order to 'stun' their husbands (Díaz, 1977). Also surprisingly, in the traditional Ayurvedic system of medicine, this plant has been popularly used by the local people in the treatment of ailments such as rabies, leprosy, and other infectious diseases (Neeraj et al., 2013).

This plant is traditionally used in Mexico as a 'hallucinogenic beverage enhancer' and Peyote ceremony: divination, prophecies, etc. 'lovesickness' (BDMTM, 2009).

In Mexico, different parts of the plant are generally used in traditional medicine to treat different types of pain (headache, muscle pains, navel pain, and neurotoxic pains), asthma, cough, dengue treatment, diabetes, fever, hemorrhoids, hits, inflammation, rheumatism, snakebite, stomachache, and strokes (García Rivas, 1990; Aguilar et al., 1994; Avilés and Juárez, 1994; BDMTM, 2009).

The leaves of *D. innoxia* are still popularly sold in the street markets of central Mexico for use in traditional folk medicine for wounds and inflammations (Argueta et al., 1994; Bye and Sosa, 2013; Benítez et al., 2018). In addition, in Mexico, the prevailing opinion is that the leaves of this plant are used in the form of cigarettes to treat asthma, and the cataplasms are used externally to the affected part which will be useful for rheumatism (BDMTM, 2009).

In Europe, especially in the Spanish traditional medical system, different parts of the plant are generally used in traditional medicine to treat asthma, burns, constipation, dermatitis, diarrhea, laxatives and rheumatism (Álvarez-Escobar, 2011; Perera, 2005; Kunkel, 1987; Mesa, 1996). All *Datura* species are extremely poisonous and potentially psychoactive, especially their seeds and flowers, which can cause respiratory depression, arrhythmias, fever, delirium, hallucinations, anticholinergic syndrome, psychosis, and death if taken internally (Ozturk, 1980). *D. innoxia* too is a strong narcotic plant but it also offers several health benefits; notable among these are: for promoting hair and skin health; easing childbirth; fertility improvement; inducing sleep; treatment of fever; heart functions enhancements; and alleviating pains. However, the fatal dose is mentioned as 15 gms of *Datura* (15–25 seeds) and side effects on the body can last longer after a high dose. The presence of atropine, scopolamine, and hyoscyamine is mainly responsible for the intoxicating as well as medicinal characteristics of this plant (Ozturk, 1980).

45.8 PHARMACOLOGY

Studies undertaken on the determination of phytochemical compounds present in the different parts of this plant have revealed that these especially include; alkaloids (especially tropane alkaloids – hyoscyamine, scopolamine, and withanolides), carbohydrates, cardiac glycosides, crude fiber, flavonoids, lipids, phenolic compounds, protein and tannins (Ayuba et al., 2011; Fatima et al., 2015; Bagewadi et al., 2019; Sharma et al., 2021).

Recently, during some pharmacological experimental studies on the leaves and seeds of this plant, different extracts have been prepared and these are reported to show important pharmacological activities such as; anti-inflammatory, antibacterial (especially against some Gram+ bacteria, and some strains of *Vibrio cholera* and *V. parahaemolyticus*), antifungal (especially against *Fusarium solani*), antioxidant, and cytotoxic effects (especially against human colon cancer cells) (Eftekhar et al., 2005; Ramadan et al., 2007; Sharma et al., 2009; Vermillion et al., 2011; Kalim et al., 2016; Jaafar et al., 2018; Benítez et al., 2018; Al-Zharani et al., 2021).

45.9 MEDICINAL EFFECTS

Although it has some beneficial effects on health, the plant contains alkaloids such as tropane, which makes it a dangerous toxic plant for the nervous system (Shah et al., 2013; Sharma et al., 2021). In particular, toxic effects from the plant have been reported in some studies to cause negative effects on the central nervous system, disorientation, memory loss, inability to process information, visual impairment due to mydriasis, myoclonic twitching, hyperpyrexia, and respiratory and cardiovascular problems (Krenzelok, 2010; Sharma et al., 2021). Therefore, the toxicity symptoms of the plant may affect human beings seriously; the main symptoms are high fever, headache, dry mouth, hallucinations, convulsions, dry skin, fast and weak pulse, and even coma and death (Sharma et al., 2021).

There are several neuropharmacological activity studies conducted on this plant. A clinical study conducted in 2018 reports that aqueous extracts from the leaves of *D. innoxia* show anti-inflammatory activity and are thus regarded as potentially active herbal analgesic (Jaafar et al., 2018). In another study undertaken by Sharma et al. (2021), it has been reported that the extracts from the seeds of *D. innoxia* may be relatively safer in inducing sleep with anesthetic indexes.

45.10 CONCLUSION

D. innoxia is widely used in American traditional folk medicine (especially in Mexico). Several pharmacological studies have been carried out on this plant. It is a natural genetic resource with a good therapeutic potential, mainly because of its antimicrobial activity, antioxidant, anti-inflammatory, and cytotoxic effects. It is one of the popular plants used in traditional systems in different areas globally, due to its effective hallucinogenic features. Although this plant has a high toxic potential, it is regarded as a promising candidate for clinical and pharmacological studies and the development of some innovative treatments at low doses. Detailed investigations are needed in order to make it available to people at global level, especially in the Americas. There is a need to fully reveal its therapeutic potential in neuropharmacological activities as well as other diseases and to ensure its safe use in modern medicine.

LIST OF ABBREVIATIONS

D. innoxia	*Datura innoxia*
USDA	United States Department of Agriculture Natural Resources Conservation Service

REFERENCES

Aguilar, A., Camacho, J.R., Chino, S., Járquez, P., and López, M.E. 1994. Herbario Medicinal del Instituto Mexicano del Seguro Social. D.F. Instituto Mexicano del Seguro Social (IMSS), México.

Álvarez-Escobar, A. 2011. Contribución al estudio etnobotánico de la isla de Tenerife. Tesis doctoral. Departamento de Biología Vegetal (Botánica). Universidad de La Laguna, Tenerife, Spain.

Al-Zharani, M., Nasr, F.A., Alqahtani, A.S., Cordero, M.A.W., Alotaibi, A.A., Bepari, A., Alarifi, S., Daoud, A., Barnawi, I.O., and Daradka, H.M. 2021. In vitro cytotoxic evaluation and apoptotic effects of *Datura innoxia* grown in Saudi Arabia and phytochemical analysis. *Applied Sciences*, 11: 2864.

Altay, V. 2012. Mustafa Kemal Üniversitesi Tayfur Ata Sökmen Kampüsü (Hatay)'nün süs bitkileri. *Karadeniz Fen Bilimleri Dergisi*, 2(6): 11–26.

Altay, V., and Karahan, F. 2017. Ruderal vejetasyon üzerine bir ön çalışma. *Kilis 7 Aralık Üniversitesi Fen ve Mühendislik Dergisi*, 1(2): 68–77.

Altay, V., and Ozturk, M. 2012. Land degradation and halophytic plant diversity of Milleyha Wetland Ecosystem (Samandağ-Hatay), Turkey. *Pakistan Journal of Botany*, 44: 37–50.

Altay, V., Ozyıgıt, I.I., and Yarcı, C. 2010. Urban flora and ecological characteristics of the Kartal District (Istanbul): A contribution to urban ecology in Turkey. *Scientific Research and Essays*, 5(2): 183–200.

Altay, V., Ozyıgıt, I.I., and Yarcı, C. 2012. Plant communities in urban habitats of Istanbul-Turkey. *Pakistan Journal of Botany*, 44: 177–186.

Altay, V., Ozyığıt, I.I., Osma, E. et al. 2015. Enviromental relationships of the vascular flora alongside the railway tracks between Haydarpasa and Gebze (Istanbul-Kocaeli/Turkey). *Journal of Environmental Biology*, 36: 153–162.

Altay, V., Silc, U., Yarcı, C. et al. 2020. Urban vegetation of the Anatolian side of Istanbul. *Phytocoenologia*, 50(2): 101–121.

Argueta, V.A., Cano, L.M., and Rodarte, M. 1994. Atlas de plantas de la medicina tradicional mexicana Vol. I, II, III. Instituto Nacional Indigenista. Biblioteca de la Medicina Tradicional Mexicana, México.

Avilés, M., and Juárez, G. 1994. *Catálogo de plantas medicinales del jardín botánico de Cuernavaca*. Centro INAH Morelos, Mexico.

Ayuba, V.O., Ojobe, T.O., and Ayuba, S.A. 2011. Phytochemical and proximate composition of *Datura innoxia* leaf, seed, stem, pod and root. *Journal of Medicinal Plants Research*, 5: 2952–2955.

Bagewadi, Z.K., Muddapur, U.M., Madiwal, S.S., Mulla, S.I., and Khan, A. 2019. Biochemical and enzyme inhibitory attributes of methanolic leaf extract of *Datura inoxia* Mill. *Environmental Sustainability*, 2: 75–87.

Benítez, G., March-Salas, M., Villa-Kamel, A., Cháves-Jiménez, U., Hernández, J., Montes-Osuna, N.,... and Cariñanos, P. 2018. The genus *Datura* L. (Solanaceae) in Mexico and Spain-Ethnobotanical perspective at the interface of medical and illicit uses. *Journal of Ethnopharmacology*, 219: 133–151.

Benkhnigue, O., Chaachouay, N., & Zidane, L. (2023). A Floristical and Ecological Study of the Medicinal Flora Used by the Local Population of the Haouz-Rehamna Region (Middle Atlantic Morocco-4). *Acta Botanica Hungarica Akadémiai Kiadó*, *65*(3-4), 247–304. https://doi.org/10.1556/034.65.2023.3-4.3

BDMTM (Biblioteca Digital de la Medicina Tradicional Mexicana). 2009. http://www.medicinatradicionalmexicana.unam.mx/index.php. Accesses on 17.02.2017.

Bock, J.H., and Norris, D. 2015. *Forensic Plant Science*. Academic Press.

Bonde, K. 1998. The Genus *Datura*: From research subject to powerful hallucinogen. *Ethnobotanical Leaflets*, 1: Article 10.

Bye, R., and Sosa, V. 2013. Molecular phylogeny of the jimsonweed genus *Datura* (Solanaceae). *Systematic Botany*, 38: 818–829.

Bye, R., Mata, R., and Pimentel, J. 1991. Botany, Ethnobotany and Chemistry of *Datura lanosa* (Solanaceae) in Mexico 61. Anales del Instituto de Biologia de la Universidad Nacional Autónoma de México, Serie Botánica.

Calina, D., Carvalho, F. and Docea, A.O. 2021. Toxicity of psychedelic drugs. In: Tsatsakis, A.M. (ed.), *Toxicological Risk Assessment and Multi-System Health Impacts from Exposure*. Academic Press, pp. 545–556.

Chaachouay, N., Douira, A., & Zidane, L. (2021a). Herbal Medicine Used in the Treatment of Human Diseases in the Rif, Northern Morocco. *Arabian Journal for Science and Engineering*, 47, 131–153. https://doi.org/10.1007/s13369-021-05501-1

Chaachouay, N., Douira, A., & Zidane, L. (2021b). COVID-19, prevention and treatment with herbal medicine in the herbal markets of Salé Prefecture, North-Western Morocco. *European Journal of Integrative Medicine*, 42, 101285. https://doi.org/10.1016/j.eujim.2021.101285

Díaz, J.L. 1977. Ethnopharmacology of sacred psychoactive plants used by the Indians of Mexico. *Annual Review of Pharmacology and Toxicology*, 17: 647–675.

Eftekhar, F., Yousefzadi, M., and Tafakori, V. 2005. Antimicrobial activity of *Datura innoxia* and *Datura stramonium*. *Fitoterapia*, 76: 118–120.

Eskin, B., Altay, V., Ozyıgıt, I.I., and Serın, M. 2012. Urban vascular flora and ecological characteristics of the Pendik District (Istanbul-Turkey). *African Journal of Agricultural Research*, 7(4): 629–646.

Fatima, H., Khan, K., Zia, M., Ur-Rehman, T., Mirza, B., and Haq, I. 2015. Extraction optimization of medicinally ımportant metabolites from *Datura innoxia* Mill.: An in vitro biological and phytochemical investigation. *BMC Complementary Medicine and Therapies*, 15: 1–18.

Fatur, K. 2019. Sagas of the Solanaceae: Speculative ethnobotanical perspectives on the Norse berserkers. *Journal of Ethnopharmacology*, 244: 112151.

Fatur, K. 2021. Peculiar plants and fantastic fungi: An ethnobotanical study of the use of hallucinogenic plants and mushrooms in Slovenia. *Plos one*, 16(1): e0245022.

Gallego, M.J. 2012. *Datura* L. In: Castroviejo, S., Aedo, C., Laínz, M., Muñoz Garmendia, F., Nieto Feliner, G., Paiva, J., and Benedí, C., eds., *Flora Iberica 11*. CSIC, Madrid, Spain, pp. 216–224.

García Rivas, H. 1990. Plantas medicinales de Méjico. Descripciones y usos. Panoramio México DF, Mexico.

Ghatak, A., Chaturvedi, P., Paul, P., Agrawal, G.K., Rakwal, R., Kim, S.T., Weckwerth, W. and Gupta, R. 2017. Proteomics survey of Solanaceae family: Current status and challenges ahead. *Journal of Proteomics*, 169: 41–57.

Jaafar, F.R., Ajeena, S.J., and Mehdy, S.S. 2018. Anti-inflammatory ımpacts and analgesiac activity of aqueous extract *Datura innoxia* leaves against induced pain and inflammation in mice. *Journal of Entomology and Zoology Studies*, 6: 1894–1899.

Jakabová, S., Vincze, L., Farkas, Á., Kilár, F., Felinger, A., and Boros, B. 2012. Determination of tropane alkaloids atropine and scopolamine by liquid chromatography-mass spectrometry in plant organs of *Datura* species. *Journal of Chromatography A*, 1232: 295–301.

Kalim, M., Hussain, F., Ali, H., Ahmad, I., and Iqbal, M.N. 2016. Antifungal activities of methanolic extracts of *Datura inoxia*. *PSM Biological Research*, 1: 70–73.

Kerchner, A., and Farkas, Á. 2020. Worldwide poisoning potential of *Brugmansia* and *Datura*. *Forensic Toxicology*, 38: 30–41.

Kerchner, A., Darók, J., Bacskay, I., Felinger, A., Jakab, G., and Farkas, Á. 2015. Protein and alkaloid patterns of the floral nectar in some solanaceous species. *Acta Biologica Hungarica*, 66: 304–315.

Krenzelok, E.P. 2010. Aspects of *Datura* poisoning and treatment. *Clinical Toxicology*, 48: 104–110.

Kunkel, G. 1987. El Libro de las Malas Hierbas y su Importancia Socio-Económica. Cajal, Almería, Spain.

Luna-Cavazos, M., Bye, R., and Jiao, M. 2009. The origin of *Datura metel* (Solanaceae): genetic and phylogenetic evidence. *Genetic Resources and Crop Evolution*, 56(2): 263–275.

McFarland, K., Sharp, N., and Loomis, J. 2022. *Datura innoxia* Mill. Checklist of Vermont Species. Version 1.8. Vermont Center for Ecostudies. Checklist dataset https://doi.org/10.15468/nu60xi. Accessed on 30.07.2023.

Mesa, S. 1996. Estudio etnobotánico y agroecológico de la comarca de la Sierra de Mágina (Jaén). Tesis doctoral. Facultad de Ciencias Biológicas. Universidad Complutense de Madrid, Spain.

Moore, D.M. 1972. *Datura* L. In: Tutin, T.G., Heywood, V.H., Burges, N.A., Moore, D.M., Valentine, D.H., Walters, S.M., and Webb, D.A., eds., *Flora Europaea*. Cambridge University Press, Cambridge, pp. 3.

Neeraj, O.M., Ayesha, K., and Balu, A.C. 2013. Rediscovering the medicinal properties of Datura sp.: A review. *Journal of Medicinal Plants Research*, 7: 2885–2897.

Orch, H., Chaachouay, N., Douiri, E. M., Faiz, N., Zidane, L., & Douira, A. (2021). Use of medicinal plants in dermatocosmetology: An ethnobotanical study among the population of Izarène. *Jordan Journal of Pharmaceutical Sciences*, *14*(3), 323–340.

Osma, E., Ozyıgıt, I.I., Altay, V., and Serın, M. 2010. Urban vascular flora and ecological characteristics of Kadikoy district, Istanbul, Turkey. *Maejo International Journal of Science and Technology*, 4(1): 64–87.

Ozturk, M. 1980. Lecture Notes on Economic Botany. Ege University, Science Faculty Press, Izmir, Turkiye.

Ozturk, M., Altay, V., Gucel, S., and Aksoy, A. 2012. Aegean grasslands as endangered ecosystems in Turkey. *Pakistan Journal of Botany*, 44: 7–18.

Perera, J. 2005. La toponimia de La Gomera. Un estudio sobre los nombres de lugar, las voces indígenas y los nombres de plantas, animales y hongos de La Gomera. IV tomos, 25 volúmenes. AIDER, Edición digital, La Gomera, Spain.

Ramadan, M.F., Zayed, R., and El-Shamy, H. 2007. Screening of bioactive lipids and radical scavenging potential of some solanaceae plants. *Food Chemistry*, 103: 885–890.

Rätsch C. 2005. *The Encyclopedia of Psychoactive Plants: Ethnopharmacology and its Applications*. Park Street Press; Rochester.

Romera-Torres, A., Romero-González, R., Vidal, J.L.M., and Frenich, A.G. 2018. Analytical methods, occurrence and trends of tropane alkaloids and calystegines: an update. *Journal of Chromatography A*, 1564: 1–15.

Rosario, L.H., Padilla, J.O.R., Martínez, D.R., Grajales, A.M., Mercado Reyes, J.A., Feliu, G.J.V., Van Ee, B., and Siritunga, D. 2019. DNA barcoding of the solanaceae family in Puerto Rico including endangered and endemic species. *Journal of the American Society for Horticultural Science*, 144: 363–374.

Sanz-Elorza, E., Dana, E., and Sobrino, E. 2001. Aproximación al listado de plantas alóctonas invasoras reales y potenciales en España. *Lazaroa*, 22: 121–131.

Shah, V.V., Shah, N.D., and Patrekar, P.V. 2013. Medicinal plants from Solanaceae family. *Research Journal of Pharmacy and Technology*, 6: 4.

Sharma, A., Potel, V.K., and Chaturvedi, A.N. 2009. Vibriocidal activity of certain medicinal plants used in Indian folklore medicine by trials of Mohakoshal region of central India. *Indian Journal of Pharmacology*, 41: 129–133.

Sharma, M., Dhaliwal, I., Rana, K., Delta, A.K., and Kaushik, P. 2021. Phytochemistry, pharmacology, and toxicology of *Datura* species-A review. *Antioxidants*, 10: 1291.

UNODC (United National, Office on Drugs and Crime). 2023. The plant kingdom and hallucinogens (part III): Solanaceae. https://www-unodc-org.translate.goog/unodc/en/data-and-analysis/bulletin/bulletin_1970-01-01_1_page005.html?_x_tr_sl=en&_x_tr_tl=tr&_x_tr_hl=tr&_x_tr_pto=sc#s190. Accesses on 14.06.2023.

USDA (United States Department of Agriculture Natural Resources Conservation Service). 2023. *Datura innoxia* Mill. https://plants.usda.gov/home/plantProfile?symbol=DAIN2. Accesses on 30.07.2023.

Valdés, B. 2012. *Datura innoxia*. – Euro+Med Plantbase - the information resource for Euro-Mediterranean plant diversity. https://europlusmed.org/cdm_dataportal/taxon/9b474e11-bf81-4921-89a6-c82cbef0be08. Accessed on 30.07.2023.

Van Wyk, B.-E., and Wink, M. 2014. *Phytomedicines, Herbal Drugs and Poisons*. The University of Chicago Press, Chicago and London Kew Publishing Royal Botanic Gardens; Kew.

Vermillion, K., Holguin, F.O., Berhow, M.A., Richins, R.D., Redhouse, T., O'Connell, M.A., Posakony, J., Mahajan, S.S., Kelly, S.M., and Simon, J.A. 2011. Dinoxin b, a withanolide from *Datura inoxia* leaves with specific cytotoxic activities. *Journal of Natural Products*, 74: 267–271.

Widelski, J., and Kukula-Koch, W.A. 2017. Psychoactive Drugs. In: Simone, B. and Rupika, D. (eds), *Pharmacognosy*. Academic Press, pp. 363–374. www.powo.science.kew.org. Accesses on 30.07.2023.

Yetman, D., and Van Devender, T. 2002. *Mayo Etnobotany: Land, History and Traditional Knowledge in Nortwest Mexico*. University of California Press, Berkeley, USA.

Zhang, Z.Y., Lu, A., and Arcy, W.G. 1994. *Datura* L. *Flora of China*, 17: 330.

46 Golden Angel's Trumpet (*Datura stramonium* L. Solanaceae)

Noureddine Chaachouay, Fathia Benlarbi, Abdelhamid Azeroual, Abdel Khalid Essamadi, Rahmatullah Qureshi, and Lahcen Zidane

46.1 INTRODUCTION

Throughout history, people have depended on plants to address many diseases and health concerns. Medicinal plant-based traditional medicine systems have autonomously evolved in many cultures globally and have been transmitted between generations (Ansari et al., 2023; Chaachouay et al., 2022a, 2022b, 2022c; Benkhnigue et al., 2022, 2023; Mahomoodally, 2013; Orch et al., 2021). Hallucinogenic plants possess chemical components influencing the human brain, modifying perception, emotion, and awareness. These plants have been used for millennia for their medical, spiritual, and recreational properties (Passos and Mironidou-Tzouveleki, 2016; Sayin, 2016).

The use of hallucinogenic plants exhibits substantial variation throughout cultures and locales and has been an integral aspect of human history for millennia. Psychoactive plants are used in religious rituals in some cultures to elicit modified states of consciousness and spiritual encounters (Passos and Mironidou-Tzouveleki, 2016). In numerous cultures, psychoactive plants are employed for recreational intentions, frequently in conjunction with festive occasions and social groups (dos Santos et al., 2021). Psychoactive medicinal plants that are often used include cannabis, opium poppy, coca, ayahuasca, peyote, and psilocybin mushrooms. Each plant has distinct physiological and psychological properties and may be used in various ways.

Jimsonweed, or Golden Angel's trumpet, scientific name *Datura stramonium* L., is a plant species of the Solanaceae family. It is an annual perennial plant that produces enormous trumpet-shaped white or purple flowers and can reach a height of 1–1.5 meters (Blakeslee et al., 1937; Mountain, 1987). Although indigenous to South and North America, the plant has been cultivated as an ornamental specimen in numerous regions around the globe (Blakeslee et al., 1937). The hallucinogenic properties of *D. stramonium* are attributed to several alkaloids, including hyoscyamine and scopolamine (Das et al., 2012; Sreenivasa et al., 2012). These alkaloids function as anticholinergic drugs, inhibiting the neurotransmitter acetylcholine in the central and peripheral nervous systems (El Bazaoui et al., 2011; Shi et al., 2022).

Scopolamine and hyoscyamine have been observed to possess medicinal properties at modest concentrations, including the ability to alleviate pain, heal asthma, and reduce inflammation (Gaire and Subedi, 2013; Soni et al., 2012). Nevertheless, when administered in larger quantities, these alkaloids may exhibit toxic properties in both people and animals, leading to hallucinations, delirium, agitation, and potentially fatal outcomes (Shagal et al., 2012). Moreover, the alkaloids in *D. stramonium* may have anticholinergic effects on the body, such as xerostomia, visual impairment, and urine retention (Debnath and Chakraverty, 2017).

Due to its inherent toxicity, the use of *D. stramonium* for its psychedelic properties is strongly discouraged and not advised. The plant is widely regarded as a toxic invasive species in several regions globally, and unintentionally consuming its seeds or foliage may lead to severe medical issues (Freye and Freye, 2010; Trancǎ et al., 2017). It is crucial to emphasize that the use of the plant should exclusively be limited to medical or other officially sanctioned intentions and should be done under the supervision of a qualified healthcare practitioner. This chapter summarizes the findings regarding botany, ecology, traditional applications, phytochemical composition, and pharmacological and hallucinogenic properties of *D. stramonium*.

46.2 SCIENTIFIC CLASSIFICATION

D. stramonium, a toxic angiosperm often known as Golden Angel's trumpet or Devil's snare, belongs to the *Datura* genus and Daturae tribe. It is a species of the Solanaceae family, which is often referred to as the nightshade family (ITIS, 2022).

Kingdom: Plantae
Subkingdom: Viridiplantae
Infrakingdom: Streptophyta
Superdivision: Embryophyta
Division: Tracheophyta
Subdivision: Spermatophytina
Class: Magnoliopsida
Superorder: Asteranae
Order: Solanales
Family: Solanaceae
Genus: *Datura*
Species: *Datura stramonium* L.

DOI: 10.1201/9781003460336-46

46.3 VERNACULAR NAMES

Arabic: Datura, Tatura Bunj, Joz-almasal, Chdeq Jmmel, Shajratul murqad, Joz-el-mathil, Joz-dab, Joz-mahabal, Joz-mulsam, Joz-masa, Joz-mahalik, Joz-maqatal; **Chinoise**: Man tuo luo; **English**: Golden Angel's trumpet, Thorn-apple, Stink weed, Jimson weed; **French**: Herbe à la taupe, Datura officinale, Datura tatula, Datura stramoine, Stramoine commune, Pomme épineuse; **Italian**: Noce spinosa, Stramonio comune, Mezzettoni; **Spanish**: Belladonna del pobre, Burladora, Chamico azul, Berenjena del diablo, Chamico morado, Higuera del infierno, Hierba ratonera, Cornescopia espantarratones, Flor de la trompeta, Trompetilla; **Japanese**: Shiro-bana-chôsen-asagao, Datsura, Yôshu-chôsen-asagao, Dachura, Chôsen-asagao; **Dutsh**: Gewone doornappel, Doornappel; **German**: Gewöhnlicher stechapfel, Gemeiner stechapfel, Weißer stechapfel, Stechapfel; **Bhutan**: Nyangmo-throkchang, Dhaturo; **Bengali**: Sada dhutura, Dhatura; **Hindi**: Kanakbij, Dhatura; **Persian**: Tatulah, Gauz ma'sal; **Portuguese**: Erva-dos- mágicos, Estramonio, Erva-dos-bruxos, Castanheiro-do-diabo, Figueira-do-inferno, Erva-do-diabo, Figueira-brava, Mamoninho, Quinquilho, Pomo- espinhoso; **Korean**: Ji-ri-san-kkum-sae, Kkum-mu-ri; **Russian**: Durman obyknovennyi; **Urdu**: Joz maasal, Tatura, Dhatura; **Sweden**: Spikklubba; **Turkish**: Boruçiçeğ i.

46.4 SYNONYMS

There are 29 synonyms attributed to *Datura stramonium* L., listed as follows: (1) *Datura laevis* L.f.; (2) *Datura hybrida* Ten.; (3) *Datura cabanesii* P.Fourn.; (4) *Datura bernhardii* Lundstr.; (5) *Datura inermis* Juss. ex Jacq.; (6) *Datura ferocissima* Cabanès & P.Fourn.; (7) *Datura ferox* Nees; (8) *Datura capensis* Bernh.; (9) *Datura bertolonii* Parl. ex Guss.; (10) *Datura stramonium* var. *canescens* Roxb.; (11) *Datura muricata* Godr.; (12) *Datura stramonium* var. *chalybaea* W.D.J.Koch; (13) *Datura praecox* Godr.; (14) *Datura lurida* Salisb.; (15) *Datura pseudostramonium* Sieber ex Bernh.; (16) *Datura microcarpa* Godr.; (17) *Datura stramonium* f. *godronii* (Danert) Geerinck & Walravens; (18) *Stramonium spinosum* Lam.; (19) *Datura stramonium* var. *gordonii* Danert; (20) *Datura stramonium* f. *inermis* (Juss. ex Jacq.) Hupke; (21) *Stramonium foetidum* Scop.; (22) *Datura stramonium* var. *tatula* (L.) Decne.; (23) *Datura stramonium* var. *inermis* (Juss. ex Jacq.) Fernald; (24) *Stramonium vulgatum* Gaertn.; (25) *Datura stramonium* f. *tatula* (L.) B.Boivin; (26) *Datura stramonium* var. *tatula* (L.) Torr.; (27) *Datura stramonium* var. *stramonium*; (28) *Datura parviflora* Salisb.; (29) *Datura loricata* Sieber ex Bernh (The Plant List, 2013).

46.5 BOTANICAL DESCRIPTION

D. stramonium is a one-year plant that may reach a maximum height of 1 meter. The root is a fleshy and branching structure that usually has a light brown color. The length of this organism may reach 20–30 cm, while its diameter ranges from 1 to 2 cm (Gaire and Subedi, 2013; Mountain, 1987). The plant's root system is characterized by a shallow and fibrous structure, where the primary roots give birth to many lateral roots. The stem is upright, with branches, and covered with fine hairs. The leaves have an alternating arrangement, with an ovate to oblong form, and have the potential to reach a length of 20 cm and a width of 15 cm (Figure 46.1). The leaf border has uneven lobes or coarse teeth (Al-Snafi, 2017).

D. stramonium produces trumpet-shaped, sizable flowers that are either pale purple or white and emit a potent, fragrant fragrance (Figure 46.2). Developing solitary in the axils of the leaves, they can attain a maximum length of 15 cm. Tubular, five-lobed, and pubescent are the calyx (Soni et al., 2012). The corolla is cylindrical and consists of five lobes that extend and diverge. Five stamens are connected to the base of the corolla tube. The fruit of *D. stramonium* is a prickly capsule about 3–4 cm in diameter. Upon reaching maturity, the fruit divides into four compartments, each housing a multitude of tiny, ebony, kidney-shaped seeds. The seeds are small, reniform, and ebony in color (Das et al., 2012; Malami et al., 2014). Their dimensions are around 3–4 mm in length and 2–3 mm in width, exhibiting a rigid and lustrous exterior. The seeds are enclosed in a prickly, spherical capsule that opens into four compartments once they reach maturity. Every segment comprises many sources (Stella et al., 2010).

46.6 ORIGIN AND GEOGRAPHIC DISTRIBUTION

D. stramonium is a plant species belonging to the Solanaceae family. The precise provenance of this plant is still being determined. It is hypothesized that its origin may be traced back to either North America or Central America, and its global distribution can be attributed to human activities such as commerce and migration. The plant is globally distributed and is regarded as a ubiquitous weed (Mountain, 1987; Weaver et al., 1985). It is distributed in several areas, encompassing South and North America, Africa, Europe, Asia, and Australia (Blakeslee et al., 1937; Chaachouay et al., 2021a, 2021b, 2021c). is distributed extensively over North America, including the United States, Canada, Mexico, and Central America. The species is distributed over many South American nations, including Brazil, Argentina, Chile, Colombia, Ecuador, Peru, and others. The plant is distributed over the Mediterranean and in France, Germany, Spain, and other European nations. It is situated in Japan, China, India, and several other regions of Asia. The distribution of this species spans throughout many African nations, including Morocco, Algeria, Egypt, Nigeria, South Africa, and others (Blakeslee et al., 1937; Chaachouay et al., 2020a, 2020b). It is distributed ubiquitously over Australia. *D. stramonium* is distributed globally beyond the areas previously listed. As an example, it is situated in the Middle East, including countries such as Iran and Palestine. In Pakistan, the plant is used for its therapeutic properties and is often called 'Kanak'. In China, this plant is often called 'jimsonweed' and has been used in traditional medicine for several decades (De-la-Mora et al., 2015; Weaver et al., 1985).

46.7 HABITAT AND ECOLOGY

D. stramonium is a resilient plant that can thrive in several environments and ecological circumstances. It often occurs in regions that have been disrupted, such as roadsides, fields, and waste sites. However, it is also capable of thriving in undisturbed environments like woods, grasslands, and deserts (De-la-Cruz et al., 2020; Núñez-Farfán and Dirzo, 1994). The plant has a high tolerance for many soil types and pH levels, enabling it to thrive in arid and humid environments. It can withstand drought conditions and may thrive in regions with abundant rainfall. *D. stramonium* is a plant that completes its life cycle in one or two years and may reach a height of 1–2 meters. The plant has sizable, deep green foliage and notable blossoms with a trumpet-like form, which may exhibit hues of white, yellow, or purple. The plant generates a prickly seed capsule that encloses several little ebony seeds (Fornoni et al., 2004). The plant is often regarded as a weed due to its propensity to increase and encroach into agricultural areas, resulting in diminished crop productivity. Nevertheless, it has significant medical value in several societies and has been used for ages due to its psychedelic characteristics. *D. stramonium* is crucial in supplying sustenance and shelter for insect and animal species, such as bees, butterflies, moths, and birds. The plant's blossoms, which have a large, funnel-shaped structure, are pollinated by bees and other insects. Meanwhile, the plant's seeds serve as a food source for birds and small animals (Castillo et al., 2013; Valverde et al., 2003). *D. stramonium*, despite its ecological significance, is recognized for its toxicity toward several animals, including humans. The plant harbors several alkaloids that may induce hallucinations, delirium, and other detrimental consequences. Livestock, including cattle and horses, may also experience adverse effects upon ingestion of the plant, resulting in symptoms such as vertigo, hallucinations, and seizures (Shonle and Bergelson, 2000).

46.8 PHYTOCHEMISTRY

The therapeutic qualities of *D. stramonium* have been used for millennia. Phytochemical analysis has shown the presence of many bioactive chemicals in the plant, including flavonoids, alkaloids, steroids, terpenoids, and cardiac glycosides. In addition, it is distinguished by its fat, significant carbohydrate, moisture, protein, crude fiber, and ash composition (Das et al., 2012; Malami et al., 2014; Sayyed, 2014). *D. stramonium* seeds have also been utilized to produce various amino acids, including phenylalanine, tyrosine, glutamate, and alanine. The principal chemical constituents identified in *D. stramonium* are detailed in Table 46.1.

Many tropane alkaloids, including atropine, scopolamine, hyoscyamine, and tropine, are found in *D. stramonium* (Ally and Mohanlall, 2020; El Bazaoui et al., 2011; Shonle and Bergelson, 2000). These alkaloids possess hallucinogenic properties and are used in medicine for the treatment of many ailments, including spasms, nausea, and vomiting. Atropine is an anticholinergic agent that inhibits acetylcholine receptors in both the central and peripheral nervous system. Scopolamine is an anticholinergic

FIGURE 46.1 (A) The root and stem and (B) aerial parts of *D. stramonium*.

FIGURE 46.2 The flower of *D. stramonium.*

medication that works by targeting acetylcholine receptors. It is often used to reduce vomiting and nausea caused by motion sickness and anesthesia. Additionally, it may have psychedelic effects, resulting in modified states of consciousness. Hyoscyamine, a tropane alkaloid found in *D. stramonium*, has comparable characteristics to atropine and scopolamine (Shi et al., 2022). The chemical makeup of the plant might vary based on variables such as the growth environment, plant age, and harvesting season (Gaire and Subedi, 2013; Shagal et al., 2012; Soni et al., 2012). Nevertheless, one must consider the plant's toxicity and susceptibility to misuse while using it for therapeutic purposes.

46.9 ETHNOBOTANY

46.9.1 Traditional Medicine

D. stramonium has been used in traditional medicine for ages, particularly in North America, Europe, Asia, North Africa, and Australia. Nevertheless, it is crucial to acknowledge that the plant presents poisonous alkaloids and should only be used with meticulous preparation and precise dosage. Traditional therapeutic uses of *D. stramonium* include:

- **Respiratory disorders**: The plant has been used as a bronchodilator for the treatment of respiratory conditions including as asthma, coughs, and bronchitis (Pretorius and Marx, 2006).
- **Pain relief**: *D. stramonium* has been used as a sedative and analgesic for alleviating pain related to rheumatism, toothaches, and other ailments (Soni et al., 2012).
- **Gastrointestinal disorders**: The plant has been used for the treatment of gastrointestinal diseases, including diarrhea, nausea, and vomiting (Bawazeer et al., 2020; Chaachouay & Zidane, 2021; Belhaj et al., 2021; Soni et al., 2012).
- **Local anesthetic**: The plant has been used as a regional analgesic for dental treatments and to alleviate rheumatism-related pain (Adebayo and Olufayo, 2017; Chaachouay et al., 2019a, 2019b).

Nevertheless, it is critical to acknowledge that improper utilization of the alkaloids present in *D. stramonium* may result in severe adverse effects, including delirium, hallucinations, and potentially fatal consequences (Chaachouay et al., 2021b; Freye and Freye, 2010; Trancă et al., 2017). Hence, it is advisable to get guidance from a certified healthcare professional or shaman before using the plant for therapeutic intentions.

46.9.2 Industrial Applications

D. stramonium has several industrial uses. Nevertheless, it is crucial to acknowledge that the plant contains poisonous alkaloids and should only be used with meticulous preparation and precise administration. This facility is used in several industrial applications (Dixit et al., 2013; Gupta, 2020). In the textile industry, the flowers and foliage of *D. stramonium* have been utilized as a source of natural dyes. When used as a dye, the alkaloids in the plant generate hues of yellow, green, and purple. The fiber extracted from the plant has been utilized to manufacture textiles and paper (Gumisiriza, 2016). The plant's fiber is robust and long-lasting, rendering it advantageous for many applications. Furthermore, the alkaloids in *D. stramonium* exhibit insecticidal characteristics, rendering it a viable natural pesticide (Bodade et al., 2019; Saka et al., 2022). Nevertheless, given the plant's poisonous characteristics, it is advisable to use care while using it as a pesticide. *D. stramonium*

TABLE 46.1
List of Compounds Extracted from *D. stramonium*

Bioactive Compounds	References
Alkaloids: Aponorscopolamine, Apoatropine, Aposcopolamine, Hyoscyamine, Litorine, Meteloidine, Methylscopolamine, Norapoatropine, Norscopolamine, Scopoline, Tropine, Scopine, Pseudotropine 3-Acetoxytropane, 7-epoxynortropane, 3-Phenylacetoxy-6, 3 (Hydroxyacetoxy)-tropane, 3-(2-Methylbutyryloxy)-tropane, 3-Tigloyloxynortropane, 3α Tigloyloxytropane, 3 α-Hydroxy-6 β acetoxytropane, 3 β Tigloyloxytropane, 3 α Phenylacetoxytropane, 3 β Phenylacetoxytropane, 3-(20-Hydroxytropoyloxy)-tropane, 3 β-Hydroxy-6 β tigloyloxytropane, 3-Hydroxy-6-propionyloxytropane, 3-(30-Acetoxytropoyloxy)-tropane, Scopolamine 3-(20 Hydroxytropoyloxy), 3 β-Hydroxy-6 β acetoxytropane, 3,6-Dihydroxytropane 6-Hydroxyacetoxytropane, 3-Isobutyryloxy-6-hydroxytropane, 3-Hydroxy-6-methylbutyryloxytropane, 3 α-Tigloiloxy-6 β hydroxytropane, 3 β-Tigloiloxy-6 β hydroxytropane, 3,6-Diacetoxytropane 3-Propionyloxy-6-hydroxytropane, 3-Hydroxy-6-isobutyryloxytropane, 3-Hydroxy-6-(2-methylbutyryloxy)-tropane, 3-Isovaleryloxy-6-hydroxytropane, 3 α-Tigloyloxy-6 β acetoxytropane, 6-Hydroxyhyoscyamine, 3 β-Tigloyloxy-6 β acetoxytropane, 3-Tigloyloxy-6-propionyloxytropane, 3α Tigloyloxy-6 β isobutyryloxytropane, 3α -Tigloyloxy-7 β isobutyryloxytropane, Dihydroxy-7-tigloyloxytropane, 3,6- 7-Hydroxyhyoscyamine, 3α, 6β-Ditigloyloxy-7β-hydroxytropane, 3-Tropoyloxy-6-acetoxytropane 6-Tigloyloxyhyoscyamine, 3α-Tigloyloxy-6β -isovaleryloxy-7β – hydroxytropane, 3 β, 6 β -Ditigloyloxytropane, 3-Tigloyloxy-6-(20 methylbutyryloxy)- tropane, 3 α, 6 β-Ditigloyloxytropane, 3β, 6β-Ditigloyloxy-7β-hydroxytropane, Apotropoyloxy-6 β -hydroxytropane, 3-Tigloyloxy-6-propionyloxy-7- hydroxytropane, 3-Tigloyloxy-6 methylbutyryloxytropane, 3 α-6-(2-Methylbutyryloxy) hyoscyamine, 3-Phenylacetoxy-6-hydroxytropane, 3β-Tigloyloxy-6β-isovaleryloxy-7β- hydroxytropane	Ally and Mohanlall (2020); El Bazaoui et al. (2011); Robins et al. (1991); Shonle and Bergelson (2000)
Flavonoids: Apigenin, chrysin, luteolin, kaempferol, quercetin,	Das et al. (2012; Malami et al. (2014)
Terpenoids: Diterpenoids, triterpenoids, monoterpenoids, sesquiterpenoids	Sreenivasa et al. (2012); Velázquez-Márquez et al. (2021)
Steroids: Campesterol, solanidine, solasodine, beta-sitosterol, stigmasterol	Al-Snafi (2017); Pan et al. (2022)
Other: Carbohydrates, fat, proteins, ash, fiber, calcium, oxalate iron, potassium, phosphorous, starch, glycosides	Ally and Mohanlall (2020); Das et al. (2012); Malami et al. (2014); Shonle and Bergelson (2000)

has been investigated in bioremediation for its capacity to eliminate heavy metals from polluted soil. The plant can assimilate heavy metals, namely lead, and cadmium, so it is a promising instrument for remediating contaminated areas (Shirkhani et al., 2021). However, it is essential to use prudence while using the plant, owing to its inherent toxicity and dangerous adverse reactions. Before using the plant for any industrial applications, it is advisable to get guidance from a certified expert.

46.10 HALLUCINOGENIC EFFECTS

D. stramonium, sometimes called jimsonweed, has a lengthy record of being used as a plant with psychedelic properties. The substance consists of several alkaloids that may induce diverse psychotropic effects, such as hallucinations, delirium, and disorientation (Gupta, 2020; Passos and Mironidou-Tzouveleki, 2016). The hallucinogenic properties of *D. stramonium* may be attributed to the existence of several alkaloids, such as scopolamine, hyoscyamine, and atropine. These alkaloids are categorized as anticholinergics, indicating that they inhibit the action of the neurotransmitter acetylcholine in the brain. These consequences may manifest as changes in perception, cognitive disorientation, and extreme mental bewilderment (Chaachouay & Zidane, 2021; Mukhtar et al., 2019; Stella et al., 2010).

The effects of *D. stramonium* can vary depending on the dose, the method of consumption, and the individual's sensitivity to the plant's alkaloids. Typically, the effects of the plant are felt within 30 minutes to an hour of consumption and can last for several hours or even days (Diker et al., 2007; Şanlıdağ et al., 2014). When consumed, the alkaloids in *D. stramonium* can affect the central nervous system. They block the activity of the neurotransmitter acetylcholine, which is involved in many brain functions, including memory, attention, and perception. This can lead to various hallucinogenic effects, including an altered perception of time and space, visual and auditory hallucinations, and changes in thought patterns (Mukhtar et al., 2019; Vearrier and Greenberg, 2010).

D. stramonium often induces visual hallucinations. These photos may be characterized by their vividness, vibrant colors, and frequent display of unusual or unpleasant elements. Users may also have auditory hallucinations, such as perceiving voices or noises that are absent. The plant may induce alterations in perception, such as a skewed impression of time and space (Forrester, 2006; Gupta, 2020). Users may experience time flowing more slowly or faster than usual, or they may experience thoughts beyond their body or in another world. *D. stramonium* may also cause confusion, disorientation, and memory impairment. Users may need assistance thinking clearly or recalling recent events, and they may need help to get distracted or confused (Boumba et al., 2004; Debnath and Chakraverty, 2017). Occasionally, *D. stramonium* may have more profound consequences, such as delirium, paranoia, and aggressive conduct. The consequences of these effects may be uncertain and risky, resulting in self-inflicted damage or harm inflicted against others.

It is crucial to acknowledge that the impacts of *D. stramonium* may be exceedingly erratic and diverge significantly based on the person and the context of consumption. The plant is also very poisonous and may induce bodily symptoms, such as urine retention, elevated heart rate, impaired eyesight, and dry mouth. Nevertheless, the plant has a significant level of toxicity and might induce unforeseeable and dangerous consequences. Therefore, it should be undertaken with utmost care or wholly abstained from.

46.11 PHARMACOLOGICAL EFFECTS

D. stramonium is a plant historically used for its therapeutic properties. The plant includes a variety of alkaloids, including scopolamine, hyoscyamine, and atropine, which are responsible for its pharmacological effects. This medicinal plant has many pharmacological effects, including:

- **Anticholinergic properties**: The alkaloids included in *D. stramonium* possess anticholinergic qualities, hence exerting an inhibitory effect on acetylcholine, a neurotransmitter responsible for muscular contraction, heart rate control, and several other physiological functions (Bouziri et al., 2011; Gaire & Subedi, 2013; Spina & Taddei, 2007).
- **Analgesic properties**: *D. stramonium* has historically been used for its analgesic effects, namely in the management of pain related to dental treatments, childbirth, and other ailments (Chandan et al., 2021; Das et al., 2012; Khalili Najafabadi & Atyabi, 2004; Soni et al., 2012).
- **Anti-asthmatic activity**: The use of *D. stramonium* for the treatment of asthma is attributed to its capacity to induce relaxation in bronchial smooth muscles. This action is attributed to the alkaloids found in the plant. (Pretorius & Marx, 2006; Singh & Singh, 2013; Soni et al., 2012).
- **Antidiabetic activities**: According to an investigation, *D. stramonium*'s hydroethanolic root extract exhibits antihyperglycemic properties. Additionally, the section significantly reduced diabetic dyslipidemia (Belayneh et al., 2019).
- **Antioxidant activities**: Numerous research have shown the antioxidant effects of *D. stramonium*. Antioxidants are substances that aid in defending cells against oxidative stress, a condition that may cause damage to cells and play a role in the development of certain illnesses (Belayneh et al., 2019; Mohammed et al., 2021). Leaves contain significant quantities of phenolic chemicals and flavonoids, which are recognized for their antioxidant characteristics. The methanol extract from the leaves exhibited considerable free radical scavenging activity and showed the ability to reduce lipid peroxidation, a primary contributor to cellular damage (Ananth & Rajan, 2015; Iqbal et al., 2017; Sreenivasa et al., 2012).
- **Antispasmodic properties**: Traditionally, *D. stramonium* has been employed for its antispasmodic properties, aiding in alleviating muscle spasms and cramping. (Das et al., 2012; Shagal et al., 2012; Stella et al., 2010).
- **Antibacterial effects**: There is a suggestion that *D. stramonium* could possess antibacterial properties. However, there is little scientific data to substantiate this assertion. Studies have shown that extracts derived from *D. stramonium* have antibacterial properties against certain bacterial strains (Eftekhar et al., 2005; Mohammed et al., 2021; Reddy, 2009). Nevertheless, the underlying processes of this activity still need comprehensive comprehension. It is essential to acknowledge that while there are some encouraging findings regarding the antibacterial properties of *D. stramonium*, further study is necessary to understand its potential in this field fully (Iranbakhsh et al., 2010).
- **Anti-inflammatory properties**: Investigations indicate that *D. stramonium* may possess anti-inflammatory characteristics that might be beneficial in the treatment of inflammatory disorders like arthritis (Aboluwodi et al., 2017; Chandan et al., 2021; Nasir et al., 2022; Sonika et al., 2010).

46.12 MEDICINAL USES

D. stramonium has a historical background of being used for therapeutic purposes, but its clinical use is restricted because of its toxicity and significant adverse effects. However, *D. stramonium* has therapeutic uses, namely in the fields of ophthalmology and anesthesia (Al-Snafi, 2017; Das et al., 2012). Jimsonweed has anticholinergic characteristics that may cause pupil dilatation, a desirable effect for ophthalmic purposes such as eye tests and cataract treatment (Choudhary et al., 2021; Leffler et al., 2017). *D. stramonium* has been used as an adjunct in anesthesia to produce profound sedation, particularly in cases involving complex airway care and mechanical ventilation, in conjunction with other medications (Holzman, 1999; Orr, 1975; Rahanandeh et al., 2022). Nevertheless, its utilization is not anticipated owing to the possibility of severe adverse reactions. It is essential to emphasize that the clinical use of *D. stramonium* should only be carried out by proficient healthcare practitioners with specialized knowledge and under stringent oversight (Freye & Freye, 2010; Stella et al., 2010). Due to its possible toxicity and significant adverse effects, it is essential to exercise care when considering the plant's usage in any therapeutic context.

46.13 TOXICITY OF *D. STRAMONIUM*

D. stramonium is a plant known for its toxicity. Its fame stems from potent tropane alkaloids in all plant portions, notably scopolamine, hyoscyamine, and atropine (Chaachouay et al., 2021a). The plant's harmful characteristics are attributed to tropane alkaloids, which are present in different amounts throughout the plant. When consumed or administered externally, they may have severe and sometimes lethal consequences for human well-being. It is vital to acknowledge and comprehend the substantial hazards linked to *D. stramonium* (Gaire & Subedi, 2013; Trancă et al., 2017).

The poisonous properties of *D. stramonium* have been widely recorded, emphasizing the significant hazards it presents to those who encounter it. Consuming even small quantities of this herb may lead to various unpleasant effects. The symptoms may include intense hallucinations, profound delirium, a fast and irregular heart rate, excessive dryness of the lips and mucous membranes, and impaired eyesight. Severe poisoning might result in a comatose condition or even fatality (Bouzidi et al., 2011; Devi et al., 2011; Freye & Freye, 2010).

Although very poisonous, *D. stramonium* has a long-standing history of being used in several traditional medicinal practices and shamanic ceremonies. It has been used in some societies due to its hallucinogenic and psychotropic characteristics. Nevertheless, the inherent danger of this plant cannot be exaggerated (Alebiowu et al., 2007). Due to its unexpected effects, propensity for overdose, and significant variation in alkaloid concentration, it is considered a very dangerous narcotic (Bouziri et al., 2011). Consequently, professionals and health authorities strongly advise against using *D. stramonium* for any reason. It is advisable to use caution and avoid contact with this plant (Benouadah et al., 2018). The possible advantages it may give are insignificant compared to the significant hazards connected with its toxicity. Therefore, persons must be fully aware of the dangers and take necessary precautions to avoid poisoning and the resulting health consequences.

46.14 CONCLUSION

D. stramonium is a very poisonous plant, nevertheless, it has had significant historical and cultural importance in many ways. Traditional civilizations have used it for medical reasons to address conditions such as asthma, coughing, and spasms. Certain societies have used the hallucinogenic qualities of this substance for spiritual purposes. In addition, scientists have researched the plant to explore its possible medicinal benefits. Scopolamine is an excellent treatment for motion sickness, nausea, and vomiting. Continuing study is investigating the potential use of the plant in the fields of pharmacology and medicine.

However, it is essential to recognize and comprehend the hazards linked to *D. stramonium*. The plant's poisonous nature may result in profound health complications and perhaps fatal outcomes. Hence, it is essential to use prudence while handling this plant, and its utilization should be rigorously regulated and supervised. It is crucial to educate oneself and others on the possible hazards of *D. stramonium*, especially in areas where the plant is prevalent. It is recommended to consult a knowledgeable practitioner or herbalist with expertise in exploring the possible therapeutic or hallucinogenic advantages of *D. stramonium*. It is essential to prioritize safety by maintaining a safe distance between children, dogs, and the plant. Additionally, practice care while disposing of any portions of the plant.

BIBLIOGRAPHY

Aboluwodi, A. S., Avoseh, N. O., Lawal, A. O., Ogunwande, I. A., & Giwa, A. A. (2017). Chemical constituents and anti-inflammatory activity of essential oils of *Datura stramonium* L. *J Med Plants Stud*, *5*(1), 21–25.

Adebayo, S. F., & Olufayo, M. O. (2017). Anaesthetic effects of *Datura stramonium* Leaf on *Heterobranchus bidorsalis* Juveniles. *International Journal of Fisheries and Aquatic studies*, *5*(2), 590–593.

Alebiowu, G., Femi-Oyewo, M. N., Elujoba, A. A., & Ojo, O. S. (2007). Toxicity studies on *Datura metel* L. with reference to official *Stramonium*. *Journal of Herbal Pharmacotherapy*, *7*(1), 1–12. https://doi.org/10.1080/J157v07n01_01

Ally, F., & Mohanlall, V. (2020). An overview of tropane alkaloids from *Datura stramonium* L. *Journal of Pharmacognosy and Phytochemistry*, *9*(3), 05–13.

Al-Snafi, A. E. (2017). Medical importance of *Datura fastuosa* (syn: *Datura metel*) and *Datura stramonium* – A review. *IOSR Journal of Pharmacy*, *7*(2), 43–58.

Ananth, A., & Rajan, S. (2015). *In-vitro antioxidant activity of Datura stramonium L. leaves*.

Ansari, M. K. A., Iqbal, M., Chaachouay, N., Ansari, A. A., & Owens, G. (2023). The concept and status of medicinal and aromatic plants: History, pharmacognosy, ecology, and conservation. In *Plants as Medicine and Aromatics* (pp. 129–144). CRC Press.

Bawazeer, S., Rauf, A., & Bawazeer, S. (2020). Gastrointestinal motility, muscle relaxation, antipyretic and acute toxicity screening of amyrin type triterpenoid (daturaolone) isolated from *Datura metel* Linnaeus (Angel's trumpet) fruits. *Frontiers in Pharmacology*, *11*, 544794.

Belayneh, Y. M., Birhanu, Z., Birru, E. M., & Getenet, G. (2019). Evaluation of in vivo antidiabetic, antidyslipidemic, and in vitro antioxidant activities of hydromethanolic root extract of *Datura stramonium* L. (Solanaceae). *Journal of Experimental Pharmacology*, *11*, 29–38. https://doi.org/10.2147/JEP.S192264

Belhaj, S., Chaachouay, N., & Zidane, L. (2021). Ethnobotanical and toxicology study of medicinal plants used for the treatment of diabetes in the High Atlas Central of Morocco. *Journal of Pharmacy & Pharmacognosy Research*, *9*(5), 619–662. https://doi.org/10.56499/jppres21.1014_9.5.619

Benkhnigue, O., Chaachouay, N., Khamar, H., El Azzouzi, F., Douira, A, & Zidane, L. (2022). Ethnobotanical and ethnopharmacological study of medicinal plants used in the treatment of anemia in the region of Haouz-Rehamna (Morocco). *Journal of Pharmacy & Pharmacognosy Research*, *10*(2), 279–302. https://doi.org/10.56499/jppres21.1196_10.2.279

Benkhnigue, O., Chaachouay, N., & Zidane, L. (2023). A Floristical and Ecological Study of the Medicinal Flora Used by the Local Population of the Haouz-Rehamna Region (Middle Atlantic Morocco-4). *Acta Botanica Hungarica Akadémiai Kiadó*, *65*(3–4), 247–304. https://doi.org/10.1556/034.65.2023.3-4.3

Benouadah, Z., Mahdeb, N., & Bouzidi, A. (2018). *Evaluation of acute and sub-acute toxicity of alkaloids from Datura stramonium sp. In mice*. http://dspace.univ-setif.dz:8888/jspui/handle/123456789/2552

Blakeslee, A. F., Bergner, A. D., & Avery, A. G. (1937). Geographical distribution of chromosomal prime types in *Datura stramonium*. *Cytologia*, *2*, 1070–1093.

Bodade, A. B., Wankhade, A. S., Kale, P. D., & Chaudhari, G. N. (2019). Preparation and characterization of nanocrystaline magnesium oxide using *Datura stramonium* leaves and its antibacterial activity. *Int. J. Innov. Stud. Sci. Eng. Technol.*, *5*, 42–45.

Boumba, V. A., Mitselou, A., & Vougiouklakis, T. (2004). Fatal poisoning from ingestion of *Datura stramonium* seeds. *Veterinary and Human Toxicology*, *46*(2), 81–82.

Bouzidi, A., Mahdeb, N., & Kara, N. (2011). Toxicity studies of alkaloids of seeds of *Datura stramonium* and synthesis alkaloids in male rats. *J Med Plants Res*, *5*(15), 3421–3431.

Bouziri, A., Hamdi, A., Borgi, A., Hadj, S. B., Fitouri, Z., Menif, K., & Jaballah, N. B. (2011). *Datura stramonium* L. poisoning in a geophagous child: A case report. *International Journal of Emergency Medicine*, *4*(1), 31. https://doi.org/10.1186/1865-1380-4-31

Castillo, G., Cruz, L. L., Hernandez-Cumplido, J., Oyama, K., Flores-Ortiz, C. M., Fornoni, J., Valverde, P. L., & Nunez-Farfan, J. (2013). Geographic association and temporal variation of chemical and physical defense and leaf damage in *Datura stramonium*. *Ecological Research*, *28*, 663–672.

Chaachouay, N., Benkhnigue, O., Fadli, M., Ibaoui, H., & Zidane, L. (2019a). Ethnobotanical and ethnopharmacological studies of medicinal and aromatic plants used in the treatment of metabolic diseases in the Moroccan Rif. *Heliyon*, *5*(10), e02191. https://doi.org/10.1080/10477845.2021.1942433

Chaachouay, N., Benkhnigue, O., Fadli, M., Ibaoui, H., El Ayadi, R., & Zidane, L. (2019b). Ethnobotanical and ethnopharmacological study of medicinal and aromatic plants used in the treatment of respiratory system disorders in the Moroccan Rif. *Ethnobotany Research and Applications*, *18*(22), 1–16.

Chaachouay, N., Douira, A., Hassikou, R., Brhadda, N., Dahmani, J., Belahbib, N., Ziri, R., & Zidane, L. (2020a). Ethnobotanical and ethnopharmacological study of medicinal and aromatic plants used in the treatment of respiratory system disorders in the Moroccan Rif. PhDThesis, Département de Biologie - Université Ibn Tofail - Kénitra, Maroc, https://tel.archives-ouvertes.fr/tel-03376377

Chaachouay, N., Benkhnigue, O., & Zidane, L. (2020b). Ethnobotanical study aimed at investigating the use of medicinal plants to treat nervous system diseases in the Rif of Morocco. *Journal of Chiropractic Medicine*, *19*(1), 70–81. https://doi.org/10.1016/j.jcm.2020.02.004

Chaachouay, N., & Zidane, L. (2021). The Symbolic Efficacy of Plants in Rituals and Socio-Religious Ceremonies in Morocco, Northwest of Africa. *Journal of Religious & Theological Information*, *21*(1–2), 34–53. https://doi.org/10.1080/10477845.2021.1942433

Chaachouay, N., Benkhnigue, O., Douira, A., & Zidane, L. (2021a). Poisonous medicinal plants used in the popular pharmacopoeia of the Rif, northern Morocco. *Toxicon*, 189, 24–32. https://doi.org/10.1016/j.toxicon.2020.10.028

Chaachouay, N., Douira, A., & Zidane, L. (2021b). COVID-19, prevention and treatment with herbal medicine in the herbal markets of Salé Prefecture, North-Western Morocco. *European Journal of Integrative Medicine*, 42, 101285. https://doi.org/10.1016/j.eujim.2021.101285

Chaachouay, N., Douira, A., & Zidane, L. (2021c). Herbal Medicine Used in the Treatment of Human Diseases in the Rif, Northern Morocco. *Arabian Journal for Science and Engineering*, 47, 131–153. https://doi.org/10.1007/s13369-021-05501-1

Chaachouay, N., Benkhnigue, O., & Zidane, L. (2022a). Ethnobotanical and Ethnomedicinal study of medicinal and aromatic plants used against dermatological diseases by the people of Rif, Morocco. *Journal of Herbal Medicine*, 32, 100542. https://doi.org/10.1016/j.hermed.2022.100542

Chaachouay, N., Azeroual, A., Bencharki, B., Douira, A., & Zidane, L. (2022b). Ethnoveterinary medicines plants for animal therapy in the Rif, North of Morocco. *South African Journal of Botany*, *147*, 176–191. https://doi.org/10.1016/j.sajb.2021.12.037

Chaachouay, N., Azeroual, A., Douira, A., & Zidane, L. (2022c). Ethnoveterinary practices of medicinal plants among the Zemmour and Zayane tribes, Middle Atlas, Morocco. *South African Journal of Botany*, *151*, 826–840. https://doi.org/10.1016/j.sajb.2022.11.009

Chandan, G., Kumar, C., Chibber, P., Kumar, A., Singh, G., Satti, N. K., Gulilat, H., Saini, A. K., Bishayee, A., & Saini, R. V. (2021). Evaluation of analgesic and anti-inflammatory activities and molecular docking analysis of steroidal lactones from *Datura stramonium* L. *Phytomedicine*, *89*, 153621.

Choudhary, M., Sharma, I., Agrawal, D. C., Dhar, M. K., & Kaul, S. (2021). Neurotoxic potential of alkaloids from thorn apple (*Datura stramonium* L.): A commonly used Indian folk medicinal herb. In D. C. Agrawal & M. Dhanasekaran (Éds.), *Medicinal Herbs and Fungi: Neurotoxicity vs. Neuroprotection* (pp. 391–420). Springer. https://doi.org/10.1007/978-981-33-4141-8_16

Das, S., Kumar, P., & Basu, S. P. (2012). Phytoconstituents and therapeutic potentials of *Datura stramonium* Linn. *Journal of Drug Delivery and Therapeutics*, *2*(3), 4–7.

Debnath, T., & Chakraverty, R. (2017). Newer insights into the psychoactive and pharmacological properties of *Datura stramonium* Linn. *Journal of Addiction - Rehabilitation Medicine Journals*, *4*, 0094–0096.

De-la-Cruz, I. M., Merilä, J., Valverde, P. L., Flores-Ortiz, C. M., & Núñez-Farfán, J. (2020). Genomic and chemical evidence for local adaptation in resistance to different herbivores in *Datura stramonium*. *Evolution*, *74*(12), 2629–2643.

De-la-Mora, M., Piñero, D., & Núñez-Farfán, J. (2015). Phylogeography of specialist weevil Trichobaris soror: A seed predator of *Datura stramonium*. *Genetica*, *143*, 681–691.

Devi, M. R., Bawari, M., Paul, S. B., & Sharma, G. D. (2011). Neurotoxic and medicinal properties of *Datura stramonium*

L. – Review. *Assam University Journal of Science and Technology*, *7*(1), 139–144.

Diker, D., Markovitz, D., Rothman, M., & Sendovski, U. (2007). Coma as a presenting sign of *Datura stramonium* seed tea poisoning. *European Journal of Internal Medicine*, *18*(4), 336–338.

Dixit, S., Upadhyay, S., Singh, H., Pandey, B., Chandrashekar, K., & Verma, P. (2013). Pectin methylesterase of *Datura* species, purification, and characterization from *Datura stramonium* and its application. *Plant Signaling & Behavior*, *8*(10), e25681.

dos Santos, R. G., Bouso, J. C., Rocha, J. M., Rossi, G. N., & Hallak, J. E. (2021). The use of classic hallucinogens/psychedelics in a therapeutic context: Healthcare policy opportunities and challenges. *Risk Management and Healthcare Policy*, *14*, 901–910. https://doi.org/10.2147/RMHP.S300656

Eftekhar, F., Yousefzadi, M., & Tafakori, V. (2005). Antimicrobial activity of *Datura innoxia* and *Datura stramonium*. *Fitoterapia*, *76*(1), 118–120. https://doi.org/10.1016/j.fitote.2004.10.004

El Bazaoui, A., Bellimam, M. A., & Soulaymani, A. (2011). Nine new tropane alkaloids from *Datura stramonium* L. identified by GC/MS. *Fitoterapia*, *82*(2), 193–197.

Fornoni, J., Valverde, P. L., & Nunez-Farfan, J. (2004). Population variation in the cost and benefit of tolerance and resistance against herbivory in *Datura stramonium*. *Evolution*, *58*(8), 1696–1704.

Forrester, M. B. (2006). Jimsonweed (*Datura stramonium*) exposures in Texas, 1998–2004. *Journal of Toxicology and Environmental Health, Part A*, *69*(19), 1757–1762.

Freye, E., & Freye, E. (2010). Toxicity of *Datura stramonium*. *Pharmacology and abuse of cocaine, amphetamines, ecstasy and related designer drugs: A comprehensive review on their mode of action, treatment of abuse and intoxication*, 217–218.

Gaire, B. P., & Subedi, L. (2013). A review on the pharmacological and toxicological aspects of *Datura stramonium* L. *Journal of Integrative Medicine*, *11*(2), 73–79.

Gumisiriza, O. (2016). *Development of a bioactive fabric using dye extracts from selected indigenous medicinal plant species*. [PhD Thesis]. Busitema University.

Gupta, A. (2020). Experimental Investigation of *Datura stramonium* biodiesel. *IJARIIT*, *6*, 1–5.

Holzman, R. S. (1999). Anesthetic uses of hyoscine and atropine alkaloids in surgical Arabic book. *Anesthesiology*, *90*(6), 1795–1796. https://doi.org/10.1097/00000542-199906000-00053

Iqbal, S., Sivaraj, C., & Gunasekaran, K. (2017). Antioxidant and anticancer activities of methanol extract of seeds of *Datura stramonium* l. *Free Radicals and Antioxidants*, *7*(2), Article 2. https://doi.org/10.5530/fra.2017.2.28

Iranbakhsh, A., Ebadi, M., & Bayat, M. (2010). The inhibitory effects of plant methanolic extract of *Datura stramonium* L. and leaf explant callus against bacteria and fungi. *Global Veterinaria*, *4*(2), 149–155.

ITIS. (2022). *Integrated Taxonomic Information System (ITIS) online database, www.itis.gov, CC0* https://doi.org/10.5066/F7KH0KBK. http://itis-web-prod.snafu.cr.usgs.gov

Khalili Najafabadi, M., & Atyabi, S. M. (2004). Evaluation of analgesic effect of *Datura Stramonium* seed extract in hot plate and formalin tested male rats. *Iranian Journal of Medicinal and Aromatic Plants Research*, *20*(3), 309–322.

Leffler, C. T., Schwartz, S. G., Wainsztein, R. D., Pflugrath, A., & Peterson, E. (2017). Ophthalmology in North America: Early stories (1491–1801). *Ophthalmology and Eye Diseases*, *9*. https://doi.org/10.1177/1179172117721902

Mahomoodally, M. F. (2013). Traditional medicines in Africa: An appraisal of ten potent African medicinal plants. *Evidence-Based Complementary and Alternative Medicine, 2013*.

Malami, I., Halilu, M. E., Mathias, S., & Alhassan, M. A. (2014). Phytochemical evaluation and investigations in to sedative properties of *Datura stramonium* (Linn) seeds in experimental mice. *J Pharm Biol Sci*, *9*(1), 1–3.

Mohammed, F. S., Kına, E., Sevindik, M., Dogan, M., & Pehlivan, M. (2021). *Datura stramonium* (Solanaceae): Antioxidant and antimicrobial potentials. *Turkish Journal of Agriculture - Food Science and Technology*, *9*(4), Article 4. https://doi.org/10.24925/turjaf.v9i4.818-821.4264

Mountain, W. L. (1987). Jimsonweed, *Datura Stramonium* L. *Solanaceae Regulatory Horticulture, Weed Circular, 12*.

Mukhtar, Y., Tukur, S., & Bashir, R. A. (2019). An overview on *Datura stramonium* L.(Jimson weed): A notable psychoactive drug plant. *American Journal of Natural Sciences*, *2*(1), 1–9.

Nasir, B., Khan, A. U., Baig, M. W., Althobaiti, Y. S., Faheem, M., & Haq, I.-U. (2022). *Datura stramonium* leaf extract exhibits anti-inflammatory activity in CCL_4-induced hepatic injury model by modulating oxidative stress markers and iNOS/Nrf2 expression. *BioMed Research International*, *2022*, e1382878. https://doi.org/10.1155/2022/1382878

Núñez-Farfán, J., & Dirzo, R. (1994). Evolutionary ecology of *Datura stramonium* L. in central Mexico: Natural selection for resistance to herbivorous insects. *Evolution*, *48*(2), 423–436.

Orch, H., Chaachouay, N., Douiri, E. M., Faiz, N., Zidane, L., & Douira, A. (2021). Use of medicinal plants in dermato-cosmetology: An ethnobotanical study among the population of Izarène. *Jordan Journal of Pharmaceutical Sciences*, *14*(3), 323–340.

Orr, R. (1975). Reversal of *Datura Stramonium* delirium with physostigmine: Report of three cases. *Anesthesia & Analgesia*, *54*(1), 158.

Pan, J., Liu, Y., Wang, S.-Y., Wu, J.-T., Guo, M.-Y., Guan, W., Algradi, A. M., Wang, D.-S., Kuang, H.-X., & Yang, B.-Y. (2022). Mantuoluosides AH, new steroids isolated from the leaves of *Datura stramonium* L. *Fitoterapia*, *163*, 105339.

Passos, I. D., & Mironidou-Tzouveleki, M. (2016). Hallucinogenic plants in the Mediterranean countries. In *Neuropathology of drug addictions and substance misuse, Vol 2: Stimulants, Club and Dissociative Drugs, Hallucinogens, Steroids, Inhalants and International Aspects* (pp. 761–772). Elsevier, Academic Press.

Pretorius, E., & Marx, J. (2006). *Datura stramonium* in asthma treatment and possible effects on prenatal development. *Environmental Toxicology and Pharmacology*, *21*(3), 331–337.

Rahanandeh, M., Tizkar, B., & Rahanandeh, M. (2022). Research article: The effects of *Datura stramonium* L. seed extract on anesthesia of farmed *Cyprinus carpio* in Guilan province. *Sustainable Aquaculture and Health Management Journal*, *8*(1), 16–25. https://doi.org/10.52547/ijaah.8.1.16

Reddy, B. U. (2009). *Antimicrobial activity of Datura stramonium L. and Tylophora indica (Burm.f.) Merr.*

Robins, R. J., Parr, A. J., Bent, E. G., & Rhodes, M. J. (1991). *Studies on the biosynthesis of tropane alkaloids in Datura stramonium L. transformed root cultures*.

Saka, A., Jule, L. T., Gudata, L., Gindaba, A., Abdisa, S. S., Nagaprasad, N., & Ramaswamy, K. (2022). Green synthesis of *Datura stramonium* (Asaangira) leaves infusion for antibacterial activity through magnesium oxide

(MgO) nanoparticles. *Advances in Materials Science and Engineering*, *2022*(1), 7134991.

Şanlıdağ, B., Derinöz, O., & Yıldız, N. (2014). A case of pediatric age anticholinergic intoxication due to accidental *Datura stramonium* ingestion admitting with visual hallucination. *Turk J Pediatr*, *56*(3), 313–315.

Sayin, H. U. (2016). Psychoactive plants used during religious rituals. In *Neuropathology of drug addictions and substance misuse* (pp. 17–28). Elsevier, Academic Press.

Sayyed, A. (2014). Phytochemistry, pharmacological and traditional uses of *Datura stramonium* L. review. *Journal of Pharmacognosy and Phytochemistry*, *2*(5), 123–125.

Shagal, M. H., Modibbo, U. U., & Liman, A. B. (2012). Pharmacological justification for the ethnomedical use of *Datura stramonium* stem-bark extract in treatment of diseases caused by some pathogenic bacteria. *Int Res Pharm Pharmaco*, *2*(1), 16–19.

Shi, Z., Zou, W., Zhu, Z., Xiong, Z., Li, S., Dong, P., & Zhu, Z. (2022). Tropane alkaloids (hyoscyamine, scopolamine and atropine) from genus *Datura*: Extractions, contents, syntheses and effects. *Industrial Crops and Products*, *186*, 115283.

Shirkhani, Z., Chehregani Rad, A., & Mohsenzadeh, F. (2021). Improving Cd-phytoremediation ability of *Datura stramonium* L. by Chitosan and Chitosan nanoparticles. *Biologia*, *76*(8), 2161–2171.

Shonle, I., & Bergelson, J. (2000). Evolutionary ecology of the tropane alkaloids of *Datura stramonium* L. (Solanaceae). *Evolution*, *54*(3), 778–788.

Singh, L. R., & Singh, O. M. (2013). *Datura stramonium*: An overview of its phytochemistry and pharmacognosy. *Research Journal of Pharmacognosy and Phytochemistry*, *5*(3), 143–148.

Soni, P., Siddiqui, A. A., Dwivedi, J., & Soni, V. (2012). Pharmacological properties of *Datura stramonium* L. as a potential medicinal tree: An overview. *Asian Pacific Journal of Tropical Biomedicine*, *2*(12), 1002–1008.

Sonika, G., Manubala, R., & Deepak, J. (2010). Comparative studies on anti-inflammatory activity of *Coriandrum sativum*, *Datura stramonium* and *Azadirachta indica*. *Asian J Exp Biol Sci*, *1*(1), 151–154.

Spina, S. P., & Taddei, A. (2007). Teenagers with Jimson weed (*Datura stramonium*) poisoning. *Canadian Journal of Emergency Medicine*, *9*(6), 467–469. https://doi.org/10.1017/S1481803500015530

Sreenivasa, S., Vinay, K., & Mohan, N. R. (2012). Phytochemical analysis, antibacterial and antioxidant activity of leaf extract of *Datura stramonium*. *International Journal of Science Research*, *1*(2), 83–86.

Stella, L., Vitelli, M. R., Palazzo, E., Oliva, P., De Novellis, V., Capuano, A., Scafuro, M. A., Berrino, L., Rossi, F., & Maione, S. (2010). *Datura stramonium* intake: A report on three cases. *Journal of Psychoactive Drugs*, *42*(4), 507–512.

The Plant List. (2013). The Plant List. Version 1.1. Published on the Internet; http://www.theplantlist.org/ (accessed 1st January).

Trancă, S. D., Szabo, R., & Cociş, M. (2017). Acute poisoning due to ingestion of *Datura stramonium* – A case report. *Romanian Journal of Anaesthesia and Intensive Care*, *24*(1), 65.

Valverde, P. L., Fornoni, J., & Núñez-Farfán, J. (2003). Evolutionary ecology of *Datura stramonium*: Equal plant fitness benefits of growth and resistance against herbivory. *Journal of Evolutionary Biology*, *16*(1), 127–137.

Vearrier, D., & Greenberg, M. I. (2010). Anticholinergic delirium following *Datura stramonium* ingestion: Implications for the Internet age. *Journal of Emergencies, Trauma and Shock*, *3*(3), 303.

Velázquez-Márquez, S., De-la-Cruz, I. M., Tapia-López, R., & Núñez-Farfán, J. (2021). Tropane alkaloids and terpenes synthase genes of *Datura stramonium* (Solanaceae). *PeerJ*, *9*, e11466.

Weaver, S. E., Dirks, V. A., & Warwick, S. I. (1985). Variation and climatic adaptation in northern populations of *Datura stramonium*. *Canadian Journal of Botany*, *63*(7), 1303–1308.

47 Jimson Weed (*Brugmansia pittieri* (Saff.) Moldenke Solanaceae)

Khola Rafique and Alvina Gul

47.1 INTRODUCTION

Hallucinogenic plants have been used since ancient times, primarily for religious reasons, but eventually also for their therapeutic properties (Calina et al., 2021). Various hallucinogenic species have played an essential role in the culture and religion of both primitive tribes and more advanced civilizations, both in the past and present (Widelski and Kukula-Koch, 2017). In an ongoing battle with their highly adaptable predators, plants continually evolve to generate diverse defensive secondary metabolites (Wink, 2016). It has been believed that certain notable psychoactive compounds, like atropine, nicotine, caffeine, cocaine, and morphine, may have arisen as a result of this evolutionary struggle between plants and their predators (Fürstenberg-Hägg et al., 2013). Substances that are psychoactive, or psychotropic, have an impact on the nervous system, which in turn affects mental processes and behavior (Rätsch, 2005). Psychoactive substances include hallucinogens that encompass a range of effects, which can alter one's perception of reality, to sedative or narcotics that cause sleep, to antidepressants, anxiolytics or calmatives, and stimulants that can stimulate the mind (Van Wyk and Wink, 2014).

In the 1950s, the term hallucinogen gained popularity and was associated with substances that were believed to induce hallucinations, but it has since been discovered that this is frequently not the case when taken in smaller amounts (Baker, 2005). The primary impact of these substances is on mood, perception, and thought. When consumed in small quantities, they have minimal impact on intellectual capabilities and do not lead to any disabling effects. Moreover, these substances are not physically addictive (Kordova and Ventegodt, 2016). Typically, these impacts are described as dreamy and involve sensory alterations, and at times actual hallucinations, generally visual or auditory in nature. Emotional states are often intensified, time can seem distorted, and the experience may have mystical elements, leading to various spiritual insights and revelations. This can include a sense of ego dissolution, feeling as if having had a near-death experience, and a sense of oneness with the universe or greater understanding (Sanz and Tagliazucchi, 2018). Following a period of heightened alarm regarding these substances in the 1970s, their usage gradually declined (Frecska, 2007). During the 1990s, there was a resurgence in the use of these substances, which has been bolstered by the widespread availability of the internet (Mrazova et al., 2011). Therefore, their current usage is of considerable interest from multiple perspectives, including medicinal, ethnobotanical, social, and legal (Fatur, 2021).

The Solanaceae family is amongst the most extensive and broadly dispersed plant families containing around 2,500 species and 100 genera. These species have historically held significant cultural importance in various societies, having played a critical role in human nutrition, medicine, and religion (Ghatak et al., 2017; Fatur, 2019; Rosario et al., 2019). The Solanaceae family's nightshade genus, *Brugmansia* Pers, comprises psychoactive flowering species known for their hallucinogenic properties (Kennedy, 2014). These plants are typically woody trees or shrubs with drooping flowers and smooth fruit devoid of spines. Numerous hallucinogenic and potentially dangerous tropane alkaloids, including atropine, scopolamine, and hyoscyamine, are derived from plants such as *Brugmansia* spp. (angel's trumpet), *Atropa belladonna* (deadly nightshade), *Datura* spp. (jimson weed or loco weed), and *Hyoscyanus niger* (henbane) (Bock and Norris, 2015; Marin-Saez et al., 2019). Each part of both *Brugmansia* and *Datura* species contains atropine and scopolamine, which are known to be toxic (Romera-Torres et al., 2018; Kerchner et al., 2015; Jakabová et al., 2012).

Genus *Brugmansia*, commonly known as 'golden angel's trumpet' or 'Jimson weed', named after their sizable and fragrant flowers. This name is occasionally used interchangeably with the closely-related *Datura* genus. In recent times, the three genera *Brugmansia* Pers, *Datura* L., and *Iochroma* Benth. have come to be recognized as constituting the Datureae Tribe of the Solanaceae family (Benítez et al., 2018). *Brugmansia* is a native flowering plant species of South America. Around 7–8 species fall under the genus *Brugmansia*, including *B. aurea, B. arborea, B. candida, B. insignis, B. suaveolens, B. sanguinea, B. sp*, and *B. versicolor* (Diaz, 2015; Kerchner and Farkas, 2020), which are distributed in America, Asia, Europe, and Africa. The species called *B. pittieri* (Saff.) Moldenke (synonym *B. aurea* Lagerh. or *Datura pittieri* Saff.) is a popular and widely cultivated similar to other members of the genus. According to The Plant List (2023), 12 species have been recognized under this genus, including various hybrids and a subspecies (*B. arborea, B.* × *candida, B.* × *cubensis, B.* × *dolichocarpa, B.* × *insignis, B. longifolia, B. pittieri,*

DOI: 10.1201/9781003460336-47

B. × *rubella*, *B. sanguinea*, *B. sanguinea* subsp. *vulcanicola*, *B. suaveolens*, and *B. versicolor*).

The widespread occurrence of *Brugmansia* and *Datura* taxa across the globe, driven by their successful reproductive strategies and ornamental cultivation, has led to an increase in their toxicological significance. This effect is not limited to tropical regions but is also observed in temperate regions worldwide (Kerchner and Farkas, 2020). The purpose of this chapter is to provide a comprehensive overview of the botany, habitat, and ecology, ethnobotany, traditional uses, phytochemistry, pharmacological activities, hallucinogenic effects, and toxicity of Jimson weed. By analyzing this information, we aim to establish a connection between these factors and provide a scientific basis for future research opportunities.

47.2 SCIENTIFIC CLASSIFICATION

In addition to *Brugmansia*, there are various types of *Datura* that are commonly referred to as angel's trumpet because of their unique flowers. However, *Datura* species are more commonly recognized as thorn apples. The *Brugmansia* species are distinguished by their flower color like the yellow (or golden) angel's trumpet (*B. aurea*), red angel's trumpet (*B. sanguinea*), white angel's trumpet (*B. suaveolens*), and peach angel's trumpet (*B. versicolor*). On the other hand, *B. arborea* is commonly known as angel's trumpet without any color indication (USDA-ARS, 2015). Although similar flower colors may belong to different species, some species are intentionally selected for their ornamental value. Therefore, relying solely on this trait may lead to inaccurate identification.

Besides the above-mentioned *Brugmansia* species, the Plant List (2015) also involves *B. longifolia* and *B. pittieri* (synonym for *B. aurea*) under this genus. The taxonomical classification and synonyms of *B. pittieri* has been presented in Table 47.1 (Hassler, 2022). Various authorities like the USDA-ARS (2015), the Plant List (2015), and Missouri Botanical Garden (2015) consider several hybrids of *Brugmansia* with other species, such as *B. x candida*, as distinct taxa. *B. candida* is commonly recognized as a distinct species (Haegi, 1976), and additional hybrids may be designated with both specific and hybrid classifications, such as *B. pittieri* and *B. x pittieri*. In 2012, a comprehensive taxonomic revision of the *Brugmansia* genus was conducted and made available by Hay et al. (2012). This was due to the confusion that existed with regard to its taxonomy and nomenclature.

The *Brugmansia* species have also been categorized into two distinct groups based on their genetics and natural isolation (Shaw, 1999). The warm-growing group, also known as Brugmansia section *Brugmansia*, comprises aurea, sauveolens, insignis, and versicolor. On the other hand, the cold group, also known as Brugmansia section Sphaerocarpium, includes arborea, sanguinea, and vulcanicola (Hay et al., 2012). A variety of ornamental plants have been created, including several hybrids and cultivars. *B.* × *candida*, which is a hybrid between *B. aurea* and *B. versicolor, B.* × *flava*, a hybrid between *B. arborea* and *B. sanguinea*, and *B.* × *cubensis* (Shaw, 1999), which is a hybrid between *B. suaveolens, B. aurea*, and *B. versicolor*, are among them. Some cultivars are double-flowered, while others have variegated leaves.

TABLE 47.1
Taxonomical Classification of *Brugmansia pittieri*

Kingdom	Plantae
Subkingdom	Tracheobionta
Superdivision	Spermatophyta
Division	Magnoliophyta
Class	Magnoliopsida
Order	Solanales
Family	Solanaceae
Subfamilia	Solanoideae
Tribe	Datureae
Genus	*Brugmansia*
Species	*B. pittieri*
Synonyms	*B. affinis* (Soff.) Moldenke
	B. amesianum (R.E.Schult.) D'Arcy
	B. aurea Lagerh.
	Datura affinis Saff.
	D.aurea (Lagerh.) Saff.
	D. pittieri Saff.
	Methysticodendron amesianum R.E.Schult.

47.3 VERNACULAR NAMES

Genus *Brugmansia* commonly known as 'golden angel's trumpet' or 'Jimson weed', named after their sizable and fragrant flowers. This name is occasionally used interchangeably with the closely related *Datura* genus.

47.4 BOTANICAL DESCRIPTION

The genus *Brugmansia* of the Solanaceae family is named after Sebald J. Brugmans (1763–1819), who was a Dutch botanist, physician, and professor of natural sciences and honored for his contributions to the field. The species of this genus possess a defining characteristic of a strong, sweet scent that is most noticeable during the evening hours (Preissel and Preissel, 2002). *Brugmansia* is a genus comprising flowering species. Popularly known as 'angel's trumpets', *Brugmansia* is a woody and perennial plant (Preissel and Preissel, 2002). *Brugmansia* species are comprised of tall evergreen shrubs and petite trees that can grow to heights ranging from 3 to 11 meters. The common name for this plant, 'angel's trumpets', is derived from its impressive, large-sized, pendulous, drooping flowers, which can measure up to 14–50 cm long and 35 cm wide. The *Brugmansia*'s flowers are ideal, singular, drooping, and emit a delightful scent during the night to lure nocturnal moths, which serve as its pollinators (Alvarez, 2008). These flowers come in a range of hues, including white, yellow, orange, pink, and red (Zayed and Wink, 2004). The leaves on the stems are arranged alternately and can either be toothed or entirely smooth (Melissa, 2023).

The Solanaceae is a monophyletic group of plants, and it includes around 100 different genera and over 2500 species (Ghatak et al., 2017; Rosario et al., 2019; Fatur, 2019). This group is very important because many popular crops come from it, such as potatoes, tomatoes, chili peppers, eggplants, and more (APG III, 2009). Traditionally, the family has been divided into two subfamilies based on their characteristics. The first subfamily is called Cestroideae and it has plants with straight or bent embryos and fruits that are in the form of a capsule. The second subfamily is called Solanoideae and it has plants with curved embryos and fruits that are berry-like (Hunziker, 1979; Olmstead and Palmer, 1992; Hunziker, 2001). In the family, eight clades have been identified, with Clade Solanoideae being one of them. This clade is made up of two significant genera, *Datura* and *Brugmansia*. These genera are known for their high alkaloid content, which has been used in traditional and modern medicine (Dhanya et al., 2018).

In the past, the members belonging to the arboreous South American genus *Brugmansia* were categorized under the genus *Datura*. *Brugmansia* was historically considered a subset of genus *Datura*, it is now widely recognized as a distinct (yet closely linked) genus (Doncheva et al., 2006). According to a recent recognition, the three genus viz. *Brugmansia* (which belongs to the 4th section of the tribe Datureae), *Iochroma*, and *Datura* have been shown to compose the Datureae Tribe belonging to family Solanaceae (Benítez et al., 2018). Since the 1970s, the two genera *Brugmansia* and *Datura* have been recognized as morphologically separate (Lockwood, 1973). In addition to longevity, plant habit, and additional morphological characteristics, it is also possible to differentiate between species in the two genera based on their contents of specific alkaloids and the microscopic characteristics of their leaves (Pasiecznik, 2014). The separation between *Datura* and *Brugmansia* is also clearly evident from the chemical analysis of their constituent alkaloids (Doncheva et al., 2006).

Datura includes herbaceous species, while *Brugmansia* contains woody species (Lockwood, 1973; Bye and Sosa, 2013). In a study, it has been shown that *Brugmansia* and *Datura* have similar morphological characteristics in terms of hairy leaves that emit an unpleasant odor when crushed or damaged, as well as fragrant funnel-shaped flowers. However, *Datura* is distinguished from *Brugmansia* by its annual or biennial growing habit, as well as its upright flowers facing upwards; giving them the name 'devil's trumpet', and rounded, spiny capsules. In contrast, *Brugmansia* is a perennial plant with drooping pendulous flowers facing downwards; giving them the name 'angel's trumpet' and elongated and smooth fruits (Preissel and Preissel, 2002; Dhanya et al., 2018).

47.5 ORIGIN AND GEOGRAPHICAL DISTRIBUTION

Brugmansia, as a genus, is naturally found in subtropics of South America, primarily along the Andes from Colombia to Northern Chile. Additionally, it can also be found in Southeast Brazil (Preissel and Preissel, 2002). These are found as ornamental container plants globally, predominantly in tropical, subtropical, and temperate regions (Kim et al., 2020) and typically have become naturalized to isolated tropical regions, such as North America, Africa, Asia, and Australia (Kurniati et al., 2010). In Sri Lanka, *B. suaveolens* is the only common species prevalent under this genus. In China, cultivation of two species (*B. arborea* and *B. suaveolens*) from this genus has been reported (Plant intelligence, 2019).

47.6 HABITAT AND ECOLOGY

B. aurea or *B. pittieri* is native to Ecuador. The International Union for Conservation of Nature (IUCN) has classified it as Extinct in the Wild since March 2014, but prior to that it was considered Vulnerable (Hay, 2014). Although it has been declared extinct in its natural habitat, *B. aurea* or *B. pittieri* is still a popular ornamental plant and is widely cultivated, along with other members of the *Brugmansia* genus. It is commonly sold and grown as a garden plant and is described as a large evergreen subtropical shrub that can grow up to 6.1 m tall. In the Northern Hemisphere, its large, drooping, trumpet-shaped blooms come out in summer and autumn, and they are usually yellow or white. The pleasant fragrance is most noticeable in the evening (Anonymous, 2023).

Brugmansia plants thrive best in a well-drained, fertile soil that is kept moist. They prefer full sun to part shade, and grow well in frost-free climates. Blooms typically emerge in mid to late spring in warm areas and continue well into the fall season. In warm conditions, it's not uncommon for the blooms to continue until early winter. During colder winters, outdoor plant life requires extra protection. However, with hardy roots, plants typically bounce back with resprouts in April or May. Species from the higher elevations in *B.* section *Sphaerocarpium* prefer moderate temperatures and cool nights, and may not flower if temperatures become too hot. On the other hand, most *Brugmansias* can be easily multiplied through cuttings of 10–20 cm size obtained from branch ends during the summer season (Anonymous, 2009). Apart from the color white, the flowers are available in other shades including cream, yellow, orange, peach, pink, or red (Kim et al., 2014).

47.7 ETHNOBOTANY

The genus *Brugmansia* is of significant importance among plant divisions due to its promised medicinal or pharmacological benefits (Diaz, 2015; Kandimalla et al., 2015). It has a diverse range of uses, both medicinal and non-medicinal. It is traditionally used for treating skin infections, various types of body pain, inflammation, cough, inducing hallucinations, and even for protection against evil (Sanz-Biset and Cañigueral, 2013; Chaachouay et al., 2021a, 2021b; Benkhnigue et al., 2023). The native use has been reported to have positive effects on diabetes, inflammation, body aches, rheumatic arthritis, and more (Diaz, 2015; Kandimalla et al., 2015). Even though different parts of the *Brugmansia* genus have traditional uses, the leaves and flowers are considered the most significant for various purposes. They have been recommended for use as externally or orally, but only if they have been crushed and decocted properly in the preparation process (Bussmann et al., 2016).

In 2018, approximately 5.6 million individuals, which accounts for 2% of the population in the USA, were reported to have used hallucinogens in the previous year (Facco et al., 2019). *Burgmansia* species are often utilized by teenagers and young adults as a legal and easily accessible hallucinogen since they are commonly found in both homes and gardens (Schultes and Farnsworth, 1980). Based on the reports, it has been stated that this Solanaceae family's primary function is highly effective in managing various medical conditions such as diabetes, body pain relief, rheumatoid arthritis, antidiabetic, and inflammation (Fatima et al., 2022; Kandimalla et al., 2015).

47.8 PHARMACOLOGY

The species within the *Brugmansia* genus are found to possess a multitude of phytochemicals, such as alkaloids, phenolic compounds, steroids, anthraquinone glycoside, terpenes, tannins, triterpenes, cyanogenic glycosides, saponins, and more. These compounds have piqued the interest of researchers due to their potential pharmacological and biological activities (Cristina Geller, 2010; Nandakumar et al., 2017). So far, eight *Brugmansia* species (*B. aurea, B. arborea, B. candida, B. sanguinea, B. insignis, B. suaveolens, B.* sp., and *B. versicolor*) have been found to contain 189 compounds including 81 alkaloids, 21 sesquiterpenoids, 31 monoterpenoids, 12 flavonoids, 7 nor-isoprenoids, 12 benzenoids, 5 aliphatic aldehydes, 4 polyamines, 3 long-chain alkanes, and 13 other compounds (Algradi et al., 2021). As indicated most species in the *Brugmansia* genus are dominated by tropane alkaloids among the various phytochemicals present. The primary alkaloids found in this plant are scopolamine, atropine, and hyoscyamine. However, many studies have not explored the plant's pharmacological activities, except for one study (Cardillo et al., 2008), which discovered anticholinergic properties.

The levels of tropane alkaloids can fluctuate depending on the specific *Brugmansia* or *Datura* plant species, the time of year, and which parts of the plant are involved in the exposure (Krenzelok, 2010; Jakabová et al., 2012; Romera-Torres et al., 2018). The roots and seeds of angel's trumpet are the primary sources of the alkaloid found in the largest concentrations. On the other hand, each flower contains approximately 0.3 mg of atropine and 0.65 mg of scopolamine. When atropine and scopolamine – tropane alkaloids containing nitrogen – are ingested orally, intramuscularly, or through inhalation, they can easily pass through the blood-brain barrier. They have the ability to inhibit the production of muscarinic receptors (Sadre et al., 2022).

Based on studies, it has been indicated that flavonoids found in *B. aurea* possess antioxidant characteristics (Kim et al., 2020). According to an earlier report, the *B. aurea* plant parts collected from the University of Heidelberg-Germany Botanical Garden have varying alkaloid contents. The anthers have the highest at 1646 μg/g, followed by the flowers at 323 μg/g, nectar at 91.7 μg/g, leaves at 56.3 μg/g, and pollen at 26.3 μg/g (Detzel and Wink, 1993). The nectar and pollen of *B. aurea* contain saponins, cardiac glycosides, and cyanogenic glycoside (Detzel and Wink, 1993).

So far, eight *Brugmansia* species (*B. aurea, B. arborea, B. candida, B. sanguinea, B. insignis, B. suaveolens, B.* sp, and *B. versicolor*) have been found to contain 189 compounds including 81 alkaloids, 21 sesquiterpenoids, 31 monoterpenoids, 12 flavonoids, 7 nor-isoprenoids, 12 benzenoids, 5 aliphatic aldehydes, 4 polyamines, 3 long-chain alkanes, and 13 other compounds (Algradi et al., 2021). The tropane alkaloids (3β-Hydroxy-6β-acetoxytropane and 3β-Tigloyloxy-6β-acetoxytropane) have been identified from the leaves of *B. aurea* (Doncheva et al., 2006). The reported alkaloids identified from *B. aurea* have been shown in Table 47.2 along with their names, formulae, chemical structures, and references (Algradi et al., 2021).

Many wild plants and weeds contain tropane alkaloids as secondary metabolites, which can act as antagonists of central and peripheral muscarinic acetylcholine receptors in mammals. As a result, these substances can cause a specific toxic syndrome (Alexander, 2008). Tropane alkaloids are present in various plant families, including Brassicaceae (Cruciferae), Proteaceae, Convolvulaceae, Euphorbiaceae, Erythroxylaceae, Olacaceae, and Rhizophoraceae. However, they are most recognized for being found in the Solanaceae family (Griffin and Lin, 2000). One interesting fact about this botanical family is that it has a higher number of psychoactive plants than most other families. Some of its members are known to contain tropane alkaloids such as hyoscyamine/atropine and scopolamine, which possess mind-altering qualities (Alrashedy and Molina, 2016).

The Solanaceae family includes several genera that are abundant in tropane alkaloids, such as *Brugmansia, Datura, Anisodus, Atropa, Hyoscyamus, Scopolia, Atropanthe, Przewalskia, Cyphantera, Physochlaina, Anthotroche, Mandragora*, and *Duboisia* (Griffin and Lin, 2000). Researchers have studied a diverse range of Solanaceous plants to explore how callus, suspension, and hairy root cultures can be used to produce medically

TABLE 47.2
The Alkaloids Identified from the *Brugmansia aurea* or *B. pittieri*

Alkaloids	Compounds	Formula	Structures	References
Monosubstituted tropane alkaloids	3-(3′-Acetoxytropoyloxy)-tropane	C19H25NO4	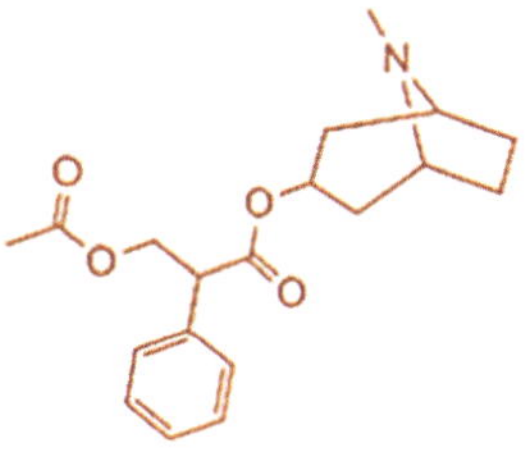	Doncheva et al., 2006
	Apoatropine	C17H21NO2	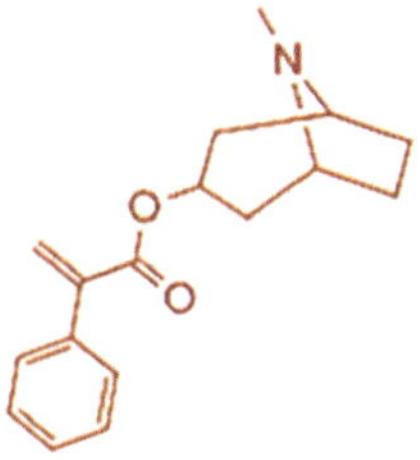	Doncheva et al., 2006; Mattioli et al., 2012
	Atropine	C17H23NO3	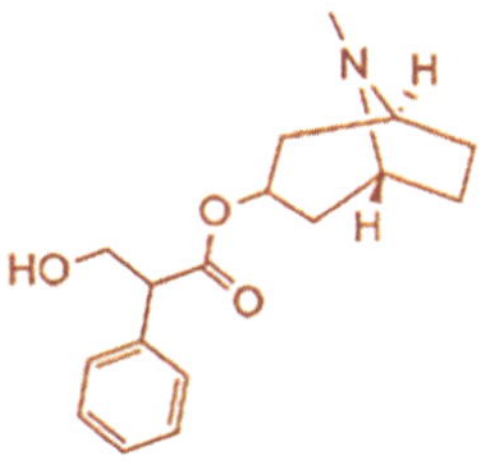	Ferreira Júnior et al., 2010; Albuquerque et al., 2015; Goldfarb et al., 2019; Singh Saroya and Singh, 2020
	Hyoscyamine	C17H23NO3	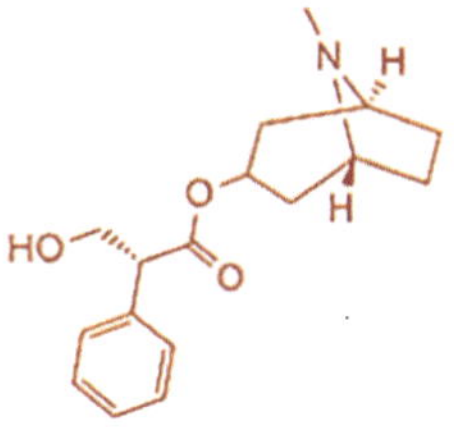	Doncheva et al., 2006
	3α-Phenylacetoxytropane	C16H21NO2	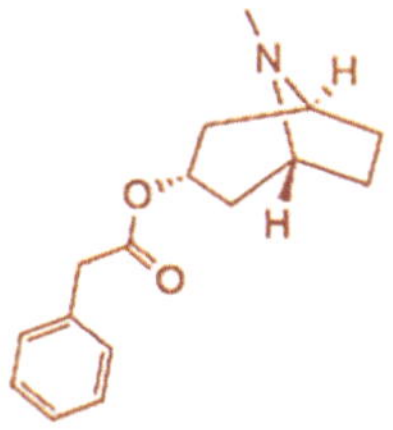	Doncheva et al., 2006
	3-(Hydroxyacetoxy)-tropane	C10H17NO3	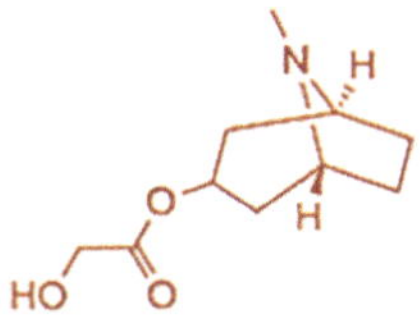	Doncheva et al., 2006
	6-Hydroxyacetoxytropane	C10H17NO3	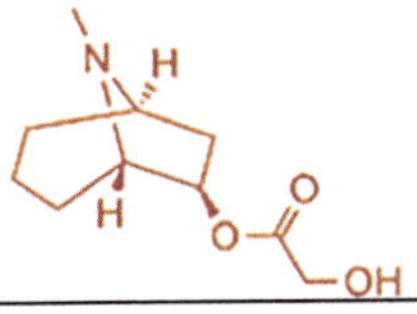	Doncheva et al., 2006

(*Continued*)

TABLE 47.2 (Continued)

Alkaloids	Compounds	Formula	Structures	References
	3α-Tigloyloxytropane	C13H21NO2	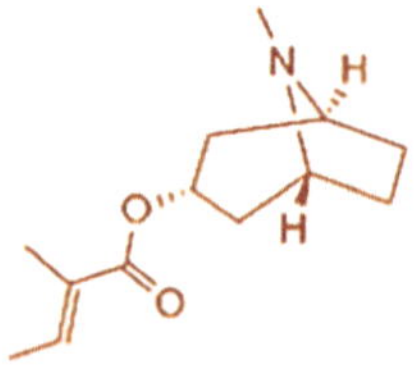	Doncheva et al., 2006
	3β-Tigloyloxytropane	C13H21NO2	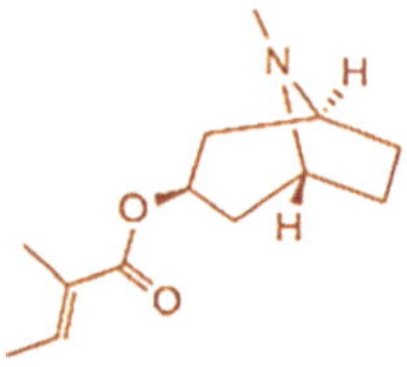	Doncheva et al., 2006
	Pseudotropine	C8H15NO	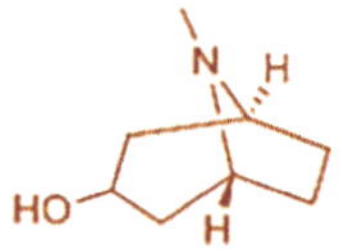	Doncheva et al., 2006
	Tropine	C8H15NO	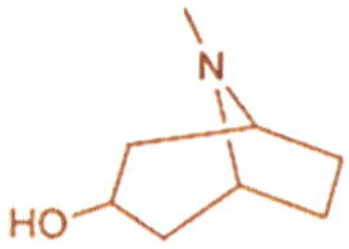	Doncheva et al., 2006
	Tropinone	C8H13NO	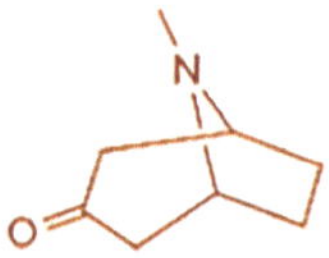	Doncheva et al., 2006
Disubstituted tropane alkaloids	3α,6β-Ditigloyloxytropane	C18H27NO4	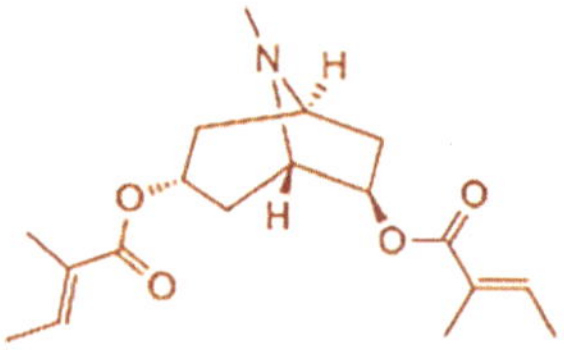	Doncheva et al., 2006
	3α-Hydroxy-6β-acetoxytropane	C10H17NO3	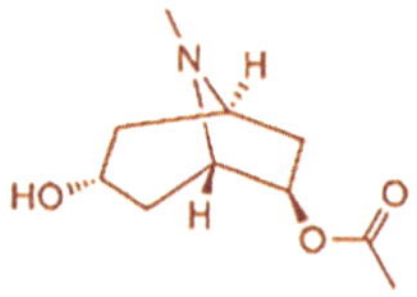	Doncheva et al., 2006
	3-Hydroxy-6-(2-methylbutyryloxy)-tropane	C13H23NO3	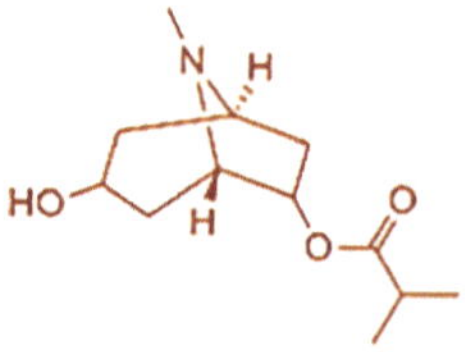	Doncheva et al., 2006
	3α-Hydroxy-6β-tigloyloxytropane	C13H21NO3	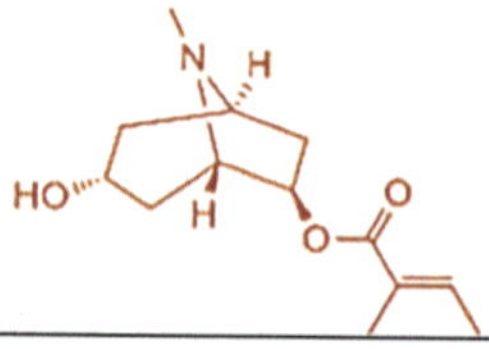	Doncheva et al., 2006

(Continued)

TABLE 47.2 (Continued)

Alkaloids	Compounds	Formula	Structures	References
	6-Hydroxyhyoscyamine	C17H23NO4	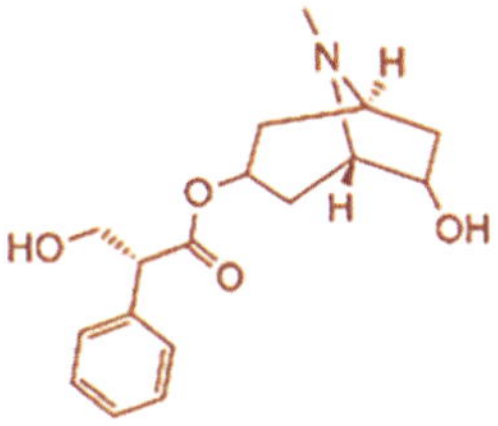	Doncheva et al., 2006
	7-Hydroxyhyoscyamine	C17H23NO4	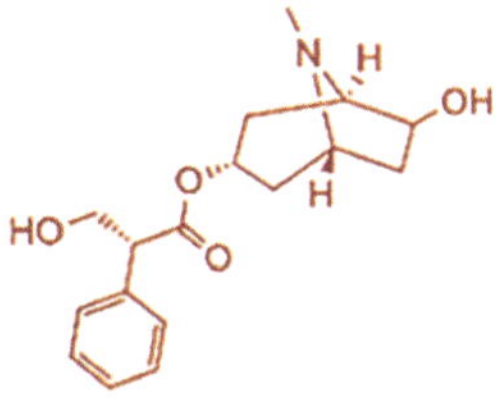	Doncheva et al., 2006
	3-Hydroxy-6-methylbutyryloxytropane	C13H23NO3	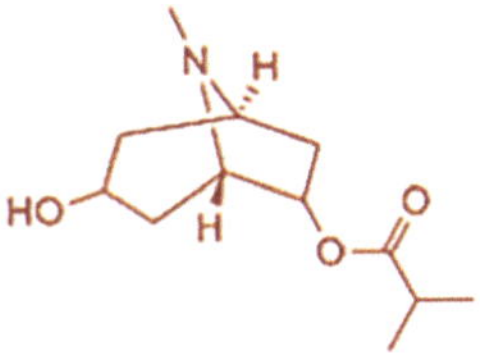	Doncheva et al., 2006
	3-Hydroxy-6-methylbutyryloxytropane	C13H23NO3	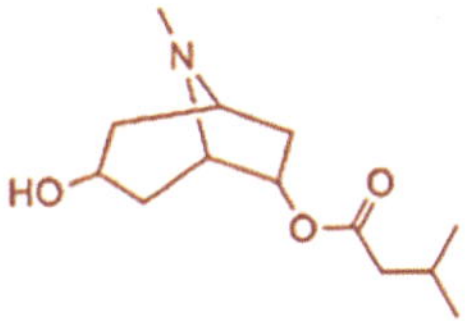	Doncheva et al., 2006
	3β-Hydroxy-6β-acetoxytropane	C10H17NO3	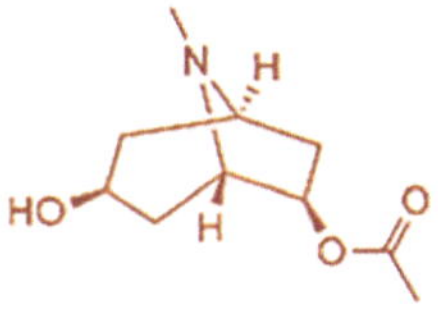	Doncheva et al., 2006
	3-Isovaleryloxy-6-hydroxytropane	C13H23NO3	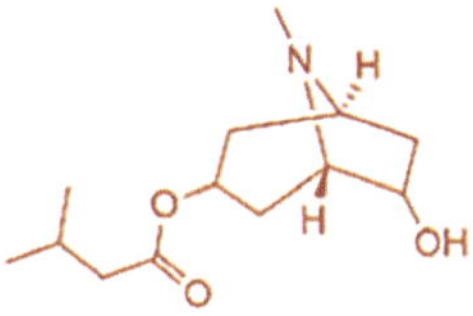	Doncheva et al., 2006
	3α-Tigloyloxy-6β-hydroxytropane	C13H21NO3	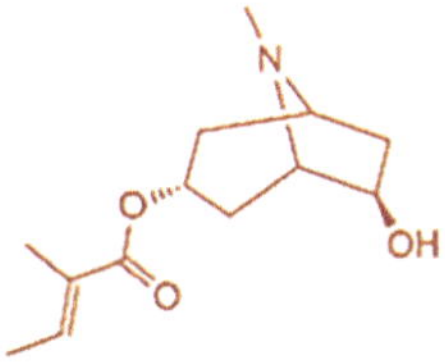	Doncheva et al., 2006
	3α-Tigloyloxy-6β-acetoxytropane	C15H23NO4	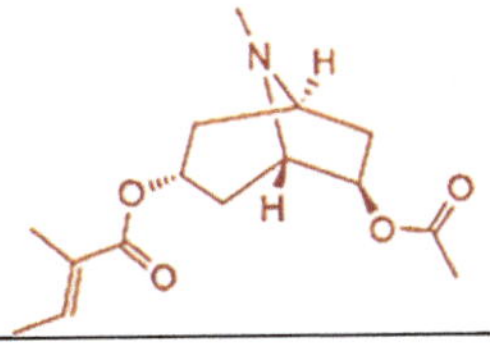	Doncheva et al., 2006

(*Continued*)

TABLE 47.2 (Continued)

Alkaloids	Compounds	Formula	Structures	References
	3-Tigloyloxy-6-propionyloxytropane	C16H25NO4	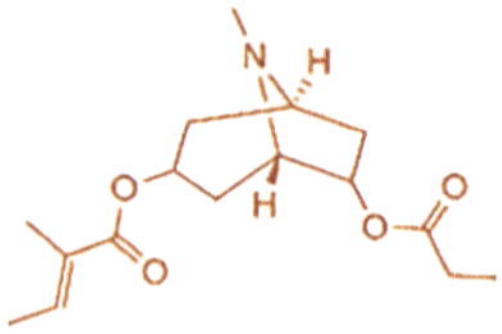	Doncheva et al., 2006
	3α-Tigloyloxy-6β-isobutyryloxytropane	C17H27NO4	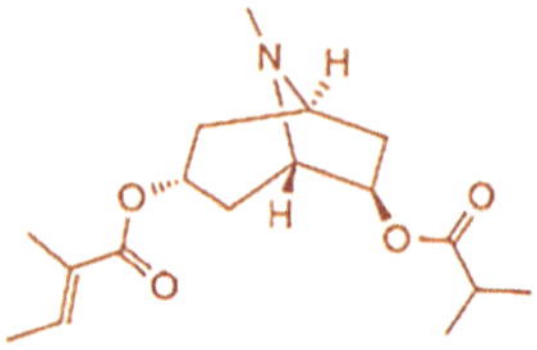	Doncheva et al., 2006
	3-Tigloyloxy-6-(2′-methylbutyryloxy)-tropane	C18H29NO4	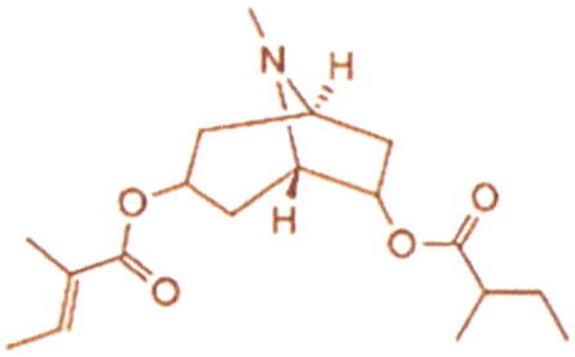	Doncheva et al., 2006
	3-Tigloyloxy-6-methylbutyryloxytropane	C18H29NO4	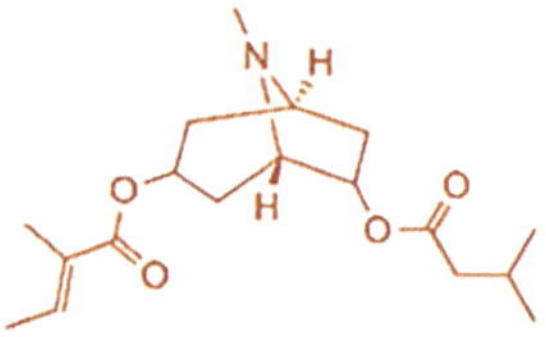	Doncheva et al., 2006
	3β-Tigloyloxy-6β-hydroxytropane	C13H21NO3	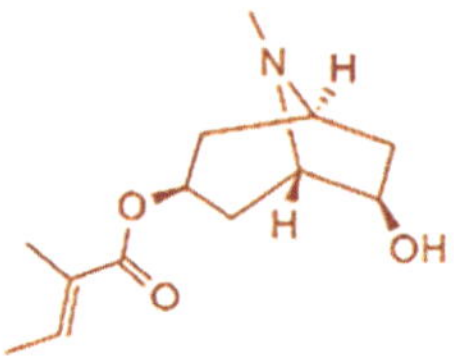	Doncheva et al., 2006
	3β-Tigloyloxy-6β-acetoxytropane	C15H23NO4	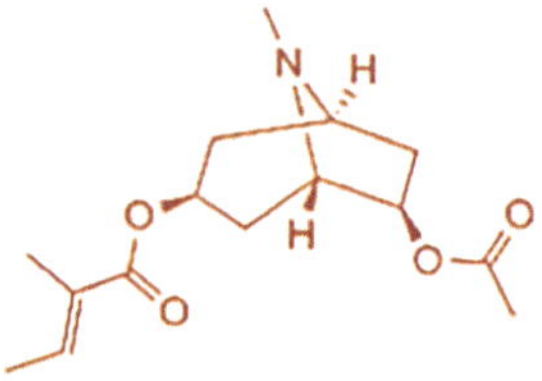	Doncheva et al., 2006
Trisubstituted tropane alkaloids	3α,6β-Ditigloyloxy-7β-hydroxytropane	C18H27NO5	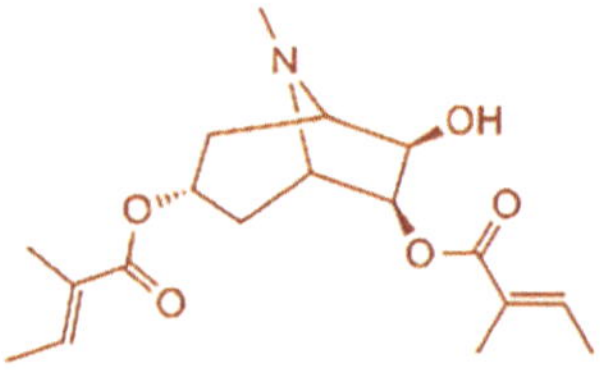	Doncheva et al., 2006
	3-Tigloyloxy-6-propionyloxy-7-hydroxytropane	C16H25NO5	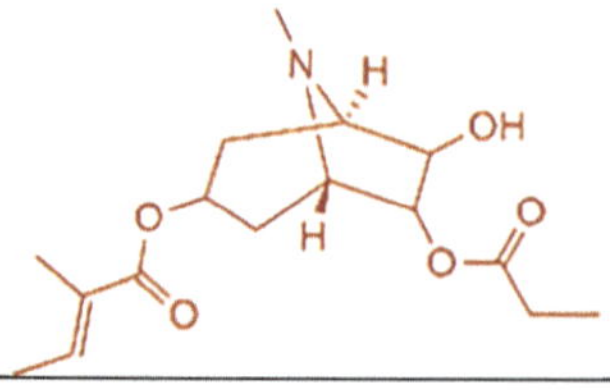	Doncheva et al., 2006

(*Continued*)

TABLE 47.2 (Continued)

Alkaloids	Compounds	Formula	Structures	References
3-substituted-6,7-epoxytropane alkaloids	Aposcopolamine	C17H19NO3	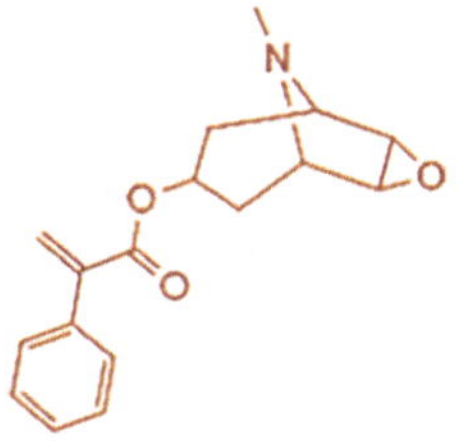	Doncheva et al., 2006; Marín-S´aez et al., 2019
	Phenylacetylscopine	C16H19NO3	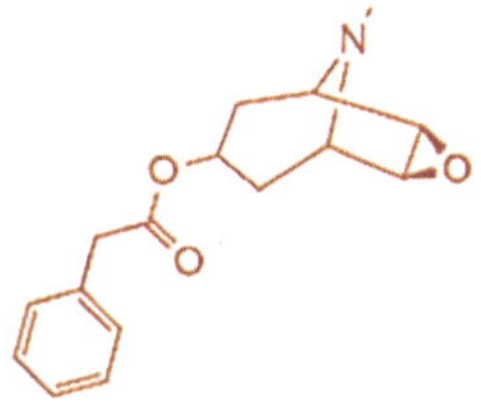	Detzel and Wink, 1993
	Scopolamine	C17H21NO4	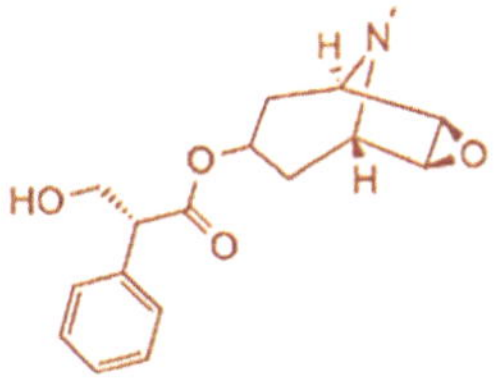	Doncheva et al., 2006
	Scopoline	C8H13NO2	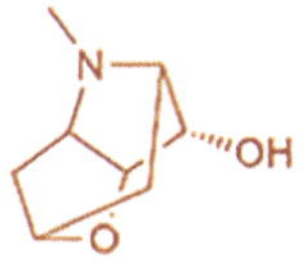	Doncheva et al., 2006; Marín-S´aez et al., 2019; Zayed and Wink, 2004
	Scopine	C8H13NO2	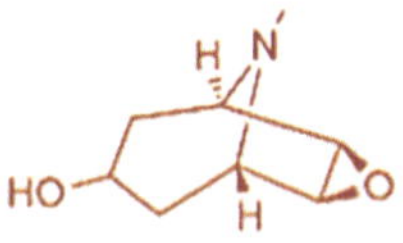	Doncheva et al., 2006; Zayed and Wink, 2004

significant alkaloids (Cardillo et al., 2013; Chandran et al., 2020). The *Brugmansia* and *Datura* species are particularly noteworthy because they have a tendency to produce significant quantities of tropane alkaloids. These plants are commonly associated with the witches of Europe and their purported flying ointments, which were believed to have psychoactive properties (Carruthers, 2015).

According to research by Li et al. (2017), alkaloids are important secondary metabolites in plant-derived medicines and have significant biological activities. The *Brugmansia* species, known for their tropane alkaloids, are a particularly rich source of these compounds (Pigatto et al., 2015), tropane alkaloids make up 40% of all isolated compounds in the genus *Brugmansia* and are present in every species. There are over 200 compounds that make up the tropane alkaloid group (Lounasmaa and Tamminen, 1993). These compounds are identified by a two-ringed structure that includes a pyrrolidine and a piperidine ring, which share a single atom of nitrogen and two atoms of carbon (Table 47.2). Most alkaloids have an amino group, which is commonly methylated. Although these substances are hallucinogenic, they are not classified as typical hallucinogens. Instead, they fall under the category of deliriants due to their potent psychoactive effects. They have a tendency to cloud the mind and produce vivid, life-like hallucinations (Díaz, 2010). According to Alam (2013), tropane alkaloids exhibit anticancer properties. Hence, it is imperative to conduct more research on this group in this context. Despite being one of the most researched chemically, 59 alkaloids have been discovered in mature flowers, immature flowers, flower nectar, fruits, corolla, and roots (Petricevich et al., 2020). The important primary known tropane alkaloids are atropine and scopolamine (Alrashedy and Molina, 2016).

Atropine, which is derived from the same biosynthetic pathway as scopolamine, acts as a muscarinic antagonist. It is commonly used as a mydriatic and cycloplegic in ophthalmology. Additionally, atropine is used for the treatment of bradycardia (Lochner and Thompson, 2016). Atropine is recognized for its ability to block the effects of acetylcholine, which can result in dilation of the pupils, blurry vision, reduced saliva production, widening of blood vessels, increased heart rate, and potential confusion (Diker et al., 2007).

Scopolamine, which is also known as hyoscine, is a type of antimuscarinic agent that can also act as a relaxant for smooth muscle. It acts as a muscarinic antagonist that has a high affinity, typically measured in nanomoles (nM). It is commonly used to treat postoperative nausea and vomiting, as well as motion sickness. Additionally, researchers often use it as a tool to induce cognitive dysfunction. However, when taken at higher doses, it can also lead to amnesia and compliance (Klinkenberg and Blokland, 2010).

The reason why scopolamine is thought to impair memory is due to its ability to quickly penetrate the blood-brain barrier and inhibit the muscarinic receptors found in the central nervous system. This inhibition of the receptors leads to a cholinergic deficit that ultimately results in memory impairment (Klinkenberg and Blokland, 2010). To model dementia, scopolamine administration has frequently been utilized as it is believed that the primary reason for age-related cognitive decline is the weakening of cholinergic neurotransmission (Bartus, 2000). Due to its ability to impair memory and cause cholinergic deficits, scopolamine has been widely employed for both preclinical and clinical evaluations of treatments intended to address cognitive impairment (Blin et al., 2009; Liem-Moolenaar et al., 2011).

Among the various species in this genus, there are several well-known ones including *B. aurea, B. arborea, B. suaveolens, B. sanguinea*, and *B. versicolor.* Historically, these plants have been used to alleviate a broad range of conditions such as rheumatic and arthritic pains, scalds, swelling, inflammations, hemorrhoids, wounds, and skin rashes. Extracts from these plants have been shown to possess anticholinergic, antiasthmatic, narcotic, spasmolytic, and anesthetic properties (Zayed and Wink, 2004). Previous studies have found that flavonoids from *B. aurea* exhibit antioxidant activity (Kim et al., 2020). This plant has been reported in previous studies to exhibit both anti-inflammatory (Kim et al., 2020; Wunder et al., 2022) and anticholinergic activities (Capasso et al., 1997). According to previous research, this herb has been found to have antidiabetic properties (Fatima et al., 2022) while related species have been associated with anti-inflammatory (Kim et al., 2020; Wunder et al., 2022), anticancer (Kumar et al., 2020), and anticholinergic (Capasso et al., 1997) properties. However, there is currently no research available on its toxicity studies.

According to a recent study (Fatima et al., 2022), it has been confirmed that a higher dose i.e. 400 mg/kg concentration of *B. aurea* leaves provides beneficial results on diabetes and other related complications such as abnormalities in hepatic and renal functions, lipid metabolism issues, and antioxidant enzyme levels. Notably, it also improves insulin sensitivity as well as resistance to a level similar to biguanides. *B. aurea*'s potential to treat diabetes may be due to the existence of several bioactive elements that display antioxidant and antidiabetic traits, as manifested by the GC-MS and LCMS analyses of the plant extract. The GC/MS analysis identified 9,12-octadecadienoic acid methyl ester and 9,12,15-octadecatrienoic acid and methyl ester as the most prevalent bioactive compounds which may be responsible for the plant's antioxidant and antidiabetic effects. Linoleic acid, also known as 9,12-octadecadienoic acid, possesses antiandrogenic, hepatoprotective, and antioxidant properties (Umeno et al., 2015). On the other hand, α-linolenic acid (ALA) or 9,12,15-octadecatrienoic acid has antiandrogenic (Sivagurunathan and Xavier, 2014), antihypocholesterolemic, antidiabetic (Erdinest et al., 2012; Sudha et al., 2013; Rao et al., 2019), and reactive oxygen species (ROS) generation inhibitory (Erdinest et al., 2012) activities. Additionally, hexadecanoic acid, 2-hydroxy-1-(hydroxymethyl) ethyl ester, exhibits anti-inflammatory and antioxidant effects (Mgbeje and Abu, 2020).

Scopolamine and atropine are frequently used in pharmaceuticals as anticholinergics, muscarinic antagonists, antispasmodics, sedatives, analgesics, and treatments for conditions like asthma, Parkinson's disease, sialorrhea, and motion sickness. Scopolamine is in higher demand and has a higher price compared to other tropane alkaloids because of its superior pharmacological activity and fewer side effects (Qiang et al., 2016).

47.9 MEDICINAL EFFECTS

According to the data collected from poison control centers across multiple countries, a considerable number of poisonings are linked to exposure to plants (Krenzelok and Mrvos, 2011). Unintentional intake of poisonous plant parts, inappropriate use of herbal products, attempted suicide, and misuse of hallucinogenic plants are the primary reasons of plant-related toxicity (Chaachouay and Zidane, 2021). The Solanaceae family species like *Brugmansia* and *Datura* are among the most commonly misused psychoactive plants (Kerchner and Farkas, 2020). The reason why species of both *Brugmansia* and *Datura* are considered highly toxic is due to the existence of tropane alkaloids such as atropine and scopolamine in all parts of the plants (Kerchner et al., 2015; Romera-Torres et al., 2018). *Brugmansia* plants are highly poisonous and can be fatal if consumed by humans, animals, and pets due to the presence of harmful toxins in all parts of the plant. Exposure to the eyes can even cause vision-related issues like mydriasis (pupil dilation) and anisocoria (unequal pupil size) (van der Donck et al., 2004). It is important to note that some countries may have regulations in place that restrict the purchase, sale, or cultivation of these plants (Preissel and Preissel, 2002).

Signs of intoxication may be enlarged pupils, disorientation, experiencing things that are not really there, dry skin and mouth, elevated blood pressure and pulse, and a rise in body temperature (Griffin and Lin, 2000; Sevketoglu et al., 2010; Goldfarb et al., 2019). Severe anticholinergic symptoms can also be caused by ingesting *Brugmansia* or *Datura* flowers and/or the nectar they produce (Disel et al., 2015). The toxic primarily alkaloids result in an anticholinergic effects by hindering the muscarinic acetylcholine receptors of the central and peripheral nervous system (Jayawickreme et al., 2019). The anticholinergic activity of atropine is responsible for causing a range of effects such as blurred vision, mydriasis, suppressed salivation, tachycardia, vasodilation, and delirium (Diker et al., 2007). Scopolamine, acts as an antimuscarinic agent as well as a smooth muscle relaxant. Anticholinergic agents like scopolamine have a wide range of applications in different medicinal fields, including narcotics, ophthalmology diagnostics, sedatives, as antispasmodics, pre- and postoperative medications, analgesics, and in the treatment of various diseases like asthma, motion sickness, and Parkinson's disease (Lukasewycz et al., 2010; Rapoport, 2010).

Kerchner and colleagues demonstrated the significance of human exposure to the Brugmansia and Datura plant species in both their native (American) and non-native (Asia, Africa, and Europe) regions in terms of toxicity. Their study focused on the possibility of contamination of various plant organs. The findings indicate that these plant species possess high toxicological importance due to their hallucinogenic property and the risk of accidental ingestion (Kerchner and Farkas, 2020). Between 2005 and 2017, there were approximately 65 cases of poisoning related to *Brugmansia* species in Hungary. In Australia, around 33 cases were reported between July 1990 and June 2000, with the flowers being the primary cause of poisoning. Additionally, between 1987 and 2006, about 68 cases of poisoning from *B. suaveolens* were reported in Taiwan (Isbister et al., 2003; Lin et al., 2009; Kerchner and Farkas, 2020).

An Australian study examining the clinical effects of poisoning due to *Brugmansia* in adult patients found that mydriasis occurred in 100% of cases, delirium in 88% of cases, and tachycardia in 33% of cases (Isbister et al., 2003). This study revealed that mydriasis lasted for an average of 29 hours, while delirium persisted for 18 hours. After ingesting the toxin, this patient began experiencing symptoms after 30 minutes. The patient's delirium showed a gradual improvement in around 15 hours, and the mydriasis reduced after approximately 24 hours since the onset. The patient had combined leaves of *Brugmansia* leaves with commonly consumed items such as 'Gotukola' (*Centella asiatica*) and 'Hathawariya' (*Asparagus racemosus*) in a drink. Interestingly, both *C. asiatica* and *A. racemosus* possess anti-cholinesterase effects that oppose the effects of *Brugmansia*, but they were unable to completely counteract the effect. However, they were not able to completely counteract the effect due to the higher level of toxicity and toxic dose of *Brugmansia* than the antagonist (Uddin et al., 2016; Arora et al., 2018).

Toxicity can result by ingesting, smoking, or absorbing the plant topically, especially through mucous membranes. Several studies have reported the intoxication cases through intake of different plant parts of *Brugmansia* and *Datura* species (Steenkamp et al., 2004; Disel et al., 2015; Le Garff et al., 2016; Lusthof et al., 2017; Trancă et al., 2017; Goldfarb et al., 2019). The extensive literature on *Brugmansia* poisonings can be attributed to their popularity as garden plants owing to their showy flowers and sweet fragrance. Engaging in gardening activities involving various *Brugmansia* taxa, which are commonly cultivated as decorative plants, is another typical way of getting intoxicated (Goldfarb et al., 2019). Gardener's pupil, also known as unilateral mydriasis, has been reported in several cases where individuals have trimmed *Brugmansia* and accidentally touched their eye with a contaminated hand (Andreola et al., 2008; Macchiaiolo et al., 2010). Children who play near *Brugmansia* plants and come into contact with certain parts of the plant, such as the flowers, may also experience signs of being intoxicated (Firestone and Sloane, 2007).

Adolescents have been reported to intentionally consume *Brugmansia* species for recreation due to the hallucinogenic and euphoric effects it produces. However, ingestion can lead to toxicity resulting in hallucinations, confusion, tachycardia, hyperthermia, urinary retention, dryness, mydriasis as well as death (Greene et al., 1996; Anonymous, 2009). There have been limited instances of accidental ingestion reported, such as a case in the year 1998 from Taiwan where the plant leaves were unintentionally added to a soup after being mistaken for an edible plant (Chang et al. 1999), and a Canadian case in 1983 where the seeds of the plant that were intended to be stored for the next year's garden were mistakenly used as spices and added into hamburgers (Canadian Disease Centre, 1984).

The degree of toxicity and levels of tropane alkaloids found in *Brugmansia* or *Datura* species can differ depending on the plant species, the time of year, maturation stage, and the specific parts of the plant involved in the exposure (Krenzelok, 2010; Romera-Torres et al., 2018). *Brugmansia* plants are deemed poisonous due to the presence of significant levels of atropine and scopolamine in various parts of the plant (Pinto et al., 2016; Kerchner and Farkas, 2020). While all parts of the plant can be toxic, the seeds are the most potent. A single blossom of *Brugmansia* comprises 0.65 mg of scopolamine and 0.3 mg of atropine. Consuming as few as ten flowers has resulted in reported fatalities due to atropine doses reaching 10 mg (Greene et al., 1996). The flowers of *Brugmansia* may contain 0.83% hyoscine, while the leaves may contain 0.4% hyoscine. As plants age, the flowers become more toxic, with levels of up to 3 mg of hyoscine (Voss, 2019). The concentration of atropine and scopolamine in the leaves of *B. pittieri* has been measured at 64.67 and 448.2 μg/g, respectively (Ciechomska et al., 2016).

Both scopolamine and atropine have also been found in the nectar of flowers belonging to *Brugmansia* and *Datura* species. Among *Brugmansia* species, *B. suaveolens* species nectar has a higher proportion of scopolamine compared to atropine (Boros et al., 2010). Small-flowered species like *D. quercifolia* and *D. stramonium* usually produce about 25 μL of nectar per flower, while the larger-flowered species like *B. suaveolens*, *D. metel*, and *D. innoxia* can produce quite a bit more, about 150 μL of nectar per flower (Farkas et al., 2011). If we consider the toxic doses of atropine and scopolamine, which typically range between 2.0 and 6.5 mg in case of oral administration (Diker et al., 2007), ingestion of the nectar of 50–100 flowers can lead to severe anticholinergic intoxication. This is especially true if we take into account the average nectar volume per flower of 100 μL in large flowering species (Farkas et al., 2011), along with an average 215 μg/mL concentration of nectar alkaloid (Boros et al., 2010). Furthermore, nectars having tropane alkaloids can be processed into toxic honeys, which can cause intoxication if ingested (Ramirez et al., 1999). Authorities have recently become concerned about the usage of *Brugmansia* and *Datura* in the form of incapacitating drugs in various sexual crimes and robbery cases. The flowers and seeds of these plants have the highest toxicological impact, but intoxication can also occur from exposure to the plant leaves and its floral nectar (Kerchner and Farkas, 2020).

47.10 CONCLUSION

This chapter reviewed the information related to botany, habitat, and ecology, ethnobotany, phytochemistry, and hallucinogenic effects of Jimson weed. Based on the information available, it has been discovered that this particular plant species possesses notable chemical components that exhibit various biological functions. Additionally, the traditional applications of certain species within this genus have been determined through their pharmacological activities, such as their ability to act as anti-inflammatories, antispasmodics, antiasthmatics, pain-relievers, anti-addictive agents, and antiprotozoals. According to the research on phytochemistry, various bioactive components have been discovered, including but not limited to tropane alkaloids, terpenes, and flavonoids. The exploration of phytochemicals in the genus Brugmansia has resulted in the identification of around 189 secondary metabolites, offering valuable insights into their potential practical applications.

Although many species of the genus *Brugmansia* have been traditionally used, their uses have not been scientifically confirmed. Additionally, the secondary metabolites of these species have not yet been identified or pharmacologically evaluated. Therefore, further research is necessary to confirm the ethnopharmacological uses, identify the chemical constituents, assess toxicity, and determine the pharmacological activities of these species (specifically *B. pittieri*). This chapter will provide useful insights for future studies on the phytochemical and pharmacological aspects of *Brugmansia* species.

ACKNOWLEDGMENTS

Our special thanks go to Prof. Dr. Munir Ozturk and Prof. Dr. Volkan Altay for their critical evaluation of this chapter.

LIST OF ABBREVIATIONS

ALA α-linolenic acid
B. *Brugmansia*
IUCN International Union for Conservation of Nature
nM nanomoles

REFERENCES

Alam E.A. 2013. In vitro cultures for the production of some anticancer agents. *Life Sci. J.*, 10: 297–310.

Albuquerque, U.P., De Medeiros, P.M., and Casas, A. 2015. Evolutionary ethnobiology. *Evol. Ethnobiol.*, 2015: 1–197.

Alexander, J. 2008. Tropane alkaloids (from *Datura* sp.) as undesirable substances in animal feed. *EFSA J*, 691: 1–55.

Algradi, A.M., Liu, Y., Yang, B.Y., and Kuang, H.X. 2021. Review on the genus *Brugmansia*: traditional usage, phytochemistry, pharmacology, and toxicity. *Journal of Ethnopharmacology*, 279: 113910.

Alrashedy, N.A., and Molina, J. 2016. The ethnobotany of psychoactive plant use: a phylogenetic perspective. *PeerJ*, 4: e2546.

Alvarez, L.M. 2008. Borrachero, Cacao Sabanero o Floripondio (Brugmansia spp.) Un Grupo de Plantas por Redescubrir en la Biodiversidad Latinoamericana. *Cultura y Droga*, 13(15): 77–93.

Andreola, B., Piovan, A., Da Dalt, L., Filippini, R., and Cappelletti, E. 2008. Unilateral mydriasis due to Angel's Trumpet. *Clinical Toxicology*, 46(4): 329–331.

Anonymous. 2009. Genus: *Brugmansia* Pers. Germplasm Resources Information Network. United States Department of Agriculture. 2009–09–01. Retrieved 2010-09-25.

Anonymous. 2023. 12 plants for evening scent. BBC Gardener's World Online. Retrieved 02 June 2023. https://www.gardenersworld.com/plants/plants-for-evening-scent.

APG III. 2009. An update of the Angiosperm Phylogeny Group classification for the orders and families of flowering plants: PG II. *Bot. J. Linn. Soc.*, 141: 399–436.

Arora, R., Kumar, R., Agarwal, A., Reeta, K.H., and Gupta, Y.K. 2018. Comparison of three different extracts of *Centella asiatica* for anti-amnesic, antioxidant and anticholinergic activities: in vitro and in vivo study. *Biomed Pharmacother.*, 105: 1344–1352.

Baker, J.R. 2005. Psychedelic sacraments. *J. Psychoactive Drugs*, 37(2): 179–187.

Bartus, R.T. 2000. On neurodegenerative diseases, models, and treatment strategies: lessons learned and lessons forgotten a generation following the cholinergic hypothesis. *Exp. Neurol.*, 163: 495–529.

Benítez, G., March-Salas, M., Villa-Kamel, A., Cháves-Jiménez, U., Hernández, J., Montes-Osuna, N., Moreno-Chocano, J. and Cariñanos, P. 2018. The genus *Datura* L.(Solanaceae) in Mexico and Spain-Ethnobotanical perspective at the interface of medical and illicit uses. *Journal of Ethnopharmacology*, 219: 133–151.

Benkhnigue, O., Chaachouay, N., & Zidane, L. (2023). A Floristical and Ecological Study of the Medicinal Flora Used by the Local Population of the Haouz-Rehamna Region (Middle Atlantic Morocco-4). *Acta Botanica Hungarica Akadémiai Kiadó*, 65(3–4): 247–304. https://doi.org/10.1556/034.65.2023.3-4.3

Blin, O., Audebert, C., Pitel, S., Kaladjian, A., Casse-Perrot, C., Zaim, M., Micallef, J., Tisne-Versailles, J., Sokoloff, P., Chopin, P., and Marien M. 2009. Effects of dimethylaminoethanol pyroglutamate (DMAE p-Glu) against memory deficits induced by scopolamine: evidence from preclinical and clinical studies. *Psychopharmacol. Berl.*, 207: 201–212.

Bock, J.H., and Norris, D. 2015. *Forensic Plant Science*. Academic Press.

Boros, B., Farkas, Á., Jakabová, S., Bacskay, I., Kilár, F., and Felinger, A. 2010. LC-MS quantitative determination of atropine and scopolamine in the floral nectar of *Datura* species. *Chromatographia*, 71(Supll 1): 43–49.

Bussmann, R.W., Zambrana, N.Y.P., Huanca, L.A.M., and Hart, R. 2016. Changing markets- medicinal plants in the markets of La Paz and El Alto, Bolivia. *J. Ethnopharmacol.*, 193: 76–95.

Bye, R., and Sosa, V. 2013. Molecular phylogeny of the jimsonweed genus *Datura* (Solanaceae). *Systematic Botany*, 38(3): 818–829.

Calina, D., Carvalho, F. and Docea, A.O. 2021. Toxicity of psychedelic drugs. In: Tsatsakis, A.M. (ed.), *Toxicological Risk Assessment and Multi-System Health Impacts from Exposure*. Academic Press, pp. 545–556.

Canadian Disease Centre. 1984. Datura Poisoning from Hamburger - Canada. *Canada Diseases Weekly Report*, 33(20): 282–283.

Capasso, A., De Feo, V., De Simone, F., and Sorrentino, L. 1997. Activity-directed isolation of spasmolytic (anti-cholinergic) alkaloids from *Brugmansia arborea* (L.) Lagerheim. *Int. J. Pharmacogn.*, 35: 43–48.

Cardillo, A.B., Talou, J.R., and Giulietti, A.M. 2008. Expression of *Brugmansia candida* hyoscyamine 6 beta-hydroxylase gene in *Saccharomyces cerevisiae* and its potential use as biocatalyst. *Microbial Cell Factories*, 7: 7–17.

Cardillo, A.B., Giulietti, A.M., Palazón, J., and Bonfill, M. 2013. Influence of hairy root ecotypes on production of tropane alkaloids in *Brugmansia candida*. *Plant Cell, Tissue and Organ Culture (PCTOC)*, 114: 305–312.

Carruthers, D.M.J. 2015. Lines of flight of the deadly nightshade: an enquiry into the properties of the magical plant, its literature and history. *Mosaic: Interdiscip. Study Lit.*, 48(2): 119–132.

Chaachouay, N., Douira, A., & Zidane, L. (2021a). Herbal Medicine Used in the Treatment of Human Diseases in the Rif, Northern Morocco. *Arabian Journal for Science and Engineering*, 47, 131–153. https://doi.org/10.1007/s13369-021-05501-1

Chaachouay, N., Douira, A., & Zidane, L. (2021b). COVID-19, prevention and treatment with herbal medicine in the herbal markets of Salé Prefecture, North-Western Morocco. *European Journal of Integrative Medicine*, 42, 101285. https://doi.org/10.1016/j.eujim.2021.101285

Chaachouay, N., & Zidane, L. (2021). The Symbolic Efficacy of Plants in Rituals and Socio-Religious Ceremonies in Morocco, Northwest of Africa. *Journal of Religious & Theological Information*, *21*(1–2), 34–53. https://doi.org/10.1080/10477845.2021.1942433

Chandran, H., Meena, M., Barupal, T., and Sharma, K. 2020. Plant tissue culture as a perpetual source for production of industrially important bioactive compounds. *Biotechnology Reports*, 26: e00450.

Chang, S.S., Wu, M.L., Deng, J.F., Lee, C.C., Chin, T.F., and Liao, S.J. 1999. Poisoning by *Datura* leaves used as edible wild vegetables. *Veterinary and Human Toxicology*, 41(4): 242–245.

Ciechomska, M., Woźniakiewicz, M., Nowak, J., Swiadek, K., Bazylewicz, B., and Kościelniak, P. 2016. Development of a microwave-assisted extraction of atropine and scopolamine from Solanaceae family plants followed by a QuEChERS clean-up procedure. *Journal of Liquid Chromatography and Related Technologies*, 39: 538–548.

Cristina Geller, F. 2010. Isolation, Structure Elucidation and Biological Investigation of Active Compounds in Cordia Americana and *Brugmansia suaveolens*. University of T¨ubingen.

Detzel, A., and Wink, M. 1993. Attraction, deterrence or intoxication of bees (*Apis mellifera*) by plant allelochemicals. *Chemoecology*, 4: 8–18.

Dhanya, C., Shabir, A.R., and Devipriya, V. 2018. Molecular systematic study of two solanaceous genera *Datura* L. and *Brugmansia* Pers. based on its sequences of nrdna. *International Journal of Advanced Research*, 6(10): 1123–1133.

Diaz, G.J. 2015. Toxicosis by plant alkaloids in humans and animals in Colombia. *Toxins*, 7(12): 5408–5416.

Díaz, J.L. 2010. Sacred plants and visionary consciousness. *Phenomenol. Cognitive Sci.*, 9(2): 159–170.

Diker, D., Markovitz, D., Rothman, M., and Sendovski, U. 2007. Coma as a presenting sign of *Datura stramonium* seed tea poisoning. *European Journal of Internal Medicine*, 18: 336–338.

Disel, N.R., Yilmaz, M., Kekec, Z., and Karanlik, M. 2015. Poisoned after dinner: dolma with *Datura stramonium*. *Turkish Journal of Emergency Medicine*, 15: 51–55.

Doncheva, T., Berkov, S., and Philipov, S. 2006. Comparative study of the alkaloids in the tribe Datureae and their chemosystematic significance. *Biochemical Systematics and Ecology*, 34: 478–488.

Erdinest, N., Shmueli, O., Grossman, Y., Ovadia, H., and Solomon, A. 2012. Anti-inflammatory effects of alpha linolenic acid on human corneal epithelial cells. *Investigative Ophthalmology & Visual Science*, 53: 4396–4406.

Facco, E., Mendozzi, L., Bona, A., Motta, A., Garegnani, M., Costantini, I., Dipasquale, O., Cecconi, P., Menotti, R., Coscioli, E. and Lipari, S. 2019. Dissociative identity as a continuum from healthy mind to psychiatric disorders: epistemological and neurophenomenological implications approached through hypnosis. *Medical Hypotheses*, 130: 109274.

Farkas, Á., Kerchner, A., Déri, H., Boros, B., and Darók, J. 2011. Nectary structure and nectar production of various *Datura* species. *International Journal of Plant Biology*, 3: 31–35.

Fatima, N., Anwar, F., Saleem, U., Khan, A., Ahmad, B., Shahzadi, I., Ahmad, H. and Ismail, T. 2022. Antidiabetic effects of *Brugmansia aurea* leaf extract by modulating the glucose levels, insulin resistance, and oxidative stress mechanism. *Frontiers in Nutrition*, 2022: 9.

Fatur, K. 2019. Sagas of the Solanaceae: Speculative ethnobotanical perspectives on the Norse berserkers. *Journal of Ethnopharmacology*, 244: 112151.

Fatur, K. 2021. Peculiar plants and fantastic fungi: An ethnobotanical study of the use of hallucinogenic plants and mushrooms in Slovenia. *Plos one*, 16(1): e0245022.

Ferreira Júnior, W.S., Cruz, M.P., Albuquerque, U.P. de, and Vieira, F.J. 2010. Are hallucinogenic plants efficacious in curing diseases?. *Bol. Latinoam. Caribe Plantas Med. Aromat.*, 2010: 9.

Firestone, D., and Sloane, C. 2007. Not your everyday anisocoria: angel's trumpet ocular toxicity. *J Emerg Med*, 33: 21–24.

Frecska, E. 2007. Therapeutic guidelines: dangers and contraindications in therapeutic applications of hallucinogens. In: Roberts, T., and Winkleman, M. (eds.), *Psychedelic Medicine: New Evidence for Hallucinogenic Substances as Treatments*. Santa Barbara: Praeger, pp. 69–95.

Fürstenberg-Hägg, J., Zagrobelny, M. and Bak, S. 2013. Plant defense against insect herbivores. *International Journal of Molecular Sciences*, 14(5): 10242–10297.

Ghatak, A., Chaturvedi, P., Paul, P., Agrawal, G.K., Rakwal, R., Kim, S.T., Weckwerth, W. and Gupta, R. 2017. Proteomics survey of Solanaceae family: current status and challenges ahead. *Journal of Proteomics*, 169: 41–57.

Goldfarb, J., Pesin, N., and Margolin, E. 2019. Gardening and dilated pupils: an interesting case of anisocoria from *Brugmansia versicolor*. *Can J Ophthalmol*, 54: e59–e61.

Greene, G.S., Patterson, S.G., and Warner, E. 1996. Ingestion of angel's trumpet: an increasingly common source of toxicity. *South Med J.*, 89: 365–369.

Griffin, W.J., and Lin, G.D. 2000. Chemotaxonomy and geographical distribution of tropane alkaloids. *Phytochemistry*, 53: 623–637.

Haegi, L. 1976. Taxonomic account of *Datura* L. (Solanaceae) in Australia with a note on *Brugmansia* Pers. *Australian Journal of Botany*, 24(3): 415–435.

Hassler, M. 2022. *Brugmansia pittieri* (Saff.) Moldenke Synonymic Checklists of the Vascular Plants of the World. In: Bánki, O. et al., Catalogue of Life Checklist. 14.3, Nov 2022.

Hay, A., Gottschalk, M., and Holguin, A. 2012. Huanduj: *Brugmansia*. London, UK: Kew Publishing.

Hay, A. 2014. *Brugmansia aurea*. IUCN Red List of Threatened Species. 2014: e.T38124A58906215.

Hunziker, A.T. 1979. South American Solanaceae: a synoptic survey. In: Hawkes, J.G., Lester, R.N. and Skelding, A.D. (eds.), *The Biology and Taxonomy of the Solanaceae*. Academic Press, London, pp. 49–86.

Hunziker, A.T. 2001. *Genera Solanacearum: The Genera Solanaceae. Illustrated, arranged According to a New System*. Gantner, Ruggell, Liechtenstein.

Isbister, G., Oakley, P., Dawson, A.H., and Whyte, I.M. 2003. Presumed Angel's trumpet (*Brugmansia*) poisoning: Clinical effects and epidemiology. *Emerg Med (Fremantle)*, 15: 376–382.

Jakabová, S., Vincze, L., Farkas, Á., Kilár, F., Felinger, A., and Boros, B. 2012. Determination of tropane alkaloids atropine and scopolamine by liquid chromatography-mass spectrometry in plant organs of *Datura* species. *J. Chromatogr. A*, 1232: 295–301.

Jayawickreme, K.P., Janaka, K.V.C. and Subasinghe, S.A.S.P. 2019. Unknowing ingestion of *Brugmansia suaveolens* leaves presenting with signs of anticholinergic toxicity: a case report. *J Med Case Reports*, 13: 322.

Kandimalla, R., Kalita, S., Choudhury, B., and Kotoky J. 2015. A review on anti-diabetic potential of genus *Solanum* (Solanaceae). *J Drug Deliv Ther.*, 5: 24–27

Kennedy, D.O. 2014. The Deliriants - The Nightshade (Solanaceae) Family. In: Kennedy, D.O. (ed.), *Plants and the Human Brain*. New York: Oxford University Press, pp. 131–137.

Kerchner, A., Darók, J., Bacskay, I., Felinger, A., Jakab, G., and Farkas, Á. 2015. Protein and alkaloid patterns of the floral nectar in some solanaceous species. *Acta Biol Hung*, 66: 304–315.

Kerchner, A., and Farkas, Á. 2020. Worldwide poisoning potential of *Brugmansia* and *Datura*. *Forensic Toxicol*, 38: 30–41.

Kim, H.G., Jang, D., Jung, Y.S., Oh, H.J., Oh, S. M., Lee, Y.G.,... and Baek, N.I. 2020. Anti-inflammatory effect of flavonoids from *Brugmansia arborea* L. flowers. *Journal of Microbiology and Biotechnology*, 30(2): 163.

Kim, Y., Kim, J., Kim, O.J., and Kim, W.C. 2014. Intoxication by angel's trumpet: case report and literature review. *BMC Research Notes*, 7(1): 553.

Klinkenberg, I., and Blokland, A. 2010. The validity of scopolamine as a pharmacological model for cognitive impairment: a review of animal behavioral studies. *Neuroscience & Biobehavioral Reviews*, 34: 1307–1350.

Kordova, P., and Ventegodt, S. 2016. Contemporary strategies in Peru for medical use of the hallucinogenic tea Ayahuasca containing DMT: In search of the optimal strategy for the use of medical hallucinogens. *Journal of Alternative Medicine Research*, 8(4): 455–470.

Krenzelok, E.P. 2010. Aspects of *Datura* poisoning and treatment. *Clinical Toxicology*, 48: 104–110.

Krenzelok, E.P., and Mrvos, R. 2011. Friends and foes in the plant world: a profile of plant ingestions and fatalities. *Clinical Toxicology*, 49: 142–149.

Kumar, S., Gupta, A., Saini, R.V., Kumar, A., Dhar, K.L. and Mahindroo, N. 2020. Immunomodulation-mediated anticancer activity of a novel compound from *Brugmansia suaveolens* leaves. *Bioorganic & Medicinal Chemistry*, 28(12): 115552.

Kurniati, H., Sumadijaya, A., Boonman, A. and Laksono, W.T. 2010. Ecology, Distribution and Bio-acoustic of Amphibians in Degraded Habitat. https://web.archive.org/web/20120426023016/http://www.biologi.lipi.go.id/bio_bidang/file_doc_bidang/Report_of_Ristek-Dikti-LIPi_Incentive_Project_2010.pdf

Le Garff, E., Delannoy, Y., Mesli, V., Hédouin, V., and Tournel, G. 2016. Forensic features of a fatal *Datura* poisoning case during a robbery. *Forensic Science International*, 261: e17–e21.

Li, C., Ying, Z., Gao, M., Wei, W., Hao, D., Xu, L.,... and Liu, J. 2017. Two new similar alkaloids from *Portulaca oleracea* L. *Natural Product Research*, 31(15): 1792–1798.

Liem-Moolenaar M., de Boer P., Timmers M., Schoemaker R.C., van Hasselt J.G., Schmidt S., and van Gerven J.M. 2011. Pharmacokinetic-pharmacodynamic relationships of central nervous system effects of scopolamine in healthy subjects. *British Journal of Clinical Pharmacology*, 71: 886–898.

Lin, T.J., Nelson, L.S., Tsai, J.L., Hung, D.Z., Hu, S.C., Chan, H.M., and Deng, J.F. 2009. Common toxidromes of plant poisonings in Taiwan toxidromes of plant poisonings T. J. Lin et al. *Clinical Toxicology*, 47: 161–168.

Lochner, M. and Thompson, A.J. 2016. The muscarinic antagonists scopolamine and atropine are competitive antagonists at 5-HT3 receptors. *Neuropharmacology*, 108: 220–228.

Lockwood T.E. 1973. Generic recognition of *Brugmansia*. *Botanical Museum leaflets, Harvard University*, 23: 273–284.

Lounasmaa, M. and Tamminen, T. 1993. The tropane alkaloids: Chemistry and Biology. In: Cordell, G.A. (ed.), *The Alkaloids*. Academic Press, New York, pp. 1–114.

Lukasewycz, S., Holman, M., Kozlowski, P., Porter, C.R., Odom, E., Bernards, C., Neil, N., and Corman, J.M. 2010. Does a perioperative belladonna and opium suppository improve postoperative pain following robotic assisted laparoscopic radical prostatectomy? Results of a single institution randomized study. *The Canadian Journal of Urology*, 17: 5377–5382.

Lusthof, K.J., Bosman, I.J., Kubat, B., and Vincenten-van Maanen, M.J. 2017. Toxicological results in a fatal and two non-fatal cases of scopolamine-facilitated robberies. *Forensic Science International*, 274: 79–82.

Macchiaiolo, M., Vignati, E., Gonfiantini, M.V., Grandin, A., Romano, M.T., Salata, M.,... and Villani, A. 2010. An unusual case of anisocoria by vegetal intoxication: a case report. *Italian Journal of Pediatrics*, 36: 1–3.

Marin-Saez, J., Romero-González, R., and Frenich, A.G. 2019. Effect of tea making and boiling processes on the degradation of tropane alkaloids in tea and pasta samples contaminated with Solanaceae seeds and coca leaf. *Food Chemistry*, 287: 265–272.

Mattioli, L., Bracci, A., Titomanlio, F., Perfumi, M., and De Feo, V. 2012. Effects of *Brugmansia arborea* extract and its secondary metabolites on morphine tolerance and dependence in mice. *Evidence-Based Complement. Altern. Med.*. https://doi.org/10.1155/2012/741925

Melissa, P. 2023. Angel's trumpet. Britannica. https://www.britannica.com/plant/angels-trumpet.

Mgbeje, B.I., and Abu, C. 2020. Chemical fingerprinting of *Nauclea latifolia*, an antidiabetic plant, using GC-MS. *Journal of Complementary Medicine Research*, 9: doi: 10.9734/jocamr/2020/v9i430148.

Mrazova, K., Navratil, T., and Pelclova, D. 2011. Use and accidental exposure to hallucinogenic agents reported to the Czech toxicological information centre from 1995 to 2008. *Subst Use Misuse.*, 46(4): 460–465.

Nandakumar, A., Vaganan, M., Sundararaju, P., and Udayakumar, R. 2017. Phytochemical analysis and nematicidal activity of ethanolic leaf extracts of *Datura metel, Datura innoxia* and *Brugmansia suaveolens* against *Meloidogyne incognita. Asian Journal of Biology*, 2: 1–11.

Olmstead, R.G. and Palmer, J.D. 1992. A chloroplast DNA phylogeny of the Solanaceae: subfamilial relationships and character evolution. *Annals of the Missouri Botanical Garden*, 79: 346–360.

Pasiecznik, N. 2014. *Brugmansia suaveolens* (white angel's trumpet). CABI Compendium. https://doi.org/10.1079/cabicompendium.107903

Petricevich, V.L., Salinas-Sánchez, D.O., Avilés-Montes, D., Sotelo-Leyva, C. and Abarca-Vargas, R. 2020. Chemical compounds, pharmacological and toxicological activity of *Brugmansia suaveolens*: a review. *Plants*, 9(9): 1161.

Pigatto, A.G., Blanco, C.C., Mentz, L.A., and Soares, G.L. 2015. Tropane alkaloids and calystegines as chemotaxonomic markers in the Solanaceae. *Anais da Academia Brasileira de Ciências*, 87: 2139–2149.

Pinto, C.F., Salinas, S., Flores-Prado, L., Echeverría, J., and Niemeyer, H.M., 2016. Sequestration of tropane alkaloids from *Brugmansia suaveolens* (Solanaceae) by the treehopper *Alchisme grossa* (Hemiptera: membracidae). *Biochemical Systematics and Ecology*, 66: 161–165.

Plant intelligence, 2019. *Brugmansia* Pers. *Natl. Key. Lab. Syst. Evol. Bot. Inst. Bot. Chinese Acad. Sci.* http://www.iplant.cn/info/Brugmansia accessed 6.19.20

Preissel, U., and Preissel, H.G. 2002. *Brugmansia* and *Datura*: *Angel's Trumpets and Thorn Apples*. Buffalo, New York: Firefly Books, pp. 106–129.

Qiang, W., Xia, K., Zhang, Q., Zeng, J., Huang, Y., Yang, C., … and Liao, Z. 2016. Functional characterization of a tropine-forming reductase gene from *Brugmansia arborea*, a woody plant species producing tropane alkaloids. *Phytochemistry*, 127: 12–22.

Ramirez, M., Rivera, E., and Ereu, C. 1999. Fifteen cases of atropine poisoning after honey ingestion. *Veterinary and Human Toxicology*, 41: 19–20.

Rao, M.R.K., Anisha, G., Prabhu, K., Shil, S., and Vijayalakshmi, N. 2019. Preliminary phytochemical and gas chromatography-mass spectrometry study of one medicinal plant *Carissa carandas*. *Drug Invitation Today*, 12: 1629–1630.

Rapoport, A. 2010. Sublingual atropine drops for the treatment of pediatric sialorrhea. *Journal of Pain and Symptom Management*, 40: 783–788.

Rätsch C. 2005. *The Encyclopedia of Psychoactive Plants: Ethnopharmacology and its Applications*. Park Street Press; Rochester.

Romera-Torres, A., Romero-González, R., Vidal, J.L.M., and Frenich, A.G. 2018. Analytical methods, occurrence and trends of tropane alkaloids and calystegines: an update. *Journal of Chromatography A*, 1564: 1–15.

Rosario, L.H., Padilla, J.O.R., Martínez, D.R., Grajales, A.M., Mercado Reyes, J.A., Feliu, G.J.V., Van Ee, B., and Siritunga, D. 2019. DNA barcoding of the solanaceae family in Puerto Rico including endangered and endemic species. *Journal of the American Society for Horticultural Science*, 144: 363–374.

Sadre, R., Anthony, T.M., Grabar, J.M., Bedewitz, M.A., Jones, A.D. and Barry, C.S. 2022. Metabolomics-guided discovery of cytochrome P450s involved in pseudotropine-dependent biosynthesis of modified tropane alkaloids. *Nature Communications*, 13(1): 3832.

Sanz, C., and Tagliazucchi, E. 2018. The experience elicited by hallucinogens presents the highest similarity to dreaming within a large database of psychoactive substance reports. *Frontiers in Neuroscience*, 12(1): 1–19.

Sanz-Biset, J., and Cañigueral, S.J. 2013. Plants as medicinal stressors, the case of depurative practices in Chazuta valley (Peruvian Amazonia). *Journal of Ethnopharmacology*, 145: 67–76.

Schultes, R.E. and Farnsworth, N.R. 1980. Ethnomedical, botanical and phytochemical aspects of natural hallucinogens. *Botanical Museum Leaflets, Harvard University*, 1980: 123–214.

Sevketoglu, E., Tatli, B., Tuğcu, B., Demirelli, Y., and Hatipoglu, S. 2010. An unusual cause of fulminant Guillain–Barré syndrome: angel's trumpet. *Pediatric Neurology*, 43: 368–370.

Shaw, J.M.H. 1999. Nomenclature Notes on *Brugmansia. The New Plantsman. Royal Horticultural Society*, 6(3): 148–151.

Singh Saroya, A., and Singh, J. 2020. Psychoactive medicinal plants and fungal neurotoxins. *Psychoactive Medicinal Plants and Fungal Neurotoxins*, https://doi.org/10.1007/978-981-15-2313-7.

Sivagurunathan A, and Xavier, B. 2014. GC-MS evaluation of bioactive compounds of *Marsilea quadrifolia* Linn. *International Journal for Pharmaceutical Research Scholars*, 3: 2277–7873.

Steenkamp, P.A., Harding, N.M., van Heerden, F.R., and van Wyk, B.E. 2004. Fatal *Datura* poisoning: identification of atropine and scopolamine by high performance liquid chromatography/photodiode array/mass spectrometry. *Forensic Science International*, 145: 31–39.

Sudha, T., Chidambarampillai, S., Mohan, V. 2013. GC-MS analysis of bioactive components of aerial parts of *Fluggea leucopyrus* Willd (Euphorbiaceae). *Journal of Applied Pharmaceutical Science*, 3: 126–130.

The Plant List, 2015. The Plant List: a working list of all plant species. Version 1.1. London, UK: Royal Botanic Gardens, Kew. http://www.theplantlist.org. Accessed on 27 April 2023.

The Plants List. 2023. *Brugmansia pittieri*. Available online: http://www.theplantlist.org/tpl1.1/record/kew-2684015. Accessed on 27 April 2023.

Trancă, S.D., Szabo, R., and Cociş, M. 2017. Acute poisoning due to ingestion of *Datura stramonium*-a case report. *Romanian Journal of Anaesthesia and Intensive Care*, 24: 65–68.

Uddin, M.S., Asaduzzaman, M., Mamun, A.A., Iqbal, M.A., Wahid, F., and Rony, R.K. (2016). Neuroprotective activity of Asparagus racemosus Linn. against ethanol-induced cognitive impairment and oxidative stress in rats brain: Auspicious for controlling the risk of Alzheimer's disease. *Journal of Alzheimers Disease & Parkinsonism*, 6(4): 1–10.

Umeno, A., Yoshino, K., Hashimoto, Y., Shichiri, M., Kataoka, M., and Yoshida, Y. 2015. Multi-biomarkers for early detection of type 2 diabetes, including 10-and 12-(Z, E)-hydroxyoctadecadienoic acids, insulin, leptin, and adiponectin. *PLoS ONE*, 2015: 10:e0130971.

USDA-ARS, 2015. Germplasm Resources Information Network (GRIN). Online Database. Beltsville, Maryland, USA: National Germplasm Resources Laboratory. https://npgsweb.ars-grin.gov/gringlobal/taxon/taxonomysearch.aspx

van der Donck, I., Mulliez, E., and Blanckaert, J. 2004. Angel's Trumpet (*Brugmansia arborea*) and mydriasis in a child - A case report. *Bulletin de la Societe Belge d'Ophtalmologie*, 292: 53–56.

Van Wyk, B.-E., and Wink, M. 2014. *Phytomedicines, Herbal Drugs And Poisons*. The University of Chicago Press, Chicago and London Kew Publishing Royal Botanic Gardens; Kew.

Voss, J.L. 2019. Guide to Poisonous Plants: College of Veterinary Medicine & Biomedical Sciences Colorado State University.

Widelski, J., and Kukula-Koch, W.A. 2017. Psychoactive Drugs. In: Simone, B. and Rupika, D. (eds), *Pharmacognosy*. Academic Press, pp. 363–374.

Wink, M. 2016. *Evolution of Secondary Plant Metabolism*. John Wiley & Sons, Ltd; Chichester.

Wunder, J., Pemp, D., Cecil, A., Mahdiani, M., Hauptstein, R., Schmalbach, K., et al. 2022. Influence of breast cancer risk factors on proliferation and DNA damage in human breast glandular tissues: role of intracellular estrogen levels, oxidative stress and estrogen biotransformation. *Archives of Toxicology*, 96: 673–687.

Zayed, R., and Wink, M. 2004. Induction of tropane alkaloid formation in transformed root cultures of *Brugmansia suaveolens* (Solanaceae). *Zeitschrift fr Naturforschung*, 59: 863–867.

48 Squirting Cucumber (*Ecballium elaterium* (L.) A. Rich. Cucurbitaceae)

Munir Ozturk, Volkan Altay, Mustafa Keskin, and Alvina Gul

48.1 INTRODUCTION

The fresh fruits of this Cucurbitaceae plant, when squeezed, squirt out a mucus-like secretion full of its seeds. Therefore, because of this feature, its vernacular name in English is known as 'squirting cucumber'. The pressure inside the fruit reaches up to approximately six bars, allowing the seeds inside to reach a speed of 36 km/h when the capsule explodes. It is one of the well-known medicinal plants in the Mediterranean and Arabic traditional medicine systems (www.wikipedia.org).

Authors have tried to provide information on various features of *Ecballium elaterium*, namely botanical aspects and vernacular names, geographical distribution and ecological features, traditional uses, and some pharmacological effects.

48.2 SCIENTIFIC CLASSIFICATION

The latest updated classification of this species according to the United States Department of Agriculture Natural Resources Conservation Service (USDA) (2023) is presented below:

Kingdom	Plantae
Subkingdom	Tracheobionta
Superdivision	Spermatophyta
Division	Magnoliophyta
Class	Liliopsida
Subclass	Dilleniidae
Order	Violales
Family	Cucurbitaceae
Genus	*Ecballium* A. Rich.
Species	*Ecballium elaterium* (L.) A. Rich.

48.3 VERNACULAR NAMES

The *E. elaterium*, widely named as 'squirting cucumber' in English, by different local names in Türkiye mainly 'acı dülek', 'cırtatan', 'eşek hıyarı', 'şeytankeleği', and 'yabani hıyar' (Gökalp, 2019).

48.4 BOTANICAL DESCRIPTION

Perennial herb; stems prostrate, stout, hispid, pale green, up to 25 cm; leaves rather thick, stiff, deep green, and scabrid above, pale green and densely hispid tomentose beneath, ovate-triangular, cordate, sinuate and undulate at the margins, rather concave above, shortly 5-lobed with the basal lobes ± erect, 4–14.5 x 3.5–17 cm; peduncles erect, hispid, 2–10 cm; male flowers in 3.5–3.7 cm racemes; petals 8–16 x 4–7 mm; female flowers often co-axillary with male flowers; peduncles 1–8 cm; petals 6–12 x 3–6 mm; fruit pendulous, 3–5 x 1.5–2.5 cm; pulp exceedingly bitter; seeds about 4 mm (Figure 48.1) (Jeffrey, 1972).

48.5 ORIGIN AND GEOGRAPHICAL DISTRIBUTION

The natural distribution of *E. elaterium* reveals that it is a herb distributed in the subtropical biome regions of the Canary Islands, Mediterranean to South European areas of Russia and North Iran. The natural distribution areas are outlined as: Albania, Algeria, Baleares, Bulgaria, Canary Islands, Cyprus, France, Greece, Iran, Italy, Lebanon, Libya, Morocco, Palestine, Portugal, South European Russia, Spain, Syria, Transcaucasus, Tunisia, Turkiye and Ukraine (www.powo.science.kew.org).

48.6 HABITAT AND ECOLOGY

The habitat preferences of *E. elaterium* in its distributional area are sandy or stony grounds near the sea and as a ruderal in disturbed habitats like roadsides, canal embankments, river banks, garbage dumps, waste places, abandoned lots, and in the vicinity of cultivation areas (Figure 48.2) (Jeffrey, 1972; Costich and Meagher, 1992; Fahn and Shimony, 2001; Rust et al., 2003; Latté, 2009). The plant in general is found on farmlands, along the roadsides, in waste areas, mowed areas, and pastures in urban and suburban places from sea levels up to 600 meters (Jeffrey, 1972).

It is one of the characteristic ruderal plants mentioned in many floristic and ecological studies conducted in some urban and rural areas (Altay and Ozturk, 2012; Altay and Karahan, 2017; Altay et al., 2010, 2012, 2015a, 2015b, 2020; Ozturk et al., 2012; Eskin et al., 2012).

From an ecological perspective, the plant generally flourishes well in the subtropical regions (www.powo.science.kew.org). Its morpho-anatomical features (especially leaves and fruits) permit it to grow well alongside such habitats. These features give important advantages to

DOI: 10.1201/9781003460336-48

FIGURE 48.1 Some morphological features of *Ecballium elaterium*: (A–C) general view of the plant; (D–F) leaves; (G–I) flower; and (J–L) fruit.

these plants in their ecological adaptations, also providing significant advantages in the habitats especially to withstand stressful ecological conditions (Benkhnigue et al., 2023; Christodoulakis et al., 2011; Ielciu et al., 2016).

48.7 ETHNOBOTANY

The fruits and roots of *E. elaterium*, a plant native to the Mediterranean region, have been popularly used in traditional folk medicine since ancient times. It has been

FIGURE 48.2 Some habitats where *Ecballium elaterium* is distributed: (A) in humus-rich forest soil; (B) on stone walls; (C, D) on roadsides and waste places in urban areas; (E, F) on meadow edges in urban areas.

evaluated as a natural source for the treatment of sinusitis, hepatitis, fever, cancer, rheumatism, hemorrhoids, liver disorders, and earache (Jodral et al., 1990; Rust et al., 2003; Janick et al., 2007; Kültür, 2007; Ielciu et al., 2016; Chaachouay et al., 2021a, 2021b). The plant is often used in traditional medicine by the locals in many countries including Turkiye as nasal drops (Ozturk, 1980) as well as for the treatment of jaundice in newborn babies (Ielciu et al., 2016). However, some studies report that the juice obtained from the fruit of this plant may have toxic effects on the respiratory, cardiac, and stomach systems, mainly the mucosa of the tissues. It is reported as dangerous if it comes into contact with the face (Chaachouay and Zidane, 2021; Ielciu et al., 2016). Therefore, if the relevant parts of this plant are taken in high doses for treatment purposes, it can irritate the large intestine and cause bloody diarrhea (Gullu and Ocal, 2016).

In Anatolian traditional folk medicine poultices and ointments prepared from the roots of this plant are used in the treatment of hemorrhoids and as a pain reliever (Ozturk, 1980). The reports depict that the fresh juice obtained by squeezing the fruits has been used in the treatment of sinusitis, jaundice, nocturia, low back pain and ear ache and inflammation (Lanfranco, 1980; Baytop, 1984; Yeşilada et al., 1988; Yeşilada, 2002; Gullu and Ocal, 2016).

E. elaterium is also a well-known plant in traditional Arabian folk medicine. It has been used especially as an intestinal softener, otitis, and brain cleanser. In Georgia, it has been evaluated for treating malaria fever. Some reports show this plant is applied to the nostrils to clear jaundice, relieve permanent headaches, and as an edema reliever in the treatment of sinusitis (Attard, 2011).

48.8 PHARMACOLOGY

The studies undertaken on the determination of phytochemical compounds present in this plant have revealed that these especially include; alkaloids, amino acids, carotenoids, cucurbitacins (B, C, D, E, G, H, I, L, and R), fatty acids, flavonoids, lipids, phenolic compounds, phytosterols, resin, soluble sugars, starch, tannins, and terpenoids (Favel et al., 1989; Jodral et al., 1990; Gök, 2023).

In numerous pharmacological experimental studies on the leaves and roots of this plant, different extracts have been prepared which are reported to show important pharmacological activities as; analgesic, antibacterial, antifungal, anti-hepatotoxic, anti-inflammatory, anti-jaundice, antimalarial, antioxidant, anti-sinusitis, antitrypsin, antiproliferative and anti-tumor or cytotoxic (di Tizio et al., 2012; Abbassi et al., 2014; Touihri et al., 2015; Touihri-Barakati et al., 2016; Gök, 2023). The plant's potential bioactive compounds are a natural genetic resource due to its ability to prevent insects and viruses that infect cucurbit plants (Ben et al., 1996; Pascual-Villalobos and Robledo, 1999). Generally, the potential positive pharmacological effects of this plant reveal that it originates from cucurbitacins as a group of bioactive compounds (Gökalp, 2019).

Adwan et al. (2011) have shown that the ethanolic extract obtained from *E. elaterium* fruit shows antimicrobial activity against *Staphylococcus aureus* and *Candida albicans*. Muftah et al. (2013), the water obtained from the fruits of *E. elaterium* produces a significant inhibition against *Escherichia coli, Pseudomonas aeruginosa, Staphylococcus aureus*, *Acinetobacter* sp., and *Klebsiella pneumonia*. Abbassi et al. (2014) have reported that extracts obtained from *E. elaterium* are effective on *Staphylococcus aureus, Enterococcus faecalis,* and *Listeria monocytogenes*. Farahani et al. (2016) report that ethanolic extracts obtained from the fruits of *E. elaterium* show strong antifungal and antibacterial activity against *Aspergillus fumigatus, Candida albicans*, and *Escherichia coli*. Felhi et al. (2017a) mention that the extract with acetone and diethyl ether obtained from *E. elaterium* fruit peels has an antibacterial effect against *Micrococcus luteus*. Felhi et al. (2017b) have also revealed that the methanol extract obtained from the fruits of *E. elaterium* is effective against *Bacillus cereus*. Hamidi et al. (2020) report that the methanol extract obtained from the fruits of *E. elaterium* in Iran has shown significant antibacterial activity against *Pseudomonas aeruginosa*.

In clinical and pharmacological experimental studies conducted on experimental animals by some researchers, it has been emphasized that extracts obtained from the fruit juice of *E. elaterium* show significant hepatoprotective bioactivity against hepatotoxicity (Agil et al., 1999; Elmhdwi et al., 2014; El Naggar et al., 2015).

Some studies point out that extracts obtained from the fruit peels and juice of this plant have high antioxidant activity (Muftah et al., 2013; Felhi et al., 2017a). It has been reported that the fruit juice of *E. elaterium* is rich in Cucurbitacin bioactive compounds and thus has a potential anti-inflammatory effect (Yeşilada et al., 1989). In another clinical trial study conducted on rats in 2017, it was reported that the methanol extract obtained from the fruits of *E. elaterium* showed a good anti-inflammatory effect (Felhi et al., 2017b).

Due to the potential anti-tumor or anti-cancer bioactivity features of the plant, it has been referred to as a powerful anti-tumor agent. Many studies have shown that its anti-tumor effect is due to the cucurbitacins found in the phytochemical content of the plant (Dinan et al., 1997; Clericuzio et al., 2004; Liu et al., 2008; Wakimoto et al., 2008; Lee et al., 2010; Adwan et al., 2011). Very recently many clinical and pharmacological experimental studies have been conducted on extracts obtained from different parts of this plant. These studies have revealed that they produce cytotoxic and strong antiproliferative effects against human hepatocarcinoma, gastric adenocarcinoma, esophagus, human colon adenocarcinoma, and fibrosarcoma cancer cell lines (Sayed and Badr, 2012; Bohlooli et al., 2012; Touihri et al., 2015; Karimi et al., 2016; Hamidi et al., 2020).

48.9 MEDICINAL EFFECTS

Touihri-Barakati et al. (2016) have reported that the seed oil obtained from the seeds of *E. elaterium* reduces the migration and proliferation of human brain cancer cell lines in a dose-dependent manner. Arslan et al. (2017) carried out experimental studies using a clinical animal model; they hypothesized that *E. elaterium* contains some components that partially exert protective effects against septicemia-associated encephalopathy by reducing pro-inflammatory cytokine accumulation in the brain and may contribute significantly to its anti-inflammatory effects during septicemia.

It has also been demonstrated in some clinical studies that bioactive compounds such as Cucurbitacin E and Cucurbitacin I isolated from *E. elaterium* have potential neuroprotective activity (Arel-Dubeau et al., 2014). Therefore, it is thought that these bioactive compounds will be effective in improving some neurological diseases such as acetylcholinesterase and butyrylcholinesterase in Alzheimer's disease and tyrosinase inhibition in Parkinson's disease and in finding potential new drugs (Orhan et al., 2016; Gökalp, 2019).

48.10 CONCLUSION

E. elaterium has been widely used in the Mediterranean traditional folk medicine. Several pharmacological studies have been carried out on this plant which reveal that it is a natural genetic resource with good therapeutic potential, mainly as an anti-tumor and anti-sinusitis as well as antimicrobial, anti-hepatotoxic, anti-inflammatory, and antioxidant activities. The chemical composition of this plant includes, in particular, cucurbitacins, which are promising potential candidates for developing some innovative treatments. Therefore, detailed investigations are needed in order to make it available to people at a global level, especially in the Mediterranean countries. There is a need to fully reveal its therapeutic potential in neurodegenerative disorders and other potential diseases and to ensure its safe use in modern medicine.

LIST OF ABBREVIATIONS

E. elaterium	*Ecballium elaterium*
USDA	United States Department of Agriculture Natural Resources Conservation Service

REFERENCES

Abbassi, F., Ayari, B., Mhamdi, B., and Toumi, L. 2014. Phenolic contents and antimicrobial activity of squirting cucumber (*Ecballium elaterium*) extracts against food-borne pathogens. *Pakistan Journal of Pharmaceutical Sciences*, 27(3): 475–480.

Adwan, G., Salameh, Y., and Adwan, K. 2011. Effect of ethanolic extract of *Ecballium elaterium* against *Staphylococcus aureus* and *Candida albicans*. *Asian Pacific Journal of Tropical Biomedicine*, 1(6): 456–460.

Agil, A., Miró, M., Jimenez, J., Aneiros, J., Caracuel, D.M., García-Granados, A., and Navarro M.C. 1999. Isolation of an anti-hepatotoxic principle from the juice of *Ecballium elaterium*. *Planta Medica*, 65(7): 673–675.

Altay, V., and Ozturk, M. 2012. Land degradation and halophytic plant diversity of Milleyha Wetland Ecosystem (Samandağ-Hatay), Turkey. *Pakistan Journal of Botany*, 44: 37–50.

Altay, V., and Karahan, F. 2017. Ruderal vejetasyon üzerine bir ön çalışma. *Kilis 7 Aralık Üniversitesi Fen ve Mühendislik Dergisi*, 1(2): 68–77.

Altay, V., Ozyıgıt, I.I. and Yarcı, C. 2010. Urban flora and ecological characteristics of the Kartal District (Istanbul): A contribution to urban ecology in Turkey. *Scientific Research and Essays*, 5(2): 183–200.

Altay, V., Ozyıgıt, I.I. and Yarcı, C. 2012. Plant communities in urban habitats of Istanbul-Turkey. *Pakistan Journal of Botany*, 44: 177–186.

Altay, V., Keskin, M. and Karahan, F. 2015a. An assessment of the plant biodiversity of Mustafa Kemal University Tayfur Sokmen Campus (Hatay-Turkey) for the view of human health. *International Journal of Scientific and Technological Research*, 1(2): 83–103.

Altay, V., Ozyigit, I.I., Osma, E., Bakir, Y., Demir, G., Severoglu, Z. et al. 2015b. Environmental relationships of the vascular flora alongside the railway tracks between Haydarpasa and Gebze (Istanbul-Kocaeli/Turkey). *Journal of Environmental Biology*, 36: 153–162.

Altay, V., Šilc, U., Yarcı, C., Kavgacı, A., Čarni, A. and Ozturk, M. 2020. Urban vegetation of the Anatolian side of Istanbul. *Phytocoenologia*, 50(2): 101–121.

Arel-Dubeau, A.M., Longpré, F., Bournival, J., Tremblay, C., Demers-Lamarche, J., Haskova, P., Attard, E., Germain, M., Martinoli, M.G. 2014. Cucurbitacin E has neuroprotective properties and autophagic modulating activities on dopaminergic neurons. *Oxidative Medicine and Cellular Longevity*, 2014: 1–15.

Arslan, D., Ekinci, A., Arici, A., Bozdemir, E., Akil, E., and Ozdemir, H. 2017. Effects of *Ecballium elaterium* on brain in a rat model of sepsis-associated encephalopathy. *Libyan Journal of Medicine*, 12(1): 1369834.

Attard, E. 2011. Use of extracts from squirting cucumber (*Ecballium elaterium*) seeds in health. In: Vinood, B., Patel, R.R., and Preedy, V.R., eds., *Nuts and Seeds in Health and Disease Prevention*, Elsevier Inc., pp. 1079–1086.

Baytop, T. 1984. Türkiye'de bitkiler ile tedavi. Istanbul Üniversitesi yayını No. 3255. Istanbul, Türkiye.

Ben, Y.D., Gal, D., Chen, M., and Rosen, D. 1996. Potential of *Aspongopus viduatus* F. (Heteroptera: Pentatomidae) as a biocontrol agent for squirting cucumber, *Ecballium elaterium* (L.) A. Rich (Cucurbitaceae). *Biological Control*, 7: 48–52.

Benkhnigue, O., Chaachouay, N., & Zidane, L. (2023). A Floristical and Ecological Study of the Medicinal Flora Used by the Local Population of the Haouz-Rehamna Region (Middle Atlantic Morocco-4). *Acta Botanica Hungarica Akadémiai Kiadó*, *65*(3–4), 247–304. https://doi.org/10.1556/034.65.2023.3-4.3

Bohlooli, S., Jafari, N., and Jahed, S. 2012. Cytotoxic effect of freeze-dried extract of *Ecballium elaterium* fruit on gastric adenocarcinoma (AGS) and esophageal squamous cell carcinoma (KYSE30) cell lines. *Journal of Gastrointestinal Cancer*, 43(4): 579–583.

Chaachouay, N., Douira, A., & Zidane, L. (2021a). Herbal Medicine Used in the Treatment of Human Diseases in the Rif, Northern Morocco. *Arabian Journal for Science and Engineering*, 47, 131–153. https://doi.org/10.1007/s13369-021-05501-1

Chaachouay, N., Douira, A., & Zidane, L. (2021b). COVID-19, prevention and treatment with herbal medicine in the herbal markets of Salé Prefecture, North-Western Morocco. *European Journal of Integrative Medicine*, 42, 101285. https://doi.org/10.1016/j.eujim.2021.101285

Chaachouay, N., & Zidane, L. (2021). The Symbolic Efficacy of Plants in Rituals and Socio-Religious Ceremonies in Morocco, Northwest of Africa. *Journal of Religious & Theological Information*, *21*(1–2), 34–53. https://doi.org/10.1080/10477845.2021.1942433

Christodoulakis, N.S., Kollia, K., and Fasseas, C. 2011. Leaf structure and histochemistry of *Ecballium elaterium* (L.) A. Rich. (squirting cucumber). *Morphology, Distribution, Functional Ecology of Plants*, 206(3): 191–197.

Costich, D.E., and Meagher, T.R. 1992. Genetic variation in *Ecballium elaterium*: Breeding system and geographic distribution. *Journal of Evolutionary Biology*, 5: 589–601.

Clericuzio, M., Mella M., Vita-Finzi P., Zem M., and Vidari G. 2004. Cucurbitane triterpenoids from *Leucopaxillus gentianeus*. *Journal of Natural Products*, 67(11): 1823–1828.

di Tizio, A., Łuczaj, Ł.J., Quave, C.L., Redžić, S., and Pieroni, A. 2012. Traditional food and herbal uses of wild plants in the ancient South-Slavic diaspora of Mundimitar/Montemitro (Southern Italy). *Journal of Ethnobiology and Ethnomedicine*, 8(1): 1–10.

Dinan, L., Whiting, P., Girault, J.P., Lafont, R., and Mugat, B. 1997. Cucurbitane-type compounds from *Hemsleya carnosiflora* antagonize ecdysteroid action in the *Drosophila melanogaster* BII cell line. *Biochemical Journal*, 53(3): 271–274.

El Naggar, E.M.B., Chalupová, M., Pražanová, G., Parák, T., Švajdlenka, E., Žemlička, M., and Suchy, P. 2015. Hepatoprotective and proapoptotic effect of *Ecballium elaterium* on CCl4-induced hepato-toxicity in rats. *Asian Pacific Journal of Tropical Medicine*, 8(7): 526–531.

Elmhdwi, M., Muftah, S., El Tumi, S., and Elslimani, F. 2014. Hepatoprotective effect of *Ecballium elaterium* fruit juice against paracetamol induced hepatotoxicity in male albino rats. *International Current Pharmaceutical Journal*, 3(5): 270–274.

Eskin, B., Altay, V., Ozyıgıt, I.I. and Serın, M. 2012. Urban vascular flora and ecological characteristics of the Pendik District (Istanbul-Turkey). *African Journal of Agricultural Research*, 7(4): 629–646.

Fahn A., and Shimony C. 2001. Nectary structure and ultrastructure of unisexual flowers of *Ecballium elaterium* (L.) A. Rich. (Cucurbitaceae) and their presumptive pollinators. *Annals of Botany*, 87: 27–33.

Farahani, Y., Amin, G., Sardari, S., and Ostad, N. 2016. Effect of *Ecballium elaterium* fruit on *Candida albicans, Aspergillus fumigatus* and *Escherchia coli*. *International Journal of Biological Research*, 4(1): 44–45.

Favel, A., Mattras, H., Coletti-Previero, M.A., Zwilling R., Robinson, E.A. and Castro, B. 1989. Protease inhibitors from *Ecballium elaterium* seeds. *International Journal of Peptide and Protein Research*, 33: 202–208.

Felhi, S., Daoud, A., Hajlaoui, H., Mnafgui, K., Gharsallah, N., and Kadri, A. 2017a. Solvent extraction effects on phytochemical constituents profiles, antioxidant and antimicrobial activities and functional group analysis of *Ecballium elaterium* seeds and peels fruits. *Journal of Food Science and Technology*, 37(3): 483–492.

Felhi, S., Saoudi, M., Daoud, A., Hajlaoui, H., Ncir, M., Chaabane, R., El Feki, A., Gharsallah, N., and Kadri, A. 2017b. Investigation of phytochemical contents, in vitro antioxidant and antibacterial behavior and in vivo anti-inflammatory potential of *Ecballium elaterium* methanol fruits extract. *Journal of Food Science and Technology*, 37(4): 558–563.

Gök, H.N. 2023. *Ecballium elaterium* (L.) A. Rich. In: Ufuk Koca Caliskan, U.K., ed., *Medicinal Plants of Turkey*. CRC Press.

Gökalp, F. 2019. A theoretical investigation of the effect of some active agents in *Ecballium elaterium* on AChE and BChE. *Academic Perspective Procedia*, 2(3): 650–654.

Gullu, İ.B., and Ocal, N. 2016. Tıbbi bir bitki olarak *Ecballium elaterium* (L.)'un tedavi alanlarının araştırılması. *Balıkesir Üniversitesi Fen Bilimleri Enstitüsü Dergisi*, 18(1): 49–57.

Hamidi, M., Ghasemi, S., Bighdilou, B., Koohi, D., and Yousefbeyk, F. 2020. Evaluation of antioxidant, antibacterial and cytotoxic activity of methanol extract from leaves and fruits of Iranian squirting cucumber (*Ecballium elaterium* (L.) A. Rich). *Research Journal of Pharmacognosy (RJP)*, 7(1): 23–29.

Ielciu, I.I., Frederich, M., Tits, M., Angenot, L., Paltinean, R., Cieckiewicz, E., ... Vlase, L. 2016. *Bryonia alba* L. and *Ecballium elaterium* (L.) A. Rich.-Two related species of the Cucurbitaceae family with important pharmaceutical potential. *Farmacia*, 64(3): 323–332.

Janick, J., Paris, H.S., and Parrish, D.C. 2007. The cucurbits of Mediterranean antiquity: Identification of taxa from ancient images and descriptions. *Annals of Botany*, 100(7): 1441–1457.

Jeffrey, C. 1972. *Ecballium* A.Rich. In: Davis, P., ed., *Flora of Turkey and the East Aegean Islands*. Edinburgh University Press, Edinburgh, Vol. 4, p. 203.

Jodral, M.M., Jimenez, J.M., Ahmad, A.M., Navarro, M.M.C., and Cabo, C.P. 1990. *Ecballium elaterium* (L.) A. Richard: II. Morphological and phytochemical studies. *Boil. Soc. Brotherina*, 63: 213–224.

Karimi, N., Bohlooli, S., and Mazani, M. 2016. Nanoliposomal formulation of *Ecballium elaterium* extract: Cytotoxic evaluation against human gastric adenocarcinoma (AGS) cell line. *Nanomedicine Research Journal*, 1(1): 9–14.

Kültür, Ş. 2007. Medicinal plants used in Kırklareli province (Turkey). *Journal of Ethnopharmacology*, 111(2): 341–364.

Lanfranco, G. 1980. Some recent communications on the folk medicine of Malta. *L-Imnara*, 3: 87.

Latté, K.P. 2009. *Ecballiun elaterium* (L.) Rich. die Eselgurke. *Zeitschrift für Phytotherapie*, 30: 148–154.

Lee, D.H., Iwanski, G.B., and Thoennissen, N.H. 2010. Cucurbitacin: Ancient compound shedding new light on cancer treatment. *The Scientific World Journal*, 10: 413–418.

Liu, T., Zhang, M., Zhang, H., Sun, C., and Deng, Y. 2008. Inhibitory effects of cucurbitacin B onlaryngeal squamous cell carcinoma. *Journal of Natural Products*, 265(10): 1225–1232.

Muftah, S., Almajberi, S., Mohamed, O., Altayar, M., Nasib, M., Attitalla, I., Al-Hejazi, I., and Elmhdwi, M. 2013. Antibacterial screening and antioxidant and free radical scavenging activity of *Ecballium elaterium*. *African Journal of Basic & Applied Sciences*, 5(6): 276–283.

Orhan, İ.E., Senol, F.S., Haznedaroglu, M.Z., Koyu, H., Erdem, S.A., Yılmaz, G., Cicek, M., Yaprak, A.E., Ari, E., Kucukboyaci, N., and Toker, G. 2016. Neurobiological evaluation of thirty-one medicinal plant extracts using microtiter enzyme assays. *Clinical Phytoscience*, 2: 9.

Ozturk, M. 1980. *Notes on Economic Botany*. Ege University, Science Faculty Press.

Ozturk, M., Altay, V., Gucel, S. and Aksoy, A. 2012. Aegean Grasslands as endangered ecosystems in Turkey. *Pakistan Journal of Botany*, 44: 7–18.

Pascual-Villalobos, M.J., and Robledo A. 1999. Antiinsect activity of plant extracts from the wild flora in southeastern Spain. *Biochemical Systematics and Ecology*, 27: 1–10.

Rust, R.W., Vaissière, B.E., and Westrich, P. 2003. Pollinator biodiversity and floral resource use in *Ecballium elaterium* (Cucurbitaceae), a Mediterranean endemic. *Apidologie*, 34(1): 29–42.

Sayed, Z.I.E., and Badr, W.H. 2012. Cucurbitacin glucosides and biological activities of the ethyl acetate fraction from ethanolic extract of egyptian *Ecballium elaterium*. *The Journal of Applied Sciences Research*, 8(2): 1252–1258.

Touihri, I., Kallech-Ziri, O., Boulila, A., Fatnassi, S., Marrakchi, N., Luis, J., and Hanchi, B. 2015. *Ecballium elaterium* (L.) A. Rich. seed oil: Chemical composition and antiproliferative effect on human colonic adenocarcinoma and fibrosarcoma cancer cell lines. *Arabian Journal of Chemistry*, 12(8): 2347–2355.

Touihri-Barakati, I., Kallech-Ziri, O., Boulila, A., Khwaldia, K., Marrakchi, N., Hanchi, B., … Luis, J. 2016. Targetting $\alpha v\beta 3$ and $\alpha 5\beta 1$ integrins with *Ecballium elaterium* (L.) A. Rich. seed oil. *Biomedicine & Pharmacotherapy*, 84: 1223–1232.

USDA (United States Department of Agriculture Natural Resources Conservation Service). 2023. *Ecballium elaterium* (L.) A. rich. https://plants.usda.gov/home/plantProfile?symbol=MIPU8. Accesses on 25.07.2023.

Wakimoto, N., Yin, D., O'Kelly, J., Haritunians, T., Karlan, B., Said, J., Xing, H., and Koeffler, H. 2008. Cucurbitacin B has a potent antiproliferative effect on breast cancer cells in vitro and in vivo. *Cancer Science*, 99(9): 1793–1797.

www.powo.science.kew.org. Accesses on 25.07.2023.www.wikipedia.org. Accesses on 25.07.2023.

Yeşilada, E. 2002. Biodiversity in Turkish folk medicine. In: Şener, B., ed., *Biodiversity: Biomolecular Aspects of Biodiversity and Innovative Utilization*. London: Kluwer Academic/Plenum Publishers, pp. 119–135.

Yeşilada, E., Tanaka S. and Sezik E. 1988. Isolation of an anti-inflammatory principle from the fruit juice of *Ecballium elaterium*. *Journal of Natural Products*, 51: 504–508.

Yeşilada, E., Tanaka, S., Tabata, M., and Sezik E. 1989. Antiinflammatory effects of the fruit juice of *Ecballium elaterium* on edem as in mice. *Phytotherapy Research*, 3(2): 75–76.

49 Ginger (*Zingiber officinale Roscoe* Zingiberaceae)

Khadija Temmante, Noureddine Chaachouay, Ouafae Benkhnigue, and Abdelhamid Azeroual

49.1 INTRODUCTION

Medicinal plants have been utilized by various cultures across the globe for centuries, fulfilling a vital function in healthcare (Ansari et al., 2023). The importance of these substances arises from their capacity to offer organic remedies for a wide range of diseases and health conditions (Chaachouay et al., 2020a, 2020b). *Zingiber officinale* Roscoe, is a well-acknowledged spice, flavor enhancer, and medicinal plant with a uniquely fragrant and tasty characteristic (Kumar Gupta & Sharma, 2014). The distinctive aroma and flavor of ginger are mainly due to a blend of bioactive substances, such as zingerone, shogaol, gingerols, and volatile oils (Ali et al., 2008a). The intense flavor of ginger is mainly attributed to non-volatile chemicals, namely phenylpropanoid derivatives, especially gingerols and shogaols (Y. Liu et al., 2019).

The ginger plant's roots and rhizomes have a long and significant history in medicine, with the ability to provide therapeutic advantages in treating several health disorders. Ginger has been studied for its potential to treat hypertension, diabetes, migraines, nausea, cardiovascular problems, Alzheimer's disease, and digestive illnesses (Krell & Stebbing, 2012; Kumar Gupta & Sharma, 2014). It demonstrates expectorant, spasmolytic, antitussive, antiallergic, and antihistaminic characteristics, successfully suppressing coughing. Moreover, studies have shown several medicinal functions of ginger, such as its impact on the movement of the gastrointestinal system, pain-relieving benefits, calming characteristics, fever-reducing effects, and ability to fight against germs.

Ginger essential oil, obtained from the plant, has shown protective properties against skin cancer in rat experiments and has proven the capacity to trigger apoptosis in ovarian cancer cells. Ginger can stimulate the production of saliva, which is known as the sialagogue effect (Munda et al., 2018; Saras, 2023). Additionally, it is considered as a mild chemical irritant. Notably, during World War I, it was used as a fever-reducing agent, namely as a suppository for horses.

The analysis of ginger essential oil indicates the existence of sesquiterpenoids, with zingiberene being the main component. Notable quantities of other sesquiterpenoids, including sesquiphellandrene, bisabolene, and farnesene, are also found, along with tiny amounts of monoterpenoids such as phellandrene, cineol, and citral (Geiger, 2005). The many characteristics of ginger make it a very valued plant that may be used in various culinary and medicinal contexts. Simultaneously, the bioactive molecules of the substance remain a topic of scientific inquiry, revealing its promise in several areas of medicine and beyond (Pakrashi & Pakrashi, 2003). This chapter delves into the classification, natural habitat, ethnobotanical uses, pharmacological properties, and therapeutic effects of Z. *officinale.*

49.2 SCIENTIFIC CLASSIFICATION

The ginger plant is classified in the following taxonomic hierarchy (GBIF, 2020):

Kingdom: Plantae
Phylum: Tracheophyta
Class: Magnoliopsida
Order: Zingiberales
Family: Zingiberaceae
Genus: *Zingiber*
Species: *Zingiber officinale* Roscoe

49.3 VERNACULAR NAMES

Arabic: Zanjabil, Skinjbir; **Tibbi**: Adrak, Zanjbeel; **English**: zinger; **Chinese**: Sheng jiang; **Nepali**: Aduwa, sutho; **German**: Gemeiner ingber/ingwer; **French**: Gingembre; **Indonesian**: Jahe; **Thai**: Khing; **Korean**: Saenggang; **Japanese**: Shokyo; **Sanskrit**: Katubhadra, Srngavera; **Hindi**: Adarakha; **Bengali**: Ada; **Punjabi**: Adi, Adrak; **Sweden**: Ingefaera; **Urdu**: Adrak; **Netherlands**: djahe, Gember; **Italian**: Zenzero; **Japanese**: Oshoga; **Spanish**: gengibre, Jengibre, jenjibre, dulcekion.

49.4 SYNONYMS

Z. officinale, commonly known as ginger, is a plant species with a long history of being cultivated and used in many civilizations around the globe. Consequently, it has amassed a multitude of synonyms and commonly used names, demonstrating its broad awareness and adaptable uses. Below are 17 other terms that may be used to refer to *Z. officinale* (The Plant List, 2013):

DOI: 10.1201/9781003460336-49

Amomum angustifolium Salisb.; **(2)** *Amomum zingiber* L.; **(3)** *Amomum zinziba* Hill; **(4)** *Curcuma longifolia* Wall.; **(5)** *Zingiber aromaticum* Noronha; **(6)** *Zingiber cholmondeleyi* (F.M.Bailey) K.Schum.; **(7)** *Zingiber majus* Rumph.; **(8)** *Zingiber missionis* Wall.; **(9)** *Zingiber officinale* var. *cholmondeleyi* F.M.Bailey; **(10)** *Zingiber officinale* f. *macrorhizonum* (Makino) M.Hiroe; **(11)** *Zingiber officinale* var. *macrorhizonum* Makino; **(12)** *Zingiber officinale* f. *rubens* (Makino) M.Hiroe; **(13)** *Zingiber officinale* var. *rubens* Makino; **(14)** *Zingiber officinale* var. *rubrum* Theilade; **(15)** *Zingiber officinale* var. *sichuanense* (Z.Y.Zhu, S.L.Zhang & S.X.Chen) Z.Y.Zhu, S.L.Zhang & S.X.Chen; **(16)** *Zingiber sichuanense* Z.Y.Zhu, S.L.Zhang & S.X.Chen; **(17)** *Zingiber zingiber* (L.) H.Karst.

49.5 BOTANICAL DESCRIPTION

Z. officinale is a perennial herbaceous plant with an erect growth habit, usually reaching a height of one to three feet (Aleem et al., 2020). The plant's stem is enveloped by a double-rowed arrangement of leaves featuring sheathing bases (Lai et al., 2022). The plant generates discernible yellowish flowers featuring violet lips arranged in a club-shaped spike; prominent yellow-green bracts accompany these flowers. It is essential to observe, however, that cultivated ginger plants rarely produce flowers (Kadirova, 2023). The commercially significant component of the ginger plant is its robust and scaly rhizomes, which are stems that grow underground. The rhizomes exhibit a branching pattern with finger-like extensions, leading to the formation of perennial individual divisions of the rhizome. The average measurements of these rhizome divisions are around 7 to 15 cm in length and 1 to 1.5 cm in width (Moghaddasi & Kashani, 2012; Ravindran & Babu, 2016). The branches that emerge from the rhizome are slanted and round, with a length ranging from 1 to 3 cm. These branches usually end in either concave marks or underdeveloped buds (Figure 49.1).

The ginger rhizome may display various colors, such as yellow, white, or red, depending on the specific type that is grown. The rhizome is enveloped with a brownish epidermis, which may exhibit variability in thickness depending on whether the ginger plant was picked at a mature or immature stage (H. Liu et al., 2020). The method commonly employed to propagate ginger plants is by utilizing rhizome cuttings, wherein each cutting contains a bud (Abbas et al., 2011). This technique enables the deliberate and dependable propagation of ginger plants, guaranteeing the ongoing cultivation of this vital and adaptable herb.

49.6 ORIGIN AND GEOGRAPHIC DISTRIBUTION

Z. officinale originates from the tropical and subtropical parts of Southeast Asia, with India being its central hub. This extraordinary plant has a significant historical background that spans many millennia, during which it was first esteemed for its gastronomic and therapeutic characteristics (Chaachouay et al., 2021a, 2021b, 2021c; da Silveira Vasconcelos et al., 2019). Ginger's extensive worldwide distribution can be attributed to its enduring popularity (Chaachouay & Zidane, 2021). Currently, it is grown in several places, demonstrating its capacity to thrive in all climates and types of soil. India continues to be a notable producer, with provinces such as Kerala, Karnataka, and Assam making considerable contributions to its production (Lai et al., 2022; Ravindran & Babu, 2016).

China, a significant ginger producer, highlights provinces such as Guangdong and Guangxi as essential cultivation regions. Ginger is a common ingredient in the traditional cuisines of Southeast Asian nations such as Thailand, Indonesia, and Vietnam. Africa cultivates ginger in Nigeria, Ghana, and Uganda, each contributing flavor characteristics (Huang et al., 2019). Jamaica and Haiti are renowned throughout the Caribbean for their exceptional ginger cultivation (Figure 49.2). This multipurpose plant is also cultivated in Latin America, Australia, and several tropical and subtropical climates across the globe (Wang et al., 2014). The flexibility and versatility of ginger have become an essential component of culinary traditions and traditional medicine systems worldwide, highlighting its extraordinary evolution from its roots in Southeast Asia to its status as a highly valued and extensively cultivated spice.

49.7 ECOLOGY

The ecological dynamics of *Z. officinale*, often known as ginger, are closely interconnected with its native environment and specific conditions for development. This adaptable botanical specimen originates from the tropical and subtropical areas of Southeast Asia, namely the Indian subcontinent, where it flourishes in hot and humid weather conditions (Koch et al., 2017; Ravindran & Babu, 2016). Ginger is an herbaceous perennial plant that usually grows to heights ranging from 1 to 3 feet. The most valuable portions of ginger are its underground rhizomes. It thrives on well-drained soils with mild acidity. Although it requires regular rainfall, it is intolerant of waterlogged conditions or cold (Govindarajan & Connell, 1983).

Ginger has allelopathic characteristics since it emits chemicals that inhibit the development of other plants, hence assisting in weed management within ginger fields. The plant's canopy offers shade and fosters microclimates that are advantageous for other plants that thrive in shaded environments. Ginger's genetic variety is enhanced by several species of ginger that may be found in woods and streambanks in their natural habitat (Khairul Alam, 2022). Gaining a comprehensive understanding of the ecology of *Z. officinale* is crucial for the effective cultivation and protection of this fragrant and medicinal plant, guaranteeing its continued flourishing in both its wild and cultivated environments.

FIGURE 49.1 (A) Aerial parts and (B) rhizome of *Z. officinale*.

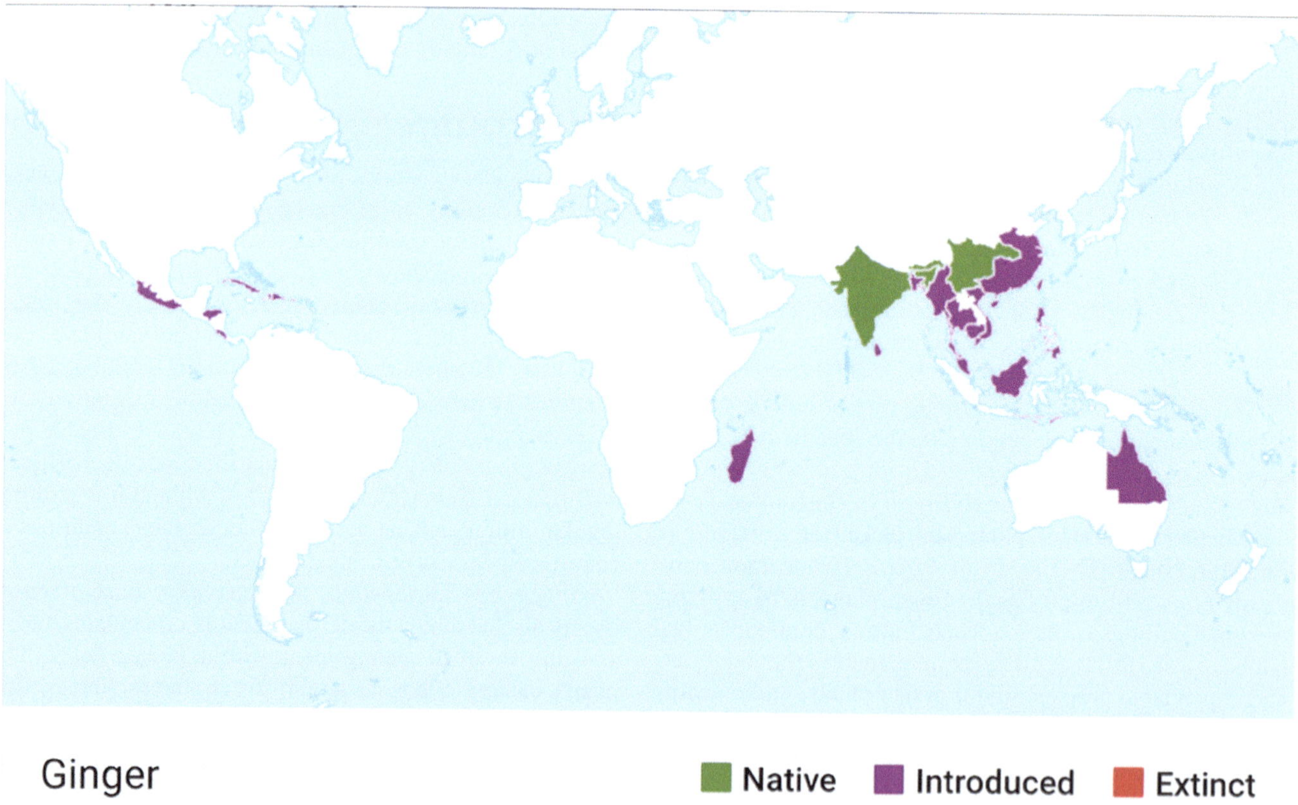

FIGURE 49.2 The geographical distribution of *Z. officinale*.

49.8 PHYTOCHEMISTRY

49.8.1 Primary Metabolites

Ginger is a prevalent spice and medicinal plant valued for its culinary and therapeutic attributes for generations. Primary metabolites are essential in carrying out the biological activities of ginger since it contains a wide range of bioactive chemicals. Primary metabolites are crucial substances that play a fundamental role in the plant's physiological activities. For ginger, these metabolites play a vital role in its growth, development, and general survival (Ali et al., 2008b).

Ginger plants mostly get their energy from carbohydrates. Ginger mainly stores carbohydrates as starch, which acts as an energy reserve for the plant throughout phases of growth and reproduction (Govindarajan & Connell, 1983). Ginger's rhizomes, the underground stems, are enriched with abundant and starchy carbohydrates.

Proteins are essential macromolecules that play a vital role in the growth and development of ginger. Enzymes, specialized proteins, participate in several metabolic activities inside the plant. Enzymes facilitate the breakdown of complex molecules into simpler ones, hence catalyzing chemical processes that aid in nutrition intake and energy generation. Amino acids are essential constituents of proteins and play a crucial role in the development and maintenance of ginger. Ginger is rich in a variety of amino acids, including several that are considered essential for human health. The presence of these amino acids enhances the overall nutritional quality of ginger (Moghaddasi & Kashani, 2012).

Ginger contains lipids, which are a primary metabolite. Fats and oils are necessary to form cell membranes and act as an energy store. The lipids present in ginger are responsible for its distinct flavor and fragrance, which enhance its appeal as a spice in culinary contexts. Ginger plants rely on nucleic acids, namely DNA and RNA, to store and transmit genetic information. These chemicals govern the plant's growth, development, and reactions to external conditions. The therapeutic benefits of ginger resin are attributed to its diverse ingredients(Y. Liu et al., 2019). These components may possess anti-inflammatory and analgesic properties.

Ginger's flavor character is influenced by primary metabolites, namely organic acids like citric acid and malic acid. Ginger's tart flavor contributes to its widespread usage in various culinary dishes and drinks. Ginger is rich in several vitamins, including vitamins C and B6, and numerous B-complex vitamins, including niacin (B3) and riboflavin (B2). Vitamin C enhances ginger's antioxidant qualities and function in bolstering the immune system. Additionally, these fundamental compounds significantly improve the total nutritional content of ginger. Water is the essential primary metabolite of ginger. It is crucial in various physiological functions, including photosynthesis, nutrition transportation, and temperature control. Ginger, while not often considered a bioactive ingredient, does include dietary fiber, which may positively affect digestion and general well-being.

Primary metabolites are essential substances that support the growth, development, and general functioning of ginger plants. Ginger is well-known for its secondary metabolites, such as gingerol and zingerone, giving it a distinct flavor and medicinal benefits. However, primary metabolites are the fundamental components that support the plant's life. The presence of these substances in ginger guarantees its life and well-being, contributing to its distinct taste as a spice and its intriguing status as a botanical specimen with a complex biochemical makeup.

49.8.2 Secondary Metabolites

Z. officinale is a plant known for its many bioactive constituents, many of which enhance its unique taste and powerful therapeutic qualities. Ginger essential oils consist of a mixture of bioactive components, such as terpenes and sesquiterpenes, which are responsible for the scent and taste of ginger (L. J. Liu et al., 1985; Marwat et al., 2015). These essential oils are often used in aromatherapy and have potential medicinal applications. Ginger has a diverse range of terpenes, including alpha-zingiberene, beta-sesquiphellandrene, beta-bisabolene, alpha-farnesene, camphene, beta-phellandrene, and curcumene. These compounds are responsible for the distinctive scent and taste of ginger (Lai et al., 2022). These chemicals have potential medicinal capabilities, such as anti-inflammatory and anti-bacterial activities. Ginger is rich in several phytonutrients, such as flavonoids and polyphenols (Ali et al., 2008b). Gingerol and its related substances belong to a broader category of phytochemicals known as glycosides. These chemicals exhibit diverse biological actions and are linked to the spicy flavor of ginger (Yousfi et al., 2021). These chemicals possess antioxidant capabilities and might potentially mitigate oxidative stress linked to chronic illnesses and the aging process. Ginger is also rich in terpene hydrocarbons such as camphene, phellandrene, and cineole (Moghaddasi & Kashani, 2012). These chemicals have the potential to enhance the scent and taste of ginger, as well as provide health advantages. Ginger is rich in phenolic compounds, with its main active ingredients being gingerol, shogaol, and paradol. These chemicals are responsible for the firm and pungent flavor of ginger (Marrelli et al., 2015; Ravindran & Babu, 2016). The main components of the volatile oil in ginger include mono- and sesquiterpenes, such as camphene, beta-phellandrene, curcumene, cineole, geranyl acetate, terpineol, terpenes, borneol, geraniol, limonene, linalool, alpha-zingiberene, beta-sesquiphellandrene, beta-bisabolene, and alpha-farnesene. Below are many essential bioactive constituents found in ginger:

- **Gingerol**: Gingerol is a critical bioactive component found in ginger. Ginger gets its distinctive pungent and robust taste from it. Gingerol has potent anti-inflammatory and antioxidant characteristics. Furthermore, it is linked to the digestive advantages and possibly analgesic properties of ginger.

- **Shogaol**: Shogaol is a derivative of gingerol that is produced when ginger is subjected to dehydration or cooking processes. It has a more intense flavor than gingerol and demonstrates comparable anti-inflammatory and antioxidant characteristics. Shogaol is thought to have a role in ginger's medicinal benefits.
- **Zingerone**: Zingerone is the compound accountable for the pleasant and earthy fragrance of ginger. It has antioxidant capabilities and may exhibit anti-inflammatory effects. Zingerone is often present in fewer amounts than gingerol and shogaol, although it still plays a role in the total bioactivity of ginger.
- **Gingerdiol**: Gingerdiol is an obscure chemical present in ginger that is regarded as a precursor to gingerol and shogaol. Ongoing research suggests that Gingerdiol has promise for several health advantages.
- **Zerumbone**: Zerumbone is a biologically active molecule belonging to the sesquiterpene class, which is present in ginger, specifically in a specific kind called *Zingiber zerumbet*. Research investigations have shown the presence of anti-inflammatory, antioxidant, and possibly anti-cancer effects.
- **Gingerenones**: Gingerenones are a collection of chemicals discovered in ginger that have attracted interest due to their potential advantages for health. They are being examined for their ability to reduce inflammation and counteract oxidative stress.
- **Gingerdiones**: Gingerdiones are a distinct category of chemicals found in ginger with anti-inflammatory actions. Continued research is being conducted to investigate the possible medicinal uses of these substances.
- **Catechins**: Ginger also has catechins, a phenolic molecule often seen in tea and other botanicals. Catechins have strong antioxidant properties and may play a role in ginger's capacity to counteract oxidative damage.
- **Kaempferol**: Kaempferol, an additional flavonoid found in ginger, possesses antioxidant characteristics. It can mitigate oxidative harm to cellular constituents, such as DNA, proteins, and lipids, by counteracting free radicals.
- **Quercetin**: Quercetin, a flavonoid, is a very effective antioxidant in ginger. Its notable characteristic is its capacity to eliminate free radicals, hence reducing oxidative stress and inflammation. Quercetin has been investigated for its possible ability to safeguard cells and tissues from harm induced by reactive oxygen species.

It is crucial to acknowledge that the makeup of ginger might differ based on variables such as ginger type, cultivation circumstances, and processing techniques. Consequently, the nutritional and bioactive composition of ginger may vary across different sources. The combination of these bioactive constituents makes ginger a helpful plant with both medical and culinary applications. Ginger has been used for ages in traditional medicine to alleviate various health conditions, such as nausea, inflammation, and digestive disorders. Continuing study is discovering more health advantages and uses for the bioactive chemicals present in ginger.

49.9 STANDARDS AND ADULTERATION

The worldwide consumption of ginger is steadily increasing, leading to a rising demand yearly. The most efficient preservation technique to fulfill this need and ensure ginger supply even during off-seasons is drying (Rubatzky & Yamaguchi, 2012). This technique not only prolongs the duration that ginger may be stored without spoiling but also hinders the proliferation of germs and prevents any chemical alterations. Dried ginger is the term used to describe fresh ginger drying at a temperature of 45°C.

Ginger meant to be consumed must feature distinct characteristics to meet rigorous quality criteria. The item should be whole and undamaged, without any indications of decay or modifications that would make it unfit for ingestion. Furthermore, it should be almost free from any harm inflicted by pests that might spoil its overall aesthetic. The ginger should also be free from any abnormal exterior moisture and, if washed, must be sufficiently dried, save for any condensation that may occur during its transfer from cold storage (Martindale & Westcott, 1892). Ginger should not have any unfamiliar smells or tastes.

Moreover, the growth and state of the ginger should be strong enough to endure transit and handling, guaranteeing its triumphant arrival at the intended location. The quality criteria are determined based on the GINGER STANDARD (CODEX STAN 218-1999). Ginger must adhere to the directives specified in the International Code of Practice for Packaging and Transport of Fresh Fruits and Vegetables (CAC/RCP 44-1995).

To ensure the integrity of ginger powder, it is crucial to maintain stringent quality standards to avoid any instances of adulteration, such as the substitution of ginger with less expensive alternatives or the utilization of ginger that has been previously employed. Official regulations permit the assessment of ash concentration in ginger as a method to authenticate its genuineness and assess its quality.

49.10 ETHNOBOTANY

The ethnobotany of *Z. officinale*, generally called ginger, is an intriguing exploration of its historical and cultural use by many people around the globe. Below is a summary of the ethnobotanical importance of ginger:

49.10.1 Culinary Delight

Ginger is a highly esteemed spice worldwide, with a unique flavor and fragrance. It is crucial in global cuisines to enhance meals with its characteristics. This adaptable root imparts a tangy and somewhat piquant flavor to savory and sweet dishes (Small & Canada, 2006). Ginger imparts complexity and dimension to marinades, stir-fries, and condiments, enhancing delectable cuisine (Nair, 2013). Furthermore, it adds a pleasant warmth to sweet treats such as gingerbread and cakes, enhancing their flavor. In addition, ginger is used in soul-warming soups, which adds richness to broths and enhances flavors (Saras, 2023). Ginger's pungent and spicy flavor contributes to the rich taste of curries, making it an essential ingredient in these recipes. Moreover, ginger plays a prominent role in drinks, providing a piquant flavor to teas and cocktails, solidifying its position as a highly valued ingredient in global cuisine.

49.10.2 Traditional Medicine

Ginger has a rich and extensive history strongly connected to traditional medical systems, which has led to its reputation as a potent medicinal agent. Throughout decades, this substance has been highly valued for its exceptional digestive characteristics, effectively relieving a wide range of health issues such as nausea, motion sickness, and gastrointestinal pain (Chaachouay et al., 2022a, 2022b, 2022c; Ravindran & Babu, 2016). The use of ginger in traditional medicine may date back millennia, originating from ancient China and India. These ancient cultures acknowledged ginger as a powerful medicinal plant, using its inherent characteristics to treat various health issues. Ginger is well recognized for its gastrointestinal advantages (Chaachouay et al., 2019a, 2019b). The substance includes bioactive components such as gingerol, which promote the formation of digestive enzymes and improve the absorption of nutrients. Ginger is a potent cure for indigestion, bloating, and gas.

The anti-emetic effects of ginger have been recognized for millennia. This cure is very reliable for alleviating morning sickness in pregnant women and has been endorsed for reducing nausea caused by chemotherapy and post-operative discomfort. Travelers have used ginger to counteract motion sickness due to its ability to stabilize the digestive system and relieve nausea. The gastrointestinal system benefits from ginger's recognized calming properties (Benkhnigue et al., 2022, 2023; Khodaie & Sadeghpoor, 2015). It can alleviate stomachaches, cramps, and pain resulting from digestive problems. Ginger helps these prevalent discomforts, whether ingested as tea or incorporated into culinary preparations. Ginger has been shown to enhance cardiovascular health by lowering cholesterol levels and blood pressure, according to some research (Orch et al., 2021). It improves cardiovascular health by facilitating improved blood flow and reducing the likelihood of heart-related ailments. The medicinal properties of ginger may be integrated into one's everyday routine. Ginger may be ingested as ginger tea, incorporated into recipes, or used as a dietary supplement, providing a flexible and easily obtainable choice for anyone trying to improve their health.

49.10.3 Ayurveda

Ginger, known as 'Shunthi' in the ancient Ayurvedic medical system, is widely esteemed for its potent therapeutic qualities and is recognized as a thermogenic plant. The medicinal uses of this substance are extensive since it has a crucial function in maintaining dosha balance, promoting digestion, and treating various illnesses (Khodaie & Sadeghpoor, 2015). Now, let's explore the complex realm of Shunthi in Ayurveda. Ayurveda acknowledges three primary doshas – Vata, Pitta, and Kapha – which correspond to distinct amalgamations of elements and regulate many facets of well-being. Ginger is highly valued for its capacity to harmonize these doshas, notably via appeasing Vata and Kapha. The warm properties of this substance counterbalance the chilly and damp characteristics often associated with these doshas, aiding in restoring bodily balance. Shunthi is revered for its potent properties in aiding digestion (Vasala, 2012). It enhances the Agni, or the metabolic process responsible for the effective decomposition and absorption of nutrients. Ginger improves digestion, preventing the buildup of undigested food particles that may cause digestive issues and disrupt the body's equilibrium. Ginger is highly recommended in Ayurveda for addressing a diverse array of health concerns. Its thermogenic characteristics make it particularly beneficial in treating cold-related ailments such as coughs, colds, and congestion (Kumar et al., 2013). Ginger's capacity to induce perspiration may aid in the elimination of toxins and the reduction of fever during periods of sickness. Ginger plays a crucial part in maintaining respiratory health. Ayurvedic treatments often recommend its use to ease respiratory conditions such as asthma and bronchitis. The anti-inflammatory effects of this substance aid in calming inflamed airways, while its warming characteristics may alleviate chest congestion. The anti-inflammatory and analgesic qualities of Shunthi are used to treat joint and muscular discomfort. Topically, it is used as poultices, while internally, it is ingested to diminish inflammation and relieve the agony linked to arthritis. Ayurvedic practitioners regard ginger as a natural enhancer of the immune system. The anti-bacterial qualities of this substance are believed to assist the body in preventing infections. It is often suggested to be used during seasonal transitions or times of susceptibility to improve general immune function. Shunthi is skillfully incorporated into Ayurvedic food and used in culinary preparations to imbue meals with its curative advantages (Ravindran & Babu, 2016). It is included in soups, stews, teas, and other herbal preparations, all meticulously designed to target specific health issues.

49.10.4 Traditional Chinese Medicine (TCM)

In the context of Traditional Chinese Medicine (TCM), ginger, which is referred to as 'Sheng Jiang', occupies a pivotal role as a medicinal herb. Due to its exceptional warming properties, it is a valuable component in TCM practices (Shahrajabian et al., 2020). Ginger is utilized to eliminate colds, enhance circulation, and treat various health ailments, such as vertigo and coughing. We examine the multifaceted function of Sheng Jiang in TCM in this article. TCM classifies foods and botanicals according to their intrinsic thermal characteristics, which may be warm, cold, neutral, heated, or mild. Ginger is categorized as a warming herb, and its therapeutic utility is predicated on its capacity to induce warmth within the body. (Goto et al., 1990). Ensuring equilibrium between Yin (calming) and Yang energies in Traditional Chinese Medicine (TCM) is essential for optimal well-being.

Ginger's thermogenic properties are valuable in eliminating cold and moisture from the body, a factor that is linked to a range of health conditions. Sheng Jiang is esteemed for its efficacy in alleviating cold symptoms, both inside and topically (Huang et al., 2019). In TCM, cold is often seen as a catalyst for stagnation and imbalance inside the body, resulting in various health problems. Ginger mitigates these effects by stimulating warmth and mobility, relieving symptoms such as chilly extremities, gastrointestinal discomfort, and stomach pain induced by cold-damp circumstances. Ginger's thermogenic effects also enhance blood circulation (Chung et al., 2003). It stimulates blood circulation and Qi (vital energy), improving bodily flow. This characteristic renders it a beneficial herb for treating ailments associated with compromised blood flow, including achy extremities, muscle rigidity, and menstrual irregularities. In TCM, ginger is renowned for its ability to aid digestion. It is essential for the digestion process, which involves the absorption of nutrients, digestion of food, and elimination of debris. Having the ability to alleviate dyspepsia, bloating, and vertigo, Sheng Jiang is a highly productive treatment for gastrointestinal distress.

49.10.5 African Remedies

Ginger is highly regarded as a botanical asset in African Traditional Medicine, which is utilized extensively to treat a wide range of health concerns. This multifunctional root is a significant component in herbal remedies and is esteemed for its effectiveness in managing conditions such as gastroesophageal reflux, colds, and migraines. Ginger holds a prominent role in African Traditional Medicine, a culture that is profoundly rooted in the utilization of plant-based remedies for medicinal purposes (Adebayo et al., 2021; Chaachouay et al., 2019b). Ginger is highly desired for its efficacy in alleviating headaches. The anti-inflammatory and analgesic characteristics of this substance provide calming relief, making it a widely favored option for relieving headaches. Ginger is often used in Africa as a cure for treating colds and flu (Moghaddasi & Kashani, 2012). The warming effects of this substance aid in relieving congestion, while its anti-bacterial characteristics are thought to support the body's defense against infections (Kemper, 1999; Ravindran & Babu, 2016). The efficacy of ginger as a gastroprotective agent is well-established in traditional medicine. It is often used to alleviate digestive pain, such as indigestion, bloating, and nausea, restoring digestive equilibrium. African Traditional Medicine prioritizes whole well-being, and ginger complements this philosophy by targeting several aspects of health.

49.10.6 Cosmetics

Z. officinale has successfully entered the cosmetics business, leading to a significant shift toward natural beauty. Ginger, renowned for its remarkable skincare advantages, has gained recognition as a valuable component in several cosmetic products (Rasool et al., 2022). The potent anti-inflammatory characteristics of this substance are very beneficial for relaxing and soothing inflamed skin. It is an excellent ingredient for treatments that reduce redness and ease disorders like acne and rosacea (Choi, 2019). In addition, ginger's abundant antioxidant content protects against premature aging by counteracting free radicals, promoting skin flexibility, and reducing the appearance of fine lines and wrinkles (Gulzar et al., 2023). This unique root, renowned for its invigorating abilities, enhances the appearance of the skin by promoting a more luminous and glowing complexion. It stimulates the flow of blood and revitalizes tired skin.

Ginger has several uses beyond skincare, including its benefits for hair care. It aids in maintaining a nourished scalp, reducing dandruff, and stimulating hair growth. Ginger's inherent allure corresponds with the growing consumer inclination toward pure and plant-based components in beauty products, making it a compelling option for those who need both effectiveness and natural aesthetics in their everyday self-care routines (Hu et al., 2023; Madankar & Nair, 2021). The increasing use of *Z. officinale* in cosmetic products highlights the long-lasting appeal of natural resources in the pursuit of vibrant, youthful, and inherently attractive skin and hair.

49.10.7 Aromatic and Flavoring Agent

Ginger surpasses its classification as a simple spice; it becomes an essential additive, enhancing a wide variety of drinks with distinct and captivating qualities. Ginger is quite versatile in beverages, enhancing the flavor and adding a vibrant touch to a wide range of drinks, including ginger tea, cocktails, and sweets. Ginger tea is highly esteemed and treasured for its ability to capture the authentic flavor of ginger (Bhatt et al., 2013; Kiyama, 2020). Usually, it infuses hot water with fresh slices or grated root of ginger. The outcome is a calming and fragrant blend with a pleasant sensation of heat and a subtle touch of spice. Ginger tea is well recognized for its many health advantages, including facilitating digestion, alleviating throat irritation, and offering solace during cold weather (Govindarajan & Connell, 1983; Wakchaure & Ganguly, 2018). Ginger tea, whether consumed straight or with honey and lemon, provides a soothing and refreshing drinking experience. Ginger's impact in the realm of mixology is revolutionary. The component is highly desired in drinks because of its pungent spiciness and fragrant characteristics (Chaachouay et al., 2022b; Marwat et al., 2015; Supartono et al., 2015). The traditional Moscow Mule is based on ginger beer, providing a stimulating and tangy blend of ginger, lime juice, and vodka. Ginger also enriches the intricacy of non-alcoholic mocktails, offering a captivating and essential component that improves the drinking experience. The diversity of rum is evident in cocktails, as it imparts a delicate warmth to a mojito and a strong impact in a dark and stormy (Bag, 2018). The use of ginger in sweets exemplifies its capacity to harmonize sweetness and delicate spiciness, creating a well-balanced taste profile. Gingerbread cookies, ginger snaps, and cakes laced with ginger are famous delicacies. The use of this ingredient in desserts adds a delightful sensation of warmth and intricacy that remains on the taste buds, enhancing the sweetness of the treat. Ginger also intensifies the flavor of ice creams, sorbets, and fruit-based sweets, creating a pleasurable juxtaposition that stimulates the taste receptors.

49.10.8 Food Preservation

Historically, ginger has played a crucial role in preserving food because of its exceptional natural anti-bacterial capabilities. This spice, renowned for its unique taste and scent, was a vital asset in keeping perishable goods without using contemporary refrigeration techniques (He et al., 2023). Ginger has been an integral part of traditional food preservation methods in many countries and locations, making a lasting impact on culinary history. Ginger is rich in bioactive chemicals such as gingerol and shogaol, which have anti-bacterial solid properties (Cai et al., 2020; Sarwar, 2014). These chemicals efficiently suppress the proliferation of bacteria, fungi, and other microbes that lead to food spoiling and deterioration. Consequently, ginger acquired the role of a natural preservative, prolonging the duration that goods may be stored and guaranteeing their suitability for eating. Pickling was a notable use of ginger in the field of food preservation. Ginger's anti-bacterial properties in pickled vegetables contribute to food conservation, ensuring its freshness and quality over prolonged periods, enabling communities to enjoy these preserved delicacies (Malik et al., 2023). In addition, ginger was used to preserve foods, enhance taste, and prevent bacterial proliferation. Ginger's ability to enhance flavor and ensure food safety has established it as an essential component in traditional methods of food preservation (Laelago Ersedo et al., 2023). While modern refrigeration has primarily supplanted the ancient practice of using ginger for food preservation, its influence remains prominent in culinary customs around the globe. The historical importance of ginger as a natural preservative emphasizes its adaptability and lasting worth beyond its position as a popular spice, showcasing its essential role in maintaining the tastes and customs of the past.

49.10.9 Cultural Symbolism

Ginger has cultural importance beyond its culinary function, interweaving many metaphorical connotations across many civilizations. It is a universally recognized icon of warmth, friendliness, and pleasant energy. Ginger is highly respected and incorporated into traditions and rituals in many cultures, adding significant meaning and importance to numerous parts of life (Chaachouay & Zidane, 2021; Miller, 2005; Saras, 2023). Ginger is often associated with creating a welcoming and cozy atmosphere in many cultures, thanks to its intense and refreshing nature. Providing ginger-flavored snacks or drinks to visitors is a traditional gesture that represents hospitality, warmly greeting them and creating a sense of warmth and coziness (Creed, 2018). In addition to its culinary qualities, ginger is renowned for its association with good energy and vigor. It stimulates both the body and mind, promoting alertness and enthusiasm. Ginger tea is often a revitalizing drink to start the day with more energy and a positive outlook. Ginger is prominently included in several worldwide rites and events. In Hindu traditions, it symbolizes purity and heavenly blessings. Chinese New Year festivities represent the arrival of good fortune and positive energy for the next year (Gurib-Fakim, 2006). Moreover, in some tribal customs, ginger is a cleaning agent during cleansing ceremonies, eliminating unfavorable points, and establishing a favorable atmosphere for beneficial encounters.

49.10.10 Historical Trade

The historical significance of ginger in commerce is evidence of its revered status as a highly desirable product throughout the age of ancient trading routes, particularly the famous Silk Road (Van der Veen & Morales, 2015). The intrinsic value of this plant, which may be used both as a spice and a powerful medicinal herb, has stimulated strong demand and the development of large trading networks. Ginger's journey as a traded commodity reached its highest point along the renowned Silk Road, an intricate network of interconnecting trade routes spanning Asia, the Middle East, and Europe (Saras, 2023). Ginger had a highly desirable status in the extensive trade between many cultures across the globe.

The allure of ginger lies in its versatile usefulness. As a spice, it imparts depth and taste to culinary creations, a highly valued characteristic in many cuisines. Simultaneously, ginger had a prominent position in traditional medical systems, where its curative properties were used to treat various illnesses, from gastrointestinal issues to respiratory diseases (Ravindran & Babu, 2016). The significant demand for ginger was generated by its dual purpose, which enticed merchants and traders who were keen to incorporate it into their cargo. Its value transcended its physical magnitude, significantly enhancing the sensory and medicinal encounters of individuals who integrated it into their routines (Pallasmaa, 2012).

Outside the domain of business, the ginger trade not only enabled the interchange of products but also the spread of other civilizations. It brought new and unique tastes and innovative medical techniques to areas along the trade routes, enhancing the local culinary traditions and healthcare practices (Chen, 2008). The recognized ancient medical systems of Ayurveda and Traditional Chinese Medicine highly value ginger's ability to integrate the body's energies and enhance general well-being. This has increased the demand for ginger in commerce (Bode & Dong, 2011). Furthermore, ginger's inherent preservation qualities were crucial in business, acting as a method to protect other perishable goods throughout the long trips along the Silk Road, ensuring their quality upon reaching their destination. Despite the changes brought about by contemporary transportation and globalization, ginger continues to have a substantial effect on culinary traditions and herbal medicine. It remains a cherished spice and natural treatment, preserving its long-standing history.

49.11 BIOLOGICAL ACTIVITIES

Z. officinale is renowned for its gastronomic and therapeutic applications, displaying a diverse array of biological properties ascribed to its many chemical constituents. *Z. officinale* exhibits many noteworthy biological functions, such as:

49.11.1 Anti-diabetic Activity

Ginger has attracted interest due to its potential as an antidiabetic agent. Research completed to March 2022 has shown many methods by which ginger exerts its beneficial benefits in diabetes management (Chaachouay et al., 2019a; Noipha & Ninla-Aesong, 2018). These methods include more excellent management of blood sugar levels via improved glycemic control, increased insulin production from pancreatic beta cells, easier glucose absorption by cells, and the translocation of GLUT4 transporters to the cell membrane (Akhani et al., 2010). In addition, ginger is thought to possess antioxidant capabilities that might counteract the oxidative stress caused by advanced glycation end products (AGEs), thereby preventing diabetes-associated issues (Mishra et al., 2012). Furthermore, ginger has shown potential in controlling the expression of genes in the liver that are involved in the metabolism of glucose, adjusting the levels of cytokines that promote inflammation, and maybe safeguarding against kidney damage that is often linked to diabetes (Chaachouay et al., 2019a; Jafri et al., 2011). Preliminary results are encouraging; however, patients with diabetes should seek guidance from healthcare experts before integrating ginger into their diabetes management regimen, since its impact may differ across individuals and potentially conflict with drugs or other therapies.

It is crucial to acknowledge that while ginger has the potential to manage diabetes, it should not be considered a replacement for traditional diabetic therapy, which may include medicines, dietary adjustments, and lifestyle changes. Before introducing ginger into their treatment plan, individuals with diabetes should seek guidance from their healthcare experts. Due to its potential medication interactions and individual variability in response, tailored counsel is crucial while using ginger. Moreover, additional study is required to comprehensively comprehend the processes and ideal doses for ginger's anti-diabetic properties.

49.11.2 Anti-Inflammatory Activity

Ginger has remarkable anti-inflammatory qualities that have been extensively studied (Ezzat et al., 2018; Mahluji et al., 2013). Gingerol, a bioactive compound in ginger, is predominantly responsible for these anti-inflammatory effects. Ginger is an effective natural remedy for inflammatory conditions because gingerol inhibits multiple pathways implicated in inflammation (Ahmed et al., 2011). It helps alleviate inflammation in disorders such as osteoarthritis and rheumatoid arthritis by inhibiting the generation of pro-inflammatory cytokines and enzymes contributing to joint discomfort and swelling (Habib et al., 2008). Ginger has an anti-inflammatory effect comparable to non-steroidal anti-inflammatory medicines (NSAIDs) but without the possible adverse effects of prolonged NSAID use, such as gastrointestinal discomfort.

Further, the anti-inflammatory properties of ginger transcend joint health. It can alleviate systemic inflammation, contributing to chronic ailments, including cardiovascular disease and specific forms of malignancy (Anosike et al., 2009; Habib et al., 2008). The antioxidants included in ginger enhance its anti-inflammatory effects by counteracting free radicals and reducing oxidative stress, which is often associated with inflammation. Ginger's incorporation into one's diet may be very beneficial for persons aiming to naturally regulate inflammation and mitigate the likelihood of inflammation-related health concerns, owing to its amalgamation of anti-inflammatory processes.

49.11.3 Anti-Obesity Activity

Ginger has shown promising anti-obesity properties, providing a natural method for addressing obesity and its related health dangers (Nazish et al., 2016). The many benefits gingers have on the body's metabolism and adipose (fat) tissue are primarily responsible for this feature. Ginger's ability to raise body temperature may help with weight control (Ebrahimzadeh Attari et al., 2018). It can increase the metabolic rate and facilitate the breakdown of stored fat, which may lead to weight reduction and decreased fat buildup. Moreover, scientific studies have shown that ginger can inhibit hunger, regulate calorie consumption, and facilitate endeavors to lose weight (Han et al., 2005; Patil & Kothavale, 2022).

Moreover, ginger has anti-inflammatory characteristics that are associated with the control of obesity. Obesity is often linked to chronic inflammation, which plays a role in causing insulin resistance and other metabolic disorders (Bin-Meferij et al., 2017). Ginger's anti-inflammatory properties may enhance insulin sensitivity and lessen the effects of obesity-related problems, including type 2 diabetes and cardiovascular disease. The antioxidants in ginger also fight oxidative stress, which is increased with obesity and may result in several health problems.

49.11.4 Cardiovascular Protection

Several studies have shown the fantastic potential of ginger's anti-inflammatory qualities in protecting against heart-related conditions(Mao et al., 2019). An essential mechanism through which ginger achieves this protective effect is through its capacity to stimulate coronary artery relaxation, consequently enhancing myocardial blood flow (Nicoll & Henein, 2009). This vasodilatory function helps alleviate the burden on the cardiovascular system and may aid in preventing disorders like coronary artery disease (Hosseini et al., 2014). Moreover, empirical evidence suggests that ginger can augment vaso-protection, a vital component of cardiovascular well-being. Ginger potentially mitigates the progression of atherosclerosis, an artery-clogging condition characterized by lipid deposition, by enhancing the functionality and overall well-being of blood vessels (Belhaj et al., 2021). Consequently, there is a decreased chance of heart attacks and strokes. Furthermore, investigations on animals have shown that ginger may alleviate ethanol-induced cardiac irregularities, indicating that it may protect against alcohol-related cardiac problems.

49.11.5 Neuroprotection

Multiple studies, primarily carried out on rodent models, have shown the potential of ginger as a neuroprotective agent, providing optimism in the battle against neurodegenerative illnesses (ALmohaimeed et al., 2021; Sutalangka & Wattanathorn, 2017). Ginger extracts have shown their capacity to mitigate memory deficits, a characteristic feature of illnesses such as Alzheimer's, by acting in the degeneration of neuronal cells and breaking synaptic connections in the brain (Waggas, 2009). Ginger has a part in maintaining cognitive function and reducing the development of neurodegenerative illnesses. In addition, several ginger chemicals, particularly 6-shogaol, have been identified for their ability to protect the nervous system (Farombi et al., 2019). This bioactive compound can preserve brain health and reduce harm to neurons. Although the precise mechanisms behind ginger's neuroprotective qualities require more investigation, these findings emphasize the promise of ginger as a natural method to enhance brain function and mitigate the likelihood of neurodegenerative disorders.

49.11.6 Gastric Effect

Ginger extracts have a variety of therapeutic properties, with the most notable being their ability to relieve nausea and vomiting (Sharma & Gupta, 1998). Ginger has been demonstrated to be effective in reducing nausea caused by antituberculosis medications and antiretroviral therapy, providing a natural and potentially beneficial solution for those suffering from treatment-induced nausea (Anosike et al., 2009; Phillips et al., 1993). Ginger extract and its bioactive ingredients benefit stomach health apart from anti-nausea characteristics. Ginger has chemo-preventive properties, implying that it may have a role in lowering the risk of stomach cancer (Gupta & Sharma, 2001; Lazzini et al., 2016). Moreover, investigations on stomach cancer have shown that ginger has antineoplastic qualities, which means that it may stop the proliferation of cancer cells (Al-Yahya et al., 1989). These results highlight the potential of ginger as a preventative measure against the onset of gastric cancer as well as a treatment for gastrointestinal pain.

49.11.7 Protective Effects Against Respiratory Disorders

Ginger is a popular herb due to its fragrant characteristics and medical benefits in treating respiratory disorders such as asthma (Hasan et al., 2016; Rout et al., 2010). Several research investigations have confirmed the bronchodilatory benefits of this substance and its capacity to reduce airway hyperactivity, making it a helpful asset in maintaining respiratory health (Mao et al., 2019). Moreover, studies on fresh ginger have shown promise in the fight against the respiratory syncytial virus (RSV). It does this by inducing respiratory cells to generate the potent antiviral protein beta interferon, which is most notable when given at a dose of 300 μg/mL (Ajith et al., 2008). Ginger has been identified in therapeutic settings for its function in improving gas exchange and shortening the time of mechanical ventilation in patients suffering from acute respiratory distress syndrome (ARDS). These diverse characteristics highlight ginger's importance as a natural cure for respiratory diseases and its ability to enhance respiratory health outcomes.

49.11.8 Antioxidant Activity

The antioxidant properties of ginger are well-established and play a vital role in its biological activity. Ginger has a diverse range of bioactive chemicals, such as gingerols, shogaols, and paradols, which contribute to its potent antioxidant properties (Stoilova et al., 2007). These compounds aid in safeguarding the cells and tissues of the body against oxidative stress, a phenomenon distinguished by the detrimental consequences of free radicals (Amir et al., 2011; Ghasemzadeh et al., 2010). These unstable chemicals, known as free radicals, may harm cellular constituents, including proteins, lipids, and DNA. Antioxidants found in ginger counteract these free radicals (Bellik, 2014). Ginger prevents premature aging and several chronic illnesses, such as cancer, heart disease, and neurological disorders like Alzheimer's, by reducing oxidative damage to cells (Tohma & Gulçin, 2010).

Furthermore, ginger has antioxidant properties that may contribute to its ability to decrease inflammation. Ginger's antioxidant processes help alleviate chronic inflammation, which is a prevalent contributing component in several chronic illnesses, hence enhancing its overall health advantages. Moreover, these antioxidants may improve immunological function, bolstering the body's ability to protect itself from infections and diseases.

49.11.9 Antimicrobial Activity

The anti-bacterial activity of ginger is well-acknowledged as one of its biological qualities, which makes it a fascinating natural resource for controlling infections and maintaining good health. The *Z. officinale* plant is the source of ginger, which includes many bioactive components, including gingerol, shogaol, and zingerone. These components are responsible for the anti-bacterial properties of ginger (Bellik, 2014; Knaak et al., 2013). These substances can impede the development and reproduction of diverse microorganisms, including bacteria, fungi, and viruses (El-Wahab et al., 2009; Singh et al., 2023). Ginger has shown anti-bacterial effects against various common pathogens in the bacterial environment. It can combat bacteria such as *Escherichia coli* and *Staphylococcus aureus*, which are causative agents of multiple infections, including foodborne illnesses and skin infections (Andriyani et al., 2015; Kawai et al., 1994). Ginger's anti-bacterial activity is thought to occur from damaging bacterial cell membranes and interfering with essential cellular functions.

Moreover, ginger has anti-fungal characteristics that make it a promising natural treatment for fungal infections, notably in cases of Candidiasis (Othman et al., 2020). Ginger extracts have shown the capacity to impede the proliferation of Candida species, which are known to induce yeast infections in people. Moreover, scientific study indicates that ginger has antiviral capabilities, which can impede the reproduction of certain viruses. Further research is required to comprehend the antiviral capabilities of ginger properly. However, some data suggests that ginger may effectively restrict the transmission of viruses such as RSV.

49.11.10 Anti-Cancer Activity

Cancer continues to be a substantial health concern on a global scale, as evidenced by its increasing prevalence and considerable loss of life (Habib et al., 2008). The growing fascination with herbal therapies and natural chemicals arises from the constraints of traditional chemotherapy medications in addressing diverse cancer forms. These medications often have cytotoxic effects on normal cells, resulting in undesirable side effects (Ajith et al., 2008; Kumar Gupta & Sharma, 2014). Various research has investigated the possible anti-cancer effects of ginger, providing insight into its intriguing role in cancer therapy. One study showed that 6-gingerol can decrease cell growth. Research conducted both in laboratory settings (in vitro) and in living organisms (in vivo) has demonstrated that 6-shogaol can effectively hinder the movement of cells, the growth of cells, and the development of tumors inside tumor tissues (Mishra et al., 2012). Furthermore, a separate study discovered that 6-shogaol amplifies the anti-cancer properties of chemotherapy drugs, such as 5-fluorouracil, oxaliplatin, and irinotecan, by stimulating autophagy and death in colon cancer cells.

Z. officinale essential oil has shown significant promise in herbal remedies for preventing the growth of breast cancer cells. For MCF-7 and MDA-MB-231 cells, it demonstrated astounding 98% and 93% inhibition rates, respectively (Park et al., 2014). Moreover, this essential oil has gained recognition for its remarkable anti-bacterial, antioxidant, and anticancer characteristics (Ansari et al., 2023; Mao et al., 2019; Mukjerjee & Karati, 2022). Furthermore, studies have shown that four new diarylheptanoid glycosides have anti-apoptotic properties since they promote the survival of normal human lung epithelial cells. Together, these results highlight the intriguing role that ginger and its derivatives may play in the battle against cancer, providing promise for new treatment approaches and fewer side effects than traditional chemotherapy. Nevertheless, more investigation and clinical studies are necessary to confirm and enhance these results for efficient cancer treatment.

49.12 CLINICAL USES

Z. officinale has a variety of therapeutic uses in both conventional medicine and healthcare (Khan et al., 2016; Mukjerjee & Karati, 2022; Shahrajabian et al., 2019). Many clinical services for it exist, such as the relief of nausea and vomiting, treatment of gastrointestinal problems, help with pain and inflammation management, enhancement of the immune system, promotion of cardiovascular health, support for respiratory health, aiding in the direction of diabetes, aiding in post-operative recovery, assisting with weight control, and possible neuroprotection in neurodegenerative diseases. It is also used in cancer care to reduce symptoms, especially in treating nausea, vomiting, and various gastrointestinal disorders because of its calming and anti-inflammatory effects on the digestive system (Abascal & Yarnell, 2009; Abdalla & Abdallah, 2018). It's crucial to remember that using ginger in a clinical setting should only be done with medical experts' supervision. When integrating ginger into clinical treatment, it is essential to consider the specific requirements of each patient, their medical history, and any possible drug interactions. Because ginger may have varied effects on different people, it is essential to carefully check doses in therapeutic settings to ensure safe and effective usage.

49.13 TOXICITY OF GINGER

Ginger is considered a benign plant substance when ingested in modest quantities as a seasoning, taste booster, or nutritional adjunct. It has a profound historical background in its use in gastronomy and traditional medicine (Chaachouay et al., 2021c). Nevertheless, the overindulgence or incorrect ingestion of ginger might result in significant negative consequences. Firstly, allergic responses to ginger are few but feasible, exhibiting symptoms that span from dermatological problems such as rashes and pruritus to severe manifestations like edema and, in exceptional instances, anaphylaxis (Bischoff & Crowe, 2005). If any indications of an allergic response arise after using ginger, it is imperative to seek emergency medical assistance.

Excessive use of ginger, particularly in its uncooked state or as a nutritional supplement, may lead to gastrointestinal discomfort, such as heartburn, indigestion, or an unsettled stomach. Pre-cooking or processing ginger before consumption may help alleviate these effects (Morakinyo et al., 2010). Ginger has minor inherent anticoagulant effects, which may be advantageous for some persons but may provide a potential hazard for those who are on blood-thinning drugs (Abebe, 2002). Consultation with a healthcare practitioner is necessary for prescription medicines since ginger can interact with certain pharmaceuticals such as anticoagulants, antiplatelet drugs, and diabetic treatments (Plengsuriyakarn et al., 2012).

Pregnant and lactating women should consume ginger in limited quantities since consuming excessive amounts or using ginger supplements without direction from a healthcare professional may pose potential dangers for pregnancy (Jefferies et al., 2012). Furthermore, persons with pre-existing cardiac issues should seek advice from a healthcare professional since ginger can reduce blood pressure and heart rate. Certain people may exhibit increased sensitivity to ginger, resulting in adverse consequences even with moderate intake, such as oral irritation, diarrhea, or a burning feeling (Setty & Sigal, 2005). Individuals who have concerns about the possible toxicity or interactions of ginger should see healthcare specialists for tailored recommendations. However, most individuals may safely use ginger in their diets. Practicing responsible use and a careful attitude is crucial to appreciating the advantages of ginger while minimizing potential dangers entirely.

49.14 CONCLUSION AND PERSPECTIVES

Ginger is a very adaptable component with many uses in several sectors, such as culinary arts, cosmetics, beverage manufacturing, and the more significant food business. The botanical makeup of *Z. officinale* contains many bioactive chemicals, including shogaol, gingerols, paradols, and gingerol. Examining different ginger extracts presents novel remedies for farmers by offering alternatives to chemical pesticides, which are acknowledged to have adverse impacts on human health, ecological equilibrium, and the environment.

Ginger has significant potential health advantages, which span a wide range of medicinal characteristics. Ginger has shown promising efficacy in combating a range of diseases, displaying antioxidant properties that fight oxidative stress, anti-inflammatory capabilities that reduce inflammation, antimicrobial actions that inhibit the growth of harmful microorganisms, anti-diabetic potential in managing blood sugar levels, anti-cancer properties with potential anti-tumor effects, antiviral activity to combat viral infections, and neuroprotective qualities that support the nervous system. Nevertheless, it is crucial to highlight the need to conduct human clinical studies to validate these results and promote the progress of pharmaceutical applications derived from the bioactive chemicals found in ginger.

Furthermore, in light of the growing difficulties presented by antibiotic-resistant microorganisms, ginger, along with other plant antimicrobials, is becoming increasingly recognized as a viable substitute for conventional antibiotics. The increasing use of plant-based antimicrobials highlights the need to decrease our dependence on antibiotics, especially considering the growing problem of antibiotic resistance. To fully use the capabilities of ginger in tackling these intricate problems, it is crucial to accumulate extensive understanding and documentation of this botanical species. This information may be used as the basis for developing improved ginger cultivars that can adjust to changing climatic conditions and other essential elements, assuring the sustainable production and usage of this unique botanical resource.

LIST OF ABBREVIATIONS

ARDS acute respiratory distress syndrome
CAC coronary artery calcification
DNA Deoxyribonucleic acid
MCF-7 *Michigan Cancer Foundation-7*
NSAIDs non-steroidal anti-inflammatory drugs
RCP Résumé des caractéristiques du produit
RNA Ribonucleic acid
RSV respiratory syncytial virus
TCM Traditional Chinese Medicine

REFERENCES

Abascal, K., & Yarnell, E. (2009). Clinical uses of *Zingiber officinale* (Ginger). *Alternative and Complementary Therapies*, *15*(5), 231–237. https://doi.org/10.1089/act.2009.15501

Abbas, M. S., Taha, H. S., Aly, U. I., El-Shabrawi, H. M., & Gaber, E.-S. I. (2011). In vitro propagation of ginger (*Zingiber officinale* Rosco). *Journal of Genetic Engineering and Biotechnology*, *9*(2), 165–172. https://doi.org/10.1016/j.jgeb.2011.11.002

Abdalla, W., & Abdallah, E. (2018). Anti-bacterial activity of ginger (*Zingiber Officinale* Rosc.) rhizome: A mini review. *International Journal of Pharmacognosy & Chinese Medicine*, *2*. https://doi.org/10.23880/ipcm-16000142

Abebe, W. (2002). Herbal medication: Potential for adverse interactions with analgesic drugs. *Journal of Clinical Pharmacy and Therapeutics*, *27*(6), 391–401. https://doi.org/10.1046/j.1365-2710.2002.00444.x

Adebayo, S. A., Amoo, S. O., Mokgehle, S. N., & Aremu, A. O. (2021). Ethnomedicinal uses, biological activities, phytochemistry and conservation of African Ginger (*Siphonochilus aethiopicus*): A commercially important and endangered medicinal plant. *Journal of Ethnopharmacology*, *266*, 113459. https://doi.org/10.1016/j.jep.2020.113459

Ahmed, K., Shaheen, G., & Asif, H. M. (2011). *Zingiber officinale* Roscoe (pharmacological activity). *Journal of Medicinal Plants Research*, *5*(3), 344–348.

Ajith, T. A., Aswathy, M. S., & Hema, U. (2008). Protective effect of *Zingiber officinale* roscoe against anti-cancer drug doxorubicin-induced acute nephrotoxicity. *Food and Chemical Toxicology*, *46*(9), 3178–3181.

Akhani, S. P., Vishwakarma, S. L., & Goyal, R. K. (2010). Anti-diabetic activity of *Zingiber officinale* in streptozotocin-induced type I diabetic rats. *Journal of Pharmacy and Pharmacology*, *56*(1), 101–105. https://doi.org/10.1211/0022357022403

Aleem, M., Khan, M., Shakshaz, F., Akbari, N., & Anwar, D. (2020). Botany, phytochemistry and antimicrobial activity of Ginger (*Zingiber officinale*): A review. *International Journal of Herbal Medicine*, *8*, 36–49. https://doi.org/10.22271/flora.2020.v8.i6a.705

Ali, B. H., Blunden, G., Tanira, M. O., & Nemmar, A. (2008a). Some phytochemical, pharmacological and toxicological properties of Ginger (*Zingiber officinale* Roscoe): A review of recent research. *Food and Chemical Toxicology*, *46*(2), 409–420.

Ali, B. H., Blunden, G., Tanira, M. O., & Nemmar, A. (2008b). Some phytochemical, pharmacological and toxicological properties of Ginger (*Zingiber officinale* Roscoe): A review of recent research. *Food and Chemical Toxicology*, *46*(2), 409–420.

ALmohaimeed, H. M., Mohammedsaleh, Z. M., Batawi, A. H., Balgoon, M. J., Ramadan, O. I., Baz, H. A., Al Jaouni, S., & Ayuob, N. N. (2021). Synergistic anti-inflammatory and neuroprotective effects of cinnamomum cassia and *Zingiber officinale* alleviate diabetes-induced hippocampal changes in male albino rats: Structural and molecular evidence. *Frontiers in Cell and Developmental Biology*, *9*, 727049.

Al-Yahya, M. A., Rafatullah, S., Mossa, J. S., Ageel, A. M., Parmar, N. S., & Tariq, M. (1989). Gastroprotective activity of ginger *zingiber officinale* rosc., in albino rats. *The American Journal of Chinese Medicine*, *17*(1–2), 51–56.

Amir, M., Khan, A., Mujeeb, M., Ahmad, A., Usmani, S., & Akhtar, M. (2011). Phytochemical analysis and in vitro antioxidant activity of *Zingiber officinale*. *Free Radicals and Antioxidants*, *1*(4), 75–81. https://doi.org/10.5530/ax.2011.4.12

Andriyani, R., Budiati, T. A., & Pudjiraharti, S. (2015). Effect of extraction method on total flavonoid, total phenolic content, antioxidant and anti-bacterial activity of *Zingiberis officinale* rhizome. *Procedia Chemistry*, *16*, 149–154. https://doi.org/10.1016/j.proche.2015.12.023

Anosike, C. A., Obidoa, O., Ezeanyika, L. U., & Nwuba, M. M. (2009). Anti-inflammatory and anti-ulcerogenic activity of the ethanol extract of Ginger (*Zingiber officinale*). *African Journal of Biochemistry Research*, *3*(12), 379–384.

Ansari, M. K. A., Iqbal, M., Chaachouay, N., Ansari, A. A., & Owens, G. (2023). The concept and status of medicinal and aromatic plants: History, pharmacognosy, ecology, and conservation. In *Plants as Medicine and Aromatics* (pp. 129–144). Taylor & Francis Group, CRC Press.

Bag, B. (2018). Ginger processing in India (*Zingiber officinale*): A review. *International Journal of Current Microbiology and Applied Sciences*, *7*, 1639–1651. https://doi.org/10.20546/ijcmas.2018.704.185

Belhaj, S., Chaachouay, N., & Zidane, L. (2021). Ethnobotanical and toxicology study of medicinal plants used for the treatment of diabetes in the High Atlas Central of Morocco. *Journal of Pharmacy & Pharmacognosy Research*, *9*(5), 619–662. https://doi.org/10.56499/jppres21.1014_9.5.619

Bellik, Y. (2014). Total antioxidant activity and antimicrobial potency of the essential oil and oleoresin of *Zingiber officinale* Roscoe. *Asian Pacific Journal of Tropical Disease*, *4*(1), 40–44. https://doi.org/10.1016/S2222-1808(14)60311-X

Benkhnigue, O., Chaachouay, N., Khamar, H., El Azzouzi, F., Douira, A, & Zidane, L. (2022). Ethnobotanical and ethnopharmacological study of medicinal plants used in the treatment of anemia in the region of Haouz-Rehamna (Morocco). *Journal of Pharmacy & Pharmacognosy Research*, *10*(2), 279–302. https://doi.org/10.56499/jppres21.1196_10.2.279

Benkhnigue, O., Chaachouay, N., & Zidane, L. (2023). A Floristical and Ecological Study of the Medicinal Flora Used by the Local Population of the Haouz-Rehamna Region (Middle Atlantic Morocco-4). *Acta Botanica Hungarica Akadémiai Kiadó*, *65*(3–4), 247–304. https://doi.org/10.1556/034.65.2023.3-4.3

Bhatt, N., Waly, M., Musthafa, M. E., & Ali, A. (2013). Ginger: A functional herb. In *Food as Medicine* (pp. 51–71). Nova Science Publishers, Inc., NY, USA.

Bin-Meferij, M. M., Shati, A. A., Eid, R. A., & El-Kott, A. F. (2017). Anti-obesity and anti-hepatosteatosis effects of dietary *Zingiber officinale* extract in male obese rats. *International Journal of Pharmacology*, *13*(6), 620–627.

Bischoff, S., & Crowe, S. E. (2005). Gastrointestinal food allergy: New insights into pathophysiology and clinical perspectives. *Gastroenterology*, *128*(4), 1089–1113.

Bode, A. M., & Dong, Z. (2011). The Amazing and Mighty Ginger. In I. F. F. Benzie & S. Wachtel-Galor (Éds.), *Herbal Medicine: Biomolecular and Clinical Aspects* (2nd éd.). CRC Press/Taylor & Francis. http://www.ncbi.nlm.nih.gov/books/NBK92775/

Cai, L., Wang, Y., & Cao, A. (2020). The physiochemical and preservation properties of fish sarcoplasmic protein/chitosan composite films containing ginger essential oil emulsions. *Journal of Food Process Engineering*, *43*(10), e13495. https://doi.org/10.1111/jfpe.13495

Chaachouay, N., Benkhnigue, O., Fadli, M., Ibaoui, H., & Zidane, L. (2019a). Ethnobotanical and ethnopharmacological studies of medicinal and aromatic plants used in the treatment of metabolic diseases in the Moroccan Rif. *Heliyon*, *5*(10), e02191. https://doi.org/10.1080/10477845.2021.1942433

Chaachouay, N., Benkhnigue, O., Fadli, M., Ibaoui, H., El Ayadi, R., & Zidane, L. (2019b). Ethnobotanical and ethnopharmacological study of medicinal and aromatic plants used in the treatment of respiratory system disorders in the Moroccan Rif. *Ethnobotany Research and Applications*, *18*(22), 1–16.

Chaachouay, N., Douira, A., Hassikou, R., Brhadda, N., Dahmani, J., Belahbib, N., Ziri, R., & Zidane, L. (2020a). Ethnobotanical and ethnopharmacological study of medicinal and aromatic plants used in the treatment of respiratory system disorders in the Moroccan Rif. PhDThesis, Département de Biologie - Université Ibn Tofail - Kénitra, Maroc, https://tel.archives-ouvertes.fr/tel-03376377

Chaachouay, N., Benkhnigue, O., & Zidane, L. (2020b). Ethnobotanical study aimed at investigating the use of medicinal plants to treat nervous system diseases in the Rif of Morocco. *Journal of Chiropractic Medicine*, *19*(1), 70–81. https://doi.org/10.1016/j.jcm.2020.02.004

Chaachouay, N., & Zidane, L. (2021). The Symbolic Efficacy of Plants in Rituals and Socio-Religious Ceremonies in Morocco, Northwest of Africa. *Journal of Religious & Theological Information*, *21*(1–2), 34–53. https://doi.org/10.1080/10477845.2021.1942433

Chaachouay, N., Benkhnigue, O., Douira, A., & Zidane, L. (2021a). Poisonous medicinal plants used in the popular pharmacopoeia of the Rif, northern Morocco. *Toxicon*, 189, 24–32. https://doi.org/10.1016/j.toxicon.2020.10.028

Chaachouay, N., Douira, A., & Zidane, L. (2021b). COVID-19, prevention and treatment with herbal medicine in the herbal markets of Salé Prefecture, North-Western Morocco. *European Journal of Integrative Medicine*, 42, 101285. https://doi.org/10.1016/j.eujim.2021.101285

Chaachouay, N., Douira, A., & Zidane, L. (2021c). Herbal Medicine Used in the Treatment of Human Diseases in the Rif, Northern Morocco. *Arabian Journal for Science and Engineering*, 47, 131–153. https://doi.org/10.1007/s13369-021-05501-1

Chaachouay, N., Benkhnigue, O., & Zidane, L. (2022a). Ethnobotanical and Ethnomedicinal study of medicinal and aromatic plants used against dermatological diseases by the people of Rif, Morocco. *Journal of Herbal Medicine*, 32, 100542. https://doi.org/10.1016/j.hermed.2022.100542

Chaachouay, N., Azeroual, A., Bencharki, B., Douira, A., & Zidane, L. (2022b). Ethnoveterinary medicines plants for animal therapy in the Rif, North of Morocco. *South African Journal of Botany*, *147*, 176–191. https://doi.org/10.1016/j.sajb.2021.12.037

Chaachouay, N., Azeroual, A., Douira, A., & Zidane, L. (2022c). Ethnoveterinary practices of medicinal plants among the Zemmour and Zayane tribes, Middle Atlas, Morocco. *South African Journal of Botany*, *151*, 826–840. https://doi.org/10.1016/j.sajb.2022.11.009

Chen, N. N. (2008). *Food, medicine, and the quest for good health.* Columbia University Press. https://doi.org/10.7312/chen13484

Choi, J.-S. (2019). Processed gingers: Current and prospective use in food, cosmetic, and pharmaceutical industry. *Recent Patents on Food, Nutrition & Agriculture*, *10*(1), 20–26. https://doi.org/10.2174/2212798410666180806150142

Chung, W. Y., Yow, C. M. N., & Benzie, I. F. F. (2003). Assessment of membrane protection by traditional Chinese medicines using a flow cytometric technique: Preliminary findings. *Redox Report*, *8*(1), 31–33. https://doi.org/10.1179/135100003125001215

Creed, B. (2018). Ginger snaps: The monstrous feminine as femme animale. In *She-wolf* (pp. 180–195). Manchester University Press. https://www.manchesterhive.com/display/9780719098192/9780719098192.00014.xml

da Silveira Vasconcelos, M., Mota, E. F., Gomes-Rochette, N. F., Nunes-Pinheiro, D. C. S., Nabavi, S. M., & de Melo, D. F. (2019). Chapter 3.18—Ginger (*Zingiber officinale* Roscoe). In S. M. Nabavi & A. S. Silva (Éds.), *Nonvitamin and nonmineral nutritional supplements* (pp. 235–239). Academic Press. https://doi.org/10.1016/B978-0-12-812491-8.00034-5

Ebrahimzadeh Attari, V., Malek Mahdavi, A., Javadivala, Z., Mahluji, S., Zununi Vahed, S., & Ostadrahimi, A. (2018). A systematic review of the anti-obesity and weight lowering effect of Ginger (*Zingiber officinale Roscoe*) and its mechanisms of action. *Phytotherapy Research*, *32*(4), 577–585. https://doi.org/10.1002/ptr.5986

El-Wahab, A. A., El-Adawi, H., & El-Demellawy, M. (2009). In vitro study of the antiviral activity of *Zingiber officinale*. *Planta Medica*, *75*(9), PF7. https://doi.org/10.1055/s-0029-1234649

Ezzat, S. M., Ezzat, M. I., Okba, M. M., Menze, E. T., & Abdel-Naim, A. B. (2018). The hidden mechanism beyond Ginger (*Zingiber officinale* Rosc.) potent in vivo and in vitro anti-inflammatory activity. *Journal of Ethnopharmacology*, *214*, 113–123.

Farombi, E. O., Abolaji, A. O., Adetuyi, B. O., Awosanya, O., & Fabusoro, M. (2019). Neuroprotective role of 6-Gingerol-rich fraction of *Zingiber officinale* (Ginger) against acrylonitrile-induced neurotoxicity in male Wistar rats. *Journal of Basic and Clinical Physiology and Pharmacology*, *30*(3), 20180114. https://doi.org/10.1515/jbcpp-2018-0114

GBIF. (2020). *GBIF—the Global Biodiversity Information Facility—Is an international network and data infrastructure funded by the world's governments and aimed at providing anyone, anywhere, open access to data about all types of life on Earth.* https://www.gbif.org/species/3001509

Geiger, J. L. (2005). The essential oil of Ginger, *Zingiber officinale*, and anaesthesia. *International Journal of Aromatherapy*, *15*(1), 7–14.

Ghasemzadeh, A., Jaafar, H. Z. E., & Rahmat, A. (2010). Antioxidant activities, total phenolics and flavonoids content in two varieties of Malaysia young ginger (*Zingiber officinale* Roscoe). *Molecules*, *15*(6), Article 6. https://doi.org/10.3390/molecules15064324

Goto, C., Kasuya, S., Koga, K., Ohtomo, H., & Kagei, N. (1990). Lethal efficacy of extract from*Zingiber officinale* (traditional Chinese Medicine) or [6]-shogaol and [6]-gingerol inAnisakis larvae in vitro. *Parasitology Research*, *76*(8), 653–656. https://doi.org/10.1007/BF00931082

Govindarajan, V. S., & Connell, D. W. (1983). Ginger – Chemistry, technology, and quality evaluation: Part 1. *C R C Critical Reviews in Food Science and Nutrition*, *17*(1), 1–96. https://doi.org/10.1080/10408398209527343

Gulzar, R., Afzaal, M., Saeed, F., Samar, N., Shahbaz, A., Ateeq, H., Farooq, M. U., Akram, N., Asghar, A., Rasheed, A., & Teferi Asres, D. (2023). Bio valorization and industrial applications of ginger waste: A review. *International Journal of Food Properties*, *26*(2), 2772–2780. https://doi.org/10.1080/10942912.2023.2254014

Gupta, Y. K., & Sharma, M. (2001). Reversal of pyrogallol-induced delay in gastric emptying in rats by Ginger (*Zingiber officinale*). *Methods and Findings in Experimental and Clinical Pharmacology*, *23*(9), 501–503.

Gurib-Fakim, A. (2006). Medicinal plants: Traditions of yesterday and drugs of tomorrow. *Molecular Aspects of Medicine*, *27*(1), 1–93.

Habib, S. H. M., Makpol, S., Hamid, N. A. A., Das, S., Ngah, W. Z. W., & Yusof, Y. A. M. (2008). Ginger extract (*Zingiber officinale*) has anti-cancer and anti-inflammatory effects on ethionine-induced hepatoma rats. *Clinics*, *63*(6), 807–813.

Han, L.-K., Gong, X.-J., Kawano, S., Saito, M., Kimura, Y., & Okuda, H. (2005). Anti-obesity actions of *Zingiber officinale* Roscoe. *Yakugaku zasshi: Journal of the Pharmaceutical Society of Japan*, *125*(2), 213–217.

Hasan, I. H., El-Desouky, M. A., Hozayen, W. G., & Abd el Aziz, G. M. (2016). Protective effect of *Zingiber officinale* against CCl4-induced liver fibrosis is mediated through downregulating the TGF-ß1/Smad3 and NF-κB/IκB pathways. *Pharmacology*, *97*(1–2), 1–9.

He, J., Hadidi, M., Yang, S., Khan, M. R., Zhang, W., & Cong, X. (2023). Natural food preservation with ginger essential oil: Biological properties and delivery systems. *Food Research International*, *173*, 113221. https://doi.org/10.1016/j.foodres.2023.113221

Hosseini, A., Shafiee-Nick, R., & Mousavi, S. H. (2014). Combination of Nigella sativa with Glycyrrhiza glabra and *Zingiber officinale* augments their protective effects on doxorubicin-induced toxicity in h9c2 cells. *Iranian Journal of Basic Medical Sciences*, *17*(12), 993.

Hu, W., Yu, A., Wang, S., Bai, Q., Tang, H., Yang, B., Wang, M., & Kuang, H. (2023). Extraction, purification, structural characteristics, biological activities, and applications of the polysaccharides from *Zingiber officinale* Roscoe. (Ginger): A review. *Molecules*, *28*(9), Article 9. https://doi.org/10.3390/molecules28093855

Huang, Z., Xie, L., Wang, H., Zhong, J., Li, Y., Liu, J., Ou, Z., Liang, X., Li, Y., Huang, H., Lin, Z., Zhang, K., Zhang, L., & Zheng, X. (2019). Geographic distribution and impacts of climate change on the suitable habitats of *Zingiber* species in China. *Industrial Crops and Products*, *138*, 111429. https://doi.org/10.1016/j.indcrop.2019.05.078

Jafri, S. A., Abass, S., & Qasim, M. (2011). Hypoglycemic effect of Ginger (*Zingiber officinale*) in alloxan induced diabetic rats (*Rattus norvagicus*). *Pakistan Veterinary Journal*, *31*(2), 160–162.

Jefferies, S., Healy, B., Weatherall, M., Beasley, R., & Shirtcliffe, P. (2012). What risks do women face when seeking advice during pregnancy from pharmacies and natural health retailers? *The New Zealand Medical Journal (Online)*, *125*(1367). https://search.proquest.com/openview/ec5347da6fe808032ce0282b05e07020/1?pq-origsite=gscholar&cbl=1056335

Kadirova, D. (2023). *Zingiber officinale*. L botaic description of. *World Bulletin of Public Health*, *20*, 76–78.

Kawai, T., Kinoshita, K., Koyama, K., & Takahashi, K. (1994). Anti-emetic principles of *Magnolia obovata* bark and *Zingiber officinale* rhizome. *Planta Medica*, *60*(1), 17–20. https://doi.org/10.1055/s-2006-959399

Kemper, K. J. (1999). *Ginger (Zingiber officinale). Longwood Herbal Task Force*, *3*(1), 1–18.

Khairul Alam, M. (2022). Traditional ecological knowledge on non-wood forest products management and biodiversity conservation: A focus on Chittagong Hill Tracts (CHTs), Bangladesh. In A. Z. M. M. Rashid, N. A. Khan, & M. Hossain (Éds.), *Non-Wood Forest Products of Asia* (Vol. 25, pp. 41–69). Springer International Publishing. https://doi.org/10.1007/978-3-030-99313-9_3

Khan, S., Pandotra, P., Qazi, A. K., Lone, S. A., Muzafar, M., Gupta, A. P., & Gupta, S. (2016). Chapter 25—Medicinal and nutritional qualities of *Zingiber officinale*. In R. R. Watson & V. R. Preedy (Éds.), *Fruits, Vegetables, and Herbs* (p. 525–550). Academic Press. https://doi.org/10.1016/B978-0-12-802972-5.00025-1

Khodaie, L., & Sadeghpoor, O. (2015). Ginger from ancient times to the new outlook. *Jundishapur Journal of Natural Pharmaceutical Products*, *10*(1), e18402.

Kiyama, R. (2020). Nutritional implications of Ginger: Chemistry, biological activities and signaling pathways. *The Journal of Nutritional Biochemistry*, *86*, 108486. https://doi.org/10.1016/j.jnutbio.2020.108486

Knaak, N., da Silva, L. D., Andreis, T. F., & Fiuza, L. M. (2013). Chemical characterization and anti-fungal activity of plant extracts and essential oils on the Bipolaris oryzae and Gerlachia oryzae phytopathogens. *Australasian Plant Pathology*, *42*(4), 469–475. https://doi.org/10.1007/s13313-013-0220-4

Koch, W., Kukula-Koch, W., Marzec, Z., Kasperek, E., Wyszogrodzka-Koma, L., Szwerc, W., & Asakawa, Y. (2017). Application of chromatographic and spectroscopic methods towards the quality assessment of ginger (*Zingiber officinale*) rhizomes from ecological plantations. *International Journal of Molecular Sciences*, *18*(2), Article 2. https://doi.org/10.3390/ijms18020452

Krell, J., & Stebbing, J. (2012). Ginger: The root of cancer therapy? *The Lancet Oncology*, *13*(3), 235–236.

Kumar Gupta, S., & Sharma, A. (2014). Medicinal properties of *Zingiber officinale* Roscoe – A review. *Journal of Pharmaceutical and Biological Sciences*, *9*, 124–129.

Kumar, K. M. P., Asish, G. R., Sabu, M., & Balachandran, I. (2013). Significance of gingers (Zingiberaceae) in Indian system of medicine - Ayurveda: An overview. *Ancient Science of Life*, *32*(4), 253–261. https://doi.org/10.4103/0257-7941.131989

Laelago Ersedo, T., Teka, T. A., Fikreyesus Forsido, S., Dessalegn, E., Adebo, J. A., Tamiru, M., & Astatkie, T. (2023). Food flavor enhancement, preservation, and bio-functionality of Ginger (*Zingiber officinale*): A review. *International Journal of Food Properties*, *26*(1), 928–951. https://doi.org/10.1080/10942912.2023.2194576

Lai, W., Yang, S., Lin, X., Zhang, X., Huang, Y., Zhou, J., Fu, C., Li, R., & Zhang, Z. (2022). *Zingiber officinale*: A systematic review of botany, phytochemistry and pharmacology of gut microbiota-related gastrointestinal benefits. *The American Journal of Chinese Medicine*, *50*(04), 1007–1042. https://doi.org/10.1142/S0192415X22500410

Lazzini, S., Polinelli, W., Riva, A., Morazzoni, P., & Bombardelli, E. (2016). The effect of Ginger (*Zingiber officinalis*) and artichoke (*Cynara cardunculus*) extract supplementation on gastric motility: A pilot randomized study in healthy volunteers. *European Review for Medical & Pharmacological Sciences*, *20*(1). https://search.ebscohost.com/login.aspx?direct=true&profile=ehost&scope=site&authtype=crawler&jrnl=11283602&AN=112542118&h=ZvXwlkMiKijxXdLRouhYpY%2Beg5k3EGf%2BoZju0YbKkFs46LW962n4FtccIOYJ87jOkKNfsCg6LKG8UJQ7xBiWiA%3D%3D&crl=c

Liu, H., Specht, C. D., Zhao, T., & Liao, J. (2020). Morphological anatomy of leaf and rhizome in *Zingiber officinale* roscoe, with emphasis on secretory structures. *HortScience*, *55*(2), 204–207. https://doi.org/10.21273/HORTSCI14555-19

Liu, L. J., Pan, J. Y., Yin, X. H., & Lu, Y. S. (1985). Quantitative determination of glycyrrhizin in Glycyrrhiza and its preparations by maero-reticular resin. *Chin Trad Herb Drugs*, *16*, 202–206.

Liu, Y., Liu, J., & Zhang, Y. (2019). Research progress on chemical constituents of *Zingiber officinale* Roscoe. *BioMed Research International*, *2019*. https://www.hindawi.com/journals/bmri/2019/5370823/abs/

Madankar, C. S., & Nair, A. (2021). Studies on extraction, microencapsulation and potential applications of ginger oleoresin. *Journal of Scientific & Industrial Research*, *80*(08), Article 08. https://doi.org/10.56042/jsir.v80i08.42736

Mahluji, S., Ostadrahimi, A., Mobasseri, M., Attari, V. E., & Payahoo, L. (2013). Anti-inflammatory effects of *Zingiber officinale* in type 2 diabetic patients. *Advanced Pharmaceutical Bulletin*, *3*(2), 273.

Malik, A., Shimpy, & Kumar, M. (2023). Advancements in Ginger drying technologies. *Journal of Stored Products Research*, *100*, 102058. https://doi.org/10.1016/j.jspr.2022.102058

Mao, Q.-Q., Xu, X.-Y., Cao, S.-Y., Gan, R.-Y., Corke, H., Beta, T., & Li, H.-B. (2019). Bioactive compounds and bioactivities of Ginger (*Zingiber officinale* Roscoe). *Foods*, *8*(6), 185.

Marrelli, M., Menichini, F., & Conforti, F. (2015). A comparative study of *Zingiber officinale* Roscoe pulp and peel: Phytochemical composition and evaluation of antitumour activity. *Natural Product Research*, *29*(21), 2045–2049. https://doi.org/10.1080/14786419.2015.1020491

Martindale, W., & Westcott, W. W. (1892). *The Extra Pharmacopœia*. H.K. Lewis.

Marwat, S. K., Shoaib, M., Khan, E. A., & Ullah, H. (2015). Phytochemistry and bioactivities of quranic plant, Zanjabil-Ginger (*Zingiber officinale* Roscoe): A review. *Environmental Science*, *15*(5), 707–713.

Miller, A. (2005). « The Hair that Wasn't There Before »: Demystifying monstrosity and menstruation in « Ginger Snaps » and « Ginger Snaps Unleashed ». *Western Folklore*, *64*(3/4), 281–303.

Mishra, R. K., Kumar, A., & Kumar, A. (2012). Pharmacological activity of *Zingiber officinale*. *International Journal of Pharmaceutical and Chemical Sciences*, *1*(3), 1073–1078.

Moghaddasi, M. S., & Kashani, H. H. (2012). Ginger (*Zingiber officinale*): A review. *Journal of Medicinal Plants Research*, *6*(26), 4255–4258. https://doi.org/10.5897/JMPR011.787

Morakinyo, A. O., Achema, P. U., & Adegoke, O. A. (2010). Effect of *Zingiber officinale* (Ginger) on sodium arsenite-induced reproductive toxicity in male rats. *African Journal of Biomedical Research*, *13*(1), 39–45.

Mukjerjee, S., & Karati, D. (2022). A mechanistic view on phytochemistry, pharmacognostic properties, and pharmacological activities of phytocompounds present in *Zingiber officinale*: A comprehensive review. *Pharmacological Research – Modern Chinese Medicine*, *5*, 100173. https://doi.org/10.1016/j.prmcm.2022.100173

Munda, S., Dutta, S., Haldar, S., & Lal, M. (2018). Chemical analysis and therapeutic uses of ginger (*Zingiber officinale* Rosc.) Essential oil: A review. *Journal of Essential Oil Bearing Plants*, *21*(4), 994–1002. https://doi.org/10.1080/0972060X.2018.1524794

Nair, K. P. P. (2013). *The Agronomy and Economy of Turmeric and Ginger: The Invaluable Medicinal Spice Crops*. Newnes.

Nazish, I., Ansari, S. H., Arora, P., & Ahmad, A. (2016). Anti-obesity activity of *Zingiber officinale*. *Pharmacognosy Journal*, *8*(5). https://phcogj.com/article/188?qt-sidebar_tabs=2

Nicoll, R., & Henein, M. Y. (2009). Ginger (*Zingiber officinale* Roscoe): A hot remedy for cardiovascular disease? *International Journal of Cardiology*, *131*(3), 408–409.

Noipha, K., & Ninla-Aesong, P. (2018). Anti-diabetic activity of *Zingiber officinale* roscoe rhizome extract: An in vitro study. *HAYATI Journal of Biosciences*, *25*(4), 160–160.

Orch, H., Chaachouay, N., Douiri, E. M., Faiz, N., Zidane, L., & Douira, A. (2021). Use of medicinal plants in dermato-cosmetology: An ethnobotanical study among the population of Izarène. *Jordan Journal of Pharmaceutical Sciences*, 323–340.

Othman, A., Azizan, N. A., Shambely, A., & Ayunie, N. (2020). *Anti-bacterial Activity of Zingiber officinale against Foodborne Pathogens 1*.

Pakrashi, S. C., & Pakrashi, A. (2003). *Ginger: A Versatile Healing Herb*. Vedams eBooks (P) Ltd. https://books.google.com/books?hl=fr&lr=&id=zfxLvxEvAh0C&oi=fnd&pg=PR7&dq=+In+essence,+ginger%27s+multifaceted+properties+make+it+a+valuable+plant+with+a+wide+range+of+culinary+and+therapeutic+applications&ots=zYjfKxRqdD&sig=2jsI_m28e6X4zr979VIXWeLORAM

Pallasmaa, J. (2012). *The Eyes of the Skin: Architecture and the Senses*. John Wiley & Sons.

Park, G. H., Park, J. H., Song, H. M., Eo, H. J., Kim, M. K., Lee, J. W., Lee, M. H., Cho, K.-H., Lee, J. R., Cho, H. J., & Jeong, J. B. (2014). Anti-cancer activity of Ginger (*Zingiber officinale*) leaf through the expression of activating transcription factor 3 in human colorectal cancer cells. *BMC Complementary and Alternative Medicine*, *14*(1), 408. https://doi.org/10.1186/1472-6882-14-408

Patil, S. B., & Kothavale, S. D. (2022). Anti-obesity activity of *Zingiber officinale*. *Research Journal of Pharmacognosy and Phytochemistry*, *14*(4), 252–256.

Phillips, S., Hutchinson, S., & Ruggier, R. (1993). *Zingiber officinale* does not affect gastric emptying rate.: A randomised, placebo-controlled, crossover trial. *Anaesthesia*, *48*(5), 393–395. https://doi.org/10.1111/j.1365-2044.1993.tb07011.x

Plengsuriyakarn, T., Viyanant, V., Eursitthichai, V., Tesana, S., Chaijaroenkul, W., Itharat, A., & Na-Bangchang, K. (2012). Cytotoxicity, toxicity, and anti-cancer activity of *Zingiber officinale* Roscoe against cholangiocarcinoma. *Asian Pacific Journal of Cancer Prevention*, *13*(9), 4597–4606.

Rasool, N., Saeed, Z., Pervaiz, M., Ali, F., Younas, U., Bashir, R., Bukhari, S. M., Mahmood Khan, R. R., Jelani, S., & Sikandar, R. (2022). Evaluation of essential oil extracted from Ginger, cinnamon and lemon for therapeutic and biological activities. *Biocatalysis and Agricultural Biotechnology*, *44*, 102470. https://doi.org/10.1016/j.bcab.2022.102470

Ravindran, P. N., & Babu, K. N. (2016). *Ginger: The Genus Zingiber*. CRC Press.

Rout, S., Dutta, S., & Rath, B. (2010). Evaluation of broncho-protective effect of *Zingiber officinale* Roscoe in histamine induced broncho-spasm. *Research Journal of Pharmacy and Technology*, *3*(2), 589–591.

Rubatzky, V. E., & Yamaguchi, M. (2012). *World Vegetables: Principles, Production, and Nutritive Values*. Springer Science & Business Media.

Saras, T. (2023). *The Wonders of Ginger: Unveiling the Health Benefits and Culinary Magic*. Tiram Media. https://books.google.com/books?hl=fr&lr=&id=MtPQEAAAQBAJ&oi=fnd&pg=PA1&dq=+In+essence,+ginger%27s+multifaceted+properties+make+it+a+valuable+plant+with+a+wide+range+of+culinary+and+therapeutic+applications&ots=wSa6YwcgSo&sig=9rbYqbczqAuT7E7WCZNYlW4n05I

Sarwar, M. T. (2014). Preparation of chitosan and its application on ginger paste and fruit juice as food preservative. *Journal of Food and Nutrition Sciences*, *2*(6), 243. https://doi.org/10.11648/j.jfns.20140206.11

Setty, A. R., & Sigal, L. H. (2005). Herbal medications commonly used in the practice of rheumatology: Mechanisms of action, efficacy, and side effects. *Seminars in Arthritis and Rheumatism*, *34*(6), 773–784. https://www.sciencedirect.com/science/article/pii/S0049017205000120

Shahrajabian, M. H., Sun, W., & Cheng, Q. (2019). Clinical aspects and health benefits of Ginger (*Zingiber officinale*) in both traditional Chinese Medicine and modern industry. *Acta Agriculturae Scandinavica, Section B — Soil & Plant Science*, *69*(6), 546–556. https://doi.org/10.1080/09064710.2019.1606930

Shahrajabian, M. H., Sun, W., Shen, H., & Cheng, Q. (2020). Chinese herbal medicine for SARS and SARS-CoV-2 treatment and prevention, encouraging using herbal Medicine for COVID-19 outbreak. *Acta Agriculturae Scandinavica, Section B—Soil & Plant Science*, *70*(5), 1–7.

Sharma, S. S., & Gupta, Y. K. (1998). Reversal of cisplatin-induced delay in gastric emptying in rats by Ginger (*Zingiber officinale*). *Journal of Ethnopharmacology*, *62*(1), 49–55.

Singh, H., Fairs, G., & Syarhabil, M. (2023). *Anti-fungal activity of Capsicum frutescence and Zingiber officinale against key post-harvest pathogens in citrus*. In *International Conference on Biomedical Engineering and Technology, IPCBEE*. *11*, 1–6.

Small, E., & Canada, N. R. C. (2006). *Culinary Herbs*. NRC Research Press.

Stoilova, I., Krastanov, A., Stoyanova, A., Denev, P., & Gargova, S. (2007). Antioxidant activity of a ginger extract (*Zingiber officinale*). *Food Chemistry*, *102*(3), 764–770. https://doi.org/10.1016/j.foodchem.2006.06.023

Supartono, W., Sukartiko, A. C., Yuliando, H., & Kristanti, N. E. (2015). Possibility of some indigenous spices as flavor agent of green tea. *Agriculture and Agricultural Science Procedia*, *3*, 62–66. https://doi.org/10.1016/j.aaspro.2015.01.014

Sutalangka, C., & Wattanathorn, J. (2017). Neuroprotective and cognitive-enhancing effects of the combined extract of Cyperus rotundus and *Zingiber officinale*. *BMC Complementary and Alternative Medicine*, *17*(1), 135. https://doi.org/10.1186/s12906-017-1632-4

The Plant List. (2013). *The Plant List. Version 1.1*. Published on the Internet; http://www.theplantlist.org/ (accessed 1st January).

Tohma, H. S., & Gulçin, I. (2010). Antioxidant and radical scavenging activity of aerial parts and roots of Turkish liquorice (*Glycyrrhiza glabra* L.). *International Journal of Food Properties*, *13*(4), 657–671.

Van der Veen, M., & Morales, J. (2015). The Roman and Islamic spice trade: New archaeological evidence. *Journal of Ethnopharmacology*, *167*, 54–63. https://doi.org/10.1016/j.jep.2014.09.036

Vasala, P. A. (2012). 18—Ginger. In K. V. Peter (Éd.), *Handbook of Herbs and Spices* (2nd Edition, pp. 319–335). Woodhead Publishing. https://doi.org/10.1533/9780857095671.319

Waggas, A. M. (2009). Neuroprotective evaluation of extract of ginger (*Zingiber officinale*) root in monosodium glutamate-induced toxicity in different brain areas male albino rats. *Pakistan Journal of Biological Sciences: PJBS*, *12*(3), 201–212.

Wakchaure, R., & Ganguly, S. (2018). *Phytochemistry and Pharmacological Properties of Ginger (Zingiber officinale)*. Lenin Media Private Limited, Delhi, India, 97–103.

Wang, L., Gao, F., Xu, K., & Li, X. (2014). Natural occurrence of mixploid Ginger (*Zingiber officinale* Rosc.) in China and its morphological variations. *Scientia Horticulturae*, *172*, 54–60. https://doi.org/10.1016/j.scienta.2014.03.043

Yousfi, F., Abrigach, F., Petrovic, J. D., Sokovic, M., & Ramdani, M. (2021). Phytochemical screening and evaluation of the antioxidant and anti-bacterial potential of *Zingiber officinale* extracts. *South African Journal of Botany*, *142*, 433–440. https://doi.org/10.1016/j.sajb.2021.07.010

50 Lesser Galangal (*Alpinia galanga* (L.) Willd. Zingiberaceae)

Noureddine Chaachouay, Ouafae Benkhnigue, Khadija Temmante, Mohd Kafeel Ahmad Ansari, Rahmatullah Qureshi, Abdelhamid Azeroual, Bouchaib Bencharki, and Lahcen Zidane

INTRODUCTION

Throughout history, several traditional communities worldwide have relied on plant-derived treatments for therapeutic concerns (Ansari et al., 2023; Chaachouay et al., 2020a, 2020b; Orch et al., 2021). In recent times, there has been a surge in the popularity of plant-based treatments as a viable substitute for synthetic chemicals. Recently, there has been a significant increase in the use of herbal medicine worldwide, as shown by studies conducted by Azaizeh et al. (2006); and Byeon et al. (2019). The increasing popularity of herbal medicine may be attributed to its natural nature, which appeals to those who choose natural cures over synthetic pharmaceuticals (Chaachouay et al., 2019a, 2019b; De Smet, 2002). Furthermore, many individuals are drawn to plant-based medicine since they believe it reduces the likelihood of adverse reactions.

Alpinia galanga (L.) Willd. is a perennial herbaceous plant from the Zingiberaceae family that has garnered interest in herbal medicine (Dissanayake et al., 2013; Ravindran et al., 2012). The plant species is indigenous to Southeast Asia and extensively grown in Thailand, Indonesia, Malaysia, and India (Hanh et al., 2014; Rangsiruji et al., 2000). The plant can reach a height of 2 meters and has elongated, broad leaves grouped in an alternating pattern along the stem (Chouni & Paul, 2018). The rhizomes, which are fleshy underground stems, are the most critical component of the plant. They are used in a variety of traditional medicinal and culinary preparations. The rhizomes possess a robust and fragrant taste with a hint of spiciness. They are used as a seasoning in several Southeast Asian culinary traditions, notably Thai and Indonesian cuisine (Eram et al., 2019a; Ravindran et al., 2012).

A. galanga has been used in traditional medicine to address various health conditions, including respiratory, digestive, and rheumatic disorders. The rhizome is often used as a culinary condiment because of its characteristic smell and spice, characterized by its fragrant ginger-like perfume and pungent, hot, and spicy flavor (Kaushik et al., 2011; Tang et al., 2018).The plant has been used to treat various illnesses, including fevers, colds, and coughs. It is thought to have antibacterial, anti-inflammatory, and antioxidant properties. (Verma et al., 2011).

As a natural cure for various health problems, *A. galanga* is becoming increasingly well-liked in the West. It comes in multiple forms, including extracts, teas, and capsules (Ramanunny et al., 2022). It is crucial to remember that *A. galanga* should be used carefully and under the supervision of a healthcare professional, just like any herbal treatment (Chaachouay et al., 2021a, 2021b, 2021c). From this perspective, this chapter provides a detailed discussion of *A. galanga*, emphasizing its taxonomy, ecology, ethnobotany, phytochemistry, pharmacology, and medicinal applications.

TAXONOMIC HIERARCHY

Alpinia galanga (L.) Willd. (Galangal, or greater galangal), is a monocotyledonous rhizomatous plant belonging to the Zingiberaceae family and *Alpinia* genus (ITIS, 2022).

Kingdom: Plantae
Subkingdom: Viridiplantae
Infrakingdom: Streptophyta
Superdivision: Embryophyta
Division: Tracheophyta
Subdivision: Spermatophytina
Class: Magnoliopsida
Superorder: Lilianae
Order: Zingiberales
Family: Zingiberaceae
Genus: *Alpinia*
Species: *Alpinia galanga* (L.) Willd.

VERNACULAR NAMES

Arabic: El Khodenjal, Khulanjan-El-kabir, Khulanjan-El-qasbi; **Amharic**: Baroch koda; **English**: Great galangal, Galanga major, Java galangal, Lengkuas, Blue ginger, Siamese ginger; **Bengali**: Kulanjan, Barakulanjar, Sugandhavacha, Bomb; **Czech**: Galgán veliký, Galgán větší, Galgán obecný; **Chinese**: Hung-dau-kau, Da guo liang jiang; **Danois**: Stor galanga; **Dutch**: Lengoewas, Grote galanga, Galigaan; **Italian**: Galanga maggiore; **German**: Galgantalpinie, Galgant, Großer galgant, Galanga, Siamingwer; **French**: Souchet odorant, Souchet long, Galanga

DOI: 10.1201/9781003460336-50

d'Inde, Galanga; **Hindi**: Sapheda- pana-ki-jhad, Bara khulanjan, Khulanjan; **Japanese**: Ryouga; **Maltese**: Kolinji, Arattha, Pera-ratta, Chitta-ratta; **Spanish**: Garengal, Calanga; **Tamil**: Peria-reta, Perarattai; **Thai**: Khaa, Ginza, Dok kha; **Turkish**: Hint zerdeçalı, Büyük zerdeçal; **Korean**: Daehwang; **Persan**: Khusravedurue- kalan, Khurduwara; **Portuguese**: Gengibre tailandés, Gengibre do Laos; **Urdu**: Khulanjan; **Russian**: Al'piniia galanga; **Vietnamese**: Một loại gừng, Cao khương hương, Riềng nếp, Sơn nại, Riềng ấm.

SYNONYMS

According to the data from The Plant List (2013), *Alpinia galanga* (L.) Willd. is a plant species that has 21 different scientific names and synonyms that all pertain to the same species. The following terms are synonymous with *Alpinia galanga* (L.) Willd.: **(1)** *Hellenia alba* (Retz.) Willd.; **(2)** *Heritiera alba* Retz.; **(3)** *Amomum medium* Lour.; **(4)** *Alpinia bifida* Warb.; **(5)** *Alpinia pyramidata* Blume; **(6)** *Alpinia alba* (Retz.) Roscoe; **(7)** *Alpinia galanga* var. *galanga*; **(8)** *Alpinia carnea* Griff.; **(9)** *Alpinia viridiflora* Griff.; **(10)** *Alpinia rheedei* Wight; **(11)** *Languas galanga* (L.) Stuntz; **(12)** *Amomum galanga* (L.) Lour.; **(13)** *Galanga officinalis* Salisb.; **(14)** *Galanga major* Garsault; **(15)** *Languas vulgare* J.Koenig; **(16)** *Zingiber sylvestre* Gaertn.; **(17)** *Maranta galanga* L.; **(18)** *Languas pyramidata* (Blume) Merr.; **(19)** *Alpinia galanga* var. *pyramidata* (Blume) K.Schum.; **(20)** *Zingiber medium* Stokes; and **(21)** *Zingiber galanga* (L.) Stokes.

BOTANICAL DESCRIPTION

A. galanga is an herbaceous perennial plant that may grow up to 2 meters tall and has long, lance-shaped leaves around 20–40 cm long and 5–10 cm wide. The rhizomes of *A. galanga* are widely used for culinary and medicinal purposes, making them the most often utilized portion of the plant (Akbar, 2020; Verma et al., 2011). The rhizomes have an asymmetrical form and a brownish color, and they have an intense and pleasant scent. Typically, their length ranges from 2 to 4 cm, and their width from 1 to 2 cm (Figure 50.1).

The foliage has a deep green hue, a lustrous appearance, and a sleek surface (Rangsiruji et al., 2000). Their maximum length may reach 60 cm, while their breadth can extend up to 10 cm. The plant produces diminutive, pale flowers that form clusters at the terminus of elongated stalks (Chouni & Paul, 2018; Eram et al., 2019b). The blooms possess a pleasant aroma and have a cylindrical form with six petals. The flowers exhibit continuous blooming throughout the year, with the highest concentration of blossoms occurring during the summer season. The plant's calyx is tubular and remains connected even after blooming, measuring 6–10 mm. The length of the corolla tube ranges from 6 to 10 mm, while the lobes are around 1.6 to 1.8 cm in size and have an oblong shape. The plant has subulate or linear purple lateral staminodes that vary in size from 2 to 10 mm. The plant capsule is elongated, measuring roughly 1–1.5 cm in length and around 7 mm in width. When the substance is deprived of moisture, it undergoes a color change to brown or red and exhibits a thin and smooth texture. The specific fruit is a spheroid to ellipsoid capsule measuring 1–1.5 cm in diameter. The capsule's core is somewhat constricted, and the fully developed fruit may have either a brown or dark red coloration. The capsule contains various seeds, often ranging from 3 to 6 (Y.-Q. Zhou et al., 2018).

ORIGIN AND GEOGRAPHIC DISTRIBUTION

A. galanga is a perennial herbaceous plant classified in the ginger family. The plant is indigenous to Southeast Asia, particularly Indonesia, Malaysia, and Thailand. The species is also distributed in other regions of Asia, including Cambodia, Vietnam, India, Thailand, Sri Lanka, Laos, and the Philippines (Akbar, 2020; Rangsiruji et al., 2000). Galangal is now dispersed worldwide, especially in tropical and subtropical areas of Africa, the Americas, the Caribbean, and Australia, due to its extensive cultivation. *A. galanga* is cultivated in these regions for medicinal and gastronomic use (Chouni & Paul, 2018). It is widely used in traditional medicine and as a culinary condiment in many cuisines, such as Indonesian, Malaysian, and Thai (Y.-Q. Zhou et al., 2018).

ETYMOLOGY

The species *Alpinia* was designated as a tribute to Prospero Alpini, a pre-Linnaean botanist who served as the director of the Padua Botanical Garden (De Judicibus, 2011). The particular descriptor galangal is taken from the medieval Latin term 'galanga', which in turn comes from the Arabic word 'Khalanǧān' or the Persian word 'Qulanjan'. These words might be derived from the Chinese term 'gao liang jiang'. In India, the plant is called kulanjan in Hindi, Kholinjan in Urdu, and kulanja in Sanskrit. These names are derived from the same root as galangal (Ravindran et al., 2012).

HABITAT AND ECOLOGY

A. galanga thrives in tropical and subtropical areas, particularly on well-drained soils that are abundant in organic matter. It prefers flourishing in areas with partial shade, such as the lower layer of wood, while it is also capable of thriving in direct sunlight. It often thrives close to aquatic environments, such as rivers and ponds (Hanh et al., 2014; Rangsiruji et al., 2000). *A. galanga* is a vital plant with both medicinal and culinary uses. The rhizomes of this plant are used to cure various conditions, including fever, cough, and stomach issues. In addition, they are used in traditional gastronomy as a means of enhancing taste and as a condiment, particularly in Thai, Indonesian, and Malaysian culinary traditions (Verma et al., 2011). The plant has insecticidal characteristics and is used as a natural pesticide in some regions of the globe (Dissanayake

FIGURE 50.1 The fresh (A) and dried rhizomes (B) of *A. galanga*.

et al., 2013). The plant's foliage and stems include essential oils that exhibit antibacterial and antifungal characteristics, making it a promising natural therapeutic agent (Chouni & Paul, 2018). *A. galanga* is a host plant for several insect species, such as the Galanga Swallowtail butterfly (*Papilio aristolochiae*), which consumes its leaves. In addition, bees, butterflies, and other insects serve as pollinators for the plant. *A. galanga* plays a vital role in forest ecosystems within its natural environment, serving as a crucial source of habitat and sustenance for species (Khairullah et al., 2020). Additionally, it is used in forestry initiatives aimed at rejuvenating deteriorated land and fostering biodiversity.

ETHNOBOTANY

Traditional Medicine

A. galanga has an extensive history of use in phytotherapy, particularly in Southeast Asia. The plant effectively alleviates various diseases, including gastrointestinal illnesses, respiratory ailments, and dermatological issues (Khairullah et al., 2020). *A. galanga* has long been used to treat digestive ailments, such as diarrhea, gas, vomiting, and indigestion (Chaachouay & Zidane, 2021; Ramanunny et al., 2022). It is believed to enhance digestion and mitigate inflammation in the gastrointestinal system. In addition, the plant has been used to treat respiratory ailments such as bronchitis, asthma, and coughs. It is believed to have anti-inflammatory properties that may help reduce inflammation in the respiratory airways, leading to easier breathing (Verma et al., 2011). *A. galanga* has been used to treat skin illnesses such as dermatitis, eczema, and acne. It is said to possess anti-inflammatory and antibacterial qualities that aid in the alleviation and regeneration of the skin (Ravindran et al., 2012). The plant has furthermore been used as a natural analgesic, namely for migraines, dysmenorrhea, and arthritic conditions. It is said to possess analgesic qualities that aid in reducing pain and inflammation (Zhou et al., 2018).

Culinary Uses

A. galanga is a plant extensively used in the culinary traditions of Southeast Asia. The rhizome of the plant is often used as a spice, with a strong and pungent taste reminiscent of ginger, although with a subtle citrus undertone and a somewhat sweet flavor (Chudiwal et al., 2010; Das et al., 2020). The rhizome of *A. galanga* is often used as a seasoning in various culinary preparations, including curries, soups, stews, and marinades. It imparts a distinctive taste and scent to the meal, making it a sought-after spice in Southeast Asian gastronomy. In some nations, such as Indonesia, Malaysia, and Thailand, the rhizome of *A. galanga* is grated or minced and combined with other components to create a condiment or dipping sauce (Ravindran et al., 2012). For instance, in Thailand, it is often used to prepare a piquant condiment known as Nam Prik Pla too. *A. galanga* may be used in the preparation of a pungent infusion and is considered to possess therapeutic attributes (Eram et al., 2019a). The rhizome is often immersed in boiling water and then sweetened with honey or sugar (Khairullah et al., 2020). The rhizome of *A. galanga* is pickled and consumed as a snack or used as a condiment for rice meals in some regions of Southeast Asia. *A. galanga* has several applications in animal healthcare, such as its utilization as medicine for cattle. Additionally, Arabs use it to augment the vitality of their horses (Eram et al., 2019b). While it is included in several combination treatments, its standalone usage is sporadic.

Cosmetic Uses

A. galanga has long been used as a customary component in body care goods, including topical skincare items and body soap, in several Asian countries (Tungmunnithum et al., 2020). Typically, these goods are created by families using uncomplicated extraction techniques, sometimes using hot water as the solvent. Presently, indigenous populations in China, Thailand, and other regions of Asia use the extract derived from *A. galanga* to create conditioner, shampoo, and several other cosmetic items (Chudiwal et al., 2010). Extraction processes often include solvents such as water and ethanol. Various groups utilize various preparation techniques and prefer employing the leaves and rhizomes of the plant (Lo et al., 2013). Although most cosmetics and cosmeceuticals are produced locally for distribution or trade within communities, the extract is also marketed as a separate component.

Other Uses

A. galanga essential oil is used as a natural insect repellant. The oil comprises cineole, camphor, and eucalyptol chemicals, recognized for their insect-repellent properties against mosquitoes, flies, and ants (Yang & Eilerman, 1999). The oil may be directly administered to the skin or dispersed throughout a space to repel insects. Due to its natural nature, it serves as a safe substitute for artificial insect repellents, which may have adverse consequences. The rhizome of *A. galanga* is used as a livestock feed addition, namely in the poultry sector (Lo et al., 2013; Sahoo et al., 2020). The rhizome harbors bioactive compounds such as phenolic acids and flavonoids, which possess antibacterial and antioxidant characteristics (Chouni & Paul, 2018). Incorporating it into the diet of poultry may promote their general health and well-being, as well as improve the quality of their meat and eggs. Moreover, it functions as an inherent preservative, effectively prolonging the durability of poultry products.

HALLUCINOGENIC EFFECTS

A. galanga is a botanical species used in herbal medicine to address several ailments, including gastrointestinal problems, dysmenorrhea, and respiratory infections (Eram et al., 2019b; Verma et al., 2011). While reports are indicating the potential psychedelic and hallucinatory properties of this plant, there is less scientific data to substantiate these assertions. *A. galanga* has been used in spiritual rituals, such as sorcery and witchcraft, by some traditional civilizations since it elicits modified states of consciousness. Nevertheless, more investigation is required to substantiate these assertions (Boufous et al., 2017; Bourgeois et al., 2014). It is imperative to underscore that the ingestion of psychoactive plants may result in detrimental consequences such as cardiovascular issues, neurological impairment, hallucinations, and mood disorders (Belhaj et al., 2021; dos Santos et al., 2021). Hence, it is advisable to abstain from ingesting *A. galanga* or any other mind-altering botanical specimen without the guidance of a competent healthcare practitioner (Chaachouay et al., 2022a, 2022b, 2022c; Qureshi et al., 1992; Zhou et al., 2018).

PHYTOCHIMESTRY

A. galanga is a perennial plant extensively used in traditional medicine systems due to its therapeutic properties. Phytochemistry is the scientific study of the chemical makeup of plants, specifically focusing on their secondary metabolites. This field of research offers vital knowledge on the possible medicinal use of plants. *A. galanga* is rich in several bioactive substances, including flavonoids, essential oils, terpenoids, alkaloids, and phenolic compounds (Hamad et al., 2016; Ravindran et al., 2012; Singh et al., 2020). These substances have many biological effects, including antibacterial, anti-inflammatory, antioxidant, anticancer, and hepatoprotective activities. The essential oil of *A. galanga* contains various bioactive compounds such as eucalyptol, terpinen-4-ol, neryl acetate, 2-methylpropyl acetate, citronellyl acetate, bornyl acetate, eugenyl acetate, geranyl acetate, trans-β-farnesene, α-terpineol, borneol, limonene, 1,8-cineole, camphor, camphene, methyleugenol, α-bergamotene, α-humulene, α -copaene, β-pinene, pentadecane, sabinene, α-pinene, β-caryophyllene, β-sesquiphellandrene, caryophyllene oxide, β-bisabolene, γ-terpinene, linalool, carveol, chavicol, chavicol acetate, terpinolene, myrcene, tridecane, and α-thujene (Charles et al., 1992; De Pooter et al., 1985; Hamad et al., 2016; Kurniawan et al., 2012; Wu et al., 2014). Phenolic compounds such caffeic acid, protocatechuic acid, syringic acid, coumaric acid, gallic acid, (+)-catechin, catechol, protocatechuic acid, kaempferol, salicylic acid, sinapic acid, apigenin 7 glucoside, quercetin, trans-ferulic acid, resveratrol, rutin trihydrate, trans-cinnamic acid, *p*- isorhamnetin, naringenin, 3′-demethoxycrataegusanoid E, (R)-4-(1-methoxypropyl) phenol, (S)-3-(4-hydroxy-3-methoxyphenyl) propane-1,2-diyldiacetate, and (R)-3-(4-hydroxy-3-methoxyphenyl) propane-1,2-diyldiacetate have also been identified in *A. galanga* (Aljobair, 2022; Barik et al., 1987; Liu et al., 2023; Mahae & Chaiseri, 2009; S. V. Nampoothiri et al., 2015; C. Zhou et al., 2021). Flavonoids such galangin, apigenin, alpinone, catechin, kaempferol, quercetin 3-methyl ether, isorhamnetin, galangin 3-methyl ether, quercetin, luteolin, pinobanksin 3-cinnamate, myricetin, pinobanksin 3-acetate, kaempferide, pinocembrin, isorhamnetin, kumatakenin, and isokaempferide have been identified in *A. galanga* (Bian et al., 2014; Jaju et al., 2009; Lin et al., 2020; Tao et al., 2006; Tungmunnithum et al., 2020). Terpenoids such as hedytriol, 1,8-cineole, (R)-4-hydroxy-4-(2-methylethyl)-cyclohex-2-en-1-one, guaia-4,10,11-triol, 2-O-b-D-glucopyranosyl-(+)-angelicoidenol, and beta-sitosterol have also been identified in *A. galanga* (Hamad et al., 2016; Morita & Itokawa, 1986; Nam Hoang et al., 2021).

Alkaloids such as galangin, galangaline, 1'-acetoxychavicol acetate, galanginol, alpiniamide, alpinetin, pinostrobin, and ethyl *p*-methoxycinnamate have also been identified in *A. galanga* (S. Nampoothiri et al., 2017; Subash et al., 2013). To summaries, *A. galanga* is an abundant reservoir of bioactive compounds that has diverse pharmacological properties. Further inquiries are necessary to comprehensively comprehend the therapeutic capabilities and mechanisms of action of these bioactive compounds.

PHARMACOLOGY

Investigators have recently started showing considerable interest in *A. galanga*. The focus surrounding this plant is mainly attributed to its possible therapeutic capabilities, which have been extensively studied in pharmacological investigations(Chudiwal et al., 2010; Eram et al., 2019a; Kaushik et al., 2011; Khairullah et al., 2020; Khattak et al., 2005). These studies examine the bioactive chemicals found in the plant and explore their potential for use in therapy, including:

Anti-Inflammatory Properties

Multiple studies have investigated the potential of *A. galanga* as an anti-inflammatory agent in various rheumatological illnesses. During a specific study, it was shown that the water-soluble component of the alcoholic extract obtained from *A. galanga* had a beneficial effect on chronic arthritis in albino rats. The anti-inflammatory effects were comparable to those of β-methazone (Chouni & Paul, 2018; Eram et al., 2019a; Ravindran et al., 2012). Furthermore, a very potent extract derived from the rhizome of galangal effectively alleviated the discomfort caused by inflammation in the knee joints due to rheumatism. The rhizome extract was shown to have specific components that preferentially inhibit the activity of T-helper cells, which play a role in the inflammatory response (Baldo et al., 2016). Furthermore, derivatives of chavicol, such as ACA and HCA, have shown notable antioxidant properties and have been found to enhance cell death while decreasing cytokine secretion by T-helper cells. Moreover, HCA has been shown to inhibit inflammatory immunological diseases resulting from excessive cytokine production (Ghosh et al., 2011; Kaushik et al., 2011; Khattak et al., 2005; Subash et al., 2016).

Antioxidant Activity

A comparative analysis was conducted to assess the composition and antioxidant activity of 50% ethanol in the water extract of *A. galanga*, in comparison to a water extract and essential oil. The results confirmed the antioxidant activity of the *A. galanga* extract (Malik et al., 2016; Nag et al., 2021; Ravindran et al., 2012). The antioxidant capabilities were evaluated using the oxygen radical absorbance capacity (ORAC) and 2, 2-diphenyl-1-picrylhydrazyl (DPPH) methods. The results suggest that, out of the three extracts, the ethanolic extract has shown the most notable capacity to eliminate DPPH free radicals and displayed the most excellent ORAC value (Mahae & Chaiseri, 2009; Tachakittirungrod & Chowwanapoonpohn, 2007). Furthermore, it has been documented that the rhizomes of *A. galanga* contain antioxidant-exhibiting compounds, including 1'-acetoxy chavicol acetate and its derivatives (El-Hadidy et al., 2020; Köse et al., 2015). Different assays have been used to assess the antioxidant properties and overall phenolic content of methanol extracts from *A. galanga*, in contrast to other techniques (Eram et al., 2019a; Johnley et al., 2020; Tang et al., 2018). The leaves and flowers had the highest chelating and β-carotene bleaching capacities compared to other sections of the plant, suggesting its potential as a valuable natural source of antioxidants for dietary purposes (Tian et al., 2022).

Antiulcer Activity

The efficacy of *A. galanga* extract in treating ulcers was evaluated using rats deliberately given stomach ulcers (Chouni & Paul, 2018; Ravindran et al., 2012). An substantial reduction in the intensity of stomach mucosal injury caused by pyloric ligation and hypothermic restraint stress was seen when the individual was administered a dosage of 500 mg/kg (Mandal et al., 2019). In previous investigations, researchers have found two potent antiulcer compounds, namely 1'-acetoxy-chavicol acetate and 1'-acetoxy-eugenol acetate, in the extract. The compounds, as mentioned earlier, were shown to impede the development of ethanol-induced damage to the stomach lining in rats, as reported by Johnley et al. (2020), Mitsui et al. (1976), and Noro et al. (1988). The findings indicate that the extract of *A. galanga* and its components may be therapeutic for treating ulcers.

Antidiabetic Activity

The hypoglycemic effects of the rhizome of *A. galanga* on rabbits were established by observing a decrease in their blood glucose levels. A substantial reduction in the blood glucose levels of control rabbits was observed after the ingestion of the pulverized rhizome as well as its methanol and aqueous extracts (Akhtar et al., 2002). Furthermore, the methanolic extract obtained from the above-ground portions of *A. galanga* has shown efficacy in controlling blood glucose levels and improving lipid profiles in rats with both standard and diabetic glycemic levels (Verma et al., 2015). Moreover, studies have shown that the methanolic extracts obtained from *A. galanga* exhibit significant anti-diabetic properties in laboratory tests. These extracts can effectively inhibit the process of hemoglobin glycosylation, as well as the activities of α-amylase and α-glucosidase, in a concentration-dependent manner (Chouni & Paul, 2018; Malik et al., 2016; Ravindran et al., 2012).

Hypolipidemic Effect

Testing conducted in a controlled laboratory environment demonstrated that the ethanolic extract derived from the rhizome of *A. galanga* exhibited hypolipidemic action. Administering the section orally at a daily dosage of 20 mg effectively decreased the levels of total triglycerides, phospholipids, and cholesterol in both the bloodstream and tissues (Achuthan & Padikkala, 1997; Chouni & Paul, 2018). During four weeks, the medication resulted in an elevation in blood high-density lipoproteins (HDL) levels in white Wistar rats that were given a high-cholesterol diet. The extract included galangin, quercetin, and kaempferol, inhibiting fatty acid synthase. These compounds had strong reversible inhibitory effects on the enzyme (Iyer et al., 2013; Tirawanchai et al., 2019). The study results indicate that the inhibitory effects of the *A. galanga* extract on fatty acid synthase vary from the mechanisms seen in previously described inhibitors (Ravindran et al., 2012).

Anticancer Activity

The rhizome extract of *A. galanga* has shown notable anticancer properties against lung and breast cancer cell lines, with IC50 values of 7.8 μM and 23.9 μM, respectively (Itokawa et al., 1987; Ravindran et al., 2012). Using bioassay and cytotoxicity tests, scientists have shown that 1′S-1′-acetoxy-chavicol acetate (ACA) is the most potent cytotoxic chemical in A. galanga. This molecule is well recognized for its ability to prevent ulcers, limit tumor growth, and reverse the development of cancer caused by exposure to chemicals (Chouni & Paul, 2018; Chudiwal et al., 2010; El-Hadidy et al., 2020; Samarghandian et al., 2014; Tian et al., 2022). Moreover, the extracted phenylpropanoids, namely 1-acetoxy-chavicol acetate and 1-acetoxy-eugenol acetate, have been shown to possess anticancer activities. These compounds have demonstrated effectiveness against sarcoma 180 ascites in mice. Furthermore, research has shown that galanin may induce apoptosis in hepatocellular carcinoma cells via the mitochondrial route, resulting in various effects such as inhibiting cell proliferation, inducing nuclear fragmentation, and disrupting mitochondrial membrane potential (Itokawa et al., 1987; Suja & Chinnaswamy, 2008; Tian et al., 2022).

Antiallergic Activity

The extract derived from the rhizome inhibited the release of β-hexosaminidase when dissolved in water and acetone. This enzyme is recognized as a trigger for IgE-mediated degranulation in RBL-2h3 cells (Matsuda et al., 2003). After the extract was isolated, nine phenylpropanoids and p-hydroxy benzaldehyde were found. I′-S-I′-acetoxy-eugenol acetate was one of these substances that showed the most inhibitory action, having I.C. (50) values between 15 and 19 mM (Tewtrakul & Subhadhirasakul, 2007). These chemicals have demonstrated the ability to inhibit passive cutaneous anaphylaxis responses in mice and reduce the generation of TNF-α and IL-4 elicited by the antigen IgE. These cytokines are involved in the delayed phase of type 1 allergic reactions in RBL-2H3 cells (Kojima-Yuasa & Matsui-Yuasa, 2020). The extract derived from the galangal rhizome shown significant efficacy in inhibiting hypersensitive responses. Hence, galangal can hinder or cure allergic reactions, encompassing anaphylaxis, allergic rhinitis, asthma, and autoimmune conditions like rheumatoid arthritis and ulcerative colitis. Additionally, it may exhibit analgesic properties (Eram et al., 2019a; Yasuhara et al., 2009).

Other Properties

An alcoholic solution derived from *A. galanga* has been shown to efficiently inhibit the growth of zoonotic dermatophytes such as *Microsporium cannas*, *Microsporium gypseum*, and *Trichophyton spp* (Nirmala, 2022; Trakranrungsie et al., 2008). Galangal oil exhibits antibacterial effects against gram-positive bacteria, yeast, and some dermatophytes, with terpene-4-ol serving as the predominant active component (Malik et al., 2016; Raveesha et al., 2021; Tachakittirungrod & Chowwanapoonpohn, 2007; Tian et al., 2022). ACA, one of the small molecular components extracted from *A. galanga* oil, has shown effectiveness against seven dermatophytic fungus and significant activity against *Trichophyton spp* (Scheffer et al., 1981). Furthermore, (E)-8,17-epoxylabd-12-ene-15,16-dial, a diterpene molecule, was revealed to dramatically boost the efficiency of chalcone and quercetin in combating *Candida albicans*, a fungal infection (Chouni & Paul, 2018; Ravindran et al., 2012). The zone of inhibition experiment has shown the antifungal properties of the DCM fractions and n-Hex of *A. galanga.* Furthermore, the particles of *A. galanga* exhibited elevated concentrations of flavonoid phenolic compounds and showed a potent ability to scavenge free radicals (Köse et al., 2015).

Moreover, studies have shown that the chloroform extract of *A. galanga* has potent inhibition against *Giardia intestinalis* (Chudiwal et al., 2010; Sawangjaroen et al., 2005). However, the rhizome of *A. galanga* has shown the capacity to inhibit the action of human immunodeficiency virus 1, human cytomegalovirus, and hepatitis C virus when extracted with methanol. Furthermore, it has been shown that administering a dosage of 500 mg/kg of *A. galanga* extract to rats has gastric antisecretory, antiulcer, and cytoprotective properties. This leads to a considerable reduction in the extent of stomach mucosal damage produced by pyloric ligation and hypothermic restraint stress (Kojima-Yuasa & Matsui-Yuasa, 2020; Raveesha et al., 2021; Tian et al., 2022).

Moreover, the extract derived from *A. galanga* has immunomodulatory activities. The polysaccharide component of the section has a substantial stimulatory effect on the reticuloendothelial system in mice, leading to a rise in the amount of peritoneal exudate and spleen cells (Ravindran et al., 2012; Tang et al., 2018). The rhizome contains flavonols, namely galangin, and kaempferol, that can hinder the functioning of tyrosinase and cyclo-oxygenase enzymes. The inhibition is most likely due to the capacity of these flavanols to form complexes with copper and other divalent cations found in enzymes (Tian et al., 2022). The inhibitory effect of galangin on mycophenolate activity is seen, but both diphenols are inhibited by galangin and kaempferol. The rhizome includes galangin and quercetin, which have antioxidant and radical-scavenging characteristics. They also control enzyme activity and reduce the harmful effects of chemicals on genes (Tungmunnithum et al., 2020).

The results of these experiments indicate that *A. galanga* has noteworthy therapeutic characteristics and might be used as a natural substitute for manufactured drugs. Consequently, this plant is gaining more significance in traditional medicine and natural medicines, necessitating more excellent investigation to comprehend its prospective applications and advantages comprehensively.

MEDICINAL USES

A. galanga has medicinal properties that are advantageous for a range of health disorders, such as diabetes, renal illnesses, rheumatic pains, bronchitis, lumbago, discomfort in the chest, catarrhal affections, sore throat, and tubercular glands (Benkhnigue et al., 2022, 2023; Khattak et al., 2005). It is often used to facilitate digestion and provide prompt alleviation of pain, specifically for angina, myocardial infarction, and cholecystitis. *A. galanga* has shown efficacy in treating dyspepsia, flatulence, vomiting, and gastric illness (Verma et al., 2011). Recent inquiries have shown this flavonoid has anticancer properties by regulating the aryl hydrocarbon receptor and impeding the detoxifying enzyme CYP1A1 (Jirovetz et al., 2003). Moreover, it hinders the cancer-causing characteristics of meals that are excessively heated or char-grilled and acts as a potent safeguard for the body's inherent antioxidant glutathione (Kaushik et al., 2011). Galangin effectively protects against heart disease by having an antioxidative effect on endothelial cells and retaining other protective antioxidants such as vitamins C and E (Chouni & Paul, 2018). Furthermore, it hinders the cyclo-oxygenase enzyme family, resulting in a widespread reduction of inflammation and a significant safeguarding impact on the heart by preventing platelet aggregation (Zhou et al., 2018). These are just a few examples of the many uses of this distinctive flavonoid, which has been discovered to possess qualities such as anti-mutagenic, anti-inflammatory, antiviral, antioxidant, cancer-preventive, and aromatase inhibitor (Ravindran et al., 2012).

CONCLUSION

A. galanga is a plant that has garnered scientific interest for its potential to exhibit various pharmacological and therapeutic properties. Multiple chemical compounds found in the plant have been recognized to display diverse pharmacological and therapeutic characteristics. Galangal, a prominent constituent of *A. galanga*, has several biological features. Furthermore, it has been shown to possess anti-inflammatory properties and may assist in enhancing digestion. Additional chemical constituents found in *A. galanga* that have been examined include flavonoids, terpenoids, and phenolic compounds. These chemicals have been discovered to exhibit antifungal, antibacterial, and antiviral characteristics, contributing to the prevention of chronic illnesses and the promotion of general health. The statement implies that while substantial findings have been achieved via the study of *A. galanga*, more investigation and assessment are required to ascertain and differentiate the different chemicals in the plant. By doing so, researchers could enhance their understanding of the possible uses of *A. galanga* in advancing human well-being. The identification of the many substances present in *A. galanga* may lead scientists to uncover new applications for the plant, especially in the development of novel medications and therapies. Given its potential to enhance human health and well-being, *A. galanga* is a crucial area of focus for future scientific investigations. Hence, the assertion suggests the need for ongoing research into *A. galanga* to comprehend its prospective advantages and make significant contributions to the advancement of medical understanding and therapy alternatives.

LIST OF ABBREVIATIONS

ACA	1′S -1′-acetoxy chavicol acetate
DCM	Dichloromethane fraction
DPPH	2, 2-diphenyl-1-picrylhydrazyl
HDL	High-density lipoproteins
IgE	Immunoglobulin E
IL	Interleukins
ORAC	Oxygen radical absorbance capacity
RBL	Rat Basophilic Leukemia cells
TNF	Tumor Necrosis Factor

REFERENCES

Achuthan, C. R., & Padikkala, J. (1997). Hypolipidemic effect of *Alpinia galanga* (Rasna) and *Kaempferia galanga* (Kachoori). *Indian Journal of Clinical Biochemistry*, *12*, 55–58.

Akbar, S. (2020). *Handbook of 200 medicinal plants: A comprehensive review of their traditional medical uses and scientific justifications. Springer Cham*, 1–2156. https://doi.org/10.1007/978-3-030-16807-0

Akhtar, M. S., Khan, M. A., & Malik, M. T. (2002). Hypoglycaemic activity of *Alpinia galanga* rhizome and its extracts in rabbits. *Fitoterapia*, *73*(7–8), 623–628.

Aljobair, M. O. (2022). Chemical composition, antimicrobial properties, and antioxidant activity of galangal rhizome. *Food Science and Technology*, *42*, e45622

Ansari, M. K. A., Iqbal, M., Chaachouay, N., Ansari, A. A., & Owens, G. (2023). The concept and status of medicinal and aromatic plants: History, pharmacognosy, ecology, and conservation. In *Plants as Medicine and Aromatics* (pp. 129–144). Taylor and Francis, CRC Press.

Azaizeh, H., Saad, B., Khalil, K., & Said, O. (2006). The state of the art of traditional Arab herbal medicine in the Eastern region of the Mediterranean: A review. *Evidence-Based Complementary and Alternative Medicine*, *3*(2), 229–235.

Baldo, D. E. B., Serrano, J. E., Diomerl, C., & Baldo, E. B. (2016). Screening for intestinal anti-inflammatory activity of *Alpinia galanga* against acetic acid-induced colitis in mice (Mus musculus). *Journal of Medicinal Plants Studies*, *4*(1), 72–77.

Barik, B. R., Kundu, A. B., & Dey, A. K. (1987). Two phenolic constituents from *Alpinia galanga* rhizomes. *Phytochemistry*, *26*(7), 2126–2127.

Belhaj, S., Chaachouay, N., & Zidane, L. (2021). Ethnobotanical and toxicology study of medicinal plants used for the treatment of diabetes in the High Atlas Central of Morocco. *Journal of Pharmacy & Pharmacognosy Research*, *9*(5), 619–662. https://doi.org/10.56499/jppres21.1014_9.5.619

Benkhnigue, O., Chaachouay, N., Khamar, H., El Azzouzi, F., Douira, A, & Zidane, L. (2022). Ethnobotanical and ethnopharmacological study of medicinal plants used in the treatment of anemia in the region of Haouz-Rehamna (Morocco). *Journal of Pharmacy & Pharmacognosy Research*, *10*(2), 279–302. https://doi.org/10.56499/jppres21.1196_10.2.279

Benkhnigue, O., Chaachouay, N., & Zidane, L. (2023). A Floristical and Ecological Study of the Medicinal Flora Used by the Local Population of the Haouz-Rehamna Region (Middle Atlantic Morocco-4). *Acta Botanica Hungarica Akadémiai Kiadó*, *65*(3–4), 247–304. https://doi.org/10.1556/034.65.2023.3-4.3

Bian, M.-Q., Wang, H.-Q., Kang, J., Chen, R.-Y., Yang, Y.-F., & Wu, H.-Z. (2014). Flavonoids from the seeds of *Alpinia galanga* Willd. *Yao xue xue bao= Acta Pharmaceutica Sinica*, *49*(3), 359–362.

Boufous, H., Marhoume, F., Chait, A., & Bagri, A. (2017). Ethnopharmacological survey of medicinal plants with hallucinogenic effect and used against pain, inflammatory diseases, diabetes and urinary lithiasis in Zagora 'Morocco'. *J Intercult Ethnopharmacol*, *6*, 342–350.

Bourgeois, J. A., Parthasarathi, U., & Hategan, A. (2014). Taking the spice route: Psychoactive properties of culinary spices. *Curr Psychiatry*, *13*(4), 21–32.

Byeon, J. C., Ahn, J. B., Jang, W. S., Lee, S.-E., Choi, J.-S., & Park, J.-S. (2019). Recent formulation approaches to oral delivery of herbal medicines. *Journal of Pharmaceutical Investigation*, *49*(1), 17–26.

Chaachouay, N., Benkhnigue, O., Fadli, M., Ibaoui, H., & Zidane, L. (2019a). Ethnobotanical and ethnopharmacological studies of medicinal and aromatic plants used in the treatment of metabolic diseases in the Moroccan Rif. *Heliyon*, *5*(10), e02191. https://doi.org/10.1080/10477845.2021.1942433

Chaachouay, N., Benkhnigue, O., Fadli, M., Ibaoui, H., El Ayadi, R., & Zidane, L. (2019b). Ethnobotanical and ethnopharmacological study of medicinal and aromatic plants used in the treatment of respiratory system disorders in the Moroccan Rif. *Ethnobotany Research and Applications*, *18*(22), 1–16.

Chaachouay, N., Douira, A., Hassikou, R., Brhadda, N., Dahmani, J., Belahbib, N., Ziri, R., & Zidane, L. (2020a). Ethnobotanical and ethnopharmacological study of medicinal and aromatic plants used in the treatment of respiratory system disorders in the Moroccan Rif. PhDThesis, Département de Biologie - Université Ibn Tofail - Kénitra, Maroc, https://tel.archives-ouvertes.fr/tel-03376377

Chaachouay, N., Benkhnigue, O., & Zidane, L. (2020b). Ethnobotanical study aimed at investigating the use of medicinal plants to treat nervous system diseases in the Rif of Morocco. *Journal of chiropractic medicine*, *19*(1), 70–81. https://doi.org/10.1016/j.jcm.2020.02.004

Chaachouay, N., & Zidane, L. (2021). The Symbolic Efficacy of Plants in Rituals and Socio-Religious Ceremonies in Morocco, Northwest of Africa. *Journal of Religious & Theological Information*, *21*(1–2), 34–53. https://doi.org/10.1080/10477845.2021.1942433

Chaachouay, N., Benkhnigue, O., Douira, A., & Zidane, L. (2021a). Poisonous medicinal plants used in the popular pharmacopoeia of the Rif, northern Morocco. *Toxicon*, 189, 24–32. https://doi.org/10.1016/j.toxicon.2020.10.028

Chaachouay, N., Douira, A., & Zidane, L. (2021b). COVID-19, prevention and treatment with herbal medicine in the herbal markets of Salé Prefecture, North-Western Morocco. *European Journal of Integrative Medicine*, 42, 101285. https://doi.org/10.1016/j.eujim.2021.101285

Chaachouay, N., Douira, A., & Zidane, L. (2021c). Herbal Medicine Used in the Treatment of Human Diseases in the Rif, Northern Morocco. *Arabian Journal for Science and Engineering*, 47, 131–153. https://doi.org/10.1007/s13369-021-05501-1

Chaachouay, N., Benkhnigue, O., & Zidane, L. (2022a). Ethnobotanical and Ethnomedicinal study of medicinal and aromatic plants used against dermatological diseases by the people of Rif, Morocco. *Journal of Herbal Medicine*, 32, 100542. https://doi.org/10.1016/j.hermed.2022.100542

Chaachouay, N., Azeroual, A., Bencharki, B., Douira, A., & Zidane, L. (2022b). Ethnoveterinary medicines plants for animal therapy in the Rif, North of Morocco. *South African Journal of Botany*, *147*, 176–191. https://doi.org/10.1016/j.sajb.2021.12.037

Chaachouay, N., Azeroual, A., Douira, A., & Zidane, L. (2022c). Ethnoveterinary practices of medicinal plants among the Zemmour and Zayane tribes, Middle Atlas, Morocco. *South African Journal of Botany*, *151*, 826–840. https://doi.org/10.1016/j.sajb.2022.11.009

Charles, D. J., Simon, J. E., & Singh, N. K. (1992). The essential oil of *Alpinia galanga* Willd. *Journal of Essential Oil Research*, *4*(1), 81–82.

Chouni, A., & Paul, S. (2018). A review on phytochemical and pharmacological potential of *Alpinia galanga. Pharmacognosy Journal*, *10*(1), 9–15.

Chudiwal, A. K., Jain, D. P., & Somani, R. S. (2010). *Alpinia galanga Willd.–An overview on phyto-pharmacological properties*, *1*(2), 143–149.

Das, G., Patra, J. K., Gonçalves, S., Romano, A., Gutiérrez-Grijalva, E. P., Heredia, J. B., Talukdar, A. D., Shome, S., & Shin, H.-S. (2020). Galangal, the multipotent super spices: A comprehensive review. *Trends in Food Science & Technology*, *101*, 50–62.

De Judicibus, M. (2011). *Botanical notebook*. UoM Custom Book Centre.

De Pooter, H. L., Omar, M. N., Coolsaet, B. A., & Schamp, N. M. (1985). The essential oil of greater galanga (*Alpinia galanga*) from Malaysia. *Phytochemistry*, *24*(1), 93–96.

De Smet, P. A. (2002). Herbal remedies. *New England Journal of Medicine*, *347*(25), 2046–2056.
Dissanayake, R., Alwis, L., & Marasingh, P. (2013). *Taxonomical Identification of Maha Aratta (Alpinia galanga (L.) Sw). Proceedings of the Research Symposium of Uva Wellassa University* 22–25.
dos Santos, R. G., Bouso, J. C., Rocha, J. M., Rossi, G. N., & Hallak, J. E. (2021). The use of classic hallucinogens/psychedelics in a therapeutic context: Healthcare policy opportunities and challenges. *Risk Management and Healthcare Policy*, *14*, 901–910. https://doi.org/10.2147/RMHP.S300656
El-Hadidy, E. M., Rashad, N. G., & Ali, M. Y. (2020). Theoretical study, antioxidant activity, and anticancer studies of Galangal (*Alpinia galangal*). *Academic Journal of Current Research*, *7*(8), 101–145.
Eram, S., Mujahid, M., Bagga, P., Ansari, V. A., Ahmad, M. A., Kumar, A., & Akhter, M. S. (2019a). A review on phytopharmacological activity of *Alpinia galanga. International Journal of Pharmacy and Pharmaceutical Sciences*, *11*(3), 6–11.
Eram, S., Mujahid, M., Bagga, P., Ansari, V. A., Ahmad, M. A., Kumar, A., & Akhter, M. S. (2019b). A review on phytopharmacological activity of *Alpinia galanga. International Journal of Pharmacy and Pharmaceutical Sciences*, *11*(3), 6–11.
Ghosh, A., Banerjee, M., & Bhattacharyya, N. (2011). Anti-inflammatory activity of root of *Alpinia galanga* willd. *Chronicles of Young Scientists*, *2*(3), 139–139.
Hamad, A., Alifah, A., Permadi, A., & Hartanti, D. (2016). Chemical constituents and antibacterial activities of crude extract and essential oils of *Alpinia galanga* and *Zingiber officinale*. *International Food Research Journal*, *23*(2), 837.
Hanh, N. P., Binh, N. Q., & Adhikari, B. S. (2014). Distribution of *Alpinia* (Zingiberaceae) and their use pattern in Vietnam. *J. Biodivers. Endanger Species*, *2*, 121.
ITIS. (2022). *Integrated Taxonomic Information System (ITIS) online database, www.itis.gov, CC0* https://doi.org/10.5066/F7KH0KBK. http://itis-web-prod.snafu.cr.usgs.gov
Itokawa, H., Morita, H., Sumitomo, T., Totsuka, N., & Takeya, K. (1987). Antitumour principles from *Alpinia galanga*. *Planta Medica*, *53*(01), 32–33.
Iyer, D., Sharma, B. K., & Patil, U. K. (2013). Isolation of bioactive phytoconstituent from *Alpinia galanga* L. with antihyperlipidemic activity. *Journal of Dietary Supplements*, *10*(4), 309–317.
Jaju, S. B., Indurwade, N. H., Sakarkar, D. M., Fuloria, N. K., Ali, M. D., Das, S., & Basu, S. P. (2009). Galangoflavonoid isolated from rhizome of *Alpinia galanga* (L) Sw (Zingiberaceae). *Tropical Journal of Pharmaceutical Research*, *8*(6), 545–550.
Jirovetz, L., Buchbauer, G., Shafi, M. P., & Leela, N. K. (2003). Analysis of the essential oils of the leaves, stems, rhizomes and roots of the medicinal plant *Alpinia galanga* from southern India. *Acta Pharmaceutica-Zagreb*, *53*(2), 73–82.
Johnley, I. I. R., Somasundaram, G., Salwe, K. J., & Manimekalai, K. (2020). Anti ulcerogenic and antioxidant activity of *Alpinia galanga* rhizomes aqueous extract in indomethacin induced gastric mucosal damage in wistar albino rats. *Biomedical and Pharmacology Journal*, *13*(1), 55–60.
Kaushik, D., Yadav, J., Kaushik, P., Sacher, D., & Rani, R. (2011). Current pharmacological and phytochemical studies of the plant *Alpinia galanga*. *Zhong Xi Yi Jie He Xue Bao*, *9*(10), 1061–1065.
Khairullah, A. R., Solikhah, T. I., Ansori, A. N. M., Fadholly, A., Ram, S. C., Ansharieta, R., Widodo, A., Riwu, K. H. P., Putri, N., & Proboningrat, A. (2020). A review of an important medicinal plant: *Alpinia galanga* (L.) willd. *Systematic Reviews in Pharmacy*, *11*(10), 387–395.
Khattak, S., Shah, H. U., Ahmad, W., & Ahmad, M. (2005). Biological effects of indigenous medicinal plants Curcuma longa and *Alpinia galanga*. *Fitoterapia*, *76*(2), 254–257.
Kojima-Yuasa, A., & Matsui-Yuasa, I. (2020). Pharmacological effects of 1′-Acetoxychavicol Acetate, a major constituent in the rhizomes of *Alpinia galanga* and *Alpinia conchigera*. *Journal of Medicinal Food*, *23*(5), 465–475.
Köse, L. P., Gülçin, İ., Gören, A. C., Namiesnik, J., Martinez-Ayala, A. L., & Gorinstein, S. (2015). L.C.–MS/MS analysis, antioxidant and anticholinergic properties of galanga (*Alpinia officinarum* Hance) rhizomes. *Industrial Crops and Products*, *74*, 712–721. https://doi.org/10.1016/j.indcrop.2015.05.034
Kurniawan, D. W., Wijayanto, B. A., & Sobri, I. (2012). Formulation and effectiveness of antiseptic hand gel preparations essential oils galanga (*Alpinia galanga*). *Asian Journal of Pharmaceutical & Biological Research (AJPBR)*, *2*(4), 245–249.
Lin, K., Wang, Y., Gong, J., Tan, Y., Deng, T., & Wei, N. (2020). Protective effects of total flavonoids from *Alpinia officinarum* rhizoma against ethanol-induced gastric ulcer in vivo and in vitro. *Pharmaceutical Biology*, *58*(1), 854–862.
Liu, P., Wu, S.-L., Wang, T., Zhang, X.-M., & Geng, C.-A. (2023). Four new phenolic compounds from the fruits of *Alpinia galanga. Phytochemistry Letters*, *55*, 75–79. https://doi.org/10.1016/j.phytol.2023.04.001
Lo, C.-Y., Liu, P.-L., Lin, L.-C., Chen, Y.-T., Hseu, Y.-C., Wen, Z.-H., & Wang, H.-M. (2013). Antimelanoma and antityrosinase from *Alpinia galangal* constituents. *The Scientific World Journal*, *2013*.
Mahae, N., & Chaiseri, S. (2009). Antioxidant activities and antioxidative components in extracts of *Alpinia galanga* (L.) Sw. *Agriculture and Natural Resources*, *43*(2), 358–369.
Malik, T., Pandey, D. K., Roy, P., & Okram, A. (2016). Evaluation of phytochemicals, antioxidant, antibacterial and antidiabetic potential of *Alpinia galanga* and *Eryngium foetidum* plants of Manipur (India). *Pharmacognosy Journal*, *8*(5), 459–464
Mandal, S. K., Dawn, S., & Bose, A. (2019). Antiulcer agents: A pharmacological update of the past ten years. *Asian J Pharm Clin Res*, *12*(8), 37–41.
Matsuda, H., Morikawa, T., Managi, H., & Yoshikawa, M. (2003). Antiallergic principles from *Alpinia galanga*: Structural requirements of phenylpropanoids for inhibition of degranulation and release of TNF-α and IL-4 in RBL-2H3 cells. *Bioorganic & Medicinal Chemistry Letters*, *13*(19), 3197–3202.
Mitsui, S., Kobayashi, S., Nagahori, H., & Ogiso, A. (1976). Constituents from seeds of *Alpinia galanga* Wild. And their antiulcer activities. *Chemical and* pharmaceutical bulletin, *24*(10), 2377–2382.
Morita, H., & Itokawa, H. (1986). New diterpenes from *Alpinia galanga* Wild. *Chemistry Letters*, *15*(7), 1205–1208.
Nag, A., Banerjee, R., Goswami, P., Bandyopadhyay, M., & Mukherjee, A. (2021). Antioxidant and antigenotoxic properties of *Alpinia galanga*, *Curcuma amada*, and *Curcuma caesia*. *Asian Pacific Journal of Tropical Biomedicine*, *11*(8), 363.
Nam Hoang, N., Kodama, T., Nwet Win, N., Minh Do, K., Abe, I., & Morita, H. (2021). A new monoterpene from the rhizomes of *Alpinia galanga* and its anti-VPR activity. *Chemistry & Biodiversity*, *18*(10), e2100401.

Nampoothiri, S., Esakkidurai, T., & Pitchumani, K. (2017). Evaluation of antidiabetic, anti-inflammatory and LDL oxidation inhibitory potential of *Alpinia galanga* and *Alpinia calcarata* – An in vitro study. *Trends in Phytochemical Research*, *1*(4), 227–234.

Nampoothiri, S. V., Esakkidurai, T., & Pitchumani, K. (2015). Identification and quantification of phenolic compounds in *Alpinia galanga* and *Alpinia calcarata* and its relation to free radical quenching properties: A comparative study. *Journal of Herbs, Spices & Medicinal Plants*, *21*(2), 140–147.

Nirmala, R. (2022). *Pharmaceutical Properties of Some Common Spices* (Vol. 1). Thanuj International Publishers.

Noro, T., Sekiya, T., Katoh, M., Oda, Y., Miyase, T., Kuroyanagi, M., Ueno, A., & Fukushima, S. (1988). Inhibitors of xanthine oxidase from *Alpinia galanga*. *Chemical and Pharmaceutical Bulletin*, *36*(1), 244–248.

Orch, H., Chaachouay, N., Douiri, E. M., Faiz, N., Zidane, L., & Douira, A. (2021). Use of medicinal plants in dermatocosmetology: An ethnobotanical study among the population of Izarène. *Jordan Journal of Pharmaceutical Sciences*, *14*(3), 323–340.

Qureshi, S., Shah, A. H., & Ageel, A. M. (1992). Toxicity studies on *Alpinia galanga* and *Curcuma longa*. *Planta Medica*, *58*(02), 124–127.

Ramanunny, A. K., Wadhwa, S., Gulati, M., Vishwas, S., Khursheed, R., Paudel, K. R., Gupta, S., Porwal, O., Alshahrani, S. M., & Jha, N. K. (2022). Journey of *Alpinia galanga* from kitchen spice to nutraceutical to folk medicine to nanomedicine. *Journal of Ethnopharmacology*, *291*, 115144.

Rangsiruji, A., Newman, M. F., & Cronk, Q. C. (2000). Origin and relationships of *Alpinia galanga* (Zingiberaceae) based on molecular data. *Edinburgh Journal of Botany*, *57*(1), 9–37.

Raveesha, H. R., Bharath, H. L., Vasudha, D. R., Sushma, B. K., Pratibha, S., & Dhananjaya, N. (2021). Antibacterial and antiproliferation activity of green synthesized nanoparticles from rhizome extract of *Alpinia galangal* (L.) Wild. *Inorganic Chemistry Communications*, *132*, 108854.

Ravindran, P. N., Pillai, G. S., Balachandran, I., & Divakaran, M. (2012). Galangal. Handbook of Herbs and Spices (Second edition) Volume 2 Woodhead Publishing Series in Food Science, Technology and Nutrition,(p303-308) Elsevier.

Sahoo, S., Singh, S., Sahoo, A., Sahoo, B. C., Jena, S., Kar, B., & Nayak, S. (2020). Molecular and phytochemical stability of long term micropropagated greater galanga (*Alpinia galanga*) revealed suitable for industrial applications. *Industrial Crops and Products*, *148*, 112274. https://doi.org/10.1016/j.indcrop.2020.112274

Samarghandian, S., Hadjzadeh, M.-A.-R., Afshari, J. T., & Hosseini, M. (2014). Antiproliferative activity and induction of apoptotic by ethanolic extract of *Alpinia galanga* rhizhome in human breast carcinoma cell line. *BMC complementary and alternative medicine*, *14*(1), 1–9.

Sawangjaroen, N., Subhadhirasakul, S., Phongpaichit, S., Siripanth, C., Jamjaroen, K., & Sawangjaroen, K. (2005). The in vitro anti-giardial activity of extracts from plants that are used for self-medication by AIDS patients in southern Thailand. *Parasitology Research*, *95*, 17–21.

Scheffer, J. J. C., Gani, A., & Svendsen, A. B. (1981). Antifungal activity of *Alpinia glanga*. *Planta medica*, *42*(06), 140–141.

Singh, S., Sahoo, B. C., Kar, S. K., Sahoo, A., Nayak, S., Kar, B., & Sahoo, S. (2020). Chemical constituents analysis of *Alpinia galanga* and *Alpinia calcarata*. *Research Journal of Pharmacy and Technology*, *13*(10), 4735–4739.

Subash, K. R., Bhaarathi, G. M., Rao, N. J., & Cheriyan, B. V. (2013). Phytochemical screening and acute toxicity study of ethanolic extract of *Alpinia galanga* in rodents. *International Journal of Medical Research & Health Sciences*, *1*(2), 93–100.

Subash, K. R., Prakash, G. B., Reddy, K. V. C., Manjunath, K., & Rao, K. U. (2016). Anti-inflammatory activity of ethanolic extract of *Alpinia galanga* in carrageenan induced pleurisy rats. *National Journal of Physiology, Pharmacy and Pharmacology*, *6*(5), 468.

Suja, S., & Chinnaswamy, P. (2008). Inhibition of in vitro cytotoxic effect evoked by *Alpinia galanga* and *Alpinia officinarum* on PC-3 cell line. *Ancient Science of Life*, *27*(4), 33.

Tachakittirungrod, S., & Chowwanapoonpohn, S. (2007). Comparison of antioxidant and antimicrobial activities of essential oils from Hyptis suaveolens and *Alpinia galanga* growing in Northern Thailand. *CMU J Nat Sci*, *6*(1), 31–41.

Tang, X., Xu, C., Yagiz, Y., Simonne, A., & Marshall, M. R. (2018). Phytochemical profiles, and antimicrobial and antioxidant activities of greater galangal [*Alpinia galanga* (Linn.) Swartz.] flowers. *Food Chemistry*, *255*, 300–308.

Tao, L., Wang, Z.-T., Zhu, E.-Y., Lu, Y.-H., & Wei, D.-Z. (2006). HPLC analysis of bioactive flavonoids from the rhizome of *Alpinia officinarum*. *South African Journal of Botany*, *72*(1), 163–166.

Tewtrakul, S., & Subhadhirasakul, S. (2007). Antiallergic activity of some selected plants in the Zingiberaceae family. *Journal of Ethnopharmacology*, *109*(3), 535–538.

The Plant List. (2013). *The Plant List. Version 1.1*. Published on the Internet; http://www.theplantlist.org/ (accessed 1st January).

Tian, Y., Jia, X., Wang, Q., Lu, T., Deng, G., Tian, M., & Zhou, Y. (2022). Antioxidant, antibacterial, enzyme inhibitory, and anticancer activities and chemical composition of *Alpinia galanga* flower essential oil. *Pharmaceuticals*, *15*(9), 1069.

Tirawanchai, N., Homongkol, P., Chansriniyom, C., Somkasetrin, A., Jantaravinid, J., Kengkoom, K., & Ampawong, S. (2019). Lipid-lowering effect of *Phyllanthus embilica* and *Alpinia galanga* extracts on HepG2 cell line. *PharmaNutrition*, *9*, 100153.

Trakranrungsie, N., Chatchawanchonteera, A., & Khunkitti, W. (2008). Ethnoveterinary study for antidermatophytic activity of Piper betle, *Alpinia galanga* and *Allium ascalonicum* extracts in vitro. *Research in Veterinary Science*, *84*(1), 80–84.

Tungmunnithum, D., Tanaka, N., Uehara, A., & Iwashina, T. (2020). Flavonoids profile, taxonomic data, history of cosmetic uses, antioxidant and anti-aging potential of *alpinia galanga* (L.) willd. *Cosmetics*, *7*(4), 89.

Verma, R. K., Garima, M., Pradeep, S., Jha, K. K., & Khosa, R. L. (2011). *Alpinia galanga* – an important medicinal plant: A review. *Der Pharmacia Sinica*, *2*(1), 142–154.

Verma, R. K., Mishra, G., Singh, P., Jha, K. K., & Khosa, R. L. (2015). Antidiabetic activity of methanolic extract of *Alpinia galanga* Linn. Aerial parts in streptozotocin induced diabetic rats. *Ayu*, *36*(1), 91.

Wu, Y., Wang, Y., Li, Z., Wang, C., Wei, J., Li, X., Wang, P., Zhou, Z., Huang, D., & Deng, Z. (2014). Composition of the essential oil from *Alpinia galanga* rhizomes and its bioactivity on Lasioderma serricorne. *Bulletin of Insectology*, *67*(2), 247–254.

Yang, X., & Eilerman, R. G. (1999). Pungent Principal of *Alpinia galangal* (L.) Swartz and Its Applications. *Journal of Agricultural and Food Chemistry*, *47*(4), 1657–1662. https://doi.org/10.1021/jf9808224

Yasuhara, T., Manse, Y., Morimoto, T., Qilong, W., Matsuda, H., Yoshikawa, M., & Muraoka, O. (2009). Acetoxybenzhydrols as highly active and stable analogues of 1′ S-1′-acetoxychavicol, a potent antiallergic principal from *Alpinia galanga. Bioorganic & medicinal chemistry letters*, *19*(11), 2944–2946.

Zhou, C., Li, C., Siva, S., Cui, H., & Lin, L. (2021). Chemical composition, antibacterial activity and study of the interaction mechanisms of the main compounds present in the *Alpinia galanga* rhizomes essential oil. *Industrial Crops and Products*, *165*, 113441.

Zhou, Y.-Q., Liu, H., He, M.-X., Wang, R., Zeng, Q.-Q., Wang, Y., Ye, W.-C., & Zhang, Q.-W. (2018). A review of the botany, phytochemical, and pharmacological properties of galangal. *In Handbook of Food Bioengineering, Natural and Artificial Flavoring Agents and Food Dyes*, Academic Press, Elsevier, 351–396. https://doi.org/10.1016/B978-0-12-811518-3.00011-9.

Index

Pages in *italics* refer to figures and pages in **bold** refer to tables.

C

H

L

M

N

O

T

U

V